Orbital Properties of the Planets and Dwarf Planets

NAME	SEMIMAJOR AXIS, IN AU	SEMIMAJOR AXIS, IN 10^6 km	SIDEREAL PERIOD	SYNODIC PERIOD, IN DAYS	AVERAGE ORBITAL SPEED, IN km/s	ORBITAL ECCENTRICITY	INCLINATION OF ORBIT, IN DEGREES
Planet							
Mercury	0.3871	57.9	87.97 days	115.9	47.9	0.206	7.00
Venus	0.7233	108.2	224.7 days	583.9	35.0	0.007	3.39
Earth	1.0	149.6	365.3 days	—	29.8	0.017	0.00
Mars	1.523	227.9	687.0 days	780.0	24.1	0.093	1.85
Jupiter	5.202	778.3	11.86 years	398.9	13.1	0.048	1.31
Saturn	9.539	1,427	29.46 years	378.1	9.6	0.056	2.49
Uranus	19.19	2,870	84.01 years	369.7	6.8	0.046	0.77
Neptune	30.06	4,497	164.8 years	367.5	5.4	0.010	1.77
Dwarf Planet							
Ceres	2.77	413.7	4.60 years	466.7	17.9	0.080	10.59
Pluto	39.53	5,900	248.5 years	366.7	4.7	0.248	17.15
Haumea	43.1	6,450	283 years	366.5	4.5	0.195	28.22
Makemake	45.8	6,850	310 years	366.4	4.4	0.159	28.96
Eris	67.7	10,210	560 years	365.9	3.4	0.442	44.19

Physical Properties of the Planets and Dwarf Planets

NAME	EQUATORIAL DIAMETER, IN km	EQUATORIAL DIAMETER, IN EARTH DIAMETERS	MASS, IN EARTH MASSES	DENSITY, RELATIVE TO WATER	ACCELERATION OF GRAVITY AT SURFACE, RELATIVE TO EARTH	ESCAPE VELOCITY, IN km/s	SIDEREAL ROTATION PERIOD, IN DAYS	ALBEDO	AXIAL TILT, IN DEGREES
Planet									
Mercury	4,878	0.38	0.055	5.4	0.38	4.3	58.65	0.11	2
Venus	12,104	0.95	0.815	5.2	0.91	10.4	243.02	0.65	177.3*
Earth	12,756	1.0	1.0	5.5	1.0	11.2	1.00	0.37	23.5
Mars	6,787	0.53	0.107	3.9	0.38	5.0	1.03	0.15	25.2
Jupiter	142,980	11.2	317.9	1.3	2.54	59.6	0.41	0.52	3.1
Saturn	120,540	9.5	95.2	0.7	1.08	35.6	0.44	0.47	26.7
Uranus	51,120	4.0	14.5	1.3	0.91	21.3	0.72	0.40	97.9*
Neptune	49,530	3.9	17.1	1.6	1.19	23.8	0.67	0.35	29.6
Dwarf Planet									
Ceres	950	0.074	0.00016	2.1	0.03	0.5	0.38	0.09	4
Pluto	2,300	0.18	0.002	2.0	0.06	1.2	6.39	0.6	122.5*
Haumea	1,400	0.11	0.0007	3	0.04	0.8	0.16	0.8	—
Makemake	1,500	0.12	0.0007	2	0.05	0.8	—	0.8	—
Eris	2,400	0.19	0.0025	2	0.06	1.3	—	0.86	—

*Retrograde rotation

Astronomy

Journey to the Cosmic Frontier

Sixth Edition

John D. Fix
University of Alabama in Huntsville

ASTRONOMY: JOURNEY TO THE COSMIC FRONTIER, SIXTH EDITION

Published by McGraw-Hill, a business unit of The McGraw-Hill Companies, Inc., 1221 Avenue of the Americas, New York, NY 10020.

 This book is printed on recycled, acid-free paper containing 10% postconsumer waste.

1 2 3 4 5 6 7 8 9 0 WDQ/WDQ 10 9 8 7 6 5 4 3 2 1 0

ISBN 978–0–07–351218–1
MHID 0–07–351218–4

Vice President & Editor-in-Chief: *Marty Lange*
Vice President EDP/Central Publishing Services: *Kimberly Meriwether David*
Publisher: *Ryan Blankenship*
Sponsoring Editor: *Debra B. Hash*
Senior Marketing Manager: *Lisa Nicks*
Senior Project Manager: *Lisa A. Bruflodt*
Design Coordinator: *Margarite Reynolds*
Senior Photo Research Coordinator: *Lori Hancock*
USE Cover Image Credit: *NASA, ESA, and the Hubble SM4 ERO Team*
Senior Production Supervisor: *Sue Culbertson*
Compositor: *S4Carlisle Publishing Services*
Typeface: *10.75/12 Garamond*
Printer: *World Color/Dubuque*

Library of Congress Cataloging-in-Publication Data

Fix, John D.
Astronomy : journey to the cosmic frontier / John D. Fix. – 6th ed.
p. cm.
Includes bibliographical references and index.
ISBN 978-0-07-351218-1
1. Astronomy–Textbooks. I. Title.
QB43.3.F59 2010
520--dc22

2009050731

www.mhhe.com

For Cynthia

Foreword

A layman challenges astronomers with the assertion: "If every object in the universe except the Earth, the Sun, and the Moon were eliminated, most people wouldn't know the difference." There is, undeniably, a measure of truth to this assertion. It resembles the assertion that humans are interested only in food, shelter, and procreation. But the full truth is much richer and more complex. Even the most primitive peoples marvel at the dazzling beauty of the night sky, identify stable patterns in the arrangement of easily identified stars, and note the movement of a few bright points of light, the planets, on the star field. They also derive spiritual inspiration from this scene. More advanced civilizations have shared this inspiration and gone beyond it to seek a scientific understanding of the grand scheme of the universe.

Even the most sophisticated modern astronomers are motivated by a primal awe of their subject. These astronomers then attempt rational explanations of its infinite detail, piece by piece, and they carry along with them the whole or nearly the whole of humanity. For example, what could be more esoteric or more remote from everyday experience than a supernova, a galaxy, the Big Bang, or a black hole? But these astronomical concepts have become a part of popular culture and language.

Astronomy takes its place with art, music, literature, drama, and religion as an inspiring subject in the minds and hearts of sensitive and thoughtful individuals. It contributes to lifting the human spirit above the break-even level of bare survival. We would be much the poorer without it. Is astronomy of practical importance? In a restricted sense, yes; it plays a key role in navigation, timekeeping, and the manifestation of physical principles at work in complex systems. But the grandeur and enormous physical scale of the universe and the realization of our tiny part in it are the aspects of astronomy that enrich our lives and permeate our culture.

Join Professor Fix and his professional colleagues in this great intellectual adventure of exploration and discovery.

James A. Van Allen
1914–2006

Preface

As James Van Allen wrote in his foreword to this book, astronomy permeates our culture. Of all the sciences, astronomy is the one that generates the most public interest. There are hundreds of thousands of amateur astronomers, two monthly astronomy magazines with healthy circulation, and television specials about important astronomical discoveries. The demotion of Pluto from planet to dwarf planet generated headlines and editorials around the world. Part of the public interest in astronomy is surely due to the dramatic scope of the science. Part, I am sure, is because nonprofessionals not only can understand astronomical discoveries but also can make some of those discoveries. Amateur astronomers regularly carry out important astronomical observations, often with telescopes they have made themselves.

The Goals of Astronomy: Journey to the Cosmic Frontier

I wrote this book as a text for an introductory course in astronomy for college students. I have taught such a course for many years at the University of Iowa and the University of Alabama in Huntsville. One of my main goals for those courses, and one of my main goals in this book, is to provide my students with a broad enough, deep enough background in astronomy that they will be able to follow current developments years after they finish my course. This book is current with recent developments such as the cosmological discoveries of the *WMAP* satellite and the results from the Mars rovers. But I want my students to continue to learn about astronomy long after these discoveries have been succeeded by newer, even more exciting, ones. I hope that years from now my students, and the readers of this book, will be able to read and watch stories about astronomy with confidence that they know what is going on and why the story is important. I can guarantee that future astronomical discoveries will occur at least as often as they do today, and I want my students to be prepared to enjoy future discoveries.

I hope that all the explanations and descriptions in the book will not obscure the awe and sense of wonder that all astronomers feel when they pause in their work and think about the beauty of the universe. People have felt that awe since prehistory and our wonderment has increased as we understand more about the order and underlying structure of the universe. If this book helps its readers to value both the sheer beauty of planets, stars, and galaxies and the equally beautiful principles that organize the universe, it will be a success.

I would be grateful for any suggestions and advice for improving this book. If you have any ideas to offer, please contact me at the Department of Physics, University of Alabama in Huntsville, Huntsville, Alabama, 35899, or by e-mail at fixj@uah.edu.

What's New?

Content Updates and Additions As stated, one of the goals of this text is to keep students up to date on current astronomical events and discoveries. In doing so, many new topics have been added to the sixth edition, and several topics from previous editions have been updated. Some of these include:

New Topics

- Fermi gamma-ray telescope (Chapter 6)
- Recent Moon missions (Chapter 9)
- Results from the *Messenger* mission to Mercury (Chapter 10)
- The discovery of ice by the *Phoenix* Mars lander (Chapter 11)
- *Mars Reconnaissance Orbiter* images of the Martian surface (Chapter 11)
- Trans-Neptunian Objects and the history of the solar system (Chapter 13)
- *Voyagers*' encounter with the termination shock of the solar wind (Chapter 17)
- Bright gamma ray burst of March 2008 (Chapter 20)
- VLBI determination of the size and rotation of the Milky Way (Chapter 22)
- Future collision between the Milky Way and M31 (Chapter 25)
- The *Corot* and *Kepler* missions to detect transits by planets orbiting other stars (Chapter 27)

Updated and Revised Topics

- Diffraction (Chapter 6)
- Large telescopes of the future (Chapter 6)
- Information on future eclipses (Chapter 9)
- The loss of Venus's water (Chapter 10)
- Discoveries by the *Spirit* and *Opportunity* rovers on Mars (Chapter 11)
- The climate history of Mars (Chapter 11)
- Three newly defined dwarf planets (Chapter 13)
- Icy plumes from Enceladus (Chapter 14)
- Possible future collisions of Apollo asteroids with Earth (Chapter 15)
- Brown dwarfs (Chapter 16)

- Massive black holes at the centers of galaxies (Chapter 23)
- Planets orbiting other stars (Chapter 27)

New and Updated Images Including images from *Hubble, Spitzer, Spirit, Opportunity, Cassini, Huygens, Mars Global Surveyor,* and *Phoenix.*

Pedagogical Features

Electronic Media Integration To help better grasp key concepts, this interactive icon has been placed near figures and selections where students can gain additional understanding through the interactives on the *Astronomy* Online Learning Center.

To help better understand key concepts, this animation icon has been placed near figures and sections where students can explore additional information on the *Astronomy* Online Learning Center.

Chapter Introduction Every chapter begins with an introduction designed to give the historical and scientific setting for the chapter material. The overview previews the chapter's contents and what you can expect to learn from reading the chapter. After reading the introduction, browse through the chapter, paying particular attention to the topic headings and illustrations so that you get a feel for the kinds of ideas included within the chapter. Also included in the chapter introduction are questions to explore while reading the text.

Worked Examples Boxes This book, like my course, presumes that many of its readers are not science majors and may not have had a college-level science or mathematics course. The book provides a complete description of current astronomical knowledge, neither at an extremely technical level nor at a level that fails to communicate the quantitative nature of physical science. I have used equations where they are relevant, but follow the equations with boxes containing one or more worked examples. The examples in the boxes show how and when to use each equation and tell why the equation is important.

Historical Emphasis Throughout the book I have emphasized the historical development of astronomy to show that astronomy, like other sciences, advances through the efforts of many scientists and to show how our present ideas developed. In the main body of the text there are many comparisons of what was once known about a particular phenomenon to what we now know about it. These historical comparisons are used to illustrate the cycle of observation, hypothesis, and further observation, which is the essence of the scientific method of discovery.

Equations 4.1 and 4.2

Sidereal and Synodic Periods

Equations 4.1 and 4.2 can be used to calculate the synodic period of a planet from its sidereal period or vice versa. Suppose there were a superior planet with a synodic period of 1.5 years. For $S = 1.5$ years and $P_{Earth} = 1$ year, Equation 4.1 is

$$\frac{1}{P} = \frac{1}{(1\ \text{yr})} - \frac{1}{(1.5\ \text{yr})} = \frac{(3-2)}{(3\ \text{yr})} = \frac{1}{(3\ \text{yr})}$$

Thus, P, the sidereal period of the planet, is 3 years. This is the hypothetical planet described in Figure 4.6. As a second example, suppose there were an inferior planet with a sidereal period of 0.25 years. For $P = 0.25$ years and $P_{Earth} = 1$ year, Equation 4.2 is

$$\frac{1}{(0.25\ \text{yr})} = \frac{1}{(1\ \text{yr})} + \frac{1}{S}$$

Rearranging this equation to solve for $1/S$ gives

$$\frac{1}{S} = \frac{1}{(0.25\ \text{yr})} - \frac{1}{(1\ \text{yr})} = \frac{4}{(1\ \text{yr})} - \frac{1}{(1\ \text{yr})} = \frac{3}{(1\ \text{yr})}$$

for which $S = 1/3$ year.

Running Summaries

Important concepts and facts are summarized in the body of the chapter immediately after the concept is introduced.

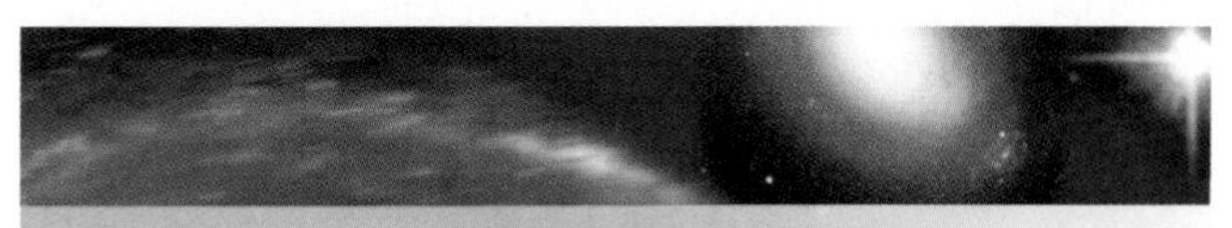

The epicyclic model perfected by Ptolemy used combinations of circular motions to reproduce the motions of the planets. The model could predict the positions of celestial objects with such accuracy that it was used for nearly 1500 years.

Planetary Data Boxes

These boxes include summaries of planetary data making this information easy to access.

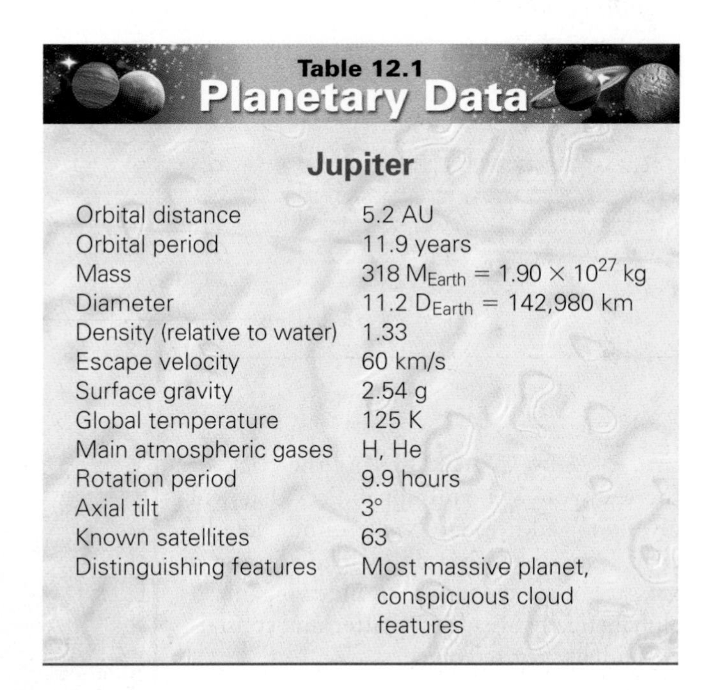

Table 12.1 Planetary Data

Jupiter	
Orbital distance	5.2 AU
Orbital period	11.9 years
Mass	318 M_{Earth} = 1.90×10^{27} kg
Diameter	11.2 D_{Earth} = 142,980 km
Density (relative to water)	1.33
Escape velocity	60 km/s
Surface gravity	2.54 g
Global temperature	125 K
Main atmospheric gases	H, He
Rotation period	9.9 hours
Axial tilt	3°
Known satellites	63
Distinguishing features	Most massive planet, conspicuous cloud features

End of Chapter Material

Chapter Summary highlights the key topics of the chapter.

Key Terms listed here are defined in the text and in the end-of-book glossary.

Conceptual Questions require qualitative verbal answers.

Problems, involving numerical calculations, test the reader's mastery of the equations.

Figure-Based Questions require the reader to extract the answer from a particular graph or figure in the chapter.

Group Activities encourage interaction between students as they work in groups to discuss different viewpoints on chapter-related issues or to complete small group projects.

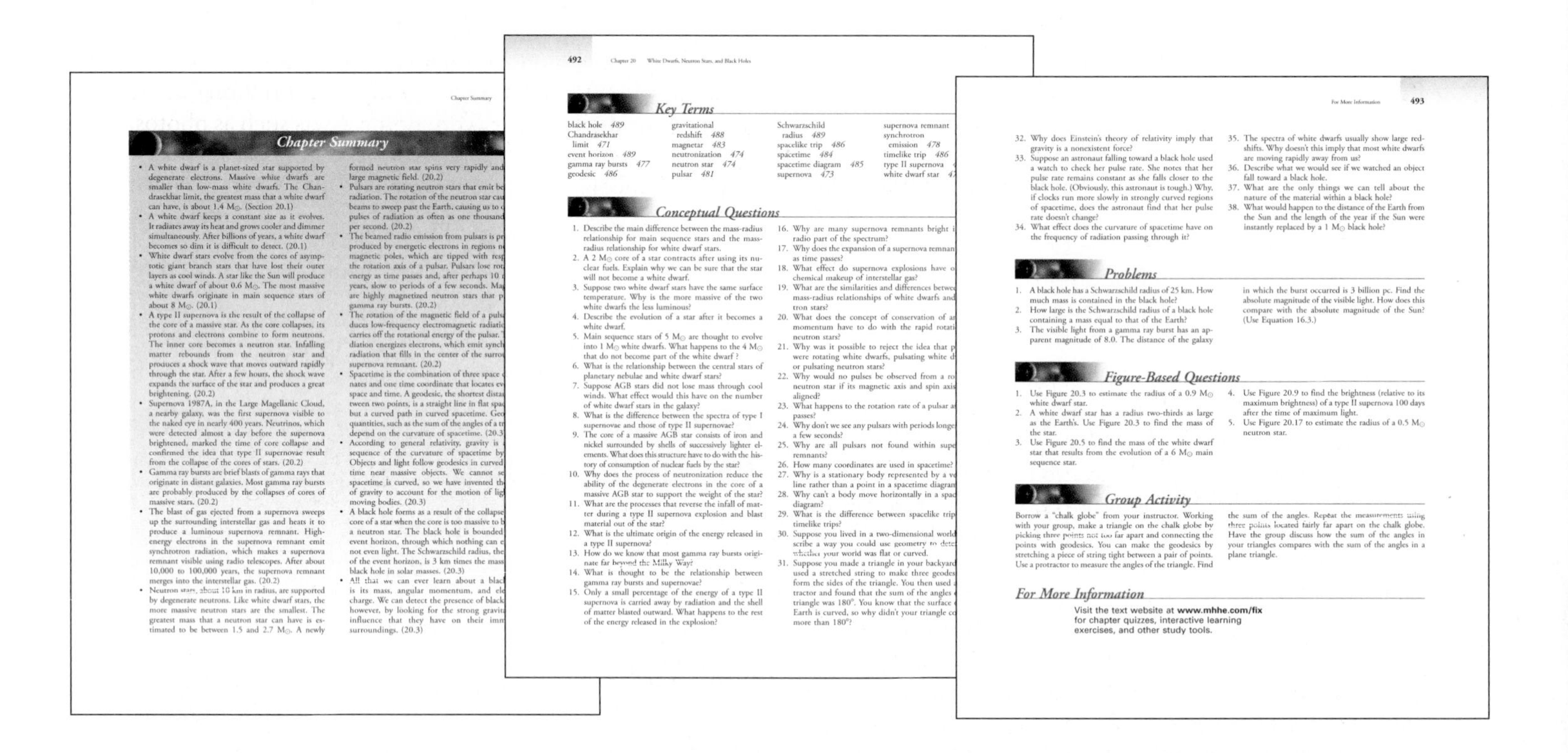

End-of-Text Material

At the back of the text you will find appendices that will give you additional background details, charts, and extensive tables. There is also a glossary of all key terms, an index organized alphabetically by subject matter, and constellation maps for reference.

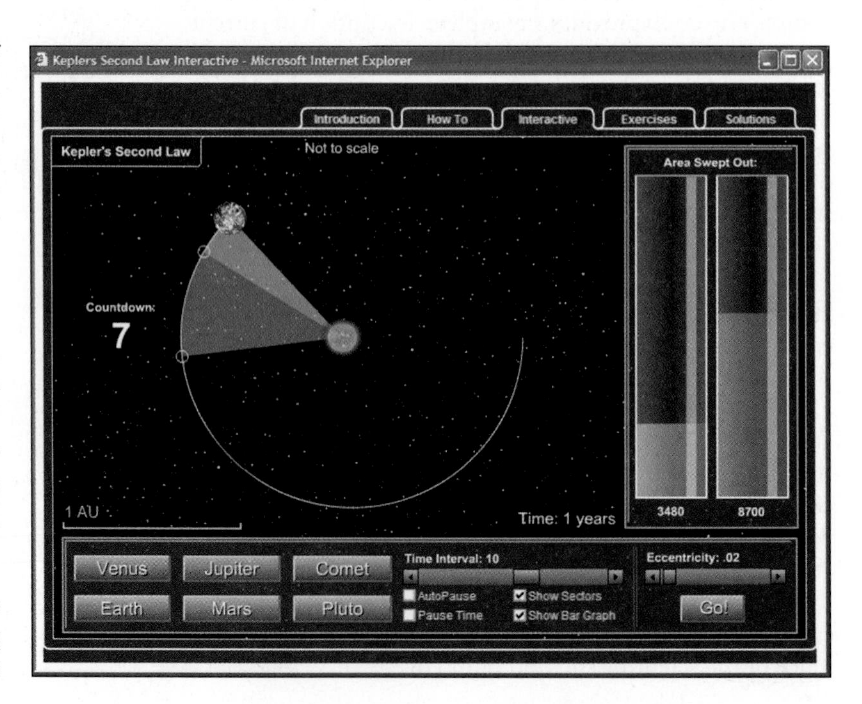

Supplements

McGraw-Hill offers various tools and technology products to support *Astronomy: Journey to the Cosmic Frontier, Sixth Edition.* Instructors can obtain teaching aids by calling the Customer Service Department at 800-338-3987 or contacting your local McGraw-Hill sales representative.

Interactives McGraw-Hill is proud to bring you an assortment of 23 outstanding Interactives like no other. Each Interactive is programmed in Flash for a stronger visual appeal. These Interactives offer a fresh and dynamic method to teach the astronomy basics. Each Interactive allows users to manipulate parameters and gain a better understanding of topics such as blackbody radiation, the Bohr model, a solar system builder, retrograde motion, cosmology, and the H-R diagram by watching the effect of these manipulations. Each Interactive includes an analysis tool (interactive model), a tutorial describing its function, content describing its principle themes, related exercises, and solutions to the exercises. Plus, users can jump between these exercises and analysis tools with just the click of the mouse. These Interactives are located on the *Astronomy* Online Learning Center.

Starry Night College Introducing *Starry Night College*! A new, online format that is now available to package with the Fix: *Astronomy* text. *Starry Night College* makes it easy to teach astronomy. The ideal companion to your astronomy textbook, *Starry Night* gives you powerful, accurate, customizable tools to inspire and engage your students. *Starry Night College* includes more than 45 computer exercises, complete with dazzling interactive simulations, all fully mapped to Fix: *Astronomy.* The step-by-step exercises in *Starry Night College* will lead your students to a solid understanding of the universe. *Starry Night College* includes:

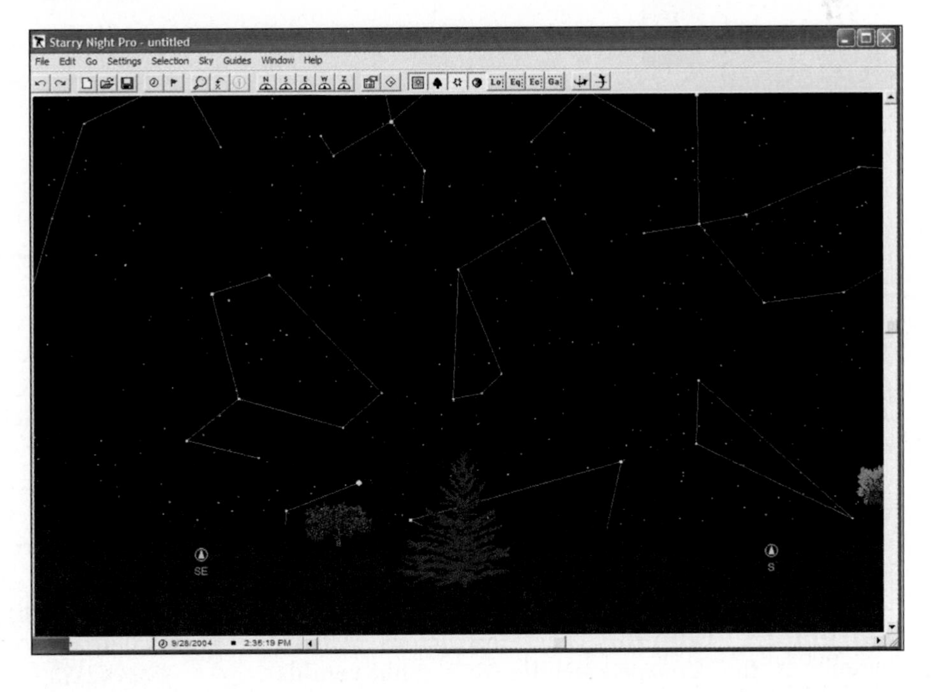

- Award-winning, customized, Starry Night Planetarium Software
- Professor's Guide
- Astronomy Companion Guide (PDF)
- 45+ integrated computer exercises
- Library of preassembled, built-in simulations
- Create movies of your customized simulations and export images
- Student Worksheets and Answer Keys
- Exercises mapped to the Fix: *Astronomy* text
- Links, references, and suggestions for online access to additional content and resources
- And much more!

www.mhhe.com/fix McGraw-Hill offers a wealth of online features and study aids that greatly enhance the astronomy teaching and learning experience. The design of the Fix website makes it easy for students to take full advantage of the following tools:

- **Interactive student technology:** Includes 23 outstanding Astronomy Interactives, Animations, and Constellation Quizzes.
- **Text-specific features:** Includes Multiple Choice Quizzes, Conceptual Questions, Problems, Figure-Based Questions, and Crossword Puzzles.
- **General astronomy features:** Includes Planetarium Activities, Group Activities, Astronomy Timeline, Astronomy Links Library, Astronomy Picture of the Day, and Further Readings.
- **Additional instructor resources:** Includes Instructor's Manual and PowerPoint Presentation.

Presentation Center

Build instructional materials wherever, whenever, and however you want! Presentation Center is an online digital library containing assets such as photos, artwork, animations, PowerPoint® presentations, and other types of media that can be used to create customized lectures, visually enhanced tests and quizzes, compelling course websites, or attractive printed support materials.

Access to your book, access to all books! The Presentation Center library includes thousands of assets from many McGraw-Hill titles. This ever-growing resource gives instructors the power to utilize assets specific to an adopted textbook as well as content from all other books in the library.

Nothing could be easier! Accessed from the instructor side of your textbook's website, Presentation Center's dynamic search engine allows you to explore by discipline, course, textbook chapter, asset type, or keyword. Simply

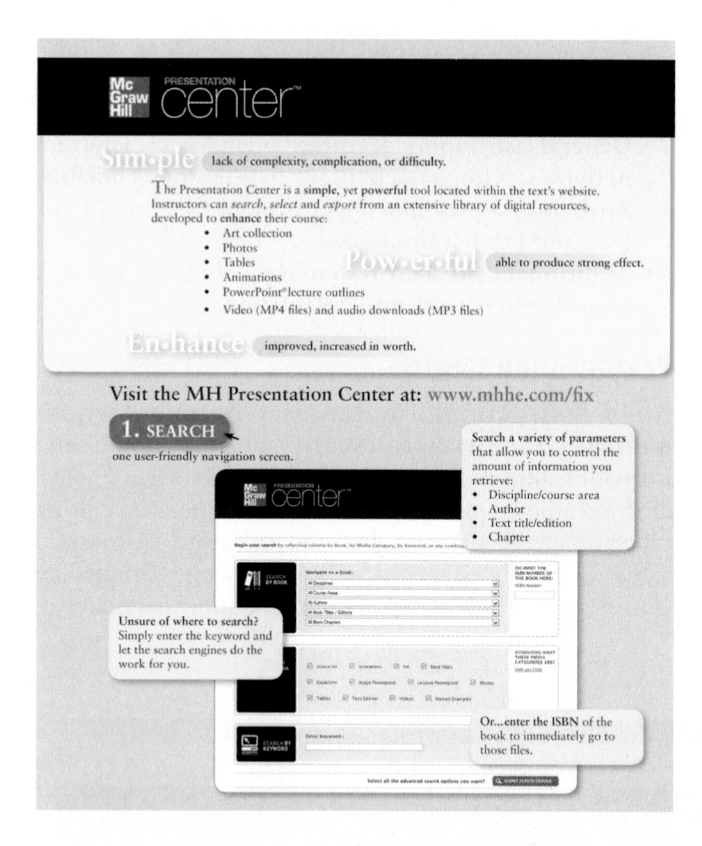

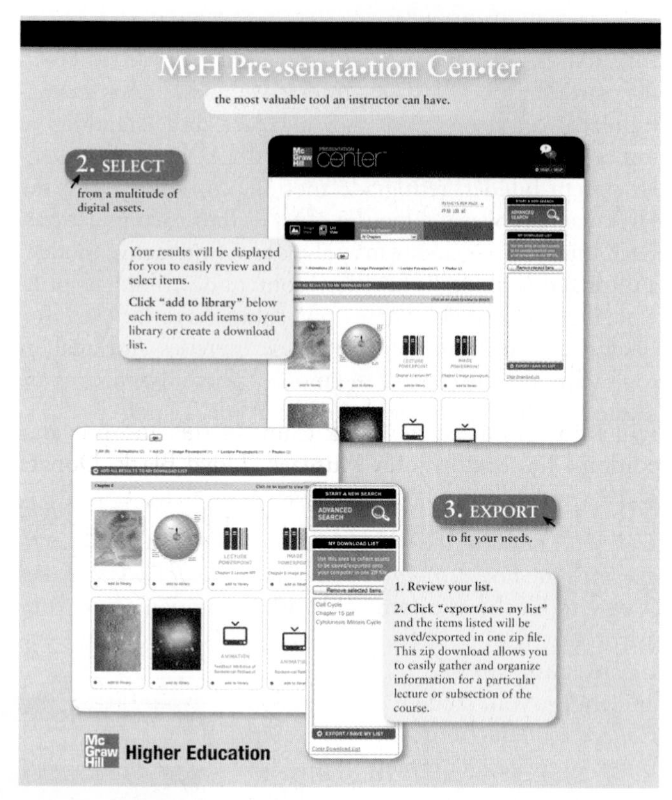

browse, select, and download the files you need to build engaging course materials. All assets are copyrighted by McGraw-Hill Higher Education but can be used by instructors for classroom purposes.

Computerized Test Bank Online A comprehensive bank of test questions in multiple-choice format is provided within the computerized test bank powered by McGraw-Hill's flexible electronic testing program—EZ Test Online (www.eztestonline.com). EZ Test Online allows you to create paper and online tests or quizzes in this easy-to-use program.

Instructor's Manual The Instructor's Manual is found on the *Astronomy* website (www.mhhe.com/fix) and on the Instructor's Testing and Resource CD, and can be accessed only by instructors.

Classroom Performance System and Questions The Classroom Performance System (CPS) brings interactivity into the classroom/lecture hall. CPS is a wireless response system that gives an instructor immediate feedback from every student in the class. Each CPS unit comes with up to 512 individual response pads and an appropriate number of corresponding receiver units. The wireless response pads are essentially remotes that are easy to use and engage students. The CPS system allows instructors to create their own questions or use the McGraw/Hill-provided astronomy questions.

Electronic Book If you or your students are ready for an alternative version of the traditional textbook, McGraw-Hill has partnered with CourseSmart to bring you innovative and inexpensive electronic textbooks. Students can save up to 50% off the cost of a print book, reduce their impact on the environment, and gain access to powerful web tools for learning including full text search, notes and highlighting, and email tools for sharing notes between classmates. eBooks from McGraw-Hill are smart, interactive, searchable, and portable.

To review comp copies or to purchase an eBook, go to www.CourseSmart.com

Acknowledgments

I am grateful to many people who helped in the development and production of this book. Perhaps the most important contributors to the book are the more than ten thousand University of Iowa and University of Alabama in Huntsville students who took my beginning astronomy classes. They taught me how to teach introductory astronomy and showed me the ingredients a good textbook must have. I owe a large debt to my fellow astronomers: Andrea Cox, Ken Gayley, Larry Molnar, Bob Mutel, John Neff, Stan Shawhan, Steve Spangler, and James Van Allen. Their help ranged from years' worth of lunchtime discussions about astronomy teaching to expert advice on difficult sections of this book. I am also grateful to Cynthia and Stephen Fix, who were the first reviewers of the first draft of the book.

Many students and teachers have pointed out errors in earlier editions of the book and offered suggestions for improving it. I found the comments of John Broderick, Wynne Colvert, Anne Cowley, Bill Keel, Kathy Rajnak, Jeremy Tatum, and Virginia Trimble especially helpful. My colleague Larry Molnar compiled a lengthy, detailed list of comments and suggestions that improved both the book and my understanding of a number of topics. I owe Larry a special expression of thanks.

I also want to thank the dozens of people at McGraw-Hill who comprised the book teams of this and previous editions. Jim Smith, John Murdzek, Lloyd Black, Donata Dettbarn, Colleen Fitzpatrick, Mary Sheehan, Lori Sheil, J. P. Lenny, Daryl Bruflodt, Brian Loehr, Liz Recker, and Wendy Langerud have been my sponsoring and developmental editors. Jim began the project and the many others saw it to completion. Their advice and comments have nearly always been right on target. I am especially grateful to John Murdzek, who not only gave me his own advice but also digested and distilled reviewers' comments so that I could benefit as much as possible from the advice of the many reviewers of the various drafts of the first edition. Debra Hash, Todd Turner, and Lisa Nicks (marketing managers), Donna Nemmers and Gloria Schiesl (project managers), Stuart Paterson, Rick Noel, and John Joran (designers), Lori Hancock (photo research coordinator), and Sandy Ludovissy and Sherry Kane (production supervisors) were also key members of the book teams.

Reviewers of the Sixth Edition

Special thanks and appreciation goes out to reviewers of the sixth edition. Their contributions, constructive suggestions, new ideas, and invaluable advice played an important role in the development of this sixth edition and its supplements. These reviewers include:

Gerald Cecil University of North Carolina
Chad L. Davies Gordon College
Ryan E. Droste Trident Technical College
James Eickmeyer Cuesta College
William A. Hollerman University of Louisiana at Lafayette
Joseph S. Miller UC Santa Cruz
Alice L. Newman California State University Dominquez Hills
Kelly A. Page Harper College

Reviewers of Previous Editions

The following astronomers and physicists reviewed previous editions of the book. Their comments and advice greatly improved the readability, accuracy, and currency of the book.

Robert H. Allen University of Wisconsin, La Crosse
Arthur L. Alt University of Great Falls
Parviz Ansari Seton Hall University
Keith M. Ashman Baker University
Leonard B. Auerbach Temple University
William G. Bagnuolo, Jr. Georgia State University
Gordon Baird University of Mississippi
Thomas J. Balonek Colgate University
Timothy Barker Wheaton College
Nadine G. Barlow Northern Arizona University
Peter A. Becker George Mason University
Ray Benge Tarrant County Junior College, NE Campus
David Bennum University of Nevada–Reno
John Berryman Palm Beach Community College
Suketu Bhavsar University of Kentucky
William E. Blass The University of Tennessee, Knoxville
Luca Bombelli University of Mississippi
Bernard Bopp University of Toledo
James M. Borgwald Lincoln University
Mark Boryta Mount San Antonio College
Richard Bowman Bridgewater College
Michael J. Bozack Auburn University
Elizabeth Bozyan University of Rhode Island
Jane K. Breun Madison Area Technical College
Michael Briley University of Wisconsin–Oshkosh
David H. Bush Eastern Michigan University
Ron Canterna University of Wyoming
Michael Carini Western Kentucky University
Steve Cederbloom Mt. Union College
Stan Celestian Glendale Community College
Joan Centrella Drexel University
Harold C. Connolly, Jr. Kingsborough College—CUNY
Larry Corrado University of Wisconsin Center (Manitowoc)
Anne Cowley Arizona State University
George W. Crawford Southern Methodist University
Mike Crenshaw Georgia State University–Atlanta
Michael Crescimanno Berea College
Charles Curry University of California, Berkeley
Steven Dahlberg Concordia College
Bruce Daniel Pittsburg State University
Norman Derby Bennington College
Don DeYoung Grace College
Andrea K. Dobson Whitman College
James Dull Albertson College of Idaho
Robert A. Egler North Carolina State University
Larry W. Esposito University of Colorado
S. R. Federman University of Toledo
Michael Fisher Ohio Northern University
Barbara A. Gage Prince George's Community College
James M. Gelb University of Texas at Arlington
Edward S. Ginsburg University of Massachusetts, Boston
David Griffiths Oregon State University
William Roy Hall Allegheny County Community College—South Campus
Padmanabh Harihar University of Massachusetts–Lowell
Charles Hartley Hartwick College
Dieter Hartmann Clemson University
Paul Heckert Western Carolina University
Esmail Hejazifar Wilmington College
William Herbst Wesleyan University
Richard Herr University of Delaware
Susan Hoban University of Maryland, Baltimore
William A. Hollerman University of Louisiana at Lafayette
Jeffrey Lynn Hopkins Midlands Technical College

David Hufnagel Johnson County Community College
Phillip James University of Toledo
Ken Janes Boston University
Scott B. Johnson Idaho State University
Adam Johnston Weber State University
Terry Jones University of Minnesota
Ron Kaitchuck Ball State University
Steve Kawaler Iowa State University
Douglas M. Kelly University of Wyoming
Charles Kerton Iowa State University
Eric Kincanon Gonzaga University
Harold P. Klein SETI Institute
William Koch Johnson County Community College
Kurtis J. Koll Cameron University
Robbie F. Kouri Our Lady of the Lake University
Louis Krause Clemson University
David J. Kriegler Creighton University
Christina Lacey University of South Carolina, Columbia
Claud H. Lacy University of Arkansas
Neil Lark University of the Pacific
Ana Marie Larson University of Washington
Stephen P. Lattanzio Orange Coast College
Henry J. Leckenby University of Wisconsin, Oshkosh
Paul D. Lee Louisiana State University
Eugene Levin York College of the City, University of New York
Ira Wayne Lewis Texas Tech University
David Loebbaka University of Tennessee at Martin
Edwin Loh Michigan State University
William R. Luebke Modesto Junior College
Donald G. Luttermoser East Tennessee State University
Michael J. Lysak San Bernardino Valley College
James MacDonald University of Delaware
R. M. MacQueen Rhodes College
Loris Magnani University of Georgia
Bruce Margon University of Washington
Philip Matheson Utah Valley State College
William E. McCorkle West Liberty State College
Marles L. McCurdy Tarrant County Junior College, Northeast Campus
Linda McDonald North Park University
José Mena-Worth University of Nebraska–Kearney
William Millar Grand Rapids Community College
J. Scott Miller University of Louisville
J. Ward Moody Brigham Young University
Mary Anne Moore Saint Joseph's University
Lee Mundy University of Maryland
Andrew M. Munro Kirkwood Community College at Swinburne University of Technology
B. N. Narahari Achar University of Memphis
Gerald Newsom Ohio State University
Melvyn Jay Oremland Pace University
Terry Oswalt Florida Institute of Technology
Ralph L. Patton St. Mary's University
James Pierce Minnesota State University–Mankato
Lawrence Pinsky University of Houston
Judith L. Pipher University of Rochester
Errol Pomerance Castleton State College
C. W. Price Millersville University
Richard Rees Westfield State College
David Reid Eastern Michigan University
David Roberts Bowdoin College
James Roberts University of North Texas
Charles W. Rogers Southwestern Oklahoma State University
Michael Rulison Oglethorpe University
Dwight Russell Baylor University
Barbara Ryden Ohio State University
Alberto C. Sadun University of Colorado at Denver and Health Sciences Center
Ronald G. Samec Bob Jones University
Anuj Sarma Eastern Kentucky University
Gary D. Schmidt University of Arizona
Ann Schmiedekamp Penn State University–Abington
James Schombert University of Oregon
Charles Schweighauser Sangamon State University
Richard Sears University of Michigan
Larry Sessions Metropolitan State College of Denver
Peter Shull Oklahoma State University
Ashley Shultz Fort Lewis College
Caroline Simpson Florida International University
Paul P. Sipiera Harper College
Michael Sitko University of Cincinnati
Tammy Smecker-Hane University of California, Irvine
Alexander G. Smith University of Florida
George F. Spagna Randolf-Macon College
Joseph L. Spradley Wheaton College (IL)
Donald R. Sprowl Louisiana College
John Stanford Georgia Perimeter College
Michael Stewart San Antonio College
Jack W. Sulentic University of Alabama
Michael E. Summers George Mason University
Jeffrey J. Sundquist Palm Beach Community College
Timothy D. Swindle University of Arizona
F. D. Talbert San Diego State University
Paul V. Temple Mohave Community College
Donald Terndrup Ohio State University
Yervant Terzian Cornell University
Charles R. Tolbert University of Virginia
Virginia Trimble University of California, Irvine
Peter Usher Pennsylvania State University
Patricia C. Vener Loyola College, Maryland
James R. Webb Florida International University
John Wernegreen Eastern Kentucky University
Walter Wesley Moorhead State University
Daniel P. Whitmire University of Louisiana at Lafayette
Dan Wilkins University of Nebraska–Omaha
Roger A. Windhorst Arizona State University
Mark Winslow Independence Community College
J. Wayne Wooten Pensacola Junior College

All of the reviewers were very helpful, but I especially want to thank James Borgwald, Steve Cederbloom, Harold Connelly, Eric Kincanon, Paul D. Lee, William Millar, David Roberts, Alexander G. Smith, Michael Stewart, and Timothy D. Swindle for their unusually valuable and thoughtful comments.

Contents in Brief

Contents

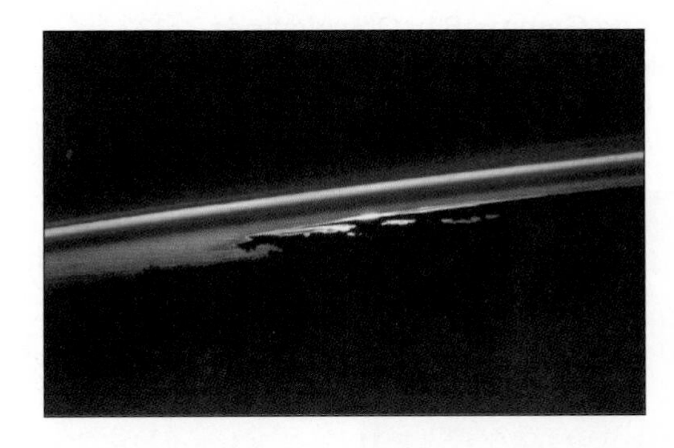

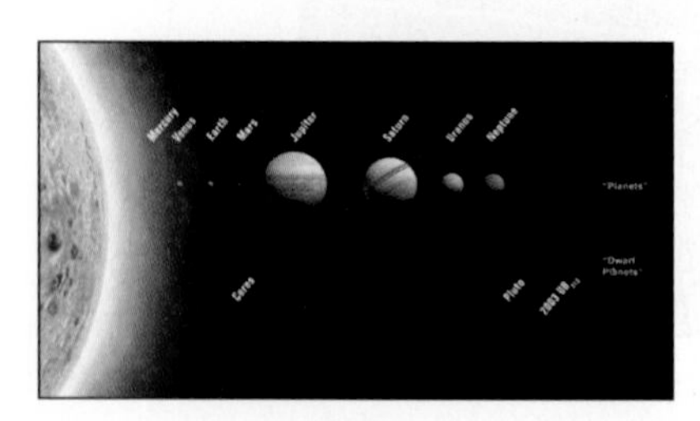

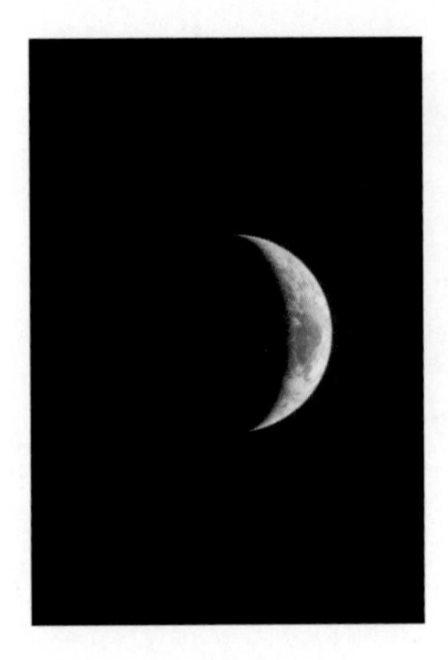

chapter **8**

The Earth, 153

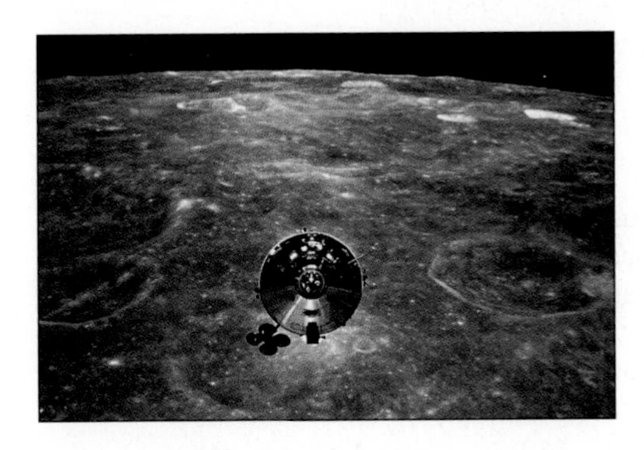

chapter **9**

The Moon, 181

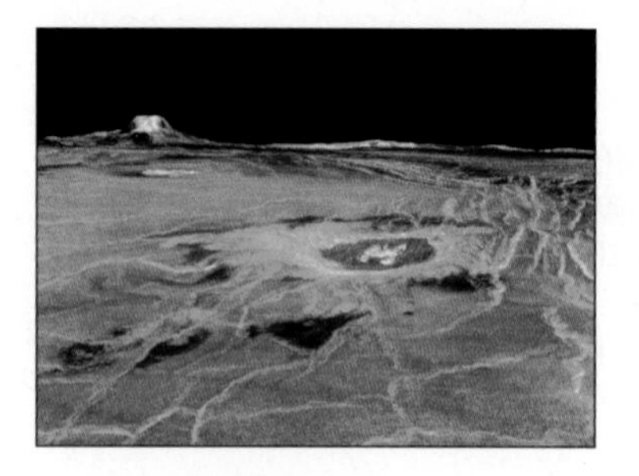

chapter **10**

Mercury and Venus, 207

chapter **11**

Mars, 233

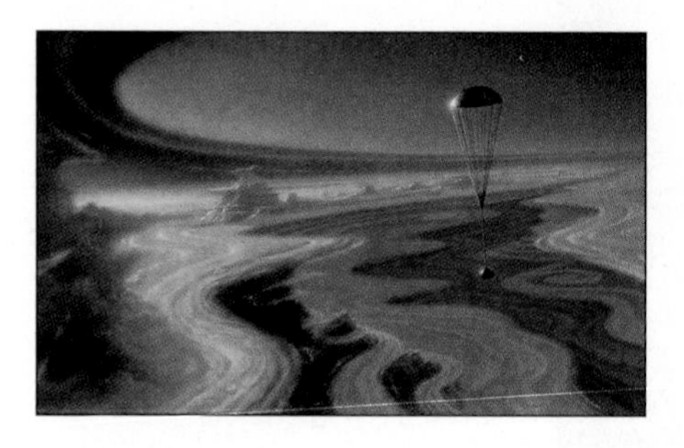

chapter **12**

Jupiter and Saturn, 267

chapter **13**

The Outer Solar System, 293

chapter **14**

Satellites, 315

chapter **15**

Small Solar System Bodies, 345

Part 3 Journey to the Stars

chapter **16**

Basic Properties of Stars, 375

chapter **17**

The Sun, 399

chapter **18**

The Formation of Stars and Planets, 425

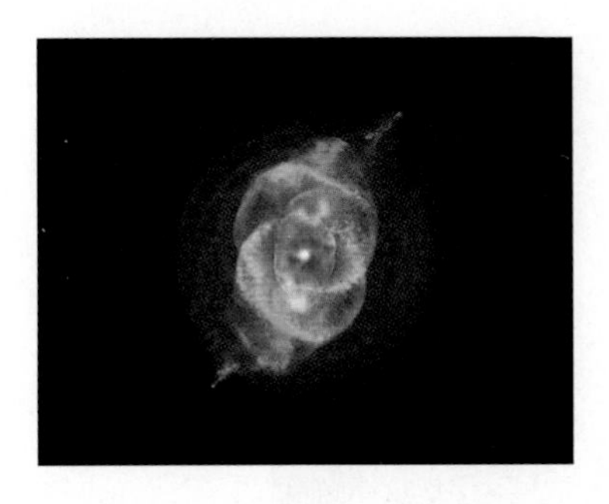

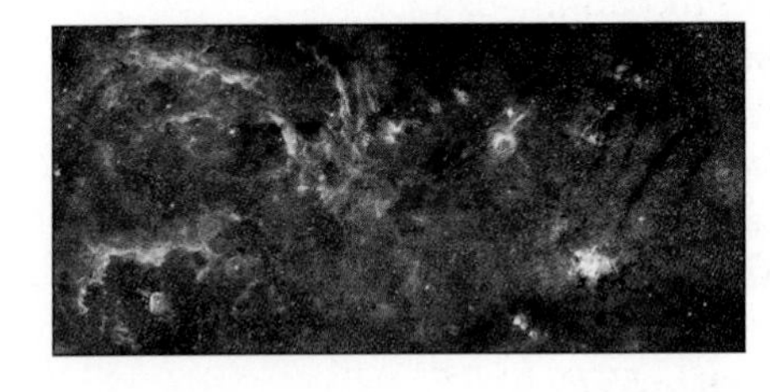

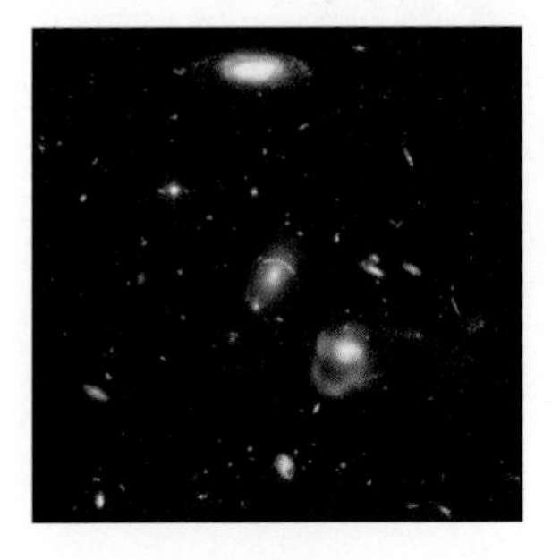

Astronomy

Part 1 The Journey Begins

1

The lights begin to twinkle from

the rocks:

The long day wanes; the slow

moon climbs; the deep

Moans round with many voices.

Come, my friends,

'Tis not too late to seek a newer

world.

Push off, and sitting well in

order smite

The sounding furrows; for my purpose holds

To sail beyond the sunset, and the baths

Of all the western stars, until I die.

Alfred, Lord Tennyson
Ulysses

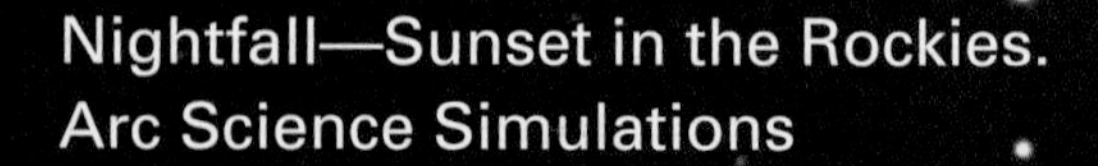

Nightfall—Sunset in the Rockies.
Arc Science Simulations

www.mhhe.com/fix

Journey's Start

The story of astronomy is one of journeys in space and time. Some of the journeys are voyages made by people or the spacecraft they have built. These journeys have expanded the realm of human experience from the Earth to the Moon and then throughout the solar system. Other journeys are trips of the mind. From prehistoric observers noting the phases of the Moon to modern astronomers measuring light from remote galaxies, the distances that our mental journeys take us have grown steadily. Our understanding has developed to the point where we can now calculate the conditions in the center of a star. We can describe the universe as it looked 10 billion years ago and predict what the solar system will be like when the Sun expands to become a giant star.

This book is about these journeys and the knowledge we have gained from them. By retracing the journeys we can discover not only the knowledge gained but also how the scientific process works. Our present knowledge, however, is far from perfect. People of today and tomorrow must undertake still more voyages of discovery.

An important step in the scientific process often involves making observations, so we'll begin with an imaginary trip to an observatory for a night of stargazing. During the night we'll see how the sky looks and how it changes through the night. We will begin to see some of the patterns that can be found in the motions of the celestial bodies. However, other patterns of motion can't be discovered by observing on a single night. Later chapters describe these patterns more fully and tell how to understand them.

Questions to Explore

- How can we describe where something is on the Earth?
- How can we describe where something is in the sky?
- How can we describe how big something looks in the sky?
- How does the appearance of the sky change during the night?

1.1 EVENING

Imagine it is a warm evening in early September. We have just arrived at the Riverside Observatory for a night of stargazing. It is 8 P.M., and the Sun has just gone down. We are sitting on the lawn behind the observatory building. The observatory grounds are surrounded by rolling cornfields, so we have a clear view of the sky in all directions.

As we wait for the sky to darken, we look in the twilight for three **planets** low in the western sky this particular evening. We search without success for Mercury. Mercury is always hard to see, and tonight it isn't high enough in the sky to be visible. Even in twilight, however, we can see two very bright lights in the southwestern sky, shown in Figure 1.1. The brighter light is Venus, the fainter is Jupiter. Venus is so bright it's hard to believe it isn't an airplane or some other artificial light. In fact, when Venus is so bright in the evening sky astronomers often get calls from people who think they've seen a UFO.

As time passes the sky gets darker. At first only the brightest **stars** are visible, but soon we can see hundreds of stars. It is a very clear, still night. The Moon isn't up yet, so the sky is very dark except in the north, where we can see a glow from the lights of Iowa City, about 13 kilometers (8 miles) away. Almost straight overhead is one of the brightest stars in the sky, Vega. Vega, in the **constellation** Lyra, is one of the stars in the summer triangle along with Deneb (in Cygnus) and Altair (in Aquila) (Figure 1.2).

Constellations are regions of the sky containing distinctive patterns of stars. Some of the patterns, such as Cygnus the Swan, remind us of the people, animals, or things for which they are named. For others, such as Aquila the Eagle, some imagination is needed to see what they are supposed to represent. Now that it is very dark we can see a pale band of light, almost ghostlike, that rises from the horizon in the southwest and passes almost overhead before it sinks to the eastern horizon, as shown in Figure 1.3. This is the **Milky Way,** which is best seen in summer or early fall. When we look closely, we can see dark streaks, or rifts, in the fabric of the Milky Way.

We turn toward the north and spot the familiar Big Dipper, which is part of the constellation Ursa Major. The Big Dipper, as shown in Figure 1.4, appears to be suspended by its handle, almost as though it were hanging on a rack of kitchen utensils. We find the pointer stars, which are now the two lowest stars in the bowl of the Big Dipper. Using these stars, we look outward from the bowl and find Polaris, the North Star. As generations of Boy Scouts and Girl Scouts have done before, we can now orient ourselves. Facing Polaris, we are looking due north. South is behind us, east on our right, and west on our left.

FIGURE 1.2
The Summer Triangle
The stars Vega, Altair, and Deneb are nearly overhead in early evening in late summer. The constellations in which they are found are Lyra, Aquila, and Cygnus, respectively.

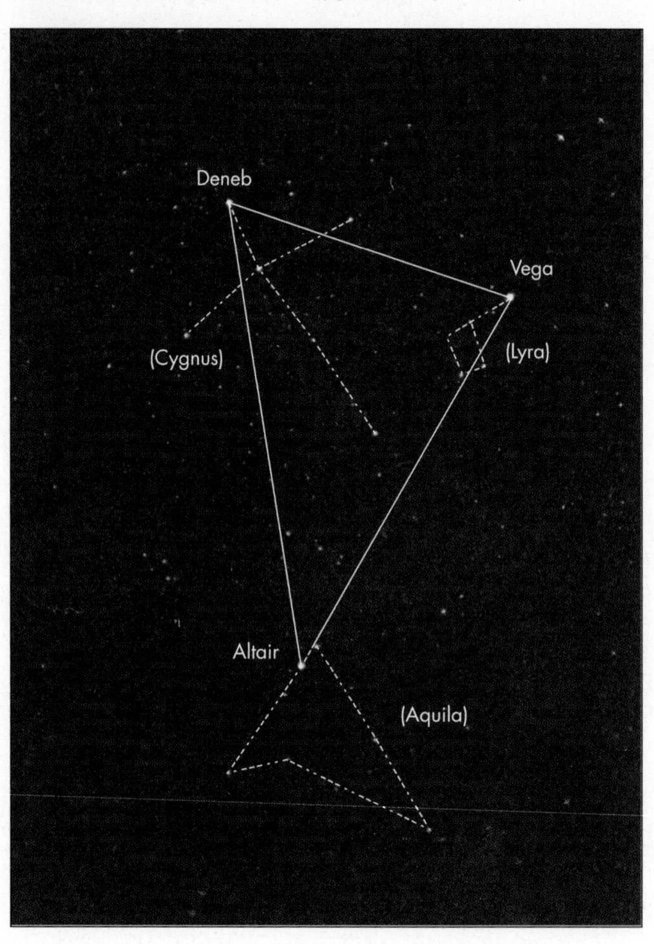

FIGURE 1.1
Venus and Jupiter Near the Western Horizon
Venus is brighter and slightly lower than Jupiter.

FIGURE 1.3
Wide Angle View of the Sky Early in the Evening in Late Summer
The Milky Way stretches across the sky from northeast to southwest.

FIGURE 1.4
Northern Constellations
The Big Dipper, on the left, has two pointer stars that point in the direction of Polaris, the North Star. The glow near the horizon is due to the lights of Iowa City, Iowa.

1.2 THE MOVING SKY

When we look back toward Venus and Jupiter, we notice they are closer to the horizon than they were when we first spotted them. At first it's hard to be sure they are really moving. To tell for sure, we move slightly so that Jupiter is exactly lined up with a telephone wire. As Figure 1.5 shows, it only takes a minute or two to be sure that Venus and Jupiter are sinking toward the western horizon. We also notice that a star just to the right of Jupiter also is sinking westward. The movement we have seen is part of the great westward wheeling of the sky. As the stars in the west sink toward the horizon, stars in the east climb higher.

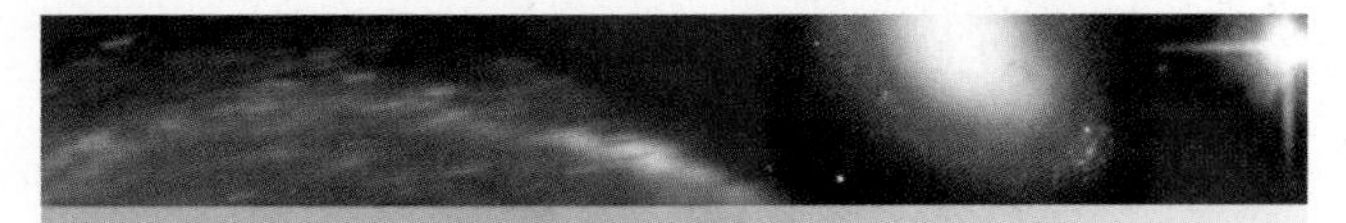

The stars and planets appear to move westward during the night. Their motion is rapid enough that it is possible to detect it in only a minute or so.

We decide to find some constellations. When we try to point some out, however, we have trouble telling each other where to look. Instructions such as "higher," "more to the south," and so on are some help, but what we really need is a more definite way to locate things in the sky. Astronomers use several ways of locating objects in the sky. Before we try to use some of them, however, it will be helpful to review how we locate ourselves on the Earth.

FIGURE 1.5
The Western Motion of the Sky
A, Jupiter, Venus, and a star seen behind telephone wires. **B,** A few minutes later, the planets and the star appear to have moved closer to the western horizon.

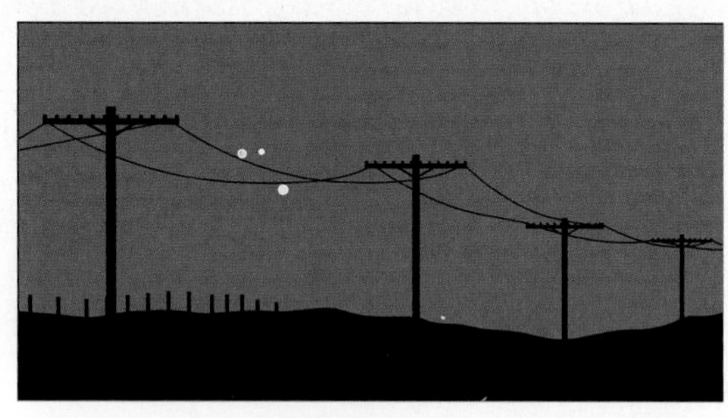

A Looking west

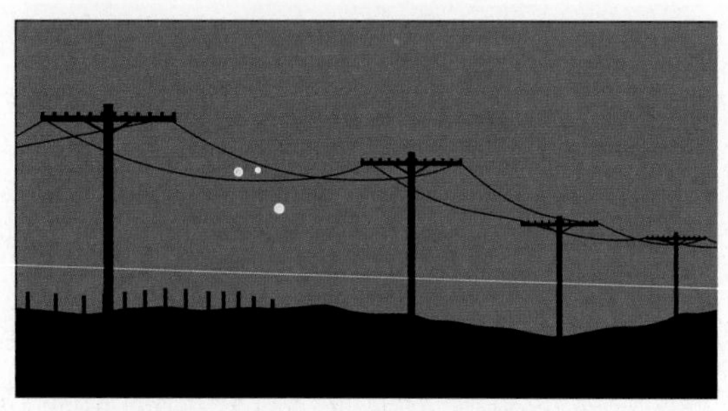

B Looking west a few minutes later

1.3 THE TERRESTRIAL COORDINATE SYSTEM

If I asked you where we are, one way you could answer would be by giving me a detailed, verbal description like the one illustrated in Figure 1.6. You could say that we are standing on the grounds of the Riverside Observatory. The observatory is located about 13 km south of Iowa City, where the University of Iowa is located. Iowa City itself is in eastern Iowa, about 80 km from the Mississippi River. The state of Iowa is in the central United States and almost in the center of the North American continent. North America stretches from just above the equator to nearly the North Pole of the Earth.

Obviously, unless you were talking to a creature from another planet, you wouldn't really have to describe your location so elaborately. However, the usefulness of a verbal description of your location depends on how familiar your listener is with your surroundings. If you try to tell a stranger how to find your home, a verbal description may be almost useless. You might be more successful if you gave the stranger the street address of your home. When we use street addresses we are using a form of **coordinate system. Coordinates** are a set of numbers that can be used to locate something. Thus, 1715 7th Avenue combines a pair of numbers (1715 and 7) that locate a particular house.

Another set of coordinates could be used to locate a particular seat in a classroom by giving the row and

FIGURE 1.6
The Location of the Stargazing Session
We are on the grounds of the Riverside Observatory, which is south of Iowa City, in the state of Iowa, near the middle of the North American continent.

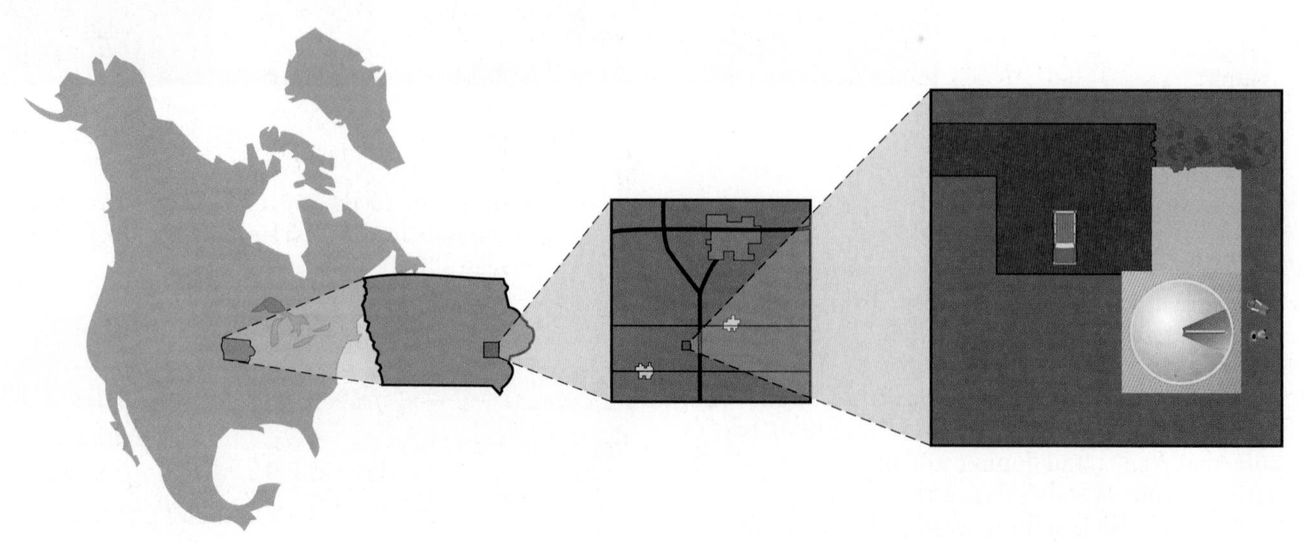

FIGURE 1.7
Longitude and Latitude
Longitude is an angle measured around the equator from the point where the prime meridian intersects the equator. Latitude is an angle measured north or south from the equator.

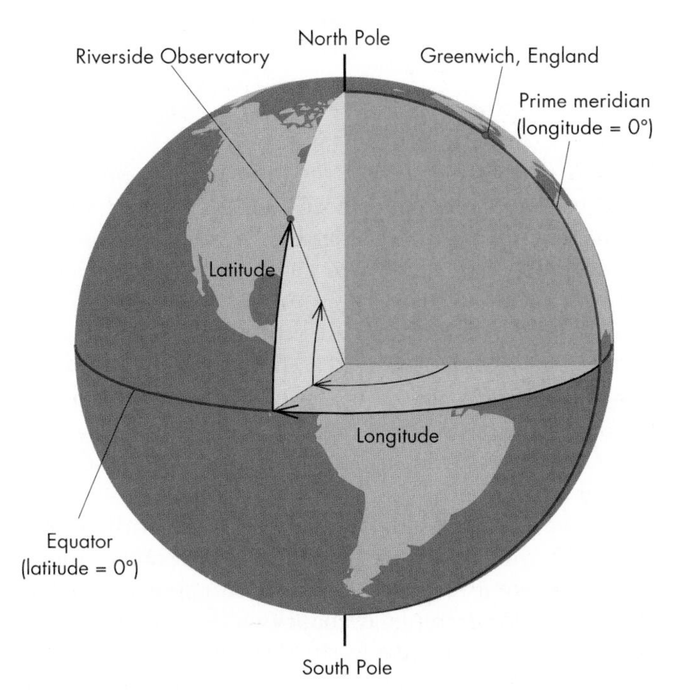

column numbers, such as "third row from the front, fourth seat from the left." For a coordinate system to be useful, everyone must agree about its **zero point** (where you start counting), and in which direction you count (left to right or right to left). If this is understood by everyone, all you would need to say is "4, 7" and everyone would know you were referring to the seventh seat from the left in the fourth row.

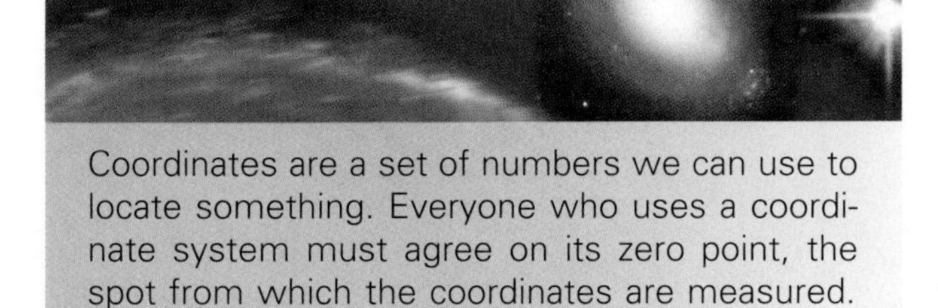

Coordinates are a set of numbers we can use to locate something. Everyone who uses a coordinate system must agree on its zero point, the spot from which the coordinates are measured.

When we make up a coordinate system for locating places on Earth, we also must agree on a zero point and the direction in which we measure. In the most commonly used terrestrial coordinate system, the zero point is a point on the Earth's **equator.** The Earth rotates about its axis, an imaginary rod that passes through its surface at the North and South Poles, as shown in Figure 1.7. The imaginary line that circles the Earth midway between the poles is the equator. The equator is an example of a **great circle,** which is a circle that divides a sphere into two equal parts. The equator divides the Earth into northern and southern hemispheres.

The location of each place on the Earth can be specified uniquely using latitude and longitude, both of which are defined using the Earth's equator. **Latitude** is the angle between the equator and a geographical location. **Longitude** is the angle, east or west, around the equator to the point nearest the location. The zero point for the coordinate system could have been chosen anywhere on the equator. For historical reasons, the zero point is on the equator due south of the former location of the Royal Observatory in Greenwich, England (Figure 1.8). The longitude line passing through Greenwich is called the

FIGURE 1.8
The Prime Meridian
A brass strip set in the ground marks a small portion of the prime meridian, which runs through the building that used to house the Royal Observatory in Greenwich, England. The roof and walls of the building can be opened so that the telescope inside can be used to measure the times that stars pass overhead.

prime meridian or the Greenwich meridian. It is crucial that everyone using this coordinate system agree about the zero point for measuring latitude and longitude.

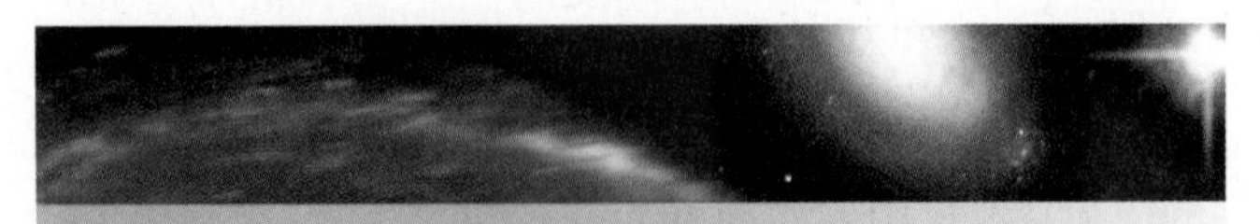

The terrestrial coordinate system is used to locate places on the Earth. The coordinates the system uses are latitude, which describes distance north or south of the equator, and longitude, which describes distance east or west from the zero point. The zero point is located on the equator, directly south of Greenwich, England.

1.4 THE CELESTIAL SPHERE

Using coordinates resembling latitude and longitude to locate stars in the sky is a natural thing to do because the sky seems to stretch above our heads like the inside of an inverted bowl, as shown in Figure 1.9. People in many cultures have referred to the sky as the "bowl of night." Another term often used to describe the appearance of the sky is **celestial sphere.** This imaginary celestial sphere completely surrounds the Earth. We see only the hemisphere above our horizon. Throughout history, the notion that the Earth is surrounded by a sphere to which the stars, Moon, Sun, and planets are attached often has been taken quite literally. Today, although we realize that the distances of the celestial bodies range enormously, we use the fiction of the celestial sphere as a convenient way of describing celestial phenomena as they appear to us.

1.5 ANGLES

To help us locate the stars on the celestial sphere, we make use of angles. The most commonly used system for measuring angles uses **degrees.** A degree is 1⁄360 of a circle. In other words, if you face a particular direction and spin around one full time, you have rotated through 360°. As Figure 1.10 shows, you have a handy angle measuring instrument on the end of your arm. At arm's length, your finger is about 2° across, your fist is about 10° across the knuckles, and your outstretched hand is about 20° across from the tip of your thumb to the tip of your little finger.

As you use your hand to measure some of the things in the sky, you find that the constellations range from a few degrees to a few tens of degrees in angular size. The Big Dipper, for example, is about two fists, or 20° across. If the Sun or Moon were in the sky, you would find that they are smaller, only about half a degree in angular size (Figure 1.11). Often, the angle an astronomer needs to measure is so small that a degree is too big to use. When this is the case, the degree is divided into smaller parts. Sometimes a

FIGURE 1.9
The Bowl of Night
The sky has the appearance of the inside of an inverted bowl over the Earth. The bowl continues beneath the horizon to form the celestial sphere.

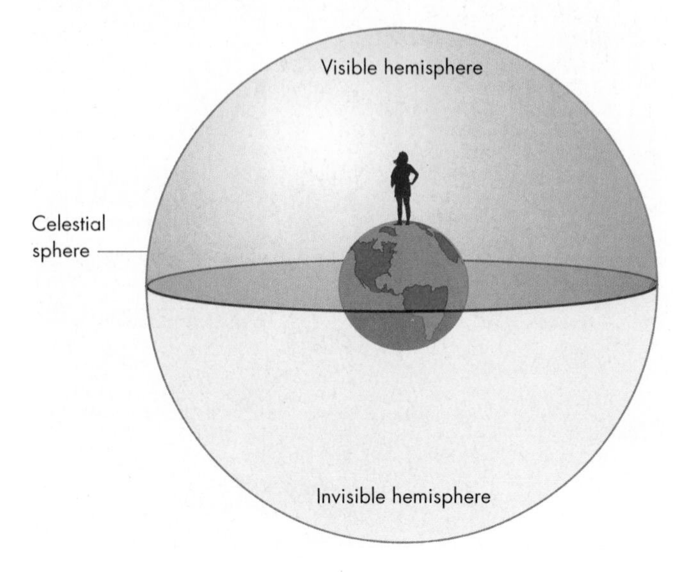

FIGURE 1.10
Measuring Angles
When held at arm's length, the fist is about 10° across. A finger is about 2° across.

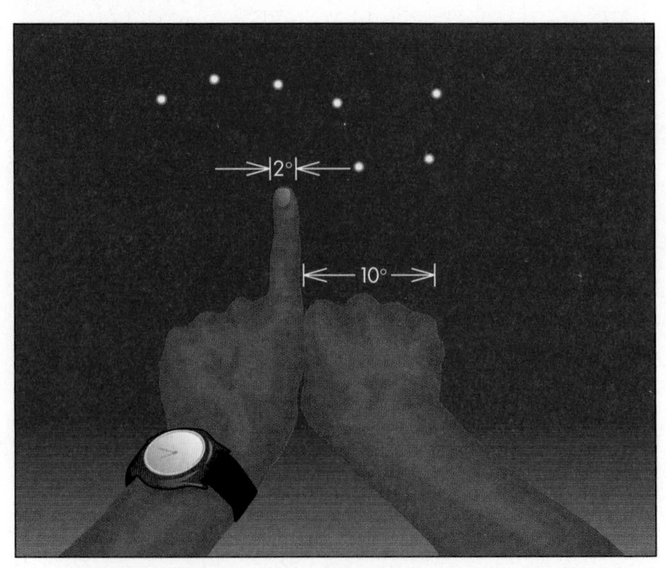

FIGURE 1.11
The Angular Size of the Moon
The angular diameter of the Moon is only about ½°, which is smaller than the width of your finger when you hold it at arm's length. The Sun also has an angular size of ½°.

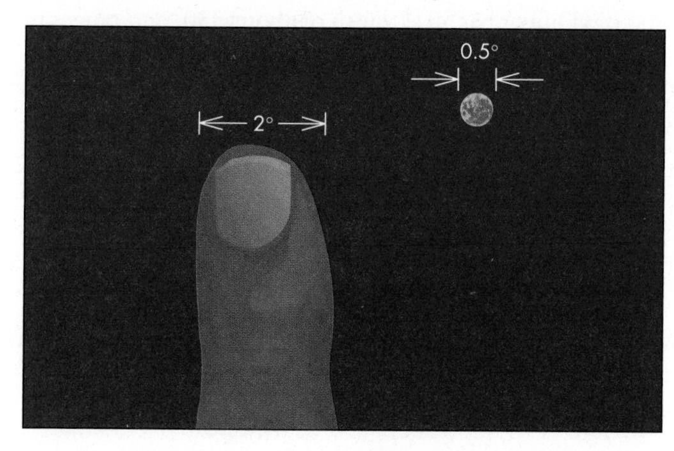

FIGURE 1.12
One Second of Arc
A penny at a distance of 4 km (2.5 miles) has an angular diameter of 1 second of arc.

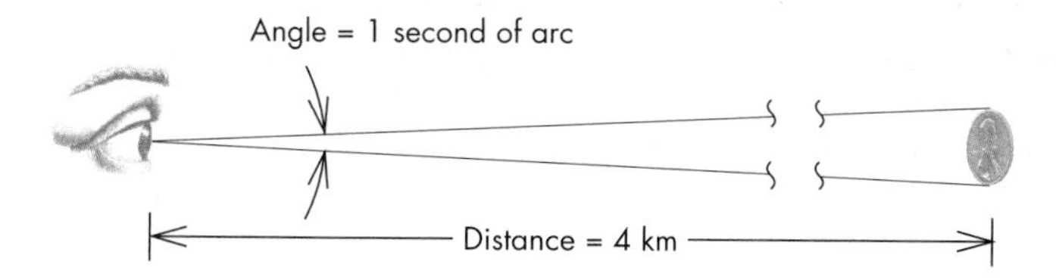

minute of arc, defined as 1/60 of a degree, is used. Some of the planets have angular sizes almost as large as a minute of arc. For even smaller angles it is often convenient to use a **second of arc.** One second of arc is equal to 1/60 of a minute of arc. As Figure 1.12 shows, a second of arc is so small that it equals the angular diameter of a penny at a distance of 4 km (2.5 miles). The angular sizes of stars are all smaller than 1 second of arc.

The sky above us looks like the inside of a bowl and is sometimes called the celestial sphere. To describe locations and sizes of things we see in the sky we use angles. In astronomy, angles are often measured in degrees, minutes, and seconds of arc.

1.6 THE HORIZON SYSTEM

One coordinate system we use to locate things in the sky has three features in common with the latitude and longitude system that we use to locate ourselves on the Earth. First, both systems are based on great circles. For the celestial coordinate system, the great circle bisects the celestial sphere just as the equator bisects the surface of the Earth. Second, in both cases one of the coordinates, such as longitude, is measured along the great circle. The other, such as latitude, is measured either above or below the great circle. Third, the zero point for the coordinates could have been chosen anywhere on the great circle. The choice is completely arbitrary, but everyone who uses the system must agree on it.

This system is called the **horizon system.** The great circle on which it is based is the horizon. Usually, when we speak of the horizon, we mean only the irregular line marking the meeting of the sky and the Earth. Astronomers, however, refer to the celestial horizon, which is defined with respect to the **zenith,** the point on the celestial sphere directly over your head. Figure 1.13 shows that the **celestial horizon** is the great circle located 90° from the zenith. It divides the celestial sphere into an upper (visible) half, and a lower (not visible) half. The only times we are likely to see the celestial horizon are when we are at sea or in the middle of a vast plain.

The two coordinates used in the horizon system are altitude and azimuth, shown in Figure 1.14. **Altitude** is the angular distance above the celestial horizon and

FIGURE 1.13
The Zenith and the Celestial Horizon
The zenith is the point on the celestial sphere directly overhead. All places on the celestial horizon are 90° away from the zenith.

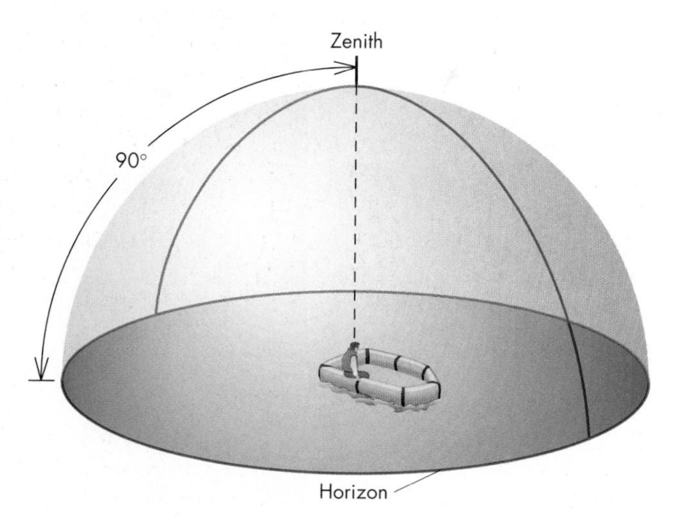

FIGURE 1.14
The Horizon System
Azimuth is the angle measured eastward around the horizon from north. Altitude is the angle measured upward from the horizon. The altitudes and azimuths of Venus, Polaris, and the Big Dipper are shown.

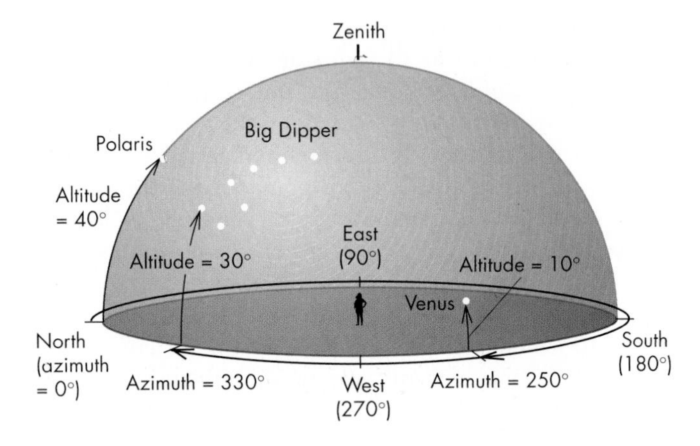

corresponds to latitude in the terrestrial coordinate system. The horizon has an altitude of 0° and the zenith has an altitude of 90°. **Azimuth** is the angular distance measured from north eastward around the celestial horizon to the point directly below the chosen location on the celestial sphere. Azimuth corresponds to longitude in the terrestrial coordinate system. The azimuth of the east point on the celestial horizon is 90°. That of the south point is 180°, and the west point 270°.

As you try out the horizon system, you find that Venus, low in the southwestern sky, has an azimuth of about 250° and an altitude of only about 10°. In the north, the Big Dipper has an azimuth of about 330° and ranges in altitude from about 30° to 50°. Polaris, straight north, has an azimuth of 0° and an altitude of about 40°, as shown in Figure 1.15. The main advantage of the horizon system is its simplicity. As we take turns pointing out the stars and constellations, we can use altitude and azimuth to tell each other exactly where to look.

The horizon system is a convenient coordinate system to use to locate stars and other celestial objects. In the horizon system, altitude is angular distance from the celestial horizon. Azimuth locates the place on the celestial horizon just below the star.

The main disadvantage of the horizon system is that it doesn't work when two people aren't together. If you left me a note telling the coordinates of a particular star and I didn't read the note until sometime later, the star would be in a different position. The altitude and azimuth in the note would no longer be correct. If you telephoned me from another state or country to describe an

FIGURE 1.15
The Altitudes of Polaris and the Big Dipper

FIGURE 1.16
Orion Rising in the East in the Early Morning in Late Summer

exciting observation, the coordinates again would be useless. Your zenith moves among the stars as you travel and as time passes. This means that the altitude and azimuth of a star depend on the time and the location where the observation was made.

Even though the day was warm, the night is growing cool, so we decide to go inside the observatory for awhile. We can come back outside later to see what new objects the apparent motion of the celestial sphere has brought into view.

1.7 CIRCUMPOLAR MOTION

When we finally get back outside it is well after midnight. Jupiter and Venus have set hours ago. Vega, which was overhead when we went inside, is now about two-thirds of the way to the western horizon. We turn and see another familiar constellation, Orion, climbing over the horizon directly in the east (Figure 1.16). Right above Orion, at an altitude of about 35°, are the Pleiades, a small cluster of stars otherwise known as the seven sisters. Many other stars and constellations have risen while we were inside. A very red object is just rising in the northeast. This is Mars. Although Mars is bright, several stars are brighter still.

We turn north and quickly find Polaris, still 40° above the horizon and directly north. The Big Dipper, however, is no longer to the left of Polaris. Now, as shown in Figure 1.17, the Big Dipper is directly below

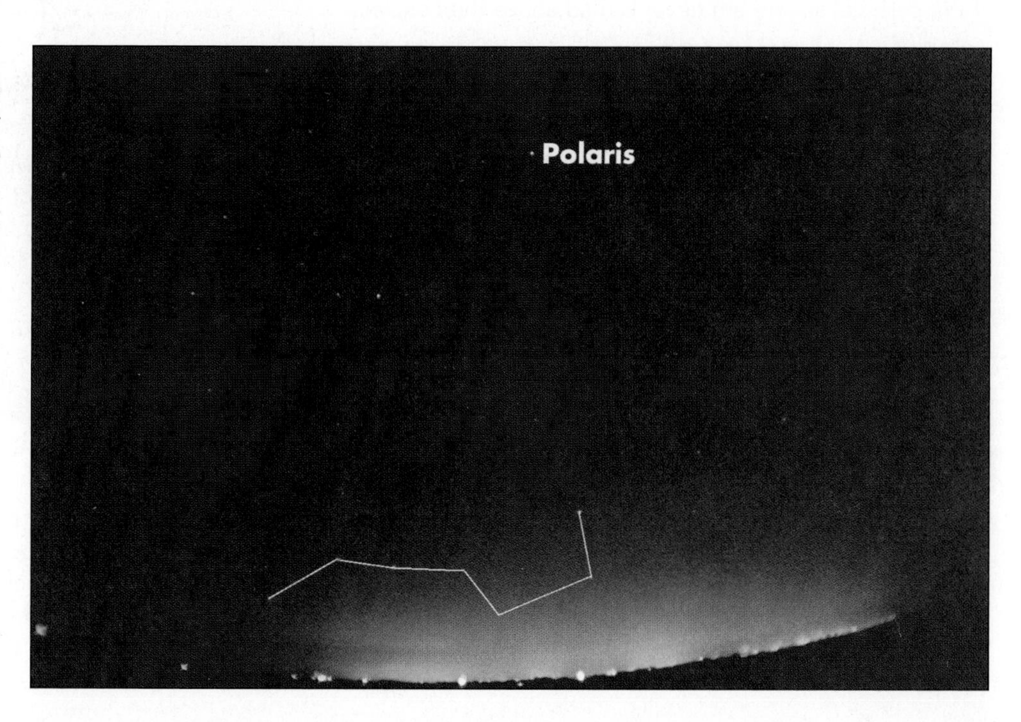

FIGURE 1.17
The Big Dipper and Polaris
At a latitude of 40°, the Big Dipper doesn't set but swings close to the horizon and then rises higher. In the southern United States, the Big Dipper rises and sets.

FIGURE 1.18
A Time Exposure of the Northern Sky
The circumpolar stars never set but move on circles centered near Polaris.

Polaris and is lying along the northern horizon. It is clear that the Big Dipper isn't going to set at all. It is as low in the sky as it can get and is beginning to climb again. Its motion is counterclockwise on a circle that is centered near Polaris. We can see that any star or constellation closer to Polaris than the Big Dipper also moves in a circular path around Polaris. The apparent motion of these **circumpolar stars,** which never set, is shown in Figure 1.18. Stars farther from Polaris also move on circles, but their circles cross the horizon so these stars rise and set each day (Figure 1.19). The size of the circumpolar zone and which stars are circumpolar depend on latitude. In northern Canada, most of the sky is circumpolar. In the southern United States, the circumpolar zone is so small that the Big Dipper rises and sets.

FIGURE 1.19
The Apparent Rotation of the Celestial Sphere
Beyond the circumpolar region, the motions of stars carry them below the horizon. Stars in the extreme southern part of the celestial sphere are never carried above the horizon.

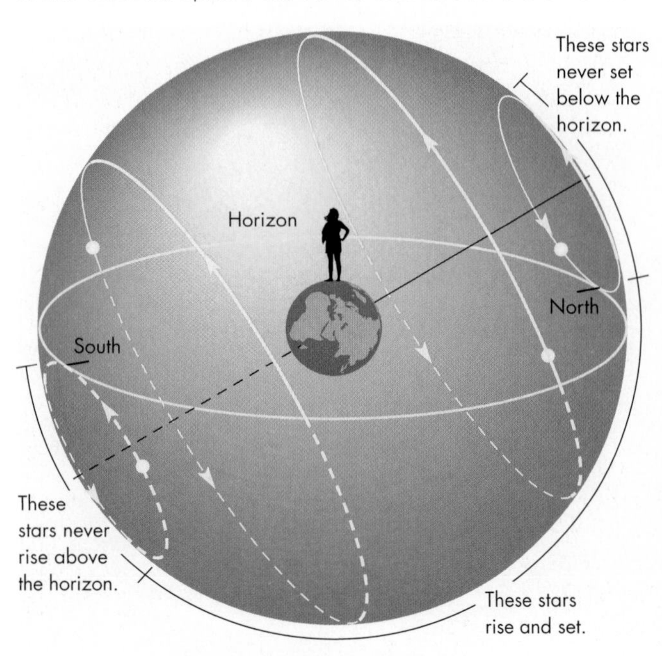
These stars never set below the horizon.
Horizon
North
South
These stars never rise above the horizon.
These stars rise and set.

Most stars rise and set. However, stars in the region centered near Polaris move counterclockwise on circles that never carry them below the horizon. This region is called the circumpolar zone. Stars outside the circumpolar zone move on circles that carry them below the horizon.

We get out a small telescope and some binoculars and begin to look at some galaxies, nebulae, and clusters of stars. Time passes quickly. After a while, we notice a faint glow on the eastern horizon. A few minutes later, the Moon begins to rise (Figure 1.20). We see that it is a beautiful crescent. Because the Moon was full about 10 days ago, it now is waning, its bright portion growing smaller as it shrinks toward new Moon.

As we observe the many objects in the sky, the hours pass. Eventually we notice that the sky is growing brighter. It gets harder and harder to see the fainter stars. Soon only the Moon can be seen. The sky continues to brighten until, at 6:30 A.M., the Sun rises. We pack away the telescope and head for the car to return to town.

FIGURE 1.20
The Waning Crescent Moon Rising in the East
The horns of the Moon point away from the Sun, which has not yet risen.

Chapter Summary

- The stars, planets, and Moon appear to move westward during the night. This westward motion can be detected in about a minute. (Section 1.2)
- Coordinates are a set of numbers that uniquely locate a particular place. Everyone who uses a coordinate system must agree on its zero point and the way the measurements are made. (1.3)
- We can locate ourselves on the Earth using the terrestrial coordinate system. Latitude describes angular distance north or south of the equator and longitude describes angular distance east or west from the prime meridian. (1.3)
- Positions and sizes of stars and other celestial objects in the sky are conveniently described using angles. Angles are often measured in degrees, minutes, and seconds of arc. (1.5)
- A convenient coordinate system for locating stars and other celestial objects is the horizon system. Altitude is the angular distance of the star above the celestial horizon. Azimuth locates the point on the horizon below the star. (1.6)
- Most stars rise and set, but stars in the circumpolar zone centered near Polaris always remain above the horizon. These stars appear to wheel counterclockwise about Polaris. Stars not located in the circumpolar zone move on circles that carry them below the horizon. (1.7)

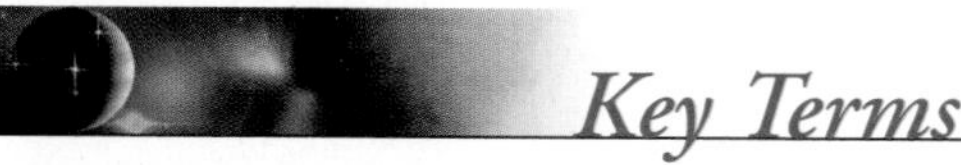

Key Terms

altitude *9*
azimuth *10*
celestial horizon *9*
celestial sphere *8*
circumpolar stars *12*
constellation *4*
coordinates *6*
coordinate system *6*
degree *8*
equator *7*
great circle *7*
horizon system *9*
latitude *7*
longitude *7*
Milky Way *4*
minute of arc *9*
planet *4*
prime meridian *8*
second of arc *9*
star *4*
zenith *9*
zero point *7*

Conceptual Questions

1. Why is it more difficult to notice the westward motions of stars that are high in the sky than those that are near the horizon?
2. Describe a coordinate system that could be used to locate uniquely the seats in the classrooms of a multistory building with one classroom on each floor. How many coordinates are needed in this case?
3. What is the latitude of the North Pole of the Earth? Why is it impossible to give the longitude of the Earth's North Pole?
4. What range of azimuth would include all the locations on the horizon at which stars can rise? Would this question make sense to an observer at the North Pole or South Pole?

Problems

1. A cluster of stars has an angular size of 1.4°. What is the size of the cluster in minutes of arc? In seconds of arc?
2. About how many hands held at arm's length would be needed to measure completely around the horizon?

Figure-Based Questions

1. Using Figure 1.7, estimate the longitude and latitude of the Riverside Observatory. Give your answers in degrees.
2. Using Figure 1.14, find the altitude and azimuth of a star located halfway between the zenith and the horizon in the northwest.
3. Using Figure 1.14, find the altitude and azimuth of a star rising in the southeast.

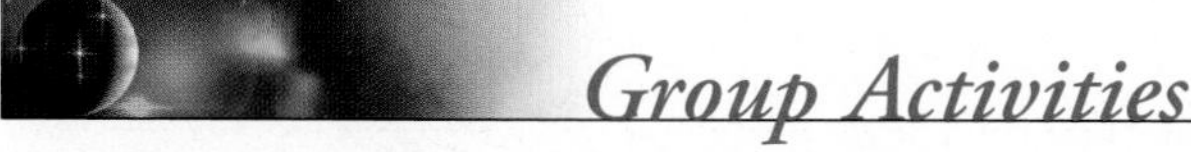

Group Activities

1. Working as a group, design a coordinate system that could uniquely locate the seats in a basketball arena. Discuss different ways to choose the zero point of the coordinate system.
2. Have one member of your group pick a star in the sky and state the altitude and azimuth of that star. See if the other members of the group can locate the chosen star in the sky. Continue with different group members picking altitudes and azimuths until everyone in the group has mastered the horizon system.

For More Information

Visit the text website at **www.mhhe.com/fix** for chapter quizzes, interactive learning exercises, and other study tools.

2

I pray the gods to quit me
of my toils,
To close the watch I keep
this livelong year;
For as a watch-dog lying,
not at rest,
Propped on one arm, upon
the palace roof
Of Atreus' race, too long,
too well I know
The starry conclave of the
midnight sky,
Too well, the splendours of the firmament,
The lords of light, whose kingly aspect
shows—
What time they set or climb
the sky in turn—
The year's divisions, bringing frost or fire.

Aeschylus
Agamemnon

Sunrise from Earth orbit by the crew of the STS-47
Space Shuttle Mission.

www.mhhe.com/fix

Patterns in the Sky

The world has always been an uncertain place. Within the fabric of unpredictable events, however, are reliable, reassuring natural patterns. And of all the aspects of nature, the most predictable and reliable are the patterns of the celestial bodies—Sun, Moon, stars, and planets. People have followed these patterns for much longer than the span of recorded history. The Moon's phases are among the earliest recorded events. The world is dotted with prehistoric observatories used to follow the annual changes in sunrise and sunset.

Today, people are probably less aware of the celestial patterns than they have been for thousands of years. The distractions of modern life conspire to keep us indoors at night. Most of us live in urban areas, where outdoor lighting and smog have led to poorer and poorer observing conditions. Very few of us have had the opportunity or interest to take regular notice of the night sky. The patterns that were so apparent to people of antiquity are now known only to a relative few.

Yet the importance of these patterns to the development of astronomy can hardly be overemphasized. After people first noticed the patterns, they began to seek their causes. Proposing explanations for observed patterns has always been a key step in the progress of science. From the beginning of recorded history, the central problem of astronomy was the explanation of the motions of the Sun, Moon, stars, and planets. The explanations that resulted eventually led, in the sixteenth century, to a radically different view of the world. The realization of the true nature of the Earth and its relation to the other planets and stars had profound spiritual and philosophical consequences.

Questions to Explore

- How do the apparent paths of stars depend on where we are on the Earth?
- How does the Sun's apparent motion differ from that of the stars?
- How do we use the Sun's motion to keep track of time?
- How are the phases of the Moon related to the apparent motion of the Moon?
- How are the apparent motions of the planets different from those of the Sun and the Moon?

Later chapters of this book deal with the evolution of modern explanations of the patterns of celestial motion. This chapter concentrates on describing the motions themselves. The motions of the stars at night have been described briefly in Chapter 1. In this chapter our focus is on a fuller description of the motions of the stars as well as the patterns of motions of the Sun, Moon, and planets. The major emphasis is on the observations that must be made to detect the motions and to deduce their patterns.

2.1 THE DAILY MOTION OF THE SKY

As the observations of Chapter 1 showed, the **diurnal,** or daily, motion of the stars is westward and can be detected in only about a minute. The rate at which a star appears to move across the sky can be as large as 1° every 4 minutes. The path that a star appears to follow is called its **diurnal circle.** Usually, you can see only part of the diurnal circle of a star because the rest is below the horizon. However, stars near Polaris lie in the **north circumpolar region,** in which the entire diurnal circle lies above the horizon. Figure 2.1 is a time exposure taken while the camera pointed at Polaris. The picture shows that the diurnal circles are all centered at a point on the celestial sphere near Polaris.

FIGURE 2.2
The Earth and the Celestial Sphere
The Earth's rotation axis, extended outward, intersects the celestial sphere at the north and south celestial poles. The Earth's equator, extended outward, intersects the celestial sphere along the celestial equator.

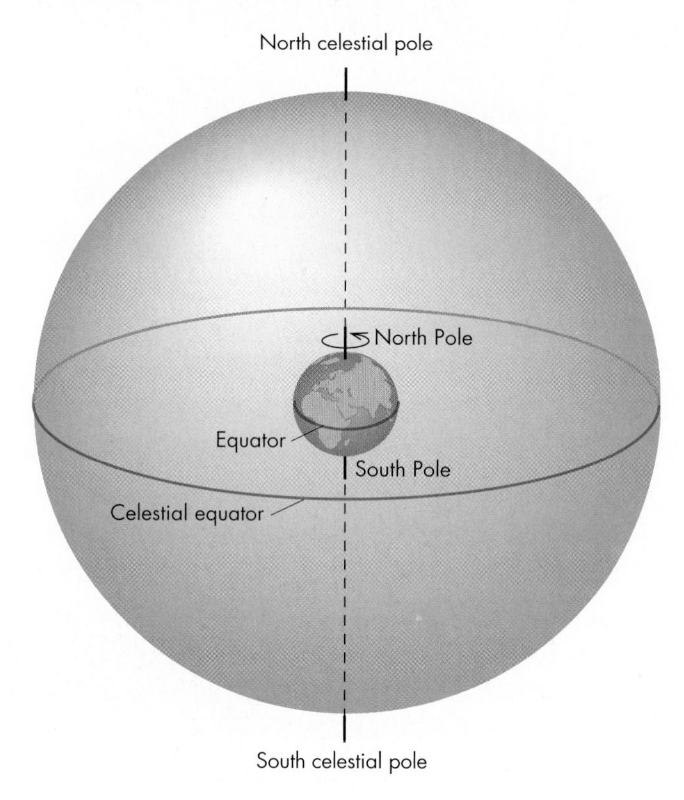

FIGURE 2.1
A Time Exposure of the Northern Part of the Sky
The tracks of the stars are partial circles centered near the North Star, Polaris. The dome in the foreground houses the Canada-France-Hawaii 3.5-meter telescope on the summit of Mauna Kea, Hawaii. The yellow lines are the trails of motor vehicle lights during the time exposure. Note how low Polaris is when viewed from Mauna Kea.

That point, about ¾° from Polaris, is called the **north celestial pole.** The north celestial pole is the point on the celestial sphere directly above the Earth's North Pole (Figure 2.2). In other words, if the Earth's rotation axis were extended into space, it would meet the celestial sphere at the north celestial pole. Extended above the Earth's South Pole, it would meet the celestial sphere at the **south celestial pole.** If there were a bright star directly at the north celestial pole, its diurnal circle would be a point. The star would remain in exactly the same place day and night. When we face north, the motion of the stars on their diurnal circles carries them counterclockwise around the north celestial pole (Figure 2.3). When we face south, the diurnal motion is clockwise. In either case the motions of the stars seem to be caused by the rotation of the celestial sphere about an axis that passes through the celestial poles.

Two important circles on the celestial sphere are the **meridian** and the **celestial equator.** As Figure 2.4 shows, the meridian passes through both celestial poles and your zenith. The half of the meridian that is above the horizon is a semicircle that divides the sky into eastern and western

FIGURE 2.3
Diurnal Circles
Whether we face north or south, the stars appear to move westward on their diurnal paths.

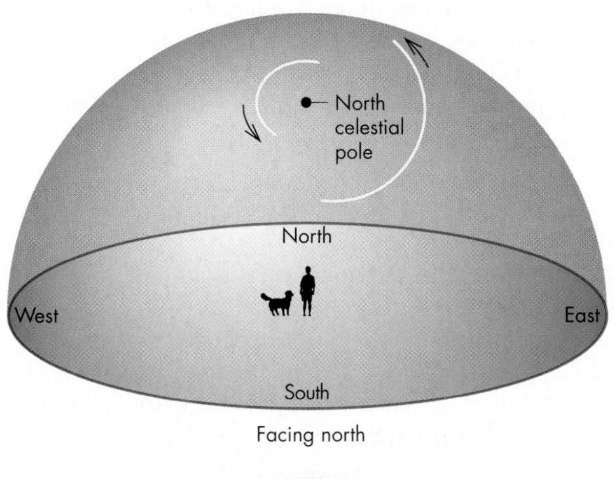

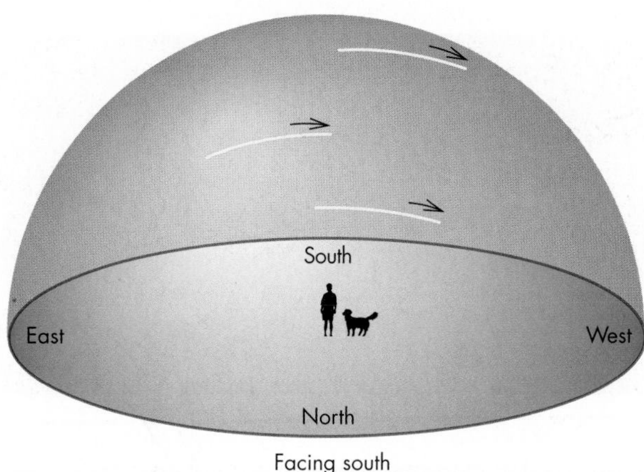

FIGURE 2.4
The Meridian and the Celestial Equator
The meridian is the circle on the celestial sphere that passes from the south celestial pole to the north celestial pole and through the observer's zenith. The part of the meridian that is below the horizon is shown as a dashed line. The north and south points on the observer's horizon lie where the meridian crosses the horizon. The celestial equator is the circle midway between the north and south celestial poles. It divides the celestial sphere into northern and southern halves.

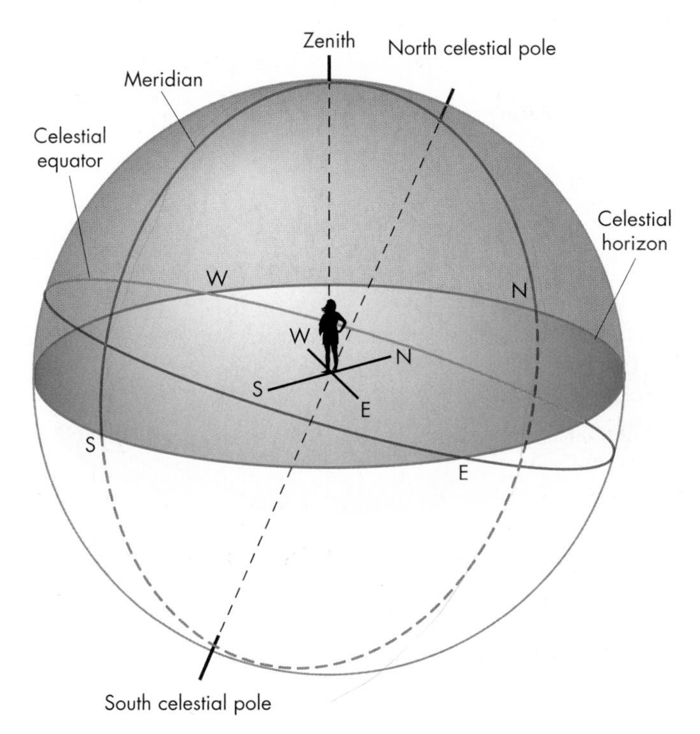

halves. As time passes, the stars move with respect to your meridian. A star rises steadily higher in the eastern sky until it crosses the meridian, then it begins to sink toward the western horizon.

The apparent daily motion of a star is westward on a circle centered on the north celestial pole. The motion appears counterclockwise if we face north and clockwise if we face south. The pattern of diurnal motions makes it look as if the celestial sphere is rotating on an axis passing through the celestial poles.

The celestial equator also is shown in Figure 2.4. Every point on the celestial equator is 90° from both celestial poles. Just as the Earth's equator divides the Earth's surface into northern and southern hemispheres, the celestial equator divides the sky into northern and southern halves. A star that lies on the celestial equator rises due east and sets due west.

Like the horizon, the celestial equator is used as the basis of a coordinate system, the **equatorial system.** The equatorial system is very similar to the terrestrial coordinate system, which is based on the Earth's equator. The two coordinates that give the position of a star in the equatorial system are shown in Figure 2.5. One is **declination,** a north-south coordinate equal to the angular distance of a star from the celestial equator. The other is **right ascension,** the angular distance measured eastward along the celestial equator from the **vernal equinox** to the point on the celestial equator nearest the star's position. The vernal equinox is the position of the Sun on the celestial sphere on the first day of spring. The vernal equinox moves with the stars as the celestial sphere rotates, so the right ascension and declination of a star remain the same throughout a night, just as the longitude and latitude of a city remain

FIGURE 2.5
The Equatorial Coordinate System
Declination is the angular distance of a star north or south of the celestial equator. Right ascension is measured from the vernal equinox eastward around the celestial equator to the point on the equator nearest the star. Local hour angle is measured westward around the celestial equator from the meridian to the point nearest the star. During a night, the declination and right ascension of a star remain constant while its hour angle changes as the star appears to move with respect to the meridian.

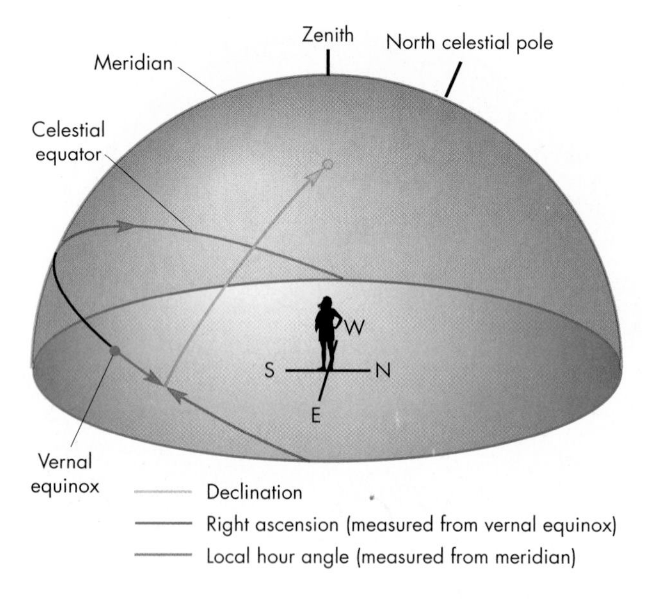

the same as the Earth rotates. Declination is measured in degrees, minutes, and seconds of arc while right ascension is measured in hours, minutes, and seconds. One hour of right ascension corresponds to 15°.

Because the equatorial coordinates of a star don't depend on the time or place where observations are made, they can be used to arrange the stars in order in catalogues. Catalogues are used by astronomers to find a particular star to observe. The fact that right ascension and declination don't vary with time or place is an advantage in this case. It means, however, that these coordinates don't directly tell you where to look to find a star. To actually locate a star in the sky, astronomers must find the star's **local hour angle.** This is the angular distance westward around the celestial equator from the meridian to the point on the equator nearest the star. The name "hour angle" suggests that there is a relationship between that coordinate and time. In fact, there is. The local hour angle of a star is zero when the star crosses the meridian and increases steadily until it next crosses the meridian. Local hour angle tells how long it has been since the star last crossed the meridian. The relationship between local hour angle and time is explored further later in this chapter.

To use right ascension and declination to find a star, an astronomer needs to know where on the celestial sphere the vernal equinox is located. To do this, special clocks, **sidereal clocks,** are used to keep track of the local hour angle of the vernal equinox. A sidereal clock reads 0^h (zero hours) when the vernal equinox crosses the meridian. The sidereal clock reaches 24^h when the vernal equinox returns to the meridian. Once the vernal equinox is located, the

FIGURE 2.6
Latitude and the Altitude of the North Celestial Pole
The altitude of the north celestial pole is the same angle as the observer's latitude. The more northerly the observer, the higher the north celestial pole is in the sky. The Little Dipper, in which Polaris is located, is also shown.

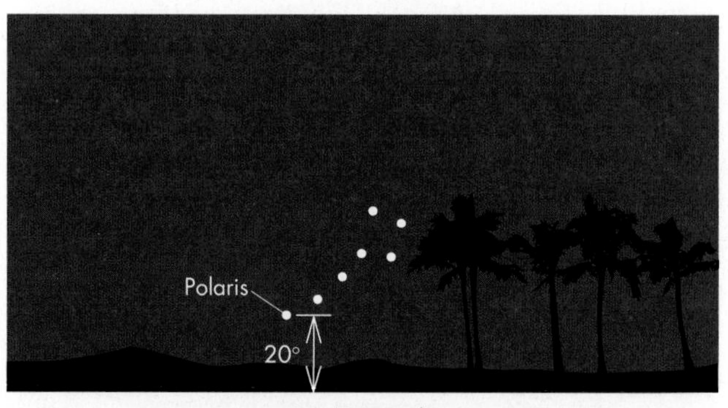

20° N Latitude

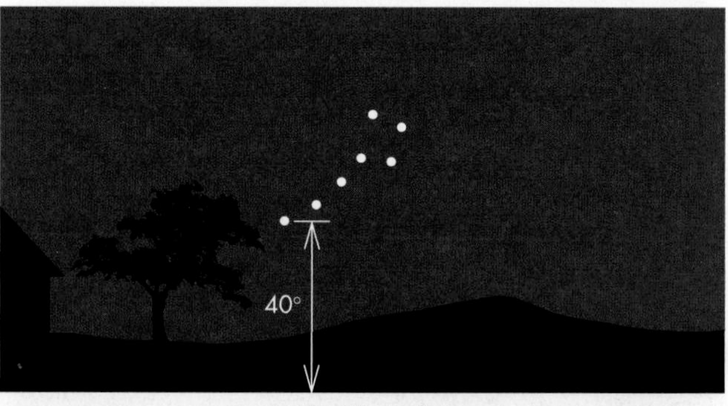

40° N Latitude

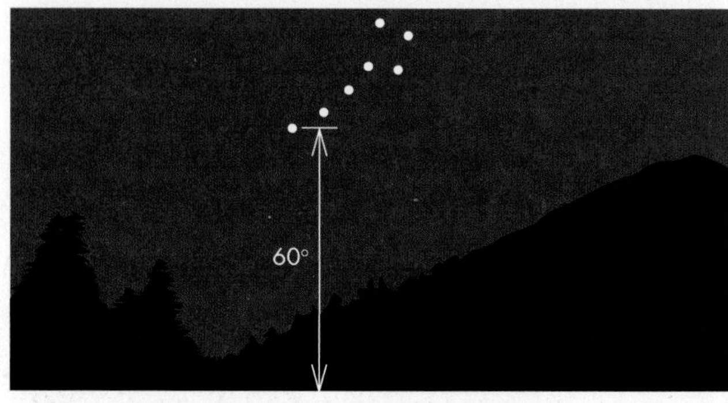

60° N Latitude

right ascension and declination of a star can be used to find the star. The relationship between right ascension and hour angle is shown in Figure 2.5.

Right ascension (or hour angle) and declination are used to locate stars in the equatorial system, which resembles the terrestrial system of longitude and latitude.

So far, the night sky has been described as it would appear to someone at the Riverside Observatory. Actually, the sky would look about the same for other stargazers in most of the United States. For those observers, only a relatively small part of the sky is contained in the north circumpolar cap, and the north celestial pole is about one-third of the way to halfway from the horizon to the zenith.

However, the appearance of the sky and the motions of the stars depend on the latitude where you are observing. The north celestial pole has an altitude equal to the observer's latitude (Figure 2.6). This means that you can use Polaris not only to find which direction is north but also to find your latitude. If you travel northward to higher latitudes, the north celestial pole climbs higher in the sky. As this happens, the circumpolar region grows in size. On reaching the terrestrial pole, you would see the north celestial pole directly overhead. Both your latitude and the altitude of the north celestial pole would be 90°. The celestial equator would lie along the horizon. Only the northern half of the celestial sphere could ever be seen, but that half would always be above the horizon. The diurnal circles of the stars, as shown in Figure 2.7, would be parallel to the horizon.

By traveling south toward the equator, you would see different changes in the sky. Polaris would sink behind you and the circumpolar region would grow smaller. At the terrestrial equator, the north celestial pole would lie on the horizon. A panoramic view of star trails at the equator is shown in Figure 2.8. Stars near the celestial

FIGURE 2.7
Observing at the North Pole
The north celestial pole is at the zenith; the celestial equator is on the horizon. The diurnal circles of stars are parallel to the horizon.

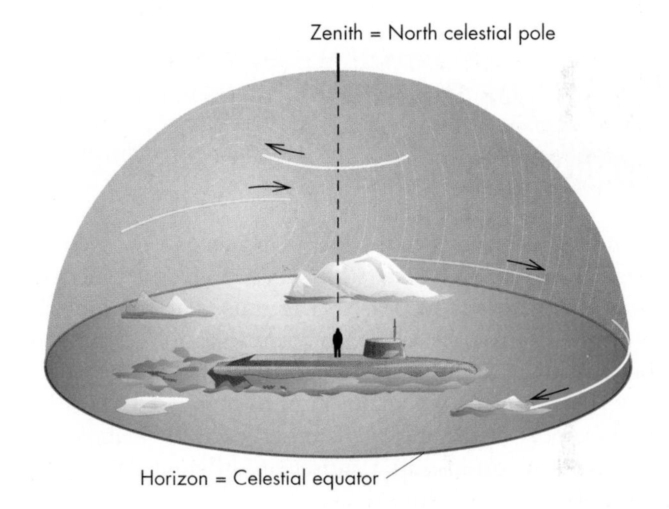

FIGURE 2.8
Star Trails at the Equator
This panoramic view shows star trails as seen from Kenya. South is at the left, west at the center, and north at the right. Star trails near the celestial poles are semicircles whereas those near the celestial equator set straight down in the west.

equator rise straight up from the horizon, pass nearly overhead, and set straight down in the west. The south celestial pole would be on the southern horizon. Half of every diurnal circle would be above the horizon and each star would spend half its time above the horizon. At the Earth's equator there is no circumpolar region. Every star in both the northern and southern hemisphere can be seen half the time. If you continued south, the north celestial pole would sink below the horizon and the south celestial pole would rise, bringing with it the south circumpolar region. Unfortunately for navigators in the southern hemisphere, there is no southern counterpart of Polaris to make it easy to find which way is south and to determine latitude. The changes in the sky as you continued south would be like those of your northern trip except that the stars and constellations would probably be unfamiliar to those of us who have spent most of our lives in the northern hemisphere.

2.2 THE APPARENT MOTION OF THE SUN

Different constellations at different times of the year

Although the Sun, Moon, and planets share the westerly diurnal motion of the stars, they have individual motions of their own with respect to the stars. For the Sun the motion with respect to the stars isn't usually very obvious because there is no starry background visible during the day. One way to see that the Sun changes its position with respect to the stars is to observe the stars above the western horizon just after sunset on different days of the year. (Observing before dawn above the eastern horizon will also work, but the time of day is less popular.)

Imagine facing west just after sunset in early September when the sky first gets dark enough to see the stars, just as in the imaginary trip to Riverside Observatory in Chapter 1. The lowest constellation you could see above the place on the horizon where the Sun had just set is Virgo, as shown in Figure 2.9. Above Virgo, and slightly to the south, would be Libra. Now repeat the observation a month later, in October. Virgo isn't visible. Instead Libra is just above the horizon. Above Libra and slightly to the south is Scorpius. In November, Scorpius is the lowest visible constellation above the place where the Sun has set. Above Scorpius in the sky are Sagittarius and Capricornus.

Throughout the fall, the constellation just east of the Sun, the one we can see just after the Sun sets, changes at the rate of one constellation per month, as shown in Figure 2.10. If we regard the stars' positions on the celestial sphere as fixed, then it appears that the Sun moves steadily eastward among the constellations. Between September and October the Sun moves into Virgo,

FIGURE 2.9
The Constellations as They Appear Above the Western Horizon Just After Sunset in Early September

then by November it moves into Libra. By December it moves eastward into Scorpius, making Sagittarius the constellation low in the southwestern sky after sunset. During a year, the Sun appears to move through the **zodiacal constellations:** Virgo, Libra, Scorpius, Sagittarius, Capricornus, Aquarius, Pisces, Aries, Taurus, Gemini, Cancer, and Leo. The length of time it takes for the Sun to move through the constellations and return to the same spot on the celestial sphere is defined as the **year.** Because there are 360° in a circle and 365 days per year, the Sun moves with respect to the stars at a rate of slightly less than 1° per day.

The path that the Sun follows among the stars is called the **ecliptic.** Although the annual motion of the Sun is primarily eastward, the ecliptic is tilted, or inclined, to the celestial equator, as shown in Figure 2.11. Thus the Sun's movement on the ecliptic also carries it north and south of the celestial equator during the year. The Sun spends about half of the year in the northern hemisphere of the celestial sphere and half in the southern hemisphere. The angle between the equator and the ecliptic is 23.5°, so the Sun's declination varies from +23.5° (north of the celestial equator) to −23.5° (south of the celestial equator) during the year. When the Sun has a declination of +23.5°, it is directly overhead at a latitude of 23.5° north. This latitude is called the Tropic of Cancer. The corresponding southern latitude, 23.5° south, is called the Tropic of Capricorn.

The point on the ecliptic where the Sun's declination is most northerly is called the **summer solstice.** The point where it is most southerly is the **winter solstice.**

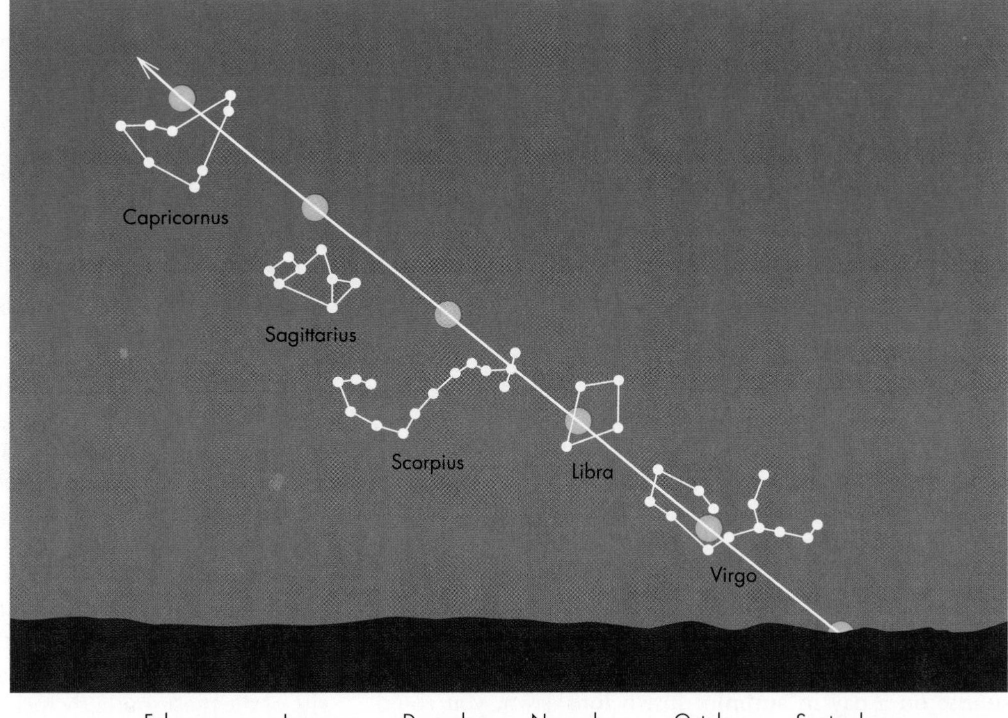

FIGURE 2.10
The Apparent Path of the Sun During Autumn
The Sun appears to move eastward, relative to the stars, at the rate of one constellation per month.

FIGURE 2.11
The Ecliptic
The Sun appears to move eastward along the ecliptic relative to the stars. The ecliptic is inclined 23.5° with respect to the celestial equator. The solstices occur when the Sun is farthest north or south of the celestial equator. The equinoxes occur when the Sun crosses the celestial equator.

ANIMATION *The Sun's motion north and south in the sky as the seasons change*

The points where the Sun crosses the celestial equator and has zero declination are called the **autumnal equinox** and the vernal equinox. Recall that the vernal equinox, one of the points where the ecliptic and the celestial equator cross, is the zero point from which right ascension is measured in the equatorial coordinate system (Figure 2.5).

The Seasons

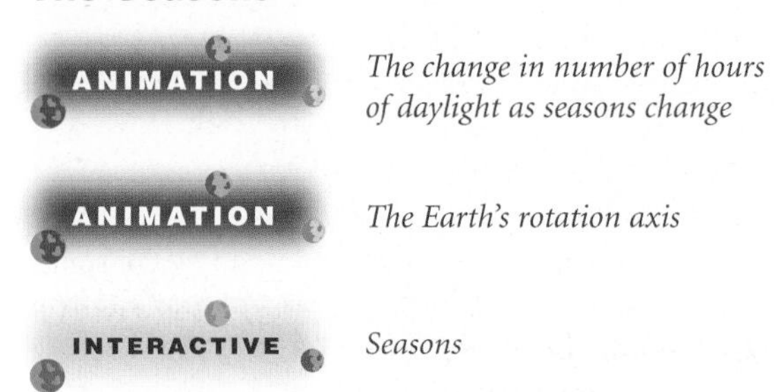

The change in number of hours of daylight as seasons change

The Earth's rotation axis

Seasons

The changing declination of the Sun affects the point on the horizon where the Sun rises (the azimuth of sunrise) and the duration of daylight. When it is north of the celestial equator, the Sun rises in the northeast and daylight lasts for more than half the day for people in the northern hemisphere. When south of the equator, the Sun rises in the southeast and day is shorter than night. On the summer solstice, the declination of the Sun is +23.5°. At a latitude of 40°, the azimuth of sunrise is 59° (Figure 2.12). At the winter solstice, the azimuth of sunrise is 121°. This means that at a latitude of 40° the direction of sunrise changes by 62°, more than one-third of the way from due north to due south, between the

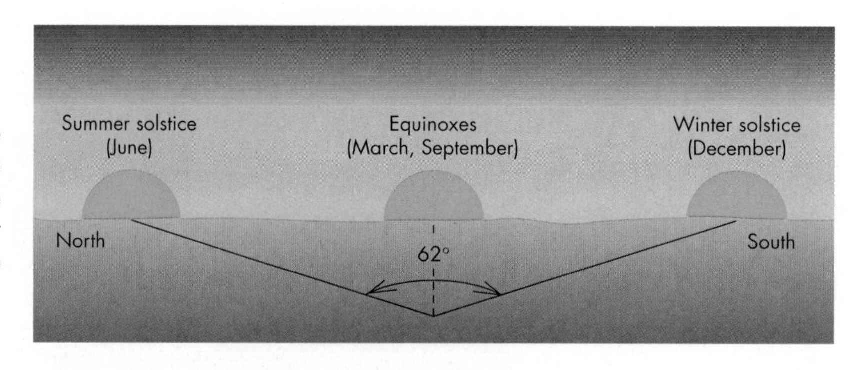

FIGURE 2.12
The Annual Variation in the Azimuth of Sunrise
For a latitude of 40°, the azimuth of sunrise increases by more than 60° between the summer solstice and the winter solstice. The annual variation in azimuth is even greater for more northerly latitudes, but is only 47° at the equator.

summer and winter solstices. The annual migration of sunrise northward and southward along the horizon was noticed by people in prehistoric times. We know this because they built monuments that helped them use the annual changes in the azimuth of sunrise and sunset to determine when the solstices occur. In other words, they kept track of the seasons.

To see how this might work, imagine that you were given a pile of rocks, a willing coworker, and an assignment to set up a system for determining the approximate date of the winter solstice. You could begin by marking the position of sunrise on a day in autumn. Just before dawn, you could stand at a chosen location and direct your assistant (who would be carrying a rock) to move back and forth in azimuth until the rock was aligned with the rising Sun. Your assistant would then place the rock to mark the azimuth of sunrise. If the exercise were repeated day after day, the two of you would build up a group of rocks, as shown in Figure 2.13, which would extend farther south as the weeks passed.

FIGURE 2.13
Determining the Solstice
Your assistant, aligned with the rising Sun, places a rock to mark the azimuth of sunrise.

One day, however, you would notice that the newly placed rock was north of the one from the previous day. Because the most southerly sunrise occurs on the winter solstice, you would know that the winter solstice had occurred on the previous day. If you piled up all the other rocks on the site of the most southerly rock, you would have a monument that could be used in later years to find the date of the winter solstice. A similar procedure could be used in the spring to locate the azimuth of the most northerly sunrise, which happens when the Sun is at the summer solstice. People at many times and many places did just this. The earliest known solar observatory, at Nabta in southern Egypt, is 7000 years old. In the Americas, a 4200 year-old temple in Peru is aligned to the azimuth of sunrise on the December solstice. Probably the most famous prehistoric solar observatory is at Stonehenge, in England, shown in Figure 2.14. At Stonehenge there are a number of alignments of standing stones to mark sunrise at

FIGURE 2.14
Stonehenge, a Prehistoric Solar Observatory
As viewed from the center of Stonehenge on the summer solstice, the rising Sun is aligned with the heel stone, which lies outside the circle of standing stones.

FIGURE 2.15
The Fajada Butte Sun Dagger on the Summer Solstice
A shaft of light passing between sandstone slabs produces the "sun dagger" that bisects the spiral petroglyph. Other distinctive patterns are produced at the winter solstice and at the equinoxes.

the solstices. The most famous of these is marked by the heel stone, over which the Sun rises on the summer solstice. Northwestern Europe is dotted with other stone monuments from which sunrises on the solstices and equinoxes could have been observed.

In North America there are solar monuments of a different sort. These are the sun daggers, built by the Native Americans as long as a thousand years ago. The most famous sun dagger is one built by the Anasazi at Fajada Butte in Chaco Canyon, New Mexico (Figure 2.15). The sun dagger is formed when narrow shafts of sunlight shine through the gaps between three sandstone slabs. At the solstices and equinoxes, the shafts (daggers) of sunlight create highly distinctive patterns on spiral petroglyphs carved on the rock wall. Early people must have been strongly motivated to expend the enormous effort required to build monuments such as Stonehenge or the sun daggers. Their motivation may have been religious or, perhaps, related to practical matters such as the timing of harvests and animal migration. Whatever the reason, it is obvious that they considered marking the seasons to be important.

Even after recorded history began, people built monuments to mark the azimuths of sunrise and sunset at the solstices and equinoxes. A dramatic early example is the Temple of Amen-Ra at Karnak, Egypt. The temple, shown in Figure 2.16, was begun more than 4000 years ago. It is built around a series of doorways and halls of columns, which define an axis almost 600 meters (about 2000 feet) long. From the innermost chamber of the temple, only a small part of the northwestern horizon can be seen. The distant doorways framed the setting Sun on midsummer's day (the summer solstice). Equinoctial alignments, in which the main axis of a building is aligned east-west,

FIGURE 2.16
The Temple of Amen-Ra at Karnak, Egypt
The temple was aligned so that its main axis points to the position of sunset on the summer solstice.

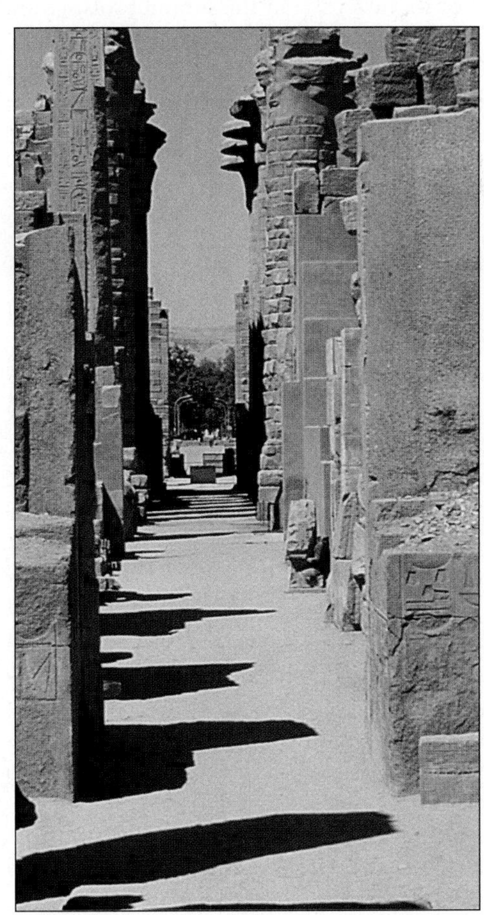

are common in temples and churches from the Temple in Jerusalem to St. Peter's Cathedral in Rome.

As the azimuth of sunrise changes, so do the altitude of the Sun at noon and the length of time that the Sun spends above the horizon. The lower the altitude of the Sun, the more widely spread out is the sunlight that strikes the ground. In the northern hemisphere, the altitude of the Sun at noon is highest at the summer solstice and lowest at the winter solstice. For a latitude of 40° (that of Philadelphia and Denver), the noonday Sun has an altitude of 73.5° at the summer solstice and only 26.5° at the winter solstice. Daylight lasts for only a little more than 9 hours at the winter solstice, but 15 hours at the summer solstice. Also, the Earth's atmosphere absorbs more sunlight when the Sun is low in the sky because sunlight's path through the atmosphere is longer. As a result of the length of the day, the spreading out of sunlight, and absorption by the atmosphere, only one-sixth as much solar energy falls on a horizontal patch of ground at the winter solstice as at the summer solstice. For latitudes more northerly than 40°, the seasonal differences are even larger. It isn't surprising that the seasons often correspond to very different temperature conditions or that other aspects of the weather should change with the seasons. The annual variation in the amount of sunlight is particularly strong above the Arctic and Antarctic Circles, at 66.5° north and south latitudes. On the Arctic Circle, the Sun never sets at the summer solstice and never rises at the winter solstice. Closer to the poles, the period of winter darkness and the period of continuous summer sunlight both increase in length until they both reach 6 months at the poles.

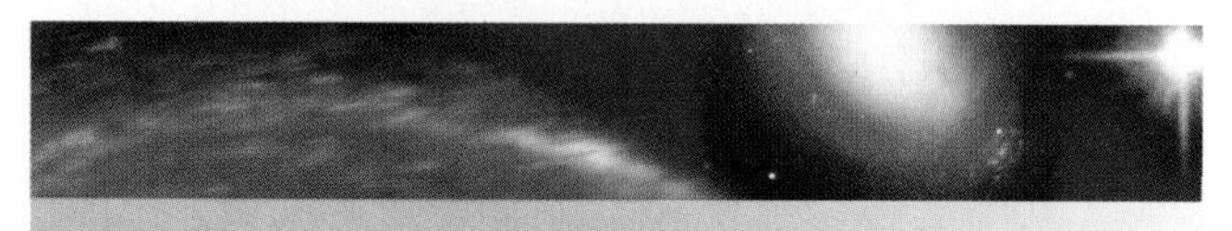

The Sun appears to move eastward among the stars on the ecliptic, which is inclined with respect to the celestial equator. The annual changes in the declination of the Sun cause changes in the rising and setting directions of the Sun and its altitude at noon. These changes are responsible for seasonal differences in temperature.

Time

People have wanted to keep track of time since prehistory. Sundials, water clocks, and other devices have been used for thousands of years to coordinate the many activities of the day and night. In some religions, the beginning of each lunar month is a very important event. Mariners have long known that tides are related to the phases of the Moon. The agricultural necessity of keeping track of the passage of the year has been recognized at least since the time of the earliest civilizations.

Two of our basic units of time, the day and the year, are both related to the motion of the Sun. The **solar day** is the amount of time that passes between successive appearances of the Sun on the meridian—in other words, the amount of time from high noon to high noon. The average length of the solar day is defined as 24 hours, and is the basis for ordinary (civil) timekeeping.

Astronomers, on the other hand, often find it convenient to use the **sidereal day,** which is the length of time it takes for a star to return to the meridian. The sidereal day is obviously related to the positions of stars in the sky. If an astronomer wants to observe a given star as it crosses the meridian, the observations need to be carried out at intervals of a sidereal day. We have already seen that the Sun moves with respect to the stars, so the solar day and the sidereal day have different lengths. Suppose the Sun and a star cross the meridian at the same moment. Before either crosses the meridian again, the Sun has moved eastward along its annual path on the ecliptic. Thus, as Figure 2.17 shows, the star returns to the meridian first, ending the sidereal day slightly earlier than the solar day ends. The length of the sidereal day is $23^h\,56^m\,4^s$ (or 23 hours, 56 minutes, and 4 seconds).

A complication in timekeeping arises because the rate at which the Sun moves along the ecliptic varies slightly throughout the year. Because of this, the length of the solar day changes during the year as well. The longest solar day is about 1 minute longer than the shortest solar day. Differences of this size, although small, accumulate during the parts of the year for which

FIGURE 2.17
The Sidereal and Solar Days
A, The Sun and a star cross the meridian at the same time. **B,** The following day the Sun and the star cross the meridian at different times. Because the Sun appears to move eastward relative to the stars, it recrosses the meridian later than the star. Thus, the solar day is longer than the sidereal day.

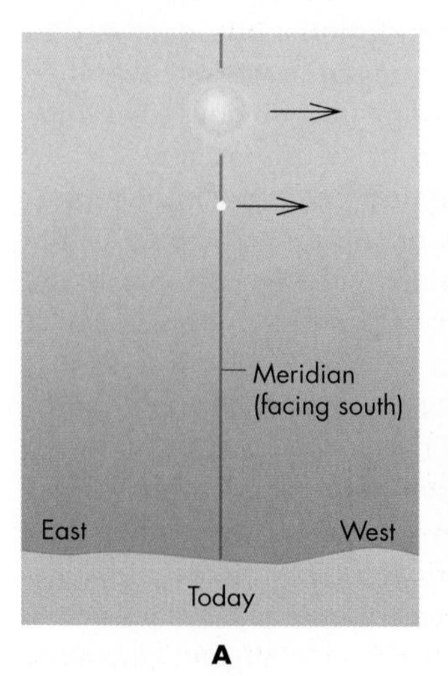

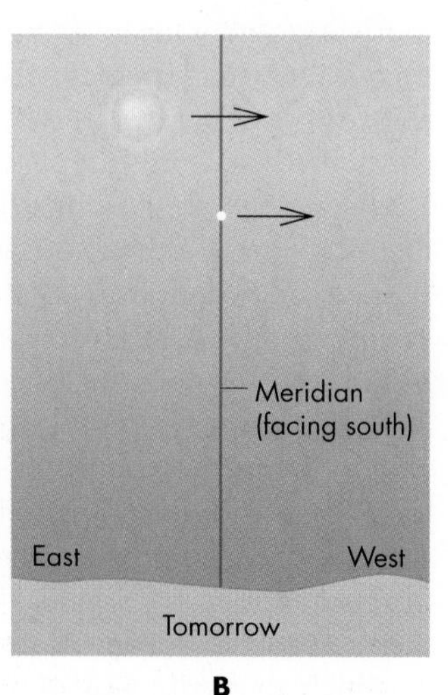

FIGURE 2.18
Time Zones in North America
The time zones are roughly 15° wide and centered at longitudes of 60° W, 75° W, and so on. Political boundaries, however, often cause the outlines of time zones to be irregular.

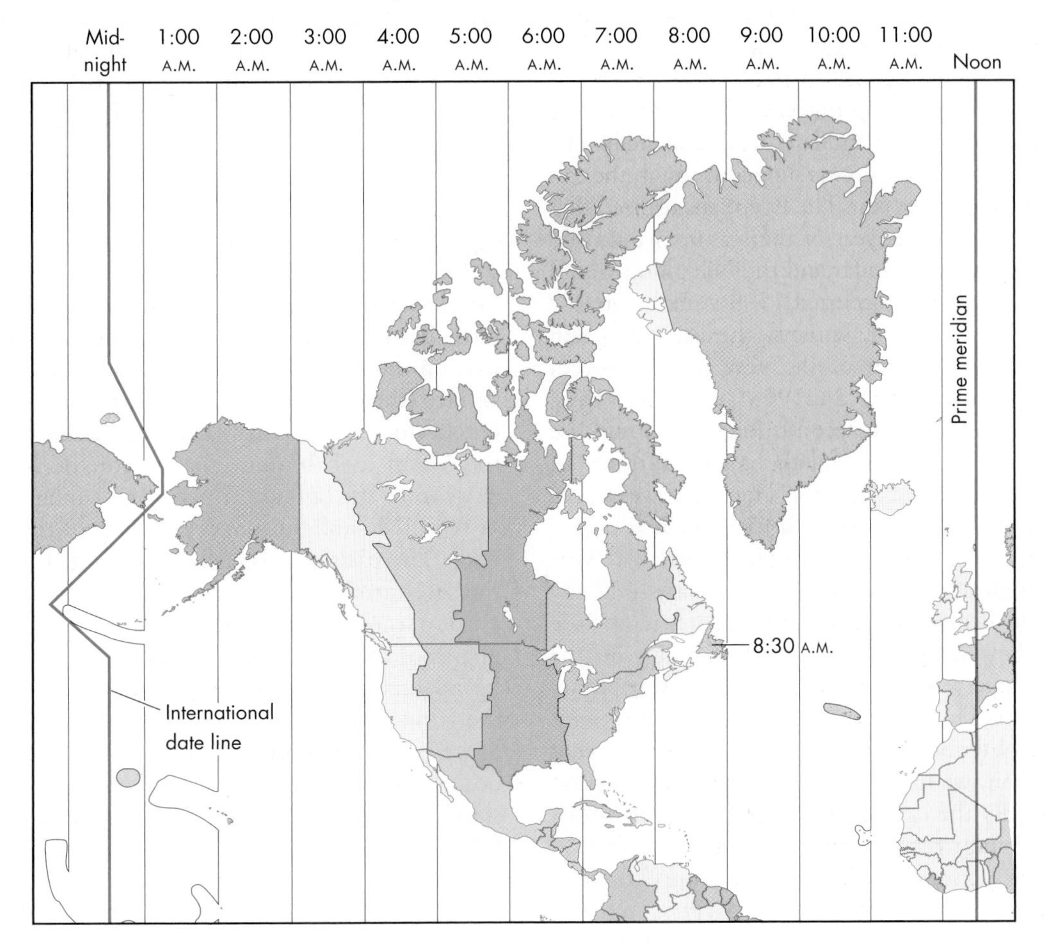

the day is longer or shorter than average. Suppose you had a clock that kept time according to the average length of the solar day. That is, the clock measured 24 hours as the average length of time it took the Sun to return to the meridian. Such a clock keeps **mean solar time,** or average solar time. (There are such clocks. You probably are wearing one on your wrist because we run our civilization on mean solar time.) On the other hand, a clock that kept time according to the actual position of the Sun in the sky would keep **apparent solar time.** It would show the hour angle of the Sun plus 12 hours, since the solar day starts at midnight.

A sundial keeps apparent solar time. A mechanical clock that kept apparent solar time would have to run at a variable rate during the year. The difference between mean and apparent solar time can be as great as 16 minutes.

A further complication with timekeeping is that the moment when the Sun crosses the meridian depends on a person's longitude. This means that if everyone kept time according to when the Sun crossed his or her meridian, the only people whose watches would agree would be those located due north or south of each other. To keep the correct time, it would be necessary to reset your watch continually as you moved east or west. This became a problem when railroads made rapid travel possible. The problem was overcome by the introduction of time zones, which are regions on the Earth, roughly 15° wide in longitude, in which everyone keeps the same **standard time** (Figure 2.18). For most of the people in a time zone, the Sun doesn't actually cross the meridian when their watches read noon. However, because the difference is usually a half hour or less, most people don't even notice where the Sun is at noon. They are quite happy to sacrifice the accuracy with which their watches can tell the Sun's position for the convenience of having the same time across the time zone.

The **tropical year** is the length of time it takes the Sun to return to the vernal equinox. This is the unit of time associated with the annual cycle of the seasons. The length of the tropical year (usually referred to simply as the year) was found by the ancient Egyptians to be roughly 365 days. The Egyptians used a calendar in which the year was divided into 12 months of 30 days each with an extra 5 days added to bring the total to 365. However, they recognized that the actual length of the year was somewhat greater than 365 days. The way they discovered that the year is longer than 365 days was that, over centuries, annual events such as the Nile flood occurred later and later in the calendar. The average rate at which annual events slipped through the calendar was about ¼ day per year. The Egyptians realized that after about 1500 years, the year of the seasons would slip all the way through the calendar and the Nile flood would return to the date where it occurred 1500 years before.

Since the time of the Egyptians, the accuracy with which we know the length of the year has improved. We now know that it is 365.242199 days or 365 days, 5 hours, 48 minutes, and 46 seconds long. We could use calendars in which each year had the correct number of days, hours, minutes, and seconds. The price we would have to pay would be to start years at various times of day rather than at midnight. Suppose that a particular year began at midnight. It would end (and January 1 of the next year would begin) 365.242199 days later, at a little before six in the morning. For the rest of that year, the date would change just before six in the morning rather than at midnight. The following year and each of its days would begin slightly before noon. Rather than accept that kind of confusion, we instead use a calendar in which no given year is really the right length of time, but in which the average of many years is correct. Some years are too short, others are too long. We do this by adding extra days to **leap years,** which occur about every 4 years. Actually, an extra day every 4 years would result in a calendar in which the average year was 365.25 days, which is a little longer than the tropical year. To shorten the average year slightly, century years are not leap years unless they can be evenly divided by 400. Thus, 1900 was not a leap year, but 2000 was. This eliminates 3 days every 400 years. Using this pattern of leap years, the average length of the calendar year is almost exactly equal to what it needs to be to keep annual events occurring at the same time of year. This procedure is the basis for the Gregorian calendar used in most of the western world.

2.3 THE PHASES OF THE MOON

The monthly pattern of the Moon's phases is among the most dramatic regular cycles in the sky. When sunlight falls on the Moon, it illuminates only half of the Moon's surface (the day side of the Moon). Some of this sunlight is reflected to the Earth, making the illuminated part of the Moon look bright. The rest of the Moon is turned away from the Sun and receives no sunlight. The absence of reflected sunlight makes this part of the Moon appear dark. The phase of the Moon depends on how much of the side turned toward us is illuminated by the Sun.

You can do a simple experiment to investigate the phases of the Moon. In an otherwise darkened room, set up a lamp. Seat yourself on a stool perhaps 10 feet from the lamp. Then, as shown in Figure 2.19, have a friend move around you holding a ball while you watch how the bright part of the ball changes in shape. Your friend should start so that the ball is nearly on the opposite side of you from the lamp. In this position, the bright hemisphere of the ball is almost entirely visible to you. The ball is at **full phase.** As your friend circles toward the lamp, less of the bright hemisphere of the ball can be seen. The phase is now **gibbous.** When your friend has moved a quarter of the way around you, only half of the bright hemisphere

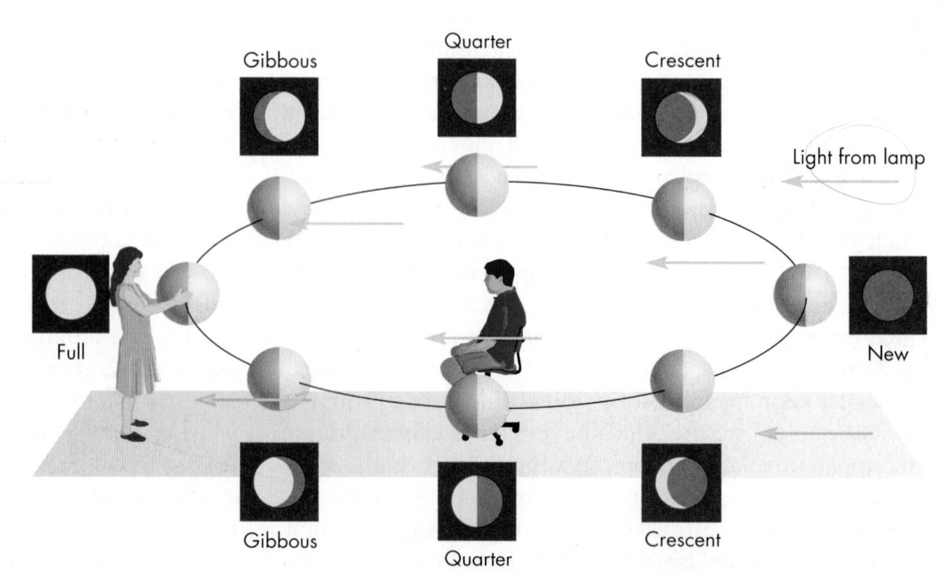

FIGURE 2.19
The Phases of the Moon
As a ball illuminated from the right is moved around an observer, the observer sees the ball go through its cycle of phases. The inset drawings show how the ball looks to the observer at each position.

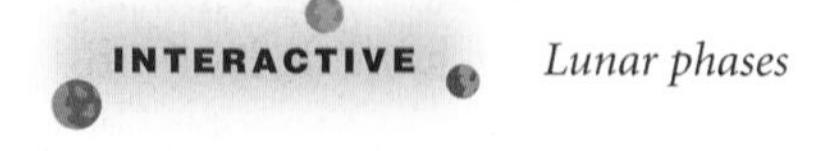

Lunar phases

FIGURE 2.20
A Prehistoric Astronomical Record
The scratches on this bone may follow the phases of the Moon.

FIGURE 2.21
The Sun and the Moon at Waxing Crescent Phase
The waxing crescent Moon is a little east of the Sun and sets shortly after sunset.

Looking west in the late afternoon on a day just after new Moon

FIGURE 2.22
The Sun and the Moon at First Quarter Phase
The first quarter Moon crosses the meridian approximately at sunset.

Looking west just before sunset

of the ball can be seen. The ball is at **quarter phase.** By the time your friend has moved nearly between you and the lamp, the phase of the ball has diminished from **crescent** to **new,** at which point almost none of the bright hemisphere can be seen. As your friend continues to circle, the phase changes from new to crescent to quarter to gibbous and back to full again.

Thus we see that the Moon's phases are determined by the position of the Moon relative to the Earth and the Sun. When the Moon is opposite the Sun in the sky so that we are nearly between the Moon and Sun, the Moon's phase is full. When the Moon is nearly between us and the Sun so that the Moon and Sun are close to each other in the sky, the Moon's phase is new. People may have recorded the Moon's phases for more than 30,000 years. Writer-archaeologist Alexander Marshack has examined bone fragments found at prehistoric sites in Europe, Asia, and Africa. Early people engraved lines and other markings on these fragments. One of these fragments is shown in Figure 2.20. Marshack suggested that the pattern of the markings follows the phases of the Moon. He has proposed that Stone Age people made records of the lunar phases. No one knows why early people made such records, but Marshack suggests that the marks may have been "remembering marks," or memory aids for the telling and retelling of stories about the Moon and its phases.

Because the phase of the Moon depends on where it is in the sky relative to the Sun, the Moon's phase is also related to the time of moonrise and moonset. Figure 2.21 shows the Moon and Sun in the sky in late afternoon on a day just after new Moon. The Moon is at **waxing crescent** phase. (Waxing means that its bright part is growing larger.) The horns of the crescent Moon point away from the Sun. The Moon is only about 15° east of the Sun and will set about an hour after sunset. Although the crescent Moon can be seen in daylight, it is most easily seen just after sunset, low in the western sky. Figure 2.22 shows the Moon and Sun at first quarter Moon. The Moon and Sun are about 90° apart. The Moon will set about midnight (6 hours after sunset) and will be conspicuous in the evening sky. When the Moon is full, it is opposite the Sun in the sky. It rises at sunset and is visible all night long. After full Moon, the Moon rises later each night until it rises just before sunrise and is a **waning crescent.** (Waning means that the bright part is shrinking.)

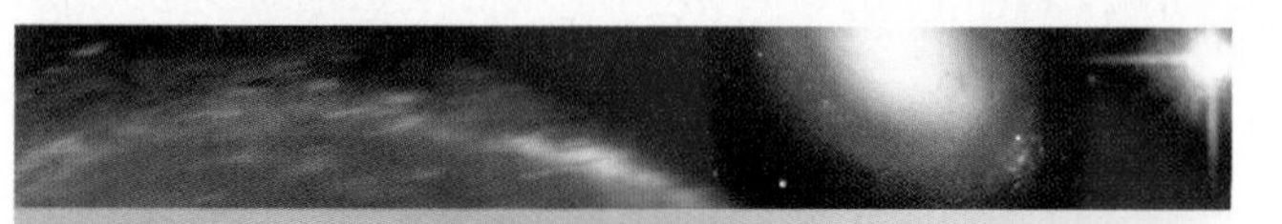

The phases of the Moon occur because we see different portions of the Moon's sunlit hemisphere during a month. The phase of the Moon depends on its location in the sky relative to the Sun.

2.4 THE MOTION OF THE MOON

The motion of the Moon among the stars is similar to that of the Sun. The Moon moves generally eastward relative to the stars, at a rate that varies throughout a month but averages about one lunar diameter per hour, or 13° per day. This means that moonrise and moonset occur nearly an hour later each day. If the Moon is near a bright star, the motion can be detected in an hour or less, as shown in Figure 2.23. The length of time it takes for the Moon to return to the same place among the stars is about 27.3 days, the **sidereal month.** Because the Sun also moves eastward among the stars, it takes more than a sidereal month for the Moon to lap the Sun and return to the same position in the sky relative to the Sun (Figure 2.24). The length of time required for the

FIGURE 2.23
The Motion of the Moon
A, The Moon is seen against a backdrop of stars. **B,** One hour later, the Moon appears to have moved eastward relative to the stars by about its own angular diameter. Both the Moon and the stars appear to move westward during that hour.

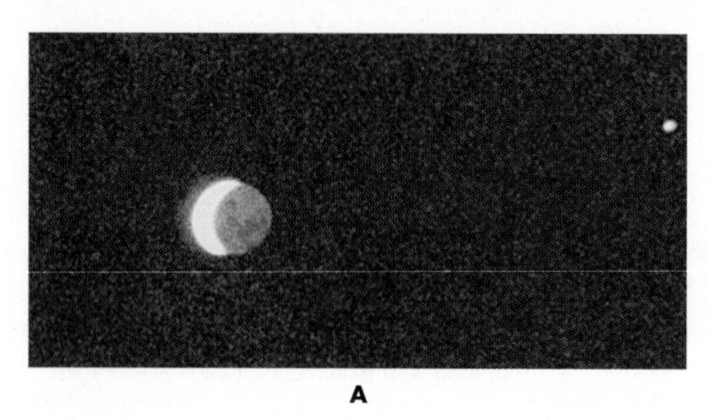

A

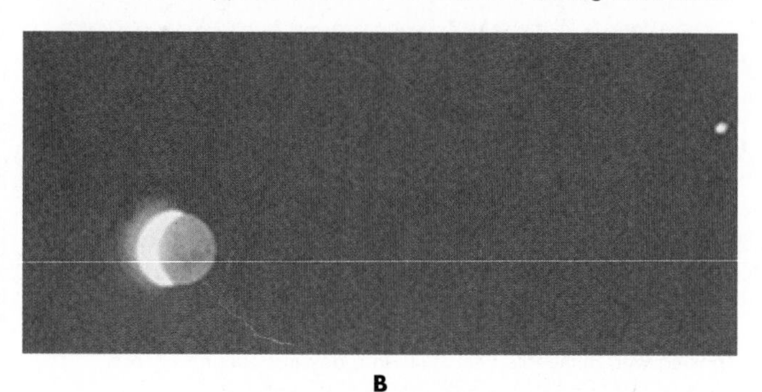

B

FIGURE 2.24
The Sidereal and Synodic Months
Because the Sun appears to move eastward with respect to the stars, it takes the Moon longer to return to the same position relative to the Sun (and the same phase) than it does to return to the same position among the stars. **A,** At new Moon, the Sun and Moon are close together in the sky. **B,** After 27.3 days, one sidereal month, the Moon returns to the same place among the stars, but the Sun has moved eastward. The Moon shows a waning crescent phase. **C,** After 29.5 days, one synodic month, the Moon catches up with the Sun and returns to new phase. The sizes of the Sun and Moon are exaggerated in this figure. The arrows indicate the eastward motion of the Sun and Moon among the stars.

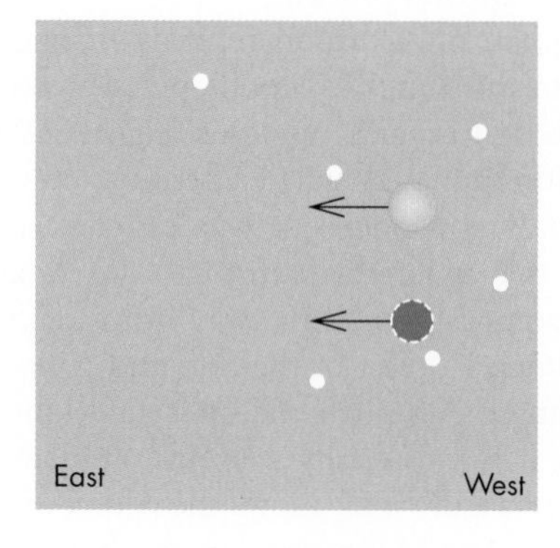

A

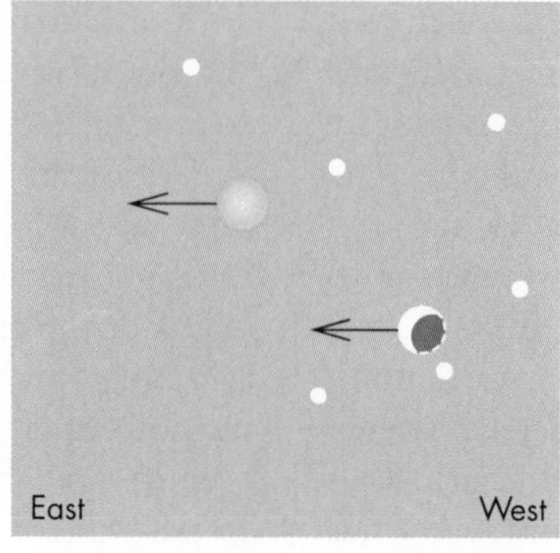

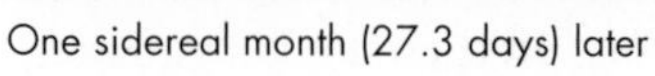

One sidereal month (27.3 days) later

B

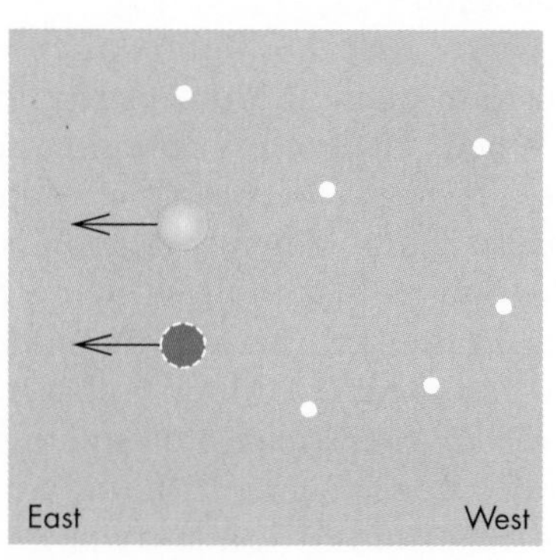

One synodic month (29.5 days) later

C

FIGURE 2.25
The Moon's Path
The Moon's path is tilted by about 5° with respect to the ecliptic.

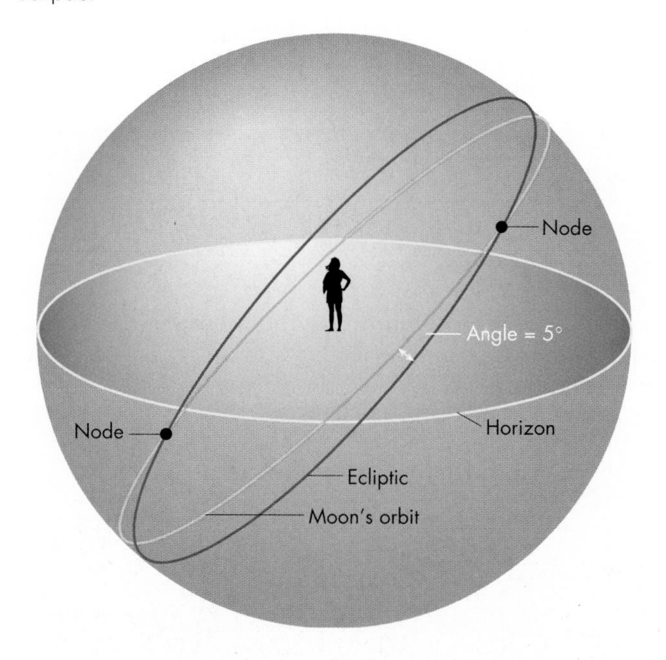

Moon to return to the same position relative to the Sun is called the **synodic month** and is about 29.5 days in length. Because the synodic month describes the relationship between the positions of the Moon and Sun in the sky, it is also related to the time of moonrise and the phases of the Moon.

The Moon's path doesn't quite coincide with the ecliptic. The angle by which the two differ, the inclination of the Moon's **orbit,** is about 5°. Figure 2.25 shows the path of the Moon and the ecliptic. The two great circles intersect at two opposite points, the **nodes** of the Moon's orbit.

The Moon's path doesn't stay fixed but rather slides around the ecliptic like one hoop slipping around another (Figure 2.26). Another way to describe the slippage is to say that the nodes of the Moon's orbit (and the imaginary line through the Earth that connects them) move around the ecliptic. It takes 18.6 years for a node to complete a trip around the ecliptic. The motion of the Moon's nodes has an important consequence for the timing of eclipses, which are described in Chapter 9.

The Moon moves eastward among the stars, returning to the same place with respect to the stars after a sidereal month (27.3 days) and the same place with respect to the Sun after a synodic month (29.5 days). Its path is inclined with respect to the ecliptic by about 5°.

2.5 THE MOTIONS OF THE PLANETS

The westerly diurnal motions of the stars and the eastward motions of the Sun and Moon relative to the stars are reasonably simple to describe. The motions of the planets, however, are much more complicated. The word "planet" is derived from the Greek word for wanderer. It was originally applied to all of the celestial objects that move with respect to the stars. These objects included the Sun and Moon as well as Mercury, Venus, Mars, Jupiter, and Saturn. The remaining planets, Uranus, and Neptune were unknown to the ancients and the similarity between the Earth and the planets was not understood.

FIGURE 2.26
The Movement of the Moon's Nodes
It takes 18.6 years for the Moon's orbit to slide completely around the ecliptic.

The movement of the Moon's nodes

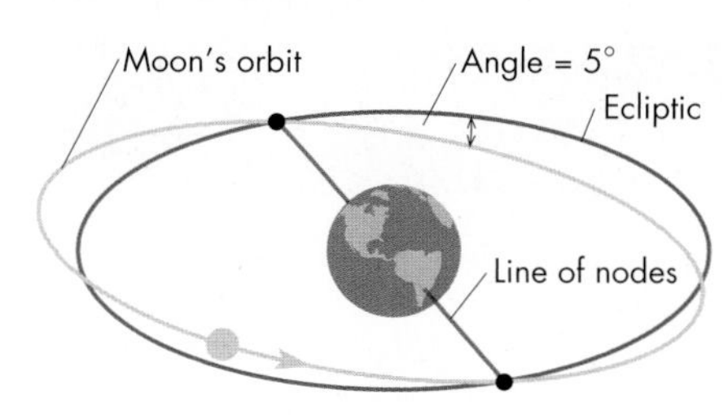

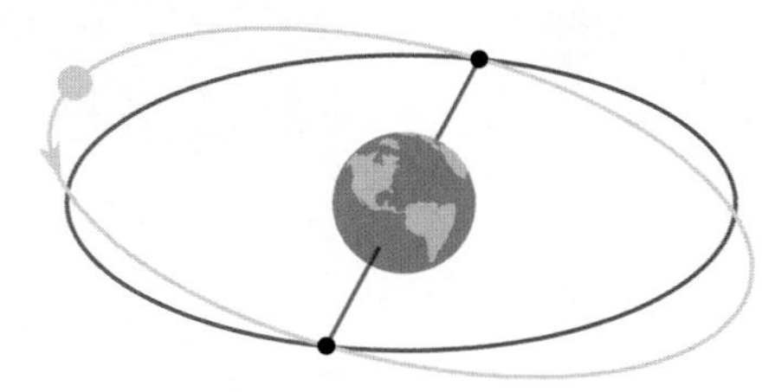

Approximately 2 years later

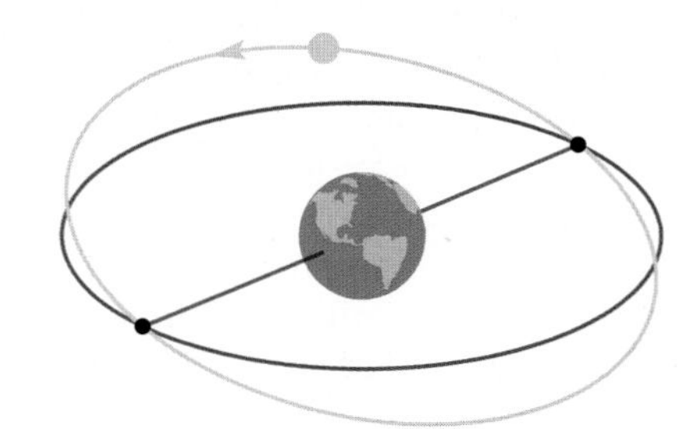

Approximately 2 years later still

FIGURE 2.27
The Retrograde Motion of Mars
The position of Mars relative to the starry background is shown for a 6-month period of time. Initially, Mars moves eastward (prograde) with respect to the stars. Its prograde motion then stops, and Mars moves westward (retrograde) with respect to the stars. Finally, its retrograde motion stops and it resumes its normal prograde motion. Mars undergoes an episode of retrograde motion every 26 months.

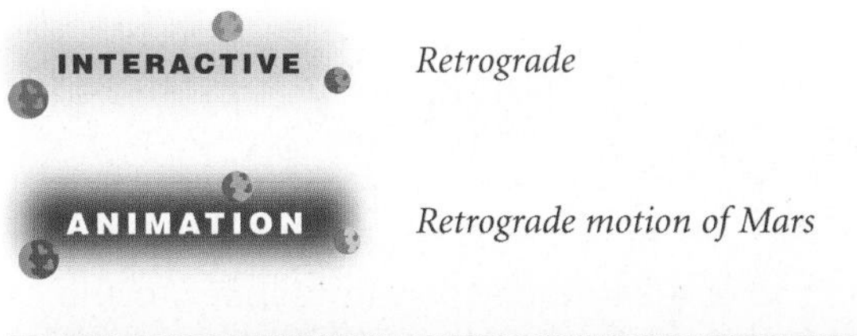

INTERACTIVE *Retrograde*

ANIMATION *Retrograde motion of Mars*

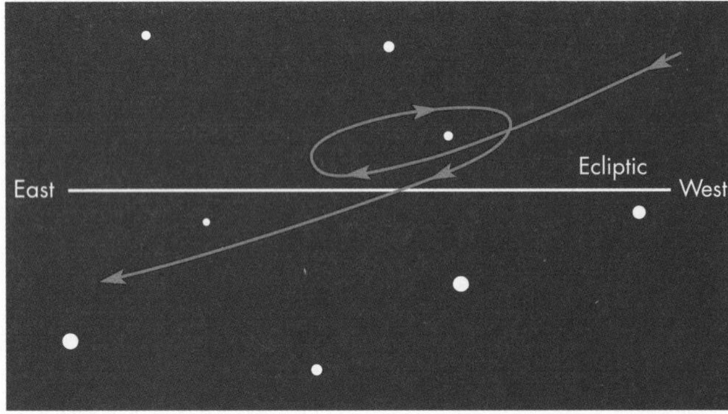

FIGURE 2.28
Three Successive Retrograde Loops for Mars
Mars underwent retrograde motion during the summer of 1984, the summer of 1986, and the fall of 1988. The starfield against which the retrograde loops took place was different for the three episodes of retrograde motion. Also, the loops differed in shape and orientation.

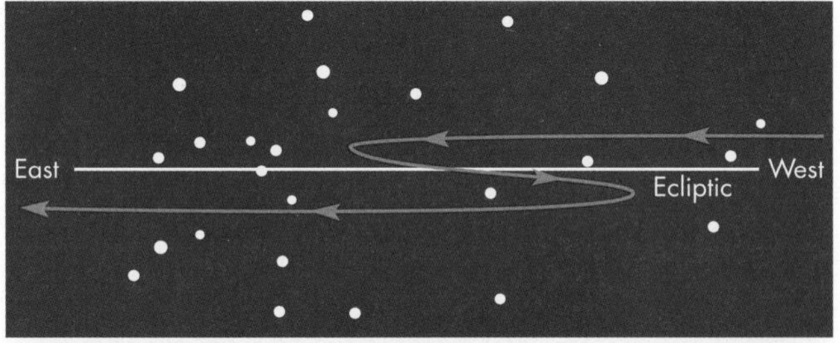

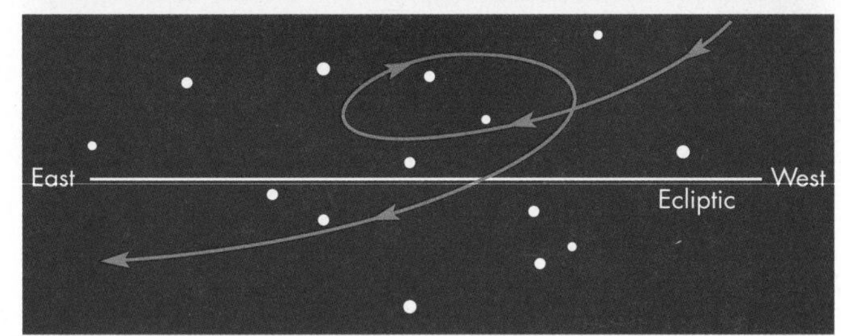

Most of the time, a planet appears to move eastward with respect to the stars. The rate varies greatly from planet to planet. When a planet moves eastward, its motion is said to be **prograde,** or **direct.** At regular intervals, however, a planet appears to reverse its direction of motion and, for a time, moves toward the west with respect to the stars. During this time its motion is said to be **retrograde.** An example of retrograde motion for Mars is shown in Figure 2.27. Retrograde motion occurs for a planet only when it is in a specific arrangement in the sky relative to the Sun. For Mercury and Venus, retrograde motion can take place only when the planet appears to pass near the Sun in the sky. The times when a planet is nearly aligned with the Sun are called **conjunctions.** (For Mercury and Venus, retrograde motion takes place only at every other conjunction.) For the other planets, retrograde motion happens when the planet is opposite the Sun in the sky—that is, when it is at **opposition.**

Episodes of retrograde motion occur at intervals of time known as the **synodic period** of a planet. Synodic periods range from about 4 months for Mercury to about 26 months for Mars. Despite these regularities, the motion of a planet is quite complex. For example, the retrograde loop of a planet differs in appearance from one episode of retrograde motion to the next (as shown in Figure 2.28) and occurs at a different position among the stars. The difficulty in describing the motions of the planets and in accounting for their complex behavior stimulated much of the best astronomical thinking for more than 2000 years. This work, which extended from the era of the ancient Greeks to the time of Isaac Newton, is described in Chapters 3, 4, and 5.

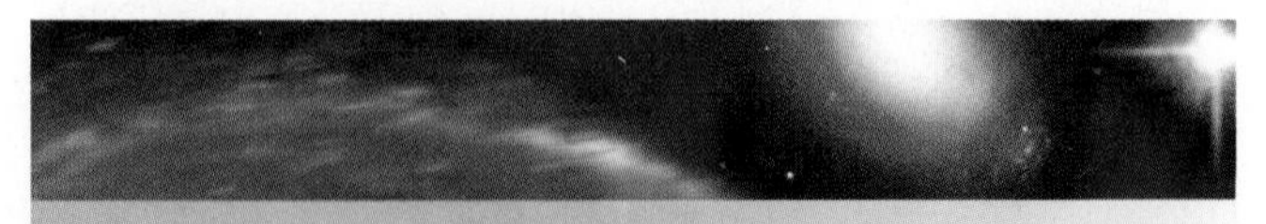

Unlike the Sun and Moon, the planets periodically reverse their usual eastward motion and undergo westward, retrograde motion with respect to the stars. The path of a planet relative to the stars is quite complex.

Chapter Summary

- The daily, or diurnal, westward paths of stars are circles centered on the north celestial pole. The diurnal circles grow smaller for stars that are nearer the north celestial pole. The celestial equator is a line on the celestial sphere that divides it into northern and southern hemispheres. (Section 2.1)
- The equatorial system is used to locate stars on the celestial sphere. This system resembles the terrestrial system of longitude and latitude. In the equatorial system, declination describes the angular distance of a star north or south of the celestial equator. Right ascension describes the east-west location of a star on the celestial sphere. (2.1)
- The Sun appears to move eastward among the stars on a path called the ecliptic, and its motion is repeated annually. The ecliptic is inclined with respect to the celestial equator, so the declination of the Sun varies during the year. (2.2)
- Changes in the declination of the Sun produce an annual pattern of change in its rising and setting points as well as change in its altitude at noon. These changes are responsible for the seasons. (2.2)
- Apparent solar time is reckoned by the position of the Sun in the sky. The variable rate of the Sun's motion on the ecliptic causes changes in the length of the solar day throughout the year. (2.2)
- The phases of the Moon happen because the part of the Moon's illuminated hemisphere that we see varies throughout a synodic month. (2.3)
- The Moon appears to move eastward among the stars, returning to the same place after a sidereal month. (2.4)
- The planets usually appear to move eastward among the stars. However, at regular intervals they appear to move westward during periods of retrograde motion. The motions of the planets among the stars are quite complex. (2.5)

Key Terms

apparent solar time *27*
autumnal equinox *23*
celestial equator *18*
conjunction *32*
crescent phase *29*
declination *19*
direct motion *32*
diurnal *18*
diurnal circle *18*
ecliptic *22*
equatorial system *19*
full phase *28*
gibbous phase *28*
leap years *28*
local hour angle *20*
mean solar time *27*
meridian *18*
new phase *29*
nodes *31*
north celestial pole *18*
north circumpolar region *18*
opposition *32*
orbit *31*
prograde motion *32*
quarter phase *29*
retrograde motion *32*
right ascension *19*
sidereal clocks *20*
sidereal day *26*
sidereal month *30*
solar day *26*
south celestial pole *18*
standard time *27*
summer solstice *22*
synodic month *31*
synodic period *32*
tropical year *27*
vernal equinox *19*
waning crescent *30*
waxing crescent *29*
winter solstice *22*
year *22*
zodiacal constellations *22*

Conceptual Questions

1. Explain why diurnal motion is counterclockwise for a star we observe toward the north but clockwise for a star we observe toward the south.
2. Suppose an observer in the northern hemisphere determines that the diurnal motion of a star keeps it above the horizon for 16 hours. Is the star in the northern or southern hemisphere of the celestial sphere? Does the star rise in the northeast or the southeast? How much time elapses between the time the star rises and the time it crosses the meridian?
3. What is the local hour angle of a star at the moment it crosses the meridian?
4. The equatorial coordinate system is very similar to the terrestrial coordinate system. Which terrestrial coordinate is the counterpart of right ascension? Which terrestrial coordinate is the counterpart of declination?
5. An observer, working at the time of the summer solstice, notes that the Sun circles about the sky at a constant altitude (23.5°). The observations are interrupted by a bear. What color is the bear?
6. What is the orientation of the celestial equator for observers at the Earth's equator?
7. At what latitude is the altitude of the south celestial pole greatest?

8. Suppose an observer finds that Aries is the constellation just above the horizon as the stars fade at sunrise. What constellation would be seen just above the horizon at sunrise 1 month later? How about 1 month later still?
9. Suppose the ecliptic weren't tilted with respect to the celestial equator. How would the azimuth of sunrise vary during a year? How would the length of day and night vary throughout a year?
10. Suppose the ecliptic were tilted by 40° rather than 23.5° with respect to the celestial equator. What effect would this have on the variation of the azimuth of sunrise during a year?
11. Suppose you and your roommate had built a monument with piles of rocks to mark the azimuths of sunrise at the solstices. How could you determine where to place a pile of rocks to mark the azimuth of sunrise at the equinoxes? (Note, there are several correct answers to this question.)
12. Describe why it would be difficult to use sidereal time for civil timekeeping.
13. Why would it be difficult to build a wristwatch that keeps apparent solar time?
14. The Julian calendar, instituted by Julius Caesar in 46 B.C. and replaced by the modern Gregorian calendar beginning in 1582, averaged 365.25 days in length. How did annual events, such as the vernal equinox, move through the calendar while the Julian system was in effect?
15. Suppose the Moon moved westward rather than eastward among the stars. Would the sidereal month be longer or shorter than the synodic month? Explain.

Figure-Based Questions

1. Estimate the right ascension and declination of the star shown in Figure 2.5.
2. Use Figure 2.18 to find the time in Mexico City, Mexico, when it is 5 P.M. in Washington, D.C. Ignore any complications that might be caused by daylight savings time.
3. Suppose an astronaut on the Moon watches the Earth throughout a month. Use Figure 2.19 to answer the following questions about what the astronaut would observe. When we see new Moon, what phase of the Earth does the astronaut see? What phases does the astronaut see when we see the lunar phases of first quarter, waxing gibbous, full, and waning crescent?

Group Activities

1. Divide your group into two subgroups. After a few minutes of preparation, have a short debate about the advantages and disadvantages of mean solar time versus apparent solar time for civil timekeeping. Be sure to take account of practicality, convenience, and universality in your arguments.
2. With a partner, mark the azimuth of the rising or setting Sun (your choice), as shown in Figure 2.13, for at least a week. Find out how many days it takes for you to be absolutely sure that the azimuth of sunrise (or sunset) is changing.

For More Information

Visit the text website at **www.mhhe.com/fix** for chapter quizzes, interactive learning exercises, and other study tools.

3

Well do I know that I am mortal, a creature of one day.
But if my mind follows the winding paths of the stars
Then my feet no longer rest on earth, but standing by Zeus himself
I take my fill of ambrosia, the divine dish.

Claudius Ptolemy

Stonehenge, a prehistoric observatory.

www.mhhe.com/fix

Ancient Astronomy

In a few minutes to a few months, a person can detect the motions of the stars, Sun, Moon, and planets. To go further and build a detailed description of the celestial motions requires far more time and effort. For example, many years of observations of solstices and equinoxes were needed to determine accurately the length of the year. This discovery, however, led to a valuable, reliable calendar for keeping track of annual events. Centuries of observing were needed to discover the regular pattern of eclipses that made eclipse prediction possible. Although astronomical knowledge was passed orally from one generation to the next, the invention of writing made it easier for records to be handed down from one astronomer to the next. Astronomical knowledge could then be accumulated by many people over many generations.

After detailed descriptions of the celestial phenomena had been accumulated, several developments became possible. First, astronomers learned to predict future events. For example, after enough observations of the retrograde motion of Mars had been accumulated, astronomers identified patterns. These patterns made it possible to predict when the next retrograde episode would occur, where in the sky it would be seen, and what shape the retrograde loop would have. Even more important than prediction, however, was explanation. After the celestial patterns became well known, it was possible to explore the reasons that the celestial bodies appeared to move as they did. The interwoven cycle of observation, explanation, and prediction is the heart of modern science. Science began when observational records became reliable enough that explanations could be tested against them. This happened in the region around

Questions to Explore

- What was different about the Greek view of the heavens compared with that of their predecessors?
- How did the Greeks determine the size and shape of the Earth nearly 2000 years before Columbus?
- Why did the Greeks reject the idea that the Earth moves around the Sun?
- What was Ptolemy's model of planetary motion and why was it the standard for 1500 years?

the eastern Mediterranean between 2000 and 3000 years ago.

By modern standards, progress in astronomy occurred very slowly. Sometimes there were centuries between significant astronomical discoveries. There also were many false starts and missed opportunities. Eventually, however, the ancient astronomers developed a workable "model" of the relationship of the Earth to the Sun, Moon, planets, and stars. The model could be used with unprecedented accuracy to predict the positions of the celestial bodies. This model, the *Ptolemaic* or *geocentric* model, was used almost without change for more than 1000 years.

This chapter describes astronomy from the earliest recorded observations of the Mesopotamians to the completion of the geocentric model. Our focus is twofold. One focus is the interaction between observations and explanations (or theories). The other focus is the cumulative nature of progress in astronomy. That is, the way that many individual astronomers contribute, often over long periods of time, to the development of new ideas about the universe. The cumulative nature of astronomical discovery has continued, right up to the present time, as the principal way progress in astronomy occurs.

3.1 MESOPOTAMIAN ASTRONOMY

The first astronomers to make long-term written records (mostly lost or destroyed) of their observations were the Mesopotamians, who began to build observatories about 6000 years ago. One of these observatories, or *ziggurats*, is shown in Figure 3.1. Ziggurats were pyramidlike towers with seven terraces, each representing one of the wandering celestial bodies—Sun, Moon, Mercury, Venus, Mars, Jupiter, and Saturn. From the tops of the ziggurats the astronomers made careful observations of the movements of the celestial bodies. The astronomers were the timekeepers of their communities. They were responsible for keeping track of the yearly cycle on which the welfare of their agricultural community was so dependent. Mesopotamian astronomers were able to do this because they had discovered that the group of stars that rises just before sunrise changes with the seasons. The predawn rising of the constellation they called "The Bull of Heaven" marked the beginning of spring. Other constellations were used to mark the start of the other seasons. Eventually, 8 constellations were added to the original 4 to form the zodiac. The 12 Mesopotamian zodiacal constellations, with some changes of boundary, are the same 12 constellations we use today to mark the annual passage of the Sun among the stars. The early Mesopotamian astronomers also named many of the brighter stars and distinguished between the fixed stars and the planets.

The Mesopotamians originated the idea of dividing a circle into 360° and then further subdividing each degree into 60 parts called *minutes* and each minute into 60 parts called *seconds*. The small circle (°) we still use to denote *degree* symbolized the Sun. A full circle, 360°, symbolized the annual motion of the Sun, which takes about 360 days.

A basic unit of time for the Mesopotamians was the month. Their years consisted of a whole number of months.

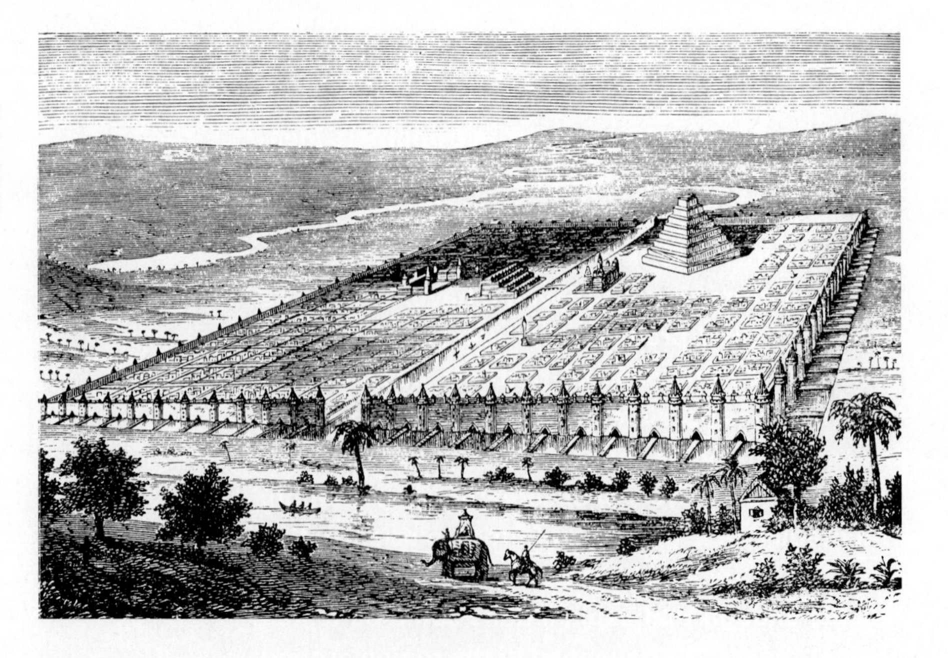

FIGURE 3.1
A Ziggurat
The Mesopotamians used ziggurats as observatories. Each of the seven terraced levels represented one of the wandering celestial objects—the Sun, Moon, and planets.

This was a problem because the actual length of a year, about 365¼ days, couldn't be evenly divided by the length of the synodic month, about 29½ days. Instead, there were about 12⅓ months in a year. The Mesopotamians solved the problem by having two kinds of years, normal ones with 12 months, and others with 13 months. With the proper mix of 12-month and 13-month years, it was possible to make the average year have the correct length even though individual years were either too short or too long. This use of "leap months" is the way calendars have been constructed whenever it was considered important to have whole numbers of months in a year and for certain months to occur at certain seasons. The modern Hebrew calendar, for example, uses leap months.

By about 500 B.C. the Babylonians (one of the many cultures in the long history of Mesopotamian civilization) were able to use the long record of observations to see regularities and patterns in the motions of the Moon and planets. They repeatedly determined the dates of the various configurations of the Sun, the Earth, and the planets shown in Figure 3.2. From these dates, the Babylonians found each planet's synodic period—the time needed for the planet to go from a particular configuration, such as opposition, back to that same configuration. The Babylonians also discovered and measured the longer intervals of time required for a planet to return to the same position with respect to both the Sun and the stars. This repetition occurs when a multiple of the synodic period of a planet is equal to nearly a whole number of years. For example, the synodic period of Mars is 780 days, so Mars is high in the sky at midnight and opposite the Sun at intervals of 780 days, or about 2 years and 2 months. Because the constellation on the meridian at midnight changes throughout the year, if Mars is in Taurus at one opposition, it will be in Cancer the next. However, the time required for 22 synodic periods of Mars is almost exactly 47 years. After 47 years, Mars returns to the same position among the stars and is in the same relationship to the Earth and the Sun. That is, Mars rises and sets at the same time, crosses the meridian at the same time, moves with respect to the stars at the same rate, and moves in the same direction as it did 47 years earlier.

This discovery gave the Babylonians a simple, practical method for predicting the positions of the planets. If you want to know where Mars will be found next month, just look at records from 47 years before the date on which you need to find the planet. It will be in nearly the same spot. The frequent and accurate observations the Babylonians made ensured that the records needed to determine future planetary positions would always be available.

One of the main reasons that the Babylonians were so interested in the positions of the Sun, Moon, and planets was their belief in **astrology.** Astrology is a pseudoscience involving the belief that the positions of the celestial objects influence events on the Earth. This was not an unreasonable assumption for the Babylonians, who had seen the relationship between the Sun's motion and seasonal events. The Babylonians believed that the influence of the Sun, Moon, and planets extended to natural events, such as floods and earthquakes. It became extremely important for them to be able to predict the positions of the celestial bodies in order to forecast natural events that might have considerable effects on political and military affairs. Only much later did people begin to believe that the positions of the celestial objects can influence events in the lives of individuals and to predict those influences using horoscopes. With our present understanding of the causes of natural events, it would seem that the belief in astrology should have died out long ago. However, some deep human feelings must be tapped by astrology, because it still thrives today. No scientific basis for astrology exists, and no controlled study has ever shown the slightest indication that the predictions of astrologers are any better than clever guesswork guided by knowledge of human nature.

FIGURE 3.2
Planetary Configurations
Conjunction occurs when the planet and the Sun are seen in the same direction from the Earth. This can happen when the planet passes between the Earth and the Sun (inferior conjunction) or when the planet lies on the opposite side of the Sun from the Earth (superior conjunction). When the planet and the Sun are opposite each other as seen from the Earth, opposition occurs. At quadrature, the planet and the Sun are separated by 90° as seen from the Earth. Greatest elongation occurs when the angular distance between the planet and the Sun is largest. Only for Mercury and Venus is greatest elongation less than 180°.

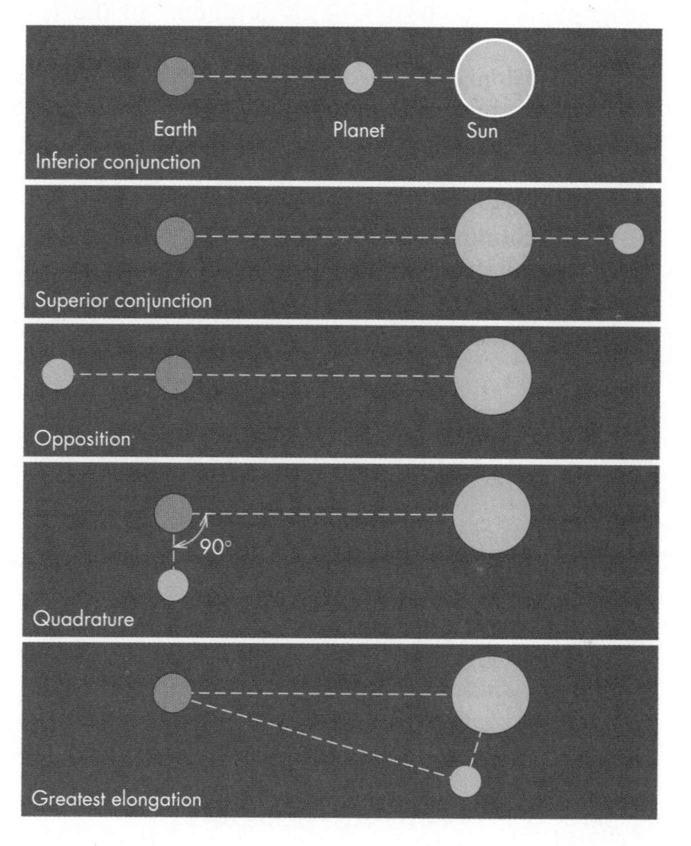

Eventually the Babylonians developed mathematical descriptions of planetary motions. They were able to express the position of a planet as the sum of a series of regular cycles, which took different lengths of time to complete. The Babylonians used the mathematical expressions to produce almanacs that, like our own almanacs, gave the positions of the planets for key dates, such as the beginning of each month. Apparently, the Babylonians never tried to explain the motions of the planets using a physical model of the solar system. They treated each planet separately and never devised a general system to account for the nature of planetary motions. However, the Babylonians' ability to predict celestial events marked the beginning of astronomy in the scientific sense.

The Babylonians applied the same kind of analysis to the motion of the Moon that they did to the planets and, by about 2700 years ago, were able to predict some lunar eclipses. They discovered that eclipses occur in series that repeat at intervals of 223 months (18.6 years). Knowing this interval of time, called the *saros,* the Babylonians were able to consult old records and predict whether an eclipse would occur at full Moon during a particular month. Eclipses are described further in Chapter 9.

The Babylonians used their lengthy records of observations to develop methods for predicting the positions of the Sun, Moon, and planets. The methods were based on their discovery of repeating patterns in the motions of celestial objects.

3.2 EGYPTIAN ASTRONOMY

While the Mesopotamians were building their vast record of astronomical observations, astronomers were also at work in Egypt. However, the Egyptian emphasis in astronomy was somewhat different from that of the Mesopotamians. As far as we know, the Egyptians developed and used astronomy entirely for practical purposes, such as developing a calendar to be used for predicting the Nile flood (see Chapter 2) and in aligning temples and monuments. Many Egyptian temples and monuments, such as the Great Pyramid, were aligned with respect to the cardinal points—north, south, east, and west. The Egyptians must have used astronomical methods to align their buildings, but no records have survived to tell us how they did it. Other buildings, such as the temple of Amen-Ra at Karnak (Chapter 2), were aligned with respect to the rising or setting of the Sun at the time of the summer solstice. Some temples were aligned with the rising points of bright stars, but no one knows the reasons they were built to do so.

Apparently, the Egyptian picture of the universe remained entirely descriptive and mythological during the long course of their civilization. Although they may have developed an alternative and more scientific view leading to plausible explanations, we have no evidence that they did. We actually know very little about Egyptian astronomy because so little of what the ancient Egyptians wrote has survived. We have a few manuscripts that tell us about their mathematics, but almost no descriptions of their science and engineering has survived. The Egyptians' influence on the Greek astronomers, which some historians believe may have been considerable, is also something that is relatively unknown. It is possible that much of what we usually consider Greek astronomy was based on much earlier work done in Egypt.

3.3 EARLY GREEK ASTRONOMY

The Greeks are generally acknowledged as the first people to raise astronomy from the level of prediction to that of explanation and understanding. The study of Greek astronomy is often frustrating, however, because such a large portion of what the Greeks wrote about astronomy has been lost. We know of many important works on astronomy only through descriptions of them by other authors. Sometimes these descriptions consist only of a tantalizing sentence or two. Although we know the general contributions of many Greek astronomers, we often are ignorant about the details of what they did.

People today often underestimate the timescales involved in ancient history. Although we tend to think of the many Greek astronomers as if they were contemporaries of each other, the interval between the first and last of the great astronomers of Greek civilization was nearly twice as long as the span of time that separates us from Copernicus. Figure 3.3 is a chronology of the development of Greek astronomy. The development of Greek astronomy also took place across a wide region, as shown in Figure 3.4. Many of the most notable Greek astronomers lived in Greek cities in Asia Minor or in colonies in Italy and Sicily. After Alexander the Great (356–323 B.C.) conquered Egypt in 332 B.C., he founded the city of Alexandria, which became a leading center of Greek civilization. Much of the significant work of later Greek astronomers was carried out in or near Alexandria.

The Astronomers of Miletus

The earliest Greek references to astronomy are found in the works of the poets Homer and Hesiod, who give practical astronomical advice for navigation and for agricultural

FIGURE 3.3
A Timeline of World Events and Famous People (left) and Notable Greek Astronomers (right)
The pictures portray (from top to bottom) Sappho, Alexander, Archimedes, and Ptolemy.

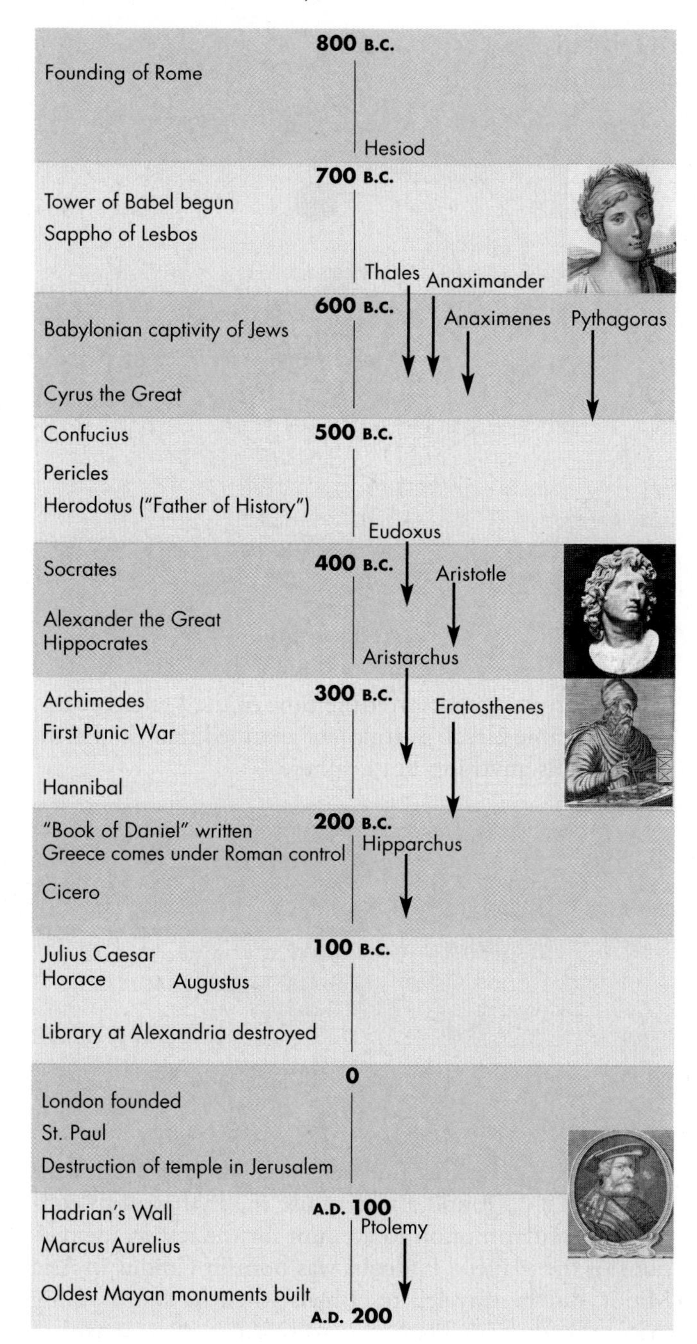

activities. For example, Hesiod, who lived about 2700 years ago, tells us

> Then, when Orion and Sirius are come to the middle of the sky, and the rosy-fingered Dawn confronts Arcturus then, Perses, cut off all your grapes, and bring them home with you.

FIGURE 3.4
Greek Astronomical Centers
Contributions to the development of astronomy were made by Greek astronomers living in many cities in the eastern Mediterranean.

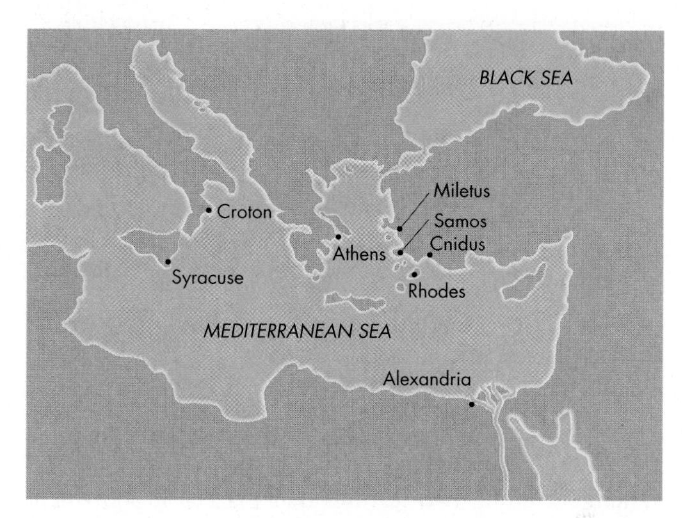

Other astronomical events tell when to do other things, such as plowing and pruning.

About a century after Hesiod, the city of Miletus became the home of a remarkable trio of astronomers. The first of the Milesian astronomers was Thales (c.624–547 B.C.), a merchant who was said to have traveled to Egypt to gather ancient knowledge. The Egyptians taught him how to find the distance to a ship at sea by observing it from two points on the shore and how to calculate the height of a pyramid by measuring the length of its shadow. The most famous story about Thales (reported more than a century later by the historian Herodotus) is that he predicted a total solar eclipse. The eclipse took place during a battle between the Lydians and the Persians in 585 B.C. The two armies were so awestruck by the eclipse that they put down their arms and ended the battle. It seems impossible that Thales could actually have predicted the eclipse. At that time not even the Babylonian astronomers could predict solar eclipses, but they could predict eclipses of the Moon. Perhaps Thales had heard of the 18.6-year cycle of lunar eclipses and had mistakenly applied the cycle to solar eclipses.

Anaximander (611–547 B.C.), a contemporary of Thales, attempted to give physical explanations for the celestial objects and their motions. He described the Earth as a cylinder floating freely in space. The Sun, Moon, and stars are hollow, fire-filled wheels with many holes from which their light emerges. This model may seem naive, but it was the forerunner of later Greek attempts to provide nonmythical explanations of celestial phenomena and explain the heavens in terms of earthly counterparts.

The third Milesian astronomer was Anaximenes (585–526 B.C.), who also gave mechanical explanations of celestial phenomena. Anaximenes believed that the stars were fixed like nail heads to a solid, crystalline vault surrounding the Earth. This idea gradually grew into the concept of the celestial sphere.

The astronomers of Miletus were the first to use mechanical explanations of celestial phenomena. The ideas of one of them, Anaximenes, eventually grew into the concept of the celestial sphere.

Pythagoras and His Students

Pythagoras (c.582–500 B.C.) and his students carried scientific model making much further than the Milesian astronomers had. They developed a model for the solar system including two features that marked a permanent break from previous thinking about the universe. First, the model assumed that the celestial bodies move in circular paths. Second, the model introduced the idea that most celestial objects, including the Earth, are spheres.

Pythagoras was born on the island of Samos. When he was about 50, he moved to Croton, in southern Italy, and founded a combination religious sect and philosophical school. Pythagoras himself left no writings, so it is difficult to determine which of the many things attributed to him were actually done by him rather than by one of his students. The Pythagoreans rejected the idea that the Earth is flat and adopted the notion of a spherical Earth. They also adopted the completely novel idea that the Earth isn't stationary but moves about an unseen, central fire each 24 hours, as shown in Figure 3.5. The central fire can't be seen because another object, a counter earth, also revolves each 24 hours and always remains between the Earth and the central fire. The Pythagoreans believed that with the planets, Sun, Moon, and Earth all whirling around, a great deal of sound is produced. The sound each body makes has a pitch that depends on its distance from the center. The combination of these sounds produces celestial "harmony." You may wonder why we don't hear the music of the spheres. The Pythagorean answer was that we always hear the sound as a constant background noise, and because we have no silence with which to compare it, we are unaware of the perpetual harmony of the spheres.

Despite the obvious problems of the Pythagorean model of the solar system, the model represented an important turning point in the development of Greek astronomy. The idea that planetary motion consists of motion on perfect circles took root and grew for 600 years, culminating in the very successful geocentric model of Ptolemy. The spherical shape of the celestial bodies also was quickly accepted. After the time of the Pythagoreans, no respectable Greek astronomer assumed that the Earth, at least, was anything but a sphere.

FIGURE 3.5
The Pythagorean Model of the Solar System
The Earth moves about a central fire, which cannot be seen because a "counter earth" is always in the way.

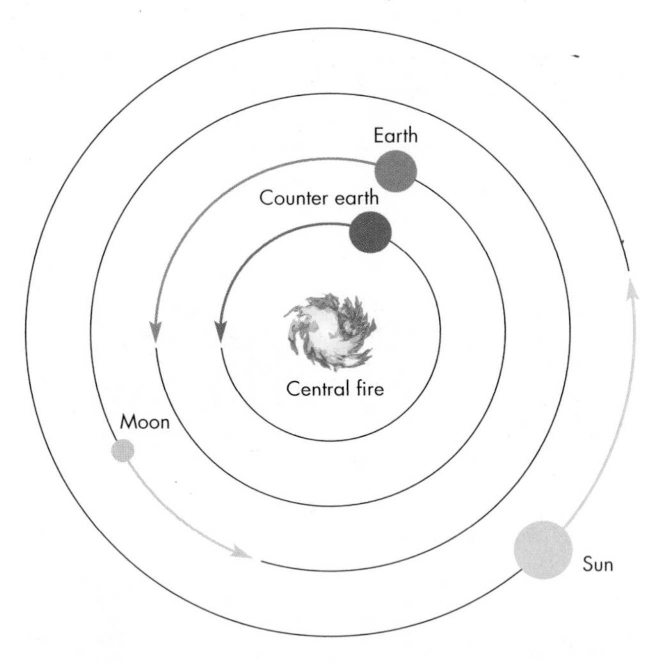

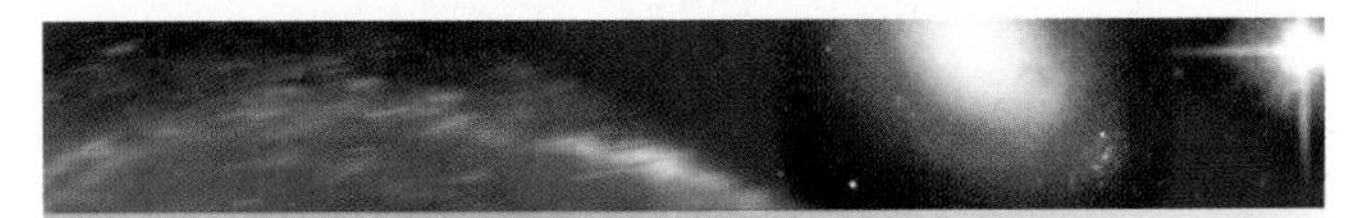

The Pythagoreans originated the idea that the universe consists of spherical bodies moving in perfect circular paths.

Eudoxus

About 200 years after the time of Pythagoras, Eudoxus (408–355 B.C.) found a way to use the Pythagorean concept of circular motion to account for the retrograde motions of the planets. Eudoxus was born in Cnidus in Asia Minor, but he traveled to Athens when he was a young man to study with Plato (427–347 B.C.). Plato is said to have proposed that planetary motion should be explained using a combination of several circular motions. This was exactly the approach that Eudoxus took. In Eudoxus's system the motions of the planets, the Sun, and the Moon were explained by the rotation of sets of spheres that have common centers. The motions of the stars were explained by the daily rotation of an outer sphere. The Sun's motion required two spheres. One of the spheres rotated westward about the Earth once a day. The other sphere rotated

FIGURE 3.6
Eudoxus's Model for the Motion of Mars
The inner two spheres, which rotate with the synodic period of Mars, produce a figure-eight motion. The figure-eight motion is stretched eastward along the ecliptic by the rotation of the second sphere. The outer sphere causes the daily westward motion of Mars.

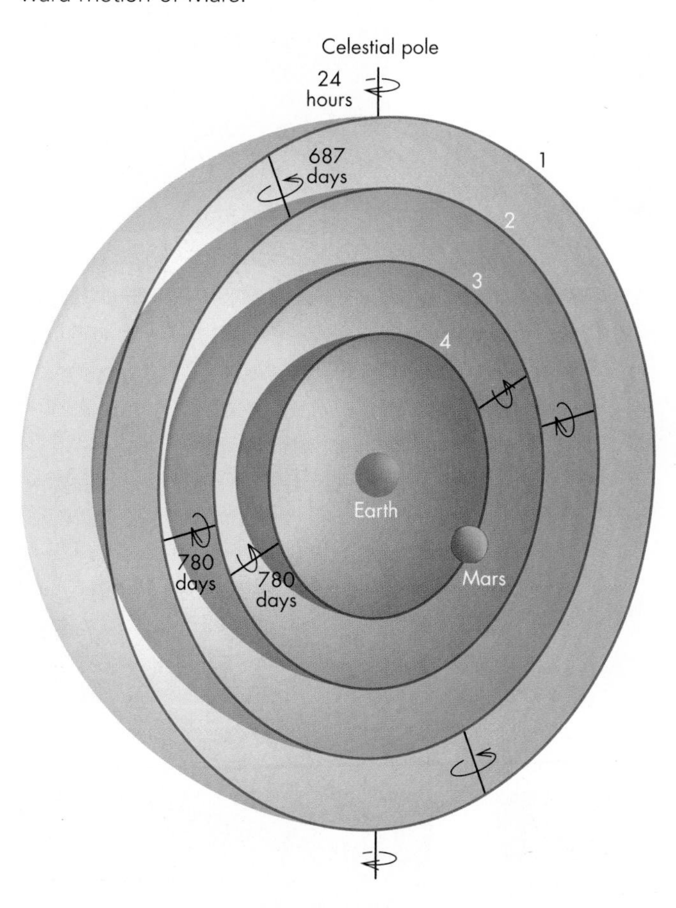

eastward once per year and carried the Sun as it went through its annual motion around the ecliptic.

Four spheres, shown in Figure 3.6, are needed to account for the motion of each planet. The inner two spheres (3 and 4) work together to produce a cyclic figure-eight motion. The spheres rotate in the same length of time, the synodic period of the planet, but in opposite directions. One sphere is tilted slightly with respect to the other. If the spheres weren't tilted with respect to each other, their motions would exactly cancel out and would have no effect on the motion of the planet. The tilt means that although they cancel out on average, the planet wanders back and forth in a figure-eight motion. The second sphere rotates eastward. When motion on the figure eight is also eastward, the planet moves more rapidly than normal. When motion on the figure eight is westward, the planet more than compensates for the normal eastward movement, and it moves westward in retrograde motion. The combined actions of spheres 2, 3, and 4 strongly resemble the observed motion of a planet. The outer sphere rotates westward in about 24 hours and reproduces the diurnal motion of the planet.

This ingenious scheme roughly imitates the observed motions of the planets, but the theory has several shortcomings. First, if you use Eudoxus's spheres to explain retrograde motion, then every retrograde loop for a planet should look exactly the same. In reality, the retrograde loops for a given planet show a striking variety of appearances. Second, although the system produces reasonable results for some planets, it doesn't work very well for others. Finally, in Eudoxus's system, a planet always stays at the same distance from the Earth. If the planet, like the stars, has a constant intrinsic brightness, then it should always look equally bright to us. However, a constant level of brightness does not agree with observations. Long ago, astronomers realized that both Mars and Venus are much brighter at some times than at others. Although the system of Eudoxus didn't work as well as later geocentric systems, it was the brilliant beginning of a long line of attempts using perfect circular motion to explain the solar system.

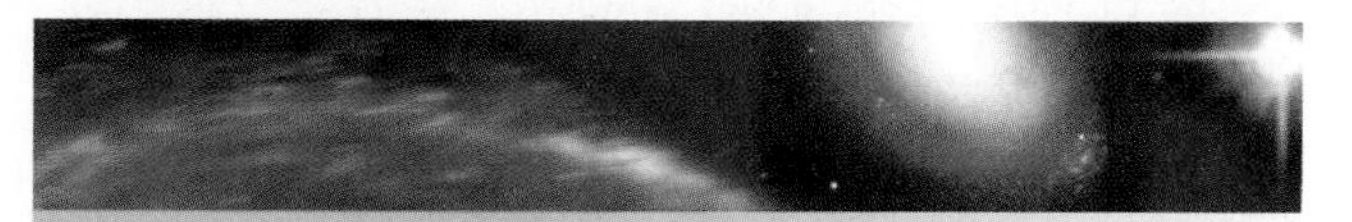

Eudoxus invented a model of the universe based on the rotation of spheres. The oppositely directed motions of two spheres produced a figure-eight motion for a planet. When combined with an eastward rotation of a third sphere, the figure-eight motion resembled the retrograde motions of planets.

Aristotle

Like Eudoxus, Aristotle (384–322 B.C.) was a student of Plato. Aristotle accepted the model of Eudoxus, but he modified it to include even more concentric spheres. Aristotle argued that all celestial motions must be circular because circular motion is the only kind of motion that can be repeated forever and is suitable for unchanging bodies such as the Moon, Sun, and stars.

Aristotle also held the Pythagorean view that the celestial bodies are spheres. He supported this claim with several proofs that showed the Earth is a sphere. One of the proofs proposes that falling objects move toward the center of the Earth. For a sphere, this means that bodies everywhere fall straight down (perpendicular to the surface). As Figure 3.7 shows, if the Earth had another shape, such as a cube, for example, this would not be true everywhere. Another proof argues that the shadow of the Earth, the shape of which can be seen on the Moon during the early and late stages of a total lunar eclipse, is always

FIGURE 3.7
One of Aristotle's Proofs that the Earth Is a Sphere
Aristotle argued that falling bodies move toward the center of the Earth. Only if the Earth is a sphere can that motion always be straight downward, perpendicular to the Earth's surface.

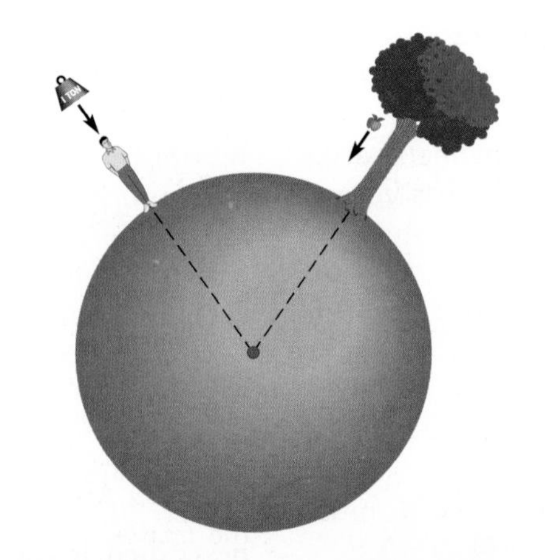

FIGURE 3.8
Another of Aristotle's Proofs that the Earth Is a Sphere
A, The Earth's shadow on the Moon during lunar eclipses is always round. **B,** If the Earth were a flat disk, there would be some eclipses in which the Earth would cast a flat shadow on the Moon.

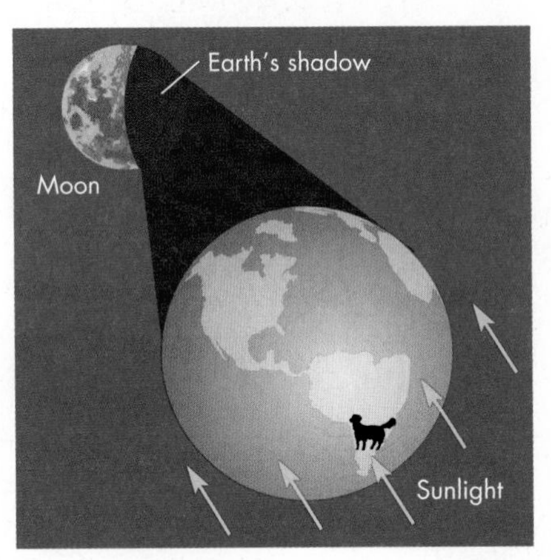

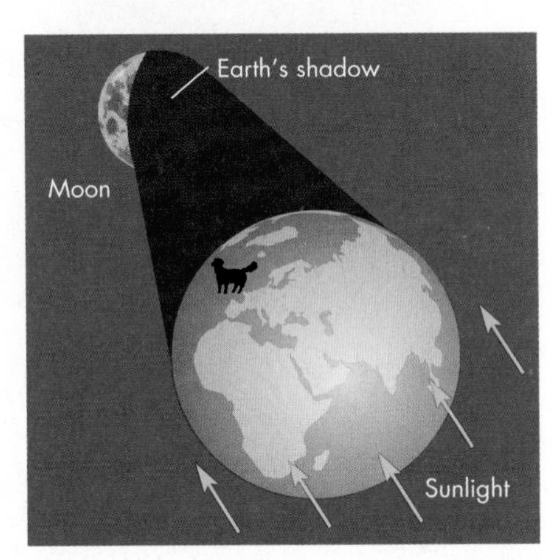

A Spherical Earth

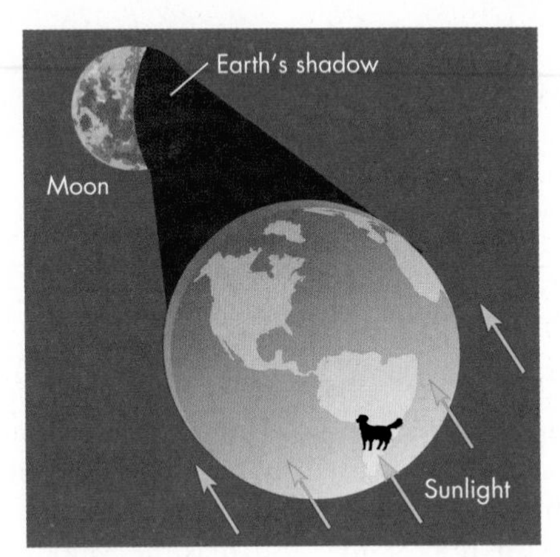

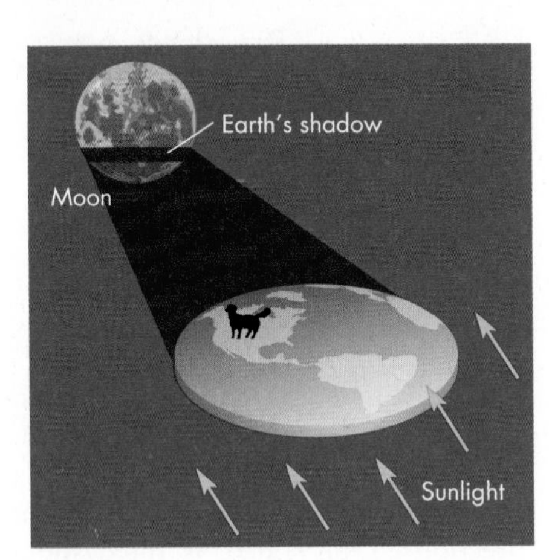

B Flat Earth

circular. This could only be true if the Earth were a sphere. Figure 3.8 shows that if the Earth were a disk, for example, there would be some occasions when the flat side of the Earth would produce a straight shadow on the Moon. In his third argument regarding the Earth's shape, Aristotle pointed out that the Earth must be a sphere because some stars can be seen in Egypt but not in Greece. If the Earth were flat, the same stars could be seen from any part of the world. Aristotle also believed that the other bodies in the universe must also be spheres. He argued that the Moon must be a sphere in order to show the phases that it does. If one celestial object is spherical, he argued, the others must be also.

Aristotle was not primarily an astronomer. His thoughts about astronomy are scattered throughout his works in science and philosophy. In these works he developed many ideas about science that remained tremendously influential for nearly 2000 years. One of his most influential ideas, which he developed from an earlier proposal by Empedocles (c.494–434 B.C.), was that the world is composed of the four elements—earth, air, fire, and water. Each of these elements has its natural place to which it moves. The heavier elements, earth and water, naturally move downward toward the center of the Earth while air and fire naturally move upward. Stones fall toward the Earth because they are seeking their natural place. The tendency of objects to seek their natural place had an impact on Aristotle's thinking about the location of the Earth in the universe. He argued that because all the heavy material had accumulated in the Earth, it is unreasonable to suppose that the heaviest part of the universe is in motion. Therefore, Aristotle concluded that the Earth must be motionless at the center of the universe.

Aristotle taught that all celestial motion must be circular and that celestial bodies were spheres. He argued that the Earth must be motionless at the center of the universe.

3.4 LATER GREEK ASTRONOMY

By Aristotle's time, the Greek astronomers had developed a view of the cosmos that was generally consistent with the observed patterns of celestial motion. In the period that followed, the Greeks began to apply geometrical principles to astronomy. The results were spectacular. During the next five centuries the Greek astronomers made important astronomical discoveries about the shapes, sizes, and distances of the solar system bodies. They also developed a new and much more accurate model for the motions of the Sun, Moon, and planets. During the later, more mathematical, period of Greek astronomy, four astronomers—Aristarchus, Eratosthenes, Hipparchus, and Ptolemy—stand out above the rest.

Aristarchus

Shortly after Euclid (c.300 B.C.) wrote his famous book on geometry, Aristarchus of Samos (c.310–230 B.C.) applied geometry to the study of astronomy. Aristarchus showed that it is possible to use geometry to find the distance to the Moon and the relative distances and sizes of the Sun and Moon. However, he apparently thought of these calculations merely as geometrical exercises and never carried them out.

The method Aristarchus described for determining the distance to the Moon required finding both the Moon's angular diameter and its linear diameter. The Moon's angular diameter can be measured directly. To find the Moon's linear diameter, Aristarchus proposed using lunar eclipses to give the relative sizes of the Moon and Earth, whose size was known. The essence of the method is shown in Figure 3.9, which shows the Earth and Moon as well as the shadow of the Earth. If the actual shape of the Earth's shadow were a cylinder rather than a cone, the method would be trivial, because the Earth's shadow would have the same diameter as the Earth itself. In that case, the relative sizes of the Earth and Moon could be found by determining how many lunar diameters the Moon moves while it is completely within the Earth's shadow. This can be found by comparing the length of time between the beginning of the eclipse and the end of the total eclipse to how long it takes the shadow of the Earth to completely cover the Moon. However, it is possible to take into account the actual shape of the converging shadow of the Earth. Once the size and angular diameter of the Moon are known, its distance can be found as well.

The distance to the Moon, the linear diameter of the Moon, and the Moon's angular diameter comprise two sides and an angle of a long, skinny triangle, like that shown in Figure 3.10. The small angle equation relates the distance to an object (d), its linear diameter (D), and its angular diameter (θ)

$$\theta = 206{,}265 \times \frac{D}{d} \tag{3.1}$$

when θ is measured in seconds of arc (see box for Equation 3.1). The number 206,265 is the number of seconds of arc in one radian, where one radian is an angle of $360°/2\pi = 57.3°$.

Equation 3.1

The Small Angle Equation

Equation 3.1 can be used to calculate the angular diameter of a body if its linear diameter and distance are known. For example, Venus has a linear diameter of 12,100 km. What is its angular diameter when it is at a distance of 5×10^7 km? (This is approximately the distance between Earth and Venus when they are closest to one another.)

Using 12,100 km for D and 5×10^7 km for d, Equation 3.1 gives

$$\theta = 206{,}265 \times \frac{12{,}100 \text{ km}}{5 \times 10^7 \text{ km}}$$

$$= 50 \text{ seconds of arc}$$

FIGURE 3.9
The Moon, the Earth, and the Earth's Shadow During an Eclipse of the Moon
By measuring the relative sizes of the Moon and the Earth's shadow at the Moon's distance from the Earth, it is possible to find the distance to the Moon in terms of the diameter of the Earth. The figure is not to scale. The Earth and Moon are shown much too large relative to the distance between them.

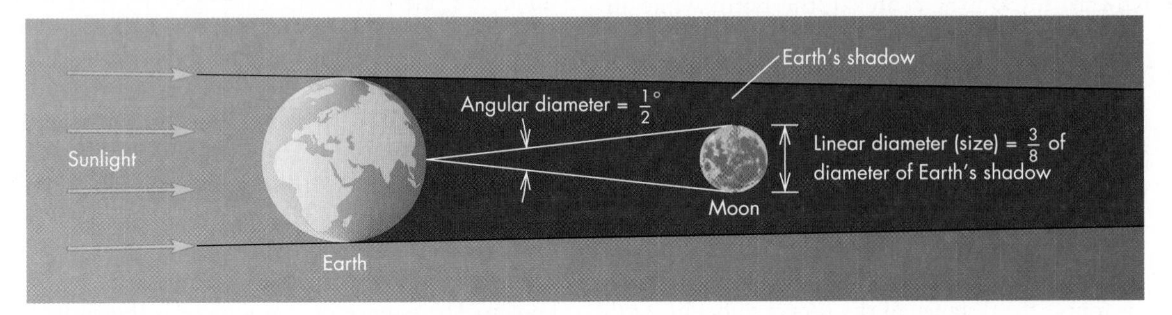

FIGURE 3.10
The Small Angle Equation
The small angle equation relates angular diameter, linear diameter, and distance. Given any two of these, the third may be found.

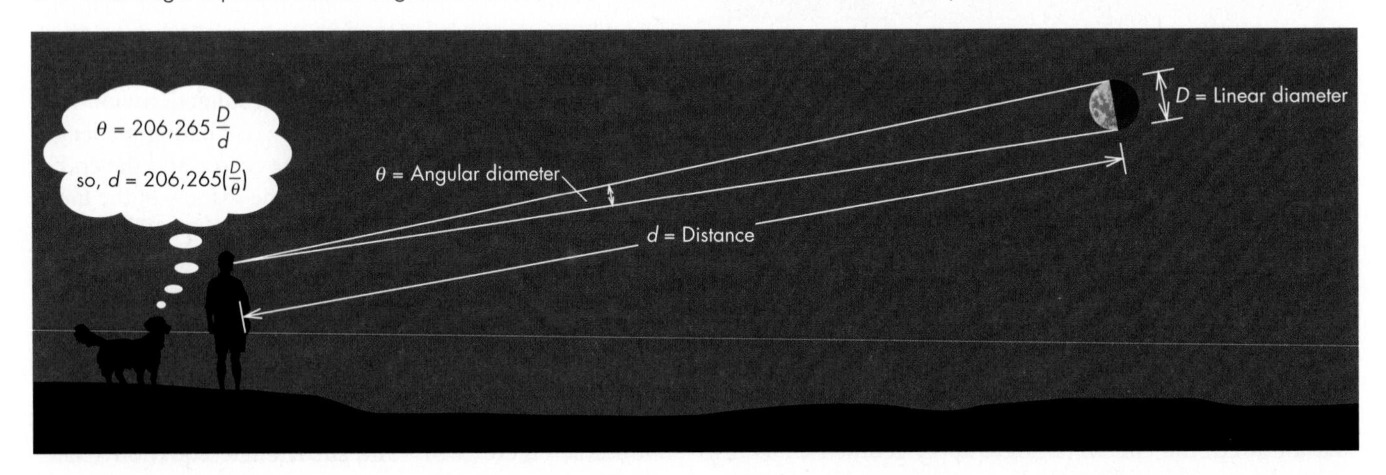

The small angle equation can be rewritten to solve for distance, *d*. Thus, when the angular diameter and linear diameter of an object are known, its distance is given by

$$d = 206{,}265 \times \frac{D}{\theta} \tag{3.2}$$

(See box for Equation 3.2.) The small angle equation is an approximation that becomes more accurate as the angle θ gets smaller and the triangle gets skinnier. Fortunately, nearly all of the angles measured in astronomy are small enough that the small angle equation can be used to relate the actual sizes of objects to their angular sizes.

It is important to realize that the angular size of an object alone isn't enough to tell us whether an object is large or small. That is, a small object that is close to us can have the same angular size as a very distant but very large object. Nevertheless, the small angle equation is consistent with our intuitive idea that for a group of objects at the

Distance, Linear Diameter, and Angular Diameter

Equation 3.2 can be used to calculate the distance to a body if its linear diameter and angular diameter are known. In the case of the Moon, the linear diameter, *D*, is 3480 km. The angular diameter of the Moon is 0.5°. In order to use Equation 3.2, the angular diameter of the Moon, θ, must be converted from degrees to seconds of arc. There are 3600 seconds of arc in a degree so 0.5° = 1800 seconds of arc. Using these values for *D* and θ, the distance of the Moon is given by

$$d = 206{,}265 \times \frac{3480 \text{ km}}{1800} = 400{,}000 \text{ km}$$

same distance, the one with the largest angular size must be the largest in actual size.

The small angle equation relates the angular size of an object, its actual size, and its distance. The equation can be used to find any of the three if the other two are known.

The method of Aristarchus was later improved and carefully carried out by Hipparchus (c.190–125 B.C.). Hipparchus found the distance of the Moon to be 59 Earth radii, which is remarkably close to the actual value of about 60 Earth radii. Because astronomers already knew the size of the Earth reasonably well, Hipparchus's measurement placed the Moon several hundred thousand miles from the Earth.

Aristarchus also showed that it might be possible to find the ratio of the distance to the Sun to the distance of the Moon by observing the angle between the Sun and Moon at the time of first or third quarter Moon. Because the Moon shines by reflected sunlight, only the half of the Moon facing the Sun is lit. At quarter Moon, as shown in Figure 3.11, we see just half of the bright part of the Moon. If we were standing on the Moon at the time of quarter Moon, the angle between the Sun and Earth would be 90°. However, there are an infinite number of triangles involving the Earth, Moon, and Sun in which that angle would be 90°. The shapes of these triangles depend on the ratio of the distance to the Sun to that of the Moon. Triangles in which the Sun is much more distant than the Moon are very skinny, as Figure 3.12 shows. In such a triangle, the angle *A*, the angular distance between the Sun and the Moon, approaches 90°. On the other hand, triangles in which the distances to the Sun and Moon are comparable are stubby ones in which the angle *A* is much less than 90°. By determining the value of the angle *A*, the relative distances to the Sun and Moon can be found. Aristarchus estimated the angle *A* to be 87°, in which case

FIGURE 3.11
The Positions of the Earth, Moon, and Sun When the Moon Is at Quarter Phase
At the Moon, the angle between the Sun and the Earth is 90°. Angle *A* is the angle between the Sun and Moon, as seen from the Earth.

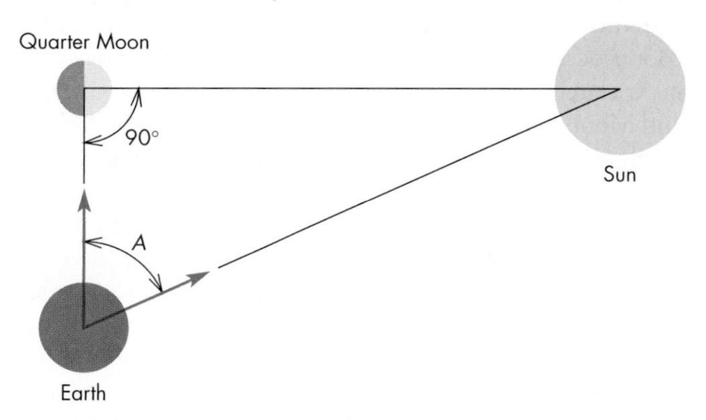

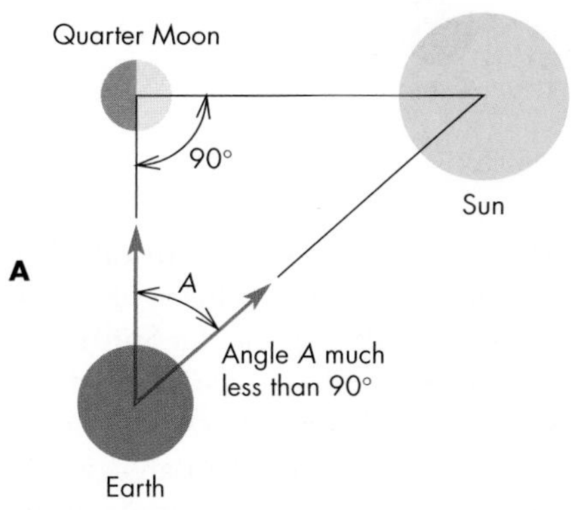

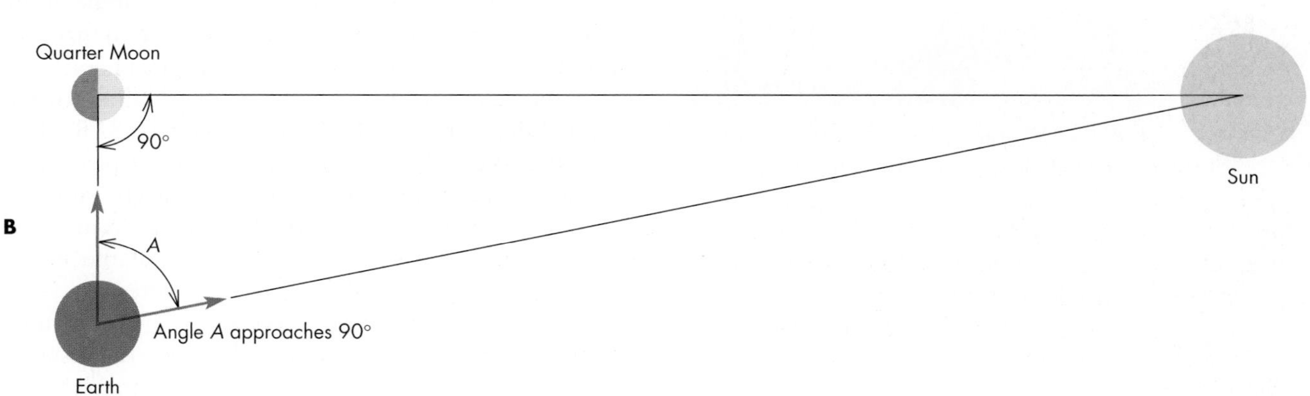

FIGURE 3.12
The Earth, Moon, and Sun at Quarter Moon
A, If the Sun and Moon are at about the same distance from the Earth, the angle *A* is much less than 90°. **B,** However, if the Sun is much more distant than the Moon, angle *A* approaches 90°.

the Sun must be 19 times as distant as the Moon. This put the Sun at a distance of several million miles. Aristarchus's result that the Sun is 19 times as distant as the Moon was accepted as a fact for nearly 2000 years.

The method of Aristarchus is geometrically sound but practically impossible to carry out. In the first place it is extremely difficult to tell the precise moment when quarter Moon occurs. In addition, it would have been impossible in Aristarchus's time to measure the angle between the Sun and Moon with enough accuracy to make the method work. Angle A is actually about 89.83°, making the distance to the Sun about 20 times as large as Aristarchus's estimate. Nevertheless, Aristarchus's methods showed for the first time that the universe is enormously larger than the Earth.

Aristarchus also calculated that the Sun has a diameter about seven times as large as the Earth. Perhaps it was the immense size of the Sun that led him to propose that the Sun is stationary and that the Earth orbits the Sun in a circular path. Aristarchus also proposed that the Earth spins on its axis once a day to produce the apparent diurnal motions of the celestial objects. Aristarchus's remarkably modern view of the solar system was almost completely disregarded by other Greek astronomers. One reason was that they found it impossible to believe the Earth could be spinning rapidly on its axis and moving at great speed about the Sun without perceptible consequences. They argued that there should be a great wind rushing in the direction opposite to the Earth's rotation. A stone dropped from a cliff should fall to the west as the Earth rushed away to the east. Objections like these were only overcome in the time of Galileo and Newton 2000 years later. An additional objection to Aristarchus's theory was that the heavens should change in appearance during a year if the Earth moved about the Sun (Figure 3.13). Stars should be brighter and farther apart when we are close to them and dimmer and closer together when the Earth is far from them. The changeless nature of the stars and the constant shapes and sizes of the constellations can be explained if they are so distant that the changing position of the Earth is completely unimportant. The Greeks, however, were not prepared to accept such an enormous universe. They imagined the celestial sphere to be just beyond Saturn.

Aristarchus applied geometry to astronomy and showed it was possible to find the Moon's distance and the relative distances to the Sun and Moon. He showed that the universe is immensely larger than the Earth. He also proposed that the Sun is stationary at the center of the universe. However, his proposal was not adopted by the other Greek astronomers.

FIGURE 3.13
An Objection to the Revolution of the Earth
Greek astronomers argued that if the Earth orbited the Sun, the brightnesses of the stars and the shapes and sizes of the constellations should vary throughout the year.

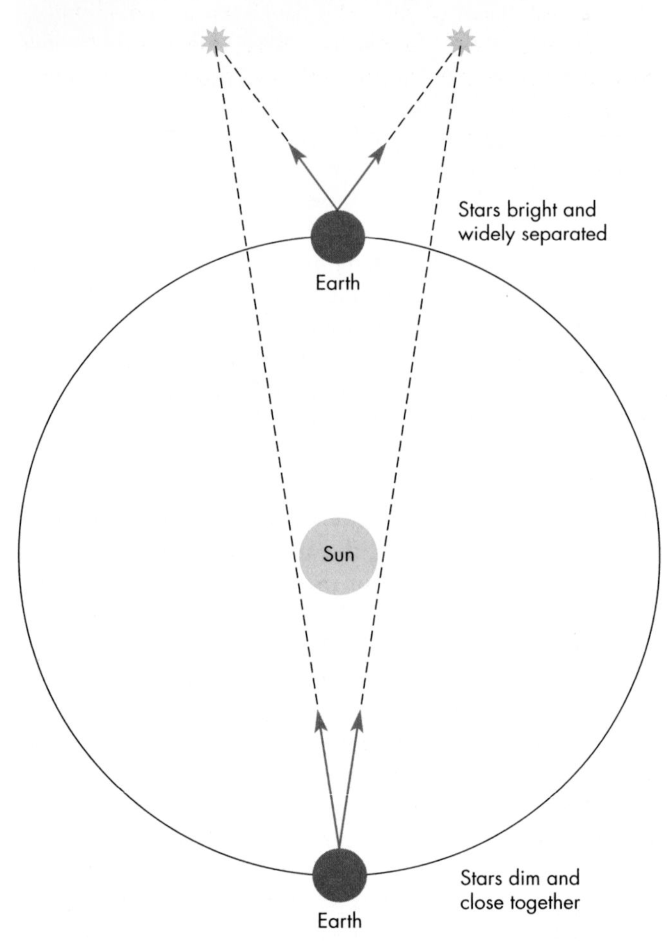

Eratosthenes

Eratosthenes (c.276–195 B.C.) used geometry to calculate the size of the Earth. His method is shown in Figure 3.14. The difference in the noonday altitude of the Sun is the same as the difference in latitude of the two places from which the Sun is being observed. Eratosthenes, who was the director of the great library at Alexandria, sent an assistant to Syene, in southern Egypt, to observe the altitude of the Sun at the summer solstice. The assistant reported that sunlight reached the bottom of a deep vertical pit at noon. This meant that the Sun was directly overhead. Eratosthenes measured the Sun's altitude at Alexandria to be 82.8°. This meant that the difference in latitude of the two cities is 7.2°, or $\frac{1}{50}$ of a circle. Thus, if Syene is directly south of Alexandria (as is nearly the case) the distance from Syene to Alexandria is $\frac{1}{50}$ of the circumference of the Earth. Eratosthenes's result was almost precisely correct.

The accuracy of Eratosthenes's result is often obscured by uncertainty about how precisely he knew the distance between Syene and Alexandria. In the time of Eratosthenes,

FIGURE 3.14
Eratosthenes's Method for Finding the Circumference of the Earth
The difference in latitude between Syene and Alexandria is the same as the difference in the altitude of the Sun at noon at those two places on the same day. Because the difference in latitude is 7.2°, 1⁄50 of a full circle, the linear distance between Syene and Alexandria is 1⁄50 of the circumference of the Earth.

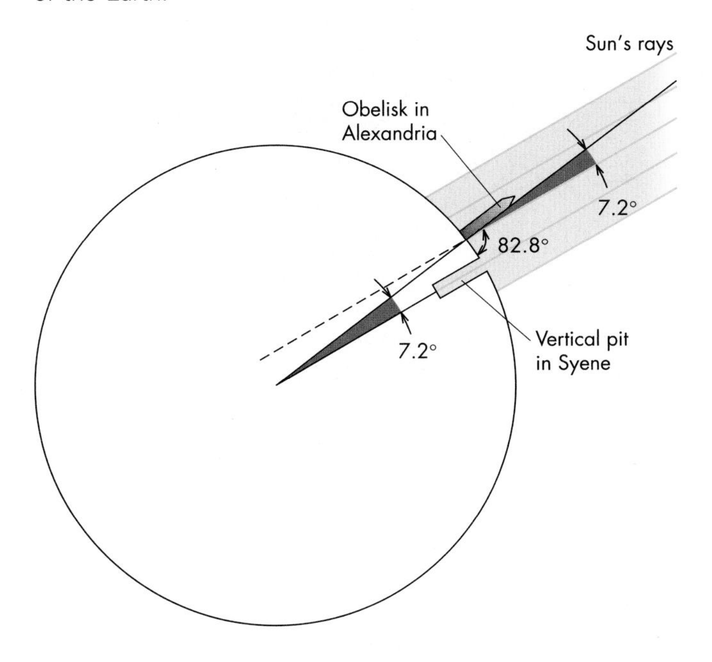

the best estimate of this distance was 5000 stadia, which was derived from the time it took the king's messengers to travel between the two cities. Several kinds of stadia were used in the ancient world. Depending on which of them Eratosthenes was referring to, his estimate for the circumference of the Earth differed from the true value of 40,100 km (25,000 miles) by somewhere between a few percent and about 15%.

However, the accuracy of Eratosthenes's result isn't really the point. Eratosthenes's method was sound and his astronomical measurements apparently were accurate. He knew that the distance from Syene to Alexandria was 1⁄50 of the way around the Earth. The main limitation on his ability to find the circumference of the Earth was his dependence on other people's measurement of the distance between the two cities.

Eratosthenes used the relationship between latitude and the altitude of the midday Sun to find the difference in latitude between Syene and Alexandria. He found that the circumference of the Earth is 50 times as large as the distance between Syene and Alexandria, a surprisingly accurate result for his time.

Hipparchus

Hipparchus, who did most of his work on the island of Rhodes, has often been called the greatest astronomer of antiquity. Yet we have very little direct knowledge of his life and works. None of his major works survive, so most of what we know is based on reports of other astronomers who lived centuries later. Among his many contributions to astronomy were the refinement of Aristarchus's method of finding the distances to the Sun and Moon, an improved determination of the length of the year, and extensive observations and theories of the motions of the Sun and Moon.

In 134 B.C., Hipparchus discovered a new star (that is, one that increased in brightness enough to become visible to the naked eye for the first time). This discovery prompted him to begin the task of making what was probably the earliest systematic catalog of the brighter stars. Although his catalog has been lost, descriptions of it suggest he gave positions and an indication of brightness for about 850 stars. A representation of Hipparchus's catalog has recently been identified on a celestial globe on the shoulders of Atlas in a statue in Naples, Italy that dates to the second century A.D. After completing his catalog, Hipparchus compared his position for the bright star Spica with the positions measured by other Greek astronomers about 170 years earlier. He found that his position and the earlier ones differed by about 2°. Other stars that he checked showed the same shift. Figure 3.15 shows that the differences were due to **precession,** the slow circular motion of the north and south celestial poles among the stars. As the pole moves, the celestial equator moves as well. The equinoxes (the points where the ecliptic meets the equator) slide around the ecliptic, as shown in Figure 3.16. As this happens, coordinates such as right ascension and declination, which are measured from the position of the vernal equinox, change steadily. Babylonian astronomers had already noticed that the positions of the equinoxes among the stars were different at different times, but they had not followed up on the discovery. Hipparchus estimated that precession shifts the vernal equinox by about 1° every 78 years and completes a full cycle of 360° in 28,000 years. This value is within about 10% of the actual length of the precession period (26,000 years).

Precession is a slow movement of the celestial poles with respect to the stars. It causes the coordinates of stars to change with time. Hipparchus discovered precession when he compared his measurements of the positions of stars with positions measured by earlier Greek astronomers.

FIGURE 3.15
Precession of the Earth's Polar Axis
The direction among the stars toward which the Earth's polar axis points changes with time as the Earth precesses like a top. It takes 26,000 years for a precession cycle to be completed.

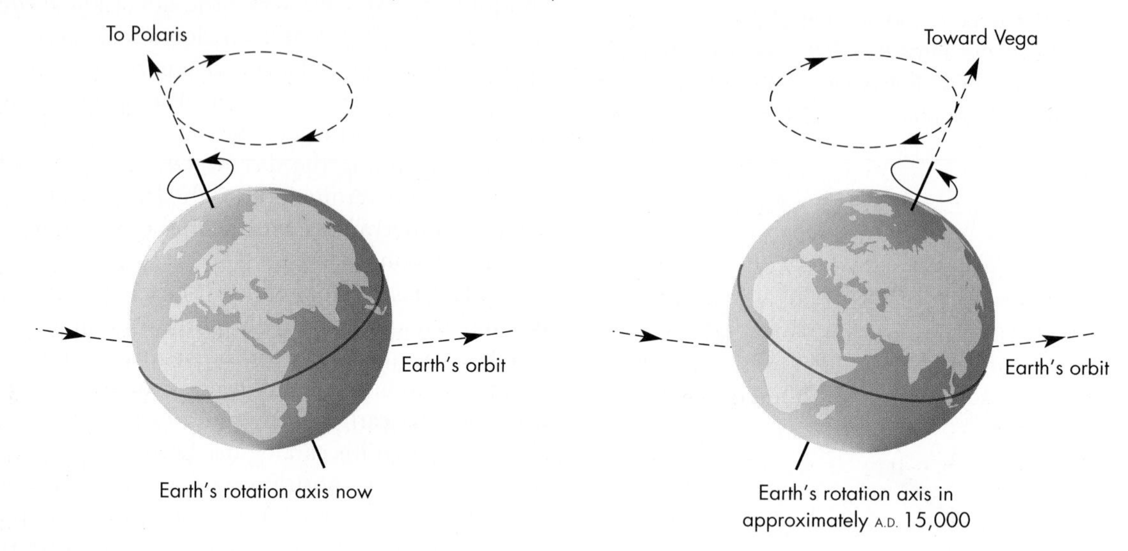

FIGURE 3.16
Precession and Celestial Coordinates
As the north celestial pole moves, so do the celestial equator and the vernal equinox. This means that the coordinates of stars, which are measured from the vernal equinox, change with time.

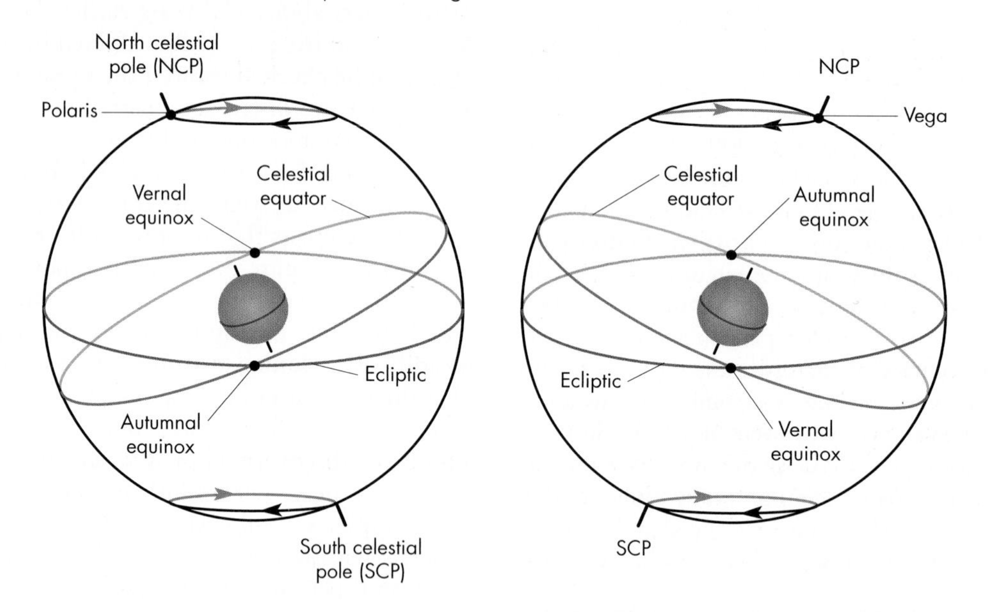

Ptolemy

Claudius Ptolemaeus, or Ptolemy, who worked in Alexandria from about A.D. 127–151, was the last of the great astronomers of antiquity. His major accomplishment was the completion of a system that could accurately account for the motions of the planets using a combination of regular circular motions. Although he is best known as an astronomer, Ptolemy also wrote important works on geography, music, and optics. As a geographer, he invented terrestrial latitude and longitude (with a zero-point that was different from the modern system) and gave these coordinates for eight thousand places. He was the first to orient maps with north at the top and east at the right.

Ptolemy's book on astronomy, usually referred to as the *Almagest,* after the Greek word μεγιστη (*megiste,* "the greatest"), is a detailed summary of his picture of the universe. This book contains improved methods for finding the distances to the Sun and Moon, and it also presents a catalog of more than a thousand stars giving their celestial coordinates as well as their brightnesses. Ptolemy's method of describing the brightnesses of stars, called the *magnitude system,* is still in use. Whether Ptolemy's catalog and magnitude system represented his own observations and ideas or were based primarily on those of Hipparchus is a matter of controversy among historians and astronomers. Ptolemy's most influential contribution to astronomy, however, was his geocentric model of the solar system.

Ptolemy's Model of the Solar System Although the system of Eudoxus could produce motions resembling those of the planets, it really couldn't stand a close comparison to those observed motions. Ptolemy's system, in contrast, was able to match closely the motion of each planet and predict its future position.

Ptolemy didn't invent the system named after him, he only completed it. Other astronomers, including Hipparchus, had been developing the idea for centuries. The system is **geocentric,** which means the Earth is stationary at the center. To justify the assumption that the Earth is stationary, Ptolemy used essentially the same arguments that had been used by Aristotle, namely that the Earth is too heavy to move and there would be drastic effects if it rotated on its axis or revolved around the Sun.

In its simplest form, Ptolemy's system proposes that the path of a planet is a combination of motions on two circular paths. As shown in Figure 3.17, the planet moves on an **epicycle,** which in turn moves along a path called the **deferent.** The combination of these paths produces a looping motion that allows the planet to move generally eastward (prograde), but with regular periods of westward (retrograde) movement. During the retrograde motion, the planet is nearer the Earth and brighter than normal, in agreement with brightness observations. Ptolemy adjusted three factors: the sizes of the epicycle and deferent, the speeds with which the planet moves along the epicycle and deferent, and the tilt of the epicycle, to give the best match to the motion of each planet. The entire system, including all the planets known in antiquity, is shown in Figure 3.18.

The simple system, however, can't quite match the observed positions of the planets. Ptolemy found that he needed to modify the system in two ways. First, he displaced the center of each deferent from the Earth, as shown in Figure 3.19. As a result, the Earth isn't truly at the center of the motion of any planet. Second, he proposed that the motion of the center of the epicycle around the deferent is uniform only if viewed from a spot called the **equant,** also shown in Figure 3.19. This meant that the planet didn't orbit at uniform speed on the deferent when viewed from the center or from the Earth. In the end, then, Ptolemy's system wasn't quite geocentric, and it did not involve uniform circular motion.

However, the system worked. The system was able to predict the motions of the planets with sufficient accuracy that with minor modifications it remained in regular use for nearly 1500 years. Whether it would have survived the challenges of other Greek astronomers is an interesting question. The system didn't have to. Shortly after Ptolemy lived, classical civilization began to unravel. Progress in astronomy (and practically everything else) came to a halt. No one seriously questioned the epicyclic model for more than 1000 years.

The epicyclic model perfected by Ptolemy used combinations of circular motions to reproduce the motions of the planets. The model could predict the positions of celestial objects with such accuracy that it was used for nearly 1500 years.

FIGURE 3.17
Ptolemy's Model of Motion of a Planet
A, A planet moves on an epicycle, which itself moves on a deferent. **B,** As viewed from the Earth, the combination of these motions mimics the prograde and retrograde motions of the planets.

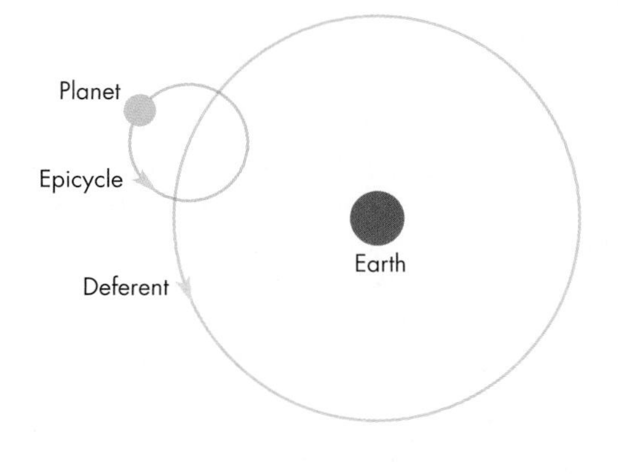

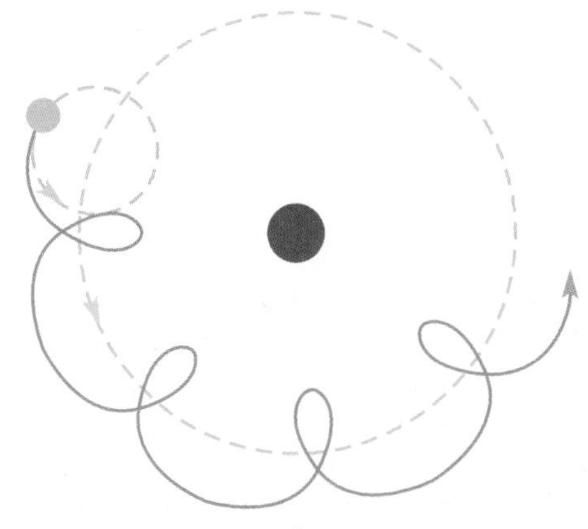

FIGURE 3.18
The Complete Geocentric Model
Venus and Mercury always move so that the centers of their epicycles lie on the Earth-Sun line.

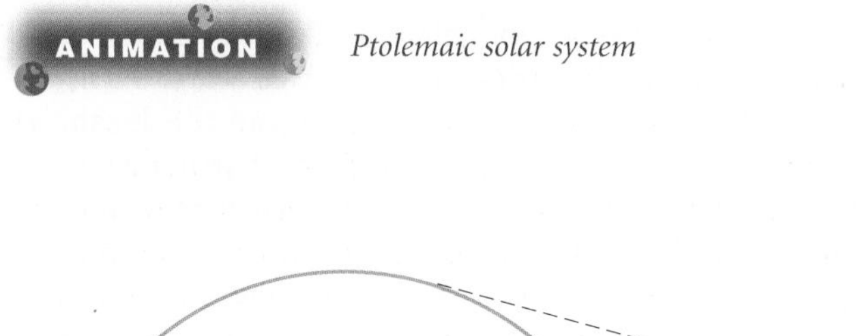

Ptolemaic solar system

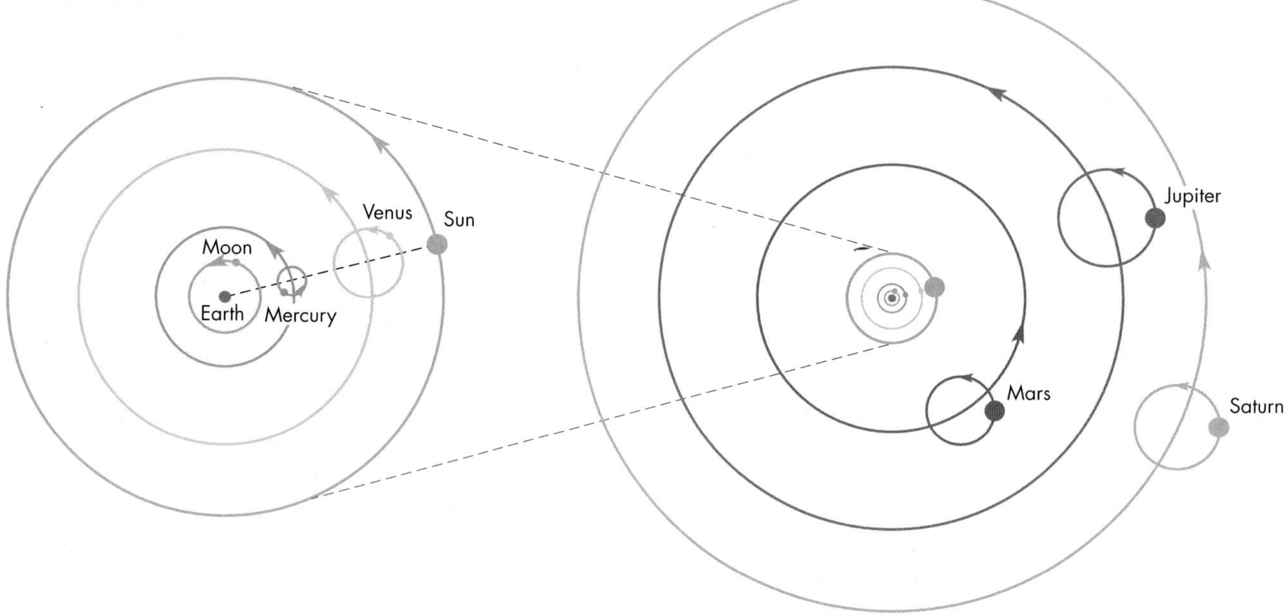

FIGURE 3.19
The "Real" Ptolemaic System
The Earth is not at the center of the deferent of the planet. Also, the planet moves so the motion of the center of its epicycle is uniform only if viewed from the equant. Viewed from either the center of the deferent or the Earth, the motion on the deferent is variable in speed.

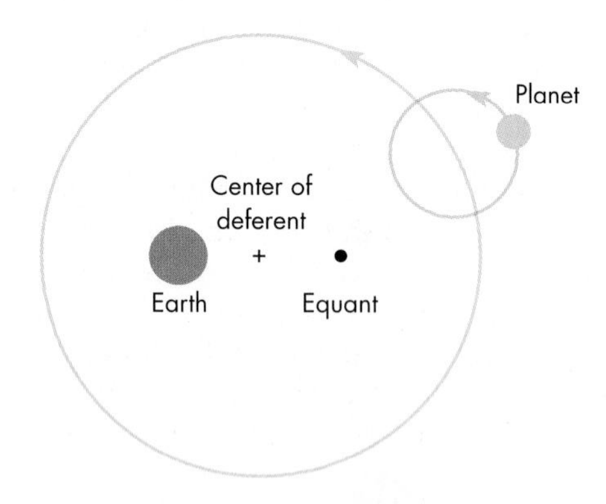

Thus, over the course of millennia, the ancient astronomers used careful observations of the heavens to develop a sophisticated model that could explain all the celestial motions they could see. The Greeks, in particular, had great success in applying mathematics to the study of astronomy. A comparable period of astronomical activity would not be seen again until the sixteenth century.

CHINESE AND MESOAMERICAN ASTRONOMY

Although it was the ancient astronomers of the eastern Mediterranean and the Near East whose work formed the foundation on which our present view of the universe is built, there were many other places in the world where ancient people made significant astronomical discoveries. Two places where astronomers were particularly active were China and Mesoamerica.

Chinese Astronomy

The Chinese made and recorded observations of eclipses at least as long ago as the thirteenth century B.C. Records survive for about 900 solar eclipses and 600 lunar eclipses that were carefully recorded during a period of about 2600 years. The Chinese also recorded other conspicuous astronomical events such as meteor showers, "guest stars" (exploding stars that suddenly appeared and then gradually faded from sight), and comets. Among these are observations of Comet Halley on many of its appearances dating back to 466 B.C. Ancient Chinese observations of comets, as well as those of guest stars, have been of great value to today's astronomers in their attempts to understand comets and exploding stars. The extensive observational records enabled Chinese astronomers, like those of Mesopotamia and Greece, to determine accurately the length of the year, to discover precession, and to use eclipse cycles to predict when eclipses would occur.

Mesoamerican Astronomy

Astronomy was also important in Mesoamerica, especially to the Mayan civilization, which flourished in Mexico and Guatemala between about 1000 and 2000 years ago. The planet Venus, which was identified with important gods of both the Maya and the Aztecs, was extremely important to Mesoamericans. Significant cultural events, such as battles, were timed to coincide with particular positions of Venus in the sky. The Maya also prepared extensive tables of risings of Venus and of lunar and solar eclipses, which they were able to predict in advance. The extremely complex Mayan calendar kept track not only of the Sun and Moon but also of the rising and setting of Venus. It seems quite unlikely that we will ever know the full extent of Mesoamerican astronomical knowledge. In contrast to the situation in China, where a very large amount of written information about astronomy and many other subjects was preserved, only a few Mesoamerican books survive. The rest were deliberately destroyed by the Spanish following their conquest of Mexico and Central America.

Chapter Summary

- The Babylonians accumulated records of astronomical observations for many centuries. The records enabled them to see repeated patterns in the motions of the celestial objects. They used the patterns to predict the positions of the Moon and planets. (Section 3.1)
- Mechanical explanations of astronomical phenomena were first used by the astronomers of Miletus. Their ideas about the stars eventually developed into the concept of the celestial sphere. (3.3)
- The idea that the celestial bodies were spheres and that they moved on perfectly circular paths originated with Pythagoras and his students. (3.3)
- Eudoxus devised a model in which the motion of a planet was produced by the rotation of several spheres. The opposite rotations of two spheres produced a figure-eight motion of a planet. The eastward rotation of a third sphere stretched out the figure eight and produced a motion resembling the observed retrograde motions of the planets. (3.3)
- Aristotle argued that all the celestial bodies were spheres and moved on circular paths. He believed that the Earth was motionless in the center of the universe. (3.3)
- The small angle equation describes the relationship of the angular diameter of an object to its linear diameter and its distance. If any two of these variables are known, the third can be found. (3.4)
- Aristarchus showed that it is possible to use geometry to find the distance of the Moon and the relative distances of the Moon and Sun. By doing so he showed that the universe is enormous compared with the size of the Earth. Aristarchus proposed that the Sun, not the Earth, is the center of the universe, but this idea was not accepted by most other Greek astronomers. (3.4)
- Eratosthenes found the difference in the altitude of the noonday Sun at Syene and Alexandria. He realized that this is the same as the difference in latitude of the two cities. This allowed him to find the ratio of the circumference of the Earth to the distance between Syene and Alexandria. (3.4)
- The celestial coordinates of stars change with time because of precession, the slow circular shifting of the celestial poles with respect to the stars. Hipparchus discovered precession when he compared his measurements of stellar positions with those of earlier Greek astronomers. (3.4)
- Greek astronomy culminated with the geocentric system of Ptolemy. In Ptolemy's model the retrograde motion of a planet was produced by the combination of two circular motions. A planet moved in a circle on an epicycle, which itself moved on a deferent. The system could predict accurately the positions of the planets and was in use for nearly 1500 years. (3.4)

Key Terms

astrology *39*
deferent *51*
epicycle *51*
equant *51*
geocentric *51*
precession *49*

Conceptual Questions

1. Suppose you lived on a planet where the month was 25 days long and the year was 330 days long. Invent a calendar (similar to that of the Mesopotamians) in which each year has a whole number of months and the average length of the year is 330 days.
2. Describe a way in which Eudoxus's model could have explained the motion of the Sun, both during a day and throughout a year.
3. Suppose the Earth had a shape like a football with the long axis lying in the plane of the Earth's equator. What effect would this shape have on the constancy of the appearance of the Earth's shadow on the Moon during eclipses?
4. Greek astronomers rejected Aristarchus's idea that the Earth moved around the Sun. They argued that the separations of stars would change during a year if the Earth moved. Describe an alternative explanation for the unchanging shapes of the constellations that works for a moving Earth.
5. In Ptolemy's model of the solar system, what range of phases could be exhibited by the planet Mars? How about the planet Venus?
6. Can Mars ever be at opposition in Ptolemy's model? How about Venus?
7. Suppose the motion of a planet on its epicycle was in the opposite direction to the motion of the epicycle on the deferent (that is, clockwise versus counterclockwise). Would retrograde motion occur for the planet?
8. In Ptolemy's model, the motion of the epicycle of a planet is at a constant rate when viewed from the equant. When does the epicycle move fastest as viewed from the center of the deferent? From the Earth?

Problems

1. The synodic period of Jupiter is 399 days. What is the interval of time between successive times that Jupiter appears in opposition at the same position among the stars? (This problem will have to be done by trial and error.)
2. What is the angular diameter of an object that has a linear diameter of 30 meters and a distance of 1000 meters?
3. What is the distance of an object that has an angular diameter of 100 seconds of arc and a linear diameter of 30 meters?
4. What is the linear diameter of an object that has an angular diameter of 20 seconds of arc and a distance of 50,000 meters?
5. Suppose the distance to an object is tripled. What happens to its angular diameter?
6. Planets A and B have the same angular diameter. Planet A has a linear diameter that is 25% larger than that of planet B. What is the ratio of the distance of planet A to the distance of planet B?
7. Suppose the angle between the Sun and Moon (as viewed from the Earth) at quarter moon is 75°. Use a protractor to find the ratio of the distances to the Sun and Moon using the method of Aristarchus.
8. Suppose Eratosthenes had found that the difference in the altitude of the Sun at Syene and Alexandria was 15°. What value would he have found for the resulting circumference of the Earth?

Figure-Based Questions

1. According to Figure 3.18, which planet has the epicycle with the smallest angular size as viewed from the Earth? (The angular size of the epicycle gives the size of the retrograde loop for a planet.)
2. Using a protractor and Figure 3.13, find the angular separation between the two stars when the Earth is near the stars and when the Earth is on the opposite side of the Sun. How far away from the Sun (in terms of the Earth's distance) would the stars have to be in order for the difference in angular separation to be only 1°?

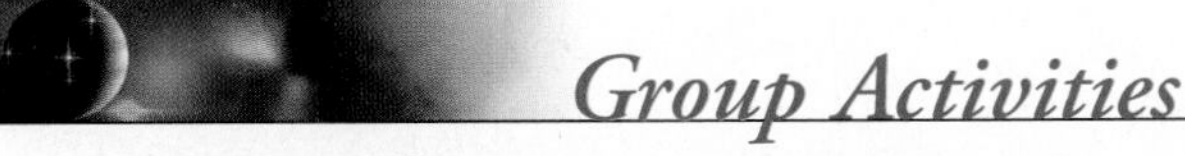

Group Activities

1. Have your partner stand 50 feet away from you. Using your hand as an angle measuring device (see Figure 1.10), find the angular size of your partner. Calculate the height of your partner using the small angle equation (Equation 3.1). Have your partner make the same measurement and calculation for your height. Compare your calculated heights with your actual heights and discuss possible reasons for any differences.
2. Have your partner move an unknown distance (less than 100 feet) away from you. Using your hand, measure the angular size of your partner. With the known height of your partner, calculate the distance of your partner and then measure the actual distance (pacing it off is probably good enough). Have your partner make the same measurement and calculation for your distance. Discuss possible reasons for any differences between the calculated and actual distances.
3. Tie a ball to a piece of rope. Then, while walking in a circle about your partner, whirl the ball around your head. Walk as fast as you can and whirl the ball as slowly as you can. Be careful not to hit your partner with the ball! Ask your partner to verify that retrograde motion of the ball occurs for this combination of motion on the deferent (the path you walk) and epicycle (the path of the ball around your head). Switch places with your partner and repeat the experiment.

For More Information

Visit the text website at **www.mhhe.com/fix** for chapter quizzes, interactive learning exercises, and other study tools.

4

Whatsoever is discovered by men of a later day must none the less be referred to the ancients, because it was the work of a great mind to dispel for the first time the darkness involving nature, and he contributed most to the discovery who hoped that the discovery could be made.

Seneca
Quaestiones Naturales

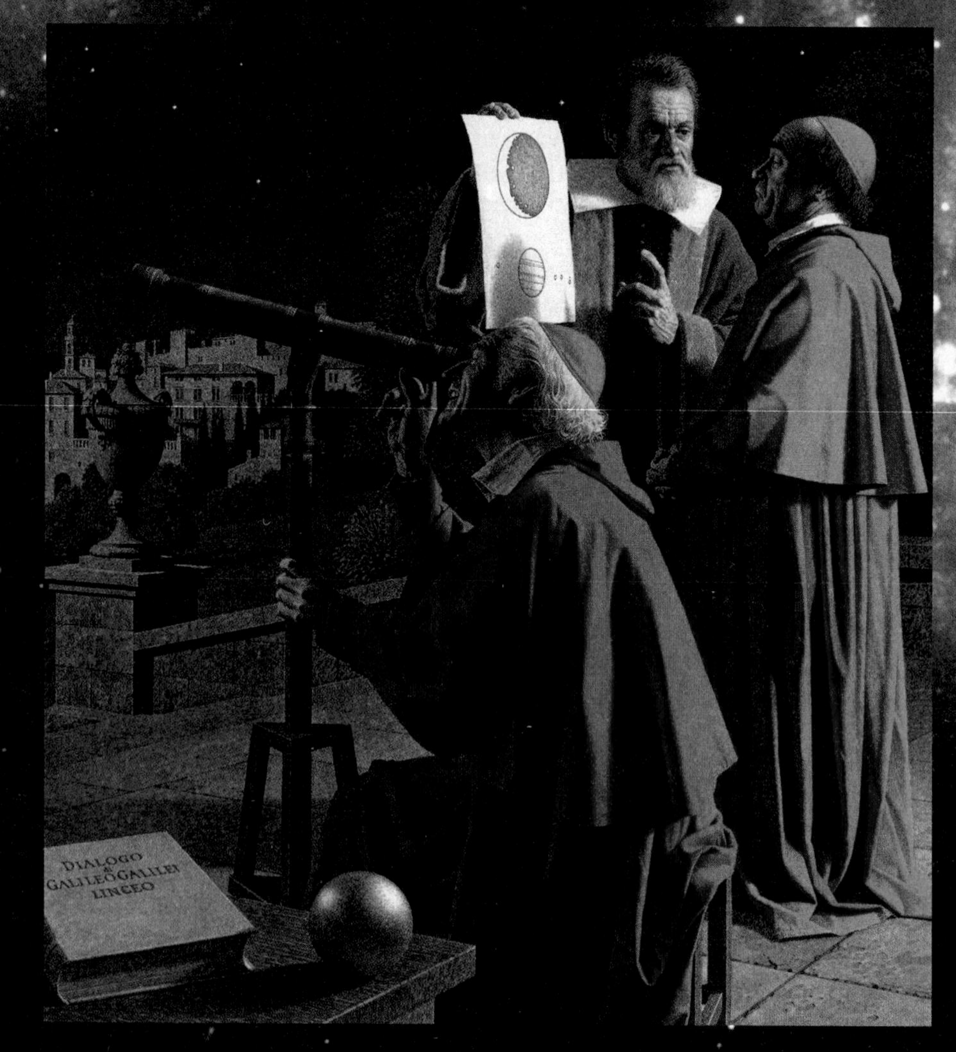

Galileo by Jean-Leon Huens. Viewing the Moon through Galileo's telescope

Renaissance Astronomy

One of the reasons for the longevity of the Ptolemaic model of the solar system is that it seems so reasonable. Our senses give us no clue that the Earth is in motion. How could the Earth be spinning eastward at the speed of a jet airplane without giving us any feeling that we are moving? Is it really possible that the Earth beneath our feet could be hurtling around the Sun at 70,000 miles per hour? On a calm evening it seems much more plausible that the stars are really moving—that it is the heavens that are in motion while we rest quietly at the center.

Even so, it is remarkable that the Ptolemaic model of the solar system dominated astronomy for more than 1000 years. But during that time, the spirit of scientific inquiry that had driven the Greeks to their many astronomical discoveries was almost absent from the western world. No theories were proposed as alternatives to the established view that the Earth was motionless at the center of the universe. Observations, when they were made at all, were intended to refine the accepted model of the solar system. None was accurate or novel enough to contradict Ptolemy's model or challenge its fundamental assumptions. Aristotle's theory of motion and arguments for a stationary Earth also reigned supreme.

The revolution that placed the Sun in the center of the solar system and relegated the Earth to the status of one among many planets was the work of many astronomers and philosophers. Like most revolutions, the overthrow of Ptolemy and Aristotle began as a trickle of objections and doubts about the authority of the ancients. It ended as a flood that washed away the established order and swept an entirely new view of the so

Questions to Explore

- What happened to the knowledge of the ancient astronomers after the fall of the Roman Empire?
- How is it possible for the heliocentric model of Copernicus to account for the diurnal motion of the celestial objects, the annual motion of the Sun, and the complicated motions of the planets?
- What does the heliocentric model have to say about the orbital distances and periods of the planets?
- Why did Tycho Brahe, the great observational astronomer, reject the heliocentric model of the solar system?
- How were Tycho's observations used by Kepler to produce his laws of planetary motion?
- What are the shapes of planetary orbits? How does the speed of a planet's revolution depend on its distance from the Sun?
- How did the telescopic observations of Galileo support the heliocentric model of the solar system?
- What finally convinced the world that the Earth revolves around the Sun rather than the other way around?

ar system into place. This chapter is about that revolution and its principal actors—Copernicus, Tycho, Kepler, and Galileo—remarkable scientists who each picked up the torch of the revolution from his predecessor and carried it forward. Their work covered more than a century, at the end of which the heliocentric revolution was nearly complete.

4.1 ASTRONOMY AFTER PTOLEMY

After the time of Claudius Ptolemy, the astronomy of the Greek and Roman world, like nearly every other aspect of western civilization, went into a long decline. In the first stage of the decline, scientific activity became less frequent and less innovative. Little new information was discovered. In the second stage of the decline, astronomical knowledge not only ceased to grow, but actually diminished. By the seventh century, Ptolemy was completely unknown in the West and Aristotle was known only through a few small works on logic.

Part of the reason for the loss of Greek astronomical knowledge can be attributed to the antagonism of the early Christian church to many of the features of Greek astronomy. Many church leaders thought that ideas such as the sphericity of the Earth contradicted descriptions of the universe found in the Scriptures. Rather than accept the spherical shape of the Earth and the celestial sphere, some Christian scholars such as Lactantius and Kosmas argued that the Sun, after sunset, traveled around the horizon toward the north and then east to rise again in the morning. Not all of the Christian scholars were as ready to reject Greek astronomy. Some tried to reconcile the scientific and scriptural ideas about the universe.

Islamic Astronomy

Although the astronomical works of the Islamic world from the seventh to the fifteenth centuries were usually written in Arabic, many of the Islamic astronomers weren't Arabs. They worked in many different cities from Spain to central Asia. The Islamic astronomers improved many observing techniques and made careful observations. They made major improvements in instruments, such as the sextant and astrolabe, used in navigation. They also improved the ways in which planetary positions were calculated. Many of the Arabic names for the stars (such as Betelgeuse, Altair, and Aldebaran) are still in use today. However, the main contribution the Islamic astronomers made to the development of astronomy was the preservation of the astronomy of the ancient Greeks. (Islamic scholars sought out knowledge from all the ancient civilizations they touched.) The Islamic astronomers translated the works of the ancient astronomers, including Ptolemy, from Greek into Arabic. When the Christian world rediscovered Greek astronomy it was through the Arabic versions.

The Islamic astronomers were the first to build observatories that were organized and equipped like their modern counterparts. The observatories had instruments to measure the positions of the celestial objects, timekeeping devices to mark the time of observations, and libraries of as many as 400,000 manuscripts. Observatories were located in many of the important Islamic cities, such as Damascus, Cairo, Toledo, Baghdad, and Samarkand, and were run by several astronomers and their students. The observatory at Samarkand was built by Ulugh Beg (1394–1449). Beg compiled a star catalog that was unique among the works of Islamic astronomers in that it was based on his own observations rather than an updating of the positions given by Ptolemy.

The observations of the Islamic astronomers were often more accurate than those of the ancients. However, the Islamic astronomers' reason for observing was not to challenge existing theories or test new ones. Instead, they wished to verify, with improved accuracy, the work of earlier astronomers. The model they used was always that of Aristotle and Ptolemy. The Islamic astronomers collected, translated, and commented on the works of the ancient astronomers, supplemented the ancient works with new observations, and then passed their astronomy on to Christian Europe. In doing so, Islam exerted a powerful influence on the rebirth of western astronomy.

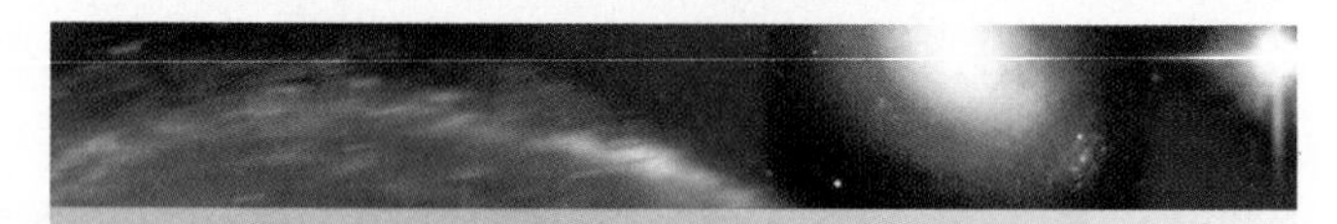

After the time of Ptolemy, western civilization suffered a long decline. The knowledge of the ancient astronomers might have been lost had it not been for Islamic astronomers, who preserved ancient writings and translated them into Arabic.

4.2 THE REBIRTH OF ASTRONOMY IN EUROPE

The rebirth of western astronomy began when scholars discovered and translated Arabic astronomical works and Islamic translations of ancient works into Latin. One of the earliest of the translations was made by John of Seville, who, in the middle of the twelfth century, translated *Jawami,* which had been written 300 years earlier by the Islamic astronomer al-Farghani. *Jawami* was a simple introduction to Ptolemy's *Almagest.* A "school" of translation arose at Toledo, which had been one of the great cities of Islamic Spain, but which was captured by the Christians in 1085. One of the scholars who settled in Toledo was Gerhard of Cremona. In the late twelfth century he translated Ptolemy's *Almagest* and works of Aristotle, Archimedes, and others into Latin. Most of the ancient works preserved by Islam had been translated and were available to scholars by the early thirteenth century. About 200 years later there was another sudden increase in translated works when many Greek manuscripts were brought west from Byzantium. These works were of great value to scholars because they were closer to the original works than were the retranslations of Islamic texts.

The development of astronomy in Europe was also linked to the birth and growth of universities. During the eleventh century students began to gather at places where scholars lectured on the wisdom of the ancients. By 1200, universities had been officially established at Bologna, Oxford, Paris, and elsewhere. Eventually, university scholarship evolved from reciting and commenting on the ancient works to genuine criticism and analysis. Even Aristotle was questioned. The Parisian scholar, Nicole Oresme (c.1330–1382) pointed out that often we perceive only relative motion. He argued that when two people riding in boats pass each other in the middle of the sea, neither can tell which boat is moving and which is still or whether both are moving. In the same way, the daily motion of the stars can be equally well explained by a westward rotation of the celestial sphere (Aristotle's explanation) or an eastward rotation of the Earth. Aristotle said that if the Earth rotated, falling objects would land west of the spot above which they had been dropped. Oresme countered that argument by pointing out that when an arrow is shot into the air, the arrow also moves eastward along with the air and the ground.

Jean Buridan (c.1300–1358) of Paris, who was the teacher of Oresme, accounted for the motion of projectiles (such as arrows) with a new theory of motion that was completely contradictory to the ideas of Aristotle. Aristotle taught that unless an object was moved by an external push or pull it could only remain at rest or fall directly to Earth. Why, then, doesn't an arrow fall directly to the ground as soon as it is released by a bow? Aristotle's answer was that the air disturbed by the arrow gives the arrow a push. Probably not even Aristotle took this explanation very seriously. Buridan suggested that the bow gives the arrow an **impetus,** a force that pushes the arrow along even after it has left the bow. The impetus is steadily decreased by the resistance of the air and the arrow slows down until gravity wins and pulls it to the ground. Although the impetus theory may not seem like a great improvement over the ideas of Aristotle, there is one very significant difference between the two. According to Aristotle, rest was the only natural state of matter. With the impetus theory, motion also was a natural state and permanent unless something resisted the motion.

Despite the work of Oresme, Buridan, and others, the Earth remained firmly at the center of the medieval **universe.** A vivid description of this universe can be found in Dante's *Divine Comedy,* written around 1300. Figure 4.1 shows Dante's universe, in which there is a central Earth surrounded by the spheres of the Moon, Sun, and planets. Beyond them is the sphere of the fixed stars, an outermost crystalline sphere, and, finally, paradise.

FIGURE 4.1
The Universe as Described in Dante's *Divine Comedy*
Dante's picture was typical of medieval views of the universe. The Earth is located at the center surrounded by the spheres of the Sun, Moon, planets, and stars.

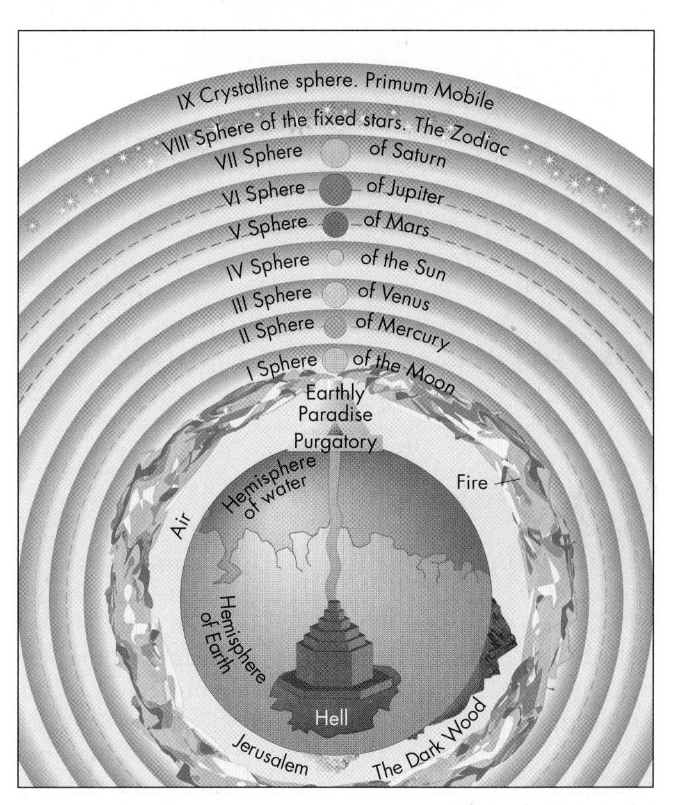

Astronomical observations, which had been neglected for centuries except in the Islamic world, began to be carried out again. Roger Bacon (c.1220–1292) was among the first to recommend experimentation as the best way to acquire knowledge. In the 1400s, this idea was taken up by astronomers such as Georg Purbach (1423–1461) and his student Johannes Regiomontanus (1436–1476). These astronomers' observations showed that the existing astronomical tables were badly in error. After Purbach's death, Regiomontanus opened a printing office and began to use the recently invented art of printing to produce astronomy books, almanacs, and tables of predictions of the positions of the Sun, Moon, and planets. After Regiomontanus, the pace of observations and the invention of new observing instruments swiftly increased. Only a few centuries earlier, scholars had studied and commented on science as they found it. By the fifteenth century, the most noted astronomers were active investigators, making observations that could be compared with the predictions of the ancient model of the universe.

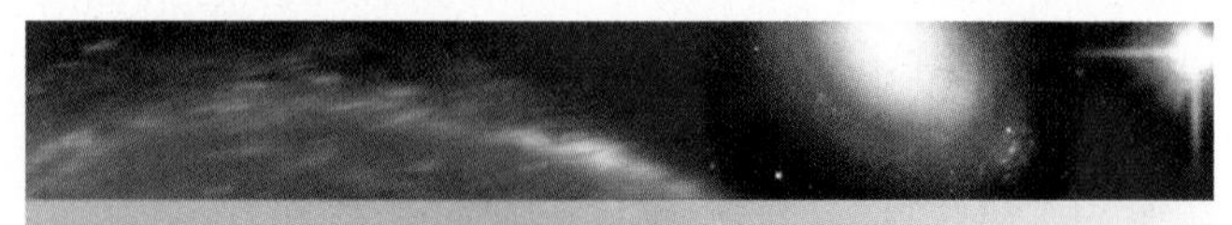

By the fifteenth century, the astronomy of the ancients had been rediscovered by Europeans. Astronomers had begun to observe again and to test hypotheses against observations. The geocentric model of Ptolemy was accepted by nearly all astronomers. Some astronomers, however, had growing doubts about Aristotle's theory of motion and his arguments that the Earth must be motionless.

4.3 COPERNICUS

Nicholas Copernicus (1473–1543), the astronomer who proposed the first fully developed heliocentric model of the solar system, was born in Torun, in what is now Poland. Copernicus, who lived at the height of the Renaissance, was a full participant in the intellectual fervor of the time. The timeline in Figure 4.2 shows some of Copernicus's contemporaries and the events of his time and that of his successors Tycho, Kepler, and Galileo. Copernicus's mother died when he was very young, and his father died when he was ten. Young Nicholas was adopted by an uncle who a few years later became Bishop of Ermland, a small, independent state.

In his late teens, Copernicus studied for several years at the University of Cracow. While there, Copernicus took an astronomy class in which he learned the details of the Ptolemaic system. When he was still in his early twenties he was made an official of his uncle's cathedral in Frauenberg. The position provided him with an income for the rest of his life. After obtaining the appointment, Copernicus traveled to northern Italy to study church law. During his 6 years as a student at Bologna, Padua, and Ferrara, he also studied medicine, astronomy, Greek, philosophy, and mathematics. By the time he returned to Ermland (which he later described as the remotest corner of the world), he was as well educated as anyone of his time.

Copernicus led a remarkably full life. He was a well-respected physician and, for a time, medical attendant to the Duke of Prussia. On several occasions he was the administrator of Ermland and its representative to the Prussian parliament. He was fluent in many languages and translated ancient literary works from Greek to Latin. In his book on economics he developed an early theory of capitalism and was the first to show that "bad money drives out good," which later became known as Gresham's law. He also published a book on trigonometry. However, astronomy was his main interest. Shortly after Copernicus returned from Italy he wrote an outline of a Sun-centered theory of the solar system and he

FIGURE 4.2
A Timeline of Notable People and Events (left) and the Four Most Prominent Renaissance Astronomers (right)
The pictures portray (top to bottom) Copernicus, Tycho, Kepler, and Galileo.

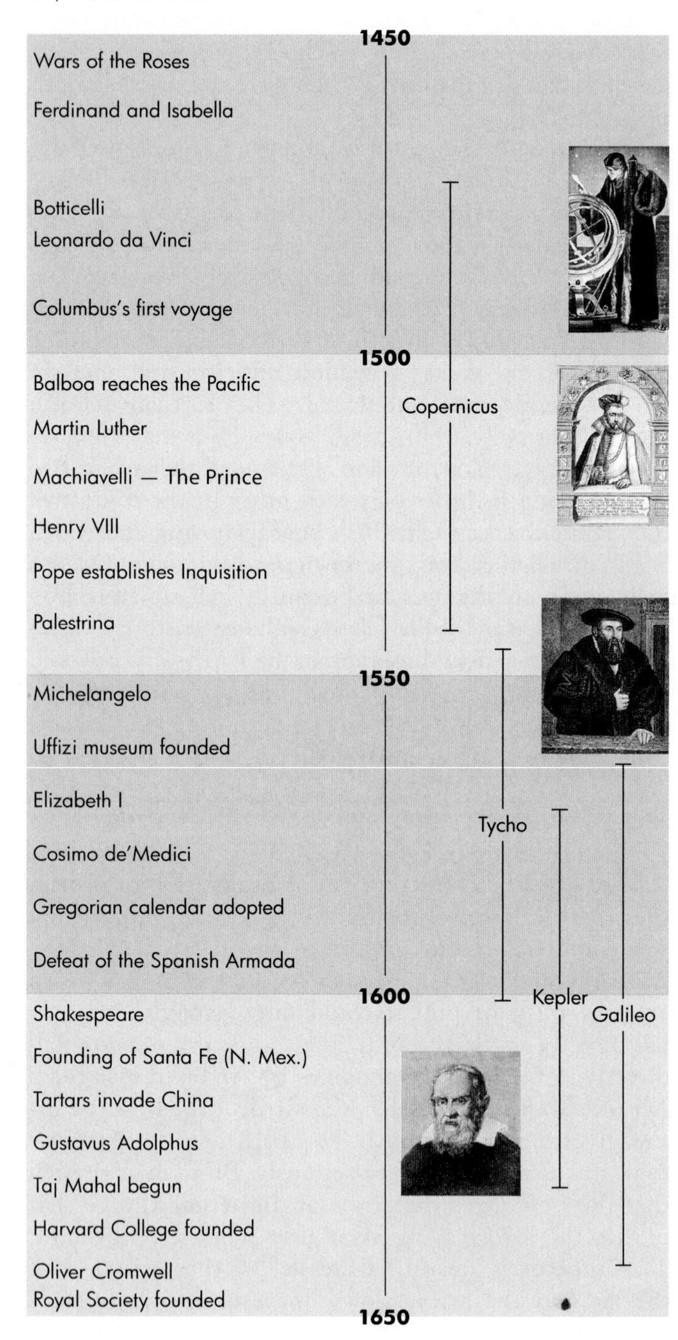

distributed it to friends. His theory probably never would have been published had his outline not been read years later by a German astronomer, Georg Rheticus (1514–1576). Very impressed, Rheticus traveled to Ermland to persuade Copernicus to publish his heliocentric theory. At that time, Copernicus was completing a book he had been writing for more than 20 years. The book, *De Revolutionibus Orbium Caelestium* (*On the Revolutions of the Heavenly Orbs*), was published in Latin, the language of scholars, in 1543, two months before Copernicus's death. According to his friend, Bishop Giese, Copernicus received the first printed copy of his book and "saw his completed work only at his last breath upon the day that he died."

The Heliocentric Model

Copernicus's search for a new model of the universe seems to have been motivated by two objections to the Ptolemaic system. One objection was that the Ptolemaic model, even in updated form, gave positions of the Sun, Moon, and planets that did not quite agree with the best available observations. The second objection was that Copernicus, like nearly everyone else, accepted Plato's idea that all celestial motions must consist of combinations of uniform motions on circles. The equants that the Ptolemaic model required did not correspond to uniform motion. Copernicus wrote, "Having become aware of these defects, I often considered whether there could perhaps be found a more reasonable arrangement of circles, from which every apparent inequality would be derived and in which everything would move uniformly about its proper center, as the rule of absolute motion requires."

The "more reasonable arrangement" that Copernicus devised put the Sun at the center of the motions of the planets, one of which was the Earth. Only the Moon was left to orbit the Earth. Copernicus pointed out that this model was not in conflict with any observations and could explain the basic motions of the celestial objects at least as well as Ptolemy's model. The eastward rotation of the Earth produces the apparent westward diurnal motion of the celestial sphere. The annual motion of the Sun against the background of the stars is caused by the Earth's annual orbit around the Sun, as shown in Figure 4.3. Finally, the retrograde motions of the planets can be explained without the use of epicycles. As shown in Figure 4.4, retrograde motion occurs whenever one planet catches up with and passes another as they both orbit the Sun. If a planet were stationary, then the Earth's motion would cause its direction to shift back and forth each year. When the orbital motion of the planet is included, the oscillation due to the Earth's motion is added to the general eastward motion of the planet, resulting in regularly spaced intervals of retrograde motion.

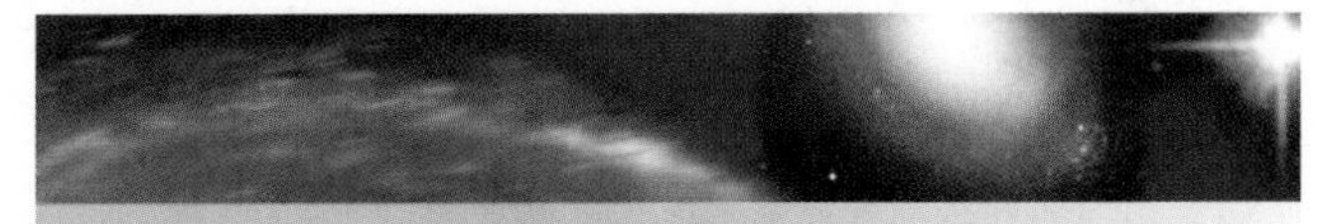

In the **heliocentric** model of Copernicus, the rotation of the Earth explains the diurnal motion of celestial objects. The Earth's revolution about the Sun explains the annual motion of the Sun. Retrograde motion of the planets occurs naturally whenever the Earth passes or is passed by another planet.

FIGURE 4.3
The Position of the Sun Against the Backdrop of Stars Throughout the Year
The zodiacal constellation behind the Sun changes as the Earth moves in its orbit.

ANIMATION *The position of the Sun against the backdrop of stars throughout the year*

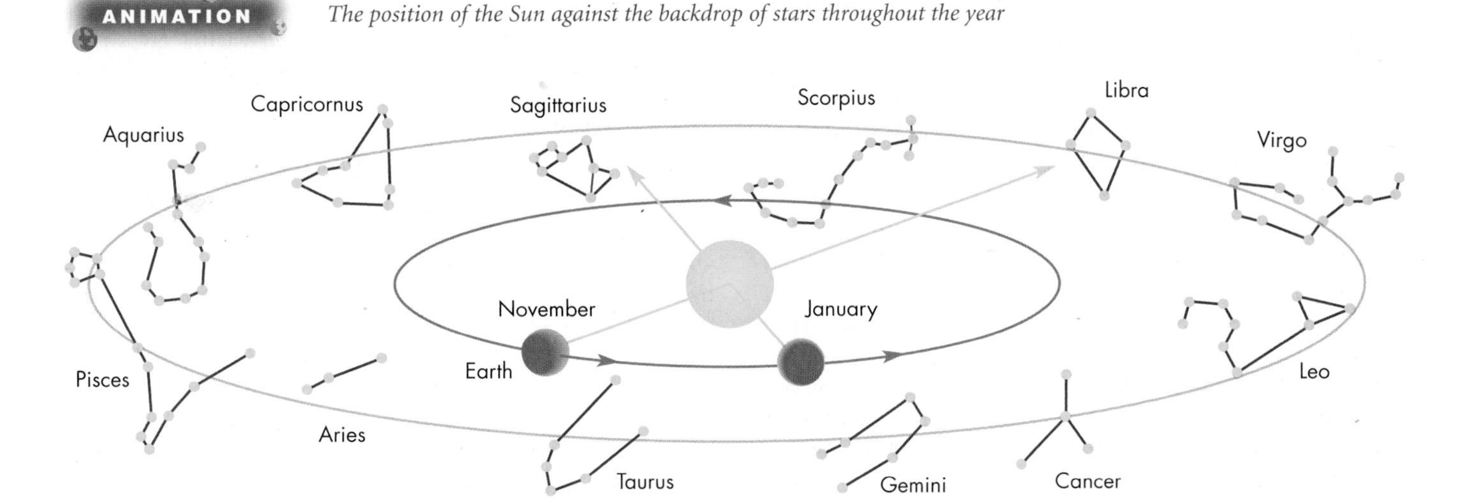

FIGURE 4.4
The Retrograde Motion of Mars According to the Heliocentric Model
A, If Mars were stationary, the Earth's orbital motion would cause Mars to appear to shift back and forth among the stars during a year. **B,** If Mars moved and the Earth were stationary, Mars would move steadily eastward among the stars. **C,** Because both the Earth and Mars are in orbit, the actual motion of Mars among the stars is the combination of **A** and **B.** The bottom three figures show Mars's apparent motion among the stars for **A,** the Earth in motion, **B,** Mars in motion, and **C,** both planets in motion.

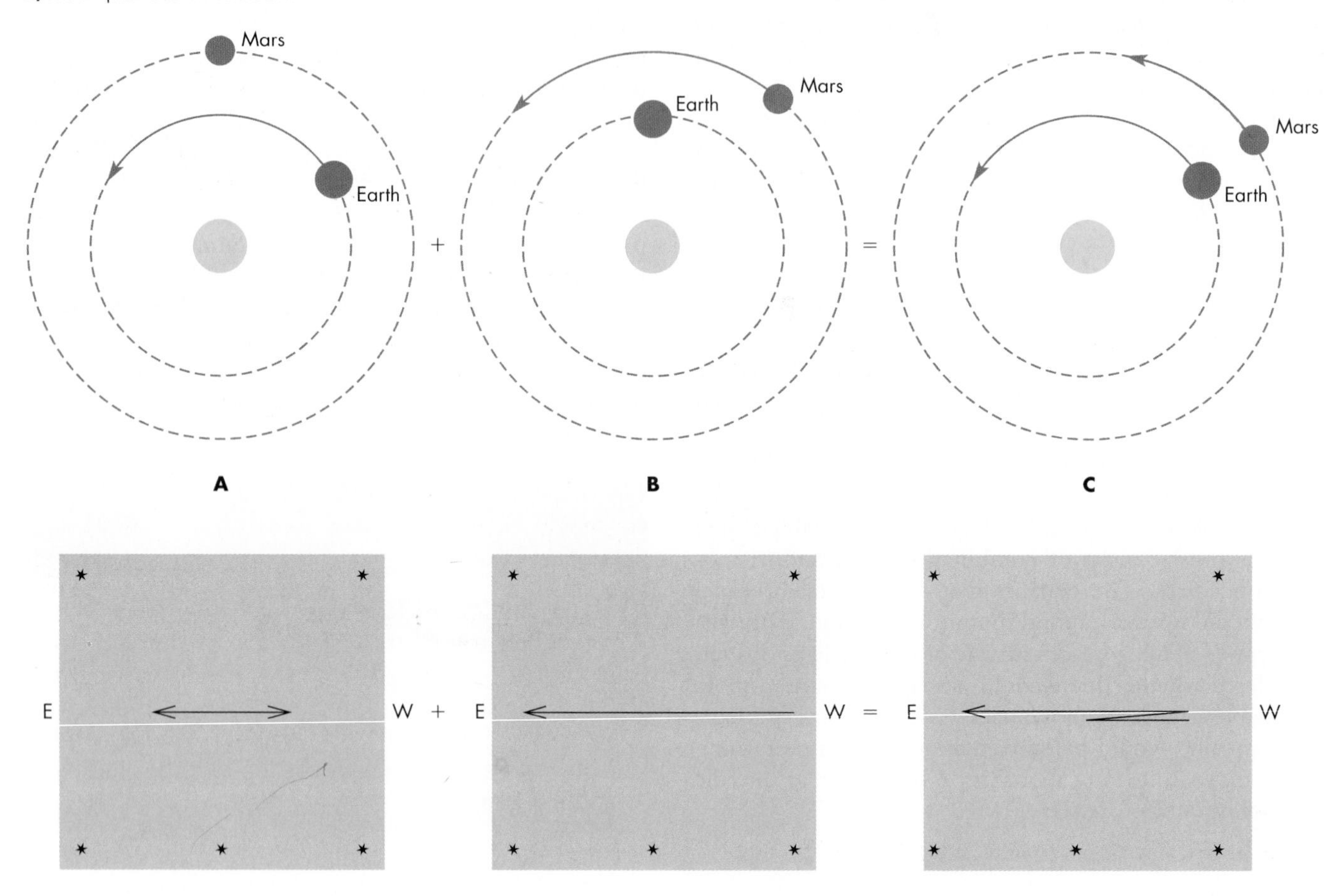

Each time the Earth passes or is passed by another planet, both the Earth and the other planet show a westward retrograde motion when viewed from the other. Episodes of retrograde motion are separated by the length of time it takes the faster-moving planet to lap the other. This length of time is the synodic period of the planet already described in Chapter 2. The length of time it takes a planet to complete one orbit about the Sun is its **sidereal period.** The relationship between the synodic and sidereal periods of a planet can be understood by looking at the analogous case of two runners on a track. Suppose, as shown in Figure 4.5, that one runner can run a lap in 4 minutes, the other runner in 5 minutes. To see how long it will take the first runner to lap the second, we calculate each runner's rate, which is just the inverse of how long it takes the runner to complete a lap. The first runner's rate is 0.25 laps per minute or 15 laps per hour. The second runner's rate is 0.2 laps per minute or 12 laps per hour. Thus, the first runner will complete 3 more laps per hour than the second runner and will lap the slower runner three times per hour, or once every 20 minutes.

For a planet, the rate of orbital motion is the inverse of the planet's sidereal period. Subtracting the rate of the slower planet from the rate of the faster planet gives the rate at which the faster planet catches up with and laps the slower. The inverse of this rate is the synodic period of the two planets. Figure 4.6 shows the relationship between the sidereal and synodic periods (as viewed from Earth) of a hypothetical planet with a sidereal (orbital) period of 3 years. The synodic period can be calculated for any two planets by calculating the inverse of the difference in their rates of orbital motion. That is,

$$1/S = 1/P_1 - 1/P_2 \tag{4.1}$$

where S is the synodic period, P_1 is the sidereal period of the inner, faster moving planet, and P_2 is the sidereal period of the outer, slower moving planet. Copernicus used

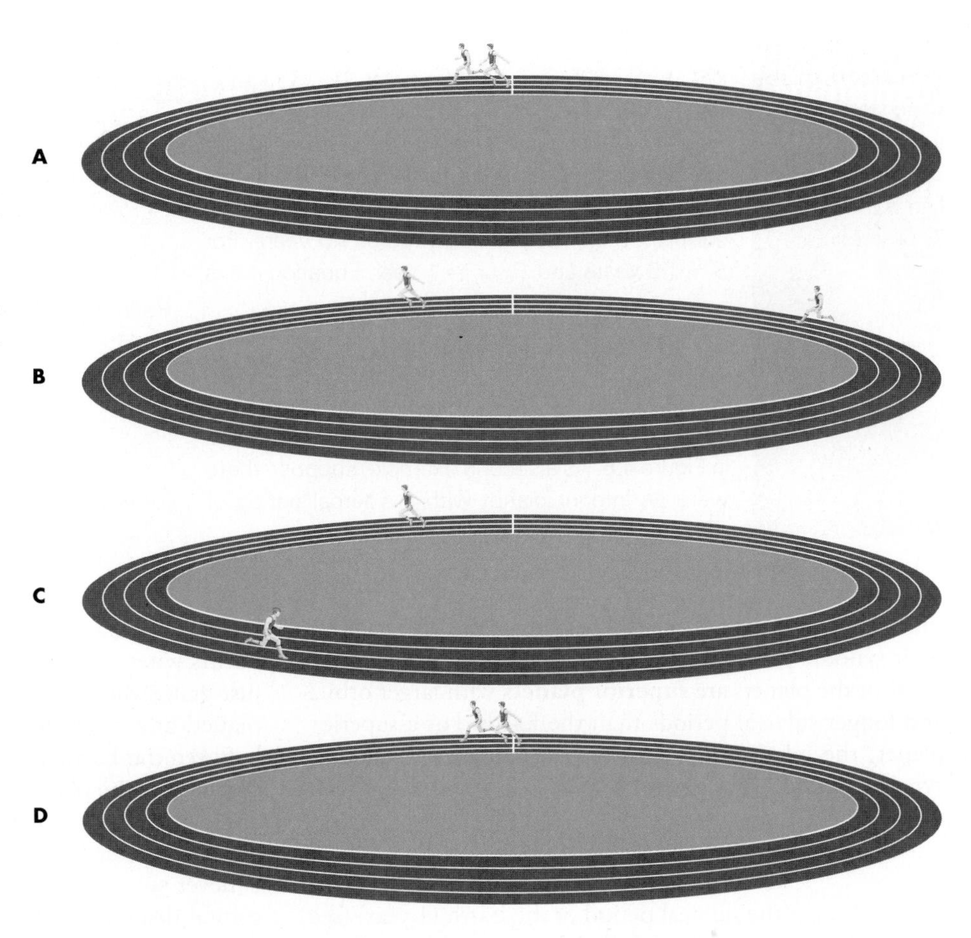

FIGURE 4.5
Two Runners of a Track
The faster runner completes a lap in 4 minutes, the slower runner in 5 minutes. **A,** The runners begin. **B,** After one lap, the faster runner is well ahead. **C,** After three laps, the faster runner is more than a half lap ahead. **D,** The faster runner passes the slower when the faster runner has completed five laps and the slower runner has completed four laps.

FIGURE 4.6
The Synodic Period of a Planet with an Orbital Period of 3 Years
The Earth needs 1.5 years to catch up with the planet. During that time, the planet has moved halfway around the Sun from its starting point.

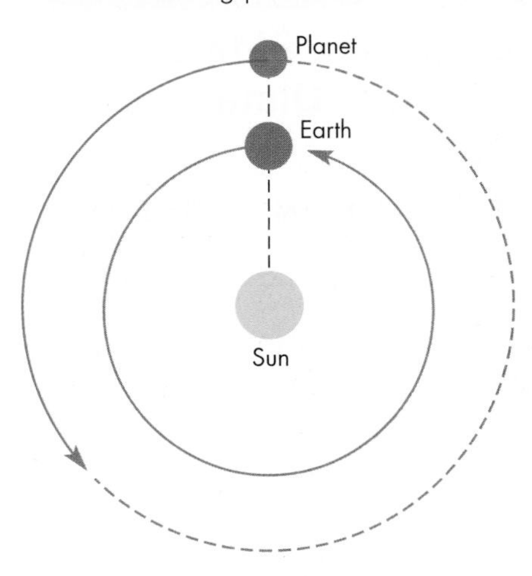

A Sidereal period of Earth = 1 year
Sidereal period of planet = 3 years

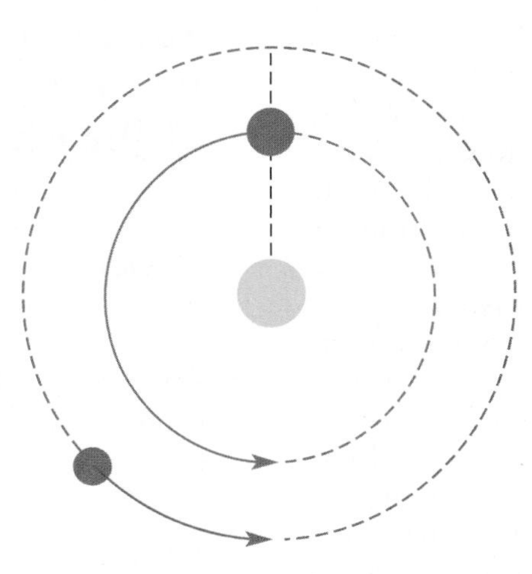

B One year later:

- Earth has completed one orbit.
- Planet has completed $\frac{1}{3}$ orbit.

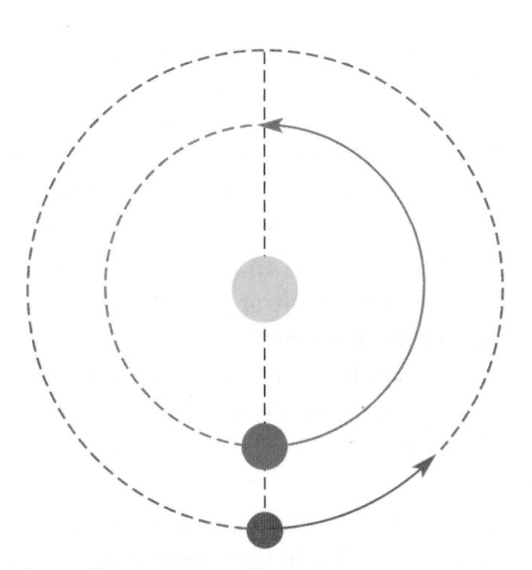

C Half a year later still:

- Earth has completed $1\frac{1}{2}$ orbits.
- Planet has completed $\frac{1}{2}$ orbit ($\frac{1}{2}$ sidereal period of planet).

Sidereal and Synodic Periods

Equations 4.2 and 4.3 can be used to calculate the synodic period of a planet from its sidereal period or vice versa. Suppose there were a superior planet with a synodic period of 1.5 years. For $S = 1.5$ years and $P_{Earth} = 1$ year, Equation 4.2 is

$$\frac{1}{P} = \frac{1}{(1\text{ yr})} - \frac{1}{(1.5\text{ yr})} = \frac{(3-2)}{(3\text{ yr})} = \frac{1}{(3\text{ yr})}$$

Thus, P, the sidereal period of the planet, is 3 years. This is the hypothetical planet described in Figure 4.6. As a second example, suppose there were an inferior planet with a sidereal period of 0.25 years. For $P = 0.25$ years and $P_{Earth} = 1$ year, Equation 4.3 is

$$\frac{1}{(0.25\text{ yr})} = \frac{1}{(1\text{ yr})} + \frac{1}{S}$$

Rearranging this equation to solve for $1/S$ gives

$$\frac{1}{S} = \frac{1}{(0.25\text{ yr})} - \frac{1}{(1\text{ yr})} = \frac{4}{(1\text{ yr})} - \frac{1}{(1\text{ yr})} = \frac{3}{(1\text{ yr})}$$

for which $S = 1/3$ year.

this method to find the sidereal periods of the planets from their synodic periods and the sidereal period of the Earth. Most of the planets are **superior planets** with larger orbits and longer sidereal periods than the Earth. For a superior planet, the relationship between sidereal period (P) and synodic period (S) is given by

$$1/P = 1/P_{Earth} - 1/S \quad (4.2)$$

where P_{Earth} is the sidereal period of the Earth (1 year) (see box for Equations 4.2 and 4.3). Mercury and Venus have smaller orbits and shorter sidereal periods than the Earth and are called **inferior planets.** For an inferior planet, the sidereal period is given by

$$1/P = 1/P_{Earth} + 1/S \quad (4.3)$$

Copernicus found that the sidereal periods of the planets ranged from ¼ year for Mercury to almost 30 years for Saturn. The synodic and sidereal periods of the planets are given in Table 4.1.

Copernicus also showed that his system made it possible to use geometry and observations to determine all the planetary distances in terms of the Earth's orbital distance, which is defined as the **astronomical unit,** or **AU.** In contrast, the distances of the planets in Ptolemy's system were arbitrary. The size of the deferent of a planet could be changed without changing the appearance of the planet's motion as seen from the Earth, provided that the size of the epicycle was changed in the same proportion. Even the order of the planets was unknowable in Ptolemy's system, although they were usually arranged in order of the average speed with which they moved relative to the stars. For the heliocentric system, as Figure 4.7 shows, the maximum angular distance between the Sun and an inferior planet depends on the size of its orbit relative to the Earth's orbital distance. **Greatest elongation,** the greatest angular distance between the Sun and the planet, occurs when the line of sight from the Earth to the planet just grazes the planet's orbit. This happens when the Sun, planet, and Earth form a right triangle with the distance between the Earth and Sun as the hypotenuse. The orbital distance of the planet, d, can then be found by measuring angle θ, the angular distance between the planet and the Sun at greatest elongation. For example, the planet Venus is never seen more than 47° from the Sun, giving Venus's orbital distance as 0.72 AU.

A slightly more complicated geometrical exercise was used by Copernicus to determine the orbital distances of the superior planets. The orbital distances and periods

The Synodic Periods, Sidereal Periods, and Orbital Distances of the Planets

PLANET	SYNODIC PERIOD (DAYS)	SIDEREAL PERIOD (DAYS)	ORBITAL DISTANCE (AU)
Mercury	116	88	0.39
Venus	584	225	0.72
Earth	—	365	1.0
Mars	780	687	1.52
Jupiter	399	4332	5.2
Saturn	378	10,750	9.6
Uranus	370	30,590	19.2
Neptune	367.5	59,800	30.1

FIGURE 4.7
Greatest Elongation, the Maximum Angular Distance from the Sun for an Inferior Planet
When the planet is seen at a maximum angular distance from the Sun, the Earth, Sun, and planet form a right triangle in which the distance between the Earth and the Sun is the hypotenuse. By measuring the angle θ, the relative distances of the planet and the Earth from the Sun can be found.

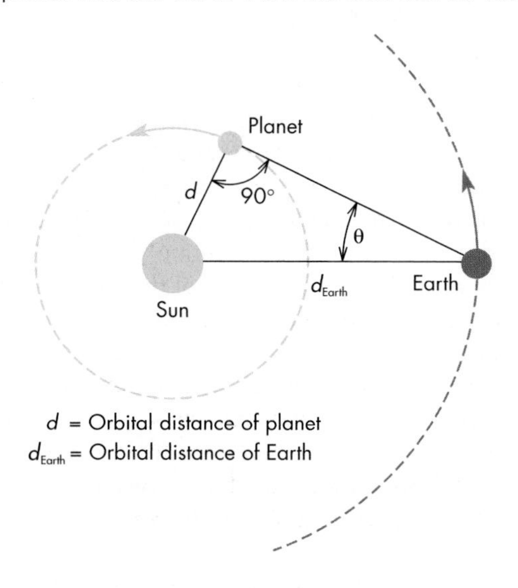

FIGURE 4.8
Epicycles in the Solar System Model of Copernicus
Mars's epicycle is needed to account for regular deviations in its orbital speed and distance from the Sun. In Copernicus's model, retrograde motion occurs with or without the epicycle.

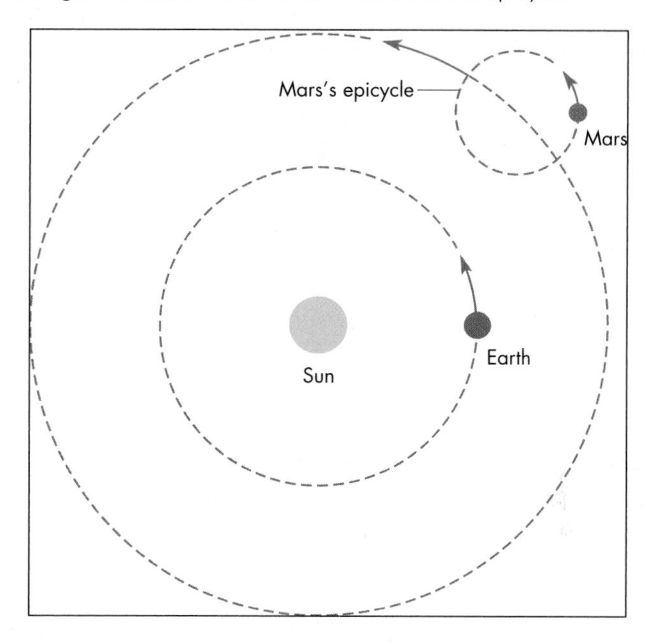

of the planets are given in Table 4.1. Notice that orbital period increases with increasing orbital distance.

The distances of the planets are arbitrary in the Ptolemaic model. In Copernicus's model, however, the orbital distances of the planets can be unambiguously determined through observations and the use of geometry.

Although Copernicus presented a revolutionary picture of the solar system, he was still very much influenced by the work of the ancient astronomers. He required that all of the solar system motions consist of combinations of circular motions. To account for variations in the speed with which the planets and the Sun moved among the stars, he had to use epicycles, as shown in Figure 4.8. The epicycles were *not* intended to account for retrograde motion, which occurs naturally when the Earth orbits the Sun as one of the planets. Copernicus hoped that the advantages of his theory compared with that of Ptolemy would be overwhelmingly obvious to other astronomers. This was not the case, however. The heliocentric theory was not immediately accepted by other astronomers. Critics of the heliocentric solar system noted that the astronomical evidence was completely neutral. There were no observations that could be explained by Copernicus but not by Ptolemy. Copernicus claimed that his theory was simpler and more elegant than the geocentric one. Yet after Copernicus had added the epicycles that he needed to account precisely for the motions of the planets, his model had as many circles (or more) than geocentric models. Another criticism of Copernicus was the old complaint that a moving Earth was in conflict with Aristotle's theory of motion. Despite growing doubts about its validity, Aristotle's theory of motion was still generally believed. Finally, Copernicus's critics argued that there would be visible consequences if the Earth were not motionless at the center of the universe.

One such consequence involves the rising and setting of stars. First, choose two stars so that one star rises at the same time that the other is setting. Then wait 6 months. If the Earth is at the center of the universe, the roles of the two stars will be reversed. On the other hand, Figure 4.9 shows that if the Earth is not in the center of the universe, as one star rises the other will have already risen or not yet come up. The observations available at the time showed that when the first of these stars is rising, the other is setting. When the second is rising, the first is setting. These observations apparently require the Earth to be at the center of the celestial sphere. However, as Copernicus pointed out, the observations of his day didn't actually say that one of the two stars rises exactly when the other sets. The observations were only accurate enough, about 0.1°, to

FIGURE 4.9
The Rising and Setting of a Pair of Stars
At a certain time of year, star *A* is rising just as star *B* is setting. If the Earth orbits the Sun, then 6 months later the stars will no longer be opposite each other in the sky. The rising of one will not coincide with the setting of the other. This effect is unnoticeable if the Earth's orbit is very small compared with the distances of the stars.

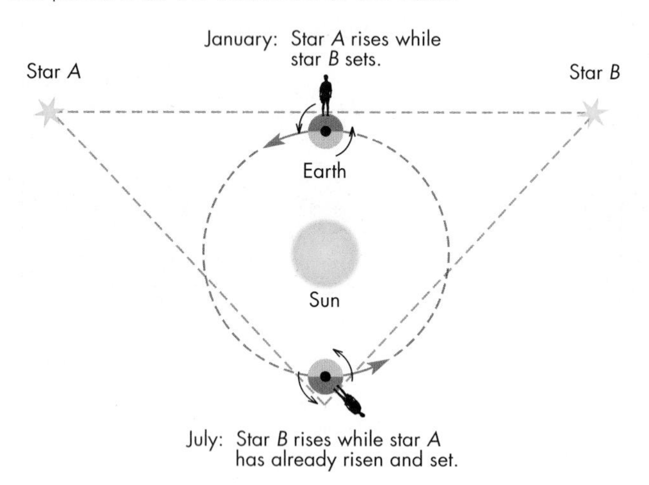

show that this was roughly true. Because of the limited accuracy, the observations only required that the Earth be near the center of the celestial sphere. The Earth could still orbit the Sun provided that the Earth's orbit is small with respect to the distance to the sphere of the stars. Copernicus argued that this was the case. He said that there were no detectable consequences of the motion of the Earth because the stars were much more distant than the Sun, Moon, or planets. In fact, the heliocentric model demanded that the stars have distances of at least 1000 AU.

TYCHO BRAHE—THE GREAT OBSERVER

De Revolutionibus attracted considerable attention as soon as it was published, and Copernicus became widely recognized as a great astronomer. However, what won the respect of other astronomers was mainly the detailed mathematical results about the positions of the Sun, Moon, and planets that make up most of *De Revolutionibus*. The heliocentric model was either ignored or disbelieved by most astronomers. The first powerful evidence in favor of Copernicus's theory came about as a consequence of a lifetime of observations made by the greatest of the pretelescopic observers, Tycho Brahe (1546–1601). Given that his observations supported Copernicus's model, it is ironic that Tycho actively opposed the heliocentric model. Because of his great reputation among his contemporaries, his opposition delayed the general acceptance of the ideas of Copernicus.

Tycho Brahe, a Danish nobleman, was born 3 years after the death of Copernicus. As a boy he became intensely interested in astronomy. While a law student at the University of Leipzig, he sneaked out at night to observe the skies while his tutor slept. When he was 26 he wrote a book describing his observations of an extremely bright new star, or "nova," which had appeared the previous year. The book established Tycho's reputation as an astronomer. A few years later, to keep Tycho from leaving Denmark, King Frederick II gave him the island of Hveen, near Copenhagen, with all its tenants and rents.

The king also instructed his chancellor to pay for the construction of a combination residence-observatory. Tycho set up a shop to build astronomical instruments and trained assistants to help him observe. Using these instruments, he regularly obtained observations that were at least twice as accurate as those of any earlier astronomer. When he took special efforts, his data were accurate to 1 minute of arc (about 3% of the angular diameter of the Moon). His accuracy came from his extreme care in constructing equipment and making observations. Whenever possible, Tycho used metal rather than wood in his instruments, thus avoiding the problems caused by warping. He also made his instruments (some of which were rather like protractors) very large to be able to mark off small fractions of a degree. One of these instruments is shown in Figure 4.10. Tycho took

FIGURE 4.10
Tycho Brahe and One of His Large Astronomical Instruments, the Mural Quadrant

FIGURE 4.11
The Parallax of a Star
A consequence of the Earth's motion about the Sun is that a nearby star is seen in slightly different directions when the Earth is on opposite sides of the Sun.

ANIMATION *The parallax of a star*

INTERACTIVE *Stellar parallax*

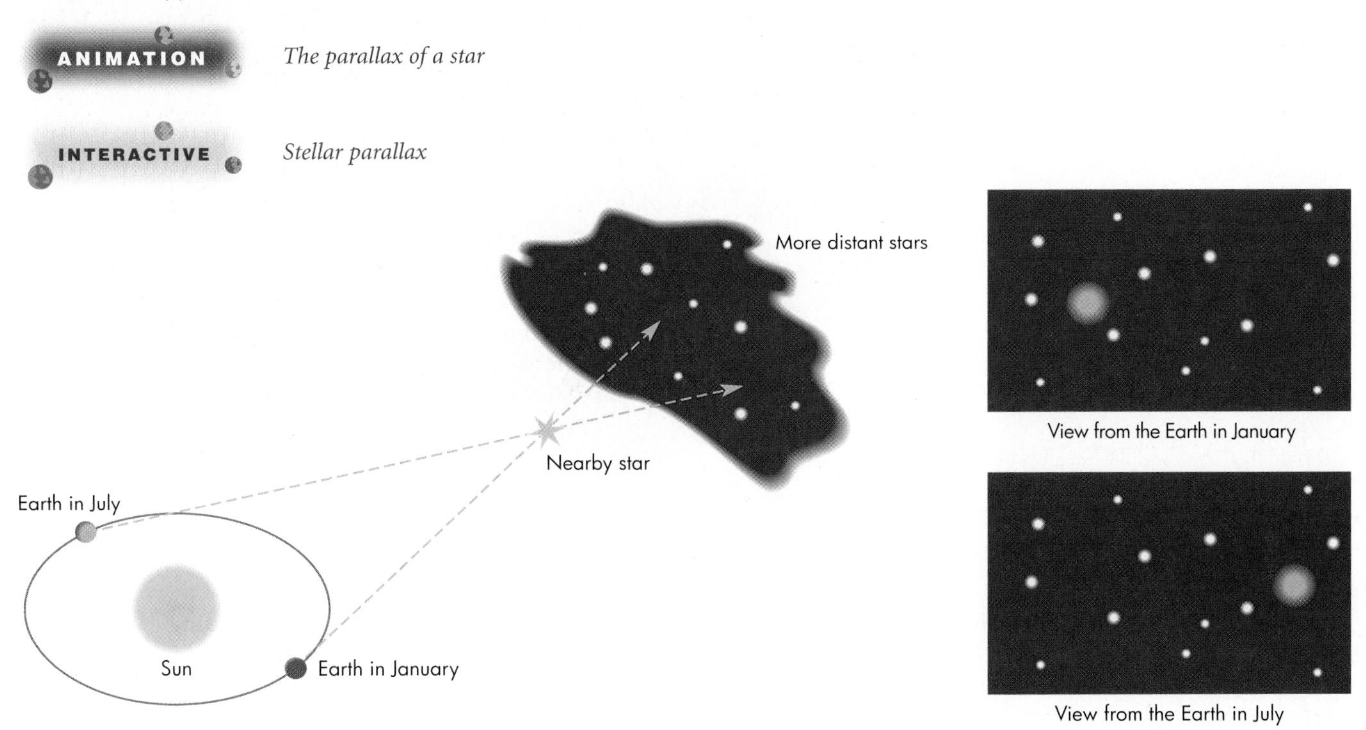

great pains to position and orient his instruments and preferred to use those instruments that were in permanent mountings. He was the first scientist to understand the value of multiple observations of the same event. He often had several assistants make simultaneous measurements, which could then be compared in order to check accuracy.

Unlike most previous observers, who had observed the Sun, Moon, and planets only on special occasions such as solstices and oppositions, Tycho carried out regular observations over a period of many years. Tycho and his assistants compiled about ten times as many observations as all earlier Renaissance astronomers combined. The scope of his data was especially important in that it replaced the error-ridden body of ancient observations on which previous astronomical theorists had to rely.

Tycho Brahe was able to make observations more accurate than those of any astronomer who had come before him. He carried out regular observations of the Sun, Moon, and planets for many years. His observations were a benchmark against which astronomical theories could be tested.

Tycho's accurate observations led him to reject the heliocentric model of the solar system. Tycho looked very carefully for **stellar parallax,** the annual apparent change in the position of a star caused by the motion of the Earth around the Sun. Figure 4.11 shows that because of orbital motion, we view the star from positions that are separated by as much as 2 AU. This changes the direction in which the star is seen. When Tycho failed to detect parallax, he was forced to adopt one of two conclusions. Either the stars were so far away that parallax was too small for even Tycho to measure, or the Earth didn't move. He found that the stars would have to be at least 7000 AU away for him not to detect parallax. Tycho might have been able to accept such immense distances except that he thought he could measure the angular sizes of the stars. The stars appeared to be several minutes of arc in angular diameter (this is actually an illusion). If they were 7000 AU away, they would have to be several AU in size, vastly larger than the Sun. He couldn't believe the stars were so big and concluded that the Earth was stationary.

Tycho eventually arrived at a model that is a compromise between those of Ptolemy and Copernicus. In Tycho's model, shown in Figure 4.12, the Sun and Moon orbit the Earth. All of the other planets orbit the Sun. Thus, the Earth is stationary, but the great regularities in the planetary motions, which are part of Copernicus's model, can be

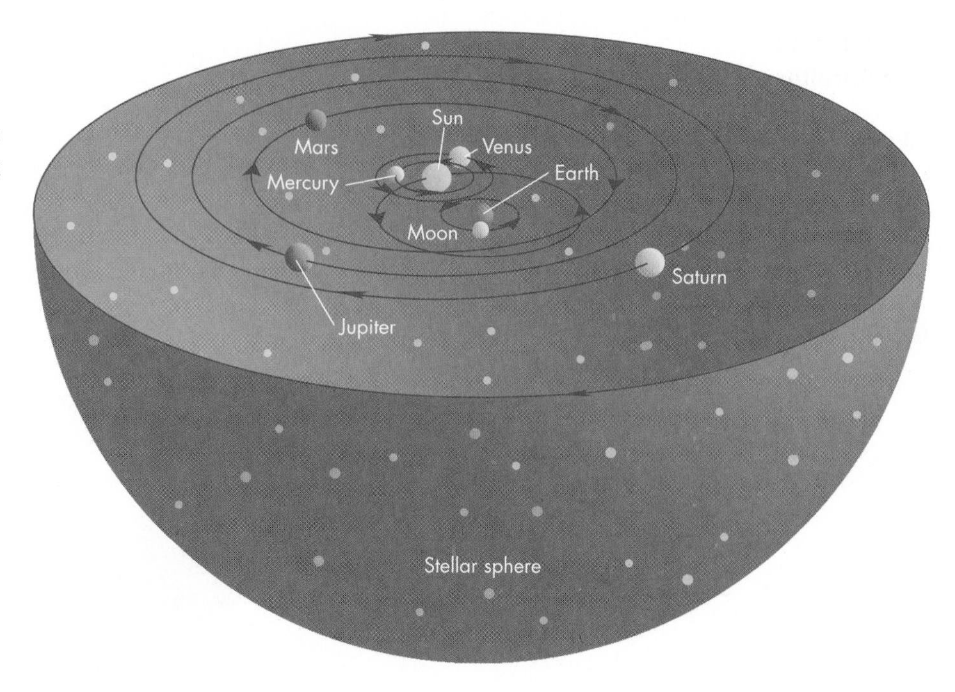

FIGURE 4.12
Tycho's Model of the Solar System
The Sun and Moon orbit the Earth while the other planets orbit about the Sun.

retained. Tycho's model gained wide acceptance as a third alternative and was in general use for about a century after he died.

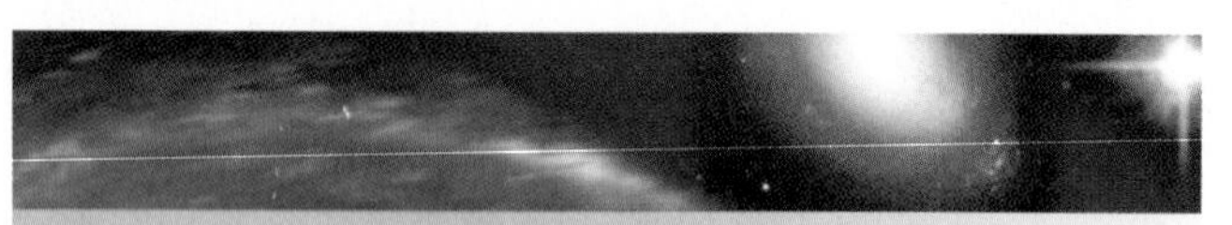

Tycho's inability to detect stellar parallax led him to reject Copernicus's model of the solar system. He proposed an alternative model in which the Sun and Moon orbit the Earth and the other planets orbit the Sun.

In 1597 Tycho had a falling out with his new royal patron, Christian IV, and left Denmark. After wandering Europe for several years he took up a new position in Prague. Although it was a personal tragedy for Tycho to leave Hveen, if he had not moved to Prague, he probably would not have met Johannes Kepler (1571–1630), who became his colleague for the last 2 years of Tycho's life. When Tycho died in 1601, he left his collection of 30 years' worth of observations in the hands of Kepler. Using these data, Kepler solved the problem of planetary motion.

4.5 KEPLER AND PLANETARY ORBITS

By the time he joined Tycho's staff, Kepler was nearly 30 years old. He had recently published a very speculative book on the orbits of the planets. The book was widely read by other astronomers including Tycho, who recognized Kepler's genius and invited him to visit. At Tycho's death, Kepler was appointed to replace Tycho as Imperial Mathematician and was put in charge of Tycho's data.

Kepler's main discoveries about the nature of the planetary orbits were made while he was trying to account for Tycho's observations of Mars. Kepler at first boasted that the task would take him only a few days. It turned out to occupy most of his effort for almost a decade. To account for Mars's motion against the background of the stars, both the orbits of the Earth and Mars had to be found. Kepler's first approach to the problem was strictly Copernican. He tried out combinations of circular orbits and circular epicycles for both planets and hoped that he could find a combination that would accurately match Tycho's data. He was able to find many solutions that would have been good enough to match the observations of Mars that had been available before Tycho. However, he could not match Tycho's data. Eventually Kepler, who had great faith in the accuracy of Tycho's observations, concluded that no adequate combination of circles could be found. He began to consider other orbital shapes and abandoned the 2000-year-old requirement of perfect circular orbits.

Kepler experimented with ovals for a time and then found that he could match Tycho's data almost exactly if Mars and the Earth moved on elliptical orbits with speeds that varied with their distances from the Sun. About 10 years later Kepler discovered a mathematical relationship between the distances of the planets and their periods of revolution. His discoveries are summarized in **Kepler's laws of planetary motion.**

FIGURE 4.13
Three Ellipses
For every point on an ellipse, the sum of the distances to the foci, *F* and *G*, is the same. The line drawn through the foci is the major axis of the ellipse. The semimajor axis is half the major axis. Ellipses **A**, **B**, and **C** all have the same major axis. Ellipse **B**, for which the separation of the foci is smaller than for ellipse **A**, is less eccentric than ellipse **A**. Ellipse **C**, for which the separation of the foci is larger than for ellipse **A**, is more eccentric than ellipse **A**.

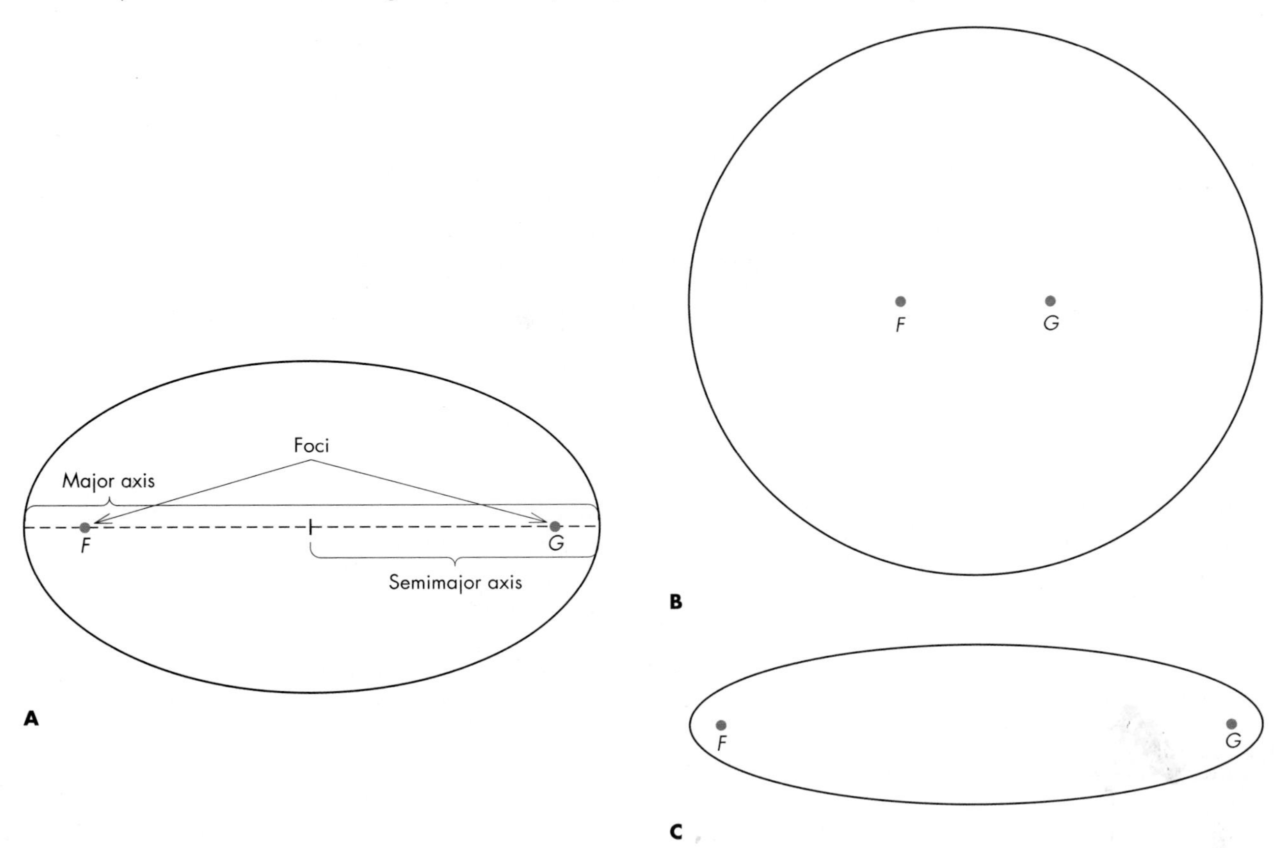

Kepler's First Law

Kepler's first law of planetary motion states that a planet moves on an elliptical orbit with the Sun at one focus. An **ellipse** is a closed curve for which the sum of the distances from two fixed points (the **foci**) is the same for every point on the curve. In Figure 4.13, *F* and *G* are the foci of the ellipse. If a piece of string has its ends secured at *F* and *G*, an ellipse can be drawn by pushing a pencil against the string until it is tight and then tracing out all of the points for which the string is taut (Figure 4.14). The shape of the ellipse can be changed by changing either the length of the string or the distance between the foci.

The maximum diameter of the ellipse (the line that passes through the foci) is called the **major axis.** Half of the major axis is defined as the **semimajor axis.** The **eccentricity** of the ellipse is the ratio of the distance between the foci to the length of the major axis. The eccentricity of an ellipse can range from 0 to 1. An ellipse with zero eccentricity is a circle, for which the two foci are located at the same point—the center of the circle. As the eccentricity of an ellipse approaches 1, the shape of the ellipse becomes nearly a straight line. In the case of planetary orbits, the Sun is located at one of the foci. Nothing, however, is located at the other focus. At **perihelion,** the planet is nearest the Sun. At **aphelion,** the planet is farthest from the Sun.

Kepler's Second Law

Kepler's second law of planetary motion states that a planet moves so that an imaginary line connecting the planet to the Sun sweeps out equal areas in equal intervals of time, as shown in Figure 4.15. This means that angular momentum, which will be discussed in Chapter 5, remains constant as a planet orbits the Sun. The second law, which is often called the law of equal areas, says that for a particular planet, the product of distance from the Sun and transverse velocity is constant. **Transverse velocity** is the part of orbital speed perpendicular to the line between the planet and the Sun. At perihelion and aphelion, transverse velocity and orbital speed are the same, as shown in Figure 4.15. This means that a planet moves fastest when it is nearest

FIGURE 4.14
Drawing an Ellipse
The ellipse consists of all the points that can be reached by a fixed length of string attached at points *F* and *G*.

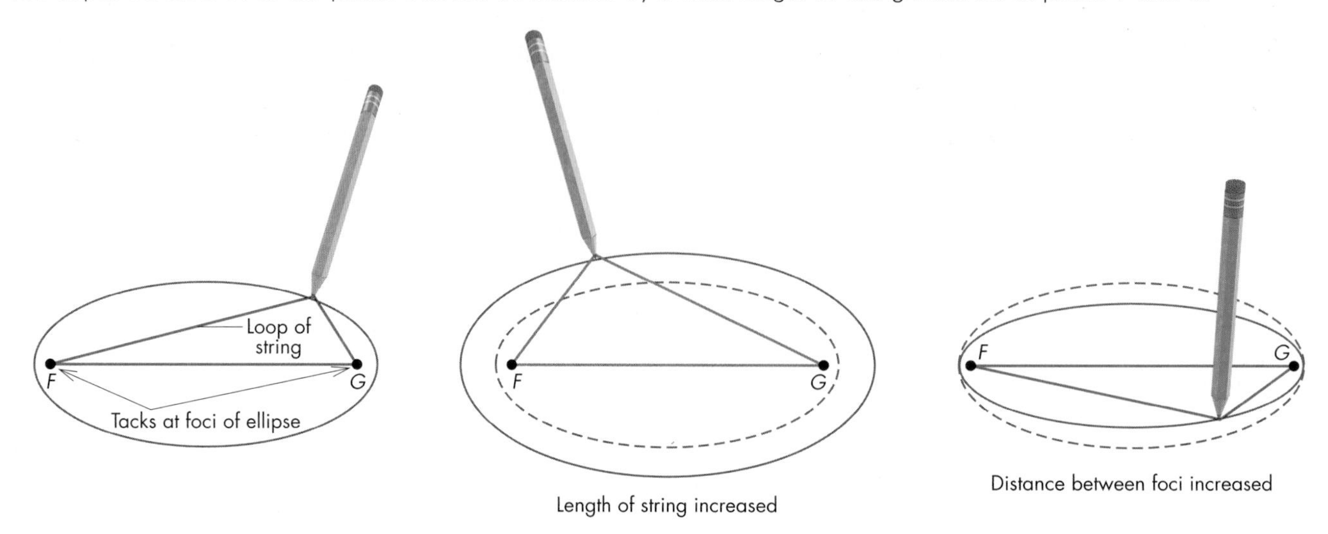

FIGURE 4.15
Kepler's Second Law
The points from *A* to *L* show the planet's position at equal intervals of time. The area of each region swept out by the planet is the same, as can be verified by counting the tiny equal-area squares in two of the regions. The arrows indicate transverse velocity.

ANIMATION *Kepler's Laws*

INTERACTIVE *Kepler's Second Law*

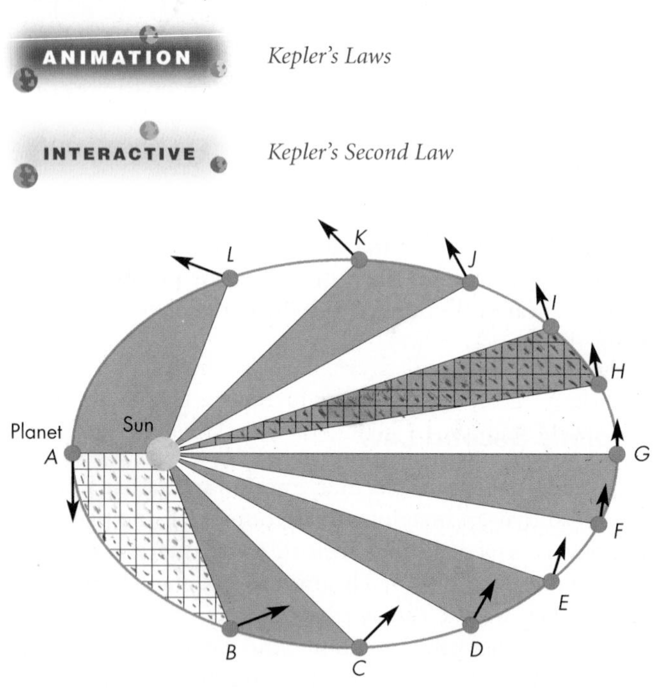

the Sun, at perihelion, and most slowly when it is farthest from the Sun, at aphelion. For Mars, which has an orbital eccentricity of 0.09, aphelion is 18% farther from the Sun than perihelion. Thus, Mars moves 18% faster at perihelion than at aphelion. The equants and eccentrics in Ptolemy's model and the epicycles in Copernicus's model were the results of efforts to account for the changing orbital distances and speeds of the planets. These devices, however, could only approximate the true motions of the planets. With Kepler's first and second laws it was possible, for the first time, to predict planetary positions that were as accurate as the observations themselves. In fact Kepler eventually used his laws of planetary motion to produce the *Rudolphine Tables,* from which planetary positions could be computed. The tables were of unprecedented accuracy and were in general use for a century.

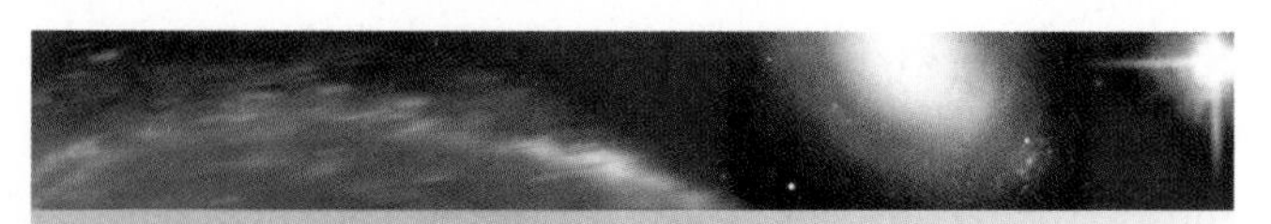

According to Kepler's first law, the orbits of the planets are ellipses with the Sun at one focus. Kepler's second law says that a planet moves so that the product of its transverse velocity and distance from the Sun is constant. Thus, the planet moves fastest at perihelion and slowest at aphelion.

Kepler's Third Law

INTERACTIVE *Kepler's Third Law*

Kepler's third law of planetary motion is very different from the first two, which describe the motions of individual planets. The third law relates the motions of all of the planets to each other. This law states that the squares

Kepler's Third Law

Equation 4.5 can be used to calculate the orbital period of a planet from its orbital distance or vice versa. To use Equation 4.5, orbital period must be given in years and orbital distance in AU. Suppose a planet has an orbital distance, a, of 3.5 AU. For a = 3.5 AU, Equation 4.5 is

$$P^2 = a^3 = (3.5\ \text{AU})^3 = 42.88$$

The square root of 42.88 is 6.55, so the orbital period, P, of the planet is 6.55 years.

Now, suppose a planet has an orbital period of 15 years. For P = 15 years, Equation 4.5 gives

$$a^3 = P^2 = (15\ \text{yr})^2 = 225$$

The cube root of 225 is 6.1, so the orbital distance of the planet is 6.1 AU.

of the sidereal periods of the planets, P, are proportional to the cubes of the semimajor axes of their orbits, a. That is,

$$P^2 \propto a^3 \tag{4.4}$$

where the symbol "$\propto$" means "proportional to."

If the periods are measured in years and the semimajor axes (which are the same as the average distances from the Sun) are measured in AU, then Kepler's third law becomes

$$P^2 = a^3 \tag{4.5}$$

The third law suggested that there was a universal underlying principle that governed the motions of the planets. The nature of that underlying principle, gravity, was discovered by Isaac Newton later in the seventeenth century.

Kepler's third law describes the relationship between the sidereal period of a planet and its average distance from the Sun. It implies that there is an underlying principle that governs the orbital motions of the planets.

4.6 GALILEO AND THE TELESCOPE

It is likely that Kepler's discoveries alone eventually would have resulted in the abandonment of the geocentric model of the solar system. The triumph of the heliocentric model, however, was also aided by a series of observations made by Galileo Galilei (1564–1642) shortly after the telescope was invented. Unlike the observations of Tycho, which were more accurate versions of measurements that had been made for thousands of years, Galileo's were totally different from any made before. With his telescope he observed remarkable phenomena that no astronomer could have seen with the unaided eye.

Galileo is often credited with inventing the telescope and being the first to turn it to the heavens. Neither is true. The telescope was invented in the early 1600s by an unknown Dutch lens maker who discovered that two lenses could be combined to magnify distant objects. Early reports about the telescope mention that it showed the night sky had many stars that were too faint to be seen by the unaided eye. In the summer of 1609, Galileo heard about the telescope. He quickly made one and began making astronomical observations. Within a year he had made important observations of the stars, Moon, Sun, and planets. All of these observations were made with small telescopes that do not compare favorably to a good pair of modern binoculars.

The Observations

Stars Like earlier users of the telescope, Galileo found previously unknown stars wherever he pointed his telescope. When he looked at the Milky Way he found that its glow was the combined radiance of myriad stars, so faint and so close together that they could not be distinguished by the unaided eye. He also noted that unlike the Moon, Sun, and planets, which were magnified when viewed through the telescope, the stars remained the same size. He quickly realized that the angular sizes of stars had been enormously overestimated by pretelescopic observers. Tycho's objection to Copernicus's model—that the angular sizes of stars were too large for them to be very distant—was based on faulty data. Thus, although Galileo's observations of stars did not prove the heliocentric model, his observations did at least dismantle one of its main counterarguments.

The Sun and Moon Before Galileo observed the Sun and Moon, they and other celestial objects were generally thought to be perfect, smooth spheres. Galileo found that the Moon had mountains, valleys, and plains like the

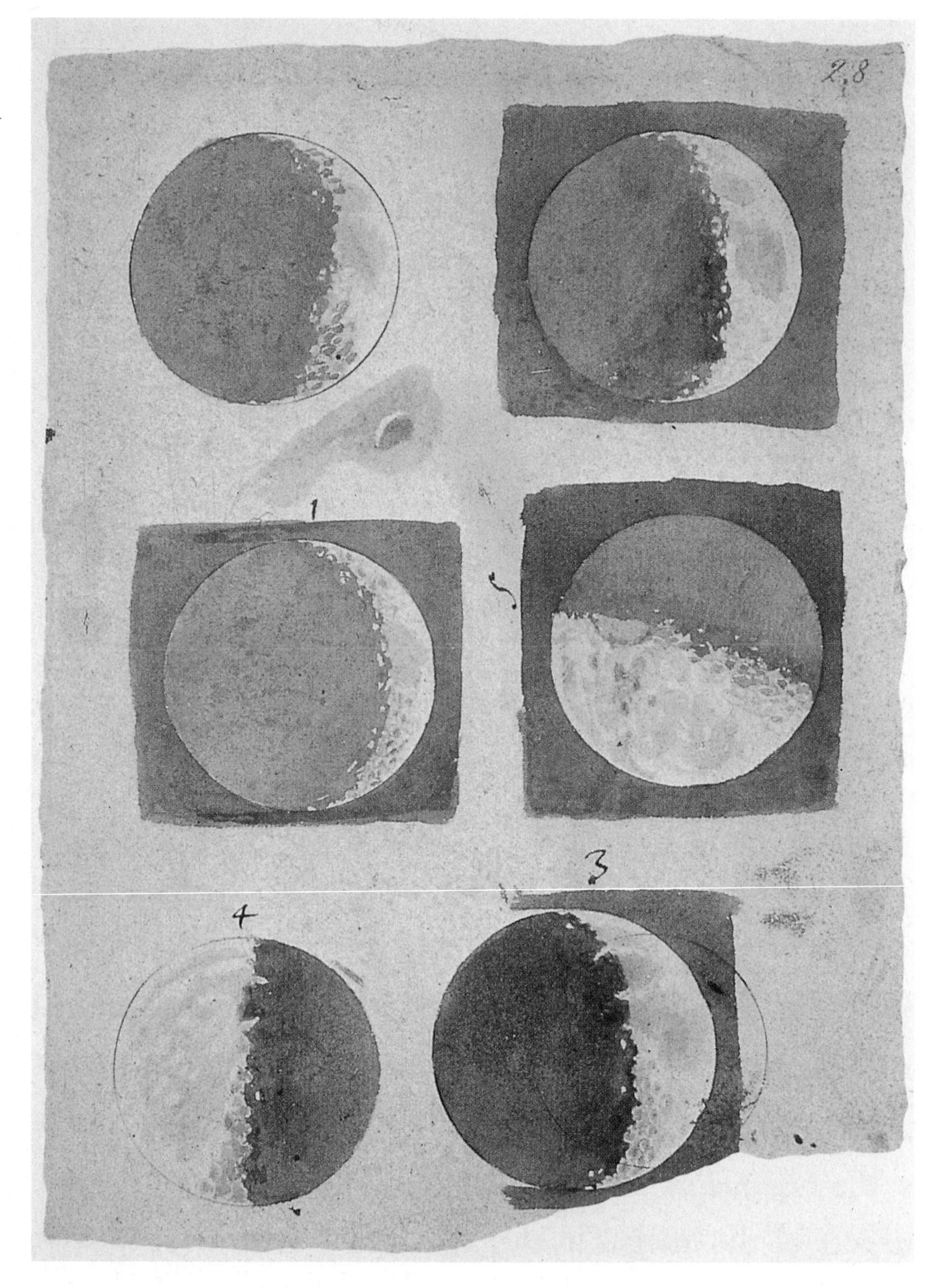

FIGURE 4.16
Some of Galileo's Drawings of the Moon
The sketches show the roughness of the lunar surface.

Earth. Some of Galileo's sketches of the Moon are shown in Figure 4.16. The similarity of the features of the Moon and the Earth suggested that the Earth, like the Moon, was a celestial object. Galileo also found that the Sun was imperfect. Figure 4.17 shows that its surface is dotted with dark spots that move across its disk. Galileo correctly interpreted the motion of sunspots as the rotation of the Sun about its axis. By observing the same group of sunspots from one day to the next, he found that the Sun rotated once in about 4 weeks. If the Sun rotates, he argued, why not the Earth?

Jupiter's Satellites Even more powerful support for the heliocentric model came when Galileo observed Jupiter. He immediately saw that Jupiter was accompanied by several smaller points of light. He followed Jupiter for several days and saw that the points shared Jupiter's motion against the stars, but, as shown in Figure 4.18, they changed their arrangement from night to night. Galileo realized that the points of light were four satellites orbiting Jupiter in the same way that the Moon orbits the Earth. The Galilean satellites of Jupiter (as they are now called) clearly demonstrated that the Earth was not at the center

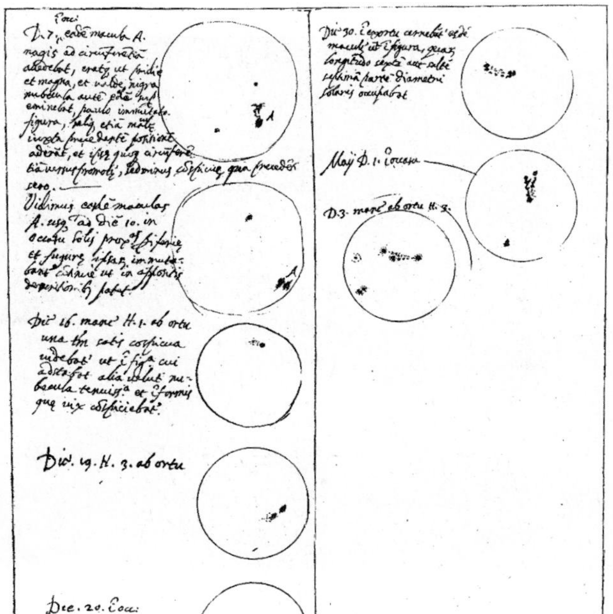

FIGURE 4.17
Galileo's Drawings of the Sun on Different Days
Galileo interpreted the changing positions of the sunspots as due to the Sun's rotation.

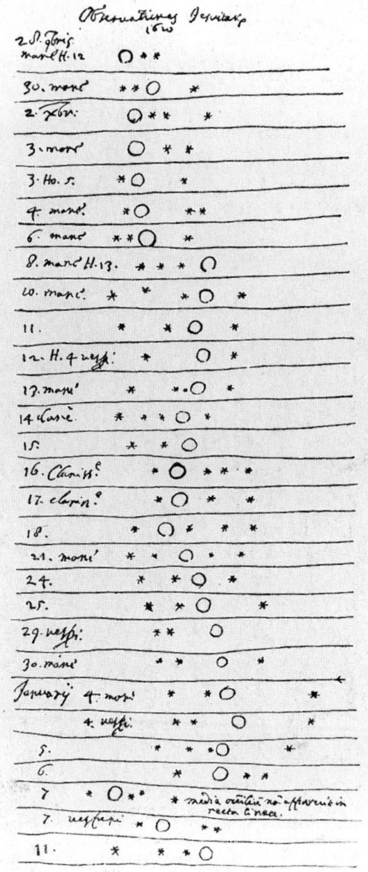

FIGURE 4.18
Early Telescopic Observations of Jupiter and Its Satellites
This drawing by Galileo is his copy of observations made by a Jesuit astronomer at about the same time Galileo himself was observing Jupiter.

of all motion in the universe. The satellites orbited Jupiter just as Copernicus and Kepler had said that the planets orbit the Sun.

The Phases of Venus The geocentric and heliocentric models of the solar system make very different predictions about the phases that Venus should be able to display if it shines because it reflects sunlight toward the Earth. The predictions of the two models are illustrated in Figure 4.19. In the Ptolemaic system, Venus is always between the Earth and the Sun. This means that we see mostly the dark hemisphere of Venus—the one turned away from the Sun. The only phases Venus can show are new and crescent. In contrast, in the Copernican system, Venus can be either between the Earth and Sun or on the opposite side of the Sun. When Venus is on the opposite side of the Sun, the hemisphere we see is the bright one—the one turned toward the Sun. At that time, we see a nearly full Venus.

Thus, in the Copernican system, Venus can show the full range of phases from new to full. Galileo found that Venus not only shows the entire range of phases, but also has a larger angular size at new phase than at full phase (also shown in Figure 4.19). This agrees exactly with the predictions of the heliocentric model. Although the Ptolemaic system can't account for Galileo's observations of the phases of Venus, Tycho's model—in which Venus orbits the Sun, which in turn orbits the Earth—can.

Galileo's observations severely damaged the Ptolemaic model and forced many confirmed geocentrists to switch their allegiance to Tycho's model. Tycho's model remained viable until about 1700, after which it was usually mentioned only to ridicule it. The model fell from favor not because new observations showed it to be incorrect, but because it lacked a comprehensive organizing principle. After it became clear that gravity could account for the motions of the planets in the heliocentric model, the lack

FIGURE 4.19
The Phases of Venus According to A, the Ptolemaic System, and B, the Copernican System
In **A,** Venus is always between the Earth and Sun and can be seen only at new to crescent phase. In **B,** all phases are possible. (However, new and full phase are difficult to see because Venus appears very near the Sun in the sky at those times.) Venus is closer to the Earth and appears larger at crescent phase than it does at gibbous phase.

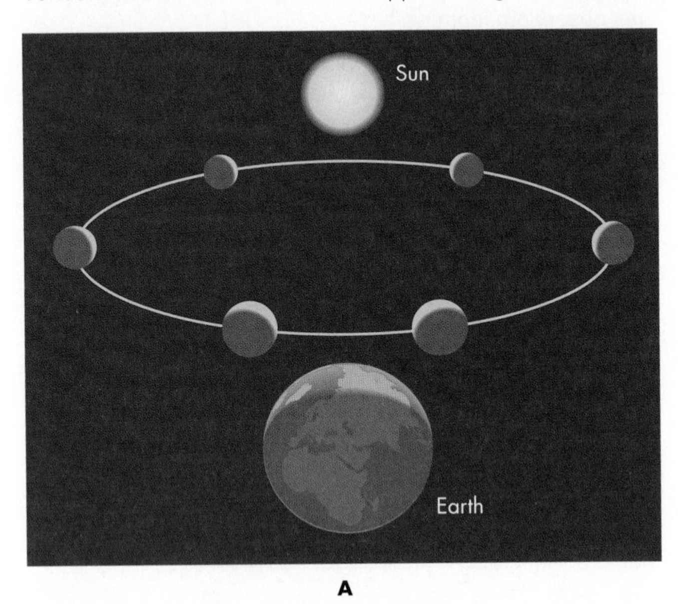

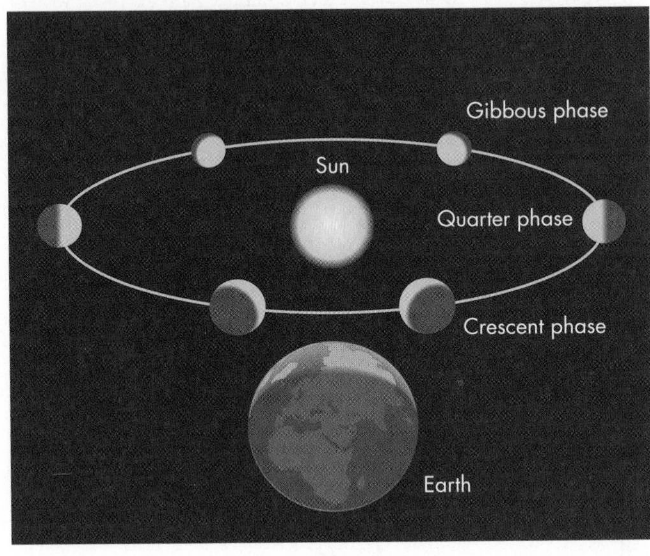

of a corresponding principle to account for the bizarre structure of Tycho's solar system became an obvious and serious weakness.

Galileo's telescopic observations of the satellites of Jupiter showed that there were at least some celestial objects that did not orbit the Earth. His observation that Venus showed a full range of phases could not be accounted for by the model of Ptolemy.

The *Dialogue*

After 1600 there was a growing tendency among astronomers to regard the heliocentric model as a realistic picture rather than a mathematical hypothesis. This posed a problem for church officials because a rotating, revolving Earth was in apparent contradiction to a number of biblical passages. In 1616 Pope Paul V asked his theological experts for a judgment on the heliocentric system. The experts replied that it was clearly erroneous and heretical to contend that the Sun was the center of the solar system or that the Earth rotated or revolved. This opinion was made into an official decree and sent throughout the Christian world. Perhaps the strong response of the church to the Copernican system, which challenged the traditional view of the physical world, was partially due to the simultaneous theological challenge of the Protestant Reformation. Galileo, who was an outspoken Copernican, was summoned by Cardinal Robert Bellarmine in 1616 and told what he must do to conform to the decree. He was instructed no longer to hold, teach, or defend the Copernican system either orally or in writing.

For many years Galileo found other scientific problems on which to work and more or less held his tongue about the heliocentric model. In 1624 Paul V was succeeded by Urban VIII, who had earlier been a supporter of Galileo's work. Galileo spoke with Urban and was told that he could write about Copernican theory so long as he treated it only as a hypothesis. With this encouragement, Galileo set to work on a major book on the structure of the universe. The book *Dialogue Concerning the Two Chief World Systems, the Ptolemaic and the Copernican* was published in 1632. It is significant that the book was published in Italian, so that it could reach a much wider audience than if it had been published in Latin. The *Dialogue* takes the form of a conversation among Salviati, an advocate of the heliocentric system, Simplicio, a staunch disciple of Aristotle and Ptolemy, and Sagredo, who is neutral. Simplicio repeatedly presents what seem to be sound arguments for the geocentric point of view. Salviati then demolishes Simplicio's arguments, making him appear foolish. Although Galileo had Salviati say that his arguments were strictly hypothetical, it was clear to readers that the *Dialogue* was a powerful demonstration of the superiority of the heliocentric system. To make things worse, the book concludes with the foolish Simplicio stating the Pope's own position, that the power of God could not be limited by scientific theories.

The book provoked a storm. Galileo was brought before the Inquisition and charged with disobeying the order of 1616 that he not advocate the Copernican theory in print. He was convicted of that charge and saved himself from being convicted of heresy only by denying the truth of the Copernican theory and stating that he cursed and detested all of his mistakes and heresies. The *Dialogue* was placed on the index of banned books, as *De Revolutionibus* had been 16 years earlier. Although he was 69 years old, Galileo was condemned to an indefinite term of imprisonment and spent the remaining 9 years of his life under house arrest. Galileo's fate had a chilling effect on astronomers in Italy, who stopped discussing the heliocentric model. In other countries the effect of the trial of Galileo was considerably weaker. Books that supported the Copernican system were published without much difficulty. Also, despite early opposition from Protestant leaders, including Martin Luther, by the time of Galileo's death, most Protestant astronomers were Copernicans. Religious opposition to the ideas of Copernicus and Kepler, like scientific opposition, gradually faded away. The Church's ban on the publication of the *Dialogue* was lifted in 1822 and Galileo was finally cleared by the Church in 1993.

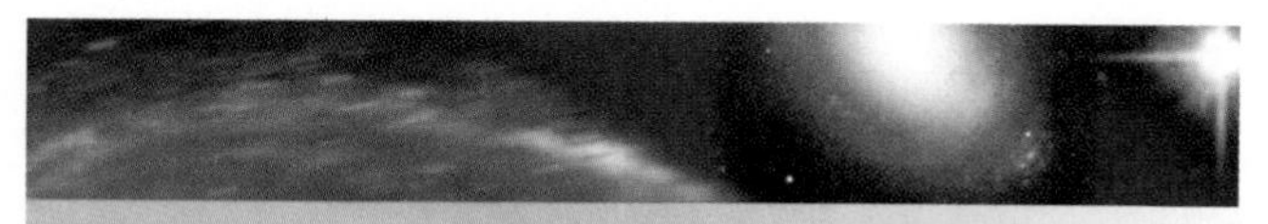

Galileo's book *Dialogue Concerning the Two Chief World Systems* was a powerful argument for the heliocentric model of the solar system. The book resulted in the Church's persecution of Galileo.

Despite the triumph of the heliocentric model, evidence that the Earth rotates and revolves was hard for astronomers to find. In fact, the first real proof that the Earth revolves around the Sun was not discovered for almost 200 years after Copernicus first questioned the model of Ptolemy. The proof came long after the heliocentric theory had become generally accepted. Thus the heliocentrists prevailed not because they were able to prove that the Earth revolved and rotated but because they were able to convince others that their ideas were more plausible than those of Ptolemy and Aristotle. They were able to show that the heliocentric model was simpler and more elegant than geocentric models and that it could be made to account for the motions of the planets with unprecedented accuracy.

Chapter Summary

- The civilization of Greece and Rome began to decline shortly after the time of Ptolemy. Fortunately, the discoveries of the ancient astronomers were preserved by Islamic astronomers, who found them and translated them into Arabic. (Section 4.1)
- Astronomical knowledge was gradually reacquired in Western Europe. By the fifteenth century, the level of knowledge matched or exceeded that at the time of Ptolemy. The geocentric model of Ptolemy was almost universally accepted as the correct description of the solar system. (4.2)
- In the early sixteenth century, Copernicus proposed that the Sun rather than the Earth is the center of the solar system. In the heliocentric model, the daily and annual patterns of celestial motion are explained by the rotation and revolution of the Earth. Retrograde motion of the planets occurs whenever the Earth passes or is passed by another planet. (4.3)
- In the model of Copernicus, the orbital distances of the planets can be found through observations and geometry. In contrast, the geocentric model makes no specific predictions about the relative distances of the planets. (4.3)
- Through his care in building and using astronomical instruments, Tycho Brahe was able to make observations of unparalleled accuracy. His regular observations of the Sun, Moon, and planets covered many years. His data replaced the ancient observations that earlier theorists had been using for centuries. (4.4)
- Tycho was unable to detect stellar parallax and thus rejected the model of Copernicus. Tycho proposed a model in which the Earth is orbited by the Sun and Moon but all of the other planets move about the Sun. (4.4)
- Using Tycho's data, Kepler was able to discover the laws of planetary motion. His first law says that the planets move on elliptical paths with the Sun at one focus. The second law says that a planet moves so that a line drawn between the planet and the Sun sweeps out equal areas in equal amounts of time. This means that the product of transverse velocity and distance from the Sun remains constant as a planet moves about the Sun. The planet moves fastest when it is nearest the Sun. (4.5)
- Kepler's third law says that the square of the sidereal period of a planet is proportional to the cube of its

average distance from the Sun. The third law implies that there is a common principle that governs the orbital motions of the planets. (4.5)
- Galileo's telescopic observations provided strong support for the heliocentric model. In particular, his observations that Venus shows all the phases from new to full could not be explained by Ptolemy's model of the solar system. (4.6)
- Galileo summarized his arguments for the heliocentric model in his book *Dialogue Concerning the Two Chief World Systems*. The book put Galileo in conflict with church authorities and resulted in his persecution. (4.6)

Key Terms

aphelion *69*
astronomical unit (AU) *64*
eccentricity *69*
ellipse *69*
foci *69*
greatest elongation *64*
heliocentric *61*
impetus *59*
inferior planet *64*
Kepler's laws of planetary motion *68*
major axis *69*
perihelion *69*
semimajor axis *69*
sidereal period *62*
stellar parallax *67*
superior planet *64*
transverse velocity *69*
universe *59*

Conceptual Questions

1. We see Mars undergo retrograde motion when the Earth passes Mars. How does the motion of the Earth look to hypothetical Martian astronomers at that time?
2. The outer planets (Uranus and Neptune) have synodic periods that are only slightly longer than a year. Explain why this is so.
3. What prediction did the Ptolemaic system make about the range of phases that could be seen for Mars, Jupiter, and Saturn? What was the prediction of the Copernican system for these planets? Could observations of the phases shown by these planets be used to show the superiority of either system?

Problems

1. Suppose an inferior planet has a sidereal period of 0.6 years. Find the synodic period of the planet. What would be the synodic period of the Earth when observed from the planet?
2. What is the sidereal period of a planet that has a synodic period of 0.7 years?
3. Suppose a superior planet has a sidereal period of 1.4 years. What would be the synodic period of the planet?
4. What are the possible sidereal periods of a planet that has a synodic period of 1.5 years?
5. What is the synodic period of Saturn when observed from Jupiter?
6. What is the synodic period of Mercury when viewed from Venus?
7. A hypothetical planet is 70° from the Sun at greatest elongation. Draw the triangle showing the positions of the Sun, planet, and Earth at greatest elongation. Measure the sides of the triangle to find the distance of the planet from the Sun in AU.
8. When viewed from Mars, what is the angle between the Earth and the Sun when Earth is at greatest elongation?
9. Use the small angle equation to confirm Tycho's conclusion that if a star had an angular diameter of 1 minute of arc and a distance of 7000 AU it would be much larger than the Sun.
10. Draw an ellipse for which the distance between the foci is 10 cm and the length of the major axis is 20 cm. What is the eccentricity of the ellipse? If a planet orbits on the ellipse, what is its perihelion distance? What is its aphelion distance?
11. How far apart are the foci for an ellipse that has a major axis of 10 cm and an eccentricity of 0.0?
12. Mercury has an orbital eccentricity of 0.21. Find the perihelion and aphelion distances of Mercury. What is the ratio of Mercury's orbital speed at perihelion to that at aphelion?
13. What is the sidereal period of a hypothetical planet with an orbital semimajor axis of 0.7 AU?
14. What is the sidereal period of a hypothetical planet with an orbital semimajor axis of 3.5 AU?
15. What is the orbital semimajor axis of a hypothetical planet with a sidereal period of 0.7 years?
16. What is the orbital semimajor axis of a hypothetical planet with a sidereal period of 25 years?

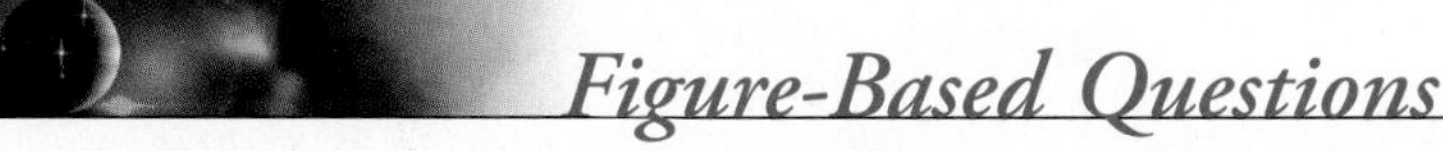

Figure-Based Questions

1. Using Figure 4.7, measure the orbital distances of the Earth and the planet. Use a protractor to measure the angle, θ. Compare the ratio of the distances with the trigonometric sine of θ.
2. Measure the eccentricities of the three ellipses shown in Figure 4.13.
3. Count the squares and partial squares in Figure 4.15 to confirm that the area in the two regions is the same. How many squares are in each region?

Group Activities

1. Go to a running track with a partner. Run slowly around the track while your partner runs quickly around the track (or vice versa). Carefully watch the background against which you see your partner as you both run. Have your partner observe your motion against the background as well. Raise your hand whenever you see your partner move in a retrograde direction against the background. Have your partner raise his or her hand whenever he or she sees you moving retrograde. Compare the times at which you and your partner see each other in retrograde motion.
2. Divide your group into two subgroups. Choose one group to be "heliocentrists" and the other group "geocentrists." After a few minutes of preparation, have a short debate about which group can best explain the daily motion of the sky, the annual motion of the Sun, and the retrograde motions of the planets. Discuss what might constitute firm evidence that one theory is better than the other.

For More Information

Visit the text website at **www.mhhe.com/fix** for chapter quizzes, interactive learning exercises, and other study tools.

5

I derive from the celestial phenomena the forces of gravity with which bodies tend to the sun and the several planets. Then from these forces, by other propositions which are also mathematical, I deduce the motions of the planets, the comets, the moon, and the sea.

Isaac Newton
Principia

Gravity and Motion

The flight of the Voyager 2 spacecraft to the outer planets and beyond helped expand our knowledge of the solar system. In order for *Voyager* to reach the many planets and satellites it visited, the team of scientists and engineers who controlled the path of the spacecraft had to take careful account of all the gravitational pulls at work as well as the effect of every firing of the *Voyager* thrusters. After launch in 1977, the scientists and engineers fired small thrusters to correct its path so that it would, in 1979, arrive at precisely the correct time and place to flash through the satellite system of Jupiter. Moreover, they aimed Voyager so that the gravitational pull of Jupiter would deflect it by just the correct amount to send it speeding on to Saturn. More small corrections were needed prior to its arrival at Saturn in 1980 so that it would miss Saturn's rings, which would have destroyed *Voyager*, and yet pass close enough to Saturn to be flung outward to Uranus. At Uranus, in 1986, the spacecraft was again deflected toward Neptune, which it encountered in 1989. Were it not for the gravitational boosts Voyager 2 received along the way, it would not have reached Neptune until the year 2010.

It is remarkable that the gravitational influences of the Sun, planets, and satellites on the path of *Voyager* and the consequences of each small thruster firing were calculated using scientific principles more than 300 years old. Isaac Newton's law of gravitation was used to find each gravitational force acting on *Voyager*. The effect of each gravitational force and each thruster firing on the motion of *Voyager* was calculated using Newton's laws of motion. Although the calculations were carried out using computers, the underlying principles were exactly the same as those given by Newton in the 1600s. Using these

Questions to Explore

- How are forces related to motion?
- What are momentum and angular momentum? What do they have to do with motion and the orbits of the planets?
- What factors control the gravitational forces objects exert on each other?
- How can the Sun's gravitational attraction account for the orbital motion of the planets?
- What forms of energy are important for an orbiting planet?
- What is meant by escape velocity?
- Why do tides occur and where can the strongest tides be found? What effect do tides have on the Earth and its oceans?

principles, the *Voyager* team was able to plan, at each planetary encounter, the spacecraft's arrival to within a few seconds and within about 15 km (10 miles) of its target. This is impressive accuracy considering the 12-year duration of the mission and the many billions of kilometers that *Voyager* traveled.

This chapter describes the development of Newton's laws of motion and his law of gravitation. We focus on the way these laws, used together, explain why the planets move as described by Kepler's laws of planetary motion. Another major theme is the way Newton showed that the same laws are at work on the Earth as well as in the heavens.

5.1 FORCE AND MOTION BEFORE NEWTON

Kepler's three laws of planetary motion give an accurate *description* of the orbits of the planets. But they do not *explain* why the planets move as they do. Before people could really understand planetary motion, they had to learn more about motion itself and the nature of the force that controls the orbits of the planets. In both cases, the development of the necessary understanding had two phases. In the first phase, a number of scientists made contributions that meandered in the right direction. In the second phase, Isaac Newton solved the problem by himself.

Galileo's Experiments

Our modern ideas about motion began to develop when Galileo turned his attention to physics and conducted experiments with moving objects. The role of experiment in science is so familiar to us today that it is hard for us to recognize the novelty of Galileo's experimental approach to physics. Although earlier ideas in physics often were based on observation, experimental investigations had seldom been attempted. When Galileo began to study moving bodies in about 1600, there had been little progress in understanding the physics of motion for about 200 years. The most widely accepted theory of motion, the impetus theory, had been developed during the fourteenth century. Impetus theory shared with Aristotle the idea that motion could continue only so long as a force, external or self-contained, was at work.

Galileo's father had intended for him to become a physician. But while he was studying medicine, he heard a lecture on mathematics and became fascinated by the subject. Galileo thought of mathematics as the best means of understanding nature. The application of mathematics to experiments was, for Galileo, a way to avoid having to rely on possibly erroneous common knowledge. A law of nature, he thought, must be describable through mathematics.

When Galileo began his experiments, it had long been accepted that heavy objects fell faster than lighter ones. It was commonly believed that objects fell with speeds that were proportional to their weights. Galileo investigated this idea not by dropping objects of different weights but by rolling balls down inclined planes (Figure 5.1). The rolling balls moved much less rapidly than dropped ones so their motion could be studied more easily. One of the reasons Galileo was able to learn so much was that he was among the first scientists to use accurate clocks (water clocks and pendula) to measure time in his experiments. Galileo found that the distance

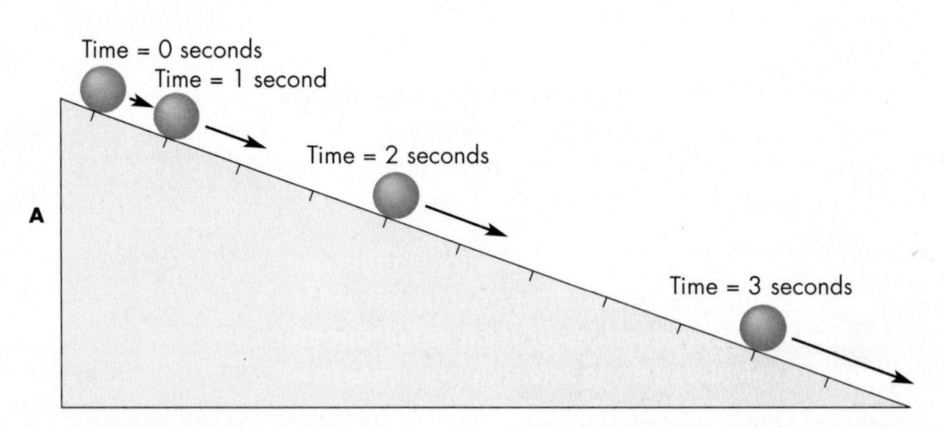

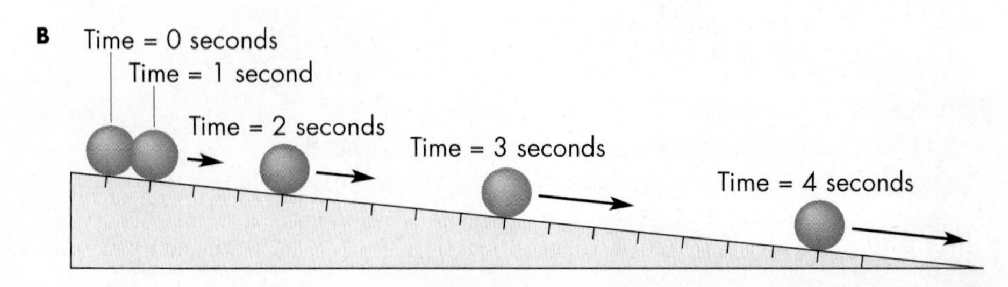

FIGURE 5.1
Two Balls Rolling Down Inclined Planes
While the ball on the steeper plane, **A,** accelerates more rapidly, both balls move so that the distance traveled increases in proportion to the square of the time that they have been moving.

FIGURE 5.2
A Lunar Astronaut Showing That a Hammer and a Feather Fall with the Same Acceleration
This occurs only when there is no air resistance because air resistance affects the feather more than the hammer.

a ball travels is proportional to the square of the time it has been in motion. This is true no matter what the slope of the plane or the weight of the ball. Galileo reasoned that the rate at which an object increased its speed while it fell applied to all objects. The reason that heavy objects reached greater speeds was that the resistance of the air was less important for them. In the absence of air resistance, all objects would fall in exactly the same way. The experiment sketched in Figure 5.2, carried out by *Apollo 15* astronaut David Scott in 1971, shows that Galileo was right. However, the story that Galileo dropped light balls and heavy balls from the Leaning Tower of Pisa is a distorted description of an experiment actually carried out by one of Galileo's critics, who dropped the balls to demonstrate that the heavy ball did, indeed, reach the ground first.

Galileo's experiments with rolling bodies also led him to conclude that motion is a natural state. That is, once an object is set in motion it will forever remain in motion unless something stops it. In the impetus theory, an object continued to move only until its impetus, which pushed it along, was used up. For Galileo, nothing was required to keep the object moving—only stopping the object required a force. Forces that slow and stop moving objects are common on the Earth. Air resistance, friction between sliding or rolling objects and the ground, and the drag felt by objects moving through water are so universal that we almost never experience motion in their absence. Galileo's experiments and his analysis allowed him to see what motion would be like if friction weren't present.

Descartes and Inertial Motion

Galileo's ideas were carried forward another step by René Descartes (1596–1650), the French philosopher. Descartes believed that all changes in motion resulted from collision with particles called "corpuscles." In the absence of such collisions, an object at rest remains at rest. An object in motion continues to move at the same speed in the same direction. This is a clear description of **inertial motion,** the motion of an object when there are no unbalanced forces acting on it. Descartes's contribution was to state that not only would an object remain in motion unless something acted on it, but that *its motion would continue in a straight line.* Self-sustaining circular motion, the basis of both Ptolemy's and Copernicus's models of the solar system, was not possible. Any departure from straight line motion at constant speed requires the presence of some kind of unbalanced force. This applies to falling objects, which increase their speed as they fall, as well as to the motions of the planets, which continually change their speeds and the directions in which they move.

5.2 PLANETARY MOTION BEFORE NEWTON

In the Universe of Aristotle and Ptolemy, the motion of a planet was part of a cosmic clocklike mechanism that drove all celestial motions. The planets, Moon, Sun, and stars were located on crystalline shells that filled space. The rotation of the outermost shell, that of the stars, drove the shell next to it, which was one of the shells that accounted for the motion of Saturn. That shell in turn drove the next shell inward and so on until the innermost shell, one of those that carried the Moon, was reached. Copernicus and his successors destroyed this explanation of planetary motion, but offered nothing to replace it. For Copernicus, perpetual circular motion was "natural," requiring no explanation. Kepler's discovery that the planets move on elliptical orbits with changing speeds strongly implied the presence of forces continually acting to modify the motions of the planets.

Kepler's Explanation

Kepler himself suggested that the motion of a planet could be accounted for by the combination of two forces connecting the Sun and the planet. The first of these forces, which he called the *anima motrix,* was a series of rays that emanated from the Sun and rotated with it. When these rays passed the planet, they pushed against it, propelling it around in its orbit (Figure 5.3). In other words, Kepler believed that the force that kept the planet circling the Sun acted in the direction in which the planet was already moving. The second force, magnetism, accounted for the ellipticity of the planet's orbit. In Kepler's theory, the Sun's south magnetic pole was buried at the center of the Sun, its north magnetic pole distributed over the Sun's surface. For a planet (like the Earth) with its magnetic axis tipped with respect to its orbit, the force between the Earth and Sun would be attractive when the south magnetic pole is tipped toward the Sun, repulsive when the north magnetic pole is inclined toward the Sun. Kepler's mechanical model of the solar system wasn't taken very seriously by other astronomers, but it was important in that it relied on forces that depended only on the position of the planet relative to the Sun.

Robert Hooke

The crucial connection between celestial and terrestrial motions was made by Robert Hooke (1635–1703), a brilliant English scientist who was a contemporary and countryman of an even greater thinker, Isaac Newton (1642–1727). Hooke understood inertial motion and realized that the planets would fly away in straight line motion if they weren't acted upon by forces. What required explanation wasn't that the planets were moving, but rather that they continually changed the direction in which they moved. Because the motion of a planet is continually deflected toward the Sun, there must be an attractive force between the Sun and the planet. Hooke showed that a **central force,** a force acting toward the center of motion, could lead to orbital motion using a demonstration he presented to the Royal Society in 1666.

Hooke suspended a heavy weight from a wire so that the weight was free to swing in any direction (Figure 5.4).

FIGURE 5.3
Kepler's Theory of Planetary Motion
Rays emanating from the rotating Sun brush past a planet and push it along in its orbit.

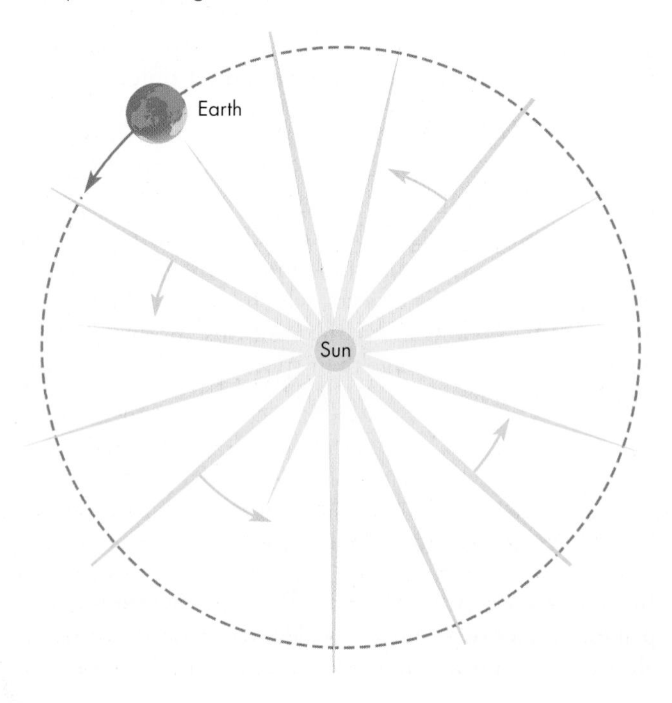

FIGURE 5.4
Robert Hooke's Demonstration of the Role of a Central Force in Planetary Motion
If a pendulum is pulled back and released, **A,** the force acting on it pulls it back toward the center and sets up the familiar oscillating motion of a pendulum. If the weight is pushed sideways when released, **B,** it moves in a rounded path resembling a planetary orbit. The only force acting on it is the central force.

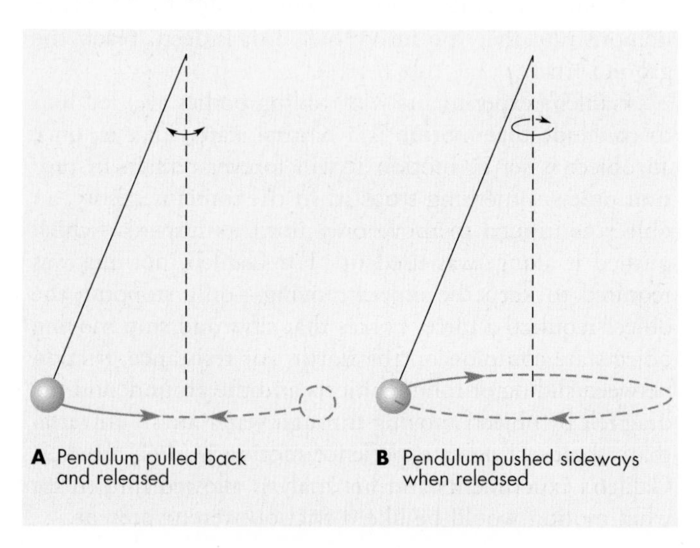

When the weight was pulled from its resting position, the force acting on it pulled it back toward the resting position. When released, the weight swung back and forth like an ordinary pendulum. However, if the weight was given a push to one side when it was released, it swung in an orbitlike motion about the resting position. With the correct strength and direction of initial push, the weight could follow a circular path or an elongated path similar to an ellipse. All the while, the force acting on the weight pulled toward the center of motion. The central force couldn't actually pull the weight to the center. It could only deflect the motion from a straight path and produce an orbitlike path. By analogy, a planet must be continually pulled toward the Sun.

Hooke also proposed that the force attracting the planet to the Sun was the same force that causes objects to fall to the Earth; this force is **gravity.** He proposed that the gravity of a celestial body attracts other objects to its center and that the strength of gravity decreases with increasing distance from the center of the celestial body. Hooke tried to verify this proposition by measuring the Earth's gravitational force in mines and on mountaintops, but the variation was too small for him to detect. Hooke never determined the way gravity depends on distance and the consequences of that dependence for planetary motion. Isaac Newton solved those problems.

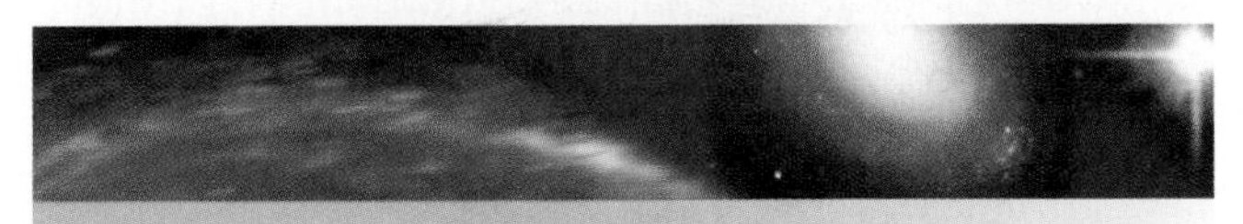

The scientists who preceded Isaac Newton had developed the concept of inertial motion—that an object will move at a constant speed in a straight line unless an unbalanced force acts on it. They had also learned that the force responsible for planetary motion was a central force directed toward the Sun.

5.3 ISAAC NEWTON

Isaac Newton, who pulled together the scattered ideas and discoveries about motion and synthesized the laws of motion and gravity, was born in the village of Woolsthorpe, England. After an early education in small schools in and near his village, Newton entered Trinity College, Cambridge, at the age of 18. As an undergraduate, Newton read and studied the works of Copernicus, Kepler, Galileo, Descartes, and other prominent astronomers, mathematicians, and natural philosophers. Shortly after he earned his bachelor of arts degree in 1665, an eruption of the plague caused Cambridge to be shut down, so Newton remained at his family farm at Woolsthorpe for most of the next 2 years. During this remarkable period he made fundamental discoveries in optics, discovered the law of universal gravitation, and invented differential and integral calculus. When Cambridge reopened, Newton returned to begin graduate study. Two years later he was appointed Lucasian Professor when Isaac Barrow, with whom Newton had studied, resigned so that Newton could take the position.

Newton often needed a challenge to get him to complete or publish his work. It has been said that there were two steps in each of Newton's discoveries. First, Newton discovered something, then the rest of the world had to discover what Newton had done. His work on gravitation, for instance, remained unpublished for about 20 years. Newton's discoveries about gravity, motion, and the orbits of the planets were finally presented in *Philosophiae Naturalis Principia Mathematica* (*Mathematical Principles of Natural Philosophy*), perhaps the greatest scientific book ever written and nearly always referred to as the *Principia.*

The Laws of Motion

The fundamental description and explanation for the motions of objects, whether terrestrial or celestial, is contained in Newton's three laws of motion.

The Law of Inertia Newton's first law states (in modern language): *An object remains at rest or continues in motion at constant velocity unless it is acted on by an unbalanced external force.* Because the property of an object that describes its resistance to changes in motion is its **inertia,** the first law is often called the law of inertia. Notice that it is essentially the same as the description of inertial motion given earlier by Descartes.

Velocity describes not only the speed of an object but also the direction in which it moves. A change in either speed or direction is a change in velocity. Because velocity has both magnitude and direction, it is a **vector. Acceleration,** also a vector, is the rate velocity changes with time. Thus, a falling body accelerates as it falls, and a bobsled accelerates when it goes around a curve, even if its speed doesn't change. A list of the accelerators on a car includes not only the gas pedal and the brakes but also the steering wheel.

Another important property of a body is its **mass,** which is a measure of the inertia of a body—its resistance to acceleration. Mass also describes the amount of material in the body. The mass of something remains the same even if it changes shape, is compressed, or is sent into space. When the mass (m) of an object is multiplied by its velocity (v), the product is defined as the object's **momentum,** p. That is, $p = mv$. Because it is defined in terms of velocity, momentum has direction as well as magnitude and is a vector. This means that two identical balls, moving at the same speed but in different directions, have different momenta. Newton's first law of motion, then, says that unless a force acts on an object, its momentum remains constant.

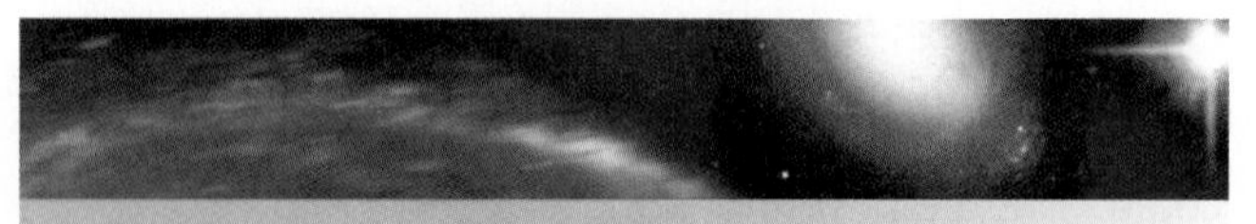

Newton's law of inertia says that the momentum of an object does not change unless a force acts on it. Momentum is the product of mass and velocity, which depends not only on the speed of an object but also on the direction in which it moves.

The Law of Force We all have a general idea about what a **force** is. It is some sort of push or pull. Newton's second law states: *When a net force acts on an object, it produces a change in the momentum of the object in the direction in which the force acts.* In other words, forces can be detected by observing the changes in motion they cause. Whenever we see an object begin to move or see its motion change speed or direction, we know that a net force must be acting on the object.

Because both mass and velocity determine the momentum of a body, a change in either or both can produce a change in momentum. A rocket, which ejects hot gases from its engine, is an example of something that changes its momentum in part because of changes in its mass. Usually, however, the mass of an object remains constant while a force acts on it, so the change in its momentum appears as a change in velocity—an acceleration. In that case, Newton's law of force takes its most familiar form

$$F = ma \tag{5.1}$$

where F is force, m is mass, and a is acceleration (see box for Equation 5.1).

Thus, we can measure the strength of a force by measuring how it accelerates a known mass. Alternatively, we can find mass by applying a known force. In fact, this is the way that masses are nearly always measured. We do this intuitively when we estimate how massive an object is by seeing how much it moves when we give it a shove or kick. Force, like velocity and momentum, is a vector and has direction as well as magnitude. Two forces are different—even if they have the same strength—if they act in different directions. It is possible for an object to be subjected to two or more forces that are applied in opposite directions and cancel each other out. In that case, even if the magnitudes of the forces are great, the momentum of the object will remain constant. Only if there is an unbalanced force will acceleration occur. In a tug-of-war contest, oppositely directed forces are applied to the rope. If the forces are exactly equal in strength, the rope remains at rest, as in Figure 5.5, A. If the forces differ in strength, the rope moves, as in Figure 5.5, B. It is the difference between the strength of the two forces, the net force, which determines the change of motion of the rope.

Newton's law of force relates a net force acting on an object to the change in momentum of the object. Usually, a change in momentum implies an acceleration, so the law of force takes the form $F = ma$.

The Law of Action and Reaction Newton's third law of motion, sometimes called the law of action and reaction, states: *When one body exerts a force on a second body, the second body also exerts a force on the first. These forces are equal in strength, but opposite in direction.* There

Newton's Law of Force

Equation 5.1, Newton's law of force, can be used to calculate the force, F, acting on a body of mass, m, that has an acceleration, a. Suppose a body having $m = 5$ kg has an acceleration of 3 m/s^2, which means 3 m/s per s, or an increase in speed of 3 m/s each second. Using these values for m and a, Equation 5.1 becomes

$$F = ma = 5 \text{ kg} \times 3 \text{ m/s}^2 = 15 \text{ kg} \frac{\text{m}}{\text{s}^2} = 15 \text{ N}$$

where the force is given in Newtons (N).

Equation 5.1 can be rewritten as $a = F/m$, which gives the acceleration of a body of mass, m, acted on by a force, F. Suppose the mass of the body is 4 kg and the force acting on it is 10 N, then Newton's law of force gives

$$a = F/m = 10 \text{ N}/4 \text{ kg} = 2.5 \frac{\text{m}}{\text{s}^2}$$

A body accelerating at 2.5 m/s^2 would reach a speed of 100 m/s (about 220 miles per hour) in 40 seconds.

FIGURE 5.5
The Balance of Forces
In **A,** the two teams exert forces equal in magnitude but opposite in direction. In **B,** the team on the right is stronger so that there is a net force on the rope. The rope accelerates in the direction of the net force.

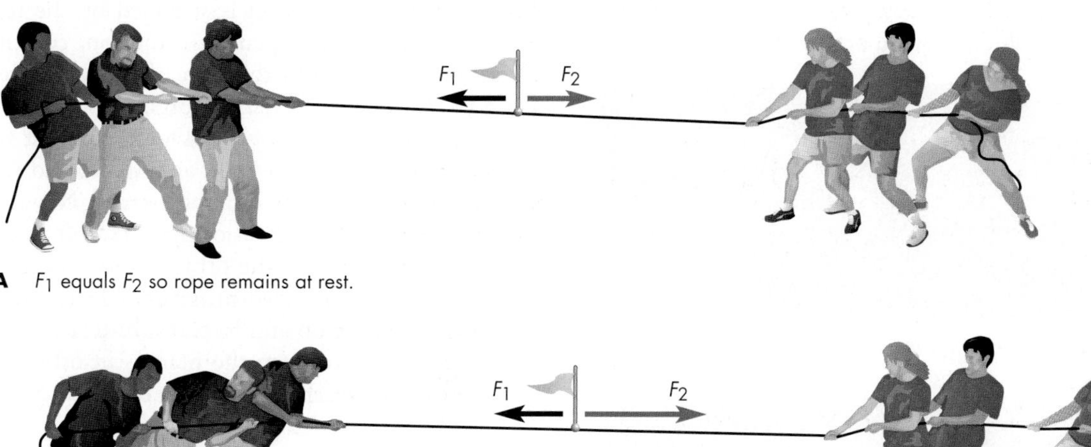

A F_1 equals F_2 so rope remains at rest.

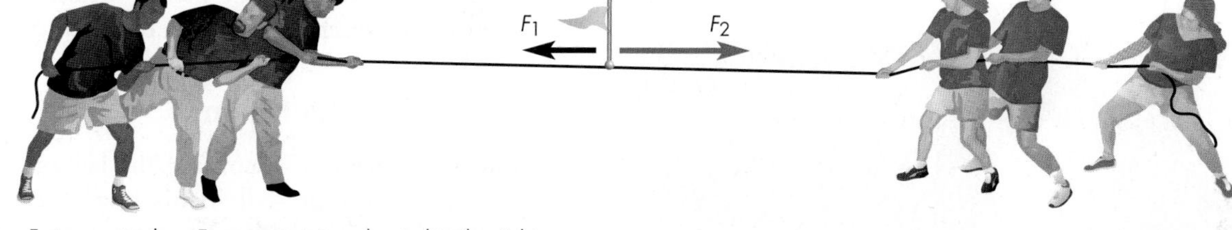

B F_2 is greater than F_1 so rope is accelerated to the right.

is never a single, isolated force in nature. If we look closely, we can always find its oppositely directed counterpart. The third law is especially easy to see when the forces involved are applied through collisions. When billiard balls strike each other, each is deflected from its original path. The forces that the balls exert on each other have the same strength but act in opposite directions. Because the balls have about the same mass, their accelerations and the resulting changes in speed and direction have the same magnitude. When a car and a truck collide, each feels a force of the same strength during the collision and the momentum of each changes by the same amount. However, the more massive truck experiences a much smaller change in velocity than does the car. Newton's law of action and reaction must be kept in mind when climbing from a boat to the dock. The boat exerts a force on the person climbing out and pushes the person toward the dock. The equally strong but oppositely directed force exerted on the boat by the person pushes the boat away from the dock.

The law of action and reaction is the basis of rocket propulsion. The exhaust gases are subjected to a force that ejects them from the back of the rocket. The exhaust gases exert an equal force on the rocket, which accelerates the rocket until the fuel and exhaust gases are used up. It is important to notice that if the rocket and its exhaust gas are considered together, there are no unbalanced forces. The force of the exhaust gas on the rocket and the force of the rocket on the exhaust are equal and opposite.

Thus, forces occur only in equal but opposite pairs and the changes in momentum they cause are oppositely directed and exactly cancel out. Total momentum is conserved, even when forces are acting. If we see the momentum of a body changing, we can be sure that there is another body whose momentum is changing in the opposite way, leaving the total momentum the same. A good example of this occurs when a moving billiard ball *squarely* strikes a stationary ball (Figure 5.6). The moving ball is brought completely to rest by the force exerted by the stationary ball. The stationary ball is accelerated to the same speed the moving ball had. Even though the momentum of each ball changes, the total momentum remains constant. It is important to realize that although forces always occur in pairs, the forces in the pair are applied to different bodies. Even though the two forces act in opposite directions, each of them can cause the body it acts on to accelerate.

Newton's law of action and reaction says that forces always occur in pairs. Whenever one body exerts a force on a second body, the first body feels a force exerted on it by the second body. The two forces are equal in magnitude but opposite in direction.

FIGURE 5.6
Newton's Law of Action and Reaction at Work
When a rolling billiard ball squarely strikes a stationary one, the force exerted by the moving ball sets the stationary ball in motion. The equal but opposite force exerted by the stationary ball completely stops the moving ball.

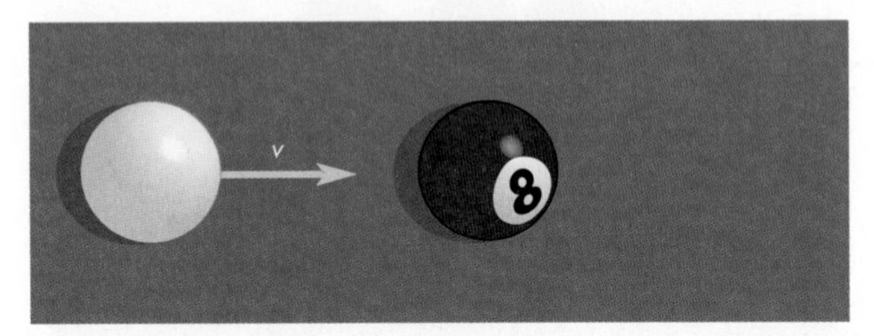

Before collision: Cue ball moves with velocity, *v*; eight ball is stationary.

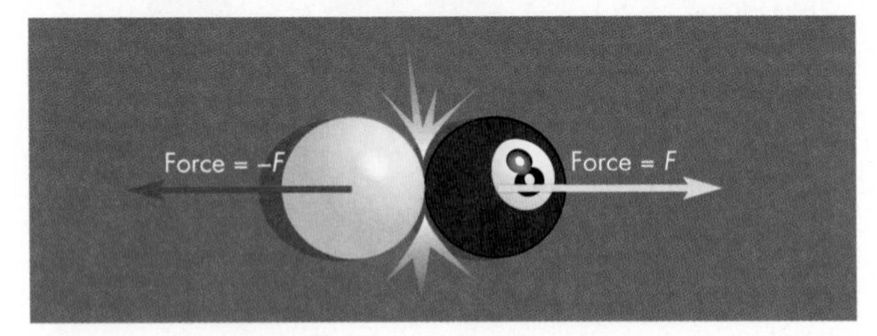

Collision: Cue ball exerts force (*F*) on eight ball; eight ball exerts equal but oppositely directed force (–*F*) on cue ball.

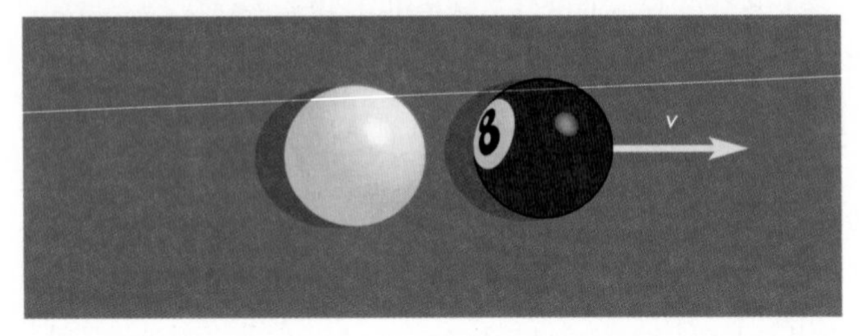

After collision: Eight ball moves with velocity, *v*; cue ball is stationary.

Gravity

The *Principia* also presented Newton's mathematical description of gravity. Newton proved that gravity is responsible for the motions of both planets and falling objects near the Earth.

Newton's Analysis

A key step in Newton's discovery of the way gravity works was finding how to calculate the centripetal acceleration of a body in a circular orbit. **Centripetal acceleration** is the acceleration toward the center of the circle. This causes the path of the orbiting object to continually bend away from the straight line path it would follow if no force were acting. In order for there to be centripetal acceleration, there must be a **centripetal force,** which acts toward the center of the circle. Figure 5.7 shows how centripetal force can lead to a circular orbit. In the case of a rock spun in a circular path on the end of a rope, the force producing the centripetal acceleration is supplied by the tension in the rope. In the case of planetary motion, centripetal acceleration is produced by the force of gravity exerted by the Sun. The mathematical description of centripetal acceleration was first published by the Dutch scientist Christian Huygens in 1673, but Newton apparently had solved the problem himself in 1666. Huygens and Newton found that the centripetal acceleration required for circular motion can be calculated if the orbital speed and distance of the body are known. Newton next used the equation for centripetal acceleration and Kepler's third law to show that the force required to keep the planets in orbit must vary with the inverse square of distance from the Sun.

The Apple and the Moon

The story that Newton discovered the law of gravitation after watching an apple fall in his mother's garden has been repeated so often and in so many forms that it has become a myth. Yet that story was told by Newton to a friend who later wrote Newton's biography. Perhaps while watching the apple fall he began to consider whether the Earth's gravity extended far beyond the Earth—even to the Moon. If so, could gravity keep an object in orbit about the Earth?

Newton discovered the answer using a famous thought experiment. Imagine firing a projectile horizontally from a cannon located on the top of a high mountain, as shown in Figure 5.8. Also imagine that there is no air resistance. In the case of any real cannon, the projectile would soon curve downward and hit the ground. But imagine a much more powerful cannon. That cannon could make the projectile travel a third of the way around the world before it fell to the ground. A still more powerful cannon could send the projectile halfway around the world or, in fact, all the way around the world. In that case, the projectile would return to its starting point traveling as fast as ever. It would keep going and circle the Earth as a satellite.

If a projectile could orbit the Earth, continually falling because of Earth's gravity, why not the Moon? Did the Earth's gravity, like the force acting on the planets, vary with the inverse square of distance? Newton found the answer by comparing the acceleration of a falling body, such as an apple, with the centripetal acceleration of the Moon. Because the Moon orbits at a distance of 60 Earth radii, the acceleration of the Moon should be $60^2 = 3600$ times smaller than the acceleration of gravity at the Earth's surface. Newton used the Moon's orbital velocity and distance to find its centripetal acceleration. He then compared this acceleration with gravity at the Earth's surface and found that the Moon's centripetal acceleration was "pretty nearly" 3600 times smaller. Thus Newton found that the Moon is

FIGURE 5.7
Centripetal Force and Circular Motion
In **A,** a ball moving on a round billiard table collides with the cushion at three points. At each collision, the ball experiences a force directed toward the center of the table. In **B,** the path of the ball yields more frequent collisions with the cushions, producing a more circular path. **C,** If the ball were in constant contact with the cushion, it would experience a continuous central force and move in a circular path.

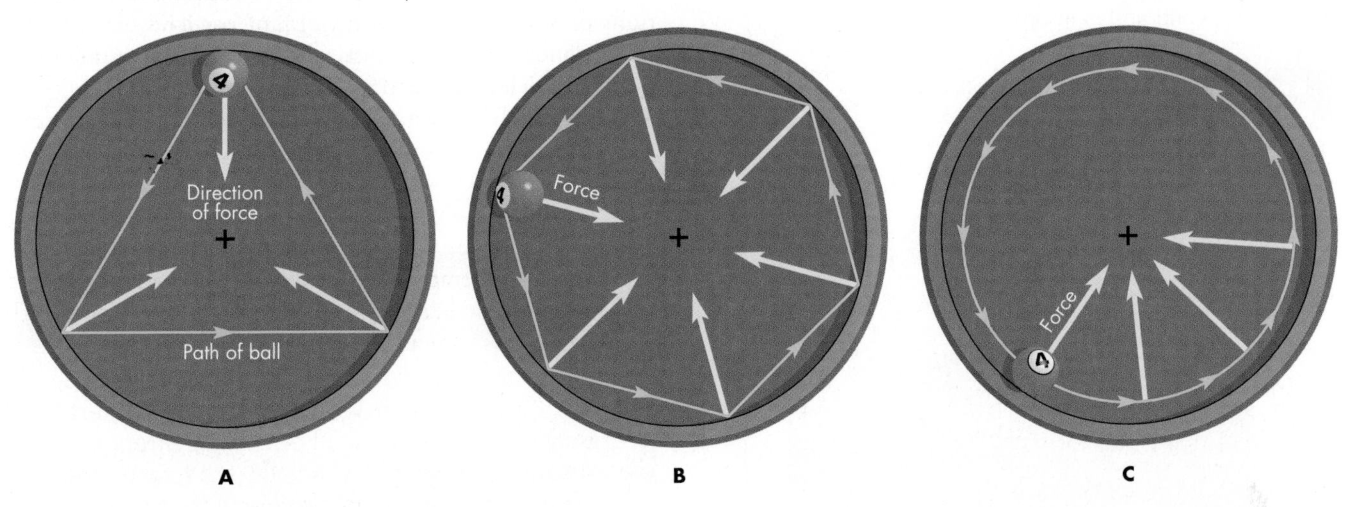

held in its orbit by the force of gravitational attraction to the Earth. By implication, the force that holds the planets in orbit is the gravitational attraction of the Sun.

The Law of Gravitation

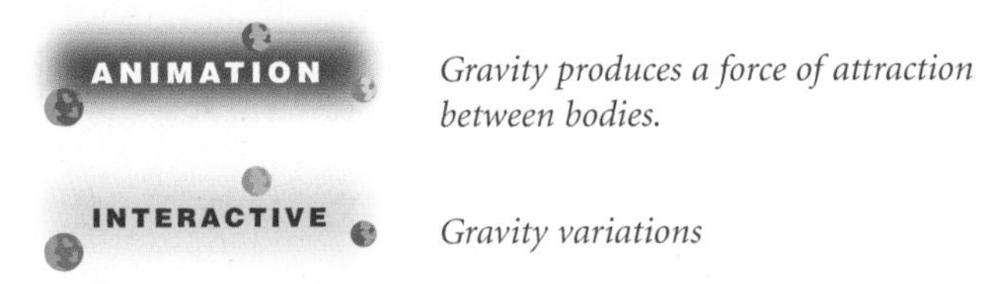

Gravity produces a force of attraction between bodies.

Gravity variations

The law of universal gravitation states that every two particles of matter in the universe attract each other with a force that depends on the product of their masses and the inverse square of the distance between them. Mathematically,

$$F_G = \frac{GMm}{d^2} \tag{5.2}$$

where m and M are the masses of the two particles of matter, d is the distance between them, and G is a number called the gravitational constant (see box for Equation 5.2).

For spherical bodies, which the Sun, Moon, and planets approximate, the distance to be used in the law of gravitation is the distance between the centers of the bodies.

Newton's Law of Gravity

Equation 5.2, Newton's law of gravity, can be used to calculate the force, F, exerted by two bodies on each other if their masses and the distance between them are known. For example, suppose the masses, M and m, of two bodies are each 1000 kg (2200 pounds) and the distance between them, d, is 10 m. Using these values for the masses and distance and knowing that G, the gravitational constant, has a value of 6.67×10^{-11} Nm²/kg², Equation 5.2 gives

$$F_G = \frac{GMm}{d^2} = \frac{6.67 \times 10^{-11} \times 1000 \times 1000}{10^2}$$
$$= 6.67 \times 10^{-7}\ \text{N}$$

where the force is given in Newtons (N). This weak force is applied to each 1000 kg mass and is capable of accelerating the mass at 6.67×10^{-10} m/s². It would be hard to detect such a small acceleration.

FIGURE 5.8
Newton's Thought Experiment on Orbital Motion
As the speed given the projectile by the cannon increases, the projectile moves farther around the Earth before falling to the ground. At a certain speed, the projectile falls completely around the world and is in orbit.

ANIMATION *Newton's cannon*

To show that this is true, Newton had to invent a way to add up the gravitational forces exerted by each particle of matter in a spherical planet (Figure 5.9). The particles exert individual gravitational forces differing in strength and the direction in which they act. Newton was able to use his invention, calculus, to show that the weaker attractions of the more numerous, distant particles in a sphere just balance the stronger attractions of the less numerous, nearer particles so that the entire sphere attracts other objects as though all of its mass were concentrated at a point at its center.

The law of gravitation says that every pair of particles of matter, no matter how far apart, exert a gravitational force on each other. The strength of the force is proportional to the product of their masses and inversely proportional to the square of the distance between the particles. A spherical body attracts other bodies as though its mass were concentrated at its center.

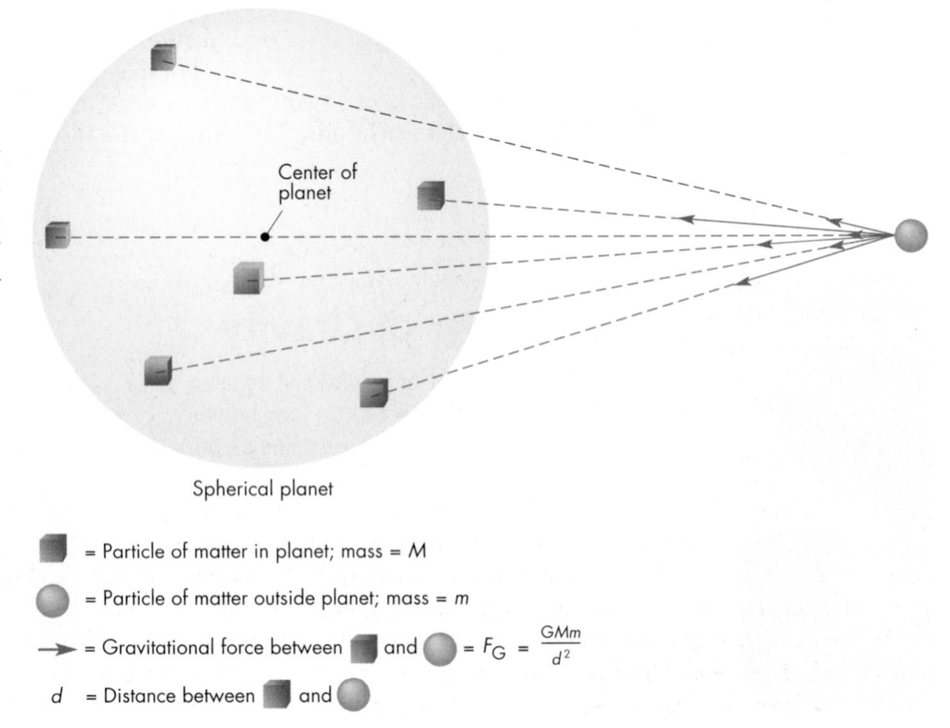

FIGURE 5.9
The Gravitational Force Exerted by a Spherical Body
The gravitational forces due to the various parts of the body act in different directions and have different strengths. Their sum is the same as the attraction that the entire body would have if its mass were concentrated at its center.

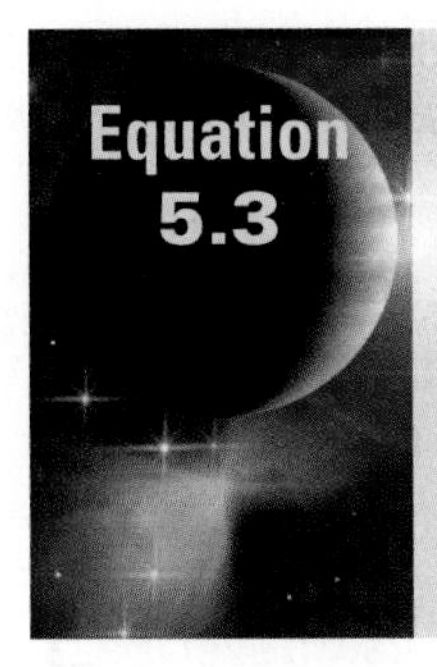

Weight

Equation 5.3 can be used to calculate how the weight of a body changes as its distance from the center of the Earth changes. Suppose, for example, that a body moved from the Earth's surface to a distance of 10 Earth radii (R_{Earth}) while its mass remained the same. Thus, m, G, and M in Equation 5.3 remain the same while R increases from R_{Earth} to 10 R_{Earth}. Because R becomes 10 times as large, R^2 becomes 100 times as large. R^2 appears in the denominator of Equation 5.3 so the weight of the body falls by a factor of 100.

Weight and Mass When we talk about **weight** we usually mean the gravitational force attracting an object to the Earth. That is,

$$W = \frac{GM_{Earth}m}{R^2} \qquad (5.3)$$

where m is the mass of the object, M_{Earth} is the mass of the Earth, and R is the distance of the object from the center of the Earth (see box for Equation 5.3). Equation 5.3 is a form of Newton's second law, $F = ma$. That is, W is a force and m is mass, so the quantity $= GM_{Earth}/R^2$ is an acceleration called the **acceleration of gravity.** Near the surface of the Earth, the value of GM_{Earth}/R^2 is 9.8 m/s^2 (32 ft/s^2) and is usually given the symbol g. Thus, $W = mg$ is another way to write Equation 5.3.

The mass of a body is always the same, but its weight can vary with its location. Near the Earth's surface, the law of gravity indicates that a body's weight decreases with increasing altitude (and distance from the Earth's center). Weight can also refer to the gravitational force attracting a body to another celestial body such as a planet or satellite. On the Moon, the acceleration of gravity depends on the mass and radius of the Moon and has the value of 1.6 m/s^2 (5.3 ft/s^2), about ⅙ g. This means that a person who weighs 120 pounds on Earth would weigh only 20 pounds on the Moon, even though the person's mass would be the same in both places.

We are aware of weight only when it is balanced by other forces. For instance, when we are standing, the ground exerts a force on the bottoms of our feet. This force balances our weight and keeps us from falling downward. When there is no balancing force, such as when a person is in orbit or skydiving, the person's weight results in acceleration and free fall. This condition is incorrectly called weightlessness.

Elliptical Orbits

More than a decade after Newton discovered the law of gravitation, Robert Hooke wrote to him asking what shape the orbit of a planet would be if the force of attraction to the Sun varied with the inverse square of distance. Newton rose to the challenge and showed that the resulting orbit would be an ellipse with the Sun at one focus. This result firmly established that the force exerted between the Sun and the planets was gravity. Newton also showed that the planets would obey Kepler's second law and that Kepler's third law is a consequence of gravity depending on the inverse square of distance from the Sun.

The Meaning of Kepler's Second Law Newton showed that any body orbiting another body under the force of gravity moves so that it sweeps out equal areas in equal amounts of time. Applied to the motion of a planet, this is Kepler's second law. The rate at which the planet sweeps out area is given by the product of its transverse velocity and its distance from the Sun. Thus, Kepler's second law says, in effect, that the product of the transverse velocity and distance of a planet remains constant as a planet orbits the Sun. The law of equal areas is a consequence of the conservation of **angular momentum,** the momentum of a body associated with its rotation or revolution. The angular momentum of an orbiting (revolving) body is proportional to the product of its orbital distance and its transverse velocity. As the planet approaches the Sun and its orbital distance decreases, its transverse velocity increases so that angular momentum remains constant.

The angular momentum of a rotating body has to be calculated by adding up the angular momenta of each of the particles of mass it contains. When this calculation is done, it turns out that angular momentum is proportional to the product of the size of the body and the speed at which it rotates. Just as an orbiting body moves faster as it approaches the point about which it is revolving, if a rotating body shrinks in size, it rotates faster. A familiar example of the conservation of angular momentum by a rotating body is a spinning ice skater (Figure 5.10). The skater pulls in his arms to spin faster and then stretches out again to slow down.

Angular momentum can never be destroyed, it can only be redistributed. For example, if two asteroids collide, the fragments of the collision have the same total angular momentum about the Sun as did the two original asteroids. An astronaut trying to slow down a spinning disk in space would find that in doing so he would acquire angular momentum about the center of the disk and be pushed in the direction in which the disk was turning when he grabbed it. The conservation of angular momentum is an important concept that we will return to many times in this book. It

FIGURE 5.10
The Conservation of Angular Momentum
When the skater wishes to spin rapidly, she pulls in her arms, reducing her "size." To slow down her spin, she stretches out again.

has important consequences for the rotation and revolution of planets, stars, galaxies, and other astronomical bodies.

Angular momentum is the momentum of a body associated with its rotation or revolution. Angular momentum can be transferred from one object to another or redistributed, but it can never be destroyed.

Kepler's Third Law Revised Newton was able to use his laws of motion and the law of gravitation to derive a version of Kepler's third law that is much more general than Kepler's original version. Newton's version is

$$P^2 = \frac{4\pi^2 a^3}{G(m + M)} \tag{5.4}$$

where P is orbital period, a is the average distance between the orbiting bodies, and m and M are their masses. (See box for Equation 5.4.)

In the solar system, the masses of the planets are much smaller than the mass of the Sun so that $m + M$ is only slightly larger than the mass of the Sun and almost the same for each planet. This is why Kepler found simply that P^2/a^3 is the same for each planet. Newton's version of Kepler's third law is extremely valuable for astronomers because it allows them to find the masses of astronomical bodies. For example, to calculate the mass of Jupiter an astronomer can study the orbit of one of Jupiter's satellites.

Equation 5.4

Geostationary Satellites

We can use Equation 5.4, Newton's version of Kepler's third law, to find the distance at which an artificial satellite would have an orbital period equal to the Earth's rotation period, 86,400 seconds. First, we solve Equation 5.5 for the cube of orbital distance, giving

$$a^3 = \frac{P^2 G(m + M)}{4\pi^2}$$

Because the mass of the satellite is so much smaller than the mass of the Earth (5.98×10^{24} kg), it can be ignored. Substituting the values of G (6.67×10^{-11} m^3/kg s^2) and orbital period, P (8.64×10^4 s), the value of the cube of the orbital distance is

$$a^3 = (8.64 \times 10^4 \text{ s})^2(6.67 \times 10^{-11} \text{ m}^3/\text{kg s}^2)(5.98 \times 10^{24} \text{ kg})/[4(3.14^2)] = 7.5 \times 10^{22} \text{ m}^3$$

Taking the cube root of a^3 gives

$$a = 4.22 \times 10^7 \text{ m} = 42{,}200 \text{ km}$$

Thus, a satellite orbiting 42,200 km from the center of the Earth, or about 35,800 km (22,400 miles) from the Earth's surface, has an orbital period equal to the rotation period of the Earth. If such a satellite is in orbit above the Earth's equator, it will hover above the same spot on the Earth and is called a geostationary satellite. Communication satellites, including those used for transmission of satellite television signals, are placed in geostationary orbit to make it easy to point dishes in their direction.

The orbital period and distance of the satellite can be used in Kepler's third law to find the sum of the mass of Jupiter and the mass of the satellite. Because the mass of Jupiter is much greater than the mass of the satellite, the result is approximately the mass of Jupiter. Similarly, it is possible to find the mass of a galaxy by observing the orbits of the stars it contains or find the mass of the stars in a binary star system by measuring their orbital separation and period.

Newton showed that a planet attracted to the Sun by the force of gravity will orbit on an ellipse with the Sun at one focus. The planet will move so as to conserve angular momentum, moving fastest when it is nearest the Sun. He also showed that Kepler's third law could be generalized to relate the period, masses, and separation of any two orbiting bodies.

Force at a Distance

Although gravity was able to account for many solar system phenomena, Newton was never comfortable with the way gravity works. Somehow, through invisible means, two widely separated bodies can exert forces on each other. Momentum is mysteriously transmitted from one to the other. Newton saw this idea of "force at a distance" as a flaw in his theory and repeatedly tried to discover an explanation of how gravity works. He wrote:

> That gravity should be innate, inherent, and essential to matter, so that one body may act upon another at a distance through a vacuum, without the mediation of anything else, by and through which their action and force may be conveyed from one to another, is to me so great an absurdity that I believe no man who has in philosophical matters a competent faculty of thinking can ever fall into it.

Newton failed to explain how gravity can act at a distance, and so did everyone else until Albert Einstein developed his general theory of relativity more than 200 years later. The theory of general relativity, and some of its most spectacular consequences, such as black holes, are described in Chapter 20.

5.4 ORBTIAL ENERGY AND SPEED

INTERACTIVE *Orbital velocity*

Another way to understand the changing speed of an orbiting body involves the energy of the body. For an orbiting body, two forms of energy are important. One of these is **kinetic energy,** or energy of motion. The kinetic energy of a body increases as its speed increases. This means that the kinetic energy of a planet is greatest when it moves fastest at perihelion and least when it moves slowest at aphelion. The other important form of energy is **gravitational potential energy.** Potential energy is stored energy. A coiled spring has potential energy, as does an object subject to the gravitational attraction of another body. If the object falls, it accelerates and picks up speed and kinetic energy. The farther an object has to fall, the more potential energy it has while suspended. This, in turn, means that the object will pick up more kinetic energy when it falls. The gravitational potential energy of a planet increases with distance from the Sun. The total energy of a planet—the sum of its kinetic and gravitational potential energies—remains constant as it orbits. When it approaches the Sun, its gravitational energy becomes smaller, but its kinetic energy increases.

Orbital Speed

A body in a circular orbit about the Sun moves at a constant speed, **circular speed,** given by

$$v_c = \sqrt{GM_\odot/d} \qquad (5.5)$$

where $M_\odot$ is the mass of the Sun and d is the distance of the body from the Sun. The average orbital speed for an elliptical orbit is approximately the same as the circular speed for the semimajor axis of that orbit.

At 1 AU from the Sun, the circular speed is 30 km/s, the Earth's average orbital speed. For the other planets, average orbital speed falls with increasing distance from the Sun. Neptune, the most distant and slowest moving planet, has an average orbital speed only about 1/5 that of

Equation 5.5

Circular Speed and Distance from the Sun

Equation 5.5 can be used to calculate how circular speed varies with distance from the Sun. For all bodies orbiting the Sun, G and $M_\odot$ are the same, so circular speed depends only on distance from the Sun. Suppose we wish to compare the circular speeds at 1 AU and at 20 AU. Because circular speed falls as the inverse square root of distance, the circular speed at 20 AU must be $\sqrt{1/20} = 1/4.5 = 0.22$ times as large as the circular speed at 1 AU. Because the circular speed at 1 AU is 30 km/s, the circular speed at 20 AU is 30 km/s × 0.22 = 6.6 km/s.

FIGURE 5.11
The Elliptical Orbit of a Planet
At aphelion, the planet moves too slowly to remain in a circular orbit and falls inward. At perihelion, the planet moves too rapidly for a circular orbit and draws away from the Sun.

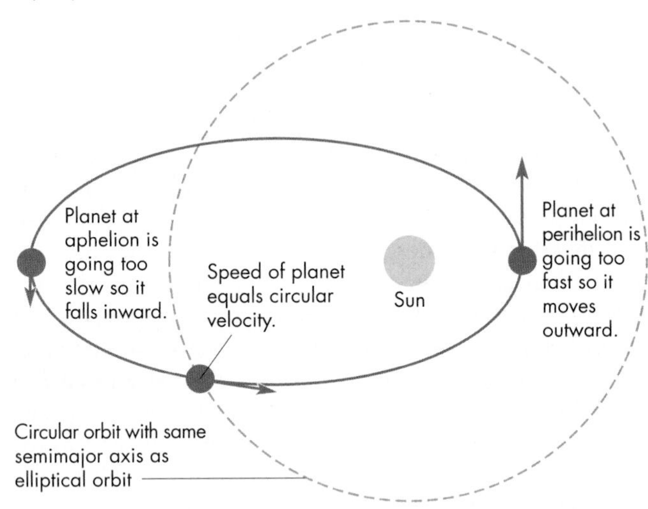
Planet at aphelion is going too slow so it falls inward.
Speed of planet equals circular velocity.
Sun
Planet at perihelion is going too fast so it moves outward.
Circular orbit with same semimajor axis as elliptical orbit

the Earth. Nevertheless, Neptune moves quite rapidly compared with the speeds of even supersonic aircraft. Its typical speed of 5.4 km/s is greater than 12,000 mph (16 times the speed of sound near the Earth's surface).

When a planet is in the part of its orbit farthest from the Sun, it moves more slowly than it would have to move to maintain a circular orbit at that distance, so it is pulled inward by gravity, as shown in Figure 5.11. When the planet reaches its average distance it is moving at a speed nearly equal to the circular speed, but is already headed inward and moves still nearer the Sun. The planet picks up speed as it falls inward. When the planet reaches perihelion, it is moving faster than the circular speed at that distance and is able to move outward against the pull of gravity. However, gravity slows it down as it moves outward so that its speed falls until it reaches aphelion again and begins to fall back inward.

For hypothetical planets having the same semimajor axis but different orbital eccentricities, the more elongated the orbit, the greater the difference between maximum and minimum speeds. The dwarf planet Pluto, for instance, moves at 6.1 km/s when it is at perihelion, 30 AU from the Sun, but only 3.7 km/s when it is 50 AU from the Sun at aphelion. Halley's comet, which has a much more eccentric orbit than Pluto, ranges between 0.6 and 35 AU from the Sun during an orbit. Its orbital speed varies between 54 km/s and 0.9 km/s.

Circular speed, the orbital speed of a body in a circular orbit, is approximately the same as the average orbital speed of a body on an elliptical orbit having a semimajor axis as large as the radius of the circular orbit. Circular speed falls with increasing distance from the Sun.

Escape Velocity

INTERACTIVE

Escape velocity

Imagine taking an elliptical orbit and stretching it until its aphelion distance extends infinitely far. The semimajor axis of the orbit would also become infinite. An object on such a path would move away from the Sun to an indefinite distance, never returning. The object would have escaped from the Sun. The speed of the object, at any point on its path, is called **escape velocity,** since it is traveling at precisely the speed required for it to escape. An object moving faster than escape velocity will also escape. Escape velocity for escape from the solar system is given by

$$v_e = \sqrt{\frac{2GM_\odot}{d}} \tag{5.6}$$

where $M_\odot$ is the Sun's mass and d is distance from the Sun.

Equation 5.6

Escape from the Earth

Escape velocity from the surface of the Earth is found from Equation 5.6, the equation for escape velocity, using the mass of the Earth (5.98×10^{24} kg) and the radius of the Earth (6.38×10^{6} m), instead of the mass of the Sun and distance from the Sun.

$$V_e = \sqrt{\frac{2GM_{Earth}}{R_{Earth}}}$$

$$= \sqrt{\frac{2(6.67 \times 10^{-11}\ m^3/kg\ s^2)(5.98 \times 10^{24}\ kg)}{6.38 \times 10^{6}\ m}}$$

$$= 1.12 \times 10^{4}\ m/s = 11.2\ km/s$$

It is easiest to achieve this speed by launching a spacecraft toward the east from a position near the equator. In this way, it is possible to take advantage of the rotation of the Earth, which is about 0.5 km/s toward the east at the equator. This means that only 10.7 km/s must be supplied by the launch vehicle. Launches from latitudes far from the equator do not take advantage of the Earth's rotation. Launches toward the west actually have to overcome the Earth's eastward rotation, making them more difficult. The escape velocities for each of the planets are given in Appendix 6.

At any given distance from the Sun, escape velocity is equal to exactly $\sqrt{2}$ times the circular velocity. That is, if the Earth were traveling $\sqrt{2}$ times as fast as it is (about 12 km/s faster), it would move away from the Sun and never return. As it moved, its speed would fall but never reach zero. Four spacecraft launched from the Earth, *Pioneer 10, Pioneer 11, Voyager 1,* and *Voyager 2,* have exceeded escape velocity from the solar system and cannot return. Instead, they will recede forever from the Earth and Sun, eventually reaching the distances of the stars. The concept of escape velocity applies not only to bodies moving about the Sun, but to objects moving under the gravitational influence of any other body. Thus, to send a probe to another planet, we must accelerate the probe to a speed at least as large as the escape velocity of the Earth.

The Four Kinds of Trajectories

The four kinds of trajectories a body moving near the Sun (or any other gravitating body) can follow are known as **conic sections** because they are the four curves that can be made by slicing through a cone (Figure 5.12). The four conic sections (and the four kinds of trajectories) are the **circle,** the **ellipse,** the **parabola,** and the **hyperbola.** A body moving at escape velocity follows a parabola. If a body has a speed greater than escape velocity, its path is a hyperbola. Bodies moving slower than escape velocity follow circles or ellipses.

The relationship between speed and the type of path on which an object moves is true whether it is traveling toward or away from the Sun. An interstellar traveler approaching the solar system from some other part of the galaxy would follow a parabolic or a hyperbolic path. Thus it would always have a speed at least as great as escape velocity. It would move rapidly through the solar system and recede again into interstellar space. To go into orbit about the Sun it would have to reduce its speed to less than escape velocity. Unless it could do so, it couldn't remain in the solar system.

If the speed of an object is equal to escape velocity, it will move away from the body attracting it on a parabolic trajectory and never return. At a given distance from the Sun, the escape velocity is $\sqrt{2}$ times the circular velocity at that distance. A body moving faster than escape velocity travels on a hyperbolic trajectory.

FIGURE 5.12
Conic Sections
The four kinds of curves that can be produced by slicing a cone are the circle, ellipse, parabola, and hyperbola. These are also the paths that a body can follow when moving under the gravitational attraction of a second body.

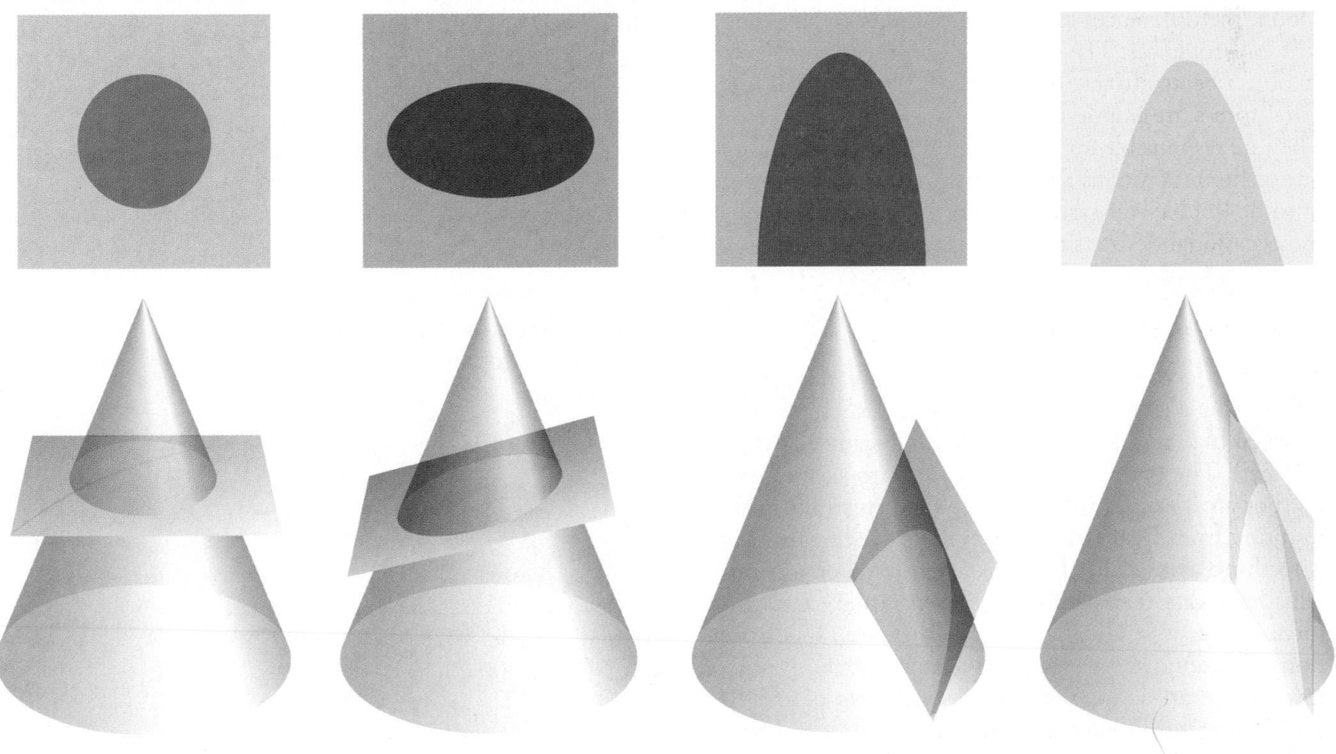

FIGURE 5.13
High and Low Tides at Bar Harbor
The same scene is shown at high and low tide. The structure of the shoreline of Bar Harbor magnifies the difference between the water levels at high tide, **A,** and low tide, **B.**

A

B

5.5 TIDES

ANIMATION *The origin of tides*

Whenever one object is gravitationally attracted to another, the various parts of the object feel gravitational forces differing in strength and direction. If the object being attracted is small, like a person or a spacecraft, the differences are extremely tiny. However, for a larger object, such as the Moon, the differences can be large enough to have important consequences. The differences in gravity that occur in an object being attracted by another are called **tidal forces.** Tidal forces are important in many astronomical settings. They are responsible for such diverse effects as volcanoes on one of Jupiter's moons and the distorted shapes of the stars in some double-star systems. Tides are also responsible for the occasional destruction of galaxies by their neighbors.

Differences in Gravity

The water level at Bar Harbor, in Maine, varies spectacularly during a day, as shown in Figure 5.13. The difference between high and low water levels is typically about 20 m (60 feet). The dramatic flows of water into and out of Bar Harbor are a result of tides. **Tides** are distortions in shape resulting from tidal forces. One way to see how tides arise is to think about what happens when one body begins to fall toward another because of their mutual gravitational force. As it falls, all of its parts accelerate. The gravitational force and the acceleration it causes are both pointed toward the attractor. Because various parts of the body are at different distances and directions from the attractor, gravitational force and acceleration vary in both magnitude and direction. The arrows in Figure 5.14 represent the magnitude and direction of force within a falling body.

The part of the falling body nearest to the attractor has the greatest acceleration and speeds up most rapidly. The part farthest from the attractor has the smallest acceleration. If the body is deformable, the difference in acceleration between its nearest and farthest parts makes it stretch along the direction it is falling. Essentially, the near side is accelerated away from the center, which is accelerated away from the far side. The nearer it gets to the attracting body, the larger the deformation becomes. Places on its surface midway between its nearest and farthest parts accelerate inward relative to the center so the body becomes thinner perpendicular to the direction of its fall. The same thing happens to orbiting bodies. The center of an orbiting body feels a gravitational force just great

FIGURE 5.14
Differences in Gravitational Force
The arrows represent forces differing in both direction and magnitude.

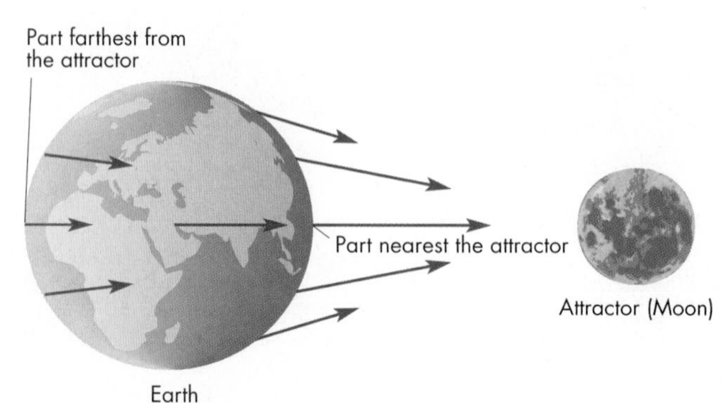

FIGURE 5.15
The Tidal Distortions of the Earth and the Moon
Both the Earth and the Moon are elongated along a line drawn between the two bodies. The amount of tidal distortion is greatly exaggerated in this figure.

ANIMATION *Tidal forces*

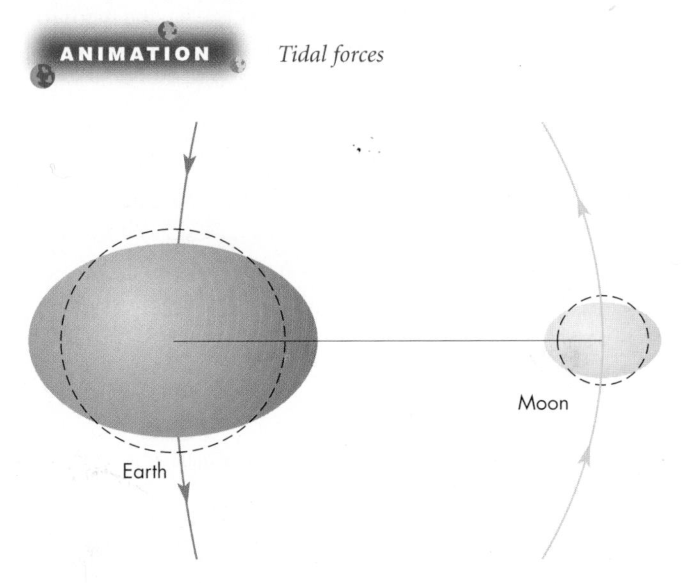

enough to keep it orbiting at the proper speed. Other points in the orbiting body feel gravitational forces different in amount or direction from that acting on the center. Thus the orbiting body experiences tidal forces, accelerations, and distortions like those of a falling body. The tidal distortions of the Earth and Moon are shown in exaggerated form in Figure 5.15.

The strength of the tidal force a body feels depends on how much the force of gravity varies from the side of the object nearest the attractor to the side farthest from the attractor. Tides are strong when the body is near a massive attractor, where the gravitational force itself is strong. A large body feels stronger tides than a small one since its greater size allows a greater decrease in gravity from its front side to its back. Thus, the strongest tidal forces are felt by fairly large bodies located close to massive attractors.

Tidal forces are differences in gravitational force that occur because the parts of a body are at different distances and in different directions from the object attracting them. Tidal forces can produce tides, which are distortions of the body when it is stretched in the direction of the attracting object. The strongest tidal forces are felt by sizable bodies located close to massive attractors.

The effect of tides can be seen by noting the differences in acceleration, including direction, at various points on and in the falling or orbiting object. The differences, measured relative to the acceleration at the center of the object, depend on whether the acceleration is larger or smaller than that at the center and whether the acceleration points in the same direction as the acceleration at the center. Some of the differences in acceleration are shown in Figure 5.16. At some locations the tidal acceleration is parallel to the surface. At other locations it is perpendicular, pointing upward at some places, downward at others. At most locations, however, the tidal acceleration has both horizontal and vertical components.

Solid Earth Tides

The Earth feels tidal forces because of the Moon's gravity. The solid Earth itself distorts in response to tidal forces, but because the Earth's interior is very rigid, the distortion is much less than that which would occur if the Earth were a liquid and easily deformed. The maximum tidal distortion of the solid Earth is only about 20 cm. Tidal forces not only raise and lower the Earth's surface, but also tilt it. The amount of tilt can be determined by careful measurement of the positions of stars. The bulges move around the Earth as it spins on its axis.

Ocean Tides

While tides in the solid Earth are difficult to observe, tides in the ocean can be quite obvious. The water in

FIGURE 5.16
Tidal Acceleration
Tidal accelerations are found by comparing the size and direction of the gravitational attraction at a point in a body to the acceleration at its center (see Figure 5.14). At some points the tidal acceleration is horizontal, at others it is vertical. At most points, however, the tidal acceleration has both vertical and horizontal components.

ANIMATION *Tidal forces*

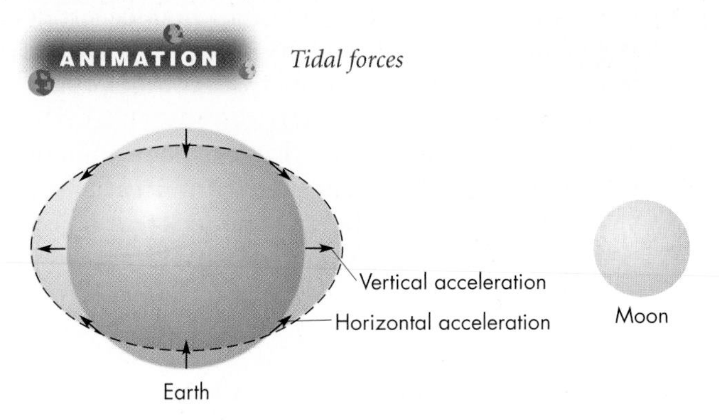

oceans and other bodies feels a tidal acceleration. The vertical part of the acceleration is unimportant relative to the horizontal part, which causes water to flow. The tidal force is quite weak, typically only about one ten millionth as strong as the gravity of the Earth. Thus a typical molecule of ocean water moves only a short distance in response to tidal acceleration. The collective motions of the water molecules, however, produce a wave that moves around the Earth, leading the Moon, at a speed of about 400 meters per second (1000 mph). In deep parts of the oceans, far from land, the tidal bulges are only about a meter in height. High tide occurs slightly after the Moon crosses the meridian and again, 12 hours 25 minutes later, when the Moon is on the opposite side of the Earth. The presence of landmasses and the structure of their shorelines causes high and low tides to be much more extreme near the shore. In coastal areas, water takes time to ebb and flow, causing a delay of high tide following the Moon's meridian crossing time. This delay can be as long as 12 hours.

The Sun also causes tides. The Sun is much more massive than the Moon, but because of the Sun's much greater distance, the solar tidal acceleration is only about half of the lunar tidal acceleration. When the Sun and Moon are aligned (full or new Moon), the total tidal acceleration is the sum of the tidal accelerations of the Sun and Moon. This leads to strong **spring tides** in which low tides are unusually low and high tides are unusually high (Figure 5.17). Spring tides have nothing to do with the season of the year. At first or third quarter Moon, the two tidal accelerations of the Sun and Moon partially cancel each other to produce unusually weak **neap tides.** The total tidal acceleration is about three times as great during a spring tide as during a neap tide.

Tides have obvious and important consequences for seafarers. Other consequences of tides, involving changes in the rotation of the Earth and the orbit of the Moon, are described in Chapter 9.

FIGURE 5.17
Spring and Neap Tides
When the Sun and Moon lie in the same or in opposite directions as seen from the Earth, the tidal bulges they cause coincide, producing a large *spring* tide, **A.** When the directions to the Sun and Moon are perpendicular, the tides of the Sun and Moon cancel to some extent, producing a weak *neap* tide, **B.**

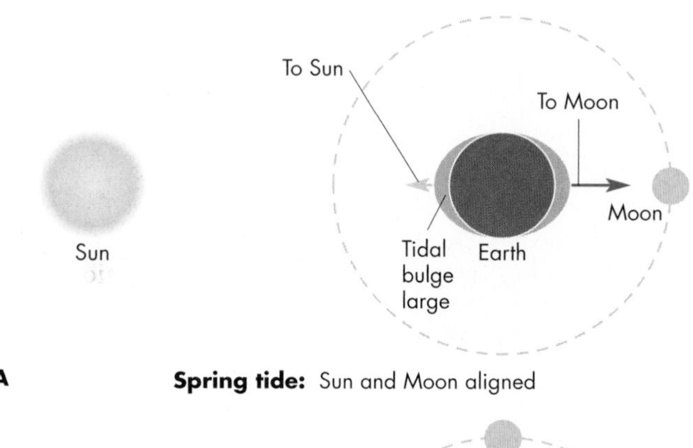

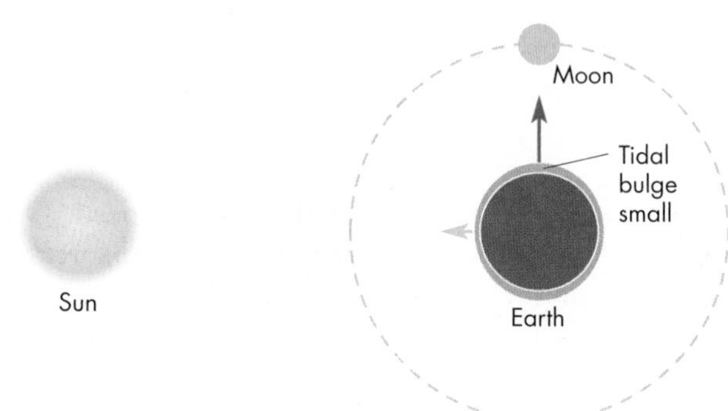

Chapter Summary

- By Newton's time, work by Galileo, Kepler, Descartes, Hooke, and others had led to an understanding of inertial motion and the need for a central force to account for the curved paths of the planets around the Sun. (Section 5.2)
- Newton's law of inertia says that the momentum of an object remains constant unless it is acted on by an unbalanced force. The momentum of an object is the product of its mass and velocity. (5.3)
- Newton's law of force says that when a force acts on a body, it causes a change in its momentum in the direction in which the force is applied. Usually, the law of force is expressed as $F = ma$. (5.3)
- Newton's law of action and reaction says that when one body exerts a force on a second body, the second body exerts a force on the first equal in magnitude but opposite in direction. (5.3)

- Newton's law of universal gravitation states that every two particles of matter attract each other with a force that depends on the product of their masses and varies inversely with the square of the distance between them. The law of gravitation is expressed as $F_G = GMm/d^2$. (5.3)
- Angular momentum is the momentum associated with rotation or revolution. Angular momentum can be transferred from one object to another or redistributed. However, it is always conserved. (5.3)
- Newton found that orbiting objects attracted to the Sun by the force of gravity obey Kepler's laws. He discovered a general form of Kepler's third law relating the separation, orbital period, and masses of any two orbiting bodies. This relationship is the basis of most astronomical mass determinations. (5.3)
- Circular speed is the orbital speed of a body that moves in a circular orbit. The circular speed at a given orbital distance has approximately the same value as the average orbital speed for an elliptical orbit having a semimajor axis equal to the circular distance. (5.4)
- A body moving at escape velocity leaves the solar system on a parabolic path and never returns. At any distance from the Sun, escape velocity is $\sqrt{2}$ times the circular speed at that distance. A body moving faster than escape velocity travels on a hyperbolic part. (5.4)
- Tidal forces are differences in gravitational force experienced by different parts of a body. Tidal forces stretch the body in the direction toward and away from the attracting object. The strongest tidal forces are felt by large bodies near massive attractors. (5.5)

Key Terms

acceleration *83*
acceleration of gravity *89*
angular momentum *89*
central force *82*
centripetal acceleration *86*
centripetal force *86*
circle *93*
circular speed *91*
conic section *93*
ellipse *93*
escape velocity *92*
force *84*
gravitational potential energy *91*
gravity *83*
hyperbola *93*
inertia *83*
inertial motion *81*
kinetic energy *91*
mass *83*
momentum *83*
neap tide *96*
parabola *93*
spring tide *96*
tidal forces *94*
tides *94*
vector *83*
velocity *83*
weight *89*

Conceptual Questions

1. Compare the way advocates of the impetus theory and the way Galileo accounted for the motion of a body moving at constant speed in a given direction.
2. What would Kepler have considered to be the path of a planet if it were influenced by the *anima motrix* but was not affected by the magnetism of the Sun?
3. In what sense does an orbiting satellite accelerate even if its speed remains constant?
4. Two forces, oppositely directed, act on a body. The force acting toward the right is twice as strong as the force acting toward the left. Describe the motion of the body.
5. Suppose a person is stranded on a frozen lake of perfectly smooth (frictionless) ice. Describe a method for the person to use to reach the shore of the frozen lake.
6. Suppose a satellite is given a speed 10% larger than circular velocity. What would be the shape of the trajectory of the body?
7. The tidal force that the Earth exerts on the Moon is stronger than that exerted by the Moon on the Earth. Explain why this is so.
8. The tidal acceleration due to the Moon is twice as large as the tidal acceleration due to the Sun. Show how this leads to a total tidal acceleration about three times as large for spring tides as for neap tides.

Problems

1. A body takes 10 seconds to go from rest to a speed of 50 m/s. What is the magnitude of the average acceleration of the body?
2. How would the momentum of a body change if its mass remained constant but its velocity tripled?
3. Suppose object A is four times as massive as object B. How does the acceleration of object A compare to that of object B if the same force is applied to each object?
4. What are the magnitude and direction of the net force acting on a body if it is acted on by a force of 50 N upward and a force of 30 N downward?
5. A body has a mass of 10 kg. A force of 20 N pulls the body to the north. A force of 40 N pulls the body to the south. What are the magnitude and direction of the acceleration of the body?
6. What is the acceleration of a 4 kg body to which an unbalanced 18 N force is applied?
7. Identical forces are applied to two objects. The acceleration of A is 20 times the acceleration of B. How does the mass of A compare with the mass of B?
8. Suppose an astronaut in space pushes a piece of equipment away from her. The astronaut is ten times as massive as the equipment. The equipment is accelerated to a velocity of 4 m/s toward Polaris. How fast and in what direction will the astronaut move as a result of pushing the equipment?
9. How would the gravitational force between two bodies change if the mass of each were doubled and the distance between them were also doubled?
10. What is the strength of the gravitational force between the Earth and the Moon?
11. The distance between two bodies is cut in half. What happens to the force of gravity between the two bodies?
12. What would happen to the weight of a body if it were lifted to a distance of two Earth radii above the ground?
13. What is the weight of a 50 kg person on the Earth's surface?
14. The acceleration of gravity on Mars is 0.38 Earth gravities. How much would a dog that weighs 80 N (18 pounds) on Earth weigh on Mars?
15. An asteroid is three times as far from the Sun at aphelion than at perihelion. At perihelion the orbital speed of the asteroid is 25 km/s. What is the orbital speed of the asteroid at aphelion?
16. Suppose the distance between two stars doubled while the mass of each star doubled as well. What would happen to the orbital period of the two stars?
17. A planet in a hypothetical planetary system orbits its parent star with an orbital period of 5 years and an orbital distance of 2 AU. How does the mass of the star compare to the mass of the Sun?
18. Suppose the Sun were twice as massive as it actually is. What would be the orbital period of a planet at a distance of 10 AU from the Sun?
19. The circular speed at 1 AU from the Sun is 30 km/s. What is the circular speed at 20 AU? At 0.2 AU?
20. What is the circular speed at a distance of 5 AU from a star having a mass of 20 $M_\odot$?
21. Neptune orbits the Sun in an almost circular orbit at a speed of 5.4 km/s. How fast would a space probe at Neptune's distance have to move to escape from the solar system?
22. What is the escape velocity from the Sun at a distance of 1000 AU?

Group Activity

Have two members of your group put on their roller skates (or ice skates, if possible). Have the two skaters stand close together facing each other. Have one skater firmly but not violently push the other skater away. Note what happens to the two skaters and compare the results with the prediction of Newton's third law of motion. If possible, do the experiment with two skaters of about equal size and again with two skaters of very different sizes.

For More Information

Visit the text website at **www.mhhe.com/fix** for chapter quizzes, interactive learning exercises, and other study tools.

6

I had a dream, which was
not all a dream.
The bright sun was
extinguished, and the stars
Did wander darkling in
the eternal space,
Rayless, and pathless, and
the icy earth
Swung blind
and blackening in
the moonless air.

Lord Byron
Darkness

www.mhhe.com/fix

Light and Telescopes

Imagine a universe without any kind of electromagnetic radiation. Light is a kind of electromagnetic radiation, so everything in that imaginary universe would be dark. The differences between that universe and ours, however, would be much more profound. For example, radios and X-ray machines wouldn't work because radio waves and X rays are kinds of electromagnetic radiation. Because energy is transported by the emission and absorption of radiation, energy flow in that imaginary universe would be negligible. For example, stars would not heat their planets. Because many atomic and nuclear processes emit and absorb radiation, matter would behave very differently there, too.

Astronomers depend on electromagnetic radiation for nearly everything they know about the universe. The amount of information obtained from cosmic rays, neutrinos, and meteorites is tiny compared with that discovered using telescopes. Perhaps more than any other kind of scientists, astronomers are intensely interested in the properties of electromagnetic radiation, how it is produced, how it travels from its source to the Earth, and how it can be gathered and measured. This chapter is about light and the tools that astronomers have created to decipher as much of light's message as they can.

Questions to Explore

- How can we describe the different kinds of electromagnetic radiation?
- How do telescopes work?
- How do astronomers measure the light from celestial objects?
- Where are some of the best places to build an observatory?
- Why is it so important for astronomers to have observatories in space?
- What is radio astronomy? How do radio telescopes work?

6.1 WAVES

A **wave** is a disturbance that moves through a material medium or through empty space. It is important to realize that in a wave it is the disturbance that moves forward, not the medium. Figure 6.1 shows that if a string is shaken up and down at one end, a disturbance propagates down the string. The oscillation is a wave and has properties similar to those of electromagnetic radiation.

Figure 6.1 shows that the **wavelength** of a wave, usually symbolized by the Greek letter lambda, λ, is the distance between successive crests of the wave. For example, waves in the ocean or in a lake typically have wavelengths that are measured in meters or tens of meters. The **frequency,** f, of a wave is the rate at which wavecrests pass a fixed point. Frequency is usually given in number of wavecrests per second, or Hertz (Hz). For waves at the beach, for example, the frequency might be about 0.2 Hz, which means that 0.2 of a wave passes each second, and 5 seconds elapse between successive passing crests.

FIGURE 6.1
A Wave
A wave is produced when the end of a stretched string is shaken up and down. The wavelength, λ, of a wave is the distance between successive wavecrests.

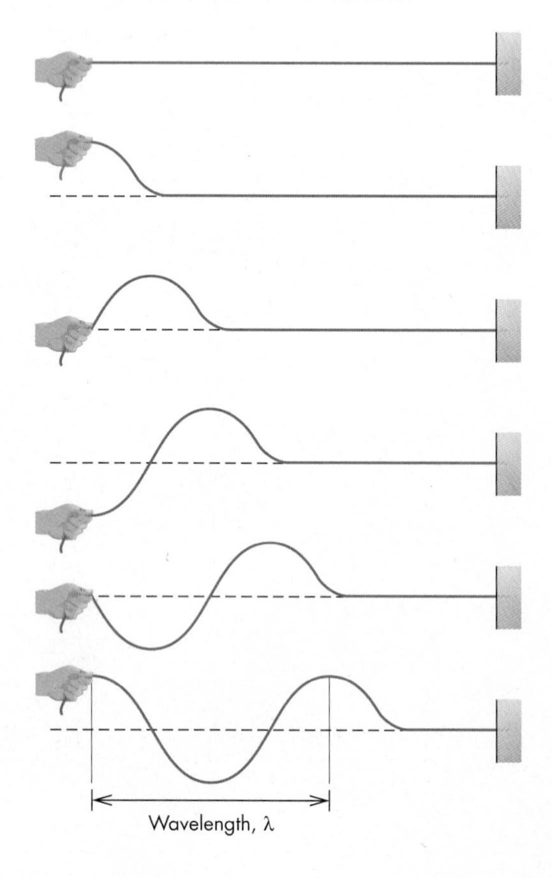

The **energy flux** of a wave is the rate at which the wave carries energy through a given area. For example, energy flux describes the amount of energy with which water waves strike a square foot of seawall each second. The higher the energy flux, the more powerful the wave.

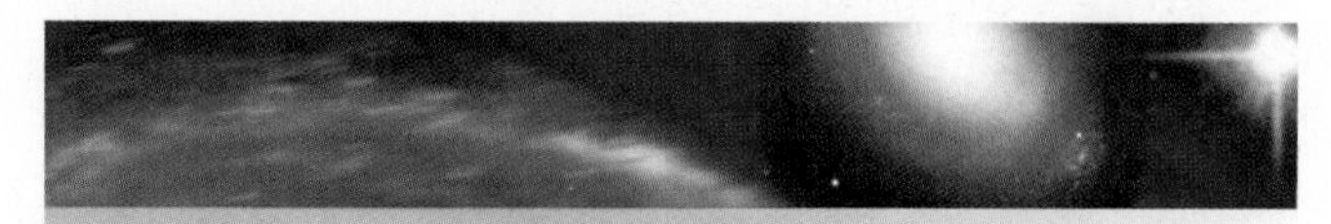

A wave is a disturbance that moves through a medium or through empty space. A wave is described by its length (wavelength), rate at which its crests pass (frequency), and the energy flux that the wave carries.

6.2 ELECTROMAGNETIC WAVES

In about 1860, the British physicist James Clerk Maxwell found a set of four equations, now known as Maxwell's equations, that summarized all of the laws of electricity and magnetism discovered experimentally by other scientists. Maxwell also found that the equations predicted the existence of electromagnetic waves. An **electromagnetic wave** is an electric and magnetic disturbance that travels through space and transparent materials. Electromagnetic waves are composed of oscillating electric and magnetic fields, as shown in Figure 6.2. When Maxwell calculated the speed of the electromagnetic waves, he found that they moved at the speed of light. What Maxwell had discovered was that **light** is a form of electromagnetic wave.

Whenever an electrically charged particle or body accelerates, it produces electric and magnetic field disturbances. If the charged particle oscillates back and forth with a certain frequency, the electric and magnetic field disturbances oscillate as a wave at that same frequency. Electromagnetic waves move outward from the accelerating charge at the speed of light, c, which is 3×10^8 m/s, or 186,000 miles per second. As the waves move outward from their source, the flux of energy they carry falls steadily. Imagine shells that surround the source of electromagnetic

FIGURE 6.2
An Electromagnetic Wave
An electromagnetic wave consists of an oscillating magnetic and electric field moving at the speed of light.

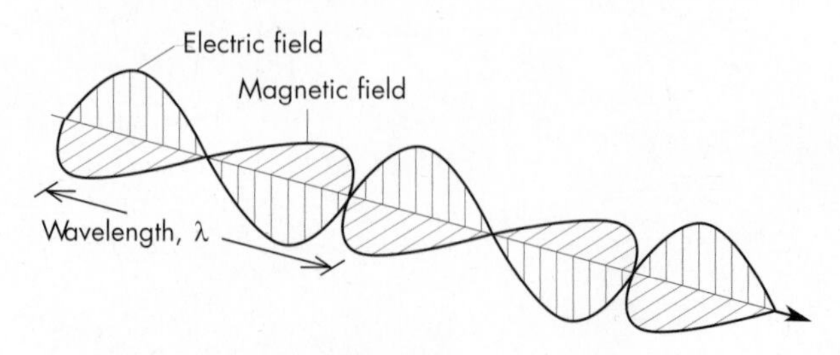

FIGURE 6.3
Electromagnetic Flux
The flux of electromagnetic waves decreases with the square of the distance from the source of the waves. The rate at which energy passes through the three shells is the same, but because the outer shells have greater area than the inner shell, the flux of the electromagnetic wave energy is less for the outer shells than for the inner shell.

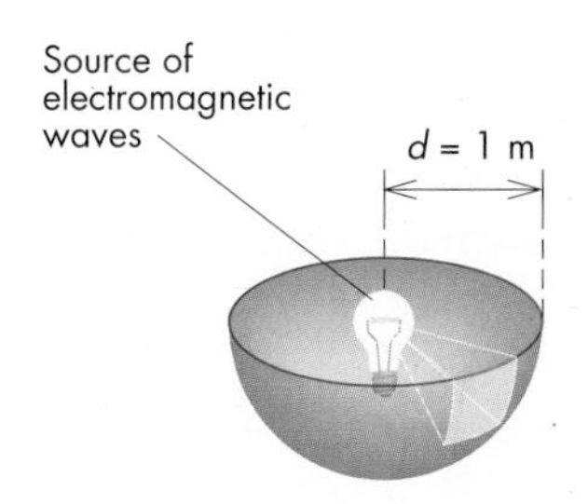

Electromagnetic flux $(F) = \frac{\text{Energy}}{\text{Area}} = \frac{E}{4\pi d^2}$

Distance (d) = 1 meter, so

$F = \frac{E}{4\pi 1^2} = \frac{E}{4\pi}$

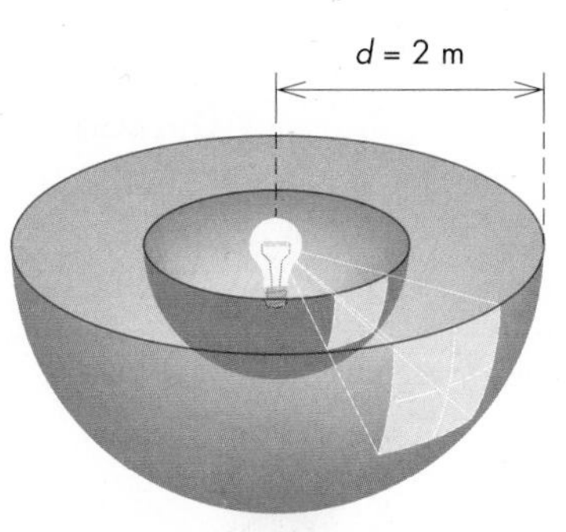

Distance (d) = 2 meters, so

Flux $(F) = \frac{E}{4\pi d^2} = \frac{E}{4\pi 2^2} = \frac{E}{16\pi}$

Notice: Flux (F) decreases as distance (d) increases because energy (E) remains the same but area of the sphere $(4\pi d^2)$ increases.

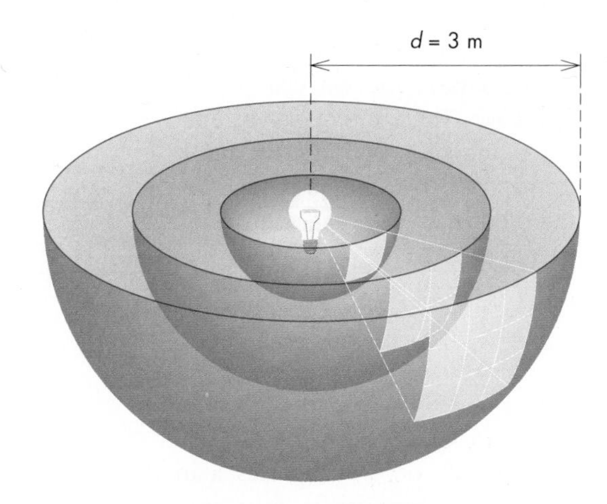

Distance (d) = 3 meters, so

Flux $(F) = \frac{E}{4\pi 3^2} = \frac{E}{36\pi}$

waves at different distances, as shown in Figure 6.3. The electromagnetic energy passing through each of the shells each second is the same; it is just the rate at which wave energy is produced at the source. However, because flux is the energy per unit area per unit time, the larger the area the smaller the flux. The area of a shell is given by $4\pi d^2$, where d is the distance from the source of electromagnetic waves, so the flux is *inversely* proportional to the square of the distance. In other words, the flux of energy, F, falls with the square of distance, d, as

$$F = \frac{E}{4\pi d^2} \qquad (6.1)$$

where E is the energy carried by the waves. Conversely, the flux increases as the square of the distance decreases. The greater the flux, the brighter the source of radiation appears to be.

The Spectrum

All electromagnetic waves travel at the same speed, the speed of light. Electromagnetic waves differ, however, in wavelength and frequency. The product of the wavelength and frequency of an electromagnetic wave is the speed of light. Figure 6.4 shows why this is so. At the beginning of a 1 second interval of time, a wavecrest of a wave with

Flux and Distance

Equation 6.1 can be used to calculate the brightness (flux of energy) of the Sun at various places in the solar system. For example, what is the flux of solar energy at the Earth's orbital distance, $d = 1.5 \times 10^{11}$ m? To solve this problem, we need to know that the rate at which energy is carried outward from the Sun by electromagnetic waves is $E = 3.83 \times 10^{26}$ watts. Using these values in Equation 6.1 gives

$$F = \frac{E}{4\pi d^2}$$
$$= \frac{3.83 \times 10^{26} \text{ watts}}{4 \times 3.14 \times (1.5 \times 10^{11} \text{ m})^2}$$
$$= \frac{3.83 \times 10^{26} \text{ watts}}{2.83 \times 10^{23} \text{ m}^2}$$
$$= 1.35 \times 10^3 \text{ watts/m}^2$$
$$= 1350 \text{ watts/m}^2$$

Thus, the solar energy striking each square meter of area at the Earth's distance is enough to power more than twenty 60-watt lightbulbs if it could be converted to electricity.

FIGURE 6.4
The Relationship Between Speed, Wavelength, and Frequency for a Wave
The wave has a frequency of 4 Hz, so four wavecrests pass a mark each second. The distance between wavecrests is the wavelength of the wave. Thus, the speed of the wave (the speed of light for electromagnetic waves) is equal to the product of its wavelength and frequency.

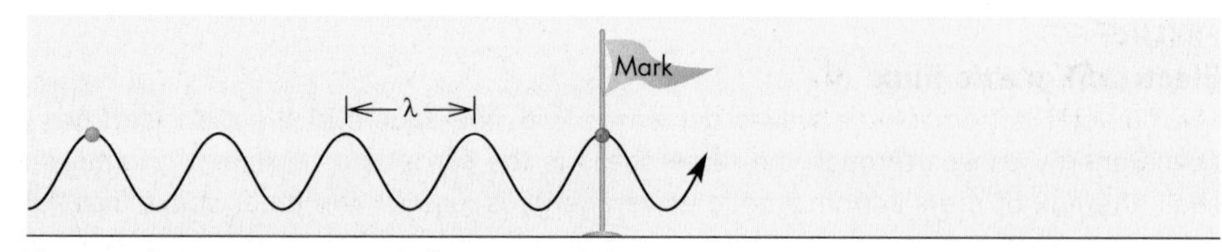

The crest of a wave passes a mark; frequency = 4 Hz.

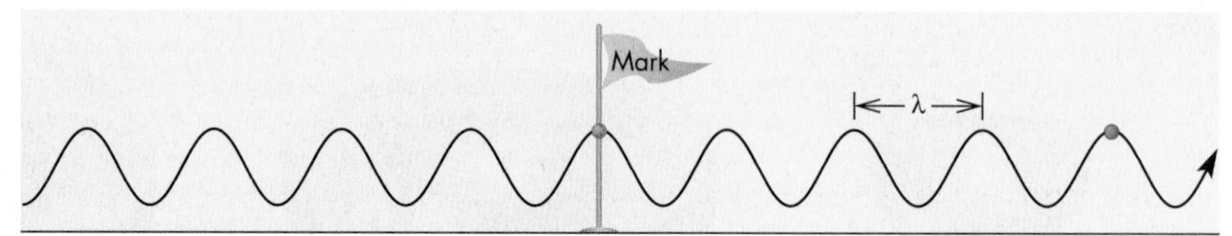

One second later: The wavecrest is four wavelengths away.

a frequency of 4 Hz passes a mark. One second later, the same wavecrest is four wavelengths away because four wavecrests pass the mark each second. The distance the wavecrest travels in 1 second, which is the speed of the wave, is the frequency (number of waves per second) times the wavelength. Because frequency and wavelength are related to each other by

$$c = \lambda f \tag{6.2}$$

electromagnetic waves with low frequencies have large wavelengths and those with high frequencies have small wavelengths.

Figure 6.5 shows the **electromagnetic spectrum** and the wavelength and frequency regions into which the spectrum has been divided. Note, for example, that radio waves cover a wide range of low frequencies but that some have wavelengths that are much larger than the Earth. Visible light, on the other hand, covers a narrow range of high frequencies that are associated with extremely small wavelengths. The smallest wavelengths and highest frequencies, however, belong to the **gamma rays,** which can be produced in nuclear reactions.

Electromagnetic waves are oscillating electric and magnetic fields produced by accelerating electric charges. All electromagnetic waves travel at the speed of light. What distinguishes them from each other are their wavelengths and frequencies. Waves with small wavelengths, such as gamma rays, have high frequencies, whereas waves with large wavelengths, such as radio waves, have low frequencies.

Frequency and Wavelength

Equation 6.2 can be used to calculate the wavelength of electromagnetic radiation given its frequency, or it can be used to calculate its frequency given its wavelength. For example, what is the frequency of an electromagnetic wave that has a wavelength, $\lambda = 5 \times 10^{-7}$ m? (Electromagnetic radiation with this wavelength is visible light.) To solve this problem we need to know that the speed of light is $c = 3 \times 10^{8}$ m/s.

We also need to rewrite Equation 6.2 to solve for frequency, *f*. This is

$$f = \frac{c}{\lambda} = \frac{3 \times 10^{8}\ \text{m/s}}{5 \times 10^{-7}\ \text{m}} = 6 \times 10^{14}\ \text{Hz}$$

This means that 600,000 billion electromagnetic wavecrests pass a point each second.

FIGURE 6.5
The Electromagnetic Spectrum
The electromagnetic spectrum ranges from radio waves, which have long wavelengths and low frequencies, to gamma rays, which have short wavelengths and high frequencies. The visible part of the spectrum is shown on an expanded scale, using nanometers (nm), where 1 nm = 10^{-9} m.

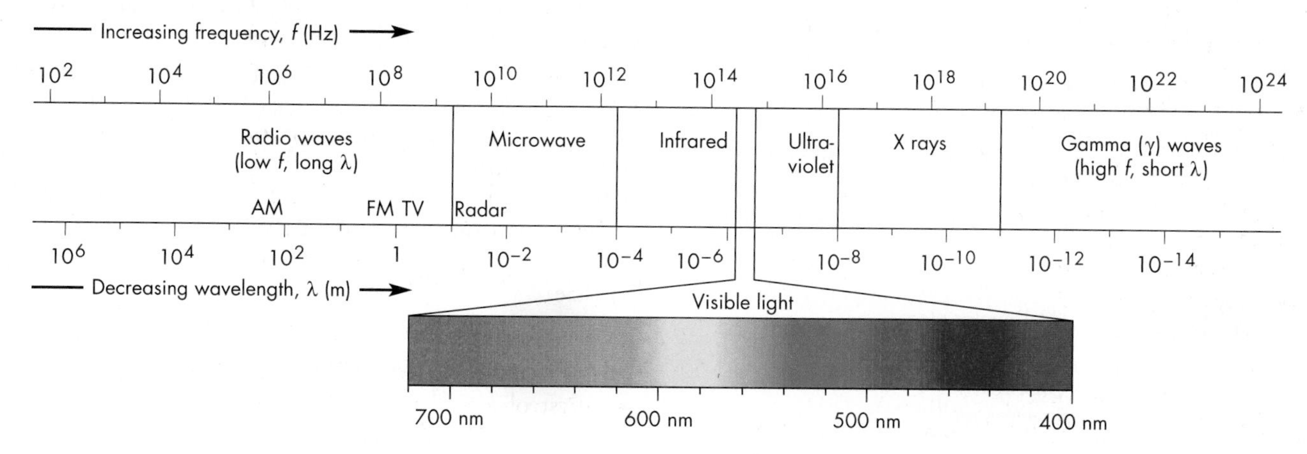

The Doppler Effect

The **Doppler effect** is a change in the frequency and wavelength of a wave that is emitted by or reflected from a body in motion with respect to the observer who measures the waves. When the observer and the source of the waves are moving toward each other, as in Figure 6.6, wavecrests arrive more often than they would if there were no motion. Thus, the observed frequency is higher and the wavelength is shorter because of the relative motion of the source and observer. On the other hand, if the source and observer are moving apart, the observer measures a lower frequency and a longer wavelength than if there were no motion. A familiar example of the Doppler effect is the abrupt drop in pitch (frequency) of the siren when an ambulance passes. While the ambulance is approaching, the observer hears its siren Doppler shifted to higher frequency. After it passes and begins to move away, the observer hears the siren Doppler shifted to a lower frequency. The Doppler effect occurs for all waves, including electromagnetic waves.

The change in wavelength produced by the Doppler effect depends on the speed with which the observer and source are approaching each other or moving apart. For electromagnetic waves, the change in wavelength due to the Doppler effect is given by

$$\frac{\Delta\lambda}{\lambda_0} = \frac{v_r}{c} \tag{6.3}$$

where $\Delta\lambda$ is the difference between the observed wavelength and λ_0, the wavelength that would be observed

The Doppler Equation

Equation 6.3, often called the Doppler equation, can be used to calculate the speed at which a source of electromagnetic radiation is moving toward or away from an observer. Suppose that the wavelength of a radio wave is 10.0 cm when there is no motion. Now, suppose the wavelength is measured to be 10.001 cm. The change in wavelength, $\Delta\lambda$ is 0.001 cm. Using 10 cm for λ_0, and $c = 3 \times 10^5$ km/s, the relative speed of the source and observer is

$$v_r = \frac{c\Delta\lambda}{\lambda_0} = 3 \times 10^5 \text{ km/s} \frac{0.001 \text{ cm}}{10 \text{ cm}} = 30 \text{ km/s}$$

The fact that the speed is positive means that the source and the observer are moving away from one another.

Suppose instead that the radio wave is observed to have a wavelength of 9.999 cm. This time, the change in wavelength is −0.001 cm. Using the Doppler equation, the speed is −30 km/s. The negative sign means that the source and observer are moving toward each other.

FIGURE 6.6
The Doppler Effect
Wavecrests spread out from a source of waves. The circles mark the crests of the waves emitted at the points indicated by dots. For the observer toward whom the source is moving, the wavecrests are closer together than they would be if the source were at rest. The observer sees a shorter wavelength and higher frequency. For the observer away from whom the source is moving, the wavecrests arrive farther apart. This observer sees a longer wavelength and lower frequency. It doesn't matter whether the observer or source is moving. Only the relative motion of the two is important.

ANIMATION *The Doppler effect*

INTERACTIVE *Doppler shift*

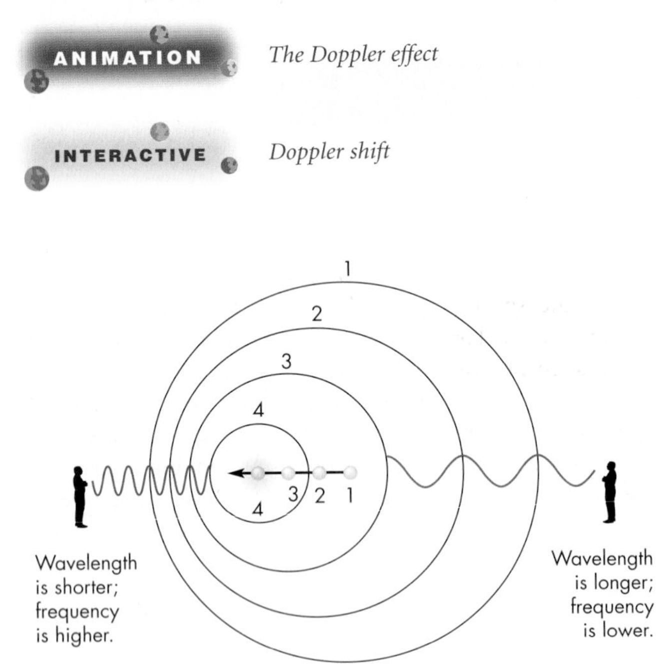

if there were no motion. The variable v_r is the speed with which the source and observer approach or recede from each other, and c is the speed of light (see box for Equation 6.3).

The Doppler effect is a change in the measured wavelength and frequency of a wave. This change happens when the source and observer of the wave are moving toward or away from each other. The change in wavelength is proportional to the speed of the source relative to the observer.

Photons

During the second half of the seventeenth century, Isaac Newton argued that light was made up of a stream of particles (what we now call **photons**). However, other scientists, including Christian Huygens and Robert Hooke, argued that light was made up of waves. Because of Newton's towering reputation, the particle theory won out and was accepted for about a century. After 1800, however, many experiments showed that light had interference properties, which could only be explained if light were composed of waves. Experiments conducted in the early part of the twentieth century demonstrated that light has a dual nature; it behaves as both a wave and a particle.

The Photoelectric Effect The photoelectric effect was first observed in the late 1800s. Researchers found that when a beam of light is directed toward a metallic surface, the light causes the metal to eject electrons, which can be collected and measured. Simply making the light brighter has no effect on the energies of the electrons; it just produces more electrons of the same energy. As Figure 6.7 shows, increasing the frequency of the light increases the energies of the ejected electrons. This result is inconsistent with the wave theory of light, in which the energy of a wave depends on its brightness. Increasing the brightness of a light should make its waves more energetic and, thus, capable of ejecting electrons with more energy. In explaining the photoelectric effect

FIGURE 6.7
The Energy of Ejected Electrons in the Photoelectric Effect
The energy of the ejected electrons depends on the frequency of the light that strikes the metal surface. Below a certain frequency, no electrons are ejected. Above that frequency, the energy of the electrons increases as frequency increases.

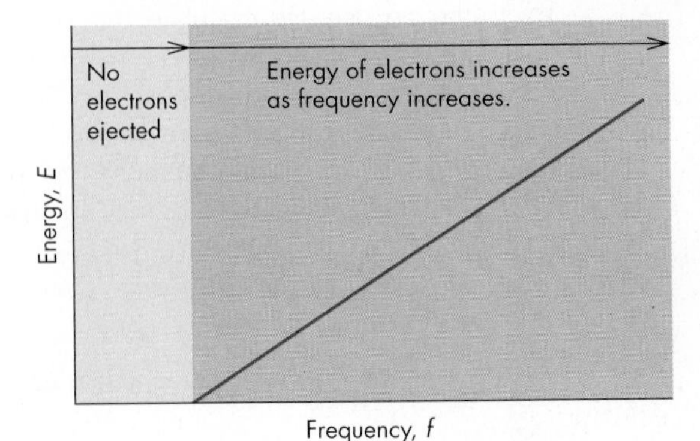

The Energy of Photons

Equation 6.4 can be used to calculate the energy of a photon given its frequency. Suppose we wish to calculate the energy of a gamma ray with frequency, $f = 10^{24}$ Hz. This frequency is about a billion times higher than that of visible light. To solve the problem, we need to know that Planck's constant is $h = 6.6 \times 10^{-34}$ J s, where J is the symbol for joule, a unit of energy, and s is the symbol for seconds. Using these values in Equation 6.4, we get

$$E = hf$$
$$= (6.6 \times 10^{-34}\ \text{J s})(10^{24}\ \text{Hz})$$
$$= 6.6 \times 10^{-10}\ \text{J}$$

Because a watt is a joule per second, a 60-watt lightbulb consumes 60 joules of energy per second. Thus it would take the energy of about 10^{11}, or 100 billion, gamma rays to keep a 60-watt bulb lit for a second. Gamma rays are the most energetic photons, so other kinds of photons carry even tinier amounts of energy.

in 1905, Einstein said that light consists of bundles of energy called photons. Unlike particles of ordinary matter, photons consist of energy only, with no accompanying mass. The energy of a photon, E, depends on its frequency as

$$E = hf = \frac{hc}{\lambda} \qquad (6.4)$$

where h is Planck's constant (named for the German physicist Max Planck).

According to Einstein's explanation of the photoelectric effect, a fixed amount of energy is required to remove each electron from the surface of the metal. Light consisting of photons with energy and frequency less than the fixed amount can't eject electrons no matter how many photons strike the surface. Increasing the brightness of the light only increases the number of photons, not the energy of each photon. On the other hand, keeping the brightness constant, but increasing the frequency increases the energy of each photon. At high enough frequencies, the photons have enough energy to remove electrons. For photons with frequency and energy above the critical amount, the energy remaining after the electron is removed provides the electron with its energy.

Light behaves both as waves and as particles. To account for some properties of light, it is necessary to think of light as a stream of massless particles called photons. The energy of a photon is proportional to its frequency.

Gamma rays, which have the highest frequencies and smallest wavelengths, have the highest energy. The radio part of the spectrum, on the other hand, with its low frequencies and large wavelengths, consists of low-energy photons. In Figure 6.8 the various regions of the electromagnetic spectrum are ranked according to their energies. Figure 6.8 is consistent with what we already know about the danger of exposure to various parts of the spectrum. Exposure to radio, **infrared,** and visible radiation, except at very high energy fluxes, is normally quite safe. Exposure to **ultraviolet** radiation is more dangerous. Exposure to **X rays** must be tightly controlled, while exposure to gamma rays can be fatal.

FIGURE 6.8
The Energy of Different Kinds of Photons
Gamma ray photons, which have the highest frequencies, have the largest energies.

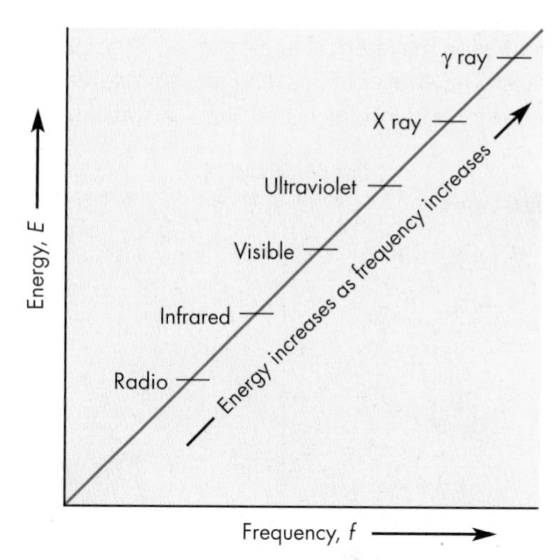

6.3 REFLECTION AND REFRACTION

Astronomers use various kinds of telescopes to measure the electromagnetic radiation from distant objects. All of these telescopes depend on reflection or refraction to bend light and bring it to a focus.

Reflection

Reflection refers to the bouncing of a wave from a surface. When a wave is reflected, its new direction depends on the direction of the incident wave and its orientation to the reflecting surface. As shown in Figure 6.9, the reflected and incident waves make equal angles to a line perpendicular to the reflecting surface at the spot where the wave strikes the surface. Tilting the surface sends the reflected wave in a different direction, but the reflected angle is still the same as the incident angle. Not all surfaces reflect equally well. The **reflectivity** of a surface describes its ability to reflect electromagnetic waves. Reflectivity ranges from 0% for a surface that reflects no light and is completely dark to 100% for a bright surface that reflects all the light that falls on it. Astronomical mirrors, like other mirrors, have surfaces made of materials with high reflectivities.

Refraction

Refraction is the bending of light when it passes from one material into another. Figure 6.10 demonstrates refraction with a meterstick partially submerged in water. The refraction of the light rays when they emerge from the water into the air makes the meterstick appear to bend where it enters the water.

Refraction occurs because the speed of light depends on the material through which it is passing. The number usually listed for the speed of light is actually correct only for light traveling through a vacuum. When light travels through any gas, solid, or liquid, it moves more slowly. The speed of light in a material is described by its **index of refraction,** the ratio of the speed of light in a vacuum to its speed in the material. For instance, light travels only about two-thirds as fast in glass as it does in a vacuum, so the index of refraction of glass is about $1/(2/3) = 1.5$. For air, the index of refraction is about 1.0003, showing that light travels only slightly slower in air than in a vacuum.

When a wave passes from one substance to another that has a different index of refraction, it is bent, or refracted. Figure 6.11 shows an analogy of the crest of a wave moving into a substance with a higher index of refraction. As each part of the wavecrest enters the new material, it slows down,

FIGURE 6.9
Reflection
When a wave is reflected from a surface, the incident and reflected waves make equal angles to a line perpendicular to the surface. The arrows in this diagram represent rays, which show the direction in which the wave is traveling.

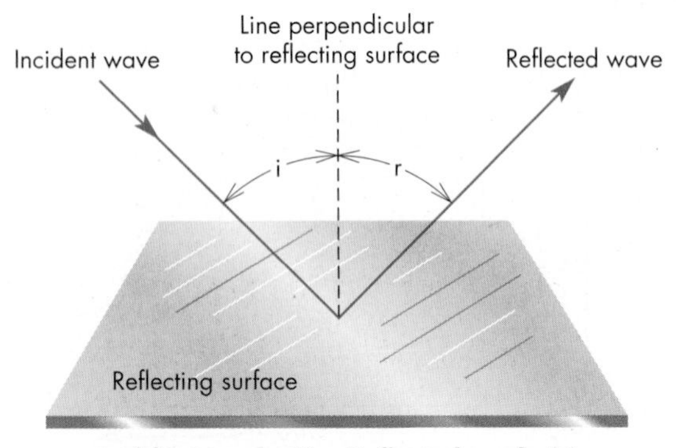

FIGURE 6.10
Refraction
Refraction makes a straight meterstick appear to bend at the point where it emerges from water.

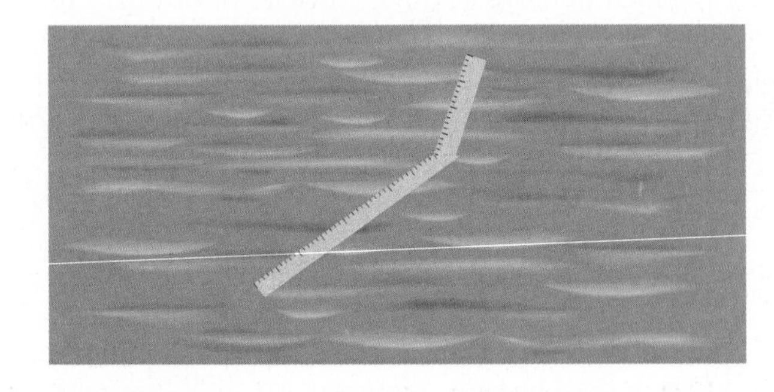

FIGURE 6.11
Refraction and the Speed of a Wave
Refraction of a wave resembles what happens when a marching band crosses from dry ground (where individual marchers move swiftly) to mud (where the marchers slow down). The slower speed in the mud causes each row of marchers to be bent, or refracted, as it crosses the boundary.

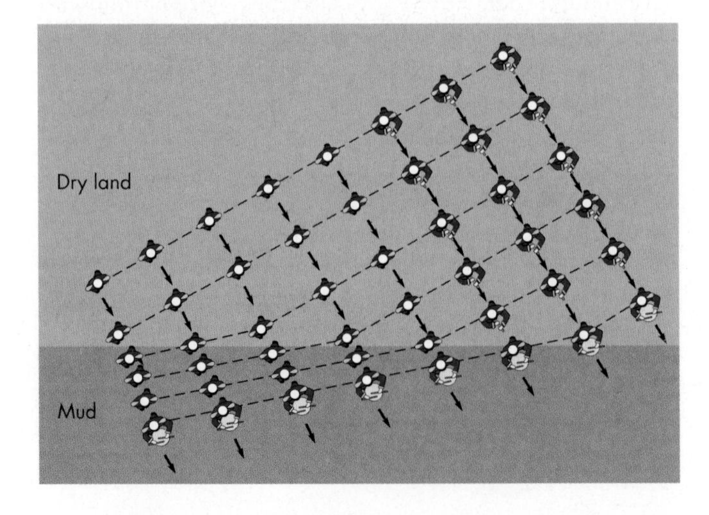

causing the direction of the wave to be bent. The greater the ratio of indices of refraction, the more a wave is bent when passing from one material to another. This process, refraction, bends the waves entering a piece of glass from air and again when the waves reemerge from the glass into the air. Figure 6.12, A, shows that if the surfaces where the waves enter the glass and where they leave the glass are parallel to each other, the directions of the incident and emerging waves will be the same. If the surfaces are not parallel, then the waves will enter and exit with different directions, as in Figure 6.12, B. This change of direction makes it possible for lenses to focus light.

FIGURE 6.12
Refraction in a Piece of Glass
Refraction bends light waves not only when the light passes from air into glass but also when it reemerges from the glass into the air. **A,** If two surfaces of the glass are parallel, the emerging wave is parallel to the incident wave. **B,** Otherwise, the emerging wave and the incident wave have different directions.

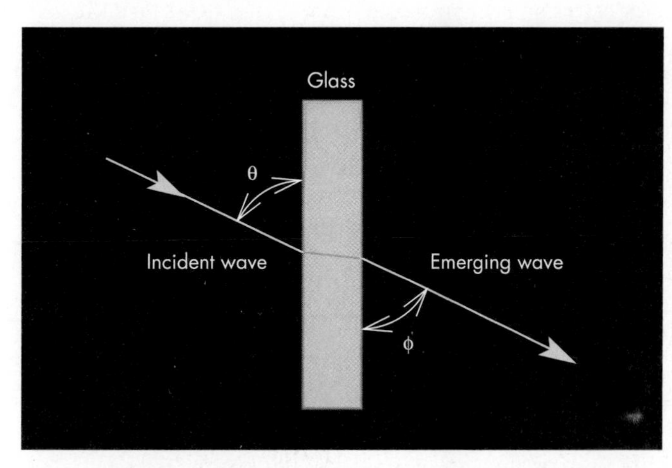

A Surfaces of glass are parallel:
- Angle θ = Angle ϕ
- Emerging wave is parallel to incident wave.

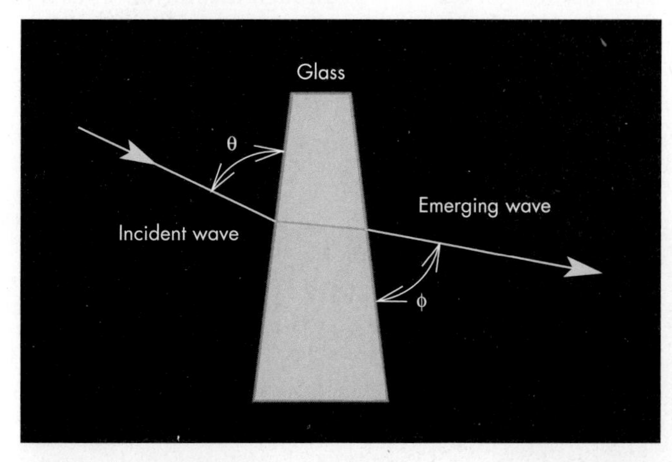

B Surfaces of glass are not parallel:
- Angle θ ≠ Angle ϕ
- Emerging wave is not parallel to incident wave.

Dispersion

When a beam of sunlight falls on a prism, as in Figure 6.13, it is separated according to wavelength to form a spectrum. The separation of light according to wavelength is called **dispersion.** Dispersion occurs because the index of refraction in glass (and other substances) depends on wavelength. Within the visible part of the spectrum, the index of refraction of violet light is greatest, so violet light is bent the most when it enters the glass and again when it emerges. Red light, on the other hand, has the smallest index of refraction in the visible spectrum and is bent the least. Light at other visible wavelengths is bent by intermediate amounts and falls between red and violet. This difference in refraction causes the light to be dispersed into the familiar visible spectrum. The same effect occurs when light is refracted in raindrops, producing the rainbow. Astronomers use dispersion to spread out light so that narrow wavelength regions can be studied in detail.

FIGURE 6.13
The Dispersion of Light by a Prism
The index of refraction of glass depends on the wavelength of light. Consequently, light of different colors is refracted by different amounts, producing the familiar spectrum of colors. Violet light is bent more than red light because the index of refraction increases as the wavelength of light decreases.

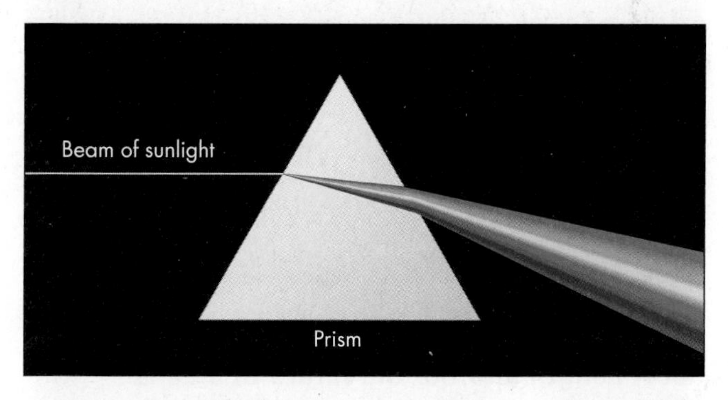

The direction of an electromagnetic wave can be changed by reflection from a surface or refraction when the wave passes from one material to another. When the amount of refraction depends on the wavelength of the light, the light will be dispersed according to its wavelength, producing a spectrum.

6.4 OPTICAL TELESCOPES

Optical telescopes use lenses and mirrors to bring light to a focus. The lens or mirror used to focus the light is called the **objective** of the telescope.

FIGURE 6.14
A Crude Lens Made from Two Prisms
Beams refracted by the two prisms are focused at the point where they cross.

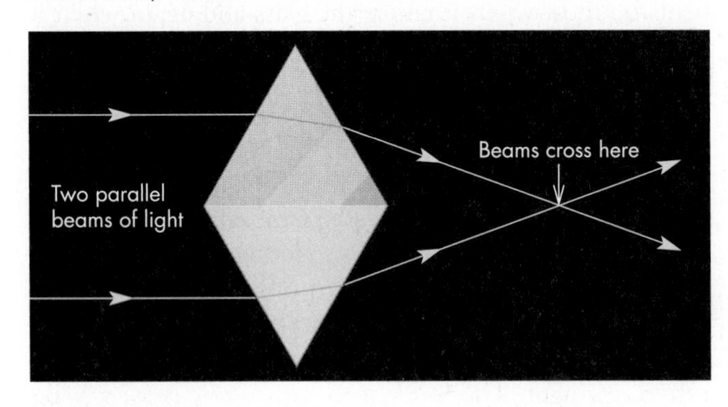

FIGURE 6.15
The Focusing of Light by a Lens
All parallel beams of light cross at the focal point of the lens, which is located one focal length behind the lens.

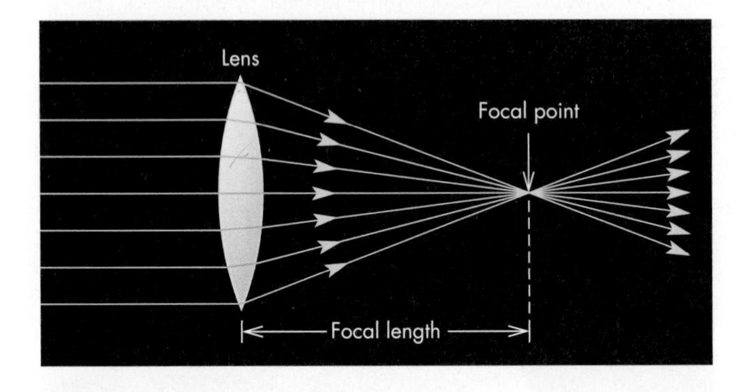

Refractors

A refracting telescope, or **refractor,** is one in which the objective is a lens. To see how a lens can focus light, consider the two prisms mounted base to base in Figure 6.14. When two parallel beams of light of the same wavelength strike the prisms, they are bent toward each other and cross. A lens can be made by smoothing out the edges of the prisms (as in Figure 6.15) so that all other parallel beams of light will cross at the same point. The **focal point** of a lens is the spot at which parallel beams of light striking the lens are brought to a focus and cross. All astronomical objects are so distant that light beams from them are essentially parallel. Thus the light from a star or planet will be brought to a focus at the focal point of a lens. The distance from a lens to its focal point is called the **focal length.** As Figure 6.16 shows, highly curved lenses have short focal lengths, and slightly curved lenses have long focal lengths. The size of a refracting telescope is the diameter of the light-collecting lens. That is, a 1-meter refracting telescope has a lens with a diameter of 1 meter.

Reflectors

A reflecting telescope, or **reflector,** is one in which the objective is a mirror. Figure 6.17 shows how flat mirrors can be used to focus two parallel beams of light. Figure 6.18 shows how rounding the mirrors into the shape of a parabola makes it possible to focus all other parallel beams of light. The focal length of the mirror is the distance from the mirror to the focal point. Just as for lenses, highly curved mirrors have short focal lengths and slightly curved mirrors have long focal lengths. The size of a reflecting telescope refers to the diameter of its objective.

Refractors Versus Reflectors The lens for a refracting telescope must be made of high-quality glass with no internal imperfections. Both the front and back surface must be ground and polished. In a reflector, on the other hand, light doesn't pass through the mirror. Thus the mirror can be made of practically anything. In fact, plastic and metal have been used to make mirrors for reflecting

FIGURE 6.16
Focal Length and the Curvature of a Lens
Highly curved lenses have short focal lengths. Slightly curved lenses have long focal lengths.

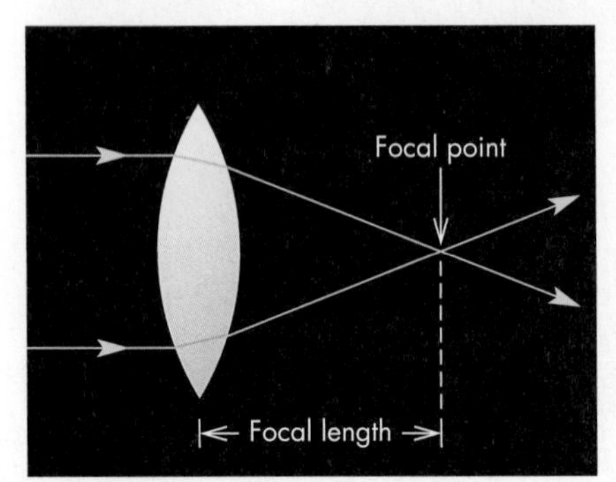

Highly curved lens: Short focal length

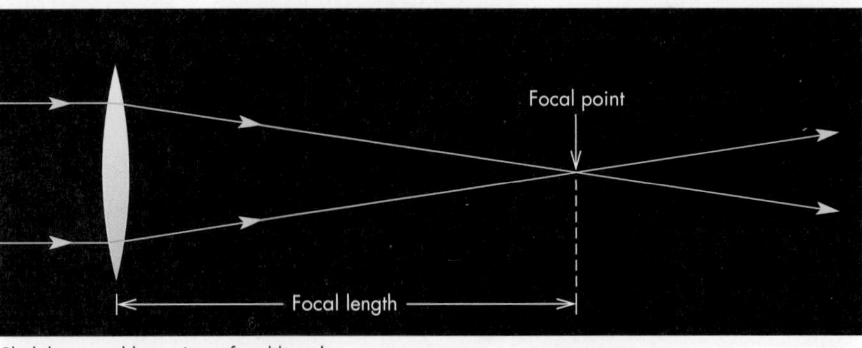

Slightly curved lens: Long focal length

FIGURE 6.17
The Focusing of Two Beams of Light by Flat Mirrors
Other beams striking the mirrors won't be reflected to the same point.

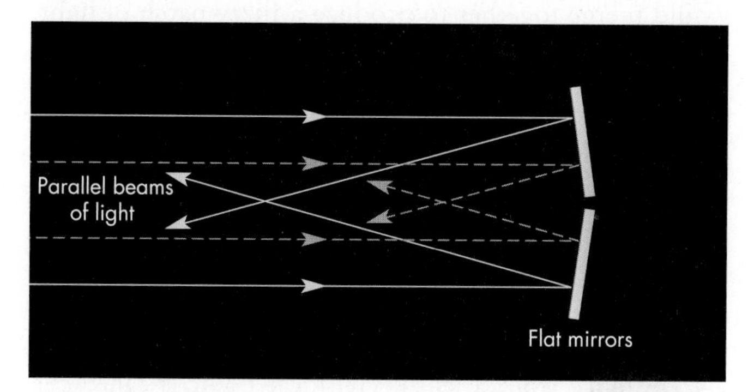

FIGURE 6.18
A Concave Parabolic Mirror
A concave mirror can bring all parallel beams that strike it to a focus at the focal point.

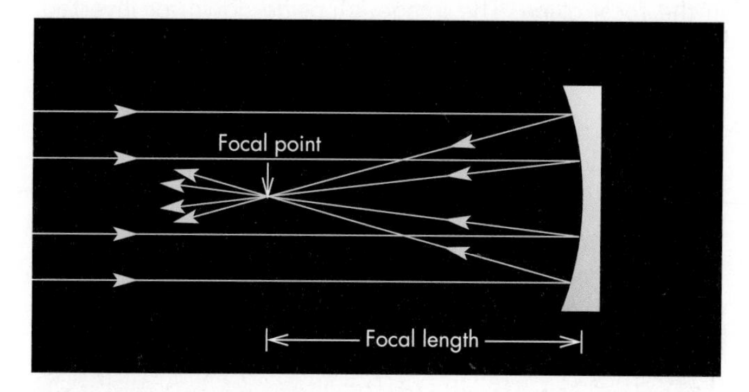

telescopes. Only one surface must be ground and polished, which makes it easier and much less expensive to produce large mirrors than it is to produce large lenses.

An additional problem is the weight of a large lens. A lens can be supported only around its rim. For a large lens, this amount of support is insufficient to keep the lens from sagging under its own weight and producing distorted images. This problem has limited the sizes of successful refracting telescopes to about 1.1 m. A mirror, on the other hand, can be supported along its entire back and has fewer problems with deformation. Even so, changes in the shape of a large mirror as the telescope is pointed in different directions must be corrected when very large reflecting telescopes are used. This correction is done by changing the pattern of support (under computer control) on the back of the telescope to obtain the sharpest possible images.

Telescopes can be made with lenses (refractors) or mirrors (reflectors). All large modern astronomical telescopes are reflectors because it is difficult to make and to support large lenses in refracting telescopes.

Forming an Image

If the source of the light is a point, the objective brings the light to focus at a single point. If the source of light is extended in angle, however, then light from different parts of the source strikes the objective from different directions. The telescope forms an image of the source of light. Figure 6.19 shows how a lens focuses light from two different points in a source of light. The **focal plane** is the surface where the objective forms the image of an extended object.

Brightness The brightness of the image depends both on the amount of light collected by the objective of the telescope and on the area of the image in the focal plane. The **light-gathering power** of a telescope is the amount of light it can collect and focus. Light-gathering power is proportional to the area of the objective. If the diameter of a circular lens or mirror is D, then the area, A, of the lens or mirror is given by

$$A = \frac{\pi D^2}{4} = \pi r^2 \qquad (6.5)$$

where r is the radius of the lens or mirror.

Equation 6.5

Light-Gathering Power

Equation 6.5 describes how light-gathering power, proportional to area, depends on the diameter of the objective of a telescope. Suppose we want to compare the light-gathering power of one of the 10-m Keck telescopes with that of a 0.2-m telescope. The area of the 10-m mirror is

$$A = \frac{\pi D^2}{4} = \frac{(3.14)(10\ \text{m})^2}{4} = \frac{314.2\ \text{m}^2}{4} = 78.5\ \text{m}^2$$

The area of the 0.2-m mirror is

$$A = \frac{\pi D^2}{4} = \frac{(3.14)(0.2\ \text{m})^2}{4} = \frac{0.126\ \text{m}^2}{4} = 0.0314\ \text{m}^2$$

Thus the 10-m telescope has $78.5/0.0314 = 2500$ times the area and light-gathering power of the 0.2-m telescope. This ratio is the same as the square of the ratio of the diameters of the two telescopes, $50^2 = 2500$.

FIGURE 6.19
The Formation of an Image
A lens forms an image in its focal plane. Light from different points in a source of light is focused at different places in the focal plane. The image is upside down, or inverted, compared with the source of light.

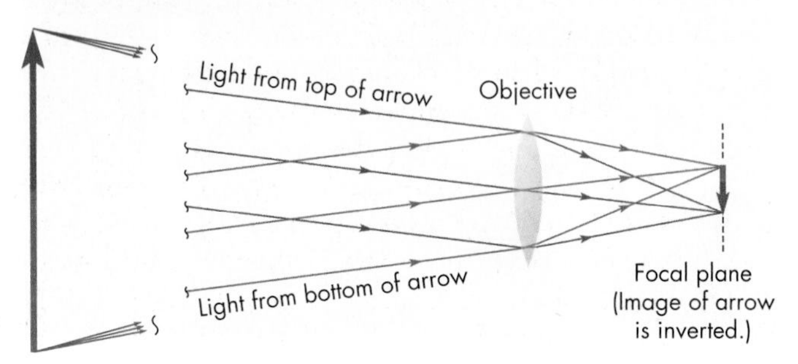

The size of an image is proportional to the focal length of the lens or mirror, so the collected light is spread out over an area proportional to the square of the focal length. The brightness of the image is then determined by the ratio of light-gathering power to image area, or the ratio of the square of diameter to the square of focal length. The ratio of focal length to diameter is often used to describe the properties of lenses or mirrors in telescopes or in cameras and is called the "speed," "f-number," or "focal ratio." An f/8 lens is one for which the focal length is eight times the lens diameter. A large f-number means that the camera or telescope will produce large images but that the images will be relatively dim compared to a telescope with a smaller f-number because the light will be spread out over a larger area.

The area of the objective determines the amount of light the telescope can gather. The brightness of an image formed by a telescope depends on the square of its focal ratio, the ratio of the focal length to the diameter of its objective.

Resolution The **resolution** of a telescope describes its ability to distinguish fine details in an image. At low resolution, the images of the individual stars in a star cluster (as an example) would be so blurred that they would merge together to produce a fuzzy patch of light. At higher resolution, the individual star images would be sharp enough that they could be distinguished from one another easily.

Every telescope blurs the images of astronomical objects because of a phenomenon called **diffraction,** which occurs whenever waves encounter an obstacle or an opening. The pattern of light and dark regions that result from diffraction depends on the size and shape of the opening. For a circular opening like the objective lens or mirror of a telescope the pattern, shown in Figure 6.20, consists of a bright central spot surrounded by a series of alternating dark rings and bright rings that grow fainter with distance from the center. The bright central spot has an angular diameter that depends on the diameter of the objective (D) and the wavelength of light (λ) according to

$$\theta = 250{,}000\ \lambda/D \qquad (6.6)$$

FIGURE 6.20
The Diffraction Pattern of a Circular Lens or Mirror
The diameter of the bright central spot is the resolution of the telescope.

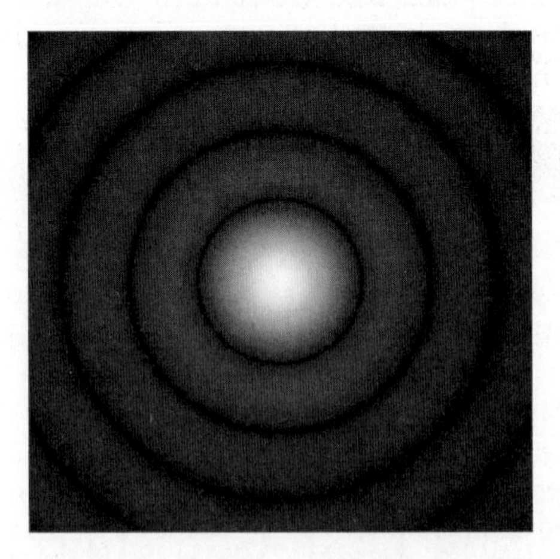

Resolution

Equation 6.6 can be used to calculate the smallest resolvable features that can be seen with a telescope of a given diameter operating at a given wavelength. For example, what is the minimum angle that can be resolved using a telescope with a diameter, $D = 2$ m, at a wavelength, $\lambda = 5 \times 10^{-7}$ m? Using these values for D and λ in Equation 6.6, we get

$$\begin{aligned}\theta &= 250{,}000\,\frac{\lambda}{D}\\ &= 250{,}000\,\frac{5 \times 10^{-7}\text{ m}}{2\text{ m}}\\ &= (2.5 \times 10^{5})(2.5 \times 10^{-7})\\ &= 6.25 \times 10^{-2} = 0.0625 \text{ seconds of arc}\end{aligned}$$

This is about 1/30,000 as large as the angular size of the Sun or Moon.

The resolution of a telescope describes its ability to distinguish fine details in an image and depends on the ratio of wavelength to the diameter of its objective. Radio telescopes, which operate at long wavelengths, usually have very poor resolution compared with optical telescopes.

where θ is in seconds of arc and D and λ must be given in the same units (see box for Equation 6.6). The angle θ is considered to the angular resolution of the telescope.

Equation 6.6 means that the longer the wavelength, the bigger the bright spot and the poorer the angular resolution. On the other hand, the bigger the objective, the smaller the bright spot and the better the angular resolution. If the angular separation of two sources of electromagnetic radiation is much less than θ, the two sources appear to be a single point. However, if two sources of light are separated by an angle larger than θ, they can be distinguished from each other.

In the visible part of the spectrum, wavelengths are so small that a resolution of a second of arc or less can be achieved with an objective of only 10 cm diameter (about 4 inches). This resolution is much better than the resolution of the human eye, which is about 1 minute of arc for most people. At longer wavelengths, particularly in the radio part of the spectrum, high resolution is much more difficult to obtain. For a resolution of 1 second of arc, a radio telescope operating at a wavelength of 10 cm would need to have a diameter of 25 km. This is almost a hundred times as large as any radio dish in the world.

Magnification One of the questions astronomers are often asked about a telescope is, "What is its magnification?" This question is difficult to answer because magnification, unlike light-gathering power, focal ratio, and resolution, is not an intrinsic property of the objective. It can be changed simply by replacing one eyepiece with another. Research telescopes, moreover, usually aren't used with eyepieces so their magnifications aren't even defined. The eyepiece is needed to convert the focused light back into parallel beams so that it can be refocused by the eye. As shown in Figure 6.21, the eyepiece is placed so that its focal point coincides with the focal point of the objective of the telescope. The magnification of the combined objective and eyepiece is given by f_o/f_e where f_o is the focal length of the objective and f_e is the focal length of the eyepiece. By using an eyepiece of very short focal length, the magnification of the telescope can be made very large. However, a very large image is quite dim, because the same amount of light makes up the image no matter what its size. Also, the field of view of the telescope, the total angle that can be seen through the eyepiece at one time, is small for very large magnifications. In practice, extremely large magnifications are not often used for observing.

Detectors

Once light has been brought to a focus, it is measured by a **detector.** The human eye is a detector, but modern astronomers use detectors that are more sensitive, can accumulate light for a long time, function in wavelength regions in which the eye is blind, and produce numerical rather than qualitative results.

Photography For more than a century, astronomers have used photography to record astronomical information. When a telescope is used for photography it operates exactly like a camera except that the objective of the telescope replaces the camera lens. The photographic film or plate is placed in the focal plane of the telescope, just as it is placed in the focal plane of the lens of a camera.

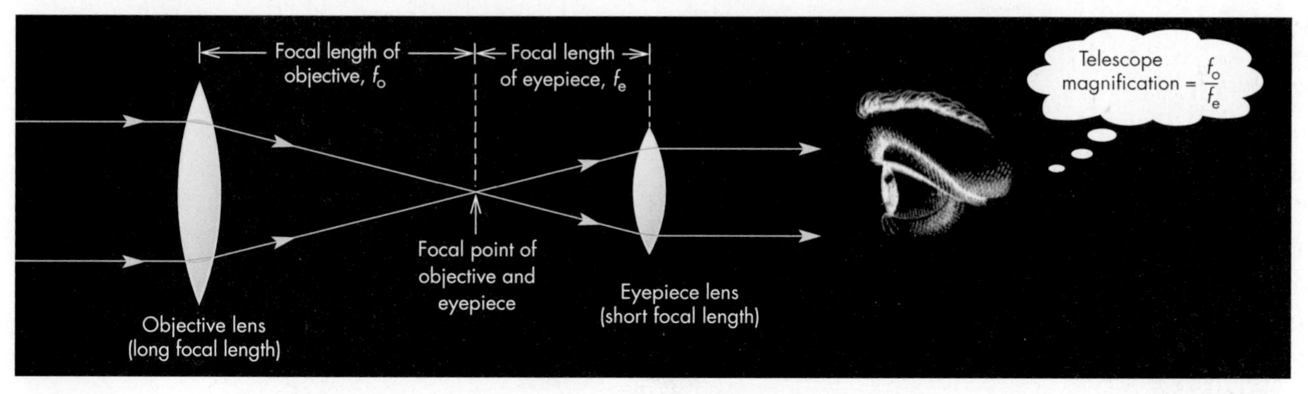

FIGURE 6.21
The Eyepiece of a Telescope
An eyepiece is needed to make visual observations with a telescope. The eyepiece is positioned so that its focal point is at the same spot as the focal point of the objective of the telescope. The ratio of the focal length of the objective to the focal length of the eyepiece determines the magnification of the telescope.

FIGURE 6.22
A CCD
CCDs may consist of millions of photosensitive detectors in an area only a few centimeters on a side.

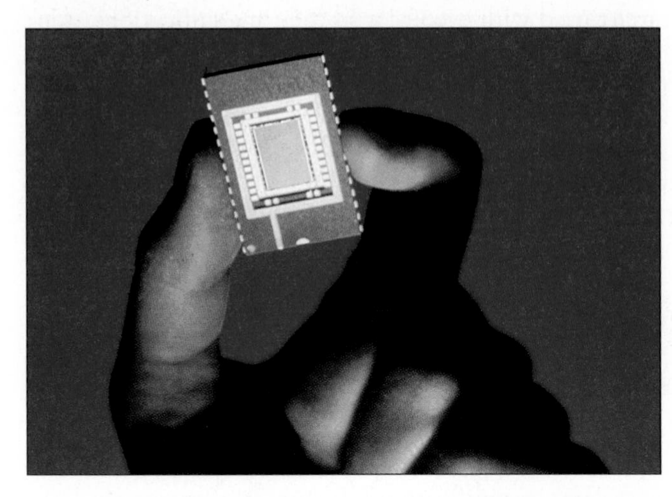

Unlike conventional photography, however, exposures for astronomical photography can be many hours long. Such long exposure times are needed to capture images of astronomical objects that may be one hundred million times too faint to be seen by the naked eye.

CCDs A significant drawback to photography is that it is not very sensitive. Only a small percentage of the photons striking a photographic plate is recorded. In the last 30 years, photography has been replaced by **charge coupled devices,** or **CCDs,** which can record images with much greater efficiency than photographic plates. CCDs have found applications in many fields other than astronomy. For example, video cameras use CCDs to capture images. CCDs have become so affordable and easily available that they are used by a rapidly growing number of amateur astronomers.

A CCD is a collection of photosensitive devices laid out in a miniaturized pattern that resembles a chessboard. Each piece of the pattern is known as a picture element, or **pixel.** CCDs used for astronomy are now as large as several thousand pixels by several thousand pixels. This means that millions of individual pixels are located in an area a few centimeters on a side (Figure 6.22). When an image is focused on a CCD, each pixel builds up an electric charge that depends on the brightness of the part of the object imaged in that pixel and the length of time the image is recorded. At the end of the exposure the amount of charge on each pixel is recorded and stored electronically so that the image can be displayed or analyzed later. The superiority of CCDs can be seen easily when images of the same region obtained with CCDs and photography are compared (Figure 6.23).

Spectroscopy One of the most powerful tools astronomers have is **spectroscopy,** through which they can study the details of the spectra of celestial bodies. In spectroscopy, light is dispersed according to wavelength. Its brightness is then measured over narrow regions of the spectrum. Using spectroscopy, astronomers can measure the shapes and strengths of the absorption and emission lines of atoms and molecules. By doing so, astronomers can learn about important properties of celestial bodies such as chemical makeup, motion, and magnetic fields.

The device that measures the spectra of celestial objects is called a **spectrograph.** An idealized spectrograph is shown in Figure 6.24. The slit of the spectrograph, located in the focal plane of the telescope, restricts the light entering the spectrograph to a small region of the sky. Without a slit, overlapping spectra from different objects would be

FIGURE 6.23
A Photographic Image, A, and a CCD Image, B, of the Same Star Field
The sensitivity of CCD arrays makes it possible to use them to obtain images of stars that are too faint to be photographed.

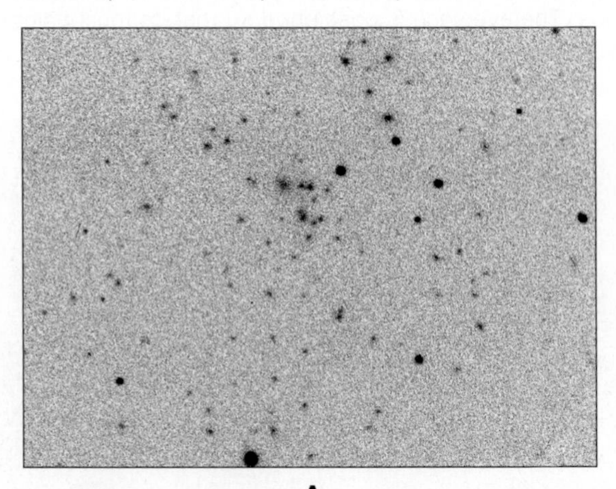

A

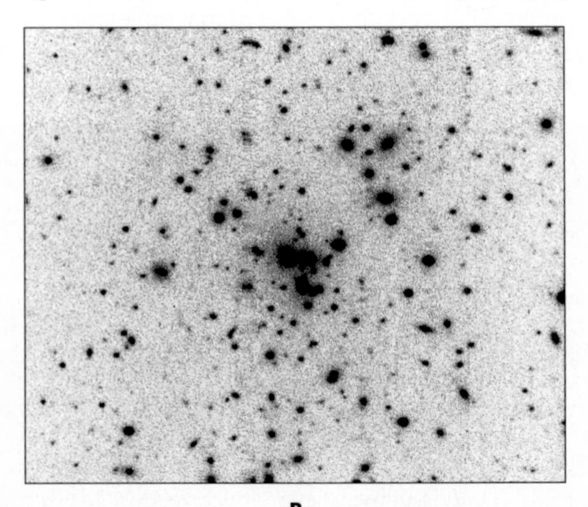

B

FIGURE 6.24
Schematic Drawing of a Spectrograph
A spectrograph is used to disperse the light from an object into a spectrum and focus the spectrum onto a detector. In this case the detector is a photographic plate.

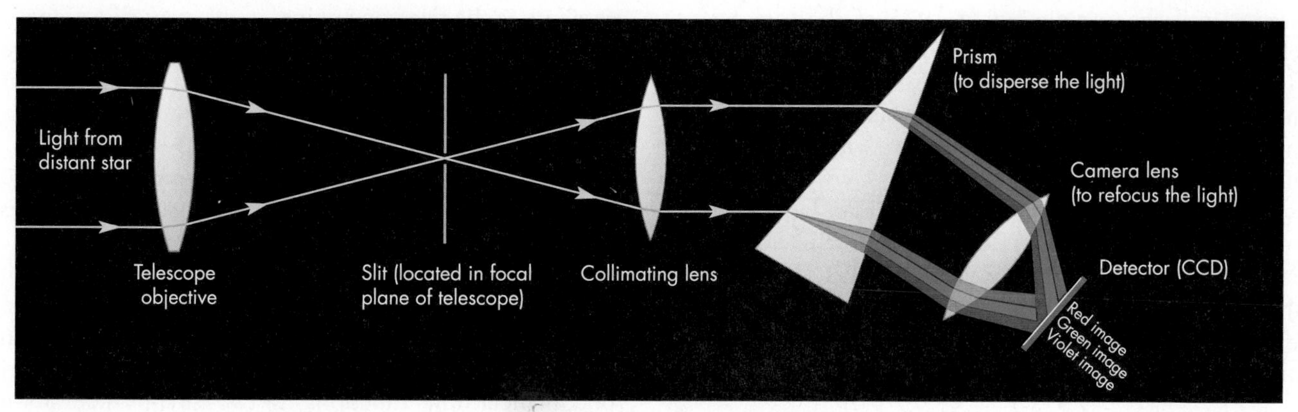

produced by the spectrograph, making the interpretation of the spectra difficult. After passing through a prism (or being dispersed by some other means), the light is refocused on a CCD. Different wavelengths are focused on different parts of the detector, producing a spectrum arranged from short to long wavelength.

Detectors are used to measure the properties of the radiation focused by a telescope. Photography and CCDs can measure the brightness of an image. Spectroscopy is used to measure the properties of light after it has been dispersed according to wavelength.

Notable Optical Telescopes

Refractors The world's largest refractor is the 40-inch (1-m) telescope of the Yerkes Observatory in Wisconsin, shown in Figure 6.25. The lens of this telescope is only slightly curved, which gives it a very long focal length. Consequently, the telescope is about 20 m (60 feet) long and needs a dome as big as those of much larger reflecting telescopes to protect it from the weather. Most large refractors, including the Yerkes telescope, were built before 1900 and are relics from a time when it was still difficult to make large reflecting telescopes.

Large Reflectors The era of the large reflecting telescope began when 1.5-meter (in 1908) and then 2.5-meter (in 1917) telescopes went into operation on Mount Wilson

FIGURE 6.25
The 40-inch Refracting Telescope of the Yerkes Observatory in Williams Bay, Wisconsin
The long focal length of the telescope is a consequence of the slight curvature of the lens.

in California. The Mount Wilson 2.5-m telescope was the world's largest optical telescope until 1948, when the 5-m-(200 inch) telescope on Mount Palomar, California, was completed. The Palomar telescope was the world's largest until the completion in 1975 of the 6-m Soviet telescope on Mt. Pastukhov. In 1991, the first of two 10-m Keck telescopes on Mauna Kea in Hawaii became the world's largest, although slightly larger telescopes in the Canary Islands and South Africa will soon become fully operational.

There are now many telescopes of 3 m or larger including a dozen larger than 8 m. Despite the growing number of reflecting telescopes with single large mirrors, the difficulty of producing very large mirrors and the problem of supporting them against their own weight have led to alternative approaches to large light-gathering power. One of these approaches is used in the twin 10-m Keck telescopes. The Keck telescopes have segmented mirrors, as shown in Figure 6.26. Each of the segments is a hexagon 1.8 m across but less than 0.1 m thick. To maintain the alignment of the segments as the telescope points in different directions, as the wind buffets the telescope, and as temperature changes cause the telescope to expand or contract, computer controlled sensors and actuators realign the segments twice each second.

Astronomers are also completing arrays of telescopes that, like the Keck telescopes, can act as single telescopes. One of these is the Very Large Telescope (VLT) in Chile. The VLT, shown in Figure 6.27, consists of four 8.2-m telescopes. When working together, the four telescopes of the VLT have the light-gathering power of a 16.4-m telescope. The Large Binocular Telescope (LBT) in Arizona, which now functions as two side-by-side 8.4-m telescopes, will soon use its telescopes together to achieve the resolution of a 23-m telescope and the light-gathering power of an 11.8-m telescope.

Even larger telescopes are now being planned. One of these is the Giant Segmented Mirror Telescope (GSMT), a 30-meter telescope based on the technology of the Keck telescopes. The GSMT may be operational as early as about 2016. A little farther in the future is the European Extremely Large Telescope (E-ELT), which is now being designed and is expected to be operational about 2018. E-ELT, shown in Figure 6.28 will be 42 m in diameter and will cost more than a billion dollars.

FIGURE 6.26
One of the Two Keck Telescopes on Mauna Kea
This telescope has a mirror made up of 36 hexagonal segments that are each 1.8 m across. The segments of the mirror of the Keck telescope have a combined collecting area equal to that of a single mirror 10 m in diameter.

FIGURE 6.27
The Very Large Telescope
The VLT has the collecting area of a single 16-m telescope and the resolution of a 200-m telescope.

FIGURE 6.28
The European Extremely Large Telescope (E-ELT)
This artists conception of the E-ELT shows a mosaic of 906 hexagonal mirror segments making up a mirror 42-m across. This is about 50% greater than the length of a college basketball court.

All of the world's largest telescopes, including the 10-m Keck telescopes, are reflectors. Many large telescopes are now being built and even larger telescopes are being planned.

6.5 OPTICAL OBSERVATORIES

The quality of observing conditions at a site depends on many factors such as the incidence of clear weather, the transparency and steadiness of the atmosphere, and the darkness of the sky at night.

The amount of clear weather varies dramatically from place to place around the world. Generally speaking, however, the clearest locations tend to be in dry or desert regions such as the southwestern United States, Chile, and Australia. Relatively cloud-free conditions can sometimes be found on mountains that rise above low-lying clouds, as on Mauna Kea in Hawaii.

The graph in Figure 6.29 shows that the Earth's atmosphere absorbs most or all of the incoming electromagnetic radiation except for the visible and radio parts of the spectrum. Note, for example, that X rays and gamma rays are completely absorbed before they reach the ground. Most infrared and ultraviolet radiation from space is absorbed, too. Even in the visible part of the spectrum, some of the light is blocked by the Earth's atmosphere. As a result, celestial objects appear fainter than they would if observed from space.

FIGURE 6.29
The Transmission of Earth's Atmosphere
The percentage of radiation that reaches the ground varies greatly with wavelength. Only in the visible, infrared, and radio wavelengths are there spectral regions in which the atmosphere is transparent.

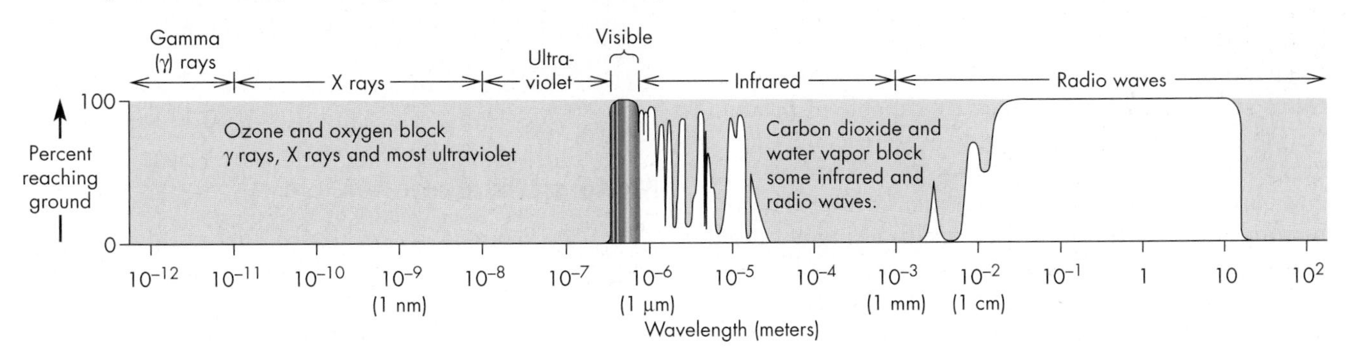

FIGURE 6.30
The United States at Night as Seen from Space
The image shows that much of the country is affected by light pollution. Notice how easy it is to trace major highways.

Dust suspended in the air can block light of all wavelengths. Air molecules themselves absorb some light, too, particularly in the violet end of the visible spectrum and in the infrared. Finally, water vapor and carbon dioxide absorb radiation, particularly in the infrared. The quest for atmospheric transparency is one of the main reasons that astronomers have built observatories on mountaintops, where much of the atmosphere lies below and the air is dry and dust free.

The light from large population centers increases the brightness of the night sky considerably. This is important because if the sky is bright, it is difficult to observe faint objects. A satellite view of the United States at night, shown in Figure 6.30, emphasizes the widespread nature of light pollution. Figure 6.30 shows clearly that dark sites within the continental United States are scarce. Large cities can brighten the sky for hundreds of kilometers around them because air molecules and dust particles scatter city light in all directions. The problem is becoming worse as increasing numbers of poorly shielded lights are put up in both urban and rural areas. It is ironic that the light that escapes upward and brightens the sky is wasted. Properly shielded and directed lighting both saves electricity and reduces light pollution.

Astronomers seldom, if ever, are able to obtain images as sharp as their telescopes are capable of producing. The culprit is called **seeing,** which is the blurring of an image caused by the unsteadiness of the atmosphere. The blurring caused by seeing is much greater than the blurring caused by the telescope itself. Seeing is caused by turbulence in which pockets of air rise and fall as well as move horizontally past the telescope. The pockets of air have different temperatures, densities, and indices of refraction. Light from a celestial object is bent each time it passes from one pocket to another. The pattern of air pockets changes rapidly, so the image changes and shimmers rapidly as well. To the naked eye, seeing is evident in the twinkling of stars, especially those near the horizon.

The effects of seeing are least noticeable when the air isn't very turbulent. This is often the case on the tops of isolated mountains, such as Mauna Kea, Hawaii, and La Palma in the Canary Islands. At a typical location near sea level, seeing blurs the images of stars so that they appear about 2 seconds of arc in diameter or larger. Good seeing produces stellar images about 1 second of arc in diameter at Kitt Peak and 0.5 seconds of arc in diameter at Mauna Kea.

Astronomers have developed techniques that can correct for the blurring effects of seeing. One of these is called **adaptive optics.** The key to adaptive optics is a flexible mirror that has actuators attached to its back. The actuators each move a small segment of the mirror up or down as often as 1000 times a second to correct for the distortion caused by atmospheric turbulence. In order for the adaptive optics system to calculate how to reshape the mirror, there must be a pointlike object, or *guidestar,* within the field of view of the telescope. The adaptive optics try to keep the image of the guidestar as small as possible. Adaptive optics systems have become steadily more sophisticated, allowing the use of fainter and fainter guidestars. As Figure 6.31 shows, adaptive optics can sharpen images dramatically. The technique permits telescopes to acquire images that are almost as sharp as the diameter of the telescope allows.

Whenever possible, observatories are located on mountains, where skies are free of clouds, the surroundings are dark, and the air is transparent. The amount of atmospheric turbulence at a site determines the seeing, the blurring of an image caused by light passing through moving pockets of air. Astronomers are using adaptive optics to correct for seeing and to produce sharp images.

Modern Observatories

There are many observatories in the world with large telescopes and up-to-date instrumentation. One is Kitt Peak National Observatory, shown in Figure 6.32. The observatory has a number of telescopes ranging in size up to 4 m. Kitt Peak is a public observatory, funded by

FIGURE 6.31
Image Sharpening Using Adaptive Optics
A, This image of a double star was obtained with the 3.6-m Canada-France-Hawaii telescope on Mauna Kea without the use of adaptive optics. Atmospheric seeing results in a blur about 0.7 seconds of arc across. There is no indication that there are really two stars rather than one. The colors in the image represent different levels of brightness. White is brightest followed by red, orange, yellow, green, and blue. **B,** This image of the same double star was obtained using the same telescope, but with the use of adaptive optics. Now two stars, separated by about 0.25 seconds of arc, clearly can be seen.

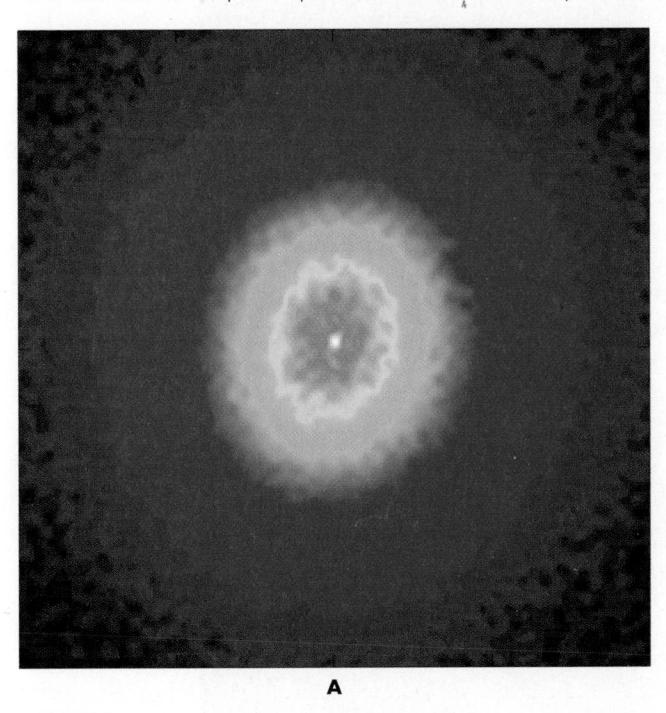
A

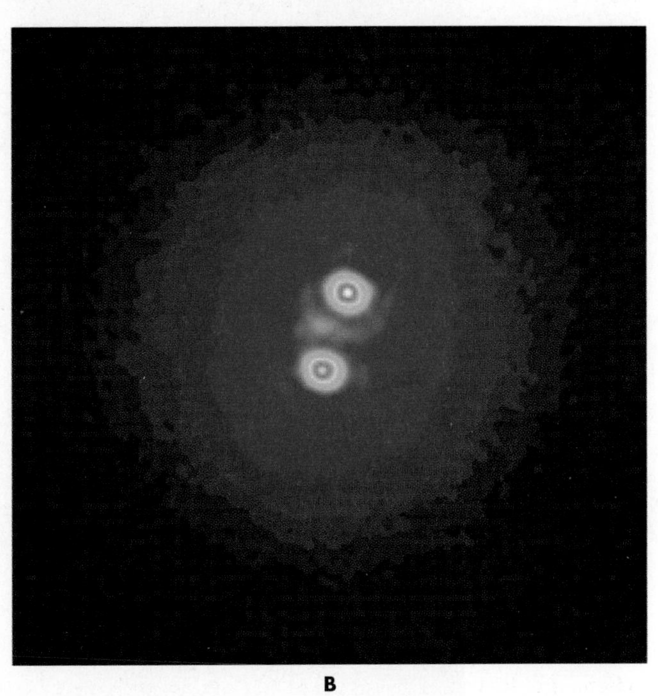
B

FIGURE 6.32
The Kitt Peak National Observatory
Located in Arizona, Kitt Peak has telescopes with mirrors as large as 4 m in diameter. Unlike private or university-owned observatories, any qualified astronomer can apply for observing time at Kitt Peak.

the U.S. government. Any qualified astronomer from the United States or anywhere else in the world can request telescope time for an observing project. The competition is intense, however, so only the very best projects are given telescope time.

Because the major observatories of Europe and North America are in the northern hemisphere, there is a large portion of the southern sky that is always below their horizons. As a result, many important astronomical objects, such as two nearby galaxies (the Large and Small Magellanic Clouds) and the center of our own galaxy, are difficult or impossible to observe from Mount Palomar, Kitt Peak, or the observatories of Europe. To gain a view of the southern skies, North American and European astronomers have built several major observatories above the 2000-m level in the Andes of northern Chile, where the observing conditions are excellent. Both the European Southern Observatory and the Cerro Tololo Inter-American Observatory (Figure 6.33) are equipped with telescopes in the 4-m class. Even larger telescopes, including the VLT (Figure 6.27) are located at other sites in Chile. Other major observatories are located in the southern hemisphere in Australia and in South Africa.

Figure 6.34 shows the summit of Mauna Kea, a dormant volcano on the island of Hawaii. Mauna Kea is generally considered the world's premier observing site. At 4200 m,

FIGURE 6.33
Cerro Tololo Inter-American Observatory
CTIO is located in a desert environment in the Andes Mountains near La Serena, Chile. This nighttime image shows the dome of the 4-m Blanco telescope. Behind the Blanco telescope are the Milky Way (right) and the Large and Small Magellanic Clouds (left).

FIGURE 6.34
Mauna Kea Observatory, Hawaii
The observatory, located at an elevation of 4200 m (14,000 feet) on Mauna Kea, a dormant volcano, is considered to be the world's premier observing site because of its great atmospheric transparency and darkness and because of excellent seeing. Haleakala, 130 km away on the island of Maui, is visible in the distance.

Mauna Kea is nearly twice as high as any other major observatory. The observatory is above almost half of the atmosphere and is extremely dry. Moreover, clouds often form below the summit at night and block the lights of Hilo and other communities on the island, making the summit one of the darkest places on Earth. Seeing is often very good, and cloud-free conditions can sometimes last for months.

There are, however, some problems with observing on Mauna Kea. First of all, bad weather on Mauna Kea can be very severe. Storms with winds of 120 mph occasionally ravage the summit. Also, astronomers do not always function well at the altitude of the summit. Some astronomers can't work at all on Mauna Kea and many others work at reduced efficiency because of the lack of oxygen. Observing must be unusually carefully planned in advance because altitude-induced mistakes are common. Conditions at the summit are severe enough that visiting astronomers usually spend only 12 hours at the observatories before descending to living and sleeping quarters at the 3000-m level.

Numerous telescopes including several radio telescopes and four optical telescopes larger than 8 m are now located on Mauna Kea.

6.6 SPACE OBSERVATORIES

By building observatories at places like Mauna Kea, astronomers have been able to significantly reduce the problems of light pollution, cloudiness, and seeing. To completely eliminate these problems, however, astronomers have put optical telescopes in orbit about the Earth. An even more important reason to put telescopes in space, however, is that the Earth's atmosphere is opaque to most of the electromagnetic spectrum. Thus astronomers have used balloons, rockets, and orbiting space observatories to carry telescopes above the atmosphere so that they can observe in the gamma-ray, X-ray, ultraviolet, and infrared parts of the spectrum.

Optical Observations

The *Hubble Space Telescope* (*HST*), shown in Figure 6.35, was launched in 1990 from the space shuttle *Discovery.* The primary purpose of the *HST* was to obtain extremely sharp images of a wide variety of objects such as planets, clusters of stars, and galaxies. From its vantage point in space, the *HST* is able to see fainter, more distant objects, too, because the extremely faint light from distant objects isn't blurred by seeing as it is when observed from the ground. The images produced by the *HST* have breathtaking clarity, as shown by the image of Mars in Figure 6.36.

Although the *HST* is still performing satisfactorily, work is under way on its replacement, the *James Webb Space Telescope* (*JWST*). The *JWST,* which is scheduled for launch in about 2013, will have a 6.5-meter segmented mirror. It will carry out visible and infrared observations from a semi-stable point 1.5 million km from the Earth in the direction opposite the Sun.

FIGURE 6.36
A *Hubble Space Telescope* Image of Mars
The image shows fine details of Mars at the beginning of summer in the northern hemisphere.

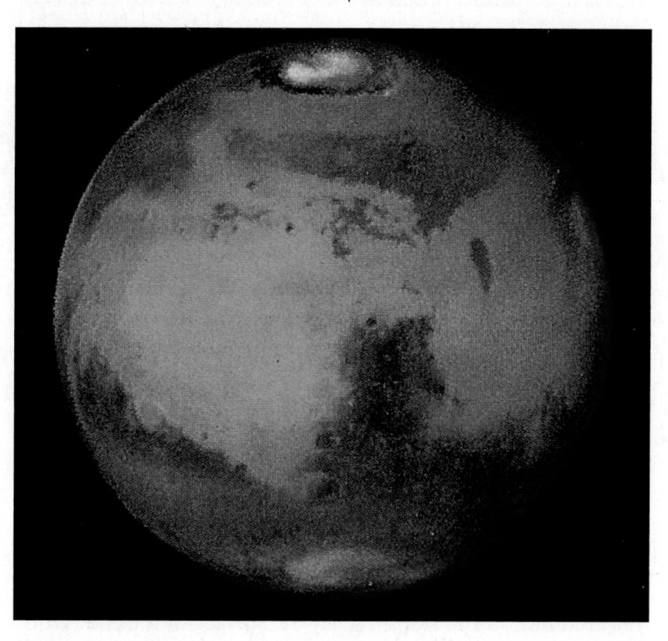

FIGURE 6.35
The *Hubble Space Telescope* (*HST*)
Launched into Earth orbit from the space shuttle *Discovery* in 1990, the *HST* has obtained high-resolution images and spectra of distant and faint objects.

Infrared and Ultraviolet Observations

Infrared and ultraviolet telescopes resemble optical telescopes except that their detectors are sensitive at longer and shorter wavelengths, respectively. Spectacular infrared images have been obtained with the *Spitzer Space Telescope* (*SST*), which was launched in 2003. A false color image of the spiral galaxy, M81, is shown in Figure 6.37. Blue indicates relatively warm parts of M81, whereas green and red indicate successively cooler parts. In reality, the human eye can't see any of the wavelengths recorded in the *SST* image of M81. The *SST* gives astronomers a new means to observe the cooler parts of the universe such as very cool stars and interstellar dust.

The *International Ultraviolet Explorer* (*IUE*) was launched in 1978 and turned off in 1996. It carried a 0.45-m reflecting telescope designed to measure ultraviolet spectra. *IUE* had a very small field of view and was incapable of surveying the entire sky. Instead, it took the ultraviolet spectra of carefully selected objects one at a time. Although there is currently no major dedicated ultraviolet telescope in space, instruments on the *HST* can carry out observation in the ultraviolet.

X-Ray and Gamma-Ray Observations

Normal imaging optics can't be used for X rays because they penetrate deeply into the reflecting surfaces of a normal telescope. In fact, if the mirror or lens is thin enough, the X rays pass right through it. However, X rays that strike a surface at *grazing incidence* (that is, almost parallel to the surface) are reflected like stones skipping on the surface of a pond. Consequently, it is possible to focus X rays using nested cylinders as shown in Figure 6.38. Grazing-incidence mirrors have been used by most space X-ray telescopes including *Einstein,* which mapped the sky in the 1970s, and *Chandra,* launched in 1999. Figure 6.39 shows a *Chandra* image of the X-ray ring around the site of a supernova that was discovered in 1987.

Gamma-ray telescopes make use of the way that gamma rays interact with matter by producing a pair of particles—an electron and a positively charged particle called a positron, which has the same mass as an electron. The paths that the electron and positron follow indicate the direction of the gamma ray that produced them. The *Compton Gamma Ray Observatory,* which orbited Earth from 1991 to 2000, carried detectors much more sensitive than those of previous satellite gamma-ray observatories. *Fermi,* a gamma-ray telescope launched by NASA in 2008 is far more powerful than any previous gamma-ray observatories. It should be able to locate thousands of sources of cosmic gamma rays during its 5-year mission. *Fermi* sees 20% of the sky at any given time and maps the entire sky every 3 hours. One of the components of *Fermi* is the Burst Monitor, which is designed to detect and measure gamma ray bursts, powerful blasts of gamma rays that reach Earth about once per day. Gamma ray bursts and theis significance are described more fully in Chapter 20.

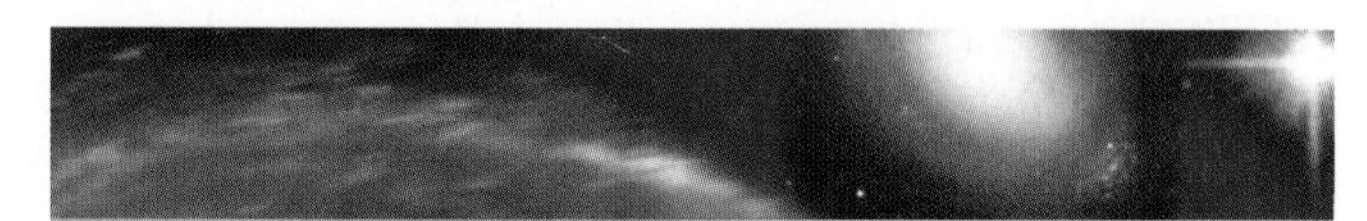

Space observatories are used to overcome atmospheric effects that limit optical telescopes that operate on the ground. Space telescopes also observe gamma rays, X rays, ultraviolet radiation, and infrared radiation, which cannot penetrate the Earth's atmosphere.

FIGURE 6.37
***Spitzer Space Telescope* Image of the Spiral Galaxy M81**
The image is a combination of images taken at three infrared wavelengths. Blue represents relatively short wavelengths at which most of the infrared radiation is due to stars. Green and red represent longer wavelengths at which most of the infrared radiation is due to interstellar dust.

FIGURE 6.38
Grazing Incidence Optics in an X-Ray Telescope
X rays pass through normal mirrors, so X-ray telescopes bring X rays to a focus by letting them graze the surfaces of nested cylindrical mirrors at shallow angles.

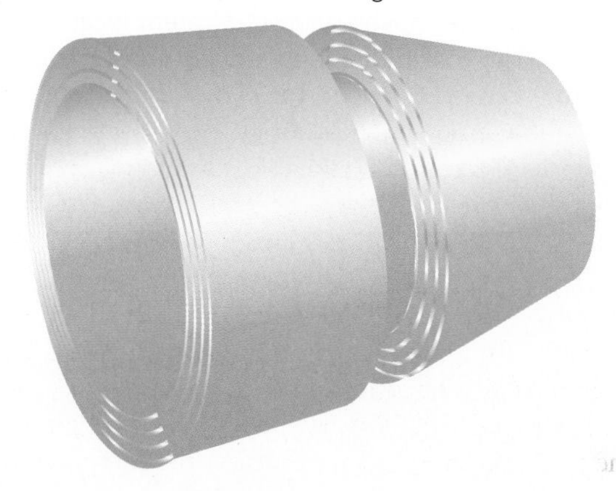

A The cylindrical mirrors of an X-ray telescope

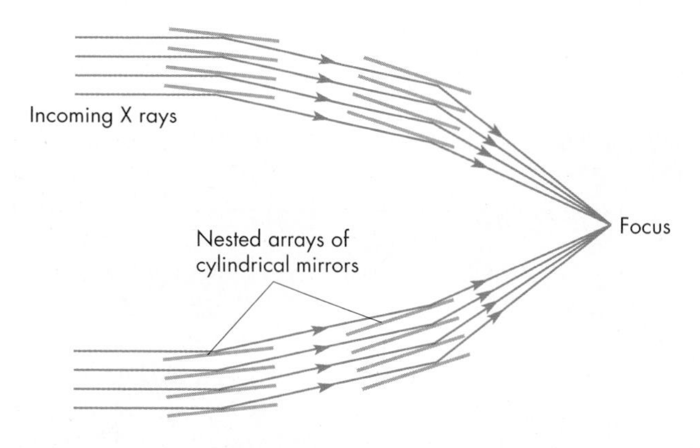

B Cross-sectional view of the grazing-incidence mirrors showing how they focus X rays

FIGURE 6.39
A *Chandra* Image of the X-Ray Ring Around SN1987A
This false-color image shows a ring of hot gas expanding outward from the site of a supernova observed in 1987. The white region indicates the brightest X-ray emission, followed by red, yellow, green, and blue. The white contour lines show the innermost ring of visible light seen in *HST* images.

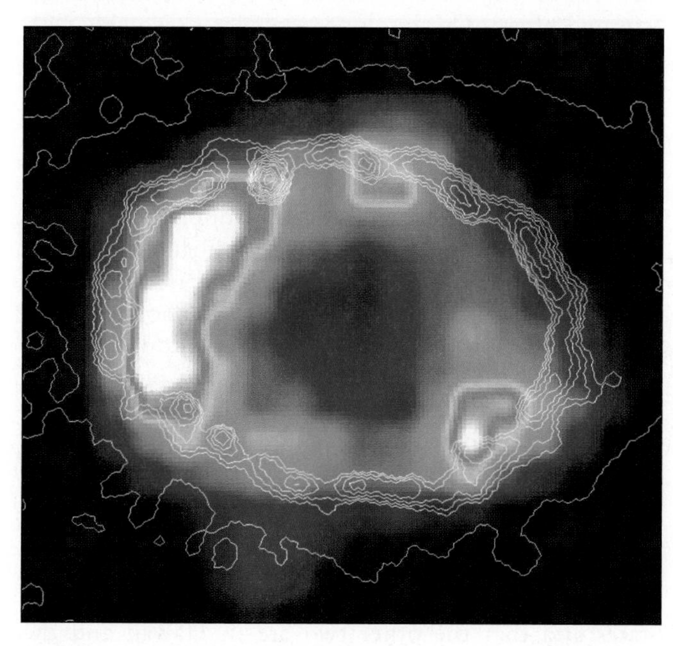

6.7 RADIO TELESCOPES

Most radio telescopes are reflectors. That is, they use metallic conducting surfaces as mirrors to reflect radio waves to a focus. To reflect radio waves with high efficiency, the surface of a radio telescope must be smooth enough that its wiggles and bumps are smaller than about 1⁄10 of the wavelength at which the telescope is designed to operate. Thus, radio telescopes operating at wavelengths longer than a meter can have surfaces with fist-sized holes or bumps in them and still be highly reflecting. However, for telescopes designed for centimeter or millimeter wavelength observations, the surface irregularities must be much smaller.

The radio waves that are brought to a focus can be measured using an antenna, which operates somewhat like an automobile radio antenna. The radio waves, which are electromagnetic field fluctuations, cause a weak electrical signal in the antenna. The signal has the same frequency as the radio waves. This signal, which can be amplified and measured, is proportional to the amount of radio radiation captured and focused by the telescope. If the signal were used to drive an audio speaker, it would sound like static.

Radio Interferometry

Because they are used at long wavelengths, even the largest radio telescopes have very poor angular resolution. The largest radio dish, at Arecibo, Puerto Rico, is 305 m across and has a circumference of about 1 km. The Arecibo telescope, shown in Figure 6.40, has great light-gathering power. Like other radio dishes, it is in constant demand for many observing programs that don't require high angular resolution. However, when operating at a wavelength of 10 cm, the resolution of the Arecibo telescope is a few minutes of arc, about the same as the human eye. If the Moon had the same pattern of bright and dark regions at radio wavelengths that it does in the visible wavelengths,

FIGURE 6.40
The 305-m Telescope of the Arecibo Observatory in Puerto Rico
The Arecibo telescope is the largest radio dish in the world. Unlike smaller radio telescopes, the Arecibo telescope can't be tilted to point in different directions. It can only observe radio sources that pass within 20° of the zenith.

FIGURE 6.41
The Very Large Array (VLA), Located Near Socorro, New Mexico
The 27 individual radio dishes of the VLA are arranged in a Y shape and can be moved among different locations on railroad tracks. The maximum separation of the VLA antennas is 40 km.

the Arecibo telescope could map out the man in the Moon, but it could not provide finer details. In order to resolve the details of galaxies as well as a small optical telescope does, it would be necessary to build a radio dish the size of Rhode Island, which is financially and technologically impossible. Instead, radio astronomers have devised other means to achieve high angular resolution.

The technique used to obtain high angular resolution is **interferometry,** in which signals from widely separated radio dishes are combined to produce the resolution of a single telescope as large as the distance between the two dishes. Sometimes the signals from the two dishes interfere destructively, or cancel each other out, when they are brought together. At other times, the two signals interfere constructively, or add together to produce a stronger signal. As time passes and the source of the radio waves moves across the sky, the signals from the two dishes alternatively interfere constructively and destructively to produce a series of interference *fringes*. The pattern of the fringes depends on the size and shape of the source of radio waves as well as the distance between the two dishes. The pattern also depends on whether the dishes lie north and south of one another, east and west of one another, or some orientation in between. To sample the pattern of fringes from a radio source at many telescope separations and orientations, modern interferometers often consist of arrays of telescopes that work in pairs. When the information from all of the pairs of antennas is combined, the observer can calculate maps (images) of the radio object.

The VLA There are many interferometric arrays in which the signals from pairs of telescopes are combined using electrical wires. Practical considerations limit the size of such arrays to tens of kilometers. The workhorse of contemporary radio astronomy is the Very Large Array (VLA) in western New Mexico. The VLA, shown in Figure 6.41, consists of a Y-shaped configuration of 27 dishes that can be moved on railroad tracks to span a region as large as 40 km across. Astronomers from around the world use the VLA to map radio objects with a resolution of about 1 second of arc. This resolution compares favorably with that of ground-based optical telescopes.

The VLBA Even larger interferometers can be created by recording the radio signals at several telescopes along with the time of observation from atomic clocks. The signals can be synchronized later and combined at a processing center. This technique is called Very Long Baseline Interferometry, or VLBI. The more telescopes that participate in a VLBI observation, the more information that is obtained about the radio object and the better the resulting map. Arrays of radio telescopes have cooperated on an informal basis for VLBI observations for decades. More recently, however, arrays dedicated to VLBI have been built. One of these is a 10-telescope array called the Very Long Baseline Array (VLBA). The map in Figure 6.42 shows that 8 of the 10 VLBA telescopes are in the continental United States and that the other two are in Hawaii and the Virgin Islands. The VLBA has a resolution of a few milliseconds of arc (a millisecond is 1⁄1000 of a second). This resolution is comparable to that of a radio dish 6000 km (about 4000 miles) in diameter and is much better than the resolution that existing optical telescopes can achieve. In 1997, the *Highly Advanced Laboratory* *for Communications and Astronomy* (*HALCA*), an 8-m diameter radio telescope in space, was combined with the VLBA to form an array. The orbit of *HALCA* carries it as far as 21,000 km (13,000 miles) from the Earth, so the VLBA plus *HALCA* has even better angular resolution than the VLBA alone. Figure 6.43 shows images of the quasar Q1156+295 obtained using the VLBA alone and using the VLBA plus *HALCA*.

Other large interferometric arrays are under construction or being planned. The Atacama Large Millimeter Array (ALMA) is now being built at an altitude of 5 km (16,000 feet) in the desert foothills of the Andes Mountains in northern Chile. ALMA will have 64 individual dishes when it is completed in about 2011. It will be capable of obtaining high resolution images in the far infrared and at millimeter wavelengths. The Square Kilometer Array (SKA), which is still in the planning stage, will have thousands of individual dishes that have a combined collecting area of a square kilometer. This will give the SKA one hundred times the light-gathering power of the VLA and more than ten times the light-gathering power of the Arecibo telescope. In terms of collecting area, the SKA will be the world's largest radio telescope. The dishes of the SKA, which will be located in Australia or South Africa, will be spread over an area 3000 km across in order to achieve high angular resolution. The SKA is scheduled for completion around 2020.

The importance of interferometry for radio astronomical imaging can be illustrated by imagining that you wanted to observe, say, Iowa with a radio telescope on the Moon. Imagine also that Iowa and the objects within it are strong emitters at radio wavelengths. Figure 6.44 shows the smallest regions resolvable using the Arecibo 305-m telescope, the VLA, and the VLBA. With the Arecibo telescope, you could distinguish eastern Iowa from western Iowa. With the VLA, you could resolve the central campus of the University of Iowa. Using the VLBA, you could resolve objects as small as desks or astronomers.

FIGURE 6.42
The Very Long Baseline Array (VLBA)
The VLBA is an interferometric array consisting of 10 radio telescopes at widely separated locations in the continental United States, the Virgin Islands, and Hawaii. Working together, the VLBA antennas achieve an angular resolution of about 1/1000 of a second of arc, about 100,000 times better than the Arecibo telescope.

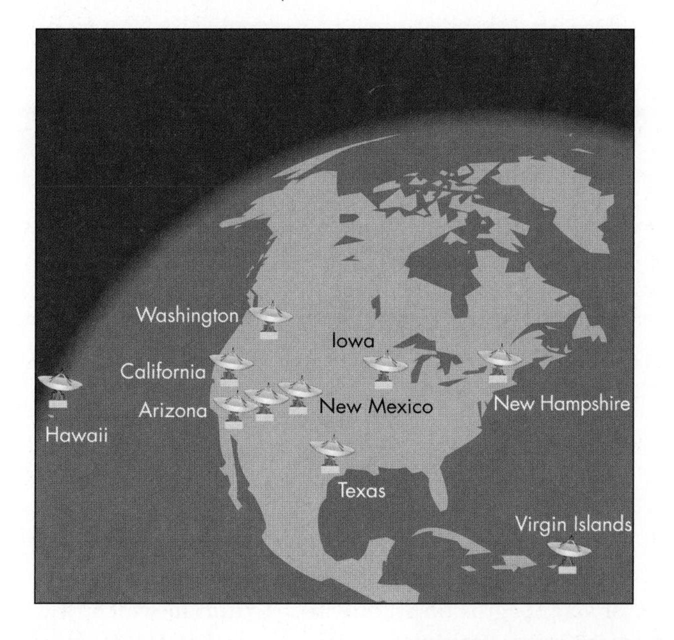

To obtain high resolution, radio astronomers use interferometers to measure the interference patterns of sources of radio radiation. The largest interferometric arrays can produce angular resolutions that surpass those available with optical telescopes.

FIGURE 6.43
High-Resolution Radio Images of the Quasar Q1156+295
The addition of HALCA to the VLBA increases the size of the array of radio telescopes and produces an image, **B,** with even better angular resolution than **A,** the image obtained with the VLBA alone. For both **A** and **B,** red indicates the brightest part of the image while surrounding colors indicate successively fainter parts of the images.

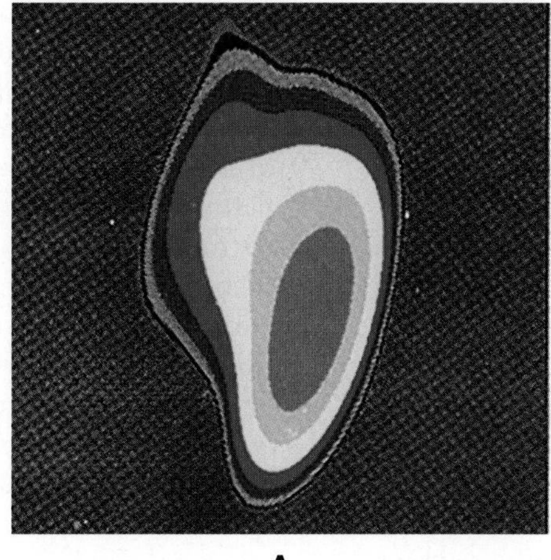

A

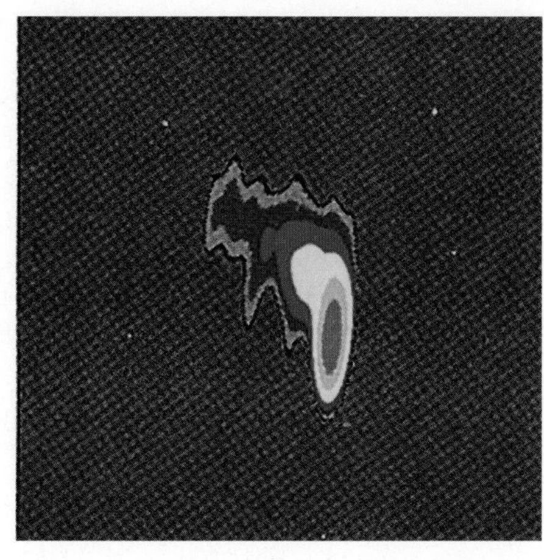

B

FIGURE 6.44
The Resolution of the Arecibo Telescope, the VLA, and the VLBA
The smallest regions that could be distinguished from the Moon using the Arecibo telescope would be the size of eastern Iowa, **A,** whereas the VLA would be able to distinguish regions the size of the University of Iowa campus, **B.** The VLBA would be able to distinguish regions as small as people or desks, **C.**

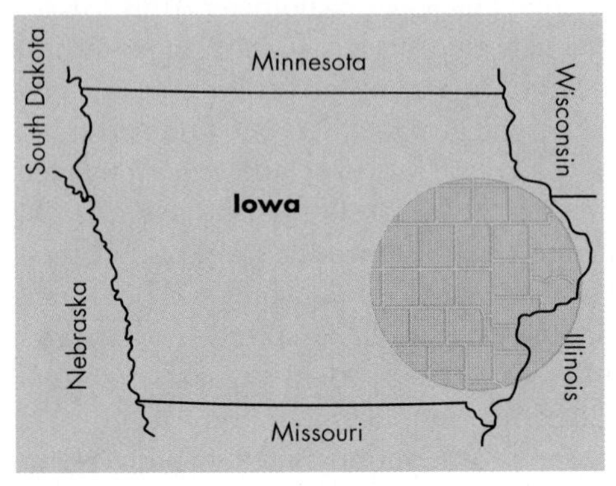

A Resolution of the Arecibo radio telescope

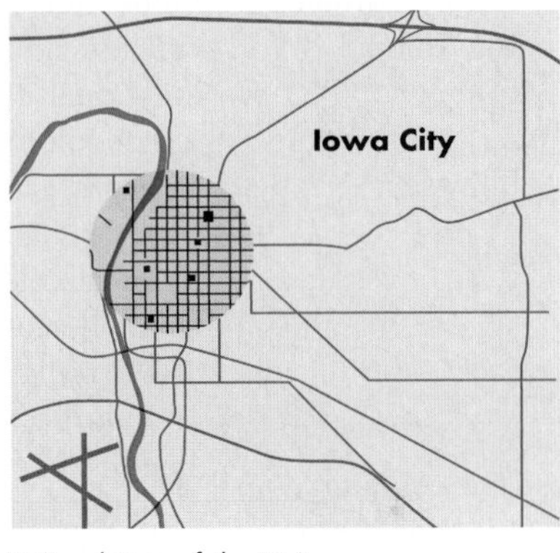

B Resolution of the VLA

C Resolution of the VLBA

Chapter Summary

- Waves are disturbances that move through space. Wavelength (the distance between crests of the wave), frequency (the rate at which waves pass), and energy describe a wave. (Section 6.1)
- Light and other forms of electromagnetic radiation are oscillating electric and magnetic fields produced by accelerating electric charges. The different forms of electromagnetic radiation all travel at the speed of light, but they differ greatly in wavelength and frequency. (6.2)
- The Doppler effect is a change in the wavelength and frequency of a wave as a consequence of motion of the source of the wave and the observer toward or away from each other. The amount of change in wavelength is proportional to the speed of the source relative to the observer. (6.2)
- Some properties of light are best explained if light is a stream of massless particles called photons. The frequency of a photon determines its energy. (6.2)
- An electromagnetic wave can be reflected from a surface or refracted when it passes from one material to another. Dispersion occurs when the amount of refraction depends on wavelength. (6.3)
- Refracting telescopes collect light using lenses. Reflecting telescopes use mirrors to bring light to a focus. Large refracting lenses are expensive to make and sag under their own weight. (6.4)
- The brightness of the image formed by a telescope depends on its focal ratio, the ratio of the focal length to the diameter of its main lens or mirror. The area of the main lens or mirror of a telescope determines its light-gathering power. (6.4)
- The resolution of a telescope, the finest detail that it can distinguish, improves as its diameter increases. Resolution worsens, however, with increasing wavelength. (6.4)
- After light has been focused by a telescope, its properties can be measured using detectors, which can determine the brightness of an image over a large range of wavelengths or produce a spectrum in which the light can be measured at many wavelengths. (6.4)
- All large modern telescopes, including the 10-m Keck telescopes, are reflectors. Other large reflectors are now being built or planned. Many existing and planned large telescopes use segmented mirrors or thin mirrors to reduce costs and improve optical performance. (6.4)
- Astronomers build observatories on mountaintops to benefit from clear skies, atmospheric transparency, good seeing, and the absence of light pollution. The best sites are those at which the blurring of images resulting from atmospheric turbulence is at a minimum. Astronomers are using adaptive optics to overcome the blurring effects of atmospheric turbulence. (6.5)

- The Earth's atmosphere blocks incoming X-ray, gamma-ray, ultraviolet, and most infrared radiation. Consequently, space observatories have been put in orbit to observe in these parts of the electromagnetic spectrum and to obtain optical images that are free of atmospheric distortion. (6.6)
- Because individual radio telescopes have poor resolution, radio astronomers use interferometers to measure the interference patterns of radio-emitting objects. Long baseline interferometers can produce much better resolution than optical telescopes. (6.7)

Key Terms

adaptive optics *118*
charge coupled device (CCD) *114*
detector *113*
diffraction *112*
dispersion *109*
Doppler effect *105*
electromagnetic spectrum *104*
electromagnetic wave *102*
energy flux *102*
focal length *110*
focal plane *111*
focal point *110*
frequency *102*
gamma ray *104*
index of refraction *108*
infrared *107*
interferometry *124*
light *102*
light-gathering power *111*
objective *110*
photon *106*
pixel *114*
reflection *108*
reflectivity *108*
reflector *110*
refraction *108*
refractor *110*
resolution *112*
seeing *118*
spectrograph *114*
spectroscopy *114*
ultraviolet *107*
wave *102*
wavelength *102*
X ray *107*

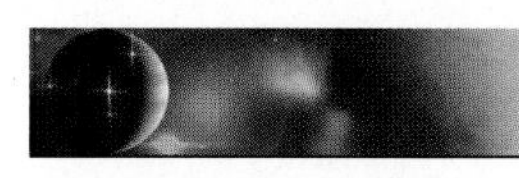

Conceptual Questions

1. Describe a type of experiment that shows that light behaves like a stream of particles and one that shows that light behaves like waves.
2. Name a substance that has a reflectivity of nearly 100% and another that has a reflectivity near 0%.
3. Draw a diagram showing how a mirror forms an image.
4. What are two advantages of using CCDs as opposed to photography?
5. List the qualities that astronomers look for when evaluating a possible observing site.
6. What are two reasons why astronomers have used space observatories?

Problems

1. Two waves have the same speed. Wave A has four times the frequency of wave B. How does the wavelength of wave A compare with that of wave B?
2. A wave has a frequency of 4 Hz and a wavelength of 6 m. What is the speed of the wave?
3. What is the wavelength of a wave that has a speed of 30 m/s and a frequency of 0.5 Hz?
4. A wave has a wavelength of 30 m. What would be the wavelength of the wave if its frequency were doubled but its speed remained constant?
5. What is the frequency of electromagnetic radiation that has a wavelength of 0.4 m? In what part of the spectrum does this radiation occur?
6. What is the wavelength of electromagnetic radiation that has a frequency of 3×10^{16} Hz? In what part of the spectrum does this radiation occur?
7. Saturn is ten times as far from the Sun as is the Earth. How does the flux of solar energy at Saturn compare with that at the Earth?
8. How close to the Sun would a spacecraft have to go to reach the distance at which the flux of solar energy is 20 times as large as the solar energy flux at the Earth?
9. Radio waves from a star are observed at a wavelength of 20.02 cm. If the star and the Earth were motionless with respect to each other, the radio waves would have a wavelength of 20 cm. Are the star and the Earth moving toward or away from each other. How fast are they moving toward or away from each other?
10. Suppose atoms at rest emit visible light with a wavelength of 500 nm. At what wavelength would the light from the atoms be observed if the atoms were moving toward the Earth at a speed of 20,000 km/s?

11. Photon A has twice the frequency of photon B. How do the energies of the two photons compare?
12. The wavelength of an electromagnetic wave is 3 m, and its frequency is 10^8 Hz. What is the index of refraction of the material through which the electromagnetic wave is passing?
13. How does the light-gathering power of a 300-m diameter radio telescope compare to that of a 50-m diameter radio telescope?
14. How much larger than a 1-m telescope would another telescope have to be in order to have 1000 times the light-gathering power?
15. Two mirrors have the same diameter. One of them has a focal ratio of f/5 and the other f/10. How do the focal lengths of the two mirrors compare?
16. A 4-m optical telescope operates at a wavelength of 5×10^{-7} m. How large would an infrared telescope operating at a wavelength of 10^{-4} m have to be to have the same resolution as the optical telescope?
17. What is the magnification of a telescope that has an objective with twice the focal length of its eyepiece?
18. About how large would an optical telescope need to be to achieve the angular resolution of the VLBA?

Figure-Based Questions

1. Use Figure 6.5 to find the frequency of a gamma ray with a wavelength of 10^{-14} m.
2. Use Figure 6.5 to find the wavelength of a radio wave with a frequency of 10^6 Hz.
3. Use Figure 6.29 to estimate the percentage of radio radiation at a wavelength of 1 cm that penetrates the Earth's atmosphere to reach the ground.

Group Activities

1. Buy or borrow five identical candles. Tie four of the candles together in a bundle. At night, find a relatively dark location. Light all five candles and wait for the candles to start burning evenly. Have one member of your group take the single candle and walk ten paces away. Have another group member take the bundle of four candles and walk away in the same direction until the rest of the group judges that the four candles together appear as bright to them as the single, closer candle. Measure how many paces the four candles are from the group. Find the ratio of the distances of the four candles and the single candle and compare the result to what you would expect on the basis of Equation 6.1.
2. Go to an asphalt parking lot early in the morning. Have one-half of the group stand on one side of the parking lot and the other half on the other side. Have the two groups observe how steady the other group appears to be when viewed across the parking lot. Repeat the experiment after the Sun has been shining on the dark parking lot for much of the day. Get the entire group together to discuss why the shimmering due to the parking lot is different at the two times of day.

For More Information

Visit the text website at **www.mhhe.com/fix** for chapter quizzes, interactive learning exercises, and other study tools.

7

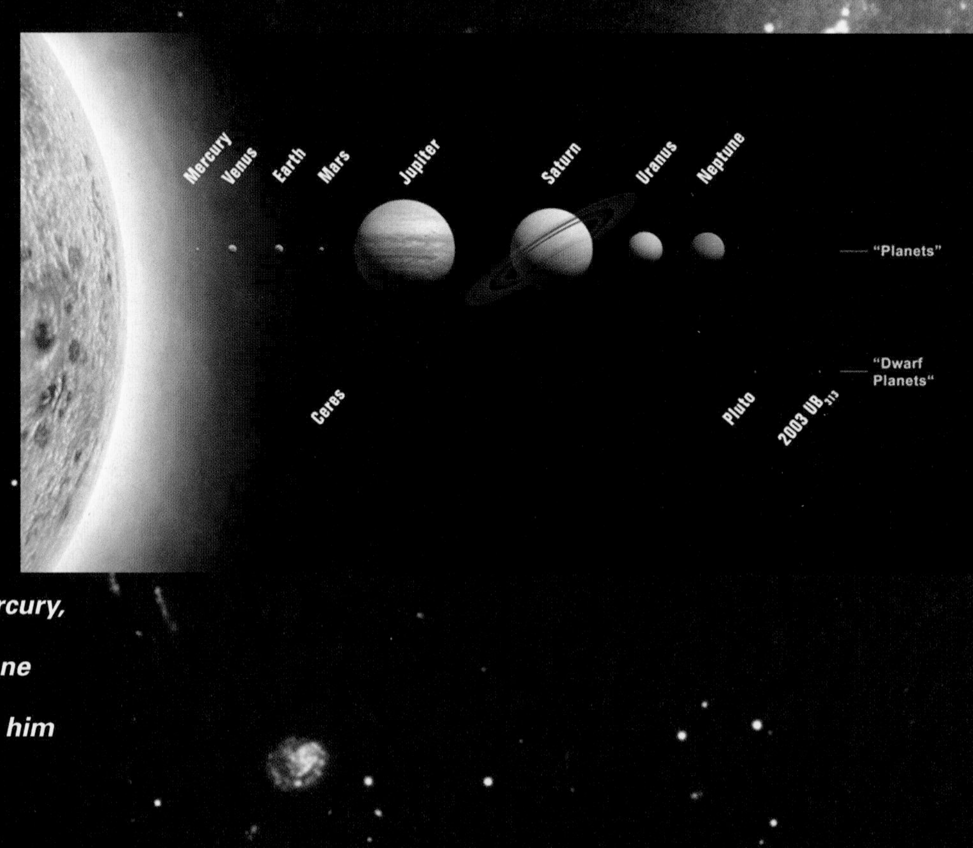

Twilight comes and

all of the

Visible planets come out.

Venus first, and then Jupiter,

Mars and Saturn

and finally

Mercury once more.

Seals bark

On the rocks. I tell Mary

How Kepler never saw Mercury,

How as he lay dying it shone

In his window, too late for him

To see.

Kenneth Rexroth
Protoplasm of Light

Overview of the Solar System

The solar system is our astronomical backyard. Because of their proximity, many solar system objects, including all of the planets, have been visited by space probes. One solar system body, the Moon, is the only body other than the Earth to have been visited by people. Because they are so close, solar system bodies often have angular sizes and brightnesses that make it possible to use novel observational techniques to study them. For instance, we can study the surfaces of nearby planets by bouncing radar pulses off them. For objects beyond the solar system, both the time required for the radio waves to reach the object and return and the weakness of the returned signals make this technique useless. Given the immense gulf of space beyond the solar system, the special nature of solar system exploration is likely to continue for a long time.

Although the solar system includes many bodies that are exotic or spectacular enough to be studied for their own sake, one of the main reasons that astronomers study the solar system so intently is in the hope that what we learn can be applied elsewhere in the universe. An important example is the relative abundances of the chemical elements. By studying meteorites and observing the Sun, astronomers have determined the abundances of the elements within the solar system far better than for the rest of the universe. Solar system abundances are then used as a standard against which to compare compositions of other objects elsewhere in the universe.

The vast amount of information that has been gathered about the solar system has shown that it contains a great diversity of objects and environments. For example, even excluding the Sun, the diameters of solar system bodies range from

Questions to Explore

- Why do some planets have atmospheres while others do not?
- Why does the pressure in the atmosphere of a planet or satellite fall with increasing height?
- What makes the surfaces of some planets hotter than others?
- Why are the interiors of planets hot?

microscopic to more than ten times the Earth's diameter. Some bodies are less dense than water, whereas others are as dense as iron. Even among the larger bodies—the planets and their satellites—the range of properties is large. Some bodies have surface temperatures as low as a few tens of kelvins, K (degrees above absolute zero), but others are as hot as 700 K (about 800° F). Some objects reflect most of the sunlight that strikes them, whereas others absorb more than 90% of the light that falls on them.

If astronomers had to start from scratch every time they studied another solar system body, the situation would be hopeless. Progress in understanding the solar system would be very slow. Fortunately, despite the great diversity of solar system objects, there are some key processes at work almost everywhere in the universe, including the solar system. For example, no matter what the surface temperature of an object, the radiation it emits is governed by laws of nature that are in effect everywhere. The laws and the equations that describe the nature of gases are the same whether the gas is deep in the atmosphere of Jupiter or in the nearly empty tail of a comet. This chapter is a summary of the contents of the solar system and a description of some important processes at work in the solar system and throughout the universe. Knowledge of these key processes makes it possible to understand and integrate what would otherwise seem to be unrelated characteristics of the host of objects in the solar system.

7.1 SOLAR SYSTEM INVENTORY

The solar system consists of a star, eight planets, five known dwarf planets, over one hundred known satellites, and a huge number of small bodies that range in size from microscopic specks of dust to asteroids. Discussion of the Sun, the dominant member of the solar system, is postponed until Chapter 17, where it will be addressed as a typical but well-studied star.

Although the planets dominate the rest of the solar system bodies in size and mass, the smaller bodies also deserve attention. The comets and asteroids, in particular, have contributed a great deal of information about the origin of the solar system. Table 7.1 summarizes key information about the planets and dwarf planets. Additional information about the planets, their satellites, and some of the comets, asteroids, and meteor swarms is given in Appendices 5 through 10.

The Abundance of the Elements

The chemical composition of a solar system body determines many of its most important properties, such as the rate at which it generates internal energy and the fraction of the sunlight striking it that it absorbs. The solar system appears to

Table 7.1

Key Properties of the Planets and Dwarf Planets

PLANET	ORBITAL DISTANCE	MASS (RELATIVE TO EARTH)	DIAMETER (RELATIVE TO EARTH)	DENSITY (RELATIVE TO WATER)
Mercury	0.39	0.055	0.38	5.4
Venus	0.72	0.82	0.95	5.2
Earth	1.0	1.0	1.0	5.5
Mars	1.52	0.11	0.53	3.9
Jupiter	5.2	318	11.2	1.3
Saturn	9.5	95	9.5	0.7
Uranus	19.2	14.5	4.0	1.3
Neptune	30.1	17.1	3.8	1.6
DWARF PLANET				
Ceres	2.77	0.00016	0.074	2.1
Pluto	39.5	0.0022	0.18	2.0
Haumea	43.3	0.0007	0.09	3.0
Makemake	45.7	0.0007	0.12	2
Eris	67.7	0.0025	0.19	2.3

FIGURE 7.1
The Relative Abundances of Atoms of the Most Common Chemical Elements
Elements other than hydrogen and helium can be shown only by using a magnified scale.

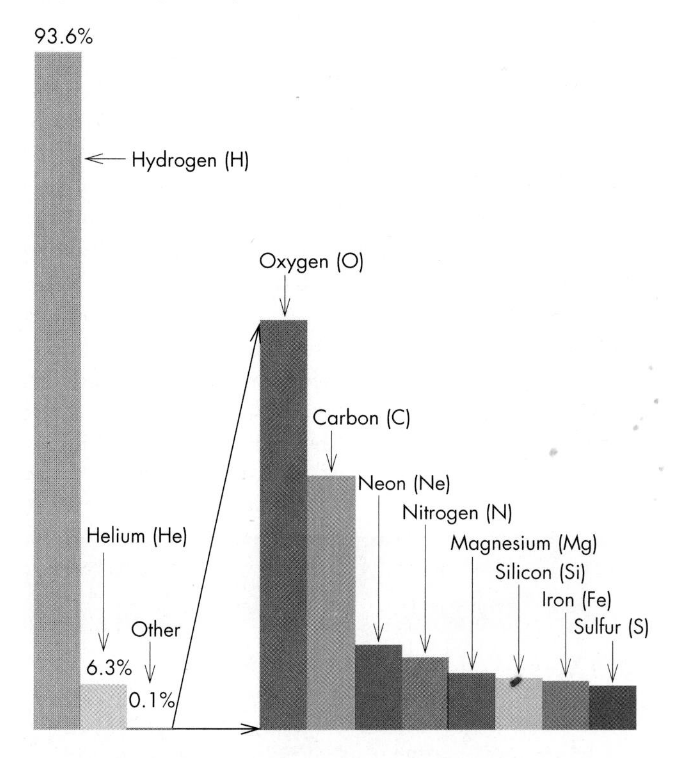

have formed from a mixture of the chemical elements that is fairly typical of the rest of the universe. Figure 7.1 shows the relative abundances of **atoms** of the most common chemical **elements.** As is the case nearly everywhere, hydrogen and helium make up more than 99% of all atoms in the solar system and in the Sun. The next most abundant elements after hydrogen and helium are oxygen, carbon, neon, nitrogen, magnesium, silicon, iron, and sulfur. Table 7.2 lists some important materials that can be formed from the most abundant chemical elements and the state (solid, liquid, gas) in which the materials are likely to be found in the solar system. These materials, under appropriate circumstances, are some of the main building materials of solar system bodies.

7.2 GASES

Most of the visible material in the universe is gaseous. Solids and liquids make up less than 1% of the total. In the solar system, gases are found in the Sun and comets, in the atmospheres of the planets and satellites, and throughout interplanetary space. Gases made up of atoms or molecules are called **neutral gases.** Gases made up of atoms that have been ionized to produce positively charged **ions** and negatively charged electrons are considered **plasmas.** Neutral gases and plasmas behave differently when there are electric and magnetic fields present, but otherwise they have the same properties.

If we want to be able to answer important questions such as how planetary atmospheres support themselves against gravity and why some planets lose their atmospheres whereas others do not, we need to know how gases behave. This behavior depends on the motions of the atoms and molecules in gases. The motions of the atoms and molecules determine pressure and temperature, two fundamental properties of a gas.

Table 7.2

Important Solar System Chemical Compounds

COMPOUND	NAME	STATE
H_2O	water	solid, liquid, gas
CO	carbon monoxide	solid, gas
CO_2	carbon dioxide	solid, gas
CH_4	methane	solid, liquid, gas
N_2	nitrogen	solid, gas
NH_3	ammonia	solid, gas
SiO_2	quartz	solid
FeO	iron oxide	solid
FeS	troilite	solid

Atoms in Motion

The atoms and molecules in a gas are in rapid motion, usually traveling short distances at high speeds between collisions with other atoms and molecules. The frequent collisions that the atoms and molecules experience change the directions and speeds of the colliding atoms. The atoms and molecules move with a considerable range of speeds, as shown in Figure 7.2 for the atmospheric nitrogen molecules

FIGURE 7.2
The Distribution of Speeds for Nitrogen Molecules at Room Temperature (70° F or 294 K) and at 500 K (440° F)
For any temperature, the most probable speed is the speed for which the relative number of molecules is greatest. Notice how much larger the relative number of molecules with speeds greater than 1 km/s is for 500 K than for 294 K.

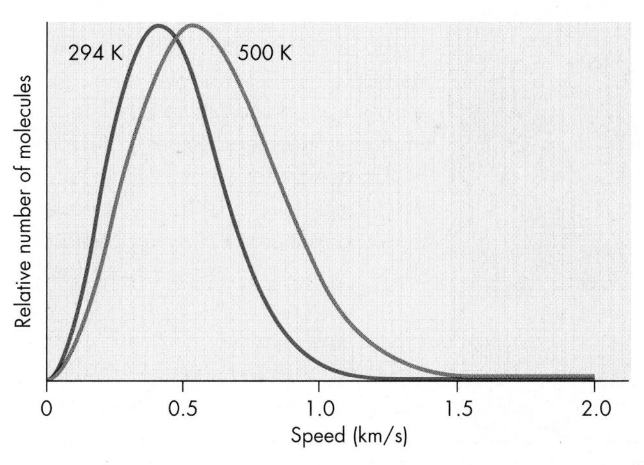

that surround you as you read this. The average speed of the nitrogen molecules is about 0.5 km/s as they travel a distance of about 10^{-6} m between collisions. Collisions occur nearly a billion times a second for each molecule. While 0.5 km/s doesn't seem like an impressive speed, expressed in different units it is 1100 miles per hour. Is this really possible? Are we constantly being bombarded by molecules traveling at such a speed? Wouldn't we be able to feel the effect of the collisions?

Speed and Temperature The answers are yes, it really is possible; yes, we are being struck at such speeds; and yes, we do feel the effect. The nitrogen molecules have extremely tiny masses, so we can't feel individual collisions. Also, the collisions occur too frequently for us to be able to isolate the effect of a single blow. The cumulative effect of the collisions is to impart energy to the skin and give us the sensation of warmth or coolness. In fact, the temperature of the gas is closely related to the speed of its atoms or molecules. More precisely, the **temperature** of a gas is proportional to the average kinetic energy of the atoms or molecules in a gas. As temperature decreases, the energies and speeds of the atoms decrease as well. Eventually, at a temperature of **absolute zero,** the motion of the atoms stops altogether. The **Kelvin temperature scale,** sometimes called the absolute temperature scale, has its zero point at absolute zero, so it is a direct indication of atomic or molecular speeds and energies. The freezing point of water is at a temperature of 273.16 K. The boiling point of water is at a temperature of 373.16 K.

Figure 7.2 also shows the distribution of speeds for nitrogen molecules at a temperature of 500 K, which is about twice room temperature (approximately 294 K). As temperature increases so does the average speed of the molecules. The number of very rapidly moving molecules rises as well. The average speed of the atoms or molecules in a gas depends not only on temperature but also on the masses of the atoms or molecules according to

$$v = \sqrt{\frac{8kT}{\pi m}} \quad (7.1)$$

where k is Boltzmann's constant (1.38×10^{-23} kg m^2/K s^2), T is temperature in kelvins, and m is the mass of the atom or molecule.

In a mixture of gases, the average speed of the lightest kind of atom or molecule is greatest. Because they are lightest, hydrogen atoms move, on average, faster than any other atoms or molecules.

Thus, the temperature of a gas describes how fast its atoms or molecules are moving and how violently they collide with each other. It is the violence of the collisions that determines how likely it is that a collision will result in the disruption of a molecule or atom. In the interiors of stars, temperature determines the likelihood that a collision will result in an energy-producing nuclear reaction. Many other important astronomical processes are controlled by the temperature of the gas in which they occur.

The average speed of the atoms or molecules in a gas increases as the temperature of the gas increases. In a mixture of gases, the lightest atoms and molecules move fastest, on average. Although most atoms move at about the average speed, a small percentage move with speeds much greater than average. Because it controls speed, temperature also controls many important processes involving collisions between the particles of which the gas is made.

Equation 7.1

Temperature, Mass, and Speed

Equation 7.1 can be used to find the average speed of atoms or molecules of a given mass and temperature. For example, hydrogen atoms at 300 K have an average speed of 2.5 km/s. Using hydrogen atoms at 300 K as a standard, we can use Equation 7.1 to find average speeds for other atoms and molecules at other temperatures. For instance, the average speed of hydrogen atoms at 600 K can be found by noting that for two gases of the same kind of atom or molecule, average speed depends only on temperature

$$v \propto \sqrt{T}$$

where the symbol $\propto$ means "proportional to." Doubling the temperature results in the average speed increasing by a factor of $\sqrt{2}$. Thus, the average speed of hydrogen at 600 K is $\sqrt{2} \times$ 2.5 km/s = 3.5 km/s. For two gases that have the same temperature but contain different kinds of atoms or molecules, average speed depends only on atomic or molecular mass

$$v \propto \sqrt{\frac{1}{m}}$$

This means that the average speed of oxygen atoms (which are 16 times as massive as hydrogen atoms) at 300 K is smaller than that of hydrogen by a factor of

$$\sqrt{\frac{1}{16}} = \frac{1}{4}$$

making the average speed of the oxygen atoms ¼ × 2.5 km/s = 0.6 km/s.

Planetary variations

Escape of Gases An example of a process controlled by temperature—a process of great importance for the evolution of planets—is the escape of planetary atmospheres into interplanetary space. Escape velocity (see Chapter 5) at the surface of the Earth is about 11 km/s. Even though the average molecule in the gas around us moves at only 0.5 km/s, a small fraction of the oxygen and nitrogen molecules move faster than escape velocity. Does this mean that these molecules can escape into space? No, they can't. Long before the molecules can reach the outer limits of the Earth's atmosphere, they strike other molecules and are deflected back toward the ground. In the lower atmosphere the molecules are too close together to permit even the most rapidly moving ones to escape. However, as height increases, the atmosphere becomes thinner and thinner. Rapidly moving atoms and molecules can travel farther before undergoing collisions. At about 1000 km above the surface, the atmosphere is so thin that collisions are rare. Above 1000 km, an atom moving faster than escape velocity is more likely to escape than to collide again. The part of the atmosphere from which atoms can escape is called the **exosphere.**

If a particular kind of gas in the exosphere had an average speed greater than escape velocity, its atoms would escape from the exosphere almost immediately. As atoms from lower in the atmosphere reached the exosphere, they too would escape quickly. All the gas of that kind would be lost from the atmosphere in a time that is very short compared with the age of the Earth. Even if the average speed of a gas is much less than escape velocity, there will be some rapidly moving atoms that can escape. Astronomers have found, as a general rule, that atoms of a gas whose average speed is greater than ⅙ of escape velocity will escape rapidly enough that the atmosphere should soon contain almost none of them. In order to predict the kinds of gases that could be retained by a planet, we need to know the escape velocity of the planet (given by its mass and radius), the temperature of its exosphere (which is about 500 K or less for most planets), and the masses of various kinds of gases. Table 7.3 gives ⅙ of the escape velocity for each planet and the Moon. The average atomic speeds for an exospheric temperature of 500 K are 3.2 km/s for hydrogen, 1.6 km/s for helium, and 0.9 km/s for nitrogen. By comparing ⅙ of the escape velocity of each planet with the speeds of these atoms, we see that Jupiter, Saturn, Uranus, and Neptune are able to retain even hydrogen, the lightest, most rapidly moving gas. The Earth and Venus are able to retain gases as heavy as nitrogen and oxygen, but not hydrogen, which is present only in tiny amounts in their atmospheres. Mars, Mercury, and the Moon have difficulty retaining even heavy gases and have thin atmospheres. Still smaller bodies, such as small satellites, asteroids, and comets have such low escape velocities that all their gases escape almost immediately. The theory of which gases should be retained by which planets is not enough to understand fully the amount and nature of every planetary atmosphere, but it does allow us to place limits on which kinds of gases should be present.

Table 7.3

One-Sixth of the Escape Velocity for Each Planet and the Moon

PLANET	$\frac{1}{6} v_{ESCAPE}$(km/s)
Mercury	0.5
Venus	1.7
Earth	1.9
Moon	0.4
Mars	0.8
Jupiter	9.9
Saturn	5.9
Uranus	3.6
Neptune	3.9

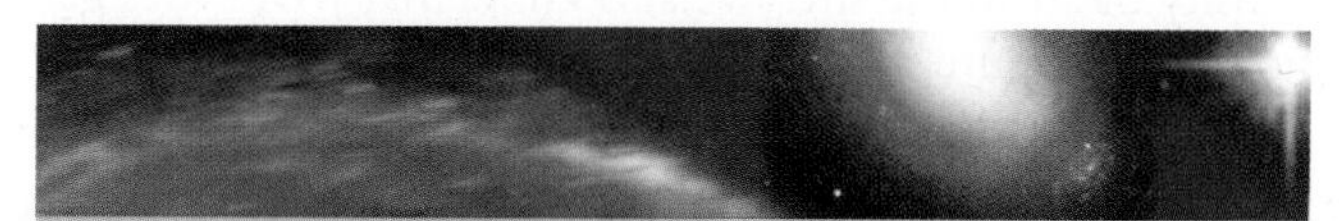

In a region of dense gas, frequent collisions prevent rapidly moving atoms and molecules from escaping. In regions where collisions are infrequent, atoms that move faster than the escape velocity can escape into space. If the average speed of a gas is greater than about ⅙ of the escape velocity, all the atoms will escape in a time that is short when compared with the age of the solar system.

Density and Pressure

The pressure of a gas describes how strongly it pushes as it tries to expand. Usually gravity or some other force acts to keep the gas from expanding. It is the balance between pressure and gravity that controls the structure of a star and the atmosphere of a planet. Pressure depends on the temperature of a gas and on its density.

The **density** of a gas describes how much of it there is in a given volume. Usually, density is defined as the mass of gas in a volume and is measured in, say, kg/m^3. Sometimes, however, density means the number of atoms or molecules in a volume of gas, in which case it is called **number density.** Density is equal to number density times the mass of a typical gas molecule or atom. The number density in the gas around you is about 10^{25} molecules per cubic meter. However, because a typical air molecule has a mass of only about 5×10^{-26} kg, the density of air is about 1 kg/m^3, or 1⁄1000 the density of water.

During each collision of gas molecules or atoms, the particles exert forces on each other that push them apart. The cumulative effect of the collisions is **pressure,** the force the gas exerts on a unit area of surface. At sea level, the average pressure of the Earth's atmosphere is about 10^5 N/m^2 or 14.7 lb/in^2. (A Newton [N] is an amount of force equal to about 0.225 lb.) This amount of pressure is defined as one atmosphere of pressure and is often used as a standard when describing the pressure of the atmospheres of other planets. Gas pressure pushes equally in all directions, which means that the pressure on one side of your hand equals the oppositely directed pressure pushing on its other side. Usually we notice pressure only when there is an imbalance, such as when air is pumped out of a can. The can is immediately crushed by the greater pressure on its outside. If there were no pressure pushing up on the bottom of your hand when you hold it out, it would be pushed downward by a force of hundreds of pounds.

Ideal Gas Law Most of us have an intuitive understanding of the way the pressure of a gas depends on its other properties. If we want to increase the pressure in an automobile tire, we pump more air into it, increasing the density of air in the tire. We also know that when tires get hot, the pressure in them increases. The same thing can happen to aerosol cans, which is why the label warns against throwing them into a fire or storing them at high temperature. Because gas pressure depends on collisions among the gas molecules or atoms, it isn't surprising that pressure depends on the rate at which collisions occur. This rate is governed by the density of the gas and the average speed of the gas particles, which depends on temperature. The relationship between pressure, density, and temperature is

$$P = nkT \qquad (7.2)$$

where P is pressure, n is number density, and T is temperature in kelvins. As before, k is Boltzmann's constant.

Notice that there are an infinite number of combinations of number density and temperature that can produce the same pressure. It is possible for a cool, dense gas to have a pressure that just balances that of a hot, thin gas. Even very thin gases can exert considerable pressure if they are hot enough.

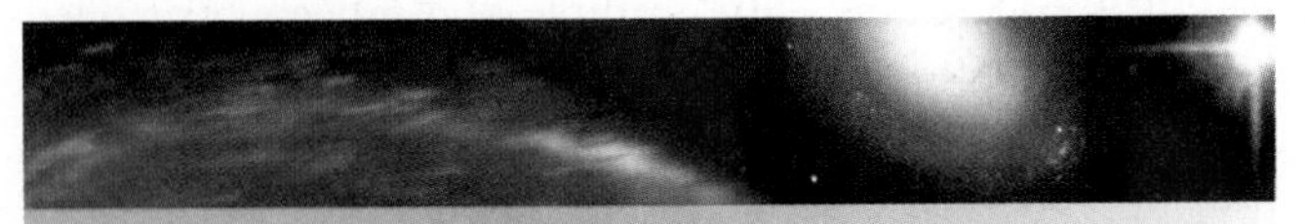

Density describes how much gas there is in a given volume. Gas pressure describes the force per unit area exerted by the gas. The pressure of a gas is proportional to the product of its density and its temperature.

Pressure and Gravity Gas pressure is what keeps the Earth's atmosphere from falling to the ground and prevents stars from collapsing inward as they would if there were no force to balance the inward pull of gravity. Figure 7.3 shows the forces that act on a cylinder of gas in the Earth's atmosphere. The cylinder has imaginary walls and contains gas that is exactly the same as the atmospheric gas surrounding the cylinder. At each point on its top, bottom, and side walls, the cylinder is pushed inward by the pressure of the surrounding gas. The gas in the cylinder pushes back with an equal force so the cylinder doesn't collapse. Gravity gives the gas in the cylinder weight, which would pull the cylinder downward if there were no balancing force. The balancing force arises because the pressure at the bottom of the cylinder is greater than the pressure at the top of the cylinder. In order to balance the force of gravity, the pressure in a region of gas must decline with increasing height. When this is the case, the gas is said to be in **hydrostatic equilibrium.** The pressure at the surface of a planet is produced by the weight of the entire atmosphere of the planet. For the Earth, the pressure at sea level is 14.7 lb/in^2, meaning that there are 14.7 pounds of air above every square inch of surface.

In the Earth's atmosphere, the pressure is halved with every increase of 5.5 km in height. This means that 5.5 km above sea level, the gas pressure is half of the pressure at sea level. At an altitude of 11 km, the pressure is one quarter the pressure at sea level. This is the cruising altitude of large

Pressure, Temperature, and Density

Equation 7.2 can be used to find the pressure of a gas if its temperature and number density are given. For example, at $T = 300$ K (81° F), the air around us has a density of about $n = 2.4 \times 10^{25}$ molecules per cubic meter. Using Boltzmann's constant k = 1.38×10^{-23} kg m^2/K s^2, or 1.38×10^{-23} N m/K, the pressure is

$$P = nkT$$
$$= (2.44 \times 10^{25}/\text{m}^3)\ (300\ \text{K})\ (1.38 \times 10^{-23}\ \text{Nm/K})$$
$$= 1.0 \times 10^5\ \text{N/m}^2 = 1\ \text{atmosphere}$$

Because one Newton equals 0.22 pounds, the atmosphere around us exerts a pressure of 22,000 pounds per square meter.

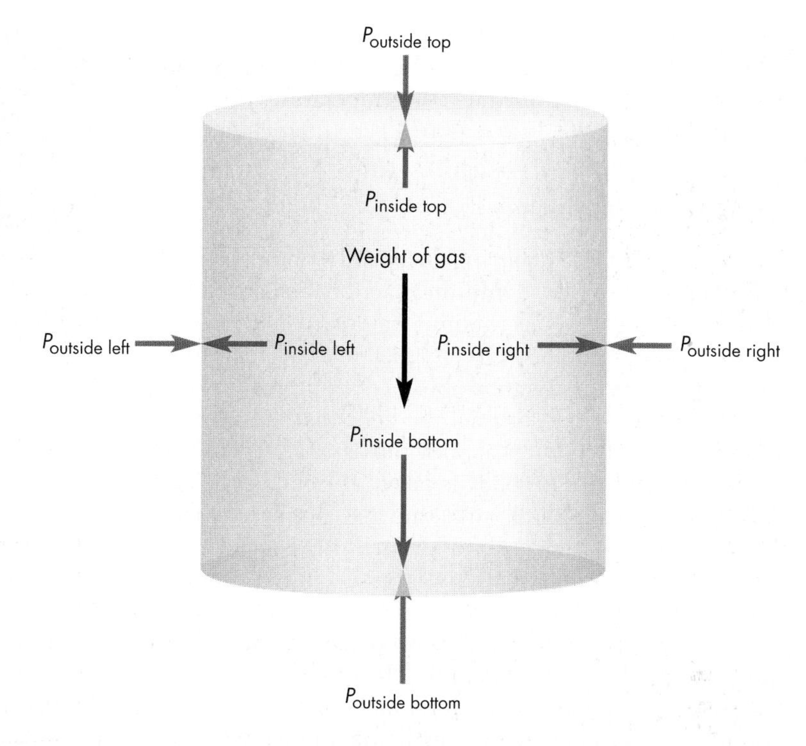

FIGURE 7.3
Pressure Forces and Gravity Acting on a Cylinder of Gas in Earth's Atmosphere
At each point on the imaginary walls of the cylinder, the inside and outside pressures exactly balance. To balance the weight of the cylinder of gas, the pressure at the bottom of the cylinder is greater than the pressure at the top.

commercial jets and is the reason why passenger cabins are pressurized. At 16.5 km, the pressure is only ⅛ of the sea level pressure. Figure 7.4 shows the decline of pressure with height for the Earth's atmosphere. Because pressure declines with height, the atmosphere gradually blends into space rather than ending abruptly when pressure runs out. The rate at which pressure drops with height depends on the acceleration of gravity, the temperature of the gas, and the masses of the gas atoms or molecules. Generally a hot atmosphere has a pressure that falls more slowly as height increases than does a cool atmosphere. Also, an atmosphere in which gravity is weak or an atmosphere that is made up of light atoms or molecules has a pressure that falls slowly as height increases.

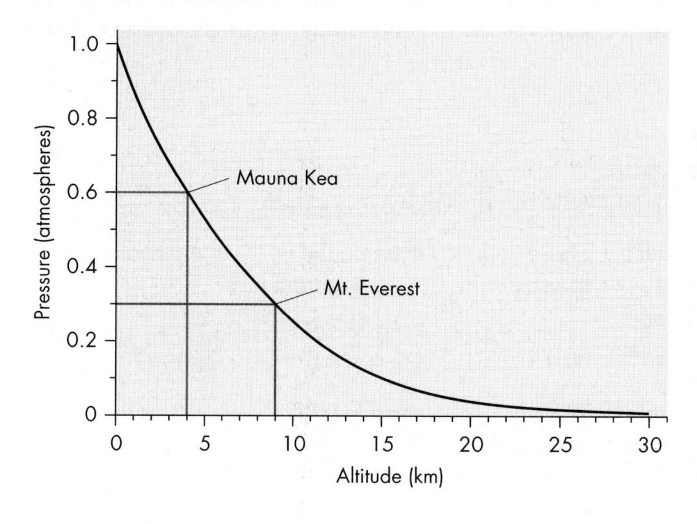

FIGURE 7.4
Atmospheric Pressure Drops with Increasing Altitude Above Sea Level
The atmospheric pressure at Mauna Kea Observatory (altitude 4 km) is about 60% that at sea level while the pressure at the summit of Mt. Everest (altitude 9 km) is only about 30% that of sea level.

When a gas is in hydrostatic equilibrium, the drop of gas pressure as height increases just balances the downward pull of gravity. The rate at which pressure drops as height increases depends on the gravity and temperature of the atmosphere as well as the masses of the gas atoms or molecules present.

7.3 RADIATION AND MATTER

Interactions between radiation and matter determine many important properties of everything from planets to stars. For example, solar system bodies are heated by energy absorbed from sunlight and cooled by radiation emitted back into space. We need to know how electromagnetic radiation is absorbed and emitted by matter in order to understand how this heating and cooling occurs. It is the balance between the absorption and emission of radiation by planets, satellites, and other bodies that controls their surface temperatures. This balance explains the steady drop of temperature with increasing distance from the Sun and accounts for the wide range of temperatures (700 K to about 30 K, or 800° F to −400° F) in the solar system.

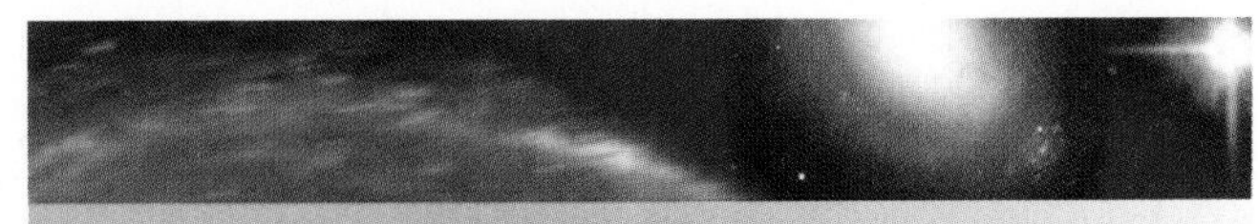

A blackbody is a (hypothetical) object that absorbs all of the light that falls on it. A blackbody also emits electromagnetic radiation. The brightness of the radiation that a blackbody emits is the greatest that can be emitted by an object of that size and temperature.

Blackbody Radiation

All of the light that falls on an opaque object is either reflected or absorbed. Only the fraction of light that is absorbed heats the object. Some materials, such as snow, are very poor absorbers. Others, such as coal, reflect very little light and are very good absorbers. A perfect absorber would be one that absorbs all the light that falls on it. A **blackbody** is a hypothetical body that is a perfect absorber at all wavelengths. This body is called a blackbody because it reflects no light at all and, thus, is completely black. Materials that are good absorbers are also good emitters of radiation, so a blackbody is a better emitter than any other object.

No material, not even a lump of coal, can perfectly absorb light of all wavelengths. However, it is possible to make an excellent approximation of a blackbody by using a hollow sphere with a small hole in it (Figure 7.5). The inside of the sphere is made rough and blackened. When light shines into the hole, it strikes the opposite side of the cavity. Most of the light is absorbed, but the rest is reflected. The reflected light strikes various points on the inside of the sphere, where most of it is absorbed. Only a small amount of the remaining light is reflected. After several reflections—long before it can be reflected back out of the hole—essentially all of the light has been absorbed. If 90% of the light is absorbed each time it strikes the wall, only 0.01% is left after four reflections. The hole in the sphere acts like a blackbody in that all the radiation that passes through the hole is absorbed. Electromagnetic radiation also is emitted from the inside of the sphere through the hole. This radiation has the greatest brightness that can be produced by any object with the same size and temperature as the hole. In other words, the hole emits radiation with maximum efficiency. The light emitted by the hole (or any other blackbody) is called **blackbody radiation.**

Temperature and Spectral Shape Blackbody radiation has a spectrum with a distinctive shape that depends only on the temperature of the blackbody. The blackbody spectrum for a temperature of 6000 K (about the same as the surface of the Sun) is shown in Figure 7.6. Brightness rises fairly abruptly with increasing wavelength until a maximum is reached and then declines gradually as wavelength continues to increase. The shape of the spectrum of a blackbody is sometimes called a blackbody distribution. The blackbody distributions for temperatures of 7500 K and 4500 K are also shown in Figure 7.6.

Wien's Law The wavelength at which a blackbody is brightest becomes shorter as temperature increases. Because blackbodies get "bluer" as they get hotter and "redder" as they cool, the color of a blackbody can be used to find its temperature. **Wien's law,** the relationship between the temperature of a blackbody, T, and the wavelength at which the emission from the blackbody is brightest, λ_{max}, is given by

$$\lambda_{max} = \frac{\text{constant}}{T} \tag{7.3}$$

When T is given in kelvins and λ_{max} in meters, this constant has the value 2.9×10^{-3} m K.

If we can find the brightest part of the spectrum of a blackbody, we can use Wien's law to find its temperature. Most other objects are good enough emitters and absorbers

Wien's Law

If the temperature of a body is known, Wien's law (Equation 7.3) can be used to find the wavelength at which it emits most brightly. This would be important if an astronomer were designing an experiment to search for radiation from a body for which the temperature was known. For example, the temperature of the visible surface of the Sun is about 6000 K. Using Equation 7.3, the wavelength at which the Sun is brightest is

$$\lambda_{max} = 2.9 \times 10^{-3} \text{ m K}/6000 \text{ K}$$
$$= 4.8 \times 10^{-7} \text{ m} = 480 \text{ nm}$$

This is in the middle of the visible part of the spectrum.

of radiation that Wien's law can be used to estimate their temperatures.

Stars hotter than about 10,000 K have λ_{max} shorter than 300 nm. Although they may emit plenty of radiation in the visible part of the spectrum, they are brightest in the ultra-violet, which we can't see from the ground. This means that the visual brightness of a hot star considerably under-estimates its actual brightness. At the center of the Sun, where the temperature is 14 million K, $\lambda_{max} = 2 \times 10^{-10}$ m. This is in the X-ray part of the spectrum. Cooler bodies, such as planets and people, also radiate. Your temperature is about 300 K so the radiation you are emitting has λ_{max} = 10 μm and falls in the infrared (1 μm is called a micrometer, or micron, and is equal to a millionth of a meter). Virtually none of the radiation you emit is visible. However, the infrared radiation that you are emitting can be detected, even at night, by infrared sensors. Even the coolest material in the universe, at temperatures of only a few kelvins, emits in the far infrared and radio parts of the spectrum.

FIGURE 7.5
The Design of a Blackbody
If the inner surface of the ball is darkened and roughened, essentially all the light that enters the hole in the ball is absorbed before it can be reflected back out of the hole.

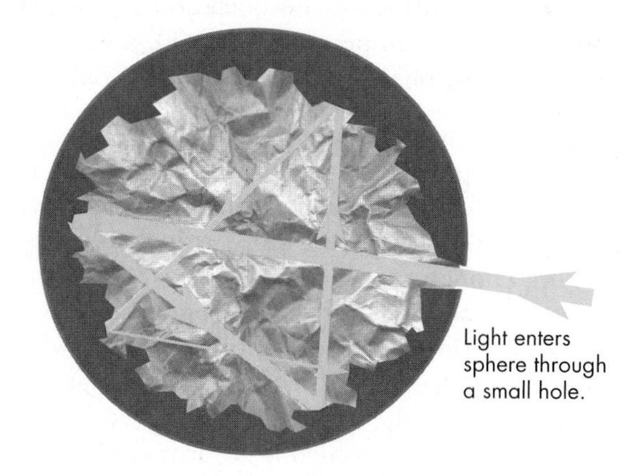

The Stefan-Boltzmann Law As the spectra of Figure 7.6 show, as the temperature of a blackbody increases it grows brighter at every wavelength. The total rate at which energy is radiated by a blackbody is the sum of the rates of emission at all wavelengths.

This varies with temperature according to the **Stefan-Boltzmann law**

$$E = \sigma T^4 \qquad (7.4)$$

where E is the rate at which energy is emitted per unit area and σ is a constant (see box for Equation 7.4). Thus, a hot object radiates more total energy than a cool one of the same size. Figure 7.7, a graph of the Stefan-Boltzmann law, shows the rate of emission for the range of temperatures encountered in the solar system.

FIGURE 7.6
The Spectra of Radiation Emitted by Blackbodies with Temperatures of 4500 K, 6000 K, and 7500 K
A hotter blackbody is brighter than a cool one at every wavelength. The visible part of the spectrum (divided here into its constituent colors) is bordered at short wavelengths by the ultraviolet and at long wavelengths by the infrared.

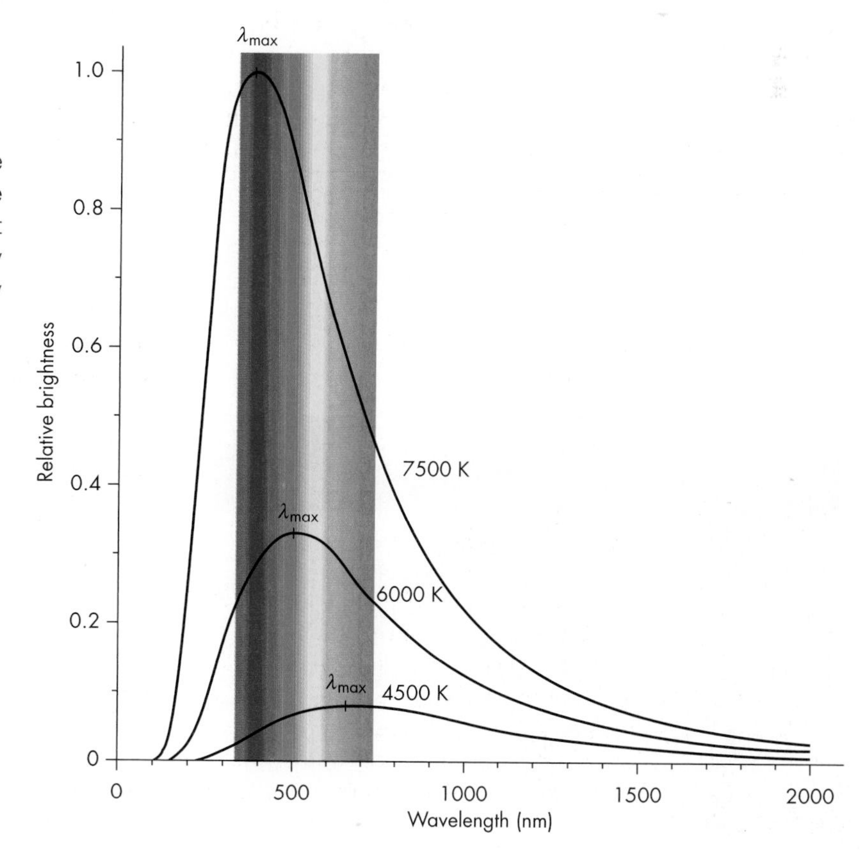

The Stefan-Boltzmann Law

Equation 7.4 can be used to find how the rate at which a body emits radiation depends on its temperature. Because the rate at which energy is emitted depends on the fourth power of temperature, the total energy emitted by a body is a sensitive function of temperature. If the temperature of a blackbody doubles, its energy output becomes $2^4 = 16$ times as large. For example, the surface temperature of the Sun is about 20 times hotter than a person's skin. How does the rate at which a square meter of the Sun's surface emits radiant energy, $E_{\odot}$, compare to the rate at which a square meter of skin emits radiant energy, E_S? To find out, we need to compare the Stefan-Boltzmann law for the two bodies.

$$\frac{E_{\odot}}{E_s} = \frac{\sigma T_{\odot}^4}{\sigma T_s^4}$$
$$= (T_{\odot}/T_s)^4$$
$$= (20)^4$$
$$= 160{,}000$$

FIGURE 7.7
The Stefan-Boltzmann Law—The Dependence of the Total Rate of Radiation by a Blackbody on Temperature
The rate of radiation increases as the fourth power of temperature.

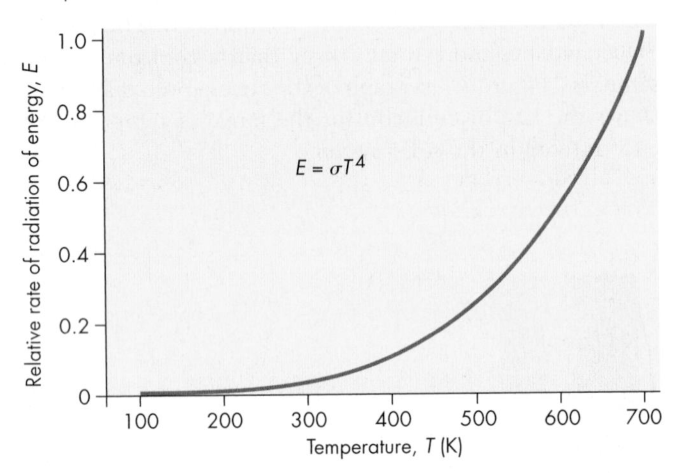

The spectrum of radiation emitted by a blackbody depends only on the temperature of the blackbody. The wavelength at which a blackbody is brightest is inversely proportional to the temperature of the blackbody, so blackbody radiation grows "bluer" with increasing temperature. Summed over all wavelengths, the rate at which a blackbody emits radiation increases with the fourth power of temperature.

Thermal Equilibrium

If a solar system body absorbs solar energy at a greater rate than it radiates energy away, it grows hotter. As its temperature increases, the rate at which it radiates energy also increases. Its temperature stops rising only when it becomes hot enough that it emits energy at the same rate that it absorbs energy. Similarly, if a body radiates more energy than it absorbs, it cools until the rate at which it emits falls to the level at which it absorbs solar energy. In either case, when the rate at which the body radiates energy equals the rate at which it absorbs radiant energy, the body is in **thermal equilibrium.** Its temperature then remains the same unless either its absorption or emission changes. This is strictly true only for bodies that are heated solely by sunlight and have no internal sources of energy, such as radioactive heating. Bodies that have internal sources are hotter than they would be if they were heated by sunlight alone. However, the temperature of such a body is still the one at which the rate of energy radiated away just equals the rate at which energy reaches the surface from internal sources plus the rate at which sunlight is absorbed.

Temperatures in the Solar System

The temperatures of bodies in the solar system (except for the Sun) can be estimated by assuming that they are blackbodies in thermal equilibrium. That is, they emit radiation at the same rate at which they absorb solar energy. Under these circumstances, the temperature of the body depends only on its distance from the Sun.

Dependence on Distance Because the brightness of sunlight falls as the square of distance from the Sun, temperatures in the solar system generally decrease with distance as well. The brightness of solar radiation at the Earth's distance from the Sun (1 AU) is given by the **solar constant,** the rate at which energy strikes a surface directly facing the Sun. The value of the solar constant is 1370 watts per square meter. (The solar constant is actually misnamed. Sunlight varies on both short and long time scales.) At other distances from the Sun, the brightness of solar radiation is equal to the solar constant divided by the square of distance from the Sun in AU. (See Equation 6.1.) This means that the brightness of sunlight is about six times as bright at Mercury's distance (0.4 AU) as at the Earth. At the distance of Neptune (30 AU), sunlight is only about 0.1% as bright as it is at 1 AU.

Because the brightness of sunlight falls with increasing distance from the Sun, the rate at which solar system bodies absorb sunlight falls as well. As a result, the temperature at which energy emitted equals energy absorbed declines with increasing distance. Near the Sun, where sunlight is very bright, a body absorbs solar energy at a high rate. In order to balance the rate at which it absorbs solar energy, the body must also emit radiation at a high rate. This requires a high temperature. Far from the Sun, sunlight is dim and the rate at which solar energy is absorbed is low. The rate of absorption can be balanced by emission at a low temperature. Figure 7.8 shows how blackbody temperature varies with distance from the Sun throughout the solar system. Temperature falls from about 440 K at Mercury's average distance to 280 K at 1 AU, to 50 K at Neptune's distance, and to only about 10 K at 1000 AU in the comet swarm that surrounds the planetary system.

Temperature and Albedo Of course, real solar system objects aren't blackbodies. They neither absorb perfectly nor emit with perfect efficiency. The fraction of light striking a body that gets reflected is defined as the body's **albedo,** A. The fraction that the body absorbs is 1-A. A blackbody has an albedo of 0.0 because it absorbs all the light that falls on it. The albedo of coal dust is 0.03, whereas most rocks have albedos between about 0.1 and 0.2. Dirty ice has an albedo of about 0.5 and fresh ice nearly 1. For the planets, albedos range from 0.11 for Mercury (showing that the planet absorbs 89% of the light that strikes it) to 0.65 for Venus. Many satellites, asteroids, and comets are even darker than Mercury and reflect only a small percentage of the light that strikes them. Others, however, are more highly reflective than Venus.

Because bodies with high albedos absorb less sunlight than those with low albedos, bodies with high albedos are cooler at a given distance from the Sun. An ice- or snow-covered planet or one with a thick cloud cover should be cooler than one covered with dark, rocky material. The temperatures of the planets and the Moon, including the effect of albedo, are compared with their actual temperatures in Table 7.4. The calculated and actual temperatures are similar for Mercury, the Moon, Mars, and Uranus, but the actual temperature of Venus is much higher than the calculated temperature. The temperatures of the Earth and the giant planets, Jupiter, Saturn, and Neptune, are also somewhat greater than can be accounted for when considering only distance from the Sun and albedo. The discrepancies in temperature are caused by difficulty in radiating away energy (Venus and Earth) and significant internal energy sources (Jupiter, Saturn, and Neptune). These important factors are described more fully in the chapters on the individual planets.

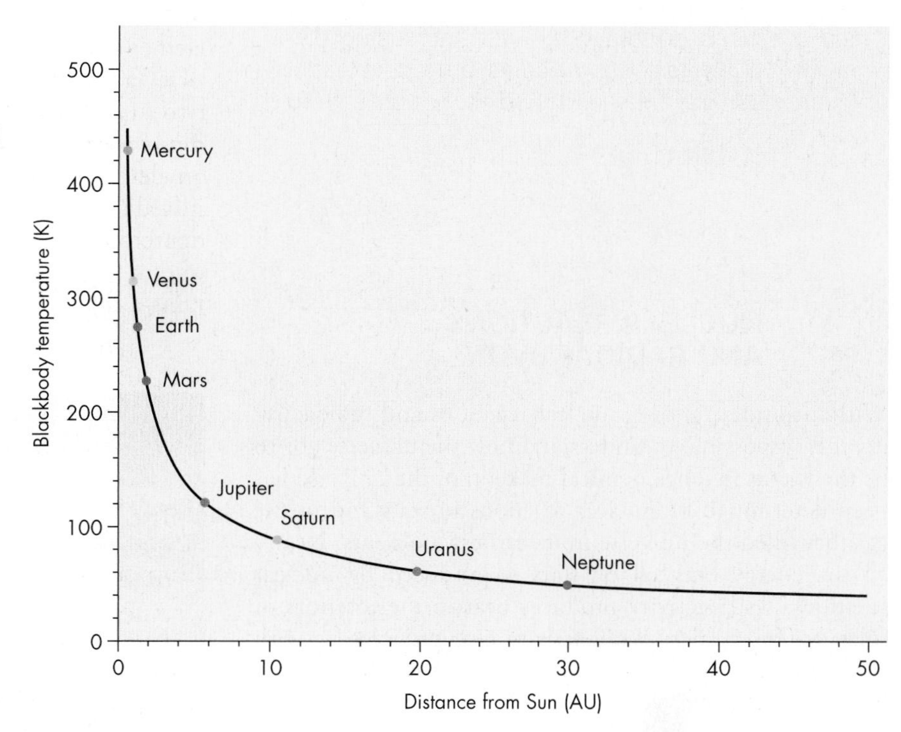

FIGURE 7.8
How Blackbody Temperature Depends on Distance from the Sun
The blackbody temperature is indicated for the average distance of each planet.

Table 7.4

The Calculated and Actual Temperatures of the Planets and the Moon

PLANET	CALCULATED TEMPERATURE (K)	APPROXIMATE AVERAGE GLOBAL TEMPERATURE (K)
Mercury	435	400
Venus	252	730
Earth	248	280
Moon	270	280
Mars	217	210
Jupiter	102	125
Saturn	77	95
Uranus	56	60
Neptune	46	60

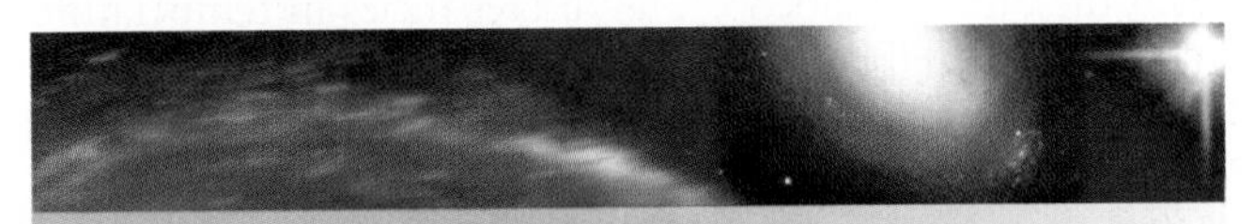

A body in thermal equilibrium has a temperature that permits it to emit radiant energy at the same rate at which it absorbs radiant energy that falls on it. Because the brightness of sunlight diminishes with increasing distance from the Sun, temperatures generally fall with increasing distance. Highly reflecting bodies are cooler, at a given distance, than dark bodies, which absorb sunlight well.

7.4 NUCLEAR REACTIONS AND RADIOACTIVITY

Without understanding nuclear reactions and radioactivity, it is impossible to understand how the universe got to be the way it is. The chemical makeup of the universe has been determined by nuclear reactions in stars and in the gas that filled the universe in its earliest moments. Nearly all the energy emitted by stars is produced by nuclear reactions. Radioactivity produces heat in the interiors of solar system bodies. Radioactivity also provides us with tools for investigating the ages of the few solar system bodies from which we have samples.

Nuclei and Nuclear Particles

Nuclei are positively charged concentrations of matter found at the centers of atoms. The **nucleus** of an atom is surrounded by one or more negatively charged particles called **electrons.** Nearly all the mass in an atom is located in its nucleus. A nucleus is made up of two kinds of particles. These are **protons,** which have a positive charge equal in magnitude to the negative charge of an electron, and **neutrons,** which have no electric charge at all. The mass of a proton is tiny, 1.67×10^{-27} kg. A neutron is about 0.1% more massive. Protons and neutrons are almost 2000 times as massive as an electron.

The number of protons in a nucleus is the **atomic number** of the nucleus. The chemical behavior of an atom is determined by the way its electrons interact with the electrons of other atoms. Because the number of protons in the nucleus is equal to the number of electrons in an atom, atomic number determines the chemical behavior of the atom and specifies which element it is. Thus, a carbon atom always has a nucleus with six protons. An iron atom always has 26 protons in its nucleus.

Although all atoms of a given chemical element have the same number of nuclear protons, they can have different numbers of neutrons. Nuclei that are composed of the same element but differ in their numbers of neutrons are called **isotopes.** For example, hydrogen, the lightest element, always has a single proton. However, there are three isotopes of hydrogen that have zero, one, and two neutrons. Some of the heavier elements, such as lead, have more than a dozen isotopes. The number of protons and neutrons in a nucleus is called its **mass number.** To distinguish between different isotopes of an element, the isotopes are written with mass number as a superscript preceding the chemical symbol of the element. Hydrogen that has only a single nuclear particle (a proton) is written ^{1}H, whereas hydrogen with two nuclear particles (a proton and a neutron) is called **deuterium** and is written ^{2}H. The various isotopes of an element behave alike chemically and are usually well mixed in nature. Thus, both normal hydrogen atoms and deuterons combine with oxygen to form water. About one in every 3400 water molecules in the oceans contains a deuteron.

The nucleus of an atom, which is composed of neutrons and protons, contains most of the mass of the atom. The number of protons in a nucleus determines what chemical element it is. Nuclei that are of the same element but differ in numbers of neutrons are called isotopes.

Nuclear Reactions

Nuclear reactions can occur when a nucleus is struck by another nucleus or by a particle such as a neutron. If the kinetic energy of the collision is low, the nuclei usually just bounce away from each other. At higher energies, however, the nuclei can react to form one or more new nuclei. In nuclear reactions, elements are transformed into other elements.

Fission and Fusion Two kinds of nuclear reactions, fission and fusion, are of great importance because of the energy that can be derived from them. In **fission,** a heavy nucleus is split into two less-massive nuclei. For example, if the nucleus of the uranium atom, ^{235}U, reacts with a neutron, it splits into two particles, each of which has about half the mass of ^{235}U. Some nuclei undergo fission spontaneously, without being struck by another particle. In **fusion,** two light nuclei combine to produce a more massive nucleus. The most common fusion reaction in the universe is the combining of two protons to form ^2H. During this fusion reaction, one of the protons changes into a neutron. This is the first step in a process in which hydrogen is converted into helium within stars.

Nuclear Energy The reason that energy can be obtained through fission and fusion reactions is that matter is converted to energy during the reaction (Figure 7.9). According to Einstein's special relativity, mass and energy are different forms of the same thing and can be changed back and forth according to the famous formula

$$E = mc^2 \quad (7.5)$$

where E is energy, m is mass, and c is the speed of light.

Energy is released in a nuclear reaction if the particles and nuclei that react are more massive than those produced by the reaction. For example, although a ^{12}C (carbon) nucleus and a ^4He (helium) nucleus contain the same number of protons and neutrons as an ^{16}O (oxygen) nucleus, the combined masses of the carbon and helium nuclei are greater than that of the oxygen nucleus. Thus, when ^{12}C and ^4He are fused together to form ^{16}O, energy is released. In general, for those elements that are lighter than iron, fusion reactions produce energy whereas fission reactions only occur if energy is supplied. For elements more massive than iron, fission reactions produce energy whereas fusion reactions require energy input. The amount of energy released in a nuclear reaction, although small in absolute terms, is extremely large compared with the energy released in a chemical reaction. The fission of a heavy element such as uranium releases about 100 million times as much energy per atom as does a chemical reaction. Fusion reactions

FIGURE 7.9
Energy Production in Nuclear Reactions
For fusion, **A,** and fission, **B,** the bars indicate the masses of the particles before and after the nuclear reaction. The small decrease of mass in both cases results in the release of energy.

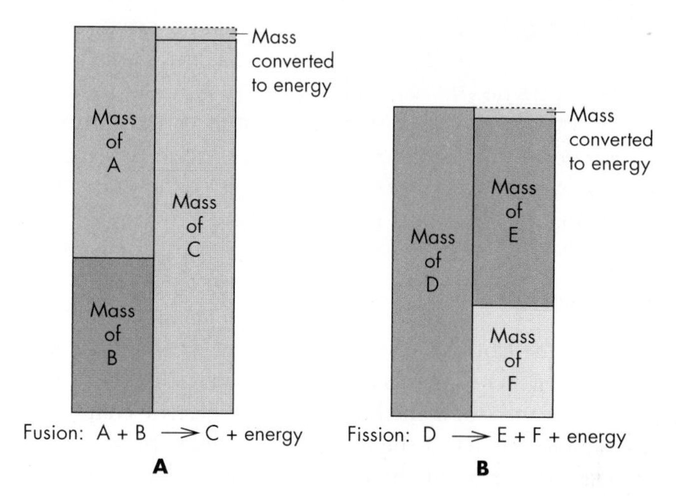

Equation 7.5

Energy and Mass

Equation 7.5 can be used to calculate the amount of energy that would be released if an object were transformed into energy. For example, the amount of energy equivalent to the mass of a 1 g (= 10^{-3} kg) paper clip is given by

$$E = mc^2$$
$$= 10^{-3}\,\text{kg}\left(3 \times 10^8\,\frac{\text{m}}{\text{s}}\right)^2$$
$$= 9 \times 10^{13}\,\text{kg}\,\frac{\text{m}^2}{\text{s}^2} = 9 \times 10^{13}\,\text{J}$$

where a joule (J) is equal to a watt second. This is enough energy to keep a 100-watt lightbulb lit for 30,000 years.

involving light elements are also very effective in extracting energy from nuclear fuel.

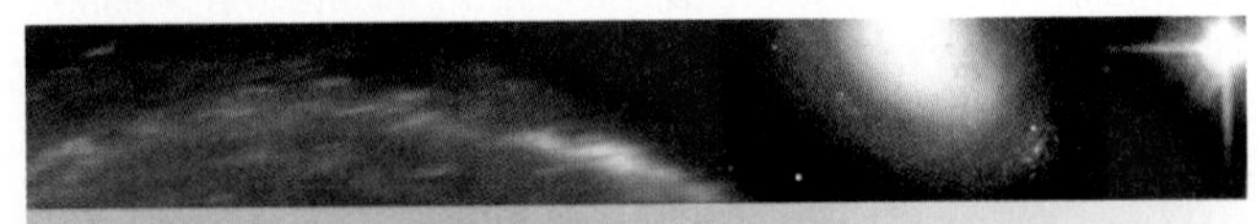

Fission occurs when a heavy nucleus splits to form two less-massive nuclei. In fusion, light nuclei combine to form a heavier one. Nuclear reactions can produce energy by transforming matter into energy. This occurs when the products of the reaction are less massive than the particles and nuclei that take part in the reaction.

Radioactivity

Many of the isotopes of the elements are **radioactive** and decay into other elements by emitting a particle that carries away mass and, for some kinds of decay, electric charge. In any radioactive decay, the masses of the decay products are less than the mass of the original nucleus. The lost mass is converted into energy.

Half-life Suppose 4000 ^{13}N (nitrogen) nuclei were produced in a nuclear physics laboratory. ^{13}N is unstable and decays into ^{13}C at a very predictable rate. In the first 10 minutes after the nuclei of ^{13}N were produced, 2000 of them would decay, leaving 2000 nuclei yet to decay. In the next 10 minutes, 1000 of these would decay, leaving 1000 nuclei after 20 minutes. After 30 minutes, 500 would be left and after 40 minutes, 250 would be left. The number of ^{13}N nuclei left is plotted against time in Figure 7.10, which shows that the number of remaining nuclei gradually approaches zero. The **half-life** of an isotope is the length of time required for the isotope to decline in number by a factor of two. In the case of ^{13}N, the half-life is 10 minutes. For any sample of radioactive material, the number of half-lives that have elapsed gives the fraction of surviving nuclei. After 5 half-lives, only $\frac{1}{32}$ of the original nuclei are left. After 10 half-lives, only about $\frac{1}{1000}$ remain. After 20 half-lives, only a millionth of the nuclei are left. As a practical matter, after 20 or so half-lives, the original sample of radioactive nuclei is essentially gone.

Radioactive Dating The predictable rate of decay of radioactive isotopes makes it possible to use them to find ages for objects that contain them. The most famous isotope used for radioactive dating is ^{14}C, which decays to ^{14}N with a half-life of 5730 years. ^{14}C is produced by nuclear reactions when energetic particles from space strike ^{12}C atoms in the Earth's upper atmosphere. About one in every 10^{12} atoms of carbon in the Earth's atmosphere is ^{14}C. ^{14}C behaves the same chemically as the more abundant carbon isotope, ^{12}C, and is incorporated in living plants and animals in the same way as ^{12}C. Every kilogram of carbon in a living plant or animal contains about the same fraction

FIGURE 7.10
The Number of Remaining ^{13}N Nuclei Declines with Time
^{13}N is an unstable isotope with a half-life of 10 minutes. After 1 half-life, only half of the original 4000 nuclei remain. During each succeeding half-life, the number of remaining nuclei drops by half.

The number of remaining ^{13}N nuclei declines with time.

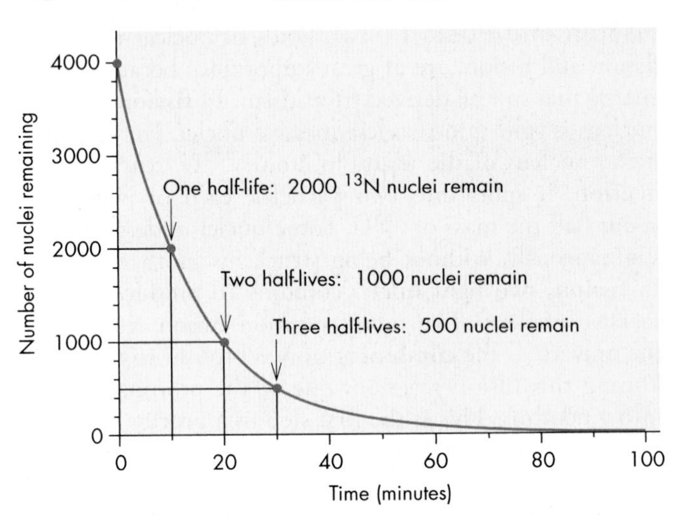

of ^{14}C as does the Earth's atmosphere. When animals and plants die, the ^{14}C in them continues to decay and, because it is no longer resupplied, begins to diminish relative to ^{12}C. A measurement of the relative amounts of the two carbon isotopes can show the length of time that has passed since the plant or animal died. For example, suppose a sample of carbon from the bones of a prehistoric human is found to contain only 0.25 atoms of ^{14}C for every 10^{12} atoms of ^{12}C. This is only one quarter of the abundance of ^{14}C found in the atmosphere and in living plants and animals. The reason is that three quarters of the original ^{14}C in the bones has decayed. This requires two half-lives, or 11,460 years, indicating that the bones are approximately 11,000 or 12,000 years old. ^{14}C dating is a reasonably accurate way to find the ages of the remains of plants and animals that lived within the past 50,000 years (about 9 half-lives).

The half-life of ^{14}C is too short for it to be useful in dating most solar system bodies, which are much older than 50,000 years. The dating of solar system materials relies on isotopes with longer half-lives. Radioactive dating, however, can only be done for bodies from which it has been possible to obtain a sample for analysis. One isotope that has been used for dating Earth rocks, lunar samples, and meteorites is ^{40}K (potassium), which decays to ^{40}Ar (argon) with a half-life of 1.3 billion years. As long as an object is molten, the gaseous argon produced by decays of radioactive potassium can bubble out of the molten material and escape. When the object solidifies and forms rocks, potassium is incorporated in the rocks. When argon is produced, it is trapped within the rock. As time passes, the amount of ^{40}K in the rock falls and the amount of ^{40}Ar increases. By measuring the relative amounts of the two isotopes, an astronomer or geologist can determine the length of time since the rock solidified.

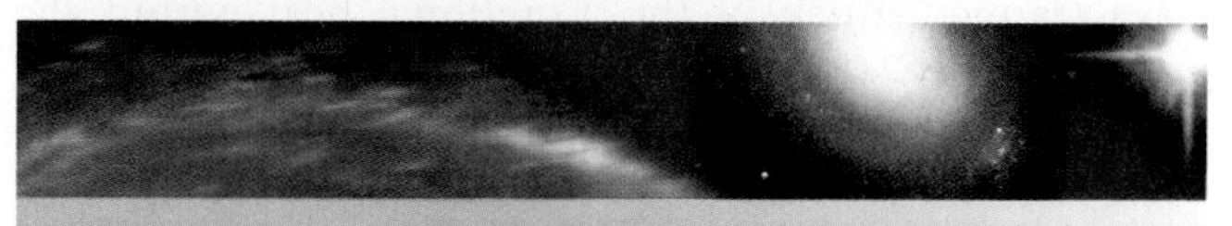

Unstable isotopes decay into other elements. The rate at which an unstable isotope decays is described by its half-life, the time needed for half of the nuclei to decay. The predictable rate of radioactive decay makes it possible to use unstable isotopes to determine the ages of the objects that contain them.

7.5 INTERNAL HEAT IN PLANETS

Radioactive Heating

The energy released by the decay of radioactive isotopes is one of the main sources of heat in the interiors of many planets and satellites. To be an important source of energy an isotope must be fairly abundant, must release significant energy with each decay, and must have a fairly short half-life so that the probability of decay is reasonably large for each nucleus. However, if an isotope has too short a half-life, all of it would have decayed long ago and there wouldn't be enough left today to produce significant heat. Isotopes that aren't important today might have been important sources of energy long ago when they were abundant.

Radioactive heating in the solar system today is mostly from decays of isotopes of K (potassium), Th (thorium), and U (uranium). The energy production rates of these isotopes are very small. It would take all the radioactive isotopes in a cube of rock 1 km on a side to light a 60-watt lightbulb. When the solar system formed about 4.6 billion years ago, each of these isotopes was more abundant and produced more energy. Figure 7.11 shows how the rate of energy released by K and U has declined with time. Some astronomers have suggested that when the solar system formed, it contained almost as much ^{26}Al as ^{40}K. No natural ^{26}Al remains today, because it has a half-life (700,000 years) thousands of times shorter than the age of the solar system. The relatively short half-life of ^{26}Al, however, would have made it an intense source of heat while it was present in the solar system. Figure 7.11 also shows an estimate of the heating rate from radioactive decays of ^{26}Al.

Accretional Heating

Astronomers think that the planets formed by **accretion,** the gradual accumulation from small pieces of matter. The significant generation of heat that occurred during the accretion process is referred to as **accretional heating.** When a small object falls onto the surface of a larger one, it hits the surface with a speed at least as great as the escape velocity of the larger body. The energy it releases when it strikes is equal to its kinetic energy, which increases as the square of the speed at which it strikes. Escape velocity, and thus energy of impact, is greater for more massive planets. To calculate the total energy released as a planet or satellite accreted, we have to add up the energy released by each piece of material as it was added, taking account of the change of escape velocity as the planet or satellite grew. When this calculation is done, the result is that the total energy released during formation, per kilogram of material, increases with the mass of the body. The amount of energy produced per kilogram of material (relative to the amount produced per kilogram as the Earth accreted) is shown for a number of solar system objects in Figure 7.12. Bodies less massive than the Moon produced even less energy per kilogram than did the Moon. For the more massive bodies, such as Jupiter, the amount of energy was immense. However, for the smaller asteroids and satellites, accretional heating was not very important.

FIGURE 7.11
The Rate of Energy Released as Heat by Radioactive Isotopes of Aluminum, Potassium, and Uranium During the Last 4.5 Billion Years
Notice that the rates of energy released by ^{40}K and ^{235}U have declined considerably over the past 4.5 billion years. ^{238}U has declined relatively little because of its long half-life. ^{26}Al, because of its short (700,000 year) half-life, may have been a very important source of energy shortly after the solar system formed, but it is nearly absent from the solar system today.

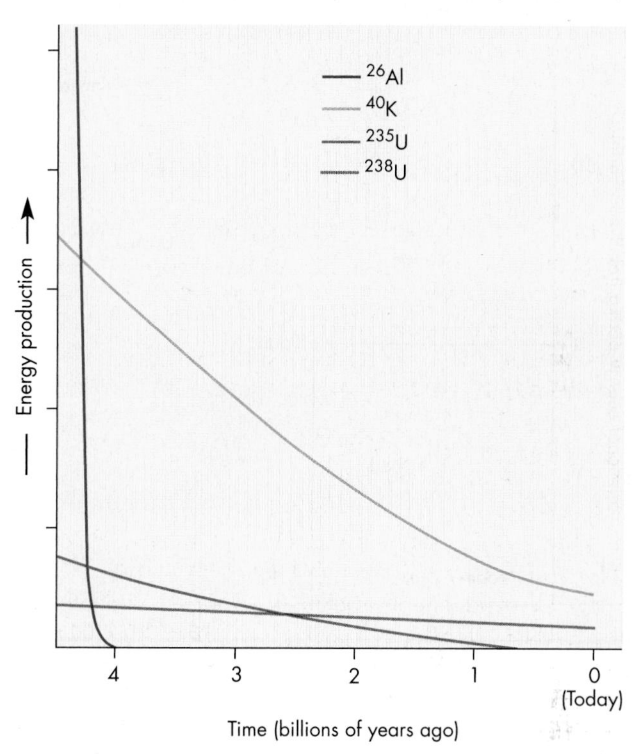

FIGURE 7.12
The Amount of Accretional Energy Per Kilogram Produced During Formation for the Moon and Five of the Planets
The accretional energy per kilogram is shown relative to the accretional energy per kilogram for the case of the Earth. The amount of energy per kilogram increases with the mass of the planet or satellite. This is an example of a graph using logarithmic scales to show several powers of ten in a single graph. (For more information about using logarithmic graphs, see Appendix 16.)

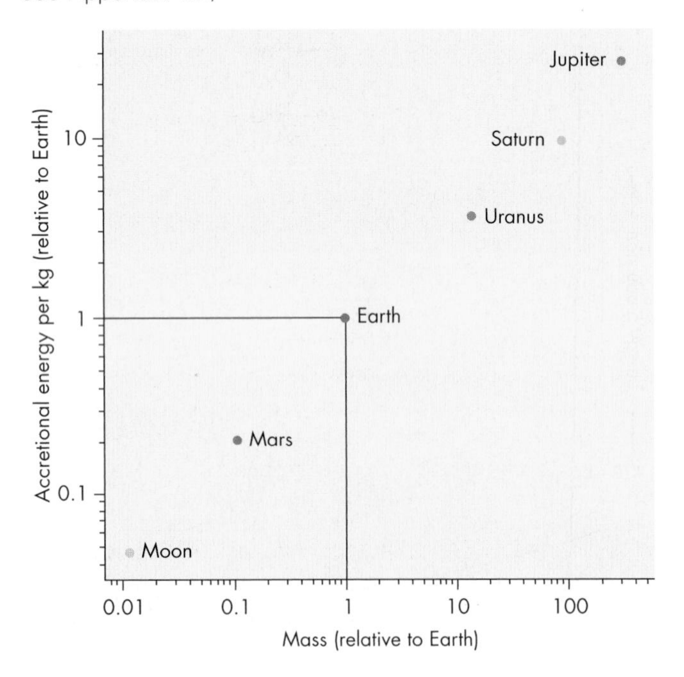

The rate at which accretional heating was released as planets and other objects formed depended on how rapid the accretion process was. Slow accumulation would have permitted accretional heat to escape without raising temperatures very much. Rapid accretion would have trapped much of the accretional heat within the forming body.

Radioactive decays are a major source of heat in the interiors of solar system bodies. The rate of radioactive heating has diminished considerably since the beginning of the solar system because the amount of radioactive material has decreased steadily. An early source of heating was the release of kinetic energy as the bodies accreted. Accretional heating was larger for more massive bodies.

Flow of Heat

Heat always flows from hot regions to cooler ones. The three processes by which heat can be transported from one place to another are **radiative transfer, conduction,** and **convection** (Figure 7.13). Radiative transfer takes place when electromagnetic radiation carries energy from a hot region to a cooler region, where it is absorbed. Radiative

FIGURE 7.13
The Three Processes by Which Energy Is Transported from Place to Place
A, Radiative transfer, in which electromagnetic radiation carries energy from the hotter to the cooler region. The absorption of sunlight by dark objects is an example of radiative transfer. **B,** Conduction, in which atoms, ions, and electrons collide with each other to transport energy. When a metal rod is heated, conduction carries energy to the unheated parts of the rod. **C,** Convection, in which rising and falling cells and currents of gas or liquid carry energy. Sailplanes rely on convection currents to gain altitude.

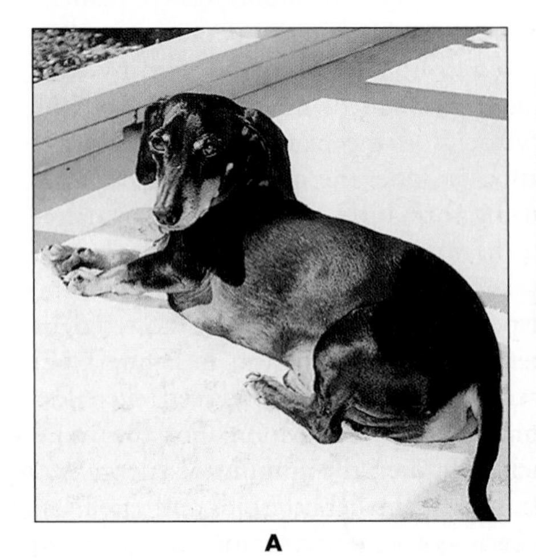

A

B

C

FIGURE 7.14
The Rate at Which Heat Flows Depends on the Rate at Which Temperature Changes with Position
If temperature is nearly uniform, little heat flows. For large temperature differences, the flow of heat is large.

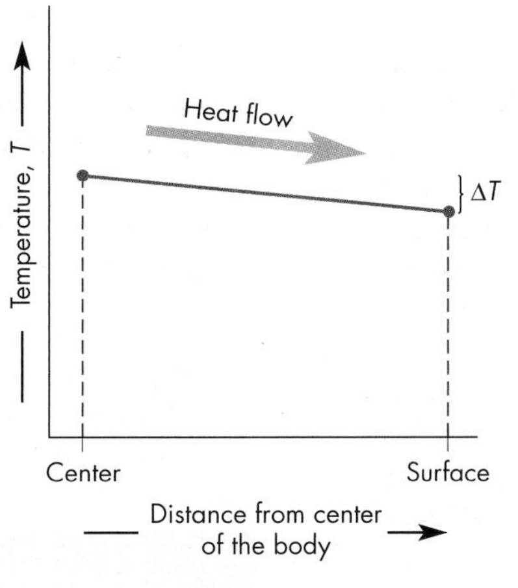

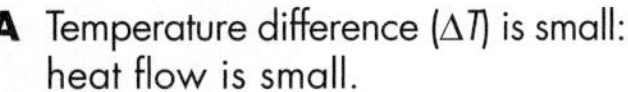

A Temperature difference (ΔT) is small: heat flow is small.

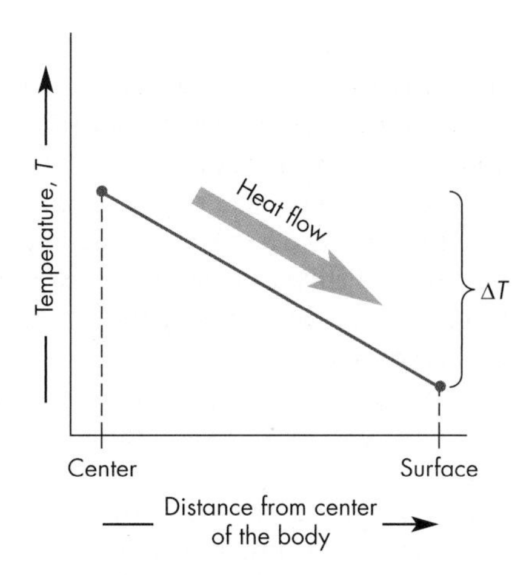

B Temperature difference (ΔT) is large: heat flow is large.

transfer is most efficient in a vacuum or in a hot gas at low density, where photons can travel a long distance between absorptions. The warmth of sunlight is due to the radiative transfer of energy between the Sun and the Earth.

Conduction occurs by collisions among the particles in a gas, liquid, or solid. The collisions gradually carry energy from hot places to cooler ones because they tend to transfer energy from hotter particles to cooler particles. The heat you feel when stirring a cup of coffee with a metal spoon has been transported from the coffee to your fingers by conduction through collisions among electrons in the spoon.

In convection, energy is carried by pockets or currents of a gas or liquid that rise from warm regions toward cool ones. Convection often occurs in a gas or liquid heated from below, such as a pan of water at a rolling boil on a stove. Heat is carried from the bottom of the pan to the surface and into the air by rolling currents of hot water.

Radiative transfer is often the dominant energy transport process in gases, such as in the atmospheres of planets and in stars. It is unimportant in the interiors of planets and satellites. Whether conduction or convection is the dominant mechanism for heat transport depends on whether the material is able to flow (which it must do if convection is to occur) and how much heat needs to be carried. The situation is different for different planets and from place to place within a single planet or satellite.

The heat generated within a body is expended in two ways. It can heat the material at the point where it is generated or it can contribute to the flow of heat toward cooler regions. The rate at which heat flows depends on how rapidly temperature declines from the center of the body to its surface (Figure 7.14). The amount of time required for the temperatures within a solar system body to become steady depends primarily on the size of the body. Large bodies take longer to cool off (or heat up) than do small bodies. This idea is consistent with our everyday experiences. Given a choice between picking up a steel cannon ball or a steel ball bearing, each of which had recently been taken out of the same oven, no one would choose the cannon ball. In the solar system, 100-km size asteroids cooled shortly after they were formed. Jupiter, however, still has a hot interior because it has retained much of the energy it accumulated when it accreted.

The flow of energy occurs through conduction, convection, and radiative transfer. The rate of heat flow depends on how rapidly temperature declines from center to surface within a body. Generally, small bodies cool more rapidly than large ones.

7.6 PLANETOLOGY

Understanding the important processes at work in solar system bodies, including those processes described in this chapter, is essential to **planetology,** the comparative study of the Earth and the other planets. Before the beginning of space exploration, scientists tended to focus on the unique properties of each planet, including the Earth, rather than on the processes that planets share. Space probes have given us a wealth of information about the atmospheres, surfaces, and interiors of the planets (and their satellites). We now know enough about other planets to try to understand their properties in terms of common processes at work in bodies of different mass, diameter, chemical composition, and proximity to the Sun. Take, for example, the relative amounts of atmosphere of the Earth, the Moon, and Mars. We can account for the relatively thick atmosphere of Earth, the thin atmosphere of Mars, and the nearly nonexistent atmosphere of the Moon by examining the process through which a planet loses atmospheric gas. As described in Section 7.2, atmospheric gas has more difficulty escaping from a massive planet, such as the Earth, than from a less-massive one, such as Mars or the Moon.

As another example, there is considerable evidence that volcanic activity is ongoing on Earth and Venus but ceased long ago on Mars, Mercury, and the Moon. Astronomers attempt to account for the different volcanic histories of the inner planets by studying how the rate at which the interior of a planet cools off and solidifies depends on the planet's size and the material from which it is made. As described in Section 7.5, the smaller planets (Mars, Mercury, and the Moon) would be expected to cool off and lose the ability to produce volcanic eruptions long before their larger counterparts (Earth and Venus).

The crucial feature of planetology is that it emphasizes trying to understand a given planet not in terms of itself alone but through comparisons with the other planets. This approach extends to the Earth, as well. Through planetology, we study the Earth as one of a large number of planets and satellites rather than as a single isolated body. By trying to account for the reasons why other planets resemble the Earth in some respects and are so different in others, we have learned a great deal about the Earth. As T. S. Eliot wrote,

> We shall not cease from exploration
> And the end of all our exploring
> Will be to arrive where we started
> And know the place for the first time.

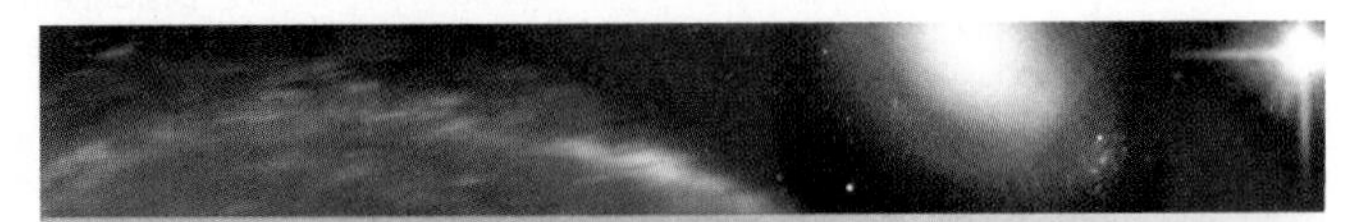

Planetology is the comparative study of the Earth and other planets. By trying to account for the similarities and differences of the planets in terms of common processes at work in them, we can learn to better understand each individual planet.

Chapter Summary

- As temperature increases, so does the average speed of the atoms or molecules in a gas. Average speed declines with increasing atomic mass. A small fraction of the atoms in a gas move much faster than the average speed. (Section 7.2)
- Atoms can't escape from the lower parts of the atmosphere of a planet because they collide with other particles after traveling a short distance. In the thin outer atmosphere, however, a rapidly moving atom is unlikely to collide and be reflected back by another atom before escaping. If the average atomic speed there is greater than about ⅙ escape velocity, all of that kind of atom will escape relatively quickly. (7.2)
- Density is the mass of gas per unit volume. Gas pressure is the force per unit area exerted by the gas. Gas pressure is proportional to the product of density and temperature. (7.2)
- Hydrostatic equilibrium occurs when the decrease of gas pressure with height balances the downward pull of gravity. Pressure falls gradually with height until the atmosphere blends into interplanetary space. (7.2)
- A blackbody is one that absorbs all of the electromagnetic radiation that falls on it. A heated blackbody emits electromagnetic radiation at the maximum rate possible for a body of a given temperature and size. (7.3)
- A blackbody emits electromagnetic radiation with a characteristic spectrum that depends only on the temperature of the blackbody. The wavelength at which the spectrum is brightest grows shorter with increasing temperature. The total rate at which a blackbody emits radiation varies with the fourth power of its temperature. (7.3)
- If the temperature of a body reaches the point where the body emits radiation at the same rate at which it absorbs radiant energy falling on it, the body is said to be in thermal equilibrium. The decline of solar radiation with increasing distance from the Sun leads to temperatures that decline with increasing distance. Highly reflecting bodies are cooler at a given distance than dark bodies, which absorb sunlight well. (7.3)
- Nearly all of the mass of an atom is contained in its nucleus, which is composed of neutrons and protons. The number of protons in a nucleus determines what element it is. Nuclei that are of the same element and differ only in numbers of neutrons are called isotopes. (7.4)

- In some nuclear reactions, matter is transformed into energy. This happens when the particles that react are more massive than the products of the reaction. The two types of nuclear reactions are fission, in which a massive nucleus splits to produce less-massive nuclei, and fusion, in which light nuclei combine to form a heavier one. (7.4)
- Radioactive decay occurs when an unstable isotope decays to form another nucleus. The rate of radioactive decay is described by its half-life, the time needed for half of the nuclei to decay. Radioactive decay is useful for determining the ages of material containing unstable isotopes because the decay occurs at a predictable rate. (7.4)
- For most solar system bodies, radioactive decays are the major source of internal heat. As the amount of radioactive material in the solar system has declined, so has the rate of radioactive heating. Solar system bodies also were heated by the release of kinetic energy as they accreted. However, accretional heating was significant primarily for larger bodies. (7.4)
- Conduction, convection, and radiative transfer are the processes that carry energy from place to place. Within a body, the rate at which heat flows is determined by the decrease of temperature from center to surface. Small bodies cool more rapidly than large ones. (7.5)
- Planetology is the comparative study of the Earth and other planets. Planetology emphasizes trying to understand an individual planet by accounting for the ways in which it is similar to and different from other planets. Different planets show how the same processes work in bodies of different mass, diameter, composition, and distance from the Sun. (7.6)

Key Terms

absolute zero *134*
accretion *145*
accretional heating *145*
albedo *141*
atom *133*
atomic number *142*
blackbody *138*
blackbody radiation *138*
conduction *146*
convection *146*
density *135*
deuterium *142*
electron *142*
element *133*
exosphere *135*
fission *143*
fusion *143*
half-life *144*
hydrostatic equilibrium *136*
ion *133*
isotope *142*
Kelvin temperature scale *134*
mass number *142*
neutral gas *133*
neutron *142*
nucleus *142*
number density *135*
planetology *147*
plasma *133*
pressure *136*
proton *142*
radiative transfer *146*
radioactive *144*
solar constant *140*
Stefan-Boltzmann law *139*
temperature *134*
thermal equilibrium *140*
Wien's law *138*

Conceptual Questions

1. Discuss why atoms can escape from the exosphere but not the lower parts of the atmosphere of a planet.
2. Explain why a massive, cool planet is more likely to have a thick atmosphere than is a less-massive, hot planet.
3. Why does atmospheric pressure decrease with increasing altitude?
4. Describe what happens to the brightness and color of a blackbody as its temperature increases.
5. Suppose the temperature of a body is higher than the temperature at which it would be in thermal equilibrium. Describe how and why the temperature of the body will change with time.
6. A body absorbs 30% of the light that strikes it. What is the albedo of the body? What happens to the light striking the body that isn't absorbed?

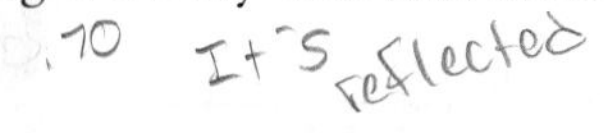

7. One atom has 14 protons and 12 neutrons in its nucleus. Another atom has 12 protons and 12 neutrons. Are these atoms different elements or different isotopes of the same element?
8. Four helium nuclei are more massive than a single oxygen nucleus. Would energy be released if an oxygen nucleus were broken up into four helium nuclei?
9. Explain how radioactive dating methods are used to determine the ages of objects.
10. Why has radioactive heating in the planets declined since the time that the planets formed?
11. What heat transfer process is likely to be most important in (a) a vacuum, (b) an opaque solid, (c) a geyser?

Problems

1. The average speed of the atoms in a gas is 3 km/s. If the temperature of the gas is tripled, what would be the average speed of the atoms?
2. Hydrogen atoms at 300 K have an average speed of 2.5 km/s. At what temperature would hydrogen atoms have an average speed of 1.0 km/s?
3. A gas contains a mixture of hydrogen atoms and neon atoms (which contain 10 protons and 10 neutrons). How does the average speed of the neon atoms compare with the average speed of the hydrogen atoms?
4. Suppose a satellite has an escape velocity of 1.5 km/s. Oxygen atoms in the exosphere of the satellite have a temperature of 300 K and an average speed of 0.6 km/s. To what temperature would the exosphere have to cool to become low enough that oxygen atoms couldn't escape?
5. What happens to the pressure of a gas if both its temperature and density double?
6. What would have to happen to the temperature of a gas in order for its density to double and its pressure to fall to 40% of its original value?
7. At what altitude in the Earth's atmosphere is the pressure equal to $\frac{1}{32}$ of the pressure at sea level?
8. At a depth of 1 km in the ocean the pressure is 10^7 N/m^2. What is the weight of the water that is in a cylinder 1 km high that has a cross-sectional area of 1 m^2?
9. The atmospheric pressure at sea level is 10^5 N/m^2. Estimate the surface area of your body and use that area to calculate the total force pushing inward on your skin.
10. A blackbody is brightest at a wavelength of 800 nm. At what wavelength would the blackbody be brightest if its temperature were tripled?
11. The wavelength at which blackbody A is brightest is four times the wavelength at which blackbody B is brightest. How does the temperature of blackbody A compare with the temperature of blackbody B?
12. A blackbody has a temperature of 5 K. At what wavelength is it brightest?
13. Suppose the temperature of a blackbody increases by 20%. What will happen to the total rate at which it radiates energy?
14. Two blackbodies have the same size. One of them radiates energy at 10 million times the rate of the other. How do the temperatures of the blackbodies compare?
15. How does the brightness of sunlight at Saturn's distance compare with the brightness at 1 AU?
16. Two satellites of the same planet have albedos of 0.8 and 0.3. Which of the satellites will have a higher surface temperature? (Assume that the satellites are identical except in albedo.)
17. Atoms of an unstable isotope are made in a nuclear physics laboratory. After 6 minutes only $\frac{1}{4}$ of the original atoms are left. What is the half-life of the isotope?
18. The half-life of an unstable isotope is 10 years. After what length of time would there be less than 1% of the original atoms of the isotope remaining?
19. What fraction of the original ^{14}C remains after 100,000 years?

Figure-Based Questions

1. Suppose a gas contains a mixture of the elements hydrogen and helium as shown in Figure 7.1. What fraction of the mass of the gas would be due to the helium atoms it contains?
2. Use Figure 7.4 to find the pressure at an altitude of 20 km.
3. Use Figure 7.6 to estimate the relative amounts of infrared radiation emitted by blackbodies of 7500 K, 6000 K, and 4500 K.
4. Suppose there were a planet between Mars and Jupiter at a distance of 4 AU from the Sun. Use Figure 7.8 to estimate the temperature of the planet.
5. Use Figure 7.11 to find the isotope with the greatest heat production (a) 4.5 billion years ago, and (b) 2 billion years ago.
6. Use Figure 7.12 to compare the accretional energy per kg for the Earth and for Mars. What implications might this comparison have for the early histories of the Earth and Mars?

Group Activity

Take a car ride with your partner. Before you leave, measure the air pressure in the tires. Measure the pressure again after riding for a half an hour or so at moderate to high speed. Discuss with your partner how to account for the difference in air pressure in the tires before and after the car was driven.

For More Information

Visit the text website at **www.mhhe.com/fix** for chapter quizzes, interactive learning exercises, and other study tools.

8

could not sleep for thinking
of the sky,
he unending sky, with all its
million suns
Vhich turn their planets
everlastingly
n nothing, where the
fire-haired comet runs.
f I could sail that nothing,
should cross
ilence and emptiness
with dark stars passing,
hen, in the darkness,
ee a point of gloss
urn to a glow, and glare,
and keep amassing,
And rage into a sun with
wandering planets
And drop behind;
and then, as I proceed,
ee his last light upon his last moon's granites
ie to dark that would be night indeed.
ight where my soul might sail a million years
n nothing, not even Death, not even tears.

ohn Masefield
Could Not Sleep for Thinking of the Sky

ww.mhhe.com/fix

The Earth

The study of the Earth is called geology, not astronomy. So why is there a chapter about the Earth in this astronomy textbook? The reason is that the planet Earth is an astronomical body just as Mars or Mercury or Neptune are. This statement seems obvious to us, but it is only since the time of Copernicus, about 500 years ago, that the Earth has been considered one of the bodies that orbit the Sun. Before that time, the Earth was considered to have a unique status. Before Copernicus, a chapter on the properties of the Earth would have had no place in an astronomy book.

Today, a description of the Earth is vitally important to the study of astronomy. This is because the Earth is by far the best investigated and, in most respects, the best understood planet. Although we don't know everything about the Earth, we are even more ignorant about the other planets. It will be a long time before we know as much about the other planets as we now know about the Earth. However, we don't have to start from scratch in trying to understand another planet. Much of what has been learned about the Earth can be applied to other planets as well. The converse is equally important: What we learn about other planets often applies to the Earth. That is, the rocks and minerals of the Earth are similar to those of other rocky planets. The high and low temperature zones in the atmosphere of some other planet are caused by the same kinds of processes that are at work in our own atmosphere. We can't expect another planet to be just like the Earth, but we can use what we know about the Earth as a starting point or a standard for comparison. Sometimes the differences will be great.

Questions to Explore

- How do we know that the planet Earth really revolves and rotates?
- What is the structure of the interior of the Earth?
- What causes mountain ranges, volcanos, and earthquakes?
- Where did the Earth's oceans and atmosphere come from and how have they changed with time?
- Why does air temperature vary with altitude?
- What is the greenhouse effect and how does it influence the temperature of the Earth?

This chapter is about the Earth as a planet. It focuses on those aspects of the Earth that are useful in understanding other planets as well.

8.1 ROTATION AND REVOLUTION

In the Copernican model of the solar system, the Earth (Table 8.1) revolves around the Sun and rotates about its polar axis. For almost 200 years after the death of Copernicus, there was no real evidence that the Earth was a revolving, rotating planet. Today, however, we have compelling evidence of the Earth's revolution and rotation.

Aberration

For several centuries after Copernicus died, astronomers (including Tycho Brahe) tried to verify that the Earth orbited the Sun by searching for stellar parallax (see Chapter 4, Section 4). Ironically, the first evidence of the Earth's revolution resulted from an unsuccessful search for stellar parallax in the 1720s by the English astronomers Samuel Molyneux and James Bradley. They discovered the **aberration of starlight,** an apparent shift in the directions of all the stars that is caused by the motion of the Earth. Molyneux and Bradley used carefully mounted vertical telescopes to observe the positions of stars that pass nearly overhead in London. They found that all the stars in a particular part of the sky appeared to shift back and forth together in an annual cycle. The total shift was about 40 seconds of arc.

An explanation for the shifts occurred to Bradley when he was a passenger on a boat in the Thames. He noticed that a windblown streamer on the mast changed direction each time the ship turned. The direction of the wind hadn't changed, it was the motion of the ship that caused the streamer to change direction. Bradley reasoned that the shifting positions of the stars were due to the changing direction in which the Earth moved around the Sun. To see how the direction of a star depends on the motion of the Earth, imagine that you are using an umbrella to keep dry during a rainstorm. Assume that the rain is falling vertically. As long as you aren't moving, you need to point the umbrella straight up, the direction from which the raindrops are falling. If you begin to move, however, you need to tilt the umbrella forward to stay dry as shown in Figure 8.1. The faster you move, the more the raindrops seem to move toward you and the more you have to tilt the umbrella. Figure 8.2 shows the Earth moving around the Sun and the orientation of a telescope pointed at a star that lies perpendicular to the Earth's orbital plane. To point at the star's apparent position, the telescope must be tilted slightly away from the star's true position. The tilt is in the direction in which the Earth is moving. In the case of an umbrella and raindrops, we can walk or run almost as fast as the raindrops are falling, so the amount of tilt can be quite large. In contrast, the Earth's speed, 30 km/s, is only about 0.01% as fast as the speed of light. Thus, for the aberration of starlight the tilt is very small, about 20 seconds of arc each way or a total of about 40 seconds of arc.

After Bradley published his results in 1729, there was no longer any reason for astronomers to cling to the geocentric model of the universe, with its stationary Earth.

FIGURE 8.1
Moving in a Rainstorm
If the rain is falling straight down, people who are standing still hold their umbrellas directly over their heads. People who are walking or running must tilt their umbrellas in the direction in which they are moving in order to stay dry.

Table 8.1
Planetary Data

Earth

Orbital distance	1.0 AU
Orbital period	1.0 years
Mass	1.0 M_{Earth} = 5.974 × 10^{24} Kg
Diameter	1.0 D_{Earth} = 12,756 km
Density (relative to water)	5.52
Escape velocity	11 km/s
Surface gravity	1.0 g
Global temperature	280 K
Main atmospheric gases	O_2, N_2
Rotation period	1 day
Axial tilt	23°
Known satellites	1
Distinguishing features	Life, liquid water

Aberration is the apparent change in the position of a star whenever the Earth's motion carries it in any direction except directly toward or away from the star. Aberration is analogous to tilting an umbrella when moving about during a rainstorm. The discovery of aberration was the first direct evidence for the revolution of the Earth.

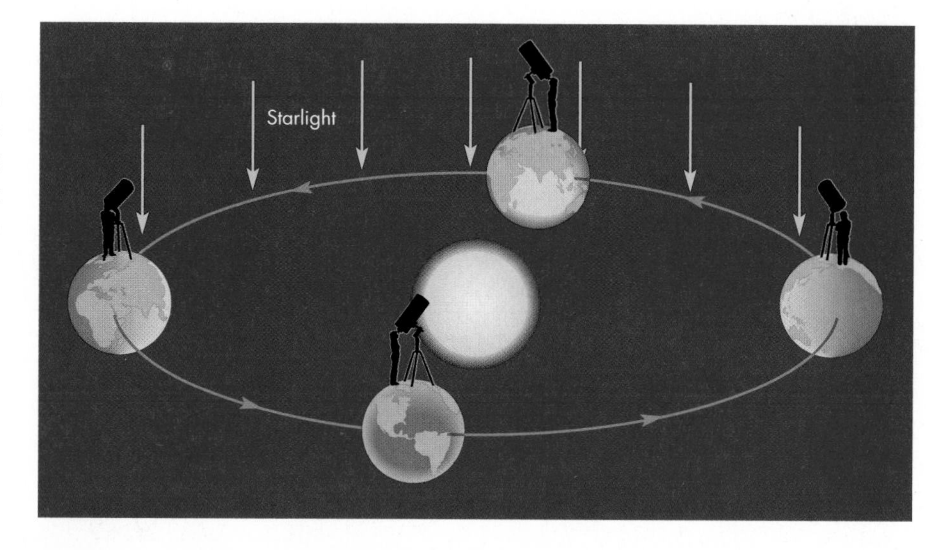

FIGURE 8.2
The Aberration of Starlight
As the Earth moves perpendicular to the direction of a star, the telescope must be tilted slightly forward to point in the apparent direction of the light from the star. The amount of tilt shown here is greatly exaggerated.

The Foucault Pendulum

Jean Foucault used a pendulum in Paris in 1851 to demonstrate that the Earth rotates. Foucault pendulums, such as the one shown in Figure 8.3, can be found today in many science museums and planetariums. If such a pendulum, free to rotate at its pivot point, is set swinging, the direction of its swing appears to change. The explanation is most easily seen for a hypothetical pendulum suspended over one of the Earth's poles. As the pendulum swings, its momentum keeps it swinging in the same arc. Figure 8.4 shows, however, that the Earth moves under the pendulum so the pendulum appears to change the direction of its swing. Its direction of swing completes a full circle in one sidereal day, $23^h\ 56^m$. Away from the poles, the period of time needed for the direction of swing to rotate through a full circle grows longer as the equator is approached. At the equator, the Foucault pendulum never changes the direction of its swing.

FIGURE 8.3
A Foucault Pendulum
As the pendulum swings back and forth, the Earth rotates underneath it so that the direction of the pendulum's swing appears to change.

A Foucault pendulum

"Centrifugal Force"

When a car turns a corner, a force seems to push its passengers against the side of the car. In reality, there is no force pushing the passengers against the side of the car. What really happens is that the inertia of the passengers causes them to continue moving in a straight line while the car is turning. As the car turns, it gets in the passengers' way and pushes against them. The fictitious force that people use to explain what happens to them when a car turns a corner is often referred to as *centrifugal force.*

Because the Earth is rotating, we experience effects similar to those of the passengers in the car. Our inertia causes us to continue moving in a straight line as the Earth rotates under us. We are constantly being hurled away from the Earth by our inertia but pulled inward by gravity. As a result, people (and everything else) weigh less than they would if the Earth weren't rotating. The inertia of the Earth itself causes it to be deformed from a

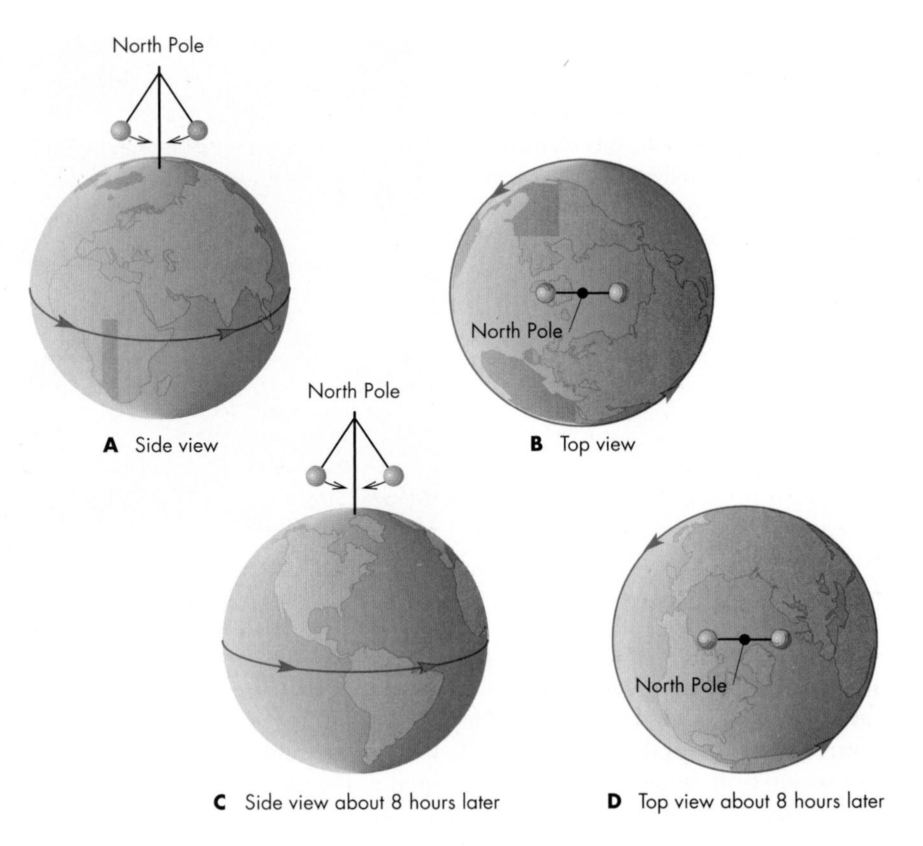

FIGURE 8.4
The Earth Moves Under a Foucault Pendulum
A Foucault pendulum above the North Pole of the Earth would seem to rotate the direction in which it swings because the Earth turns beneath the pendulum. The direction of the pendulum's swing would complete a full circle in $23^h\ 56^m$, one sidereal day.

spherical shape. The consequences of inertia are greatest at the equator, where the speed at which the Earth's surface rotates is largest. The effect falls to zero at the poles. Because of its rotation, the Earth has an **oblate** shape, as shown in Figure 8.5. The Earth is about 0.3% (or 40 km) larger in equatorial diameter than it is in polar diameter. Other planets are also deformed by their rotations. The amount of oblateness of a planet depends on its size and rotation rate as well as the kind of matter it contains and the way matter is distributed inside the planet.

Because a person at the equator is farther from the center of the Earth than a person at the poles, gravity is weaker at the equator. The lower gravitational attraction of the Earth at the equator combines with the inertial reduction of weight to lower a person's weight by about 0.5% at the equator compared with the same person's weight at the poles. For latitudes in the continental United States, weight is about 0.3% more than at the equator. Thus a person who weighs 200 pounds could lose about ⅔ of a pound by moving from the United States to Ecuador.

Coriolis Effect

The **Coriolis effect** is another fictitious force caused by the Earth's rotation. The Coriolis effect appears to curve the paths of moving objects, winds, and water currents

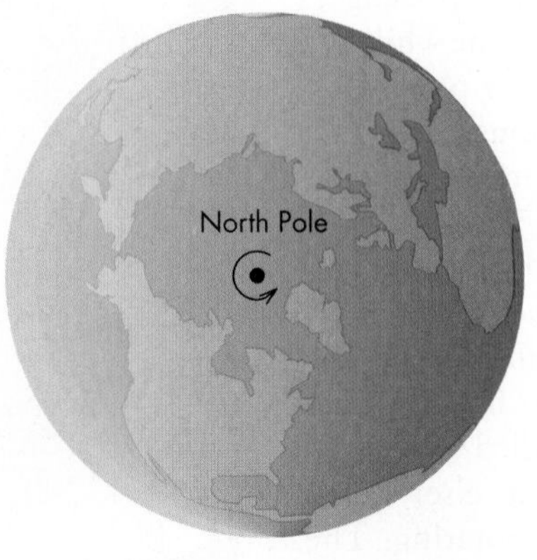

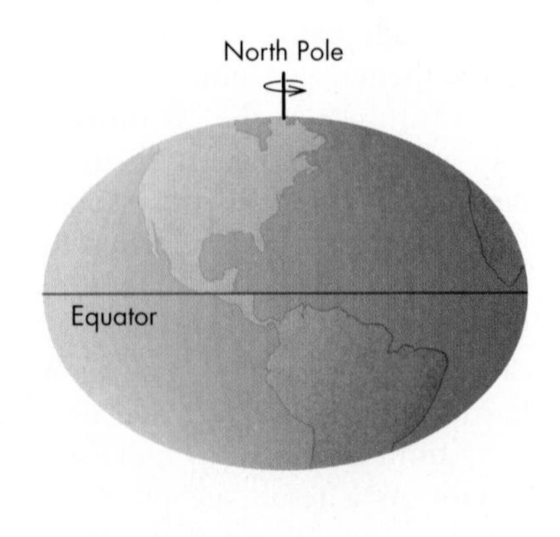

FIGURE 8.5
The Oblate Earth
The Earth's rotation distorts the shape of the planet from a sphere to an oblate spheroid. **A,** Seen from above the North Pole, the Earth appears round. **B,** From a position above the equator it appears elongated in the direction of the equator. The amount of deformation is greatly exaggerated in the figure. In reality, the Earth has an equatorial diameter only 40 km greater than its polar diameter.

to the right in the northern hemisphere and to the left in the southern hemisphere. Suppose a projectile is fired straight north from a cannon located on the equator. In addition to its northward motion, the projectile is also moving 1675 km/hr eastward because of the Earth's rotation. However, 10° north of the equator, the Earth is turning eastward at only about 1650 km/hr. When the projectile reaches that latitude it will be moving eastward 25 km/hr faster than the countryside over which it passes. If its path is plotted on a map, it will curve eastward as shown in Figure 8.6. Any projectile fired away from the equator moves faster than the land over which it passes and curves eastward. A projectile fired toward the equator moves eastward more slowly than the landscape and curves westward. The Coriolis effect must be taken into account when aiming artillery shells. For instance, a shell fired southward at 3000 km/hr in the United States would curve westward by about 1 km by the time it traveled 60 km to the south.

The real significance of the Coriolis effect is its influence on the motions of currents of air or water. For example, air flows inward toward a low-pressure region in the atmosphere. If that low-pressure region is in the northern hemisphere, as in Figure 8.7, air flowing in from the south will curve eastward. Air from the north will curve westward, setting up a circulation pattern that is counterclockwise when seen from above (as on weather maps and satellite photographs). The effect is just the opposite in the southern hemisphere. The air flowing outward from a high-pressure region causes a clockwise circulation pattern in the northern hemisphere, counterclockwise in the southern. These patterns are familiar to anyone who has experienced a hurricane, an intense low-pressure region (Figure 8.8). Elsewhere in the solar system, we would expect spiral-like circulation patterns to be common only for rapidly rotating planets, where the Coriolis effect is large.

FIGURE 8.6
The Coriolis Effect
A projectile fired northward from the equator appears to veer eastward because it retains the eastward speed of the equator, which is greater than the eastward speeds north or south of the equator. Because of the Coriolis effect, projectiles and currents of air and water veer to the right in the northern hemisphere, **A,** and to the left in the southern hemisphere, **B.**

ANIMATION *The Coriolis effect*

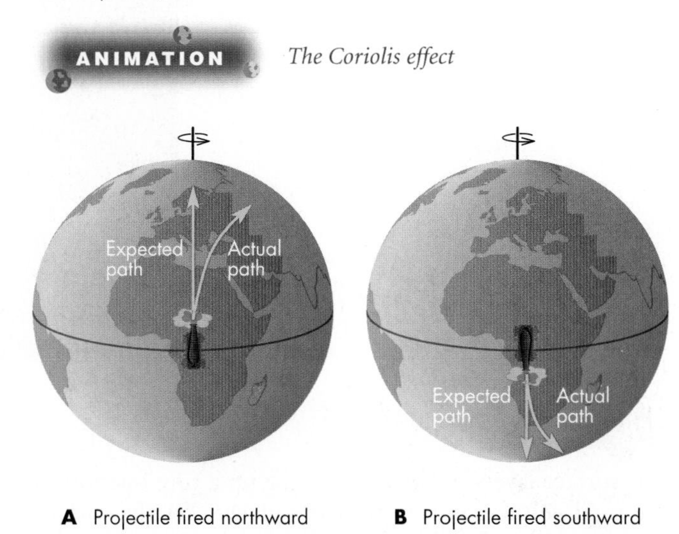

A Projectile fired northward **B** Projectile fired southward

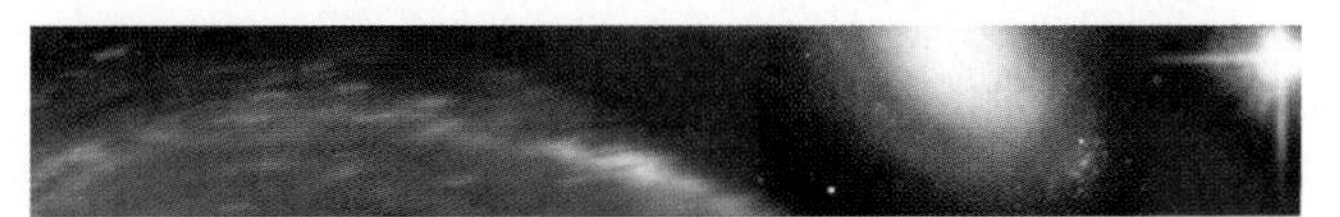

Rotation causes the Earth to have an oblate shape. The Coriolis effect causes moving objects and air masses to appear to be deflected from straight line paths. It is responsible for spiral-like circulation patterns in the atmosphere and oceans.

FIGURE 8.7
The Circulation Pattern Around Low-Pressure and High-Pressure Regions
A, In the northern hemisphere, air flowing into a low-pressure region veers to the right, setting up a counterclockwise rotation pattern. **B,** High-pressure regions, from which air flows outward, are accompanied by clockwise rotation patterns in the northern hemisphere.

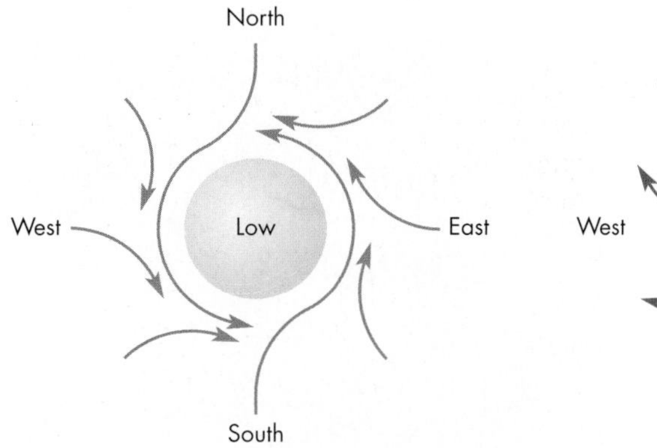

A Low-pressure region in northern hemisphere

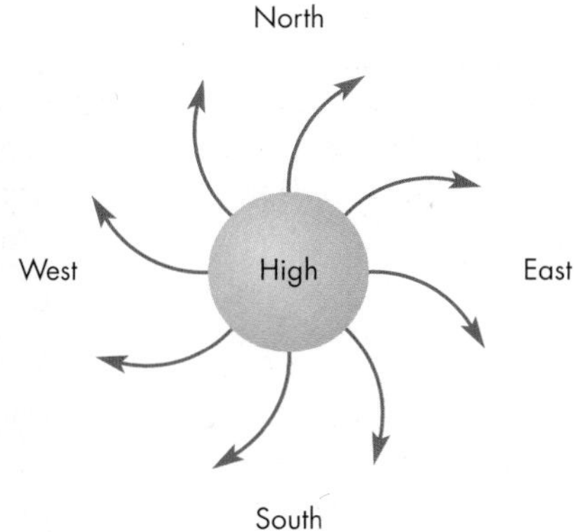

B High-pressure region in northern hemisphere

FIGURE 8.8
The Cloud Patterns in Hurricane Linda in 1997
The spiral pattern of cloud features in a hurricane is a consequence of the Coriolis effect. A hurricane is an intense low-pressure region with a strong counterclockwise wind pattern. The pattern of winds stretches cloud features into the observed spiral shape.

8.2 SURFACE

The surface layers of the Earth make up a tiny fraction of the total volume and mass of the planet. Yet the surface layers are the only part of the Earth that we can directly sample. The surface layers give vital information about the age of the Earth, its chemical makeup, and the dynamic processes that are at work in it.

The Age of the Earth and Its Rocks

The length of time over which significant changes in the Earth occur are usually very long when compared with the lifetime of a person or with all of human history. As far as we know, the first people to begin to comprehend the great age of the Earth were the Greeks. About 2500 years ago, Xenophanes examined beds of rock in which there was horizontal layering called stratification, as shown in Figure 8.9. He reasoned that such rocks originated as layers of sediments on the seafloor. Given the thickness of the sedimentary layers in some places, he concluded that the Earth was very old. A little later, the historian Herodotus observed that every flood of the Nile laid down a thin layer of sediment in the Nile delta. He reasoned that it would have taken thousands of years for similar floods to have produced the delta.

FIGURE 8.9
Stratification in the Walls of the Grand Canyon
Beds of rock showing horizontal layering formed as layers of sediment on the ocean floor.

These early efforts to determine the age of the Earth were eventually forgotten and replaced by attempts to date the beginning of the Earth on the basis of the sequence of biblical events. The most famous of these efforts was by Archbishop James Ussher of Ireland in 1654. He concluded that the Earth was created in 4004 B.C. A more scientific approach was taken about a century later by Comte de Buffon of France. He guessed that the interior of the Earth was iron and calculated how long it would have taken to cool from a molten state. The result was 75,000 years. This was already in conflict with many geologists, who had concluded that millions of years would have been necessary to lay down the entire sequence of stratified rocks that had been found.

Early in this century, the discovery of radioactivity led to measurements of the ages of rocks. The initial results surprised even geologists, for they showed that some rocks were not millions, but hundreds of millions or billions of years old. Throughout the twentieth century, discoveries of more and more ancient rocks have pushed the record for "the Earth's oldest rock" back to about 4 billion years.

The oldest rocks yet found are younger than the Earth because even older rocks have been eroded, covered up, or folded into the Earth's interior. It is likely that the Earth and the other planets are about as old as the oldest meteorites (see Chapter 15, Section 2), about 4.6 billion years. Such an immensity of time is practically impossible to comprehend. Suppose the history of the Earth were represented by a single day on the hands of the clock shown in Figure 8.10. Each hour would represent 190 million years and each minute 3 million years. Each second would represent 50,000 years, or nearly 1000 human lifetimes. All of recorded history would fill only 1/10 second. Just as the universe is far bigger than we can really understand, it is also far older. Intervals of time like millions and billions of years are routinely used by geologists and astronomers. Even astronomers are occasionally startled by the realization of how far beyond human experience such quantities of time really are.

Minerals and Rocks

didn't talk about

Minerals A **mineral** is a solid chemical compound. There are an enormous number of ways that the more than 100 chemical elements could combine to form minerals. The most common minerals in the Earth are **silicates,** which combine silicon and oxygen, the two most abundant elements in the Earth's crust. Four oxygen atoms bond to each silicon atom to form a unit. A vast number of different minerals can be built by connecting these units in chains and rings. In these minerals, atoms such as sodium, calcium, iron, and others can be located in the spaces between the silicon-oxygen units. An experienced geologist can distinguish hundreds of different minerals on the basis of differences in hardness, color, density, and many other qualities.

FIGURE 8.10
The Age of the Earth Represented by a Single Day
The actual age of the Earth is 4.6 billion years, so each hour represents 190 million years, each minute represents 3 million years, and each second represents 50 thousand years. If the day representing the history of the Earth began at midnight, then life began at about 5 A.M., the dinosaurs became extinct at about 11:40 P.M., and recorded human history began at about 11:59:59.9 P.M., only a tenth of a second ago.

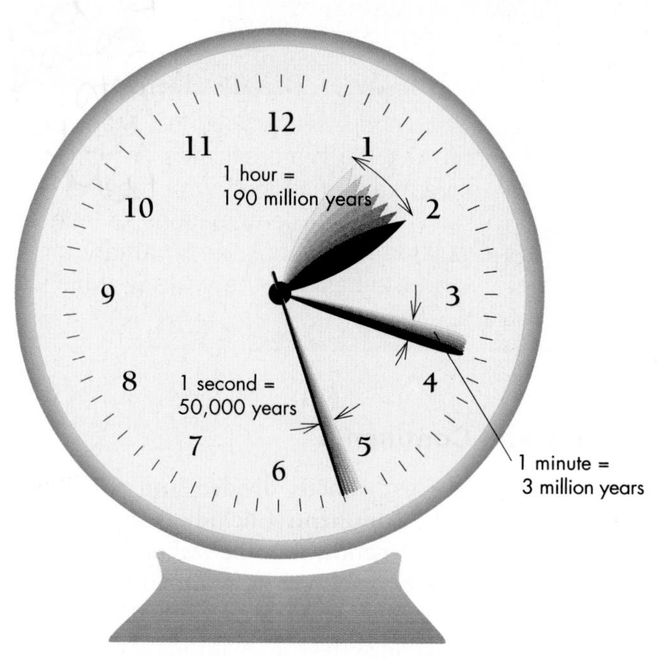

Rocks A **rock** is a solid aggregate of grains of one or more minerals. Rocks can be grouped on the basis of their chemical makeup or their texture and appearance. But the most common system of grouping rocks is on the basis of their origin. The three major categories are igneous, sedimentary, and metamorphic.

Igneous Rocks **Igneous rocks** are those that have solidified from molten material. Molten rock is called **magma** when it is beneath the surface of the Earth and **lava** when it flows on the Earth's surface. One of the most common igneous rocks is **basalt,** the characteristic rock of the Earth's crust under the oceans. Basalts are also common on the Moon, Mars, and Venus.

Sedimentary Rocks A **sedimentary rock** is one that formed by the accumulation of small mineral grains carried by wind, water, or ice to the spot where they were deposited. The grains then became cemented together to form a rock. Limestone, sandstone, and shale are common sedimentary rocks. Sedimentary rocks are usually deposited as relatively thin horizontal layers. This deposition has made it possible to use them as a kind of geological clock, as described earlier in this section.

Metamorphic Rocks **Metamorphic rocks** are rocks that originally formed as igneous or sedimentary rocks. They were then transformed into new forms by high temperatures and pressure either by being buried deeply or by contact with molten material near the surface. Limestone, for example, becomes marble after metamorphosis.

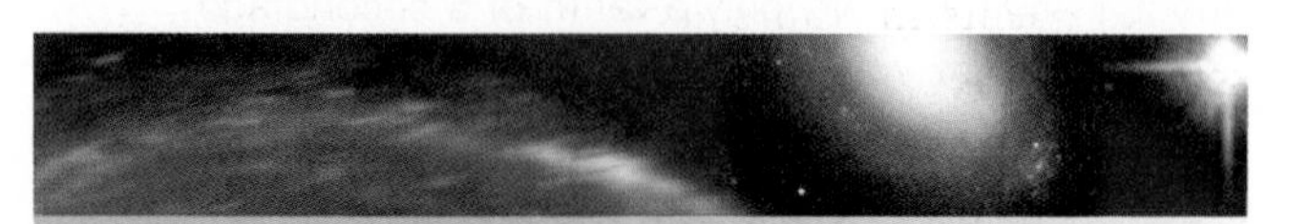

Minerals are solid chemical compounds. Many of the minerals in the Earth are silicates, in which silicon and oxygen are combined with other elements. Rocks, which are combinations of minerals, are classified as igneous, sedimentary, or metamorphic on the basis of the ways in which they formed.

Oceans and Continents

The Earth's surface is distinctly divided into two major parts. These are the continents (including surrounding shallows known as the continental shelves) and the ocean floors. Extremes of elevation, such as mountains and ocean trenches, make up a small fraction of the Earth's surface. Most of the continental surface lies within a few hundred meters of sea level. Most of the ocean floor lies between about 3 km and 5 km below the ocean surface. The transition from continental shelf to deep ocean is usually rather abrupt, as is shown in Figure 8.11.

The central parts of the continents, the continental shields, are ancient. Rocks from many of them are older than 2 billion years. Away from the continental shields, the rocks are generally younger. Major mountain regions tend to be near the margins of the continents. The ocean floor is dominated by broad, abyssal plains covered by layers of sediments. Figure 8.11 shows that abyssal plains are split by midocean ridges that extend for tens of thousands of kilometers. The ridges are the youngest parts of the oceans, and there are no ocean rocks older than about 200 million years.

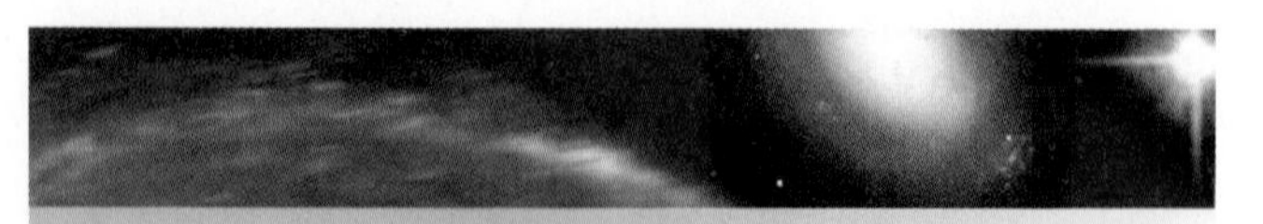

The surface of the Earth is divided into continents (including continental shelves) and oceans. The rocks of the continents are ancient, but oceanic rocks are younger than a few hundred million years.

FIGURE 8.11
The Floor of the North Atlantic Ocean
At the edge of the continental shelf, which borders the continents, the ocean floor drops off rapidly to the abyssal plain. The mid-Atlantic ridge, a submerged mountain range, runs generally north to south along the middle of the abyssal plain.

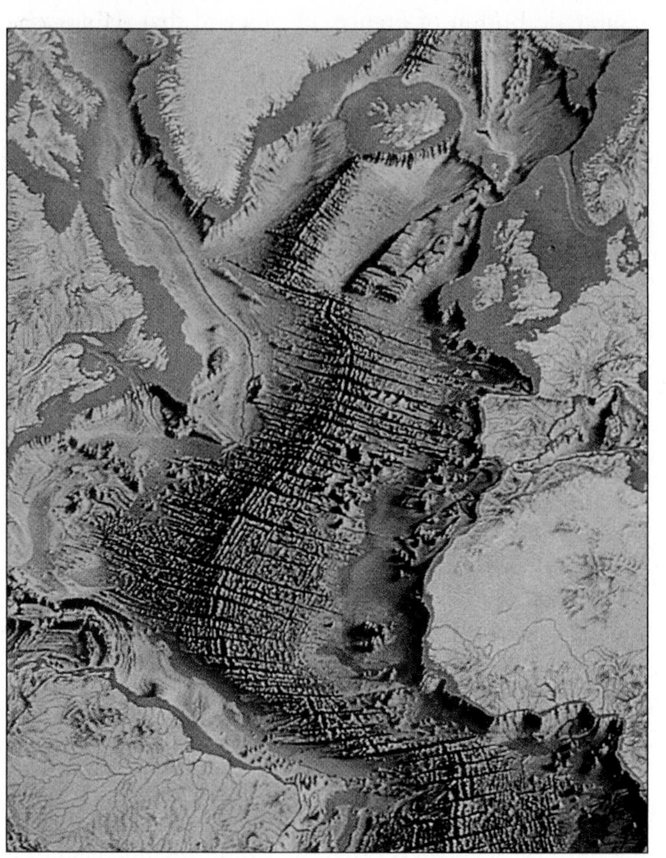

8.3 INTERIOR

The Earth's interior is as inaccessible to us as another galaxy. Miners in South Africa have penetrated about 4 km into the Earth, only about 0.06% of the distance to the center. Holes have been drilled deeper into the Earth, but the deepest of these is only about 15 km. Even so, we know a great deal about the interior of the Earth. Information from heat-flow measurements, studies of seismic waves, and other sources, such as volcanic rocks that come from great depths, has made it possible to piece together a comprehensive picture of the interior.

Probing the Interior

Internal Density and Pressure The average density of the Earth, its mass divided by its volume, is 5500 kg/m^3, or 5.5 times the density of water. By comparison, the densities of most surface rocks are in the range of 2000 to 3500 kg/m^3. This means that the density of the Earth's interior must be even greater than 5500 kg/m^3.

The interior pressure of the Earth is caused by the weight of the overlying matter, just as the pressure in the Earth's atmosphere is caused by the weight of overlying air. Pressure increases steadily with depth to a central pressure of about 3.5 megabars (Mb). One Mb is one million times the atmospheric pressure at sea level. The average pressure in the Earth's interior is about 1 Mb, so most of the material in the Earth is subject to very high pressure.

Consequences of High Pressure There are several important consequences of the great pressures inside the Earth and other planets. One of these is compression, or an increase in the density of material when it is subjected to crushing pressures. Sometimes the density of a rock increases abruptly as pressure increases. The reason for the abrupt change is that the structure of the minerals that make up the rock changes to a more compact, denser form.

Pressure also determines whether a substance is liquid (molten) or solid at a given temperature. Raising the pressure can cause a molten material to resolidify, resulting in a transition from liquid to solid as depth increases.

The pressure inside the Earth increases with depth and has a typical value of about one million atmospheres. Such enormous pressure crushes the material inside the Earth, causing its density to increase. High pressure can also produce changes from one mineral to another and resolidify molten material.

Heat Flow and Temperature Nearly all of the energy that reaches the surface of the Earth is sunlight, which penetrates the atmosphere and is absorbed by the ground. Only about 1/5000 as much energy comes from the interior of the Earth. The internal heat is due to radioactive decays and heat remaining from the era of the Earth's formation. Heat flow, the rate at which heat reaches the surface from the interior, has been measured at many locations on the continents as well as on the ocean floor. The surface layers of the continents are mostly granite, which contains a relatively large amount of radioactive uranium, thorium, and potassium. The "hot" rock in the upper 10 km or so of the Earth contributes somewhat less than half of the heat flow on the continents. The rest comes from deeper layers of the planet. The basaltic rock of the ocean floor contains only about 1/6 as much radioactive material as the continental granite. Nevertheless, the heat flow in the ocean floor is about as great as that of the continents. Most of the heat in the ocean floor comes from the relatively hot, young rock that makes up the ocean bottom.

Because heat flows outward through the Earth's surface, the interior must be hotter than the surface. Measurements in mines show that as you move inward the temperature increases by about 20 to 30 K per km. Direct evidence that temperature continues to increase below the depth of the deepest drillings comes from volcanos. Volcanic magmas, which come from depths of many kilometers, reach the surface at temperatures that exceed 1000 K.

The Earth's Magnetic Field The reason that a compass works has been known since 1600, when William Gilbert proposed that the Earth acts like a big magnet whose field aligns the small magnet used as a compass needle. The magnetic field of the Earth is very much like that which would exist if there were a powerful permanent bar magnet located near the center of the Earth. As shown in Figure 8.12, the magnetic axis is inclined by about 11° to the Earth's rotation axis, so magnetic and geographic north aren't exactly the same at most locations on the Earth's surface.

Although a bar magnet seems to be a good analogy for the Earth's magnetic field, the Earth can't really be a bar magnet. When magnetic materials are heated, they lose their permanent magnetism at a temperature of about 800 K. Because all but the outer 10 to 20 km of the Earth is hotter than this, there can't be a permanent magnet near the center of the planet. Instead, the Earth's magnetic field is probably the result of a **dynamo,** in which the rotation of the Earth causes electric currents to flow in liquid, electrically conducting portions of the planet's interior. The electric currents, in turn, produce a magnetic field in the same way that a current passing through a copper wire produces an electromagnet. The dynamo model for the Earth's magnetic field requires that somewhere within the Earth there must be large scale fluid motions of conducting (probably metallic) material. Planets that lack either relatively rapid rotation or electrically conducting fluid in their interiors cannot produce magnetic fields through dynamos.

The dynamo model is supported by evidence of changes in the Earth's magnetic field. The most dramatic changes are reversals of the direction of the magnetic field on time-scales of hundreds of thousands of years or less. The evidence for such changes is found in the alignment of magnetic materials within lava beds. When lava cools, the magnetic materials within the lava become aligned with the direction of the Earth's magnetic field. In lava beds where lava flows have recurred over long periods of time, the alignment of the magnetic materials in the

FIGURE 8.12
The Earth's Magnetic Field
The magnetic field of the Earth resembles that of a bar magnet tipped by about 11° with respect to the Earth's spin axis.

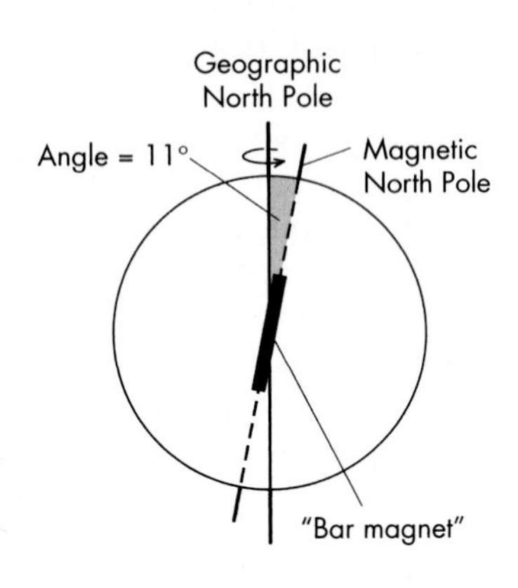

A Geographic North Pole vs. magnetic North Pole

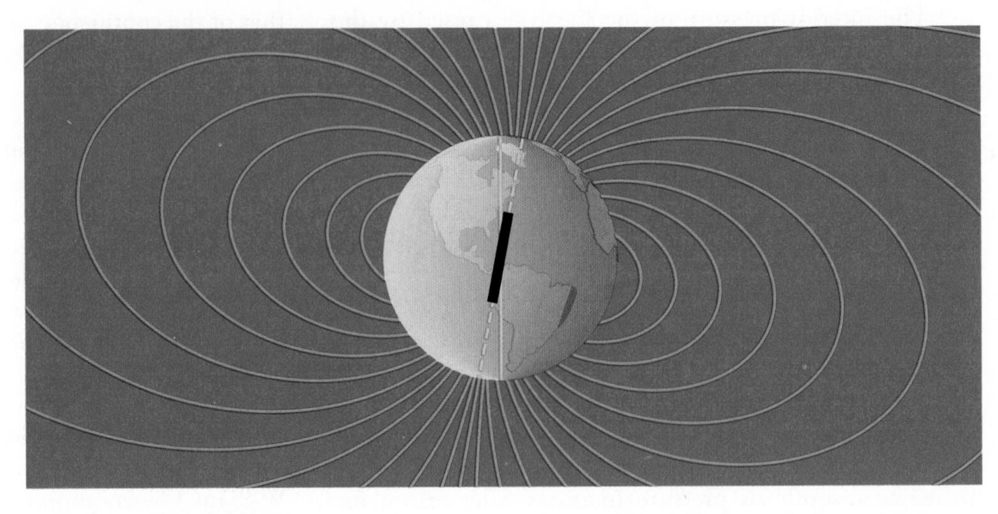

B Earth's magnetic field lines

FIGURE 8.13
Evidence of Magnetic Field Reversals
The magnetic materials in lava are aligned with the Earth's magnetic field. When the lava solidifies, it records the direction of the field. In lava beds, where fresh lava flows have covered up older lava many times, the orientation of the magnetic materials shows abrupt changes with depth. The thicknesses of layers with the same magnetic orientation show that field reversals occur on a timescale of hundreds of thousands of years.

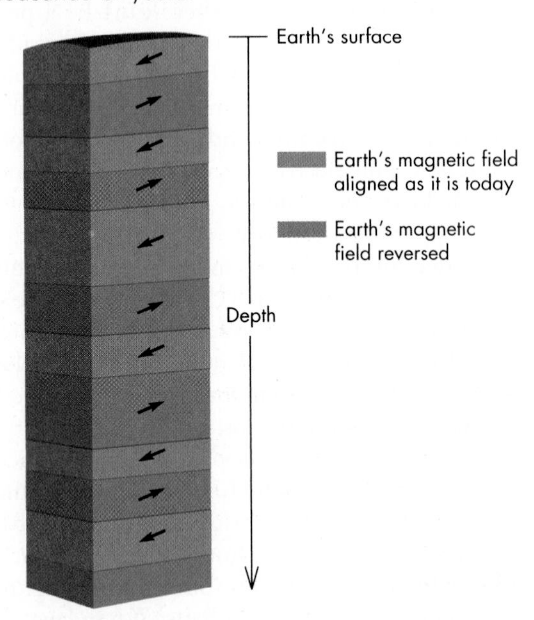

rock switches back and forth with depth, as shown in Figure 8.13. The rock record shows that as recently as 30,000 years ago the Earth's north magnetic pole was located near the present south magnetic pole. Although the reason that there are reversals in the Earth's magnetic field is not well understood, one possibility is that the reversals occur when the pattern of fluid motions within the Earth undergoes a change.

The Earth's magnetic field resembles that of a bar magnet. However, it is actually produced by currents of conducting material in the Earth. The direction of the Earth's magnetic field undergoes reversals on a timescale of hundreds of thousands of years.

Seismology Earthquakes occur when sections of the Earth's crust move suddenly with respect to each other. Nearly a million earthquakes are recorded each year with sensitive measuring devices called **seismometers.** Most earthquakes are too weak to be felt even by people who are near the epicenter of the quake. However, hundreds of earthquakes capable of damaging buildings occur each year. An earthquake capable of destroying a city occurs once every few years. The most powerful earthquakes release as much energy as thousands of nuclear bombs.

Although they are destructive, earthquakes are useful to geologists because they produce **seismic waves,** which

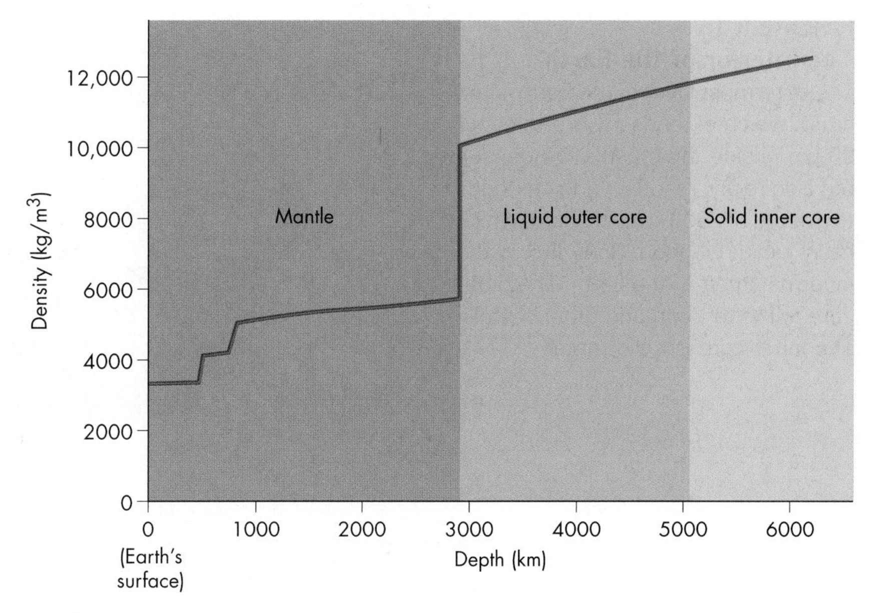

FIGURE 8.14
Density Versus Depth in the Earth
Regions in which density changes rapidly as depth increases correspond to zones in which the properties of the materials of the Earth change rapidly with depth.

travel at great speed through the Earth. Seismic waves give us information about composition and density at different depths within the Earth. When an earthquake occurs, seismic waves radiate out in all directions in the Earth. Seismic waves that reach a point far from the epicenter of the earthquake must travel deep within the Earth. Thus, the times required for seismic waves to travel to different distances from the epicenter of an earthquake show how the speed of the seismic waves varies with depth in the Earth. Because the speed of seismic waves depends on the density of the rock the waves are passing through, the arrival times of seismic waves give important information about the way that density changes with depth beneath the Earth's surface.

The density profile of the Earth's interior is derived from seismic studies and is shown in Figure 8.14. At some depths there are abrupt changes in density, indicating changes in the chemical makeup, mineral structure, or physical state of the Earth at those depths. Generally, though, density increases from the surface to the center of the Earth.

Information from seismic waves allows us to determine the density at all depths in the Earth. Zones in which density changes rapidly with depth are regions in which there are changes in chemical composition, mineral structure, or physical state.

Internal Structure

The differentiation of the Earth's core

Crust, Mantle, and Core The information from seismic studies has led to a model of the Earth in which the interior is divided into the **crust, mantle,** and **core,** as shown in Figure 8.15. The boundary between the crust and the mantle is marked by an abrupt increase in the speeds of seismic waves. The thickness of the crust varies greatly, as shown in Figure 8.16. Beneath the oceans it is only about 10 km thick, but it ranges from about 35 km to more than 50 km under the continents.

The upper mantle, like the crust, is solid and rigid. This region, which extends down to about 70 km, together with the crust is called the **lithosphere.** At a depth of about 70 km, temperature is high enough that the material of the mantle is partially melted. The mantle becomes completely solid again at a depth of about 250 km where increasing pressure raises the melting temperature enough to end the zone of partial melting. The partially melted zone between depths of 70 km and 250 km of the mantle is called the **aesthenosphere.** It is somewhat plastic, or puttylike, and flows relatively easily. It is likely that the molten basalt that reaches the surface as volcanic lavas comes from the aesthenosphere. Below the aesthenosphere, in the solid part of the mantle, there are zones in which the speeds of seismic waves increase rapidly. These zones are thought to be regions of phase transitions in which pressure compresses the rock to form denser minerals.

The Earth's metallic core begins at a depth of about 2900 km, where there is an abrupt increase in density. The outer core is liquid. However, the central core is

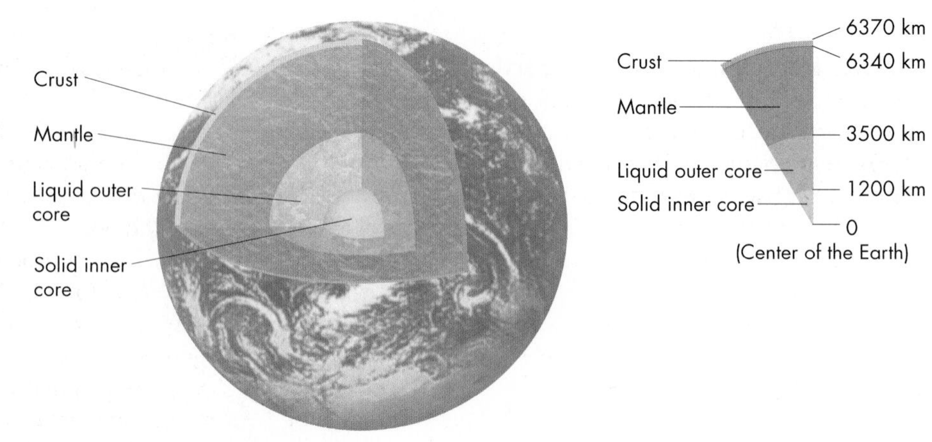

FIGURE 8.15
The Interior of the Earth
The outermost layer of the Earth is the crust, which extends inward to about 30 km. Inside the crust is the denser, but also rocky, mantle, which extends to a depth of 2900 km. The core, which has a radius of about 3500 km, is the innermost part of the Earth. The outer core is molten iron and other metals. The inner core is solid metal.

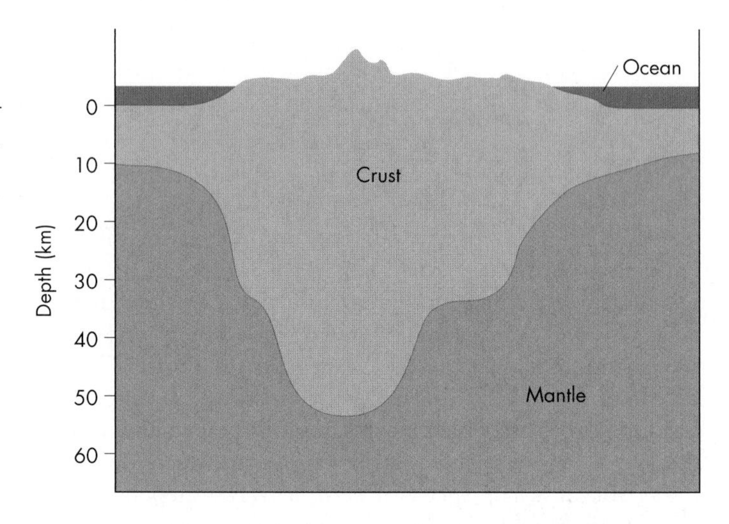

FIGURE 8.16
The Crust of the Earth
The crust of the Earth is much thicker beneath the continents than beneath the oceans. Beneath mountain ranges, the continental crust is especially thick.

subjected to great enough pressure that the metallic material is solid. The identification of the core as metallic is based on the high density of the core, which is too great to be rock under high pressure. It is likely that the metal is mostly iron and nickel and that these metals are mixed with small amounts of lighter elements such as sulfur, silicon, and oxygen.

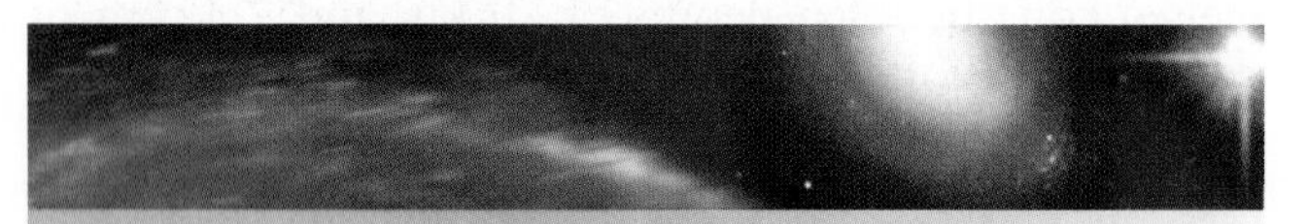

The three major layers of the Earth are the crust, the mantle, and the core. The crust, made of relatively light rock, sits atop the denser rock of the mantle. The core is mostly metal. The upper 70 km of the crust and mantle make up the rigid lithosphere. Under the lithosphere is the aesthenosphere, a plastic, partially melted zone about 200 km thick.

Plate Tectonics

Plate tectonics caused by convection

Until the last 50 years or so, most geologists thought that the continents and oceans were stable, permanent features of the planet. This point of view has been overturned with the development of the concept of **plate tectonics.** Plate tectonics brings together and unifies many observed characteristics of the Earth's outer layers and presents a picture of a dynamic, changing planet.

Continental Drift One important aspect of plate tectonics is the movement of segments of the continental crust over long periods of time. As long ago as 1620, Francis Bacon noticed that the eastern and western shores of the Atlantic Ocean were parallel and could be fitted together rather snugly. After Bacon, the idea of continental drift was revived many times. In the 1920s Alfred Wegener, a German meteorologist, supported continental drift by noting the strong similarity of rocks and fossils on opposite shores of the Atlantic. Wegener proposed that the present continents had been parts of a supercontinent, Pangaea, which began to break apart about 200 million years ago, as shown in Figure 8.17. A little later, Arthur Holmes, a British geologist, proposed that the mechanism that produces continental drift is convection currents beneath the crust. Ascending currents under a continent diverge near the surface and drag the continental crust apart. A new ocean develops in the gap left behind.

Seafloor Spreading Despite the work of scientists such as Wegener, the concept of continental drift was not quickly accepted. The evidence that eventually compelled its acceptance came from mapping and magnetic studies of the ocean floor. By 1960 geologists had learned that there is a system of connecting ridges on the midocean floor. These ridges, which are somewhat like the seam on a baseball, extend for 60,000 km. The ridge system is shown in Figure 8.18. Harry Hess, an American geologist, suggested that the ridges occur where the ocean floor cracks apart. Magma from the interior oozes out of the crack to form a new ocean bottom, which then spreads apart as the process continues. This process is called **seafloor spreading.** The continents, frozen into sections of the lithosphere called **plates,** are carried along as the plates are pushed apart by seafloor spreading.

When magma solidifies to form a new ocean bottom, the magnetic material it contains becomes aligned with the Earth's magnetic field. As the rock is split apart and pushed away from the ridge system, it acts like magnetic tape, preserving the direction of the magnetic field

FIGURE 8.17
The Motion of the Continents During the Past 540 Million Years
Pieces of the present continents gradually came together to form the supercontinent Pangaea about 200 million years ago. The disruption of Pangaea eventually led to the present configuration of the continents.

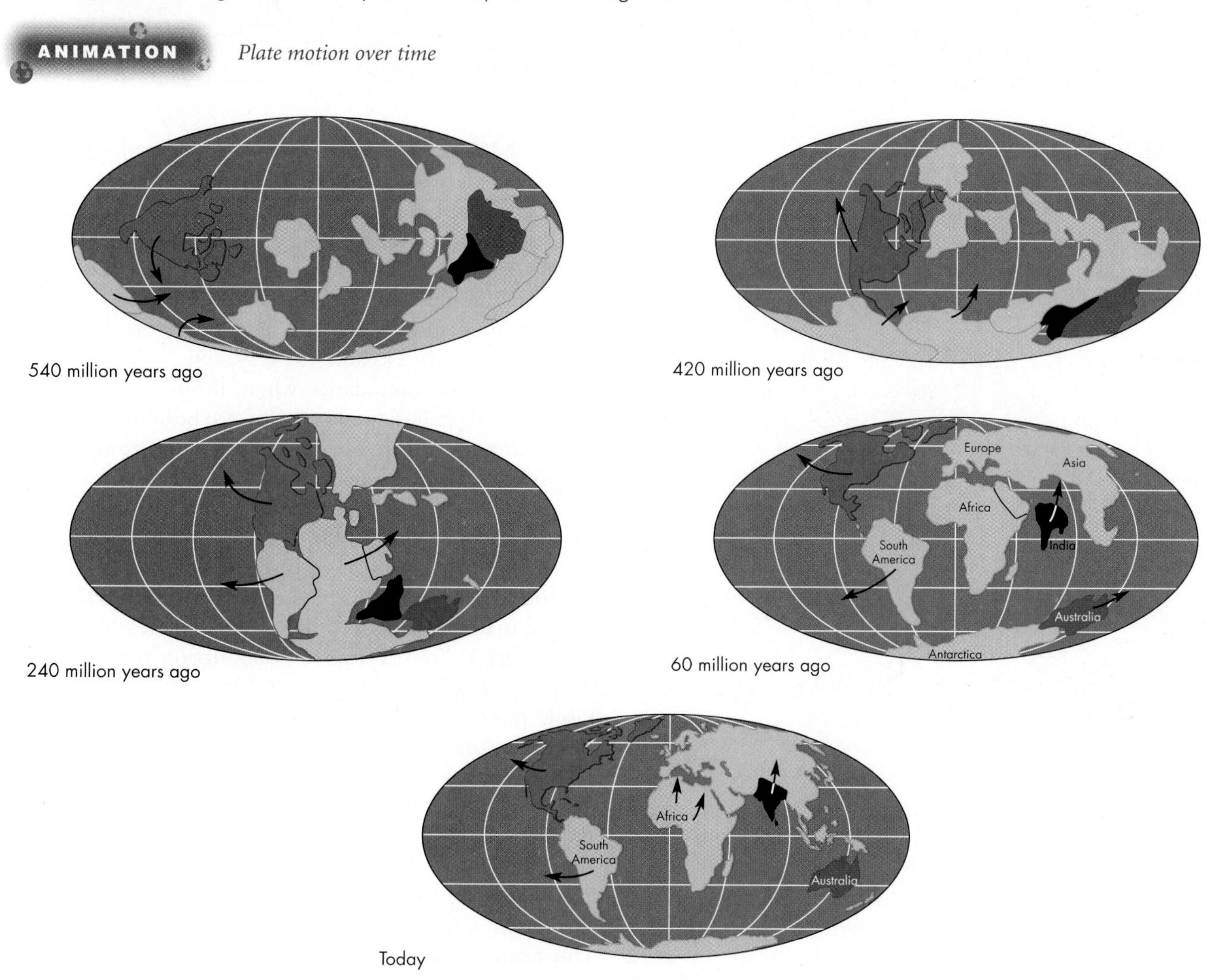

FIGURE 8.18
The Midocean Ridge System
The ocean bottom has a worldwide system of connecting mountain ranges, which stretch for 60,000 km around the Earth. Most of the ridge system is far beneath the surface of the ocean, but the ridges emerge above the surface in places such as Iceland. (Bruce C. Heezen and Marie Tharp, "World Ocean Floor," copyright by Marie Tharp, 1977)

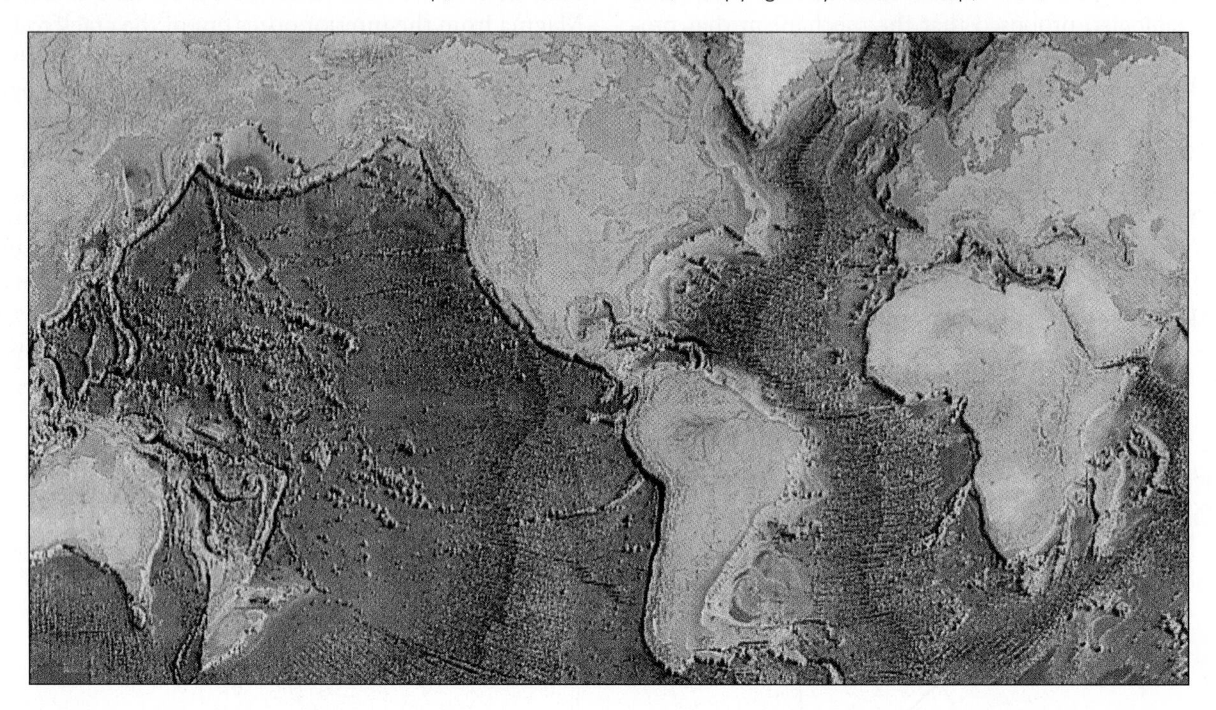

FIGURE 8.19
Seafloor Spreading
New crust forms at the midocean ridges and is imprinted with the magnetic field orientation of the Earth at the time it forms. When the crust is pushed apart by newer crust, it forms bands of magnetically imprinted rock on the ocean floor.

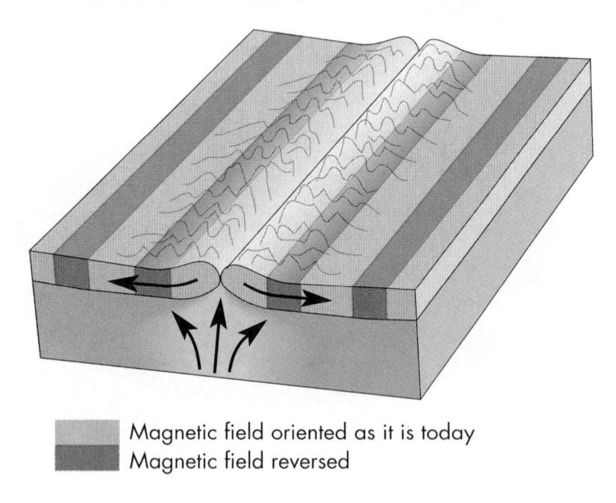

when it formed. Because the Earth's magnetic field reverses from time to time, the ridge system should be flanked by symmetrical stripes of rock with alternating magnetic field alignment, as shown in Figure 8.19. This is exactly what sensitive measurements of the magnetization of the ocean floor have shown.

Plate Motion The dozen or so lithospheric plates, shown in Figure 8.20, are dragged along by currents in the aesthenosphere, in which the heated, partially melted rock flows like a liquid at speeds of centimeters per year. Figure 8.21 shows that there are three kinds of boundaries between the plates. A **zone of divergence** is a boundary along which plates grow and separate. These boundaries correspond to the oceanic ridge systems.

If there are boundaries where lithospheric plates are growing and separating, there must also be boundaries where plates come together. These boundaries are **zones of convergence,** also shown in Figure 8.21. When plates collide, either one of them overrides the other or they buckle to form mountains. When overriding occurs, the leading edge of the overridden plate is destroyed by being driven down into the mantle and melted. This process is called **subduction.** The melted material becomes a source of magma, which reaches the surface to form volcanoes. Subduction zones appear as deep trenches in the ocean floor. One example is the Peru-Chile trench, which runs along most of the western edge of South America. The cycle of formation and destruction of the ocean floor is shown in Figure 8.22.

When two plates with continents at their leading edges collide, the thicker, lighter continental material isn't subducted. Instead, the collision of the two plates buckles the continental crust and thickens it to produce enormous mountain ranges, such as the Himalayas. Older mountain ranges in the middle of continents mark places where

FIGURE 8.20
The Crustal Plates of the Earth
There are about a dozen plates, some of which consist only of ocean bottom and some of which contain pieces of continents. The plates are in slow but constant motion with respect to each other. The arrows indicate the directions in which the plates are moving with respect to one another.

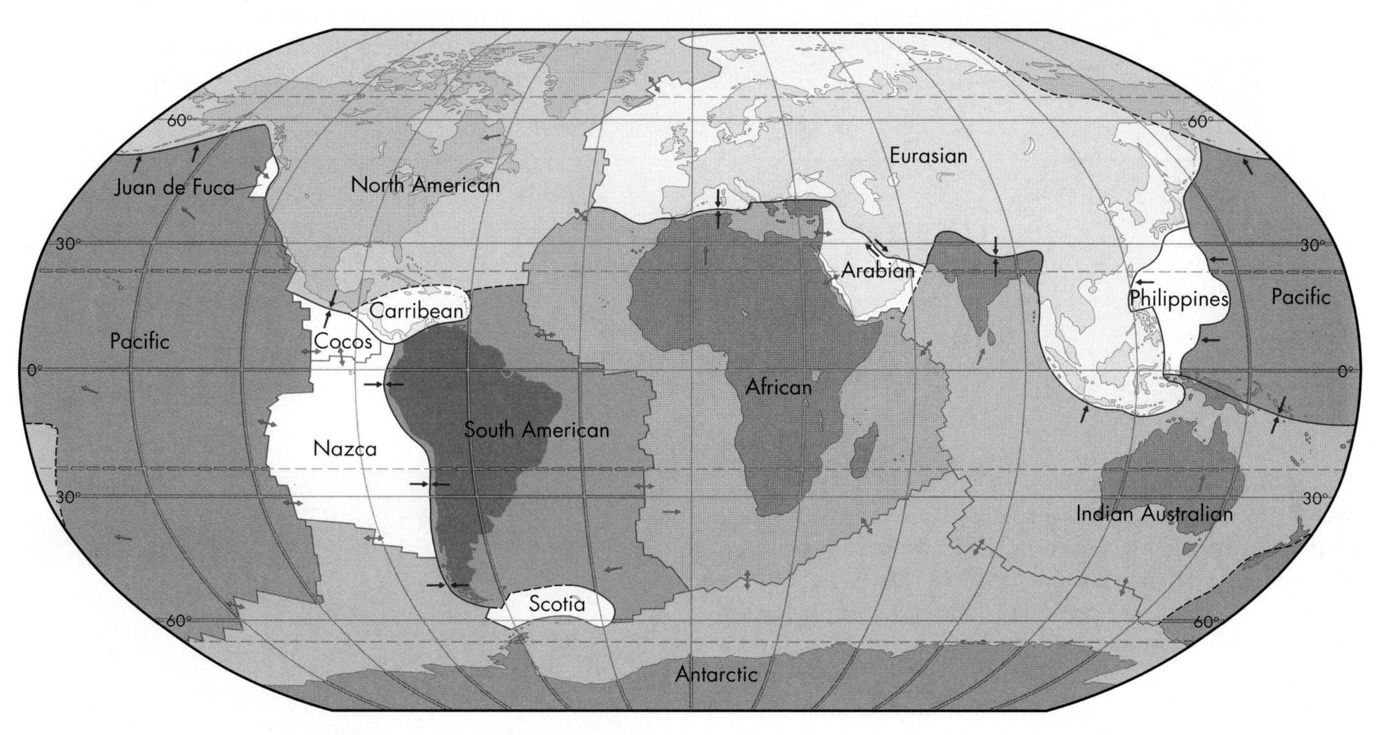

FIGURE 8.21
The Three Kinds of Plate Boundaries
A, A zone of divergence is a plate boundary at which plates are pushed apart. This occurs at the midocean ridges. **B,** In a zone of convergence, plates come together. When this happens one of the plates may override the other and push it down into the Earth's interior. **C,** At a transform fault, plates move past each other.

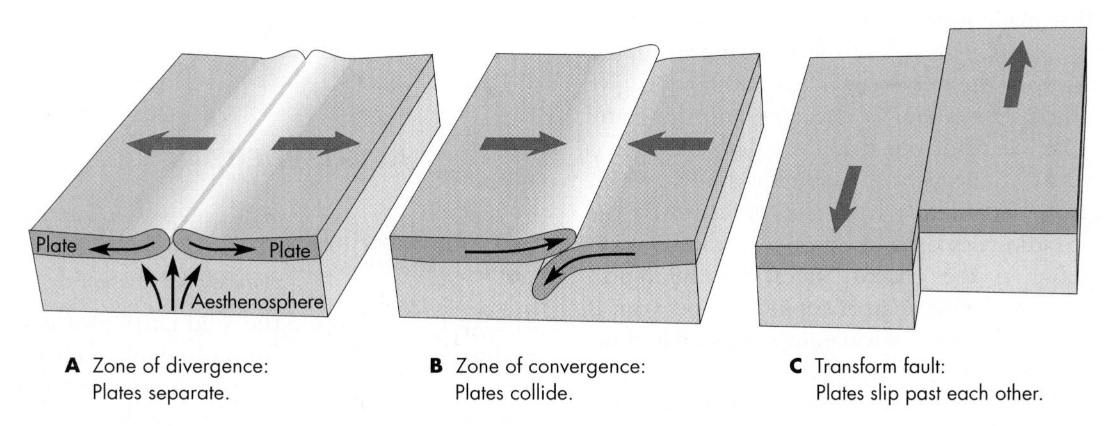

plate convergence has occurred in the past. For example, the Appalachian Mountains may be the site where an ancient ocean closed up when Africa and North America converged about 375 million years ago. Although continental crust is continually formed and destroyed, the amount remains about constant. The continental crust is pushed about and formed into new combinations and configurations, but nevertheless some continental crust is quite ancient and permanent. The oceanic crust, in contrast, is formed and destroyed on a timescale of hundreds of millions of years and is always young.

The third type of plate boundary, the **transform fault,** occurs when two plates slip past each other, as shown in

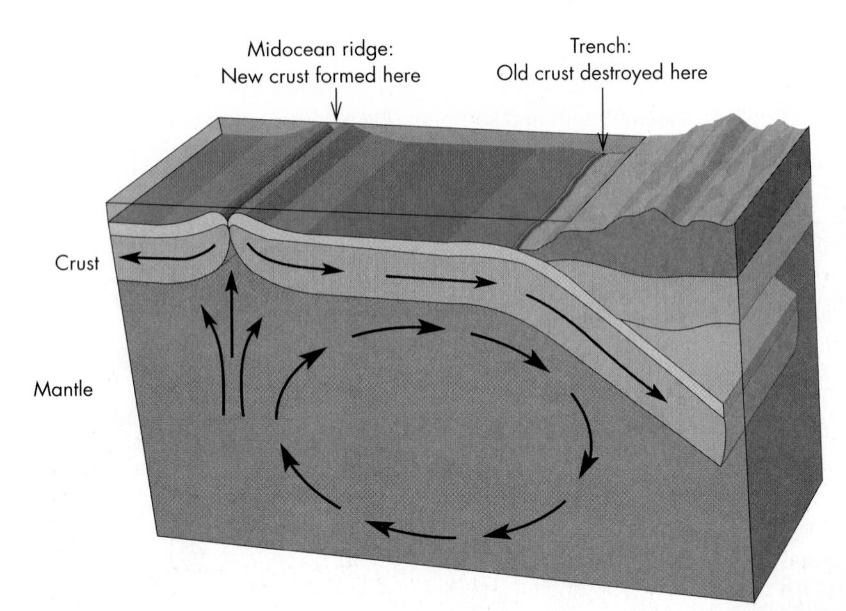

FIGURE 8.22
The Cycle of Formation and Destruction of the Earth's Crust
New crust is formed in the midocean ridges and destroyed in trenches at zones of convergence. The rate at which these two processes go on is nearly the same, so the amount of the Earth's crust remains roughly constant over time.

Figure 8.21. When this happens, lithospheric material is neither formed nor destroyed. The plate movement along a transform fault isn't uniform. Instead, friction between the plates slows the movement of the plates as pressure builds. Eventually, the pressure is released abruptly as the plates quickly move many centimeters or meters past each other, producing an earthquake. Earthquakes are common along transform faults such as the San Andreas fault, which marks the boundary between the North American plate and the Pacific plate. The Pacific plate is moving northward with respect to the North American plate. In about 25 million years, this movement will carry Los Angeles into contact with San Francisco.

Rate of Crustal Motion The magnetic record on the ocean floor first made it possible to find the rate at which plate tectonics occurs. The ages of the magnetic strips and their distances from the midocean ridges combine to give the rate at which the seafloor is spreading and the plates are moving apart. It turns out that the Atlantic Ocean is widening at 2 to 3 cm per year, about the rate at which fingernails grow. This motion has been confirmed by the discovery that radio telescopes in Europe, on the Eurasian plate, and those in the United States, on the North American plate, are moving apart at about 2 cm per year. If we extrapolate its present rate of widening backward in time, we can estimate that the Atlantic began to form about 200 million years ago. It is important to realize that although the Earth's surface is widening in some places, it is becoming narrower in others. Approximately the same amount of crust is created each year in seafloor spreading as is destroyed in regions where plates collide and subduction occurs.

Astronomers have inspected other solar system bodies for telltale signs of plate tectonics such as long trenches and mountain chains like the midocean ridge system. None of the other planets and satellites that have been examined show evidence of active plate tectonics.

Plate tectonics describes the Earth's outer layers as dynamic and changing. New crust is produced when molten material reaches the surface at oceanic ridge systems. The continents ride atop plates that are pushed apart by the arrival of new material. The boundaries where plates rub past each other or collide are marked by earthquakes, volcanic activity, trenches, and mountain building. The plates move about at speeds of a few centimeters per year.

ATMOSPHERE

The origin of the Earth's atmosphere by volcanos, comet impacts, and planetesimal collision

The layer of gas around the solid Earth provides the oxygen we breathe, protects us from ultraviolet radiation from the Sun, and acts as insulation to prevent extreme temperature swings from day to night. Our benign atmosphere is unique in the solar system. Nevertheless, it shares many properties with the atmospheres of other planets.

Composition

Near the surface, the Earth's atmosphere contains about 78% nitrogen, 21% oxygen, 1% argon, and traces of water vapor, carbon dioxide, and many other gases. Although nitrogen and oxygen predominate throughout the atmosphere, the atomic and molecular makeup of the atmosphere changes with altitude.

Thermal Structure

Although density and pressure fall steadily with altitude, temperature does not. Temperature alternately decreases and increases with height. The temperature profile of the atmosphere has led to the definition of several layers that describe the atmosphere's structure. The temperature profile of the atmosphere is shown, along with the locations of atmospheric layers, in Figure 8.23. Notice that the boundaries between adjacent atmospheric layers occur where atmospheric temperature reaches a maximum or minimum. In general, temperature maxima occur where atmospheric gases absorb sunlight well or radiate energy away poorly. Temperature minima occur where atmospheric gases absorb sunlight poorly or radiate energy away well.

Atmospheric Layers Most of the gas in the Earth's atmosphere is in the lowest region, the **troposphere.** On average, the troposphere extends from the ground to an altitude of about 10 km. The ground absorbs sunlight and heats the air that is in contact with it. This heating, plus the absorption of infrared solar radiation by water vapor, is concentrated in the lower troposphere. The rate of heating falls swiftly with altitude and so does temperature. Near the ground, the temperature drops by 6.5 K for every kilometer of increasing altitude. Whenever temperature decreases rapidly with altitude, convection (see Chapter 7, Section 5) occurs, as it does in the troposphere. The consequences of convection in the troposphere include cloud formation and turbulence. The water vapor in a rising column of tropospheric gas is chilled as it rises, so it condenses as water droplets or crystals. Because water condenses before reaching the top of the troposphere, it is more or less trapped in the troposphere.

The drop of temperature with height gradually becomes less steep until, at the tropopause, it becomes so small that convection ceases. The **tropopause** is the boundary between the troposphere and the **stratosphere.** The stratosphere is very cold at its base (about 200 K, or −100° F), but temperature rises steadily until it reaches a maximum at

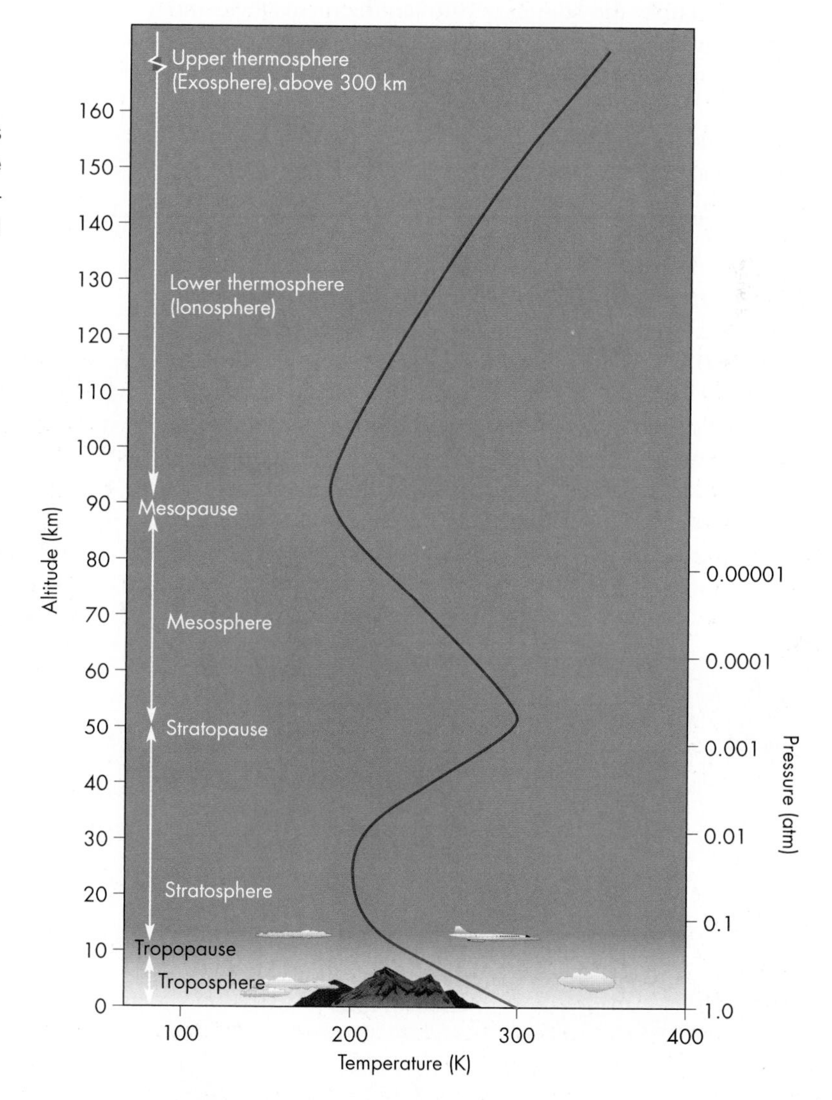

FIGURE 8.23
The Temperature Structure of Earth's Atmosphere
The Earth's atmosphere contains warm regions and cool regions. The warm regions are those in which solar radiation is absorbed. Cool regions do not absorb much solar radiation or cool themselves by radiating well.

an altitude of about 50 km. In the lower stratosphere there are often high-speed winds, known as the jet stream, that influence the speeds of jet aircraft.

Temperature increases with altitude in the stratosphere because of heating due to the absorption of solar radiation. Ultraviolet sunlight is absorbed by O_2 molecules, which are broken up into oxygen atoms. The oxygen atoms react with other O_2 molecules to produce **ozone,** O_3, which is effective at absorbing ultraviolet solar radiation. The heating produced by ozone absorption is greatest at about 50 km, where the temperature is 300 K (about the same as the temperature at the ground). The ability of the ozone layer to block potentially lethal solar ultraviolet radiation is vital to the continued existence of life on Earth.

The temperature maximum at an altitude of 50 km marks the **stratopause,** the boundary between the stratosphere and the **mesosphere,** a layer in which temperature falls with increasing height. Temperature declines with height in the mesosphere both because the absorption of ultraviolet sunlight by ozone becomes less effective and because cooling due to the emission of infrared radiation by carbon dioxide becomes stronger. Temperature drops until it reaches a minimum of 180 K (−135° F) at about 90 km. This region is the **mesopause,** the boundary between the mesosphere and the **thermosphere.** In the thermosphere, temperature climbs steadily, mostly because of the absorption of ultraviolet sunlight by nitrogen and oxygen atoms. The solar radiation that heats the thermosphere is relatively short-wavelength ultraviolet radiation, which is completely absorbed in the thermosphere and doesn't penetrate far enough into the atmosphere to heat lower-lying layers. Atoms are ionized by the radiation they absorb, so that a small fraction (about 1/500th or smaller) of the gas in the lower thermosphere is made of electrons and ions. The lower thermosphere, from 90 to 300 km, is the **ionosphere.** At some frequencies, radio waves are reflected by the ionosphere, making it possible for AM radio stations to be received thousands of km from the transmitter. Above the ionosphere is the exosphere. As described in Chapter 7 (Section 2), collisions among the particles in the exosphere are so infrequent that a rapidly moving atom or ion can escape into space.

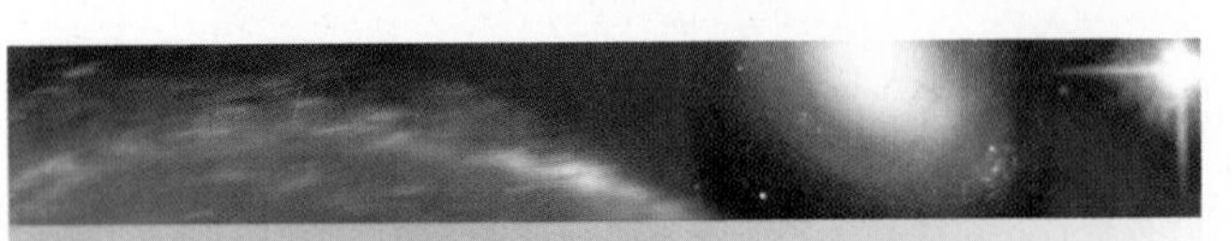

The layers of Earth's atmosphere are defined by temperature variations with altitude. Warm regions are those in which solar radiation is absorbed. Cold regions are those in which absorption of sunlight is small or in which cooling is effective. The troposphere, the lowest layer of the atmosphere, is heated from the ground. Convection in the troposphere causes weather.

Greenhouse Effect The **greenhouse effect** is the warming of a planet that occurs because its atmosphere prevents infrared radiation from easily escaping into space. The surface of a planet is heated by absorbing sunlight (mostly in the visible part of the spectrum). The surface, radiating like a blackbody, emits infrared radiation. The warmer the surface, the greater the rate at which it emits infrared radiation. The temperature of the planet is the one at which the planet is in thermal equilibrium (Chapter 7, Section 3) and absorbed solar energy just balances emitted infrared radiation.

If the gases in the atmosphere of the planet are effective absorbers of infrared radiation, they absorb some of the infrared energy emitted by the ground and lower atmosphere. The absorption reduces the fraction of the infrared radiation that escapes into space after it is emitted by the planet. The temperature of the surface and lower atmosphere rises, and the amount of infrared radiation increases, until the infrared energy escaping through the atmosphere just balances the solar energy absorbed by the planet. As Figure 8.24 shows, the net effect of the presence of the atmosphere is to make the planet warmer than it would be with no atmosphere. Generally speaking, the thicker the atmosphere of a planet, the stronger the greenhouse effect and the larger the increase in temperature.

The Earth's atmosphere produces a greenhouse effect that, so far, has been beneficial for the Earth's environment. Without the greenhouse effect the Earth's average temperature would be about 266 K, about 14 K (25° F) cooler than it actually is. The Earth would then probably have much larger polar regions and smaller habitable zones than it actually does.

The greenhouse effect takes place when the atmosphere of a planet absorbs the infrared radiation emitted by the planet's surface, raising the planet's temperature. Without the greenhouse effect, the Earth would be about 14 K (25° F) cooler than it is.

Human Changes

Natural changes in the Earth's atmosphere are probably going on today, but these changes are significant only over thousands or millions of years. Changes induced by human activity have been much more rapid and are occurring at an accelerating pace. None of the changes caused by human activity seems likely to be beneficial for human beings.

FIGURE 8.24
The Greenhouse Effect
The surface of the Earth is warmed by absorbing sunlight and cools itself by emitting infrared radiation. **A,** In the absence of an atmosphere, the infrared radiation emitted by the Earth would freely escape into space and keep the Earth cool. **B,** Atmospheric gases such as carbon dioxide and water vapor partially block the escape of infrared radiation. The temperature of the Earth and the rate at which its surface emits infrared radiation rise until the rate at which infrared energy escapes through the atmosphere equals the rate at which solar energy is absorbed.

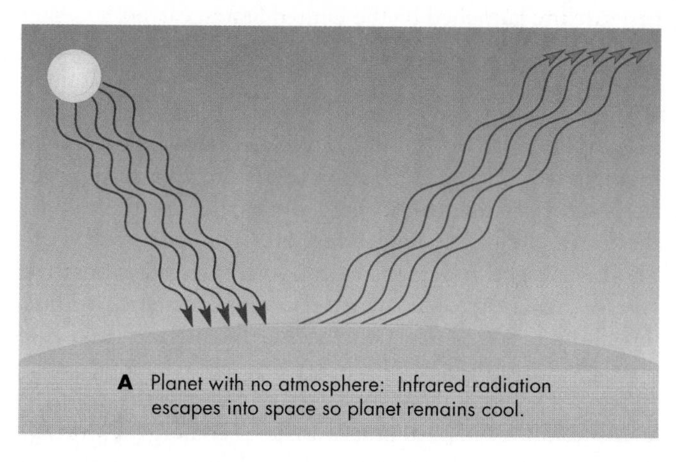

A Planet with no atmosphere: Infrared radiation escapes into space so planet remains cool.

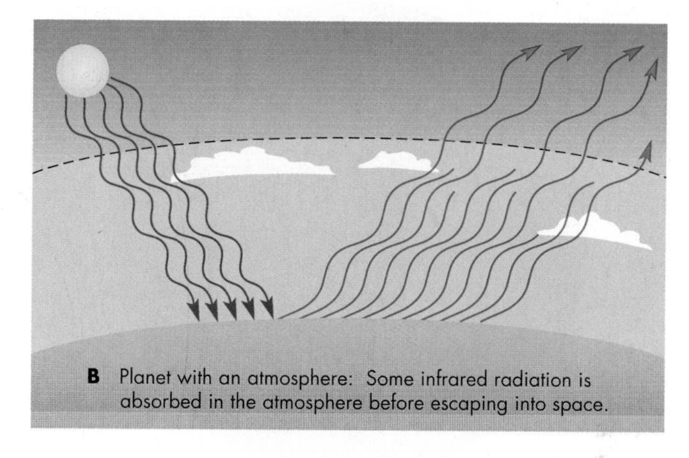

B Planet with an atmosphere: Some infrared radiation is absorbed in the atmosphere before escaping into space.

Carbon Dioxide Water vapor and carbon dioxide are the principal "greenhouse gases," or absorbers of infrared radiation, in the Earth's atmosphere. One of the main products of the burning of fossil fuels is carbon dioxide. Carbon dioxide is also released into the atmosphere when forest areas are cleared and the vegetation either is burned or decays. Since the industrial revolution the rate at which carbon dioxide has entered the atmosphere has increased steadily. Some of the carbon dioxide has been absorbed into the oceans, but enough has remained in the atmosphere to increase the concentration of CO_2 by about 15% in the last century. It is clear that the improved greenhouse effect caused by increased CO_2 will lead to increases in the Earth's average temperature. However, scientists are still unsure of the magnitude of the change, how soon it will become significant, and how severe regional consequences will be. The reason for this uncertainty is that the Earth's atmosphere and its interaction with the oceans and surface are complicated and not yet well enough understood to make predictions with complete assurance. For example, warming the Earth should produce changes in the extent of cloud cover, which may in turn partially offset the effect of increased CO_2. In any event, the problem of the greenhouse effect is now recognized as a serious threat to our present climate.

Aerosols **Aerosols** are liquid droplets and solids suspended in the atmosphere. The general effect of aerosols is to reflect sunlight and increase the Earth's albedo (see Chapter 7, Section 3). Thus aerosols act to reduce the temperature of the Earth. Volcanic eruptions, such as the eruption of Mount St. Helens shown in Figure 8.25, are a major source of atmospheric aerosols. It has long been noted that cool periods often follow large eruptions, such as the 1982 eruption of El Chichon in Mexico. Atmospheric aerosols are also produced by agricultural and industrial activity. The level of this activity has increased to the point where it is comparable to the natural production of aerosols. The consequences of atmospheric aerosols are not even as well understood as those of increased CO_2. In addition to reflecting sunlight,

FIGURE 8.25
The Eruption of Mount St. Helens in May 1980
The aerosols injected into the atmosphere by volcanos reflect sunlight back into space and reduce the solar heating of the Earth. Major volcanic eruptions produce enough aerosols to reduce global temperatures for months or years.

FIGURE 8.26
The Antarctic Ozone Hole in October 2009
The color in this image indicates the relative amounts of ozone above different regions of the Earth's surface. The purple region above Antarctica shows only about one-third of its normal amount of ozone.

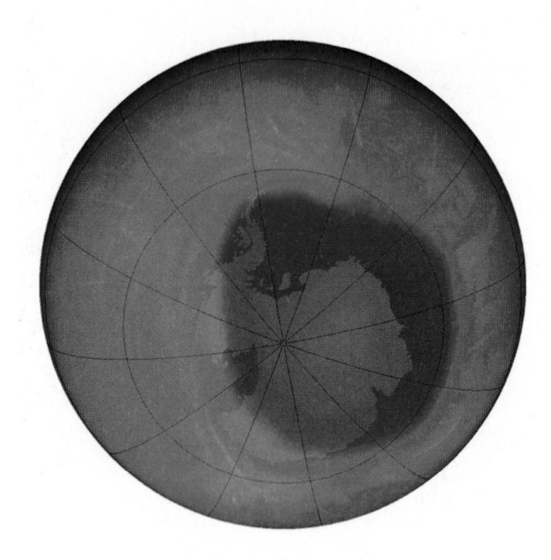

aerosols help water vapor form clouds, which also reflect sunlight. It is possible that increased aerosol production may cancel a significant part of the greenhouse effect. However, either effect by itself could have serious consequences.

Ozone Chlorofluorocarbons, such as the freons commonly used as coolants in refrigerators and air conditioners and once used as propellants in hair sprays and deodorants, gradually spread upward to the ozone layer. Once they reach the ozone layer, chlorofluorocarbons are extremely effective in destroying ozone molecules. The reduction in ozone content depends on many atmospheric processes and varies greatly by region and season. The most extreme reduction of ozone occurs over Antarctica during southern spring, as shown in Figure 8.26. A reduction in the amount of upper atmospheric ozone reduces the atmosphere's ability to block solar ultraviolet radiation. In populated parts of the planet, an increased amount of ultraviolet radiation reaching the ground will lead to increased skin cancers and possible damage to crops. Even if materials that can destroy ozone were eliminated today, decades would be needed for their influence on the ozone layer to fade away. However, efforts to control the release of chlorofluorocarbons have resulted in a reduction in the rate at which they are reaching the ozone layer.

8.5 MAGNETOSPHERE

The domain of the Earth doesn't end with the exosphere. The outermost part of the Earth's atmosphere is the **magnetosphere.** The magnetosphere is the part of the Earth's atmosphere dominated by the Earth's magnetic field and filled by a very thin plasma of protons and electrons.

Van Allen Belts

Before the 1950s it was believed that space surrounding the Earth was nearly a vacuum. In 1958, *Explorer 1,* the first satellite launched by the United States, carried a Geiger counter built by James Van Allen. The Geiger counter showed that the region beyond the atmosphere contained an unexpectedly large number of energetic charged particles. The particles reside in the **Van Allen belts,** shown in Figure 8.27. The Van Allen belts are two doughnut-shaped regions in which charged particles are trapped by the Earth's magnetic field. The outer Van Allen belt, containing energetic electrons, lies about 15,000 km above the surface. The inner belt, which contains energetic protons, lies about 2500 km above the Earth's surface. Like the outer belt, the inner belt doesn't have sharp boundaries. Instead, the number of energetic protons gradually diminishes both outward and inward. The energetic protons in the inner belt constitute a hazard both to astronauts and electronic equipment, which has a reduced lifetime in the inner belt because of radiation damage. Fortunately, nearly all of the missions involving human crews have been carried out in the relatively safe region within 400 km of the surface.

Structure of Magnetosphere

Interaction with Solar Wind The **solar wind** is a hot plasma that flows outward from the Sun. The wind passes the Earth at speeds of between 400 and 700 km/s (about 1.6 million miles per hour). As the solar wind approaches the Earth's magnetosphere, it slows rapidly. The **bow shock** (shown in Figure 8.28) where the slowing takes place resembles the shock front produced by a supersonic airplane. Inside the bow shock is the **magnetopause,** which marks the outer boundary of the magnetosphere. The solar wind flows around the Earth in the region just outside the magnetopause. However, the solar wind distorts the shape of the magnetosphere. The solar wind pressure pushes the magnetopause to within about 10 Earth radii (65,000 km) on the day side of the Earth. On the night side of the Earth, the solar wind stretches the magnetosphere into a **magnetotail,** which extends millions of kilometers into space.

Aurorae Although the Earth's magnetic field keeps solar wind particles from entering the magnetosphere, some of the energy carried by the solar wind is transferred to charged particles in the magnetosphere. This happens mainly in the magnetotail, where the solar wind creates enormous currents of energetic electrons that flow into the Earth's upper atmosphere in regions about 20° from the north and south magnetic poles. At about 100 km

FIGURE 8.27
The Van Allen Belts
The Earth's magnetic field traps charged particles in belts, or shells, that surround the Earth. The belts were discovered by James Van Allen with a detector on *Explorer 1,* the first U.S. satellite.

A Cross-sectional view of the Van Allen belts

B Cut-away view showing that the belts are actually shells

FIGURE 8.28
The Earth's Magnetosphere
The bow shock occurs where the solar wind is slowed as it approaches the Earth. The outer boundary of the magnetosphere is the magnetopause. The magnetosphere is compressed by the solar wind on the sunward side and pushed back on the opposite side to form the magnetotail, which extends for millions of kilometers.

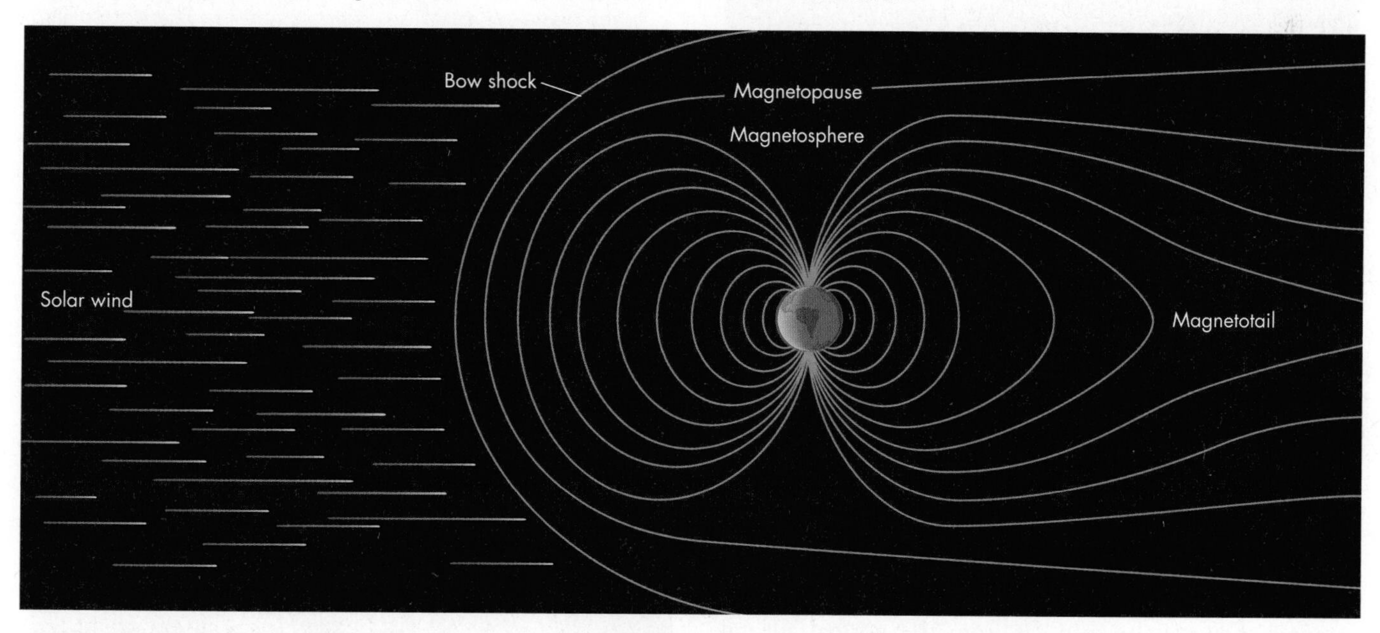

above the surface, the energetic electrons strike hydrogen, nitrogen, and oxygen atoms and molecules, causing them to emit red, green, and blue light in much the same way that light is produced in a fluorescent bulb. This pulsing, shimmering glow is the **aurora borealis** (northern lights) and **aurora australis** (southern lights). Occasional outbursts on the Sun produce large increases in the number and energy of solar wind particles streaming past the

FIGURE 8.29
An Aurora Borealis as Seen from the Ground
Aurorae are produced when the solar wind causes magnetospheric currents that strike atoms and molecules in the Earth's upper atmosphere, causing them to emit light in a glowing, shimmering pattern.

FIGURE 8.30
An Aurora as Seen from Space
This image shows the aurora australis.

FIGURE 8.31
Global Auroral Images
From the ground, only a small portion of an auroral display can be seen at once. Images taken from satellites, such as *Dynamics Explorer 1*, show the entire auroral region. The auroral display is centered approximately on the magnetic pole.

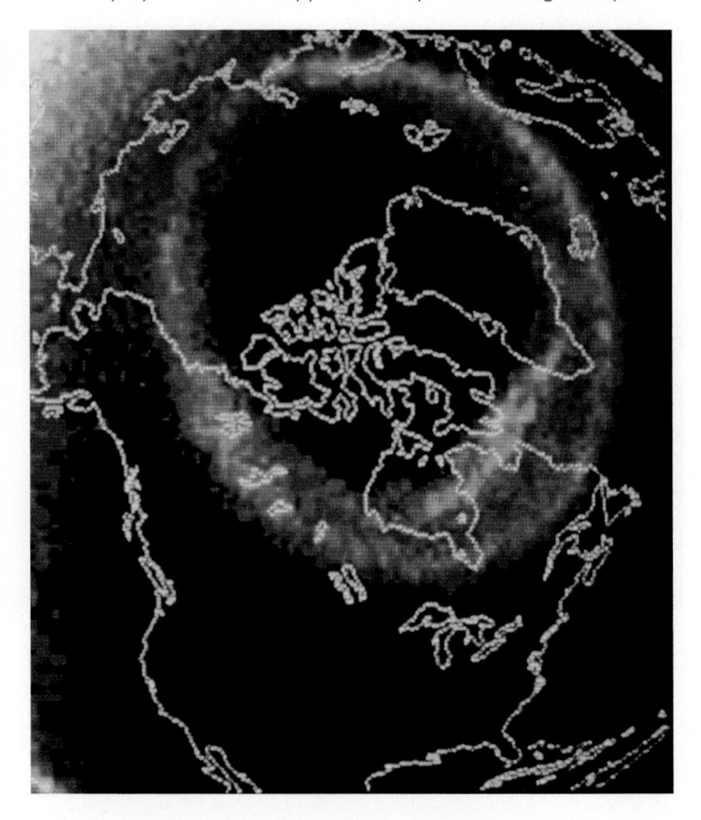

Earth. These solar wind enhancements disturb the magnetosphere even more than usual and generate such strong magnetospheric currents that the auroral region expands. Aurorae then can be seen far from the normal auroral region, sometimes as far south as Florida.

From the ground or from low-orbit spacecraft like the space shuttle, it is possible to see only a small part of an auroral display, such as those shown in Figures 8.29 and 8.30. Satellite images, such as the one shown in Figure 8.31, show the entire auroral region and can be studied by geophysicists trying to understand the complex interactions between the magnetosphere and the solar wind.

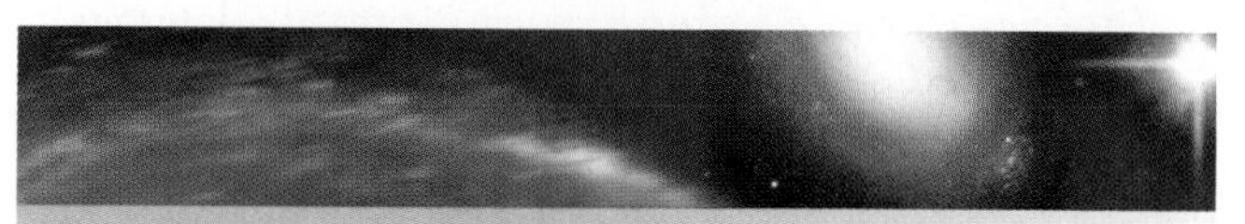

The magnetosphere is a region about the Earth that is dominated by Earth's magnetic field. Within the magnetosphere are regions of trapped electrons and ions known as the Van Allen belts. The solar wind creates currents of electrons in the magnetosphere. These currents strike the upper atmosphere near the magnetic poles, exciting atmospheric atoms and molecules and producing aurorae.

8.6 EVOLUTION OF THE EARTH

The earliest stages of the development of the Earth will be discussed in more detail in Chapter 18, which describes the origin of the stars and planets. Briefly, astronomers think the Earth formed by the accumulation of infalling bodies, some of which were larger and more massive than the Moon or Mercury. This accumulation occurred about 4.6 billion years ago. Energy released in the impacts kept the growing Earth molten throughout.

Development of the Core

At first, the chemical composition of the Earth probably was more or less uniform from center to surface. Gradually, however, **differentiation,** the gravitational separation of the interior of a planet according to density, began to occur. Dense materials such as iron, nickel, and other metals accumulated into molten drops and sank to the center of the Earth to form the core. The least dense materials were displaced and rose to the surface where they solidified to form the crust of the Earth. Since the differentiation of the Earth, its interior has changed relatively little. The Earth is too large and well insulated to have cooled much, even though the rate of radioactive heating has steadily declined.

One way that the Earth's interior has changed since differentiation, however, is that the core has cooled enough that it has begun to solidify from the center outward. Solidification began at the center, despite higher temperatures there, because of the great pressure at the Earth's center. One consequence of the gradual solidification of the core is that the inner, solid core rotates slightly faster than the outer, liquid core and the outer layers of the Earth. The inner core rotates in the same direction as the rest of the Earth, but completes its daily rotation about two-thirds of a second faster than the outer layers. Over a year's time the inner core rotates about one degree of longitude more than the outer core, mantle, and crust. This means that the inner core makes a complete rotation with respect to the rest of the Earth about once every 400 years.

The inner core has grown over the past billion years to a diameter of 2400 km (1500 miles). In comparison, the inner and outer core together have a diameter of nearly 7000 km (4350 miles) and the Earth's diameter is 12,700 km (7900 miles). Although the inner core occupies only 0.7% of the volume of the Earth, its great density (13 times the density of water) means that it is more massive than the Moon.

Oceans

Water in the primitive Earth was chemically locked up in minerals. Gradually, water was released from the molten interior and reached the surface along with lava in volcanic eruptions. At the surface, water entered the atmosphere as steam, but then cooled and fell as rain. We don't know how rapidly the oceans grew in depth. However, there must have been widespread surface water by no later than 3.5 billion years ago, when fossils of early marine organisms were formed.

Although the Earth has had oceans for billions of years, the water in today's oceans has been there for a far shorter time. Once water reaches the Earth's surface, it doesn't remain there permanently. Water is incorporated in sediments on the ocean floor and then carried into the interior by subduction in zones of convergence. However, volcanic activity recycles the subducted water back to the surface. The timescale for recycling of water is probably hundreds of millions of years.

Atmosphere

When the Earth formed, it probably had a **primeval atmosphere,** which has been almost entirely lost, possibly as a result of violent impacts between the Earth and large infalling bodies. The present atmosphere is a **secondary atmosphere,** which gradually developed as gas was released from the interior during volcanic eruptions. Another source of the secondary atmosphere may have been impacts of icy bodies similar to comets. The release of gas from the interior is known as **outgassing.** The gas emerging from the interior of the Earth was composed mainly of hydrogen, water vapor, carbon dioxide, and nitrogen. The hydrogen escaped into space. Water mostly condensed as rain and formed the oceans. Nearly all of the outgassed carbon dioxide was dissolved in the oceans as weak carbonic acid, which then reacted with silicate rocks to produce carbonate rocks such as limestone. Thus the early secondary atmosphere was mainly nitrogen with small amounts of carbon dioxide and water vapor.

The large amount of oxygen in the atmosphere today is mostly the result of photosynthesis by plants, which make organic matter and oxygen from carbon dioxide, water vapor, and sunlight. Apparently, O_2 in the atmosphere was relatively scarce until about 2 billion years ago. After that time it increased to its present level. As oxygen molecules became more abundant, they were broken up by solar ultraviolet radiation and then combined with other oxygen molecules to produce ozone. This change made it possible for life to emerge onto land. Without the protection of stratospheric ozone, the only safe habitats on Earth would have been in the oceans or beneath Earth's surface.

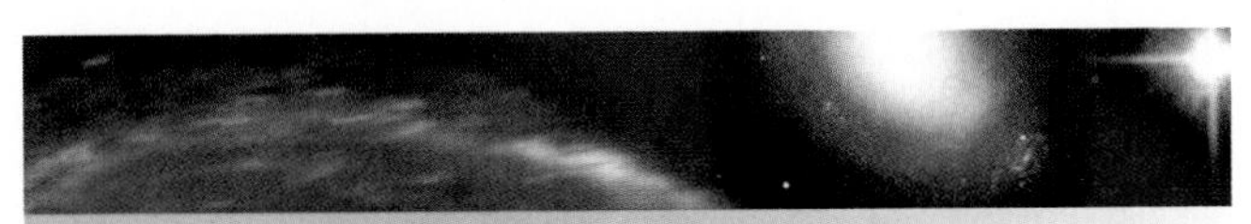

The Earth formed by the accumulation of relatively large bodies. The energy released during the impacts kept the Earth melted and permitted the heavy materials, such as iron, to sink to the center as a core. The oceans and atmosphere probably formed from the escape of gases from the interior of the Earth. The outgassed carbon dioxide mostly dissolved in the oceans and then formed carbonaceous rocks. The oxygen present in the atmosphere today is mostly the result of photosynthesis by plants.

Chapter Summary

- As the Earth moves in its orbit, the positions of stars appear to shift back and forth just as the apparent direction of falling raindrops changes as we move back and forth during a rainstorm. This effect, the aberration of starlight, was the first direct evidence for the revolution of the Earth. (Section 8.1)
- The rotation of the Earth causes the planet to have an oblate shape an equatorial bulge. The Coriolis effect, also caused by the Earth's rotation, causes apparent deflections of moving objects and currents of air and water. The Coriolis effect is responsible for spiral patterns in atmospheric circulation. (8.1)
- Minerals are solid chemical compounds. Rocks are assemblages of minerals. We classify rocks according to whether they were formed from cooling magma (igneous) or collected sediments (sedimentary), or were transformed by high pressure and temperature (metamorphic). (8.2)
- The Earth's surface is divided into ocean basins and continents. The continents contain very old rocks, but the ocean basins are relatively young. (8.2)
- Pressure increases as depth within the Earth increases. A typical value of pressure in the Earth is one million atmospheres. Such high pressures have an effect on the density, mineral structure, and molten state of rock. (8.3)
- The magnetic field of the Earth resembles that of a bar magnet, but it is actually produced by flows of electrically conducting material in the interior. The magnetic field direction reverses on a timescale of hundreds of thousands of years or less. (8.3)
- The study of seismic waves has led to knowledge of the variation of density with increasing depth in the Earth. Zones in which density changes rapidly as depth increases correspond to regions of changing mineral structure, molten state, or chemical composition. (8.3)
- The layers of the interior are the crust, which is made of lighter rock; the mantle, which is made of denser rock; and the core, which is composed mostly of metal. The crust and outer mantle make up the lithosphere, a rigid shell. Beneath the lithosphere is the aesthenosphere, a partially melted, plastic zone. (8.3)
- Plate tectonics describes the Earth's crust as divided into about a dozen moving plates. New crust is produced at oceanic ridges, where magma reaches the surface. The plates are driven apart by the spreading of the seafloor. When plates collide or move past each other, earthquakes, volcanoes, mountain building, and ocean trenches result. (8.3)
- Atmospheric layers are defined by changes in temperature with altitude. Warm layers are those in which solar radiation is absorbed. Cool layers absorb sunlight poorly or radiate away energy well. The lowest layer, the troposphere, is heated by the ground. Convection in the troposphere causes weather. (8.4)

- The greenhouse effect insulates the Earth by making it difficult for infrared radiation to escape into space. The gases mainly responsible for the absorption of infrared radiation are carbon dioxide and water vapor. Without the greenhouse effect, the Earth would be about 14 K (25° F) cooler. (8.4)
- The Earth's magnetic field dominates the motion of electrons and ions in the magnetosphere. Within the magnetosphere are regions of trapped electrons and ions known as the Van Allen belts. The solar wind creates currents of energetic electrons that strike the upper atmosphere causing the aurorae. (8.5)
- The Earth formed by the accumulation of infalling bodies, some of which were as large as the Moon or Mercury. Energy from the impacts of these bodies kept the Earth molten and permitted heavy materials to sink to the center to form the core. The oceans and atmosphere accumulated from gas that emerged from the interior in volcanic eruptions. There would be much more carbon dioxide in the present atmosphere if it had not dissolved in the oceans and formed carbonaceous rocks. The oxygen in the atmosphere is mostly the result of photosynthesis by plants. (8.6)

Key Terms

aberration of starlight *154*
aerosol *171*
aesthenosphere *163*
aurora australis *173*
aurora borealis *173*
basalt *159*
bow shock *172*
core *163*
Coriolis effect *156*
crust *163*
differentiation *175*
dynamo *161*
greenhouse effect *170*
igneous rock *159*
ionosphere *170*
lava *159*
lithosphere *163*
magma *159*
magnetopause *172*
magnetosphere *172*
magnetotail *172*
mantle *163*
mesopause *170*
mesosphere *170*
metamorphic rock *160*
mineral *159*
oblate *156*
outgassing *175*
ozone *170*
plates *165*
plate tectonics *164*
primeval atmosphere *175*
rock *159*
seafloor spreading *165*
secondary atmosphere *175*
sedimentary rock *159*
seismic wave *162*
seismometer *162*
silicates *159*
solar wind *172*
stratopause *170*
stratosphere *169*
subduction *166*
thermosphere *170*
transform fault *167*
tropopause *169*
troposphere *169*
Van Allen belts *172*
zone of convergence *166*
zone of divergence *166*

Conceptual Questions

1. Describe how a telescope would have to be tilted throughout a year to point at a star whose direction is in the Earth's orbital plane.
2. What do you think would happen to the size of the shift due to aberration if the speed of light were only half as large as it actually is?
3. Which way do the winds circulate around an intense low-pressure region in the southern hemisphere? Why?
4. Suppose the Earth's atmosphere had no ozone. What consequences would the absence of ozone have for the temperature structure of the atmosphere?
5. Why are observers more likely to see aurorae in Canada than in Florida?

Problems

1. If a person weighed 200 pounds at the South Pole, what would he or she weigh at the equator? How would the person's mass at the equator compare to his or her mass at the pole?
2. Suppose the history of the Earth were represented by a year. How many days ago would all of the continents have been merged together as the supercontinent Pangaea? At what date and time during the year would the birth of Copernicus have taken place?
3. Geologists have proposed drilling downward through the crust into the mantle. Where should they drill in order to minimize the depth of the hole? If they succeed, what percentage of the distance to the center of the Earth will they have drilled?
4. Only about $1/5000$ as much energy comes from the interior of the Earth as is absorbed at the surface from sunlight. If no energy reached the Earth from the Sun (or any other celestial object), how would the temperature of the Earth's surface compare to its actual temperature? (To solve this problem, you need to use the Stefan-Boltzmann law, given in Chapter 7.)
5. What fraction of the volume of the Earth is occupied by its core?
6. Near one of the midocean ridges the rock on a strip of ocean bottom shows the same magnetic orientation. The strip is 10 km wide and formed during the period of time between two magnetic reversals that were separated by 200,000 years. How rapidly is the ocean bottom moving with respect to the midocean ridge?
7. Europe and North America are moving apart at 2 cm per year. Find (or estimate) the width of the North Atlantic Ocean and determine how long ago the North Atlantic began to form.
8. How much colder is it likely to be at the top of a 6000-meter mountain than it is at the bottom?
9. Suppose greenhouse gases are introduced into the atmosphere of a planet. The effect of the gases is to cut in half the probability that infrared radiation emitted by the planet can escape into space. What would happen to the amount of infrared energy emitted by the surface of the planet? By approximately what fraction would the temperature of the planet need to increase to bring about the change in the amount of infrared energy emitted? (As in question 4, you need the Stefan-Boltzmann equation to solve this problem.)

Figure-Based Questions

1. Use Figure 8.14 to find the density of the mantle at a depth of 2000 km. Also find the density of the core at a depth of 5000 km. How do these densities compare with the densities of typical surface rocks?
2. Using Figure 8.20, find the type of plate boundary that lies between the Indian Australian and Antarctic plates. What kind of plate boundary lies between the Eurasian plate and the Arabian plates? Between the Indian Australian plate and the Eurasian plate?
3. Use Figure 8.23 to find the temperature of the Earth's atmosphere at the altitude where pressure is 0.001 atmosphere. Find the pressure where the temperature is 250 K. (There are multiple correct answers to this part of the question.)

Group Activity

During a steady rain, go outside with your partner and an umbrella for each of you. Have your partner stand still and note the direction your partner has to tilt her or his umbrella to keep from getting wet. Walk rapidly through the rain and note the direction you have to tilt your umbrella to keep dry. Walk even faster and note the direction of tilt. Compare the tilts you had to use with that of your stationary partner. Discuss the results in terms of the aberration of starlight.

For More Information

Visit the text website at **www.mhhe.com/fix** for chapter quizzes, interactive learning exercises, and other study tools.

9

I can see the moonlight shining

on my couch.

Can it be that frost has fallen?

I lift my head and watch the

mountain moon,

Then my head droops in

meditation of earth.

Li Po
Still Night Thoughts

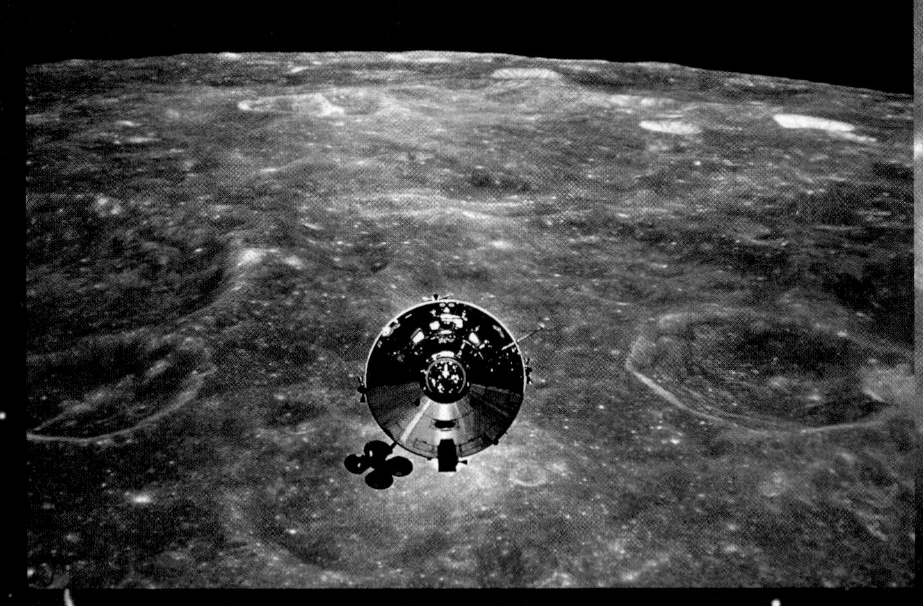

Apollo Command Module Above the Moon.

www.mhhe.com/fix

The Moon

Except for the Sun, the Moon affects us more than any other celestial body. The ways in which it affects us range from literary to physical. The Moon has always attracted the interest of poets. The beauty of the Moon, the subtlety of moonlight, and the regularity of the Moon's phases are celebrated in almost countless poems and songs. Perhaps these qualities are what intrigued prehistoric people and caused them to record their observations of the Moon as long as 30,000 years ago.

A more tangible influence of the Moon is felt during eclipses and through tides. Lunar eclipses are fun and interesting to watch. Solar eclipses actually affect our surroundings. During total solar eclipses, temperature drops and daylight dims. Eclipses have ended battles and brought disaster to military expeditions. The Moon is mostly responsible for tides, which distort the Earth and affect activities such as deep-sea fishing and navigation.

This chapter is about the Moon's influence on us. In part, this means physical effects like tides and eclipses. But it also means the influence of the Moon on our understanding of the solar system. The Moon is the only world, other than the Earth, where people have been. It is the only celestial body from which materials have been selected and brought back to Earth for analysis. What we have learned about the similarities and differences between the Earth and the Moon has been important for understanding the planets in general. The Moon preserves a much better record of the early stages of its development than the Earth does, so the exploration of the Moon has been vital to our ideas about when and how the planets were formed.

Questions to Explore

- Why does the Moon keep the same face turned toward the Earth?
- What causes eclipses of the Sun and the Moon?
- How can we predict eclipses?
- How do tides affect the Earth and the Moon?
- What caused the craters, maria, and other surface features of the Moon?
- What have samples from the Moon told us about the Moon and the origin of the solar system?
- What do we know about the interior of the Moon?
- How did the Moon form?

REVOLUTION AND ROTATION

The monthly motion of the Moon and the cycle of its phases are described in Chapter 2, Section 4. Briefly, relative to the stars, the Moon moves eastward by about one diameter per hour or 13° per day. This eastward motion is much more rapid than the eastward motion of the Sun, so the Moon moves to the east with respect to the Sun as well. The times of moonrise and moonset both grow later throughout the month.

The Month

The time required for the Moon to go through its cycle of phases is 29.53 days and is defined as the synodic month. Thus, after one synodic month, the Moon returns to about the same position in the sky relative to the Sun. During a month, however, the Sun moves eastward relative to the stars, so when the synodic month is completed, the Moon has passed its previous position among the stars. The length of time for the Moon to return to the same position relative to the stars, the sidereal month, is 27.32 days, about 2 days shorter than the synodic month. The sidereal month is essentially the length of time for the Moon to complete one orbit about the Earth. The reason for the difference between the two kinds of months is that the Earth moves in its own orbit during a month. As Figure 9.1 shows, the Earth's orbital motion makes the Sun appear to move eastward among the stars by about 1° per day. During a sidereal month, the Sun appears to move eastward by about 27°, so it takes the Moon about 2 days longer to "catch up" with the moving Sun than it does to orbit the Earth once and complete its motion relative to the stars. If the Earth were much farther from the Sun than it is and orbited very slowly, the sidereal and synodic months would be almost equal in length.

Synchronous Rotation

As the Moon revolves about the Earth, it also rotates on its axis with a period that is the same as its orbital period. Because the Moon keeps the same face turned toward the Earth, many people think it doesn't rotate. Actually, if it didn't rotate, we would see the entire surface of the Moon each month. Figure 9.2 shows what we would see if the Moon didn't rotate and what we actually see as a consequence of the Moon's rotation. A planet, satellite, or star that revolves and rotates in the same length of time is said to be in **synchronous rotation.** Like the Moon, many other solar system satellites rotate synchronously. The explanation for the synchronous rotations of the Moon and other satellites is to be found in the tidal interactions they have with the planet about which they revolve. The way that tidal interactions work is described later in this chapter.

If people inhabited the Moon, they would see consequences of the Moon's synchronous rotation that are very different from those seen from the Earth. Although the

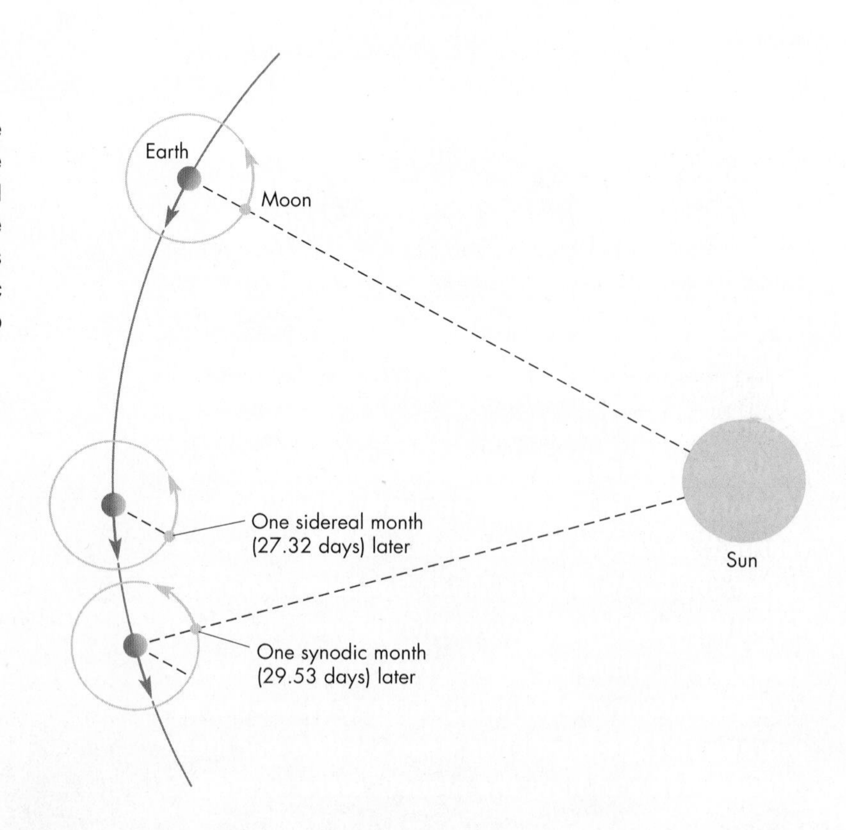

FIGURE 9.1
The Synodic and Sidereal Month
The Moon takes 27.32 days to revolve about the Earth and complete a sidereal month. Because the Earth moves in its orbit during a sidereal month, the Moon must travel more than one complete trip about the Earth before it returns to the same phase and completes a synodic month. The synodic month takes 29.53 days to complete.

FIGURE 9.2
The Rotation of the Moon
A, If the Moon didn't rotate, it would present both faces to the Earth during a month. **B,** We see only one face of the Moon because the Moon rotates synchronously. That is, its rotation period equals its period of revolution.

ANIMATION *The rotation of the Moon*

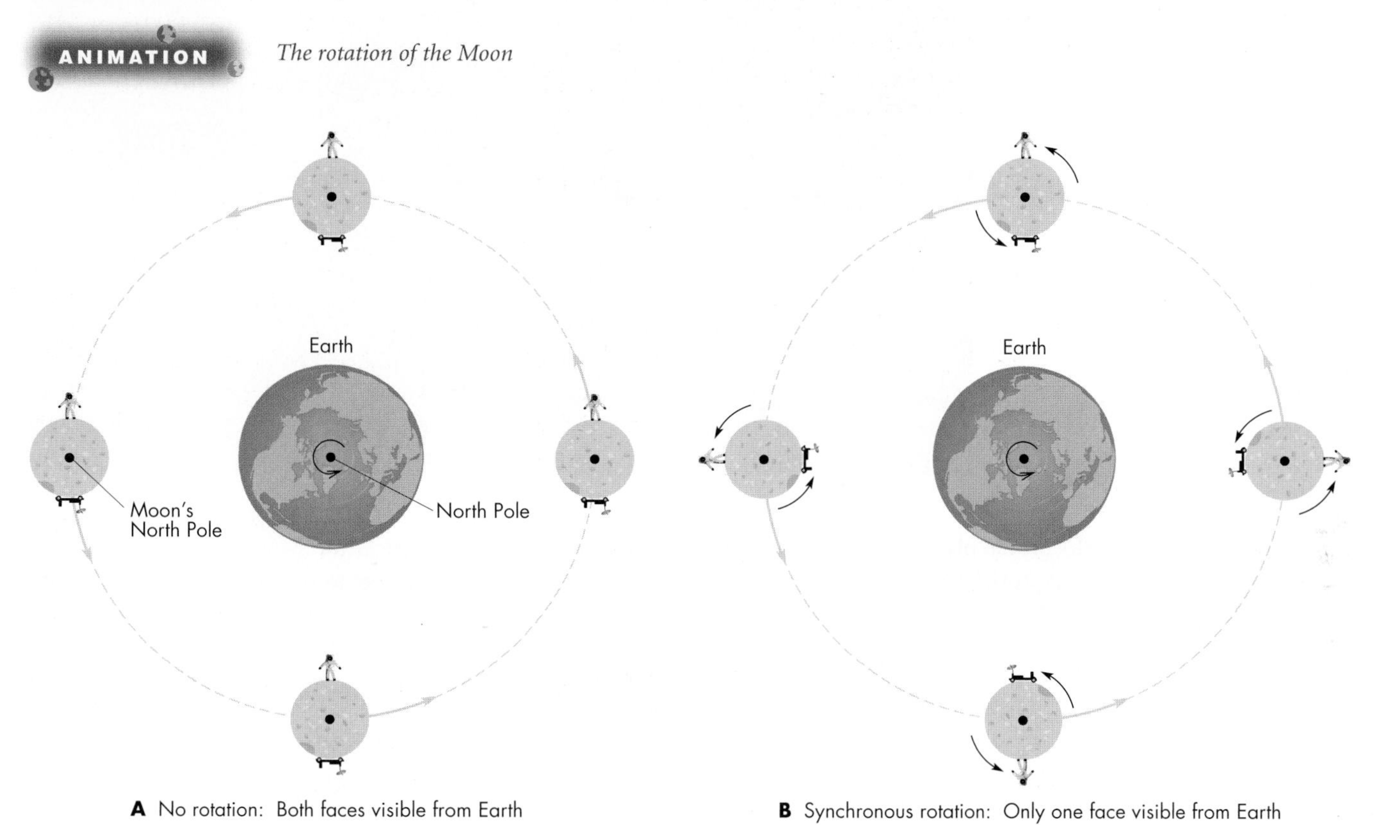

colonists would see the Earth move among the stars about once per "lunar day" (the same as our synodic month) and go through the same cycle of phases we see for the Moon, the Earth would hang nearly motionless in the sky. For colonists living at the center of what we call the near side of the Moon, the Earth always would be near the zenith. From colonies near the edge of the part of the Moon we can see, the Earth would always hover near the lunar horizon, as shown in Figure 9.3. Colonists living on the far side of the Moon would have to go on expeditions to see the Earth, because the Earth could never be seen in the sky over their homes.

The Moon rotates and revolves in the same length of time, the sidereal month. This synchronous rotation keeps the same face of the Moon turned toward the Earth.

The Orbit and Size of the Moon

The Moon's orbit about the Earth is elliptical. The average distance from the center of the Earth to the center of the Moon is 384,400 km but the distance can be as small as 356,400 km (at perigee) and as large as 406,700 km (at apogee). The best measurements of the Moon's distance from the Earth have been found by measuring the time required for laser signals aimed at the Moon to bounce back to the Earth from reflectors left on the Moon during the *Apollo* landings. Using this method the distance to the Moon can be found to an accuracy of a few centimeters. The measurements give not only the Moon's distance but also provide data for accurate calculations of the Moon's orbital shape and variations in the orbital shape due to the gravitational attraction of the Sun.

Figure 9.4 shows that the orbit of the Moon lies almost in the ecliptic. The angle between the two is only about 5°. Among the solar system satellites, the Moon's orbit is quite unusual. Most large satellites orbit nearly in the equatorial plane of their parent planet. The plane of the Moon's orbit

FIGURE 9.3
The Earth Seen from Lunar Orbit
Because the Moon keeps the same face turned toward the Earth, the Earth hovers in the black lunar sky. During a month, the Earth goes through a full cycle of phases, but remains in nearly the same spot in the sky.

is much closer to the Earth's orbital plane (the ecliptic plane) than it is to the Earth's equatorial plane. The two points where the orbit of the Moon passes through the ecliptic plane are called the nodes of the Moon's orbit. The **ascending node** of the Moon's orbit is the point where the Moon moves northward across the ecliptic. The **descending node** is where the Moon moves southward across the ecliptic. Although it is customary to say that the Moon orbits the Earth, this is really not correct. The gravitational force exerted on the Moon by the Sun is more than twice as great as the gravitational force exerted on the Moon by the Earth. As a result, the Moon and Earth orbit the Sun together, in paths that weave in and out with respect to one another.

Once the Moon's distance from the Earth and its angular diameter are known, the small angle equation (Equation 3.1) can be used to calculate its size. Using 31 minutes of arc (about ½°) for the angular diameter of the Moon and 384,400 km for its distance, the Moon's diameter is found to be 3476 km, 27% as large as the Earth. Some important properties of the Moon are summarized in Table 9.1.

FIGURE 9.4
The Tilt of the Moon's Orbit
The Moon's orbit lies nearly in the ecliptic, but is tipped by 5°. This means that the Moon passes through the ecliptic twice during each orbit about the Earth. The points in the Moon's orbit where this happens are called nodes. The line of nodes is an imaginary line connecting the two nodes.

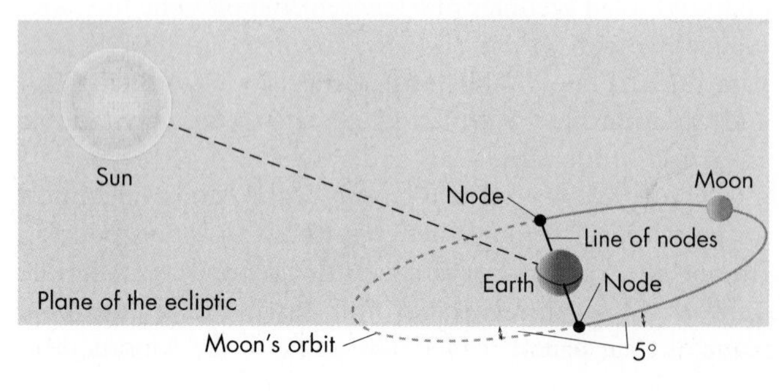

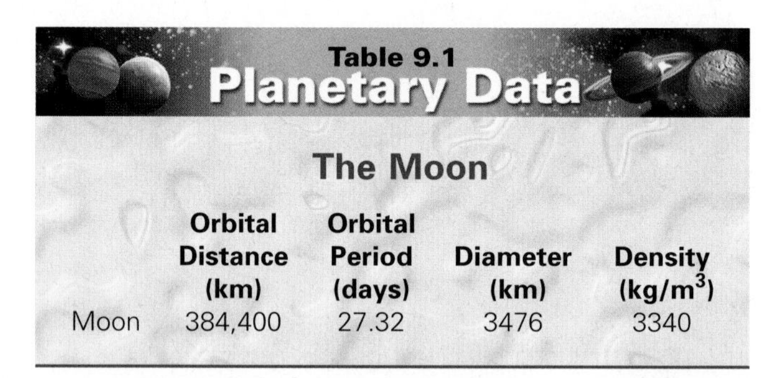

Table 9.1
Planetary Data

The Moon

	Orbital Distance (km)	Orbital Period (days)	Diameter (km)	Density (kg/m^3)
Moon	384,400	27.32	3476	3340

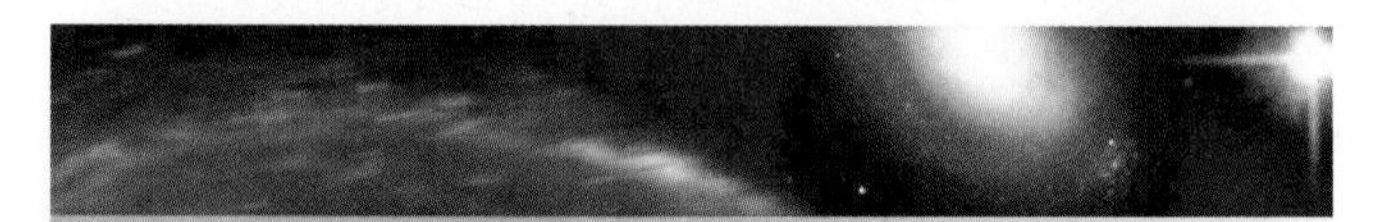

The Moon's orbit about the Earth is an ellipse lying nearly in the ecliptic plane. The angular size and distance of the Moon give its diameter, which is about 27% as large as the Earth's.

Eclipses

Eclipses

A total **eclipse** of the Sun begins gradually as the silhouetted Moon begins to intrude on the Sun's disk. The Sun appears to be slowly eaten away from west to east as more than an hour passes. During the partial phase of the eclipse, the sky darkens as though evening were approaching. Your senses tell you that something is wrong because the shadows are shorter than the long shadows of late evening. Birds flock to their evening roosts and there is a drop in temperature. As the crescent of sunlight grows smaller, the planets and bright stars become visible in the darkening sky.

As the Moon blocks all but the last sliver of sunlight, the sliver breaks into many bright spots of light called **Baily's beads,** shown in Figure 9.5. This happens because sunlight can be seen longer when it shines through valleys on the rim of the Moon. Lunar mountains cause the gaps between the bright beads. The Sun sinks behind the Moon and the beads disappear one by one. In the growing darkness, the last bead seems to shine with special brilliance in an effect called the **diamond ring,** shown in Figure 9.6.

Within a second the diamond ring disappears and the total phase of the eclipse has begun. All that can be seen of the Sun is the pale glow of the **corona,** which is the tenuous outer layer of the Sun (Figure 9.7). The corona is usually lost in the glare of the Sun's disk, so it is best seen from the ground during total eclipses. At the edge of the dark disk of the Moon it is sometimes possible to see solar prominences. Figure 9.8 shows that prominences point away from the Sun like flames. Unfortunately, totality

FIGURE 9.5
Baily's Beads
Just before totality, the Sun can still be seen through valleys on the rim of the Moon. The effect, seen at the bottom of the image, resembles bright beads of light.

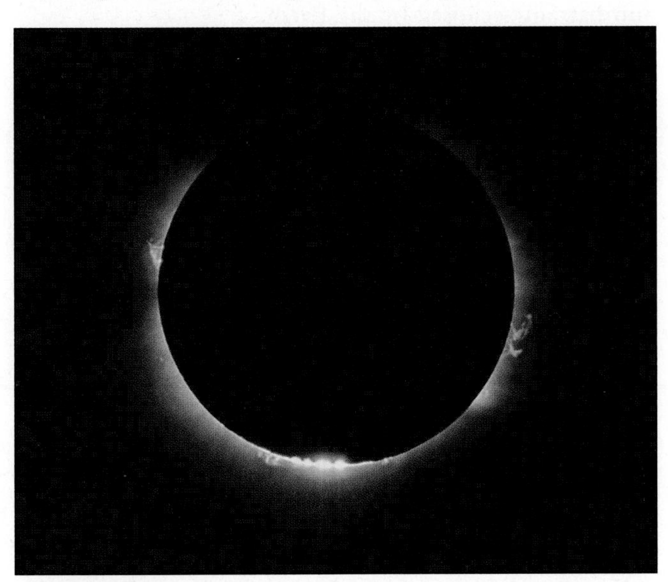

FIGURE 9.7
The Solar Corona
The tenuous outer region of the Sun's atmosphere is normally lost in the glare of the Sun's disk. However, it can be seen during total solar eclipses.

FIGURE 9.6
The Diamond Ring
The final glimpse of the Sun, just before totality, appears especially bright because the Sun is otherwise eclipsed.

FIGURE 9.8
Solar Prominences
Large arcs of glowing gas can sometimes be seen above the edge of the Moon during solar eclipses.

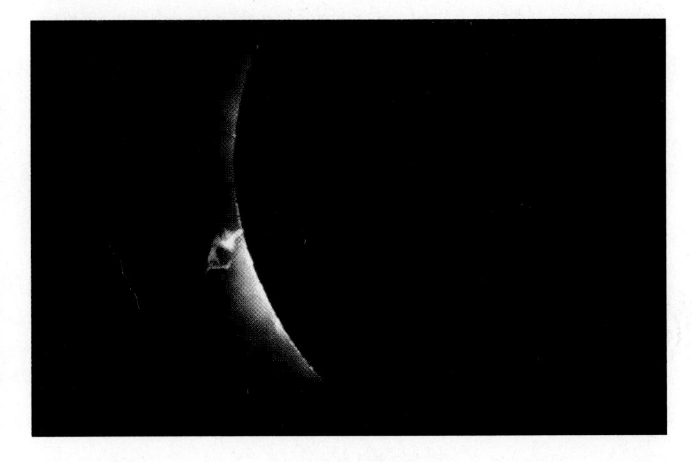

never lasts for longer than 7⅔ minutes. With a flash, the Sun reappears. Gradually, the sky brightens until the last dark sliver of the eclipse retreats from the Sun.

Eclipses as Omens

People have recorded solar and lunar eclipses for thousands of years. The oldest recorded lunar eclipse was observed more than 4000 years ago at Ur, in Mesopotamia. The oldest record of a solar eclipse was made by the Chinese a few hundred years later. Mesopotamian astronomers began keeping fairly detailed records of eclipses at least as early as 750 B.C.

Even if you understand why eclipses happen and are able to predict them, the disappearance of the Sun or Moon is dramatic and impressive. For those people who don't understand them, eclipses can evoke powerful reactions, as many historical examples demonstrate. Columbus, on his fourth voyage to the New World, was shipwrecked and marooned on the island of Jamaica. In his almanac he found that a lunar eclipse would occur on February 29, 1504. Knowing this, he threatened to make the Moon rise "inflamed with wrath." When the Moon rose dim and reddish in color, as often happens during a lunar eclipse, the native Arawaks were sufficiently impressed that they quickly brought food for Columbus and his starving men.

Plutarch, in his biography of the Athenian leader Nicias, described the consequences of a lunar eclipse that occurred in 413 B.C., just as the Athenians were about to be evacuated from Sicily during the Peloponnesian War. The Athenians regarded the eclipse as an uncertain omen. They waited a month to see if the Moon would be restored to normal the next time it was full. As a result of the delay, the entire Athenian army was either killed or captured and enslaved. Curiously, the Greeks of that time understood the cause of solar eclipses, but not lunar eclipses. They were perplexed because they could not identify the object that they thought passed in front of the Moon and blocked its light from reaching the Earth.

Eclipses and Shadows

Eclipses are produced by shadows. When a luminous body is a large one, such as the Sun, a body blocking its light casts a shadow that has two parts, as shown in Figure 9.9. The **umbra** is the inner part of the shadow. No sunlight penetrates directly into the umbra because all parts of the Sun's disk are blocked by the body. In the **penumbra,** the outer part of the shadow, only part of the Sun's disk is blocked by the body. Figure 9.10 shows the Earth's shadow. A body in the Earth's penumbral shadow is partially illuminated by the Sun. If it shines only by reflecting sunlight, it will appear dimmer than usual, but this may be hard to notice. If you were to look sunward from a location within the Earth's penumbra, you would see the Sun partially eclipsed by the Earth.

The situation is different within the Earth's umbral shadow. Here sunlight is almost completely blocked. Figure 9.11 shows that the only sunlight that enters the umbra is light refracted in passing through the Earth's atmosphere. More red sunlight than blue sunlight is refracted into the umbra so objects, such as the Moon, that enter the Earth's umbral shadow often appear red but very dim. From within the umbra, the Sun is completely eclipsed by the Earth. The penumbra and umbra are shaped differently. The penumbra extends infinitely far behind the Earth and grows broader with increasing distance from the Earth. The umbra, on the other hand, tapers to a point behind the Earth and has only a finite length. We enter the Earth's umbral shadow each night at sunset, experiencing a form of total solar eclipse. In the part of the penumbra directly behind the point of the umbra (point *C* in Figure 9.10), only the rim of the Sun can be seen. The visible part of the Sun forms an annulus (Latin for "ring").

FIGURE 9.9
The Parts of a Shadow
The inner part of the shadow is the umbra, where sunlight is completely blocked. The umbra tapers to a point behind the body that casts the shadow. The outer part of the shadow is the penumbra; sunlight is only partially blocked there. The penumbra widens as distance increases behind the body casting the shadow.

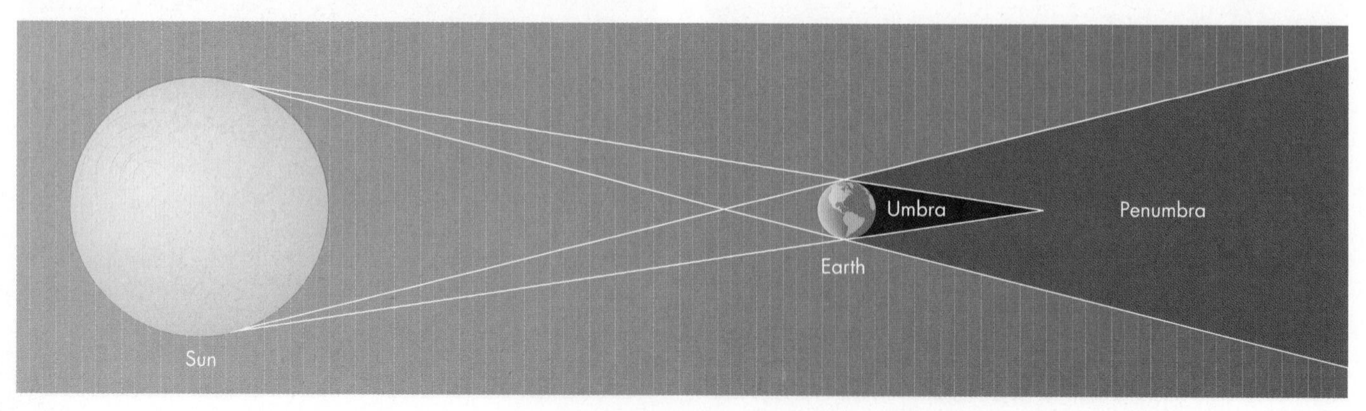

FIGURE 9.10
Eclipses in the Earth's Shadow
An object, such as an Earth satellite, which enters the Earth's penumbra receives and reflects less light and dims as it approaches the umbra, within which it is no longer visible. An observer on the satellite would see a partial eclipse of the Sun while in the penumbra but a total eclipse while in the umbra. An observer in the part of the penumbra directly behind the umbra would see an annular eclipse.

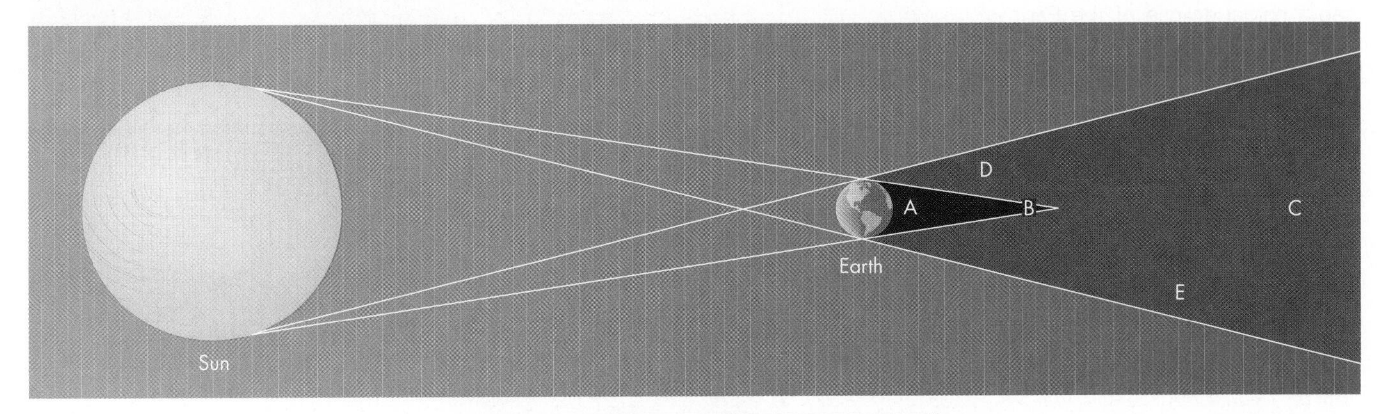

Location of observer	What the observer sees there
Point A (umbra)	Total eclipse
Point B (umbra)	Total eclipse
Point C (penumbra)	Annular eclipse
Point D (penumbra)	Partial eclipse
Point E (penumbra)	Partial eclipse

FIGURE 9.11
Sunlight Refracted into the Umbra
Light passing through the Earth's atmosphere is bent into the Earth's umbral shadow. More red light than blue light passes completely through the atmosphere into the umbra. For this reason the eclipsed Moon can be seen while in the umbra. The Moon often looks red or coppery in color during an eclipse.

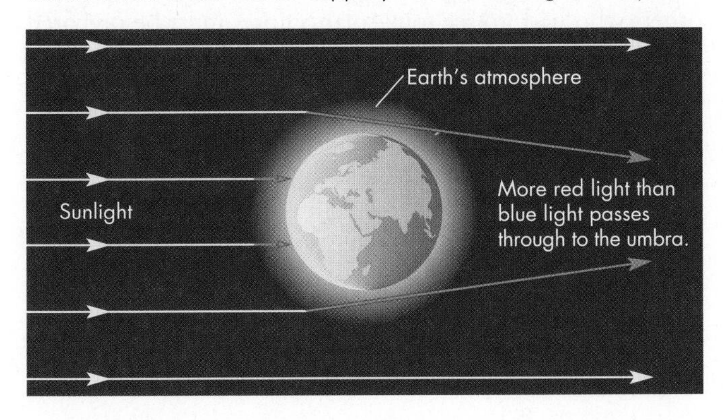

This arrangement produces what is called an **annular eclipse** of the Sun.

The Earth and Moon have shadows that consist of two parts, the umbra and the penumbra. Sunlight is partially blocked within the penumbra, but it is completely blocked within the umbra. The umbra is the region of total eclipses.

Eclipses and Lunar Phases Eclipses involving the Earth and Moon occur only when one of the two enters the shadow of the other. For this to happen, the Earth, Moon, and Sun must be in almost perfect alignment. Therefore, eclipses occur only at new or full phases of the Moon.

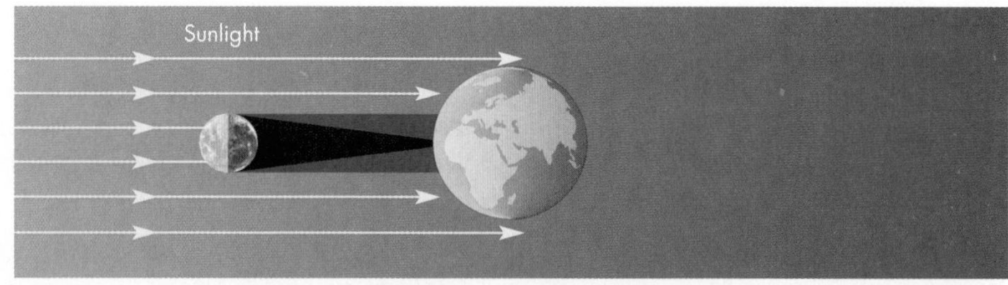

A Moon between Earth and Sun

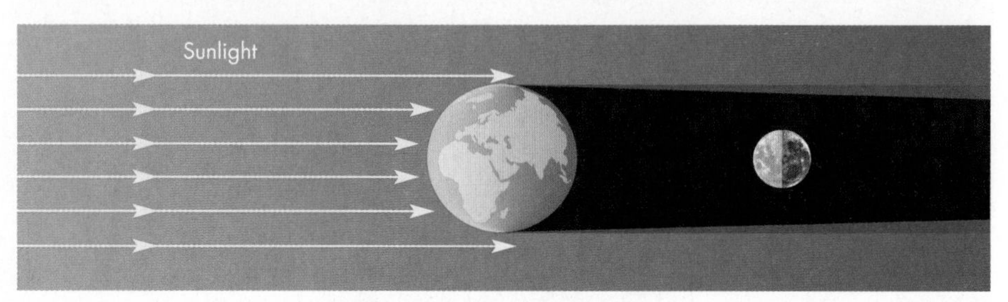

B Earth between Sun and Moon

FIGURE 9.12
Phases and Eclipses
A, When the Moon is new and is between the Earth and the Sun, we see a solar eclipse. Someone on the side of the Moon facing the Earth would see a partial eclipse of the Earth at that time. **B,** When the Earth is between the Sun and the Moon and the Moon is full, we see a lunar eclipse. An observer on the Moon would see a solar eclipse.

Figure 9.12 shows the arrangements of Earth, Moon, and Sun during lunar and solar eclipses. A solar eclipse occurs when part of the Earth enters the Moon's shadow. Thus, the Sun and Moon are in the same direction as seen from the Earth. The side of the Moon turned toward the Earth is turned away from the Sun and is dark, so the Moon's phase is new. A lunar eclipse occurs when the Moon enters the Earth's shadow. At this time the Moon and Sun are opposite each other in our sky. The side of the Moon turned toward the Earth is also turned toward the Sun. We see the illuminated part of the Moon and observe a full Moon. Knowing the dates of the new and full lunar phases is thus a prerequisite for predicting eclipses.

Umbral Shadows and Eclipse Tracks The Earth and Moon have different sizes and so do their umbral shadows. The Earth's umbral shadow averages about 1,380,000 km in length and extends far beyond the orbit of the Moon. At the Moon's average distance, the Earth's umbra is almost 9200 km across. If it could be seen projected on a screen, the umbra would be a dark circle about 1.3° across as seen from the Earth and would easily cover the entire Moon, which is only about 0.5° across. The Moon crosses the Earth's umbral shadow in about 3 hours and can be entirely within the umbra for as long as about 100 minutes. During this time, people all over the night side of the Earth can see the eclipse.

The Moon's umbral shadow averages about 371,000 km in length. It is shorter than the Earth's umbral shadow because the Moon is smaller than the Earth. The average distance between the Earth's surface and the Moon is about 378,000 km, so the Moon's umbra doesn't quite reach the Earth's surface most of the time. If the Moon's umbral shadow never reached the Earth's surface, we could never see total eclipses of the Sun. The best we could hope for would be annular solar eclipses. However, the length of the Moon's umbral shadow depends on the Moon's distance from the Sun and varies throughout the year. Also, the distance between the Earth and the Moon varies during a month because the Moon has an elliptical orbit. As a result of these factors, the Moon's umbra occasionally reaches the Earth's surface. However, even when it does reach the Earth, the umbra's diameter is never greater than about 270 km. It sweeps across the Earth as a round patch of darkness, as shown in Figure 9.13. The path of the Moon's umbral shadow across the Earth is called an **eclipse track.** People located on the eclipse track see a total solar eclipse. People located near but not on the eclipse track see a partial solar eclipse.

If the Earth didn't rotate, the Moon's shadow would sweep across the Earth at the Moon's orbital speed, about 3400 km/h, and the shadow would cover a point on the Earth's surface for no more than about 5 minutes. However, because the Earth rotates in the same direction as the Moon orbits, a point on the Earth moves in the same direction as the Moon's umbra. This lengthens the time that an eclipse can be seen from a spot on the Earth. Even so, the longest possible total eclipse lasts only 7 minutes 40 seconds. A more typical eclipse is total for only 2 or 3 minutes at a given place. A typical eclipse sweeps out a path about 10,000 km long and 150 km wide. The area covered is only 0.3% of Earth's surface area, so the chance of a given eclipse being total at a particular spot is quite small. Astronomers and others who wish to see total solar eclipses must be prepared to travel extensively, often to remote places.

During solar eclipses, the Moon's umbral shadow sweeps swiftly across the Earth. The shadow is never more than 270 km across. Less than 1% of the Earth's surface area is covered by the track of the umbral shadow during a single eclipse.

FIGURE 9.13
The Moon's Umbral Shadow
The shadow is only a few hundred kilometers across when it reaches the Earth's surface. In this image taken from the Mir space station, the Moon's umbra crosses Europe during an eclipse on August 11, 1999.

FIGURE 9.14
Eclipses and the Moon's Orbital Inclination
Because the Moon's orbit is inclined by about 5° relative to the ecliptic, the Moon is usually above or below the ecliptic when it is new or full. At least twice a year there are periods when the Moon, in its new or full phase, is near a node and eclipses can occur.

ANIMATION *Eclipses and the Moon's orbital inclination*

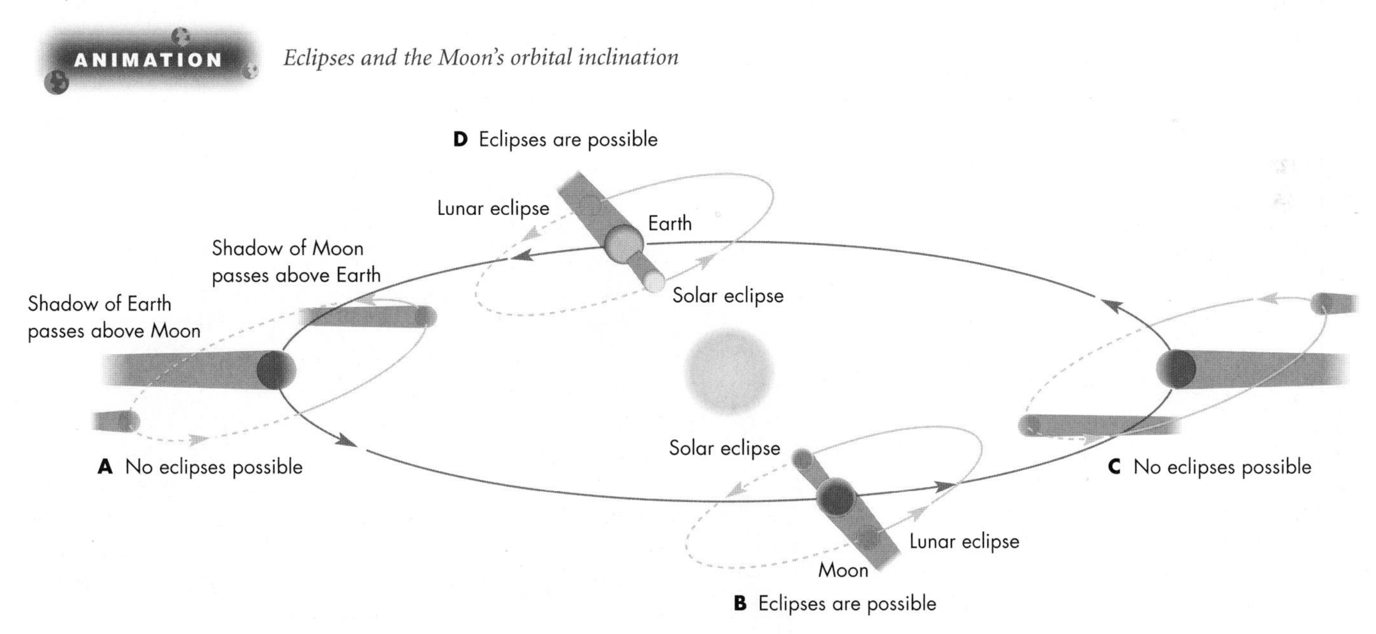

Predicting Eclipses

As shown in Figure 9.14, the inclination of the Moon's orbit means that even when the Moon is in its new or full phase it isn't necessarily in the Earth's orbital plane and directly aligned with the Sun and Earth. Only if the Moon is near one of the nodes of its orbit at the same time it is new or full can an eclipse occur. This happens when the **line of nodes,** the imaginary line connecting the two nodes of the Moon's orbit, points nearly at the Sun. The line of nodes points directly enough at the Sun for eclipses to occur during intervals of time called **eclipse seasons** which last for approximately a month. When the Moon reaches the descending node of its orbit during an eclipse season, it moves southward through the ecliptic and its umbra sweeps toward the southeast across the Earth's surface. When the Moon reaches the ascending node, its umbra sweeps toward the northeast. Each eclipse season lasts about a month, during which the alignment of the Sun, Moon, and Earth during the new or full phases of the Moon is precise enough to produce eclipses.

Eclipse seasons are separated by intervals of 5.7 months. This means that there are two of them in most years, but in some years there are three. The **eclipse year** is the interval separating successive times when the same node of the Moon's orbit is aligned with the Sun. This is just twice the amount of time separating eclipse seasons, or 346.6 days. The variable number of eclipse seasons during a year and the fact that more than two eclipses can occur in a single eclipse season mean that the number of eclipses varies from year to year. The maximum number of eclipses (counting both solar and lunar) that can take place during a year is seven. Seven eclipses occurred in 1973 and 1982, but this won't happen again until 2038.

Eclipses occur in series. The interval of time between members of a series of eclipses is determined by the coincidence that both 223 synodic months and 19 eclipse years add up to nearly the same amount of time, about 6585⅓ days = 18 years 11⅓ days. The significance of this coincidence can be seen in the following hypothetical situation. Suppose a solar eclipse occurs today. Then 18 years 11⅓ days from today a whole number (223) of synodic months will have passed, so the Moon will be at the same phase as it is today. Also, a whole number (19) of eclipse years will have passed, so the Moon will be at the same node of its orbit as it is today. Because the Moon will be in its new phase and at a node of its orbit, an eclipse will occur. Moreover, the eclipse will be similar to the one that occurs today. That is, because the Moon will be in its new phase, the eclipse will be a solar eclipse. Because the Moon will be at the same node of its orbit, its umbral shadow will sweep across the Earth in the same direction.

The length of time from one member of the series of eclipses to the next, 6585⅓ days, is called the **saros.** Because the saros exceeds a whole number of days by about ⅓ day, consecutive members of a series of solar eclipses occur ⅓ of the circumference of the Earth, or about 120° in longitude, apart. However, after three saros periods (about 54 years, 1 month), another eclipse will be seen at nearly the same longitude as the first eclipse. Figure 9.15 shows eclipse tracks for six consecutive members of the same series of solar eclipses.

Someone who is aware of the saros can predict future eclipses of the Sun and Moon from records of the dates of previous eclipses. The Babylonian astronomers recognized this by about 600 B.C., but at first they were able to use the saros only for lunar eclipses. Thales's prediction of an eclipse in 585 B.C. may have been based on knowledge of the saros.

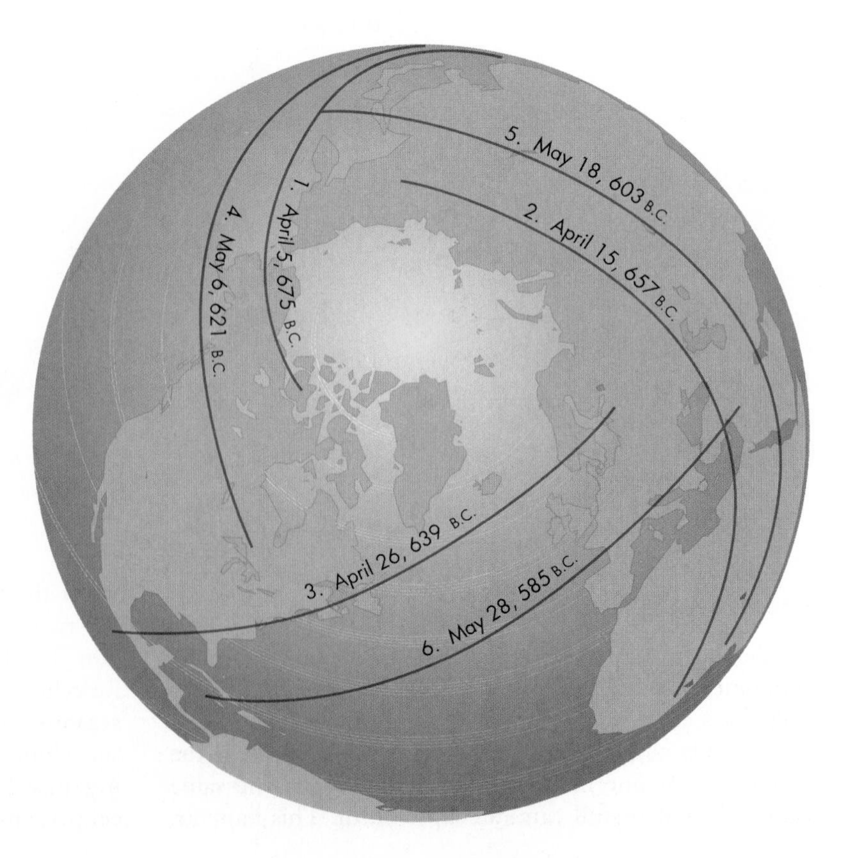

FIGURE 9.15
Eclipse Tracks
Eclipses in a saros series are very similar to each other. Notice that the six eclipses shown here are separated from each other by about 120° in longitude so that the first and the fourth, second and fifth, and third and sixth in the series have tracks that are near each other. No doubt many people in the Near East saw both eclipses two and five as partial solar eclipses.

Table 9.2

Solar and Lunar Eclipses

DATE	TYPE OF ECLIPSE	WHERE VISIBLE
January 26, 2009	Annular solar	Indian Ocean, Indonesia
February 9, 2009	Penumbral lunar	Asia, Australia, Western North America
July 7, 2009	Penumbral lunar	Australia, North and South America
July 22, 2009	Total solar	India, China
August 6, 2009	Penumbral lunar	North and South America, Europe, Africa
December 31, 2009	Partial lunar	Europe, Africa, Asia, Australia
January 15, 2010	Annular solar	Africa, Asia
June 26, 2010	Partial lunar	Asia, Australia, North America
July 11, 2010	Total solar	South Pacific, South America
December 21, 2010	Total lunar	Asia, Australia, North America, Europe
January 4, 2011	Partial solar	Europe, Africa, Asia
June 1, 2011	Partial solar	Asia, Alaska, Canada
June 15, 2011	Total lunar	South America, Europe, Africa, Asia, Australia
July 1, 2011	Partial solar	Indian Ocean
November 25, 2011	Partial solar	South Africa, Australia
December 10, 2011	Total lunar	Europe, Africa, Asia, Australia
May 20, 2012	Annular solar	Asia, Pacific Ocean, United States
June 4, 2012	Partial lunar	Asia, Australia, North America, South America
November 13, 2012	Total solar	Australia, Pacific Ocean, South America
November 28, 2012	Penumbral lunar	Europe, Asia, Australia, North America
April 25, 2013	Partial lunar	Europe, Africa, Asia, Australia
May 10, 2013	Annular solar	Australia, Pacific Ocean
May 25, 2013	Penumbral lunar	North America, South America, Africa
October 18, 2013	Penumbral lunar	North America, South America, Africa, Europe, Asia
November 3, 2013	Total solar	Atlantic Ocean, Africa

Eclipses can occur only near new or full phases of the Moon and only when the Moon is near the ecliptic plane. Eclipses occur in series separated by the saros, a length of time equal to about 18 years 11⅓ days. Knowledge of the saros enabled ancient astronomers to predict eclipses.

There are about 240 solar eclipses per century. Of these, about 35% are partial, 32% are annular, and 33% are total (or total for part of the eclipse track and annular for the rest). Solar eclipses and lunar eclipses are almost equally numerous, yet solar eclipses are thought of as rarer events. This is because only a very small fraction of the world's population is able to witness a given solar eclipse. On the other hand, about half of the people on the Earth are on the side facing the Moon during a lunar eclipse. Barring cloudy weather, each of them has the opportunity to see the Moon enter the Earth's shadow and be eclipsed. Table 9.2 gives the total and partial eclipses of the Sun and Moon that will occur between 2009 and 2013. Figure 9.16 shows the tracks of total solar eclipses visible in North America between 2009 and 2050. Notice that the next total solar eclipse visible in the contiguous United States will occur on August 21, 2017. This will be almost 40 years after the most recent total solar eclipse visible in the contiguous United States, which occurred on February 26, 1979.

9.2 TIDES

The Earth and the Moon exert tidal forces on each other. These tides, described in Chapter 5, Section 5, have had a strong influence on the history of the Earth-Moon system.

FIGURE 9.16
North American Eclipse Tracks: Total Solar Eclipses, 2001–2050
The figure shows the tracks of total solar eclipses visible in North America between 2001 and 2050. The next total solar eclipse visible in the contiguous United States will occur on August 21, 2017.

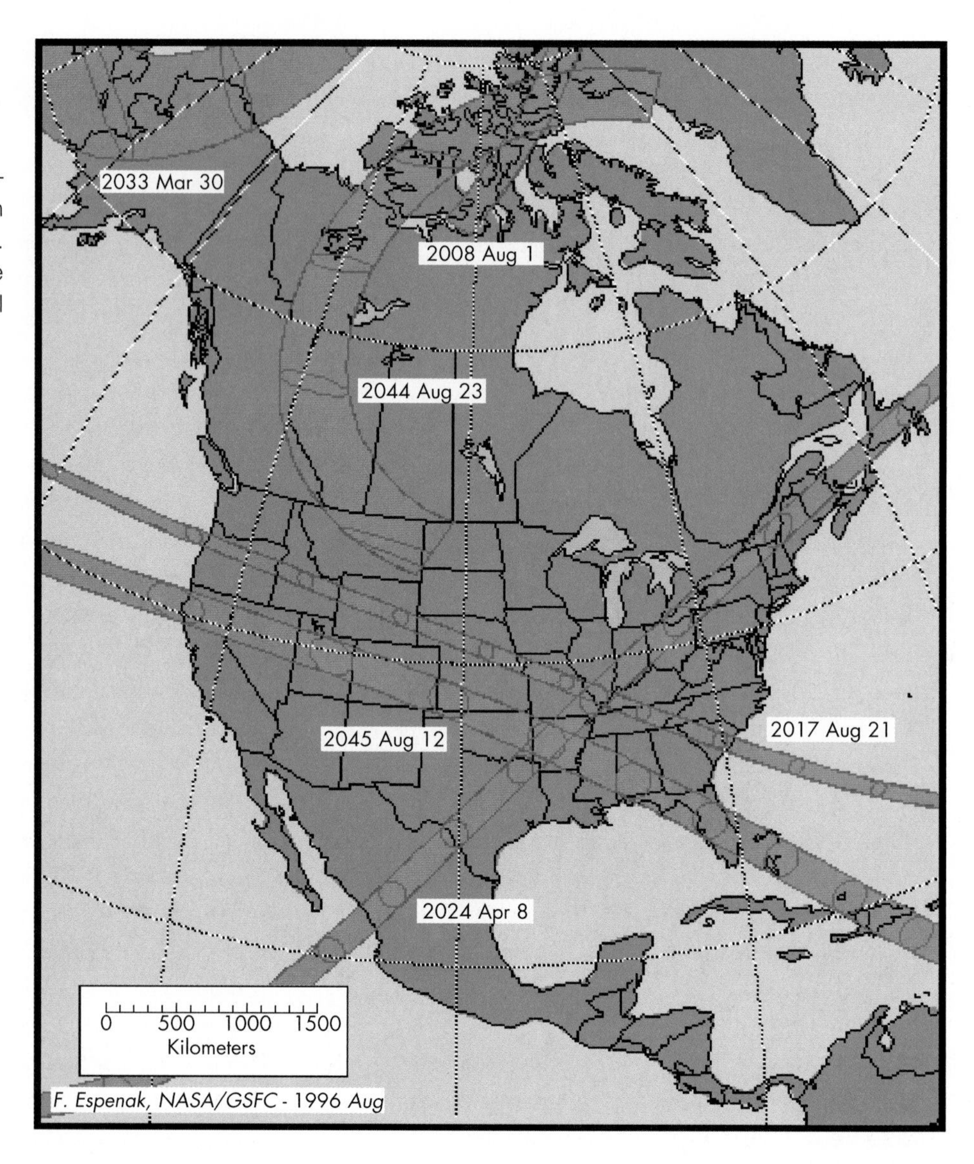

Tides and the Earth's Rotation

Tides cause the Earth's surface to rise and fall by as much as 20 cm. The flexing of the solid Earth and the tidal flow of ocean water, particularly in shallow regions above continental shelves, result in friction, which dissipates some of the energy of the Earth's rotation and the revolution of the Earth and Moon about each other. The frictional interaction between the Earth and its tidal bulges resembles what happens if someone puts his or her hand on a spinning bicycle tire. Just as the person's hand slows the spin of the tire, tidal friction slows down the Earth's spin. The friction causes the day to lengthen at a rate that is now about 0.0015 seconds per century. This may seem like a negligible rate of change, but it adds up over the course of the Earth's history. For example, studies of fossil corals have shown that days were shorter so there were more days in a year in the remote past. This result came from studying the width of the rings secreted by corals. Coral animals secrete one ring per day. The width of the ring depends on the amount of sunlight received, which in turn depends on the season of the year. The fossils show that there were about 384 days per year 270 million years ago and 407 days per year 440 million years ago. Thus, the length of the day has increased from 21.5 hours 440 million years ago to 22.8 hours 270 million years ago to 24 hours today. The average rate of change over this time was about the same as the rate at which the length of the day is now changing.

Because the Earth is much more massive than the Moon, the Moon feels stronger tides than the Earth does. Long ago tides slowed the rotation of the Moon so much that the lunar day became a month long and the Moon's rotation became synchronous with its revolution. The Moon's tidal bulges, which are much larger than the tidal bulges of the

FIGURE 9.17
Tides and the Moon's Orbit
The Earth's tidal bulge that is closer to the Moon produces a stronger acceleration on the Moon than does the more distant bulge. The effect is to pull the Moon ahead in its orbit and increase its angular momentum.

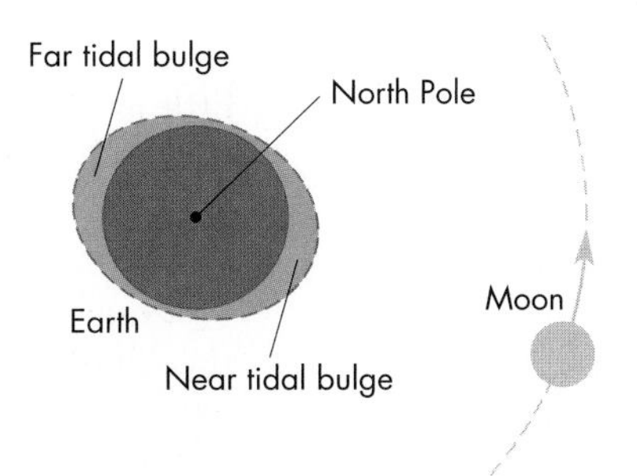

A If the Earth didn't rotate, its tidal bulges would align with the position of the Moon as it orbits Earth.

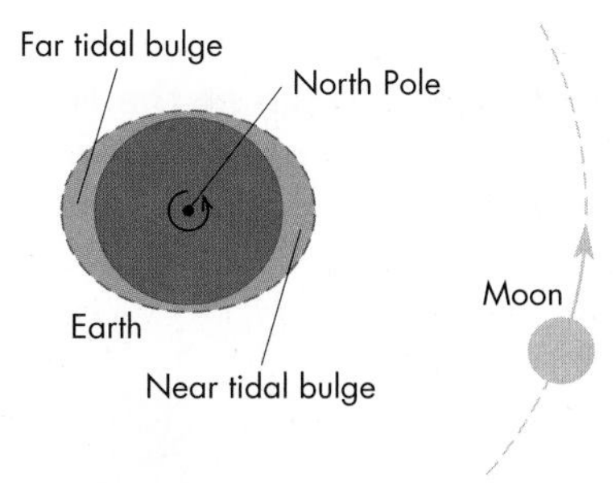

B Instead, the Earth's rotation drags its tidal bulges ahead of where they would be if the Earth didn't rotate.

solid Earth, are locked into place so that they always point toward and away from the direction of the Earth.

Tides and the Moon's Orbit

The frictional interaction between the Earth and its tidal bulges also affects the location of the tidal bulges. Once again think about someone placing his or her hand on a rotating bicycle tire. Figure 9.17 shows that just as the person's hand is pulled forward by the spinning tire, the Earth drags the tidal bulges ahead of where they would be if the Earth weren't spinning.

The Moon feels accelerations due to both of the Earth's tidal bulges. However, because the two bulges are at different distances from the Moon, the accelerations they produce are unequal. The nearer bulge produces the stronger acceleration and pulls the Moon forward in its orbit. As the Moon accelerates, its angular momentum (see Chapter 5, Section 3) increases, causing it to recede from the Earth. The rate at which the Moon is now receding from the Earth is about 4 cm per year. As the Moon recedes, the length of the lunar month increases by about 0.014 seconds per century. The fact that the Moon is receding implies that it was closer to the Earth in the past. If the Moon's distance has always increased at today's rate, then the Moon must have been in a very close orbit about the Earth only 1.6 billion years ago. However, it is more likely that tidal friction was smaller and the Moon moved away from the Earth more slowly billions of years ago than it does today.

If the present rate of recession of the Moon continues, in about 1.1 billion years the Moon will be so distant that its umbral shadow will no longer reach Earth, even under the best of circumstances. Total solar eclipses will cease to be visible from the Earth.

As both the solar day and the lunar month grow longer, a time may come in the extremely remote future when they are both equal to about 47 of our present days. If and when that happens, the Earth will rotate synchronously with the Moon. Each will keep the same face turned toward the other so the Moon will be visible from only one side of the Earth. At that time, the Moon will hang in the sky just as the Earth now does for lunar astronauts. Moonrise will be a thing of the past. Elsewhere in the solar system, synchronous rotation has already been reached by both Pluto and its satellite, Charon. Outside the solar system, many binary star systems are synchronously coupled.

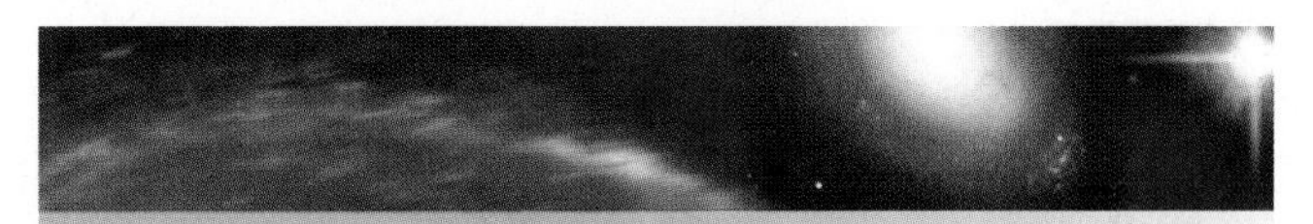

The tidal interaction between the Earth and the Moon is gradually slowing the rotation rate of the Earth and causing the Moon to grow more distant from the Earth. In the very distant future, the day and month may have the same length, causing the Earth to keep the same face turned toward the Moon.

9.3 THE MOON'S SURFACE

Observations from the Earth

The best pretelescopic sketches of the Moon were probably those of Flemish painter Jan Van Eyck, in the early 1400s, and Leonardo da Vinci, in the early 1500s. Despite da Vinci's skill, the sketches like the one in Figure 9.18 showed only a few major features and bore only a modest resemblance to modern lunar maps. Meaningful observations of the surface of the Moon began in 1609, when Galileo viewed the Moon through his telescope. Figure 9.19 shows one of Galileo's sketches of the Moon. Galileo saw two distinct types of lunar terrain. Although Galileo understood the Moon probably didn't actually have large bodies of water, he called the dark regions **maria** (or seas) and the lighter, cratered regions **terrae** (land). The smaller dark regions of the Moon are called Lacus (lake), Palus (marsh), or Sinus (bay). The names attached to the dark regions of the Moon were invented in the mid-1600s by Giovanni Riccioli. Riccioli also began the custom of naming lunar craters after notable scholars. Lunar mountains, however, are named after mountain ranges on Earth.

FIGURE 9.18
A da Vinci Sketch of the Moon
This pretelescopic sketch shows only the relatively few features that can be seen by the unaided eye.

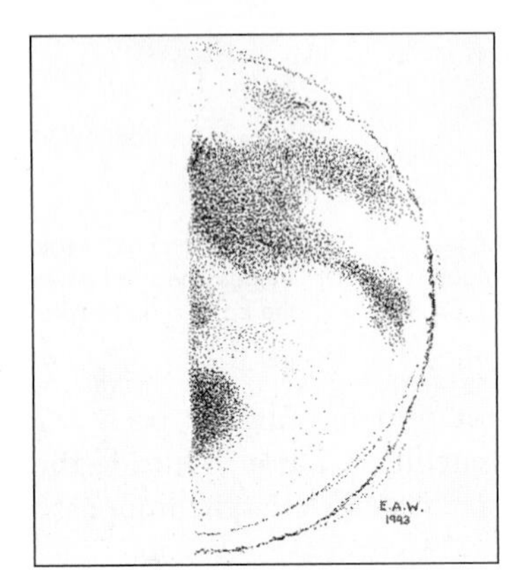

FIGURE 9.19
Galileo's Sketch of the Moon
This sketch was made using a telescope. Most of the features drawn by Galileo can be identified with features on modern maps of the Moon.

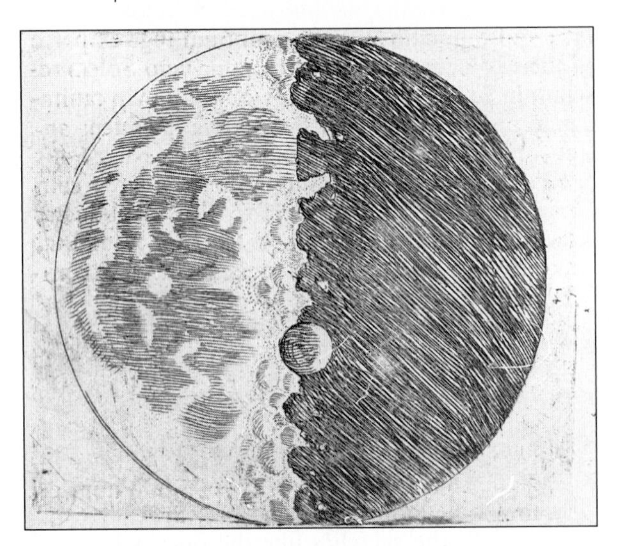

Exploration of the Moon

The exploration of the Moon by robot probes and by astronauts has provided a wealth of information about the lunar interior, surface, and atmosphere. By the time of the first robot lunar landers in 1966, there had been a number of Soviet and American flybys; hard landers, which crashed into the Moon (some of the hard landers were intended to be "soft" landers); and orbiters. Some of these missions were among the first great scientific accomplishments of the space age. *Luna 3* (1959) passed behind the Moon and returned the first images of the side that can never be seen from the Earth. The images showed that the far side of the Moon, which has almost no large maria, is surprisingly different from the near side, as shown in Figure 9.20. The *Ranger* series in 1964–65 used television cameras to obtain images of the lunar surface while approaching the Moon on crash courses. Some of the pictures were taken from as little as 500 m above the surface. The pictures showed some regions littered with large boulders and other locations that were smooth and flat enough for safe landings.

In 1966, both the USSR and the United States began series of robot landers. Some scientists had predicted that the Moon had such a thick dust layer that landings would be impossible. By landing and sending back pictures of the Moon's surface, the first lunar lander, *Luna 9,* showed that the surface of the Moon was *not* covered by an extremely thick layer of dust. The goal of the U.S. *Surveyor* program was to determine whether a human landing was possible. The *Surveyors* measured the strength and composition of the lunar soil and found that the lunar maria were quite suitable for human landings. In fact, in 1969 *Apollo 12* landed about 600 m from the *Surveyor 3* landing site. *Apollo 12* astronauts Charles Conrad and Alan Bean visited the *Surveyor 3* landing site during their exploration of the lunar surface. After the *Surveyor* program came the *Apollo* program of human landings on the Moon. Beginning with the historic landing of *Apollo 11* on July 20, 1969, and ending with the *Apollo 17* mission in December 1972, teams of astronauts carried out examinations of a variety of maria and terrae regions. They explored as far as 30 km from their landing sites, carried out numerous scientific experiments, and brought back a total of about 400 kg of carefully selected lunar rocks and soil as well as cores bored from the lunar surface layers. This priceless material has since been carefully studied for information about the nature and history of the Moon.

After the numerous *Luna* and *Apollo* missions in the 1960s and 1970s, there have been few lunar missions. *Clementine,* launched in 1994, mapped the Moon with nearly two million images. *Lunar Prospector,* which orbited the Moon in 1998 and 1999, mapped the composition of lunar surface rocks, measured magnetic and gravitational fields, and studied gases escaping from the Moon. However, lunar missions have become more frequent in recent years. China (in 2007) and India (in 2008) have both put spacecraft in lunar orbit. India also landed a probe on the Moon in 2008. The United States has announced plans for a series of manned lunar missions beginning as early as 2020. In the first mission, a four-person crew would visit the lunar surface for as long as a week. In later missions, astronauts may stay on the Moon for as long as 6 months. The lunar missions would be a prelude to a program of human exploration of the planet Mars.

Lunar Craters

As Figure 9.21 shows, **craters** are the dominant features of the lunar surface. Most, if not all, of them were produced by the impacts of meteoroids. Craters range in size from rimless depressions less than a meter in diameter to features like Clavius, more than 220 km in diameter with a high rim and a depth of several kilometers. Large craters are surrounded by vast regions littered with ejected material. The regions of ejecta contain numerous secondary craters produced by the impacts of large rocks blasted from the

FIGURE 9.20
The Surface of the Moon
A, The near side of the Moon can be mapped from the Earth. **B,** The far side is known only from images taken by spacecraft orbiting or flying past the Moon. The dark areas in these images are maria. Note the near absence of maria on the far side compared with the near side.

A

B

original impact site. Some of the youngest craters, like Copernicus (shown in Figure 9.22) and Tycho, are surrounded by bright patterns of **rays,** which are particularly conspicuous at full Moon. The nature of the rays is not completely understood. However, their appearance may simply be due to the lighter color of the ejected material compared with the older surface. Presumably, the rays eventually fade as the ejected material is darkened

FIGURE 9.21
The Cratered Surface of the Moon
The craters shown here range from less than 1 km in size to 100 km in diameter. The large crater at the left center is Eratosthenes. An even larger crater, Copernicus, is in the distance on the right.

FIGURE 9.22
The Crater Copernicus
Only relatively young lunar craters, such as Copernicus (100 km in diameter), show systems of bright rays. The large crater in the upper right is Eratosthenes.

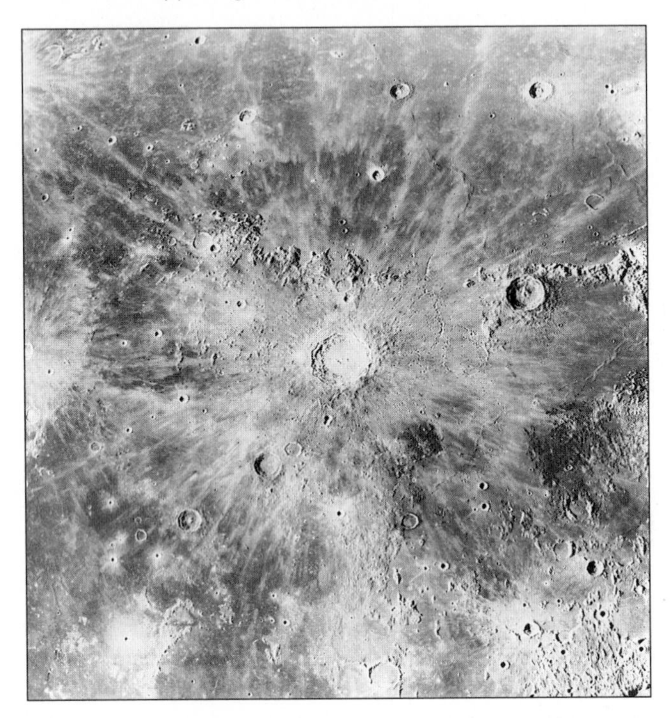

FIGURE 9.23
The Apennine Mountains
The Apennines stretch from lower left to right center across this picture. Mare Imbrium is in the upper left. Lunar mountains are portions of the rims of large craters and basins.

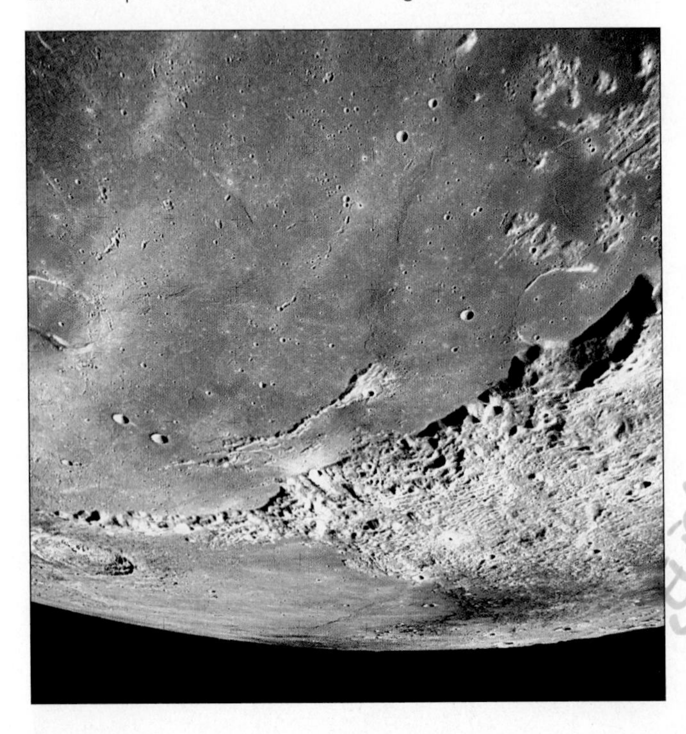

by radiation damage from the solar wind and melted or stirred up by new impacts. The lunar mountains, such as the Apennines shown in Figure 9.23, are just portions of the rims of large craters and other large impact structures. Some of the lunar mountains are nearly 8 km higher than the adjacent flat regions. Unlike the mountains of the Earth, the lunar mountains did not form as a result of the collisions of crustal plates and do not indicate internal activity.

Formation of Craters Meteoroids falling toward the Moon are accelerated by lunar gravity and strike the Moon at speeds at least as great as the Moon's escape velocity (2.4 km/s). Some impacts occur at speeds as great as several tens of kilometers per second. The extremely energetic nature of the impacts has several important consequences. First, the impacting body can melt an amount of lunar crust several times more massive than the body itself. Second, the impact generates an expanding shock wave that moves into the Moon from the point of impact and has a pressure of millions of atmospheres. Thus crater formation resembles an underground nuclear explosion. The resulting crater tends to be roughly round unless the impacting body strikes the Moon almost horizontally. As a general rule, the diameter of the crater is five to ten times the diameter of the impacting body.

The Regolith A lot of meteoritic material has accumulated on the Moon and much of the lunar surface has been pulverized as a result of the long history of impacts. The layer of surface debris is called the **regolith** ("blanket of rock"). The regolith averages several meters in thickness although its depth varies considerably. Measurements by *Apollo* astronauts showed that the regolith is almost 40 m thick in Mare Serenitatis but only a few centimeters thick on the edge of Hadley Rille (a valley), where most of the debris has slipped down the sloping side of the valley. The thickness of the regolith implies that there are probably few places where outcroppings of lunar rock are exposed. Much of the debris from impacts consists of dust-size fragments. These fragments make the Moon a very dusty place, though not as dusty as had been feared before the first robot landings on the Moon.

The surface of the Moon is heavily cratered. One effect of the impacts of infalling bodies has been to pulverize the Moon's surface, creating a layer of fine debris called the regolith.

Volcanic Features

The Maria Although cratering accounts for much of the appearance of the Moon, some lunar features are volcanic. The most conspicuous of these are the maria, which lunar samples have shown to consist of volcanic basalts. The maria formed within very large, roughly circular basins produced by impacting bodies 100 km or more in diameter. These impacts fractured the Moon to great depth. Much later, magma from a depth of hundreds of kilometers reached the surface through the fractures and flooded the impact basins. The lava flows that formed the maria wiped out most of the previous surface features and left a smooth, uncratered surface. Apparently, the lava that flooded the maria was too fluid to pile up as volcanic mountains. Instead, it spread out to fill most of the large impact basins to a depth that averages only a few hundred meters.

The intrusion of dense mantle material into the fractured crust beneath the maria produced concentrations of mass called **mascons.** A spacecraft passing over one of the mascons feels a gravitational acceleration that slightly alters its orbit. Other indications of lunar volcanic activity are mountains that may be extinct volcanos. These low, rounded hills, or domes, are found near some of the maria.

Sinuous Rilles Other lunar features that may have been caused by volcanic activity are **rilles,** which may be either sinuous (winding) or straight. **Sinuous rilles,** such as Hadley Rille shown in Figure 9.24, occur in the maria and resemble dry riverbeds. A possible explanation of the sinuous rilles involves the collapse of lava tubes. A lava tube is produced when a crust forms on a river of lava, which then drains away leaving a tunnel. Eventually, the roof of the lava tube collapses to form a sinuous rille.

FIGURE 9.24
Hadley Rille
This sinuous rille curves along Mare Imbrium at the foot of the Apennine Mountains. Sunlight illuminates the left wall of Hadley Rille, whereas the right wall is in shadow. The arrow marks the landing site of *Apollo 15.*

If there has been any recent volcanic activity on the Moon, it cannot have been very widespread, because none of the rocks collected on the Moon is less than 3 billion years old. Whether there is now any volcanic activity on the Moon is somewhat controversial. There have been numerous reports of transient lunar phenomena, such as lights in the night hemisphere of the Moon. These phenomena tend to be seen on crater rims, central peaks of craters, and the edges of maria. Although the transient phenomena, if real, have no adequate explanation, most astronomers doubt that they are really indications of volcanism.

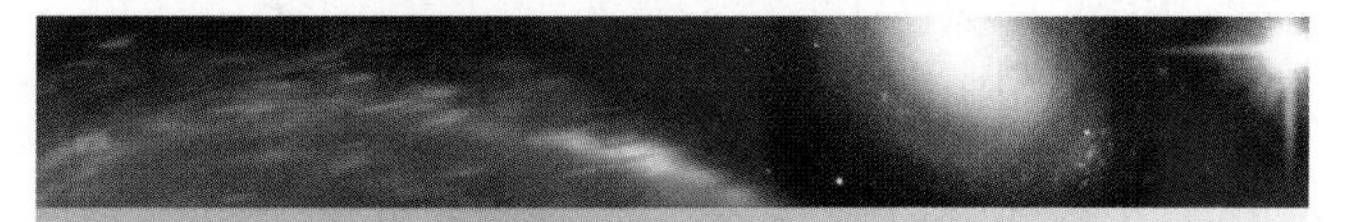

Maria, great lakes of solidified lava, are the most widespread evidence of volcanic activity on the Moon. Volcanic activity on the Moon has not been widespread for billions of years.

9.4 LUNAR SAMPLES

Chemical Composition

The lunar samples returned by six *Apollo* missions and by three Soviet sample return missions have provided detailed information about the chemical makeup of the surface layers of the Moon. Generally, the lunar rocks are remarkably similar to the silicate rocks of the Earth's crust, although there are some significant differences in detailed chemical makeup. Elements such as aluminum, calcium, chromium, titanium, and uranium, which form compounds that condense from a gas at relatively high temperatures, seem to be more abundant on the Moon than on the Earth. On the other hand, water is almost completely absent in the lunar samples. Easily vaporized elements such as sodium, chlorine, zinc, and lead are also low in abundance compared with terrestrial rocks. Iron and those elements that dissolve easily in molten metal are less abundant as well. Some of the differences in composition between the Moon and the Earth's crust are thought to represent differences in the chemical nature of the material from which the two objects formed. The similarities and differences between lunar and terrestrial rocks are some of the most important clues to understanding how the Moon formed.

Ages of Lunar Samples

The oldest lunar rocks are about 4.5 billion years old and are thought to have solidified shortly after the formation of the Moon. Most lunar rocks, however, are significantly younger. Rocks from the terrae sites are generally more than 4 billion years old, but rocks from the maria range from 4 billion to 3.2 billion years old. Apparently, magma from the lunar interior flooded the surface beginning about 4 billion years ago. Eruptions continued until the maria were completely formed by about 3.2 billion years ago. Since then, little rock formation has occurred.

Ages and Cratering

Crater density is the number of craters of different sizes on a given amount of lunar surface area. The older regions of the Moon's surface have had more time to accumulate impacts, so crater density increases with age. Crater counts using Earth-based telescopes or lunar orbiter images show that the maria have crater densities about 1⁄30 the crater densities of the terrae. The terrae are virtually completely covered with craters 4 km in diameter or larger. In fact, the terrae have reached **crater saturation.** That is, new impacts would obliterate as many craters as they would produce. The difference in crater density between the maria and the terrae tells us that the terrae are older than the maria, but how much older can't be determined using crater density alone.

History of the Lunar Surface The ages obtained for lunar samples make it possible to know just how much older terrae are than maria. Combined with crater densities for terrae and maria, the ages also make it possible to estimate the rate at which craters were formed throughout the Moon's history and to deduce how the lunar surface has changed with time. A key observation is that the terrae are only about 15% older than the maria, but they are much more heavily cratered. Apparently, the cratering rate was very much larger when the terrae were formed 4 billion years ago than when the maria formed a few hundred million years later. Perhaps the intense bombardment of the Moon 4 billion years ago was the final stage in the accumulation of the Moon. The cratering rate may have been even higher earlier than 4 billion years ago, but little of the earlier surface is left. Most of it was wiped out by impacts toward the end of the bombardment.

At the same time that the bombardment of the lunar surface was producing the terrae, impacts of especially large meteoroids produced the circular basins in which the maria now lie. This stage in the history of the lunar surface is sketched in Figures 9.25 and 9.26. The largest of the basins, and the second largest impact basin in the solar system, is South Pole-Aitken. South Pole-Aitken is as much as 9 km deep and stretches 2250 km from the south pole to nearly the equator on the Moon's far side. The impact that produced South Pole-Aitken penetrated the Moon's mantle and brought iron- and magnesium-rich mantle material to the surface. Because the interior of South Pole-Aitken is nearly filled with later basins and craters, it wasn't recognized as a giant impact basin until the 1970s. The end of the bombardment era and the formation of the last basins took place at about the same time, 3.8 billion years ago. Over the next 600 million years or so, the magma that produced the maria rose to the surface. The Moon has

FIGURE 9.25
The History of the Lunar Surface
A, The Moon's original crust formed about 4.6 billion years ago. **B,** The original crust was obliterated by impacts including those which formed the basins between 4.6 and 3.8 billion years ago. **C,** After the basins were formed, they were filled with lava between 3.8 and 3.2 billion years ago. **D,** Since that time, impacts have continued to produce lunar craters.

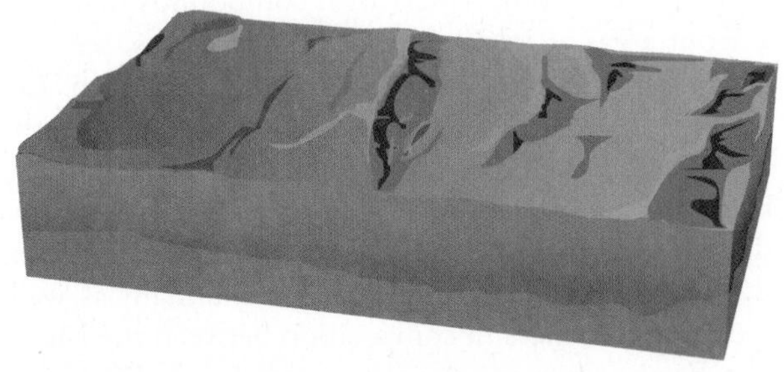

A 4.6 billion years ago

B 4.6–3.8 billion years ago

C 3.8–3.2 billion years ago

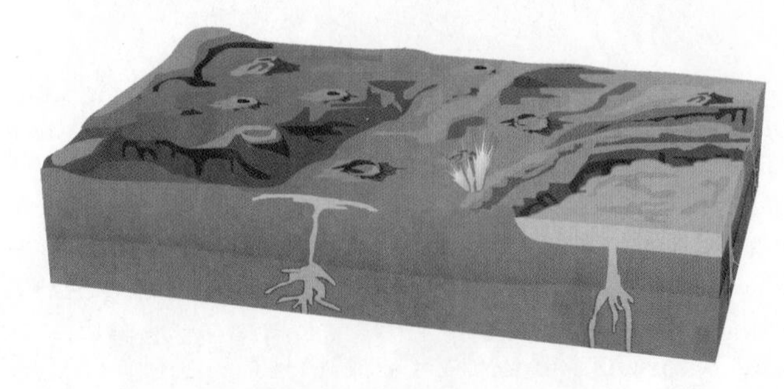

D 3.2 billion years ago to today

FIGURE 9.26
The Changing Appearance of the Moon
A, About 4 billion years ago, the basins of the Moon had been formed, but they had not yet filled with lava. **B,** By 3.2 billion years ago, lava flows had formed the smooth-floored maria. The present Moon, **C,** looks similar to the Moon of 3.2 billion years ago, but it has more heavily cratered maria.

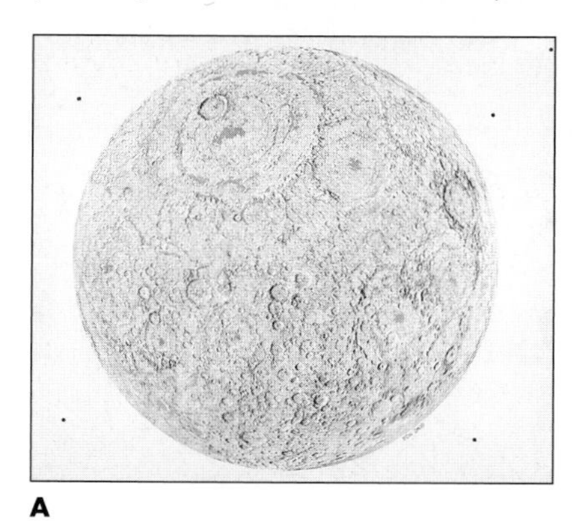
A

B

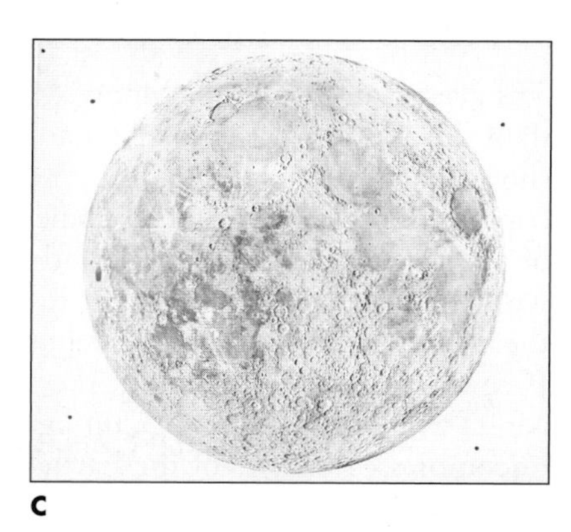
C

changed very little since the formation of the maria was completed 3.2 billion years ago. Since then, the main changes have been due to impacts that have gradually scarred the maria.

Crater Densities on Other Planets and Satellites

The general relationship between age and crater density holds for the surfaces of other planets and satellites as well as for the Moon. That is, the more heavily cratered parts of the surface of a planet or satellite must be older than the less heavily cratered areas. Because radioactive dating has not been possible for other solar system bodies, only relative ages can be found from crater densities. However, sometimes astronomers have assumed that the rate of crater formation has been the same throughout the solar system. Using this assumption, astronomers have used the lunar relationship between age and crater density to estimate the ages of other solar system bodies from Mercury to the satellites of Neptune.

Crater density, the number of craters of different sizes on a region of the Moon's surface, increases with increasing age. By relating the age of a region of the Moon to its crater density, it has been possible to learn the history of lunar cratering. Impacts were much more frequent before the maria formed between 3 and 4 billion years ago. The relationship between crater density and age determined for the Moon has been used to estimate the ages of other planets and satellites.

9.5 THE MOON'S ATMOSPHERE

Density

Even before space probes reached the Moon, it was obvious that the Moon has almost no atmosphere. Experiments left on the Moon by lunar astronauts have confirmed the extremely tenuous nature of the lunar atmosphere.

The Moon's atmosphere consists primarily of helium, neon, argon, and hydrogen. The total mass of the atmosphere is only a few thousand kilograms. If compressed to the density of water, the Moon's atmosphere would fill a cube only a little more than 1 m on a side. In contrast, if the Earth's atmosphere were compressed to the density of water it would fill a cube about 170 km (100 miles) on a side. There is so little atmosphere on the Moon that the landings in the *Apollo* program seriously contaminated it with their exhaust gases. In the extreme cold of the lunar night, however, many of the gases emitted by the spacecraft condensed to solid or liquid form, making it possible for experiments left on the lunar surface to measure the natural atmosphere of the Moon.

The thin atmosphere of the Moon poorly insulates the surface. Thus, temperature falls dramatically after sunset and rises swiftly after sunrise. The temperature on the Moon at noon can be as high as 400 K (260° F) but the temperature just before dawn is only 100 K (−280° F). On the Earth, the average difference between day and night temperatures is usually only 20 K (36° F) or less. Another consequence of the near absence of an atmosphere on the Moon is that the sky, even in the daytime, is black. The dark sky means that the planets and stars are visible even at lunar noon.

Gain and Loss of Gases

Radioactive decay in surface rocks produces atoms of helium and argon. The escape of these two gases from lunar rocks and the capture of solar wind gases, primarily hydrogen and helium, are the main sources of the atmosphere of the Moon. About 50 g of gas are captured by the Moon from the solar wind each second. However, the weak gravity of the Moon prevents the accumulation of a substantial atmosphere. The lunar escape velocity is so low that lighter gases, such as hydrogen and helium, escape into space in a matter of hours. Heavier gases are swept away from the Moon by the solar wind, or are absorbed by rocks when gas atoms strike the lunar surface. The low rate at which the Moon accumulates atmospheric gases and the ease with which it loses them combine to produce the near vacuum found at the surface of the Moon.

From time to time, the Moon's thin atmosphere is temporarily enhanced when comets strike the lunar surface. Upon impact, ice in the comet is vaporized and forms an atmosphere of water molecules. Most of the water molecules escape into space but some migrate to the lunar poles. There, on the extremely cold floors of craters that never see sunlight, the water molecules can condense as ice crystals. Over billions of years, deposits of ice, mixed with lunar soil, may have built up. Similar ice deposits on Mercury are described in Chapter 10, Section 1.

The existence of ice deposits at the north and south poles of the Moon was suggested by measurements made by the *Lunar Prospector* spacecraft. *Lunar Prospector* detected subsurface hydrogen atoms, presumably in water molecules. The rocky soil of the Moon absorbs radio waves and is poorly reflecting. Ice, in contrast, is highly reflecting at radio wavelengths. Astronomers have used the Arecibo telescope to search for ice in lunar craters by bouncing radio signals off the Moon and examining the reflected signals. The results showed that any ice in the polar regions of the Moon must be widely spread out and mixed with lunar soil rather than concentrated in thick deposits.

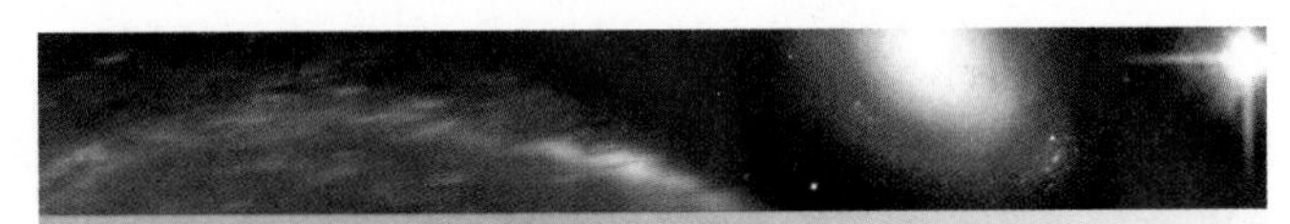

The Moon's atmosphere is extremely tenuous. The atmosphere consists of temporarily trapped solar wind gases and gases released by radioactive decays in surface rocks. Atmospheric gases escape rapidly from the Moon. Comet impacts on the Moon produce a temporary atmosphere of water molecules that may eventually condense at the lunar poles to form ice deposits.

9.6 THE MOON'S INTERIOR

Mass

The Moon's mass has been accurately determined from measurements of the orbital periods and distances of satellites in orbit about the Moon. When the period and distance of a lunar orbiter are used in Newton's version of Kepler's third law (Equation 5.4), the mass of the Moon is found to be 7.4×10^{22} kg. Thus, the Moon is only $\frac{1}{81}$ as massive as the Earth. Because of its small mass, the Moon is 81 times as far from the center of mass of the Earth-Moon system as the Earth is. In fact, the center of mass of the Earth and Moon is only 4700 km from the center of the Earth. This means it is located about 1700 km beneath the surface of the Earth, as shown in Figure 9.27.

The Density of the Moon

The average density of the Moon is 3300 kg/m^3, much less than the average density of the Earth (5500 kg/m^3). This shows that there must be significant differences in chemical composition between the two bodies. The average density of a planet or satellite depends both on the material from which it is made and the extent to which the interior of the body is compressed by the weight of overlying material. If compression were removed, the average density of the Moon would be reduced to 3200 kg/m^3, about 70% of the uncompressed density of the Earth. This density is about

FIGURE 9.27
The Earth-Moon System
In this scale model of the Earth-Moon system, both bodies appear quite small compared with the distance separating them. The center of the mass of the Earth-Moon system lies within the Earth.

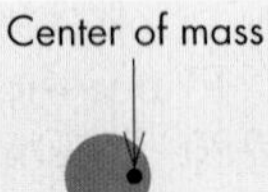

the same as the density of the crustal rocks of the Earth. *Lunar Prospector* measurements of the distribution of material in the Moon suggest that the Moon has a partially molten metallic core between 440 and 700 km across, or between 13% and 20% of the Moon's diameter. The core makes up 2 to 3% of the Moon's mass. In contrast, the Earth's core occupies about 55% of Earth's diameter and makes up about a third of Earth's mass.

Lunar Seismology and Internal Structure

Just as the study of seismic waves in the Earth has contributed greatly to our knowledge of the Earth's interior, lunar seismic studies have been used to probe the interior of the Moon. Four seismometers placed on the Moon during the *Apollo* landings recorded thousands of mild quakes, but none exceeded 2 on the Richter scale. The absence of strong moonquakes shows that the Moon's interior is inactive compared with that of the Earth. Analysis of the seismic data gives the speed of seismic waves at various depths in the Moon. The speeds imply that the crust is 65 to 150 km thick (much thicker than Earth's crust) and that the rigid lithosphere extends down to about 1000 km beneath the surface, where the Moon becomes partially molten. Figure 9.28 shows the internal structure of the Moon. Curiously, the lunar crust is generally thicker on the far side of the Moon than on the near side. Magma rising toward the lunar surface may have had greater difficulty reaching the surface on the far side, which may explain why there are so few maria on the far side of the Moon. The seismic detectors also recorded numerous meteoroid impacts, some by objects as massive as 1000 kg. Meteoroid impacts have also been seen as very brief, bright flashes on the night side of the Moon.

The *Apollo* astronauts drove heat probes several meters into the lunar soil to measure how rapidly temperature increases with depth and the rate at which heat flows out of the lunar interior. The heat flow in the Moon is ⅓ of that in the Earth. This is surprisingly large for such a small body, which should have cooled more rapidly than the larger Earth. Most of the heat is probably produced by radioactive decays in the crust, where radioactive elements are somewhat enriched compared with Earth surface rocks.

FIGURE 9.28
The Interior of the Moon
The Moon's crust and lithosphere are both much thicker than those of the Earth. The lithosphere extends down to a depth of about 1000 km, where the Moon becomes partially molten. The Moon may have a metallic core between 440 km and 700 km across.

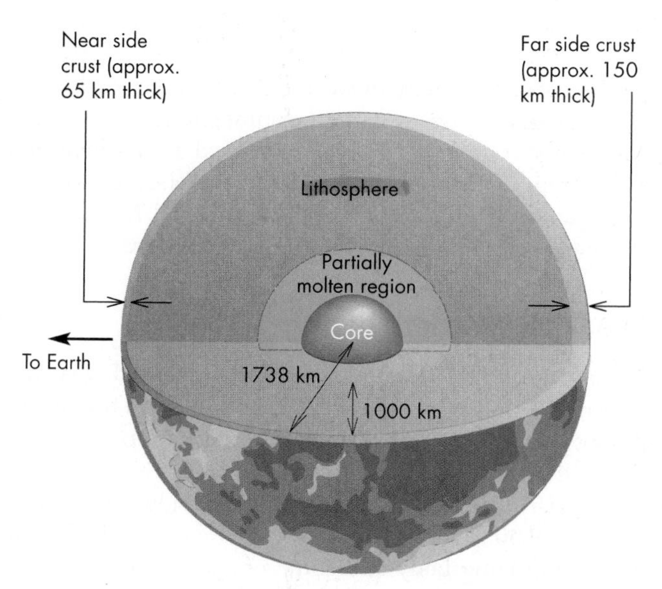

The average density of the Moon is about the same as that of the Earth's crustal rocks, and is consistent with the small size of the Moon's metallic core. Seismic measurements have shown that the Moon's crust and lithosphere are both much thicker than that of the Earth. Compared with the Earth, the Moon's interior is very inactive.

9.7 THE ORIGIN OF THE MOON

Fission, Accretion, Capture

The origin of the Moon has been a persistent problem for astronomers. Any successful theory of the origin of the Moon needs to account for the fact that the chemical composition of the Moon is similar, but not identical, to that of the Earth. In particular, the theory needs to account for the small amount of iron in the Moon when the Earth has an iron core that makes up 32% of its mass. The theory also needs to explain how the Moon began to orbit about the Earth in the first place. Until recently there were only three general theories of the origin of the Moon. All of them, however, have serious flaws. The three theories, shown in Figure 9.29, are the following:

1. **Fission theory**—The Moon was once a part of the Earth, but it broke away because of the rapid rotation of the primitive Earth. If the breakup happened after the Earth's iron sank to its core, then the reason for the lack of iron in the Moon is that the Moon formed from the outer layers of the Earth, which were already depleted of iron. One problem for the fission theory is that the parts of the Earth that should have been flung outward would have been along the equator, where the speed of rotation is fastest. Yet the Moon's orbit is inclined to the Earth's equator. A more serious problem is understanding why the primitive Earth was rotating fast enough (about once every 4 hours) to have spun off the Moon.

FIGURE 9.29
The Fission, Binary Accretion, and Capture Theories for the Formation of the Moon
Each theory can account for some properties of the Moon but can't explain other properties.

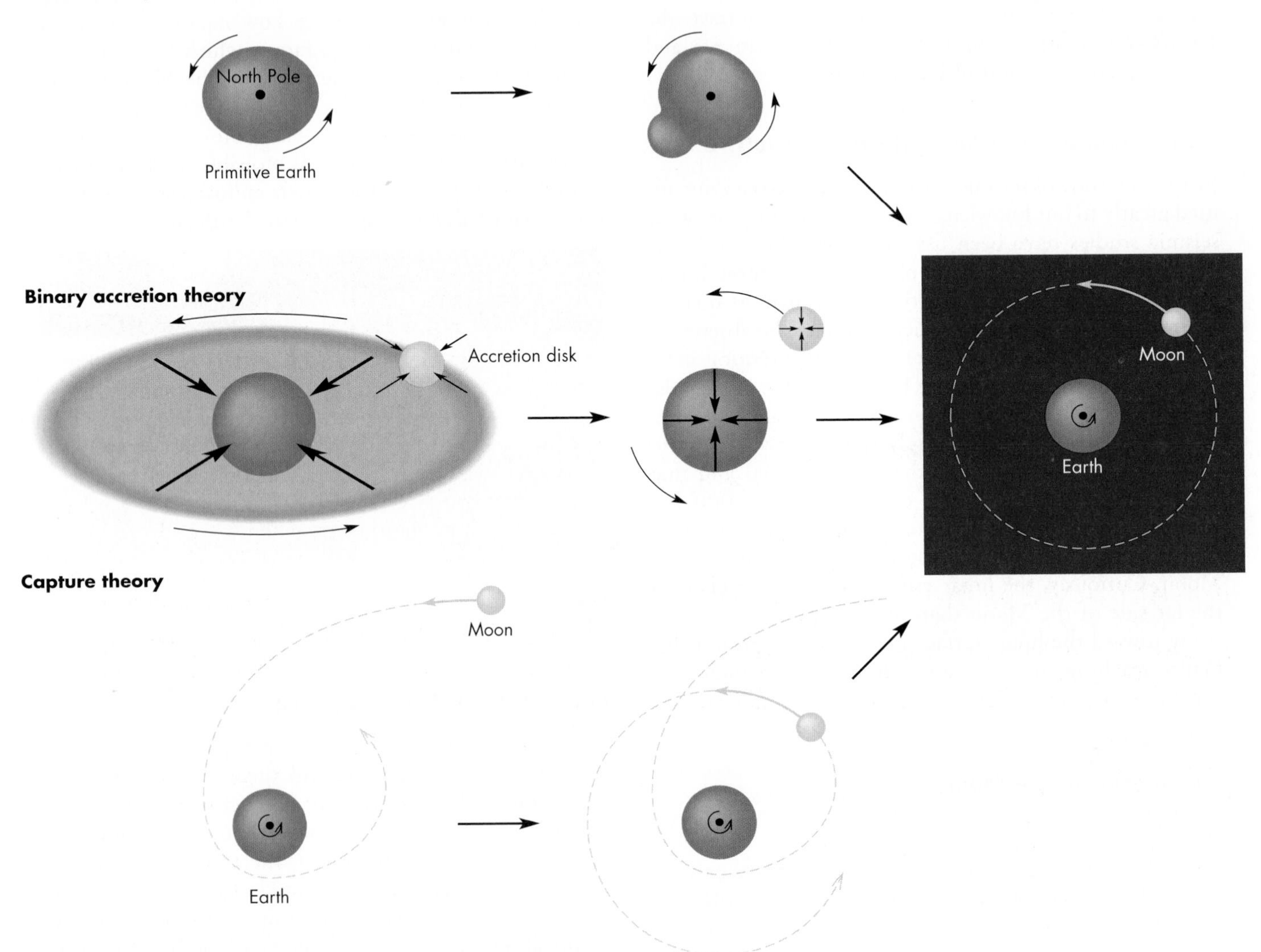

2. **Binary accretion theory**—The Moon formed in orbit about the Earth, but not out of material drawn from the Earth. Both bodies formed at about the same time out of the same swarm or cloud of material. This theory is also called the *double planet theory.* The formation of the Earth and Moon from the same swarm of material explains why the two bodies have similar composition. A major problem, however, is explaining why the abundance of iron in the Earth and the Moon is so different.
3. **Capture theory**—The Moon formed elsewhere in the inner solar system and was captured by the Earth during a close encounter. However, because the Moon would have approached the Earth at a speed in excess of escape velocity, a tremendous amount of energy would have had to be dissipated in order for the capture to occur. No mechanism for the required dissipation has been discovered. Another problem for the capture theory is that if the Moon formed somewhere else in the solar system, it is hard to understand why the chemical compositions of the Earth and Moon aren't much more different than they actually are.

The Giant Impact Theory

During the past 25 years, the **giant impact theory** has gained wide acceptance among astronomers. This theory proposes that the Moon formed from debris expelled when the Earth was struck a glancing blow by a body about as massive as Mars. The consequences of the collision are sketched in Figure 9.30. The violence of the collision vaporized most of the impacting body and some of the Earth's crust and

FIGURE 9.30
The Giant Impact Theory of the Moon's Formation
In this theory, the Earth was struck by a Mars-size body. The collision vaporized the body and part of the Earth. Some of this debris fell into orbit about the Earth and then condensed to form the Moon.

ANIMATION *The birth of the Moon*

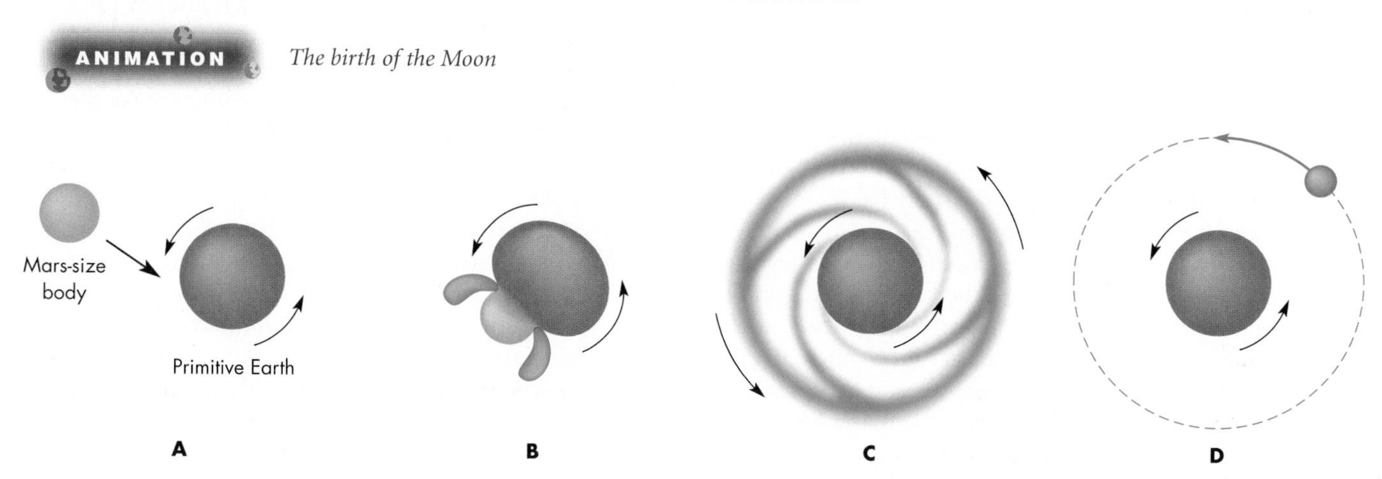

mantle. Most of the vaporized material then jetted away from the Earth in the same direction that the impacting body had been moving before the collision. Some of the ejected material fell back to Earth, some escaped into interplanetary space, and some began to orbit the Earth as a ring of hot gas. Hydrogen and other easily vaporized materials escaped from the cooling gas during this time, which accounts for the lack of water and volatiles in the Moon. After about a thousand years, solid particles began to condense from the cooling gas, and the Moon began to accumulate. The collision took place after the Earth's iron core formed, so the Moon is iron poor compared with the Earth, because only the Earth's outer layers contributed to the material from which the Moon formed. The Moon and Earth are similar in composition because a fraction of the Earth's mantle was ejected by the impact and became part of the ring of gas from which the Moon formed. The differences in composition between the Moon and the Earth's crust are accounted for by the presence of material from the impacting body in the gas from which the Moon formed. The giant impact theory may not be the ultimate answer to the mystery of the Moon's origin, but as of now it seems to account for most of what is known about the history of the Moon and its chemical composition.

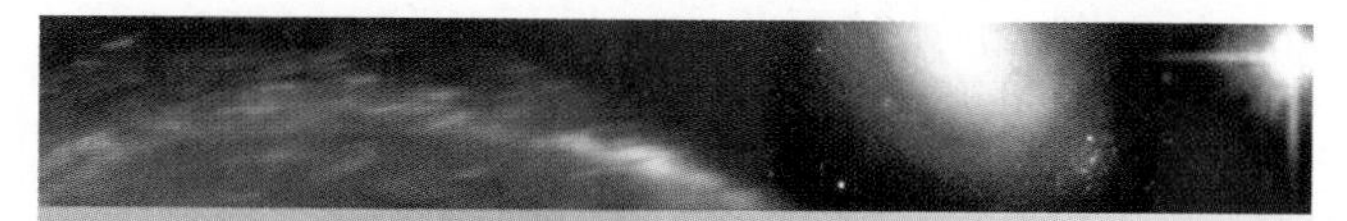

Older theories of the origin of the Moon all suffer from serious difficulties. A newer theory, in which the Moon formed from debris after a large body struck the Earth, seems better able to account for the history of the Moon and the Moon's chemistry.

Chapter Summary

- The rotation period of the Moon is the same as the period of its revolution about the Earth. This arrangement keeps the same face of the Moon turned toward the Earth. (Section 9.1)
- The distance to the Moon averages 384,400 km but can be as large as 406,700 km and as small as 356,400 km. The Moon's diameter is about ¼ that of the Earth. (9.1)
- The two parts of the shadows of the Earth and the Moon are the penumbra, within which sunlight is partially blocked, and the umbra, within which sunlight is completely blocked. Total eclipses occur within the umbral shadow. (9.1)
- The Moon's umbral shadow is only about 150 km wide where it reaches the Earth. During a solar eclipse, the Moon's umbral shadow sweeps rapidly across the Earth, producing an eclipse track that covers less than 1% of the Earth's surface area. (9.1)
- Eclipses can occur only when the Moon is in its new or full phase and when the Moon is near the ecliptic. Eclipses occur in series in which successive eclipses are separated by 18 years 11⅓ days, the saros period. Using their knowledge of the saros, ancient astronomers were able to predict eclipses. (9.1)
- Tides cause the day to lengthen and the Moon to recede slowly from the Earth. In the remote future,

the day and month may become equal in length. When that happens, the Earth will keep the same face turned toward the Moon. (9.2)

- The Moon's surface is covered with craters. The meteoroids that made the craters have pulverized the Moon's surface layers and produced a fine layer of debris. (9.3)
- The lunar maria were produced when vast volcanic flows flooded the floors of basins produced by earlier impacts. Little or no volcanic activity has taken place for more than 3 billion years. (9.3)
- Analysis of lunar samples has shown that they are similar to the surface rocks of the Earth but lack water and are deficient in easily vaporized elements, such as chlorine, zinc, and lead. Lunar rocks range in age from 3.2 billion to 4.5 billion years. The terrae are about 1 billion years older than the maria. (9.4)
- Measurements of the crater densities of regions of the Moon that have different ages have been used to show that the Moon experienced an early period of intense bombardment. The relationship between crater density and age for the Moon has been applied to other planets and satellites to estimate their ages. (9.4)
- The Moon has almost no atmosphere because any gases that enter the atmosphere escape almost immediately. These gases come from the solar wind and radioactive decays in lunar surface rocks. When comets strike the lunar surface, vaporized water molecules form an atmosphere, some of which eventually condense at the lunar poles to form ice deposits. (9.5)
- The Moon is mostly rocky with a small metallic core. Seismic studies show that the Moon's interior is virtually inactive. The crust and lithosphere of the Moon are both very thick compared with the Earth's. (9.6)
- Old theories of the origin of the Moon all suffered from serious deficiencies. A new theory, in which the Moon accumulated from debris from an impact of a large body with the Earth, accounts much more successfully for what we know about the history and composition of the Moon. (9.7)

Key Terms

annular eclipse *187*
ascending node *184*
Baily's beads *185*
binary accretion theory *202*
capture theory *202*
corona *185*
crater *194*
crater density *198*
crater saturation *198*
descending node *184*
diamond ring *185*
eclipse *184*
eclipse seasons *189*
eclipse track *188*
eclipse year *190*
fission theory *201*
giant impact theory *202*
line of nodes *189*
maria *193*
mascon *197*
penumbra *186*
rays *195*
regolith *196*
rilles *197*
saros *190*
sinuous rille *197*
synchronous rotation *182*
terrae *193*
umbra *186*

Conceptual Questions

1. Suppose the Earth orbited the Sun in 225 days. What effect would that have on the length of the Moon's synodic period? Explain your answer.
2. On June 8, 2004, the planet Venus crossed the disk of the Sun. What part of the shadow of Venus were we in when that happened?
3. When we see a total lunar eclipse, what kind of eclipse would be seen by an observer on the Moon?
4. What is the phase of the Earth, as seen from the Moon, during a total solar eclipse?
5. Describe the ways in which the chemical compositions of Earth rocks and lunar samples are similar and different.
6. How do the ages of lunar samples compare with the ages of samples from the Earth's ocean bottoms? What is the reason for the great difference in age?
7. How has the crater density of the maria changed since the maria formed?
8. The terrae have reached crater saturation. In the future, how will the crater density of the terrae change with time?
9. Examination of images of the surface of Jupiter's satellite Io has failed to find a single impact crater. What conclusion about the surface of Io can you draw from this observation?
10. Why does the Moon have so little atmosphere?
11. Summarize the evidence that the Moon does not have a large metallic core.
12. The giant impact theory of the origin of the Moon combines features of the three previous theories of the Moon's origin. Describe features held in common by the giant impact theory and each of the three earlier theories.

Problems

1. The angular diameter and distance of the Moon are 31 minutes of arc and 384,400 km. Use the small angle equation to find the linear diameter of the Moon.
2. Suppose the Moon's rotation period were half as long as its orbital period about the Earth. How long would it take the Earth to complete one trip around the sky? (Hint: One way to find the answer is to sketch the position of an observer on the Moon throughout several lunar orbits.)
3. When flight controllers on Earth spoke to *Apollo* astronauts on the Moon, there was a noticeable time delay in the communications. How long does it take a radio signal to travel to the Moon and back?
4. What is the angular diameter of the Earth as seen from the Moon?
5. If the Earth's rotation continues to slow at its present rate, how long will it take for the day to lengthen by 1 hour?
6. If the month continues to lengthen at its present rate, how long will it take for the month to reach 47 days in length?
7. The Imbrium Basin is about 700 km in diameter. Approximately how large was the impacting body that produced the Imbrium Basin?

Table-Based Questions

1. Using Table 9.2, find the date of new Moon in June 2011.
2. Using Table 9.2, find the date of full Moon in May 2013.
3. Using Table 9.2, find an eclipse season with three eclipses.
4. Using Table 9.2, find a year with three eclipse seasons.

Group Activities

1. Walk in a circle around your partner while you keep facing the same direction. (Use a distant object to be sure you keep facing the same direction and aren't rotating.) Have your partner note whether all or only one side of you can be seen during your "orbit." Discuss the results in the context of the rotation of the Moon (Figure 9.2). (Note: Your partner should turn so that he or she keeps facing you.)
2. Working with a partner, get a plastic dish of flour and some marbles. Stand over the dish of flour and take turns dropping marbles into the flour. (This may be messy, so be careful where you do the experiment.) Retrieve the marbles after every couple of impacts, taking care not to mess up the flour with your fingers or obliterate any "craters." Keep track of how many marbles have been dropped and how many craters you can identify. See how long it takes to reach crater saturation.

For More Information

Visit the text website at **www.mhhe.com/fix** for chapter quizzes, interactive learning exercises, and other study tools.

10

Now the bright morning-star,

Day's harbinger,

Comes dancing from the East . . .

John Milton
Song on May Morning

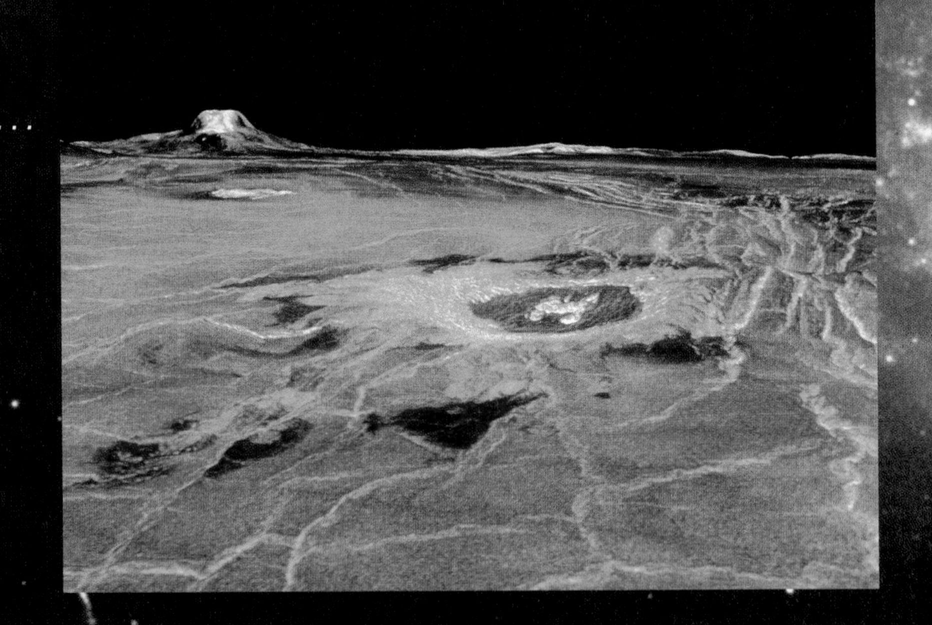

Three Dimensional View of the Surface of Venus.

Mercury and Venus

The terrestrial planets are defined as the ones that resemble the Earth in size and composition. Counting the Moon, there are five terrestrial planets, made mostly of rock and metal, that occupy the inner part of the planetary system. These planets differ considerably in important ways, such as their internal activity, the thicknesses of their atmospheres, and the presence of water on their surfaces. Such differences are due primarily to two key properties: distance from the Sun and diameter. Table 10.1 summarizes the distances, masses, and diameters of the terrestrial planets. Figure 10.1 shows the relative diameters of the terrestrial planets.

The distance from the Sun at which a planet formed determined the raw materials it contained at the start of its evolution. In turn, the mix of raw materials controlled radioactive heating within the planet, the kinds of gases that escaped from its interior into its atmosphere, the size of its metallic core, and many other properties. Distance from the Sun directly determines the amount of solar energy that strikes the top of its atmosphere or its surface.

Diameter and mass (which is closely related to diameter) determine the gravity and escape velocity of a planet. Through its influence on escape velocity, diameter controls the energy released when objects strike a planet and determines which gases can escape from the planet's atmosphere. The larger the diameter of a planet, the longer it takes the planet to cool off.

The terrestrial planets vary considerably both in diameter and distance from the Sun. This gives us the opportunity to figure out, if we can understand all the clues, how each of these factors controls the evolution of a planet. It is important and helpful to keep in mind what we know about the Earth and Moon (the celestial bodies that we know best) when learning about the other terrestrial planets. By comparing and contrasting the properties of the entire group of terrestrial planets, we can discover general principles that would be hard to figure out

Questions to Explore

- Why does Mercury show a much wider range of temperatures than the Earth does?
- What does the surface of Mercury tell us about the history of the planet?
- Why is Mercury so dense?
- How has Mercury's small diameter influenced its internal evolution?
- Why is the atmosphere of Venus so much hotter and thicker than that of the Earth?
- Does the surface of Venus show that the same kinds of internal activity have occurred on both Venus and the Earth?
- If Venus and the Earth began to evolve under similar conditions, what accounts for the large differences that exist today?

Table 10.1

Properties of the Terrestrial Planets

	DISTANCE FROM SUN (AU)	MASS (EARTH = 1)	DIAMETER (EARTH = 1)	AVERAGE DENSITY (RELATIVE TO WATER)
Mercury	0.39	0.055	0.38	5.43
Venus	0.72	0.82	0.95	5.24
Earth	1.00	1.00	1.00	5.52
Moon	1.00	0.012	0.27	3.34
Mars	1.52	0.11	0.53	3.94

FIGURE 10.1
The Relative Sizes of the Terrestrial Planets
Venus and Mercury are the second largest and fourth largest terrestrial planets, respectively.

Venus

Mercury

Moon

Mars

Earth

10.1 MERCURY

Mercury is the nearest planet to the Sun (Table 10.2). Figure 10.2 shows that because of its proximity to the Sun, Mercury is never farther than 29° from the Sun as seen from the Earth. Usually it is too close to the Sun to be seen at all with the unaided eye. When it is visible, it is low above the western horizon just after sunset or the eastern horizon just before sunrise. Thus, Mercury is always viewed in either morning or evening twilight. The brief and widely separated periods of time when the planet can be seen confused the Greek astronomers, who thought that Mercury was two separate objects. The Greeks called the planet "Hermes" when it appeared in the evening sky and "Apollo" when it appeared in the morning sky.

Mercury's Orbit

Mercury's average distance from the Sun (its semimajor axis) is 0.39 AU. Its orbit is more elliptical than that of any other planet. Mercury can be as near the Sun

Table 10.2 Planetary Data

Mercury

Orbital distance	0.39 AU
Orbital period	0.24 years
Mass	$0.055\ M_{Earth} = 3.30 \times 10^{23}$ kg
Diameter	$0.38\ D_{Earth} = 4878$ km
Density (relative to water)	5.43
Escape velocity	4.3 km/s
Surface gravity	0.38 g
Global temperature	400 K
Main atmospheric gases	Na, O, He
Rotation period	59 days
Axial tilt	2°
Known satellites	0
Distinguishing features	Greatest range of surface temperature, great cliffs on surface

FIGURE 10.2
The Greatest Elongation of Mercury
The greatest angular distance between Mercury and the Sun occurs when Mercury is at aphelion at the same time it is at its greatest western or eastern (shown here) elongation. When this happens, Mercury is 29° from the Sun.

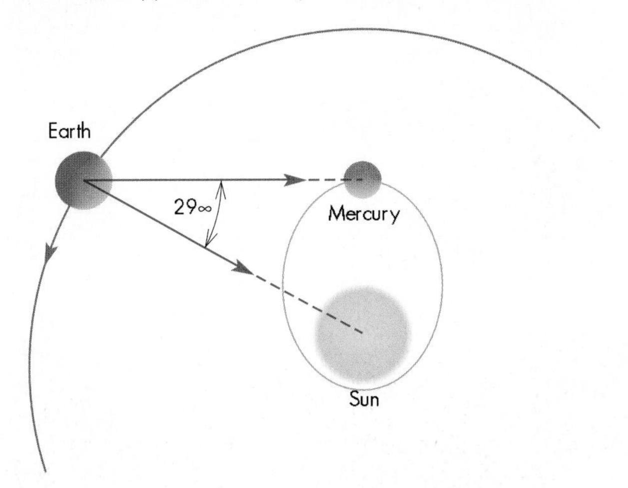

FIGURE 10.3
The Rotation of Mercury During 2 Mercury Years
On each circle representing Mercury, the dot shows the same fixed location on Mercury's surface. The dot is shown in blue during the first Mercury "year" and in orange during the second Mercury "year." During the 2 Mercury years, only a single Mercury day passes. Notice that the longitude marked with a dot and the longitude opposite the dot take turns facing the Sun when Mercury is at perihelion. Other longitudes never experience noon when Mercury is nearest the Sun.

ANIMATION *The rotation of Mercury during 2 Mercury years*

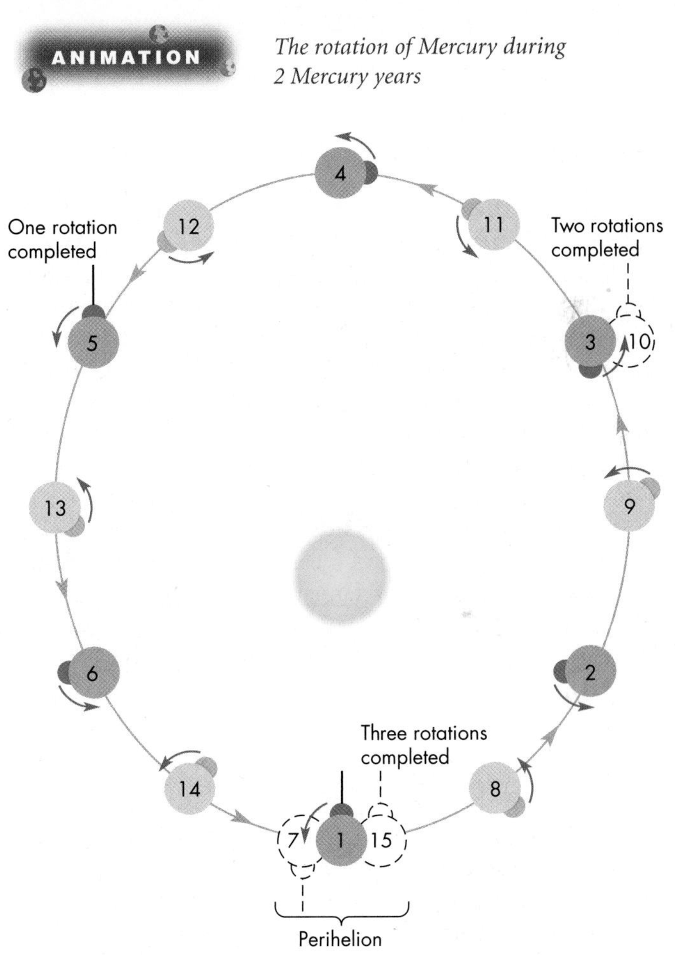

as 0.31 AU and as distant as 0.47 AU. Mercury revolves around the Sun in 87.97 Earth days. The planet rotates once every 58.64 Earth days, which is exactly two thirds of its period of revolution. If the two periods were the same, then Mercury would always keep the same face turned toward the Sun. Instead, as Figure 10.3 shows, Mercury turns slowly with respect to the Sun. A solar day lasts 2 Mercury years. Only two longitudes on Mercury ever experience noon while the planet is at perihelion. For two other longitudes, 90° away from the first two, noon occurs only at aphelion. Thus, two quadrants of Mercury (called *hot poles*) receive much more sunlight during a day and have much higher daytime temperatures than the other two quadrants (called *warm poles*). The terms *warm pole* and *hot pole* refer to regions of Mercury's surface and have nothing to do with the rotation poles of the planet.

A Day on Mercury The Sun's motion across the sky on Mercury is very different from what we see on Earth. For one thing, the rate at which the Sun crosses the sky is quite variable. When it is near the Sun, Mercury's orbital motion is fast enough to keep the same longitude of the planet facing the Sun. While this is happening, the Sun hangs nearly motionless in the sky (in fact it moves slightly eastward). On the other hand, when the planet is far from the Sun, Mercury's orbital motion is much slower than its rotation and the Sun moves nearly twice as fast as average across the sky. Seen from one of Mercury's hot poles, the Sun would appear to rise relatively quickly (though slowly by Earth standards). The Sun would slow down and hover in the sky at noon and then increase its speed as it approached the horizon in the evening. From one of Mercury's warm poles, the Sun would appear to hover near the horizon at both sunrise and sunset but move rapidly across the sky during the day.

The surface of Mercury gets very hot during the 88 Earth days that make up 1 Mercury day, but the planet cools off greatly during the 88 Earth days that make up 1 Mercury night. The temperature can be as high as 700 K at noon but falls at night to about 100 K. There are warmer places than Mercury in the solar system as well as cooler ones. However, the daily temperature range on Mercury is greater than that of any other planet or satellite in the solar system. Although most of Mercury's surface gets much too hot for any ice to exist, there may be ice deposits within craters at Mercury's poles. These deposits appear as bright spots in maps made by bouncing radar pulses off Mercury. Calculations have shown that ice in the floors and inner walls of polar craters would evaporate at a rate of only a few centimeters per billion years. Thus, polar ice deposits, once formed, could survive for much longer than Mercury has existed.

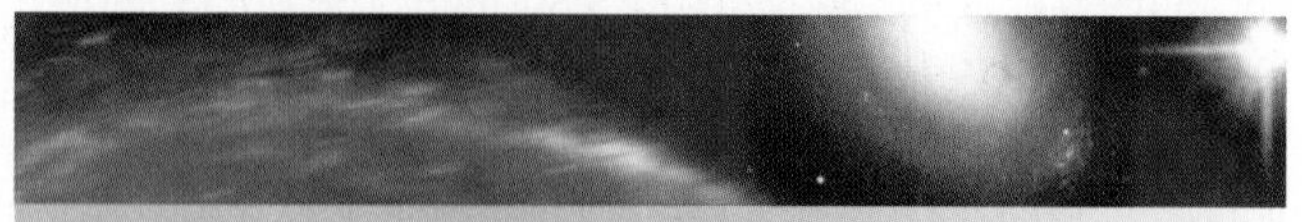

Because of Mercury's slow rotation, its solar day is as long as two of its years. The surface of Mercury gets very hot during its long days but cools greatly during nights, which are 88 Earth days long. The range of temperatures on Mercury is greater than anywhere else in the solar system.

Surface

Observations of Mercury using Earth-based telescopes have a resolution that is similar to observations of the Moon made with the unaided eye. Thus it is hardly surprising that we knew almost nothing about the surface features of Mercury until the planet was visited in 1974 and 1975 by *Mariner 10.* Because the same face of Mercury was turned toward the Sun on each of *Mariner 10*'s three encounters with the planet, only about 45% of the surface of Mercury was imaged by *Mariner 10.* The finest details that can be seen in *Mariner 10* images of Mercury are about the same as those seen on the Moon using Earth-based telescopes.

After a gap of a third of a century, Mercury was again visited by a spacecraft in January 2010, when *Messenger* flew to within 200 km of the planet's surface. *Messenger* was launched in August 2004. Before beginning to orbit Mercury, it will have flown past Earth once, Venus twice, and Mercury three times. Each flyby was designed to change *Messenger*'s path and eventually bring it close to Mercury at low enough speed that its braking rocket will be able to put it in orbit about Mercury. This will happen in March 2011. *Messenger* will orbit Mercury for a year, mapping the entire planet at much better resolution than *Mariner 10. Messenger* will also measure the chemical composition of Mercury's surface, map Mercury's magnetic field, and determine how material is distributed in Mercury's interior.

Even before it begins to orbit Mercury, *Messenger* is sending back spectacular images of the planet. Figure 10.4 is a global view of Mercury taken during the flyby of January 2008. The light-colored lines are rays from several relatively young craters. Figure 10.5 shows the rayed crater located toward the lower right of the global view of Mercury in Figure 10.4. The large ray systems on Mercury were something of a surprise to astronomers because ray systems on the Moon are thought to weather due to the accumulation of dust and become inconspicuous fairly rapidly after the impact that causes them. Why ray systems last so long on Mercury isn't yet understood.

Compared with the Earth, Mars, or Venus, Mercury's surface does not show much evidence of large-scale crustal movements or distortions. However, this makes Mercury quite interesting because its surface preserves a partial record of events that occurred early in the history of the planet. On more active planets, the record of early activity has been erased by extensive, more recent activity.

FIGURE 10.4
A Messenger Image of Mercury

FIGURE 10.5
A Rayed Crater in the Southern Hemisphere of Mercury

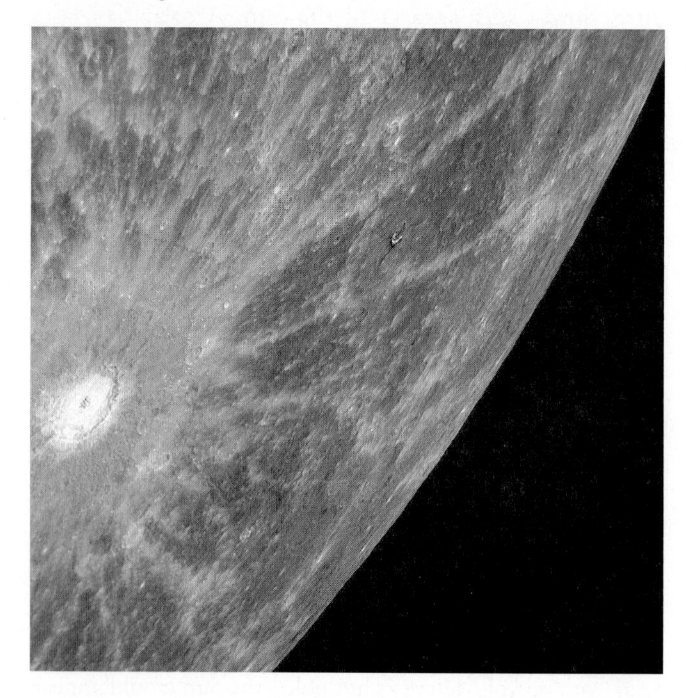

FIGURE 10.6
Caloris Planitia, an Impact Basin as Large as Mare Imbrium on the Moon
Caloris, which fills the left half of this image, was at the edge of the sunlit part of Mercury when the *Mariner* images were taken, so only about half of it can be seen. The diameter of the outermost ring of the Caloris basin is 1300 km (800 miles).

Craters Like the Moon, the surface of Mercury is dominated by impact craters and large basins. Figure 10.6 shows part of one of the basins, Caloris Planitia, about 1500 km in diameter. Caloris is about 70% as large as South Pole-Aitken, the largest lunar basin. Figure 10.7 shows an unusual structure, nicknamed Spider, located near the center of Caloris. Spider consists of a 40-km-diameter crater from which cracks and troughs radiate outward. The crater may have formed the cracks or its location near the center of the cracks may just be a coincidence. Most of the craters on Mercury probably were formed during an early period of intense bombardment. The bombardment era ended 3.8 billion years ago on the Moon and probably ended at about the same time on Mercury.

The craters on Mercury look very much like lunar craters. However, the larger gravity of Mercury affects the distance to which material can be blasted from craters. The ejecta from a typical Mercurian crater covers only about 10% as large an area as that covered by debris from a lunar crater of the same size. This means that even in the heavily cratered regions of Mercury there are smooth areas between the craters. Thus Mercury is much better than the

FIGURE 10.7
"Spider", a Curious Feature near the Center of Caloris Planitia on Mercury

Moon at preserving the record of its early bombardment because more recent craters had had a smaller effect on the previously cratered surface.

Mariner 10 and *Messenger* images have revealed over a dozen basins on Mercury. The basins are sometimes hard to see, however, because nearly all of them were almost obliterated by lava floods that took place after the basins formed. It is likely that Mercury's degraded basins represent the largest craters of an ancient, heavily cratered surface. Ancient impact craters smaller than about 500 km seem to have been almost completely eradicated, leaving only the basins to show that the ancient surface ever existed.

The surface of Mercury is dominated by impact craters. Mercury has a larger gravity than the Moon, so material ejected from craters lies closer to the crater than is the case for the Moon. The smaller size of the area affected by an impact on Mercury means that the record of very old impacts is preserved better on Mercury than on the Moon. Mercury also has badly degraded basins that appear to be the remnants of an ancient, heavily cratered surface.

Plains Although much of Mercury's surface is heavily cratered, there are also plains regions that have relatively few craters and appear to have been produced mainly by lava flows. There are two types of plains terrain on

FIGURE 10.8
Mercury's Smooth Plains
This *Messenger* image of the crater Rudaki and its surroundings shows an example of Mercury's smooth plains. Rudaki is 120 km in diameter.

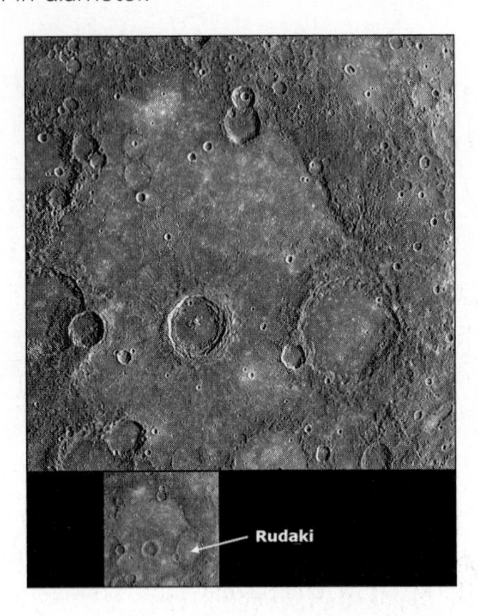

FIGURE 10.9
Beagle Scarp
Beagle Scarp is 600 km long and 1 km high. It cuts across the crater Sveinsdóttir (120 km × 220 km), indicating that it formed after the crater was formed.

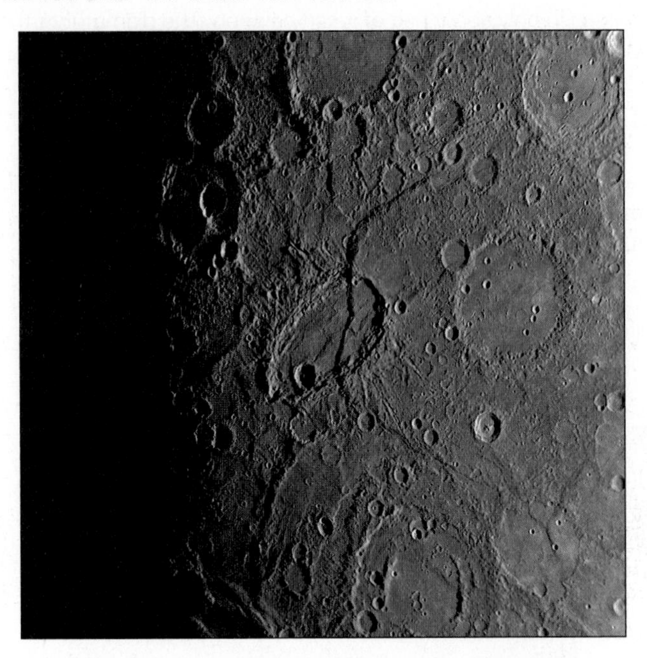

Mercury. One is the **intercrater plains,** which form a gently rolling landscape between and around large craters and basins. The intercrater plains have a high density of small craters (less than 15 km in diameter), many of which occur in clusters or in chains. These small craters seem to be secondary craters formed by the impact of debris ejected during the formation of the large craters and basins. This means that the intercrater plains already existed before most of the impacts that produced the heavily cratered terrain.

The other principal kind of plains on Mercury is the **smooth plains,** which cover almost 40% of the portion of Mercury viewed by *Mariner 10.* Figure 10.8 shows the smooth plains that fill the crater Rudaki and the region around it. The largest region of smooth plains on Mercury lies in a broad ring around Caloris basin. The smooth plains that lie within Caloris and other basins, resemble the lunar maria and, like the lunar maria, probably were formed when lava flows filled existing impact basins. The smooth plains suggest that much of Mercury was resurfaced by widespread volcanic flows after the impact that formed the Caloris basin.

FIGURE 10.10
The Early Evolution of Mercury's Surface
The graph shows the cratering rate for a billion years or so after Mercury formed. The arrows indicate the approximate time periods during which important events took place.

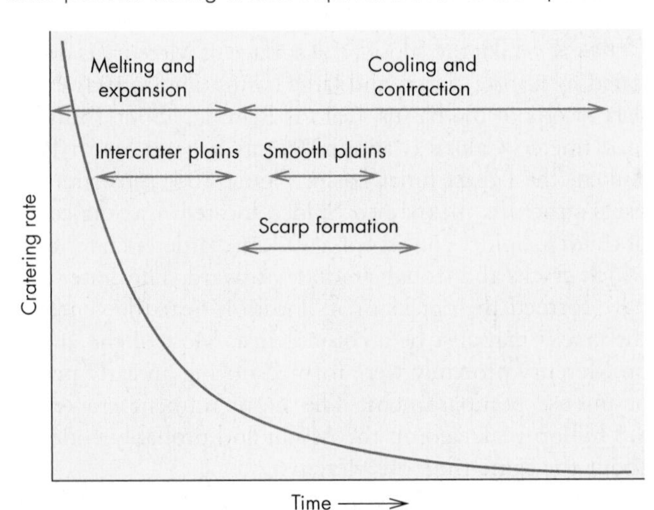

Scarps and Other Surface Features

In addition to craters and plains, there are other important surface features. The most conspicuous of these features are great **scarps,** or cliffs. Some of the scarps, such as the Beagle Scarp shown in Figure 10.9, are several kilometers high and run for hundreds of kilometers. The formation of most of the scarps must have taken place after the formation of Mercury's heavily cratered terrain because scarps cut across many Mercurian craters.

Mercury's craters, plains, and scarps provide clues about the geological history of the planet. Figure 10.10 sketches a scenario for the history of Mercury that is consistent with what we have learned about its surface features. As with the Moon, the rate of impacts on Mercury was very high just after it formed and then fell steadily over the next several hundred million years. It appears that soon after Mercury formed, its interior almost completely melted inside a relatively thin crust. As Mercury melted, it expanded by about 15 km. This expansion fractured Mercury's crust and allowed lava to reach the surface and form the intercrater plains. Eruptions of lava within and near impact basins produced Mercury's smooth plains. As the interior began to cool, Mercury shrank, and its diameter decreased by 2 km or more. The consequences of the shrinkage of Mercury after its crust had formed were spectacular. Imagine dipping a balloon in mud, waiting until the mud had hardened, and then deflating the balloon a little. The balloon, like Mercury, would then have an excess of crust, which would buckle and crack in a pattern resembling the scarps seen on Mercury's surface. After Mercury had shrunk a little, the fractures that permitted lava to reach the surface were sealed off and volcanic activity ceased. All this probably happened during the first 700 million years after Mercury formed. For the last 3.8 billion years Mercury has probably been unchanged except for the formation of craters by impacting bodies.

Mercury's surface shows a pattern of large scarps several kilometers high and hundreds of kilometers long. The scarps probably formed when Mercury shrank after an early period of expansion, which fractured the crust and allowed lava flows to produce plains regions. All this activity probably ended about 3.8 billion years ago.

Directly opposite Caloris Planitia on Mercury is a region of broken terrain covering about 500,000 sq km (about 1% of Mercury's surface). Figure 10.11 shows part of the low hills and closely spaced ridges that characterize the region. The broken terrain was probably formed when seismic waves produced by the Caloris impact were focused at the point opposite Caloris as shown in Figure 10.12. The seismic waves caused vertical shaking of as much as a kilometer. This greatly disrupted the terrain and left the region looking jumbled.

Atmosphere Like the Moon, Mercury has essentially no atmosphere. The gases known to be present in Mercury's atmosphere are (in order of decreasing abundance) sodium, oxygen, helium, potassium, and hydrogen. Other undetected gases may be present as well, but the total amount of atmospheric gas is extremely small. The hydrogen and helium atoms in Mercury's atmosphere have probably been captured from the solar wind, whereas the other gases may have been produced when meteoroids vaporized on impact with Mercury's surface. Mercury cannot have had an appreciable atmosphere at any time since its oldest cratered terrain was formed because there is no evidence of the erosion that atmospheric winds would

FIGURE 10.11
Hilly Terrain on Mercury at the Location Opposite Caloris Basin
This terrain probably formed when seismic waves caused by the Caloris impact were focused and shook Mercury's surface.

FIGURE 10.12
The Focusing of Seismic Waves
The strong seismic waves produced by the impact that formed the Caloris basin at first spread out through Mercury but were then focused toward the point on Mercury opposite Caloris.

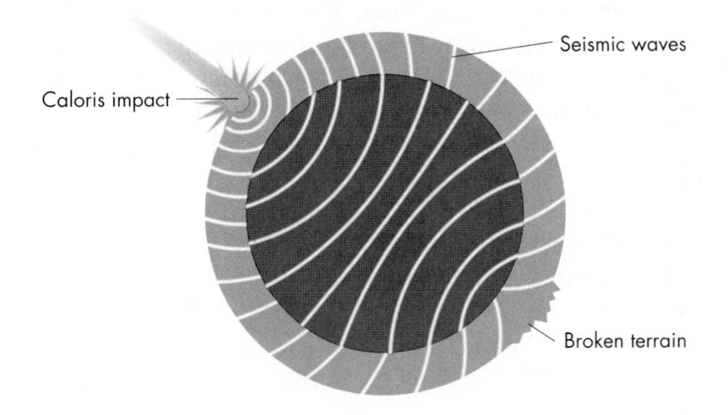

have produced. Instead, the craters of Mercury are sharp and uneroded.

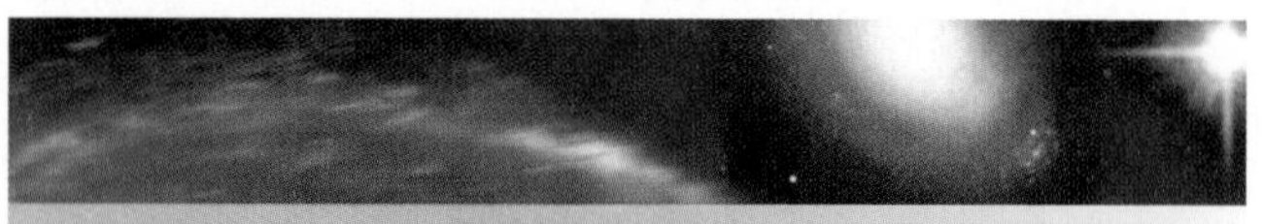

Mercury has a very tenuous atmosphere, which probably consists of gases trapped from the solar wind or produced when meteoroids vaporized as they struck Mercury's surface. The uneroded condition of old craters shows that Mercury never had a thick atmosphere.

Mercury's Interior

Much of what we know about the Earth's interior has come from seismic studies and other measurements made on the surface. Until a lander is able to make similar measurements for Mercury, our ideas about the interior of Mercury will be guesses based on a few solid pieces of information. *Messenger*'s measurements of the gravitational field of Mercury, however, will tell us about the extent to which material is concentrated toward Mercury's center.

The most remarkable property of Mercury is probably its average density, 5440 kg/m^3. This is almost as large as the average density of the Earth, 5520 kg/m^3. The large density of the Earth is partly due to high internal pressures, which compress rock and metal and make it denser. What makes Mercury's density remarkable is that Mercury is too small to have internal pressures capable of compressing its material very much. If the pressure were removed from the material in the Earth, it would expand until it attained an average density of about 4000 kg/m^3. Mercury's uncompressed density is much larger, 5300 kg/m^3. Thus Mercury must contain a large proportion of material denser than the rocky silicate material that makes up most of the Earth. The most likely possibility is iron, which is abundant in the solar system and more than twice as dense as typical rocks. If the dense material inside Mercury is iron and other metals, then Mercury is composed of about 70% metal and only about 30% silicate rock.

Before Mercury was visited by *Mariner 10* it was generally assumed that the planet's very slow rotation ruled out any prospect of a magnetic field like that of the Earth. It was surprising, then, that *Mariner 10* detected a magnetic field strength near Mercury that was much weaker than the Earth's, but much stronger than that of any other terrestrial planet. The strength of the magnetic field at Mercury's surface is 0.5% of the strength of the magnetic field at the surface of the Earth. Nevertheless, Mercury's magnetic field is strong enough to deflect the solar wind and produce a magnetosphere similar in many respects to the Earth's but only about one thirtieth as large. Mercury's magnetic field is probably produced by a dynamo similar to that at work in the core of the Earth. This suggests that Mercury's core is electrically conducting and that the metal in Mercury is concentrated in its core. If that is the case, then the metallic core of Mercury, as shown in Figure 10.13, extends outward to about three-fourths of Mercury's radius.

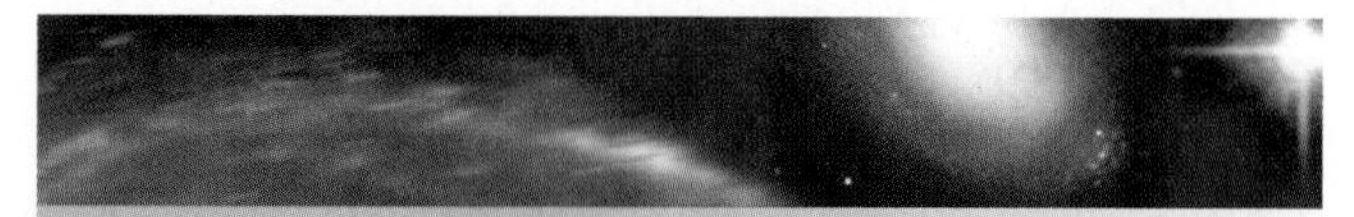

Mercury's average density is nearly as large as the Earth's. Because the material in Mercury's interior is less compressed than that of the Earth, Mercury must contain a large proportion of iron or other heavy elements in order to be so dense. Mercury has a magnetic field, suggesting that the planet has a liquid metallic core. The core reaches about three-fourths of the way to Mercury's surface.

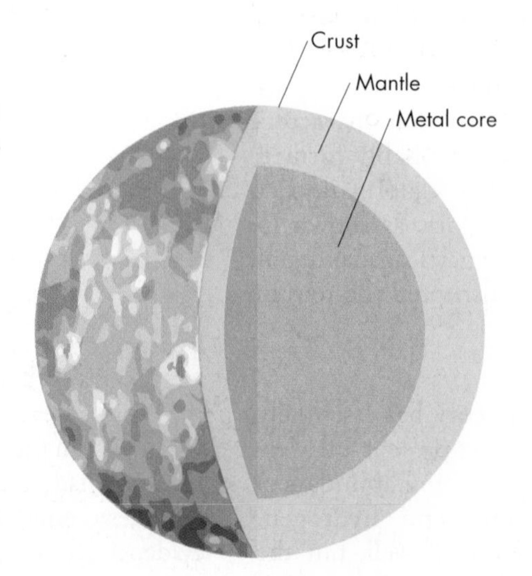

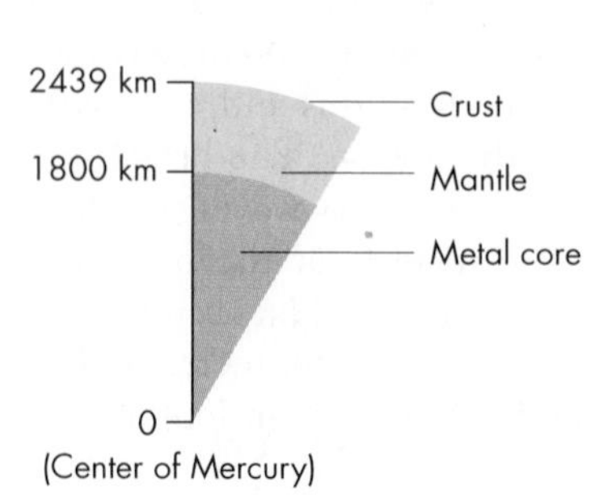

FIGURE 10.13
The Metallic Core of Mercury
Mercury's core extends to three-fourths of the planet's radius. The core makes up about 42% of the volume of Mercury.

FIGURE 10.14
A Model for the Origin of Mercury's Large Metallic Core
A, Mercury is struck by a body almost as large as itself. **B,** Both Mercury and the other body are disrupted. **C,** The silicate outer layers are ejected at high speed and mostly lost. **D,** The cores of the two bodies are also disrupted, but enough metallic material remains to form a large metallic core inside a relatively thin layer of silicate material.

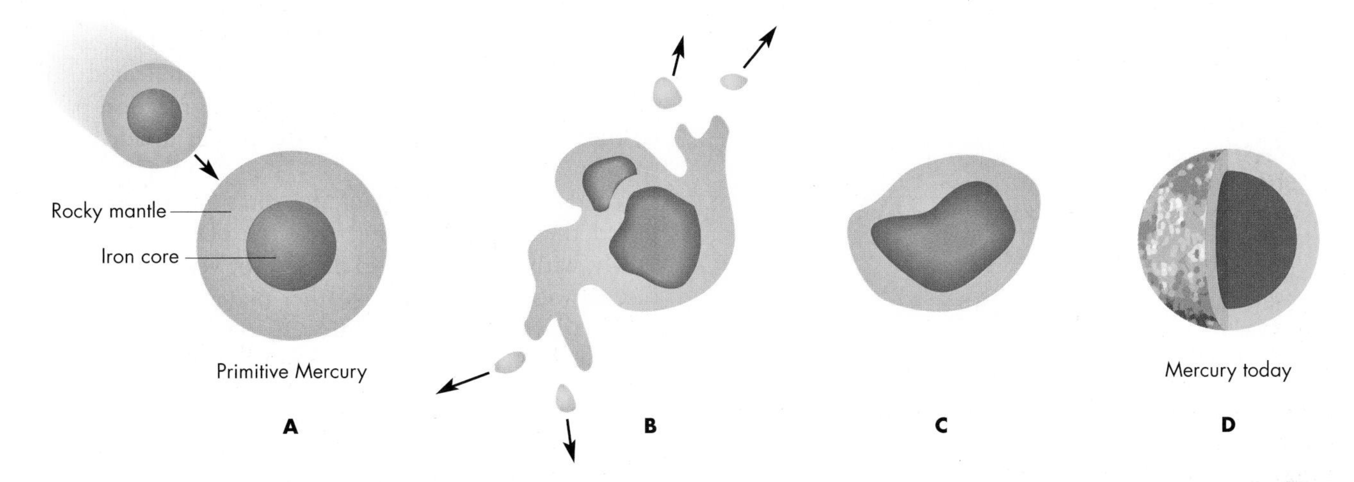

The Origin of Mercury The high iron content of Mercury has been difficult for astronomers to explain. Most attempts to calculate what kinds of materials should have accumulated to form Mercury have concluded that Mercury originally had a much higher proportion of rock than it does today. If that is the case, then Mercury must have lost a considerable fraction of its rocky material since it formed. Figure 10.14 shows how this might have happened as a result of the collision of Mercury with another planet-sized object. The collision would have had to happen after the iron in Mercury had already sunk to its center to form the core. The violence of the collision vaporized and ejected most of the rocky envelopes of both the impacting body and Mercury. The iron cores of Mercury and the impacting body would have been less disturbed by the impact and would have merged inside a much reduced rocky envelope.

The proposal that a collision between planet-sized bodies was responsible for the loss of much of Mercury's rocky envelope was an outgrowth of the giant impact theory of the origin of the Moon. Astronomers now suspect that giant impacts may have been common in the early solar system and may be responsible for other anomalous properties of planets.

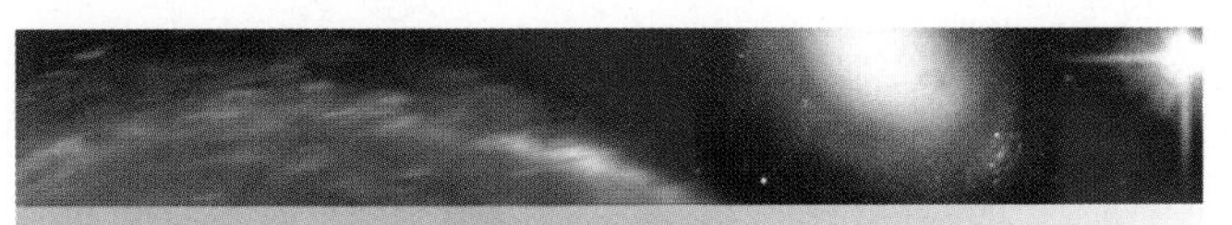

One explanation of Mercury's high iron content is that Mercury lost most of its rocky exterior after the planet formed and its iron sank to form a core. This may have happened through a collision between Mercury and another large body.

10.2 VENUS

For several months every year and a half, Venus is visible above the western horizon in the evening, more than ten times as bright as the brightest stars. During this time, people are so surprised by Venus's brilliance that they can't believe it is a natural object. Astronomers frequently get calls from people who want to know about the mysterious light in the sky or want to report a UFO sighting.

Like Mercury, Venus is most often seen in the western sky in the evening or in the eastern sky in the morning. The ancient Greeks thought that these were different objects, so they called it Phosphoros when it appeared in the morning sky and Hesperus when it appeared in the evening sky. Venus is so bright that it sometimes can be seen in the daytime if the sky is very clear and if an observer knows where to look.

When someone first views Venus through a telescope, he or she is usually a little disappointed. Venus has a nearly uniform yellowish appearance with none of the variations in brightness and color that can be seen for Mars, Jupiter, and Saturn. The only features that can be detected from the Earth are vague, dark blotches that show up in blue or ultraviolet images.

The mass, radius, and density of Venus given in Table 10.3 are all slightly smaller than those of the Earth. The Earth's orbital distance is closer to that of Venus than any other planet. Despite their apparent similarity, however, Venus and the Earth are strikingly different in many important ways. Their atmospheres, for example, have little in common. Venus's atmosphere is much hotter and denser than Earth's and has an entirely different

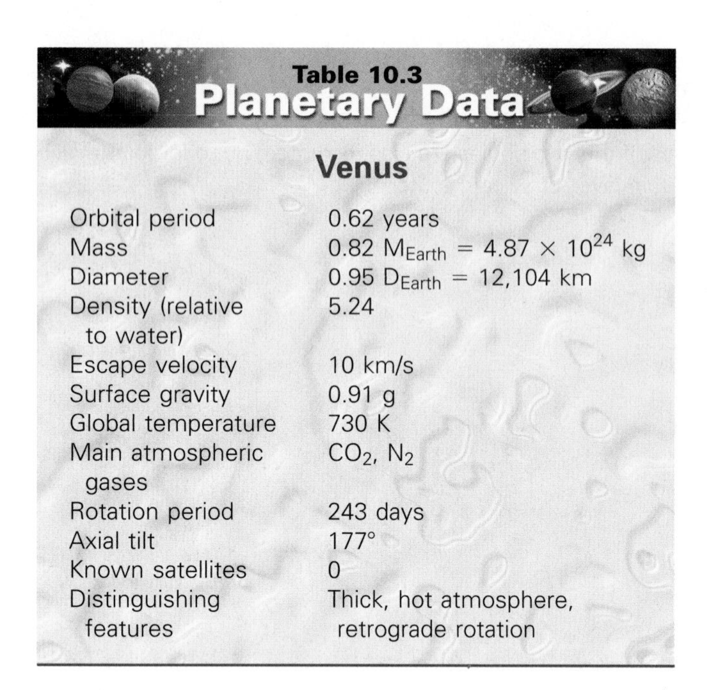

Table 10.3
Planetary Data

Venus

Orbital period	0.62 years
Mass	0.82 M_{Earth} = 4.87×10^{24} kg
Diameter	0.95 D_{Earth} = 12,104 km
Density (relative to water)	5.24
Escape velocity	10 km/s
Surface gravity	0.91 g
Global temperature	730 K
Main atmospheric gases	CO_2, N_2
Rotation period	243 days
Axial tilt	177°
Known satellites	0
Distinguishing features	Thick, hot atmosphere, retrograde rotation

chemical composition. The types of surface features on the two planets are quite different as well. The similarity of the Earth and Venus in mass, radius, and distance from the Sun makes their great differences in atmosphere and surface features surprising and important. It seems likely that the Earth and Venus formed and began to evolve along similar paths. However, relatively small early differences, particularly distance from the Sun, soon led those paths to diverge drastically. By comparing the Earth and Venus, we can see how early differences affect the evolution of Earthlike planets.

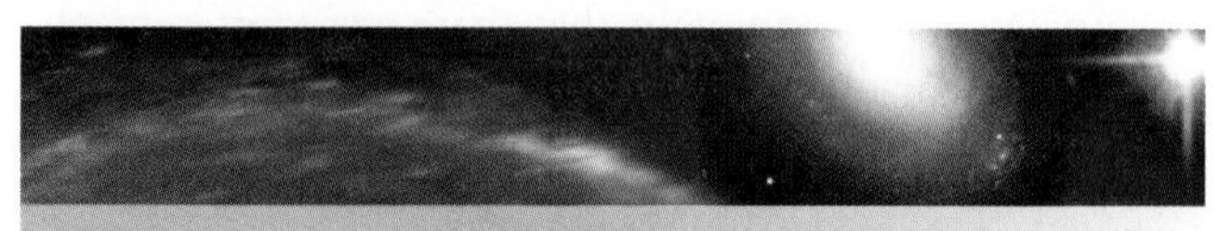

Venus is similar to the Earth in mass, radius, and density. Venus's orbit is nearer the Earth's than that of any other planet. By studying the ways in which Venus and the Earth differ, we can discover how initial differences in important properties such as distance from the Sun can affect the subsequent evolution of similar planets.

Rotation and Revolution

No reliable measurement of the rotation period of Venus was made until after the development of radar astronomy in the 1950s. Bouncing radar pulses off of Venus made it possible to use the Doppler effect to find the speed of rotation of the planet. If a radar pulse is reflected from a region of the surface of a rotating planet, as in Figure 10.15, the rotating surface produces a Doppler shift in the wavelength of the reflected pulse. The part of the planet that is rotating toward the Earth produces a Doppler shift to shorter wavelengths, whereas the part that is rotating away from the Earth produces a Doppler shift to longer wavelengths. The difference between the short and long wavelength parts of the reflected pulse gives the planet's speed of rotation. The rotation period is, then, the circumference of the planet divided by its speed of rotation.

Radar measurements of the rotation of Venus produced the surprising result that the planet rotates very slowly, with a period of 243 days. Also, Venus rotates from east to west. Almost every other planet and satellite in the solar system rotates from west to east. The westward rotation of Venus is called **retrograde rotation.**

A Venusian Day The slow retrograde rotation of Venus leads to an unusual relationship between the Venus solar day and the Venus year. Figure 10.16 shows that it takes almost half a Venus year to complete a single Venus day. The Venus year is only 1.93 Venus days, but each of these Venus days is 116.8 Earth days long. At a given spot

FIGURE 10.15
A Radar Pulse Reflected from a Rotating Planet
The pulse is broadened by the Doppler effect. This happens because part of the pulse reflected from the side of the planet turning toward the Earth is Doppler shifted to shorter wavelengths. The part of the pulse reflected from the side of the planet turning away from the Earth is Doppler shifted to longer wavelengths. The amount of broadening depends on how rapidly the planet rotates.

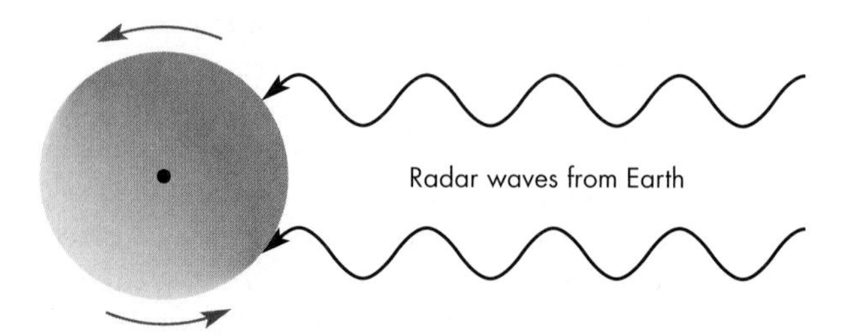

A Radar waves are bounced off a rotating planet.

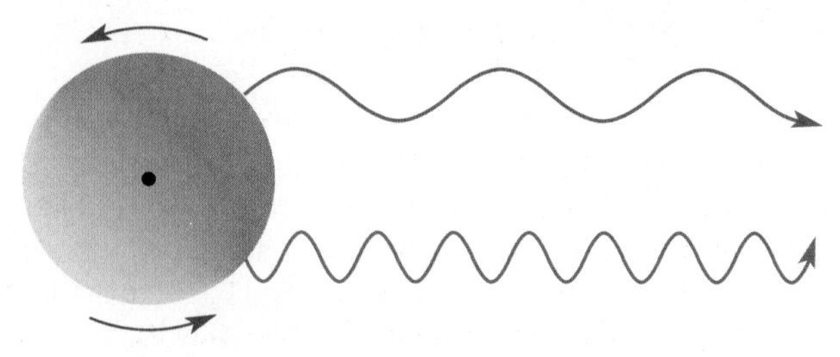

B The reflected waves are Doppler shifted.

FIGURE 10.16
A Venusian Day
The slow retrograde rotation of Venus produces days that are 116.8 Earth days long. The dot on Venus marks a particular spot on its surface. A Venusian day lasts for slightly more than half of a Venusian year.

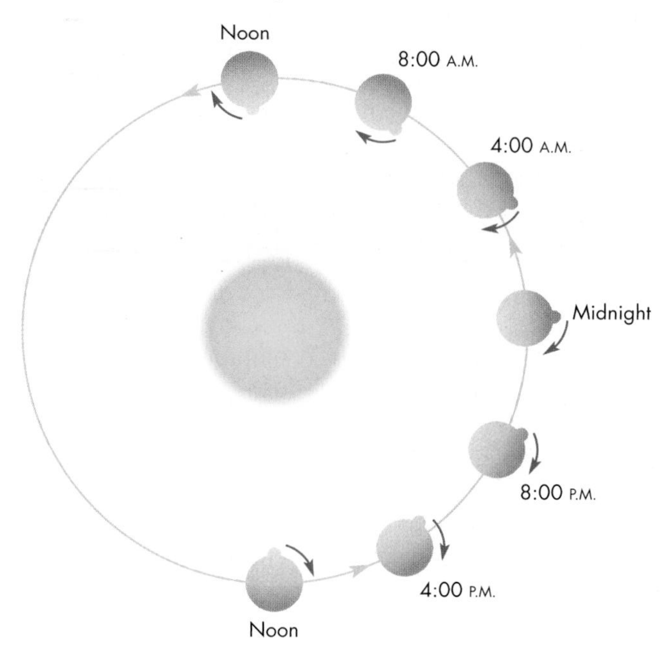

on Venus, the Sun is up for about 58.4 Earth days, and night lasts for the same length of time.

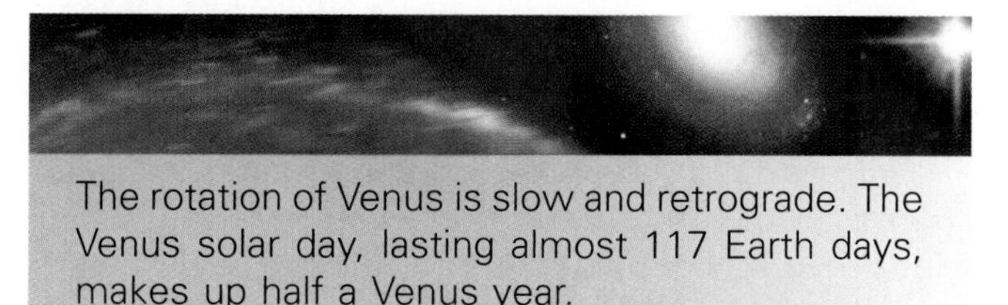

The rotation of Venus is slow and retrograde. The Venus solar day, lasting almost 117 Earth days, makes up half a Venus year.

Venus's Hostile Atmosphere

As Table 10.4 shows, the atmospheres of Venus and the Earth are different in almost every respect. The atmosphere of Venus is much hotter than the atmosphere of the Earth. The first indication that Venus has a hot atmosphere was found when radio telescopes were used to observe Venus. The unexpectedly large radio brightness of Venus implied that its surface had to be very hot, perhaps 600 or 700 K (about 600° to 800° F). This conclusion was later verified by temperature measurements carried out by spacecraft that probed Venus's atmosphere and landed on its surface. The average surface temperature is 730 K. Atmospheric probes and landers have measured the temperature of Venus's atmosphere at a number of latitudes both at night and during the day. The probes showed that Venus's surface temperature is remarkably uniform, varying by no more than a few kelvins from equator to pole and from day to night. Figure 10.17 shows that the temperature in Venus's atmosphere falls with altitude throughout the troposphere, reaching about 200 K at the top of the troposphere at 60 km. Above the tropopause, the temperature is relatively steady with increasing height, but it is

Table 10.4

The Atmospheres of Venus and the Earth

	VENUS	EARTH
Average surface temperature (K)	730	280
Surface pressure (atmospheres)	92	1
Most abundant gases	CO_2 96.5% N_2 3.5%	N_2 78% O_2 21%
Cloud composition	H_2SO_4	H_2O

FIGURE 10.17
The Temperature Profile of Venus's Atmosphere
Temperature decreases steadily from the surface of Venus to an altitude of 100 km. Although temperatures in the troposphere are the same day and night, the upper atmosphere cools dramatically at night.

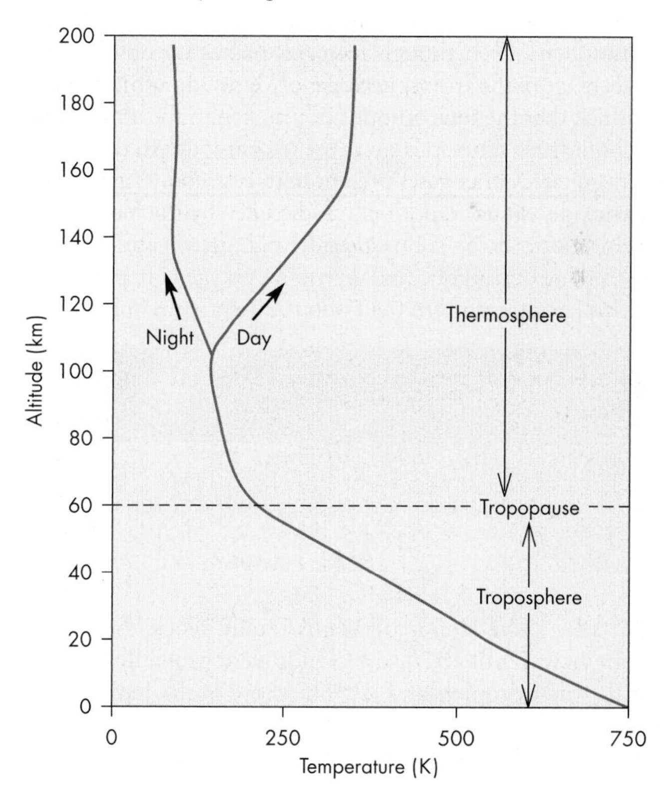

considerably colder at night than during the day. The upper atmosphere cools itself by emitting infrared radiation. During the day, this cooling is balanced by the absorption of sunlight. At sunset, however, the absorption of sunlight ceases while cooling continues. The temperature quickly falls to about half the daytime temperature as energy is emitted into space faster than it can be transported from the day side of the planet to the night side by winds.

Just as the surface temperature of Venus is very high, the pressure at Venus's surface is also extremely large, about 92 atmospheres (atm). This is about the same as the pressure 1 km deep in the Earth's oceans (beneath the zone in which submarines usually operate). The pressure declines with altitude but doesn't fall to 1 atm until 50 km above Venus's surface. The gas at the surface of Venus is so thick that it has a density about 1⁄17 as large as that of water. The Earth's atmosphere, in contrast, has a surface density only about 1⁄800 as dense as water.

Spectroscopic studies using Earth-based telescopes in the 1930s showed that Venus's atmosphere contains a considerable amount of carbon dioxide. Entry probes have confirmed that the dominant gas in Venus's atmosphere is carbon dioxide, which makes up 96.5% of all the atoms and molecules present. Essentially all of the remaining 3.5% is molecular nitrogen, the dominant gas in the Earth's atmosphere. Even though nitrogen makes up only a small percentage of the atmospheric gas of Venus, the atmosphere is so thick that the total amount of nitrogen in the atmosphere is about three times as large as the total amount in the Earth's atmosphere. Other gases present in trace amounts are carbon monoxide, water vapor, sulfur dioxide, hydrochloric acid, hydrofluoric acid, sulfuric acid, and inert gases such as helium, neon, argon, and krypton. Oxygen, if present at all, makes up no more than about 20 parts per million.

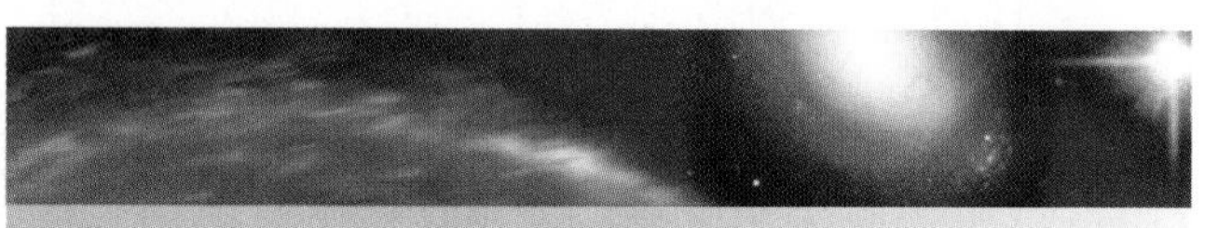

Venus has a very hot atmosphere made primarily of carbon dioxide. Its average surface pressure is 92 atm.

Clouds The clouds of Venus totally block the surface from view. Until 1970, the clouds were generally believed to be water droplets and ice, like those of the Earth. Since then, both telescopic observations and measurements by atmospheric entry probes have shown that the clouds are primarily droplets of concentrated sulfuric acid. Although most terrestrial clouds are water or ice, the Earth too has a thin layer of sulfuric acid cloud at an altitude of about 20 km. If this layer were thick rather than thin, the Earth would look very much like Venus.

Venus's cloud layer is very thick, extending from 45 to 70 km above the surface, with thin haze both above and below the main cloud layer. Compared with the Earth's clouds, though, Venus's clouds aren't very dense. Even in the densest part of Venus's cloud layer it would be possible to see for several kilometers. The enormous thickness of the clouds of Venus might suggest that virtually no sunlight could penetrate through them and reach the surface. However, the main way that the clouds block sunlight is by reflecting it rather than by absorbing it. Thus, some sunlight is scattered back and forth by cloud droplets but finally emerges through the bottom of the cloud layer. The regions below the clouds are about as bright as the Earth is on an overcast day.

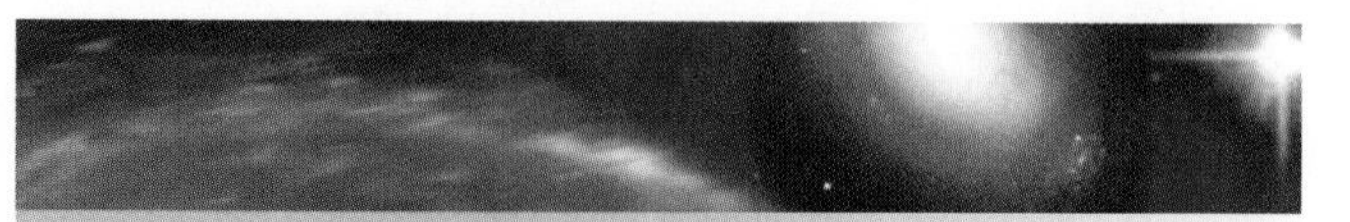

The clouds of Venus are made of droplets of concentrated sulfuric acid. A thick layer of clouds, between 45 and 70 km in altitude, completely conceals the surface of Venus. Enough sunlight penetrates the clouds to make the surface as bright as the Earth on a cloudy day.

Atmospheric Circulation The Earth's atmospheric circulation is driven by the solar radiation absorbed near the ground and is strongly influenced by the Coriolis effect. For Venus, on the other hand, only about 2.5% of the sunlight that strikes the planet is absorbed at the ground. About ten times as much solar energy is absorbed within and above the highest cloud layer, causing the circulation of Venus's atmosphere to be driven from above. Also, the extremely slow rotation of Venus means the Coriolis effect is negligible. Because of these differences, atmospheric circulation is completely different for Venus and the Earth.

The entire atmosphere of Venus rotates westward faster than the solid planet itself rotates westward. The wind speed is about 350 km/hr at the top of the clouds near the equator. Below the clouds, the wind speed drops steadily, reaching speeds of only 0.3 to 1 m/s (about 3 km/hr) at the surface. Westward wind speeds also diminish from the equator toward the poles. Although the main atmospheric circulation is westward, there must be weaker north-south winds as well. The atmosphere absorbs more sunlight at the equator than at the poles, yet the polar regions are almost as hot as the equator. For this to happen, the atmosphere must transport heat poleward. As Figure 10.18 illustrates, in both the northern and southern hemispheres of Venus warm gas rises over the equator, flows poleward, sinks over the pole, and then flows toward the equator at the surface. Each of these loops is called a Hadley cell. Earth's atmosphere also has Hadley cells, but there are three circulation cells per hemisphere rather than Venus's single cell. Astronomers think the reason Venus's atmosphere rotates faster than the solid planet is that air moving toward the equator near the surface picks up angular momentum from the

FIGURE 10.18
The Atmospheric Circulation Pattern of Venus
Warm air rises at the equator, travels poleward, cools, and sinks at the pole. Air then returns to the equator along the surface.

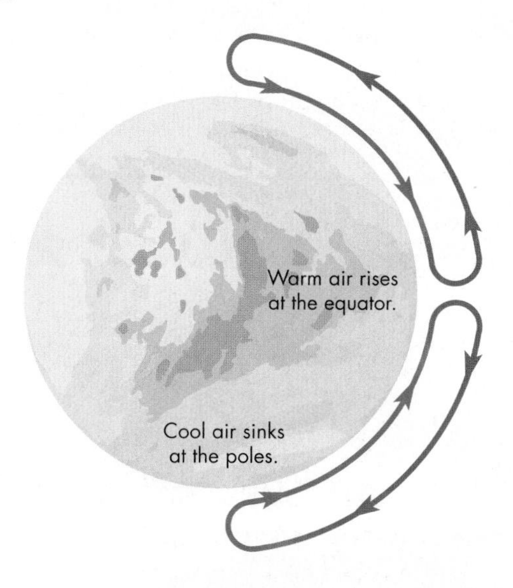

Warm air rises at the equator.
Cool air sinks at the poles.

solid planet. This extra angular momentum leads to faster rotation of the atmosphere.

Greenhouse Effect About twice as much sunlight strikes Venus each second as strikes the Earth. However, Venus reflects about 80% of the sunlight striking it, whereas the Earth reflects only about 30%. This means that the Earth actually absorbs more solar energy than Venus does. Why, then, is Venus so much hotter than the Earth?

The explanation is that Venus has a greenhouse effect like the one that slightly warms the Earth (Chapter 8, Section 4). For Venus, however, the effect is much greater. About 2.5% of the sunlight that strikes Venus eventually reaches the ground and is absorbed there. Because of the large amounts of carbon dioxide and other gases in Venus's atmosphere, infrared radiation is unable to escape easily from the surface or lower atmosphere into space and cool the planet. The absorbed solar energy is trapped near Venus's surface and produces the very high surface temperature.

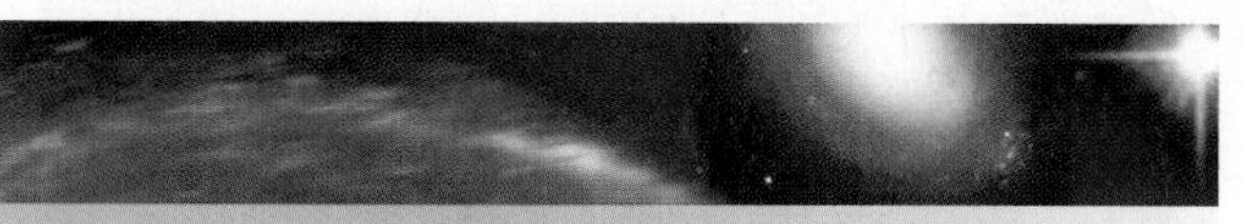

The high surface temperature of Venus is caused by the greenhouse effect. The thick atmosphere makes it difficult for infrared radiation from the ground and lower atmosphere to escape into space and cool Venus.

Surface

Because the surface of Venus is hidden by clouds, the only two ways to observe the surface are to land a remote probe on Venus or to examine the planet using radar, which can penetrate the clouds.

Images from the Surface Despite the high temperature and pressure on the surface of Venus, the Soviet Union succeeded in landing a series of spacecraft, the *Venera* series, at a variety of locations on the planet. Four of the *Venera* landers returned pictures of the places where they landed. *Venera 9* and *Venera 10* each returned only the image of a strip running from the horizon to the ground and back up to the opposite horizon. Later, *Venera 13* and *Venera 14* each sent back color images of the full 360° scene around the landing site. The images of the surface of Venus returned by *Venera 13* and *Venera 14* are shown in Figure 10.19.

The images from *Venera 13* and *Venera 14* show a flat surface with outcroppings of bedrock. All of the images show a regolith of broken rock. The color images show an orange world, colored not by the nature of the surface material than by the orange light that filters through the thick atmosphere. If the surface of Venus could be viewed in white light, it would appear dark gray. Several astronomers have noted that the general appearance of the surface of Venus is more like Earth's ocean bottoms than Earth's desert regions.

The early *Venera* landers also determined the abundances of major chemical elements in surface samples by measuring the X rays emitted when the samples were irradiated by radioactive sources. The results showed that the surface of Venus is covered with silicate rocks that resemble the common igneous basalts of the Earth.

Radar Mapping Although many maps of portions of Venus have been made using radar transmitters and receivers on the Earth, the most complete coverage of the planet has come from much less powerful radar equipment in orbit around Venus. The orbiters transmit pulses of radio radiation toward the surface. The time it takes for a radar pulse aimed at the surface to return to the orbiter shows the distance from the spacecraft to the surface and, thus, the altitude of the surface beneath the spacecraft. The radar reflectivity of the surface, which depends on its chemical composition and roughness, can be found from the strength of the reflected radar pulse. Images obtained by the *Magellan* spacecraft, which explored Venus in the 1990s, resolved features as small as the size of a football field. *Magellan* produced high-resolution images of almost the entire surface of Venus.

Venus Geography Figure 10.20 shows a radar map of the elevations on the surface of Venus. As Figure 10.20

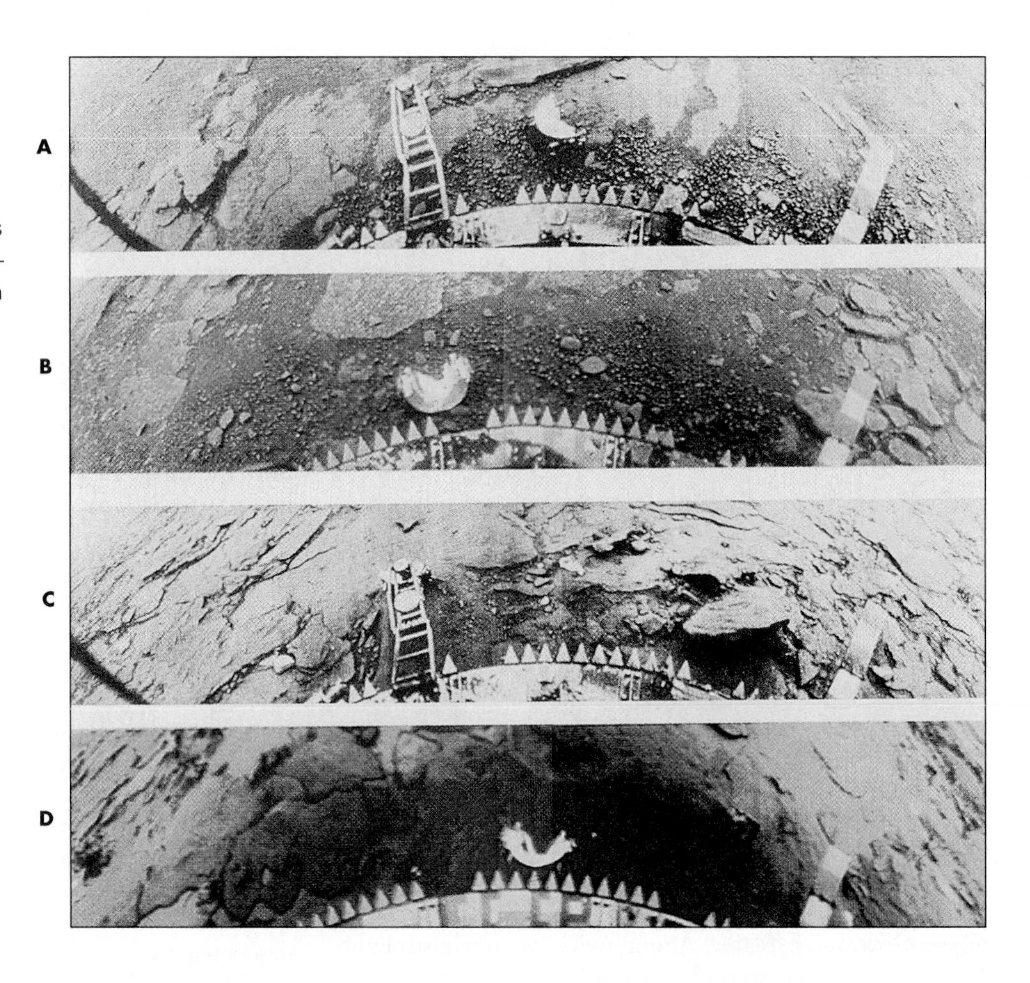

FIGURE 10.19
Panoramic Views at the Landing Sites of the *Venera 13* and *Venera 14* Spacecraft
The *Venera 13* panorama is shown in **A** and **B,** and the *Venera 14* panorama is shown in **C** and **D.**

FIGURE 10.20
A Map of Surface Elevations of Venus
Red and yellow regions show Venusian high spots, whereas blue regions are lower than the average elevation. This map greatly exaggerates the sizes of regions that lie near the poles, such as Ishtar. The names of conspicuous regions are also shown.

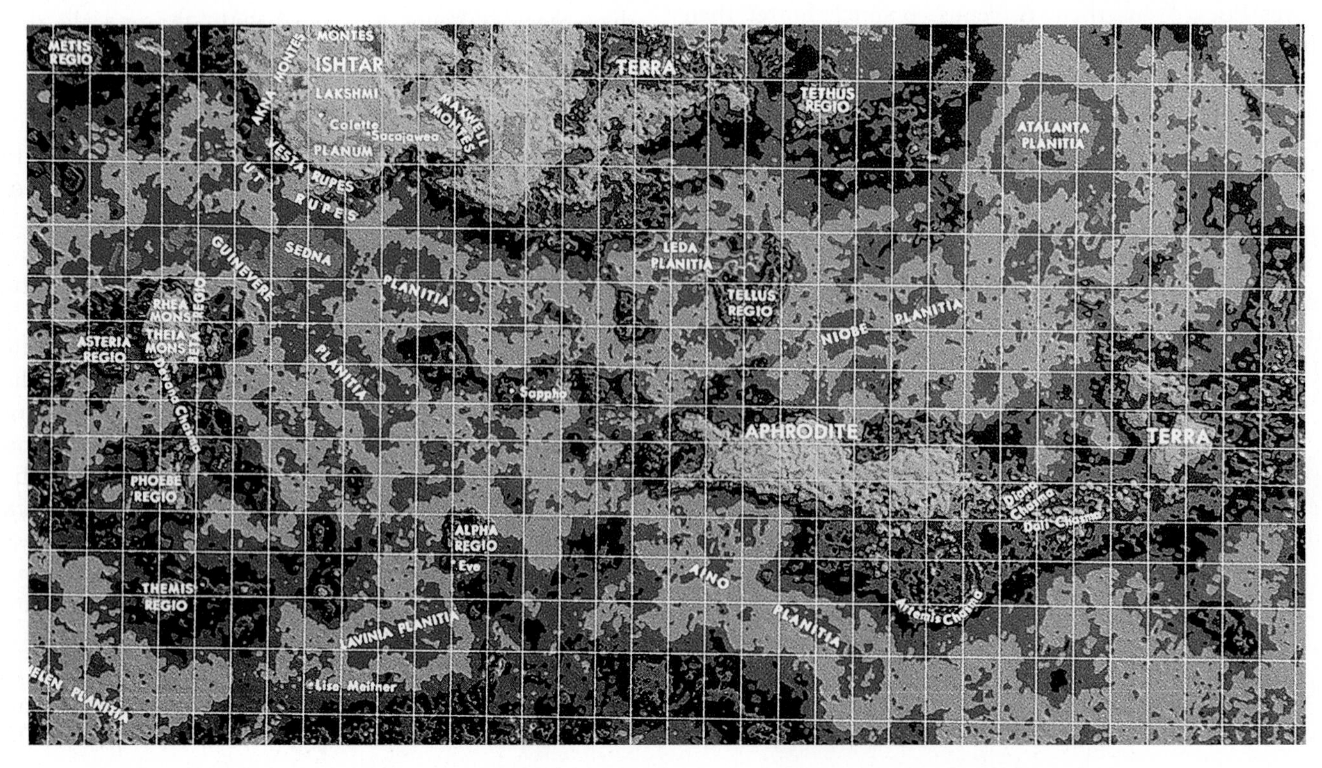

FIGURE 10.21
One Hemisphere of Venus
Red and yellow represent high regions, whereas blue regions are low. Ishtar Terra is at the top of the map. Beta Regio is the roundish high region near the center of the map.

illustrates, most of Venus's surface is rolling plain with little vertical relief. The radar maps also show that Venus has two Australia-sized highland regions—Ishtar Terra in the north polar region and Aphrodite Terra near the equator. Ishtar Terra is also shown in Figure 10.21. Ishtar Terra and Aphrodite Terra are about 6 km (4 miles) above the plain, with peaks reaching about 12 km (8 miles). The difference in elevation between the highest and lowest points on Venus is about 15 km (9 miles). This is 75% of the 20-km (12-mile) difference between the peak of Mount Everest and the deepest ocean trenches on the Earth.

The radar reflectivity of Venus is shown in Figure 10.22. Figure 10.23 shows the radar image of Venus from a point above the eastern end of Aphrodite Terra. Rough, fractured regions and recent lava flows appear bright in the radar images. Notice that many of the features in the radar map can also be found in the elevation map, but other features of the radar map represent reflectivity differences that do not correspond to elevation differences. Venus's surface has a great variety of features, many of which have no counterparts anywhere else in the solar system. Some of these unique features resulted from impacts and others have a volcanic origin.

Craters Fewer than 1000 impact craters have been found on Venus. In fact, impact craters are completely

FIGURE 10.22
The Radar Reflectivity Map of Venus
Generally, fractured regions appear bright, and smooth regions dark. Some of the features of the radar reflectivity map can be found in the elevation map (Figure 10.20).

Venus

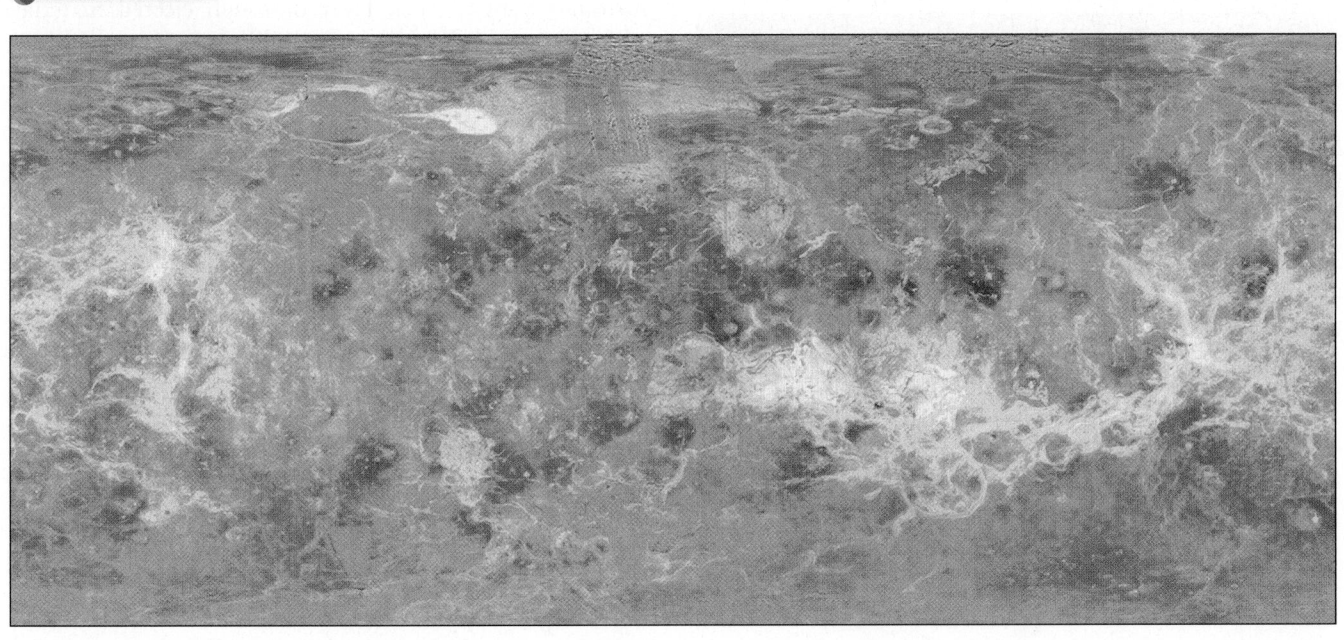

FIGURE 10.23
The Radar View of Venus as Seen from Above Eastern Aphrodite Terra
Notice the network of fractures and faults that runs generally east to west across Aphrodite Terra.

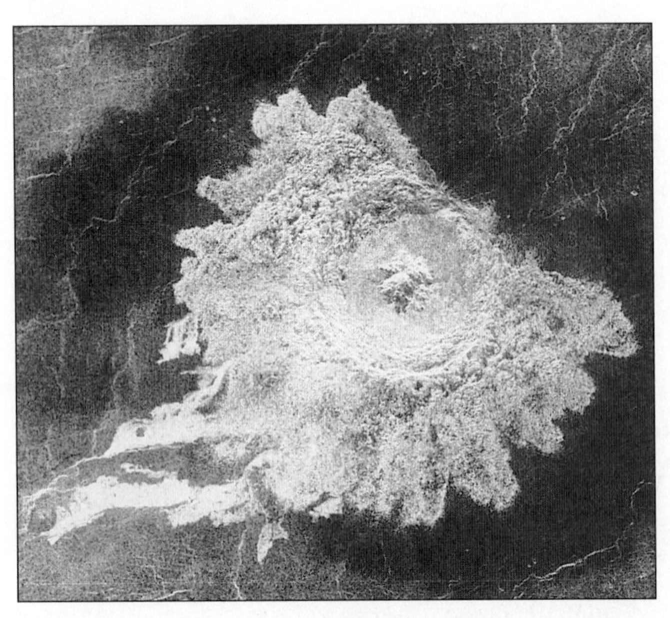

FIGURE 10.24
Aurelia Crater
This crater is 30 km in diameter and is surrounded by a thick layer of ejected material.

missing from regions as large as 5 million square kilometers (2 million square miles). Most of the impact craters on Venus are tens of km in diameter. Aurelia, a typical crater 30 km in diameter, is shown in Figure 10.24. Aurelia is surrounded by a thick layer of rough ejected material, which appears bright because it reflects radar well. Many of the impact craters of Venus have flat, smooth floors and seem to have been flooded with lava that may lie close to the surface of Venus.

The atmosphere of Venus seems to have had a strong effect on meteoroids passing through it. Meteoroids smaller than a few hundred meters across, capable of producing relatively small craters, apparently break up when they encounter Venus's thick atmosphere. This has resulted in few craters smaller than 25 km in diameter and a complete absence of craters smaller than 3 km.

The surface of Venus is dotted with dark splotches like those shown in Figure 10.25. The splotches may be related to the effect of the atmosphere on incoming meteoroids. When a meteoroid passes through the atmosphere, it creates a shock wave that may strike the surface with enough force to pulverize surface rocks. In the case of the upper splotch in Figure 10.25, the meteoroid struck the surface

FIGURE 10.25
Dark Splotches on Venus
This *Magellan* image shows two of the many dark splotches that dot the surface of Venus. Splotches were probably formed when atmospheric shock waves from meteoroids struck Venus's surface and broke surface material into small particles, which reflect radio waves poorly and appear dark in radar maps.

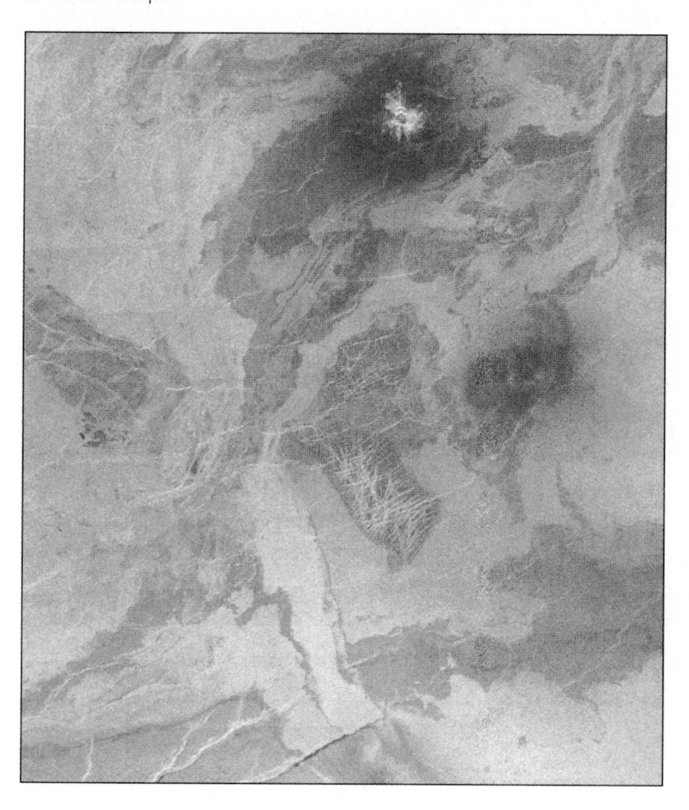

and made a crater surrounded by a splotch 30 to 50 km in diameter. In the case of the lower splotch, the meteoroid apparently disintegrated before striking the surface. The triple-crater Stein, shown in Figure 10.26, may have resulted from the breakup of a single meteoroid as it passed through Venus's atmosphere. All three craters are named for the American author and poet Gertrude Stein. Almost all of the surface features of Venus have been named for real or mythological women.

The most likely explanation for the small number of large impact craters on Venus is volcanic resurfacing. It appears that about a half billion years ago vast planetwide flows of lava completely erased all of the impact craters that existed on Venus at that time. To erase all preexisting impact craters, even large ones, the surface of Venus must have been covered with lava to a depth of several kilometers. Although such a dramatic event may sound impossible, models of the interior evolution of Venus have shown that Venus may undergo relatively brief, intense periods of volcanic activity every few hundred million years.

FIGURE 10.26
The Triple-Crater Stein
This triple-crater may have resulted when a meteoroid broke into three or more pieces while passing through Venus's atmosphere.

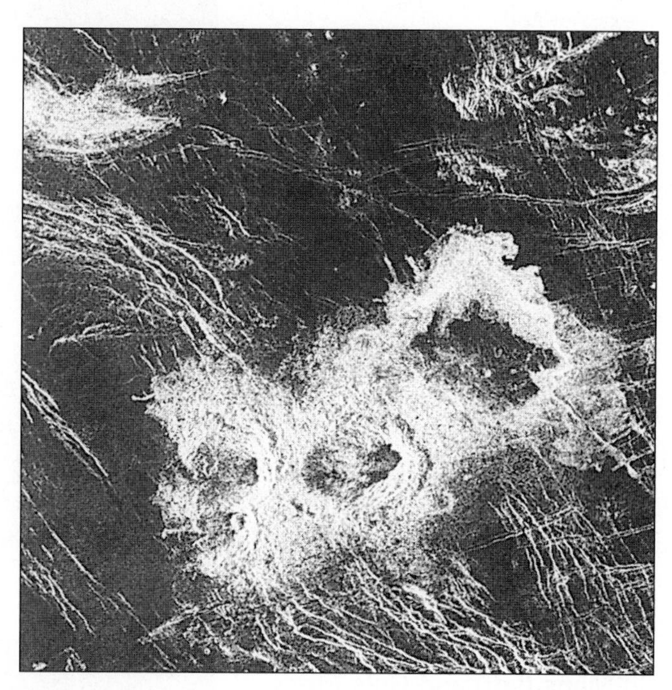

In between these episodes volcanic activity goes on at a much lower rate, as is the case at the present time.

There are fewer than 1000 impact craters on Venus. Small meteoroids are destroyed by the thick atmosphere of Venus before they reach the ground, leading to a scarcity of small craters. Venus appears to have been resurfaced by lava flows about a half billion years ago.

Volcanic Features Except for Jupiter's satellite Io (Chapter 14), Venus shows more evidence of volcanic activity than any other body in the solar system. There are nearly 500 volcanos on Venus larger than 20 km in diameter and tens of thousands of smaller ones. Figure 10.27 shows what Venus's second tallest volcano, 8-km-high Maat Mons, would look like if we could view it from the surface. Many of the volcanos of Venus appear to be **shield volcanos,** large cones with gentle slopes. Shield volcanos are built up by flows of very fluid lava. Gula Mons, shown with its slightly shorter neighbor, Sif Mons, in Figure 10.28, is about 300 km in diameter but only 4 km high. Many of

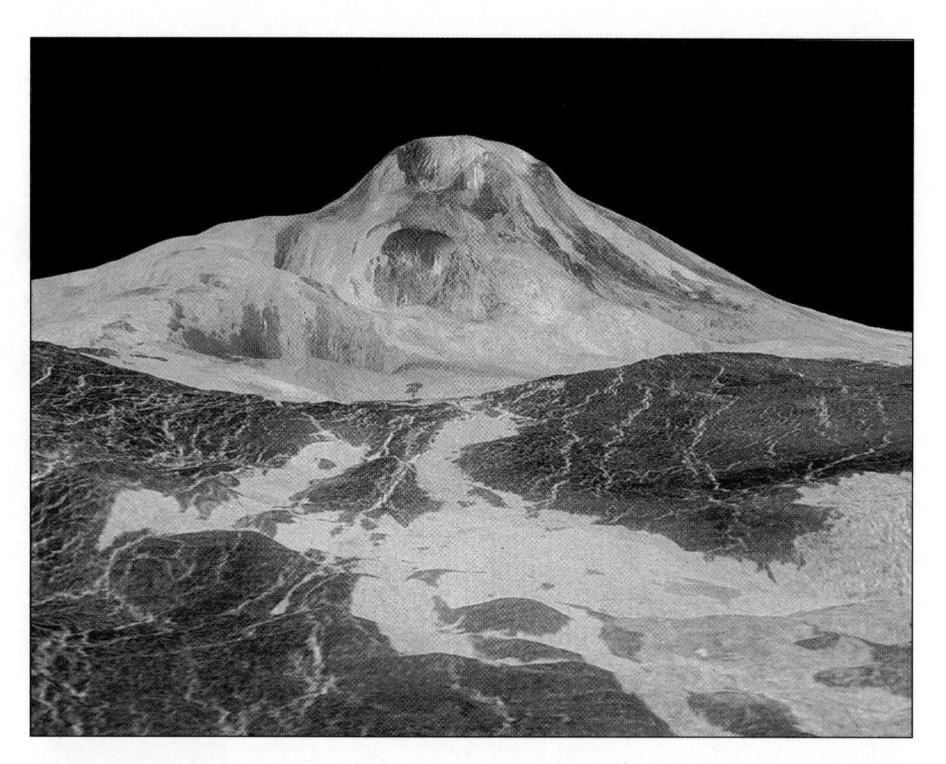

FIGURE 10.27
Venusian Volcano Maat Mons
This view of the 8-km-high volcano was synthesized from *Magellan* radar images. The vertical scale of Maat Mons is exaggerated.

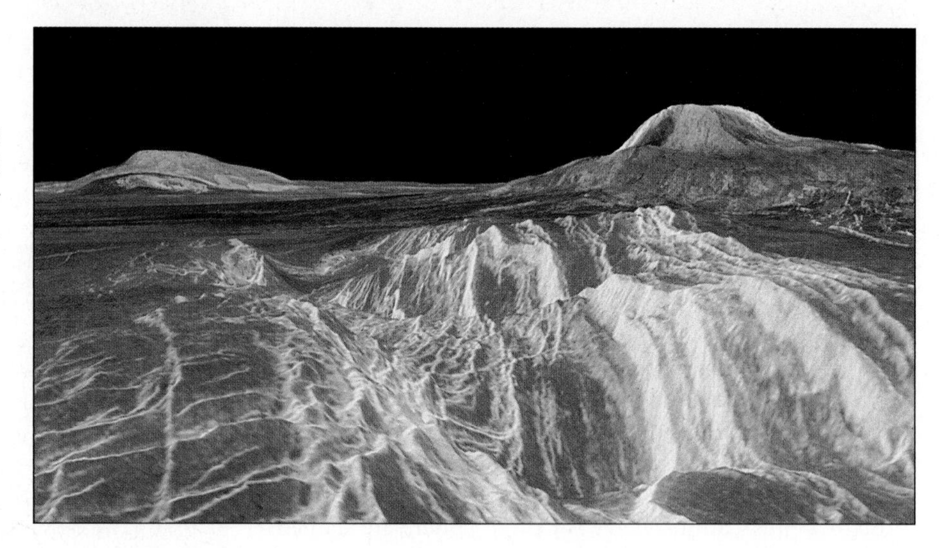

FIGURE 10.28
Sif Mons (left) and Gula Mons (right)
This panoramic view, synthesized from *Magellan* images, exaggerates vertical distances, making both volcanos appear taller than they would appear in reality.

the smaller volcanos on Venus have such flattened shapes that they have been nicknamed "pancakes." These steep-sided domes, like the ones shown in Figure 10.29, are typically 15 to 25 km in diameter, but only 1 or 2 km high. They form when thick, sticky lava oozes out in large puddles as the lava cools. Similar but smaller volcanic domes are common in some regions of California.

Two other unique kinds of features on Venus probably also formed when molten rock rose toward the surface. **Arachnoids,** like those shown in Figure 10.30, are circular regions tens of kilometers in diameter that are interconnected by a web of fractures. (The name "arachnoid" is derived from the name Arachne, a woman in Greek mythology who was turned into a spider for defeating a goddess in a weaving contest.) Arachnoids probably formed when magma rose toward the surface, causing it to bulge and fracture. *Magellan* also found hundreds of large, circular **coronae.** The coronae, such as Fotla shown in Figure 10.31, can be as large as several hundred kilometers in diameter. They consist of a collapsed dome surrounded by a trench and an outer ring. Like arachnoids, they appear to have formed when magma caused the surface to bulge, although in the case of coronae, widespread volcanic eruptions seem to have occurred as molten material reached the surface.

FIGURE 10.29
Pancake Domes
This view of pancake domes was synthesized from *Magellan* images. The pancake-shaped volcanos are typically 25 km in diameter but only 1 or 2 km high. The pancake domes were formed from thick, sticky lava.

FIGURE 10.30
Arachnoids
These volcanic structures consist of smooth, circular regions connected by webs of bright fractures. Arachnoids probably formed when magma pushed the surface upward, fracturing it.

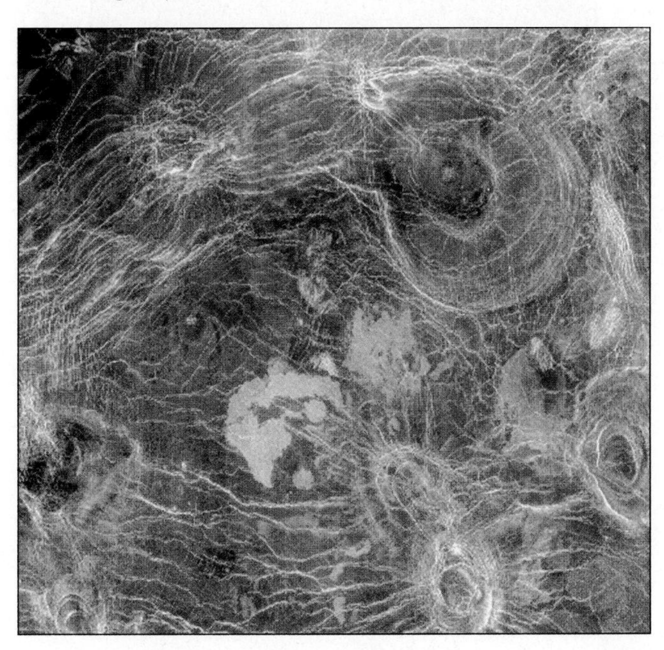

FIGURE 10.31
Fotla Corona
Coronae are large (in the case of Fotla, 200 km in diameter), circular, flattened volcanic domes surrounded by deep trenches.

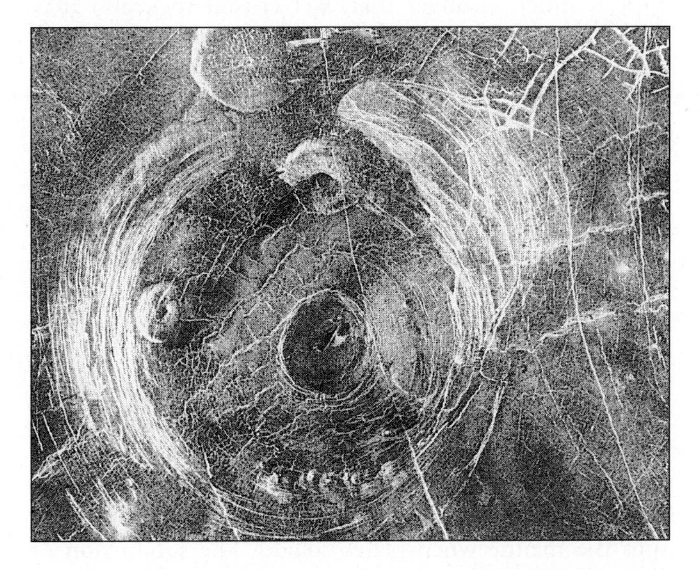

Comparison with the Earth's Surface The surface of Venus shows evidence of widespread and possibly ongoing internal activity. In addition to volcanos, arachnoids, and coronae, there are also mountain ranges, belts of ridges and grooves, and regions of intersecting ridges and troughs, such as those in Figure 10.32. Venus definitely has undergone distortions of its crust, but the crust does not seem to have broken into plates like the Earth's crust has. None of the *Magellan* images shows features like the Earth's midocean ridges, where new crust is formed, or linear trenches, where crust is destroyed when one plate is forced beneath another.

The absence of midocean ridges and trench systems indicates that the internal activity of Venus does not

FIGURE 10.32
A Region of Intersecting Ridges and Troughs in Ishtar Terra
Regions such as this probably formed when Venus's crust was compressed and buckled as part of its surface sank. The region shown is about 100 km (60 miles) across.

FIGURE 10.33
Trenches of the Earth and Venus
A, The purple region marks a deep ocean trench on the outside of the Aleutian Island chain in Alaska. **B,** Deep blue and purple mark part of the trench surrounding Artemis Corona on Venus. The horizontal and vertical scales of the two images are the same, showing that the two regions have similar structure.

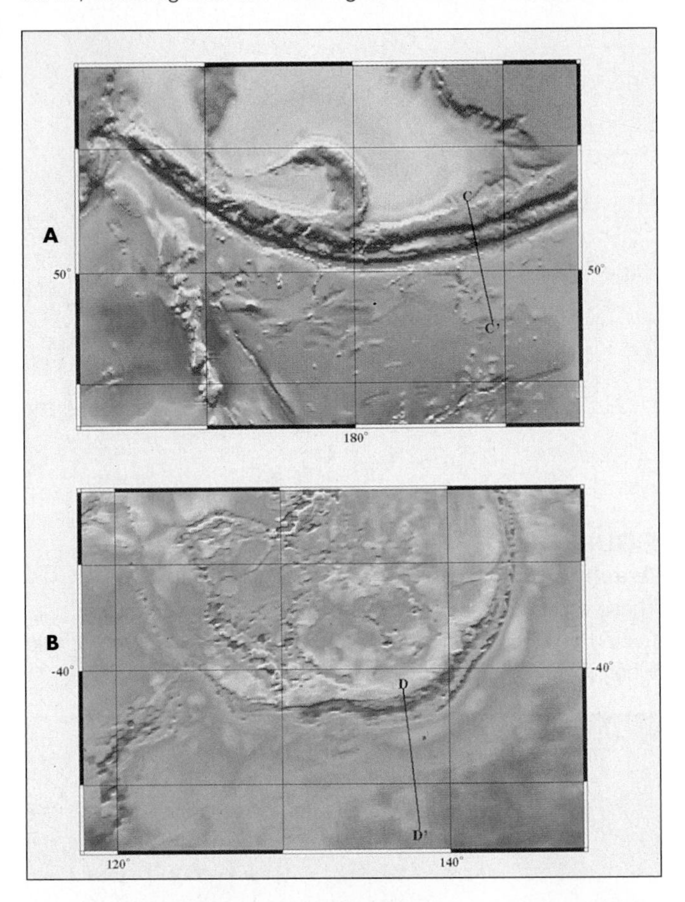

result in horizontal motion in the crust. This may be related to the near absence of water on Venus. Dry crustal rock is much stronger than wet crustal rock and may have been too strong to have broken into crustal plates, a key feature of plate tectonics. Instead, the movement of the crust of Venus is mostly vertical. Rising plumes of magma bulge the crust of Venus upward, stretching the crust and forming large regions of cracks, grooves, and ridges. Volcanos, coronae, and arachnoids form above the plumes. This also happens on the Earth in places like the Hawaiian Islands, where volcanos have formed above a long-lasting plume in the mantle. On Venus, especially large plumes may have resulted in highland regions, such as Aphrodite Terra. In parts of the crust that are sinking rather than rising, the crust is compressed and buckles into mountain ranges like the ones surrounding the western end of Ishtar Terra.

The absence of plate motion on Venus means that there are no regions where crustal material is forced back into the mantle when plates collide. The subduction of crustal material may, however, take place at the edges of Venus's numerous coronae. As shown in Figure 10.33, these coronae resemble portions of the trenches found at plate boundaries on the Earth. Figure 10.34 shows one possible scenario to explain the structure of coronae. In the first stage, a corona forms as a bulge above a mantle plume. The weight of the volcanic material erupting onto the surface next causes the crust to break. The broken edges of the crust then sink into the mantle. The trench surrounding a corona is thought to be where the crust begins to dip into the mantle.

Venus shows widespread evidence of volcanic activity and crustal distortion. However, there is no evidence that plate tectonics has occurred on Venus. The vertical motion of Venus's crust in response to rising and falling plumes of magma in the mantle can account for many of the features seen on Venus. The dry crustal rock of Venus may be too strong to be broken into crustal plates and enable plate tectonics to occur.

Interior

The widespread volcanic activity and crustal deformation on Venus is hardly surprising, given the similarity of Venus to the Earth. The overall similarity in size and density of the two planets implies that they formed from similar materials

FIGURE 10.34
A Model of the Formation of Coronae
A, Lava pours out of a volcanic vent to form the central bulge of a corona. **B,** The weight of the volcanic material breaks the crust beneath the corona. **C,** The edges of the broken crust sink into the interior of Venus, producing the trench that surrounds the corona.

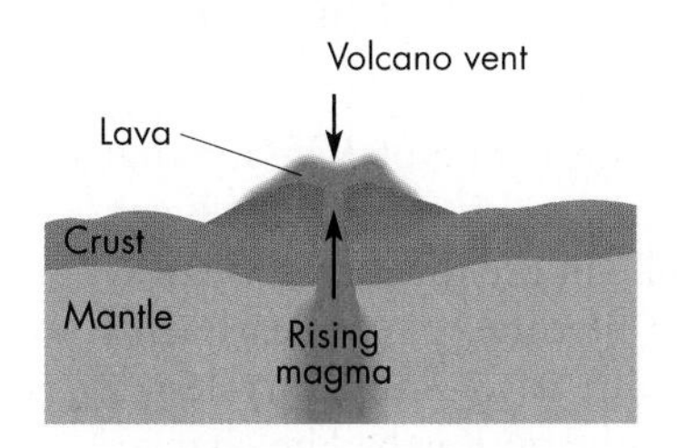

A Lava forms central bulge of the corona.

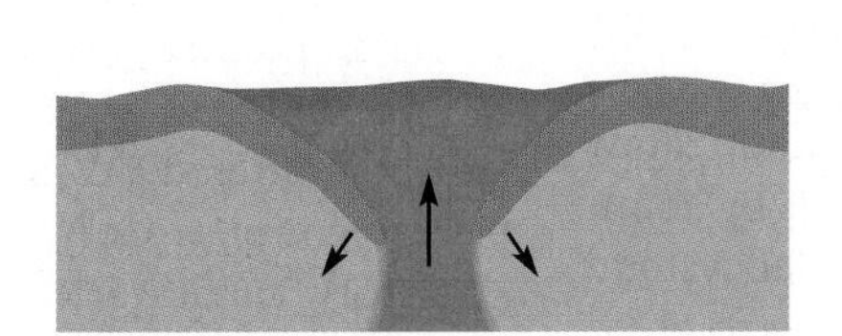

B Crust breaks under the weight of the central bulge.

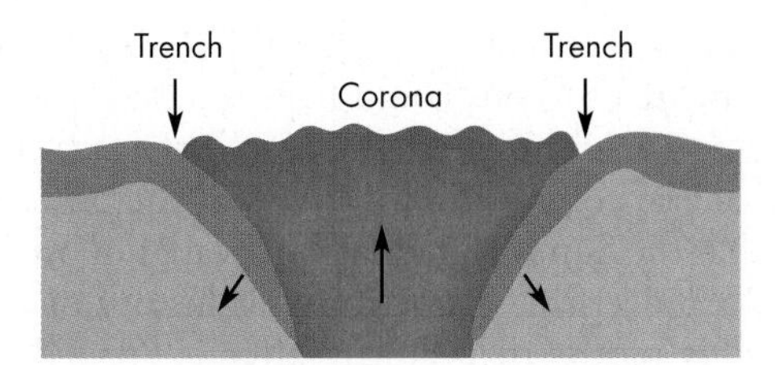

C Trench forms when broken crust sinks into interior of Venus.

and have similar internal structures. There are no seismic data to probe the interior of Venus, but it is thought that Venus has a liquid metal core extending halfway to the surface. Although Venus probably has a metal core, it has no measurable magnetic field. This is likely to be a result of Venus's extremely slow rotation, which prevents the planet from having an Earthlike dynamo.

Evolution of Venus

The internal evolution of Venus probably has paralleled the Earth's internal evolution, except for differences in the mode of tectonic activity. However, the evolution of the atmosphere and climate of Venus must have been very different from that of the Earth.

Evolution of the Atmosphere The present atmospheres of both Venus and the Earth are thought to have formed through the release of gases from their interiors. The three principal gases that resulted from outgassing on the Earth are carbon dioxide, nitrogen, and water vapor. Only nitrogen has remained primarily in the atmosphere. Nearly all of the Earth's outgassed carbon dioxide is locked up in carbonate minerals that form in the oceans and become rocks on the ocean floor. These rocks are eventually subducted into the interior, where the carbon dioxide is released from the rock and recycled back into the atmosphere in volcanic venting. Most of the Earth's outgassed water now forms the oceans.

For Venus, the principal atmospheric gases are carbon dioxide and nitrogen, both of which presumably were released from the interior. The total amounts of outgassed carbon dioxide and nitrogen are remarkably similar for the two planets. If the carbon dioxide now locked up in the Earth's rocks were released into the atmosphere, the result would be an atmosphere strongly resembling that of Venus in thickness and composition.

There is a small amount of atmospheric water vapor on Venus. Venus's surface temperature, however, is far too hot for there to be any oceans or other bodies of liquid water. The total amount of water in Venus's atmosphere, if converted to liquid form, would cover Venus to a depth of only about 3 cm. Why Venus has so much less water than the Earth is not clear, but is probably a result of the loss of most of Venus's water. This possibility is supported by the isotopic makeup of Venus's water. Most water molecules contain two atoms of hydrogen that have only a single proton in their nuclei. A small percentage of water molecules, however, have a hydrogen atom with both a neutron and a proton in its nucleus. Such hydrogen atoms are called deuterium. When water molecules reach the outer parts of a planet's atmosphere they are split into their constituent atoms by ultraviolet radiation from the Sun. The hydrogen atoms can then be lost to space (see Chapter 7 for more details). Deuterium, being heavier than ordinary hydrogen, is lost more slowly. Venus has about 120 times as much deuterium, relative to hydrogen, than does the Earth. This implies that Venus has lost a considerable amount of water, though just how much water has been lost is still uncertain. The more rapid loss of water from Venus's atmosphere may be due to the absence of a global magnetic field. The Earth's magnetic field creates a bubble (the magnetosphere) that keeps the solar wind from striking the upper atmosphere. This doesn't happen on Venus, where the solar wind may have stripped away the upper atmosphere and contributed to the great loss of water.

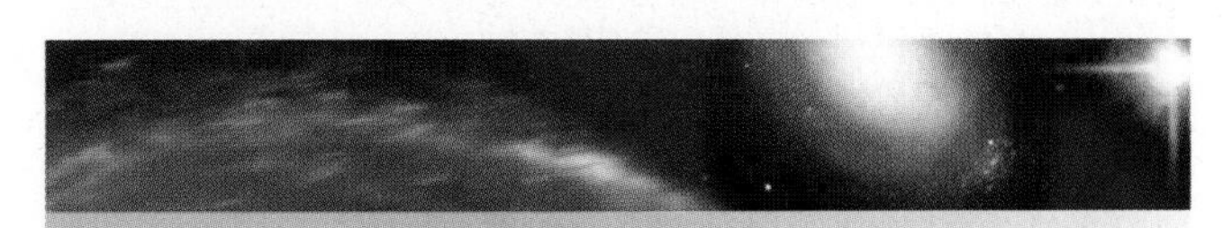

Venus and the Earth have outgassed similar amounts of nitrogen, carbon dioxide, and probably water vapor. Essentially all of the outgassed nitrogen and carbon dioxide remain in Venus's atmosphere. However, nearly all of the outgassed water vapor has been removed from Venus's atmosphere.

Evolution of Venus's Climate Venus's atmosphere is now extremely different from the Earth's, mainly because Venus lacks oceans that could absorb atmospheric carbon dioxide to form carbonaceous rocks. But did the differences arise as soon as Venus began to develop an atmosphere or later, after a period during which Venus might have had a more Earthlike climate? Changes in the climate of Venus could have resulted from changes in the amount of sunlight falling on Venus or from varying atmospheric composition. Models of the Sun's evolution show that the Sun has gradually increased in brightness by about 25% over the 4.6-billion-year history of the solar system. This means that shortly after it formed, Venus received an amount of sunlight only about 25% larger than the Earth now receives. Venus might have been cool enough to have had oceans.

However, as the temperature on a water-covered planet rises, so does the humidity and the amount of water vapor in its atmosphere. When the temperature reaches a critical level, water vapor begins to be effective in blocking the escape of infrared radiation from the planet. At this point feedback begins to occur. As the greenhouse effect develops, it warms the planet. This in turn raises the amount of water vapor, which increases the efficiency of the greenhouse, and so on. The planet's temperature quickly increases to 500 K or higher. To make matters worse, as the oceans dwindle, outgassed carbon dioxide can no longer be deposited as carbonaceous rocks and instead accumulates in the atmosphere, increasing the effectiveness of the greenhouse.

Whether Venus ever had cooler surface temperatures depends on how quickly its atmosphere began to develop. If enough water vapor were outgassed while the Sun was still relatively dim, oceans could have accumulated quickly and prevented a greenhouse from occurring right away. Eventually, though, Venus's atmosphere underwent what was probably a sudden increase in temperature and reached the hostile conditions that exist today. The possible significance of the history of Venus's climate has been summarized by Thomas Donahue and James Pollack:

> The picture now emerging is one of a Venus that began to evolve along a path similar to that of Earth but suffered a catastrophe in the form of a runaway greenhouse early in its lifetime. . . . Clearly this story carries a lesson with it for humanity lest it trigger a similar catastrophe on Earth. The chances seem to be very small that such an event can happen, but prudence suggests that the probability be evaluated with the case of Venus in mind.

The climates of Venus and the Earth probably were similar at first. However, the somewhat higher temperature of Venus resulted in more water vapor in its atmosphere. This inhibited the cooling of Venus by infrared radiation and quickly led to a much hotter climate. Carbon dioxide entering the atmosphere could not be stored away in carbonaceous rocks. Instead, it accumulated to produce the very thick atmosphere of Venus today.

Chapter Summary

- Because of its slow rotation, Mercury's solar day is twice as long as its year. Although daytime temperatures are very high, temperature plummets during the long nights. The range of surface temperatures is greater for Mercury than for any other planet or satellite. (Section 10.1)
- The dominant features of Mercury's surface are impact craters. Because of Mercury's relatively large gravity, debris from impacts was not thrown as far as for lunar craters. Thus, young craters have not been effective at erasing old craters. (10.1)
- A pattern of great scarps shows that Mercury's crust has been altered since it formed. This probably happened when Mercury shrank, fracturing the crust. All internal activity in Mercury probably ended about 3.8 billion years ago. (10.1)
- Mercury's atmosphere is extremely tenuous. It consists of gases trapped from the solar wind or released when meteoroids struck Mercury. (10.1)
- The average density of Mercury is almost as large as that of the Earth. This shows that Mercury contains a large proportion of heavy elements, the most abundant of which is probably iron. (10.1)

- Mercury's high iron content may have resulted from the loss of most of its rocky exterior. This may have happened when Mercury collided with another body almost as large as Mercury. (10.1)
- Venus is an especially important planet to understand, because it is similar to the Earth in many ways. By comparing the Earth and Venus, we can see how small initial differences in the two planets have led to significant differences today. (10.2)
- Unlike the other terrestrial planets, Venus has a retrograde rotation. The rotation of Venus is very slow, resulting in a solar day that is 117 Earth days long. (10.2)
- The lower atmosphere of Venus is as hot as an oven. The large amount of carbon dioxide in Venus's atmosphere results in a surface pressure 92 times as large as that of the Earth. (10.2)
- The thick cloud layer of Venus is made of concentrated sulfuric acid droplets. The main cloud layer is 25 km thick. Nevertheless, enough sunlight penetrates the clouds to make Venus's surface as bright as the Earth's on a cloudy day. (10.2)
- The thick atmosphere of Venus produces a strong greenhouse effect that is responsible for Venus's high atmospheric temperature. (10.2)
- Venus's surface has many impact craters. However, there are no small impact craters because small meteoroids are destroyed in the dense atmosphere before they reach the ground. Venus appears to have undergone planetwide lava flooding about a half billion years ago. (10.2)
- The surface of Venus shows abundant evidence of volcanic activity and crustal distortion. However, surface features seem to have been formed through vertical motion in Venus's crust rather than plate tectonics. A possible reason that plate tectonics is inhibited on Venus may be that the dry crustal rock of Venus is too strong to be broken into plates. (10.2)
- Similar amounts of nitrogen and carbon dioxide have been released from the interiors of the Earth and Venus. In the case of Venus, almost all of the outgassed nitrogen and carbon dioxide remains in the atmosphere. Whatever water that was released from Venus's interior, however, has been removed from the atmosphere. (10.2)
- The climates of Venus and the Earth were probably quite similar at first. However, the warmer temperature of Venus's atmosphere caused it to be more humid than that of the Earth. The additional water vapor partially blocked the escape of infrared radiation and produced a greenhouse effect, further warming Venus. The atmosphere quickly became much hotter, evaporating surface water. Without oceans, carbon dioxide could not be deposited as carbonaceous rocks and therefore entered the atmosphere, resulting in the thick, hot atmosphere of Venus today. (10.2)

Key Terms

arachnoid *224*
coronae *224*
intercrater plain *212*
retrograde rotation *216*
scarp *212*
shield volcano *223*
smooth plain *212*

Conceptual Questions

1. What are two reasons why Mercury's surface gets so cold at night?
2. Describe the way in which the scarps of Mercury were probably produced.
3. What is the principal reason that astronomers think Mercury has a large metallic core?
4. How do the Moon and Mercury differ with respect to crater density, the relative number of small craters, and the proximity of ejected material to the impact site? How can these differences be accounted for?
5. If the cloud layer of Venus is so thick, how is it possible for any sunlight to reach the surface of the planet?
6. What makes the greenhouse effect for Venus so much more effective than the Earth's greenhouse effect? (You might want to review the description of the greenhouse effect in Chapter 8.)
7. Why does the surface of Venus appear orange in the *Venera* images?
8. What evidence do we have that most of the surface of Venus has been flooded with lava during the past 800 million years?
9. If Venus had plate tectonics, what kinds of features should be visible in the *Magellan* images?

10. Without plate tectonics, how could the crustal deformations of Venus have been produced?
11. Why does Venus have so much carbon dioxide in its atmosphere as compared with the Earth?
12. The atmospheres of Venus and the Earth may originally have been quite similar. What may have happened to cause their more recent evolution to diverge so strongly?

Problems

1. Use Kepler's second law (Chapter 4) to calculate the ratio of Mercury's orbital speed at perihelion to its speed at aphelion.
2. Suppose Mercury's rotation period were four-fifths of its orbital period. How many Mercury days would there be in a Mercury year?
3. Use Equation 4.1 to find the synodic period of Venus and the Earth. What does the synodic period of Venus have to do with the interval of time between appearances of Venus in the western sky in the evening?

Group Activity

Have one member of your group be the Sun, one member Mercury, and one member Earth. Have Mercury move about the Sun in a circle and Earth in a bigger circle. Have Mercury complete an orbit in about half the time Earth completes an orbit (this isn't actually the correct ratio of orbital periods). Ask Earth to verify that Mercury's angular distance from the Sun (its elongation) increases and decreases. Have Earth call out when Mercury's elongation is greatest. The rest of the group should note where in their orbits both Mercury and Earth are when this happens and how many orbits each has completed by the next time greatest elongation occurs. Have the entire group discuss the relationship between the interval of time between successive greatest elongations of Mercury and the synodic period of Mercury.

For More Information

Visit the text website at **www.mhhe.com/fix** for chapter quizzes, interactive learning exercises, and other study tools.

11

That life inhabits Mars now is the only rational deduction from the observations in our possession; the only one which is warranted by the facts.

Percival Lowell, 1907

First Light by Pat Rawlings. Early morning fog obscures the floor of Noctis Labyrinthus, a Martian canyon 6 km (4 miles) deep.

www.mhhe.com/fix

Mars

Whereas Venus excited people's imagination because they could see so little of its surface, Mars was exciting because its surface showed so much. Many of Mars's features seemed to have counterparts on Earth. Mars has bright polar caps (Figures 11.1 and 11.2) that wax and wane with the Martian seasons. Clouds can often be seen in Mars's atmosphere, and patterns of dark markings on the surface come and go with the seasons, much like the seasonal patterns of vegetation on the Earth. Even the Martian day is similar in length to a day on the Earth. (The Martian day is about 40 minutes longer.)

In 1877 Mars and the Earth passed unusually close to each other, enabling Asaph Hall of the U.S. Naval Observatory to discover Phobos and Deimos, the two tiny moons of Mars. Also in 1877 Giovanni Schiaparelli, director of the Milan Observatory, carried out extensive observations of Mars. He was able to distinguish features on Mars that were about the size of Ohio. Schiaparelli also reported thin, dark lines crisscrossing the surface of Mars. Similar features had been reported earlier, but not in the number and complexity seen by Schiaparelli. He called the features "canali," which translates as channels. However, Schiaparelli's observations raised the possibility that the "canali" might be canals dug by intelligent beings, a very provocative idea.

By the end of the 1880s a number of astronomers reported seeing the canals, although others were unable to see them at all. The public became fascinated by the prospect of life on Mars. Novels about Martians became popular and large prizes were offered for establishing communication with the Martians. Bright flashes on Mars were interpreted as attempts

Questions to Explore

- In what ways is Mars similar to the Earth and in what ways is it different?
- What is the relationship between internal activity and Mars's spectacular surface features such as its gigantic volcanos and immense canyon system?
- What can we learn about the history of Mars by studying its surface features?
- Has liquid water ever existed on Mars and, if so, where is that water today?
- Has Mars's atmosphere always been as thin, dry, and cold as it is now?
- Is there any evidence of present or past life on Mars?

FIGURE 11.1
A Series of Mars Photographs Taken in 1907 at Lowell Observatory
Bright polar caps are seen at the top and bottom of the pictures. North is at the bottom and east is on the left side of each picture. Surface markings move from right to left during Mars's 24.6-hour rotation period.

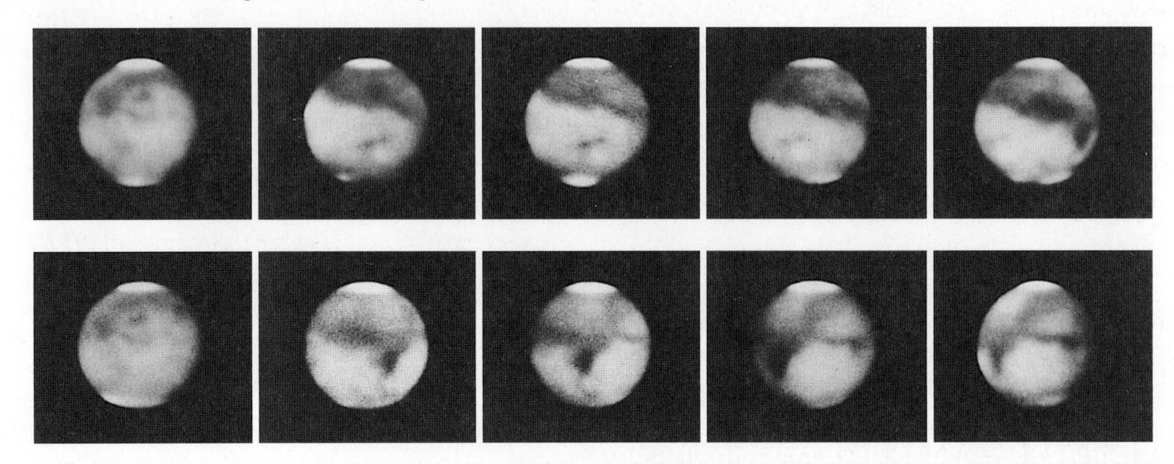

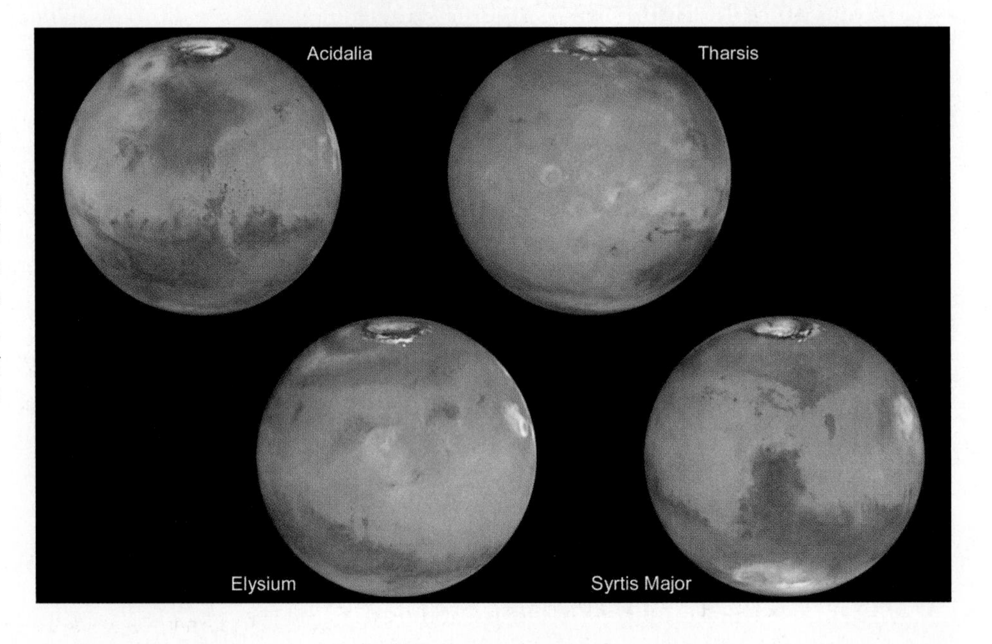

FIGURE 11.2
***Hubble Space Telescope* Images of Mars**
These images, taken in 1999, show Mars at four longitudes that cover the entire Martian surface. The images are a modern version of Figure 11.1 and show the northern polar cap, bright and dark surface markings, and wispy white clouds. The images were taken as summer was beginning in Mars's northern hemisphere.

to signal the Earth. An Austrian suggested constructing a set of giant geometrical symbols (made of kerosene-filled trenches) in the Sahara Desert and then igniting the kerosene to catch the Martians' attention. Charles Cos of France suggested using giant mirrors to focus sunlight and burn messages into the Martian deserts or perhaps ignite vegetation. (This seems to be a questionable way to make ourselves known to a civilization capable of constructing a global network of canals.)

The canals of Mars also caught the interest of Percival Lowell, the man who became the most famous and dedicated advocate of the idea that Mars was inhabited. Lowell was a Boston aristocrat. His brother was the president of Harvard University and his sister was the poet Amy Lowell. When Lowell learned in 1893 that failing eyesight had forced Schiaparelli to give up his studies of Mars, he decided to dedicate himself to planetary studies. To carry out his studies he built an observatory (still an active center of planetary research) in Flagstaff, Arizona. He and his staff observed Mars systematically and routinely for more than 20 years. His observations convinced him that the canals of Mars were real. A gifted writer and speaker, Lowell used his talents to popularize his idea that Mars was an "abode of life" inhabited by intelligent creatures striving to survive on a planet becoming an ever more hostile desert.

Lowell was so successful in this that the American public became convinced there was a Martian civilization. In 1907 even the conservative *Wall Street Journal* decided that the most important event of the year was the presentation of "the proof afforded by astronomical observations that conscious, intelligent human life exists upon the planet Mars."

After Lowell died in 1916 his influence faded quickly. New observations with better telescopes demonstrated that Lowell's Martian canals didn't really exist. The new observations reinforced the view that the Martian environment was too hostile for higher organisms. Nevertheless, the idea of Mars as Earthlike, only drier and with a thinner atmosphere, remained the standard picture of Mars until *Mariner 4* flew past the planet in 1965. Novels and movies continued to feature civilizations on Mars. Popular science books, as late as the 1950s, took it for granted that plant life existed on Mars. Even scientists thought in terms of an Earth model and described the Martian environment as "the Gobi Desert raised into the stratosphere."

In the space age there has been a revolution in our understanding of Mars. The first flybys saw only craters, so many astronomers adopted a Moonlike model for Mars. However, more complete observations of Mars have shown it to be a much more dynamic, complex body than the Moon. The Mars of Percival Lowell may be no more, but the Mars we know today is also interesting and intriguing.

This chapter is about Mars as we understand it today, a planet with a hostile and forbidding environment, but showing evidence of a much more benign early environment. The focus is on how the history of Mars can be read in the atmosphere and surface we see today. That history gives us yet another example of how terrestrial planets evolve and helps us understand our own planet.

11.1 THE EXPLORATION OF MARS

As the twentieth century progressed, the idea of Mars as a habitable planet was gradually discarded. New observations showed that Mars's atmosphere is far thinner than that of the Earth, that carbon dioxide is the principal gas, and that oxygen and water vapor are nearly absent. Any notion of Mars as an "abode of life" was shattered when the first Mars probe, *Mariner 4,* flew past Mars in July 1965. The 22 images it returned showed a heavily cratered surface, much like that of our Moon (Figure 11.3). Several hundred images radioed back to the Earth from the *Mariner 6* and *7* flybys in 1969 confirmed the new picture of Mars as a dead planet. They showed that the atmospheric pressure of Mars is less than 1% that of the Earth at sea level and that the polar caps are primarily frozen carbon dioxide (dry ice) rather than water.

Mariner 4, 6, and *7* returned pictures of only a small percentage of the total surface of Mars. In 1971 *Mariner 9* went into orbit around Mars with the mission of imaging the entire surface of the planet. When *Mariner 9* arrived, a dust storm covered the entire planet. The only visible features were a cluster of dark spots. As the dust settled, the spots proved to be immense volcanos with peaks so high they poked above the surrounding dust. Other pictures showed canyon systems that dwarfed our Grand Canyon and widespread channels that looked like dry riverbeds. Like the earlier *Mariner 4* pictures, those from *Mariner 9* changed astronomers' views about Mars. The pictures showed that Mars must have a very interesting, complicated history. Even better pictures of Mars were returned by the two *Viking* orbiters that arrived at Mars in 1976. Each *Viking* orbiter was accompanied by a lander that safely touched down on Mars's surface and, for several years, returned images of its surroundings, meteorological data, and seismic measurements. Each lander also carried equipment to examine the soil of Mars for microbial life.

After a gap of almost 20 years, Mars exploration resumed again in 1997. In July 1997, *Pathfinder* landed on Mars. Two days after landing, *Pathfinder* released *Sojourner,*

FIGURE 11.3
The Surface of Mars as Seen from *Mariner 4* in July 1965
The image shows a cratered surface very much like that of the Moon.

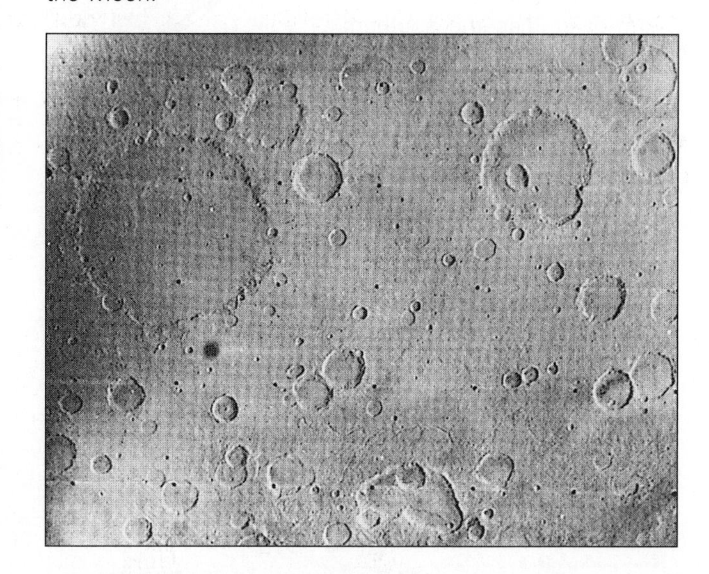

a small six-wheeled rover that explored the Martian surface in the immediate vicinity of *Pathfinder*. *Sojourner* carried a spectrometer to measure the chemical composition of the soils and rocks around the landing site. In September 1997, *Mars Global Surveyor* went in orbit around Mars with the mission of mapping Mars with much better resolution than ever before. *Mars Global Surveyor*, which stopped transmitting in 2006, was able to resolve features on Mars as small as 1.4 m, about the size of a bicycle. In 2003–2004, a number of missions reached Mars, including the *Spirit* and *Opportunity* rovers *Mars Express*. *Mars Reconnaissance Orbiter* went in orbit about Mars in 2006 and *Phoenix* landed near Mars's north polar cap in May 2008. Some important properties of Mars are summarized in Table 11.1.

Table 11.1
Planetary Data

Mars

Orbital period	1.88 years
Mass	$0.11\ M_{Earth} = 6.42 \times 10^{23}$ kg
Diameter	$0.53\ D_{Earth} = 6794$ km
Density (relative to water)	3.94
Escape velocity	5.0 km/s
Surface gravity	0.38 g
Global temperature	210 K
Main atmospheric gases	CO_2, N_2, Ar
Rotation period	24.6 hours
Axial tilt	25°
Known satellites	2
Distinguishing features	Most "Earthlike" environment, evidence of ancient bodies of water

11.2 THE SURFACE OF MARS

The images obtained by the *Mariner* and *Viking* spacecraft showed the entire surface of Mars and gave us a better view of the geography of that planet than was available for Earth early in the twentieth century. Figure 11.4, a *Mars Global Surveyor* relief map of Mars, shows Mars to be a planet with two very different faces. An observer far above Mars's south pole would see a densely cratered planet resembling the highlands of the Moon. An observer above Mars's north pole, however, would see sparsely cratered plains resembling the lunar maria. On average, the northern plains of Mars are 5 km lower in elevation than the cratered highlands. The plains probably originated in vast volcanic flows that buried the ancient craters of the northern hemisphere. Other explanations for the smoothness of the northern plains involve the deposition of thick

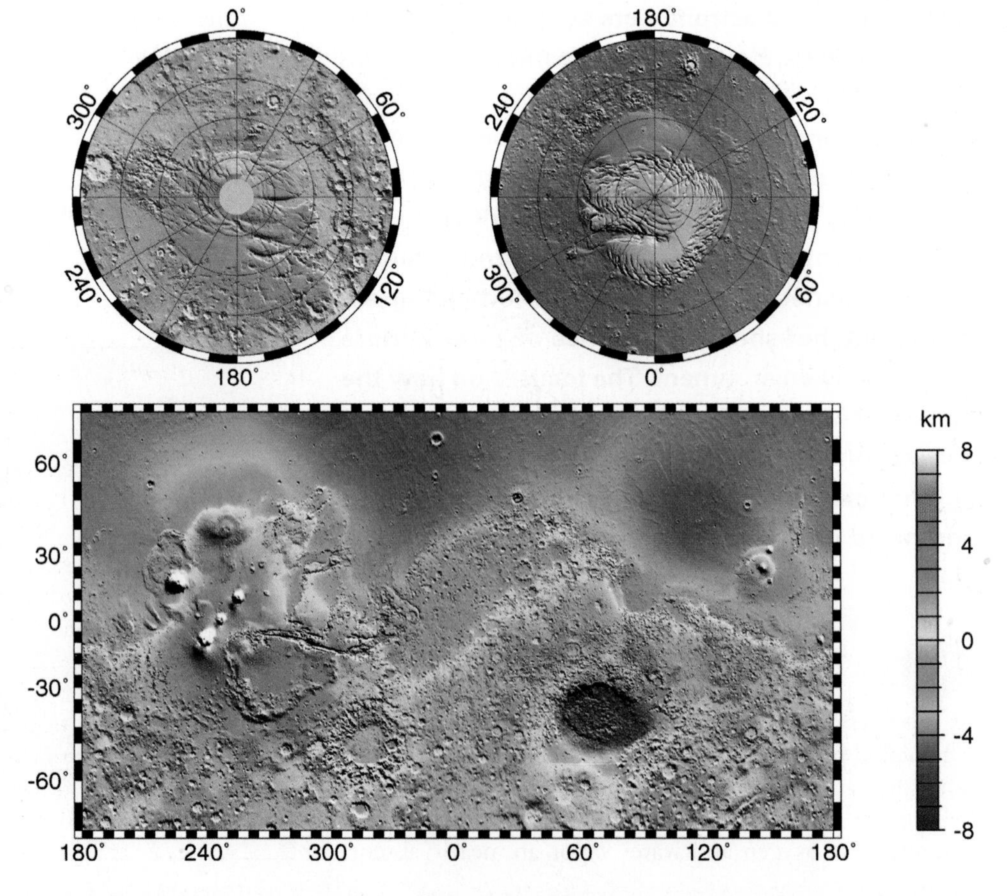

FIGURE 11.4
A Relief Map of Mars
Mars Global Surveyor obtained this relief map of Mars by bouncing laser pulses off Mars and timing how long it took them to return. White and red indicate high locations, green and blue are low regions. Notice that the northern plains are typically 5 km lower than the southern cratered terrain. The light region at center left is Tharsis and its giant volcanos. The circular low region at lower right is Hellas, the biggest impact basin in the solar system.

sediments by either wind or water. A group of astronomers have recently proposed that the reason the northern plains of Mars are smooth and low in elevation is that a large meteoroid struck Mars a glancing blow. The impact stripped away 40% of Mars's crust and left a huge elliptical basin. The northern hemisphere has many volcanos, including several gigantic ones in the region nineteenth-century observers called Tharsis.

Martian Craters

The Martian craters look like those on the Moon and Mercury except for the distinctive appearance of the surrounding ejected material. On the Moon and Mercury the layer of ejecta is thickest near the crater rim and thins out with increasing distance from the crater. This is the pattern that occurs when the ejected material is thrown upward and outward from the crater. On Mars, however, many craters are surrounded by one or more sheets of ejecta that end abruptly. As Figure 11.5 shows, the ejected material looks as if it flowed, rather than flew, away from the crater. In some cases the ejected material apparently flowed around obstacles such as preexisting craters. Clearly the ejecta produced by impacts on Mars are more fluid than those of the Moon and Mercury. One possible explanation for the fluidity of Martian ejecta is ice beneath the Martian surface. When an impact heats the Martian surface layers it melts the ice to produce water, making the ejected material a muddy mixture.

The density of craters has been used to estimate the ages of different portions of the Martian surface. As in the case of the Moon, the heavily cratered highland regions were formed during a period of heavy bombardment that took place after Mars formed 4.6 billion years ago. The more lightly cratered regions formed about 3.8 billion years ago after the period of heavy bombardment ended. Ages for the younger, more sparsely cratered parts of Mars also can be estimated from their crater densities. Most age estimates for Martian surface features are based on the assumption that the Martian cratering rate has declined over time in the same way as the lunar cratering rate.

The densely cratered regions of Mars have both rugged, heavily cratered terrain and intercrater plains. The heavily cratered regions are probably the oldest remaining part of the Martian surface. Relatively smooth intercrater plains are found within the cratered terrain. Their similarity to the lunar maria strongly suggests that the intercrater plains are the result of local volcanic flows. Mars was probably more volcanically active than the Moon during the early period of heavy bombardment. In some regions local volcanic flows filled in the craters more rapidly than new ones were formed, producing the smooth intercrater plains.

FIGURE 11.5
Yuty, a Martian Crater 20 km in Diameter
The material ejected from Yuty appears to have flowed away from the crater in several thin sheets.

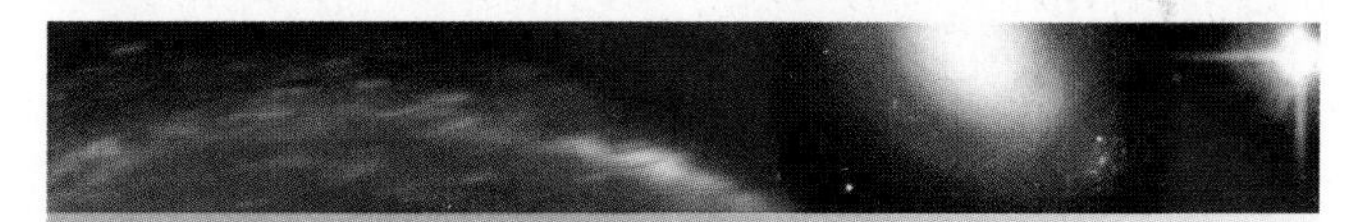

About half of the surface of Mars is covered with ancient craters. Ejected material appears to have flowed away from many craters. This suggests that there may be subsurface ice or water on Mars.

Meteorites from Mars

Although there have been proposals for a mission to return samples of the Martian soil and rocks to Earth, there are already some samples on the Earth. Debris from some Martian impact craters was blasted into space and eventually reached the Earth. About thirty meteorites, known as the SNC meteorites, show some remarkable properties. While virtually all other meteorites solidified about 4.5 billion years ago, the lavas that formed most of the SNC meteorites solidified only 1.3 billion or fewer years ago. Mars is one of the few places in the solar system believed to have had volcanic activity as recently as 1.3 billion years ago. The SNC meteorites contain inert gases identical in composition to the inert gases in the Martian atmosphere as measured by *Viking*. One of the SNC meteorites, discovered in Antarctica in 1984, was found to contain possible evidence of ancient microbial Martian life. This discovery is described more fully in Chapter 27.

But how could an SNC meteorite have been thrown from the impact site with a speed great enough to escape from Mars? One possibility is that an impact in a region with a large amount of subsurface ice might have melted and vaporized the subsurface ice. The explosive expansion of the hot steam could have supplied an additional, jetlike push to help propel debris outward.

FIGURE 11.6
Knobs and Mesas Near the Boundary Between the Cratered Highlands and the Smooth Plains
The knobs are the rims of craters that were flooded with lava.

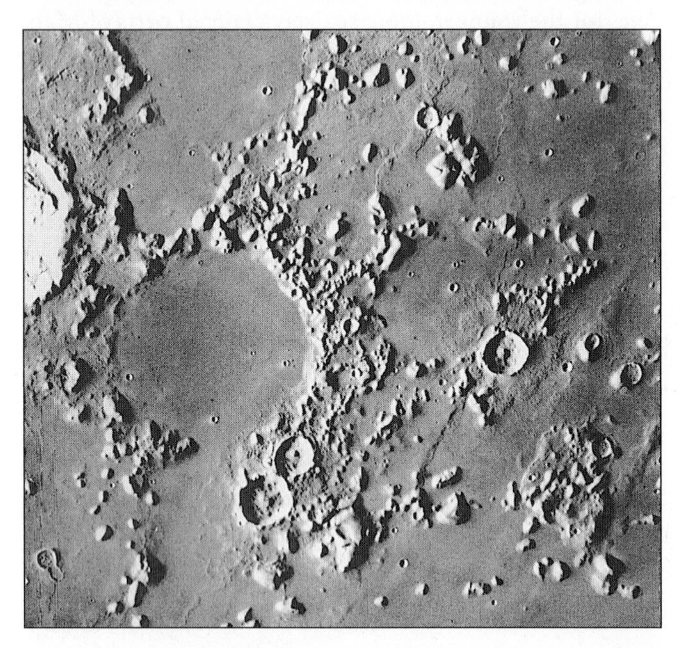

Volcanic Activity

Although there are many volcanic features in the southern, heavily cratered hemisphere of Mars, evidence of volcanic activity is even more obvious over the rest of the Martian surface. Most of Mars's northern hemisphere is covered by plains about as heavily cratered as the lunar maria.

The crater densities on the plains show that the plains were formed over a very long period of time. The youngest plains are probably much less than 1 billion years old, while the oldest were formed nearly 4 billion years ago. Episodes of widespread volcanic flows may have occurred throughout the history of Mars.

Near the boundary between the cratered highlands and the smooth plains, there are numerous examples of flattopped mesas and knobs. These mark the outlines of craters, like those in Figure 11.6, nearly completely covered by lava or other deposits. Apparently the northern hemisphere of Mars was once as heavily cratered as the southern hemisphere, but the ancient craters have been nearly obliterated by lava floods.

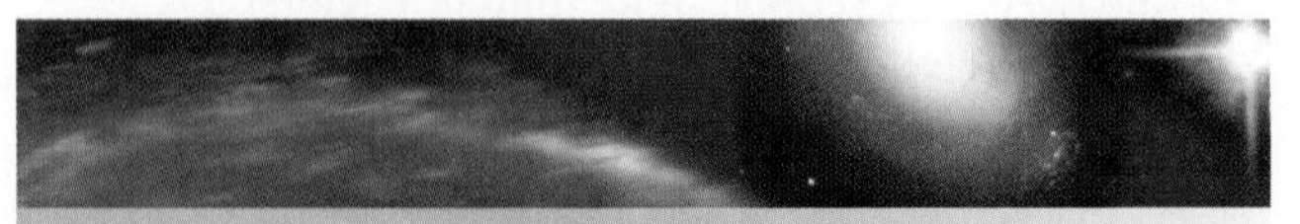

Extensive volcanic flows, particularly in Mars's northern hemisphere, have obliterated the ancient cratered terrain.

Volcanos of the Tharsis Region

Although there are many volcanos on Mars, the most spectacular by far are Arsia Mons, Pavonis Mons, Ascraeus Mons, and Olympus Mons, clustered in the Tharsis region shown in Figure 11.7. These four volcanos are located on a continent-sized elevated region called Tharsis, which rises to about 10 km above the average Martian elevation. The origin of Tharsis, the dominant surface feature of the planet, is not completely clear. One plausible way the bulge could have been produced is through the uplifting of Mars's crust by convection currents in its mantle. Another possibility is simply the accumulation of volcanic lava for several billion years.

Three of the Tharsis volcanos, Arsia Mons, Pavonis Mons, and Ascraeus Mons, are spaced about 700 km apart near the ridge of Tharsis. Their peaks rise to about 27 km (17 miles). They are shield volcanos (see Chapter 10), and

FIGURE 11.7
The Volcanos of the Tharsis Region
Olympus Mons is the large spot in the upper part of the sunlit crescent of Mars. Below it, from left to right, are Arsia Mons, Pavonis Mons, and Ascraeus Mons. Notice how large the volcanos are in comparison with the size of Mars.

FIGURE 11.8
Olympus Mons Compared with the Hawaiian Islands
Olympus Mons, which rises 24 km above the surrounding plains, is broader than the entire region covered by the main islands of the Hawaiian chain. The vertical scale of this figure is greatly exaggerated.

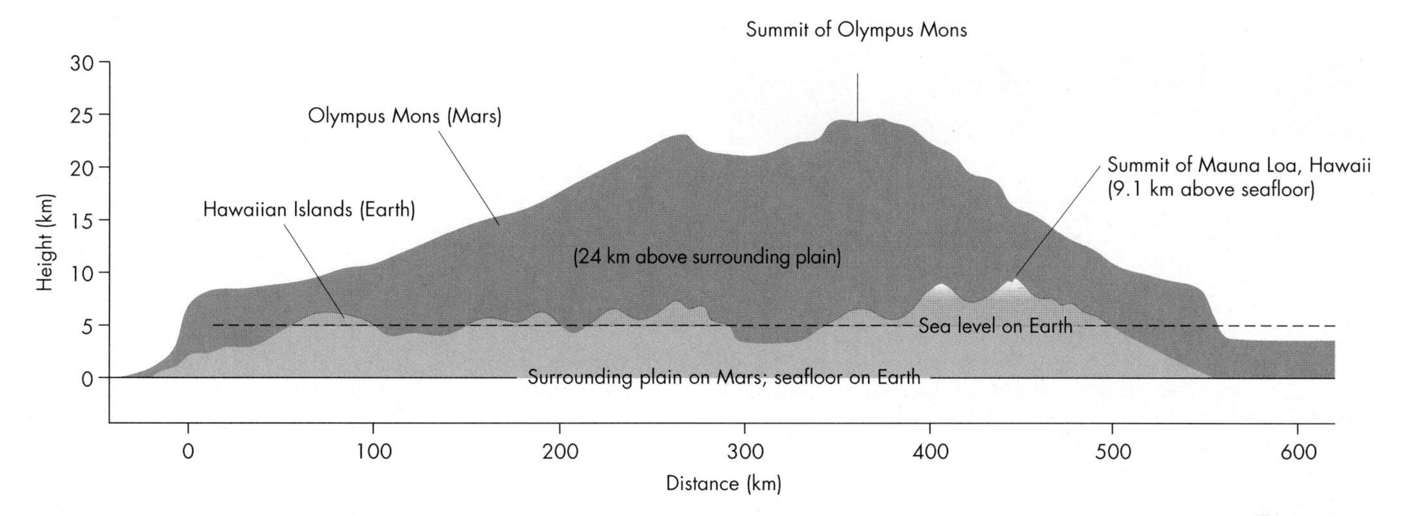

have very gentle slopes near their bases, but become steeper near their summits. Shield volcanos are built up gradually by thousands of individual lava flows. The shields of the Tharsis volcanos are about 400 km in diameter. Their calderas (summit pits) are as much as 100 km across and several kilometers deep. By way of comparison, the shield volcanos of the Hawaiian Island chain are among the largest on the Earth but are all less than about 120 km in diameter and rise less than 10 km (5.6 miles) above the ocean floor.

Olympus Mons, sketched in Figure 11.8, is located near the edge of Tharsis and also rises to 27 km. However, it does so from a lower base than the other Tharsis volcanos, which sit on a base about 10 km high. Olympus Mons towers 24 km above the surrounding plains and is the tallest mountain in the solar system. It is 550 km across. If it were centered in New York City, as in Figure 11.9, it would reach from Boston to Washington, D.C. It would cover much of New England, New York, and Pennsylvania, and extend almost 300 km out to sea. Olympus Mons is so tall that it rises above most of the Martian atmosphere. The atmospheric pressure at its summit is about 5% of the pressure at its base.

The gentle slopes of shield volcanos, like those of Tharsis and the Hawaiian chain, result from the very fluid nature of their lavas, which flow easily and cover a wide area. Like the Hawaiian volcanos, the calderas of the Tharsis volcanos have very complex shapes. This shows that their summits have undergone multiple cycles of expansion and collapse. The great sizes of the Tharsis volcanos imply that the Martian lithosphere has moved less than the Earth's lithosphere. One reason the Hawaiian volcanos are relatively small is because each of the Hawaiian volcanos was active for a limited period of time. The motion of the Pacific plate on which Hawaii is located carries each newly formed volcano northwest away from the source of the magma that built it. The oldest Hawaiian volcanos are on the northwest end of the chain; the youngest (and only active) volcanos are on the southeast. The Tharsis volcanos, however, apparently are permanently fixed over their magma sources.

Great pressure is required to lift magma from its source to the summit of a volcano. The deeper the source of the magma, the higher the pressure exerted on the magma and the higher it can rise. The fact that the four large Tharsis

FIGURE 11.9
Olympus Mons Superimposed on a Map of the Eastern United States
If Olympus Mons were centered in New York City, it would reach north to Boston, south to Washington, D.C., west to central Pennsylvania, and east 300 km (200 miles) into the Atlantic Ocean.

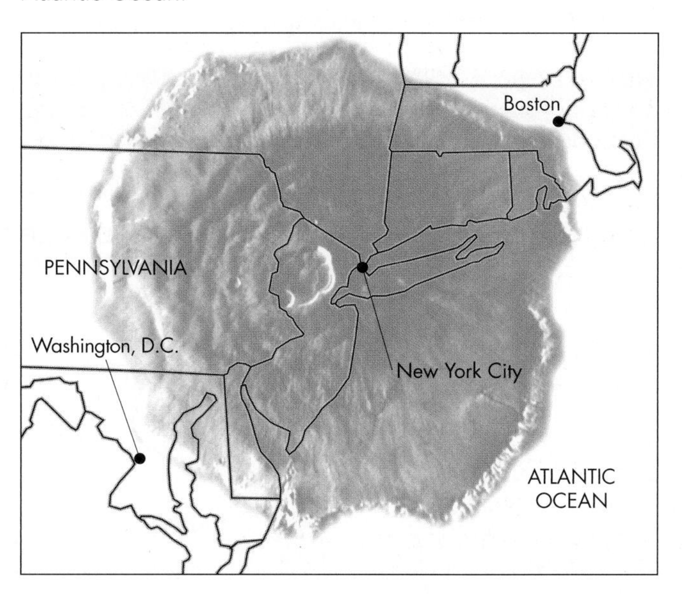

volcanos all rise to such great heights implies that their magma source, at the base of the lithosphere, is very deep, probably at a depth of about 160 km. Thus, the Martian lithosphere is probably considerably thicker than the Earth's, which averages about 100 km thick.

The largest known volcanos in the solar system are found in the Tharsis region, a continent-sized bulge on Mars. Volcanos of such immense size show that the crust of Mars is very stable, allowing the Tharsis region to remain over its magma source for billions of years. Also, the lithosphere of Mars must be very thick so that magma is lifted with sufficient pressure to reach the peaks of the volcanos.

Other Volcanos

Another kind of large Martian volcano is the **patera,** which resembles shield volcanos, but is even more gently sloping. Paterae can be very large, but not nearly as tall as shield volcanos. There are no counterparts to the Martian paterae on the Earth. One of the paterae, Alba Patera, is located at the northern edge of Tharsis and covers the largest area of any known volcano in the solar system. Although only about 6 km high, its lava flows cover a circular region about 1600 km across. It is big enough to cover the state of Texas completely. The average slope on the flanks of Alba Patera is only 0.5°, so someone climbing the volcano would barely be able to tell that he or she were walking uphill. The very gradual slope of Alba Patera and other paterae may be due to pyroclastic flows. Pyroclastic flows are clouds of volcanic debris containing a mixture of hot ash, solid rock fragments, and gases. These clouds can travel extremely rapidly even over gentle slopes, so they can flow hundreds of kilometers. A particularly disastrous example of a pyroclastic flow on the Earth occurred when Mont Pelee, in the West Indies, erupted in 1902. The flow swept upon the town of St. Pierre, destroying it and killing 30,000 people.

Mars's Elysium region, a bulge resembling Tharsis, but smaller, also contains large shield volcanos. The largest of the Elysium volcanos, Elysium Mons, is about 170 km across and rises 9 km above the surrounding plains. Some of the volcanic flows in the Elysium region are nearly free of impact craters and are among the youngest volcanic sites on the entire planet. Hundreds of additional Martian volcanos have been identified, including many ancient ones in the heavily cratered highlands.

Volcanic History

The history of volcanic activity on Mars can be read from the density of craters on volcanic flows and volcanos. Some important crater densities and the ages they imply are given in Table 11.2.

The oldest recognizable volcanic flows are the intercrater plains of the cratered highlands. They are quite heavily cratered and date from the time shortly after the impact rate declined, about 3.8 billion years ago. Evidence of any older volcanic activity is buried under younger lava flows or obliterated by impact craters. The ridged plains like Chryse Planitia, which cover Mars's northern hemisphere, are slightly younger, about 2.5 to 3.5 billion years old. Lava flows on the flanks of the volcanos in Elysium and Tharsis show a great range of ages. The most complete study of the crater densities in Elysium and Tharsis was carried out using high resolution *Mars Express* images that made it possible to see craters as small as 10 m across. The ages of lava flows on the Elysium and Tharsis volcanos range from 3.8 billion years to as little as 2.4 million years for a flow on the lower flanks of Olympus Mons.

The history of Martian volcanism shows that volcanic activity has become restricted to a smaller and smaller region of Mars as time has passed. Until about 2.5 billion years ago, Martian volcanic activity was nearly global, producing broad

Table 11.2

Ages of Martian Volcanic Regions

	CRATER DENSITY (RELATIVE TO LUNAR MARIA)	AGE ESTIMATE (BILLIONS OF YEARS)
Olympus Mons	<0.1	<0.2
Arsia Mons	0.1	0.2
Tharsis volcanic plain	0.5	1.6
Elysium volcanos	0.7	2.5
Chryse volcanic plain	1.1	3.2
Alba Patera	1.8	3.5
Cratered highlands	13.0	3.8

volcanic plains in the north and the intercrater plains in the south. After 2.5 billion years ago, volcanic activity was mostly confined to the Tharsis region. In the last billion years volcanic activity has occurred only in the volcanos near the crest of Tharsis and in Olympus Mons and in Elysium. Are there still active volcanos on Mars? No one knows for sure, but there is no reason to believe that the Tharsis volcanos have recently become extinct after billions of years of activity. No one has seen a volcanic eruption since telescopic observations of Mars began. However, we have been observing Mars for a length of time that is very brief compared with the history of Martian volcanic activity. The intervals of time between volcanic outbursts may be so long that we couldn't expect to see one in only a few hundred years of observations.

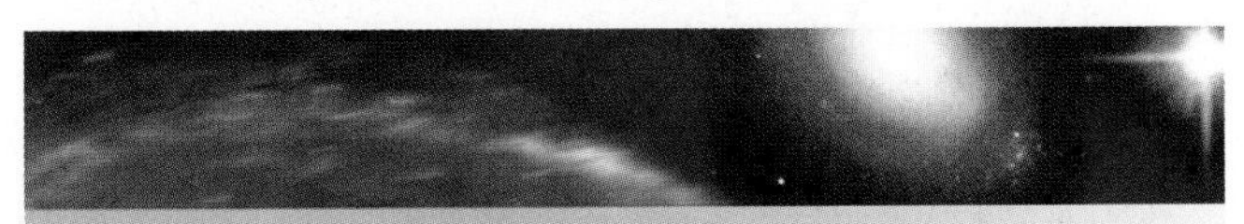

The ages of Mars's volcanic features show that Martian volcanic activity was widespread until about 2.5 billion years ago. Since that time the region of volcanic activity has become steadily more restricted.

Crustal Motion

There is ample evidence, in the form of faults and chasms, that there has been crustal motion on Mars, particularly in the region around Tharsis. Figure 11.10 shows part of the region around Tharsis where there are many faults, some 1 km wide and extending for hundreds of kilometers. Most of these cracks point toward the center of Tharsis. Apparently the fractures were a result of stresses in the Martian crust caused by the enormous weight of Tharsis. The effect is similar to the system of cracks that would develop if a weight were placed on the top crust of a pie. The most heavily fractured regions are 2.5 to 3.5 billion years old, as indicated by their relatively high crater densities. Younger regions have fewer faults. However, even some very young volcanic flows show fractures. It appears that most of the fractures formed shortly after the intense bombardment of Mars's surface ended. Since that time fracturing has taken place less and less frequently. The fact that most of the fractures were formed long before the Tharsis volcanos achieved their immense size supports the idea that the massive Tharsis bulge was at least partly caused by an uplifting of the Martian crust.

Another sign of Martian crustal motion is Valles Marineris shown in Figure 11.11. Valles Marineris is a vast system of interconnected canyons descending from Tharsis and stretching eastward for nearly a quarter of the circumference of Mars. Valles Marineris is more than 600 km wide and as much as 8 km deep in some places. Although erosion has shaped the canyons of Valles Marineris, they certainly were not cut primarily by the force of running water. Instead the individual canyons are gigantic cracks produced by faulting. Like the smaller faults near Tharsis, Valles Marineris points more or less

FIGURE 11.10
Parallel Fractures Pointing Toward Tharsis Bulge
This *Mars Global Surveyor* image shows a region about 100 km across. The parallel cracks are faults that point toward Tharsis.

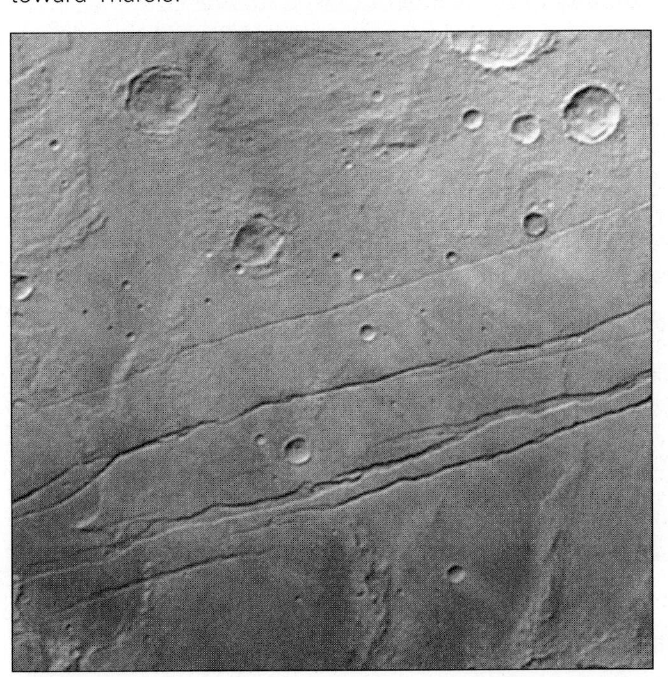

FIGURE 11.11
Valles Marineris
Valles Marineris appears as a gash at the center of this *Viking* image. The system of chasms extends for 4000 km (about a quarter of Mars's circumference) around Mars's equator.

at the center of Tharsis. In places the canyon floors look as if they are pieces of the original collapsed surface. The walls of Valles Marineris occasionally undergo landslides as large as the one 100 km wide shown in Figure 11.12. A high resolution *Mars Global Surveyor* image of cliffs and rock outcrops on the northern wall of Valles Marineris is shown in Figure 11.13.

Mars Global Surveyor found evidence that Mars had plate tectonic activity long ago. The evidence, shown in Figure 11.14, consists of magnetic field measurements that show a pattern of bands of opposite magnetic polarity. The bands resemble the magnetic strips found in regions of Earth's ocean floor where new crust has formed (see Figure 8.19). The magnetic bands suggest that Mars also

FIGURE 11.12
Capri Chasma, Part of Valles Marineris
Part of the 2-km-high wall has been eroded by landslides that extend outward from the chasm wall for 20 to 30 km.

FIGURE 11.13
A View of the Northern Wall of Valles Marineris
This *Mars Global Surveyor* image shows a region about 8 km across. The resolution is about 10 m (the size of a house). The most striking features of the image are sheer cliffs a kilometer high and layering in the cliff wall.

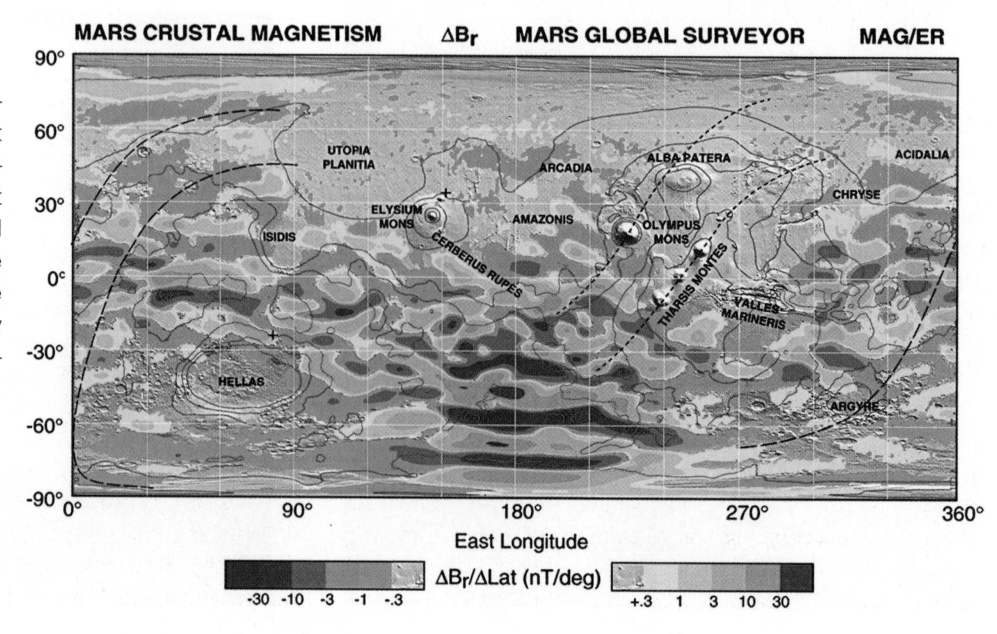

FIGURE 11.14
Magnetic Bands on Mars
The *Mars Global Surveyor* magnetic measurements show that much of Mars's southern hemisphere consists of east-west bands produced when crustal motion created new surface material. Blue and red indicate different magnetic polarity. Grey indicates regions of low magnetic field strength.

had moving crustal plates and a varying magnetic field. Plate tectonic activity on Mars, however, seems to have ended long ago. Notice that regions that have been heated by impacts (Hellas) and volcanic flows (Tharsis, Elysium) have lost the magnetic field stripes imposed during the era of Martian plate tectonics.

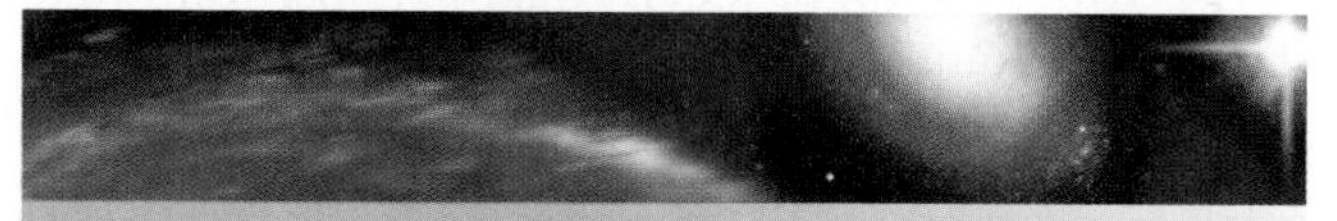

Crustal motion on Mars has produced many faults as well as the immense Valles Marineris canyon system. The faults and chasms may be fractures that resulted from the weight of Tharsis. Mars may have experienced an early period of plate tectonic activity.

The Martian Channels

One startling result of the orbiter images of Mars was the discovery of numerous channels apparently cut by running water. This was surprising because the pressure of Mars's atmosphere is too low for liquid water to exist, except very briefly, today. If a pool of water were placed on Mars it would quickly evaporate or freeze, leaving no liquid water at all. The channels of Mars can be grouped into two types: **runoff channels** and **outflow channels.** The Martian channels discovered by Mars orbiters have nothing to do with the canali reported by Schiaparelli, Lowell, and others. The channels are far too narrow to be seen from the Earth.

River Systems

The runoff channels consist of branching networks of tributaries feeding into larger and larger valleys. Typical systems are hundreds of kilometers in length and cover tens of thousands of square kilometers. Figure 11.15 shows a very extensive system of tributaries that look like they were formed by the collection of widespread rainfall. Most of the Martian runoff channel systems, however, show some significant differences from the river systems on the Earth. First, there are broad spaces between tributaries. These spaces are too lightly eroded to have had much liquid water running across them. Second, the large distances between individual networks are inconsistent with widespread rainfall. A better suggestion for the origin of most of the runoff channels is the release of underground water, which flowed downhill collecting into larger streams as it flowed. In any case, for rivers to have existed the Martian climate must have been very different in the past. *Mars Global Surveyor and Mars Reconnaissance Orbiter* produced images of Martian river valleys showing unprecedented detail. Figure 11.16 shows part of the Nanedi Vallis network of valleys at a resolution of 1 m.

FIGURE 11.15
Closely Spaced Runoff Channels
Runoff channels as densely distributed as this are unusual on Mars. Rainfall may have been responsible for the channel system. The scene is about 200 km across. Water in the channels flowed generally toward the bottom of the image.

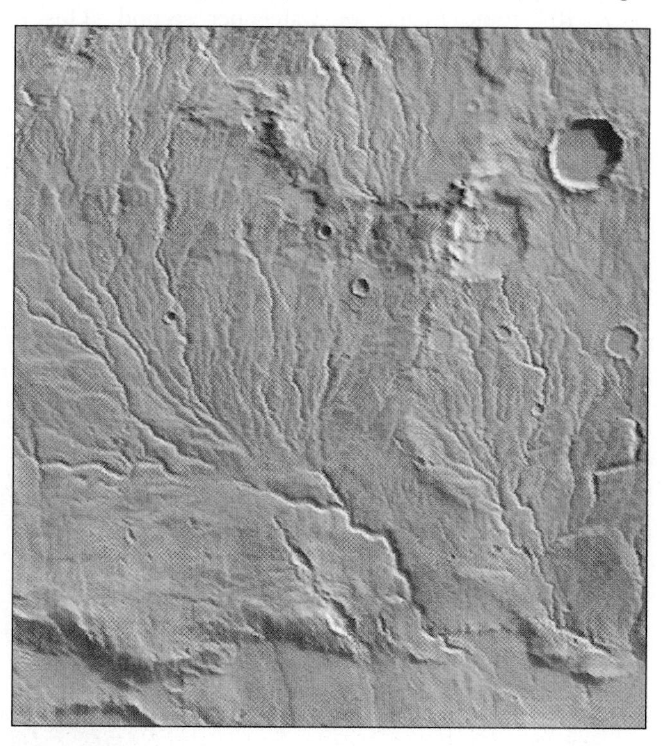

How long ago the rivers flowed can be estimated from the age of the terrain where the channel networks are seen. The networks are found almost everywhere in the heavily cratered uplands but are very scarce on the sparsely cratered plains. This suggests that most of the river systems were formed before lava flows produced the volcanic plains 3.5 to 4 billion years ago.

Ancient Floods

Outflow channels, concentrated in the equatorial regions of Mars, are probably the results of catastrophic floods. What triggered the floods may have been the sudden release of underground water or the sudden melting of underground ice. The released water then flowed rapidly downhill, cutting the channels. Unlike runoff channels, outflow channels have few tributaries and their widths and depths don't change much from beginning to end. Figure 11.17 shows three of the outflow channels of Mars. Where the outflow channels emerge onto the smooth plains, the plains show many indications of flooding such as shallow channels, elongated islands, as in Figure 11.18, and places where temporary damming of the flow occurred. *Pathfinder* landed in Ares Vallis about 100 km from the region shown in Figure 11.18.

Some of the floods must have been spectacular, because the channels cut by them are tens of kilometers wide and several kilometers deep, and extend for hundreds of kilometers. The maximum flow rates for the floods can be estimated from the widths and depths of the channels they produced. The estimated flow rates range up to 3×10^8 cubic meters of water per second. This is about 10,000 times as great as the average flow of the Mississippi River and 3000 times as great as the average flow of the Amazon River. The total amount of water that flowed through the channels can be estimated from the amount of erosion the floods produced. In the Chryse region the total volume of eroded material is about 5×10^6 cubic km. If the flood were 60% water and 40% sediment, then 7.5×10^6 cubic km of water would have been involved. The water required to account for the outflow channels

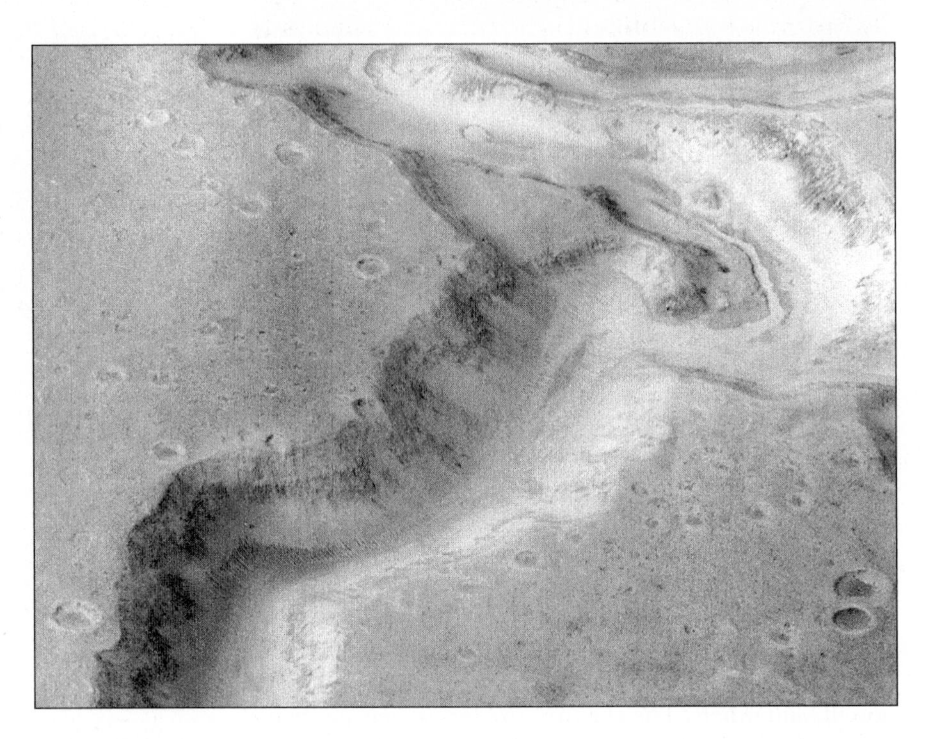

FIGURE 11.16
A High Resolution View of a Martian River Valley
This *Mars Reconnaissance Orbiter* image shows a 7-km-wide portion of Nanedi Vallis. The terraces and channels suggest that the valley was cut by running water.

FIGURE 11.17
Three Outflow Channels
This *Mars Global Surveyor* image shows an 800-km-wide region east of Hellas. The three outflow channels are, from left to right, Dao Vallis, Niger Vallis, and Harmahlis Vallis. The channels are about 1 km deep and 10 km to 40 km across. Water in the channels flowed toward the bottom of the image.

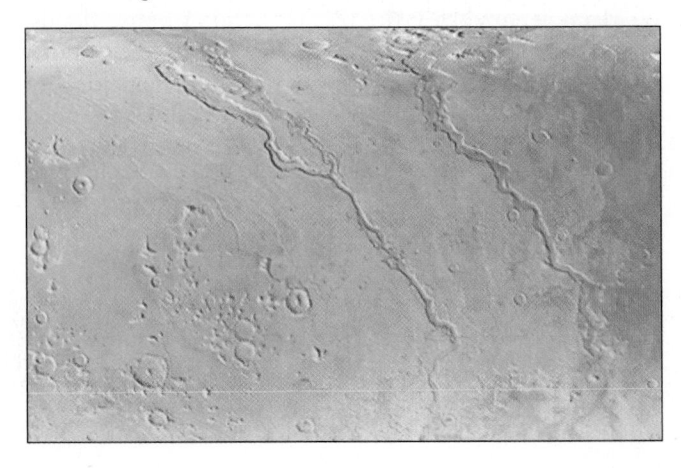

FIGURE 11.18
"Islands" Located Where Ares Vallis Emerges onto Chryse Planitia
The islands were formed when water flowing from the south (lower left) was diverted around craters. The large island at bottom center is about 15 km across and 40 km long.

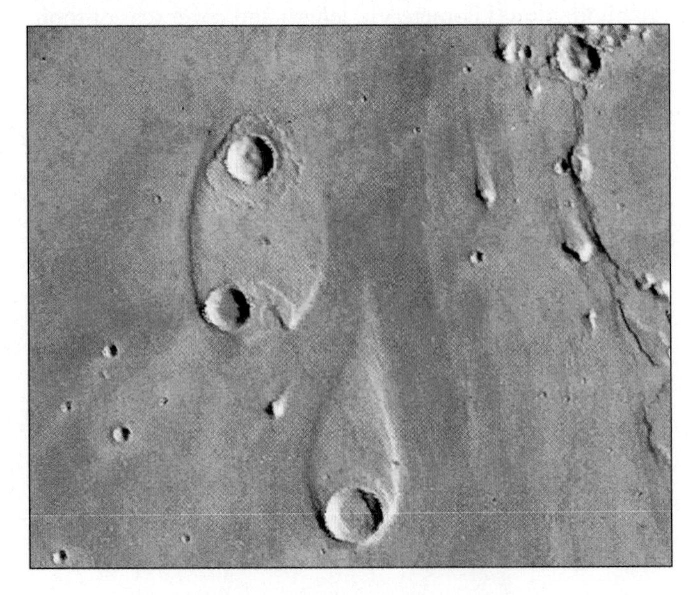

of Chryse alone would have been enough to cover Mars to a depth of 50 m!

There is no widely accepted explanation of why the water that cut an outflow channel collected so suddenly. Two possible mechanisms are the volcanic melting of underground ice and the eruption of water under high pressure from underground springs. Whatever the mechanism, it must have operated over a very long period of time, because the outflow channels show a great range of ages. Some of the outflow channels in the cratered uplands may be as much as 3.9 billion years old. Others, found on sparsely cratered lava flows, may be less than 1 billion years old. It is possible that outflow channels could be produced even in the present Martian climate if enough underground water were suddenly released. A sudden surge of water could produce considerable erosion before the water froze or evaporated. Before *Mars Global Surveyor* and *Mars Reconnaissance Orbiter* returned high resolution images of Mars, the youngest Martian channels that had been discovered were hundreds of millions to billions of years old. *Mars Global Surveyor* and *Mars Reconnaissance Orbiter,* however, imaged some small channels, like the ones shown in Figure 11.19, that may have formed as recently as a few years ago. The channels, or gullies, originate in layers near the tops of the rims of craters and flow downward into the crater for a few hundred meters. Many of the gullies end in debris fields. The gullies may have formed when subsurface heating melted local pockets of ice. The water produced by the melting was prevented from reaching the inner slopes of the crater, however, by blocks of ice. When enough pressure built up, the ice dams burst, allowing the pent up water to rush out and carve gullies. Shortly after it emerged, the water evaporated.

Some of the gullies are very sharp in appearance and do not seem to have been filled in by the dust that blows everywhere on Mars. This suggests that the minifloods that formed such gullies may have taken place only a few years ago. Repeated images of Mars show that some gullies have changed in appearance, possibly as a result of new flows of water. Figure 11.20 show *Mars Global Surveyor* images of the same scene taken 4 years apart. The later image shows a fresh deposit inside an existing gully, showing that water may still reach the surface of Mars and form or modify small channels.

The surface of Mars shows two kinds of channels, both almost certainly cut by running water. Runoff channels look like river systems and were cut by the collection of underground water or rainfall. Outflow channels were cut by catastrophic floods produced by the sudden release of large amounts of underground water. Some small outflow channels, or gullies, may be very young.

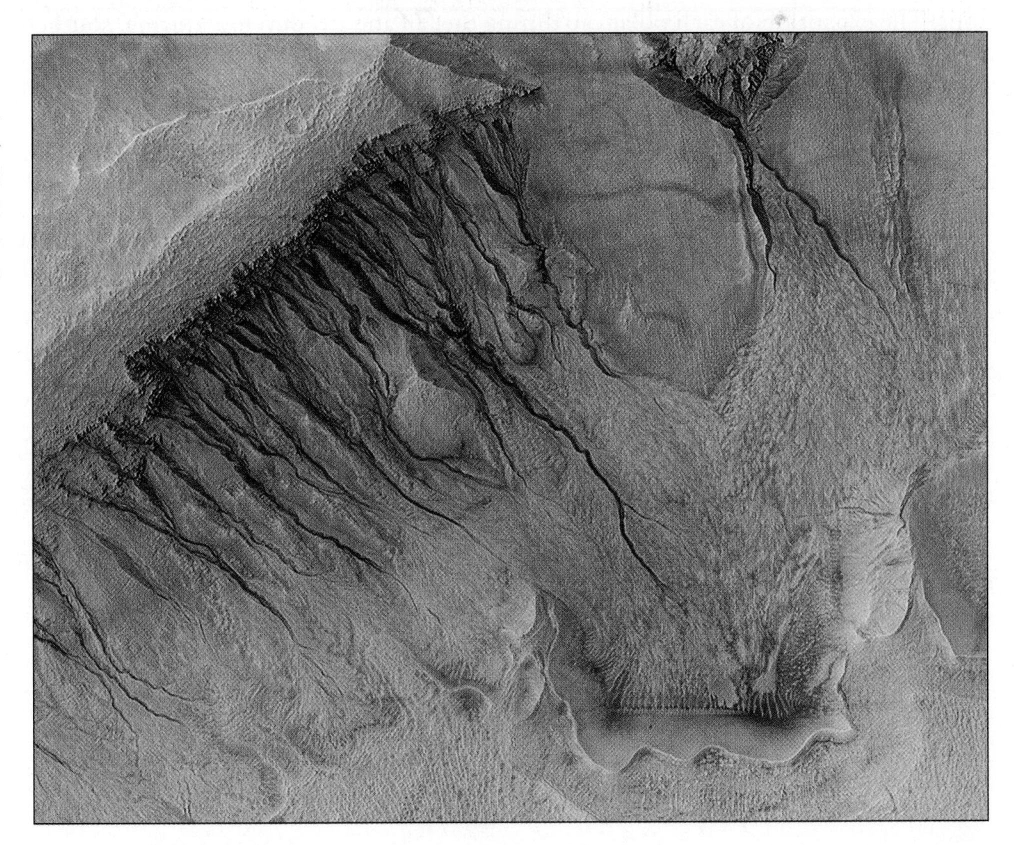

FIGURE 11.19
Young Martian Channels
The numerous small channels shown in this *Mars Reconnaissance Orbiter* image originated near the top of the crater wall and stretch downhill for hundreds of meters. The scene is about 5 km across.

FIGURE 11.20
Possible Evidence of a Recent Flow of Water
Both the left and right images show the same small Martian crater. Between 2001 and 2005 water may have flowed in the marked channel and left a bright deposit. The fresh deposit is about 500 m long.

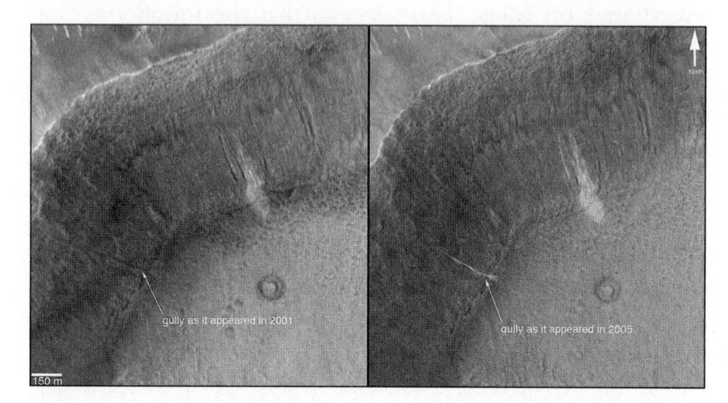

11.3 THE POLAR REGIONS

The nature of the Martian surface changes dramatically at latitudes greater than about 70°, where the polar caps have helped shape the surface.

Seasonal Caps

Both polar caps grow and shrink with the Martian seasons. The expansion of each polar cap during the fall and winter occurs because atmospheric carbon dioxide forms a frost of dry ice wherever the surface temperature drops to about 150 K. Because of Mars's orbital eccentricity the seasons in the southern hemisphere are more extreme than in the northern hemisphere. The eccentricity of Mars's orbit is much larger than that of the Earth. The Earth is only about 3% nearer the Sun at perihelion than at aphelion. For Mars the difference is about 20%. This means that the amount of sunlight reaching the surface of Mars is much greater at perihelion than at aphelion. As Figure 11.21 shows, Mars's southern hemisphere is tipped toward the Sun at perihelion and away from the Sun at aphelion. Thus, in the southern hemisphere of Mars, summer occurs when Mars is much nearer the Sun than in winter, so the temperature difference between summer and winter is large. In Mars's northern hemisphere, on the other hand, winter occurs when Mars is near the Sun so northern seasonal changes are relatively mild. During the long, cold southern winters, the seasonal polar cap extends to about 55° S latitude. In the milder northern winter, the northern cap extends to about 65° N latitude. Mars's northern polar cap in northern summer is shown in Figure 11.22. Near the poles the seasonal cap may be a meter or more in thickness. However, near the edge of the seasonal cap only a very thin frost forms.

Residual Caps

The permanent ice cap in the southern hemisphere is about 350 km across and seems to be made primarily of carbon dioxide, although water ice may also be present. The northern residual cap has a diameter of about 1000 km and is made primarily of water ice. The southern residual cap contains carbon dioxide while the northern does not because the summer temperatures in the northern polar regions aren't cold enough to keep carbon dioxide frozen. How can the southern residual cap be colder than the northern cap when the southern summers are warmer than the northern summers? The answer may lie in the planetwide dust storms that often develop during southern summer when the northern polar cap is forming and growing. Dust is captured in the growing northern cap, making it dirtier and darker than the southern cap. Because it is darker, the northern cap absorbs sunlight better and is warmer than the southern cap.

Layered Deposits

Within about 10° of each pole, the surface of Mars is covered by a thick layer of sediment. The sedimentary layers lie on top of heavily cratered terrain in the southern

FIGURE 11.21
The Effect of Mars's Orbital Eccentricity on the Severity of the Seasons
Mars is nearest the Sun in southern summer and farthest from the Sun in southern winter. Thus, the southern seasonal variations are greater than those in the northern hemisphere, which is nearest the Sun in winter and farthest from the Sun in summer.

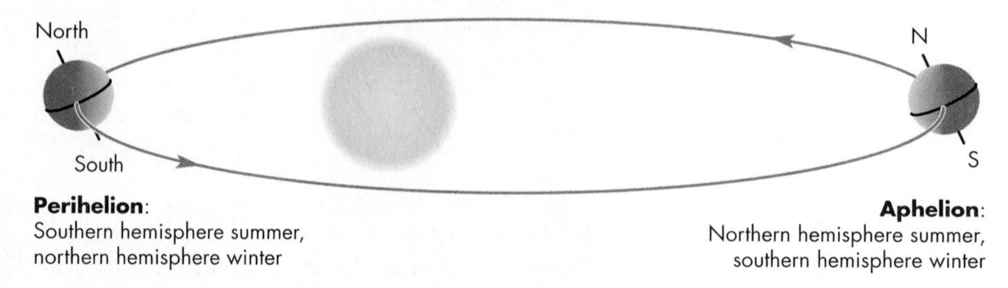

FIGURE 11.22
Mars's Northern Polar Cap During Northern Summer
The permanent polar cap is made of ice and dirt and is about 1000 km across. During northern winter the cap collects a thin pinkish layer of carbon dioxide frost. Sand dunes form a dark band that encircles the polar cap.

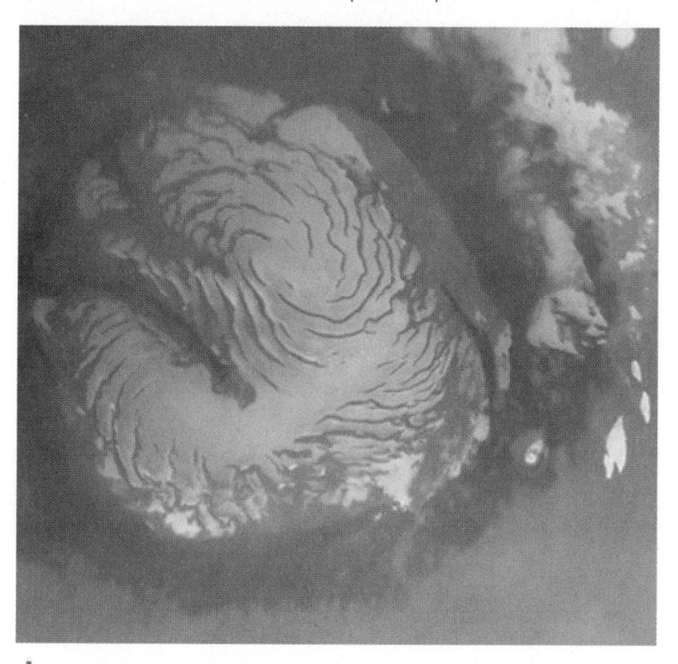

Each pole of Mars has a permanent cap. The larger northern cap is made mostly of water ice while the southern cap contains mostly frozen carbon dioxide. During winter the polar caps grow as thin layers of dry ice form about the permanent cap.

hemisphere and the volcanic plains of the northern hemisphere. The sedimentary layers are most easily seen on ice- and frost-free slopes. Individual layers of sediment range in thickness from about 50 m down to 10 m. The total thickness of the deposits is several kilometers. The deposits are essentially free of impact craters so they must be very young.

The presence of sedimentary layers tens of meters thick in the Martian polar regions is probably related to periodic variations in the orientation of Mars's polar axis. Just as the Earth's spin axis precesses, so does that of Mars, but with a period of 51,000 years rather than the Earth's 26,000-year period. In about 25,000 years the northern hemisphere of Mars will be tilted toward the Sun at perihelion, and the roles of the northern and southern polar caps will be reversed. The polar deposits are accumulations of both ice and dust. The layering may reflect variations in the relative amount of dust accumulated with the ice. A layer of sediment is probably now building at the north pole of Mars because it is the north polar cap that grows while the global dust storms are taking place. In 25,000 years it will be the southern cap that builds while the dust storms lift dust into the atmosphere. The length of Mars's precession period is consistent with the time it would take to build up layers as thick as the layers of deposits are observed to be. We can estimate how rapidly individual layers could be built up by calculating how much Martian dust is raised into the atmosphere during a major dust storm. If most of that dust were deposited at the polar caps, the sediments could grow by about 1 mm per year. This means that it would take tens of thousands of years to build up layers tens of meters thick.

The polar regions of Mars are covered by thick layers of sediments deposited as windblown dust from other parts of Mars. Individual layers of sediment probably are built up over tens of thousands of years. The layering is probably related to the precession of Mars's polar axis.

11.4 THE EXPLORATION OF MARS'S SURFACE

The *Mariner, Viking, Mars Global Surveyor, Mars Express,* and *Mars Reconnaissance Orbiter* images of Mars have produced an abundance of information about the processes that have shaped the Martian surface. However, those views were remote ones. To complement the orbiter images, six spacecraft—two *Viking* landers, *Pathfinder, Spirit, Opportunity,* and *Phoenix*—have successfully landed on Mars and returned pictures and measurements from the Martian surface.

The *Viking* Landers

In the summer of 1976, landing modules separated from each *Viking* orbiter and landed on Mars. The landing sites were sparsely cratered volcanic plains in Mars's northern hemisphere. *Viking 1* landed about 3000 km east of the Tharsis bulge in Chryse Planitia at 22° north latitude. *Viking 2* landed in Utopia Planitia at 48° north latitude—almost half a world away in longitude from the *Viking 1* landing site. The landing sites were chosen with the safety of the landers in mind. It might have been more exciting to touch down in Valles Marineris, on top of Olympus Mons, or in one of the Martian channels. However, steep slopes and debris-covered areas had to be avoided to reduce the chance that the landers would tip over on landing.

To survey the landing sites, each *Viking* lander carried a weather station, seismometer, chemical analysis equipment, cameras, and a biology package to look for signs of Martian life. Each lander was equipped with a radio dish to receive instructions from the Earth and to transmit data back to Earth.

***Viking* Views of the Surface of Mars** Within minutes of the time it touched down, Lander 1 took its first picture of the Martian surface. This picture, a view of the ground beneath the lander, is shown in Figure 11.23. The image shows a sandy surface littered with rocks as large as 10 cm across. Later images, such as Figure 11.24 shows the landscape around Lander 1. The region of Chryse Planitia doesn't look particularly alien. In fact, it strongly resembles rocky deserts of the Earth. The gently rolling landscape is littered with rocks of all sizes. Most of the rocks probably were ejected from impact craters near the landing site. The rocks have rough surfaces, pitted by the impacts of windblown sand and dust. Drifts of sand-sized particles can be seen in the spaces between

FIGURE 11.23
The First *Viking* Lander Image of the Surface of Mars
Rocks as large as about 10 cm are scattered in fine sand or dust.

FIGURE 11.24
Part of the Landscape near Lander 1
The many rocks in the scene were probably ejected from nearby impact craters. Drifts of sand can also be seen.

FIGURE 11.25
The Landscape near Lander 2 in Summer and Winter
The winter scene (bottom) shows a thin frost of water ice.

the larger rocks. One unusual feature of the *Viking* images from Mars's surface is the pinkish color of the sky (see Figures 11.25 and 11.26). The color is due to reddish dust suspended in the Martian atmosphere. Without the dust, the sky of Mars would be nearly black because there is so little atmospheric gas.

The region around the Lander 2 site looks much like Chryse Planitia, but rockier and less hilly. The greater number of rocks around the Lander 2 site is probably due to its proximity to Mie, an impact crater 100 km in diameter and about 180 km away. No drifts of sand can be seen near the lander, but several channels, perhaps a meter wide and 10 cm deep, wind through the scene. The origin of these small troughs and how they are related to various features seen in the orbiter pictures is not clear. Due to the smoothness of the underlying surface near Lander 2, some geologists have proposed that the rocks sit on a broad layer of sediments like those within Mars's polar regions. The only major change seen at either *Viking* landing site was the appearance of frost around Lander 2 during the Martian winter (Figure 11.25). For about 100 days, dust particles covered by water ice sank to the surface around the lander, producing a thin layer of water frost.

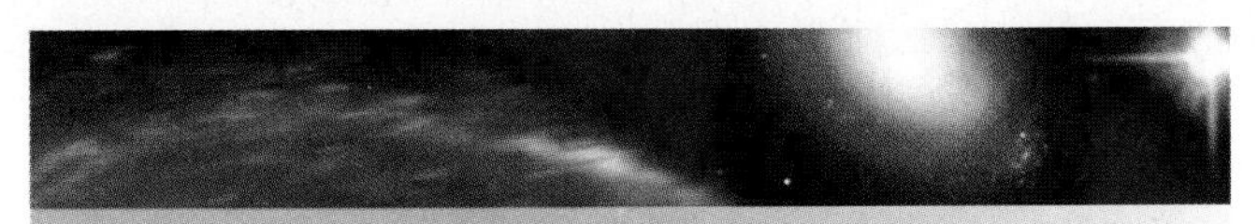

Pictures from the *Viking* landers show a rolling landscape littered with rocks ejected from nearby craters. The surface rocks have been eroded by windblown dust and sand.

Soil Analysis The *Viking* landers carried several experiments for determining the chemical composition (but not the mineralogical composition) of the fine debris at the landing sites. The soil had nearly the same chemical composition at both sites, apparently because global dust storms lift loose surface material from all parts of the planet, mix them together, and redeposit the mixture. The soil is rich in iron and silicon. The highly oxidized iron compounds give Mars its reddish, rusty color. Each lander was also equipped with a scoop mounted on a maneuverable arm used to bring soil samples into the lander for analysis. The top picture in Figure 11.25 shows a series of trenches about 15 cm deep and 30 cm wide dug by the scoop.

Pathfinder

Pathfinder was designed to be able to land and right itself in terrain the *Viking* landers could not have survived. On July 4, 1997, *Pathfinder* landed in Ares Vallis, the site of catastrophic Martian floods 1 to 3 billion years ago. A parachute and rockets brought *Pathfinder* within 20 m of the surface. Wrapped in airbags, it then fell, striking the surface at a speed of 14 m/s (30 miles per hour). It bounced at least a dozen times before coming to rest.

A panorama of the scene around the *Pathfinder* landing site is shown in Figure 11.26. The landscape shows abundant evidence that floods once roared through Ares Vallis. Rocks all around the landing site are tilted and stacked toward the northeast, the direction in which water rushed through Ares Vallis. Many of the rocks are rounded and appear to have been shaped by running water. In fact, the floods may have carried rocks and boulders to the landing site from hundreds of kilometers away in the ancient southern highlands. A pair of hills (the "Twin Peaks") about 1 km from the landing site appear to have gullies and terraces cut by repeated floods. Dune fields and sandbars lie among the rocky debris dropped by floodwaters. The floodwaters in Ares Vallis must have been hundreds of meters deep and traveled at about 70 m/s (150 miles per hour). The rate at which water flowed through Ares Vallis at the height of the floods was about 1000 times the rate of flow of the Amazon River.

FIGURE 11.26
The *Pathfinder* Landing Site
This 360° panorama shows the view around the spot in Ares Vallis where *Pathfinder* landed in July 1997. *Pathfinder* appears in the foreground. The *Sojourner* rover can be seen inspecting the rock nicknamed "Yogi" about 4.5 m (15 feet) from *Pathfinder*. Many of the rocks near *Pathfinder* were carried there by ancient Martian floods. Other rocks were ejected by the impacts that formed craters near or in Ares Vallis.

FIGURE 11.27
Endurance Crater
This panoramic view covers 180° from a viewpoint on the rim of Endurance Crater, an impact feature located about 1 km from the *Opportunity* landing site. Endurance Crater is about 130 m across and 20 m deep. Part of *Opportunity* can be seen in the lower right corner of the image.

Sojourner A more detailed inspection of the landing site was carried out by *Sojourner,* a six-wheeled rover that rolled down a ramp from *Pathfinder* to the surface about a day after the landing. *Sojourner* can be seen examining a rock nicknamed "Yogi" in Figure 11.26. Although *Sojourner* crept across the surface of Mars at a maximum speed of less than a meter per minute, it was able to cover quite a bit of ground before *Pathfinder* stopped sending back signals on September 27, 1997.

In addition to its cameras, *Sojourner* carried an instrument called the Alpha Proton X-ray Spectrometer (APXS). The APXS was able to measure the abundances of the chemical elements in a rock or soil sample by recording the energies of protons and X rays given off by different kinds of atoms. Astronomers were surprised that some of the rocks measured by the APXS resemble Earth rocks that have solidified and then been remelted by volcanic heat. When this happens, low-density rocks like the ones measured by the APXS form at the top of the melted material as it solidifies. This suggests that early in the history of Mars large regions melted thoroughly enough to separate into heavy and light layers.

Spirit and *Opportunity*

In January 2004, the *Spirit* and *Opportunity* rovers bounced to a landing and began to explore the surface of Mars. Unlike *Sojourner,* which traveled only a few tens of meters away from the *Pathfinder* lander, *Spirit* and *Opportunity* both traveled many kilometers from their landers. The rovers were equipped with panoramic cameras to send back high resolution images of the landscape, microimagers to obtain close-ups of soil and rocks, instrumentation to perform chemical and mineral analyses, and drills to bore into rocks to expose their interiors. Both rovers returned spectacular images such as the panorama of Endurance Crater shown in Figure 11.27. Figure 11.28, a remarkably detailed image obtained by *Mars Reconnaissance Orbiter,* shows *Opportunity* on the edge of Victoria Crater. The image, which resolves features as small as 0.8 m across, also shows tracks produced by the rover.

The landing sites for *Spirit* and *Opportunity* were selected in the hope of obtaining unambiguous evidence that parts of Mars's surface were once covered with water. *Spirit* landed in Gusev Crater, into which runs an 850-km-long sinuous channel. Gusev appears to have once been a lake bed. *Opportunity* was sent to Meridiani Planum, one of the few spots on Mars that is covered by the mineral hematite. Hematite is an iron compound that usually forms in the presence of liquid water. The most compelling evidence for water came from *Opportunity.* Figure 11.29 is a close-up image that shows the surface is covered with hematite spherules a few mm across. The spherules were formed when water filtered through cracks in surface rocks. Figure 11.30 shows a layered outcrop that was visited by *Opportunity.*

FIGURE 11.28
***Opportunity* at Victoria Crater**
This image obtained by *Mars Reconnaissance Orbiter* shows the *Opportunity* rover as it investigates Victoria Crater. *Opportunity* traveled for over two and a half years to reach Victoria Crater. The image resolves details as small as 0.8 meters (3 feet) across, including *Opportunity,* the shadow of its camera mast, and tracks made by the rover. The inset in the upper left shows the entire image of Victoria Crater. Victoria Crater is about 0.75 km in diameter.

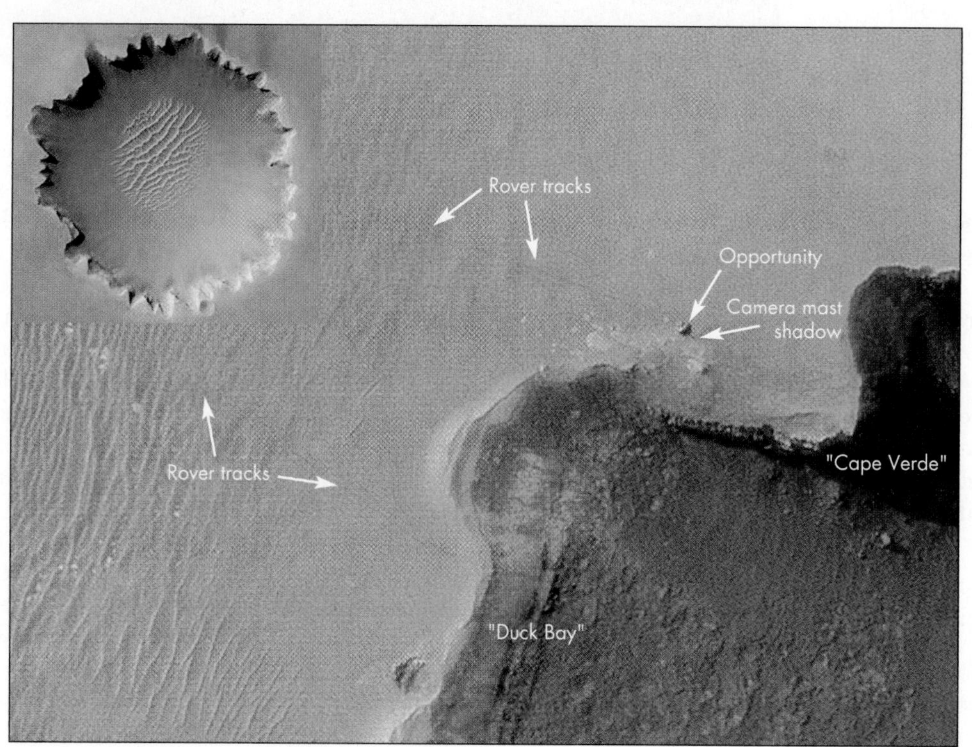

FIGURE 11.29
A Close-up of the Surface of Mars
This image shows a part of the Martian surface about 20 cm across. The reddish bedrock is sprinkled with hematite spherules a few mm in diameter. Hematite spherules were also found embedded in rocks at the *Opportunity* landing site. The large round feature to the left of center is a shallow hole bored by *Opportunity's* drill in order to measure the composition of the rock.

FIGURE 11.30
The *Opportunity* Landing Site
This view from *Opportunity* shows the lander that brought it to the Martian surface in the upper-right portion of the scene. *Opportunity* landed in a shallow crater. The impact that formed the crater exposed a small outcrop of layered bedrock that can be seen on the left side of the image. *Opportunity* itself fills the bottom of the image.

As shown in Figure 11.31, the layers in the outcrop are fossilized ripples that resemble those found on Earth on beaches and in riverbeds. The layers form when water gently flows across a lake or sea bottom. Other evidence found by *Opportunity* includes outcrops with holes that formed when crystals dissolved away, sulfate salts that precipitate out of seawater, and minerals that need liquid water to form. The evidence conclusively shows that water at least several inches deep (and probably much deeper) covered Meridiani Planum for a considerable period of time.

Phoenix

In May 2008, *Phoenix* touched down at 68° north latitude near Mars's north polar cap. Figure 11.32 shows *Phoenix* and its parachute descending through the Martian atmosphere. The mission of *Phoenix* was to search for ice beneath the surface of Mars and to investigate whether the environment at the interface between the ice and the soil had ever been suitable for life. Figure 11.33 shows the terrain where *Phoenix* landed. The landscape is flat and littered with small rocks. The most notable feature of the landscape is a pattern of polygons separated by shallow troughs. Polygonal terrain is also found in regions of permafrost on Earth, where it is caused by the seasonal expansion and contraction of subsurface ice.

Scientists expected that there might be many centimeters of soil covering subsurface ice. It turned out, however, that the covering of soil was very thin at the spot where *Phoenix* landed. Figure 11.34 shows a patch of ice exposed beneath the lander when its landing

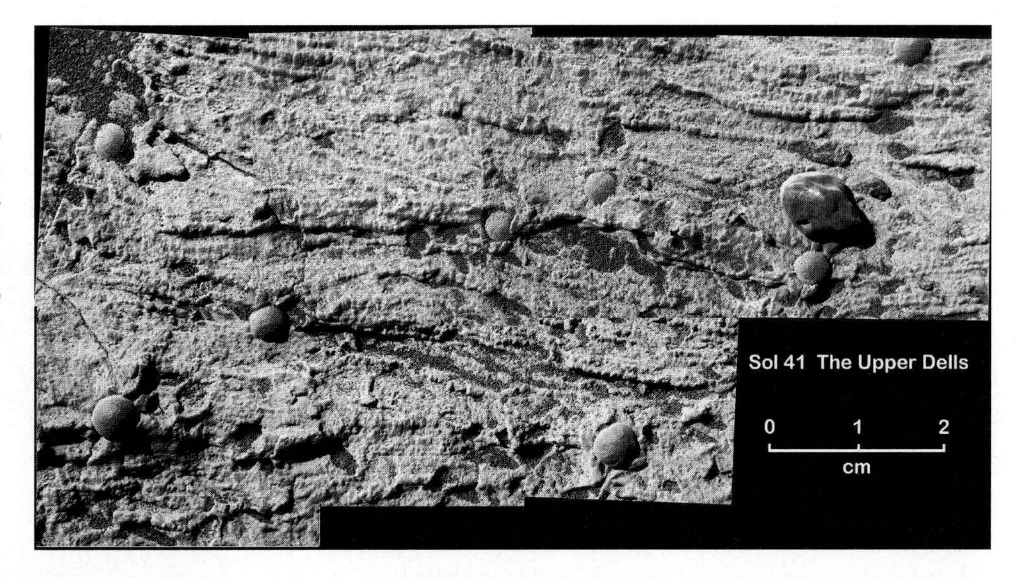

FIGURE 11.31
Evidence for Water on Ancient Mars
The layered rock in this image is part of the outcrop shown in Figure 11.28. The wavy layers resemble those formed on Earth in shallow, gently flowing water. The image shows a part of the rock about 15 cm across.

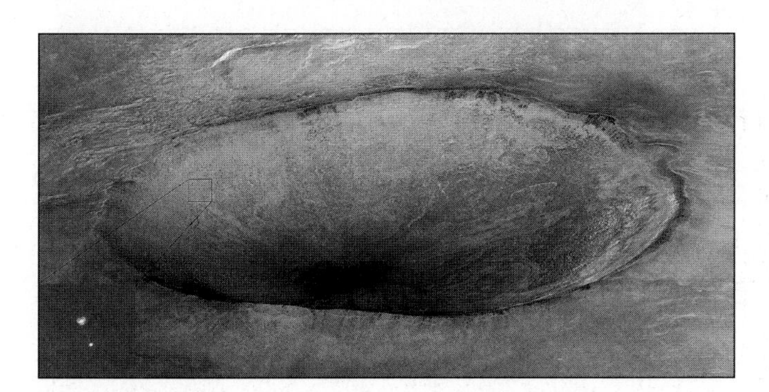

FIGURE 11.32
***Phoenix* Descending to the Martian Surface**
This *Mars Reconnaissance Orbiter* image shows *Phoenix* descending through the Martian atmosphere. The inset clearly shows both *Phoenix* and its parachute. The crater Heimdall, about 10 km across, is actually about 20 km farther away than *Phoenix*.

FIGURE 11.33
***Phoenix* on Mars**
The terrain on which *Phoenix* landed is flat and strewn with rocks. Note the polygonal pattern of the landscape. One of *Phoenix*'s solar panels and its robotic arm are seen in the foreground.

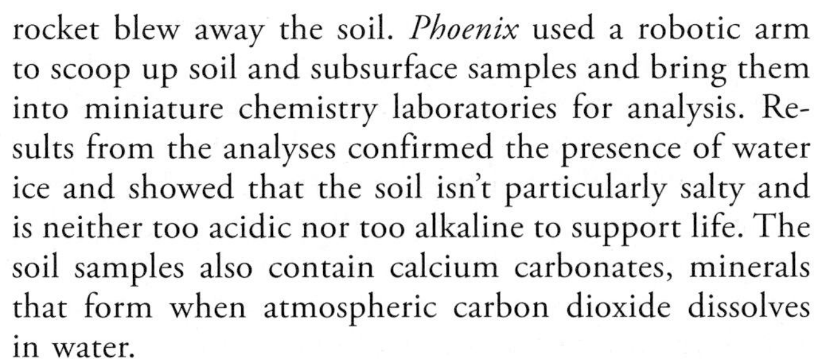

rocket blew away the soil. *Phoenix* used a robotic arm to scoop up soil and subsurface samples and bring them into miniature chemistry laboratories for analysis. Results from the analyses confirmed the presence of water ice and showed that the soil isn't particularly salty and is neither too acidic nor too alkaline to support life. The soil samples also contain calcium carbonates, minerals that form when atmospheric carbon dioxide dissolves in water.

As Martian winter approached, the amount of sunlight reaching the solar panels on *Phoenix* fell until in early November 2008, the last communication from *Phoenix* was received. Even so, *Phoenix* exceeded its planned lifetime by 2 months.

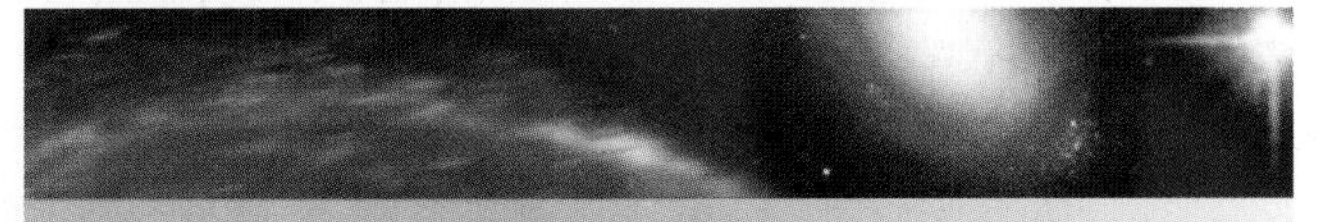

Pathfinder landed in Ares Vallis in July 1997. The images it returned showed evidence of ancient, catastrophic floods in Ares Vallis. *Sojourner,* a six-wheeled rover carried by *Pathfinder,* obtained close-up images of Martian rocks and carried out chemical analyses of Martian rocks and soil. *Spirit* and *Opportunity* traveled far from their landing sites and sent back powerful evidence of bodies of water on ancient Mars. *Phoenix* confirmed the presence of subsurface ice in Mars's polar regions.

FIGURE 11.34
Ice Beneath *Phoenix*
The large whitish object just in front of the landing pad is ice that was exposed when *Phoenix*'s landing rockets blew away a thin layer of Martian soil.

FIGURE 11.35
A Graph of Temperature Versus Altitude in the Martian Atmosphere
There is almost no ozone in Mars's atmosphere, so there is no ozone layer to absorb solar radiation and produce a warm stratopause such as occurs in the Earth's atmosphere.

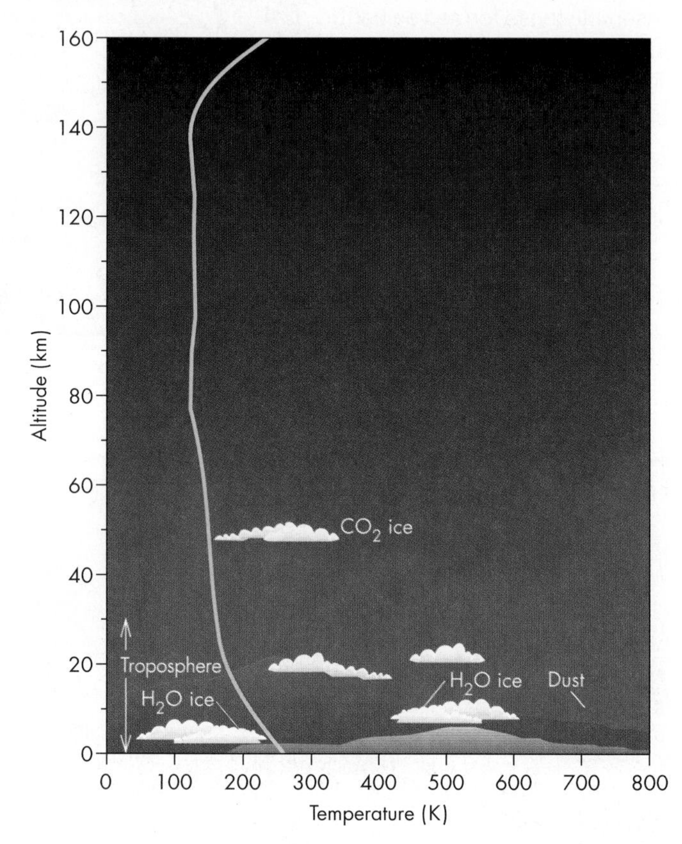

11.5 THE ATMOSPHERE OF MARS

The surface atmospheric pressure of Mars averages about 650 N/m^2 or 0.1 pounds per square inch. This is less than 1% of the Earth's surface pressure and equivalent to the atmospheric pressure 35 km (21 miles) above the ground on the Earth. Although Mars's atmosphere is much cooler and thinner than that of Venus, they are quite similar in chemical makeup. Carbon dioxide makes up about 95% of the Martian atmosphere with most of the rest divided between nitrogen and argon. The small amount of water vapor present varies with the season and time of day.

Because the atmosphere is so thin, it absorbs very little sunlight. Nearly all of the atmospheric heating occurs from contact with the ground where solar energy is absorbed. The relatively warm stratopause of the Earth's atmosphere 50 km above the surface is caused by the absorption of solar ultraviolet radiation by ozone molecules. Because ozone is nearly absent from the Martian atmosphere there is no corresponding warm layer. The graph in Figure 11.35 shows that atmospheric temperature falls more or less steadily with height. The atmosphere is unable to insulate the Martian surface, which shows large daily and seasonal temperature variations. Daily fluctuations of 60 K (100° F) are common. On a mild summer day the Martian temperature may reach the melting point of water (273 K), but a chilly, winter night is about 100 K colder.

Weather on Mars

The *Viking* and *Pathfinder* landers carried instruments to measure atmospheric temperature, pressure, wind speed, and wind direction. Martian weather data were returned for two Martian years by Lander 2 and for more than three Martian years by Lander 1, so we now have information on daily and seasonal weather cycles as well as year-to-year variations. The weather station of *Pathfinder* was able to record data much more often than the weather stations on the *Viking* landers. *Pathfinder* also could measure temperature and winds at several different heights above the ground. The *Pathfinder* measurements show that atmospheric conditions on Mars can change with surprising speed. Temperatures can drop 20° C (40° F) in a few minutes. There can also be differences of as much as 8° C (15° F) between the surface temperature and the temperature a meter above the ground. Such large temperature fluctuations show that the thin atmosphere of Mars can be extremely turbulent, with eddies and convection currents rising from the surface when it is heated by sunlight.

Apart from brief fluctuations, Martian weather is very monotonous; each day is nearly the same as the previous one. This is because Mars has little water vapor and no surface water, factors responsible for much of the weather that occurs on the Earth. Light winds

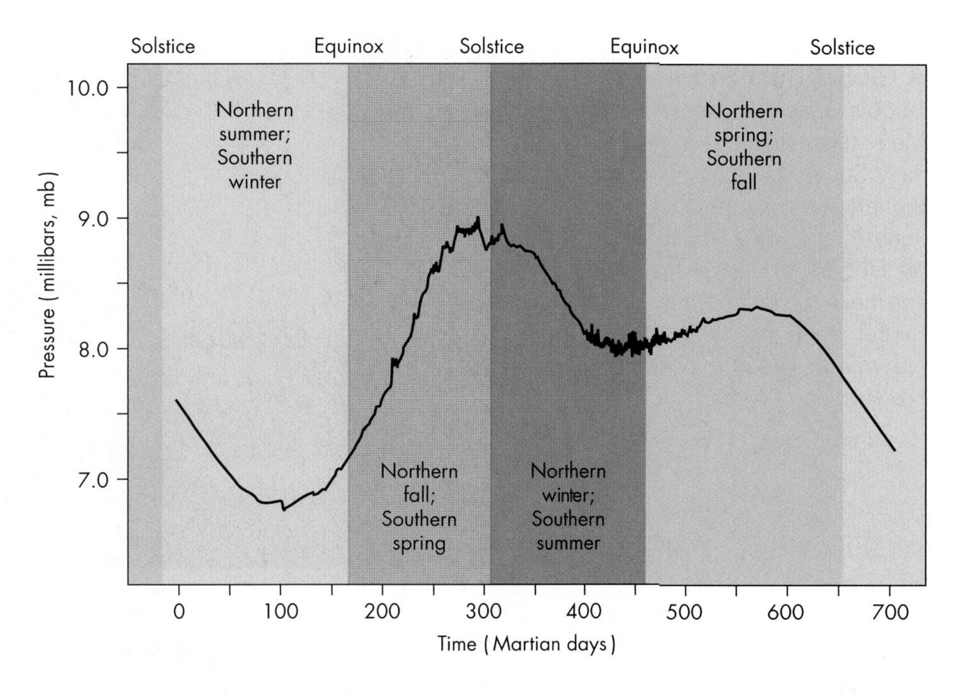

FIGURE 11.36
A Graph of Martian Atmospheric Pressure Versus Time
Atmospheric pressure is lowest in winter and summer, when much of the carbon dioxide shared by the atmosphere and polar caps is frozen in the polar caps. In spring and fall, carbon dioxide is released from the shrinking polar cap, raising atmospheric pressure.

prevailed nearly all the time at both landing sites. A surprising observation from the Martian surface was how much the atmospheric pressure changes throughout a Martian year (Figure 11.36). The changes in pressure occur because both the atmosphere and part of the polar caps of Mars are made of the same substance, carbon dioxide. The smallest atmospheric pressure is found during southern winter, when about 25% of the Martian atmosphere is condensed as dry ice in the southern polar cap. In southern spring the carbon dioxide reenters the atmosphere, raising the pressure. Pressure again falls, but not by so much, during southern summer when the smaller northern polar cap forms. Pressure rises again in southern autumn when the northern cap shrinks. Nothing remotely like this annual cycle takes place on the Earth, where the polar caps are made of frozen water rather than oxygen and nitrogen, the main atmospheric gases.

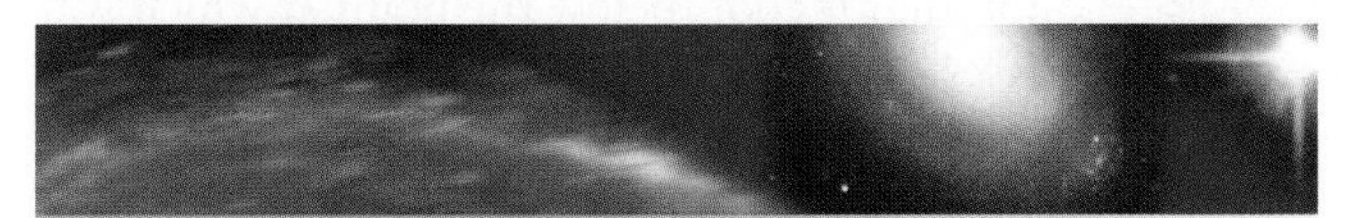

The atmosphere of Mars consists mostly of carbon dioxide. Because the atmosphere is too thin to insulate the surface, there are very large daily and seasonal temperature fluctuations. Atmospheric pressure fluctuates with the seasons as carbon dioxide is trapped in and then released from the Martian polar caps.

Atmospheric Dust

The major atmospheric events on Mars are great dust storms that fill the atmosphere with dust for weeks at a time during southern summer. Some dust storms are so large they cover Mars nearly from pole to pole as shown in Figure 11.37. Smaller, more localized dust storms occur frequently, keeping some dust in the atmosphere all the time.

Mars is closest to the Sun in late southern spring and early summer, so the greatest solar heating takes place at the beginning of summer in the south. The heating is great enough to cause strong atmospheric convection that lifts dust into the air. The lifted dust absorbs sunlight, causing more atmospheric heating and the elevation of more dust into the atmosphere. As the cycle continues a dust storm rapidly develops. Because the atmosphere is so thin, the wind speed required to lift dust from the ground is about 180 km/h, much larger than for the Earth. Such strong winds quickly spread the dust throughout most of Mars's atmosphere. (Lander 1 measured winds of nearly 100 km/h at the arrival of a dust storm.)

During an intense dust storm the Sun's disk dims by more than 99%. Shadows become much lighter as the light from the sky becomes brighter because there is more dust to scatter sunlight. The absorption of solar energy by the dust heats the atmosphere and increases atmospheric pressure during the daytime. The dust begins to settle after several weeks, and eventually, after many months, the amount of dust in the atmosphere returns to normal levels and the normal weather pattern resumes.

Dust storms often occur in southern summer, but not always. During the first year after the *Viking* landings, there

FIGURE 11.37
A Global Dust Storm
Hubble Space Telescope pictures of Mars taken in late June 2001 and early September 2001 are shown on the left and right, respectively. In the June image, small dust storms can be seen near Hellas in the lower right and near the northern polar cap. By September, the dust storms had grown and spread to obscure the entire surface of Mars.

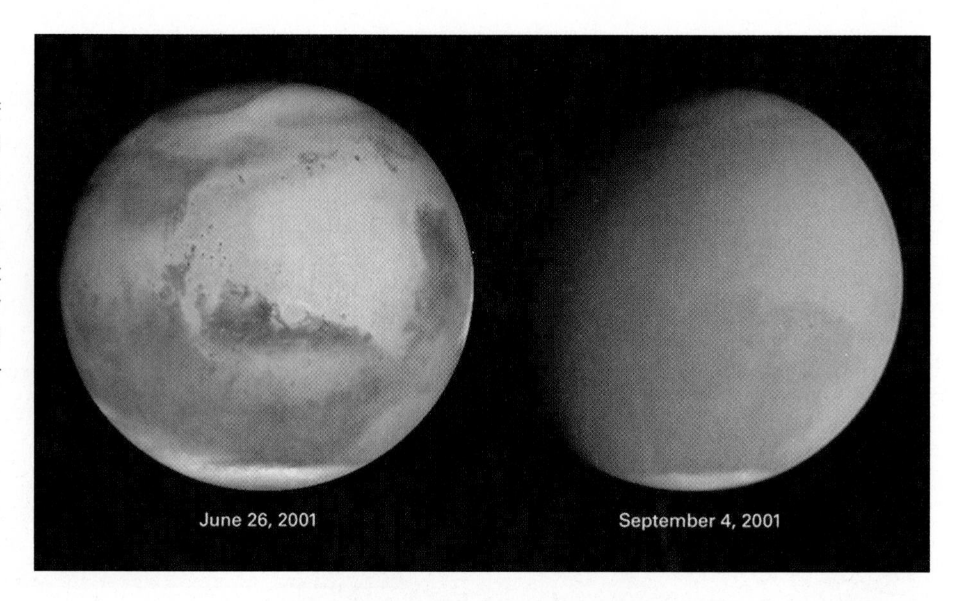

were two global dust storms. For the next two Martian years no large dust storms took place. In the fourth year the most intense storm observed by the *Viking* landers occurred. Why global dust storms happen during some Martian years and not in others is a mystery, as is the exact mechanism that begins the dust storm.

The frequent, intense dust storms that occurred after the *Viking* probes landed left an unusually great amount of dust suspended in Mars's atmosphere. The dust produced unusually warm northern summers and the pinkish color of the Martian sky. A visitor to Mars would usually experience summers 20 to 30 K cooler than the landers reported. The visitor would also see a dark blue, almost violet, sky.

Dust and wind are responsible for the variable dark markings that so fascinated Mars observers a century ago. Many of the dark markings on Mars are due to layers of dust lying atop the soil or rock. Wind direction and speed change with the Martian seasons, blowing off the dust and then blowing it back on at a different time of year. This produces the seasonal pattern of dark markings that Lowell and others wrongly interpreted as evidence for the annual growth and decay of vegetation.

Only very small dust particles can be carried for long distances by the Martian winds. Larger particles can only be picked up briefly and then dropped. This causes the larger particles to skip across the surface. The large grains remain close to the ground, so they are very effective in eroding the surface, producing new dust particles while they themselves are broken into finer and finer dust. The prevailing winds of Mars apparently have caused the larger dust particles to migrate to the polar regions. Large fields of sand dunes practically encircle the northern polar cap. The smaller dust particles settle to the surface in sedimentary layers. This is particularly evident at the poles, but also at other places where the surface layers are exposed.

Frequent dust storms keep some dust suspended in Mars's atmosphere at all times. Very large dust storms sometimes occur in southern hemisphere spring and summer and grow to cover the entire planet. The seasonal changes in surface markings are caused by blowing dust. Large dust particles are blown across the ground by Martian winds and are very effective at eroding surface rocks.

Clouds

The atmosphere is always hazy with dust, but individual clouds are relatively rare on Mars. At a given time only about 1% of the surface is cloud covered. During the winter the northern latitudes of Mars are obscured by a "polar hood," which is probably a combination of water ice and dry ice particles. Figure 11.38 shows another type of cloud frequently seen on Mars on the slopes of the large volcanos. Martian air flowing up the flanks of a volcano grows cooler. When this happens in late morning the air is moist enough that it cools below the point at which water ice condenses, forming clouds. Exactly the same thing happens on the Earth when moist air is forced upward to cross mountain ranges. The windward side of a terrestrial mountain range can be much cloudier and rainier than the leeward side. Sometimes the air in some Martian channels and crater bottoms is humid enough that surface fog forms in the early morning, as shown in Figure 11.39. When sunlight warms the ground a small amount of water vapor is released into the still cold atmosphere where it condenses as a fog. The presence of morning fog is strong evidence that the Martian surface layers contain frozen water.

FIGURE 11.38
Clouds near the Summit of the Volcano Apollinaris Patera
The clouds are thought to be made of water ice, which condenses when Martian air cools as it is pushed up the slopes of the volcano by Martian winds. Apollinaris Patera is about 5 km high.

11.6 THE INTERIOR AND EVOLUTION OF MARS

Mars is intermediate in size and mass between the largest terrestrial planets (Venus and the Earth) and the smallest (the Moon and Mercury). It is about $\frac{1}{10}$ as massive as the Earth, but twice as massive as Mercury. Thus we might expect Mars to have remained internally active longer than Mercury or the Moon, but to have cooled more rapidly than the Earth or Venus. (See Chapter 7 for a discussion of size and cooling time.) Mars's small size, relative to the Earth, means that the material in its interior does not support as much weight as the material in the interior of the Earth. This leads to smaller internal pressures and less compression of Mars's internal material. (See the discussion of pressure and gravity in Chapter 7.)

A clue about the nature of Mars's interior is the reddish color of its surface, which the *Viking* lander experiments showed is due to iron compounds. The present Martian surface originated mostly in volcanic flows, so the composition of the surface rock tells us about the composition of the mantle, where the volcanic magma originated. The iron-richness of the surface suggests that the mantle of Mars also contains a high percentage of iron. Iron-rich rock, when melted, is more fluid than iron-poor rock, which might explain the very fluid nature of the magma that produced the Martian lava flows.

FIGURE 11.39
Morning Fog in the Bottoms of Martian Craters
The picture on the left shows the scene just after sunrise. The picture on the right shows the same scene later in the morning after fog has formed. The bright bottoms in some craters may be the result of a thin fog of water vapor. Apparently sunlight warms the surface, releasing water vapor that freezes in the atmosphere to form a thin fog.

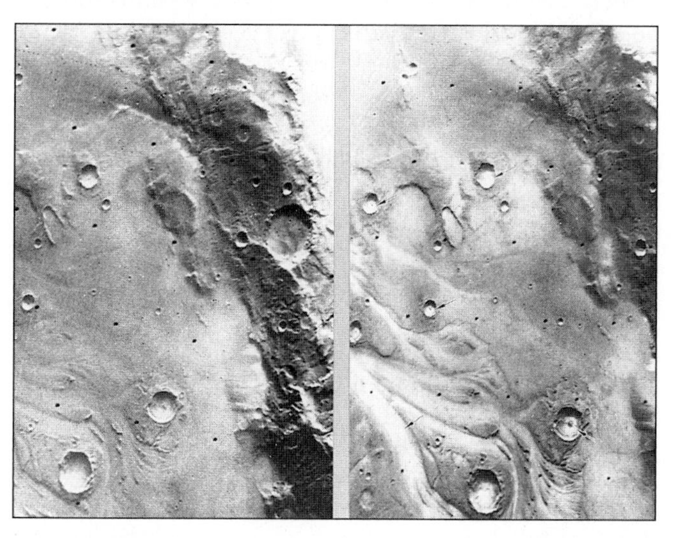

Internal Structure

The pressure at the center of Mars is about 0.4 Mb (1 Mb is equal to about a million atmospheres of pressure). Pressures this large are reached in the Earth at a depth of about 1000 km. Thus, there is little compression in Mars. Its average density is almost 3940 kg/m^3. This is larger than the average density of the Earth's crustal rocks and shows that Mars has a considerable amount of iron or other heavy elements. If Mars's core is made mostly of iron and nickel, then the core extends about 40% to 50% of the way to the surface of the planet. The outer part of Mars's core is liquid. Mars's core may also contain large amounts of sulfur. The iron in Mars's core amounts to about 15% of the mass of the planet. Including iron in the mantle, Mars is about 27% iron. The Earth is about 40% iron, so the amount of iron in Mars continues a general trend of lower iron to rock ratios with increasing distance from the Sun.

Mars's surface materials contain a considerable amount of iron, giving the planet its reddish color. Mars also has iron in its core, which may be primarily metal or may contain iron mixed with sulfur.

Thermal History

The earliest volcanic activity on Mars may have originated in a layer of molten material near the surface. This layer could have been melted by impacts while the deeper interior of Mars remained cool and solid. Radioactive decays gradually heated Mars's interior until iron melted and began to accumulate in a core, providing an additional source of heat as it sank. (For a description of accretional and radioactive heating, see Chapter 7.) Eventually the temperature in the mantle became high enough to melt silicate rock, which reached the surface to produce the smooth plains and led to the development of the large volcanos 2 to 3 billion years ago. As the planet grew warmer its radius increased by 10 km. This expansion may have pulled apart Mars's crust, producing the Valles Marineris rift system.

Mars has already begun to cool. As a result of cooling, volcanic activity has become steadily more restricted since the time the smooth plains were formed. Whether the aesthenosphere has completely disappeared isn't clear. Throughout Mars's history, cooling has occurred mostly through the loss of heat in volcanic flows and volcanos. This is very different from the case of the Earth, where most internal heat reaches the surface at the boundaries of crustal plates.

A better understanding of the thermal structure of Mars's interior could help us resolve an apparent dilemma about Mars's magnetic field. Mars rotates almost as rapidly as the Earth does and seems to have an iron-rich liquid core, yet it has, at best, a very weak global magnetic field. Magnetometers aboard *Mars Global Surveyor* detected localized magnetic fields such as those shown in Figure 11.14. The magnetic field strength in these regions was about 1⁄800 as strong as that of the Earth. Any global magnetic field must be even weaker. It is believed that the local magnetic fields on Mars may come from rocks that cooled when Mars had a relatively strong magnetic field. As they cooled, the rocks were imprinted by the global magnetic field. The local magnetic fields remain long after Mars's global magnetic field has faded away.

It was hoped that seismometers on the *Viking* landers would make it possible to use Marsquakes to probe the interior of Mars and discover the size and state of the core. However, this required two working seismometers. Unfortunately, the seismometer on Lander 2 failed to work, leaving the problem of Mars's core unsolved. The seismometer on Lander 1 measured little seismic activity, showing that Mars has a quiet interior relative to the Earth.

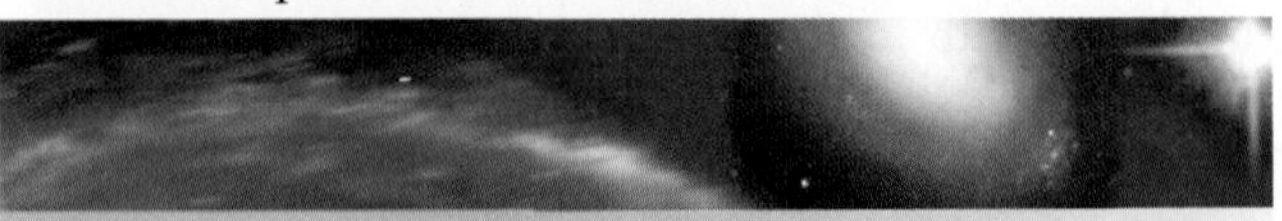

The interior of Mars grew hotter until its metallic core formed and magma reached the surface to produce volcanos and widespread volcanic flows about 2 to 3 billion years ago. During that time Mars expanded, possibly pulling its crust apart to form the Valles Marineris chasms. Since then Mars has cooled and become less volcanically active.

11.7 WATER ON MARS

Where Is Mars's Water?

Although the Martian channels and measurements from Mars's surface provide ample evidence that there was water on Mars in the remote past, the amount of water present in the past and what has happened to it are very controversial. Some of the water on Mars is directly detectable in atmospheric clouds and in the polar caps. However, if all the water in Mars's atmosphere were condensed as ice, it would make an ice cube only about 1 km on a side. (By comparison, the Earth's water could make a cube more than 1000 km on a side.) If the water in Mars's atmosphere were spread uniformly around the planet, it would cover the surface to an average depth of only eight-millionths of a meter.

The amount of water that we can see on Mars is much less than the amount thought to have been released from Mars's interior during volcanic activity. Estimates of the total amount of outgassed water yield quantities of liquid water that would have covered Mars to an average depth of perhaps 100 m. An even larger amount of water, as much as 500 m, is implied by geological evidence such as the large flood features and river valley networks. Thus Mars may have more than 1000 times as much water as we can directly detect. Where is it?

The answer is that it is probably trapped beneath the Martian surface. Some is frozen as ice in the layered terrain and deeper ground ice near the Martian poles. Even more may be permanently frozen (as permafrost) within the upper few kilometers of Mars at latitudes greater than about 40°. Throughout the history of Mars, the ground ice between the equator and 40° N latitude and 40° S latitude has been warmed enough to drive water vapor into the atmosphere. The water was deposited in the polar regions after reaching the atmosphere. Although it may seem unlikely that hundreds of meters of ice could be stored beneath the Martian surface without being detected by the *Viking* and *Pathfinder* landers, Figure 11.40 shows that deposits of permafrost can be equally invisible on Earth.

Strong evidence of the existence of buried ice on Mars was obtained by the *Mars Odyssey* spacecraft. By observing neutrons produced when cosmic rays strike Mars, *Mars Odyssey* could map the distribution of hydrogen atoms, presumably in the form of water molecules, below the surface. The hydrogen content map of Mars, shown in Figure 11.41, shows large concentrations of buried ice in a large region centered on the south pole and within the northern plains. The layer a few centimeters below the surface contains amounts of water ranging from 2% in some equatorial regions to nearly 100% at the poles. Buried ice can also be discovered on Mars using radar. Radio waves transmitted by Mars orbiters can penetrate the rocky surface layers of Mars and then reflect from layers of subsurface ice. *Mars Express* probed Mars's southern polar cap and layered deposits near the northern polar cap using radar and found

FIGURE 11.40
Permafrost in the Earth's Polar Regions
Much of the surface of Mars may conceal a thick layer of permafrost as well.

ice deposits that are as much as 3.5 km thick. *Mars Express* also found extensive ice deposits beneath smooth plains near the southern polar cap and possibly near Mars's equator. The amount of subsurface ice discovered by *Mars Express* could supply enough water to cover Mars to a depth of several tens of meters. Further measurements of subsurface ice deposits are being made with radar on *Mars Reconnaissance Orbiter*, which began to orbit Mars in 2006.

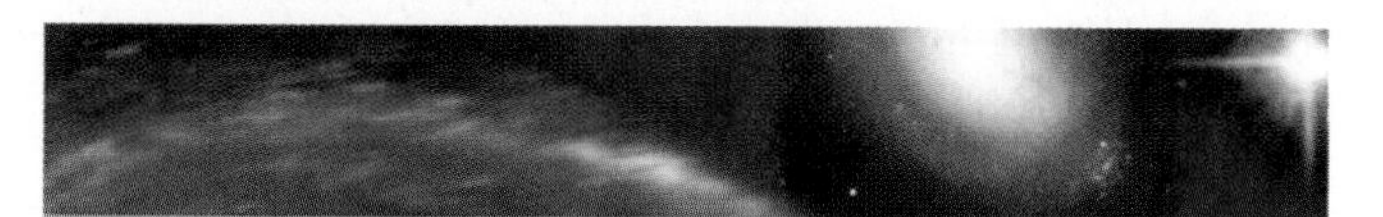

Little water can be detected on Mars today. However, Martian channels show that considerable amounts of water once existed on Mars. That water is probably trapped as ice beneath the Martian surface.

Climate History

It is highly likely that bodies of water existed on Mars in the past. The evidence pointing to bodies of water includes the runoff and outflow channels. Other evidence can be seen in images of Mars, such as Figure 11.42, that show rock layers that may have been formed from sediments in ancient Martian lakes. Layered deposition of windblown sand has also been proposed to account for the layering. The layers, which resemble sedimentary layers deposited in bodies of water on Earth, are each many meters thick and extend for many kilometers. They appear to have been deposited over long periods of time, covering up previously existing terrain. It is also possible that the northern plains of Mars are the dry bottom of an ancient ocean basin. Images have shown what may be the remains of ancient shorelines around the northern plains.

Bodies of water could not have existed on Mars unless the atmosphere was much warmer and thicker than it is today. Accounting for such an atmosphere has been a major problem for astronomers who study Mars. Several scenarios have been proposed.

1. Mars had a thick atmosphere, primarily of carbon dioxide, that produced a strong greenhouse effect. At least several atmospheres of carbon dioxide would have been required, but carbon dioxide condenses into clouds at a pressure of about 1 atmosphere. Also, if a thick atmosphere of carbon dioxide existed, then it should gradually have dissolved in Mars's oceans and lakes and then been deposited as carbonate sediments. This is the mechanism that has limited the amount of carbon dioxide in Earth's atmosphere. Thus, there should be widespread carbonate sediments on Mars today. But only small amounts of carbonates have been discovered.
2. Numerous impacts on Mars by comets and asteroids heated the atmosphere and released subsurface ice into the atmosphere. This led to rainfall, flooding, and the runoff channels. The timing of impacts required for this scenario has yet to be worked out.

FIGURE 11.41
Buried Ice on Mars
This *Mars Odyssey* map shows where subsurface hydrogen is concentrated on Mars. Most of the hydrogen is probably in the form of frozen water. The greatest concentration of ice is indicated in red, the lowest concentration in blue. Notice that the ice concentration is greatest near the poles.

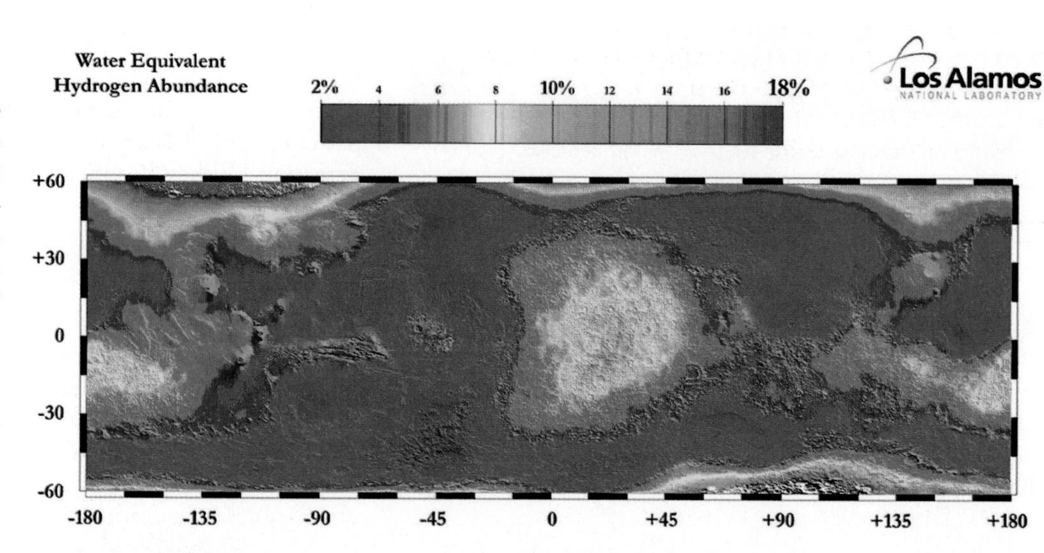

FIGURE 11.42
Layered Rocks in Meridiani Planum
This *Mars Reconnaissance Orbiter* image shows a region about 6 km across in Meridiani Planum. The rocks shown here consist of dozens of layers. The layers resemble layers of sediments deposited over time in lakes and oceans on Earth.

3. Mars had a strong greenhouse effect early in its history, but one produced by sulfur dioxide rather than carbon dioxide. Sulfur is abundant in Martian rocks and soils and was released in Martian volcanos. Like carbon dioxide, however, sulfur dioxide would have dissolved in water and been removed from the atmosphere. Perhaps there was an equilibrium between the rate at which volcanos produced sulfur dioxide and the rate at which it was removed from the atmosphere by water.
4. Methane as well as carbon dioxide contributed to the Martian greenhouse effect. Methane has been detected in the Martian atmosphere, but much less than the amount required for a strong greenhouse effect.

As yet, however, none of these scenarios or any others have been able to account fully for an early warm, wet atmosphere on Mars.

Mars once had a warm, wet atmosphere. None of the scenarios proposed so far have been able to fully account for such an atmosphere.

11.8 LIFE ON MARS

Even though life on Mars seemed unlikely, the two *Viking* landers carried life detection experiments. One of the main justifications for the biology experiments was the evidence that there is water on Mars today and that Mars may have been far wetter in the past. The presence of liquid water early in the Earth's history was crucial to the development of life on our planet.

Viking Search for Life

Each *Viking* lander conducted five experiments to investigate the possibility of life on Mars. The imagers, which returned many pictures of the landscape surrounding the landers, found no evidence of plant or animal life. Almost no one was surprised at the absence of waving vegetation or migrating herds of animals. However, there are many places on the Earth, such as parts of the Sahara Desert and Antarctica, where it would be unusual to find organisms large enough to be seen by the *Viking* imagers. Nevertheless, microorganisms inhabit these hostile places and nearly every other square meter of the surface of the Earth.

Molecular Analysis

All life on the Earth is based on the chemistry of carbon atoms. **Organic molecules** contain carbon atoms or chains of connected carbon atoms. A sensitive way to test for the presence of microbial life is to look for organic molecules, either from living organisms or the remains of dead ones. Each lander used a device known as a gas chromatograph-mass spectrometer (GCMS) to measure the number of molecules of different molecular weights in samples brought into the lander. The GCMS was also used to measure the chemical composition of the Martian atmosphere and the Martian soil.

The GCMS was tested with soil samples taken from Antarctica before it was sent to Mars. It was able to detect more than two dozen organic molecules, some present only at a level of a few parts per billion. No organic molecules were detected in any of the samples taken from the surface, from beneath the surface, or from under rocks.

Recent analyses of the GCMS results have shown that the GCMS instruments were not very sensitive to the molecules likely to be derived from organisms under Martian conditions. As many as several million organisms per cubic cm of Martian soil would not have produced a detection by the GCMS.

Biology Experiments

In view of the results of the molecular analyses carried out by the GCMS experiments, the scientists who had designed the biology detection experiments were doubtful that there would be any interesting results from their experiments. However, they were in for a major surprise.

Two of the biology experiments were wet experiments in which small Martian soil samples were wetted with liquids containing nutrients. (One of the nutrient mixtures was nicknamed "chicken soup.") Both experiments were designed to look for gases, such as carbon dioxide, methane, and oxygen, that would be released as a consequence of microorganisms metabolizing the nutrients. The results were spectacular. In one experiment a large amount of oxygen was released. The other experiment recorded a swift increase in carbon dioxide in the test chamber when the nutrient was added. At first these results seemed to indicate that biological activity had been detected. However, further experimentation showed results that would not be expected if microorganisms were responsible for the released gases. For instance, preheating a soil sample to kill the microorganisms did not reduce the amount of oxygen released. Also, when the soil sample was wetted a second time no additional carbon dioxide was produced. It could be argued that neither of the wet experiments was realistic in the sense that they used liquid water, which hasn't been stable on the surface of Mars for billions of years.

The third experiment, a dry one, searched for biological activity under conditions similar to those actually existing on Mars. Soil samples were put in a small chamber and exposed to gas resembling the Martian atmosphere. However, some of the carbon dioxide and carbon monoxide in the gas contained radioactive ^{14}C. The samples were also exposed to simulated sunlight. After the sample was incubated for 5 days the atmospheric gases were removed and the soil sample was heated to about 1000 K to incinerate and vaporize any microorganisms. Any microorganisms present in the soil sample should have incorporated radioactive carbon atoms into the organic molecules they contained. Heating the sample should have released gas containing radioactive carbon atoms. And this is exactly what happened in seven of the nine trials of the experiment. The amount of biological activity implied was small, but radioactive carbon was definitely incorporated into the Martian soil under Martian conditions. As was the case for the wet experiments, further tests raised a problem for the biological explanation of the experiment. When the soil sample was heated to a temperature of 450 K (350° F), which should have killed any organisms present, radioactive carbon was still incorporated into the soil.

The consensus among the scientists who carried out the *Viking* biology experiments was that the results showed that the soil of Mars is chemically but not biologically active. Perhaps the Martian soil contains oxygen compounds that could react with some of the *Viking* nutrients to produce water and carbon dioxide. No completely acceptable chemical explanation of the entire trio of biology experiments has yet been discovered. It is also worth noting that a minority of the *Viking* experimenters still believe that the simplest explanation is that Martian microorganisms are responsible for the *Viking* results.

It is probably not completely valid to generalize the apparently negative *Viking* biology results to the planet as a whole. After all, there are terrestrial life-forms, such as those living inside rocks and in dry valleys in the Antarctic, that could survive on Mars. Still, the winds of Mars seem to have widely redistributed its surface materials. If life existed elsewhere on Mars why weren't redistributed organic molecules detected at the *Viking* sites? It seems clear that any definitive conclusions about Martian life, at least about its existence today, will await a much more thorough exploration of Mars and the use of a much more complete set of procedures for searching for evidence of biology.

A tantalizing hint that there may be life on Mars today has come from the detection by several Earth-based and orbital experiments of methane in the atmosphere of Mars. One of these experiments on *Mars Express* found concentrations of methane in several regions near Mars's equator. While there are several explanations of the presence of methane that don't require biology, much of the methane in Earth's atmosphere is produced by microorganisms.

Surprisingly, the question of whether there was life on Mars billions of years ago might be answered before we can be absolutely sure about whether life exists on Mars today. A study of a meteorite thought to have come from Mars showed possible signs of ancient life. Further tests of the meteorite might indeed show that life was present on Mars 3.6 billion years ago. This possibility is described in more detail in Chapter 27.

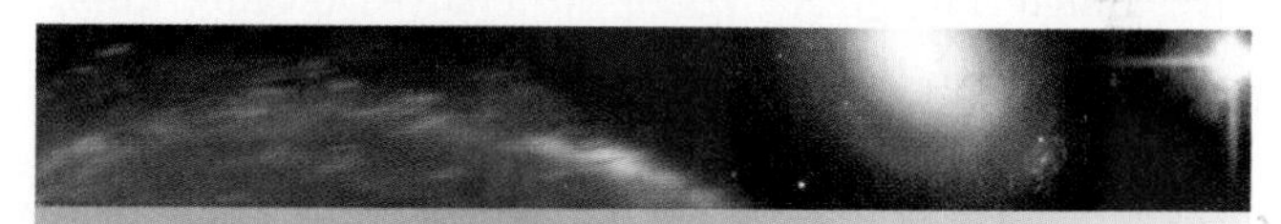

The biology experiments on the *Viking* landers produced many puzzling results, but found no conclusive evidence for biological activity.

THE EVOLUTION OF THE TERRESTRIAL PLANETS

The history of the terrestrial planets could not easily have been discovered from studies of the Earth alone. The Earth's dynamic nature has eradicated most of the evidence of its early history. On other planets, however, the early evidence is better preserved. That evidence shows that the terrestrial planets formed by accretion about 4.6 billion years ago. The impacts that formed them and the decay of their radioactive materials heated the planets enough that their internal temperatures reached the melting point of iron. When this occurred, iron and other metals sank toward the centers of the planets, releasing still more heat in the process. Light, silicate materials floated to the surface and cooled to become the planetary crusts. By the time the heavy bombardment period ended about 3.8 billion years ago, the terrestrial planets had rigid crusts upon which the craters of later impacts could be recorded.

After the bombardment ended, the terrestrial planets cooled at a rate that depended on their size. The smaller bodies, Mercury and the Moon, cooled relatively rapidly and soon became volcanically inactive. These bodies have changed very little for more than 3 billion years. Mars, somewhat larger, cooled more slowly and probably has not completely lost its volcanic activity. Venus and the Earth, the largest terrestrial planets, have cooled relatively little and remain tectonically and volcanically active, although in different ways.

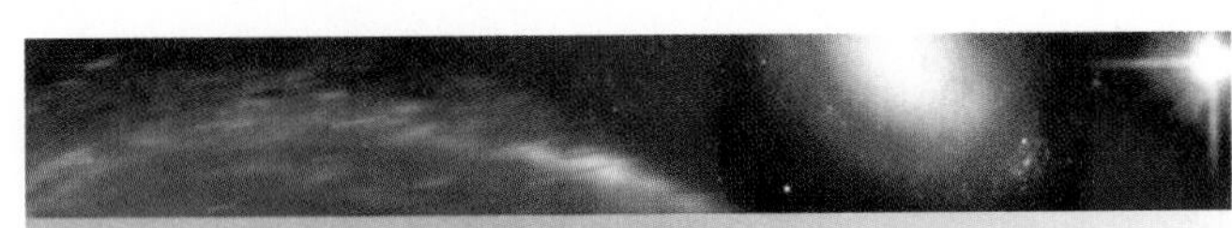

Each of the terrestrial planets contributes information to our knowledge of their origin and evolution. Shortly after formation, impacts and radioactive decays heated each planet until its interior melted. This permitted much of its iron to sink to the center to form a core. Rocky material floated to the surface to form a crust. The larger terrestrial planets have cooled slowly and remain volcanically active. The smaller terrestrial planets cooled more quickly and are now inactive.

Chapter Summary

- The terrain in the southern hemisphere of Mars is ancient and heavily cratered. The pattern of ejected material from many Martian craters shows that the ejecta were very fluid and may have contained a high proportion of water. (Section 11.2)
- Widespread volcanic flows have eradicated the ancient, cratered terrain in the northern hemisphere of Mars. (11.2)
- Mars has many volcanos. The largest Martian volcanos are located in the Tharsis region, a continent-sized bulge. The great sizes of the Tharsis volcanos imply that the Tharsis region has remained over its magma source for billions of years. The heights of the Tharsis volcanos show that the magma that formed them rose from great depths, where pressure was large enough to lift the magma to the summits of the volcanos. This shows that the lithosphere of Mars is thicker than the lithosphere of the Earth. (11.2)
- We can estimate the ages of Martian volcanos and lava flows by measuring crater densities. Volcanic activity occurred all over Mars about 2.5 billion years ago but has gradually become restricted to a smaller and smaller region of the planet. (11.2)
- The many faults and chasms on Mars show that there has been considerable crustal motion. The most spectacular example of crustal motion is the Valles Marineris canyon system. Much of the crustal motion seems to have occurred as a consequence of the weight of the Tharsis bulge. Long ago, Mars experienced a period of plate tectonic activity. (11.2)
- Two kinds of channels on Mars appear to have been cut by running water. Runoff channels, resembling terrestrial river systems, were formed when underground water or rainfall was collected from a large region. Outflow channels were cut by great floods that took place when large pools of water suddenly were produced in a local region. Small outflow channels may have been formed very recently. (11.2)
- Both the north and south poles of Mars have permanent polar caps. The larger northern cap is made primarily of water while the southern cap also contains a large percentage of carbon dioxide (dry ice). A seasonal cap of dry ice forms about each permanent polar cap in winter. (11.3)
- The polar regions of Mars are covered with thick layers of sediment formed from dust blown poleward by strong winds. It takes tens of thousands of years to accumulate a single layer tens of meters thick. The layering probably is caused by the precession of Mars's polar axis. (11.3)
- Pictures taken by the *Viking* landers show rolling countryside littered with rocks ejected from nearby impact craters. The regions around the landers resemble rocky deserts on the Earth. (11.4)
- The *Pathfinder* lander and the *Sojourner* rover explored Ares Vallis for about 3 months in 1997. *Pathfinder* images showed abundant evidence of ancient floods while *Sojourner* measured the chemical composition of rocks and soil. The *Spirit* and *Opportunity* rovers found strong evidence that water once covered the surface of Mars. (11.4)
- *Phoenix* landed near the northern polar cap of Mars. *Phoenix* found ice beneath the surface and showed that the soil of Mars would be hospitable to life. (11.4)
- The atmosphere of Mars consists primarily of carbon dioxide. The atmosphere is too thin to insulate the

surface, where daily temperature fluctuations are as much as 60 K. Atmospheric pressure is lowest in the summer and winter when carbon dioxide is trapped in one of the polar caps. Pressure is highest in the fall and spring when carbon dioxide is released into the atmosphere. (11.5)

- Frequent dust storms keep dust suspended in Mars's atmosphere at all times. Great dust storms sometimes spread a thick layer of dust over the entire planet. The seasonal changes in dark markings are caused by deposition and removal of dust by seasonal winds. Large dust particles are skipped along the surface by winds and are very effective at eroding the Martian surface. (11.5)
- Although Mars contains a smaller percentage of iron than the Earth does, it has a core made of iron or iron mixed with sulfur. The crust of Mars is rich in iron, giving Mars its reddish color. (11.6)
- After Mars formed, radioactive decays heated its interior until iron melted and accumulated in its core. Eventually the rocks in the mantle melted partially to produce magma, which flooded the surface and produced large volcanos. Mars expanded as it grew hotter, perhaps causing its crust to pull apart and produce Valles Marineris. Mars has cooled and become less active for the last several billion years. (11.6)
- Little water (in solid, liquid, or gaseous form) can be found on Mars today. However, surface features such as its channels show that Mars once had much more water than we have been able to measure. It is likely that most of Mars's water is trapped as permanently frozen, subsurface ice. (11.7)
- Early in its history, Mars may have had an atmosphere thick enough for a greenhouse effect to occur. The atmosphere may have been warm and thick enough for liquid water to exist. No fully satisfactory explanation of the early atmosphere of Mars has yet been proposed. (11.7)
- The *Viking* biology experiments showed that the Martian soil is chemically active, but probably does not harbor life. (11.8)
- The evolution of the terrestrial planets can be pieced together using information from each of them. Impacts and radioactive decays heated each planet until much of its iron sank to form a core, and light rocky material floated to the surface to become the crust. The extent to which each planet has cooled and become volcanically inactive depends on size. The smaller planets quickly became inactive while the larger planets remain tectonically and volcanically active. (11.9)

Key Terms

organic molecule *260* outflow channel *243* patera *240* runoff channel *243*

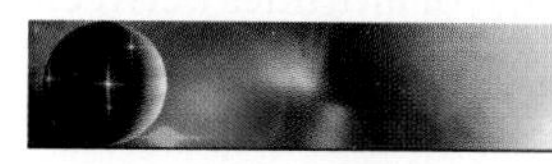

Conceptual Questions

1. What evidence for underground ice on Mars is provided by the ejecta from Martian craters?
2. How fast would a piece of rock need to be blasted outward from the surface of Mars to escape into space as a meteoroid?
3. What is the reason that the northern plains of Mars have a much smaller crater density than the cratered terrain of the southern hemisphere?
4. What evidence do we have that the volcanos of the Tharsis region are younger than most of the rest of the surface of Mars?
5. What is believed to be the reason that Alba Patera and other Martian paterae have such gentle slopes?
6. What differences in the properties of runoff channels and outflow channels show that the two types of channels originated in different ways?
7. The Earth is closest to the Sun in January, which is summer in the southern hemisphere. Why isn't the difference between northern and southern seasonal variations as extreme on the Earth as on Mars? (You should give an astronomical reason, which is part of the full explanation.)
8. Suppose Mars had such a large orbital eccentricity that the north polar region was too warm in winter for carbon dioxide to condense as dry ice. What would the annual variation in atmospheric pressure be like in that case?
9. What is the reason that the Martian surface and sky are somewhat pink?
10. Why does Mars lack a warm stratopause?
11. What is the connection between Martian wind patterns and the variable dark surface markings seen from the Earth?

12. Mars has nearly the same rotation period as the Earth, yet Mars has no detectable global magnetic field. What possible difference between the two planets could account for the absence of a Martian magnetic field?
13. In what kinds of places is water observed to exist or believed to exist on Mars today?
14. What evidence do we have that Mars was once much wetter and hotter than it is today?
15. What may have caused the thick atmosphere that Mars is believed to have had in the distant past to have disappeared?
16. What evidence did the gas chromatograph-mass spectrometers on the *Viking* landers yield about the presence of life on Mars?
17. What did the biology experiments on the *Viking* landers tell us about the presence of life on Mars?
18. What evidence of water on Mars was found by the *Opportunity* rover?

Problems

1. The distance between the Earth and Mars when the two planets are at opposition varies greatly because of the large eccentricity of Mars's orbit. The perihelion distance of a planet is given by $r_{min} = a(1 - e)$ and the aphelion distance by $r_{max} = a(1 + e)$ where a is the semimajor axis and e the orbital eccentricity. Find the smallest and largest opposition distances assuming that the Earth's orbit is a circle.
2. At an average opposition, the Earth and Mars are separated by 0.52 AU. Suppose an astronomer observes Mars at opposition and that seeing blurs the images to a resolution of 1.0 seconds of arc. What is the smallest surface feature the astronomer would be able to resolve on Mars? How does this size compare with the diameter of Mars? (You will need the small angle equation, Equation 3.1, to do this problem.)
3. An astronomer observed Mars at opposition using a telescope that has a lens 50 cm (20 inches) in diameter. The astronomer observes in visible light at a wavelength of 500 nanometers. What is the smallest feature the astronomer could resolve on Mars if the image were not affected by seeing? How does this compare to the diameter of Mars? To the diameter of Olympus Mons? (Use the equation for angular resolution, Equation 6.6.)
4. How long, deep, and wide would a terrestrial chasm have to be to have the same proportions relative to the Earth that Valles Marineris has to Mars?

Figure- and Table-Based Questions

1. The slopes of a hypothetical Martian volcano have a crater density twice that of the lunar maria. Use Table 11.2 to estimate the age of the volcano.
2. Use Table 11.2 to estimate the crater density (relative to the crater density of the lunar maria) of a part of Mars's surface that is 2 billion years old.
3. Use Figure 11.35 to find the range of altitudes between which the temperature of the Martian atmosphere is less than 200 K.

Group Activities

1. Divide your group into two subgroups. After a few minutes of preparation, have one subgroup present the evidence used in Percival Lowell's time to demonstrate the likelihood of life on Mars. Let the other subgroup huddle for a minute and then try to refute the evidence of the first subgroup based on more recent information.
2. With your group, prepare a plan for what you would need to take along in order to survive 6 months on the Martian surface. If possible, make a brief presentation of your plan in class. Invite the class to discuss whether recent discoveries of ice on Mars would have any impact on your plan.

For More Information

Visit the text website at **www.mhhe.com/fix** for chapter quizzes, interactive learning exercises, and other study tools.

12

There appear on the surface of Jupiter certain bands darker than the rest of the disc, and they do not always preserve the same form . . .

Christiaan Huygens, 1698

Into the Forge of God by Alan Gutierrez. The *Galileo* probe descends into the clouds of Jupiter.

Jupiter and Saturn

Since 1970 there has been a spectacular increase in what we know about the giant planets—Jupiter, Saturn, Uranus, and Neptune. Before the 1970s even Jupiter, the best observed giant planet, was only a distant ball on which intriguing but fuzzy details could be seen. The other giant planets were known even less well. For some of the giant planets, even fundamental information such as rotation rate was not well known. The many satellites of the giant planets were visible only as points of light. Much of what has been learned since 1970 has been the result of the enormously successful probes of the outer solar system—*Pioneers 10* and *11*, *Voyagers 1* and *2*, *Galileo*, and *Cassini*. One after another, starting in 1973, the giant planets and their satellites were examined in detail by these spacecraft.

The giant planets are much farther from the Sun than their terrestrial counterparts. As a consequence, it is much more difficult and expensive to send probes to the giant planets than to Venus or Mars. To send a probe directly to Neptune, for example, it would be necessary to place the probe in an elliptical orbit with its aphelion at Neptune's distance and its perihelion at the Earth's distance. This would require accelerating the probe until it were traveling about 13 km/s faster than the Earth. Even starting at such a high speed, the trip would take 30 years. In contrast, a probe sent to Mars must be accelerated to a speed only about 3 km/s faster than the Earth is traveling and would arrive in less than a year.

Thirty years is a long time—nearly as long as the scientific careers of the people who have experiments on the probe. Also, a huge amount of fuel must be used to accelerate probes to such high speed.

Questions to Explore

- Why are the densities of Jupiter and Saturn so low compared with the terrestrial planets?
- What makes the cloud patterns of Jupiter and Saturn so colorful and complex?
- Why are the interior structures of Jupiter and Saturn so different from those of the terrestrial planets?
- How did Jupiter and Saturn evolve to their present form?
- What is the source of the internal energy radiated away by Jupiter and Saturn?
- Why are the rings of Jupiter and Saturn so thin compared with their diameters?
- How can we account for the complex structure of the rings of Saturn?

Fortunately, there is an alternative method, both quicker and cheaper, in which "gravitational assists" are used to send a probe past multiple planets. To illustrate the usefulness of gravitational assists, Figure 12.1 shows the speed of *Voyager 2* as it moved outward through the solar system. After leaving the Earth, *Voyager 2* reached the vicinity of Jupiter and its considerable gravitational influence in 1979. Jupiter's gravity accelerated *Voyager* from about 10 km/s to a maximum speed of 27 km/s when the probe passed closest to Jupiter. *Voyager* slowed to about 20 km/s as it moved away from Jupiter, but this was twice as fast as it had been traveling before the encounter. *Voyager 2* then sped to Saturn in 1981 and Uranus in 1986. Each of these planets provided an additional gravitational assist, allowing *Voyager 2* to reach Neptune in 1989, only 12 years after launch. *Voyager 2* was actually slowed by the Neptune encounter, but still had enough speed to escape the solar system when it left Neptune. The probe is destined to become an interstellar wanderer as it passes beyond the remotest parts of the solar system. Figure 12.2 shows both the "direct" elliptical path to Neptune and the actual path taken by *Voyager 2*. Clearly the shortest path in this case is not the fastest. An arrangement of the giant planets that makes possible a four-planet mission like that of *Voyager 2* occurs only every 179 years, so scientists were extremely eager to take advantage of the window of opportunity that opened up in the late 1970s.

During a gravitational assist, the energy of a probe increases significantly. Because total energy is conserved, the additional energy for the spacecraft has to come from somewhere. But where? The answer is that it is stolen from the planet that supplies the gravitational assist. The planet's

FIGURE 12.1
The Speed of *Voyager 2* During Its Flight out of the Solar System
Gravitational assists during encounters with Jupiter, Saturn, and Uranus sped up the spacecraft, whereas the encounter with Neptune slowed it down.

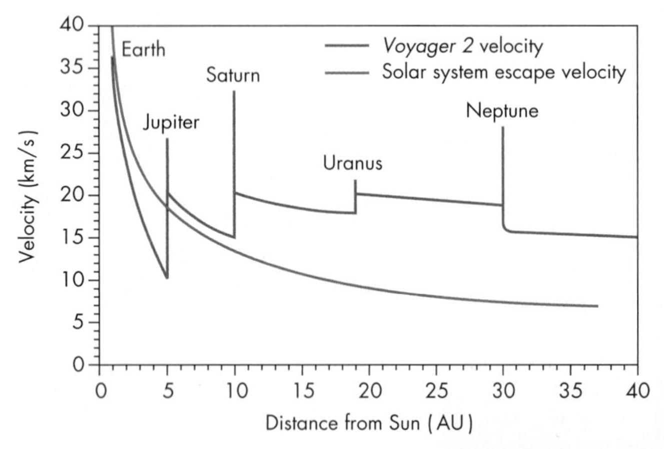

FIGURE 12.2
The Path Followed by *Voyager 2* on Its Way to Neptune
The elliptical orbit that would take a probe directly to Neptune is also shown. The path taken by *Voyager 2* is longer but faster than the elliptical path.

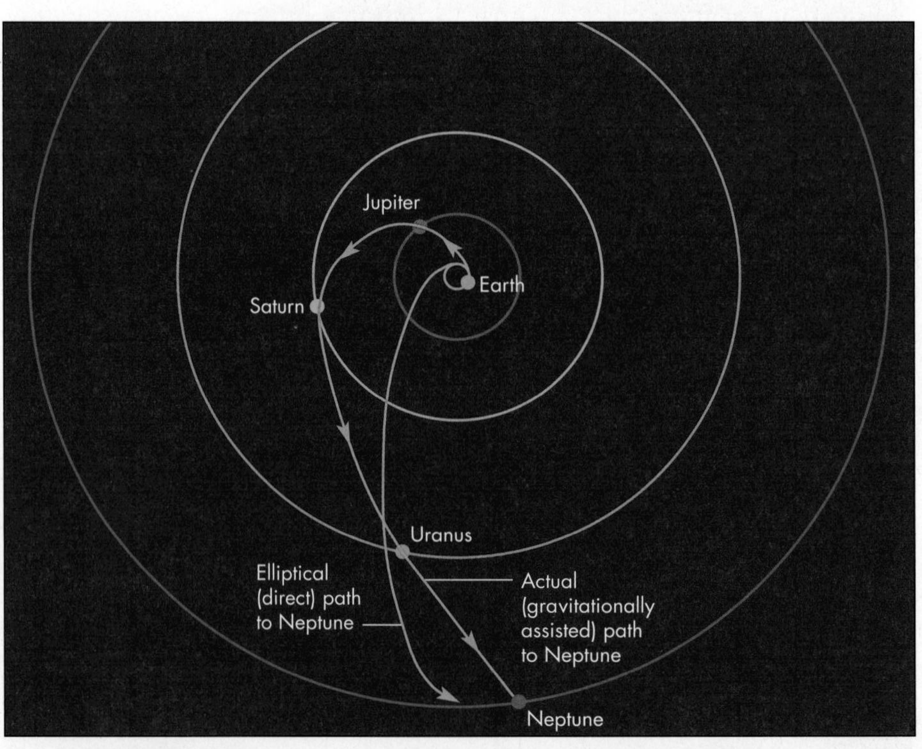

orbital speed decreases during the encounter, but only by a tiny amount.

JUPITER AND SATURN

Jupiter is the most massive planet. In fact, it makes up 70% of all the planetary matter in the solar system. It is also the largest planet; its diameter is approximately 11 times as large as the Earth's and 1⁄10 as large as the Sun's.

Saturn was the most distant planet known to the ancients. When Galileo pointed his new telescope at Saturn, he was perplexed because Saturn seemed to have ears. His sketch of Saturn is shown in Figure 12.3. Two years later he found that the ears were gone. He couldn't explain their disappearance, but guessed that they would soon return, which they did. What Galileo had seen were the spectacular rings of Saturn, as shown in Figure 12.4. The rings are very thin and were viewed edge-on at the time Galileo thought they had vanished (see Figures 12.25 and 12.27). Saturn's rings give the planet a unique appearance and a special beauty.

BASIC PROPERTIES

Some basic properties of Jupiter and Saturn are summarized in Tables 12.1 and 12.2.

FIGURE 12.3
A Sketch of Saturn Drawn by Galileo in 1610
Galileo's telescope produced an image that was too poor for him to see that Saturn had a ring. He was puzzled by the "appendages" of the planet, particularly when they later disappeared.

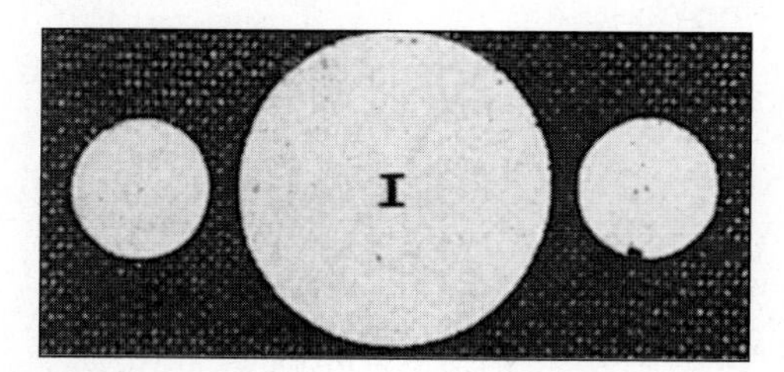

FIGURE 12.4
Saturn as Seen by the *Hubble Space Telescope*

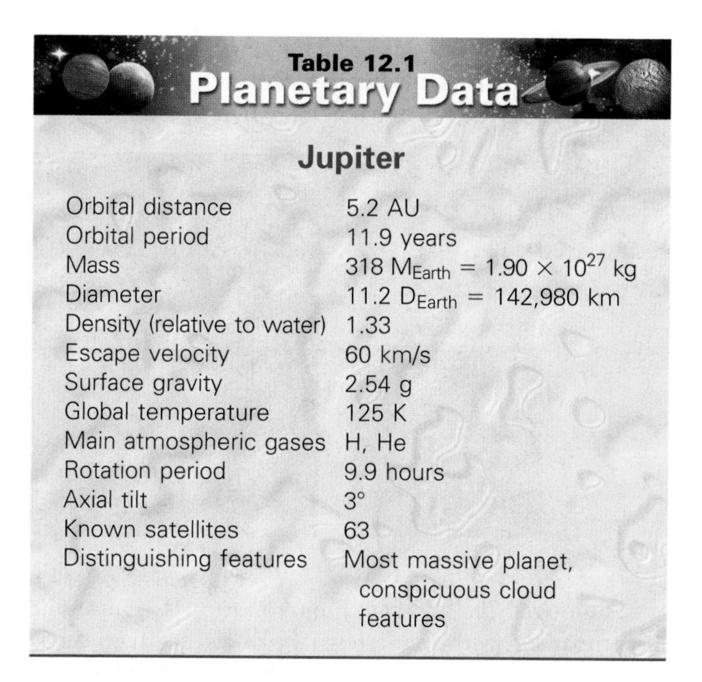

Table 12.1
Planetary Data

Jupiter	
Orbital distance	5.2 AU
Orbital period	11.9 years
Mass	318 M_{Earth} = 1.90×10^{27} kg
Diameter	11.2 D_{Earth} = 142,980 km
Density (relative to water)	1.33
Escape velocity	60 km/s
Surface gravity	2.54 g
Global temperature	125 K
Main atmospheric gases	H, He
Rotation period	9.9 hours
Axial tilt	3°
Known satellites	63
Distinguishing features	Most massive planet, conspicuous cloud features

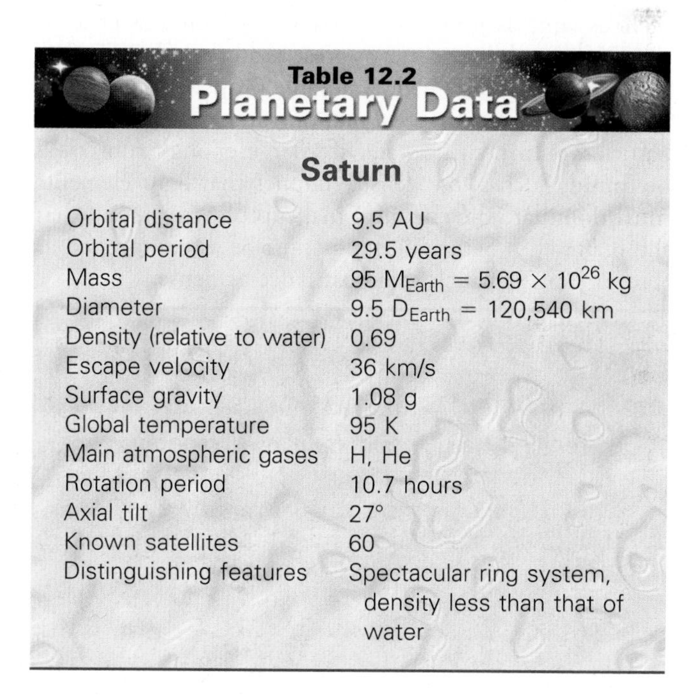

Table 12.2
Planetary Data

Saturn	
Orbital distance	9.5 AU
Orbital period	29.5 years
Mass	95 M_{Earth} = 5.69×10^{26} kg
Diameter	9.5 D_{Earth} = 120,540 km
Density (relative to water)	0.69
Escape velocity	36 km/s
Surface gravity	1.08 g
Global temperature	95 K
Main atmospheric gases	H, He
Rotation period	10.7 hours
Axial tilt	27°
Known satellites	60
Distinguishing features	Spectacular ring system, density less than that of water

Diameter, Mass, and Density

A trip once around the equator of Jupiter would be longer than a flight to the Moon. Figure 12.5 shows the relative diameters of the Earth, Jupiter, and Saturn. Notice how small the Earth is in comparison with Jupiter and Saturn. Jupiter is as massive as 318 Earths. Despite its huge size and tremendous internal pressures, the density of Jupiter is only 1330 kg/m^3 (1.33 times the density of water). This is only 25% of the Earth's density, which shows that Jupiter cannot be made of rocks and metals like the Earth and other terrestrial planets. Instead, Jupiter must contain large amounts of the lighter elements hydrogen and helium.

FIGURE 12.5
The Relative Sizes of Jupiter, Saturn, and the Earth
The Earth is 12,756 km in diameter, whereas the diameter of Jupiter is 143,000 km, and that of Saturn is 120,500 km.

Although smaller than Jupiter, Saturn is enormous in comparison with the terrestrial planets. Its diameter is more than nine times as large as the Earth's and almost 85% as large as Jupiter's. Its mass is 95 times as great as the Earth's and three times as great as all the other planets, except Jupiter, combined. A notable property of Saturn is its low density—only 69% that of water. This means that Saturn would float in water if a large enough tub could be found. Such a low density implies that light elements must dominate the chemical makeup of Saturn, and that the material in Saturn's interior can't be as compressed as that of Jupiter, which is almost twice as dense.

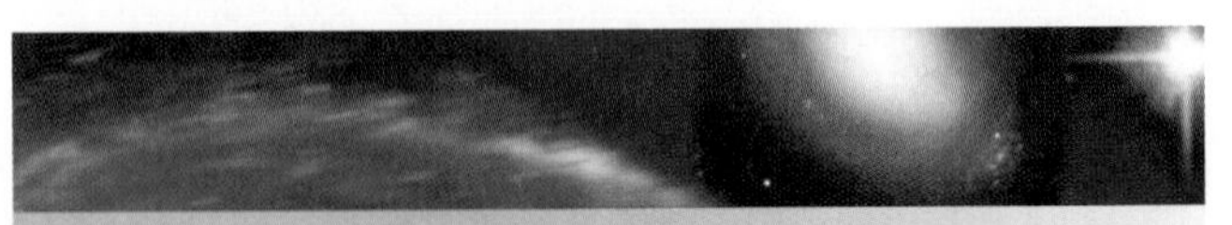

Jupiter and Saturn dominate the planetary system. They are the largest and most massive planets. Their average densities, however, are much lower than those of the terrestrial planets, indicating that they must be made mostly of the light elements hydrogen and helium.

Visual Appearance

Although the outer layers of Jupiter (and the other giant planets) consist almost entirely of transparent gas, clouds of liquid and solid droplets produce a wealth of colored features, as shown in Figure 12.6. Notice that there are alternating dark **belts** and light **zones** lying parallel to its equator. In addition, numerous bright and dark spots are visible. The largest and most famous of these is the **Great Red Spot,** an immense rotating cloud pattern that has been present at least since its discovery by Giovanni Cassini in 1665. The colors of Jupiter range from reddish-pink to blue-gray. Although Jupiter is certainly colorful, its colors are much more subdued than those of the Earth. Jupiter's red isn't as vivid as an apple and its blue isn't as bright as a mountain sky.

Unlike the terrestrial planets, Jupiter is significantly flattened. Its polar diameter is about 6% less than its

FIGURE 12.6
A *Hubble Space Telescope* Image of Jupiter
Even when viewed by *Hubble* from Earth orbit at a distance of 960 million km (520 million miles), Jupiter has a complex appearance.

The rotation of Jupiter

equatorial diameter. Jupiter's flattening is caused by a combination of its rapid rotation and large size. These result in an equatorial rotation speed of about 12.5 km/s, larger than the Earth's escape velocity and about 20% of the escape velocity of Jupiter. The inertia of the material in the equatorial regions of Jupiter lifts it outward, producing a larger equatorial than polar diameter.

Saturn is also flattened by rapid rotation. Its polar diameter is only about 90% as large as its equatorial diameter. Figure 12.4 shows that Saturn is quite muted in appearance. Saturn has a pattern of belts and zones similar to Jupiter's, but with much smaller differences in color and brightness. Fewer than a dozen bright or dark spots have ever been seen on Saturn from the Earth. The *Voyager* images, however, showed many atmospheric features that resemble those of Jupiter. Some of the cloud features seen at Saturn's high southern latitudes are shown in Figure 12.7. Overall, though, Saturn has fewer spots of all sizes than Jupiter and has no spot nearly as large as Jupiter's Great Red Spot.

Rotation and Winds

The rotation period of Jupiter or Saturn can be found by measuring how long it takes a conspicuous atmospheric feature to return to the same spot on the disk of the planet. If this is done for features near the equator of Jupiter, it is found that Jupiter rotates eastward with a rotation period of $9^h50^m30^s$. For features at higher latitudes, the rotation period is $9^h55^m41^s$. The variation of rotation period with latitude can only be explained if the part of Jupiter we see consists of clouds and gas rather than solid surface. The 5-minute difference in rotation period between the equator and high latitudes means that the clouds near the equator rotate eastward almost 1% (300 km/h) faster than those at high latitudes. This is about as fast as the speed of the Earth's jet streams. Actually there are many changes in Jupiter's rotation period between its equator and its poles. The variation of rotation period with latitude can be described as **zonal winds,** as shown in Figure 12.8. The equatorial winds are sometimes referred to as Jupiter's **equatorial jet.** Regions in which there are large changes in zonal wind speed, such as 20° N latitude or 20° S latitude, often show large rotating patterns of clouds, such as the Great Red Spot.

How fast the deep interior of Jupiter rotates can't be determined from observations of Jupiter's atmosphere. In 1955, however, astronomers found that electrons trapped in Jupiter's magnetic field emit radio waves. They also found that the emission varies with a period of $9^h55^m29^s$, the time it takes for Jupiter's magnetic field to rotate about the axis of Jupiter. Because Jupiter's magnetic field is produced in the planet's interior, the period for the radio emission is assumed to be the rotation period of the parts of Jupiter deep beneath the visible cloud layers. The internal rotation rate of Jupiter is slower than its equatorial rotation rate. This means that the equatorial region rotates eastward faster than the interior of the planet. Thus the equatorial jet blows toward the east. The zonal winds at high latitudes alternately blow eastward and westward.

Saturn rotates almost as rapidly as Jupiter and, like Jupiter, has a rotation period that varies with latitude. Near the equator the rotation period is about 10^h14^m, while high-latitude regions rotate in about 10^h40^m. The rotation

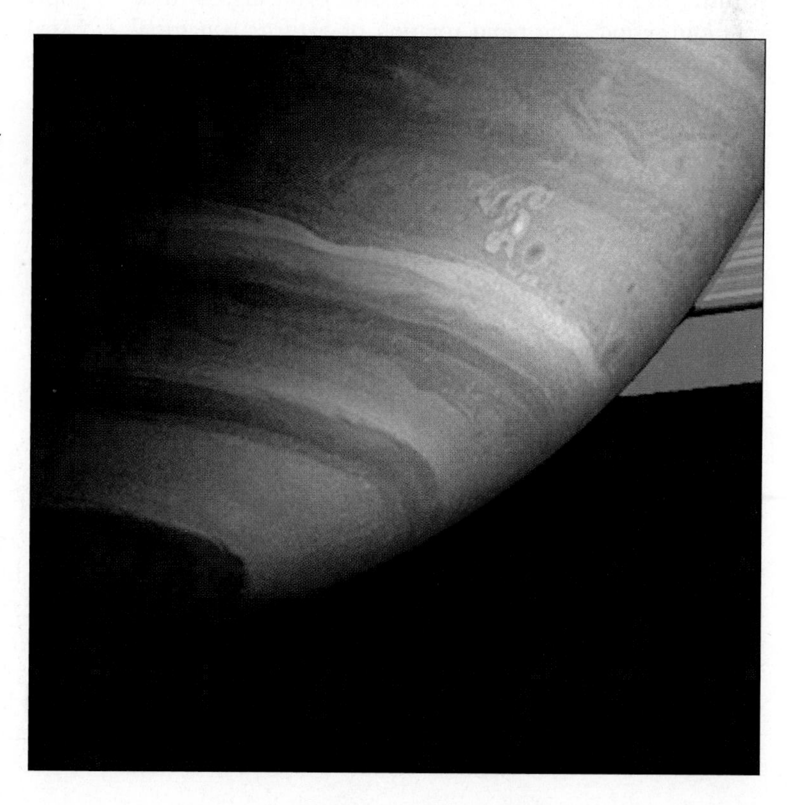

FIGURE 12.7
The Atmosphere of Saturn at High Southern Latitudes
This infrared, false-color image taken by *Cassini* shows that Saturn has many cloud features. Red regions are deep in the atmosphere, and grey and brown clouds are higher.

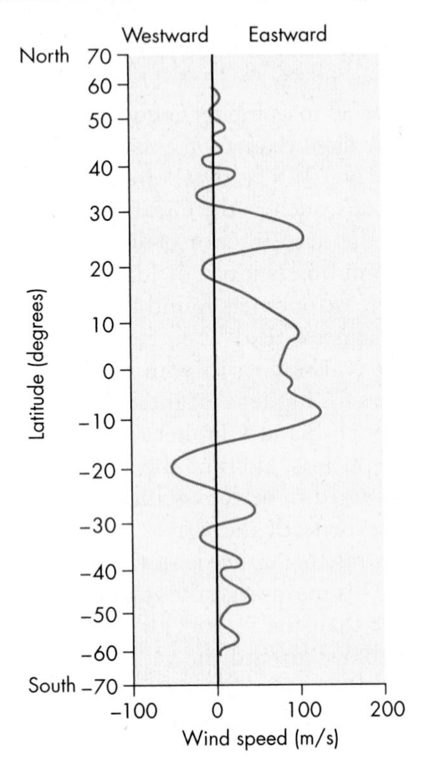

FIGURE 12.8
The Zonal Winds of Jupiter
Many of the places where wind speed is greatest or where wind speed changes fastest with latitude correspond to belts and zones in the cloud features of Jupiter.

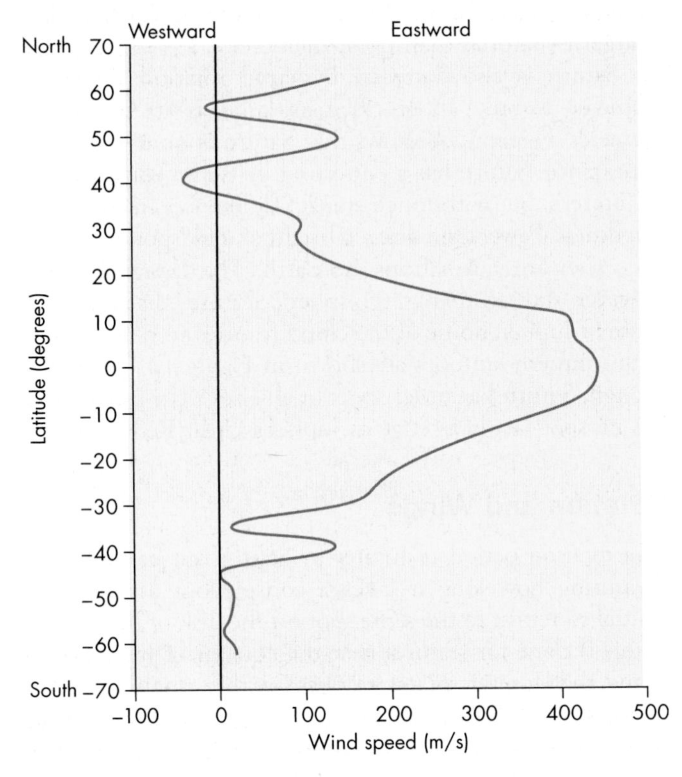

FIGURE 12.9
The Zonal Winds of Saturn
Saturn's equatorial jet is about four times as fast as Jupiter's. There are few places where abrupt changes in wind speed with latitude correspond to the belts and zones of Saturn.

period of the interior of Saturn, found from the rotation of Saturn's magnetic field, is about the same as the high latitude rotation period. The rapid rotation of Saturn's equatorial region can also be described as strong winds that blow toward the east at speeds up to 500 m/s (or 1800 km/h). Figure 12.9 shows the pattern of wind speeds with latitude for Saturn. Saturn's equatorial jet is much faster and wider than Jupiter's. On Jupiter, changes in wind speed can often be associated with boundaries between adjacent belts and zones. This is usually not the case on Saturn.

High-speed eastward or westward zonal winds are found not only in the atmospheres of Jupiter and Saturn but in the atmospheres of other planets as well. The Earth, for instance, has equatorial westward surface winds (the trade winds) and high-altitude eastward winds (the jet streams) at intermediate latitudes. Much of the upper atmosphere of Venus streams westward (the direction of Venus's rotation) at several hundred kilometers per hour. These winds on the Earth and Venus, and the zonal winds of Jupiter and Saturn, are driven by the rotation of the planet and by the transport of heat upward through the atmosphere. In the case of the Earth and Venus, the heat comes from sunlight that warms the ground and lower atmosphere. For Jupiter and Saturn the heat originates deep in the interior. Much remains to be explained about the general process that produces the zonal winds of the giant planets, and details such as the different patterns of zonal winds on Jupiter and Saturn. However, understanding the complicated zonal winds of Jupiter and Saturn might teach us something about atmospheric circulation on the Earth.

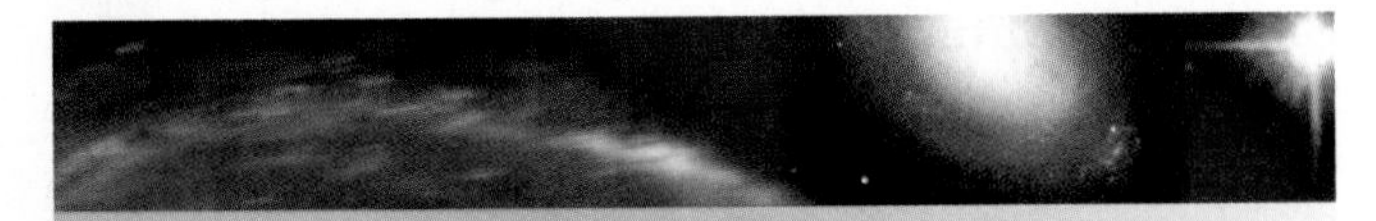

For both Jupiter and Saturn, the atmospheric rotation period varies with latitude, showing that the layer we see is clouds and gas. The variations of rotation period with latitude produce a pattern of high-speed zonal winds similar to the Earth's jet streams.

12.3 ATMOSPHERES

The atmospheres of Jupiter and Saturn are all we are ever likely to see of these planets. The information we have about conditions in the atmosphere of Jupiter was greatly increased when the *Galileo* spacecraft dropped an entry probe into Jupiter's atmosphere. *Galileo* itself entered Jupiter's atmosphere and was destroyed in 2003. On December 7, 1995, the probe entered Jupiter's atmosphere and sent back data for about an hour before succumbing to high pressure and temperature far below the visible clouds. The solid parts of Jupiter and Saturn, if they exist at all, are buried beneath thousands of kilometers of gas.

Chemical Composition

Hydrogen molecules (H_2) and helium atoms (He) make up almost 99.9% of the atmospheric gas of Jupiter. There are about 4 kg of hydrogen to every 1 kg of helium, nearly the same ratio as in the Sun and most other stars. This is also believed to be the ratio of hydrogen to helium in the material from which the solar system formed, which suggests that Jupiter captured and retained almost all of the gas in its vicinity while it was forming.

The great abundance of hydrogen dominates the chemistry of Jupiter's atmosphere and leads to the presence of hydrogen compounds in very small amounts. Methane (CH_4) and ammonia (NH_3), for instance, make up about 0.1% of Jupiter's atmosphere. Other hydrogen compounds, such as water vapor (H_2O), acetylene (C_2H_2), ethane (C_2H_6), and propane (C_2H_8), make up even smaller proportions.

The atmosphere of Saturn is similar to that of Jupiter. One important way in which they differ, however, is in the relative abundance of helium. On Saturn the ratio of hydrogen to helium is almost seven to one. Thus, helium is much less abundant in Saturn's atmosphere than in Jupiter's atmosphere. It seems unlikely that Saturn could have formed with a deficiency of helium. Instead, the low abundance of helium in Saturn's atmosphere suggests that helium has been redistributed within the planet. The importance of this possibility is described later in this chapter.

Vertical Structure

Because Jupiter seems to have no solid surface, the atmospheric layer chosen to be zero altitude is arbitrary. But pressure, which falls steadily with height, can be used to denote different vertical layers. The visible layers of the atmosphere of Jupiter have pressures similar to those near the surface of the Earth. For each 100 km increase in height, pressure decreases about ten times.

The temperature structure of Jupiter's atmosphere is similar to that of the Earth's atmosphere except that Jupiter is about 100 K cooler for a given pressure. The visible cloud tops have a temperature of 125 K. The rate at which temperature drops in the troposphere is exactly what would be expected for an atmosphere heated from below and in which energy is carried outward by convection. Just as for the Earth, the troposphere of Jupiter is a turbulent region of rising and falling columns of gas. The convection in Jupiter's atmosphere strongly influences its cloud structure and its visible appearance.

The temperature structure of Saturn's atmosphere also resembles that of Earth's. Saturn's atmosphere, however, is even cooler than Jupiter's, averaging about 150 K cooler than the Earth's atmosphere for a given pressure.

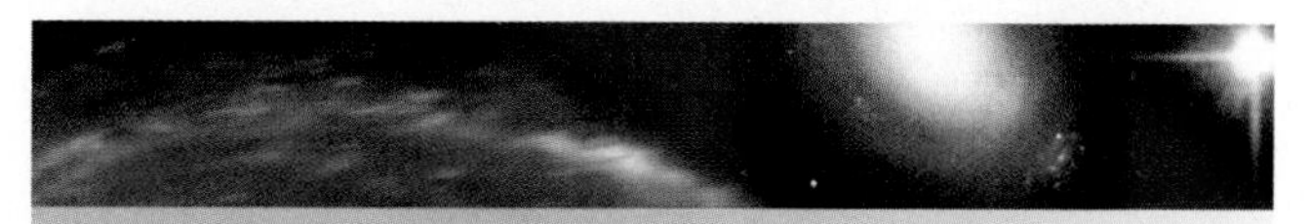

The principal gases in the atmospheres of Jupiter and Saturn are helium and molecular hydrogen. Pressure falls steadily with altitude in the atmosphere of both planets. Temperature falls with altitude in the troposphere.

Cloud Features—Jupiter

The *Voyager* and *Galileo* images reveal the planet in great detail. Figure 12.10 shows a *Galileo* image in which the smallest cloud features are only about 20 km across.

FIGURE 12.10
A High Resolution Image of Jupiter
The smallest features in this *Galileo* image are about 20 km across. The atmospheric entry probe released by *Galileo* is believed to have entered Jupiter's atmosphere in a region similar to the dark region in this image. The colors in this image are exaggerated.

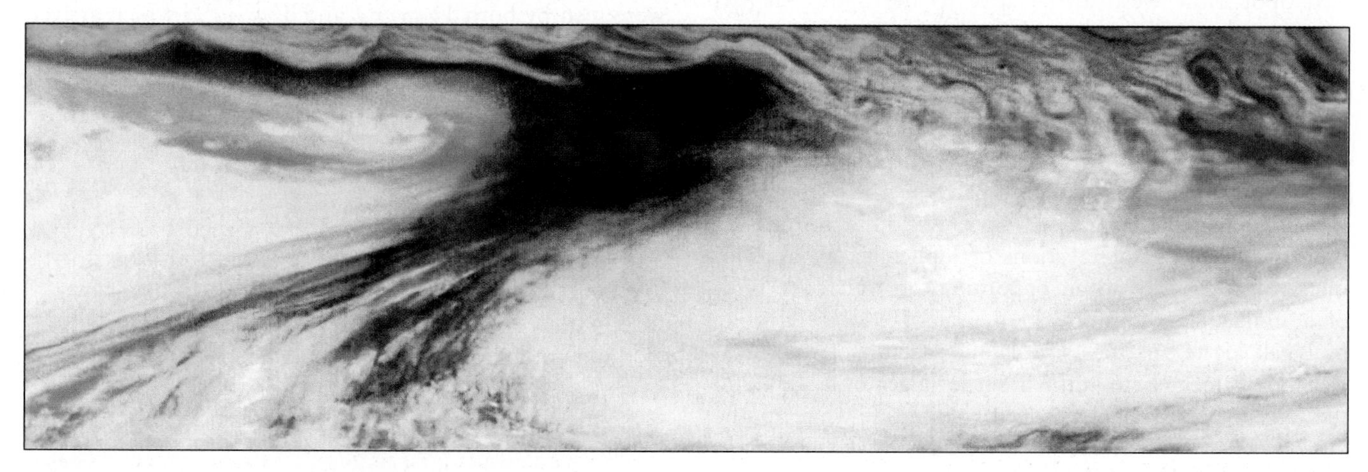

FIGURE 12.11
A *Galileo* Image of Jupiter's Great Red Spot
The Great Red Spot is a long-lived cloud feature that has been observed for more than 300 years. Gas moving from the center of the Great Red Spot to its edges is deflected by the Coriolis effect. This produces a counterclockwise rotation pattern.

FIGURE 12.12
A *Galileo* Image of White Ovals in the Atmosphere of Jupiter
Large white ovals are located near the Great Red Spot in Jupiter's southern hemisphere. Many of these swirling high-pressure regions are larger than Earth.

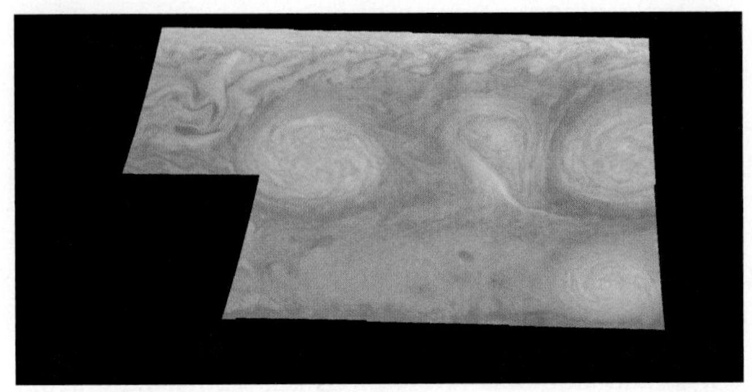

Great Red Spot Figure 12.11 shows the Great Red Spot, an atmospheric storm in Jupiter's southern hemisphere. Earth-based observations during the last 300 years have shown that the Great Red Spot varies in size and color. Sometimes it is three times the size of the Earth, and at other times it is smaller than the Earth. The Great Red Spot rotates counterclockwise, which is the pattern for a southern high-pressure region. This indicates that the Great Red Spot is a region in which pressure is relatively high and gas flows outward. These conditions are just the opposite of those in a hurricane, a low-pressure region in which air flows inward.

The interaction of the Great Red Spot with the zonal winds is complicated. The winds to the north of the Great Red Spot blow westward relative to the Spot, whereas winds to the south blow to the east. The Great Red Spot acts as an obstacle to onrushing gas, which flows past it and is deflected to the north or south. Often a small cloud feature flows around the Great Red Spot to the north, hovers at its western edge, and then either continues westward or is drawn into the circulation pattern of the Spot.

Although the Great Red Spot was fully formed by the time telescopic observations of Jupiter began, astronomers recently have had an opportunity to watch the birth of a similar cloud feature. In the late 1990s three white ovals, all located at the same latitude (south of the Great Red Spot) merged to form a single larger oval. In 2006 the large oval suddenly changed color from white to the same reddish color as the Great Red Spot. The new spot, nicknamed "Red Spot Jr.", has grown to about ⅔ as large as the Great Red Spot and has winds about as strong as those of the Great Red Spot. Astronomers speculate that a similar merger of white ovals may have formed the Great Red Spot centuries ago. A third Red Spot formed from a white oval in 2008.

Other Features Figure 12.12 shows white ovals, high-pressure regions that resemble the Great Red Spot except that they are much smaller. Notice the counterclockwise rotation pattern of the white ovals, which, like most white ovals, are located in Jupiter's southern hemisphere. White ovals last for decades rather than centuries like the Great Red Spot.

Dark brown ovals like the one shown in Figure 12.13 are common in Jupiter's northern hemisphere and last for only a few years. Unlike the Great Red Spot and the white ovals, brown ovals are regions of low pressure.

Cloud Features—Saturn

Many of the individual cloud features of Saturn appear to be formed by rising columns of warm gas. These produce local high-pressure regions with accompanying circulation patterns. Some of the storms last for only a few days; others were seen by both *Voyager 1* and *Voyager 2*, which arrived at Saturn nearly a year apart. Figure 12.14 shows some of the different kinds of cloud features seen on Saturn.

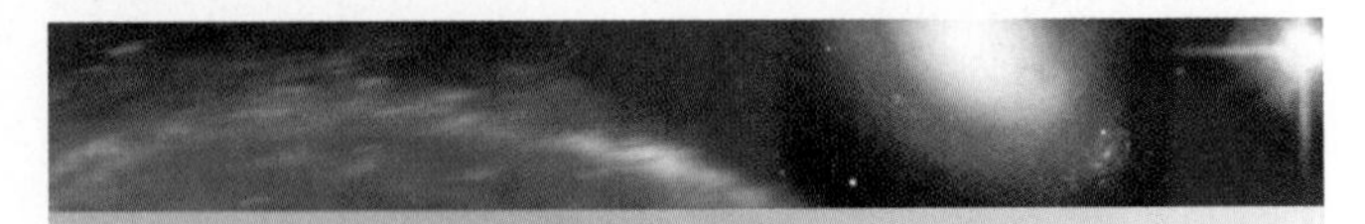

In addition to their overall banded structures, the cloud features of both Jupiter and Saturn have distinctive circulation patterns associated with high- and low-pressure regions. Jupiter's Great Red Spot is a high-pressure region that has persisted for centuries while other cloud features last only a few years.

FIGURE 12.13
A Dark Brown Oval
This Earth-sized atmospheric feature is a hole in the main cloud deck of Jupiter. Dark gases in the lower atmosphere can be seen through the hole. The colors in this *Voyager* image are exaggerated.

FIGURE 12.14
Cloud Features of Saturn
This *Cassini* image has a resolution of 100 km. The image shows numerous cloud features in Saturn's northern hemisphere.

Infrared and Radio Appearance

Our ability to interpret the cloud features of Jupiter and Saturn has been aided considerably by infrared and radio measurements of the two planets. Unlike visible images, which show the sunlight reflected by Jupiter and Saturn, infrared and radio images show the radiation the planets emit. (See the discussion of radiation and matter in Chapter 7.) Bright infrared and radio regions are relatively warm, whereas dark infrared and radio regions are cool. Comparing the infrared and visible images of Jupiter in Figure 12.15 shows that there is a strong relationship between color and temperature.

FIGURE 12.15
Infrared and Visible Images of Jupiter
The bright regions in the infrared image (left) are gaps in the cloud deck through which deeper and warmer atmospheric layers can be seen. Bright features in the visible image (right) tend to be high, cool clouds that appear dark in the infrared image. The Great Red Spot, a high cool feature, can be found toward the left side of the infrared image.

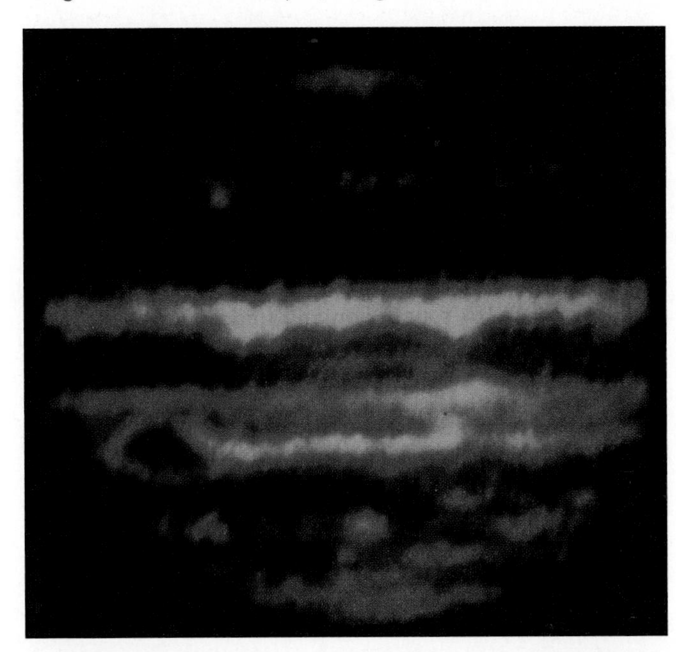

FIGURE 12.16
A Radio Image of Saturn and Its Rings
The bright regions in the radio map are relatively warm, the dark regions relatively cool. This image, obtained in February 1997, shows quite a different pattern of warm and cool regions than radio images of Saturn obtained in the 1980s.

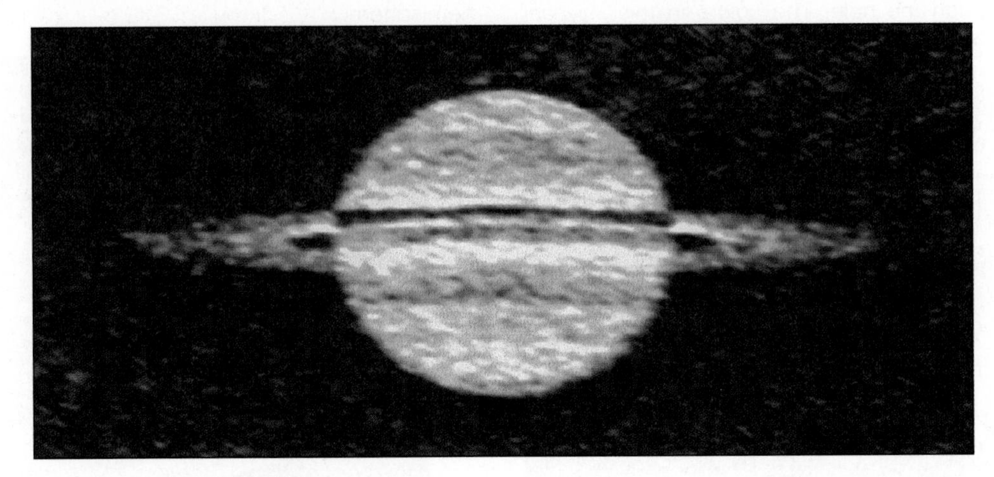

Light, pale clouds are generally cool, whereas dark clouds are warm. Thus, the belts of Jupiter, first pointed out in Figure 12.6, are warmer than the lighter zones with which they alternate. The white ovals are very cool whereas the brown ovals are warm. The Great Red Spot, on the other hand, is dark in the infrared map, indicating that it is cold even though you might expect it to be warm based on its dark color in the visible image.

The radio image of Saturn in Figure 12.16 shows a pattern of warm and cool atmospheric bands resembling those of Jupiter seen in Figure 12.15. This radio image of Saturn was obtained in 1997 and is remarkably different from the pattern of warm and cool bands seen in radio images from the 1980s. The changes in the pattern of bands suggests that Saturn's atmosphere undergoes significant temperature changes on a timescale of decades, possibly in response to Saturn's seasons.

Temperature, Depth, and Convection

The temperature variations seen on Jupiter are caused by differences in the depths of the clouds that we see. Temperature decreases with height in the troposphere, within which most of the clouds are found. This means that warm clouds lie relatively deep in the atmosphere, and cool clouds lie toward the top of the troposphere. The pattern of zones and belts is believed to reflect an organized convective pattern shown in Figure 12.17. The zones consist of rising columns of gas in which bright clouds form, just as water clouds form in rising columns of air on the Earth. Sinking columns of gas in the belts expose warmer regions lower in the atmosphere. Relatively hot spots, such as the brown ovals and some blue-gray regions, lie even deeper in the atmosphere than the dark clouds of the belts.

The reason the Great Red Spot and the white ovals are cool is that their clouds lie at the top of great rising columns

FIGURE 12.17
The Belt and Zone Structure of Jupiter's Atmosphere
The pattern of belts and zones is a result of the circulation pattern in Jupiter's atmosphere. Warm gases from the interior of Jupiter rise upward in the zones producing high, cool, light clouds. Descending gas in the belts produces low, warm, dark clouds.

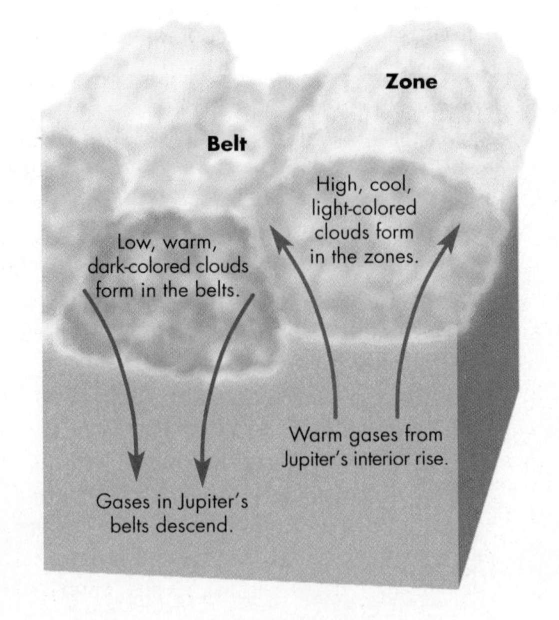
Zone
Belt
High, cool, light-colored clouds form in the zones.
Low, warm, dark-colored clouds form in the belts.
Warm gases from Jupiter's interior rise.
Gases in Jupiter's belts descend.

of gas. When these columns reach their maximum heights the gas they carry spreads outward. Under the influence of the Coriolis effect (see Chapter 8, Section 1), the outflowing gas produces an anticyclonic circulation pattern.

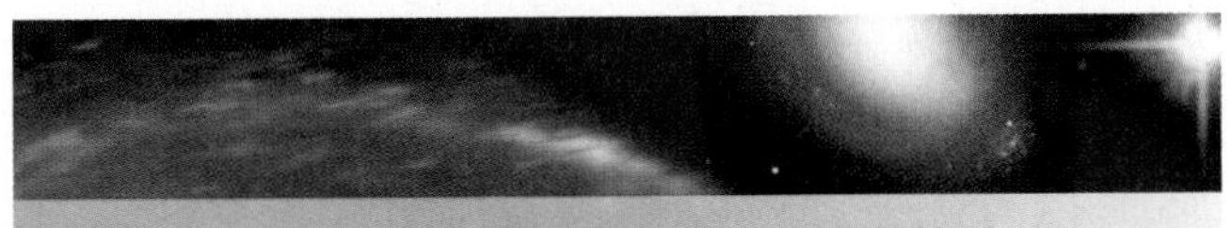

Infrared observations show that the dark atmospheric features of Jupiter are warmer and deeper than the light features. The Great Red Spot, very high and cool, is an exception to this pattern.

Atmospheric Coloration

Jupiter Many materials, including sulfur, sodium, phosphorus compounds, and complex organic molecules, have been proposed to account for the colors of Jupiter's clouds. It has been difficult to identify the molecules responsible for the colors of Jupiter, however, because many different compounds can produce the same colors. The only colored molecule that has been identified in Jupiter's clouds is phosphine (PH_3), which was detected by the *Galileo* entry probe. The way the colored compounds are formed is also unknown, although it seems likely that chemical reactions produced by the absorption of solar ultraviolet radiation may be important.

Colored molecules need to be present only in very tiny amounts in Jupiter's clouds to create the browns, blues, reds, and oranges observed. The clouds themselves are probably made primarily of ammonia, ammonium hydrosulfide (NH_4SH), and water, all of which form colorless (white) droplets and crystals. Water condenses at a higher temperature than ammonium hydrosulfide, whereas ammonia has the lowest condensation point of the three. This means that water clouds form deeper in the atmosphere than ammonium-sulfide clouds. Ammonia clouds are generally highest of all, as shown in Figure 12.18. There are clear regions between the layers of clouds. Because water vapor condenses relatively deep in the atmosphere, it is absent from the upper atmosphere of Jupiter just as it is absent from the upper parts of the Earth's atmosphere. The high, cool clouds of Jupiter are white because they consist of essentially pure ammonia crystals and lack colored compounds. They are the equivalent of cirrus clouds on the Earth.

The Great Red Spot is made up of the highest and coldest clouds on Jupiter even though it is red rather than white. A possible explanation for the dark color of the Great Red Spot involves the motions of the cloud droplets within it. Droplets that rise to the center of the Great Red Spot take months to reach the edge of the spot, where down drafts can return them to the deeper atmosphere.

FIGURE 12.18
The Cloud Layers of Jupiter
The highest clouds are made primarily of frozen ammonia. Beneath the layer of ammonia clouds is a cloud layer consisting mostly of ammonium hydrosulfide ice. The deepest cloud layer is mostly water ice.

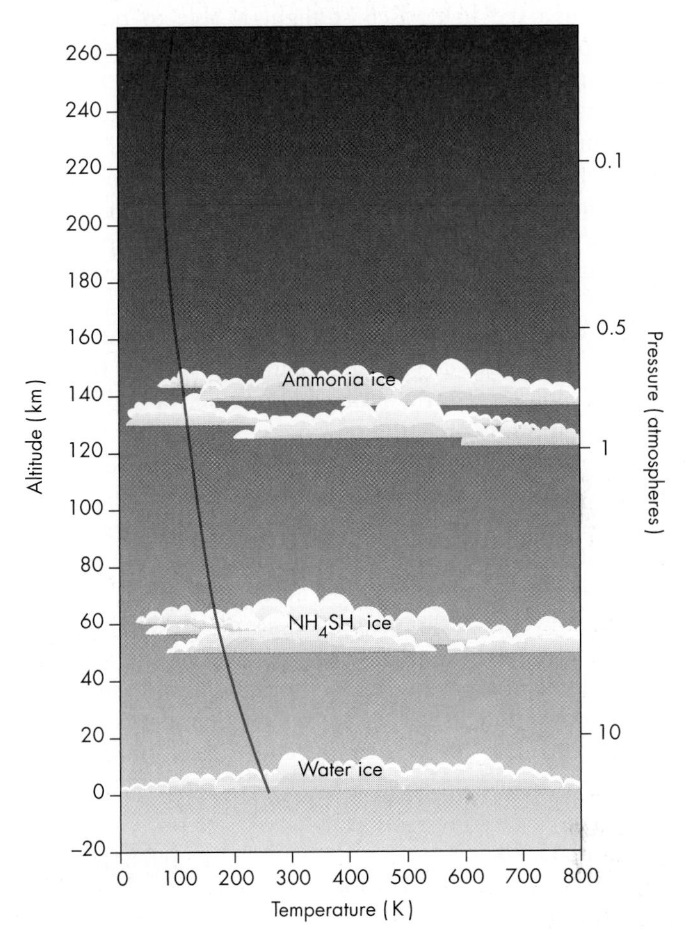

Thus the droplets are exposed to solar ultraviolet radiation for months on end and can undergo chemical reactions that might produce colored molecules. Alternatively, the updraft responsible for the Great Red Spot may be so strong that it lifts red-colored material from the lower atmosphere to heights the colored material can't normally reach.

The main layer of tan to brownish clouds is probably made of ammonia plus ammonium hydrosulfide. Again, other molecules are responsible for the coloration. The dark brown ovals are thought to be holes in the main cloud layer through which we can see deeper, warmer regions. The dark brown cloud layer seen in the ovals is too deep in the atmosphere for solar ultraviolet radiation to penetrate and cause photochemical reactions. The molecules responsible for the dark colors may be produced by lightning discharges or produced much deeper and brought upward in rising columns of gas.

The warmest, deepest parts of Jupiter's atmosphere that we can see are blue-gray regions near the equator. These may not be clouds at all. Instead, the color may be produced by the same mechanism that gives our sky its blue color. The molecules in the atmosphere of Jupiter, like those of the Earth's atmosphere, scatter blue light better than red light. Red sunlight is able to penetrate so deeply that it is absorbed by Jupiter's gases, whereas blue sunlight, which penetrates into cloud-free regions of Jupiter's atmosphere, is reflected back outward. Thus the sunlight reflected from cloud-free regions is bluish in color. Apparently the dark brown material seen in the ovals at high latitudes is absent near the equator so we can see especially deep into the atmosphere. The possible structure of the cloud layers of Jupiter is summarized in Figure 12.19. The *Galileo* entry probe apparently entered Jupiter's atmosphere in a relatively clear, almost water-free region that appears as a hot spot in infrared images. Instead of three cloud layers, the probe detected only a single layer of ammonium hydrosulfide droplets.

Figure 12.20 suggests what Jupiter's atmosphere would look like if we could float in it, supported by a balloon. If our balloon floated at a height where the pressure is about the same as at the Earth's surface, we would probably be in a relatively clear layer between decks of clouds. Above us would be towering masses of white ammonia clouds. Their shape and the patterns of light and shadow might look like our own clouds on a humid, summer afternoon. Below us the clouds would be more complex. Clouds of many colors, heights, and shapes would form and dissipate in columns of rising and falling gas. Those same strong up and down drafts would buffet the balloon and produce rapid rises and abrupt falls. Above and below us spectacular lightning flashes would light up the clouds. Occasionally we would pass over holes in the clouds and

FIGURE 12.19
The Locations of Various Colored Clouds in Jupiter's Atmosphere
Blue-gray regions may not be clouds at all, but rather holes through which we can see to great depths.

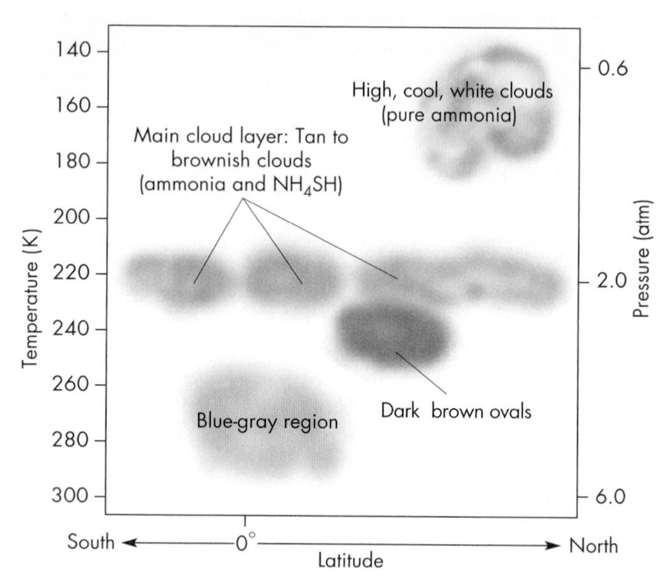

FIGURE 12.20
The View from Within Jupiter's Cloud Layers
This model, based on *Galileo* data, shows the high, light-colored clouds and reddish main cloud layer. The blue streaks are holes in the main cloud layer. Height variations are exaggerated.

stare 50 km downward into the bluish haze of the lower atmosphere. No solid or liquid surface would ever come into view.

There have been proposals to float just such a balloon (though without human observers) in Jupiter's atmosphere to carry out a long-term study of its atmospheric structure and dynamics. Someday we may be able to view pictures from Jupiter that show us incredible scenes we now can only imagine.

Saturn Saturn is twice as far from the Sun as Jupiter is and receives only 25% as much solar energy. This makes the temperature of Saturn's atmosphere cooler than Jupiter's. Saturn has layers of ammonia, ammonium hydrosulfide, and water clouds. These cloud layers are widely separated in height, with relatively clear regions in between each layer. The layers of clouds exist at about the same temperatures as the clouds on Jupiter, but they lie deeper in Saturn's atmosphere, so they are beneath much more gas.

It is difficult to explain the more subtle, uniform colors of Saturn. One possibility is that the greater depth of the cloud layers on Saturn protects them from the ultraviolet radiation that can produce chemical reactions resulting in colored molecules. Another possibility is that the ammonia-ice cloud layer of Saturn is thick enough to partially block our view of the deeper, colored regions.

The clouds of Jupiter and Saturn are made primarily of ammonia, ammonium hydrosulfide, and water, all of which are colorless. The colors seen in the two planets are produced by phosphine and other, unidentified, molecules. Both atmospheres contain several cloud layers with clear regions between them.

12.4 INTERIORS

Although little is known about the interiors of Jupiter and Saturn, several things are clear. First, the low average densities of the two planets require that their interiors consist mainly of hydrogen and helium, the lightest elements. Second, the temperatures and pressures in the deep interiors of Jupiter and Saturn must be very high. The pressure at the center of Jupiter is about 50 Mb. (Recall from Chapter 8 that the pressure at the center of the Earth is only 3.5 Mb.) The central pressure of Saturn, although not as great as that of Jupiter, is still greater than 10 Mb. Temperatures in the interiors of Jupiter and Saturn are expected to be high as well. The center of Jupiter is probably about 20,000 K and the center of Saturn is probably almost as hot.

Metallic Hydrogen

Under such extremes of temperature and pressure, hydrogen, which we normally think of as a gas, takes on other forms. As temperature and pressure increase below the clouds of Jupiter and Saturn, hydrogen gas becomes steadily denser until it eventually liquefies. There is probably no abrupt transition from gas to liquid. Instead, gas imperceptibly gives way to liquid H_2. At a pressure of 2 or 3 Mb, however, another change takes place, this time very abruptly. Electrons within the hydrogen molecules are squeezed away from the hydrogen protons and become free to flow throughout the liquid. This is how electrons behave in a metal, so this state of hydrogen is called **metallic hydrogen.** Metallic hydrogen behaves chemically like molten sodium or potassium and probably has a shiny appearance.

The electrons in metallic hydrogen can move freely through the liquid, so metallic hydrogen is a very good electrical conductor. This means that large electrical currents can flow within Jupiter and Saturn. Rapid rotation and vigorous convection within Jupiter and Saturn drive the currents, generating the planets' large magnetic fields. However, detailed models of Jupiter and Saturn's magnetic dynamos (or other planetary dynamos, for that matter) have not yet been worked out.

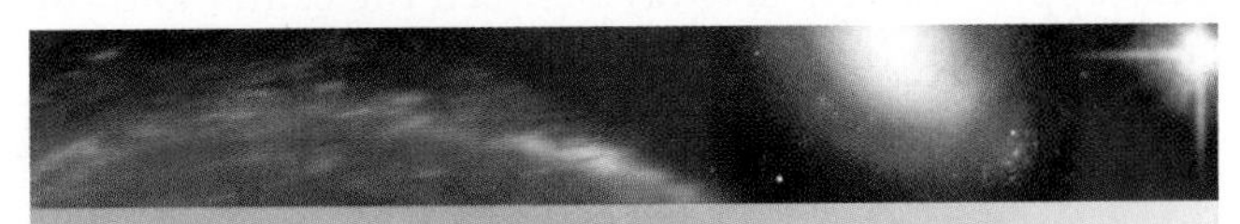

As depth increases in Jupiter and Saturn, hydrogen gradually changes from a gas to a liquid. Deep within both planets, hydrogen changes to a liquid, electrically conducting metal.

Internal Structure

The average densities of Jupiter and Saturn are too large for either planet to be made entirely of hydrogen and helium. Evidently both planets also contain some heavy materials such as rock, metal, and ice. Although it isn't completely clear how Jupiter's heavier elements are distributed within the planet, one possibility is that the rocky material and metals form a central core, whereas the icy material (ammonia, methane, and water) is mixed with the hydrogen and helium. Figure 12.21 shows a possible model of Jupiter's interior. The core contains about 10 Earth masses of rock and metal and is about as large as the Earth. Thus Jupiter, made mostly of hydrogen and helium, has about five times as much rocky material and metal as all the

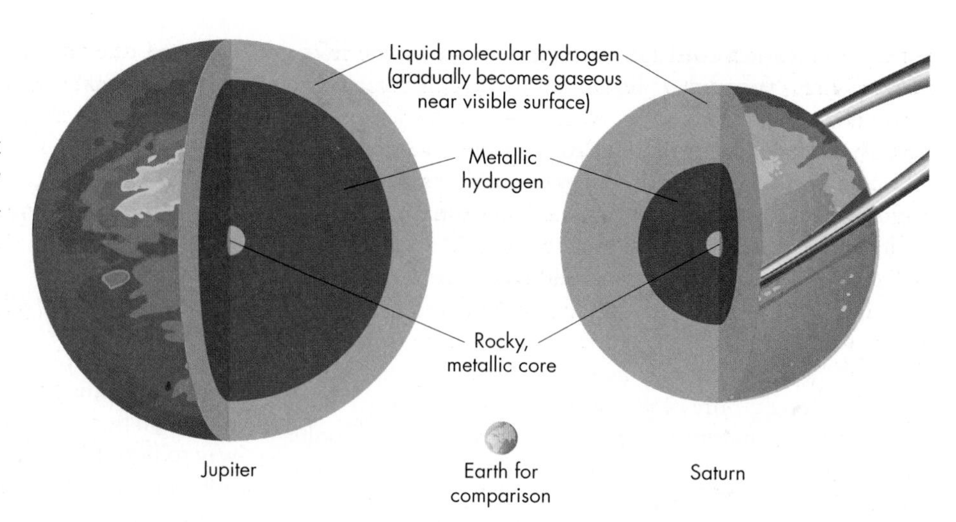

FIGURE 12.21
Models for the Interiors of Jupiter and Saturn
Rocky cores are shown, although it isn't known for certain that the heavy elements within Jupiter are concentrated in its core.

terrestrial planets put together. Outside the core is a thick shell of metallic hydrogen that extends outward to about 55,000 km, approximately 80% of the radius of Jupiter. The outer shell of Jupiter is a layer of liquid molecular hydrogen that gradually becomes gaseous near the visible surface.

The interior of Saturn is mostly hydrogen and helium as well. About 15 Earth masses of rock and ice are needed to account for the density of Saturn. The rock is believed to form the core of the planet. The internal pressure in Saturn, although lower than that in Jupiter, is still great enough to compress liquid hydrogen into metallic hydrogen. Figure 12.21 shows that this happens approximately halfway between the center and the surface of Saturn.

Both Jupiter and Saturn are too dense to be made entirely of hydrogen and helium. Instead, each contains some rock, metal, and ice. These materials are probably concentrated in the centers of Jupiter and Saturn.

Internal Energy

Astronomers discovered in the 1960s that Jupiter is unexpectedly bright in the infrared part of the spectrum. Subsequent measurements, both from the Earth and from space probes, have shown that Jupiter emits 70% more energy each second than can be accounted for by absorbed sunlight. Nearly 50% of the energy emitted by Saturn can't be accounted for as absorbed and reradiated sunlight. This means that Jupiter and Saturn are self-luminous.

If the Sun turned off abruptly, Jupiter's visible brightness, which is almost entirely due to reflected sunlight, would plummet. In the infrared, however, Jupiter's brightness would drop only by 60%. The typical atmospheric temperature of Jupiter would fall from 125 K to 100 K, and the layers of clouds might settle a little lower in Jupiter's cooler atmosphere. Other than that, nothing about Jupiter would change very much at all.

An oversimplified distinction between stars and planets is that stars are self-luminous whereas planets shine only by reflected light. Jupiter and Saturn obviously blur this distinction. Actually all planets are self-luminous to a limited extent. Radioactive decays inside the Earth, for example, produce energy that flows to the Earth's surface and is radiated into space. The difference between the Earth, and Jupiter and Saturn, however, is that the Earth's internally produced energy is tiny compared with the amount of energy it absorbs from sunlight, whereas the internal energy of Jupiter and Saturn is significant. In fact, some very faint stars produce only a little more internal energy per kilogram of matter than Jupiter and Saturn.

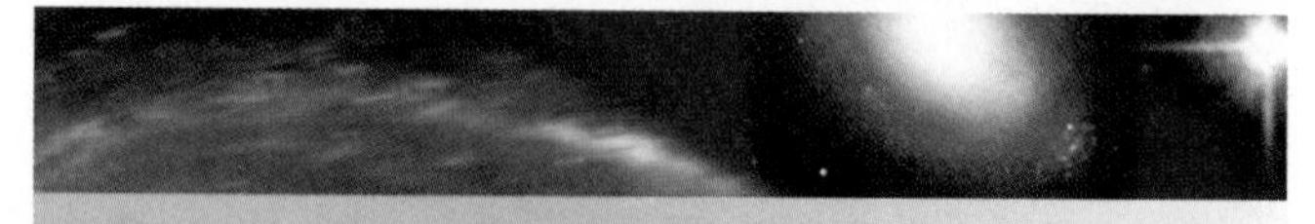

Both Jupiter and Saturn are self-luminous. Forty percent of the energy emitted by Jupiter and nearly 50% of the energy emitted by Saturn is due to internal energy sources.

12.5 THE INTERNAL ENERGY OF JUPITER AND SATURN

Just after the planetary system formed there was a brief period, perhaps 10,000 years, when Jupiter was almost 1% as bright as the Sun. However, it was only one quarter as hot as the Sun and would have appeared distinctly red in color. During that time, Jupiter was about as big as the Sun is now. If there had been an observer on the primitive Earth (not likely, because the Earth probably had a molten surface then), Jupiter would have been spectacular. Its angular size was about 25% as large as today's full Moon and its brilliance was equal to that of the full Moon. At night the landscape would have been dull red whenever Jupiter was in the sky. During that brief period of time, the solar system strongly resembled a binary star system.

Energy from Gravitational Contraction

Jupiter already contained all the mass that it has today and was collapsing under its own weight. Like any other massive body, when Jupiter shrank it released energy. To visualize how energy is released when a body contracts, imagine that the body consists of many separate layers. As each layer falls inward it is accelerated by the gravity of the layers beneath it. When it strikes the layer beneath it, it stops abruptly and its kinetic energy is converted to heat. In reality, a body collapses as a whole rather than in separate layers, but the result is the same—as the body shrinks, heat is released. The stronger the gravity of the body and the faster the rate of collapse, the greater the rate of heat production.

Jupiter shrank tenfold during the first million years of its collapse. Since that time it has shrunk by only another 40%. Today it is shrinking far too slowly for us to measure. At first the rate of collapse was so rapid that the energy released could not be transported to Jupiter's surface and radiated away fast enough. As a result, Jupiter heated up until its central temperature rose to 50,000 K. As the collapse slowed, the interior of Jupiter began to cool, and energy flowed to the surface faster than it was released by collapse. This is still the situation today. The conversion of gravitational energy is no longer rapid enough to match the energy Jupiter radiates away. Most of the energy now being radiated by Jupiter is internal heat that built up when Jupiter was collapsing rapidly.

Saturn's Internal Energy

The source of Saturn's internal energy is not as easy to explain as Jupiter's. Saturn is smaller than Jupiter, so it cooled much more quickly. As a result, the internal heat produced as Saturn shrank to its present size should have mostly dissipated by now. In fact, Saturn's luminosity should have declined to its present level about 2 billion years ago and fallen even further since then.

Saturn may be generating energy from the separation of the helium dissolved in its liquid metallic hydrogen. At temperatures above 10,000 K, helium is dissolved in metallic hydrogen. At lower temperatures it isn't. Instead, helium slowly settles inward through the metallic hydrogen, converting gravitational energy to heat. This probably doesn't happen within Jupiter because Jupiter is hot enough to keep helium dissolved in its metallic hydrogen. The interior of Saturn may be cool enough, however, that helium condenses at the top of the metallic hydrogen core and falls toward the center. As time passed, the envelope and atmosphere of Saturn should have become depleted in helium. This is exactly what measurements of the helium to hydrogen ratio in Saturn's atmosphere have shown.

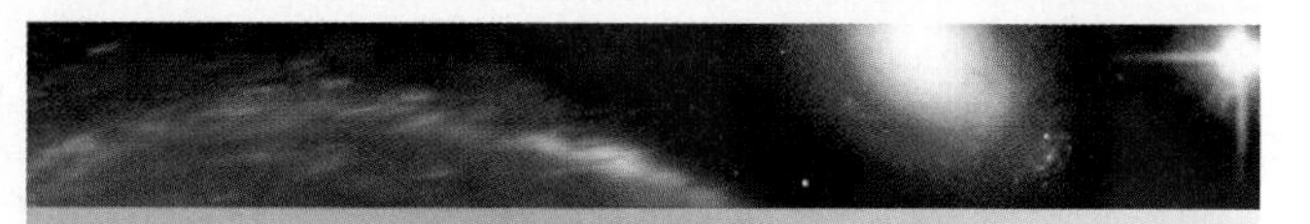

While it contracted to its present size, Jupiter converted large amounts of gravitational energy into heat. Early in its evolution Jupiter was almost 1% as bright as the Sun because of this internal heat. Heat stored from Jupiter's early contraction is the source of its present luminosity. Saturn's luminosity, on the other hand, may be due to helium that is still settling into its core.

12.6 MAGNETOSPHERES

The dynamos operating within Jupiter and Saturn produce strong magnetic fields. The magnetic field at Jupiter's cloud tops is 14 times as strong as the magnetic field at the surface of the Earth. This is especially impressive considering that Jupiter's diameter is 11 times the Earth's. Thus Jupiter's clouds are much farther from the internal dynamo that produces its magnetic field than the Earth's surface is from the Earth's dynamo. The magnetic field at Saturn's visible surface is about two-thirds as strong as the Earth's.

The magnetospheres of Jupiter and Saturn generally resemble the Earth's. They obstruct the flow of the solar wind, just as the Earth's magnetosphere does, and cause it to flow around them. The solar wind compresses the magnetic fields on the sunward sides and stretches them to great lengths on the night sides. The strong magnetic fields of Jupiter and Saturn produce magnetospheres much larger than the Earth's. Figure 12.22 shows that Jupiter's magnetosphere is ten times the diameter of the Sun. If it emitted visible light, Jupiter's magnetosphere would produce a glowing area covering four times the area of the full Moon. The magnetotail of Jupiter extends at least 650 million km behind Jupiter, well beyond the orbit of Saturn. Saturn's magnetosphere is smaller than Jupiter's, but still many millions of kilometers across. The inner magnetospheres of Jupiter and Saturn trap energetic ions and electrons in the Jovian equivalent of the Earth's Van Allen belts. When particles in these belts penetrate into the upper atmospheres of Jupiter and Saturn, they produce

FIGURE 12.22
Jupiter's Magnetosphere
The strong magnetic field of Jupiter carves a huge magnetosphere out of the solar wind. The size of the Sun is shown for comparison.

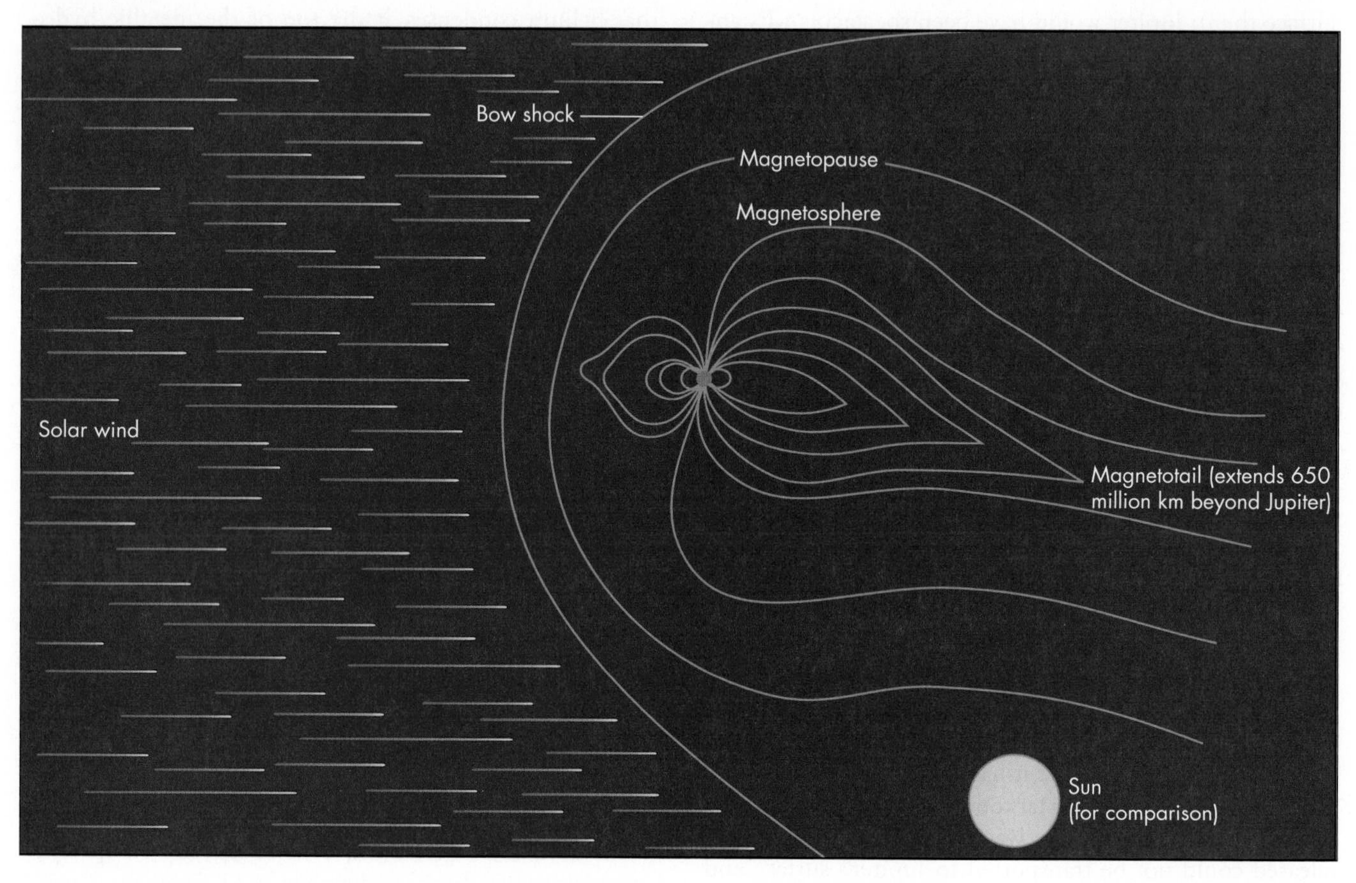

FIGURE 12.23
Aurorae and Lightning in Jupiter's Atmosphere
Aurorae are the bright arcs near the top of the image. The bright spots in the lower half of the image are lightning.

aurorae (Figure 12.23) like those seen in the Earth's polar regions, only brighter.

The ionized gas in the Earth's magnetosphere is captured from the solar wind. For Jupiter and Saturn, however, their satellites (Chapter 14) are the major source of gas. For example, volcanic eruptions on Jupiter's satellite Io eject 1000 kg (1 ton) of gas (sulfur, oxygen, sodium, and potassium) into space each second. This gas is then ionized by solar radiation and becomes trapped in Jupiter's magnetosphere. Nitrogen gas from the thick atmosphere of Saturn's satellite Titan escapes into Saturn's magnetosphere.

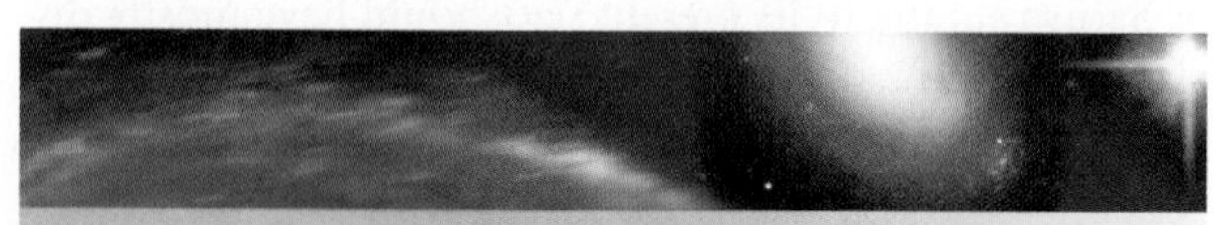

The dynamos within Jupiter and Saturn produce large magnetic fields and magnetospheres. The size of Jupiter's magnetosphere is many times larger than the Sun.

FIGURE 12.24
How the Appearance of Saturn's Rings Changes as Saturn Orbits the Sun
Saturn's rings are tilted 27° with respect to the plane of its orbit. This means that twice during each Saturn orbit, at intervals of about 15 years, there are periods of time when we see the rings edge on. At those times the extreme thinness of the rings makes them difficult to see.

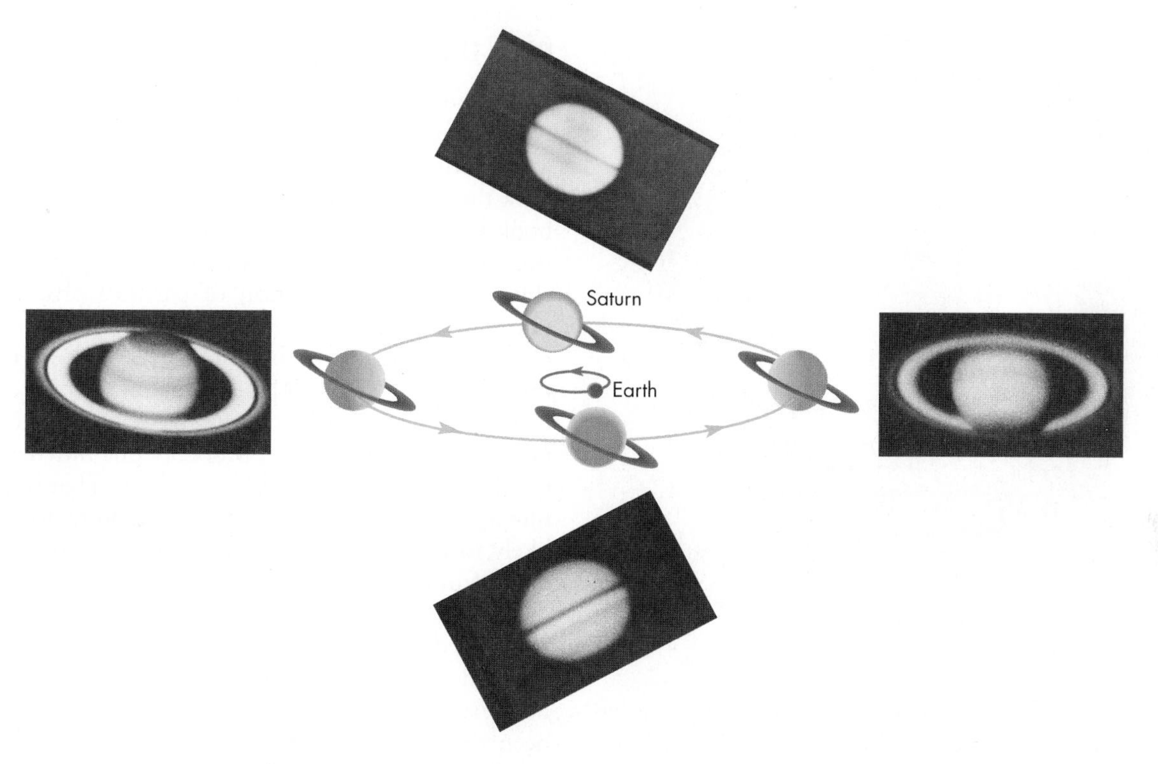

12.7 RINGS

The rings of Jupiter were discovered in 1979 during the *Voyager 1* flyby. By that time, Saturn's rings had been known for three and a half centuries. After Galileo first saw Saturn's "appendages," many other astronomers observed them without recognizing what they were. The nature of Saturn's rings was finally understood by the Dutch astronomer Christiaan Huygens in 1659, after he saw the appendages disappear in 1655 and reappear several years later. He suggested that the effect was caused by a flat ring that, when viewed edge on, was difficult to see. Figure 12.24 shows that because the rings are tipped 27° to Saturn's orbital plane, the rings appear and disappear twice during Saturn's 29.5-year orbit around the Sun. Figure 12.25 shows how the appearance of Saturn's rings changed between 1996 and 2000.

Astronomers speculated for centuries about whether the rings were solid sheets, a liquid layer, or large numbers of individually rotating particles. In 1850 Edouard Roche showed that the rings were located inside the distance where tidal forces from Saturn would tear a satellite apart. In 1859 James Clerk Maxwell pointed out that if the rings of Saturn were solid disks, they would be torn apart by their rotation. Observational confirmation that the rings of

FIGURE 12.25
***Hubble Space Telescope* Images of Saturn's Rings from 1996 to 2000**
In 1996 (lower left) Saturn's rings were seen nearly edge on. By 2000 (upper right), the rings were seen more nearly face on.

FIGURE 12.26
The Rotation of Solid Rings
Because all parts of a solid ring must rotate in the same amount of time, the outer part of the ring must move faster than the inner part.

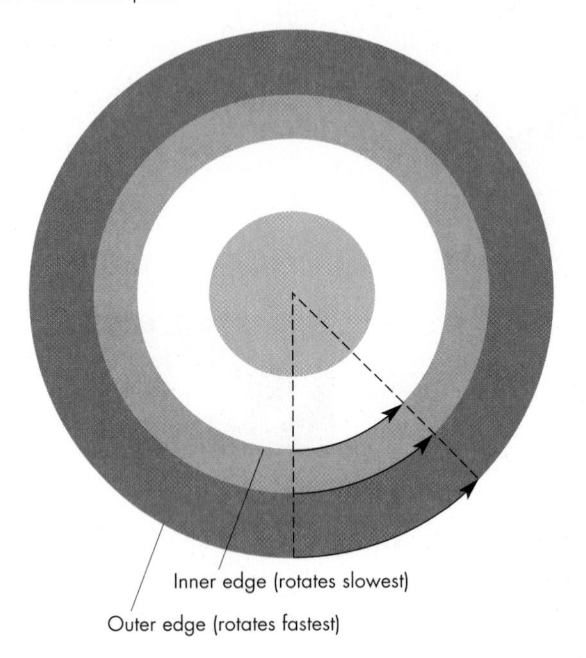

Saturn aren't solid disks came in 1895 when James Keeler measured the Doppler shift of the sunlight reflected from different parts of the rings. He found that the orbital speed of the outer edge of the rings was less than the orbital speed of the inner edge of the rings. This was consistent with rings made up of many individually orbiting particles. Figure 12.26 shows that, if the rings were solid sheets, then both the inner and outer edges would have to rotate in the same period of time. Therefore, the orbital speed of the outer edge would be greater than the orbital speed of the inner edge. In fact, the rings are made up of a myriad of individually orbiting particles that range from 1 cm to tens of meters in diameter. Small particles outnumber large ones, and most of the particles are bits of water ice or rocks covered by ice.

Ring Dimensions

As Figure 12.27 shows, the rings of Jupiter and Saturn are extremely thin. Saturn's rings are about 200,000 km across but, in most places, less than 200 m thick. To have the same ratio of width to thickness, a page of this book would have to be 100 m across. The rings are so thin because of collisions among the ring particles. The rings began as a collection of particles orbiting in the same general direction, but with individual orbits that carried them out of what would become the plane of the rings (Figure 12.28). While on the part of its orbit that passed through the ring plane, a particle was likely to collide with another particle, probably one moving through the ring plane in the opposite direction. That is, a particle moving through the ring plane from north to south was likely to collide with a particle moving through the ring plane from south to north. When a collision occurred, the north/south velocity of each particle was reduced or eliminated, leaving each particle moving in an orbit less inclined to the plane than its old orbit. Over time, the north/south motions of the entire swarm of particles have been nearly eliminated, and now their orbits lie almost exactly in the same plane.

The flattening of a collection of orbiting particles (or gas clouds) is a very general process. It occurred in the solar system before and during the formation of the planets. It is going on today in our galaxy as collisions among orbiting gas clouds are slowly flattening the shape of the Milky Way.

Figure 12.29 is a picture of Jupiter's ring taken by *Voyager 1* as it approached Jupiter. The main part of

FIGURE 12.27
The Thinness of Saturn's Rings
This *Hubble Space Telescope* image was taken in August 1995, when the Earth crossed the plane of Saturn's rings and the rings were seen edge on. Saturn's largest satellite, Titan, is above the rings at the left. Titan's shadow can be seen as a black dot on Saturn. Four other moons lie near the rings at the right. The dark stripe on Saturn is the shadow of the rings.

FIGURE 12.28
Collisions and the Flattening of Saturn's Rings
Collisions between ring particles tend to reduce the north/south component of the velocities of the particles. As a result, the rings have become very flat and thin.

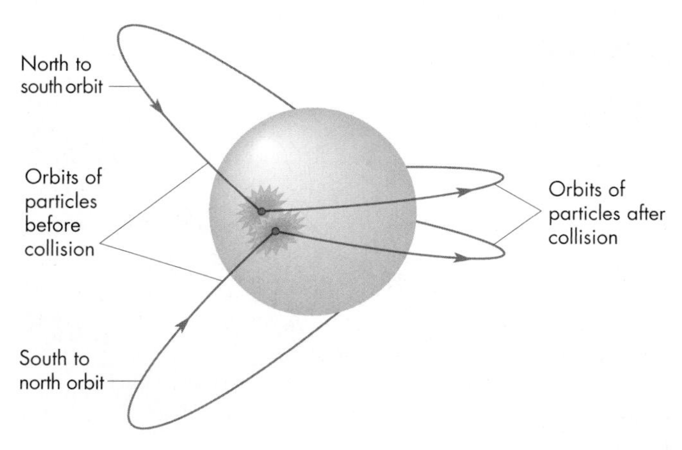

FIGURE 12.29
Jupiter's Ring
This image, taken by *Voyager 1* as it approached Jupiter, shows that the main part of the ring is very narrow.

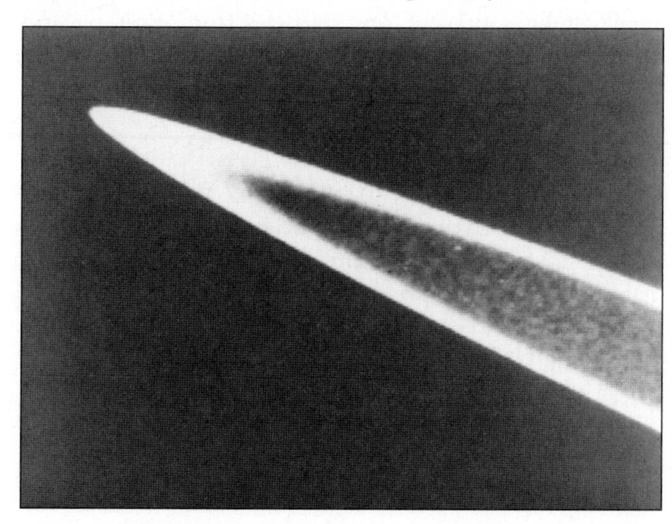

Jupiter's ring is only 7000 km wide, although there are very faint regions both inside and outside the main ring. Figure 12.30 is a picture of the ring taken after *Voyager 2* had passed by the planet. Jupiter's ring is more than 20 times as bright when looking back at the ring than when viewed while approaching. To account for this, the ring must contain many very small particles. Such small particles should last only a very short time—perhaps only a few thousand years—so the dust in Jupiter's ring must be continually resupplied. The supply of small particles probably comes from meteoroids striking three small satellites that orbit within the rings and from volcanos on Jupiter's satellite Io.

The rings of Jupiter and Saturn are made of myriad particles following individual orbits. Collisions among ring particles are responsible for the thinness of the rings.

Structure of Saturn's Rings

Structure in Saturn's rings was first discovered in 1675 when Giovanni Cassini found a gap about two-thirds of the way outward from the inner edge of the rings. This gap, known as the **Cassini division,** divided the rings into the inner B ring and outer A ring. The C ring, much dimmer than the A and B rings, lies inside the B ring and wasn't discovered until 1848.

The *Voyager* and *Cassini* images showed that Saturn's rings are much more complex than had been suspected. Figure 12.31 shows that, instead of a few gaps, there are many gaps, and instead of three large rings, there are a large number of individual, thin ringlets. Some of the ringlets and the spaces between them are only 20 to 30 km wide. The spaces between the ringlets aren't actually empty either. Instead, they are regions with fewer ring particles to reflect light. The Cassini division has some small ringlets within it.

Detailed information about the structure of Saturn's rings was obtained when the *Voyager 2* spacecraft recorded the brightness of the star δ Scorpii as it passed behind the rings. The star appeared bright when viewed through parts of the rings where there were few particles to extinguish

FIGURE 12.30
Jupiter's Ring as Seen from Behind Jupiter
The Sun is eclipsed by Jupiter in this image. The bright arcs are sunlight scattered through Jupiter's atmosphere. The ring is much brighter than when it is seen with the Sun at the viewer's back, as in Figure 12.29.

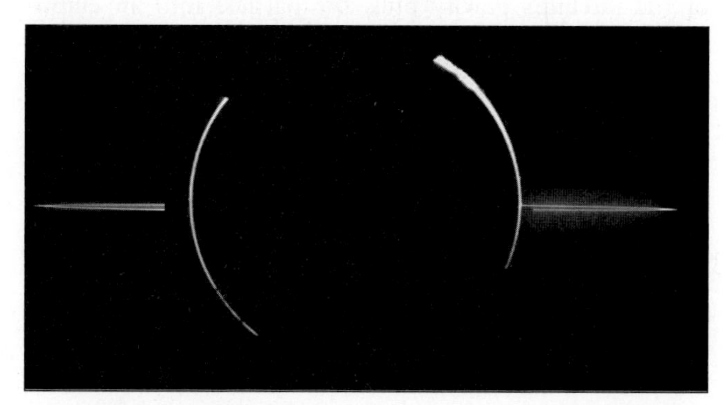

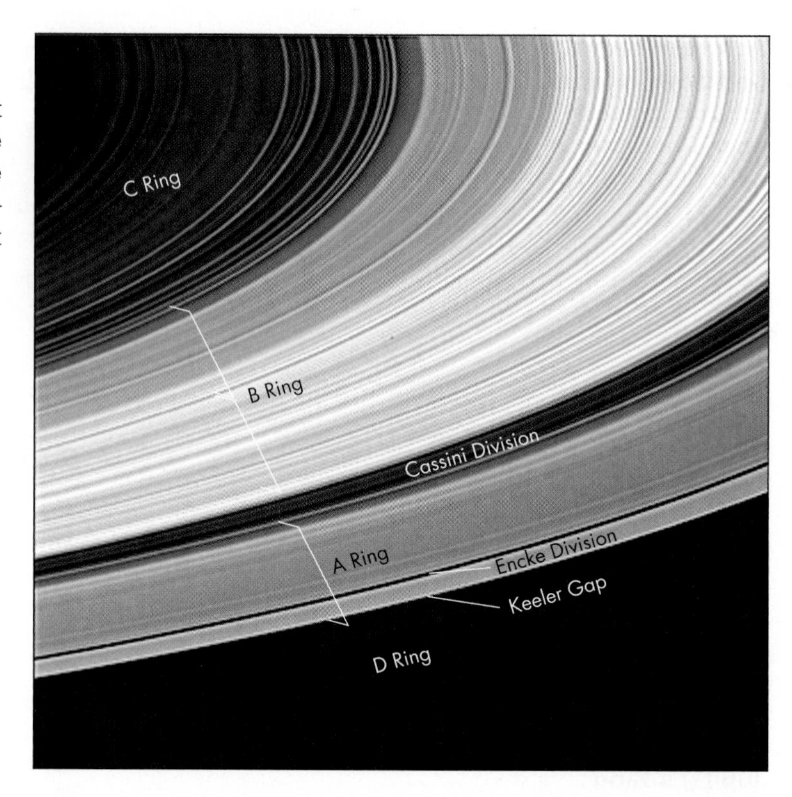

FIGURE 12.31
Saturn's Rings
This *Cassini* image shows that Saturn's rings consist of many narrow ringlets. The A and B rings, which are the brightest of Saturn's rings, are indicated on the image along with the Cassini division, which separates the A and B rings. Notice that there are bright ringlets within the Cassini division.

its light. When δ Scorpii passed behind parts of the rings where there were many particles, its brightness dropped dramatically. The measurements showed that the number of particles in the rings changes in distances as short as 100 m. The detailed structure of part of the B ring is shown in Figure 12.32. The *Voyager* images also showed that there is a very dim D ring inside the C ring, a narrow, uneven F ring outside the main system of rings, and two broad, faint rings (G and E) beyond the F ring.

Resonances

Part of the structure found in Saturn's rings is due to **resonances,** which occur at positions in the rings where the ring particles have orbital periods that are integer fractions of the orbital periods of Saturn's satellites. Suppose a ring particle has an orbital period exactly one-half as long as that of a satellite, as shown in Figure 12.33. The particle completes two orbits for every orbit of the satellite. The repetitive tug of the satellite's gravity pulls the particle into an elliptical orbit. The disturbed orbits near the resonance cause the resonant particles to collide more often with neighbors and deplete the resonant region of particles. An example of a resonance occurs at the inner edge of the Cassini division where ring particles have one-half the orbital period of the satellite Mimas. The structure of the rings, however, is far too complicated to be explained by satellite resonances alone. Most of the "gaps" between ringlets do not correspond to a resonance with any known satellite.

Satellites affect the structure of Saturn's rings in other ways as well. The satellites Pandora and Prometheus orbit just inside and just outside the very narrow F ring. These small satellites prevent ring particles from escaping and give the F ring a knotted structure. The gravitational disturbance of a satellite can also cause ring particles to bunch

FIGURE 12.32
The Detailed Structure of Saturn's B Ring
Voyager 2 showed the amount of ring material at different distances from Saturn by measuring the amount of light from a bright star that passed through the ring material. This figure shows the relative amount of material in a part of the B ring that is 2000 km wide. The abrupt changes in the amount of ring material with distance show that there is structure in the rings on scales of 10 km and smaller.

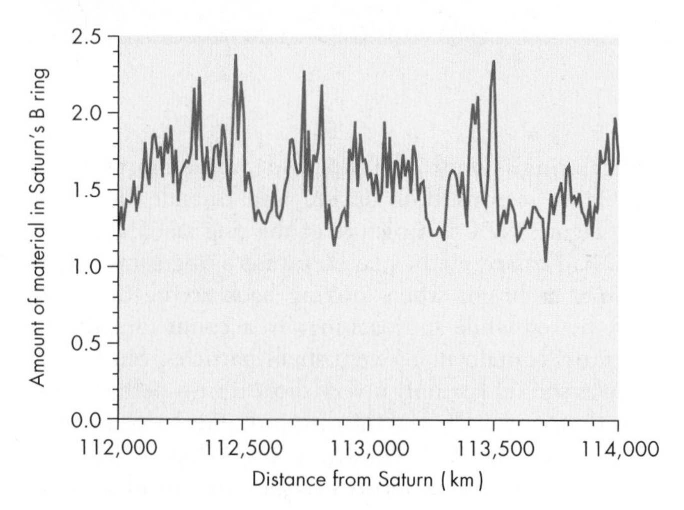

FIGURE 12.33
A 2:1 Resonance
A ring particle completes two orbits in the same time that a satellite completes one orbit. The repeated tug when the two are closest alters the particle's orbit into an elliptical one, increasing collisions with neighboring particles.

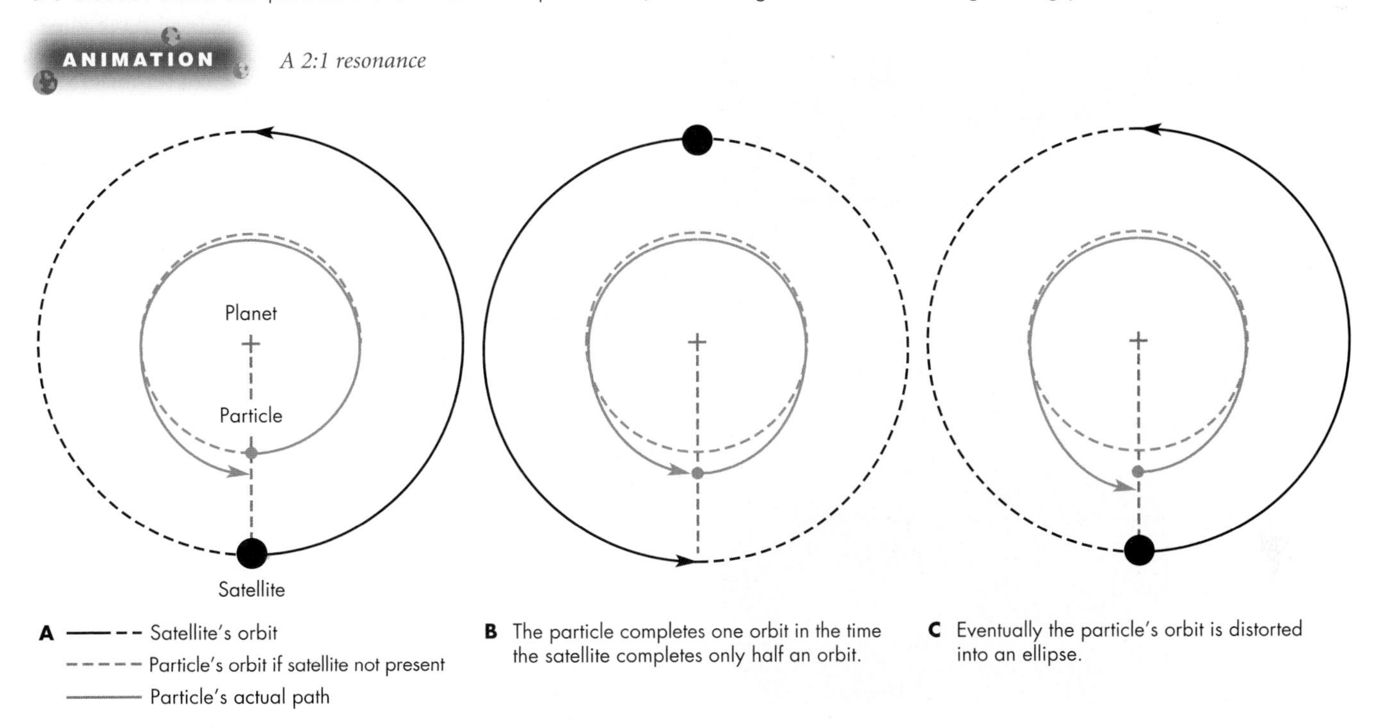
ANIMATION
A 2:1 resonance
Planet
Particle
Satellite
A Satellite's orbit
Particle's orbit if satellite not present
Particle's actual path
B The particle completes one orbit in the time the satellite completes only half an orbit.
C Eventually the particle's orbit is distorted into an ellipse.

up at different distances from Saturn. The result is a spiral density wave like the one shown in Figure 12.34.

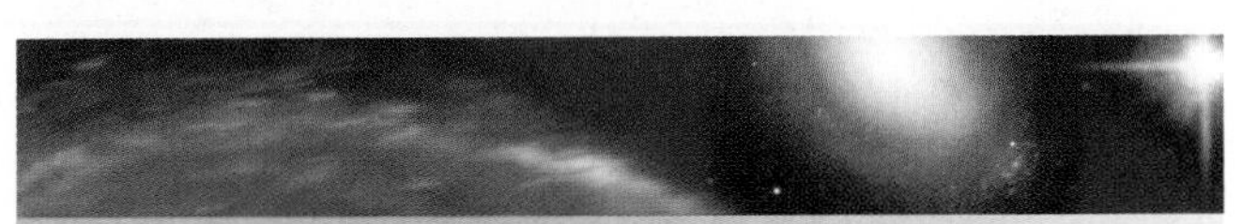

The rings of Saturn consist of many individual ringlets. Some of the spaces between ringlets can be accounted for by gravitational interactions between ring particles and Saturn's moons.

Spokes

The *Voyager* images also show short-lived features that form and dissolve in the B ring. Figure 12.35 shows several of these features, called **spokes**, which appear as dark fingers that form very quickly and stretch radially outward for tens of thousands of kilometers through the rings. The formation of a spoke possibly begins with the impact of a small meteoroid on the rings. The impact vaporizes and ionizes material in the rings, and the ionized gas quickly spreads radially across the rings. Tiny ring particles that come in contact with the ionized gas become negatively charged. They are then levitated above the rings by electromagnetic forces produced by the rotating magnetic field of Saturn. While above the rings, the tiny particles partially block sunlight from reaching the B ring beneath them. The shadow cast by the levitated particles appears as a spoke.

FIGURE 12.34
A Spiral Density Wave in Saturn's Rings
This *Cassini* image shows a spiral density wave in a 125-mile-wide region of Saturn's A ring. The density wave is a result of bunching up of ring particles at different distances from Saturn, leaving relatively empty regions in between.

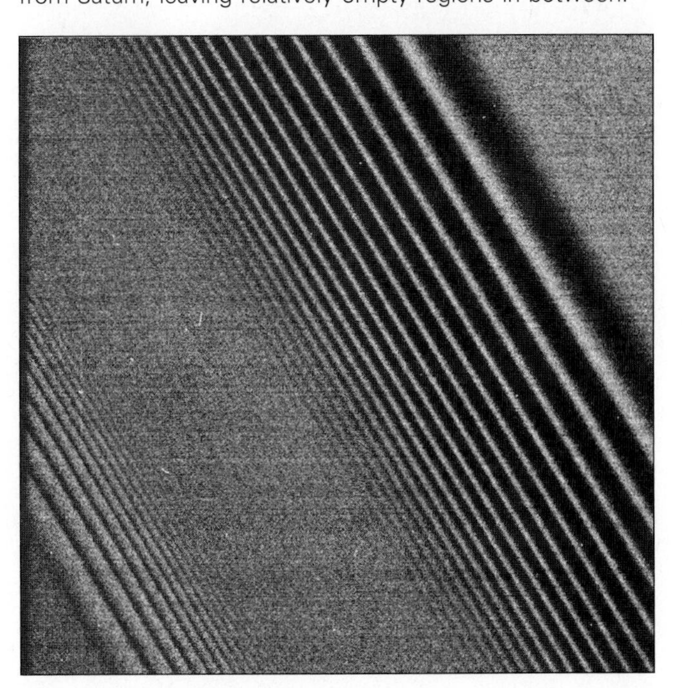

FIGURE 12.35
Ring Spokes
These dark radial features are thought to be shadows that occur when electromagnetic forces levitate small particles above the main ring.

The spokes in Saturn's rings were first seen from the Earth in the nineteenth century. They were seen in images taken by the *Voyager*'s in the early 1980s. When *Cassini* began taking images of the rings in July 2004, the spokes were gone. In late 2005, they reappeared. Clearly, there are aspects of the spokes that are not yet understood.

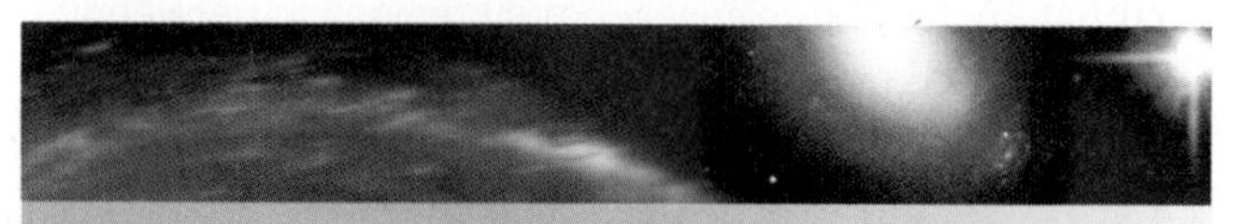

Spokes are dark, radial structures that form rapidly in Saturn's rings. The spokes may be caused by the lifting of small particles above the rings by electromagnetic forces. The particles then shadow the rings.

The Formation and Evolution of Planetary Rings

The **Roche distance** is the minimum distance at which a body held together by its own gravity can orbit a planet without being pulled apart by tidal forces. The Roche distance depends slightly on the density of the body, but as a general rule, is about 2.5 times the radius of the planet. Figure 12.36 shows that most of the material in the rings of Saturn lies inside the Roche distance. Planetary rings may have formed from primordial material that would

FIGURE 12.36
The Roche Distance
Inside the Roche distance, bodies held together by their own gravity are pulled apart by tidal forces. Most of the material in the rings of Saturn is located within the Roche distance.

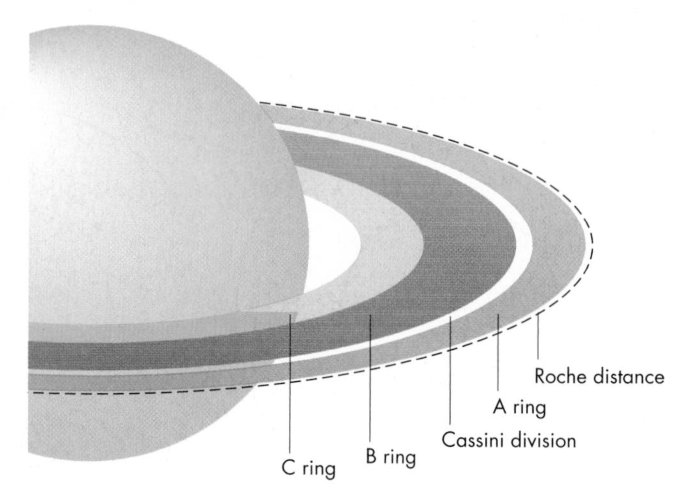

have become a satellite if it had been outside the Roche distance. However, it now appears that the length of time required for planetary rings to dissipate is relatively short. As a ring system evolves, individual ring particles work their way slowly inward or outward through the rings. If they reach the outer edge of the rings, they leave the ring system. If they move inward far enough, they encounter the tenuous outer layers of the planet's atmosphere and are destroyed. It may take a few hundred million years for a ring system to fade away. This means that the ring systems that now exist are quite young compared with the age of the solar system. Therefore it is more likely that planetary rings formed when a body, possibly a satellite, ventured within the Roche distance of a planet and was destroyed by the planet's tidal forces. The fact that we see rings around all four of the giant planets demonstrates that catastrophic destruction of meteoroids or satellites must produce new ring systems at least as often as old ring systems fade away.

Most of the material in the rings of planets lies within the Roche distance, the distance at which an object would be torn apart by tidal forces. Planetary rings probably formed relatively recently when a body, perhaps a satellite, approached a planet too closely and was destroyed. There are many features of planetary rings, such as their complex structure and their spokes, that are not yet completely understood.

Chapter Summary

- Jupiter and Saturn have much larger diameters and are much more massive than the other planets in the solar system. Their densities, however, are lower than the terrestrial planets. This implies that Jupiter and Saturn are made up primarily of the light elements, hydrogen and helium. (Section 12.2)
- The parts of Jupiter and Saturn that we can see are atmospheric clouds. For both planets, rotation period varies with latitude, producing a system of zonal winds similar to the Earth's jet stream. (12.3)
- Jupiter and Saturn have tropospheres in which pressure and temperature decrease steadily with altitude. The atmospheres of Jupiter and Saturn are made primarily of hydrogen and helium. (12.3)
- Jupiter shows many cloud features. One of these, the Great Red Spot, is an atmospheric storm that has persisted for centuries. Other features, such as white ovals, last for decades. Saturn also shows many cloud features, but not as many as Jupiter. (12.3)
- The clouds of Jupiter and Saturn are separated into several decks with clear regions in between. They consist primarily of ammonia, ammonium hydrosulfide, and water, all of which are colorless. Most of the compounds that color the clouds brown, blue, red, and orange have not yet been identified. (12.3)
- The dark cloud features of Jupiter and Saturn are warmer and lie deeper in the atmosphere than the light cloud features. The Great Red Spot is an exception, however. It is one of the coldest, highest atmospheric features on the planet. (12.3)
- In the atmospheres of Jupiter and Saturn, hydrogen exists as a molecular gas. Beneath the atmospheres, however, as the pressure increases, hydrogen gradually turns into a liquid. Even deeper within each planet, hydrogen changes abruptly to a metallic liquid. (12.4)
- Both Jupiter and Saturn are too dense to be made entirely of hydrogen and helium. They must contain other, heavier materials, such as rock, ice, and metal. It is likely that the heavy materials form the core of Saturn and possibly Jupiter. (12.4)
- Both Jupiter and Saturn are self-luminous. For each planet, about one-half of the energy emitted is derived from internal sources. (12.4)
- Just after it formed Jupiter was much larger than it is now. While it contracted to its present size, Jupiter converted large amounts of gravitational energy into heat. For a brief time during its early contraction, it was almost 1% as luminous as the Sun. Today Jupiter's luminosity is derived from energy stored from the time when it was contracting rapidly. Although Saturn also converted gravitational energy into heat as it formed, most of the planet's present internal energy has a different origin. Possibly it is due to the separation of hydrogen and helium within Saturn. (12.5)
- Jupiter and Saturn have large magnetic fields and enormous magnetospheres. Jupiter's magnetotail extends beyond the orbit of Saturn. (12.6)
- The rings of Jupiter and Saturn consist of many individually orbiting particles. Collisions among the particles have gradually brought all of the particle orbits into the same plane. Thus the rings are extremely thin compared with their diameters. (12.7)
- The rings of Saturn consist of many individual ringlets separated by gaps. Some of the gaps are caused by the gravitational influences of Saturn's satellites. (12.7)
- Spokes, dark radial features in Saturn's rings, may be the shadows of minute particles that have been lifted above the rings by electromagnetic forces. (12.7)
- Most of the material in planetary rings lies inside the Roche distance, the distance from the planet within which bodies are destroyed by tidal forces. Planetary rings were probably formed relatively recently by the destruction of a body, perhaps a satellite, which ventured inside the planet's Roche distance. (12.7)

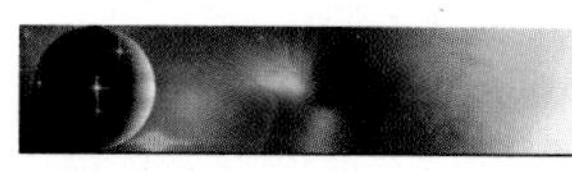

Key Terms

belts *270*
Cassini division *285*
equatorial jet *271*
Great Red Spot *270*
metallic hydrogen *279*
resonances *286*
Roche distance *288*
spokes *287*
zonal winds *271*
zones *270*

Conceptual Questions

1. Why are Jupiter and Saturn flattened?
2. Suppose there were a narrow belt near Jupiter's equator that had a rotation period of 9^h45^m. In which direction would the zonal winds blow in that belt relative to the rest of the equatorial region?
3. Why are oxygen compounds such as carbon monoxide and carbon dioxide extremely low in abundance in the atmospheres of Jupiter and Saturn?
4. How do the pressure and temperature in the visible layers of the atmosphere of Jupiter compare with those at the surface of the Earth?
5. The large sizes and rapid rotations of Jupiter and Saturn result in very large Coriolis effects in their atmospheres. How are the large Coriolis effects on Jupiter and Saturn related to the appearance of their cloud features?
6. What evidence do we have that the Great Red Spot of Jupiter is a high-pressure region with outflowing gas?
7. What is the relationship among color, temperature, and height for the atmospheric features of Jupiter and Saturn?
8. Why are there at least three separate cloud decks on Jupiter?
9. Why do the deepest parts of Jupiter's atmosphere that we can see have a bluish color?
10. What forms does hydrogen take in Jupiter from the surface to the center of the planet?
11. What evidence do we have that Jupiter and Saturn must be made of more than hydrogen and helium alone?
12. What would happen to the temperature at the Earth's surface if the Sun stopped shining? How about the atmosphere of Jupiter?
13. About 60% of the energy radiated away by Jupiter comes from its interior. What is the origin of that excess energy? What about the case of Saturn?
14. How do the size and shape of Jupiter's magnetosphere compare with that of the Earth?
15. Why are Saturn's rings conspicuous at times and very difficult to see at other times?
16. What evidence do we have that Saturn's rings aren't solid disks?
17. What did *Voyager* tell us about Saturn's rings that wasn't already known from telescopic observations from the Earth?
18. What is thought to be the nature of the spokes seen in Saturn's rings?
19. What happens to an asteroid that passes within the Roche distance of a planet?

Problems

1. What percentage of all planetary material in the solar system is contained in Jupiter and Saturn together?
2. What is the ratio of Jupiter's volume to the volume of the Earth?

Figure-Based Questions

1. Use Figure 12.1 to find how fast *Voyager 2* was moving at the slowest point in its trip through the solar system.
2. Use Figure 12.18 to find the temperature in the part of Jupiter's atmosphere where the pressure is 1 atmosphere.
3. Use Figure 12.21 to find the fraction of Saturn's radius that marks the outer boundary of the region of metallic hydrogen.
4. How many distinct "ringlets" can be identified in Figure 12.32 between 113,000 km and 113,200 km from Saturn?

Group Activity

With your group, plan a balloon flight in Jupiter's atmosphere (see Figure 12.20). Include a list of the important observations and measurements you would like to make. Discuss the hazards of the flight. If possible, make a brief presentation of your preparations in class.

For More Information

Visit the text website at **www.mhhe.com/fix** for chapter quizzes, interactive learning exercises, and other study tools.

13

On Tuesday the 13th of March, between ten and eleven in the evening, while I was examining the small stars in the neighborhood of H Geminorum, I perceived one that appeared visibly larger than the rest: being struck with its uncommon magnitude, I compared it to H Geminorum and the small star in the quartile between Auriga and Gemini, and finding it so much larger than either of them, suspected it to be a comet.

William Herschel, 1781

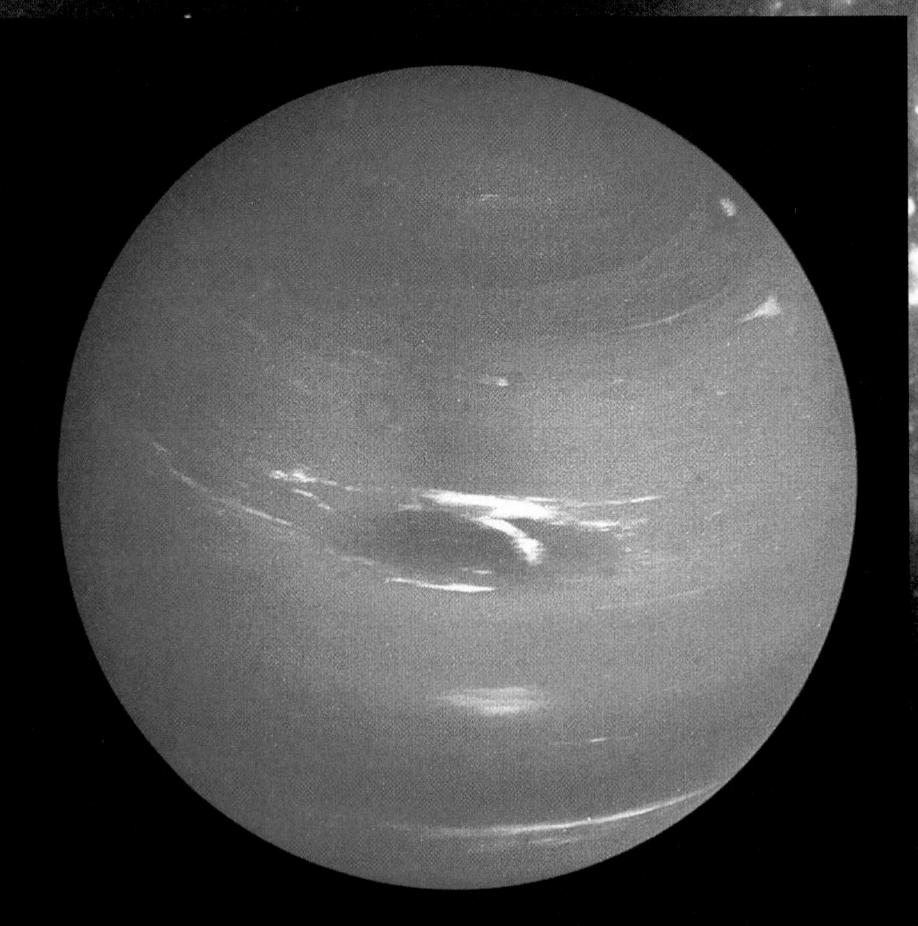

Neptune

www.mhhe.com/fix

The Outer Solar System

Beyond Saturn we find the frontier of the solar system. This is a region that is expensive and time-consuming to visit with space probes and difficult to observe because the faintness of sunlight makes outer solar system bodies much dimmer than they would be in the inner solar system. The two outer solar system planets, Uranus and Neptune, are the only ones that haven't been known since prehistoric times. Each of them has been visited only once by a space probe, whereas all of the other planets have been visited multiple times and, except for Mercury, studied in detail by one or more orbiters and landers.

Just as the geographic frontier on Earth was the place where new discoveries were made, new discoveries are being made regularly in the outer solar system. In the last 20 years astronomers have discovered over 1000 Trans-Neptunian Objects (TNOs), an entirely new class of objects that are possibly the most numerous kind of bodies in the solar system. Several TNOs are about as large or larger than Pluto and led to the reclassification of Pluto from the smallest planet to the first known TNO.

The TNOs have proved to be a very important group of bodies in that their orbital properties give astronomers important clues about the early history of the solar system. These clues suggest that the solar system was originally quite different from what we see today and evolved to its present form because of the interactions between the planets and an originally much larger population of TNOs.

This chapter will first describe what is known about Uranus and Neptune and then discuss what we are learning about the vast reaches beyond the planets.

Questions to Explore

- How did astronomers discover Uranus and Neptune?
- Why are Uranus and Neptune so different in appearance?
- Why is Neptune self-luminous whereas Uranus is not?
- What do we know about the ring systems of Uranus and Neptune?
- What kind of objects lie beyond Neptune?
- What role did Trans-Neptunian Objects possibly play in the early evolution of the solar system?

13.1 DISCOVERIES

Until 1781 astronomers believed that the orbit of Saturn marked the outer edge of the solar system. The existence of Uranus and Neptune was unknown.

Uranus

William Herschel's discovery of Uranus in March 1781 is usually described as an accident. This is a little misleading. It is true that Herschel didn't set out to find a new planet. However, the discovery is best described as serendipitous rather than accidental or lucky. William Herschel (1738–1822) was a professional musician who began a new career as an astronomer at the age of 30. He discovered Uranus in March 1781 while he was working on a project to measure the brightnesses and positions of all the naked-eye stars and the brighter stars that can be seen through a small telescope. Because Herschel was extremely thorough and because Uranus was bright enough to fall within the range of brightness he set out to study, it was inevitable that he would observe it during the course of his observations. Herschel is acknowledged as the discoverer of Uranus even though it must have been observed by many people from prehistoric times onward. Uranus is bright enough to be seen with the unaided eye but had always been mistaken for a faint star. In fact, it was recorded on at least two dozen maps of the sky made before Herschel began his work. Herschel was the first person to notice the motion of Uranus and realize that it wasn't just another faint star.

On March 13, 1781, Herschel noticed that a "star" in the constellation Taurus seemed too large to be a star. Thinking that the object might be a comet, he observed it with eyepieces of successively greater magnification and soon found that it had an angular size of about 4 seconds of arc—much larger than the angular sizes that the stars near its position appeared to have. Four nights later he observed it again and found that it had moved relative to the stars by about 1 minute of arc. Its motion relative to the stars confirmed that the object was within the solar system. Herschel, still thinking that he had found a comet, wrote to other astronomers, who soon confirmed his discovery. Because no tail could be seen and over many months the object grew neither brighter nor fainter, other astronomers eventually became convinced that Herschel had discovered a new planet rather than a comet.

Herschel named the new planet "Georgium Sidus," or George's star, after the ruling monarch of Britain, but this name soon faded from use. The name Uranus, originally proposed by the German astronomer Johann Bode, eventually became accepted. Bode proposed the name because, in Roman mythology, Uranus was the father of Saturn just as Saturn was the father of Jupiter.

The discovery of Uranus made Herschel famous. King George III granted him an annual stipend that made it possible for him to give up his musical career and become a full-time astronomer. He built a number of large telescopes. With them he carried out important studies of double stars, the Milky Way, and nebulae that are described later in this book.

Neptune

The discovery of Uranus started astronomers wondering if other undiscovered planets might lie beyond Uranus. By 1821 it had become clear that Uranus's motion about the Sun showed deviations from a simple elliptical shape. These deviations could not be accounted for by the gravitational attractions of the other known planets. The deviations were especially large for observations of Uranus that dated back to 1690, many years before Herschel's discovery. Because Uranus's orbital period is 84 years, the observations between 1690 and 1821 spanned more than one and a half orbital periods. Astronomers found that it was impossible to find a single orbit for Uranus that was consistent with both the older and newer observations. It was widely acknowledged that an additional planet was exerting a gravitational influence, or **perturbation,** on Uranus.

Fortunately, **celestial mechanics,** the aspect of astronomy that deals with the motions of celestial bodies under the influence of gravity, had matured considerably since Newton first made it possible to calculate orbits. The perturbations of planets on the motions of other planets, comets, and asteroids could be calculated. When it became clear that something was perturbing the motion of Uranus, several astronomers tried to deduce the orbit of the perturber.

One of the attempts was made by John Couch Adams, a mathematics student at Cambridge, who came up with several rather vague predictions for the location of the planet responsible for the motion of Uranus. His various calculations differed from one another by as much as 20°. In the summer of 1846, astronomers at Cambridge used his results in a fruitless search for the planet.

Meanwhile, also in 1846, a young French astronomer named Urbain Leverrier published the results of his own calculations for the orbit and position of the unknown planet. Leverrier became discouraged by his inability to get French astronomers to search for a planet at his predicted position. In September 1846, Leverrier wrote to Johann Galle at the Berlin Observatory. Galle already had an accurate map of the region of sky near Leverrier's calculated position. On September 23, 1846, the same day he got the letter from Leverrier, Galle and a student compared the map with what they saw through the telescope and almost immediately found Neptune.

To save face, British astronomers manufactured a version of what had happened in England, which resulted in Adams being given credit as the codiscoverer of Neptune. According to the British story, Adams had made an accurate prediction of the position of the unknown planet.

He sent his calculation to Sir George Airy, the Astronomer Royal of Great Britain, who didn't take him seriously. Later, when Airy saw that Adams's position was within 2° of the position published by Leverrier, he told James Challis, the director of the Cambridge Observatory, to search for the planet in the region of the sky predicted by Adams and Leverrier. By the time the search was carried out, however, Galle had already discovered Neptune.

This story ignores the fact that Adams made many predictions and that only one of his earlier attempts was close to the actual position of Neptune. It also ignores the search carried out at Cambridge in the summer of 1846. The full story was kept secret from 1846 until the Royal Greenwich Observatory's Neptune file was made public in 1998. It now seems apparent that credit for the discovery of Neptune should be given solely to Leverrier, who made a clear and accurate prediction of Neptune's position, something that Adams never really did.

The discovery of Neptune is often cited as a triumph of celestial mechanics. In reality, though, it required some luck, too. Leverrier assumed that the unknown planet had an orbit twice as large as that of Uranus, just as Uranus's orbit was twice as large as the orbit of Saturn. This meant that the missing planet had a semimajor axis of 38 AU and an orbital period of 217 years. To account for the effect of the planet on the orbit of Uranus, Leverrier calculated that the planet had an eccentricity of 0.1 and was at perihelion in 1820. In reality, Neptune has an almost perfectly circular orbit with a distance of 30 AU and a period of 164 years. Only in 1846 were the real Neptune and the "hypothetical" Neptune at nearly the same position in the sky. If the search had been carried out a few years earlier or later, Neptune would have been far away from the predicted position and might not have been found.

Given the data and the methods used by Leverrier, there are many possible orbits that could account for the perturbations on the orbit of Uranus. Leverrier found an orbit that coincidentally put the calculated planet near enough to the actual planet that Neptune was discovered.

The circumstances of the discoveries of Uranus and Neptune seem, at first glance, to be very different. Both discoveries, however, resulted from a combination of luck and hard work.

13.2 URANUS AND NEPTUNE

Prior to the *Voyager 2* flybys, Uranus and Neptune were thought to be so much alike that they were often described as "twin planets." The *Voyager* data showed, however, that they are different in a number of important ways.

Basic Properties

Some important properties of Uranus and Neptune are given in Tables 13.1 and 13.2. Uranus and Neptune have masses and diameters that fall about midway between those of the Earth and Jupiter. Each is about 4 times as large as the Earth and about one-third the diameter of Jupiter. Each is about 15 times as massive as the Earth and about 1⁄20 as massive as Jupiter. Neptune, both a little smaller and

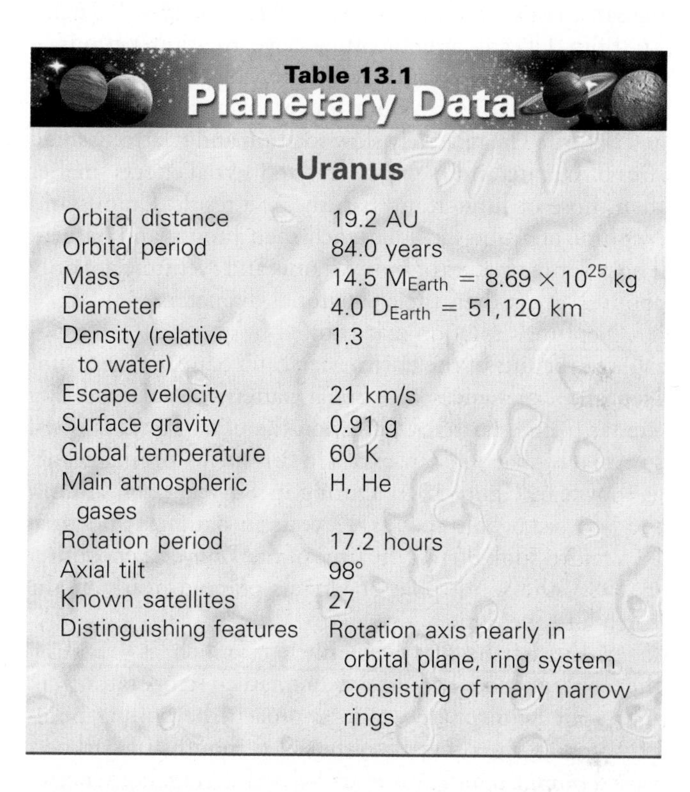

Table 13.1 Planetary Data

Uranus	
Orbital distance	19.2 AU
Orbital period	84.0 years
Mass	$14.5\ M_{Earth} = 8.69 \times 10^{25}$ kg
Diameter	$4.0\ D_{Earth} = 51{,}120$ km
Density (relative to water)	1.3
Escape velocity	21 km/s
Surface gravity	0.91 g
Global temperature	60 K
Main atmospheric gases	H, He
Rotation period	17.2 hours
Axial tilt	98°
Known satellites	27
Distinguishing features	Rotation axis nearly in orbital plane, ring system consisting of many narrow rings

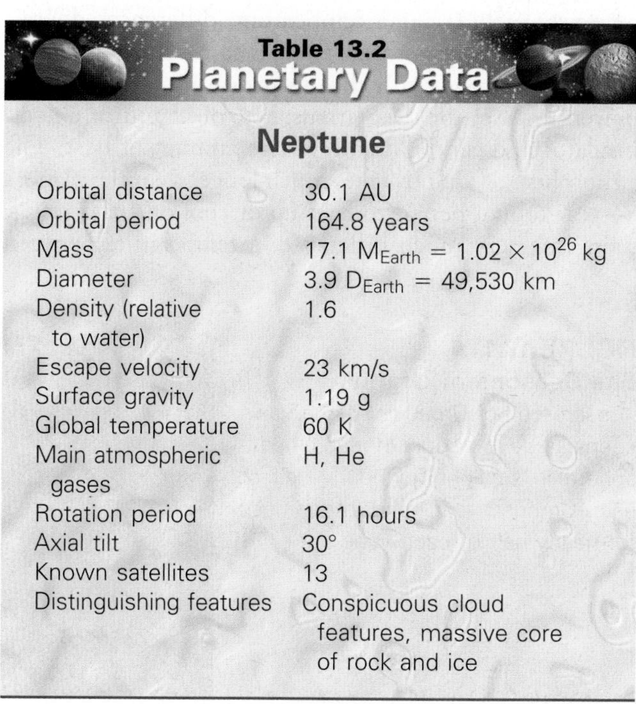

Table 13.2 Planetary Data

Neptune	
Orbital distance	30.1 AU
Orbital period	164.8 years
Mass	$17.1\ M_{Earth} = 1.02 \times 10^{26}$ kg
Diameter	$3.9\ D_{Earth} = 49{,}530$ km
Density (relative to water)	1.6
Escape velocity	23 km/s
Surface gravity	1.19 g
Global temperature	60 K
Main atmospheric gases	H, He
Rotation period	16.1 hours
Axial tilt	30°
Known satellites	13
Distinguishing features	Conspicuous cloud features, massive core of rock and ice

a little more massive than Uranus, has an average density of 1640 kg/m^3, about 20% larger than the average density of Jupiter. Uranus's average density is 1290 kg/m^3, about the same as Jupiter's. Both Uranus and Neptune are small compared with Jupiter and Saturn so their internal pressures aren't capable of compressing hydrogen and helium enough to produce their observed densities. Instead, water is likely to be a major constituent of Uranus and Neptune. In fact, each of them is only slightly larger and less dense than a planet of the same mass would be if it were made entirely of water.

Both Uranus and Neptune have rotation periods—17^h14^m for Uranus and 16^h7^m for Neptune—that are shorter than the Earth's but longer than those of Jupiter and Saturn. The relatively slow rotation and relatively small sizes of Uranus and Neptune lead to inertial effects smaller than those of Jupiter and Saturn. As a result, Uranus and Neptune are much less flattened than Jupiter and Saturn. The polar diameters of both Uranus and Neptune are only about 2% less than their equatorial diameters.

Neptune's rotation axis is tilted by about 29°, comparable to the tilts of the Earth, Mars, and Saturn. As a result, Neptune experiences a seasonal pattern of solar heating similar to the Earth's. Uranus, on the other hand, is tilted so severely that its spin axis is nearly in its orbital plane, as shown in Figure 13.1. During its 84-year orbit around the Sun, each pole spends 42 years in darkness and then 42 years in sunlight. At the time of the *Voyager 2* encounter in 1986, the south pole of Uranus pointed nearly at the Earth and the Sun.

If the Earth's axis were tilted as much as Uranus's, the consequences would be dramatic. Temperatures in the sunlit hemisphere would approach the boiling point of water. In the other hemisphere, 6 months of darkness would plunge temperatures far below freezing. For Uranus, however, such extreme temperature differences do not occur. One reason is that Uranus is so much farther from the Sun than the Earth that solar heating is only 1⁄400 as powerful. Also, the deep atmosphere of Uranus is a good insulator and can hold a very large amount of heat. The atmosphere of Uranus takes much longer than the planet's 84-year orbital period to heat up or cool off. As a result, there isn't very much difference in temperature between the hemispheres turned toward and away from the Sun. Averaged over an entire Uranian year, the polar regions actually experience more solar heating than the equator does. This, obviously, is very different from the situation for planets with smaller axial tilts.

Uranus and Neptune have similar masses, diameters, and rotation periods, but very different axial tilts. Neptune's tilt is similar to the Earth's, but Uranus's is so great that its poles are almost in its orbital plane.

Atmospheres

The *Voyager 2* images of Uranus are colorful but disappointingly uniform, as Figure 13.2 shows. The striking blue color of Uranus is caused by methane gas, which strongly absorbs red sunlight leaving only blue light to be reflected back into space. Although methane provides the color of Uranus, most of the atmosphere is made of hydrogen and helium in essentially the same proportion as in the Sun. Very few individual clouds were seen in the *Voyager 2* images and those that could be seen were only slightly different from their surroundings in brightness and color. As Uranus rotated, these cloud features were tracked to find the pattern of zonal winds. However, because the northern hemisphere of Uranus was in darkness and couldn't be observed at all and because only a few cloud features could be seen in the southern hemisphere, the zonal wind pattern of Uranus is not very well known. The trend of wind speed latitude is shown in Figure 13.3. The pattern appears to be similar to the Earth's, with eastward winds at midlatitudes and westward winds near the equator.

Since *Voyager 2* flew past Uranus in 1986, the planet has completed a quarter of an orbit. This means that the equator rather than the south pole of Uranus is turned toward the Sun. Both hemispheres now spend about half of the

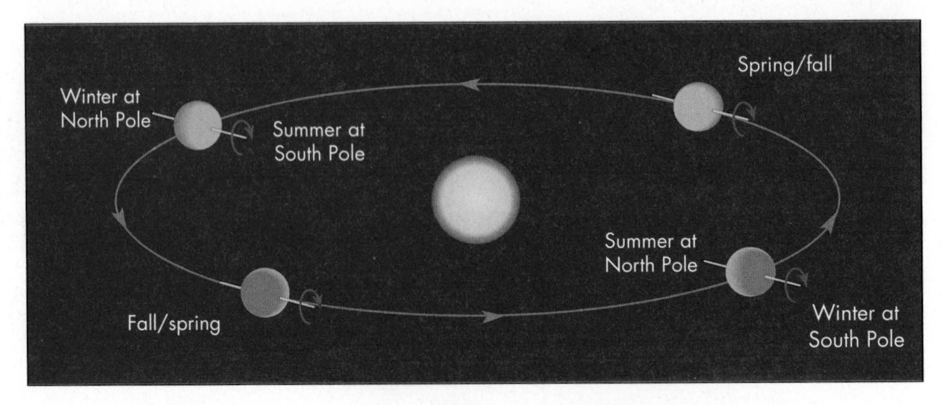

FIGURE 13.1
The Seasons of Uranus
The large tilt of Uranus's rotation axis keeps large portions of each hemisphere in sunlight for nearly half of each Uranus "year" and in darkness for nearly half of each year.

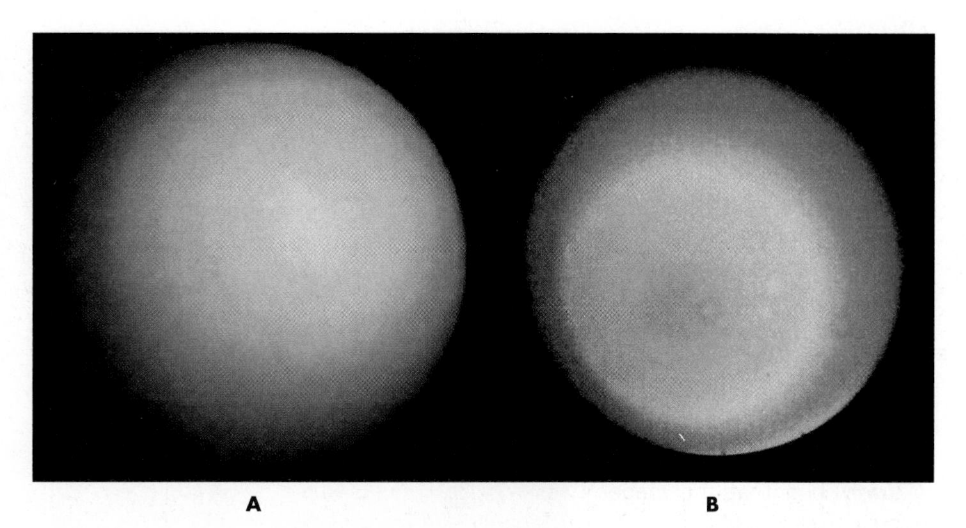

FIGURE 13.2
A *Voyager 2* Image of Uranus
A, Uranus appears as a nearly featureless blue ball. **B,** Only by enhancing the images of Uranus so that small differences in color and brightness were greatly exaggerated was it possible to find cloud features in the atmosphere of Uranus.

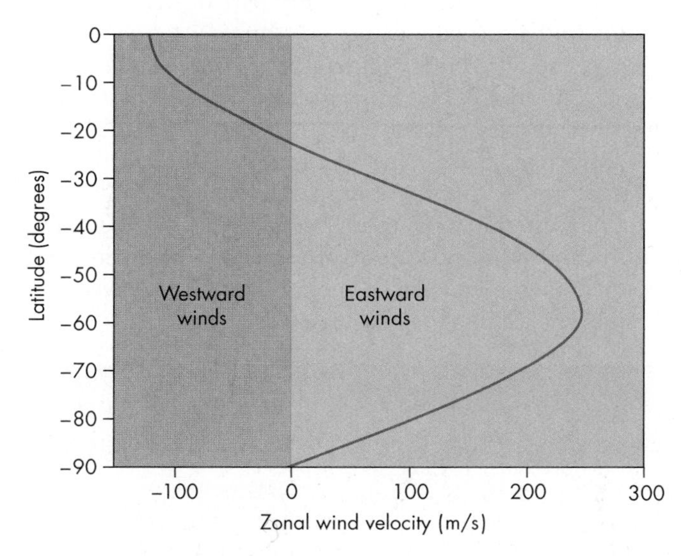

FIGURE 13.3
The Zonal Winds of Uranus
The pattern of wind speed with latitude shown here is an estimate based on measurements at the few latitudes at which clouds could be seen. Notice that Uranus's zonal wind pattern seems to resemble the Earth's, with westward winds (negative zonal velocities) near the equator and eastward winds (positive zonal velocities) at high latitudes. No information about Uranus's northern hemisphere was obtained because that hemisphere was in darkness during the *Voyager 2* flyby.

Uranus day in sunlight. Uranus's northern hemisphere is now in view. Infrared images of Uranus taken with one of the Keck telescopes using adaptive optics have shown numerous bright clouds and storms, many of them in the northern hemisphere, as shown in Figure 13.4. The reason the storms have appeared is not known, but may be part of a seasonal pattern. Astronomers speculate that near the solstices, when the hemispheres of Uranus are tipped toward or away from the Sun, the atmosphere becomes stratified, inhibiting convection and storms. During spring and fall, however, whatever inhibits convection disappears and storms occur. This suggests that storms and cloud features will become less numerous as summer approaches in Uranus's northern hemisphere. It will take decades to check this idea, however, because the next solstice won't occur until 2028.

Figure 13.5 shows the temperature profile for the troposphere of Uranus as well as for Jupiter, Saturn, and Neptune. Notice that the temperature declines with decreasing pressure just as for the tropospheres of Jupiter and Saturn. Uranus's atmosphere, however, is considerably cooler for a given pressure. A uniform layer of methane clouds lies at the level where pressure is about 1 atm. There are probably ammonia and water clouds below the methane cloud layer, but these are too deep to be visible.

Neptune is blue like Uranus, but Figure 13.6 shows atmospheric storms, a pattern of belts and zones, and bright cloud features that resemble Jupiter and Saturn more than Uranus at the time of the *Voyager 2* flyby. Neptune is blue because its atmosphere contains methane. In fact, most of the bright clouds are made of ice crystals of methane. Beneath the layer of methane clouds is a very clear atmosphere above a deeper deck of ammonia or hydrogen sulfide clouds. Figure 13.7 shows very high, bright methane clouds casting shadows on the main cloud deck 100 km below. The pattern of Neptune's zonal winds was found to be very similar to Uranus's, with weak eastward winds at intermediate latitudes and a very strong westward jet at the equator. The equatorial winds blow at speeds of more than 2200 km/h. These are the fastest winds found in any planet's atmosphere.

Neptune's Great Dark Spot, shown in Figure 13.8, was a storm system that strongly resembled Jupiter's Great Red Spot. Both were high-pressure regions and were located at about 20° S latitude. The Great Dark Spot was almost the same size as the Earth, so it was about one-half to one-third the size of the Great Red Spot. *Hubble Space Telescope* observations of Neptune showed that the Great Dark Spot had disappeared by 1994. Other *Hubble* images showed that a new great dark spot, this one in Neptune's northern hemisphere, had formed.

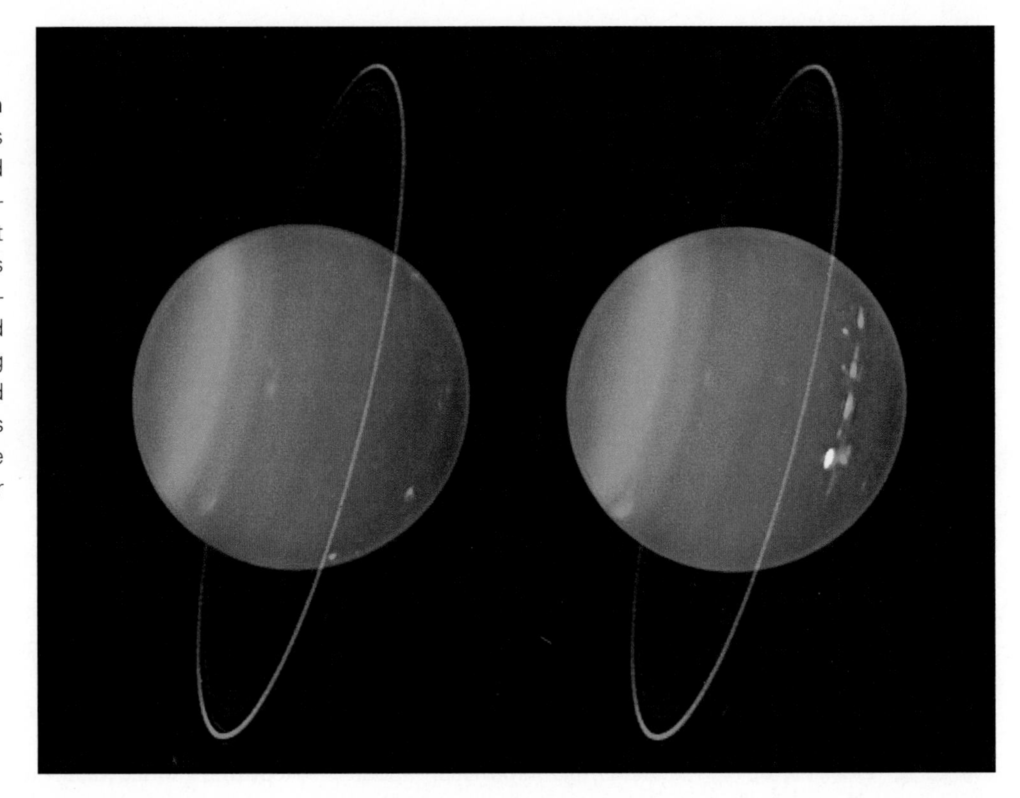

FIGURE 13.4
Storms on Uranus
These infrared images taken with one of the Keck telescopes show many bright clouds and storms in the northern hemisphere of Uranus, on the right side of the images. The images are false color, with blue indicating shorter wavelength infrared emission and red indicating longer wavelength infrared emission. The rings of Uranus appear red because they are relatively bright at the longer infrared wavelengths.

FIGURE 13.5
The Atmospheric Temperature Profiles of the Outer Planets
Pressure is used instead of altitude to mark the different depths in the atmospheres of the four planets. Like Jupiter and Saturn, Uranus and Neptune have tropospheres in which temperature falls with increasing altitude and decreasing pressure. Methane clouds lie at depths where the temperature is about 80 K and ammonia clouds lie where the temperature is about 140 K.

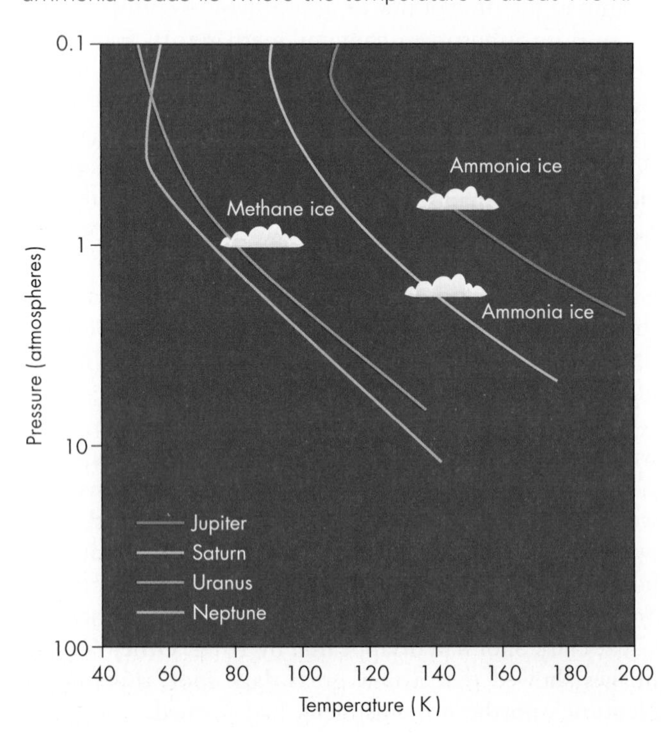

The reason that Neptune always shows atmospheric storms while Uranus does so only part of the time may be due to the extreme seasons of Uranus. Alternatively, it may be a result of the fact that Uranus's atmosphere isn't heated from below by energy produced inside Uranus. As described in more detail later in this chapter, Neptune, like Jupiter and Saturn, has a large amount of heat flowing outward from its interior into its atmosphere. Uranus does not. The internal heating of the atmospheres of Jupiter, Saturn, and Neptune leads to atmospheric convection and weather. In the absence of internally driven convection, weather on Uranus may be seasonal.

Both Uranus and Neptune appear blue in the *Voyager* images because both have methane gas in their atmospheres. Neptune shows a pattern of belts and zones as well as storms and bright clouds. In contrast, cloud features are sometimes absent from Uranus, possibly because there is no internal energy source to drive convection.

Internal Structure

There are many ways that rock, ice, and gas could be combined and distributed within Uranus and Neptune to match their average densities. Many of these possibilities could be ruled out, however, after *Voyager 2* measured the

FIGURE 13.6
A Map of Neptune's Cloud Features at the Time of the *Voyager 2* Encounter
The pattern of belts and zones resembles those of Jupiter and Saturn. Notice the bright cloud features and oval storms.

FIGURE 13.7
Cloud Shadows on Neptune
High bright clouds cast shadows on the cloud deck 100 km lower in the atmosphere. Sunlight shines from the bottom of the image toward the top.

FIGURE 13.8
Neptune's Great Dark Spot
The Great Dark Spot resembled the Great Red Spot of Jupiter in that both were located at about 20° S latitude and both covered about the same range of longitude and latitude.

ANIMATION *The rotation of Neptune*

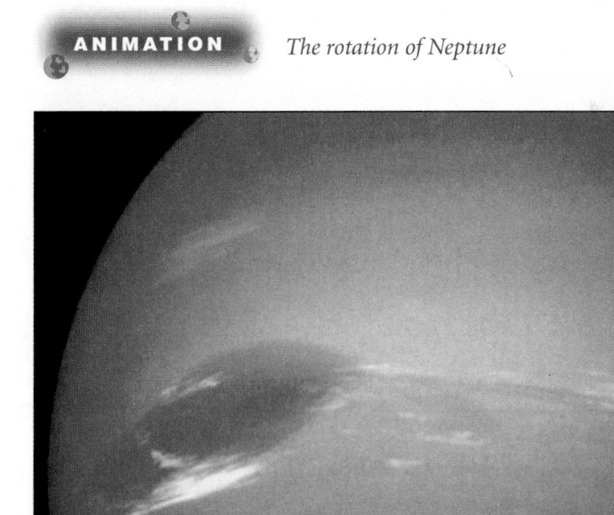

gravitational fields of Uranus and Neptune. The shapes and gravitational fields of both planets are distorted by rotation. The most obvious distortion of a planet by its rotation is that it flattens at the poles or becomes oblate. Other, more subtle distortions occur as well.

The oblateness of a planet and the severity of its other distortions depend on the distribution of material within the planet. A planet with its mass concentrated at the center will be distorted differently than a more uniform planet of the same size and mass. Images of a planet can determine its oblateness. The more subtle distortions can be measured only by tracking the accelerations experienced by a probe (like *Voyager 2*) that passes through the distorted gravitational field of the planet.

Before the *Voyager 2* encounter, most astronomers thought Uranus and Neptune had cores of molten rock and iron surrounded by a thick layer of hot liquid water, ammonia, and methane. On top of that they believed there was a gaseous layer, rich in hydrogen, which eventually merged with the atmosphere. It turned out, however, that these models were too centrally concentrated to be consistent with the *Voyager 2* measurements of the distorted gravitational fields of Uranus and Neptune. Instead, it seems that Uranus and Neptune have small, rocky cores that comprise no more than 3% of their masses. Figure 13.9 shows that about 85% of the mass of each planet is contained in a thick layer of rock and ice. These materials are liquids at the temperatures and pressures inside Uranus and Neptune and probably have the consistency of watery mud. The gaseous outer envelopes

FIGURE 13.9
A Possible Model for Uranus's Interior
The planet probably has a rocky core surrounded by a layer of liquid ice and rock. An outer envelope, rich in hydrogen and helium, merges with the atmosphere. Neptune's internal structure is probably similar to that of Uranus.

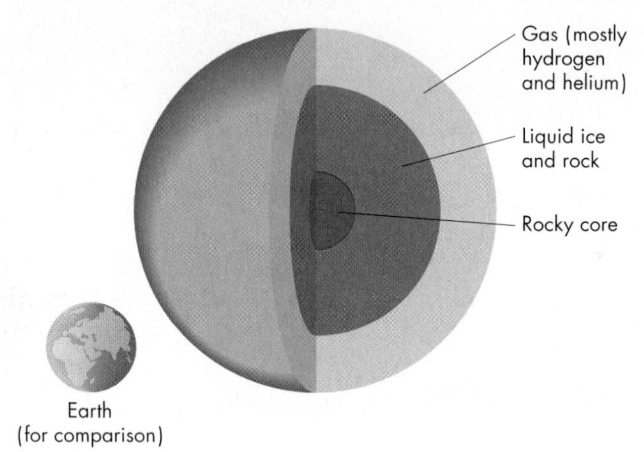

of Uranus and Neptune are made mostly of hydrogen and helium but also contain water, methane, and ammonia. The pressure at the base of the gaseous layer is only a few hundred kilobars (or a few hundred thousand atmospheres), far less than the pressure required to produce metallic hydrogen. It thus seems unlikely that either Uranus or Neptune contains hydrogen in metallic form.

If their outer gaseous envelopes were stripped away, both Uranus and Neptune would resemble the cores of Jupiter and Saturn. Without their outer layers rich in hydrogen and helium, all four planets would consist of about 10 to 15 M_{Earth} of ice and rock. The main difference between Uranus and Neptune on one hand and Jupiter and Saturn on the other is the amount of hydrogen and helium that they collected as they formed. It seems plausible that the cores of ice and rock of all four planets formed first and then attracted the gases in their vicinity. Evidently, Uranus and Neptune were unable to attract as much gas as Jupiter and Saturn. This may be because the gas had a lower density in the vicinity of proto-Uranus and Neptune than in the vicinity of proto-Jupiter and Saturn. Alternatively, it may have taken longer for the icy-rocky cores of Uranus and Neptune to form so that by the time they did, most of the gas in the solar system was gone.

Uranus and Neptune probably have small, rocky cores covered by thick layers of water and molten rock. Both have outer layers consisting mostly of hydrogen and helium, but neither are as thick as the hydrogen and helium envelopes surrounding Jupiter and Saturn.

Internal Energy Uranus and Neptune have identical atmospheric temperatures despite the fact that Neptune is farther from the Sun and absorbs only 40% as much sunlight as Uranus. The temperature of Uranus's atmosphere is consistent with heating only by absorbed sunlight, whereas Neptune is significantly warmer than it would be if it were heated by sunlight alone. Only about 40% of the energy Neptune radiates can be accounted for by absorbed and reradiated sunlight. The remaining 60% must come from internal sources.

Why does Neptune have so much heat flowing from its interior, but Uranus does not? The heat flowing out of Neptune is believed to have the same origin as the excess energy emitted by Jupiter. That is, the interior of Neptune was heated by the conversion of gravitational energy as the planet contracted as it formed. Neptune is insulated well enough that its interior is still hot and heat continues to flow out of it.

Is it possible that Uranus formed in such a way that its interior never became hot? Given the similarity of Uranus and Neptune, that doesn't seem very reasonable. Instead, it seems likely that the interior of Uranus was about as hot initially as the interior of Neptune. Could it be that all of Uranus's internal heat has already escaped? Again, the masses, diameters, compositions, and internal structures of the two planets are so similar that it seems improbable that Neptune would be well insulated, whereas Uranus would not. In fact, it may be that Uranus is a much better insulator than Neptune. One possibility is that the chemical makeup within the rocky-icy part of Uranus's interior varies with depth. A gradual increase in the rock to ice ratio with depth would suppress convection and greatly reduce the rate at which heat can flow outward.

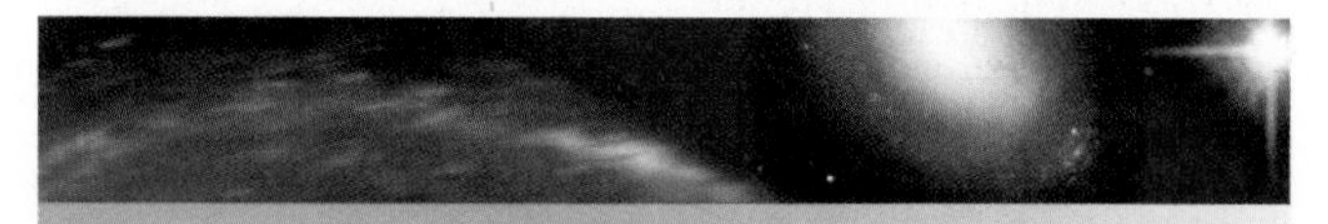

Uranus radiates away essentially no internal heat, whereas 60% of the energy emitted by Neptune comes from internal sources. It is possible that the interior of Uranus is quite warm but that the flow of energy out of Uranus is inhibited.

Magnetic Fields and Magnetospheres

The experiments on *Voyager 2* showed that Uranus's magnetic field was about 100 times stronger than the Earth's but about 100 times weaker than Jupiter's. This value agreed with the best pre-*Voyager* estimates, but everything else about the magnetic properties of Uranus was completely unexpected. For example, the angle between Uranus's magnetic axis and its rotational axis was found to be almost 60°—the largest tilt in the solar system. Figure 13.10 shows a comparison of the orientations of the

FIGURE 13.10
The Magnetic Fields of the Earth, Jupiter, Saturn, Uranus, and Neptune
The fields of the Earth, Jupiter, and Saturn are closely aligned with their rotational axes and are centered near the centers of those planets. The fields of Uranus and Neptune, in contrast, are greatly tilted with respect to their rotational axes and are centered at locations far from the centers of Uranus and Neptune.

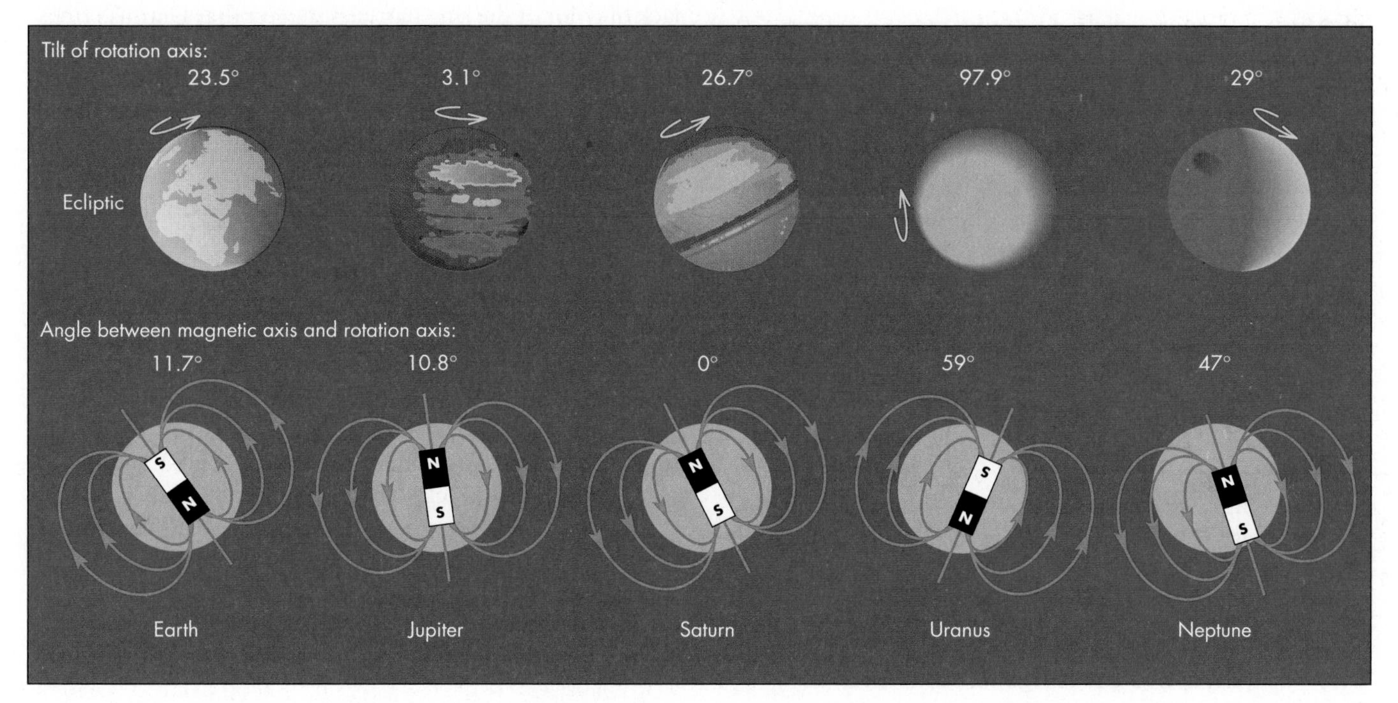

rotation and magnetic axes of the Earth, Jupiter, Saturn, Uranus, and Neptune. If the Earth's magnetic axis were tilted as much as Uranus's, the north magnetic pole would at the same latitude as Florida. As a result, a compass would be of little use to a balloonist floating in the atmosphere of Uranus, because geographic north and magnetic north would be unrelated.

Another unexpected property of Uranus's magnetic field is that it is centered about one-third of the way from the center to the surface of the planet. The Earth's magnetic field is slightly offset from the center of the planet, but not by nearly so much as in the case of Uranus. One theory proposed to account for the large tilt and offset of Uranus's magnetic field is that *Voyager 2* visited the planet while it was undergoing a reversal of its magnetic field. If this were true, the magnetic field of Uranus presumably would normally be almost aligned with the poles of rotation.

The *Voyager 2* measurements showed that Neptune's magnetic field is about as strong as Uranus's and is also greatly tipped and displaced from the center of the planet. The magnetic field of Neptune is tilted by about 47° relative to Neptune's rotation axis and the center of the magnetic field is offset by more than half of Neptune's radius.

Because it is unlikely that both Uranus and Neptune were undergoing field reversals in the late 1980s when they were visited by *Voyager 2*, there must be a different explanation for their large tilts and displacements. One possibility is that the magnetic fields of Uranus and Neptune are generated by currents within icy fluids far from the centers of the planets. The unique nature of the magnetic fields of Uranus and Neptune is probably related to the fact that their interiors are made of different material and structured differently from the other planets.

Both Uranus and Neptune have magnetic fields tipped at large angles relative to their rotation axes. Neither planet's magnetic field is centered at the center of the planet. It is probable that the magnetic fields of Uranus and Neptune are produced by currents in icy layers far from the centers of the planets.

Rings

Uranus's rings were discovered during efforts to observe a **stellar occultation.** Stellar occultations occur when a planet (Uranus, in this case) passes in front of a star as seen from the Earth. Occultations are important because an accurate timing of the interval between the disappearance and reappearance of the star can give the size of the planet. If the occultation is observed from different places on the Earth, the oblateness of the planet can be found as well. If the planet has an atmosphere, the disappearance and reappearance of the star will be gradual rather than abrupt, depending on the structure of the planet's atmosphere.

Uranus Uranus occulted a faint star on March 10, 1977. As Figure 13.11 shows, before and after the occultation there were brief, sudden, drops in the brightness of the star. When astronomers realized that the dips that occurred before the occultation matched the spacing of those that took place after the occultation, it became clear that the dips were themselves occultations caused by nine rings around Uranus. The brevity of the dips showed that the rings are very narrow. The outer ring is 100 km wide, whereas most of the rings are only about 10 km wide. Images from *Voyager 2,* such as Figure 13.12, confirmed the nine rings of Uranus and discovered two additional faint rings between the rings that were already known. *Hubble Space Telescope* observations have shown two distant, faint rings, bringing the total to thirteen.

The rings had never been seen from the Earth before the occultation because they are very dark. The particles that make up the rings are drab gray (almost black), and reflect only a small percentage of the light that strikes them. The particles that make up Saturn's rings, on the other hand, reflect about half of the light that strikes them.

Although we can't be certain, the low reflectivity and lack of color of the ring particles suggest that Uranus's rings are made mostly of carbon. One possible explanation for the origin of the carbon is that the ring particles were once made mostly of methane ice. When methane ice is exposed to energetic particles such as those in the magnetosphere of Uranus, it decomposes to produce carbon. Another possibility is that the rings formed when a carbon-rich asteroid broke up under the tidal influence of Uranus.

While *Voyager 2* was in the shadow of Uranus, it obtained an image of the ring system from forward-scattered light rather than back-scattered light. This means that the imager was looking toward the Sun rather than away from the Sun, as it was while *Voyager 2* approached Uranus. The arrangements of the Sun, ring particles, and spacecraft are shown for both forward scattering and back scattering in Figure 13.13. Figure 13.14 shows that the forward-scattered image is very different from the back-scattered images. The thin rings seen in back-scattered light are very faint in forward-scattered light, whereas many other bands and rings not seen in the back-scattered image are quite

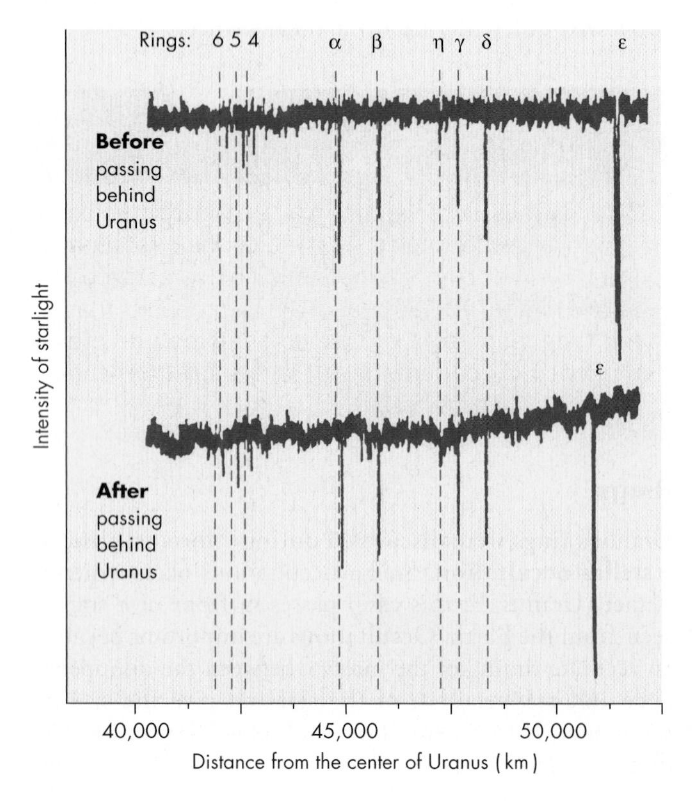

FIGURE 13.11
The First Evidence for Rings Around Uranus
The brightness of a star is shown before and after it passed behind Uranus. In both cases, the star's brightness dipped nine times as it passed behind the individual rings of Uranus. The individual rings are designated by numbers or Greek letters.

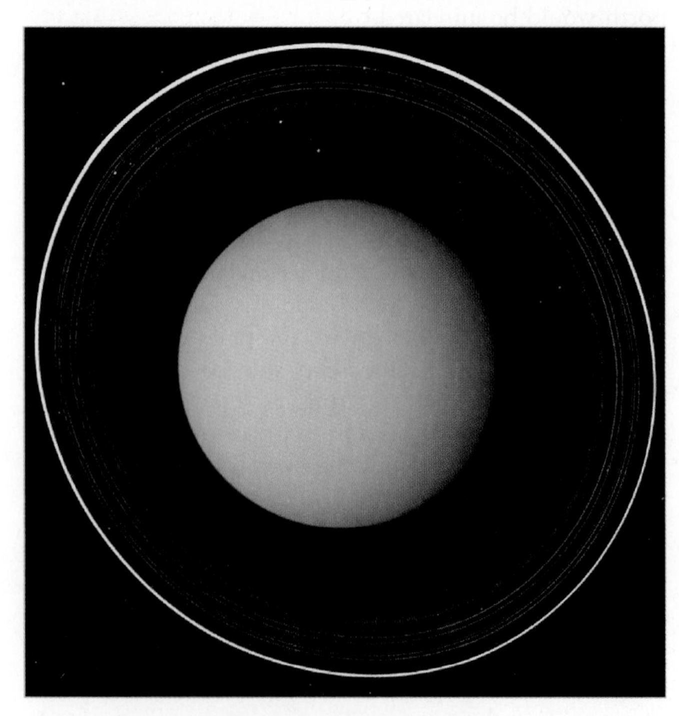

FIGURE 13.12
A *Voyager 2* Image of the Rings of Uranus
The image was taken before *Voyager 2* passed Uranus, so the Sun was behind *Voyager 2.* The points of light are Uranus's satellites and background stars.

FIGURE 13.13
Back Scattering and Forward Scattering
A, In order to observe light that is back scattered from ring particles, the spacecraft has to be roughly between the Sun and the particles. **B,** To observe forward-scattered light, the spacecraft must be beyond the rings, looking generally backward toward the Sun.

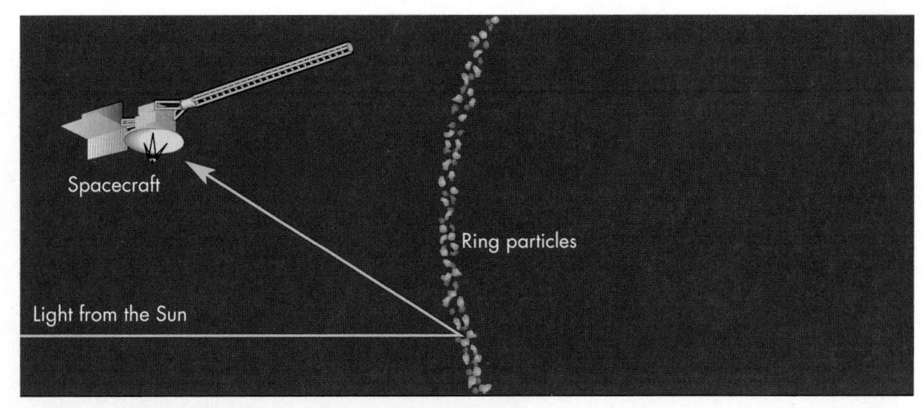

A Back-scattered light

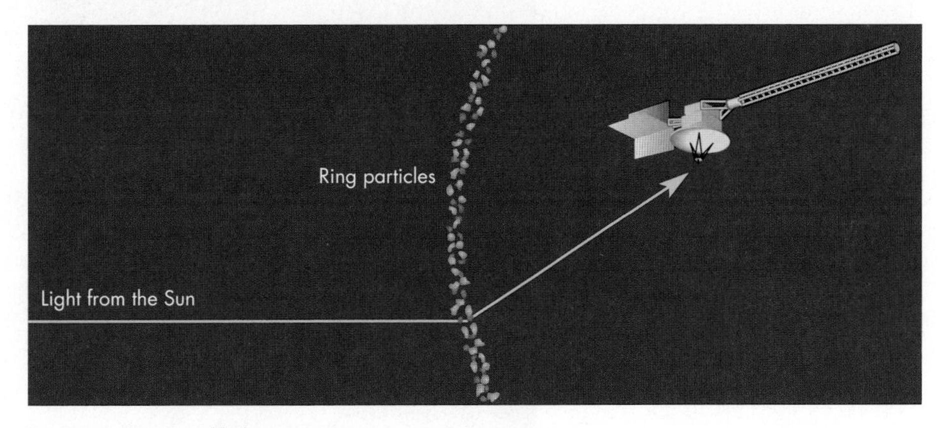

B Forward-scattered light

conspicuous in forward-scattered light. Large particles are good back scatterers but poor forward scatterers, whereas small particles are good forward scatterers but poor back scatterers. Therefore, comparing the two images in Figure 13.14 suggests that the thin rings that had been detected from the Earth were made of particles that were larger than the particles in the space between the rings. The thin rings are made up of particles that range in size from many centimeters to several meters, whereas the space between the rings is filled with fine dust.

It was somewhat surprising that there was so much fine dust in the rings of Uranus. The thin outer atmosphere of Uranus extends into the rings, so it should slow down very tiny dust particles and cause them to sink into the inner atmosphere in a few thousand years or less. Even particles as large as a meter should have been lost by this point, considering the age of Uranus. Evidently something, perhaps meteoroid impacts on undiscovered satellites within the ring system, continually replenishes the supply of ring particles.

Collisions between ring particles have the effect of slowly making the ring wider. The rings of Uranus, however, are narrow and have sharp inner and outer edges. Before the *Voyager 2* encounter, astronomers proposed that there might be small satellites orbiting just inside and just outside each ring. The repeated gravitational tugs from these "shepherding" satellites could keep ring particles from straying from the rings.

Evidence for this theory for the thinness of Uranus's rings was provided by *Voyager*'s discovery of two small satellites, Cordelia and Ophelia, which bracket the orbital distance of Uranus's ϵ ring. Cordelia and Ophelia are shown just inside and outside Uranus's ϵ ring in Figure 13.15. Searches for shepherding satellites for the other rings, however, were unsuccessful. It is not clear whether the satellites are just too small and too dim to be found or whether there is another explanation for the thinness of the remaining rings.

Neptune Attempts to detect the occultations of stars by rings around Neptune gave mixed results. About 90% of the attempts showed no occultations. In five cases, however, dips in brightness were detected. The best explanation for these results was that Neptune had rings, but that they only partially encircled Neptune. *Voyager 2* found rings, but they were extremely dark and faint when observed in back-scattered light as *Voyager 2* approached Neptune with the Sun at its back. The rings are several times brighter in forward-scattered light. The images of the rings in Figures 13.16 and 13.17 show two thin rings and a very faint, broad ring that may extend all the way down to the atmosphere of Neptune.

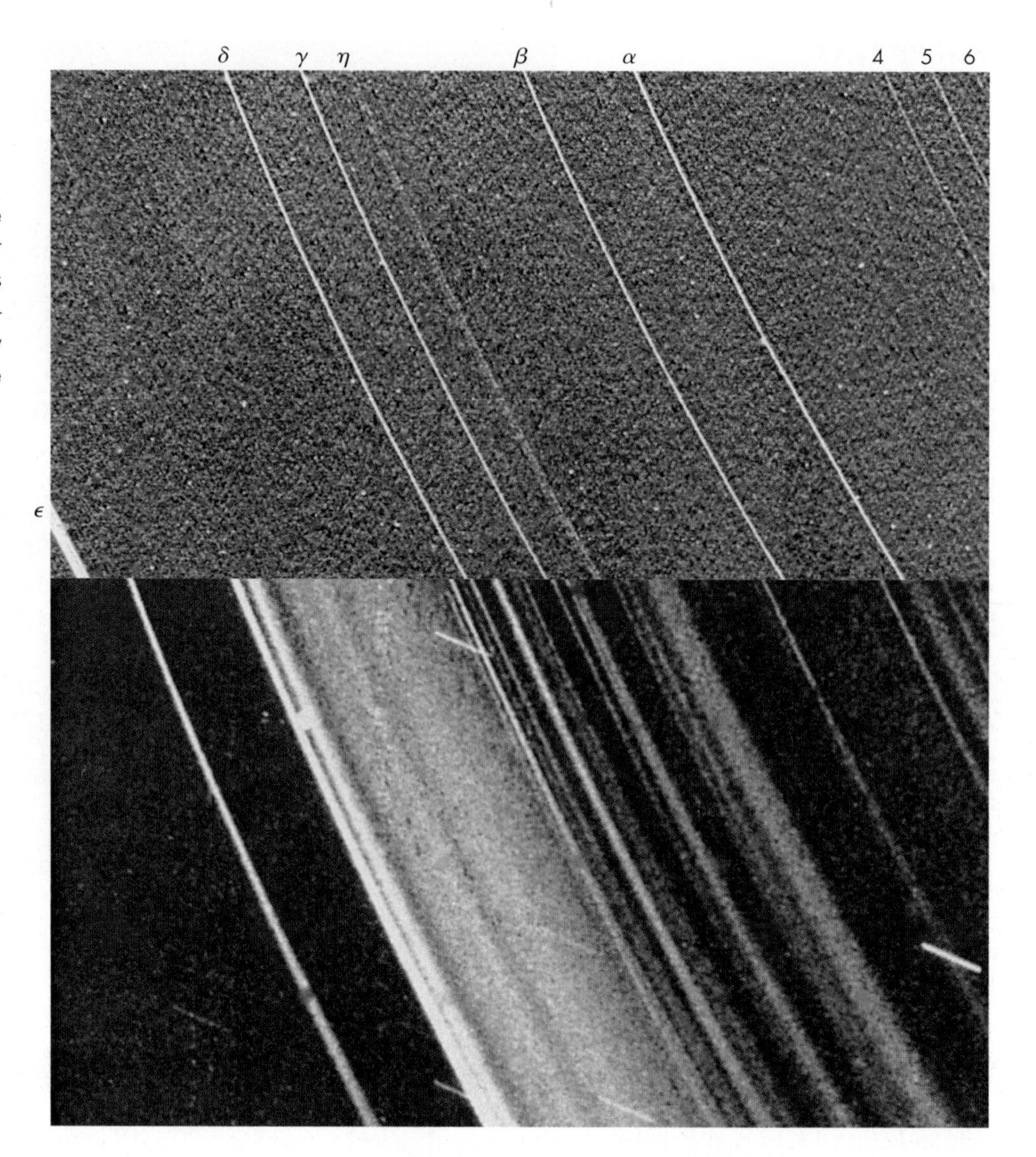

FIGURE 13.14
The Rings of Uranus in Back-Scattered (top) and Forward-Scattered (bottom) Light
The bottom half of the image was taken with the imager looking back toward Uranus and the Sun. Many forward-scattered features, produced by fine dust, cannot be seen in the back-scattered image.

The *Voyager* images also show several other things about the rings of Neptune. First, there is very little dust between the rings, which is quite different from the rings of Uranus. The rings themselves are made mostly of microscopic dust particles rather than larger chunks of matter. The total amount of material in Neptune's rings is about one ten-thousandth of that in Uranus's rings. Second, as Figure 13.18 shows, there are bright arcs in the outer ring. The three bright arcs were responsible for stellar occultations observed as many as 8 years prior to the *Voyager* encounter. The longevity of the three clumps of material requires an explanation, because the particles within a clump orbit Neptune with slightly different orbital periods. This means that a clump should spread completely around the ring in only a few years. It is likely that gravitational interactions between Neptune's satellite Galatea and the ring particles confine the particles in arcs rather than in a uniform ring.

Both Uranus and Neptune have ring systems. The rings of Uranus consist of narrow rings made of large particles. The spaces between the rings are filled with fine dust. Neptune has two thin rings and a broad inner ring, all made of microscopic particles. The clumping in Neptune's outer ring is probably caused by gravitational interaction with Neptune's satellite Galatea.

13.3 TRANS-NEPTUNIAN OBJECTS

The region beyond Neptune is occupied by a very large number of relatively small, icy bodies known as **Trans-Neptunian Objects,** or TNOs.

FIGURE 13.15
Shepherding Satellites of Uranus's Outer Ring
Gravitational forces resulting from the satellites, which orbit just inside and just outside the ϵ ring, act on individual ring particles and prevent them from spreading out to form a broad ring.

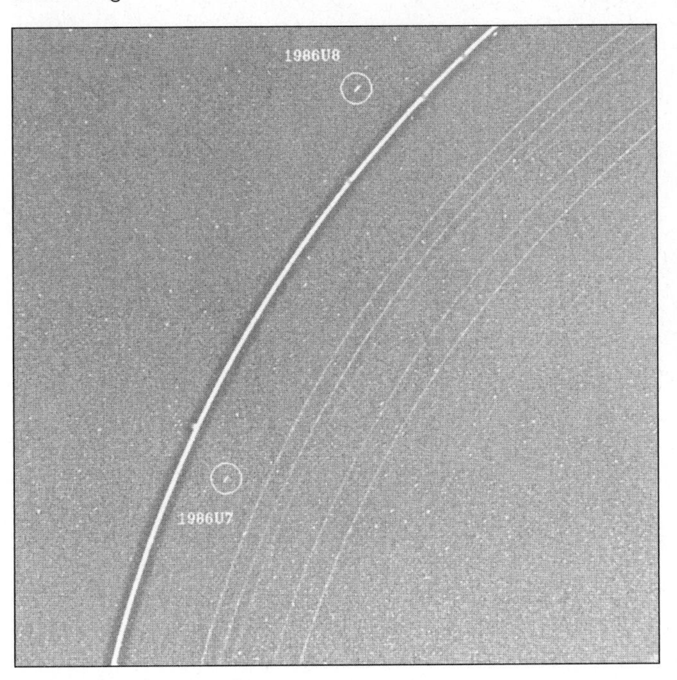

FIGURE 13.16
Neptune's Rings
This image of Neptune's rings was taken by *Voyager 2* before it passed Neptune. Only two thin, faint rings can be seen in back-scattered light.

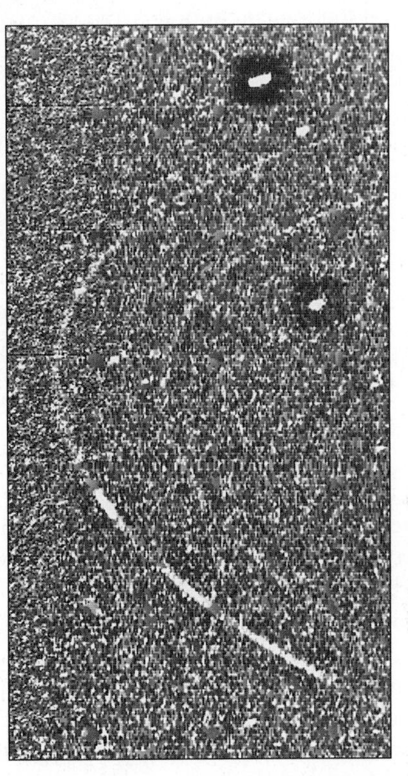

FIGURE 13.17
Neptune's Rings in Forward-Scattered Light
The greater brightness of the forward-scattered image shows that the particles in the rings of Neptune are very small. The ring system consists of two thin outer rings and a faint, thicker inner ring.

Pluto

For more than 60 years, Pluto was the only known TNO. It is by far the best-studied TNO and, although among the largest TNOs, can serve as a prototype.

Discovery Pluto was discovered as an indirect result of Leverrier's 1846 prediction of the location of Neptune. Leverrier's success prompted other astronomers to look for other discrepancies between the calculated and observed

FIGURE 13.18
Clumps in the Outer Ring of Neptune
The three bright regions in the outer ring are clumps of ring particles. One of the bright arcs is seen at the top center of this image. These clumps were responsible for the occultations of stars viewed from the Earth.

orbits of Uranus in hopes of finding yet another planet. One of these astronomers was Percival Lowell, who predicted that a planet about six times as massive as Earth with an orbital distance of 43 AU could be found in the constellation Gemini. Lowell searched for the planet from 1905 until he died in 1916. While carrying out the search, Lowell realized that it would be possible to search more efficiently using a wide-angle telescope. In 1927 Lowell's brother donated $10,000 to buy a wide-angle telescope to search for the predicted planet.

Using the new telescope, the search was resumed in 1929 by Clyde Tombaugh. Tombaugh's strategy was to take pairs of photographs of the same region of the sky at intervals of about a week. By comparing the two photographs, he could determine which objects had moved relative to the stars and thus were within the solar system. This was extremely tedious work because there were hundreds of thousands of images to compare on each pair of plates. On February 18, 1930, Tombaugh spotted a faint object that moved by the right amount (70 seconds of arc per day) to be a planet beyond Neptune. The new planet was found within 6° of the position Lowell had predicted. Pluto, the name chosen for the new planet, seems appropriate because Pluto was the Roman god of the underworld and Pluto orbits in the darkness far from the Sun. It is now known that Pluto has far too small a mass—less than that of our Moon—to have been capable of causing the perturbations in the orbit of Uranus that led to Lowell's prediction. Thus the discovery of Pluto, like those of Uranus and Neptune, was the result of fortuitous circumstances that caused an astronomer to look in the right place at the right time.

Basic Properties Almost from the time of its discovery, Pluto seemed to be an anomaly in the solar system. It was much smaller than the other planets. It was neither a gas giant nor a rocky body, rather it was a mixture of rock and ice. Its orbit was both the most eccentric and the most inclined to the plane in which the other planets orbit the Sun. Nevertheless, for over 70 years it was considered to be one of the nine planets.

Figure 13.19 shows the orbits of Neptune and Pluto. The semimajor axis of Pluto's orbit is 40 AU, but Pluto's large orbital eccentricity means that there are large variations in its distance from the Sun. At perihelion, Pluto is not quite 30 AU from the Sun and is inside the orbit of Neptune, as it was between 1979 and 1999. Even though Pluto regularly crosses Neptune's orbital distance, no collisions are possible because the two planets are in a 2:3 **orbital resonance.** This means that Neptune's orbital period is 2/3 that of Pluto. This is a stable orbital arrangement that makes it impossible for Neptune and Pluto to collide. When Pluto crosses Neptune's orbital distance, Neptune is always nearly on the other side of its orbit. The two planets pass each other only when Pluto is near aphelion and far beyond Neptune. Many of the recently discovered TNOs are also in orbital resonances with Neptune.

In the 1950s astronomers discovered that the brightness of Pluto varies with a period of 6.4 days. The brightness variations are caused by variations in reflectivity on Pluto's surface as the rotation of Pluto carries regions of different reflectivity into view. Over the next several decades, Pluto continued to vary in brightness, but the character of the variations changed. The average brightness of Pluto (corrected for changing distance from the Sun) dropped, whereas the amount that its brightness varied increased from about 10% to over 20%.

These gradual changes were interpreted as evidence that Pluto, like Uranus, rotates with its poles almost in its orbital plane. As a result, as Pluto moved in its orbit, its orientation as seen from Earth changed. When Pluto was discovered, its south pole pointed nearly at Earth, as seen in

Figure 13.20. The reason that it appeared relatively bright at that time was because it has a bright polar region. Furthermore, the reason that no brightness variations were seen was that the same hemisphere (the southern hemisphere) was constantly in view. Since its discovery, Pluto has moved far enough in its orbit that we now look down nearly on its equator. Because we aren't looking directly at its bright polar region, Pluto has dimmed. However, there are bright and dark patches near the equator, which the rotation of Pluto now brings into view, causing relatively large brightness variations. The surface markings inferred from Pluto's brightness variations were later confirmed by *Hubble Space Telescope* images, which were used to produce the map of Pluto shown in Figure 13.21. The map, which has a resolution of about 150 km, shows bright polar regions and about a dozen distinct bright and dark features. Some of the features may be impact basins and fresh craters.

FIGURE 13.19
The Orbits of Neptune and Pluto
The orbit of Pluto is eccentric enough that it passes within Neptune's orbit for about 20 years when Pluto is near perihelion. No collision between the two is possible.

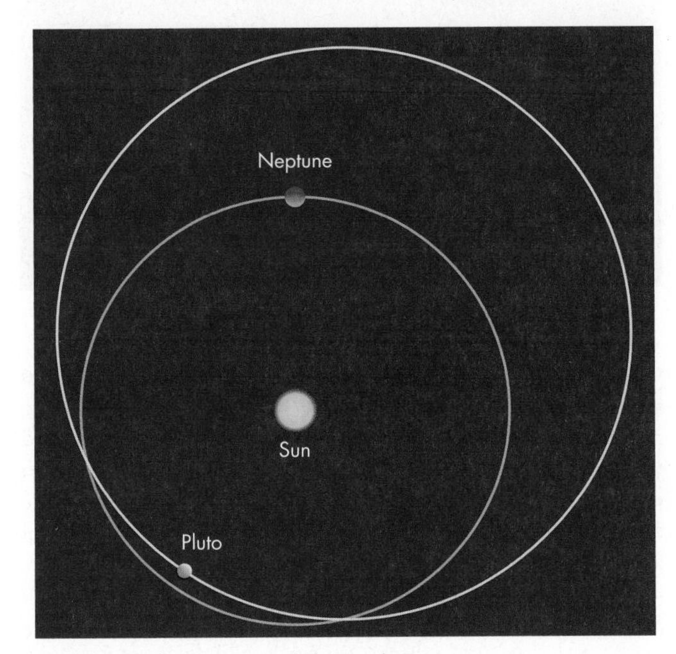

A Top view of Neptune's and Pluto's orbits

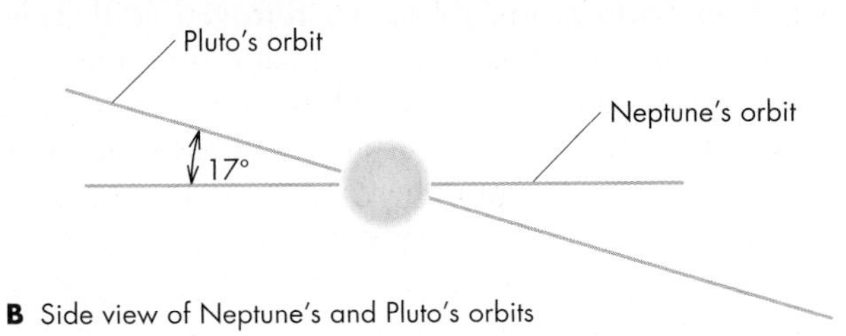

B Side view of Neptune's and Pluto's orbits

Charon and the Mass of Pluto The key to determining the mass of Pluto was the discovery that Pluto has a satellite. In 1978 James Christy, an astronomer at the U.S. Naval Observatory, was examining photographic plates of Pluto to determine its position and orbit with better precision. He noticed that the images of Pluto on several of the photographic plates were all distorted in the same direction. He quickly concluded that the distortion of Pluto's image, shown in Figure 13.22, was caused by a satellite. Figure 13.23 is a *Hubble Space Telescope* image clearly showing both Pluto and its inner satellite, which has been named Charon after the mythological boatman who ferried the dead across the River Styx to Pluto's underworld. The distance between Pluto and Charon is only 19,600 km—about one-twentieth the distance between Earth and the Moon.

The orbital period of Charon is 6.4 days, the same as Pluto's rotation period. This means that Pluto always keeps the same face turned toward Charon. Charon is probably also locked in a synchronous orbit so that it keeps the same face turned toward Pluto. None of the planets rotate synchronously with the orbit of a satellite, although satellites in synchronous orbits (like our Moon) are common.

FIGURE 13.20
The Changing View of Pluto
Pluto's axis of rotation is nearly in its orbital plane. When first discovered, the view from the Earth was of Pluto's relatively bright south polar region. We now view Pluto from nearly above its equator. Bright and dark spots produce changes in the brightness of Pluto as it rotates, but the bright polar region is less conspicuous, so Pluto is dimmer than when it was discovered.

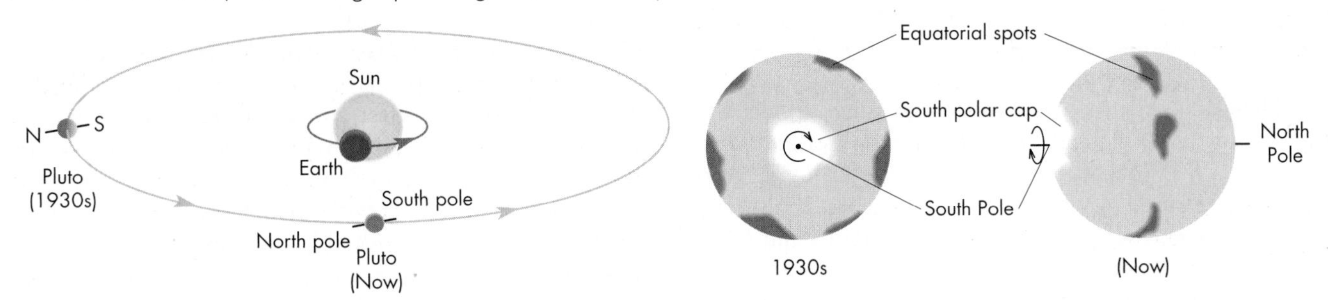

A The position of Pluto on its orbit when it was discovered (1930s) and today

B The view of Pluto from Earth

FIGURE 13.21
The Surface of Pluto
These maps of Pluto, obtained using the *Hubble Space Telescope,* show Pluto as viewed with four different longitudes at the center of its disc. Note the large bright south polar region.

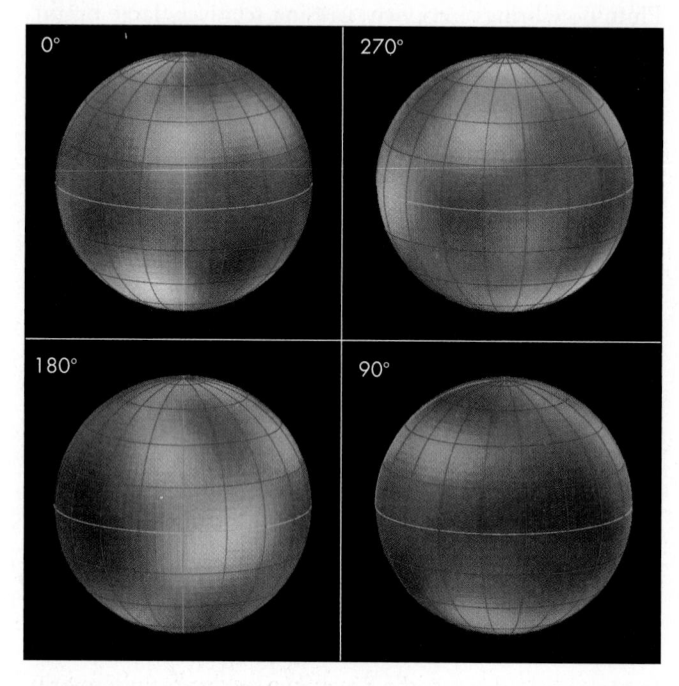

FIGURE 13.22
The Discovery of Charon
Charon appears as a bump at the upper right (arrow) of this picture of Pluto taken from the Earth.

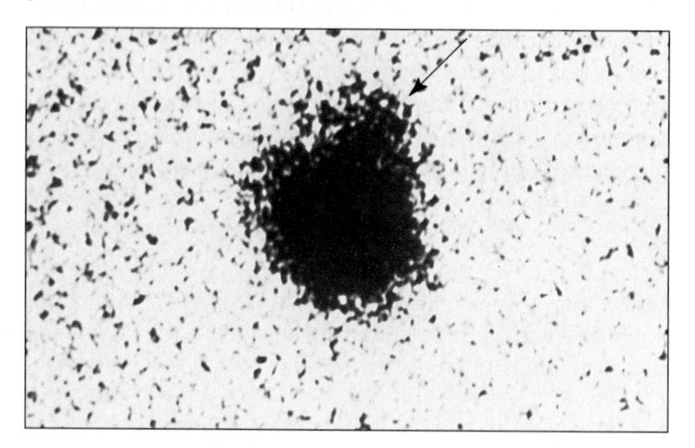

The discovery of Charon made it possible to calculate the mass of Pluto (actually, the combined masses of Pluto and Charon). When the distance and period of Charon were plugged into Newton's version of Kepler's third law (Equation 5.4), the combined masses of Pluto and Charon were found to be only 0.0022 M_{Earth}. Of this, Pluto accounts for 0.0020 M_{Earth} and Charon accounts for 0.0002 M_{Earth}. Thus, the mass of Pluto is only one-sixth of the mass of the Moon and is also smaller than the masses

FIGURE 13.23
Pluto and Charon
The high angular resolution of the *Hubble Space Telescope* makes it possible to resolve Pluto and Charon as separate bodies.

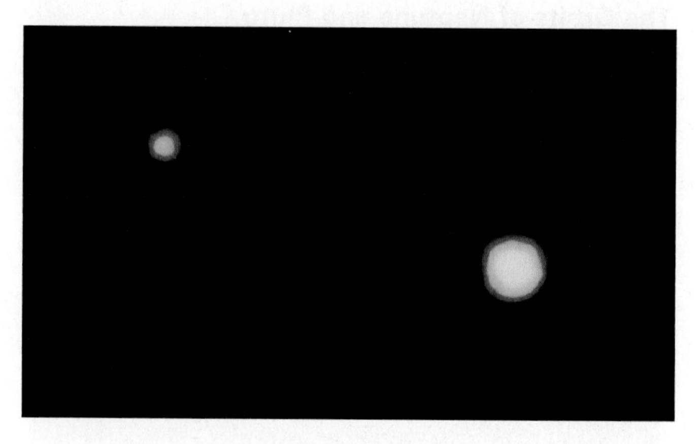

of six other solar system satellites. Pluto's mass is far too small to have an effect on any of the planets.

The density of Pluto was calculated from its mass and diameter and found to be about 2000 kg/m^3. To have a density that large, Pluto cannot be made entirely of ice and must contain a substantial amount of rocky material as well.

Surface and Atmosphere Parts of Pluto's surface are highly reflective, probably due to ice material. The icy material on Pluto is mainly frozen nitrogen, with small proportions of frozen carbon monoxide and methane. Methane gas is also present on Pluto as a very thin atmosphere with a pressure about one-hundred thousandth the pressure of Earth's atmosphere. The amount of gaseous methane that can coexist with methane ice depends on temperature, just as the maximum amount of water vapor in Earth's atmosphere depends on temperature. Because Pluto can be as close to the Sun as 30 AU and as far away as 50 AU, Pluto's temperature varies considerably during the course of a Pluto year. When it is near the Sun (as it is now), the amount of methane in Pluto's atmosphere may be many times larger than the amount of atmospheric methane when Pluto is far from the Sun.

The View from Pluto The sky from Pluto would be quite different from what we see on Earth. The thin atmosphere would scatter almost no sunlight, so the sky would be very dark (day or night) and the stars would appear pointlike and steady. During the day the Sun would appear pointlike as well, with an angular size too small for the human eye to resolve. Nevertheless, the Sun would be far brighter than the brightest stars and would shine hundreds of times more brightly than our full Moon. It would cast conspicuous shadows and provide enough light to read by, but it would provide little warmth. The ground would be heated to only about 50 K (−370° F). The feeble heating of sunlight and the small amount of energy radiated away by Pluto at night would lead to temperature variations of only a few degrees during the 6.4 day-long Pluto "day." Pluto's satellite, Charon, would hover in the sky on the side of Pluto that faces it. Charon would have an angular size of almost 4°, so it would appear about seven times as large as our Moon looks to us. It would be very dim, however, because the sunlight it reflects is so weak at Pluto's distance. Charon would go through its cycle of phases in 6.4 days as Pluto and Charon mutually orbit one another.

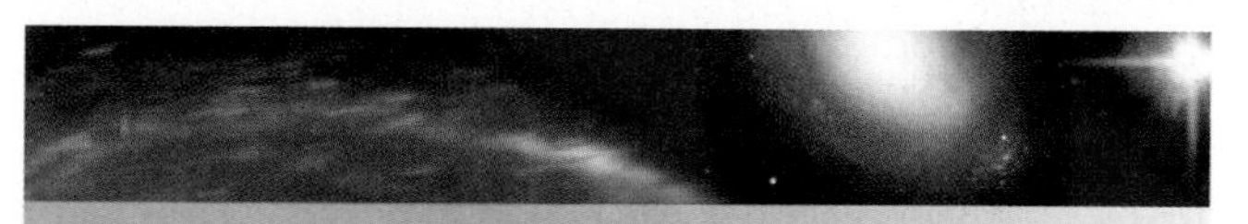

Pluto is the first of the Trans-Neptunian Objects to have been discovered and for over 60 years was the only TNO known. It was considered to be a planet until other TNOs as large or larger than Pluto were discovered. Despite crossing Neptune's orbit, Pluto is in an orbital resonance with Neptune that makes collisions impossible. Pluto is made of a mixture of rock and ice.

Other TNOs

After he discovered Pluto, Tombaugh continued to search for other planets, but without success. He and his colleagues observed most of the northern hemisphere of the sky and found numerous comets but no planets. The failure to find additional planets, even ones as small and dim as Pluto, was partly a consequence of the relative insensitivity of the photographic plates they used. In the 1990s new searches were carried out, this time using sensitive CCD arrays rather than photography. The era of new discoveries began in 1992 when David Jewitt and Jane Luu discovered the first TNO to be found since Pluto's satellite, Charon, was discovered in 1978. The body they found, 1992 QB_1, was much fainter and presumably smaller than Pluto. It has a roughly circular orbit with a semimajor axis of 44 AU. Since then over 1200 TNOs have been discovered. Astronomers estimate that there are perhaps one million TNOs larger than a few tens of kilometers in diameter, most of them yet to be discovered.

Large TNOs Although most of the TNOs that have been discovered are smaller and fainter than Pluto, there are some exceptions. About a half dozen of the known TNOs are at least 50% as large as Pluto. Eris, discovered in 2003, is actually larger than Pluto. The discovery of Eris, and the prospect that there may be many other undiscovered TNOs larger than Pluto, led to Pluto's demotion from the status of planet. After Eris was found, astronomers had several options. One was to pick an arbitrary size and declare that in order to be a planet a body must orbit the Sun and be bigger than the chosen size. If the arbitrary size were the diameter of Pluto, then Pluto would remain a planet and any larger bodies beyond Neptune would also be considered planets. Thus, there would be ten planets now and the number of planets would probably increase. If the arbitrary size were larger than Pluto (the diameter of Mercury, say) then Pluto would be dropped from the ranks of the planets. The other TNOs that have been discovered or are likely to be discovered would also be too small to be planets.

Another possibility was to adopt a physical definition of the word "planet." One definition that was offered is that, in addition to orbiting the Sun, a planet's gravity must be large enough to pull it into a spherical shape. The dividing line between bodies that are round and those that aren't is about one-half to one-third the diameter of Pluto. With this definition, Pluto would remain a planet. However, the asteroid Ceres and a handful of known TNOs would also be considered planets. By this definition, there may be dozens of other "planets" beyond Neptune.

The decision of how to define "planet" rested with the International Astronomical Union (IAU), a worldwide astronomy organization. In 2006, the IAU passed a controversial resolution that defined a planet as a body that orbits the Sun, is massive enough to have a spherical shape, isn't the satellite of a planet, and dominates the region near its orbit by clearing it of smaller bodies. A **dwarf planet** differs from a planet in that it does not dominate the region near its orbit. There may be dozens or hundreds of undiscovered TNOs that may one day be found to be dwarf planets. At the present time, however, only Ceres, Pluto, Eris, and two other TNOs, Haumea and Makemake, are considered to be dwarf planets.

The demotion of Pluto from the ranks of the planets isn't unprecedented in astronomy. After the first four asteroids were discovered between 1801 and 1807 they were generally considered to be planets (see Chapter 15 for more about asteroids). They even had astrological signs associated with them. Then, in the 1850s, when many more asteroids began to be discovered, the four (Ceres, Vesta, Juno, and Pallas) were dropped from the list of planets. The circumstances were similar to that of Pluto. Keeping them as planets would have led to too many planets.

TNO Orbits The TNOs have orbits with a wide range of distances, eccentricities, and inclinations. These orbital characteristics have allowed astronomers to divide the TNOs into groups with different kinds of orbits and different histories.

Kuiper Belt Objects (KBOs) are named for Gerard Kuiper, one of several astronomers who speculated about the existence of TNOs long before any except Pluto had actually been found. **Classical KBOs** have orbits with small inclinations and eccentricities and semimajor axes between 42 and 45 AU. Their perihelia are greater than 40 AU, so they have relatively stable orbits because they never come close enough to Neptune to suffer strong gravitational perturbations. **Resonant KBOs** are trapped in an orbital resonance with Neptune. Most of the known resonant KBOs, like Pluto, are trapped in the 2:3 resonance with Neptune. Like Pluto, they have large orbital eccentricities and inclinations. Like the classical KBOs their orbits are relatively stable. One remarkable feature of the Kuiper Belt is that it ends fairly abruptly at 48 to 50 AU from the Sun. Why there are no KBOs with larger semimajor axes is not yet understood.

Scattered disk objects have orbits larger, more eccentric, and more highly inclined than the KBOs. Eris, the largest known TNO, is a scattered disk object. Its 560-year orbit carries it as close to the Sun as 38 AU, but at aphelion it is at nearly 100 AU. Because of their large eccentricities, the perihelion distances of the scattered disk objects generally are between 30 and 38 AU, which means they pass close enough to Neptune to make their orbits unstable. At each encounter with Neptune, their semimajor axes and eccentricities change significantly. Eventually, an encounter with Neptune will cause a scattered disk object to lose angular momentum and move into the planetary system. The **Centaur** objects are a group of bodies that have encountered Neptune and entered the planetary system in the relatively recent past. The Centaurs are on elliptical orbits that cross the orbits of Saturn, Uranus, and Neptune. Within a few tens of millions of years after encountering Neptune, a Centaur is likely to suffer another planetary encounter that will eject it from the solar system or cause it to move into the inner solar system where it becomes visible as a comet.

A final group of TNOs have orbits similar to those of the scattered disk objects but with perihelion distances that are large enough they can never have an encounter with Neptune. One of these objects, Sedna, has an orbital period of 11,000 years, a perihelion distance of 76 AU, and an aphelion distance of 900 AU. Astronomers are uncertain how bodies such as Sedna could have got into their present orbits.

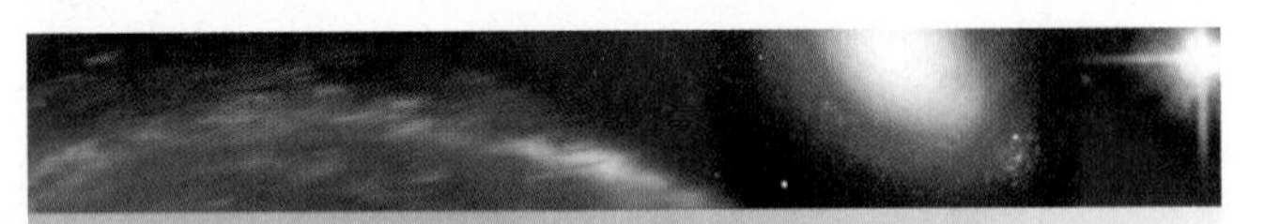

Over 1200 TNOs have been discovered since 1992. Some of them are comparable in size to Pluto. TNOs that are in orbital resonances with Neptune or have perihelion distances far enough from Neptune to escape its influence are called Kuiper Belt Objects (KBOs). Those with perihelion distances that bring them close to Neptune are scattered disk objects.

TNOs and the History of the Solar System

Astronomers suspect that the TNOs played a role in a dramatic reshaping of the solar system that may have taken place 4 billion years ago. According to a recent model of the evolution of the outer solar system, the reshaping began long after the planets formed about 4.6 billion years ago. At that time, as Figure 13.24 shows, the giant planets orbited the Sun on circular orbits that were closer together than they are today. The outermost planet, at that time Uranus, was about 17 AU from the Sun. The planets were surrounded by a disk containing about 35 M_{Earth} of icy bodies—a much more massive, more compact version of today's Kuiper Belt. Some of the smaller bodies that had formed the planets were still present in the outer solar system. (These are called planetesimals. How they accumulated to form the planets is described in Chapter 18.) Through encounters with Neptune some of them were thrown outward into larger orbits. Most, however, were deflected into orbits that took them near Jupiter, where they picked up enough speed to be thrown out of the solar system. Because most of the planetesimals that encountered Neptune were deflected inward, Neptune took orbital energy from the planetesimals and moved outward. Obviously, planets are much more massive than small, icy bodies, so during an encounter with a planetesimal Neptune moved outward very little. Over many encounters, however, the effect was significant. Uranus and Saturn also moved outward. Jupiter lost energy in its encounters with planetesimals and moved inward.

All this took place very slowly until, after about 600 million years, Jupiter and Saturn moved into a 1:2 resonance so that Jupiter orbited the Sun twice for each time Saturn orbited. Because they passed each other at the same points in their orbits, their gravitational tugs on each other stretched their orbits out, making them elliptical. Now moving on elliptical orbits, Jupiter and Saturn occasionally passed close to Uranus and Neptune, destabilizing the entire planetary system. Both Uranus and Neptune moved outward on elliptical orbits, entering the Kuiper Belt and, for about 50 to 100 million years, scattered the icy bodies in all directions. Some of the icy bodies entered the inner solar system and hit the planets in catastrophic collisions. While this was going on Neptune moved outside Uranus, becoming the most distant planet. Encounters with the icy bodies made the orbits of Uranus and Neptune more circular and left the planets in the orbits they occupy today.

This scenario is consistent with a number of known facts about the solar system. For one, the major impact basins on the Moon (which are mostly occupied by the lunar maria) seem to have been formed at about the same time about 3.9 billion years ago. This could have been the result of a lengthy, steady bombardment in

FIGURE 13.24
Interaction of KBOs and Planets
The four diagrams show how interactions between the Kuiper Belt Objects and planets led to a period of rapid change of the solar system. **A**, The solar system just after the formation of the planets. The orbits of the planets are slowly changing as they encounter debris in the solar system. The planets are surrounded by a thick Kuiper belt. **B**, About 3.9 billion years ago Jupiter and Saturn fall into a resonance that destabilizes the planetary system. **C**, Uranus and Neptune move into the Kuiper Belt scattering KBOs in all directions. Some of them strike the Moon and planets. **D**, After most of the KBOs have been ejected from the solar system, the planetary system stabilizes.

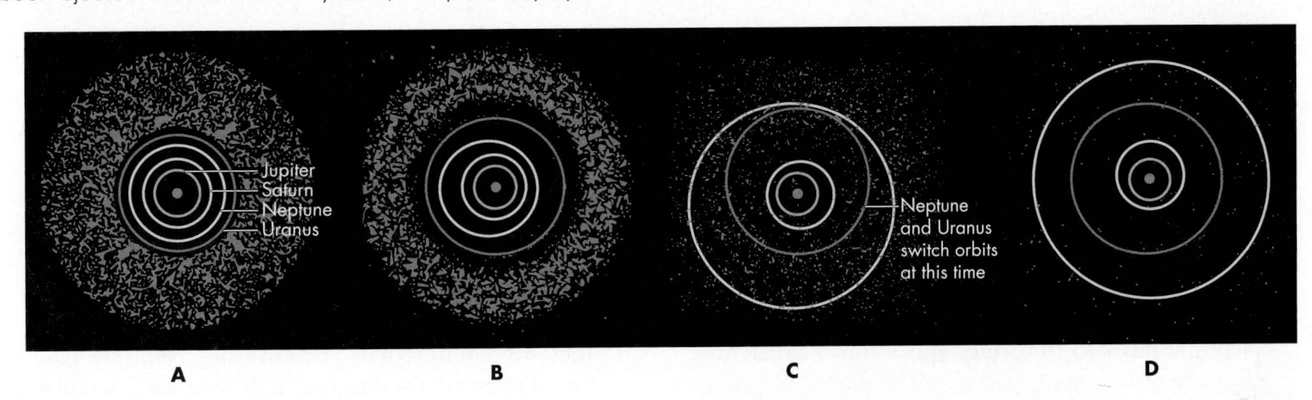

which new impacts wiped out the evidence of older ones until the bombardment stopped about 3.9 billion years ago. But it also can be explained by a brief, intense bombardment by bodies from the Kuiper Belt. The scenario can also explain the orbits of TNOs in the Kuiper Belt. As Neptune moved outward through the Kuiper Belt it could have trapped TNOs into orbital resonances, which then would have moved outward with Neptune. Pluto would have been one of the TNOs swept up in this manner. Other bodies would have been thrown outward into larger orbits and become the scattered disk objects.

In order for all this to have happened, about 99.9% of the original Kuiper Belt Objects would have to have been lost either through collisions with planets or, more often, by being ejected from the solar system. This would leave only about 0.03 M_{Earth} of TNOs today—an amount or material consistent with the estimated number of TNOs of different sizes.

There may once have been a massive disk of TNOs just outside the planetary system. Encounters with Neptune and Uranus may have scattered these bodies in all directions, causing major impacts on the Moon and other solar system bodies about 3.9 billion years ago.

The *New Horizons* space probe is now on its way to Pluto. (At the time it was launched, Pluto was still considered to be a planet.) *New Horizons* will fly past Pluto and Charon in 2015 and if all goes well will fly past another Kuiper Belt Object in 2020. Thus, we are only a few years away from a close-up examination of distant and still not well-understood members of the solar system.

Chapter Summary

- Although it is not really accurate to say that Uranus and Neptune were discovered by accident, fortuitous circumstances contributed to the discovery of each of them. (Section 13.1)
- Uranus and Neptune are so similar in mass, diameter, and rotation rate that they are often called "twin planets." Their axial tilts, however, are very different. Neptune's is similar to the Earth's, but Uranus's is so great that its pole lies almost in its orbital plane. (13.2)
- Uranus sometimes is a nearly featureless blue sphere, whereas Neptune shows banded structure, bright clouds, and storm systems. The difference between the appearances of the two planets is caused by large seasonal effects on Uranus and the absence of a source of internal energy to drive convection in its atmosphere. (13.2)
- The interiors of Uranus and Neptune probably consist of rocky cores surrounded by thick shells of a fluid mixture of rock and ice. On top of that are

relatively thin envelopes rich in hydrogen and helium. The two planets apparently were less effective than Jupiter and Saturn in gathering thick envelopes of hydrogen and helium while they formed. (13.2)

- Neptune is significantly self-luminous, whereas Uranus is not. It seems likely that Uranus, like Neptune, has a hot interior but that heat is unable to escape easily. (13.2)
- The magnetic fields of Uranus and Neptune are about one hundred times as strong as the Earth's magnetic field but about one hundred times weaker than Jupiter's. Moreover, Uranus's and Neptune's magnetic fields are tilted with respect to their rotation axes, and offset from the centers of the planets. The current theory is that the magnetic fields are generated by currents in the icy shells of Uranus and Neptune rather than in their cores. (13.2)
- The ring system of Uranus consists of thin rings of large particles separated by regions filled with fine dust. Neptune has two thin rings and a broad inner ring, all made of very small particles. The arcs in Neptune's outer ring are probably caused by the gravitational influence of Neptune's satellite, Galatea. (13.2)
- Pluto was the first Trans-Neptunian Object to be discovered and was considered to be a planet for over 70 years. Pluto is in an orbital resonance with Neptune that prevents it from ever passing near or colliding with Neptune. Pluto's density shows that it is made of a mixture of rock and ice. (13.3)
- The first TNO other than Pluto and Charon was discovered in 1992. Since then over 1200 TNOs have been discovered. One known TNO, Eris, is larger than Pluto. Kuiper Belt Objects (KBOs) are TNOs on orbits that keep them from encountering Neptune. Scattered disk objects have perihelion distances that bring them close enough to Neptune to have their orbits significantly altered. (13.3)
- Astronomers suspect that the TNOs of today are the remnants of a much more massive Kuiper Belt that once formed a disk around the planetary system. Encounters with Uranus and Neptune may have scattered the original KBOs in all directions and sent some of them into collisions with the Moon and the planets. (13.3)

Key Terms

celestial mechanics 294
Centaur 310
classical KBO 309
dwarf planet 309
Kuiper Belt Object 309
orbital resonance 306
perturbation 294
resonant KBO 310
scattered disk object 310
stellar occultation 301
Trans-Neptunian Object 304

Conceptual Questions

1. Why was it necessary for Herschel to observe Uranus on more than a single night for him to be convinced that he had discovered a new object within our solar system?
2. What was the reason that, after the discovery of Uranus, astronomers began to believe that there was a planet beyond Uranus?
3. The discovery of Neptune is usually attributed to Urbain Leverrier. What did he do to warrant this recognition?
4. In what ways are Uranus and Neptune similar and in what ways are they different?
5. Describe why the pattern of sunlight and darkness for a hypothetical inhabitant of Uranus's polar regions has little relationship to the rotational period of Uranus.
6. Why doesn't the severe tilt of the rotation axis of Uranus result in great temperature differences between the day side and the night side of the planet?
7. What may be the reason that Uranus sometimes lacks the cloud features found on Neptune?
8. Neptune is 50% farther from the Sun than Uranus is, yet their atmospheric temperatures are nearly the same. How is this possible?
9. What evidence do we have that the rock and ice in Uranus and Neptune are concentrated near the centers of those planets?
10. How do the icy-rocky cores of Neptune and Uranus compare with those of Jupiter and Saturn?
11. What is unusual about the magnetic fields of Uranus and Neptune compared with those of the other planets?
12. What evidence do we have to suggest that the magnetic fields of Uranus and Neptune don't originate in currents in the central regions of those planets?
13. How were the ring systems of Uranus and Neptune detected before they were actually seen in *Voyager 2* images?
14. How does the appearance of the rings of Uranus compare with the appearance of the rings of Saturn?
15. How do we know that there is fine dust between the rings of Uranus?
16. What is the reason that some of the predicted occultations of stars by Neptune's rings could not be detected from the Earth?

17. Why is it sometimes incorrect to say that Pluto is farther from the Sun than is Neptune?
18. How do we know the length of the rotation period of Pluto?
19. How did the discovery of Pluto's satellite, Charon, make it possible to calculate the mass of Pluto?
20. How do dwarf planets differ from planets?
21. How do KBOs and scattered disk objects differ?
22. What role may TNOs have played in reshaping the solar system?

Problems

1. Given the masses and radii of Uranus and Neptune, calculate the gravity at the "surface" of each of the planets. (Note: The "surfaces" of Uranus and Neptune are the visible cloud tops.)
2. Pluto has a temperature of about 50 K. Use Wien's law (Chapter 7) to calculate the wavelength at which Pluto is brightest.
3. What is the orbital period of a scattered disk object with perihelion at 40 AU and aphelion at 100 AU?

Figure-Based Questions

1. Use Figure 13.5 to estimate the pressure at the depths in the atmospheres of Uranus and Neptune where ammonia (NH_3) clouds lie.
2. Use Figure 13.11 to estimate the distance of Uranus's β ring from the center of the planet.

Group Activity

With your group, draw up a plan for an observatory on Pluto. Include the advantages the observatory would have compared with one on Earth and the difficulty of maintaining the observatory and its staff. If possible, make a brief presentation of your plan in class.

For More Information

Visit the text website at **www.mhhe.com/fix** for chapter quizzes, interactive learning exercises, and other study tools.

14

. . . two satellites, which revolve about Mars, whereof the innermost is distant from the center of the primary planet exactly three of its diameters, and the outermost five . . .

Jonathan Swift,
Gulliver's Travels, 1726

Jupiter's Satellite, Europa

www.mhhe.com/fix

Satellites

In 1610 Galileo discovered four satellites of Jupiter. By 1970 a total of 30 satellites were known for Mars, Jupiter, Saturn, Uranus, and Neptune. All were small and distant, so very little was understood about them.

Thanks to the *Voyager* missions and searches for faint satellites using telescopes on Earth, the number of known satellites has more than quadrupled since 1970. Moreover, high-resolution images obtained by *Viking,* the *Voyagers, Galileo,* and *Cassini* have led to detailed maps of the surfaces of many satellites. We now know more about the surface features of many satellites than we did about Mars before the *Mariner* and *Viking* missions.

The *Viking, Voyager, Galileo,* and *Cassini* missions visited only five planets, but they explored dozens of different worlds. This chapter is about many of those worlds—the satellites of Mars, Jupiter, Saturn, Uranus, and Neptune. When the Viking, *Voyager, Galileo,* and *Cassini* missions observed the satellites of the planets, they answered many important questions and raised many new ones.

Questions to Explore

- The surfaces of many satellites show evidence of eruptions. What energy sources power the eruptions and keep the satellite interiors molten?
- In what ways are worlds made of ice like those made of rock? In what ways are they different?
- Why does Io have the most intense volcanic activity in the solar system?
- Where in the outer solar system can we find "rain" and "lakes"? What kind of liquids make up "rain"?
- What evidence do we have that some satellites contain liquid water?

14.1 KINDS OF SATELLITES

With only a few exceptions, the satellites of the planets and dwarf planets can be grouped into three different classes: regular satellites, collision fragments, and captured asteroids. These three classes have different orbital properties, origins, and physical properties—such as size and shape.

Regular Satellites

The **regular satellites** form miniature solar systems around Jupiter, Saturn, and Uranus. All of them are so large and so bright that they were observed from the Earth long ago. Their orbits are regularly spaced about their parent planets (just as the orbits of the planets in our solar system are regularly spaced about the Sun). The orbits of the regular satellites are nearly circular and are almost in the equatorial planes of their parent planets. The regular satellites are thought to have originated in the same sequence of events that formed the Sun and planets. That is, the regular satellites formed in orbit about their parent planets just as the planets formed in orbit about the Sun. Studying the regular satellites may teach us something about the way in which the solar system formed.

Collision Fragments

Each of the four giant planets has many small, irregularly shaped satellites with orbits similar to those of the regular satellites. These satellites probably are the fragments of larger satellites broken apart by collisions with meteoroids. Many of these **collision fragments** are close enough to their parent planet that they are immersed in the planet's ring system. In these cases, the satellites act as shepherding satellites or interact with the ring particles in other ways.

Captured Asteroids

Both satellites of Mars, and many of the outer satellites of Jupiter, Saturn, Uranus, and Neptune may be captured asteroids or Kuiper Belt Objects. Except for the satellites of Mars, their orbits are highly elliptical and are inclined to the equatorial plane of their parent planet. In fact, many of them have orbits that are so highly inclined they move in a retrograde direction. None of the captured asteroids are very large. Neptune's satellite, Nereid, the largest, is about 400 km in diameter. These satellites are about as far from their parent planet as a satellite can be without being torn from the planet by the gravitational attraction of the Sun.

Other Satellites

Two satellites do not fall into any of the three main classes. One of these is our Moon, which probably formed following a collision between the Earth and a Mars-sized body (Chapter 9). Neptune's largest satellite, Triton, is another satellite that defies simple classification. Triton has a retrograde orbit just like some of the captured asteroid satellites. Nevertheless, it is difficult to imagine how Neptune could have captured a body as massive as Triton, although several captive scenarios have been proposed.

Most solar system satellites are either regular satellites, which formed in orbit about their parent planet; fragments of regular satellites; or captured asteroids. Two satellites—the Moon and Triton—do not fall into any of these categories.

14.2 GENERAL PROPERTIES OF SATELLITES

The *Viking, Voyager, Galileo,* and *Cassini* missions obtained reliable measurements of the diameters and the masses of many satellites. Diameters were measured directly from the images of the satellites. The strength of the gravitational force on the spacecraft while it flew past a satellite was used to calculate the mass of the satellite. From the diameters and the masses of the satellites, we can find their densities and estimate the kinds of materials from which they are made. If we know their compositions, diameters, and masses, we can begin to deduce their internal structures and their histories.

Density and Composition

The density of a planet or a satellite obviously depends on the mixture of materials from which it is made. Gases have low densities, ices have intermediate densities, and rock and metal have high densities. For planets, it is usually necessary to consider the compression that occurs because of the tremendous pressure deep in the interior. Internal compression isn't very great in satellites, however, because their internal pressures are small compared with those in larger bodies like the Earth. The satellite with the largest central pressure is Io, the innermost of the regular satellites of Jupiter. Io's central pressure is about 57 kilobars (or 57,000 atmospheres). Recall from Chapter 8, Section 3, that the central pressure in the Earth is about 3.5 megabars (or 3,500,000 atmospheres). A pressure of 57 kilobars (kb) can be found in the Earth at a depth of only 150 km. Thus,

Io's central pressure is very small compared with the Earth's and is much too small to cause significant compression.

Internal pressure and compression are even smaller for other satellites. For example, Rhea, one of the moons of Saturn, has a central pressure of about 1.5 kb, which can be found only a few kilometers deep in the Earth. The low pressures within satellites mean that we usually can ignore compression in making estimates of their composition.

Mixtures of Rock and Ice With the exceptions of Io and Europa (another of the regular satellites of Jupiter), all of the satellites of the outer planets have densities smaller than about 2000 kg/m^3. Because this is smaller than the density of any silicate mineral, the satellites must be made partly from ices, the only low-density solids likely to be abundant in the solar system. The term *ice* includes not only frozen water but also frozen methane and ammonia. Usually, however, frozen water makes up most of the icy material in a satellite. Ices have densities of about 1000 kg/m^3, whereas rocky material has a density of about 3500 kg/m^3. A mixture of 25% ice and 75% rock has a density of 2200 kg/m^3. Mixtures of 50% and 75% ice have densities of 1600 and 1200 kg/m^3, respectively. Satellites large enough that compression can't be completely ignored are a little denser for a given composition.

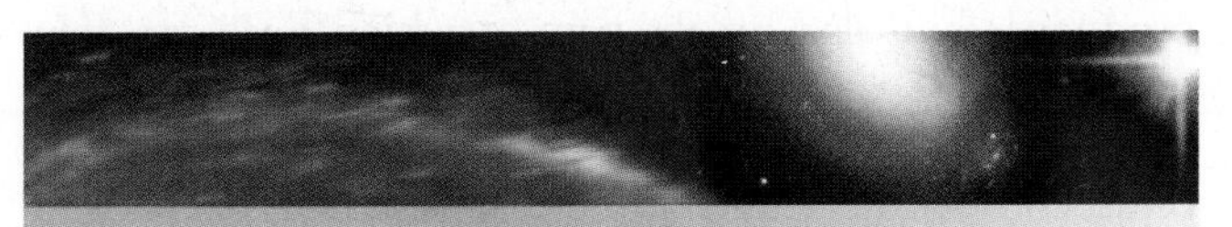

The densities of nearly all the satellites of the outer planets are less than 2000 kg/m^3. Because rocky material has a density of about 3500 kg/m^3, the low densities of the satellites show that a large portion of their material must be ice. Frozen water is probably the dominant ice in outer solar system satellites.

Internal Activity

Many astronomers expected the satellites of the outer solar system to be dead, uninteresting worlds. Instead, some show evidence of crustal motion and volcanic activity, which shows that these satellites must have had significant sources of internal energy. Three sources of internal energy for the satellites of the outer planets are gravitational energy, radioactive heating, and tidal heating. (Gravitational energy and radioactive heating are discussed in Chapter 7, Section 5.) The conversion of gravitational energy into heat was most important early in the history of the satellites as they accumulated infalling matter. In some cases, additional energy may have been released later, when heavy materials sank to form a core.

Radioactive Heating The radioactive elements in a satellite are concentrated in its rocky material rather than its ices. This means that satellites made mostly of ice have much less radioactive heating than satellites that are mostly rock. The amount of radioactive material in a satellite decreases with time because the radioactive elements decay, so the rate at which radioactive heating occurs drops as well.

Tidal Heating **Tidal heating** occurs when the shape of a body changes as the tidal forces acting on it vary in strength. A satellite's shape deviates from a sphere because of tidal forces from its parent planet. The tidal bulges on a satellite point nearly toward and away from the planet it orbits. The tidal force exerted on a satellite increases as the satellite approaches its parent planet. Similarly, the tidal force decreases as the satellite moves away from its parent planet. Thus, for an elliptical orbit like Io's, Figure 14.1 shows that the shape of the satellite is most distorted when it is nearest the planet and least distorted when it is farthest from the planet. The steadily changing shape of the satellite results in internal friction, which heats the satellite. The effect is similar to the heating that occurs

FIGURE 14.1
The Tidal Flexing of Io
The eccentricity of Io's orbit (here greatly exaggerated) and the strong dependence of tidal force on distance from Jupiter cause Io to be more elongated when it is nearest Jupiter (at perijove) than when it is most distant from Jupiter (at apojove). The cycle of stretching and relaxing produces friction and internal heating in Io.

FIGURE 14.2
The Orbital Resonance Between Io and Europa
Europa's orbital period is twice as long as Io's. Thus they are always nearest each other at the same point in their orbits. The eccentricities of the orbits of Io and Europa are greatly exaggerated in this drawing.

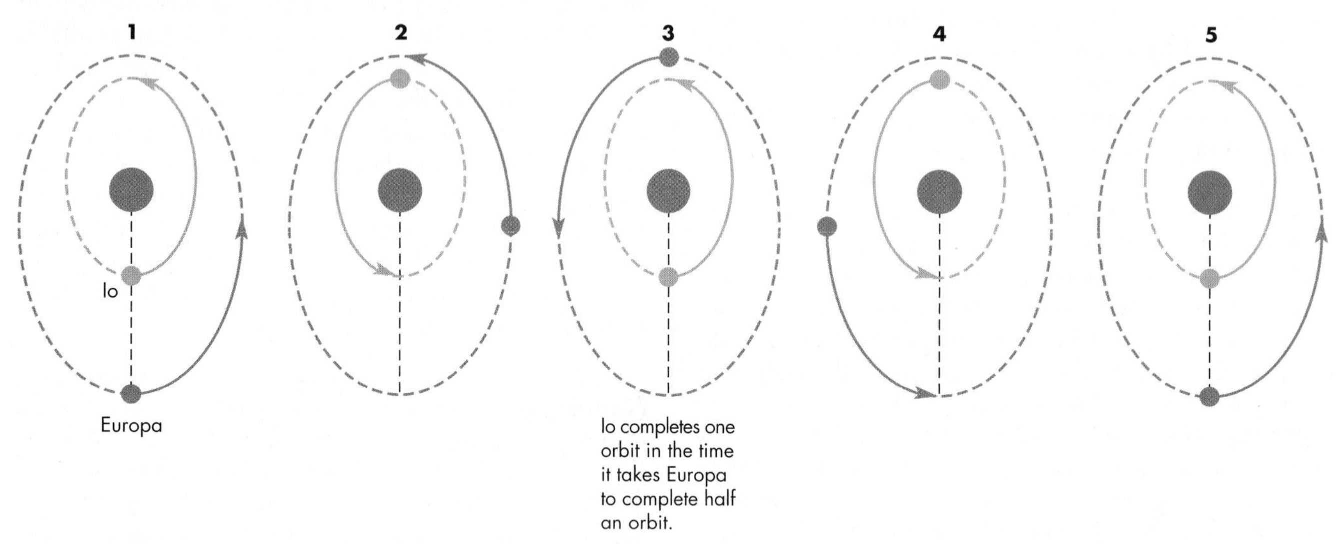
1
2
3
4
5
Io
Europa
Io completes one orbit in the time it takes Europa to complete half an orbit.

when a cold basketball is repeatedly bounced at the beginning of a game. The compression of the ball heats the air within it. This increases the pressure of the air, making the basketball harder and bouncier.

The rate of tidal heating depends on the strength of the tidal force, which increases with the size of the satellite and the mass of the planet and decreases with orbital distance. Only for large satellites that orbit close to massive planets can we expect tidal heating to be important at all.

The tidal interaction between a satellite and a planet not only generates internal heat, but also reduces the eccentricity of the satellite's orbit. In fact, an isolated satellite would soon achieve a circular orbit, in which case it would experience no changes in shape and thus no tidal heating would take place. However, the gravitational attractions of companion satellites can sometimes cause a satellite to retain an elliptical orbit despite the effect of tidal forces. For instance, in the cases of Io, Europa, and Ganymede, three of the satellites of Jupiter, gravitational interactions between the satellites keep their orbits from becoming circular. The orbital period of Io is almost exactly half of Europa's. Thus, Figure 14.2 shows that Io catches up with and passes Europa exactly once each time Europa orbits Jupiter. Moreover, the figure shows that whenever the two satellites are closest together, Io is at the point in its orbit nearest Jupiter, whereas Europa is farthest from Jupiter. The mutual gravitational interaction between Io and Europa prevents their orbits from becoming circular. The orbital period of Europa is nearly half of Ganymede's, so the eccentricity of Europa's orbit is affected by the mutual gravitational interaction between Europa and Ganymede. The consequences of tidal heating for Io and Europa are described later in this chapter.

Thermal Histories The way that temperature rises and falls with time throughout a body is referred to as its thermal history. Just how hot a satellite can become depends not only on the rate at which internal heat is produced, but also on the rate at which heat can flow to the surface and escape. The important heat transfer processes in satellites are conduction and convection. Of these, convection is much more efficient, but it occurs only when the material within a satellite is able to flow. Both rock and ice flow best when they are liquids, but they can flow while they are solids, too. The movement of the ice in glaciers is an example of solid flow.

Recall from Chapter 7, Section 5, that small bodies cool faster than large bodies because large bodies retain heat more effectively. The planetary satellites are made of different mixtures of ice and rock and have a wide range of sizes. Thus, they also have a wide variety of thermal histories. Some of them have never melted and remain almost unchanged since they formed, while others have melted, differentiated, and refrozen.

The release of gravitational energy, radioactive heating, and tidal heating can all be important sources of internal energy for satellites. The thermal history of a satellite depends not only on its heat sources but also on how rapidly heat can be carried out of the satellite by convection and conduction.

14.3 THE SATELLITES OF MARS

The remarkable thing about the quotation from *Gulliver's Travels* at the beginning of this chapter is that it was written 145 years before Phobos and Deimos, the two satellites of Mars, were discovered by Asaph Hall in 1871. At the time that Jonathan Swift wrote *Gulliver's Travels,* there were no known satellites of Venus, one of Earth, and four of Jupiter. Several astronomers, including Kepler, examined this numerical sequence and reasoned that Mars (located between the Earth and Jupiter) must have two satellites. Apparently, Swift adopted this idea and made guesses about the distances and periods of the two satellites that are almost correct. Phobos is unusual because its orbital period (7.7 hours) is shorter than the rotation period of Mars (24.6 hours). This means that an astronaut exploring Mars would see Phobos rise in the west and set in the east every 11.2 hours. Some important properties of Phobos and Deimos are given in Table 14.1.

The best images we have of Phobos and Deimos were returned by the *Viking* orbiters, *Mars Global Surveyor* and *Mars Reconnaissance Orbiter.* Figure 14.3 shows one of the *Viking* views of Phobos, which is about 28 by 20 km and is shaped roughly like a potato. Phobos's surface is marked by many craters, including Stickney, whose diameter of 10 km is almost half the average diameter of Phobos. The violence of the impact that formed Stickney produced a pattern of cracks that radiate away from Stickney. Some of these cracks, about 10 meters deep and hundreds of meters across, can be seen in Figure 14.3. The pattern of cracks suggests that the Stickney impact must have nearly demolished Phobos. *Mars Reconnaissance Orbiter* obtained high-resolution images of Phobos in 2008. Figure 14.4 shows features as small as 7 m (20 feet) in size.

Deimos, shown in Figure 14.5, is smaller than Phobos and does not have craters as large as those on Phobos. The *Viking* orbiter came to within just 23 km of the surface of Deimos in May 1977. A close-up image returned at that time is shown in Figure 14.6. The image, which has details as small as 5 m across, shows that the craters on Deimos look blurry. This is probably because they are blanketed by a layer of dust many meters thick, which softens the appearance of their rims.

Both Phobos and Deimos are believed to be captured asteroids. Mars could have captured Phobos and Deimos when it had a much thicker atmosphere than it does today. Possibly Phobos and Deimos both passed close enough to the surface of Mars that they were slowed down by the friction of atmospheric gases and captured into orbit. The

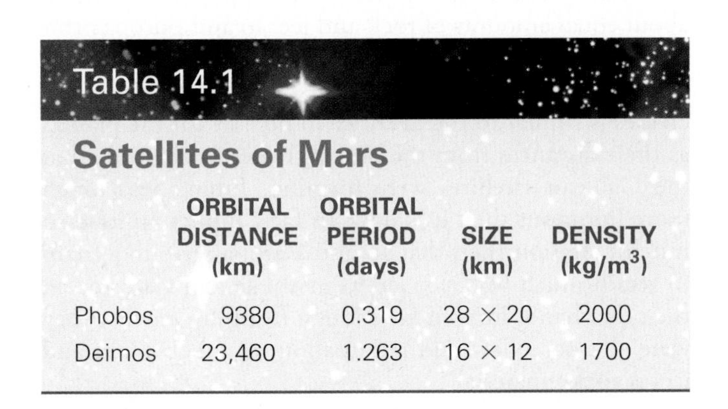

Table 14.1

Satellites of Mars

	ORBITAL DISTANCE (km)	ORBITAL PERIOD (days)	SIZE (km)	DENSITY (kg/m^3)
Phobos	9380	0.319	28 × 20	2000
Deimos	23,460	1.263	16 × 12	1700

FIGURE 14.3
Phobos, the Larger of the Two Martian Satellites
The large crater at the bottom left of the image is Stickney, which has a diameter nearly half as large as Phobos. Note the cracks that radiate away from Stickney.

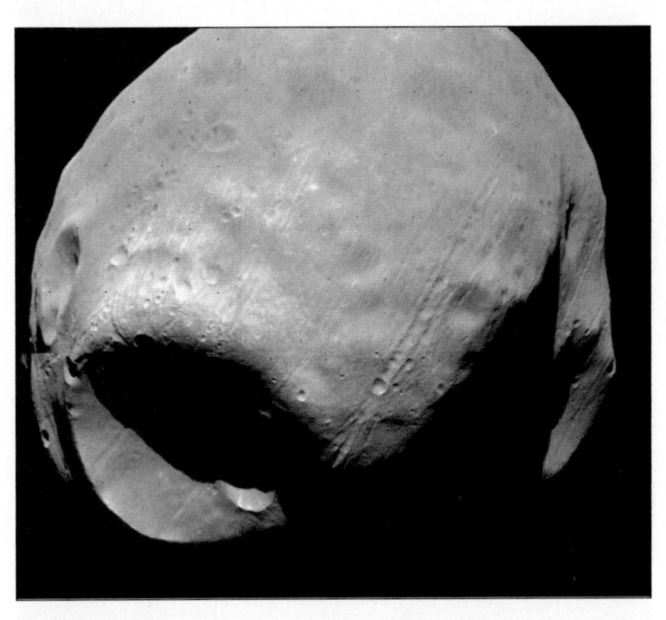

FIGURE 14.4
A *Mars Reconnaissance Orbiter* Image of Phobos
Craters having a range of sizes are visible along with numerous cracks.

FIGURE 14.5
Deimos, the Smaller of the Two Martian Satellites
The craters on Deimos have a more subdued appearance than those of Phobos. The softer appearance is probably due to a thick layer of dust on Deimos.

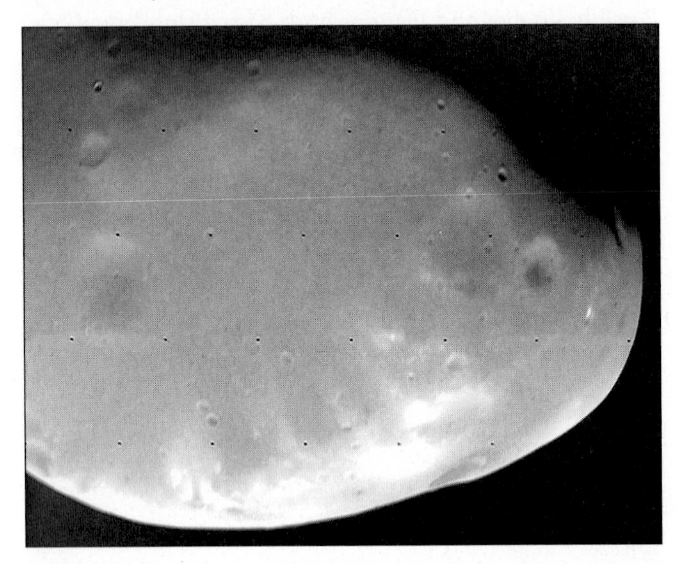

FIGURE 14.6
A Close-up View of Deimos
This image was taken by *Viking* at a distance of only 23 km above the surface of Deimos. The smallest visible craters are only about 5 m across.

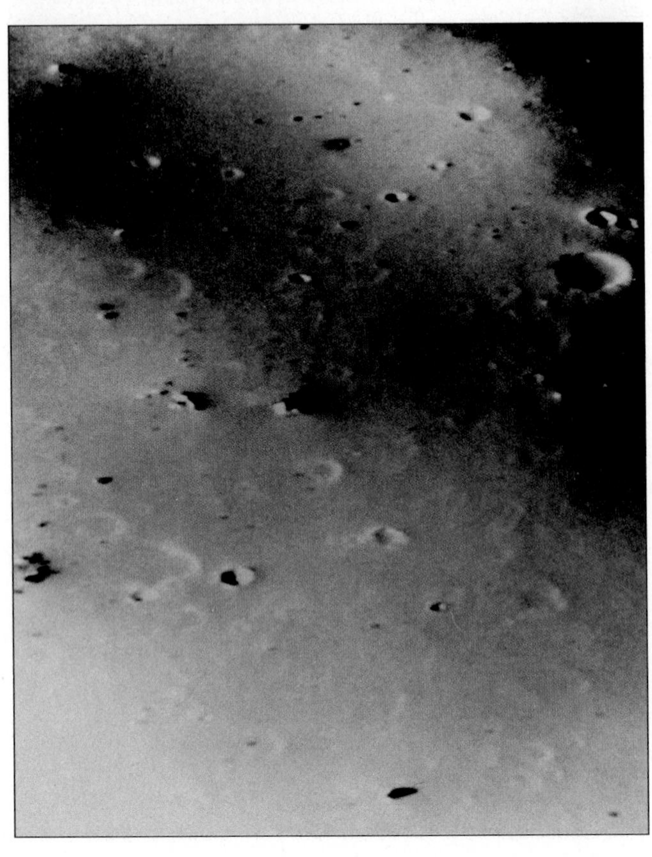

Martian atmosphere is now much too thin, and has been for billions of years, to assist in the capture of an asteroid.

Phobos, which lies within the Roche distance of Mars, is slowly spiraling inward. Within about 100 million years Phobos will either be torn apart by tides, perhaps forming a ring, or will crash into Mars.

The two Martian satellites, Phobos and Deimos, are thought to be captured asteroids. Their surfaces show many craters. The crater Stickney, on Phobos, was the result of an impact that nearly destroyed Phobos. The impact produced a system of cracks that radiate from Stickney.

14.4 THE GALILEAN SATELLITES

On January 7, 1610, Galileo pointed his new telescope at Jupiter. He saw not only Jupiter but also what he described as three "small stars." Later, he found another small star. By the end of January, Galileo had convinced himself that these were four new planets that orbited Jupiter, a larger planet. He named them, in order of distance from Jupiter, Io, Europa, Ganymede, and Callisto. They became known as the Galilean satellites. Some important properties of the Galilean satellites are given in Table 14.2.

The Galilean satellites are among the largest moons in the solar system. Compared with the Moon, Europa is slightly smaller, whereas Io is slightly larger. Somewhat larger still is Callisto, nearly the same size as Mercury. Ganymede, the largest satellite in the solar system, is about 10% larger than Mercury. All four might be considered planets if they orbited the Sun rather than Jupiter. In fact, all four are so large and bright that they could be seen from the Earth without a telescope if the glare of Jupiter didn't obscure them.

The densities of Io and Europa are 3530 kg/m^3 and 3010 kg/m^3, respectively, so they must be made mainly of rock. However, the densities of Ganymede and Callisto are both about 1900 kg/m^3, so they probably contain about equal amounts of rock and ice. Io and Europa orbit closer to Jupiter than Ganymede and Callisto do. Thus, the decrease in density as the distance from Jupiter increases is similar to the decrease in density for the planets as their distances from the Sun increase. At the time that the Galilean satellites were forming, Jupiter was much more luminous than it is now. In fact, Jupiter emitted so much radiation then that Io and Europa were too warm to retain much water or ice. Ganymede and Callisto, on the other hand, formed far enough from Jupiter that they were able to collect and retain about as much water and ice as rocky material.

Table 14.2

Galilean Satellites

	ORBITAL DISTANCE (km)	ORBITAL PERIOD (days)	DIAMETER (km)	DENSITY (kg/m^3)
Io	421,600	1.77	3640	3530
Europa	670,900	3.55	3122	3010
Ganymede	1,070,400	7.15	5262	1940
Callisto	1,882,700	16.69	4820	1830

Io

Surface Shortly before the *Voyager* flybys of Jupiter, Stanton Peale and his coworkers published a remarkable paper in which they analyzed the effect of tides on the interior of Io. Io's relatively large size, proximity to Jupiter, and forced orbital eccentricity imply that Io should experience very large tidal effects. Peale and his colleagues calculated that the tidal heating in Io is about 1000 times as great as heating due to radioactive decays. As a result, they predicted that there might be extensive volcanic activity on Io. Seldom have scientific predictions been so spectacularly confirmed.

Figure 14.7 shows one hemisphere of Io. The yellow and orange surface with its numerous dark spots resembles a pepperoni pizza. At first the significance of the dark spots wasn't recognized. Later, after *Voyager 1* had passed Jupiter, close examination of an Io image revealed a bright feature extending above the edge of Io. This proved to be a volcanic **plume,** like the one rising hundreds of kilometers above the edge of Io in Figure 14.8. Some of the dark spots in Figure 14.7 are plumes. Close-ups of other dark spots, like the one in Figure 14.9, showed that they are volcanic calderas with surrounding lava flows. Figure 14.10 shows glowing lava during a volcanic eruption on Io.

FIGURE 14.7
A *Galileo* Image of Jupiter's Satellite, Io
The surface color is produced by elemental sulfur and sulfur compounds. Many of the round, dark markings are volcanic plumes or calderas.

FIGURE 14.8
A Plume Seen on the Limb of Io
The plume extends more than 100 km above the volcano Pillan Patera.

FIGURE 14.9
Io's Volcano Zal Patera
This *Galileo* image shows Zal Patera, a 5-km-high volcano on Io. The caldera of the volcano, seen toward the top of the image, is surrounded by lava flows.

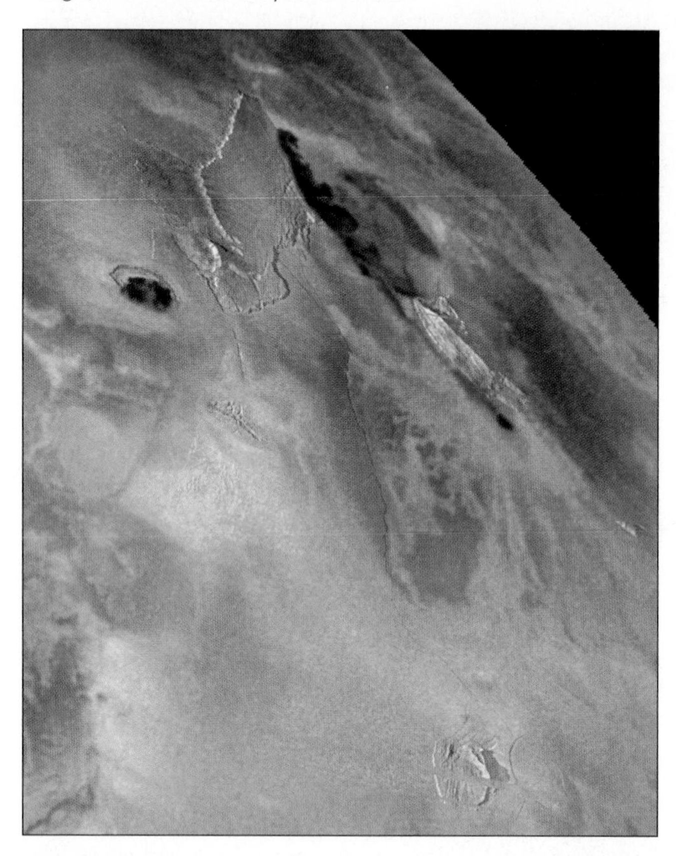

FIGURE 14.10
A Volcanic Eruption on Io
Hot, glowing lava erupts from the fissure at the left of the image.

Although there is volcanic activity on Earth, Venus, and possibly on the Moon and Mars, Io is by far the most volcanically active body in the solar system. There are about 500 active volcanos on Earth, but few of these are erupting at any given time. If a spacecraft were to fly past the Earth, it is unlikely that a major volcanic eruption would be seen. The *Voyager 1, Voyager 2,* and *Galileo* images have shown that on Io, many major eruptions are going on all the time.

Volcanic activity on Io is so intense that it rapidly changes Io's surface features. Some changes were even seen during the 4 months between the times of the *Voyager 1* and *Voyager 2* flybys. A close inspection of Figure 14.11 reveals many surface changes during the 17 years between *Voyager* and *Galileo* images of the same portion of Io. Io is continually resurfaced by volcanic eruptions. The rate at which volcanos deposit new material may be as high as 10 cm per year, averaged over the entire surface of Io. This is equivalent to about 4000 km^3 of volcanic material

FIGURE 14.11
***Voyager* and *Galileo* Images of Io**
The *Voyager* image (left) and *Galileo* image (right) show the same region of Io. Note the many changes during the 17 years between the two images. One of the most conspicuous changes is a dark lava flow from the volcano Prometheus, the bright ringed feature just to the right of the center of the images.

each year. The Earth, in contrast, is much larger than Io but the rate at which new material is deposited on Earth by volcanic eruptions is about 1000 times smaller. One consequence of the rapid deposition of volcanic material on Io is that impact craters are swiftly filled in and become undetectable. As a result, not a single impact crater has been found in any of the *Voyager* or *Galileo* images of Io. The absence of impact craters shows that Io's surface is the youngest in the solar system.

Despite its intense volcanic activity, Io doesn't have huge volcanic mountains like Olympus Mons on Mars or Mauna Loa on Earth. Apparently the lava on Io is so fluid that it flows away too fast to build up large volcanos. In some cases, lava extends hundreds of kilometers from the volcanic calderas.

The yellow, orange, and black colors of Io are characteristic of sulfur compounds such as sulfur dioxide (SO_2). Molten sulfur is very fluid, so it has been suggested that some of the longer flows on Io may be sulfur rather than silicate rock. Sulfur and SO_2 may also be responsible for powering the plumes of Io. The plumes resemble terrestrial geysers, which are powered by the change of superheated water to steam as it nears Earth's surface. Because water seems to be entirely absent from Io, some other volatile substance must take its place in driving the plumes. Sulfur and SO_2, which would be molten at depths of only a few kilometers in Io, are likely candidates.

Although sulfur compounds are present in the flows and plumes of Io, it doesn't seem likely that all of the surface materials of Io can consist of sulfur compounds. Sulfur forms brittle materials that can't be piled high enough to form tall mountains. Some of Io's mountains (which do not appear to be volcanic in origin) are more than 9 km high, requiring surface materials of considerable strength. Also, some of Io's calderas are several kilometers deep, a depth difficult to understand if the surface were entirely sulfur compounds. Figure 14.12 shows one plausible model of the surface layers of Io consisting of a mountainous silicate surface covered by layers of both sulfur and silicate lavas. In this model, the mountains of Io are spots where the silicate crust penetrates the lava layers and is exposed.

FIGURE 14.12
A Model for the Surface Layers of Io
Mountains of silicate rock extend upward through layers of sulfur and silicate lava.

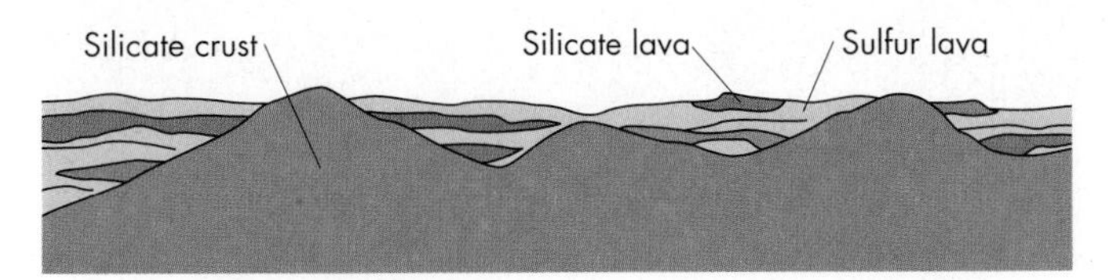

The large tidal heating experienced by Io produces extensive volcanic activity. Volcanic material is deposited continuously on the surface of Io and covers up impact craters soon after they form. Sulfur and sulfur compounds color the surface of Io and power its volcanic plumes.

Internal Structure Two possible models for Io's interior have been proposed. In one model, a thin rigid crust covers an entirely molten interior. In the other model, a thin rigid crust covers a thin molten, or partially molten, layer. Solid mantle is believed to be beneath this molten layer. In both models, Io has an iron-rich core extending out to half the radius of Io. At the present time, it is impossible to judge which of the two models is correct because both are consistent with Io's average density.

Atmosphere Before the *Pioneer* and *Voyager* encounters, Io was thought to have no atmosphere at all. Measurements carried out by the probes, however, showed that Io has a thin atmosphere consisting mainly of gaseous SO_2. During the night, the surface temperature of Io falls and a frost of SO_2 condenses on the surface, reducing the atmospheric pressure dramatically. During the day, sunlight warms the surface and vaporizes the frozen SO_2, thus restoring the atmosphere. Continuously cycling SO_2 in this way must create strong winds that blow from the relatively high pressure day side of Io to the near vacuum of the night side.

The gases that escape from Io contain atoms and ions of sulfur, sodium, potassium, and oxygen. Eventually Io's escaping atmosphere diffuses throughout Jupiter's entire magnetosphere. Some of the atoms that originated on Io may be deposited on Europa and other satellites that orbit within Jupiter's magnetosphere, and others may escape into interplanetary space.

Europa

As the *Galileo* image in Figure 14.13 shows, the surface of Europa is highly reflecting and crisscrossed by dark bands, some which stretch halfway around Europa. Overall, Europa is smoother than a billiard ball. It has no hills higher than about 1 km, no deep chasms, and few impact craters.

Europa's surface is highly reflective because it is made primarily of water ice. Europa's high density (about 3000 kg/m^3), however, suggests that only about 15% of its material can be ice or water. The rest must be rock and metal. If all the water and ice in Europa were located in a

FIGURE 14.13
The *Galileo* Image of Europa
The surface of Europa is smooth and crisscrossed by a pattern of dark bands. The bright feature at the lower right is the impact crater Pwyll, about 26 km in diameter. The dark spot at the lower left is the impact crater Callanish.

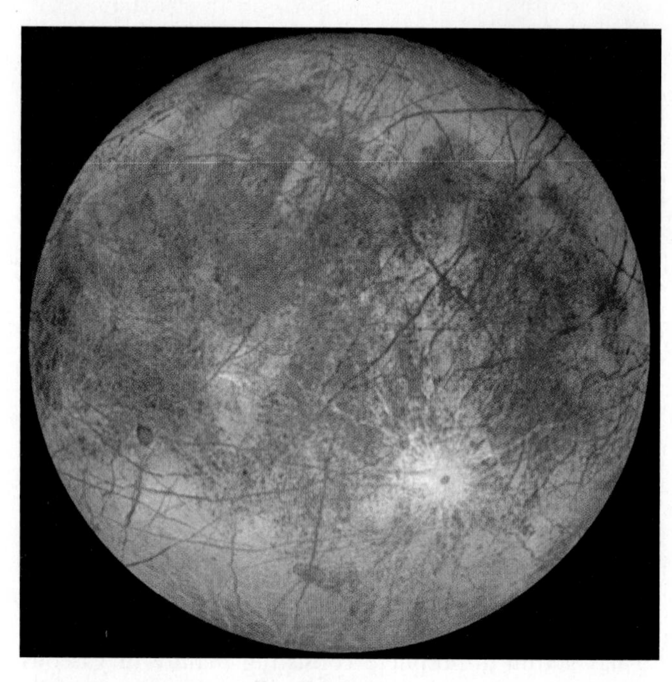

FIGURE 14.14
The Interior Structure of Europa
Europa has a metallic core and rocky mantle covered by a layer of ice. Inside the ice layer is a layer of liquid water.

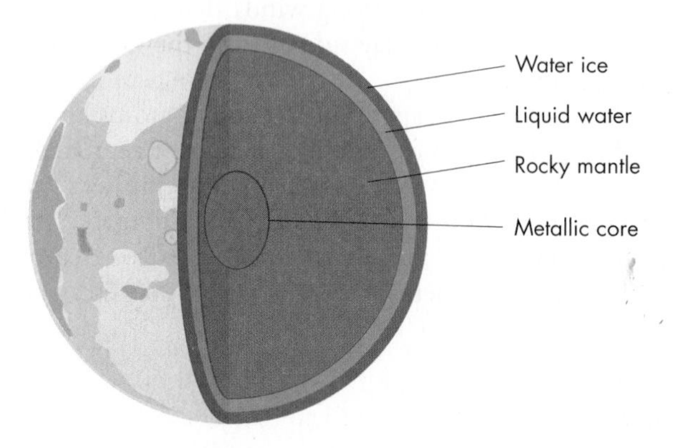

layer at its surface, that layer would be about 100 km thick, as shown in Figure 14.14. Just how much of the layer of water in the outer layers of Europa is liquid depends on whether tidal and radioactive heating have been able to keep the interior of Europa hot. Before *Galileo* began sending back images of Europa, astronomers had little convincing evidence that there was a water ocean under Europa's frozen crust. The *Galileo* images, however, give striking evidence that Europa does have liquid water near its surface and that the icy crust might be no more than a few kilometers in thickness.

The fact that no impact craters larger than 50 km were found in the *Voyager* images, and only a few craters in the 20 km size range were visible, was the first indirect evidence for liquid water on Europa. The scarcity of craters implied that the ancient surface of Europa—presumed to have been heavily cratered—must have been eroded or covered up. The most likely kind of erosive process is glacierlike flows of ice. This kind of flow could gradually reduce the heights of crater rims and fill in the central pits of craters. To be effective in eroding large craters, however, the ice layer must be relatively warm, because warm ice flows much faster than cold ice. *Galileo* observations of Pwyll, a 26-km-diameter crater shown in Figure 14.15, confirm that glacierlike flows occur. Pwyll is a young crater, but is extremely shallow.

High-resolution *Galileo* images also show that the dark bands that crisscross Europa are systems of ridges and

FIGURE 14.15
The Young Impact Crater Pwyll as Seen from *Galileo*
The white region around Pwyll is a 40-km-wide ejecta blanket blasted outward by the impact. A pattern of rays also radiates outward from Pwyll. Although Pwyll is young, it is very shallow, indicating that glacierlike flows of ice have filled in Pwyll's floor.

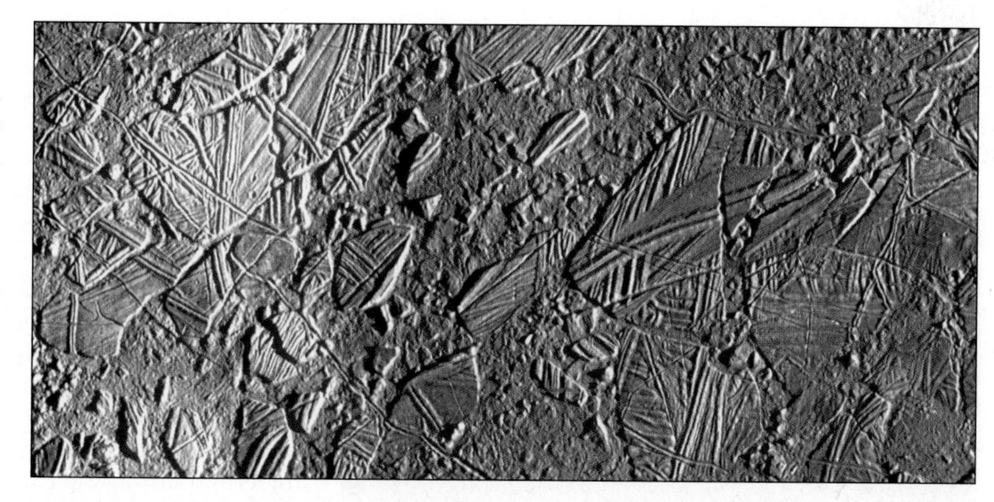

FIGURE 14.16
Ice Rafts on Europa
This *Galileo* image shows iceberglike chunks of Europa's icy crust that have drifted and rotated away from their original locations. The rafting of chunks of Europa's surface may be due to warm currents in a liquid water ocean just beneath the icy crust.

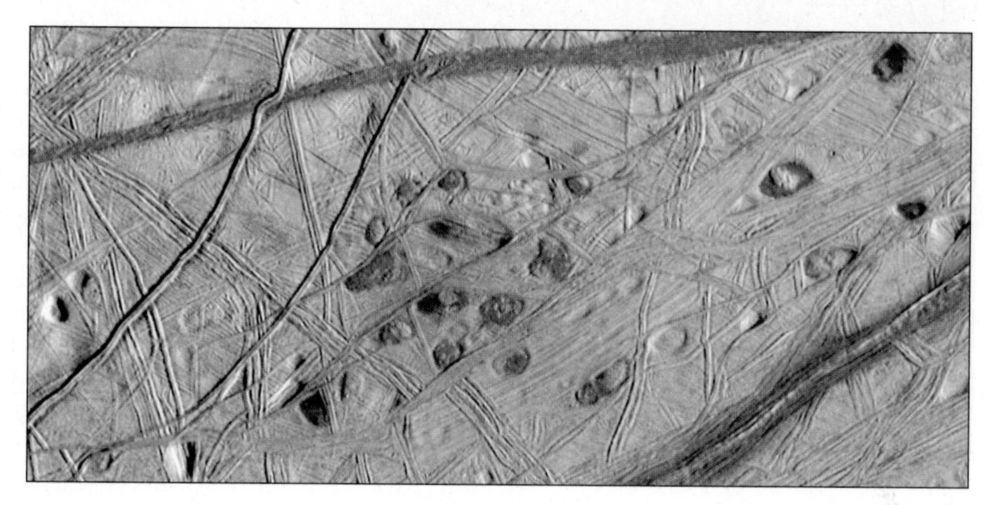

FIGURE 14.17
Europa's Lenticulae
This *Galileo* image shows a region of Europa about 100 km across. The dark spots are lenticulae, which are 100 m high ice mounds that may have formed when relatively warm blobs of ice rose through the colder surrounding ice. The warm ice in the mounds may contain chemicals from the ocean beneath the surface ice.

grooves probably caused by the movement of Europa's icy crust. The ridges and grooves may result from tidal distortion as Europa's orbit carries it closer and then farther from Jupiter during its elliptical 3.6-day orbit. The tidal distortion fractures Europa's surface. Water rises through the cracks from below and then freezes at the surface to form ridges on either side of the cracks.

Still more evidence of liquid water beneath Europa's surface is found in Figure 14.16, which shows features resembling icebergs that have broken away from their original locations and then drifted and rotated. One explanation for the rafting of chunks of Europa's surface away from their original locations is that the chunks are moved about by warm water currents beneath a layer of ice that may be as little as 1 km thick.

Figure 14.17 shows dark lenticulae, or ice mounds about 5 km across and 100 m high. The mounds may have been formed by pockets of warm ice rising through the colder surrounding ice. The probability that Europa has liquid water has focused attention on the possibility that Europa may harbor life. This possibility is discussed in Chapter 27. Astronomers are developing plans to explore Europa further. Among the ideas that have been suggested is a spacecraft that would land on Europa, melt through the icy crust, and release a submarine to explore the concealed ocean.

Europa is made mostly of rock. *Galileo* images of Europa's surface show shallow craters, grooves and ridges, and rafting of surface features that strongly suggest the surface is an icy crust covering an ocean of liquid water.

Ganymede

Ganymede's surface, shown in Figure 14.18, has two distinct types of terrain. Heavily cratered, dark regions are separated by bright regions that form intricate patterns. Figure 14.19 shows that the bright regions are bands of parallel ridges and valleys and resemble the dark bands seen on Europa. The bands run for thousands of kilometers across Ganymede. The dark terrain, also shown in Figure 14.19, appears to have been deformed by episodes

FIGURE 14.18
The Surface of Ganymede
This *Galileo* image shows that the surface of Ganymede consists of regions of ancient dark terrain separated by light-colored terrain.

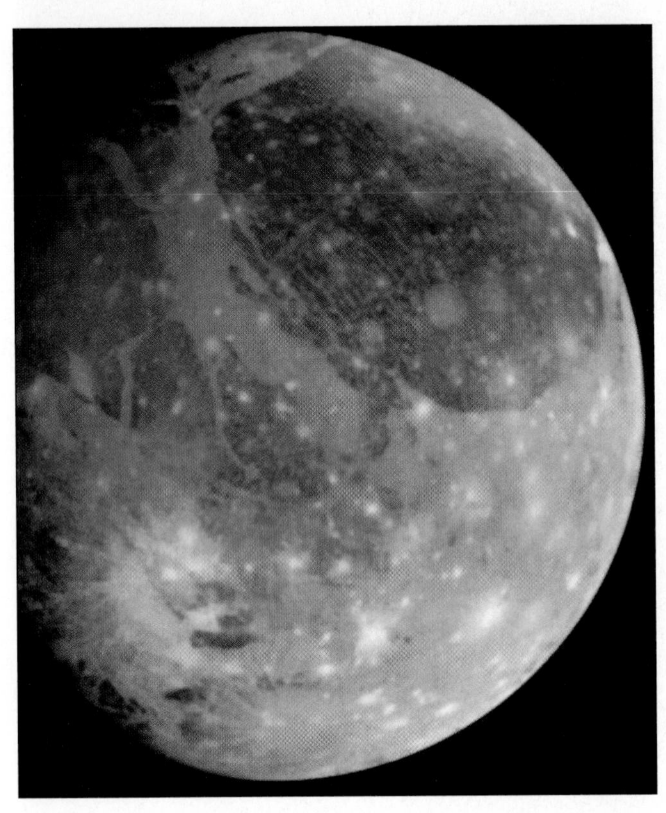

of faulting. Figure 14.20 shows what it would be like to hover over Ganymede's dark terrain. At first glance it appears that the dark regions represent ancient crust that was broken and pushed apart by the growth of the bright regions. The history of Ganymede's surface must be more complicated than this, however, because the dark regions wouldn't fit together very well if they were pushed back together. Apparently, the bright terrain destroyed ancient darker terrain as it developed.

One clue to the origin of the complex surface features of Ganymede was the discovery, based on *Galileo* images, that the bright bands are quite smooth and lower than the dark terrain by about a kilometer. This suggests that the bright bands began as troughs that were then filled in by eruptions of water from within Ganymede. Although crater densities show that the bright bands are about a billion years old, there may still be liquid water within Ganymede today. One of the findings of *Galileo* is that Ganymede has a magnetic field generated within a molten core. In addition to the permanent magnetic field, Ganymede has a secondary magnetic field that changes as Ganymede orbits Jupiter. The best explanation for the varying magnetic field is that Jupiter's magnetic field sets up electric currents in a layer of salty water buried about 150 km beneath Ganymede's surface. The buried ocean may be 5 km thick, implying that Ganymede contains about as much liquid water as does the Earth. As Ganymede orbits Jupiter, Jupiter's magnetic field changes direction and strength, changing the currents within Ganymede and, thus, the secondary magnetic field.

FIGURE 14.19
A High-Resolution *Galileo* Image of Ganymede
This image shows a region 89 by 26 km across with a resolution of 34 m. The smooth area at the center of the image is a region of bright terrain that shows parallel sets of grooves. The regions at the right and left are dark terrain that has been deformed by faulting.

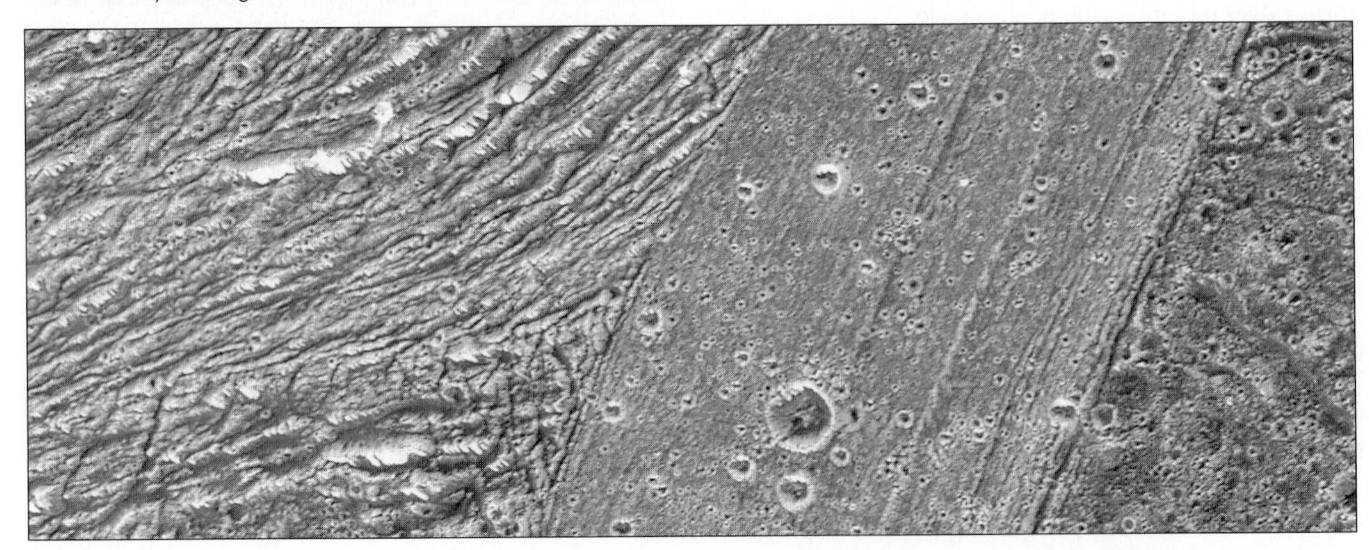

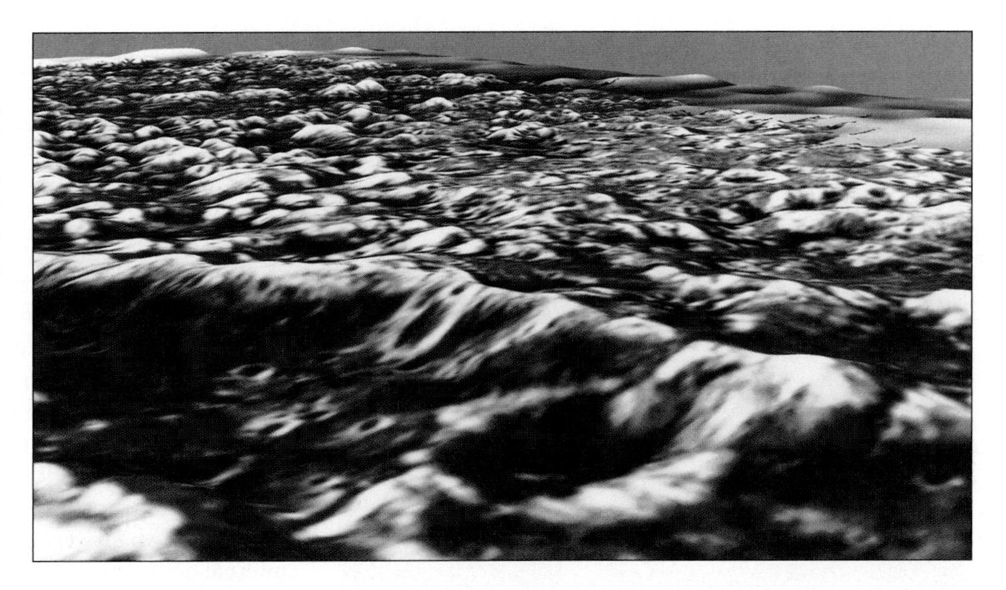

FIGURE 14.20
A View Across Ganymede
This view, reconstructed from *Galileo* images, shows what it would be like to hover above a portion of Ganymede's dark, ancient terrain. Both deep furrows and impact craters can be seen.

Radioactive heating in Ganymede's rocky material is probably sufficient to produce the molten core and deeply buried ocean layer that exist today. The flow of water to Ganymede's surface a billion years ago, however, required an additional source of heat. One possibility is tidal heating. Although it is not quite in an orbital resonance with Europa, Ganymede's orbital distance is increasing. Ganymede may have been in an orbital resonance with Europa a billion years ago and may have had an elliptical orbit at that time. If so, it would have been subject to tidal flexing and heating such as Io and Europa experience today. Tidal flexing may have cracked Ganymede's surface. The additional heating would have enlarged the buried ocean and brought it close enough to the surface for water to flow into troughs and produce the bright bands that are the scars of that more active part of Ganymede's past.

Callisto

Callisto and Ganymede are very similar in size and composition. However, Callisto's surface seems to have been almost unchanged since the time it formed. Impact craters cover virtually the entire surface of Callisto, as shown in Figure 14.21. Figure 14.22 shows that the largest impacting bodies produced multiring structures similar to those on the Moon and Mercury. A large scarp produced by the Valhalla impact is shown in Figure 14.23. Relatively young craters are white and bright, because the impacts uncovered clean ice, which has not yet been darkened by micrometeorite bombardment. Because the outer layer of Callisto is icy, Callisto's craters are shallower than those of the Moon and other rocky bodies. Not even the large multiring structures have central depressions of any great depth. Overall, Callisto shows no evidence that its surface has ever been modified by tectonics or volcanic activity. *Galileo* measurements show that, like Ganymede, Callisto has a varying magnetic field induced by the magnetic field of Jupiter. This suggests that Callisto, too, has an ocean of salty water buried deep beneath the surface. Apparently,

FIGURE 14.21
A *Galileo* Image of Callisto
Callisto's surface is heavily cratered and very ancient. The bright spots on Callisto are fresh ice, exposed when relatively recent impacts removed a thin layer of dark meteoritic debris.

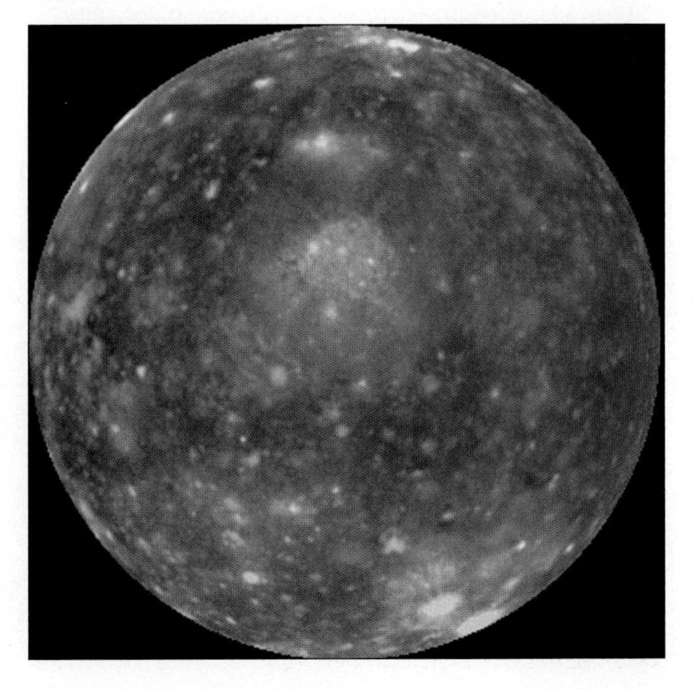

FIGURE 14.22
Valhalla, a Large, Multiringed Impact Basin on Callisto
Valhalla is similar to the large impact basins within which the lunar maria are found. Concentric ridges extend outward to a distance of about 1500 km from the center of Valhalla.

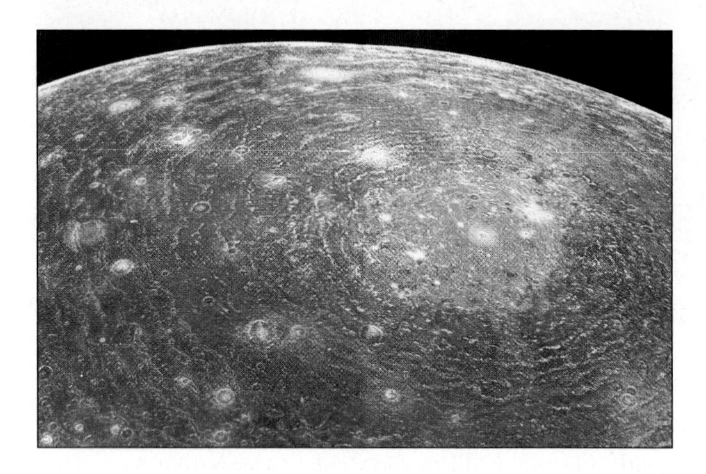

Callisto has never experienced enough tidal heating to enlarge its buried ocean and thin its outer, icy layer enough for liquid water to reach the surface.

Ganymede and Callisto are similar in diameter and mass yet different in appearance. Ganymede's surface shows evidence of considerable internal activity including eruptions of water from within. Callisto has a heavily cratered, ancient surface. Both satellites probably have thick oceans of salty water buried deep beneath the surface.

14.5 THE ICY MOONS OF SATURN

Saturn has 60 known satellites, eight of which have average diameters greater than 300 km. Some important properties of Saturn's eight largest satellites are given in Table 14.3. One of these satellites, Titan, is about as large as Ganymede. Titan is described later in this chapter. The other seven larger satellites of Saturn (Mimas, Enceladus, Tethys, Dione, Rhea, Hyperion, and Iapetus) range in diameter from 300 to 1500 km. Their densities show that they contain at least 50% ice, which, in the cold of the Saturnian system, is as hard as rock. Each of these seven satellites is heavily scarred by impact craters. Before the *Voyager* images were obtained, most astronomers assumed that bodies as small as the satellites of Saturn (except Titan) were too small to have ever had warm interiors. Little or no evidence of internal activity was expected. Not surprisingly, Hyperion, Rhea, and Mimas show little evidence of internal activity. Dione and Tethys, however, have surfaces that reveal past internal activity. The surface of Iapetus has been extensively modified by internal activity, and activity on Enceladus has been directly observed by the *Cassini* orbiter.

Hyperion

Figure 14.24 shows a *Cassini* image of Hyperion, one of the oddest looking satellites in the solar system. The depression in the center of the image may be a large impact crater, heavily eroded by later impacts. The numerous honeycomb-like pits may have resulted when dark material collected in small impact craters. The dark material absorbed sunlight and melted the ice around it, deepening and widening the crater. Sometimes the widened craters intersect. The low density of Hyperion suggests that its interior may have many large caverns. An alternative explanation is that bodies striking Hyperion's porous outer layers compress the surface into deep craters.

FIGURE 14.23
A Scarp on Callisto
This *Galileo* image covers a region 38 km across. Sunlight illuminates the scene from the left so the dark band on the right is the shadow of a scarp. The scarp was formed by the impact that produced the Valhalla basin. Notice the many impact craters that dot Callisto's surface.

Table 14.3

Major Satellites of Saturn

	ORBITAL DISTANCE (km)	ORBITAL PERIOD (days)	DIAMETER (km)	DENSITY (kg/m^3)
Mimas	185,520	0.94	400	1140
Enceladus	238,020	1.37	500	1000
Tethys	294,660	1.89	1050	1000
Dione	377,400	2.74	1120	1500
Rhea	527,040	4.52	1530	1240
Titan	1,221,850	15.95	5150	1880
Hyperion	1,481,100	21.27	300	1500
Iapetus	3,561,300	79.33	1440	1020

Rhea and Mimas

Figure 14.25 shows that the surface of Rhea is heavily cratered, with no evidence of cracks or other indications of prolonged internal activity. Rhea's crater density varies somewhat from region to region, however, which suggests that some resurfacing took place during the period of heavy bombardment more than 4 billion years ago, just after Rhea formed. Remarkably, Rhea may have a ring system consisting of three narrow rings between 800 and 1200 km above the satellite's surface.

Mimas, the second smallest of the larger icy satellites of Saturn, also has a heavily cratered surface. Figure 14.26 shows Mimas's most remarkable surface feature, Herschel, a gigantic crater that has a diameter nearly one-third as large as Mimas. It seems likely that the impact of the body that produced Herschel must have come close to destroying Mimas. As is the case for Rhea, Mimas's crater density varies from region to region, indicating that resurfacing may have taken place during the period of heavy bombardment.

FIGURE 14.24
A *Cassini* Image of Saturn's Satellite Hyperion
This false-color image emphasizes relatively small color differences. In reality, Hyperion has a slightly reddish color. The feature in the center of the image may be a relatively large, heavily eroded impact crater. Note the unusual honeycomb-like appearance of the smaller impact craters.

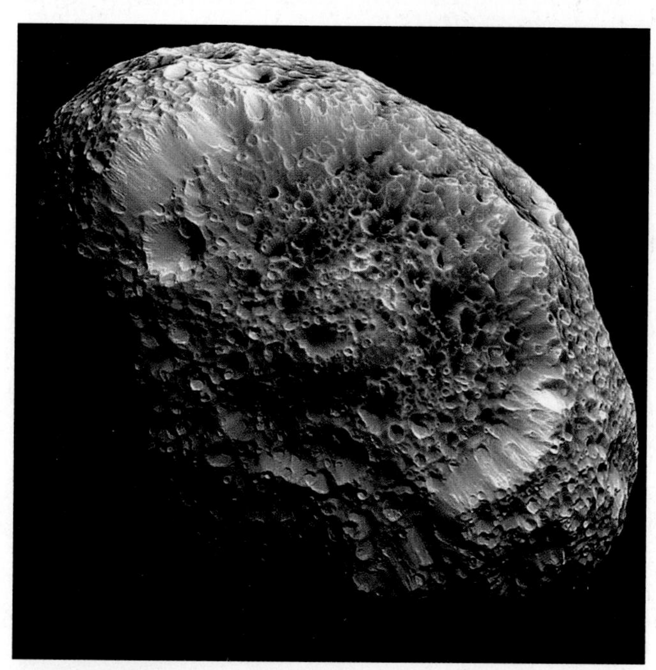

Dione and Tethys

The surfaces of Dione and Tethys, like those of Rhea and Mimas, are heavily cratered. However, both Dione and Tethys show evidence of resurfacing and internal activity.

FIGURE 14.25
A *Cassini* Image of Saturn's Satellite Rhea
Rhea's ancient surface is heavily cratered. The white regions on the inner rims of several craters are probably exposures of fresh ice.

FIGURE 14.26
A *Cassini* Image of Mimas and the Crater Herschel
Herschel is almost one-third as large as Mimas. It is thought that the impact of a body any larger than the one that produced Herschel would have fragmented Mimas.

There are extensive smooth plains on Dione and Tethys, which probably formed when watery material flowed out from their interiors. The most notable surface features of Dione are bright wispy streaks. High-resolution *Cassini* images, such as Figure 14.27, have shown that the wisps are bright icy cliffs produced by fractures.

Another indication of past internal activity on Tethys is Ithaca Chasma, a great trench that extends for about 2500 km (three-fourths of the circumference of Tethys). Ithaca Chasma, shown in Figure 14.28, is about 100 km wide and has walls several kilometers high. It is so large that it covers about 10% of Tethys's surface. One proposal to explain Ithaca Chasma is that Tethys was once a ball of liquid water with a thin icy crust. When the interior froze, the satellite expanded and cracked. Why the cracking was confined to a single trench is not clear. Expansion upon freezing may also explain the fractures of Dione.

Iapetus

Iapetus, like the other satellites of Saturn, orbits with the same face always turned toward Saturn. This means that the same hemisphere of Iapetus always leads as it orbits. The leading and trailing hemispheres of Iapetus have completely different appearances. The trailing hemisphere is bright and heavily cratered like the other icy satellites of Saturn, while the leading hemisphere is so dark that it is difficult to see craters or other surface features. Figure 14.29 shows that the boundary between the dark and bright regions of Iapetus is very sharp.

FIGURE 14.27
A *Cassini* Image of Saturn's Satellite Dione
This detailed image was obtained during a close approach to Dione in December 2004. One of Dione's wispy features runs from the upper right to center bottom of the image. *Cassini* images like this one showed that the wisps are produced by bright icy cliffs caused by fractures in Dione's surface. One of Dione's relatively smooth regions can be seen on the bottom right side of the image.

FIGURE 14.28
A *Cassini* Image of Saturn's Satellite Tethys
Notice that the lower right portion of Tethys appears smoother and less cratered than the upper left. The smoothness of the lower right part of the image is probably a result of flows of watery material (melted ices) from within Tethys. Ithaca Chasma, a great trench about 100 km wide, stretches downward from the large crater in the top.

FIGURE 14.29
A *Cassini* Image of the Two Faces of Iapetus
The leading side of Iapetus (upper right) is much darker than the trailing side. The boundary between the dark and light portions of Iapetus is quite sharp. Note the ridge circling Iapetus's equator.

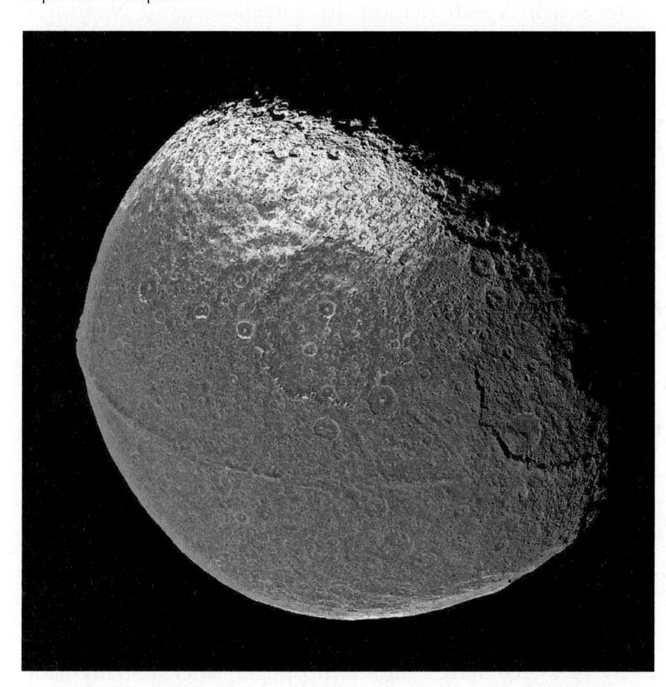

FIGURE 14.30
A *Cassini* Mosaic of Enceladus
This mosaic of 28 individual images of Enceladus shows few craters, indicating that the part of Enceladus shown here is very young. Note, however, the more heavily cratered older region at the lower left. The bluish stripes contain narrow fractures from which liquid water emerges to flood the surface and from which plumes of vapor are propelled into space. Labtayt Sulci, at the top of the image, is a canyon about 1 km deep.

Because the dark region lies on the leading hemisphere of Iapetus, it seems probable that the dark material consists of a layer of debris swept up by Iapetus as it orbits. The dark color of the swept-up material is due to carbon compounds trapped in water ice. The bright hemisphere is covered by nearly pure water ice. The absence of bright craters within the dark region shows that either the dark deposit is so thick that it is not penetrated by impacts or that dark material is deposited fast enough to cover impact craters as fast as they are formed.

Another curious feature of Iapetus is a ridge around its equator. The ridge is 20 km wide and as much as 13 km high. Possibly the ridge is the remains of a ring that once circled Iapetus and then fell into its surface around the equator.

Enceladus

Of all Saturn's satellites. Enceladus shows the greatest evidence of internal activity. High-resolution *Cassini* images like Figure 14.30 show that the surface of Enceladus has a low crater density. Even the most heavily cratered regions of Enceladus are more lightly cratered than those of other icy satellites of Saturn, indicating that global resurfacing must have taken place, perhaps within the past few tens of millions of years. The bluish stripes (nicknamed "tiger stripes" by the *Cassini* imaging team) are narrow fractures where icy material emerges from the interior of Enceladus. The icy material has flooded old craters and produced a ridged terrain that has almost no impact craters.

Icy material emerging from the tiger stripes has also given Enceladus a thin, patchy atmosphere consisting mostly of water vapor and molecular hydrogen with traces of carbon dioxide, carbon monoxide, and nitrogen molecules. At least some of the icy material escapes from the interior of Enceladus in the form of plumes or fountains that emerge from the regions of the tiger stripes. Figure 14.31 shows that sunlight scattered by icy particles makes the plumes visible above the crescent of Enceladus. The plumes are powerful enough that some material escapes from Enceladus and supplies fresh material to Saturn's broad E ring. In addition to water vapor, the plumes also contain both simple and complex carbon compounds, suggesting that Enceladus could support life.

The fact that Enceladus contains liquid water is surprising because a body as small as 500-km Enceladus should have cooled rapidly and frozen long ago. For Enceladus to have a liquid interior today, it must have a significant source

FIGURE 14.31
Icy Plumes from Enceladus
Icy material erupting from the vicinity of the "tiger stripes" on Enceladus can be seen on the left side of the image because it scatters sunlight.

of internal heat. The best possibility for explaining the internal activity of Enceladus and the other satellites of Saturn appears to be tidal heating as tides rub the sides of cracks against each other. Recall that in order for tidal heating to be effective, a satellite must be in an orbital resonance with another satellite. Otherwise tides would circularize the orbit of the satellite and tidal flexing and heating would stop. Dione has an orbital period exactly twice that of Enceladus, keeping Enceladus's orbit slightly elliptical and making tidal heating of Enceladus possible. Similarly, Tethys has an orbital period twice that of Mimas. Mimas doesn't show obvious signs of internal activity, but perhaps Mimas is too small to experience significant tidal heating. There are no other strong orbital resonances in the satellite system of Saturn today. Saturn's satellites have been slowly moving outward from the planet, however, so it is likely that orbital resonances existed among various satellites in the past.

The icy satellites of Saturn vary greatly in the extent to which they show evidence of past internal activity. At one extreme, Rhea is heavily cratered and has undergone little resurfacing. At the other extreme, Enceladus is lightly cratered and has some smooth regions that have been recently resurfaced. Plumes of water vapor are blasted into space from cracks in the surface of Enceladus. Tidal heating may explain the pattern of internal activity shown by Saturn's icy moons.

14.6 TITAN

Titan is similar to Ganymede and Callisto in both diameter and composition. Unlike the two Galilean satellites, however, Titan has a very thick atmosphere. Only four other satellites—the Moon, Io, Enceladus, and Neptune's satellite Triton—are known to have atmospheres. However, the atmosphere of Titan is much thicker than the atmospheres of the other four satellites. Observations from the Earth detected methane (CH_4) and a smaller amount of ethane (C_2H_6) in Titan's atmosphere. *Voyager 1* discovered, however, that Titan's atmosphere is mostly nitrogen (N_2). Methane makes up a few percent of the atmosphere, and other hydrocarbons, such as propane, acetylene, and ethylene, are also present in small quantities. The atmospheric pressure at Titan's surface is about 50% greater than the pressure at the surface of the Earth. Because atmospheric pressure is equal to the weight of the gas above each square meter of surface, and Titan's gravity is only 14% as large as the Earth's, there is actually about ten times as much gas above each square meter of the surface of Titan as there is above the Earth's surface. Only Venus and the giant planets have thicker atmospheres than Titan.

Origin of Titan's Atmosphere

Why does Titan have a thick atmosphere, whereas Ganymede and Callisto do not? The answer probably stems from the lower temperature at which Titan formed. The amount of gas that water traps as it freezes increases as the temperature at which it freezes decreases. The temperature at which the ices in Titan formed was low enough to trap large amounts of methane, ammonia, and nitrogen. Methane may also have been present in the form of ice. Later, when the interior of Titan warmed due to heat generated by the impacts that formed Titan as well as heat generated by radioactive decays, the gases were released from the ice. The gases then migrated to the surface, possibly through cracks like those that can be seen on Dione, Tethys, and Enceladus. Ganymede and Callisto, however, are closer to the Sun and warmer than Titan is. Also, they were heated by Jupiter's luminosity when they formed, so their ices trapped very little gas.

Atmospheric Structure

Titan's surface has a temperature of only 94 K. It is probably the only place in the solar system where a person would freeze to death while walking about in the type of space suit worn by the lunar astronauts. Other solar system bodies have colder surface temperatures, but they lack thick atmospheres to draw heat from the space suit. Titan's atmosphere is very cold and very dense. Temperature falls with altitude until a minimum is reached at the tropopause, 45 km above the surface. At higher altitudes, temperature rises again in layers where solar ultraviolet radiation is absorbed by atmospheric gas.

If the atmosphere of Titan consisted only of gases, its surface would be clearly visible. Actually, a haze of aerosol droplets and particles lies between altitudes of about 20 and 200 km. Some of the aerosols are simply droplets or flakes of condensed atmospheric gases such as methane and ethane, which are clear or gray. Titan's haze, however, is orange, so it must contain other chemicals, most of which are probably formed in reactions triggered by the absorption of solar ultraviolet radiation.

FIGURE 14.32
A View of Titan from *Huygens*
The mosaic shows Titan from an altitude of 8 km. The dark channels were cut by liquid methane. The dark region at the bottom of the image is a dry lake or ocean bottom. The white streaks may be methane or ethane fog or low-lying clouds.

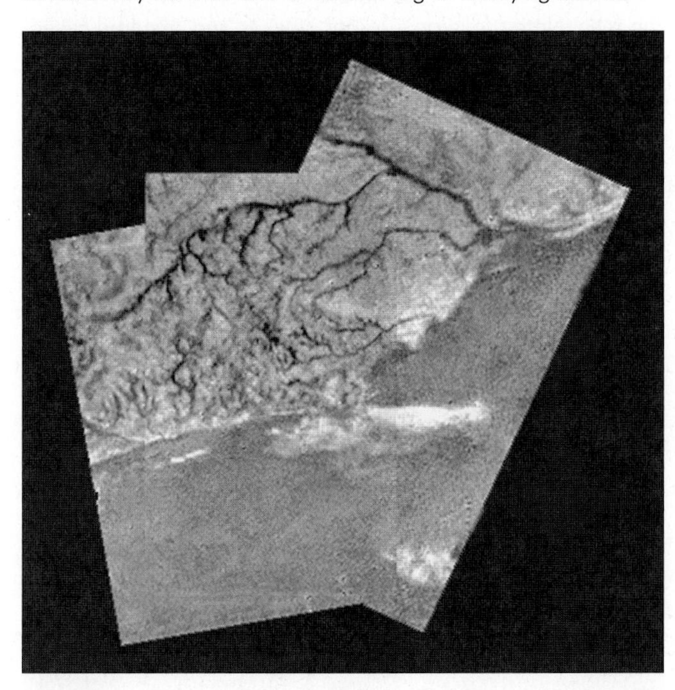

At the Surface of Titan

The *Cassini* orbiter carried with it the smaller *Huygens* spacecraft, which detached from *Cassini* and landed on Titan in January 2005. *Huygens* returned images of Titan's surface both during its descent and after landing. A view of Titan's surface from an altitude of 8 km is shown in Figure 14.32. The image shows what appears to be dark river channels fed by smaller streams. The rivers flow from relatively light-colored terrain into a darker ocean or lake basin that fills the lower part of the image. But what could have cut the river channels? Water is frozen solid at the temperature of Titan's surface. The answer is methane. We normally think of methane as a gas, but at the temperatures and pressures of Titan's lower atmosphere, methane also exists as a liquid. When droplets of methane in Titan's atmosphere grow to large enough sizes, they fall to the surface like raindrops. Other hydrocarbons that settle out of the atmosphere also reach Titan's surface and form a dark coating. Methane rain then washes the dark hydrocarbons off the highland regions, into the river channels, and then into lake or ocean basins leaving the highland regions lighter in color than the lowlands. A panoramic view of Titan's surface, including the landscape in Figure 14.32, is shown in Figure 14.33. This image also shows the boundary between

FIGURE 14.33
A Panoramic View of Titan
This view from 8 km above Titan was made by combining many individual images. The view shows the "shoreline" between light-colored highlands and darker lowlands. Details as small as 20 m can be seen. Color has been added to the image to show the reddish color of the sunlight that penetrates Titan's thick atmosphere. The scene is dark because it is only about 1⁄1000 as bright on Titan as at Earth's surface.

FIGURE 14.34
At the Surface of Titan
The view from the *Huygens* lander showed a soft surface littered with rocks probably made of frozen water. The rocks in the foreground are only inches across and are a few feet away.

FIGURE 14.35
A Radar Map of Titan's Lake Region
This false-color radar image, obtained by the *Cassini* spacecraft, shows a region near the north pole of Titan. The region is about 140 km across and shows details as small as about a half a km across. Poorly reflecting areas are shown in blue and regions that reflect radio waves well are shown in tan. Lakes of liquid, which are smooth and calm, reflect radar very poorly and are the dark features in the map. The map has been foreshortened to give the impression hovering over one end of the map. Note the large number of lakes and, in some cases, interconnecting rivers. Titan is one of only two solar system bodies to have liquid at its surface. Earth is the other.

the lighter uplands and the darker compounds collected in the lowlands. The white patches may be clouds or fog made of methane or ethane droplets.

Huygens touched down in a lowland region. The impact was relatively gentle and indicated that the surface was soft like wet sand or clay. Heat from *Huygens* warmed the surface materials and evaporated methane, which lay just below the surface. Apparently, the solid surface, probably made of water ice, lies over a reservoir of liquid methane. The view from lander is shown in Figure 14.34. The "rocks" in the foreground are only inches across and lie only a few feet from *Huygens*. The rocks are probably made of water ice.

Despite the abundant evidence of erosion by liquids on the surface of Titan, the riverbeds and lake or ocean bottoms examined by *Huygens* appeared to be dry. This surprised most scientists, who had expected large oceans and lakes of liquid methane. Radar observations by *Cassini* have mapped much more of Titan's surface than could be seen by *Huygens*. The maps show that bodies of liquid on Titan are confined to the regions within 15° of Titan's north and south poles. Figure 14.35 shows that the north

Table 14.4

Major Satellites of Uranus

	ORBITAL DISTANCE (km)	ORBITAL PERIOD (days)	DIAMETER (km)	DENSITY (kg/m^3)
Miranda	129,390	1.41	470	1200
Ariel	191,020	2.52	1160	1670
Umbriel	266,300	4.14	1170	1400
Titania	435,910	8.71	1580	1710
Oberon	583,520	13.46	1520	1630

polar region has many lakes with interconnecting rivers. The lakes, which probably contain methane and ethane, are smooth and calm. They appear dark because they reflect radio waves poorly. The reason that lakes only exist near Titan's poles is not well understood. Perhaps liquid methane erupts from Titan's interior only from time to time. Because methane is destroyed in Titan's atmosphere on a timescale of tens of millions of years, the amount of liquid gradually dwindles. At the present time Titan may be in a dry period between eruptions or eruptions may have ceased and the time when lakes can exist on Titan may be coming to an end.

The most notable characteristic of Titan is its nitrogen atmosphere, thicker than the Earth's atmosphere and much colder. Methane, one of Titan's atmospheric gases, exists as rain in the atmosphere and forms lakes on Titan's surface.

14.7 SATELLITES OF URANUS

Uranus has more than two dozen known satellites, of which only five—Miranda, Ariel, Umbriel, Titania, and Oberon—are large enough and bright enough to have been discovered from the Earth before the *Voyager 2* flyby. Most of the remaining satellites were discovered by the *Voyager 2* imaging team and range in diameter from only 25 km up to approximately 150 km.

Although the five largest satellites of Uranus can be seen from the Earth, they appear only as points of light. Nothing about the appearances of their surfaces was known before the arrival of *Voyager 2.* It was generally believed that they were heavily cratered balls of rock and ice like Saturn's satellites, Rhea and Mimas. The *Voyager* images showed, however, that each of the Uranian satellites has been more active than Saturnian satellites of comparable size. Some important properties of the larger satellites of Uranus are given in Table 14.4.

Oberon, Titania, Umbriel, and Ariel

At the time of the *Voyager 2* encounter, the south pole of Uranus pointed almost directly at the Sun. This meant that the southern hemispheres of Uranus and its satellites were illuminated by the Sun, but their northern hemispheres were dark. As a result, we know nothing about the appearances of the northern hemispheres of Uranus's satellites.

All of the larger satellites of Uranus have features that indicate great internal activity. Umbriel (Figure 14.36) and Oberon (Figure 14.37) have heavily cratered, ancient surfaces. They also have large cracks, which may

FIGURE 14.36
A *Voyager* Image of Umbriel
Scarps and cracks can be seen in Umbriel's heavily cratered surface.

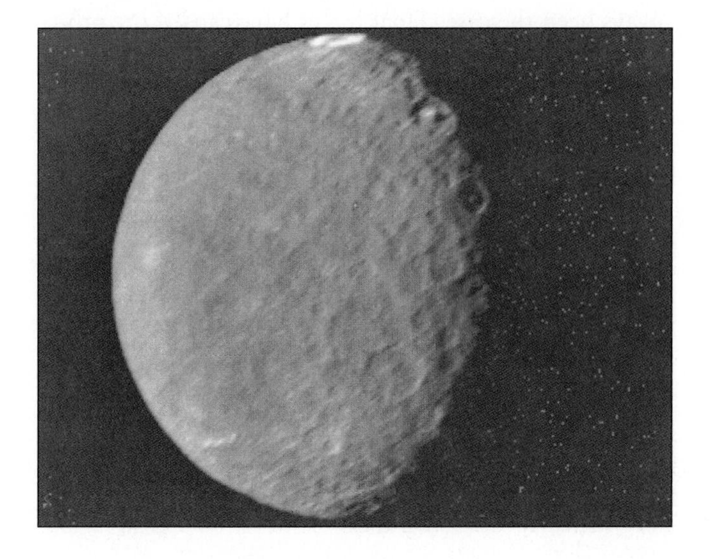

FIGURE 14.37
A *Voyager* Image of Oberon
Several chasms cross the cratered terrain.

FIGURE 14.38
Uranus's Satellite Titania
Titania has systems of deep cracks. Its surface is less heavily cratered than those of Umbriel and Oberon, which suggests that resurfacing has taken place.

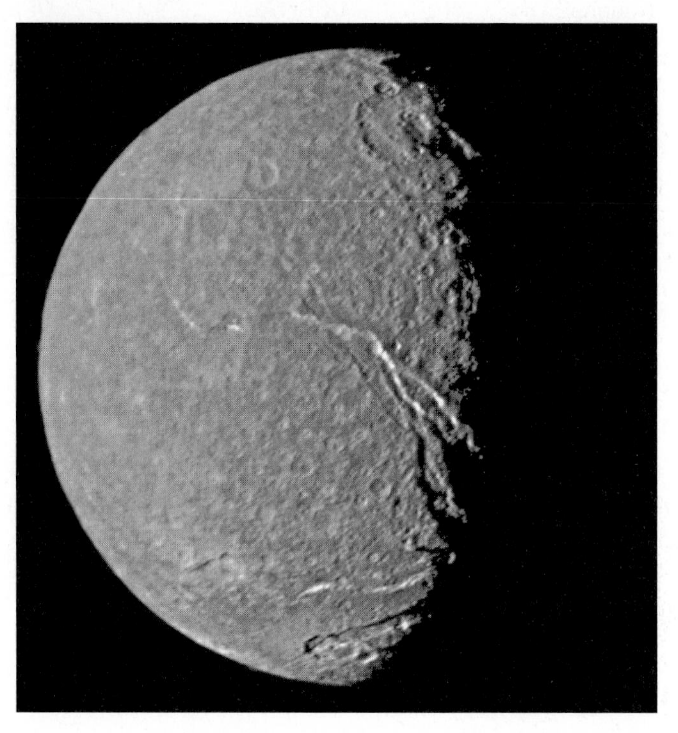

have formed when their interiors expanded as the liquid water within them froze. Even more evidence of internal activity can be seen on Titania (Figure 14.38) and Ariel (Figure 14.39). Both of these satellites are more lightly cratered than are Umbriel and Oberon, indicating that Titania and Ariel have been resurfaced, probably by flows of water from geysers and cracks. Some regions of Ariel are almost craterless, which suggests that resurfacing has occurred quite recently. Titania and Ariel also have systems of deep cracks, which show that they too expanded after their surfaces solidified. The internal activity of Uranus's satellites is probably due to tidal heating. Although tidal heating is now small, this hasn't always been the case. Tidal forces have caused the orbits of the satellites to evolve outwards (just as in the case of the Moon). They have moved in and out of orbital resonances that increased orbital eccentricities and tidal heating just as the resonance between Io and Europa causes tidal heating in those satellites.

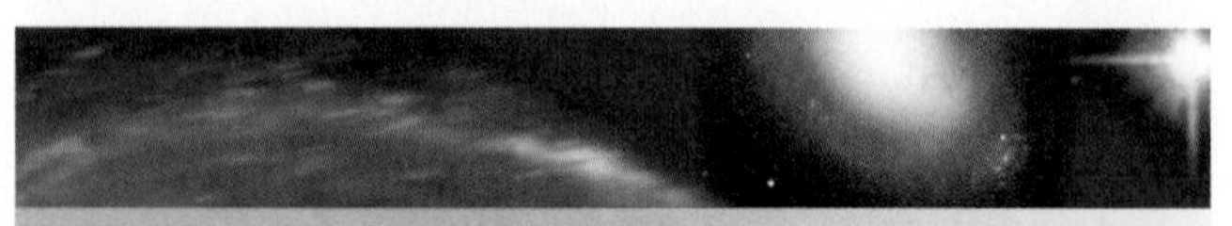

Each of Uranus's large satellites shows more evidence of internal activity than Saturnian satellites of comparable size. Past tidal heating is thought to be responsible for the intense activity of Uranus's satellites.

FIGURE 14.39
Uranus's Satellite Ariel
Ariel's surface is heavily fractured.

FIGURE 14.40
A *Voyager* Mosaic of Miranda
Miranda's surface is a chaotic assemblage of several distinct types of terrain including rolling, cratered terrain, fractures, and coronae (systems of parallel ridges and troughs).

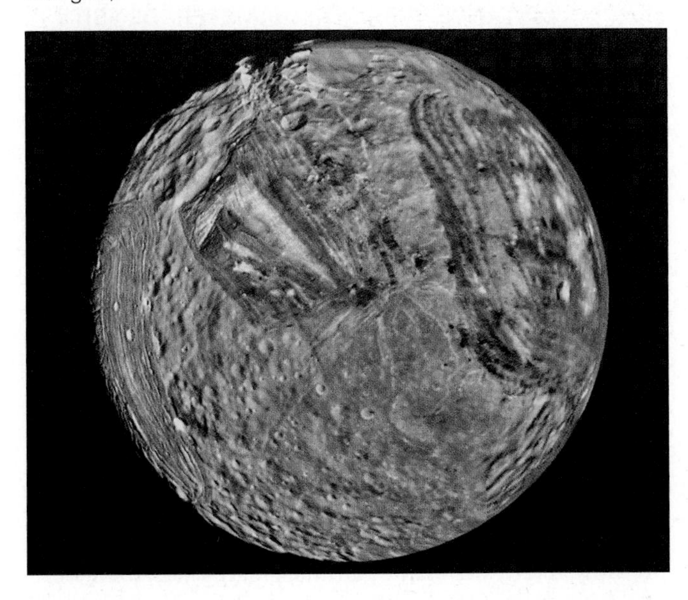

Miranda

Miranda

The surface of Miranda shows far more evidence of internal activity than any of the other satellites of Uranus. *Voyager 2* passed within 28,300 km of Miranda, so its images show details as small as 1 km. As Figure 14.40 illustrates, Miranda has a bewildering variety of types of terrain. Much of the surface is rolling, cratered terrain, but in places the cratered terrain is broken by fractures. Several very strange-looking features, called **coronae,** consist of parallel ridges and troughs that produce the striped appearance shown in Figure 14.41. The coronae have very sharp boundaries. In places, the edges of the coronae are marked by trenches with steep cliffs on either side. One of these, shown in Figure 14.42, is 20 km high. In the weak gravity of Miranda, it would take nearly 10 minutes to fall from the top of this great cliff to the bottom.

The appearance of Miranda has not been adequately explained, although a number of possibilities have been proposed. One theory is that Miranda became so warm that it began to differentiate. Its rocky material began to sink to its core, and its icy material began to rise to the surface. Before the process could be completed, however, Miranda cooled. According to this theory, we see it suspended as a partially differentiated body. Another theory is that Miranda was shattered one or more times by large

FIGURE 14.41
Coronae on Miranda
Coronae consist of parallel ridges and valleys within sharp boundaries. In some places, they make almost right-angle turns.

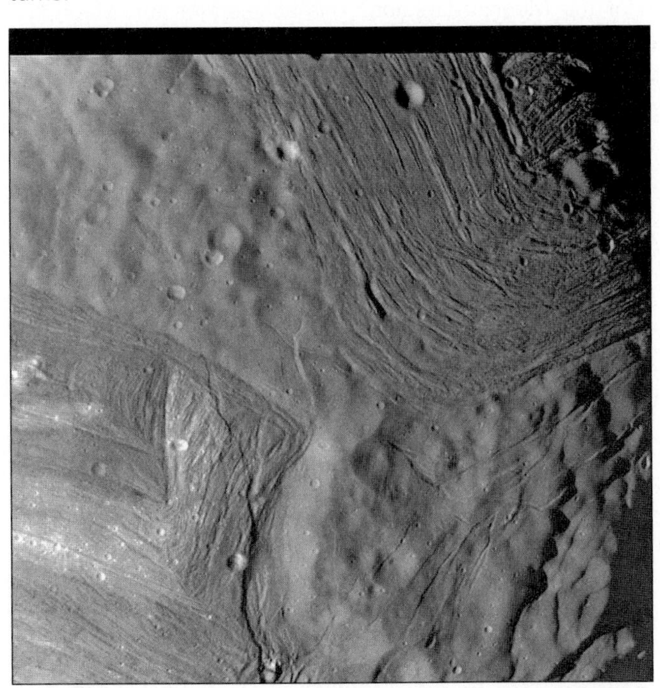

FIGURE 14.42
A 20-km-High Cliff on the Terminator of Miranda
Sunlight from the bottom of the image illuminates the face of the cliff at the upper right.

impacting bodies. As it was pulled back together again by gravity, some pieces that had been part of the deep interior became part of the new surface. This gave Miranda the random, chaotic appearance that we see today. Although it is fair to say that we don't fully understand any of the larger satellites of the outer solar system, Miranda is probably the most poorly understood of all.

Miranda has several distinct types of surface terrain that appear to have been randomly thrown together. A satisfactory explanation for Miranda's unusual appearance has not yet been found.

14.8 SATELLITES OF NEPTUNE

Neptune has 13 known satellites, six of which were discovered in images taken during the *Voyager* flyby in 1989. Only one of the satellites, Triton, is larger than about 400 km in diameter. The *Voyager* images showed that Triton is an internally active body with a complex surface. Some important properties of Triton are given in Table 14.5.

Triton

Triton is slightly smaller than Europa, the smallest Galilean satellite. However, Triton's density, 2050 kg/m^3, is larger than any of the other outer solar system satellites, except for Io and Europa. A density this large indicates that Triton is made of about 75% rock and 25% ice, so Triton's radioactive heating may be greater than that of satellites containing a higher proportion of ice.

Table 14.5

Triton

	ORBITAL DISTANCE (km)	ORBITAL PERIOD (days)	DIAMETER (km)	DENSITY (kg/m^3)
Triton	354,800	5.877	2700	2050

Orbit and Evolution Of all the large satellites in the solar system, Triton is the only one with a retrograde orbit about its parent planet. The most probable explanation is that Triton wasn't always a satellite of Neptune but was captured into its retrograde orbit after forming elsewhere. This could have happened if Triton struck and destroyed one of Neptune's original satellites or, more likely, was half of a binary pair when it encountered Neptune. Its companion was ejected during the encounter and Triton fell into orbit about Neptune.

Triton is about as far from Neptune as the Moon is from the Earth. Like our Moon, it is massive enough to produce tides within Neptune. The consequence of those tides for Triton, however, is quite different than for the Moon. Figure 14.43 shows that lunar tides produce tidal

FIGURE 14.43
The Effect of Tides on the Moon and Triton
A, The rotation of the Earth pulls the Earth's tidal bulge ahead of the Moon's direction. This accelerates the Moon and causes it to spiral slowly away from the Earth. **B,** Because of Triton's retrograde orbit, the tidal bulge it raises on Neptune lags behind Triton's direction. This causes Triton to spiral inward toward Neptune.

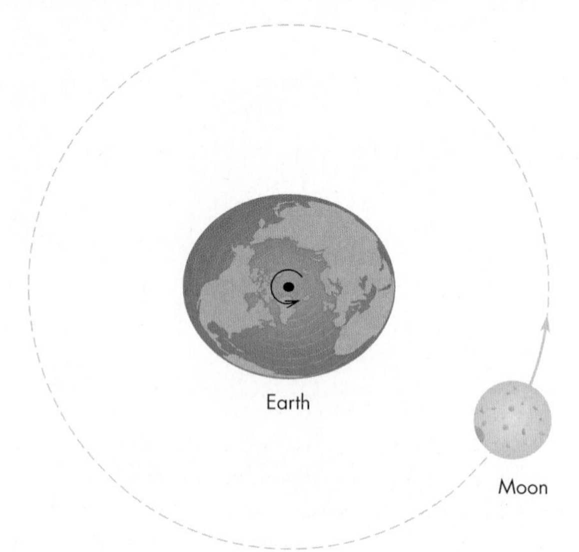

A The Earth's tidal bulge precedes the Moon so the Moon spirals slowly away from the Earth.

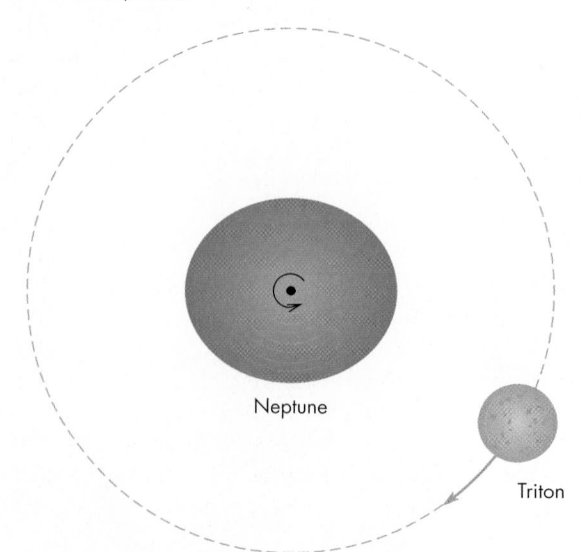

B Neptune's tidal bulge lags behind Triton so Triton spirals slowly inward.

FIGURE 14.44
The Extreme Seasons on Triton
Neptune's axis is tipped 30° from its orbital plane. Triton's orbit and its spin axis are tipped by another 23° with respect to the axis of Neptune. At times the Sun can be overhead as much as 50° north or south of Triton's equator, leading to very strong seasonal effects.

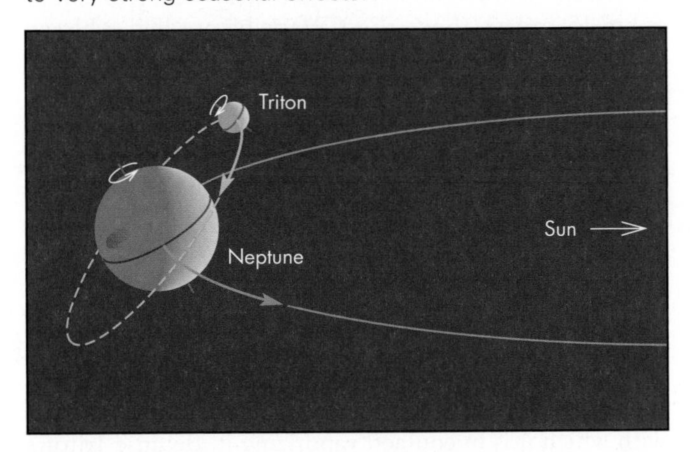

FIGURE 14.45
A *Voyager* Mosaic of Triton
Triton's surface shows large cracks and regions of pits and dimples but relatively few impact craters. This suggests that Triton has been extensively resurfaced.

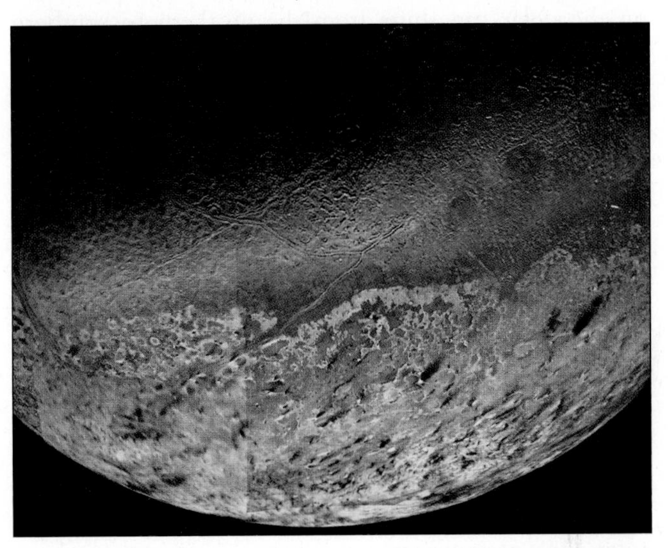

bulges that precede the Moon in its orbit. The tidal bulges accelerate the Moon, causing it to spiral slowly away from the Earth, as described in Chapter 9, Section 2. Neptune rotates in the opposite direction from the orbital motion of Triton, so the tidal bulges on Neptune lag behind the position of Triton. As a result, tidal forces retard Triton's motion and produce a slow inward spiral. Eventually, Triton will fall inside the Roche distance of Neptune and be shattered by tidal forces. Triton is moving inward so slowly, however, that billions of years will pass before it is destroyed.

Neptune's equator is tilted 30° from its orbital plane and Triton's orbit is tilted another 23° with respect to Neptune's equator. As Figure 14.44 shows, the combination of these two tilts can produce situations in which the Sun is directly overhead at Triton's 50° N or S latitude. If this occurred on Earth, there would be times when the Sun would be directly overhead at the southern tip of South America. This would leave virtually all of Europe, all of Canada, and most of the northern United States in total darkness. For Triton such extreme seasonal effects occur in a cycle about 680 years long. At the present time, the region within 40° of Triton's north pole is in complete darkness, as it has been for decades and as it will continue to be for many years.

Atmosphere Triton's very thin atmosphere contains nitrogen (N_2) and a much smaller amount of methane (CH_4). The surface pressure is only 14 millionths of the pressure at the surface of the Earth. Nevertheless, scattered clouds of solid nitrogen have been detected.

Surface Features and Internal Activity Figure 14.45 shows that Triton shows considerable evidence of resurfacing and internal activity. Relatively few impact craters are visible. The most heavily cratered regions of Triton have crater densities like those of the maria on the Moon. Figure 14.46 shows a very smooth region that resembles a frozen lake. Smooth regions were probably formed by flows of watery material (melted ices) from

FIGURE 14.46
Smooth Regions on Triton
These regions appear to be frozen lakes formed when watery material emerged from within Triton.

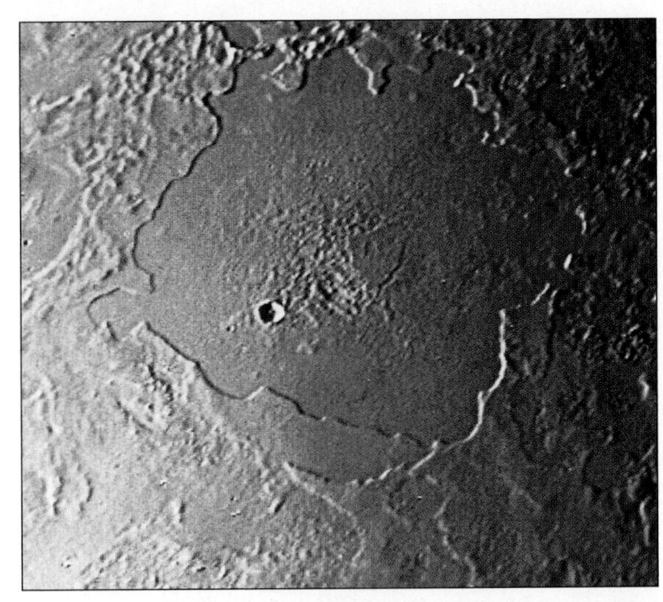

Triton's interior. Other regions are dominated by pits and dimples that produce a "cantaloupe-like" appearance. Large cracks crisscross Triton. Almost the entire visible part of the southern hemisphere is covered by a thin polar cap of nitrogen and methane ice. The material that makes up the polar cap shifts from the northern to southern hemisphere and then back again each time Triton goes through one of its 680-year seasonal cycles.

Radioactive heating and tidal heating are probably sufficient to maintain liquid water ice and other liquids within Triton's interior. The crust is water ice with thin coatings of nitrogen and methane ices. Eruptions of liquid from the interior may continue to occur even today.

Plumes and Geysers Figure 14.47 shows a few of the many dark, elongated features seen in the images of Triton. These features, also seen in Figure 14.45, were produced by plumes. These local eruptions of nitrogen gas propelled dark particles into Triton's thin atmosphere. The sooty particles were then blown downwind to produce a dark streak. At least four geyserlike eruptions were detected in the *Voyager* images of Triton. Dark material can be seen rising to about 8 km (5 miles) above the surface before being blown downwind for more than 100 km. The dark streaks are found mainly in the southern polar cap, which means that they must be young surface features because the ones that we see today will be covered up by fresh polar ices the next time that the southern polar cap forms. This means that geyser activity must be fairly common on Triton.

One way in which the geysers might be produced requires a 1-meter thick layer of transparent nitrogen ice that lies above a layer of darker material. Sunlight passes through the nitrogen ice and heats the dark material. As the dark material warms, it also warms the frozen nitrogen with which it is in contact, vaporizing it. Because Triton's surface is only 37 K (the coldest place yet found in the solar system), an increase of a few degrees in temperature produces a large amount of nitrogen gas. When the pressure of the nitrogen gas rises far enough, it ruptures the overlying ice and launches a plume of nitrogen gas mixed with dark material.

FIGURE 14.47
Plumes on Triton
The dark, elongated markings are caused by sooty material that erupts from Triton's surface and is blown downwind by Triton's thin atmosphere.

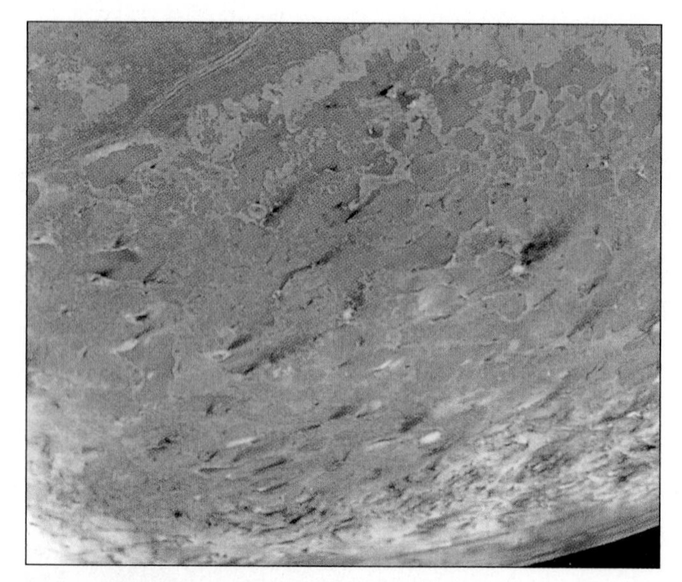

Triton, the only large satellite with a retrograde orbit, probably was captured by Neptune. Radioactive and tidal heating have led to considerable resurfacing of Triton. Local eruptions of gas from beneath Triton's surface produce geyserlike plumes of soot, which leave streaks of dark material more than 100 km long.

Chapter Summary

- Most of the satellites in the solar system fall into one of three categories: regular satellites that formed about their parent planets, collisional fragments of regular satellites, and captured asteroids. The Moon and Triton do not fit into any of these three categories. (Section 14.1)
- Nearly all of the satellites of the outer planets have densities less than 2000 kg/m^3. This is much lower than the density of rocky material, 3500 kg/m^3, so ices must make up a large fraction of the material of most satellites. Most of the ice in the outer solar system is probably frozen water. (14.2)

- There are three sources of internal heat for satellites: the conversion of gravitational energy into heat during formation, radioactive heating, and tidal heating. The thermal history of a satellite depends not only on these sources of energy, but also on how easily heat is transported to its surface by convection and conduction. (14.2)
- Phobos and Deimos, the satellites of Mars, are believed to be captured asteroids. Their surfaces are pocked by craters. Cracks on Phobos radiate from the crater Stickney, which was produced by an impact nearly violent enough to destroy Phobos. (14.3)
- The intense tidal heating of Io makes it the most volcanically active body in the solar system. Volcanic material is deposited so rapidly on Io's surface that all evidence of impact cratering has been covered up. Sulfur and its compounds provide the yellow and red colors of Io's surface and may also power the plumes that propel material hundreds of kilometers above Io's surface. (14.4)
- Europa has a very smooth, icy surface even though it is made mostly of rock. Europa's surface has been smoothed by glacierlike flows and, probably, by flows of water from the interior. A thick ocean of water probably exists below its icy crust. (14.4)
- Ganymede's surface shows evidence of past internal activity and crustal movement. Callisto's heavily cratered surface shows no evidence of any internal activity. Both Ganymede and Callisto may have oceans of water far beneath the surface. (14.4)
- Saturn's icy satellites show evidence of a wide range of internal activity. At one extreme is Rhea, which has a heavily cratered surface, while at the other extreme is Enceladus, which has been extensively resurfaced. Tidal heating may explain why Rhea seems to have experienced little or no internal activity while Enceladus seems to have experienced so much. (14.5)
- Saturn's largest satellite is Titan. Its atmosphere contains mostly nitrogen and is thicker than the Earth's. Titan is cold enough that atmospheric methane can condense to liquid form. Titan has methane rain and lakes of liquid methane. (14.6)
- All of the larger satellites of Uranus have been more active than the Saturnian satellites of comparable size. Tidal heating may have powered this internal activity. (14.7)
- Uranus's satellite, Miranda, is very unusual in appearance. Its surface is a jumbled mixture of different types of terrain. The way in which Miranda evolved to its present form is not known. (14.7)
- Triton orbits Neptune in a retrograde direction and is slowly spiraling inward. Triton probably formed elsewhere and was captured into a retrograde orbit. The surface of Triton shows evidence of considerable resurfacing and internal activity. Plumes of dark material erupt from beneath the surface of Triton and are carried far downwind by its thin atmosphere. (14.8)

Key Terms

collision fragment *316*
coronae *337*
plume *321*
regular satellite *316*
tidal heating *317*

Conceptual Questions

1. Suppose a satellite has a diameter of 800 km and an orbit that lies nearly in the equatorial plane of its parent planet well outside of the planet's ring system. Is the satellite likely to be a regular satellite, a collision fragment, or a captured asteroid?
2. What properties of a satellite would suggest that it may be a captured asteroid?
3. How did the *Voyager, Viking, Galileo,* and *Cassini* spacecraft measure the masses of satellites?
4. What is the reason that an icy satellite has less radioactive heating than a rocky satellite does?
5. Under what conditions is tidal heating likely to be important for a satellite?
6. The density of Io is 3600 kg/m^3 while that of Callisto is 1900 kg/m^3. What do these densities tell us about the relative amounts of rock and ice in Io and Callisto?
7. What evidence is there that the surface of Io is very young?
8. What powers the internal activity of Io?
9. The surface of Europa is smooth and relatively craterfree, yet it lacks volcanos to deposit fresh material on the surface. What are possible reasons why Europa's surface is so smooth?
10. What is a possible explanation for the different types of terrain seen on Ganymede?

11. What evidence is there that the satellites of Saturn once had internal activity?
12. Iapetus always appears brighter when it is about to pass behind Saturn than when it reappears from behind Saturn. Draw a diagram to show that this effect is related to the distribution of dark and bright regions on the surface of Iapetus.
13. Compare the atmosphere of Titan to that of Earth with respect to composition, pressure, and temperature.
14. What is the reason that the atmospheric temperature of Titan increases with increasing altitude at heights above 45 km?
15. Why is Titan orange?
16. Write a description of what you might see if you visited the surface of Titan.
17. Compare the images of Miranda with those of the other satellites shown in this chapter. Find a satellite with surface features similar to the coronae of Miranda. Find a satellite with a surface that resembles the rolling, cratered terrain of Miranda.
18. Both the Moon and Triton raise tidal bulges on the planet they orbit. Why does the tidal bulge on Neptune cause Triton to spiral inward while the tidal bulge on the Earth causes the Moon to spiral outward?
19. What is believed to be the origin of the dark streaks on the surface of Triton?

Problems

1. Suppose one-third of the mass of a satellite is ice and the rest is rock. Calculate (or estimate) the average density of the satellite.
2. Suppose the orbital period of one satellite is exactly five-fourths as long as the orbital period of another. How often would the two satellites pass one another as they orbit? Give your answer in terms of the orbital period of the inner satellite.
3. The orbital period of Phobos is 7.7 hours. Show why an observer on Mars would see Phobos rise every 11.2 hours.
4. Show that the orbital periods and distances of the Galilean satellites of Jupiter are consistent with Kepler's third law.

Group Activity

With your group, draw up a plan for a mission to Europa to search for life. Be sure to include your strategies for searching for life and how you would recognize life if you found it. If possible, make a brief presentation of your plan in class.

For More Information

Visit the text website at **www.mhhe.com/fix** for chapter quizzes, interactive learning exercises, and other study tools.

15

When in your middle years
The great comet comes again
Remember me, a child,
Awake in the summer night,
Standing in my crib and
Watching that long-haired star
So many years ago.
Go out in the dark and see
Its plume over water
Dribbling on the liquid night,
And think that life and glory
Flickered on the rushing
Bloodstream for me once, and for
All who have gone before me,
Vessels of the billion-year long
River that flows now in your veins.

Kenneth Rexroth
Halley's Comet

Asteroid Impact

www.mhhe.com/fix

Small Solar System Bodies

The town of Homestead is about 20 miles west of Iowa City. Homestead, normally a quiet place, was the scene of an extraordinary happening a little more than a century ago. At about 10:15 in the evening on February 12, 1875, a small solar system body made a spectacular arrival in the vicinity of Homestead. Here is a contemporary account by a Mr. Irish:

> From the first the light of the meteor could hardly be tolerated by the naked eye turned full upon it. Several observers who were facing south at the first flash say that upon looking full at the meteor it appeared to them round, and almost motionless in the air, and as bright as the sun. . . .
>
> The observers who stood near to the line of the meteor's flight were quite overcome with fear, as it seemed to come down upon them with a rapid increase of size and brilliancy, many of them wishing for a place of safety, but not having time to seek one. In this fright animals took part, horses shying, and plunging to get away, and dogs retreating and barking with signs of fear. The meteor gave out marked flashes in its course, one more noticeable than the rest, when it had completed about two-thirds of its visible flight. All observers who stood within twelve miles of the meteor's path say that from the time they first saw it, to its end, the meteor threw down "coals" and "sparks."
>
> Thin clouds of smoke or vapor . . . would seem to burst out from the body of the

Questions to Explore

- What causes meteors? Why do meteor showers occur?
- What can meteorites tell us about objects that formed early in the history of the solar system?
- How are meteorites related to asteroids?
- How do comets produce their spectacular comas and tails?
- Where do new comets come from?
- What happens when asteroids and comets strike the Earth?

meteor like puffs of steam from the funnel of a locomotive or smoke from a cannon's mouth, and then as suddenly be drawn into the space behind it. The light of the meteor's train was principally white, edged with yellowish green throughout the greater part of its length, but near to the body of the meteor the light had a strong red tinge. The length of the train was variously estimated, but was probably about nine degrees, or from seven to twelve miles, as seen from Iowa City. . . .

From three to five minutes after the meteor had flashed out of sight, observers near the south end of its path heard an intensely loud and crashing explosion, that seemed to come from the point in the sky where they first saw it. This deafening explosion was mingled with, and followed by, a rushing, rumbling, and crashing sound that seemed to follow up the meteor's path, and at intervals, as it rolled away northward.

After the explosion, a large number of stones rained down on an area of about 50 km^2 around Homestead. About 100 meteorites, weighing a total of 250 kg, were quickly picked up. The largest of them weighed 35 kg. Professor Gustavus Hinrichs of the University of Iowa rushed to the scene and collected several of the meteorites, which are now in the collections of universities and museums.

The incident at Homestead was the result of the encounter between the Earth and a desk-sized chunk of interplanetary debris. Such bodies, although far smaller and less massive than the planets, can affect us more strongly than any objects except the Sun and the Moon. This chapter is about small interplanetary bodies and the ways that they can influence us, both physically and intellectually.

15.1 METEORS

Figure 15.1 shows a **meteor,** a bright streak of light produced when a piece of interplanetary debris moves rapidly through the Earth's atmosphere. The piece of debris is usually referred to as a **meteoroid.** Meteors occur almost continuously. On any clear night it is possible to see a meteor by looking at the sky for 5 or 10 minutes. To be able to see a meteor, someone must be within a few hundred kilometers of it. This means that from a given spot on the Earth's surface it is possible to see only a tiny fraction of all the meteors that occur each day. For 8 or 10 meteors to be visible each hour at a given spot, tens of millions of them must occur each day.

The Meteor Phenomenon

The meteoroids that strike the Earth's atmosphere arrive at speeds that range from about 11 km/s (the Earth's escape velocity) to 72 km/s (about 160,000 mph). Heat generated by friction between atmospheric gas and the meteoroid melts the surface of the meteoroid and then vaporizes it, reducing

FIGURE 15.1
A Meteor
A typical meteor is a brief, bright streak of light that takes place when a small piece of interplanetary debris enters the Earth's atmosphere at great speed. Several meteors can be seen each hour on a clear, dark night.

its size as well as heating and charring it. Atmospheric gas and the vaporized meteoroid material are heated to the point where they glow, producing the meteor. A typical meteoroid begins to glow when it is about 90 km above the ground, and it is completely vaporized by the time it reaches a height of about 80 km. How far the meteoroid penetrates into the atmosphere depends on its mass, its speed when it strikes the atmosphere, and the angle at which it strikes.

Sizes of Meteoroids

Although the glowing region of gas that we see as a meteor can be many meters across, the meteoroid responsible for the meteor is much smaller. Bright meteors can be produced by meteoroids that have masses of about 1 gram and are less than 1 cm in diameter. Faint meteors, produced by less massive meteoroids, are much more numerous than bright ones. A few thousand times each day the atmosphere is struck by a meteoroid big enough (several centimeters across) to produce a spectacularly bright meteor called a **fireball.** Some fireballs are accompanied by thunderlike rumbling and produce luminous trains of glowing gas that can last for an hour.

Although the meteoroids responsible for ordinary meteors are entirely consumed high in the atmosphere, this is not necessarily the case for fireball meteoroids. Air resistance slows them at the same time that it strips away their outer layers. Before a meteoroid larger than the size of a fist can be completely destroyed, the atmosphere slows it to **terminal velocity,** the speed at which air resistance equals the force due to the Earth's gravity. The remaining portion of the meteoroid hits the ground at a few hundred meters per second (in other words, a few times as fast as a well-thrown baseball). The portion of the meteoroid that survives passage through the atmosphere and reaches the ground is called a **meteorite.** Sometimes, as in the case of the Homestead meteor, the meteoroid disintegrates in the air, causing a shower of meteorites.

The height at which a meteoroid is slowed to terminal velocity depends on the mass and the initial speed of the meteoroid. Typically, this occurs tens of kilometers above the ground. A very large meteoroid (larger than the size of an automobile) may strike the ground before it can be slowed to terminal velocity. In such cases an impact crater may be formed.

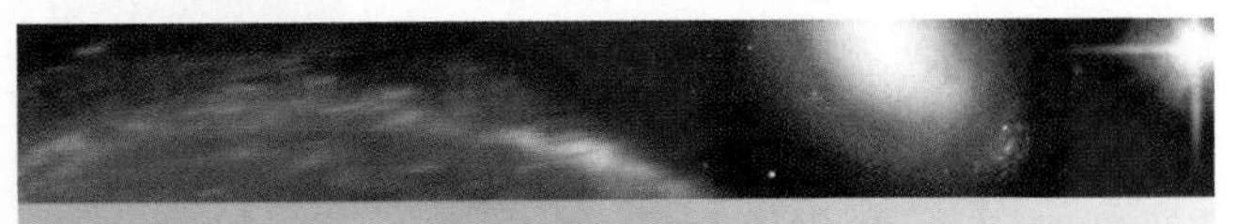

A meteor is the flash of light that occurs when a small piece of interplanetary matter called a meteoroid strikes the Earth's atmosphere at such high speed that the atmosphere heats up until it glows. Most meteoroids are smaller than 1 cm, but a few are large enough that portions of them reach the ground. The part of a meteoroid that reaches the ground is called a meteorite.

Meteor Showers

About ten times a year, the rate at which meteors can be seen rises dramatically. The rate rises until it reaches a maximum of 15 to 100 per hour and then declines again over several days. These events are called **meteor showers.** Table 15.1 describes some of the major meteor showers. The dates of meteor showers are very predictable, because they recur on nearly the same date each year. The maximum meteor rate during a shower is much harder to predict. Occasionally, spectacular meteor showers occur. For example, in the early morning of November 13, 1833, people all over the eastern part of the United States witnessed a meteor shower in which at least 100,000 meteors per hour could be seen. This amazing shower prompted many preachers to announce that the end of the world was at hand. More recently, on November 17, 1966, observers in the southwestern United States were treated to a display that peaked at a rate of more than 100 meteors per second. During a 40-minute period, 60,000 meteors were seen.

Table 15.1

Major Meteor Showers

SHOWER	DATE OF MAXIMUM	DURATION (days)	RATE (number/hour)	PARENT OBJECT
Quadrantid	Jan. 3	0.4	80	
Lyrid	Apr. 22	1	15	Comet Thatcher
Eta Aquarid	May 4	6	60	Comet Halley
Delta Aquarid	Jul. 29	8	30	
Perseid	Aug. 12	3	100	Comet Swift-Tuttle
Orionid	Oct. 21	2	30	Comet Halley
Leonid	Nov. 16	2	20	Comet Tempel-Tuttle
Geminid	Dec. 13	3	90	Asteroid Phaethon

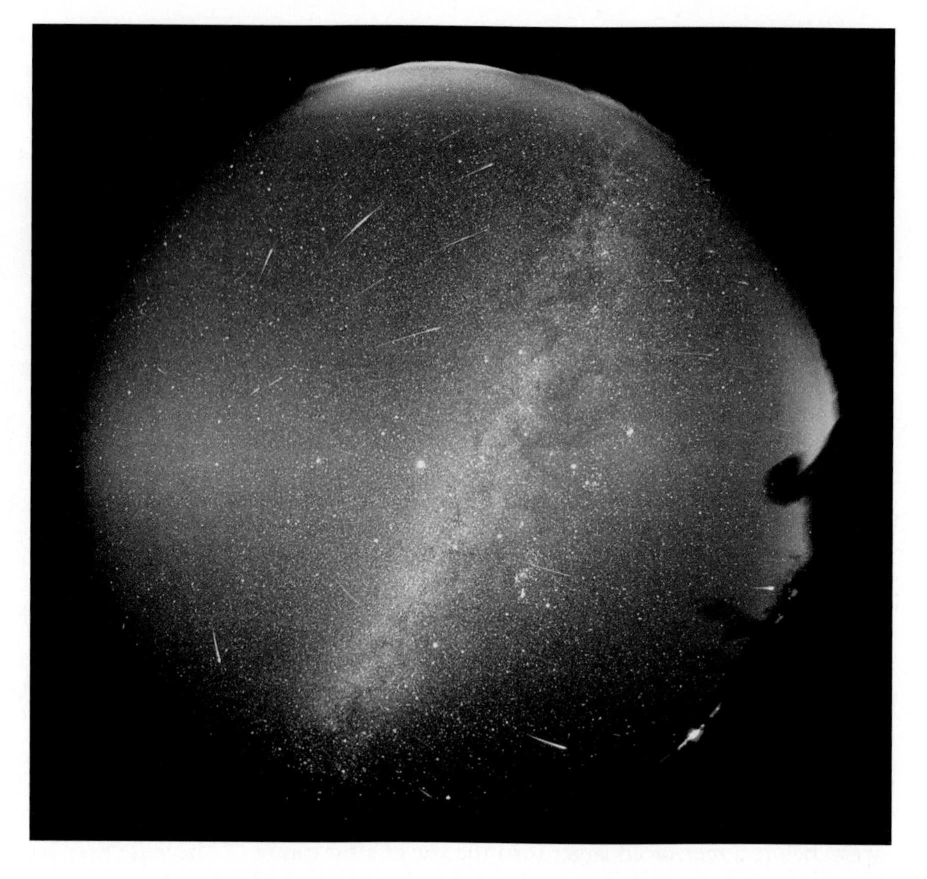

FIGURE 15.2
The Leonid Meteor Shower in 2001
During meteor showers the rate of meteors can rise to hundreds or thousands per hour. The radiant of this meteor shower, in the left center of the image, can be found by extending the meteor trails until they meet.

A meteor shower

Radiants and Meteor Orbits As Denison Olmstead, a Yale professor, watched the November 1833 shower, he noted that the meteors appeared to be coming from a point within the constellation Leo. The spot in the sky from which the meteors in a shower seem to originate is called the **radiant** (Figure 15.2). Meteor showers are named for the constellation within which the radiant is located, so the great shower of November 1833 was an occurrence of the Leonid meteor shower.

Olmstead realized that the radiant point was caused by perspective. Just as snowflakes seem to diverge from a point in front of us when we drive through a snowstorm, the paths of individual meteors appear to diverge from the radiant. In reality, as shown in Figure 15.3, the meteoroids that produce a meteor shower are moving along parallel paths through the solar system when they strike the Earth. A meteor shower, then, occurs when the Earth collides with a swarm of meteoroids that orbit the Sun on nearly identical paths.

Figure 15.4 shows that a given meteor shower recurs about the same date each year because the Earth crosses the orbit of the meteoroids responsible for the shower on approximately that same date each year. If the meteoroids are smoothly distributed around their orbit, the meteor rate will remain stable from year to year. However, if the Earth encounters a relatively dense clump of meteoroids, a spectacular shower will result. The Earth moves

FIGURE 15.3
The 1997 Leonid Meteor Shower as Seen from Space
This image, taken by the *MSX* satellite, shows many meteors moving along parallel paths through the Earth's atmosphere.

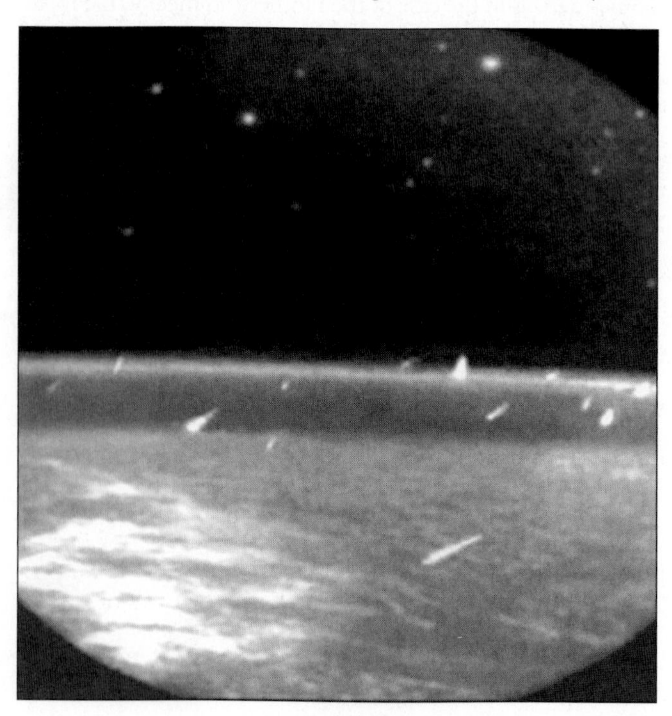

FIGURE 15.4
Why Meteor Showers Occur
Meteor showers occur when the Earth intersects the orbit of a meteoroid swarm. Because the Earth returns to the region of intersection at the same point in its orbit each year, a given meteor shower occurs on approximately the same date each year.

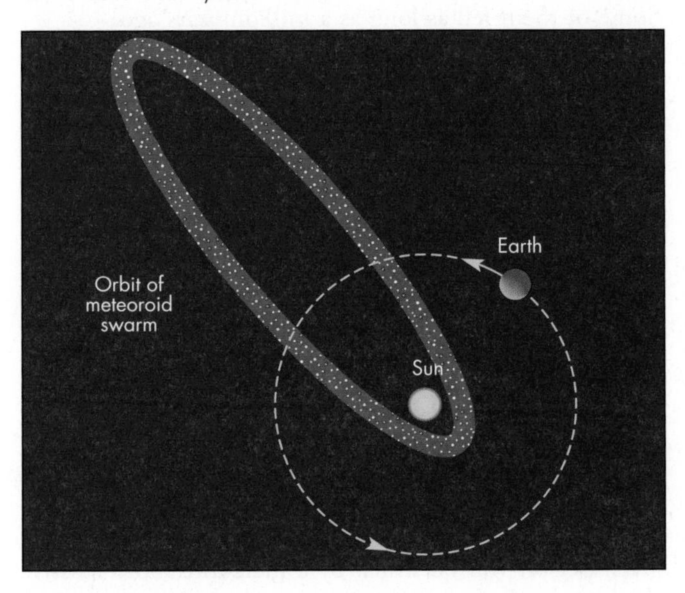

about 2.5 million km along its orbit each day. Because meteor showers usually last for several days (though not at their peak rates), the meteoroid swarms that cause meteor showers must be several millions of kilometers across.

Meteor showers are temporary increases of the rate at which meteors are observed. Showers recur on approximately the same date from year to year. Meteor showers happen when the Earth runs into a swarm of meteoroids that share a common orbit about the Sun.

Micrometeorites

Although most meteoroids are destroyed in the atmosphere, those smaller than about 50 millionths of a meter in diameter are slowed down without becoming hot enough to vaporize. This happens at altitudes near 100 km. These meteoroids are called **micrometeorites** and they slowly settle through the atmosphere until they reach the ground. Astronomers have collected many micrometeorites using high-altitude aircraft. One of these is shown in Figure 15.5. They tend to be very porous, fragile objects that look somewhat like dustballs.

FIGURE 15.5
A Micrometeorite
Most micrometeorites are fragile, fluffy objects that originated in comets. The micrometeorite shown here is about 10^{-5} m across or about one-tenth as big as the thickness of this page.

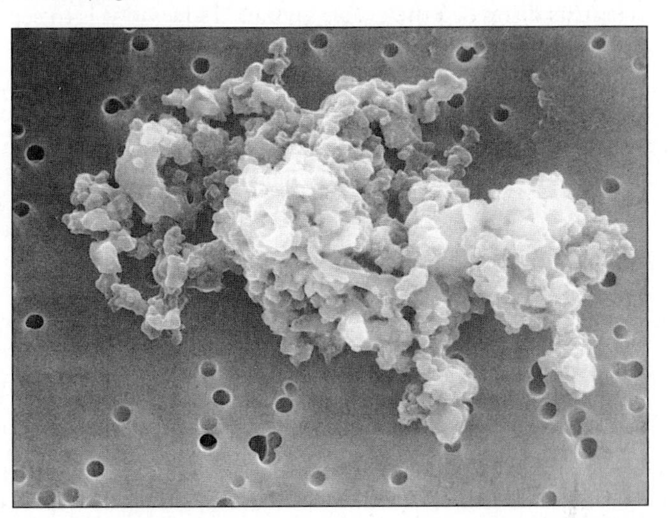

15.2 METEORITES

Until lunar samples were returned to the Earth by the Apollo astronauts, meteorites were the only pieces of material from the rest of the universe that could be analyzed in a laboratory. Important ideas about the development of the solar system were based on the evidence available in meteorites. Yet the conclusion that meteorites are cosmic intruders is surprisingly recent.

Even in prehistory, people recognized that meteorites were unusual objects. In some cases, they also realized that meteorites fell from the sky. In some cultures, meteorites were revered as holy objects. At many Greek and Roman temples, meteorites were worshiped as gods fallen from the heavens. Meteorites were also venerated in Japan, China, India, and Africa. A very common belief about meteorites was that they were omens or that they possessed supernatural powers. For example, after the fall of the Ensisheim (Germany) meteorite in 1492, the Emperor Maximillian was advised by his council that the meteorite was a favorable omen for his wars with the French and the Turks. The meteorite was hung in a local church.

Many Native American cultures treated meteorites with great respect, often burying them in special graves or crypts. Before going to battle, warriors of the Clackamas tribe of Oregon bathed their faces and dipped their arrows in water that had collected in an iron meteorite. Before the development of iron metallurgy, meteorites were an important source of nearly pure metal. They were used for diverse items such as buttons, ear ornaments, and swords.

Although reports of stones falling from the sky were common, most eighteenth-century scientists were skeptical that such events really happened. They even doubted the reliability of eyewitnesses. A meteorite shower that took place in Barbotan, France, in 1790 was witnessed by the mayor and the city council. Nevertheless, French scientists dismissed the event, saying, "How sad it is to see a whole municipality attempt to lend credibility, through a formal deposition, to folktales that arouse the pity not only of physicists but of all sensible people." In the United States, in 1803, President Thomas Jefferson is reported to have doubted that stones could fall from the sky, although his main objection was to the idea that the stones could have formed in the atmosphere.

The first scientist to contradict this skepticism about meteorites was the German physicist E. E. F. Chladni. After collecting many reports of meteorite falls, he concluded in 1794 that meteorites were the surviving bodies of fireballs. He also concluded that the meteorites were extraterrestrial. His views were ridiculed by his contemporaries. After all, no less an authority than Isaac Newton had written that interplanetary space was completely empty. For Newton, it was impossible that meteoroids could exist in space before striking the Earth. Even those who believed that meteorites were real thought that they were ordinary stones that had been swept up by whirlwinds or ejected from volcanos.

In the decade after Chladni published his conclusions about meteorites, there were several well-observed falls, which convinced many other scientists that meteorites were real. The most influential of these falls took place in April 1803 in l'Aigle, France, when about 3000 stones fell in a 40 km^2 region. The French Academy of Sciences dispatched physicist Edouard Biot to investigate the event. His careful investigation convinced nearly all skeptics that meteorites really do fall from the sky. Within about 30 years, even the extraterrestrial origin of meteorites became widely accepted.

How Often Do They Fall?

Meteorite falls are quite rare. To collect one meteorite larger than about 0.1 kg each year, it would be necessary to patrol a region of about 1000 km^2 (about the size of the Washington, D.C., metropolitan area). Larger meteorites are even less common. To collect a 10 kg meteorite each year, it would be necessary to monitor an area of more than 1 million km^2. Because the annual sprinkling of meteorites is spread around the globe, it is hardly surprising that most meteorites are never recovered. In fact, until a few decades ago, the typical harvest of meteorites was only a few dozen per year for the entire planet.

The number of collected meteorites has risen dramatically during the past 20 years, however, thanks to the discovery that the Antarctic ice sheet acts as a natural collector of meteorites. The Antarctic ice sheet is much thicker at the center of the continent than at the coast. Ice flows toward the coast, carrying with it any meteorites that have fallen into the ice. In some places, the flow of the ice is stopped by rock barriers beneath the surface. As the ice evaporates, the meteorites it carries emerge at the surface, where they can be collected in large numbers. Tens of thousands of meteorites have been found in Antarctica. Some of them fell as long as a million years ago.

The discovery that meteorites can be readily found in Antarctic ice fields, where they have been preserved for up to millions of years, has greatly increased the number of meteorites that have been collected.

Meteorite Insurance? Despite rumors to the contrary, there isn't a single documented case of a person being killed by a meteorite. In fact, the chances that a given person will be struck by a meteorite more massive than 0.1 kg during a given year are about 1 in 10 billion. However, an Alabama woman was slightly injured by a meteorite in 1954. Her injuries might have been worse had the meteorite not been slowed by the roof and ceiling of her house before striking her. There are also a few reports of meteorites striking and killing animals. For instance, in 1860 a horse was struck and killed in Ohio, and a dog was killed in Egypt in 1911.

Kinds of Meteorites

Differentiation in asteroids and their subsequent breakup by collision to form iron and stony bodies

Meteorites can be divided into three broad categories: stony meteorites, iron meteorites, and stony-iron meteorites. **Stony meteorites,** such as the one shown in Figure 15.6, make up about 94% of all the meteorites that fall to the Earth. Most stony meteorites are **chondrites,** named for the **chondrules** they contain. Chondrules (Figure 15.7) are spheres of silicate rock only millimeters in size. Many of them are glassy. They appear to have been liquid drops that solidified, crystallized, and were then incorporated into chondrites. The origin of chondrules is still quite mysterious. They may have formed directly from the gas cloud that preceded the solar system. Perhaps shock waves or lightning discharges within the gas cloud triggered their formation. Another possibility is that they resulted from liquid droplets sprayed outward during impacts between early solar system bodies.

FIGURE 15.6
A Stony Meteorite
The stony meteorite shown here was part of the Allende meteorite, which fell in Mexico in 1969. This piece is about 15 cm across. After a few years of weathering, stony meteorites are often mistaken for terrestrial rocks.

FIGURE 15.7
Chondrules Within a Chondrite
Chondrites are the most common kind of stony meteorite. This close-up of a piece of the Allende meteorite shows the chondrules as light-colored, round inclusions a few millimeters in diameter.

Carbonaceous chondrites are an important class of chondrites. Because they contain considerable amounts of water (bound in minerals) and other compounds that would have been driven off or decomposed if they had been heated, the carbonaceous chondrites are believed to be samples of early solar system material that have never been altered by high temperatures or pressures.

Carbonaceous chondrites are named for their relatively high content of carbon, much of which is in the form of organic compounds. Among the organic compounds are about 20 amino acids, the building blocks of proteins. However, there is no indication that life developed from the amino acids within carbonaceous chondrites. Instead, the presence of amino acids in carbonaceous chondrites demonstrates that the mixture of materials in the early solar system, in the right environment of temperature and pressure, could yield the basic building blocks of life.

Another type of stony meteorites is the **achondrite,** which contains no chondrules. The achondrites resemble terrestrial igneous rocks. The achondrites may have been formed when chondritic material melted, which would have destroyed the chondrules that chondrites contain. Many of the achondrites contain much less iron than the chondrites. Apparently, the material from which the achondrites formed became separated from iron (and other metals) when melting took place.

Iron meteorites, such as the one shown in Figure 15.8, make up about 5% of the meteorites that fall to Earth. They are nearly pure alloys of iron and nickel. The minerals in iron meteorites occur as large crystals that can only be produced when molten metal cools very slowly. Rapid cooling would produce many small crystals instead. Large crystals suggest that the iron meteorites solidified in an environment where the temperature declined over tens of millions of years.

FIGURE 15.8
An Iron Meteorite
Etching the polished face of an iron meteorite with weak acid makes the pattern of crystals within the meteorite visible. This slab of the Edmonton, Kentucky, meteorite is about 15 cm across.

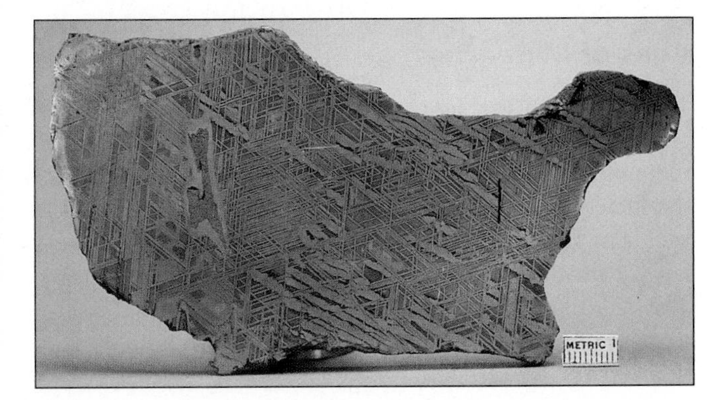

FIGURE 15.9
A Stony-Iron Meteorite
Stony-irons consist of a matrix of silicate rock within which are small pieces of metal. This polished slice of the Pavlodar, Siberia, meteorite is about 10 cm across.

Stony-iron meteorites, shown in Figure 15.9, make up the remaining 1% of the meteorites. The stony-irons are a mixture of metal and silicate rock. They appear to have been formed when molten silicates came in contact with molten metal.

Most meteorites are stones that resemble terrestrial rocks. One important type of stony meteorite is the carbonaceous chondrite, which contains water and other volatile substances. Carbonaceous chondrites may represent primitive solar system material. Two other types of meteorites are irons, which are nearly pure alloys of iron and nickel, and stony-irons, which are mixtures of stone and metal.

Ages of Meteorites

The **solidification age** of a meteorite is the amount of time that has passed since the meteorite solidified from the molten state. This happened when the parent bodies of the meteorites cooled sufficiently to solidify. Solidification ages can be measured by radioactive dating, and most fall within the narrow range of 4.55 to 4.65 billion years. The solidification ages of meteorites are a little greater than the ages of the oldest Moon rocks brought back to the Earth.

Another important moment in the history of a meteorite is described by its **cosmic ray exposure age.** This is the length of time that has passed since the meteorite broke off from a larger body. Interplanetary space is filled with **cosmic rays,** which can cause nuclear reactions when they strike meteorites. However, cosmic rays can penetrate only about 1 m into a meteorite. As long as a meteorite is buried within a parent body or a large fragment of the parent body, it can't be struck by cosmic rays, and none of the products of cosmic ray-induced nuclear reactions can build up. Typical cosmic ray exposure ages are tens of millions of years. This means that the meteorites that strike the Earth today were part of larger bodies for nearly all of their history and that these larger bodies were broken up to produce meteorite-sized objects only quite recently.

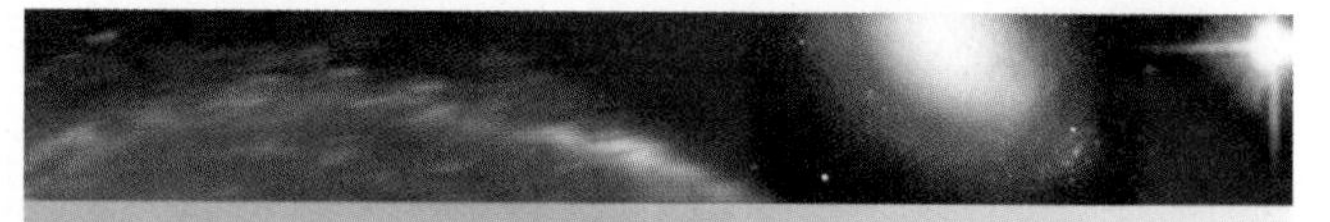

Analysis of the radioactive elements in meteorites shows that most of them solidified about 4.6 billion years ago. The oldest lunar samples date back to about the same time. The cosmic ray exposure ages of meteorites show that they were located within larger bodies until only a few tens of millions of years ago.

Parent Bodies of Meteorites

Most meteorites are quite uniform in texture and mineral structure, which suggests that they remained at fairly constant temperatures of 1000 K or more for long periods of time. During that time they were cooked as though in a pressure cooker. This changed their textures and made their mineral structures uniform. Some meteorites, however, seem to have resulted from temperatures high enough to melt part of their parent bodies. Figure 15.10 shows one of the parent bodies before and after it melted. After melting, iron and other metals sank to the bottom and rocky material rose to the top of the melted zone. The remnants of these zones are the iron meteorites and the achondrites. The stony-irons may have formed from material that lay at the boundary between the molten metal core of a parent body and its outer layer of rock. There is little evidence, however, that carbonaceous chondrites were ever heated. Instead, they may have been at or near the original surface of the parent body or perhaps were never part of a parent body.

The sizes of the parent bodies can be estimated from the very slow rate at which iron meteorites cooled once they solidified. For the iron to remain hot for tens of millions of years, it must have been well insulated. Cooling at the required rate could have taken place only within bodies larger than approximately 100 km in diameter.

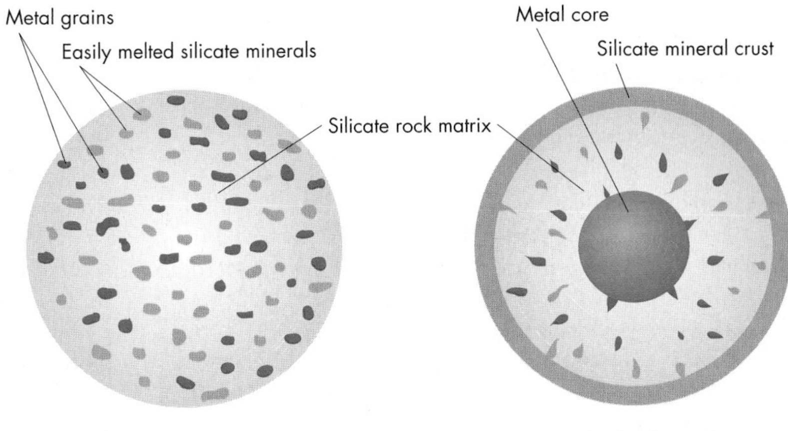

FIGURE 15.10
The Possible Evolution of a Parent Body of Meteorites
The parent body is shown in its original form on the left. It consisted of metal grains (shown as small blue regions) and easily melted silicate minerals (shown as green inclusions) in a matrix of silicate rock. When the parent body was heated, it partially melted. Molten metal sank to the center to form a core. Easily melted silicate minerals rose to form the crust.

Astronomers also have been able to learn something about where the parent bodies of the meteorites were located in the solar system and the kinds of orbits they had. A number of meteoroids have been tracked through the Earth's atmosphere accurately enough to determine that, before they hit the Earth, they were traveling on elliptical orbits with aphelia between the orbits of Mars and Jupiter.

Putting the clues together, we can conclude that the parent bodies of the meteorites must have ranged in size up to at least hundreds of kilometers in diameter. There must have been enough of them that they have been colliding with and fragmenting each other for as long as the solar system has existed. Finally, these bodies must have orbits that carry them into the region between Mars and Jupiter. We might expect to be able to find some evidence for the group of objects that produce meteorites. We do. These objects are asteroids.

Most meteorites were cooked at high temperatures over long periods of time. Also, the iron meteorites contain crystals that could have formed only if the iron had cooled from a molten state very slowly. To have cooled slowly enough, the parent bodies of the meteorites must have been at least 100 km in diameter. The orbits of meteoroids show that they came from bodies whose orbits carried them into the region between Mars and Jupiter.

15.3 ASTEROIDS

Asteroids are the multitude of rocky bodies that orbit the Sun within the planetary system and that range from a few kilometers in diameter up to nearly a thousand kilometers in diameter. These bodies are also called **minor planets.**

Discovery of Asteroids

The first asteroid was discovered by the Sicilian astronomer Giuseppe Piazzi on January 1, 1801. As he made routine observations of the positions of stars, Piazzi saw that one of the objects he thought to be a star moved with respect to the other stars. He observed the object long enough to determine that its orbit lay in the large gap between the orbits of Mars and Jupiter. Because many astronomers had speculated that there might be an undiscovered planet within that gap, Piazzi decided that he had found a new planet, which he named Ceres. The mathematician Carl Friedrich Gauss used Piazzi's observations to show that Ceres had an elliptical orbit with a semimajor axis of 2.77 AU and an orbital period of 4.62 years.

Ceres had escaped previous detection because it is too faint to be seen with the unaided eye. In fact, it is less than one one-thousandth as bright as Mars and Jupiter, the planets that orbit inside and outside its orbit. Piazzi's "planet" was certainly different from the other known planets. Ironically, Ceres, the first discovered asteroid, is now considered to be a dwarf planet because it is in orbit about the Sun, is not a satellite of a planet, and is massive enough that its gravity pulls it into a nearly spherical shape.

While looking for Ceres in 1802, Wilhelm Olbers found a second body, which he named Pallas, orbiting between Mars and Jupiter. Two more bodies with similar orbits were soon discovered—Juno in 1804 and Vesta in 1807. Now there was an abundance of bodies between Mars and Jupiter rather than a dearth. All of these bodies were so faint and so small that astronomers realized they represented a new kind of celestial object and began to refer to them as asteroids.

After the discovery of Vesta, nearly 40 years passed before another asteroid was discovered. However, when charts became available for stars below the limit of unaided vision, asteroids began to be discovered at a steady rate. One hundred ten asteroids had been found by 1870.

In the 1890s, a new technique utilizing long-exposure photographs of starfields came into practice. Asteroids, which moved with respect to the stars, showed up as streaks on the images. By the turn of the century, the number of known asteroids had risen to 450.

Over the years, ever fainter and smaller asteroids have been discovered. There are now more than 300,000 asteroids that have been observed often enough that we know their orbits. Hundreds of thousands of other asteroids have been seen but have not been well enough observed for their orbits to be determined.

Asteroid Names If you discover an asteroid and determine its orbit, you get to name it. As a result, there is a wide variety of asteroid names. Many asteroids are named after characters from different mythological traditions (Aten, Frigga, Agamemnon). Others are named for observatories (Alleghenia), cities (Chicago), benefactors (Rockefellia), relatives (Winifred), operas (Turandot), and many other people and things.

Orbits of Asteroids

The orbits of asteroids vary in shape and size, but nearly all of them orbit the Sun in the same direction as the Earth does. Compared with the planets, however, asteroids tend to have more eccentric orbits that are more inclined to the ecliptic plane.

The Asteroid Belt The overwhelming majority of known asteroids are located in the **asteroid belt,** which lies between 2.1 and 3.3 AU from the Sun. Figure 15.11 shows the positions of more than 7000 asteroids on March 7, 1997. Notice that nearly all of them were located in the asteroid belt between the orbits of Mars and Jupiter. The large number of asteroids orbiting within this region may give the impression that the asteroid belt is very crowded, with frequent collisions and near misses. Science fiction movies reinforce this impression when they portray asteroid belts as if they were as crowded and as hazardous as shopping mall parking lots. This picture is completely wrong, however. If there were a million asteroids in the asteroid belt, the average distance between nearest neighbors would be about 2 million km (more than five times the separation of the Earth and the Moon). You could spend your entire life on an asteroid without ever getting close enough to another asteroid to see it as anything larger than a point of light.

The largest known asteroids are in the asteroid belt. Pallas and Vesta are both a little larger than 500 km in diameter. Thousands of other asteroids in the main belt are larger than 10 km in diameter.

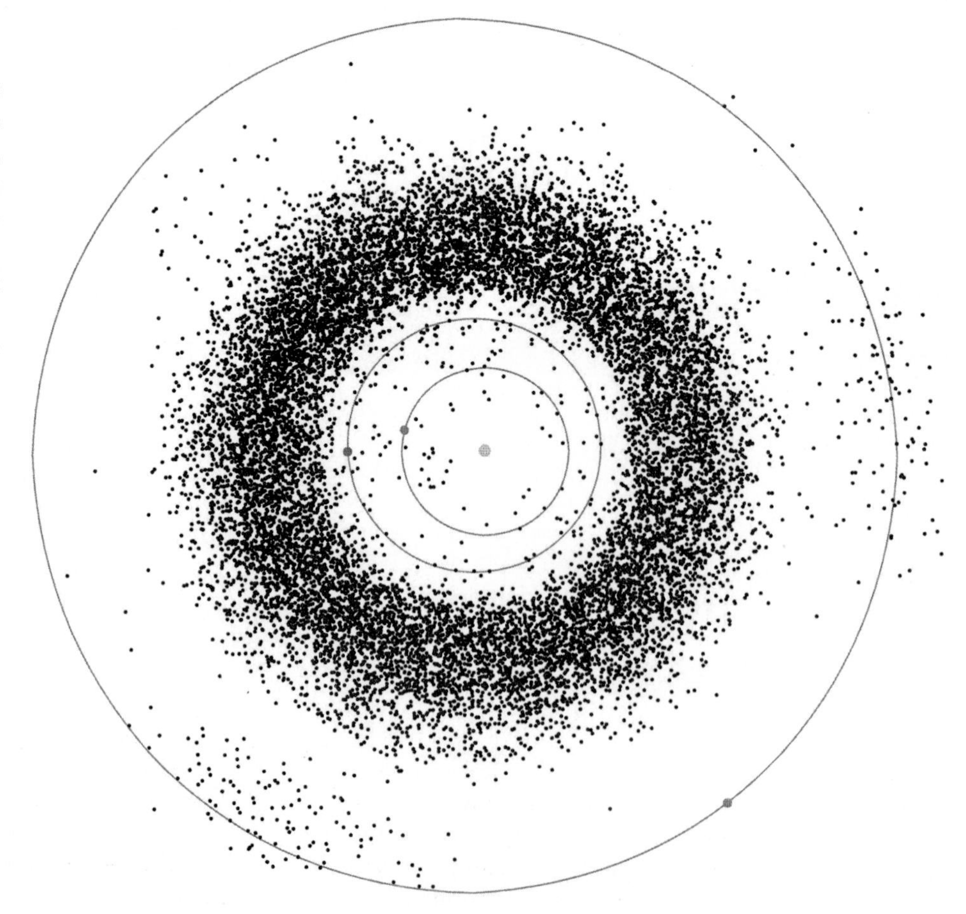

FIGURE 15.11
Asteroid Locations
The locations of more than 7000 asteroids on March 7, 1997, are shown along with the positions and orbits of the Earth, Mars, and Jupiter. Most asteroids orbit in the main belt, which lies between 2.1 and 3.3 AU. This is outside the orbit of Mars. Two clusters of asteroids, known as the Trojan asteroids, follow Jupiter's orbit but are 60° ahead of or behind Jupiter. Note that there are some asteroids inside the orbits of Mars and the Earth.

FIGURE 15.12
Asteroids Mathilde, Gaspra, and Ida
The asteroids are shown at the same scale in this composite image. Mathilde is about 60 km in diameter, Gaspra is about 12 by 20 by 11 km, and Ida is about 56 km long.

In the last decade, astronomers have obtained close-up images of a handful of main-belt asteroids when spacecraft flew past them on their way to other destinations. Images of three of these asteroids are shown in Figure 15.12. Gaspra, which has an elongated shape that measures 20 by 12 by 11 km, resembles Mars's satellite Phobos in that it is pitted with craters and shows a series of cracks that apparently resulted from a particularly violent collision with another, smaller asteroid. Ida, about 56 km in length, is elongated and heavily cratered like Gaspra. Remarkably, Ida has a satellite of its own. Ida's satellite, shown in Figure 15.13, is about 1.5 km across and orbits Ida at a distance of about 100 km. Observations from Earth have identified over 50 other asteroids that have satellites. One asteroid, Sylvia, is orbited by two satellites. A third main-belt asteroid visited by spacecraft is Mathilde, a roughly round body about 60 km across. Figure 15.12 shows that Mathilde is pocked by craters, including one 30 km across and 10 km deep. Mathilde has an albedo of only 0.04, indicating that it reflects sunlight about as poorly as a chunk of coal. Its surface materials may be similar to those of carbonaceous chondrites. As the spacecraft passed Mathilde, it was deflected slightly by the asteroid's gravity. This made it possible for astronomers to calculate Mathilde's mass and density. Mathilde's density is less than 2000 kg/m^3—about half the density of carbonaceous chondrites or other rocky materials. This suggests that Mathilde is porous. It may be a "rubble pile" fragmented by the numerous impacts it has experienced. The main-belt asteroids Steins and Annefrank have also been visited by spacecraft, but at great enough distances that images of them do not show many details.

The most detailed examination of an asteroid occurred when the *NEAR* (Near Earth Asteroid Rendezvous) *Shoemaker* spacecraft orbited Eros for nearly a

FIGURE 15.13
The Asteroid Ida and Its Satellite, Dactyl
This picture of Ida and its satellite was obtained by the *Galileo* spacecraft when it passed Ida at a distance of 11,000 km in 1993. Ida is elongated and has a cratered surface. The small object to the right of Ida is its satellite, Dactyl, about 1.5 km across. The satellite appears to be very close to Ida, but this is because it is closer to the spacecraft than Ida is. The distance between Ida and the satellite is actually about 100 km, or twice the length of Ida.

FIGURE 15.14
Asteroid Eros
This image of Eros is the result of many measurements by *NEAR Shoemaker.*

FIGURE 15.15
The Surface of Eros
This view of Eros was taken by *NEAR Shoemaker* from a distance of 50 km and shows the interior of one of Eros's largest craters. The view shows large boulders and numerous small craters. Much of the surface appears to be covered with a layer of brown dust.

FIGURE 15.16
A High-Resolution Image of Eros
This image was taken by *NEAR Shoemaker* from 130 m above the surface. The image shows a region only 6 m across. Details as small as 10 cm can be seen.

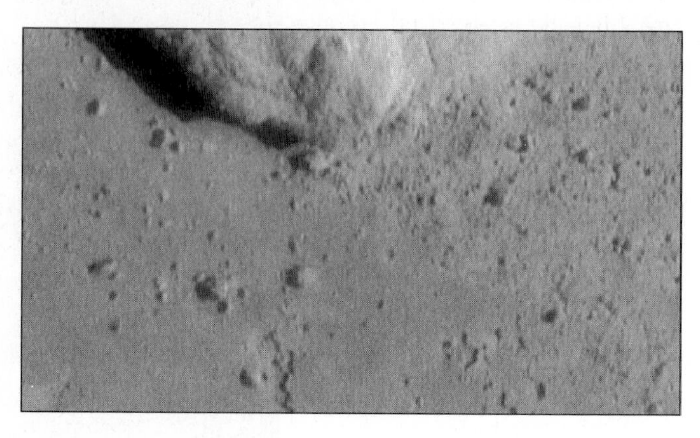

year in 2000 and 2001. Figure 15.14 shows a composite image of Eros, which is an irregularly shaped body about 40 km long that orbits between the orbits of Earth and Mars. *NEAR Shoemaker* images showed many fine details, including grooves and ridges that suggest that Eros may be a collision fragment from a larger body. Figure 15.15 shows boulders as large as 40 m across that may be debris from impacts with other asteroids. On February 12, 2001, *NEAR Shoemaker* touched down on Eros. Just before it landed, *NEAR Shoemaker* returned an image shown in Figure 15.16, in which features as small as about 10 cm (4 inches) can be seen. After it landed, *NEAR Shoemaker* used gamma rays to determine the chemical composition of Eros. Eros's composition resembles that of chondrite meteorites, suggesting that, like the chondrites, the surface of Eros may be as old as the solar system.

Another close examination of an asteroid was carried out by the Japanese spacecraft *Hayabusa,* which landed on Itokawa in 2005. *Hayabusa* collected samples of Itokawa's surface materials that will eventually be returned to Earth for analysis.

Most of the known asteroids orbit the Sun in the asteroid belt, which lies between 2.1 and 3.3 AU from the Sun. Even though there are many asteroids, they are widely spread out in this immense region of space. Spacecraft encounters have provided us with our best views of asteroids.

Asteroids at Jupiter's Distance and Beyond In addition to asteroids in the asteroid belt, Figure 15.11 shows two swarms of asteroids that share Jupiter's orbit around the Sun. These are the **Trojan asteroids,** located 60° ahead of or behind Jupiter, where the gravitational attractions of Jupiter and the Sun combine to produce regions where small bodies can have stable orbits. The orbital periods of the Trojan asteroids are the same as Jupiter's, but they have a range of orbital eccentricities and inclinations. This means that they lie 60° from Jupiter, on average, but their individual orbits cause them to wander back and forth around their average positions.

Surveys of the Trojan asteroids are much less complete than surveys of main-belt asteroids. About 3000 Trojan asteroids are known. The largest measures about 150 by 300 km and has been named Hector. Estimates suggest that there may be nearly half as many Trojan asteroids (mostly undiscovered) as asteroids in the main belt. Although the Trojan asteroid swarms are usually portrayed as curiosities of our solar system, it is more appropriate to think of them as a second asteroid belt quite unexplored compared with the main belt.

Many asteroids are not found in the asteroid belt. The Trojan asteroids share Jupiter's orbit, leading or trailing Jupiter by 60°.

Earth-crossing Asteroids Just as some asteroids travel far beyond the main belt, others penetrate deeply into the inner solar system. The **Amor asteroids,** for example, pass inside Mars's orbit and approach the Earth without ever crossing the Earth's orbit. The **Aten asteroids,** on the other hand, have orbital semimajor axes smaller than 1 AU and cross the Earth's orbital distance when they are near aphelion. All of the **Apollo asteroids** have orbits with aphelia well beyond the Earth's orbit and perihelia inside the Earth's orbit, so they are sometimes referred to as Earth-crossing asteroids (although the Atens are Earth crossing as well).

So far, only about 5000 of these asteroids have been discovered. This is partly because they are so small. The largest known Amor asteroid is only 40 km in diameter, and the largest known Apollo asteroid is 8 km in diameter. Figure 15.17 compares the sizes of some of these asteroids with the altitude at which commercial jets fly.

Even though the Earth is a relatively small target, a large fraction of the Apollo asteroids will strike the Earth and be destroyed within the next few tens of millions of years. The consequences of such an encounter are described at the end of this chapter. Near misses occur quite frequently. In 2004, for example, asteroid 2004 FH came within 45,000 km (7 Earth radii) of the Earth. It was discovered only 2 days earlier, so there was essentially no warning. The Aten asteroid Apophis will pass within 10 Earth radii in 2029 and there is a small chance it will strike Earth in 2036. It will be several years

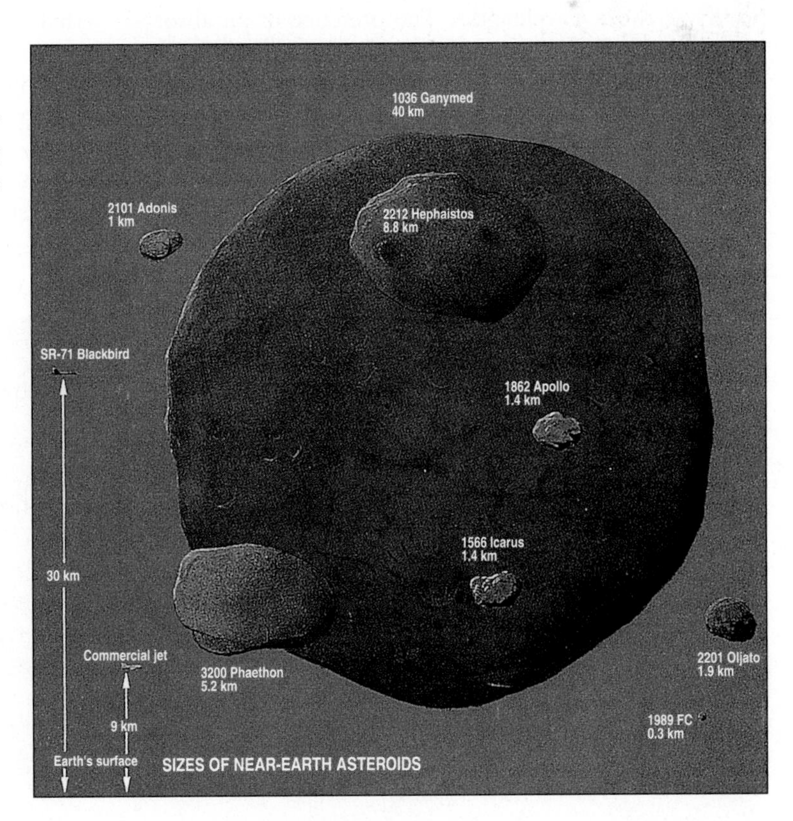

FIGURE 15.17
The Sizes of Near-Earth Asteroids
Although near-Earth asteroids are smaller than many of the asteroids in the asteroid belt, they are big compared with the altitudes at which jet airplanes fly in Earth's atmosphere.

before astronomers will be able to tell for sure whether a collision with occur in 2036.

The solar system has been in existence for much longer than tens of millions of years, so the Apollo asteroids would have all been destroyed long ago unless there was a source to resupply them. This source is the asteroid belt. Asteroids at some orbital distances can be influenced by Jupiter's gravity so that their orbits become more eccentric. Eventually, they pass close enough to Mars to be pulled even deeper into the solar system.

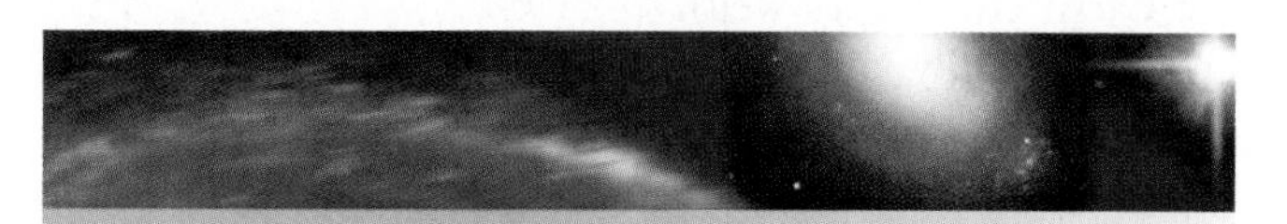

There are thousands of asteroids that have orbits that cross the Earth's orbit. A large fraction of the asteroids in Earth-crossing orbits will strike the Earth within a few tens of millions of years.

Classes of Asteroids

The asteroids have been grouped into different classes based on their **reflectance spectra,** which show how well they reflect light of different wavelengths. When sunlight falls on the surface of an asteroid, some is reflected and some is absorbed. For many minerals, the absorption of sunlight is much stronger at some wavelengths than at others. This makes the asteroid absorb well but reflect poorly at those wavelengths. The presence of an absorption feature of a particular mineral in the spectrum of an asteroid indicates that the surface layers of the asteroid contain that mineral.

About three-fourths of the asteroids are **C-type.** The C-type asteroids are very dark, so their reflectivities are only a few percent. Their spectra show no strong absorption features due to minerals. Most of the remaining asteroids are **S-type,** which show an absorption feature due to the mineral olivine. Many other classes exist, including the **M-type** asteroids, which have reflectance spectra like those of metallic iron and nickel, and the unique **V-type** asteroid Vesta, which shows a strong absorption feature due to pyroxene, a common mineral in basaltic lava flows.

Asteroids and Meteorites

Most of the meteorites that fall to the Earth are derived from main-belt asteroids. In some cases the immediate parent bodies of the meteorites are Earth-crossing asteroids that have been deflected from the asteroid belt. Most of the different classes of meteorites have reflectivity spectra that strongly resemble classes of asteroids. That is, the carbonaceous chondrites (as distinct from the much more common ordinary chondrites) look like smaller versions of the C-type asteroids, whereas the iron meteorites resemble the M-type asteroids.

There is one very important exception to the identification of asteroids as parent bodies of meteorites, however. Asteroids that look like ordinary chondrites, by far the most common kind of meteorite, seem to be rare among the large asteroids for which reflectivity spectra have been measured. Perhaps large asteroids were heated enough that the chondritic material near their surfaces was melted and ceased to be chondritic. If this is true, then only small asteroids, which were never heated very much, could retain their chondritic characteristics. Thus, smaller asteroids, for which reflectivity spectra have not yet been measured, may be the missing source of the ordinary chondrites.

One important thing to keep in mind when comparing asteroids and meteorites is that at any given time, the majority of meteorites may have come from only a few asteroids that happen to be in Earth-crossing orbits. This means that the kinds of meteorites that strike the Earth may change significantly over millions of years as the population of Earth-crossing asteroids changes. This idea is supported by the fact that there are significant differences between the Antarctic meteorites that fell to the Earth a million years ago and the ones that have fallen in the last few centuries. For instance, iron and stony-iron meteorites are approximately four times less common in Antarctic meteorites than in recent falls. This suggests that ordinary chondrite parent bodies may not be very common in the asteroid belt despite the fact that the majority of fragments striking the Earth today are chondrites. In the past and in the future, the most common kinds of meteorites may very well not be ordinary chondrites. Only if we could collect meteorites over many millions of years might we expect the collection to give a good representation of the kinds of objects in the asteroid belt.

The reflectance spectra of asteroids show that many of them resemble larger versions of different classes of meteorites. One problem for the identification of meteorites with asteroids is the absence or rarity of asteroids that resemble the ordinary chondrites, the most common kind of meteorite. It is possible that the kinds of meteorites that have fallen to Earth in the last few centuries reflect only the kinds of Earth-crossing asteroids that exist today and are not characteristic of the kinds of objects in the asteroid belt.

15.4 COMETS

A **comet** is a small icy body in orbit about the Sun. When it passes near the Sun, a comet becomes much brighter and can produce a conspicuous tail. Records of observations of comets date back to the time of the Babylonians. During most of history, however, comets were believed to be phenomena within the Earth's atmosphere rather than celestial objects. At various times, comets were thought to be reflections from high clouds or violent, fiery winds. Aristotle's view—that comets were fiery phenomena in the atmosphere—was extremely influential and generally accepted until the sixteenth century.

Comets were also considered to be omens, sometimes good and sometimes bad. Suetonius, in his biography of Julius Caesar, said that on the first day of the games given in honor of the deification of Caesar, a comet appeared in the evening sky and was visible for a week. The comet was thought to be Caesar's soul, elevated to heaven. When Halley's comet appeared in April of 1066 (Figure 15.18), it was considered an ill omen for King Harold of England. This interpretation of the comet seemed to be confirmed when Harold was killed at the Battle of Hastings and England fell to the Norman invasion. Even in the twentieth century, the appearance of Halley's comet in 1910 was thought by many to be a sign that the world was about to end.

The modern view of comets as celestial bodies began to develop when Tycho Brahe made careful observations of the position and motion of the bright comet of 1577. He showed that the comet had to be at least six times as far away from the Earth as the Moon was. He suggested that comets revolved about the Sun, possibly on elongated orbits. More than a century later, in 1682, another bright comet captured the interest of Edmund Halley. He found that the comet had the same elliptical orbit as other comets seen in 1456, 1531, and 1607. Concluding that these were all appearances of a single comet with an average orbital period of 75 years, he predicted that the comet would be seen again in 1758 or 1759. The prediction came true when the comet, now known as Comet Halley, was seen by an amateur astronomer on Christmas Night in 1758.

Anatomy of a Comet

Figure 15.19 shows Comet Halley as most people picture comets. It is a spectacular sight with a bright head and tails stretching across the sky. For the great majority of their lives, however, comets are small, dim lumps of frozen ices.

Nucleus The only part of a comet present at all times is the **nucleus.** The nucleus of a comet is an irregularly shaped, loosely packed lump of dirty ice that measures between a few hundred meters and several kilometers across. Most of the ice is frozen water, but frozen carbon monoxide, carbon dioxide, and formaldehyde are present as well. Microscopic dust particles are also trapped within the mass of ice. In some respects, the nucleus of a comet resembles a larger version of the mounds of snow and dirt plowed into piles in shopping center parking lots. When the snow evaporates or melts, most of the dirt is left behind, forming a dirty crust that makes the pile very dark.

FIGURE 15.18
Part of the Bayeux Tapestry
This tapestry was woven in about 1070 to commemorate the Norman conquest of England in 1066. Comet Halley can be seen at the top center of this section of the tapestry.

FIGURE 15.19
Comet Halley in March 1986
This picture, taken by astronomers at the European Southern Observatory in Chile, shows the comet about a month after it had passed perihelion. At that time, Comet Halley was bright and easily seen by viewers in the southern hemisphere but was difficult to observe for those in the northern hemisphere. The picture was taken in the same week that five spacecraft encountered the comet.

Like the shopping center snowpile, a comet nucleus begins to evaporate when solar energy warms it. Within about 3 AU of the Sun, water and other molecules within the comet nucleus evaporate and flow outward, carrying with them some of the dust mixed with the ice. Much of the escape of gas and dust seems to occur in spots where subsurface material can break through the crust of the nucleus to produce jets that stream outward. The gas and dust that escape from the nucleus form the coma and tails shown in Figure 15.20. Thus, the coma and tails exist only when the comet is near the Sun.

The nuclei of comets are far too small to be resolved by even the largest telescopes on Earth. They appear only as

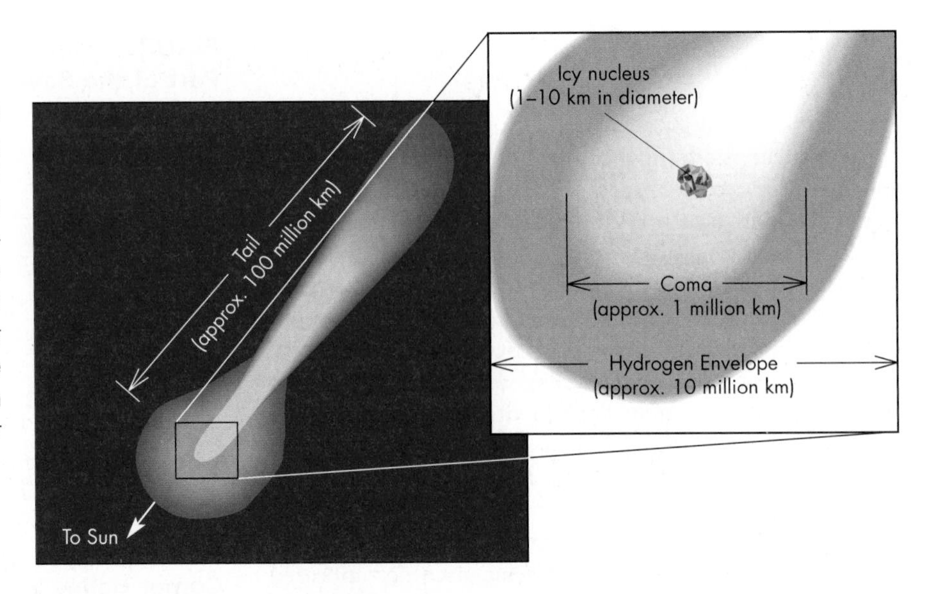

FIGURE 15.20
The Parts of a Comet
The nucleus is so small that it appears pointlike from the Earth. Surrounding the nucleus is the coma, a spherical cloud of gas and dust that extends as far as a million kilometers from the nucleus. Around the coma is an invisible cloud of hydrogen that can be several million kilometers across. The tail or tails of the comet point away from the Sun and extend as much as 100 million km (nearly 1 AU) outward into the solar system.

points of light. To learn more about comets, space probes have been sent to encounter four comets as they passed through the inner solar system. The *Giotto* spacecraft obtained images of Comet Halley in 1986 and *Deep Space 1* imaged Comet Borelly in 2001. In 2004 the *Stardust* spacecraft imaged Comet Wild 2 as it flew within 250 km of the nucleus. *Stardust* collected dust samples from the coma of Wild 2 and returned them to Earth in 2006. The most detailed examination of a comet nucleus was accomplished by the *Deep Impact* spacecraft. In July 2005, *Deep Impact* released a 372-kg copper projectile, which impacted the surface of Comet Tempel 1 at a speed of 10 km/s. The projectile produced a crater about 100 m across and ejected 250 million kg of material into space. Light emitted by the ejected material was analyzed by *Deep Impact.* In addition to ice and silicate rock, the ejected material was found to have a substantial amount of organic molecules. This suggests that comets impacting the Earth early in its history may have brought organic compounds to Earth and may have played a role in the origin of life. Observations of the motion of the plume of material ejected by the impact yielded the gravity of the nucleus and showed that Tempel 1 has a density only 60% that of water ice and, thus, must be very porous. As it approached the nucleus, the projectile obtained a series of high-resolution images. A composite of the images is shown in Figure 15.21. The nuclei of Comets Halley, Borelly, Wild 2, and Tempel 1 are generally quite similar. Their surfaces are very irregular, with craters, mountains, and rolling terrain. The nucleus of Tempel 1 differs from those of Halley, Borelly, and Wild 2 in that it shows what appear to be impact craters. The image returned by *Giotto, Deep Space 1, Stardust,* and *Deep Impact* generally confirmed the ideas astronomers had developed about the nuclei of comets.

FIGURE 15.21
The Nucleus of Comet Tempel 1
This picture is a composite of high-resolution images. The images were obtained by a projectile released by the *Deep Impact* spacecraft just before the projectile struck the nucleus. The impact occurred between the two impact craters at the bottom of the image. The nucleus of Tempel 1 is 7.6 km long and 4.9 km wide. Unlike other comet nuclei for which images have been obtained, Tempel 1 shows a number of impact craters.

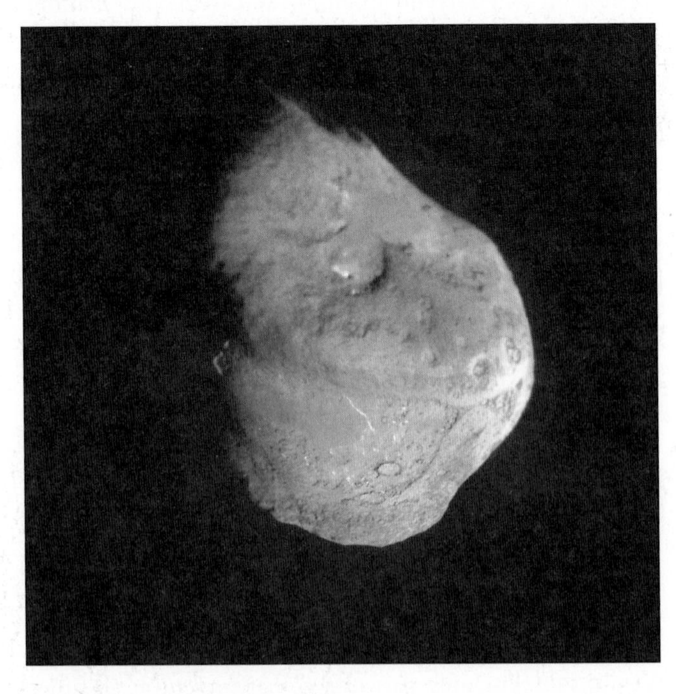

Coma The **coma** of a comet is a ball of outflowing gas and dust that surrounds the nucleus. A coma, which can be a million kilometers in diameter, is much bigger than the nucleus that produces it. Although there are probably only a few kinds of molecules released from the nucleus, these molecules are broken apart and ionized by solar radiation to produce a large variety of atoms, ions, and molecules within the coma. The coma appears bright because of a

FIGURE 15.22
The Dust and Plasma Tails of Comet West (1975)
The dust tail, on the left, is yellow and broader than the plasma tail. Plasma tails are blue and often much longer than dust tails. Plasma tails are very straight, whereas dust tails often appear as curving arcs.

combination of emission from the gas and sunlight reflected by the dust. The coma looks quite substantial, but this is very deceptive. All of the gas and dust in the coma comes from a thin outer layer of the nucleus. This small amount of material spreads out over an enormous volume of space to form the coma. The coma actually is a much better vacuum than we can make in laboratories on the Earth.

Tails Many comets show two tails. Figure 15.22 shows that one of these tails is blue, whereas the other appears white or yellow. The white or yellow tail is made of dust swept from the nucleus, so it is called the **dust tail.** A common misconception about comets is that their tails trail behind their nuclei like hair streaming behind a person who is running. Figure 15.23 shows that their tails actually point away from the Sun. The reason the dust tail points away from the Sun is that the force of sunlight pushes the dust particles outward.

The blue tail in Figure 15.22 is called the **plasma tail,** because it is made of ions and electrons. The color of the plasma tail is produced by ionized carbon monoxide, which emits strongly in the blue part of the spectrum. The plasma tails of comets interact strongly with the ionized gases in the solar wind. The interaction is so strong, in fact, that the solar wind can sweep the plasma tail outward to more than 1 AU in length.

Solar warming diminishes as a comet recedes from the Sun, so the rate at which new gas and dust are supplied to the coma and tail diminishes as well. The coma and tail shrink and fade. By the time the comet reaches a distance of 3 AU on its way outward, it looks like a snowball again. It is somewhat smaller and dirtier than it was before approaching the Sun, but otherwise it is not seriously altered.

FIGURE 15.23
The Orientation of Comet Tails
Both the dust tail and the plasma tail of a comet point away from the Sun. This means that the tails sometimes follow the nucleus and coma and sometimes precede the nucleus and coma as the comet orbits the Sun.

Orientation of comet tails

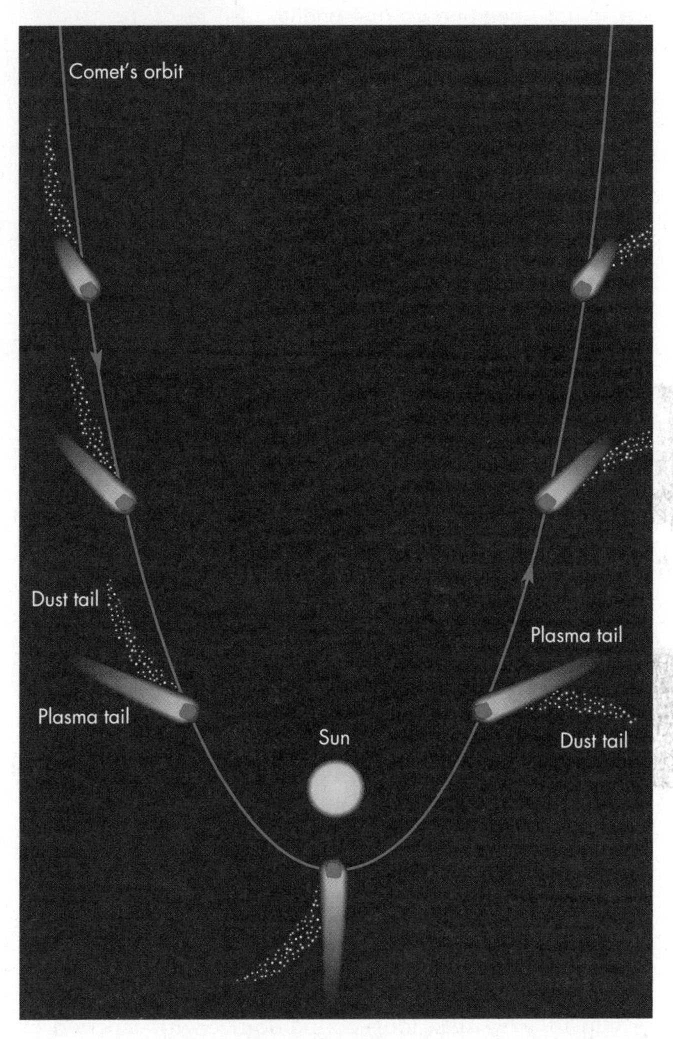

Far from the Sun, a comet consists only of a nucleus, a loosely packed chunk of water ice, other ices, and dust. The warmth of sunlight, however, drives gas and dust from the nucleus. These form the coma, the dust tail, and the plasma tail of the comet.

FIGURE 15.24
Comet Hale-Bopp
Comet Hale-Bopp was a spectacular sight during spring 1997, yet didn't pass very close to either the Earth or the Sun. If Hale-Bopp had passed through the inner solar system 4 months earlier it would have passed within 0.1 AU of Earth and would have been one of the brightest comets in history.
(© Wally Pacholka/AstroPics.com)

Comet Orbits

Since the Babylonians began keeping astronomical records, more than a thousand different comets have been observed. Only about 500 of these, however, have been observed well enough and often enough to determine their orbits with any accuracy. Astronomers use orbital information to divide comets into two groups, long-period and short-period. The **long-period comets** have orbital periods longer than 200 years, and the **short-period comets** have orbital periods shorter than 200 years. This is an arbitrary, but convenient, dividing point. Most short-period comets have periods much shorter than 200 years (in fact, half of them orbit the Sun in less than 6.5 years), and most long-period comets have periods much longer than 200 years. Comet Halley, with a period of 76 years, is a short-period comet, whereas Comet Hale-Bopp (shown in Figure 15.24), which passed through the inner solar system in 1997, is a long-period comet with a period of about 2400 years.

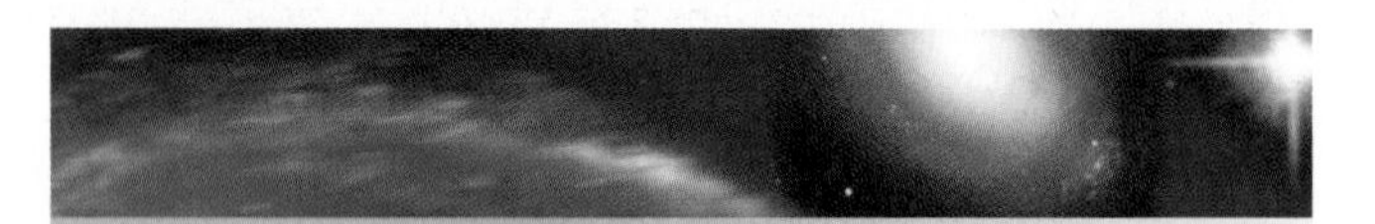

Comets are divided into two groups, the long-period and short-period comets, based on the lengths of their orbital periods. The long-period comets have orbital periods longer than 200 years, whereas the short-period comets have orbital periods shorter than 200 years.

Long-period Comets and the Oort Cloud As a long-period comet enters the planetary system, its orbit is influenced by the gravitational pulls of the planets. Therefore, to find out where the comet came from, it is important to determine the orbit it was following before it entered the planetary system. For those comets for which this has been done, the orbits are extremely large. Almost half of them, in fact, have aphelia more than 10,000 AU from the Sun (about 250 times the orbital distance of Pluto). Some have aphelia as great as 100,000 AU from the Sun. At such a distance, a comet is more than one-third of the way to the nearest stars. Orbital periods for such large orbits are tens of millions of years.

Many of the long-period comets that we see are entering the inner solar system for the first time, so they are called **new comets.** Even though astronomers observe only a few new comets each year, it is thought that there is an enormous number of comets orbiting the Sun on large, elliptical orbits. For one thing, the long orbital periods of the new comets mean that each one is visible for only a small fraction of its orbital period (about 1 year out of perhaps 10 million). Thus, for every new comet that we see each year, there are millions moving on orbits that will eventually bring them near enough to the Sun to become visible from Earth. Also, only those comets that have perihelion distances within about 3 AU are likely to be detected. There must be many more that have perihelia at 10 or 100 or 10,000 AU and never develop highly visible comas and tails. As a result of these factors, the total number of long-period comets has been estimated to

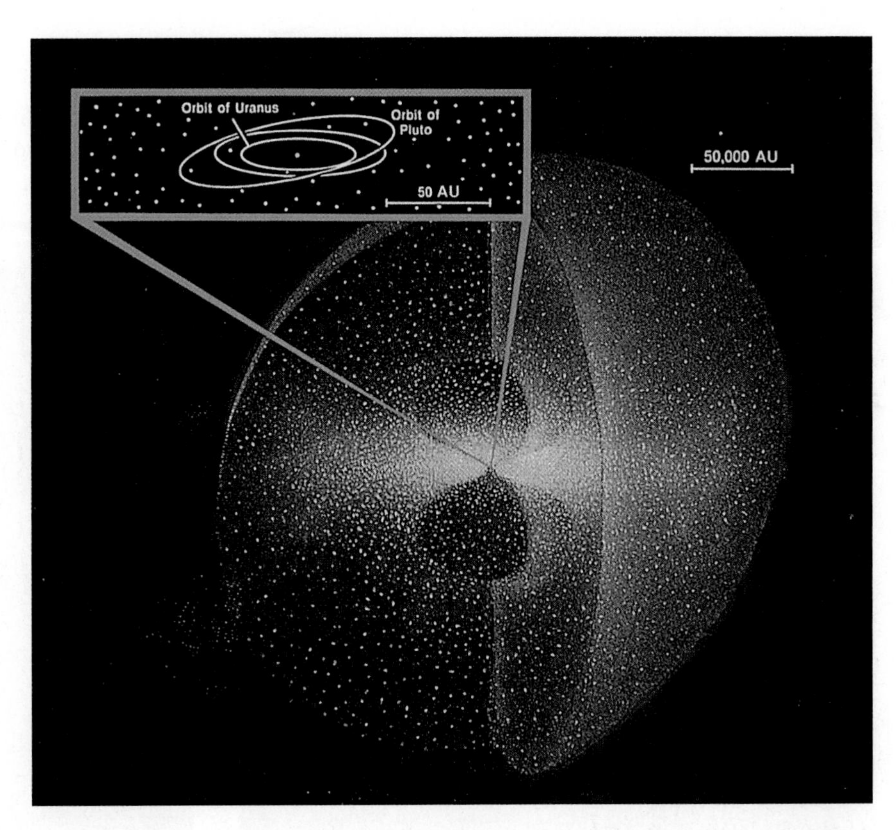

FIGURE 15.25
The Oort Cloud
The cloud of comets around the Sun extends about one-third of the way to the nearest stars. Although there may be as many as 1 trillion comets in the Oort cloud, the volume of space that the Oort cloud occupies is so immense that the comets are separated from one another by distances that are typically about 10 AU.

Oort cloud and Kuiper Belt

be at least a trillion. The total mass of the long-period comets is thought to be between 10 and 100 times the mass of the Earth. The swarm of comets shown orbiting the Sun on long-period orbits in Figure 15.25 is called the **Oort cloud** after Jan Oort, who first realized the significance of the large aphelion distances of the new comets and deduced that the comet swarm exists. In 2004, astronomers discovered Sedna, a body about half the size of the Moon and 90 AU from the Sun. Sedna's elliptical orbit carries it as far as 900 AU from the Sun. Sedna is the most distant known solar system body and may be an inner Oort cloud comet.

Gravitational forces caused by the planets have a strong effect on the orbits of new comets. About half are accelerated to higher speeds and are ejected from the solar system on hyperbolic orbits. The other half lose speed and assume less eccentric orbits. These remain long-period comets, but with periods of thousands or tens of thousands of years rather than millions of years. In effect, all of the Oort cloud comets whose orbits carry them into the inner solar system are lost to the Oort cloud. The comets on such orbits when the solar system formed disappeared billions of years ago. So why do we continue to see new comets? The answer to this question depends on other stars and on the Milky Way galaxy.

As described in Chapter 16, the stars are in motion with respect to each other. Every million years or so, a few of them pass close enough to the Sun to penetrate or graze the Oort cloud. When this happens, the star's gravity alters the motions of the comets that it passes near. Some are speeded up and escape from the solar system. Others lose a little speed and, on their next orbit, pass closer to the Sun than they ever have before, as shown in Figure 15.26. These are the new comets, which are influenced by the planets and are either ejected from the solar system or have their orbital periods shortened to thousands of years. The theory of stellar encounters is difficult to check directly, because the stars that produced the new comets we see today passed near the solar system millions of years ago. Moving on their individual paths through space, they are no longer among the near or bright stars. The comets that are propelled toward the inner solar system by the stars near us now won't arrive for millions of years. Another force disturbing the Oort cloud is the tidal force due to the Milky Way galaxy. The galactic tide is always present and supplies new comets at a more or less steady rate.

The Origin of the Short-period Comets The short-period comets don't originate in the Oort cloud but rather in the Kuiper Belt, which is discussed in Chapter 13, Section 3. The reason that the Kuiper Belt is thought to be a flattened disk is that the short-period comets have orbits that lie much closer to the ecliptic plane than do the orbits of the long-period comets. The Kuiper-Belt Objects experience gravitational perturbations due to

FIGURE 15.26
Stellar Perturbations
A, The gravitational attraction of passing stars slows down some of the comets in the Oort cloud so that, **B,** their orbits carry them into the inner solar system for the first time. The orbits of Oort cloud comets are actually much larger and more elongated than shown here.

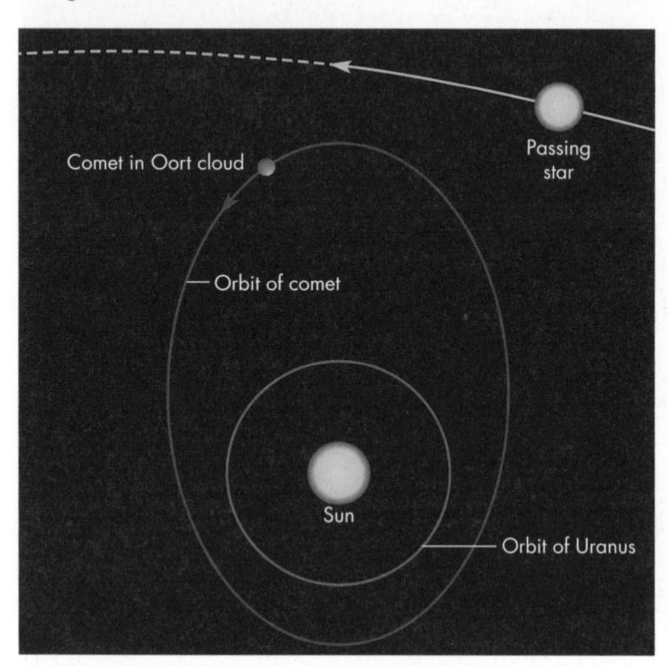

A Orbit of comet before it is slowed by passing star

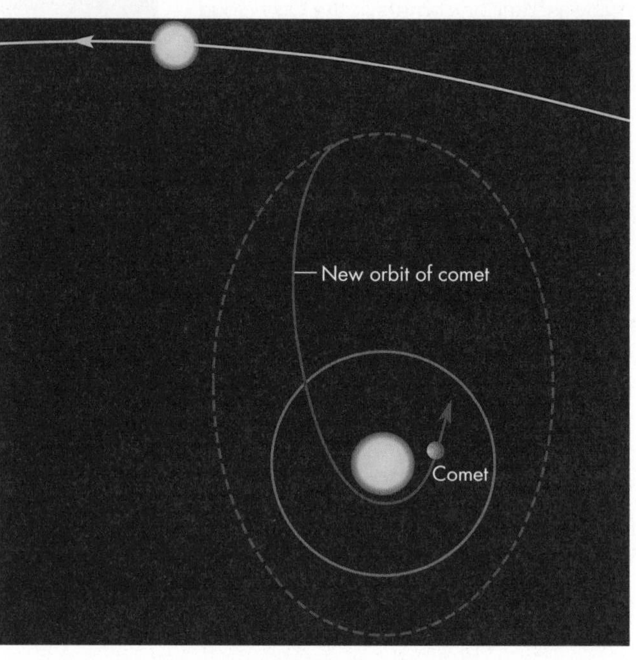

B Comet enters inner solar system after being slowed by passing star

the giant planets, and their orbits are sometimes altered enough to bring them close enough to the Sun to be seen as short-period comets.

The long-period comets have extremely elongated orbits that carry them as far as 100,000 AU from the Sun. It is estimated that 10 billion comets orbit the Sun on such orbits. This swarm of comets is called the Oort cloud. Passing stars disturb the orbits of some of the comets from the Oort cloud. These comets enter the planetary system for the first time and become visible as new comets. Short-period comets are thought to originate in the Kuiper Belt, a flattened disk of comets lying outside the orbit of Neptune.

What Happens to Comets?

We have seen how the orbits of comets in the Oort cloud and Kuiper Belt are modified so that they carry the comets into the inner planetary system. There must be processes that eliminate comets from the inner planetary system as well. Otherwise, the inner solar system would be clogged with them by now. In fact, at least three mechanisms exist that destroy comets. The first is a collision with one of the planets or the Sun. The second, erosion, is much more probable than a collision. On each passage near the Sun, a comet loses an outer layer of ice and dust. Therefore its size and perihelion distance determine how many orbits it will make before all of its icy material is gone. Comet Halley, for example, has been observed on each of the 30 times it has approached the Sun since 240 B.C., and there is reason to think that it made many more appearances before that. Although Comet Halley grows smaller on each trip near the Sun, it is large enough to survive many more orbits. The third way that a comet can be destroyed is by being broken into several pieces when it passes too near to the Sun or one of the planets. After a comet is broken into smaller pieces, erosion continues to wear away each piece until it is gone or until it strikes a planet or satellite.

Figure 15.27 shows a trail of at least 18 glowing objects—pieces of Comet Shoemaker-Levy 9, which was disrupted when it passed too close to Jupiter in 1992. The fragments of Comet Shoemaker-Levy 9 struck Jupiter and were destroyed in July 1994. The disruption of Comet Shoemaker-Levy 9 has led some astronomers to propose that the nucleus of a typical comet consists of a collection of smaller balls of ice held together by their mutual gravitational attraction. Near a planet, tidal forces pull

FIGURE 15.27
Fragments of a Comet
The glowing bodies aligned from lower right to upper left are the fragments of Comet Shoemaker-Levy 9, which was disrupted by Jupiter's gravity when it passed near that planet in 1992. The fragments struck Jupiter in July 1994.

the balls apart, forming a line of fragments. When the fragments strike a solid body, they may produce a chain of craters like the ones shown in Figure 15.28. Many crater chains have been found on the Moon, Ganymede, and Callisto.

The impacts of the fragments of Comet Shoemaker-Levy 9 on Jupiter's southern hemisphere gave astronomers their first direct glimpses of the violent collisions between planets and smaller bodies. Actually, the impacts took place just around the limb of Jupiter on the far side and were hidden from Earth. The impact sites could be seen by the *Galileo* spacecraft, which was approaching Jupiter. The first sign of an impact was the flash of a meteor as a fragment of the comet entered Jupiter's atmosphere. Within seconds the fragment exploded, producing a fireball hotter than the surface of the Sun as shown in Figure 15.29. Each time Jupiter's rotation brought a fresh impact site into view, astronomers saw a new multiringed cloud feature that, for some fragments, was as much as 25,000 km in diameter (twice the size of the Earth). One of the new cloud features is shown in Figure 15.30. Although Jupiter's winds quickly began to pull the new cloud features apart, they could be seen for more than a year after the impacts. The energy released by each fragment's impact was up to hundreds of times greater than the total energy of all of the Earth's nuclear bombs. If they had hit the Earth, the fragments would have left craters many tens of kilometers in diameter.

Meteoroid Swarms Once the icy material in the nucleus of a comet is used up, the nucleus can never produce a coma or a plasma tail again. However, even when the ice is gone, dust is left behind. The dust particles ejected from the nucleus of a comet (that form its dust tail) continue to follow the orbit of the comet for many years. If the orbit of the comet nearly crosses the orbit of the Earth, then the Earth can pass through the swarm of

FIGURE 15.28
A Crater Chain on Ganymede
This chain of craters stretches 150 km across the surface of Ganymede. The chain may have been formed when a line of comet fragments struck Ganymede. Similar crater chains are seen on the Moon and Callisto.

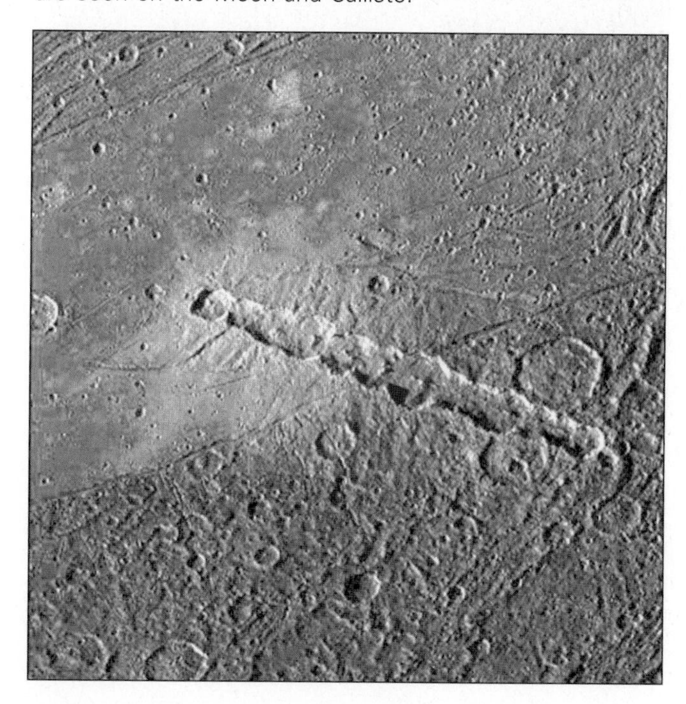

FIGURE 15.29
An Infrared Image of an Impact Fireball
At infrared wavelengths, the fireballs produced by some impacts of comet fragments were momentarily brighter than the disk of Jupiter. The glow of warm regions produced by earlier impacts can be seen to the right and above the fireball.

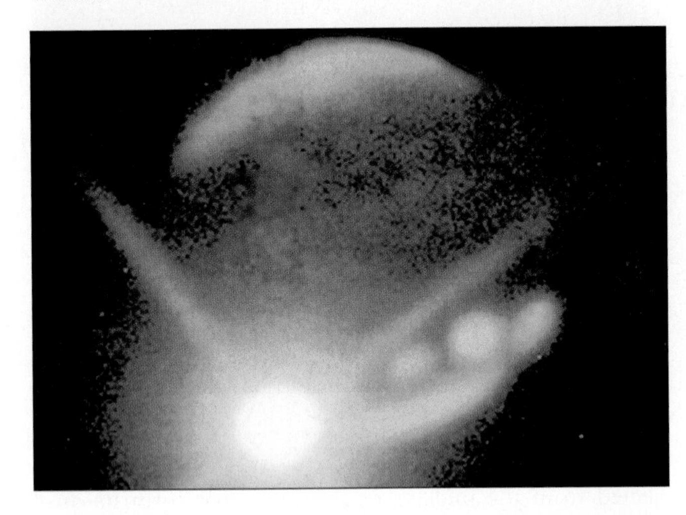

FIGURE 15.30
An Atmospheric Feature Produced by the Impact of a Comet Fragment on Jupiter
Debris from the impact of a fragment of Comet Shoemaker-Levy 9 formed a dark, multiringed feature high in the atmosphere of Jupiter. The feature was about twice as large as the Earth.

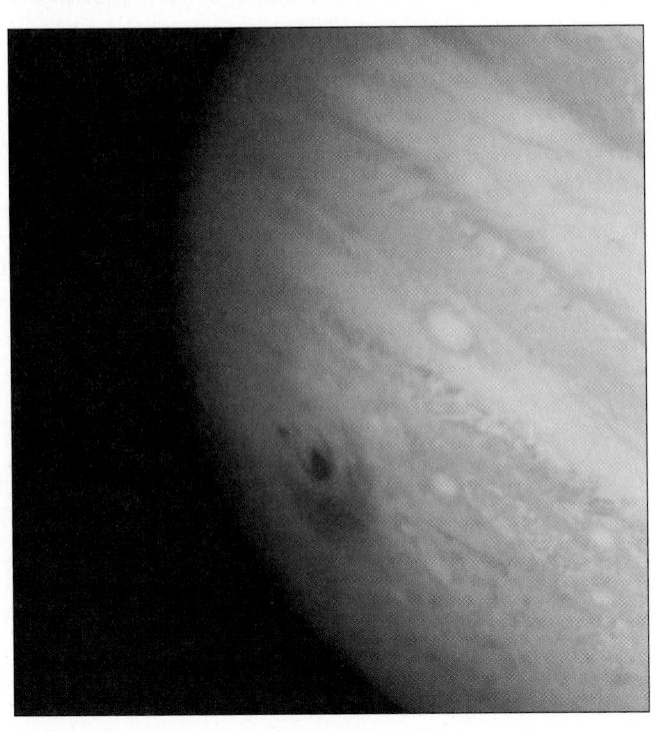

dust particles, or meteoroids, every year where the two orbits nearly intersect (see Figure 15.4). When this happens, a meteor shower occurs.

If the nucleus of the comet is also near the place where the two orbits cross, the meteor shower can be very intense. For example, the Draconid meteor shower occurs each October when the Earth passes close to the orbit of Comet Giacobini-Zinner. However, the Draconid shower varies greatly in intensity from year to year, depending on where the comet is situated in its 6.5-year orbit about the Sun. In 1946, for instance, the Earth crossed the orbit of Comet Giacobini-Zinner only 15 days after the comet had passed. The result was a spectacular meteor shower, in which 5000 meteors per hour were seen. In other years, when Comet Giacobini-Zinner was far away from the crossing point, only a few Draconid meteors per hour were seen.

Most meteor showers, however, aren't associated with any known comet. In those cases, it is thought that the meteoroid swarm was produced by a comet, but that the comet has used up all its ice and become invisible. It may be that all that is left is the swarm of dust particles. This may have happened in the case of Comet Biela, which was seen at several perihelion passages between 1826 and 1852. It was never seen again, but when the Earth crossed its orbit four times between 1872 and 1899, spectacular meteor showers (with rates as high as 50,000 per hour) were seen.

Only a small fraction of the dust particles produced by comet nuclei are destroyed by hitting the Earth or another planet. The rest eventually spread out from the comet orbit into the region near the ecliptic. Figure 15.31 shows that sunlight reflected from these dust particles produces a glow, called the **zodiacal light,** which can be seen above the horizon when the sky is dark after sunset or before sunrise.

FIGURE 15.31
The Zodiacal Light
Sunlight reflected from dust particles orbiting near the ecliptic can be seen on dark nights after sunset or before sunrise. The glow is usually very faint.

Asteroids from Comets? It's possible that something in addition to dust particles is left behind when a comet uses up all of its ice. Some comets may have solid cores of rocky material that become asteroids when the comet's icy material is gone. The Centaur Chiron and the asteroid Hidalgo, for example, have cometlike orbits. Chiron, which sometimes develops a gaseous coma, may be an extinct comet nucleus. On the other hand, Chiron is so far from the Sun (13.7 AU) that the evaporation of its ices is negligible. Therefore, Chiron may be mostly ice, so it might be better to think of Chiron as an inactive comet rather than an extinct comet nucleus. Three bodies in the asteroid belt have been found to have tails and may be comets rather than asteroids. Another possibility is that these asteroids are gradually releasing gas and dust derived from a layer of ice buried beneath a few meters of regolith. Calculations have shown that such an icy layer could survive for billions of years.

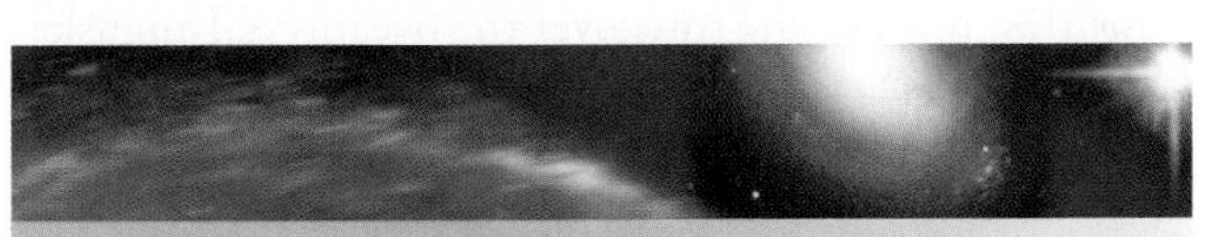

Repeated passages near the Sun eventually erode all of the icy material in the nucleus of a comet. When the ice is gone, a swarm of dust particles is left behind. These form the meteoroid swarms responsible for meteor showers. It is possible that some comet nuclei have rocky cores that become asteroids when all the ice is gone.

15.5 WHERE DID COMETS FORM?

Could the comets have originated beyond the solar system, elsewhere in the galaxy? This possibility can be ruled out by the orbits of the new comets. If comets came from interstellar space they would move on hyperbolic orbits and have speeds in excess of escape velocity. In fact, no comet has ever been found to have a hyperbolic orbit when it entered the planetary system. Thus, the comets formed within the solar system, probably at the same time that the Sun and the planets formed. The question, then, is where within the solar system did they form?

It is unlikely that comets formed in the Oort cloud itself. At such enormous distances from the Sun, the density of matter has probably never been high enough for bodies 1 km in diameter or larger to have accumulated. At the present time, it seems most likely that the comets formed near the orbit of Neptune. This hypothesis is supported by the similarity in composition between the comets and the icy satellites of the outer planets. Two different things could happen to comets orbiting near Neptune. Interactions with Neptune could reduce the size of the comet's orbit, bringing it into the inner solar system as a short-period comet. Alternatively, repeated interactions with Neptune and then Uranus could increase the speed of the comet, ejecting it from the solar system or sending it to the Oort cloud. Thus, icy bodies in the Kuiper Belt near the edge of the planetary system could both resupply the Oort cloud and provide some of the short-period comets.

Wherever they formed, comets should have a distribution of sizes. Most of the comet nuclei should be only a few kilometers in diameter, but there should also be some comets much larger than that. The recently discovered objects beyond Neptune's orbit are some of the largest comets in the Kuiper Belt. Chiron, also, may be a large comet nucleus. Perhaps even Pluto is a giant comet nucleus.

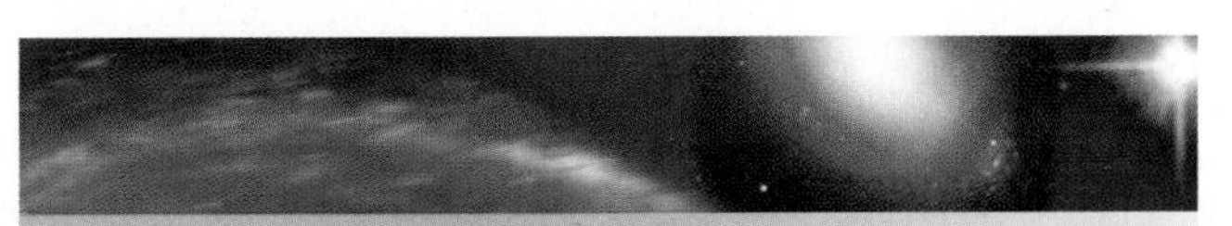

The comets probably formed at the outer edge of the planetary system, near Neptune. Many of them may still exist there as the Kuiper Belt. There may be some very large comet nuclei, of which Chiron and Pluto may be examples.

15.6 COLLISIONS WITH EARTH

Given the presence of Earth-crossing asteroids and comet nuclei that cross the orbit of the Earth, it is certain that from time to time the Earth is struck by bodies several kilometers or more in diameter and moving at speeds as great as 70 km/s.

How Often Do Collisions Occur?

As a rule of thumb, a large meteoroid that strikes the Earth produces a crater with a diameter about 10 to 20 times larger than the meteoroid. With this information and our estimates of the numbers of Earth-crossing asteroids and comets of different sizes, we can estimate how often craters of different sizes should be formed on the Earth. The result is that a crater with a 10 km diameter should be produced, on average, about once every 100,000 years. A crater 50 km in diameter should be formed once every 5 million years and a crater 100 km in diameter once every 50 million years.

Record of Impacts

The Moon, Mars, Mercury, and other bodies preserve their impact scars for very long periods of time. On Earth, impact craters are destroyed or eroded relatively quickly. About two-thirds of all impact craters are formed on the ocean bottoms. Over the course of 100 million years, the ocean floor on which a crater lies is carried to a subduction zone by plate tectonics and disappears into the interior of the Earth. On the continents, erosion, the deposition of sediments, and glaciation smooth crater rims and fill in the crater itself.

Only very young craters such as 50,000-year-old Meteor Crater in Arizona (Figure 15.32) are easy to recognize. Much older impact craters, such as 200 million-year-old Manicouagan in Canada (Figure 15.33), can sometimes be identified in photographs from aircraft or from space.

FIGURE 15.32
Meteor Crater
This impact crater in central Arizona was formed about 50,000 years ago. It is about 1 km in diameter and 200 m deep.

Still other ancient craters are completely undetectable at the surface. They can only be discovered by studying the shattered, disturbed rock beneath the surface.

Consequences of an Impact

On the morning of June 30, 1908, a meteoroid 30 to 50 m across exploded in Earth's atmosphere about 6 km above the ground near the Tunguska River in Siberia. The energy of the explosion has been estimated to be about the same as a very large hydrogen bomb. Because the event occurred in a remote area, it appears that only one person was killed by the blast. However, perhaps 60 million trees were leveled over an area of about 2000 square km (a larger area than Washington, D.C.). At 60 km from the blast, people were knocked to the ground and windows were broken. People 500 km away heard loud explosions and saw a fiery cloud.

A meteoroid as large or larger than the one that caused the Tunguska event strikes the Earth every few centuries, most of the time over the oceans. A Tunguska-like impact near a population center would cause severe damage and loss of life. Yet we have reason to believe that the Earth is occasionally struck by far larger meteoroids that cause explosions of such power that the entire Earth is affectd.

The impact of an asteroid or comet 1 km in diameter or larger would have extremely serious local and global consequences. The impact would vaporize both the impacting body and a large volume of terrestrial rock. These hot gases would expand upward and outward, somewhat like a nuclear fireball, but rising as high as tens of kilometers into the atmosphere. Somewhat deeper rock would be melted rather than vaporized. The melted rock would be splashed outward on trajectories that would carry molten droplets

FIGURE 15.33
The Manicouagan Impact Structure in Quebec
The structure is 70 km in diameter and formed about 200 million years ago. The circular lake lies just within the original walls of the crater, which eroded away long ago. The pencil-shaped object in the upper left is the tail of the space shuttle *Columbia,* from which this picture was taken in 1983.

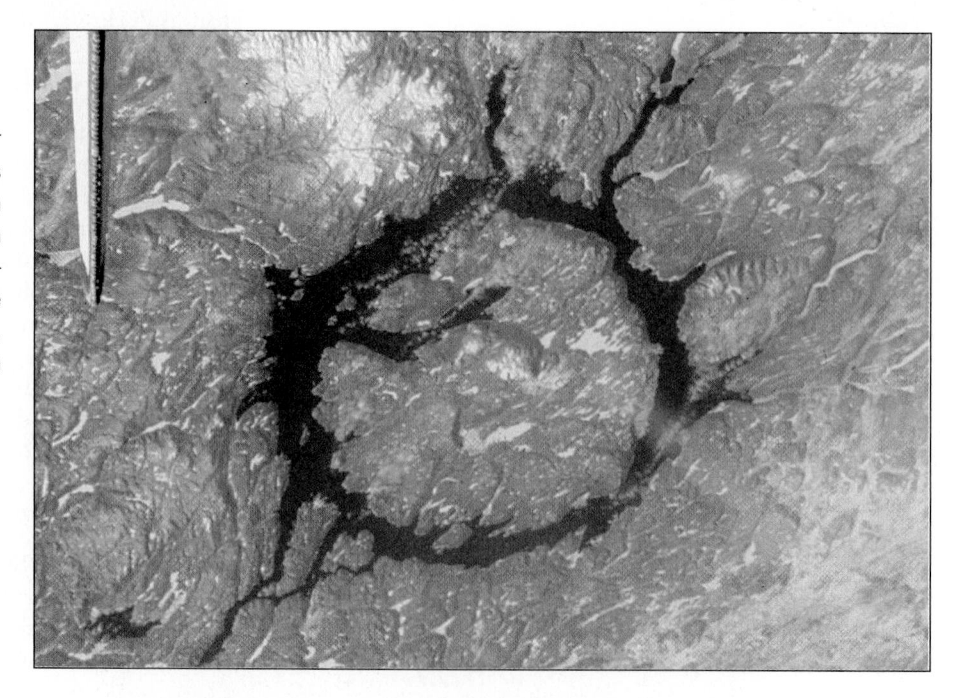

as far as several thousand kilometers from the impact site. Still deeper rock would be smashed and thrown outward as a blanket of ejecta. If the impact took place on the ocean floor, enormous waves would sweep across the oceans.

The flash of light and heat accompanying the fireball would ignite vegetation near the impact, perhaps triggering vast forest fires. The shock wave from the impact would be capable of killing anyone within perhaps a thousand kilometers of the impact site and injuring those considerably farther away. The ejected molten and solid material, of course, would obliterate everything close to the impact.

The relative importance of several possible global consequences is still uncertain. After the hot gas from the vaporized material reached the upper atmosphere, it would cool to form dust particles that would remain suspended in the atmosphere for months or years. The dust would obscure sunlight, plunging the Earth into a period of darkness and cold. Just how much dust would be formed, how long it would last, and how serious the consequences for the climate would be are uncertain. Thick clouds of smoke from forest fires could also contribute to the darkening.

Another serious consequence would be that large quantities of atmospheric nitrogen would burn within the fireball, yielding the same kinds of nitrogen oxides that are partly responsible for acid rain. Following an impact, however, the amount of nitrogen oxides produced would far exceed those formed by industrial pollution. A possible result would be global rain with the acidity of strong laboratory acids. Such rain would have devastating consequences for life on land, in lakes, and in shallow ocean waters. Still another possibility is that ejected material, blasted to heights above the atmosphere, would heat the atmosphere upon reentry, producing a thermal pulse a few hours or days in duration that would roast any unsheltered plants and animals.

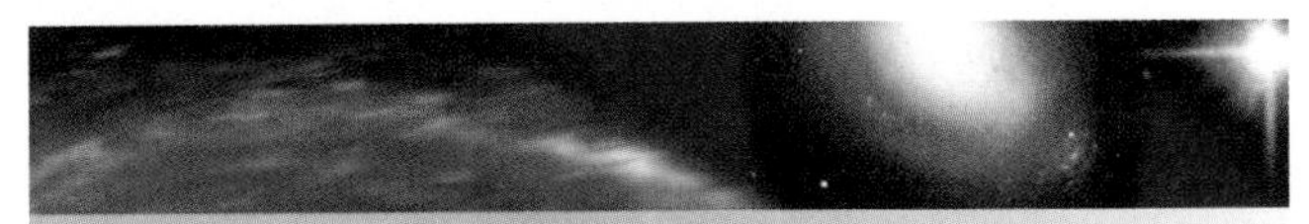

The impact of a large meteoroid on the Earth would have many severe consequences, both local and global. Some of the possible global consequences include prolonged darkness, very acidic rain, and temporary heating of the atmosphere. Molten and solid material ejected from the impact site would be propelled to great distances.

Has This Ever Happened?

We have ample evidence that large bodies have struck the Earth. Is there any reason to think that such impacts have produced global disasters? In the last 30 years or so, evidence has accumulated that answers this question with a reasonably definite yes. The evidence shows that meteoroid (or comet) impacts are sometimes associated with dramatic changes in the nature of plant and animal life on Earth.

The first indication that impacts play a role in the evolution of life on the Earth came when Luis and Walter Alvarez examined a thin layer of clay, shown in Figure 15.34, that lies at the Cretaceous-Tertiary boundary. Cretaceous and Tertiary rocks, like other rocks, are identified by the life forms that existed during the time when the rocks were deposited. The Cretaceous period ended 65 million years ago and was followed by the Tertiary period, in which the life forms were distinctly different. Boundaries between geological periods are, almost by definition, times of wholesale extinction of species. At the end of the Cretaceous period, about 60% of all species of plants and animals became extinct. The end of the Cretaceous period brought the extinction of the dinosaurs and all species of land animals

FIGURE 15.34
The Layer of Dark, Iridium-Rich Clay at the Cretaceous-Tertiary Boundary
The iridium in the clay may have resulted from the impact of an iridium-rich asteroid or comet with the Earth. Widespread extinction of life on Earth took place at the time that the layer was produced. The circle shown in the picture is a coin about as large as a quarter. The upper-left edge of the coin is touching the iridium-rich layer.

larger than goats. About half of all species of floating marine vegetation also died.

The Alvarezes and their colleagues discovered that the layer of clay at the Cretaceous-Tertiary boundary is about 30 times richer in the element iridium than are typical terrestrial rocks. This is significant because iridium is much more common in meteorites than it is in the Earth's crustal rocks. The Earth's supply of iridium is believed to have accompanied iron and other metals when they sank to form the Earth's core. The Alvarezes proposed that the iridium came from the vaporized material in a large meteoroid that struck the Earth. On impact, the iridium, the rest of the meteoroid, and some terrestrial rock were vaporized. They rose to the upper atmosphere, solidified to form dust, slowly settled, were washed from the atmosphere by rain, and were carried with the rainwater to form the sediments of the Cretaceous-Tertiary boundary. Shattered mineral grains and rock fragments in the boundary clay seem to be solid debris from the impact.

The Impact Site? Geologists have eagerly sought the site of the Cretaceous-Tertiary impact. One clue to its whereabouts is that although the shattered mineral grains are found around the world, they are most numerous in North America, suggesting that the impact may have occurred there. Currently, the best candidate is a crater centered near Chicxulub in northern Yucatán, Mexico. The location of the Chicxulub crater is shown in Figure 15.35. The Chicxulub crater has a diameter of 180 km, making it one of the largest craters on Earth, but it is buried beneath a kilometer of sedimentary rocks. The crater was formed 65 million years ago. Geological evidence from the region surrounding Chicxulub supports the idea that a violent impact occurred there. At the time of the impact, Yucatán lay beneath about 100 m of water in the Gulf of Mexico. It appears that the impact produced enormous waves. The waves swept over the coasts of the Gulf of Mexico, dragging rocks and vegetation back into the sea. This material settled to the bottom to form an unusual layer of sedimentary rock that has been found in several sites around the Gulf of Mexico. The disturbed sedimentary layers also include tiny spheres of glassy rock melted by the impact and hurled thousands of kilometers from the impact site. These glassy spheres are found as far away as Haiti.

Impacts and Extinctions We probably need to be cautious in concluding that large impacts lead to extinctions or that extinctions are caused by impacts. For one thing, the extinctions that took place at the end of the Cretaceous period were fairly gradual. The dinosaurs had been on the way out for millions of years. Perhaps the impact only accelerated the extinction of species that were doomed anyway. Also, although there is evidence that impacts may have triggered relatively minor extinctions several times in the past 100 million years, there have been other extinction events, some much more severe than the Cretaceous-Tertiary event, for which there isn't yet any compelling evidence of an impact.

Although the relationship between impacts and the extinction of species has not yet been completely worked out, it is becoming clear that the course of life on the Earth might have been quite different had it not been for the impacts of solar system debris.

FIGURE 15.35
The Chicxulub Impact Structure
The impact at Chicxulub took place 65 million years ago and produced a crater 180 km in diameter. The crater has been buried under a kilometer of sedimentary rock so that no trace of it can be seen at the surface. The impact at Chicxulub may have been responsible for the mass extinction that took place at the end of the Cretaceous period.

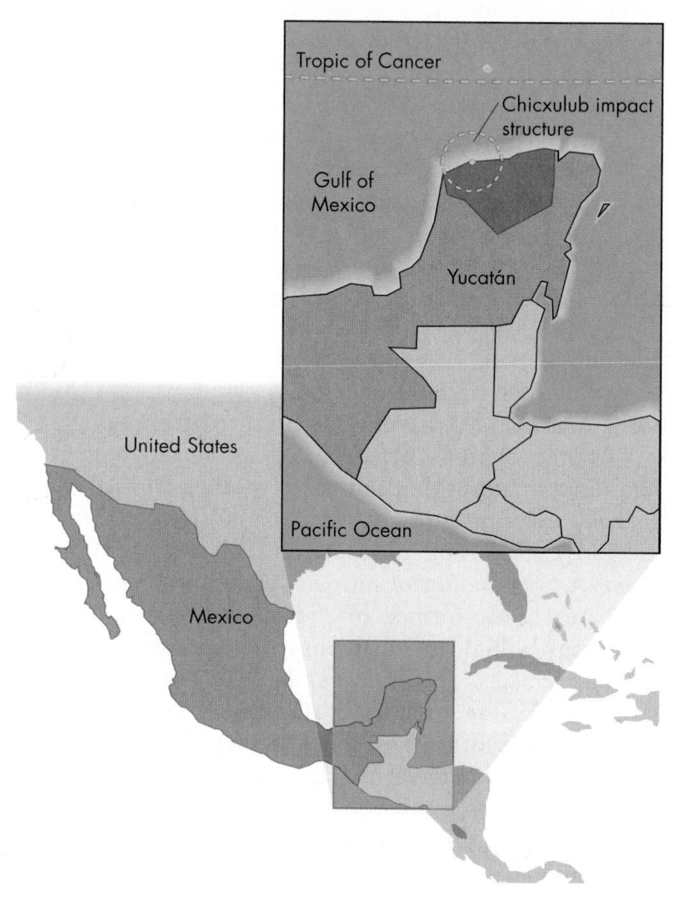

The discovery that rocks formed at the end of the Cretaceous era (a time of mass extinction) were enriched in the element iridium suggests that the Earth was struck by an asteroid at that time. Although the impact may have played a role in the Cretaceous extinctions, the nature of that role and whether extinctions and impacts are generally related is not yet clear.

Chapter Summary

- A meteor occurs when a meteoroid enters the Earth's atmosphere and vaporizes, heating itself and atmospheric gases so that they glow. Most meteoroids are no more than 1 cm in diameter. (Section 15.1)
- High meteor rates occur during meteor showers, when the Earth runs into a swarm of meteoroids. Showers take place on or close to the same date each year, when the Earth crosses the common orbit of the meteoroids. (15.1)
- The number of recovered meteorites has risen dramatically with the discovery that Antarctic ice fields collect and preserve meteorites for millions of years. (15.2)
- Meteorites are classed as stones, irons, and stony-irons. Stones resemble Earth rocks and are the most common meteorites. Carbonaceous chondrites are a type of stony meteorite and may represent unaltered material from early in the history of the solar system. Irons are alloys of iron and nickel and stony-irons are mixtures of stone and metal. (15.2)
- The radioactive elements in meteorites show that most of them solidified at almost the same time as the oldest Moon rocks, about 4.6 billion years ago. (15.2)
- Many meteorites appear to have been kept at high temperatures for a long period of time or to have cooled very slowly. To cool slowly, they must have been part of a body at least 100 km in diameter. The orbits of meteorites show that the bodies from which they came had orbits that carried them into the region between Mars and Jupiter. (15.2)
- Most of the known asteroids orbit in a belt located between the orbits of Mars and Jupiter. These asteroids are very widely spread out. (15.3)
- Many asteroids are not found in the asteroid belt. The Trojan asteroids, for example, either trail or precede Jupiter in its orbit around the Sun. (15.3)
- Many asteroids have orbits that carry them inside the orbit of the Earth. Within tens of millions of years, these asteroids are likely to be destroyed by striking the Earth. (15.3)
- The reflectance spectra of many asteroids resemble those of various kinds of meteorites. However, asteroids that resemble ordinary chondrites, the most common kind of meteorite, are very scarce. (15.3)
- The nucleus of a comet is a low-density chunk of ice and dust. Upon coming within 3 AU of the Sun, however, the nucleus is warmed enough by sunlight to release gas and dust. These flow away from the nucleus to produce the coma and the dust and plasma tails. (15.4)
- Comets with orbital periods shorter than 200 years are called short-period comets. Comets with orbital periods longer than 200 years are called long-period comets. (15.4)
- The Sun is surrounded by the Oort cloud, a swarm of comets extending as far as 100,000 AU from the Sun. There may be as many as a trillion comets in the Oort cloud. Passing stars alter the orbits of Oort cloud comets, causing some of them to enter the planetary system and become visible as new comets. Short-period comets are thought to come from the Kuiper Belt, a disk of icy bodies just beyond the orbit of Neptune. (15.4)
- A comet loses icy material each time it passes the Sun. Eventually, the ice is entirely eroded. The dust particles left behind form meteoroid swarms that produce meteor showers. Some comet nuclei may have rock cores that become asteroids once the surrounding ice is gone. (15.4)
- Comets may have formed in a region near the orbit of Neptune. The bodies recently found beyond the orbit of Neptune may be among the largest comets that still remain in that region. (15.5)
- If a large meteoroid or comet struck the Earth, there would be serious local and global consequences. The global consequences might include darkness for weeks or months, very acidic rain, and temporary heating of the atmosphere. (15.6)
- An excess of the element iridium, discovered in rocks formed at the end of the Cretaceous period 65 million years ago, suggests that an asteroid struck the Earth at that time. The consequences of the impact may have played a role in the Cretaceous extinctions. However, the way that the impact affected life on the Earth and the relationship between extinctions and impacts is not yet understood. (15.6)

Key Terms

achondrite *351*
Amor asteroid *357*
Apollo asteroid *357*
asteroid *353*
asteroid belt *354*
Aten asteroid *357*
carbonaceous chondrite *351*
chondrite *350*
chondrule *350*
coma *360*
comet *359*
cosmic ray *352*
cosmic ray exposure age *352*
C-type asteroid *358*
dust tail *361*
fireball *347*
iron meteorite *351*
long-period comet *362*
meteor *346*
meteorite *347*
meteoroid *346*
meteor shower *347*
micrometeorite *349*
minor planet *353*
M-type asteroid *358*
new comet *362*
nucleus *359*
Oort cloud *363*
plasma tail *361*
radiant *348*
reflectance spectra *358*
short-period comet *362*
solidification age *352*
stony-iron meteorite *352*
stony meteorite *350*
S-type asteroid *358*
terminal velocity *347*
Trojan asteroid *357*
V-type asteroid *358*
zodiacal light *366*

Conceptual Questions

1. Distinguish between a meteoroid, a meteor, and a meteorite.
2. Tell what happens to meteoroids with diameters of (a) 10 millionths of a meter, (b) 1 centimeter, (c) 1 meter, and (d) 10 meters beginning from the time they enter the Earth's atmosphere.
3. Explain why the meteors in a given shower appear to be diverging from a fixed point in the sky.
4. Why do meteor showers recur on the same date each year?
5. Why are carbonaceous chondrites especially important for the study of the early solar system?
6. What is the significance of the large crystals contained in iron meteorites?
7. What does the cosmic ray exposure age tell us about the history of a meteorite?
8. How do we know that the parent bodies of the meteorites were larger than about 100 km in diameter?
9. Ceres, the first asteroid to be discovered, lies between Mars and Jupiter. Why was Ceres unknown to ancient astronomers?
10. What is the relationship between asteroids and meteorites?
11. Describe what happens to the nucleus of a comet as it moves from 10 AU from the Sun to 0.5 AU from the Sun.
12. Where does the gas and the dust in the coma of a comet come from?
13. Why do the dust tail and the plasma tail of a comet point away from the Sun?
14. What evidence is there that the Sun is surrounded by a vast cloud of comets known as the Oort cloud?
15. What causes Oort cloud comets to enter the inner solar system and become visible as new comets?
16. Where do short-period comets come from?
17. What is the eventual fate of most short-period and long-period comets?
18. What is the relationship between comets and meteor showers?
19. What evidence do we have that comets are permanent solar system members rather than interlopers from other parts of the galaxy?
20. Suppose the Earth were struck by an asteroid 10 km in diameter. Describe the local and global consequences of the impact.
21. What evidence do we have that the extinction of species at the end of the Cretaceous era was associated with an asteroid or comet striking the Earth?

Problems

1. Most asteroids are located in the asteroid belt that lies between 2.1 and 3.3 AU from the Sun. What orbital periods correspond to the inside and outside of the asteroid belt?
2. What is the orbital period of one of the Trojan asteroids?
3. Icarus, an Apollo asteroid, has a perihelion distance of 0.19 AU and an aphelion distance of 1.97 AU. How often does it cross the orbit of the Earth?
4. An Oort cloud comet has an orbital period of 5 million years. What is its average distance from the Sun?

Group Activities

1. Travel with your group to a dark spot on a clear, moonless night. Divide the group into two or more subgroups and have each group sit facing a different direction. Each group should look about halfway from the horizon to the zenith and count the number of meteors each member of the group sees in 15 minutes. Average the number of meteors seen by each member of the group and multiply by four to get the meteor rate in number per hour. If possible, repeat the experiment during one of the meteor showers listed in Table 15.1. Compare the meteor rates of the two observations.
2. Go to the website http://cfa-www.harvard.edu/iau/lists/MPNames.html where you will find a list of asteroid names. Have each member of your group use the list to try to make up a sentence using only asteroid names. (For example: Rockefellia Neva Edda McDonalda Hamburga.) After you have an entry from each member of the group, have the group decide which is the best entry.

For More Information

Visit the text website at **www.mhhe.com/fix** for chapter quizzes, interactive learning exercises, and other study tools.

16

Vhen I heard the learn'd
ıstronomer,
Vhen the proofs, the figures,
vere ranged in columns
ɔefore me,
Vhen I was shown the charts
ınd diagrams, to add,
livide, and measure them
Vhen I sitting heard
he astronomer where
ıe lectured with
nuch applause in the
ecture-room,
How soon unaccountable
became tired and sick,
ill rising and gliding out I wander'd off
ɔy myself,
n the mystical moist night-air, and from
ime to time,
.ook'd up in perfect silence at the stars.

Valt Whitman
Vhen I Heard the Learn'd Astronomer

An Icy Planet Orbiting a Dim Red Star

www.mhhe.com/fix

Basic Properties of Stars

As we describe the properties of stars in this chapter, it is very easy to make them seem ordinary and commonplace. Nothing could be further from the truth. Stars are marvelous and stupendous. There are no ordinary stars.

Astronomers would like to get a closer look at stars other than the Sun, but that hardly seems possible in the foreseeable future. Even the nearest stars are much too distant to be explored by today's spacecraft or those we can realistically envision for the future. It is difficult to comprehend, even with the help of analogies, how far it is from the Earth to the stars. Suppose we compare the size of the region of space we have explored so far to the distance to the nearest star. The *Voyager* 1 space probe, traveling since the early 1970s, is now more than 100 AU, or 13 light hours, from the Sun. If that 13 light hours were reduced to 1 cm (about the size of your small fingernail), then the nearest known star, Proxima Centauri, would be 30 m away (about 100 ft).

Another way to try to understand the distances from the Earth to the stars is to compare their distances with their sizes. Suppose we use salt grains to make a scale model of our galaxy. An average grain of salt is about 0.5 mm ($^1/_{50}$ inch) in diameter. There are about as many grains of salt in a cube 2 m across as there are stars in our galaxy. If we let each salt grain represent a star like the Sun, how far apart would neighboring salt grains have to be to model the galaxy? The answer is about 15 km, roughly the size of Washington, D.C. Despite the sizes of the stars, the distances between them are so large that the universe is nearly empty. Thus, relatively crowded places like our solar system are the exception rather than the rule.

Questions to Explore

- How can we measure the distances to the stars?
- How do we know the Sun and stars are in motion with respect to each other?
- How can we describe a star's brightness?
- How do atoms in stars produce emission and absorption lines?
- What can we learn by combining information about the temperatures and luminosities of the stars?
- How can we determine the masses of stars?

Given the almost unimaginable distances to the stars, you might think we would have little chance of ever learning much about them. However, astronomers have been able to learn enough about stars that they are among the best-understood astronomical objects. We have firm ideas about how they form, their structure, how they evolve, and how they die. There are still numerous unanswered questions about stars, but many of their most important characteristics are known. This chapter is about the basic properties of stars and how astronomers have learned them. The properties described in this chapter form the basic set of facts that theories of stellar evolution must explain.

16.1 STAR NAMES

To organize the fundamental data about a star, we must agree which star we are talking about. In other words, we need a name for the star. It turns out, however, that some stars have many names—some quite new, but some very ancient. Star-naming was an activity common to many cultures. Each had its own familiar names for the brighter stars and constellations. When astronomers refer to stars by their familiar names, they usually use Greek or Latin names adopted in classical times or names given by Islamic astronomers during the period from A.D. 500 to 1000. The name *Sirius,* for example, is taken from the Latin word for "brilliant one," whereas *Altair* is derived from Arabic for "flying eagle." Because "al" is a common Arabic article, the Arabic origin for the name of a star often can be seen in the initial two letters of its name, as in Altair, Aldebaran, and Alphard.

A more systematic way of naming the brighter stars pairs a Greek letter with the name of the constellation in which the star is located. Stars in a constellation are designated in roughly descending order of brightness with α brighter than β, β brighter than γ, et cetera. Thus Sirius, the brightest star in the constellation Canis Major, is also α Canis Majoris. Stars are also named according to the catalogs in which they are listed. The designation of Sirius in the *Catalog of Bright Stars,* for example, is BS 2491, whereas it is IRAS 06429-1639 in the *Infrared Astronomy Satellite Catalog.* The existence of numerous catalogs of stars has led to a sometimes confusing plethora of names for a given star.

16.2 THE DISTANCES OF STARS

It is of fundamental importance to know the distance of a star from the Earth because distance is needed to find other important information about it, such as its intrinsic brightness and mass. Despite repeated attempts to determine stellar distances, however, the first reliable measurement was not made until the early nineteenth century.

Pretelescopic Estimates

Early calculations of the distances to stars were based on the nesting of the celestial spheres in the geocentric (Ptolemaic) model of the solar system (Chapter 3, Section 4). In the ninth century, for example, Al Fargani calculated that the region within which Mars moved extended to 8867 Earth radii. The region within which Saturn moved extended outward to 20,110 Earth radii, or about 120 million km. The stars, presumably, lay just beyond the region of Saturn. This estimate, which was actually less than the distance between the Earth and the Sun, was too small by a factor of several hundred thousand. Nevertheless, this calculation (and others like it) demonstrated that the universe of stars was immense compared with the size of the Earth.

Later Estimates

The true scale of stellar distances was first found through an ingenious method devised by James Gregory in 1668 and later used by Isaac Newton. They assumed that all stars, the Sun included, were equally luminous. If this were true, then the apparent brightnesses of stars would depend only on their distances (a nearby star would be brighter than a distant star). If the ratio of the brightness of the Sun to the brightness of a star could be measured, then their relative distances could be found as well. The apparent brightness of the Sun, however, is so much greater than that of any other star that it is difficult to make a direct comparison. Gregory proposed using a planet for comparison instead, because the brightness of the planet relative to the Sun can be calculated from its size, its distance from the Earth, and its distance from the Sun. Thus, combining the two comparisons, the relative distances of the star and Sun could be found. Using this method, Newton calculated a distance for Sirius of about 1 million AU, about twice the actual distance. Newton's calculation was in error partly because a basic assumption of this method is incorrect—that all stars are equally luminous. As a result, the method can provide only rough estimates of stellar distances.

Parallax

As early as the time of Aristotle, people realized that if the Earth orbited the Sun, the stars should change their positions as we view them from different locations. (When we refer to the "position" of a star, we usually mean the direction, relative to the other stars, in which we see it.) This concept, called stellar parallax, is introduced in Chapter 4, Section 4, and is illustrated in Figure 16.1. The nearest stars should show the largest changes of position, so distance is correlated with parallax. Many observers, including Copernicus and Tycho, looked for parallax in the pretelescopic era. Tycho's inability

FIGURE 16.1
The Parallax of a Nearby Star
The directions of the star when viewed from opposite sides of the Earth's orbit in December and June are different. The star appears to shift back and forth with respect to the background of much more distant stars. A comparison of the sky in the direction of the star at these two times of year shows that the star has shifted slightly with respect to the background stars.

ANIMATION *Parallax*

INTERACTIVE *Stellar parallax*

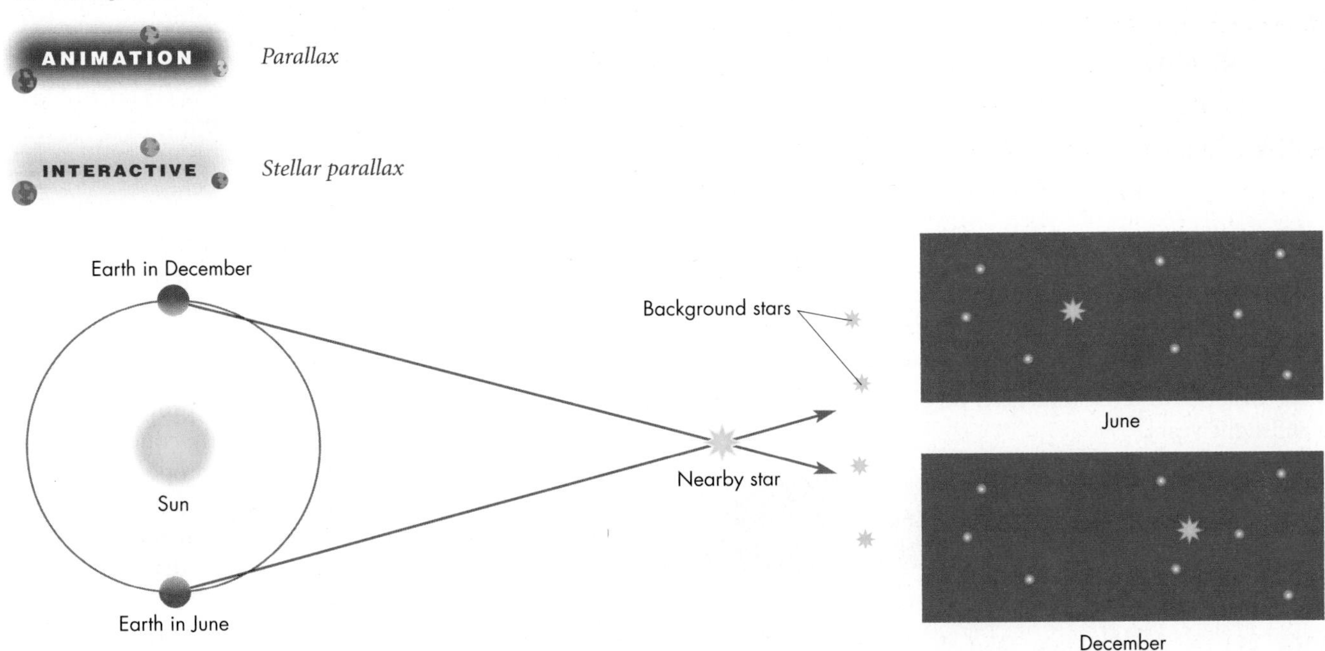

to detect parallaxes of 1 minute of arc or larger implied that the stars were at least many thousands of AU away.

After the invention of the telescope it became possible to measure angles much smaller than 1 minute of arc. In the early 1700s James Bradley attempted to measure the parallax of γ Draconis by making very careful measurements of its altitude when it crossed the meridian. His measurements of its annual variation in altitude led to his discovery of the aberration of starlight (see Chapter 8, Section 1). He failed to detect the parallax of γ Draconis, however, because he was unable to measure angles smaller than 1 second of arc.

Following Bradley's work, it became obvious to astronomers that to detect parallaxes they would need to measure the positions of stars to an accuracy much better than a second of arc. To do this, astronomers measured the position of a nearby star relative to more distant stars that lie in nearly the same direction. The more distant stars have extremely small parallaxes, so they form a constant backdrop against which the small shifts of the nearby star can be seen. Combining this method with the use of improved measuring instruments resulted in the almost simultaneous detections of stellar parallax by Friedrich Bessel, Wilhelm Struve, and Thomas Henderson in 1838. Bessel, the first to succeed, found that the parallax of 61 Cygni was 0.30″. (Recall that the symbol ″ is used for second of arc.)

The largest known stellar parallax is 0.75″ for α Centauri, a star for which Henderson made one of the first parallax measurements. This places α Centauri at a distance of 275,000 AU. The number of stars for which we have accurate parallaxes underwent a dramatic increase as a result of measurements made by the *Hipparcos* satellite. *Hipparcos,* which operated in Earth orbit between 1989 and 1993, was designed to measure stellar positions to an accuracy of about 0.001″. This made it possible to determine parallaxes to an accuracy about ten times better than ever before. Before *Hipparcos,* astronomers had found distances to an accuracy of 1% for only a few dozen stars. After *Hipparcos,* we have distances accurate to 1% for over 400 stars. Distances accurate to 5% are available for 10,000 stars. We now have reliable distances for most stars within about 200 pc of the Sun, including several important kinds of stars for which no reliable parallaxes existed before *Hipparcos*. *Hipparcos* also made highly accurate measurements of the motions and brightness of the stars.

Several units of distance have been used to describe the distances of the stars. One of the most widely used is the **light year** (9.46×10^{15} m), the distance light travels in a year. Another is the **parsec** (3.09×10^{16} m), the distance at which a star has a parallax of 1 second of arc. Because the parallax of a star is inversely proportional to its distance, the distance of a star d, given in parsecs,

Parallax and Distance

Equation 16.1 can be used to calculate the distance of a star from its parallax or, if rearranged, the parallax of a star from its distance. Suppose there were a star with a parallax of 0.2 second of arc. For $p = 0.2''$, Equation 16.1 is

$$d = 1/(0.2'') = 5 \text{ pc}$$

Alternatively, Equation 16.1 can be solved for parallax, giving

$$p\,('') = 1/d \text{ (pc)}$$

For example, if a star had a distance of 20 pc, its parallax would be

$$p = 1/(20 \text{ pc}) = 0.05''$$

and its parallax, p, given in seconds of arc, are related by Equation 16.1:

$$d \text{ (parsec)} = 1/p('') \qquad (16.1)$$

A parsec (abbreviated pc) is equal to 3.26 light years, so the distance to α Centauri can be given as either 1.3 pc or 4.3 light years.

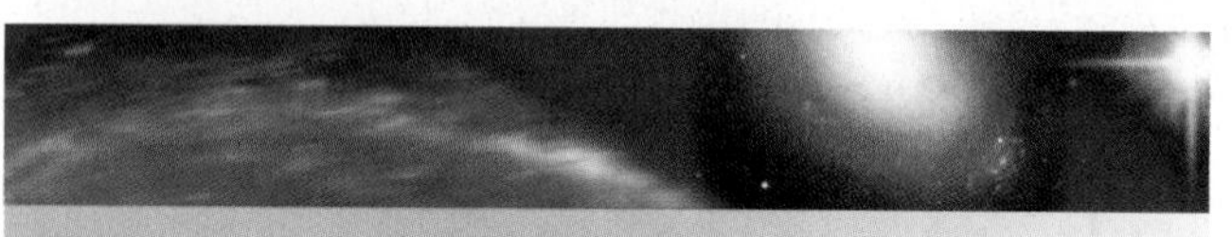

The distance to a star can be found by measuring its parallax, the apparent change in the direction of the star that results from viewing it from different places on the Earth's orbit around the Sun. All stellar parallaxes are smaller than 1 second of arc. The distance to a star, measured in parsecs, is the reciprocal of its parallax, measured in seconds of arc.

16.3 THE MOTIONS OF THE STARS

At about the same time astronomers were able to determine the distances to stars, they also discovered that the stars change their positions as time passes. This motion is too slow to be detected in the course of a year or two, so its detection depended on comparing current positions of stars with those given in older star catalogs.

Proper Motions

The stars are often referred to as "fixed" to distinguish them from the "wanderers," such as the Moon, the Sun, and the planets. This is not really correct, however, because the positions of the stars are not fixed. The positions of stars change, but they change much more slowly than those of the Sun, the Moon, and the planets. The person who discovered the position changes of the stars, called **proper motions,** was Edmund Halley. In 1718 Halley compared recent observations of the positions of stars with the positions listed by Hipparchus and Timocharis 2000 years earlier. He found that the stars Aldebaran, Sirius, and Arcturus were located 0.5° away from their positions in antiquity. Halley reasoned that because Aldebaran, Sirius, and Arcturus were among the brightest stars in the sky, they were probably among the nearest as well. If all the stars moved through space at about the same speed, then the nearest stars would change position most rapidly and would have the most conspicuous proper motions, as shown in Figure 16.2. Halley correctly concluded that many or most of the stars would show proper motions if they were observed carefully enough.

Modern observations (pairs of measurements, separated in time by years or decades) have determined the proper motions of a great number of stars, often with an accuracy of a few thousandths of a second of arc per year. Proper motions of comparable accuracy were obtained by *Hipparcos* for over 100,000 stars. The star with the largest proper motion is Barnard's Star (the second nearest known star), which has a proper motion of 10.25″ per year. Even though this is the largest proper motion, it corresponds to only 1° every 350 years. A few hundred stars have proper motions as large as 1″ per year, but most proper motions are less than this. Although the proper motions of the stars can be detected through careful measurements, these motions are extremely slow. The shapes of the constellations have looked essentially the same since the dawn of human history and will continue to look the same for many thousands of years.

The Motion of the Sun

By the time of Halley's discovery of proper motions, astronomers had considered the possibility that the Sun and the Earth move through space. If this were true, the proper motions of the stars would appear to diverge from the point toward which the Sun and the Earth were moving and to converge toward the point behind the Earth and the Sun. As astronomer Tobias Mayer put it in 1760, it would be "just as when you are walking through a wood the trees which are in front of you seem to be separated and those which are behind you seem to be joined together." The trees (or stars) that would appear to change direction most rapidly would be those that the observer was just passing. The motion of the Sun with respect to the nearby stars is referred to as the **solar motion** and is illustrated in Figure 16.3. The direction toward which the

FIGURE 16.2
The Proper Motions of Nearby and Distant Stars
A, The arrows represent the distance each star moves through space in a year. Even though the two distances are the same, the nearer star moves through a larger angle, so it shows a larger proper motion. **B** and **C,** These two photographs of the region near Barnard's star, which has the largest known proper motion, were taken 22 years apart. Arrows indicating Barnard's star show that it moved nearly 4 minutes of arc (one-eighth the angular diameter of the Moon) during that time.

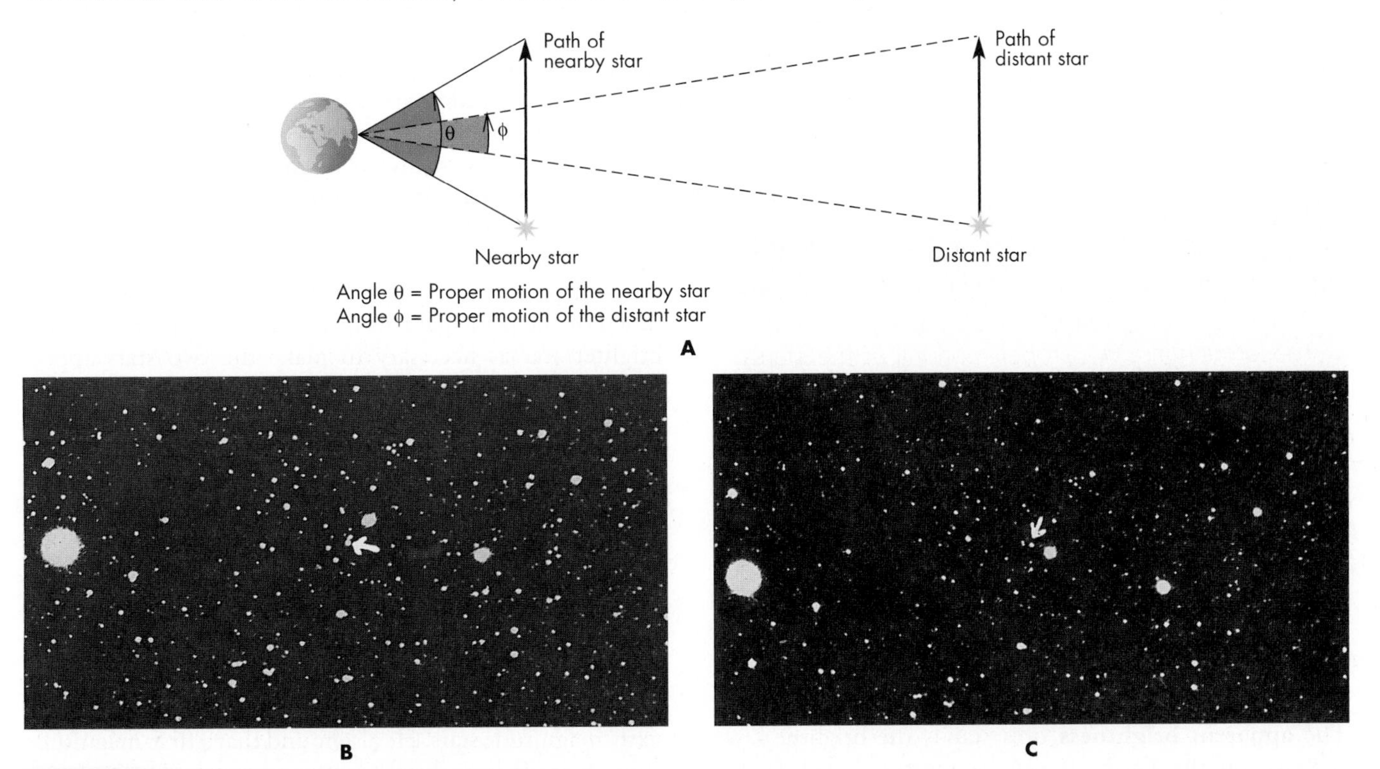

FIGURE 16.3
Consequences of the Sun's Motion
A, The stars that the Sun is moving toward appear to diverge from the apex of the Sun's motion. **B,** The stars that the Sun is moving away from appear to converge toward the antapex of the Sun's motion.

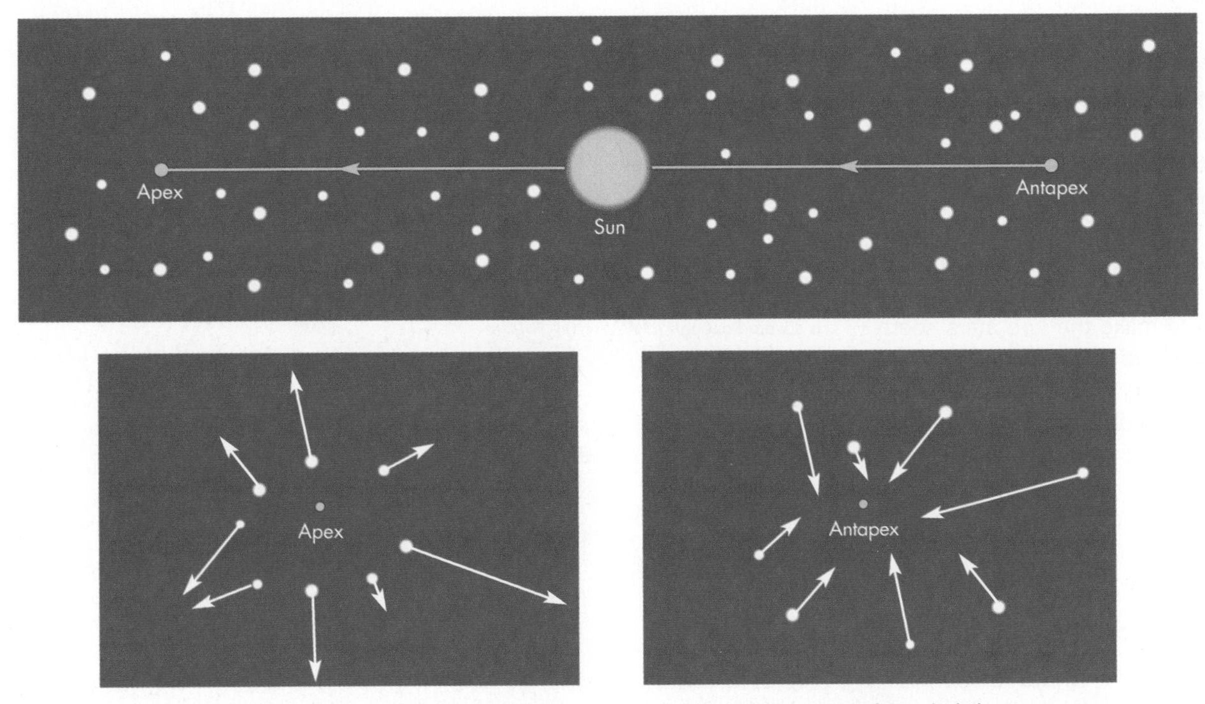

Sun is moving is called the **apex** of the Sun's motion. The direction from which it is moving is its **antapex.**

The first determination of the solar motion was made by William Herschel in 1783. He analyzed the proper motions of only seven stars (those that he felt were the most reliable) and discovered that the Sun was moving in the direction of the constellation Hercules. Modern determinations of the solar motion, which agree rather well with Herschel's original result, place the apex of the Sun's motion in Hercules. The antapex is in the constellation Columba, opposite Hercules on the celestial sphere.

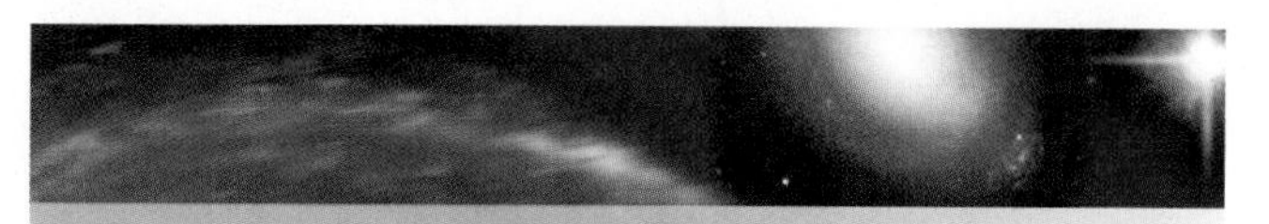

The positions of stars change gradually with time. These changes, the proper motions of the stars, are caused by their movement through space relative to the Sun. The motion of the Sun with respect to the other stars causes the stars toward which the Sun is moving to diverge and the stars away from which the Sun is moving to converge.

16.4 THE BRIGHTNESSES OF STARS

The **apparent brightness** of a star is the brightness we perceive on the Earth. It can be measured directly using telescopes and detectors as described in Chapter 6, Section 4.

Stellar Magnitudes

The first person to compile a systematic list of stellar brightnesses probably was Hipparchus. Ptolemy, whose catalog included estimates of the brightnesses of about 1000 stars, probably included Hipparchus's list along with additional stars he observed himself. Ptolemy grouped the stars into six **magnitude** groups according to their apparent brightnesses. He called the brightest stars firstmagnitude stars. The dimmest visible stars were called sixth-magnitude stars.

Measurements of stellar brightnesses continued to be only rough visual estimates until the early 1800s, when William Herschel began using two telescopes to compare the brightnesses of pairs of stars. As shown in Figure 16.4, he pointed each telescope at a star, then covered up as much of the principal lens of the telescope pointed at the brighter star as necessary to make the two stars appear equal in brightness. Recall from Chapter 6, Section 4, that the brightness of a star as viewed through a telescope depends on the area of the principal lens or mirror of the telescope. In Herschel's experiments, therefore, the ratio of the uncovered areas of the two lenses was equal to the ratio of the brightnesses of the two stars. That is, if star A appeared as bright as star B when only 10% of the lens used to observe star A was uncovered, then star A must be ten times as bright as star B.

Herschel discovered that Ptolemy's first-magnitude stars are, on average, about 100 times brighter than his sixth-magnitude stars. He also found that a first-magnitude star is as many times brighter than a second-magnitude star as a second-magnitude star is brighter than a third-magnitude star and so on. With this information, astronomers devised

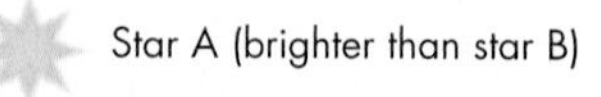
Star A (brighter than star B)

Star B
Lens of telescope pointed at star A is covered enough to make stars A and B appear equally bright:
Lens of telescope pointed at star B is uncovered:

FIGURE 16.4
Herschel's Method of Measuring the Relative Brightnesses of Stars
Herschel pointed two telescopes at two different stars (**A** and **B**), then covered enough of the lens of the telescope pointed at the brighter star (**A**) to make the two stars seem equally bright. The ratio of the uncovered areas of the two telescope lenses was proportional to the relative brightnesses of the two stars.

Magnitudes and Relative Brightnesses

Equation 16.2 describes the relationship between the difference in the magnitudes of two stars and the ratio of their brightnesses. Suppose stars 1 and 2 have magnitudes $m_1 = 7$ and $m_2 = 3$. The ratio of their brightnesses is, then,

$$\frac{b_1}{b_2} = 2.512^{m_2 - m_1} = 2.512^{3-7} = 2.512^{-4}$$

This can be solved using a calculator, giving

$$\frac{b_1}{b_2} = 0.025$$

This tells us that star 1 is only 2.5% as bright as star 2, or that star 2 is 40 times as bright as star 1.

a standard scale of magnitudes designed to match the brightnesses of Ptolemy's magnitude groups. In the standard magnitude scale, an increase of five magnitudes (such as for Ptolemy's first- and sixth-magnitude stars) corresponds to a decrease of 100 times in brightness. Each increase of one magnitude corresponds to a decrease of $\sqrt[5]{100} = 2.512$ times in brightness. Thus, a first-magnitude star is 2.512 times brighter than a second-magnitude star, a second-magnitude star is 2.512 times brighter than a third-magnitude star, and so on. The apparent brightnesses of stars, b_1 and b_2, and their magnitudes, m_1 and m_2, are related by

$$\frac{b_1}{b_2} = 2.512^{m_2 - m_1} \tag{16.2}$$

Objects brighter than the brightest stars are given negative magnitudes whereas faint objects have large positive magnitudes. This is somewhat analogous to golf, in which a low score is better than a high one. Thus, the Sun has a magnitude of -26.5, Jupiter -3, and the faintest detectable stars $+29$. More information about the magnitude scale is given in Tables 16.1 and 16.2.

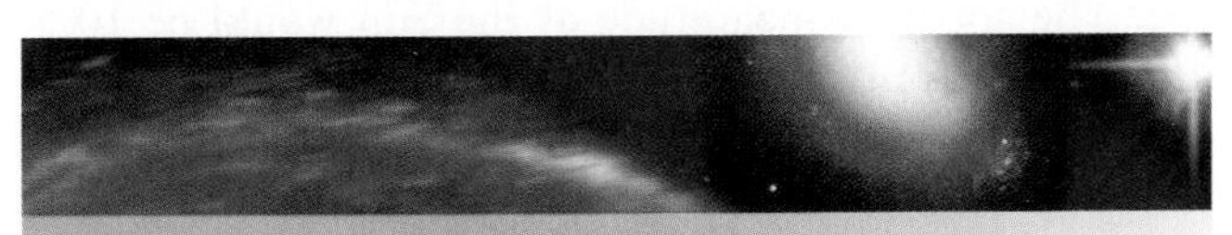

The magnitude system was devised to describe the apparent brightnesses of the stars. In the magnitude system, each additional magnitude corresponds to a decrease in brightness by a factor of 2.512.

Absolute Magnitude

The magnitude system devised by Ptolemy was based on the apparent brightnesses of stars and employs **apparent magnitude.** Apparent brightnesses are distance dependent, however, so they don't tell astronomers anything about how bright the stars really are. It is quite possible, in fact, that an intrinsically bright star would appear dimmer than an intrinsically faint star because the bright star was much farther from the Earth. To address this situation, astronomers have developed the concept of **absolute magnitude,** the apparent magnitude a star would have if it were at a distance of 10 pc.

Calculating the absolute magnitudes of stars compensates for their differences in distance. If all stars were at 10 pc, then those that are most luminous would have the brightest apparent magnitudes, whereas intrinsically faint stars would have the dimmest. This means that absolute magnitude is a measure of a star's intrinsic brightness. The intrinsic brightness of a star, summed over the entire electromagnetic spectrum, is called its **luminosity.** Luminosity and absolute magnitude both describe the

Table 16.1

Some Apparent Magnitudes

OBJECT	APPARENT MAGNITUDE
Sun	−26.5
Full Moon	−12.5
Venus (at brightest)	−4.0
Jupiter, Mars (at brightest)	−3.0
Sirius (brightest star)	−1.4
Polaris	2.0
Uranus	5.5
Naked eye limit (dark location)	6.5
Neptune	7.8
Limit with binoculars	10.0
Pluto	15.0
Limit for large telescope	24.0
Limit for *Hubble Space Telescope*	29.0

Luminosity and Absolute Magnitude

Equation 16.3 describes how the ratio of the luminosities of two stars is related to the difference in their absolute magnitudes. Suppose star 1 is 100 times as luminous as star 2. Then

$$\frac{L_1}{L_2} = 100$$

Using Equation 16.3, the magnitudes of stars 1 and 2 are related by

$$\begin{aligned} M_2 - M_1 &= 2.5 \log \frac{L_1}{L_2} \\ &= 2.5 \log (100) \\ &= 2.5 \times 2 \\ &= 5 \end{aligned}$$

Thus, the absolute magnitudes of star 2 is 5 magnitudes greater (fainter) than that of star 1.

Table 16.2

Magnitude and Brightness Ratio Difference

MAGNITUDE DIFFERENCE	BRIGHTNESS RATIO
0.0	1.0
0.1	1.1
0.2	1.2
0.3	1.3
0.4	1.45
0.5	1.6
0.7	1.9
1	2.5
2	6.3
3	16
4	40
5	100
7	630
10	10,000
15	1,000,000
20	100,000,000

intrinsic brightness of a star. The luminosities, L, and absolute magnitudes, M, of two stars are related to each other by

$$M_2 - M_1 = 2.5 \log \frac{L_1}{L_2} \qquad (16.3)$$

The stars that are actually more distant than 10 pc appear fainter than they would at 10 pc. Therefore their apparent magnitudes are higher (fainter) than their absolute magnitudes. Stars nearer than 10 pc appear to be brighter than they would at 10 pc, so their apparent magnitudes are lower (brighter) than their absolute magnitudes. For a given absolute magnitude (M), the apparent magnitude (m) of a star increases by five for every tenfold increase in distance. That is,

$$m = M + 5 \log \left(\frac{d}{10}\right) \qquad (16.4)$$

where d, the distance of the star, is measured in parsecs.

The Sun has an absolute magnitude (M) of 5. According to Equation 16.4, therefore, if the Sun were at a distance of 10 pc, it would have an apparent magnitude of 5 and be one of the faintest stars visible with the naked eye. The apparent magnitude of the Sun would be 10 at 100 pc, 15 at 1000 pc, and an undetectable 30 at 1 million pc (the distance to some of the nearer galaxies). The Sun's

Distance and Magnitude

Equation 16.4 describes how the apparent magnitude of a star with a given absolute magnitude varies with distance. Suppose a star had an absolute magnitude (M) of −5. (This star would be 10,000 times as luminous as the Sun.) What would be the apparent magnitude (m) of the star if it were at a distance (d) of 1000 pc? Equation 16.4 gives

$$\begin{aligned} m &= M + 5 \log \left(\frac{d}{10}\right) = -5 + 5 \log \left(\frac{1000}{10}\right) \\ &= -5 + 5 \log (100) \end{aligned}$$

Because the logarithm of 100 is 2, the apparent magnitude of the star is

$$m = -5 + (5 \times 2) = -5 + 10 = 5$$

Table 16.1 shows that such a star would be visible but dim if viewed without a telescope.

tremendous brilliance ($m = -26.5$), therefore, is due to its proximity to the Earth.

Just as apparent magnitude describes the apparent brightness of a star, absolute magnitude describes the luminosity, or intrinsic brightness, of a star. Apparent magnitude and absolute magnitude are related to each other through the distance to the star.

Luminosity Functions

Stellar luminosities and absolute magnitudes vary widely. The relative number of stars for each value of absolute magnitude is called a **luminosity function.** For the stars in the Sun's vicinity, the luminosity function is shown in Figure 16.5. Notice that most of the stars have absolute magnitudes greater than the Sun's. This means that most are less luminous than the Sun. In fact only about 1 in 20 stars is as bright as or brighter than the Sun. The brightest stars are extremely rare. They have absolute magnitudes brighter than −10 (not shown in Figure 16.5) and are more than 1 million times as bright as the Sun. Even though stars more luminous than the Sun are rare, they are so much more luminous than typical stars that they contribute much more to the brightness of the galaxy.

FIGURE 16.5
The Luminosity Function of the Nearby Stars
A plot of the relative number of stars versus absolute magnitude shows that fainter stars (large magnitudes) are much more numerous than brighter stars. The Sun, which has an absolute magnitude of about 5, is more luminous than the majority of stars. The brightest stars, with absolute magnitudes less than zero, are too rare to show up in the plot.

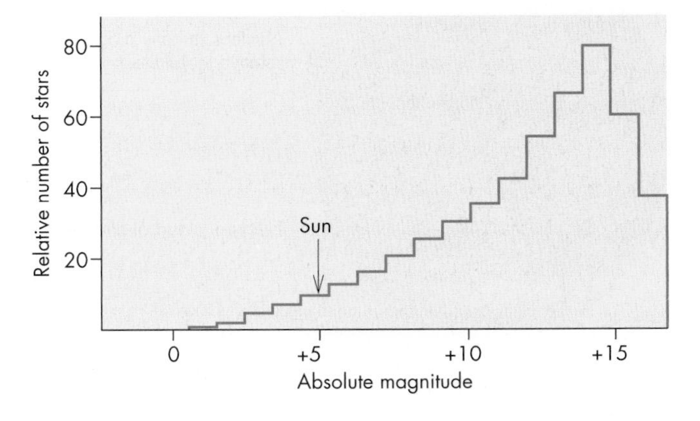

16.5 STELLAR SPECTRA

INTERACTIVE *Stellar spectroscopy*

Astronomers use stellar spectra to obtain a tremendous amount of information about stars, including their surface temperatures, chemical compositions, and the speeds with which they approach or recede from us. The techniques for recording a spectrum at the telescope are described in Chapter 6, Section 4.

The study of stellar spectra began when Isaac Newton passed sunlight through a prism and discovered that "white" sunlight actually consists of a mixture of every color. About a century and a half later, early in the nineteenth century, William Wollaston and Joseph Fraunhofer found that a more detailed examination of the solar spectrum revealed the presence of numerous dark bands or lines—that is, places in the spectrum where sunlight is unusually dim. Fraunhofer found that stars also have dark lines in their spectra. At about the same time, scientists began to study the spectra of terrestrial materials by producing glowing gases of the various elements and compounds. They found that the spectra of hot gases consist mainly of bright lines—narrow regions of the spectrum in which the emission of the glowing gas is concentrated. Furthermore, they soon found that the wavelengths of some of the dark lines in the Sun's spectrum were the same as some of the bright lines in the spectra of chemical elements. The conclusion was inescapable—the Sun contains those elements.

Atoms and Spectral Lines

Kirchhoff's Laws The conditions under which dark line, bright line, and continuous spectra can be produced were first described by Gustav Kirchhoff in the middle of the nineteenth century. The descriptions of these conditions, illustrated in Figure 16.6, are known as **Kirchhoff's laws.**

1. A hot solid, liquid, or dense gas produces a **continuous spectrum** in which emission appears at all wavelengths.
2. A thin gas, seen against a cooler background, produces a bright-line, or **emission-line,** spectrum. An emission-line spectrum consists of narrow, bright regions separated by dark regions.
3. A thin gas in front of a hotter source of continuous radiation produces a dark-line, or **absorption-line,** spectrum. An absorption-line spectrum looks the same

FIGURE 16.6
Kirchhoff's Three Kinds of Spectra
A hot solid, liquid, or dense gas emits a continuous spectrum. A dark-line (absorption) spectrum is obtained when the continuous spectrum is observed through a thin gas cooler than the object that produces the continuous spectrum. A bright-line (emission) spectrum is obtained when the thin gas is viewed against a cooler (darker) background. The bright lines in the emission spectrum correspond to the dark lines in the absorption spectrum of the same gas.

ANIMATION *Absorption*

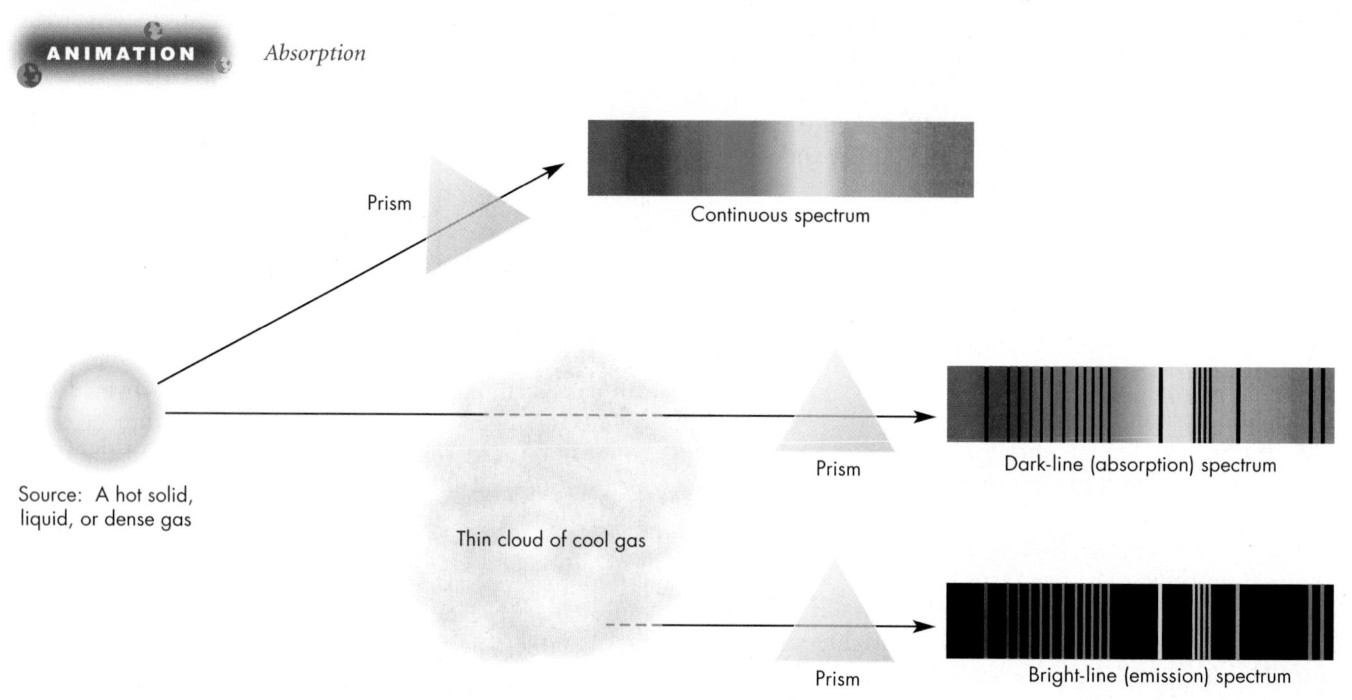

as a continuous spectrum except that there are narrow wavelength regions in which the radiation is diminished or absent. The spectra of the Sun and most stars are examples of absorption-line spectra. This shows that continuous radiation from inner stellar layers passes through relatively cool gas in the outermost layers of the star before it reaches us.

There are three things worth noting about the types of spectra shown in Figure 16.6. First, the wavelengths of the dark lines in the absorption-line spectrum of a particular gas are identical to the wavelengths of the bright lines in an emission-line spectrum of the same gas. Second, the same cloud of gas will produce an absorption-line spectrum if it is seen against a background hotter than the gas, but it will produce an emission-line spectrum if it is seen against a background cooler than the gas. Third, each chemical element or compound gives rise to its own unique pattern of dark or bright lines, so absorption-line or emission-line spectra can be used to identify particular elements or chemical compounds in a cloud of gas.

Atomic Structure The characteristics of bright-line and dark-line spectra are consequences of the structure of atoms. As shown in Figure 16.7, an atom consists of a

FIGURE 16.7
A Model of a Hydrogen Atom
The extremely tiny nucleus (which for hydrogen consists of a single proton) lies at the center of the atom. The electron that surrounds the nucleus may be thought of as a cloud of charge concentrated at a certain distance from the nucleus. Hydrogen is the simplest atom. Other atoms have more than a single electron cloud.

ANIMATION *A model of a hydrogen atom*

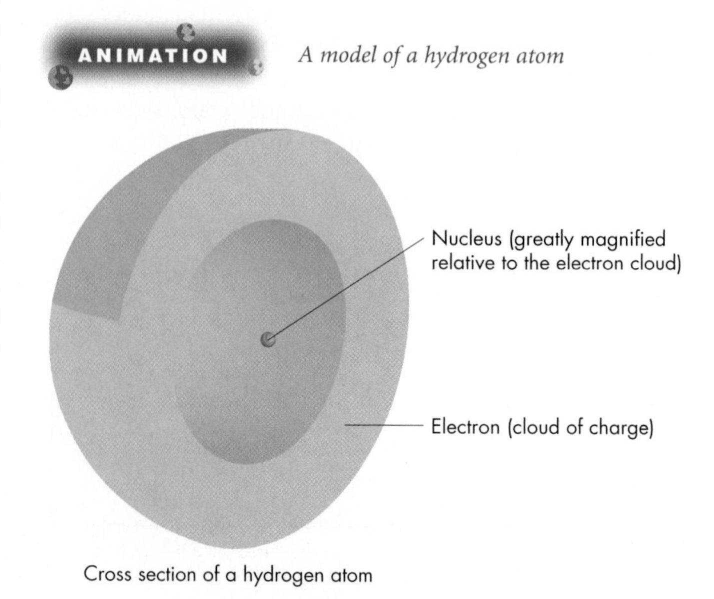

very tiny nucleus surrounded by one or more electrons. The nucleus, made up of protons (and, for most atoms, neutrons) contains most of the atom's mass and has a positive electric charge. The electrons, which have very little mass but occupy most of the atom's volume, have negative electric charge. The nucleus and the electrons are held together by electric force.

The electrons in an atom can exist only at specific distances from the nucleus. Each allowed electron distance has a different energy. For this reason, the allowed electron distances are often referred to as **energy levels.** The electron distances nearest the nucleus have lower energies than the ones farther away from the nucleus. Figure 16.8 shows the distances and energy levels for hydrogen, which is the simplest atom because it has only one electron. The innermost energy level in a hydrogen atom is only 5×10^{-11} m from the nucleus. The energy levels get closer together as their distances from the nucleus increase. The energies of the levels are very tiny. For example, the second level for hydrogen is only 1.6×10^{-18} J higher than the first energy level. This is 100 trillion times smaller than the energy needed to lift 1 cubic cm of water a distance of 1 cm. With increasing distance from the nucleus, the energy difference between successive levels becomes less and less until the energies of the outer levels gradually approach the amount of energy needed to completely separate the electron from the nucleus. Removing an electron from an atom is called **ionization.**

An electron can jump from one level to another. For one of its electrons to jump outward, however, an atom must absorb energy. For one of its electrons to jump inward, the atom must lose energy. There are several ways in which atoms may gain or lose energy. One is by colliding with other atoms, electrons, or ions. Another is by absorbing or emitting photons. Suppose an atom is bombarded with photons that have energies exactly equal to the energy difference between two of the atom's energy levels. By absorbing one of those photons, the atom's electron can jump from one energy level to a higher one. Photons with energies slightly higher or lower than the required energy can't be absorbed at all. If a continuous spectrum of photons is shined on a group of atoms of the same kind, the only photons they can absorb are those that correspond to energy differences between their energy levels. Removing the absorbed photons from a continuous spectrum produces a dark-line absorption spectrum.

If an electron is in any energy level other than the innermost, it can jump inward to a lower energy level. The level it jumps to may be the next lowest in energy or one much lower in energy. The point is, an electron can only *spontaneously* jump to levels that are *lower* in energy. Like a bowling ball on a staircase, an electron continues to jump downward in energy until it reaches the innermost energy level, called the **ground state.** With every jump, the electron emits a photon that has an energy equal to the difference in energy of the levels between which the electron jumps. Thus, the electron can emit only photons with a certain set of energies or wavelengths. The photons emitted in this way make a bright-line, or emission-line, spectrum.

The energies of the photons that can be emitted or absorbed by hydrogen atoms are shown in Figure 16.9. The upward-pointing arrows represent absorptions of energy and the downward-pointing arrows represent emissions of energy. The absorptions that begin in the lowest energy level (level 1) and extend to levels 2, 3, 4, and so on make up the **Lyman series.** These absorptions correspond to wavelengths in the ultraviolet part of the spectrum and are invisible to the human eye. The **Balmer series** comprises absorptions from level 2. They

FIGURE 16.8
Electron Distances and Energy Levels
A, The electron in a hydrogen atom can exist only at certain distances from the nucleus. **B,** Each of the possible electron distances corresponds to a different energy level.

Atomic emission and absorption

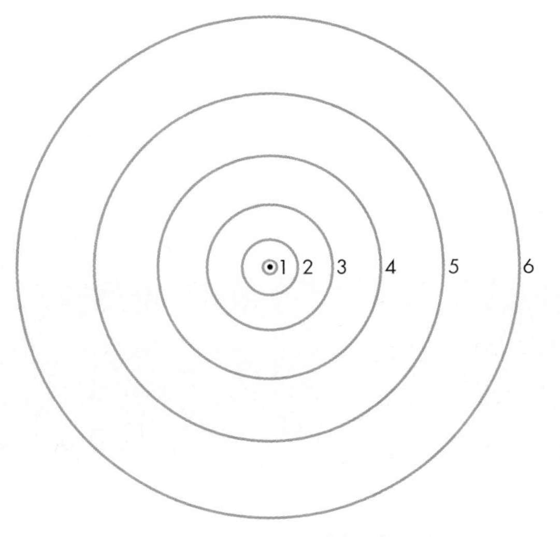

A The inner possible distances of the electron in a hydrogen atom

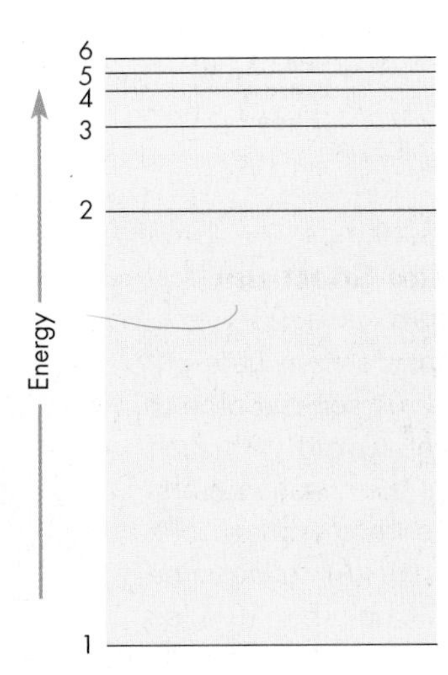

B The lowest energy levels for the hydrogen atom

have lower energies than the Lyman series and are the only hydrogen lines that fall in the visible part of the spectrum (Figure 16.10). Absorptions from levels 3 and 4 are the Paschen and Brackett series, respectively. They appear in the infrared part of the spectrum. Absorptions from still higher levels lie even farther in the infrared. Atoms of other elements have their own unique energy level structures and thus their own unique patterns of absorption and emission lines.

FIGURE 16.9
Possible Absorption and Emission Lines for the Hydrogen Atom
All absorptions that begin with the electron in the first (lowest) energy level or emissions that end with the electron in the first energy level are called Lyman lines (α, β, γ, etc.). Absorptions from or emissions to the second, third, and fourth energy levels are called Balmer, Paschen, and Brackett lines, respectively. Only the Balmer lines fall in the visible part of the spectrum.

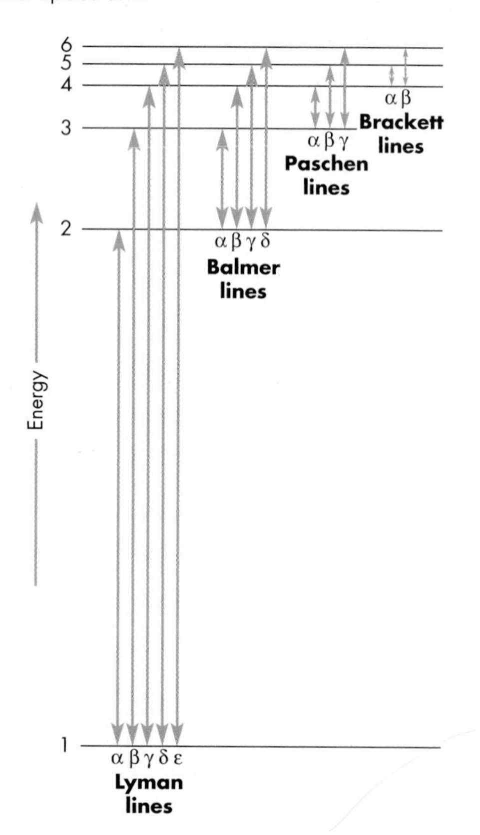

When an atom absorbs a photon, the energy carried by the photon causes an electron to jump to a higher energy level. When an atom emits a photon, one of its electrons jumps to a lower energy level. The absorption and emission of radiation by atoms produces dark-line and bright-line spectra. A given element or compound can only absorb or emit photons of certain energies. Thus the pattern of dark or bright lines in the spectrum of an element or compound is characteristic of that element.

Spectral Classification

In the early 1860s William Huggins studied the spectra of a few bright stars and was able to identify many of their lines as those of atoms of hydrogen, calcium, sodium, iron, and other terrestrial elements. Almost simultaneously, Angelo Secchi examined the spectra of several thousand stars. He discovered that many stars had spectra that did not closely resemble that of the Sun. Secchi also found that he could match most of the dark lines of other stars with those of known chemical elements.

Visual studies were replaced in the 1880s by photography, which allowed astronomers such as William Huggins and Henry Draper to make permanent records of stellar spectra. After Draper died, his widow gave his scientific instruments and a monetary contribution to Harvard Observatory to allow his work to continue. The

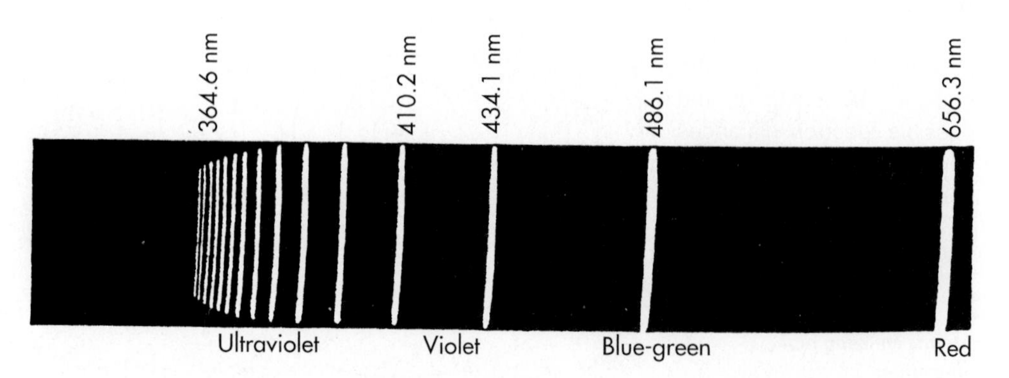

FIGURE 16.10
An Emission Spectrum of Hydrogen
All of the lines shown here are from the Balmer series. Some of the lines are marked with their wavelengths in nanometers. Astronomers sometimes use angstrom units (Å) to describe wavelength. One nm = 10 Å, so the Balmer line on the right side of the figure has a wavelength of 656.3 nm or 6563 Å.

Harvard program of stellar spectroscopy was begun by E. C. Pickering and continued by Antonia Maury and Annie J. Cannon for about 40 years. It eventually produced a catalog of spectra for about 225,000 stars. This catalog is known as the *Henry Draper Catalog* in honor of the first American to observe stellar spectra.

The Harvard astronomers devised a classification system in which stellar spectra with common features could be grouped. The groupings, or **spectral classes,** originally consisted of 15 types labeled A through O according to the strengths of the absorption lines of the Balmer series of hydrogen. Spectral class A contained the strongest Balmer lines and class O the weakest. Later, the number of spectral classes was reduced and some of the original letter designations were dropped when it was found that they represented only low-quality spectra, not separate classes.

The Harvard classification system was based entirely on the appearances of the spectra and didn't arrange the spectra according to any meaningful property of the stars producing the spectra. In 1920, however, M. N. Saha, an Indian physicist, showed that the temperature of the outer layers of a star was primarily responsible for the appearance of its spectrum. Soon afterward, Cecilia Payne, an American astronomer, found that stars of different spectral classes have essentially the same chemical composition. Cannon then reorganized the classification using the letters O, B, A, F, G, K, and M in order of decreasing temperature. Generations of astronomy students have memorized the sequence of spectral classes using the mnemonic "*O*h *B*e *A* *F*ine *G*uy (*G*irl), *K*iss *M*e."

Finer spectral classification has been achieved by dividing each spectral class into ten subclasses numbered 0 to 9. The hottest G stars, for example, are classified G0 whereas the coolest are G9. The Sun, a G2 star, has a spectrum 20% of the way from G0 to K0. Although there are some stars with such unusual spectra that they cannot be placed in one of the seven main spectral classes, the system succeeds in classifying the great majority of stellar spectra. Figure 16.11 shows examples of spectra for various spectral classes. The way the Balmer lines at 486 and 657 nanometers (abbreviated nm, equal to 10^{-9} m) grow stronger and darker from class O to class B, reach a maximum in class A, and then weaken through classes F and G can be seen clearly in Figure 16.11. Lines of other elements also appear in stellar spectra and can sometimes be stronger than the Balmer lines. These other lines also reach peak strengths at some spectral class.

FIGURE 16.11
Spectra of the Spectral Classes
Classes O and B are brightest in the blue part of the spectrum whereas classes G, K, and M are brightest in the red.

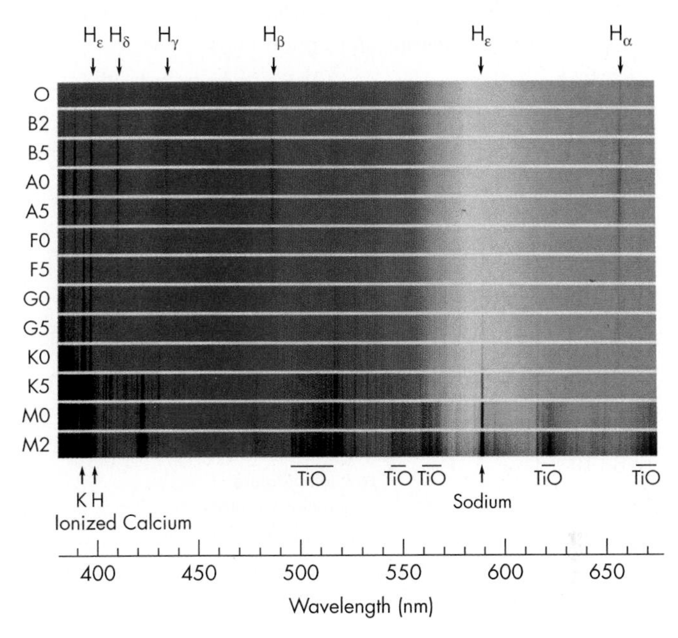

The Role of Temperature

Temperature controls the strengths of atomic absorption lines, because temperature determines the energy level in which the electrons of atoms are likely to be found. At low temperatures, more atoms are in the lowest possible energy level (the ground state) than any other energy level. As the temperature of a gas increases, however, the atoms (or other particles) it contains move more rapidly and have more kinetic energy (see Chapter 7, Section 2). The electrons of these atoms can be excited to higher energy levels either by colliding with other atoms or by absorbing radiation. The probability that an electron will be excited to a higher energy level increases as the collisions become more violent and more frequent. As a result, as temperature increases, the number of atoms in the ground state decreases and the number of atoms in excited levels increases. The strength of a particular absorption line depends on the number of atoms that have their electrons in the level from which the absorption occurs. The greater the number of atoms in a particular state, the stronger the spectral line. Thus the strengths of lines that result from absorptions from the ground state should decrease with increasing temperature whereas those of lines from excited levels should intensify.

The Balmer lines of hydrogen are produced by hydrogen atoms in which the electrons are in the first excited level (level 2 in Figure 16.9). The fraction of all hydrogen atoms that have their electrons in the first excited level increases with increasing temperature, as shown in Figure 16.12. From this we would expect the strengths of the Balmer lines to grow with increasing temperature and, in fact, they do. Notice that the Balmer lines are extremely weak in the very cool

FIGURE 16.12
The Relative Number of Hydrogen Atoms in the Second Energy Level for Various Temperatures
The relative number grows steadily with increasing temperature. Only atoms with their electrons in the second energy level can produce Balmer absorption lines.

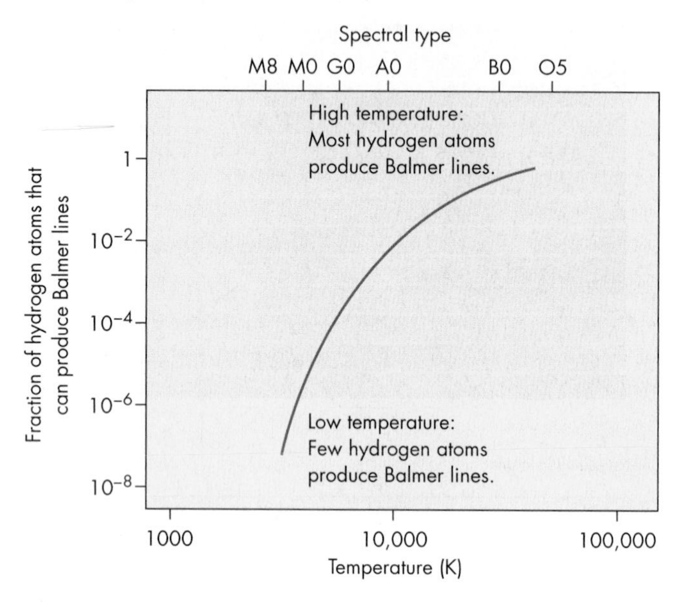

M stars but become increasingly strong as temperature increases through spectral classes K, G, F, and A. In B and O stars, however, the Balmer lines become weaker as temperature rises. To understand why this is true, we must consider collisions violent enough to completely strip the electron from a hydrogen atom and ionize it. After it becomes ionized, a hydrogen atom can no longer contribute to the strength of the Balmer lines. As temperature increases, the energy of collisions in a gas also increases. More and more of the collisions result in the ionization of a hydrogen atom, so a larger fraction of the hydrogen atoms become ionized. For temperatures above 10,000 K, the loss of hydrogen atoms through ionization becomes great enough that the strengths of the Balmer lines begin to decrease as temperature increases.

The graph in Figure 16.13 shows that the relative fraction of hydrogen (including both atoms and ions) that can produce Balmer absorptions peaks at temperatures near 10,000 K, the temperature of the A0 stars. Other elements are more easily or less easily ionized than hydrogen and have different patterns of energy levels, so the strengths of the spectral lines of the other elements peak at different temperatures than those of hydrogen. At high temperatures, absorption lines caused by ionized atoms are strong, whereas at low temperatures molecules such as CN, CH, and TiO produce spectra with huge numbers of lines.

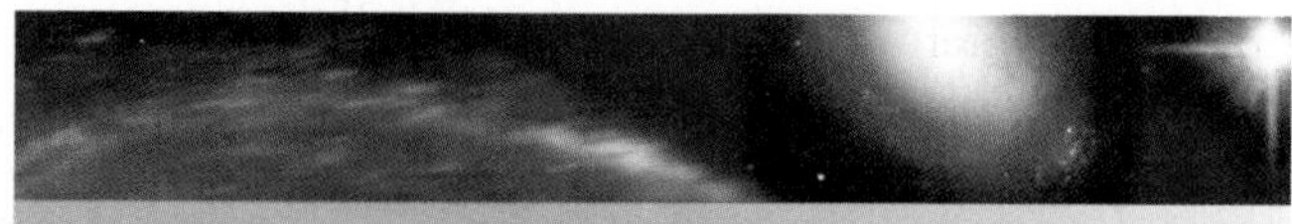

The appearance of the spectrum of a star is controlled mostly by the stellar temperature because temperature affects the fraction of the atoms of a given element that have their electrons in various energy levels. The system for classifying stellar spectra is based mainly on the strengths of the Balmer lines of hydrogen. In descending order of temperature, the classes are O, B, A, F, G, K, M.

Luminosity Class

Although temperature affects the appearance of stellar spectra more than anything else, other factors also play a role. One of these is the density of the gas in the star's outer layers, where the spectral lines are produced. Recall that the strengths of the spectral lines of an element depend on the extent to which the element is ionized. If the atoms of an element are mostly ionized, there will be few atoms left to produce absorption lines. Once an atom is ionized and an electron is freed, the electron remains free

FIGURE 16.13
The Number of Hydrogen Atoms with Their Electrons in the Second Energy Level Compared with the Total Amount of Hydrogen, Whether in Atomic or Ionized Form
Because ionizations reduce the number of hydrogen atoms as temperature increases, the number of atoms capable of producing Balmer lines peaks at 10,000 K.

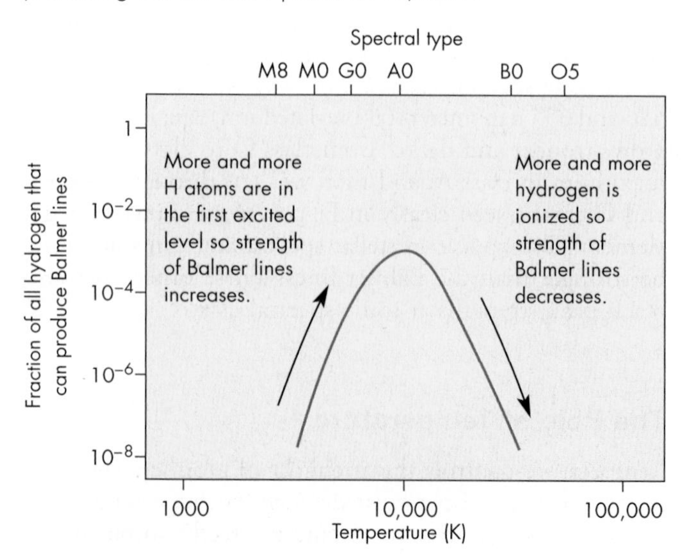

until it encounters another ion and recombines to form an atom. The number of atoms in comparison with the number of ions, then, depends on how quickly electrons and ions encounter each other and recombine. At low densities, where ions and electrons are relatively far apart, encounters and recombinations are infrequent, so a larger fraction of the atoms of an element are ionized than at high densities, where recombination takes place almost as soon as ionization occurs.

The density in the outer layers of a star depends on the gravity there. Recall from Chapter 5, Section 3 that the gravity of a body increases as its mass increases but decreases as the square of its radius increases. Two stars that have similar temperatures can have very different masses and radii and therefore different gravities, pressures, and densities in their outer layers, where spectral lines are formed. The Sun and Canopus, for example, are similar in temperature, but Canopus is 13 times as massive as the Sun and its radius is 60 times as large. The gravity in its atmosphere is 1/250 that of the Sun, resulting in lower density in its outer layers. Because of the lower density, the ionization of atoms is greater in Canopus than in the Sun. This means that for a given temperature, large luminous stars such as Canopus have weaker lines of atoms yet stronger lines of ionized elements than do smaller, less luminous stars such as the Sun.

A trained observer can distinguish the spectra of large luminous stars from smaller, less luminous ones. These distinctions are used to classify a star according to its **luminosity class.** Luminosity class is indicated by a Roman numeral that ranges from I for the largest, most luminous stars **(supergiants)** to III for large luminous stars **(giants)** to V for relatively small, low-luminosity stars **(dwarfs),** which are the most common type of stars. The Sun is a dwarf star, so its complete spectral classification is G2V.

Chemical Abundances

The strengths of individual spectral lines can be used to determine the chemical composition of a star. Figure 16.14 shows, for example, the spectra of two stars that are similar in temperature, mass, and radius but have different abundances of the element lithium. As lithium increases in abundance, absorption lines of lithium grow deeper and broader. The analysis of an enormous number of stars by many astronomers has yielded the perhaps surprising result that, although individual stars vary somewhat in chemical makeup, most have chemical compositions that are similar. Hydrogen and helium are nearly always the most abundant elements; in fact, almost all of the gas in most stars is hydrogen and helium.

Atomic abundances generally decrease as their atomic numbers increase. Thus light elements such as hydrogen, carbon, and oxygen tend to appear in greater abundances than heavy elements such as tin, lead, and gold. The relative abundances of the chemical elements for a typical star are shown in Figure 16.15. The distribution of chemical elements in stars has provided significant clues about the origin of the chemical elements both in the early universe (Chapter 26) and in stars (Chapter 19).

FIGURE 16.14
Lithium Lines in Two Stellar Spectra
For each star the spectrum is presented as a graph of brightness versus wavelength. A dip in the graph for a star represents an absorption line, which is a relatively dim spectral region. A peak represents an emission line. The smallest dips and peaks in the spectra are "noise" rather than real absorption and emission lines. The stars I-11 and I1-13 have spectra with similar strengths of calcium (Ca) and silicon (Si) lines. However, the greater abundance of lithium in I-11 makes its lithium lines much stronger than those of I1-13.

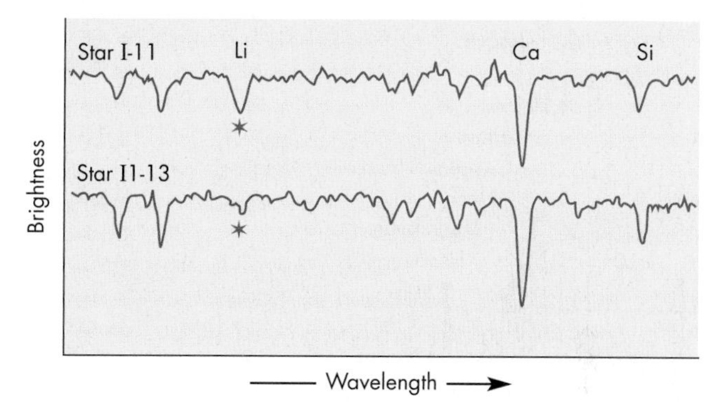

The Doppler Effect and Stellar Spectra

As described in Chapter 6, Section 2, the frequency and wavelength of a wave depend on the relative motion of the source of the wave and the observer of the wave. When the observer and the source are moving toward each other, the frequency measured by the observer is higher than the frequency emitted by the source of the waves. The observed wavelength is shorter, too. If the source and observer are moving apart, on the other hand, the observer measures a lower frequency and longer wavelength. This is called the Doppler effect.

Because of the Doppler effect, the wavelengths at which spectral lines from a star are observed depend on the **radial velocity** of the star—that part of its motion directly toward or away from the observer (Figure 16.16). The Doppler effect changes the wavelength of a line from what it would be if the star were motionless with respect to us. For the stars in our galaxy, the changes in wavelength are small (a tenth of a percent or smaller) because the radial velocities encountered in the galaxy generally are less than a few hundred kilometers per second. Radial velocity describes how rapidly a star moves toward or away from the observer, whereas proper motion is a measure of how rapidly the star moves across the line of sight. Therefore, by combining both measurements, astronomers can discover the actual path of the star through space.

The Doppler effect sometimes makes it possible for astronomers to calculate the speed with which a star rotates. If a star rotates, as shown in Figure 16.17, parts of its surface approach us at speeds that can be as large as the equatorial rotational velocity. Other parts of the star recede from us with equally large speeds. The spectrum of the entire star is a composite of the spectra emitted by all parts of its surface and includes a range of Doppler shifts. As a result, the star's spectral lines are broader than those of a star that does not rotate. If the same star were viewed from above its north or south pole, however, its rotation would not result in any motion toward or away from us, so there would be no broadening of its spectral lines. It is not possible, therefore, to determine whether a star with narrow spectral lines is rotating very slowly or oriented so that we view it from above a pole. Figure 16.18 shows the spectra of two stars that are nearly identical except that one of the stars is rotating rapidly.

FIGURE 16.15
The Relative Numbers of Atoms of Different Elements in a Typical Star
Notice that the abundance of hydrogen is almost one trillion (10^{12}) times that of gold.

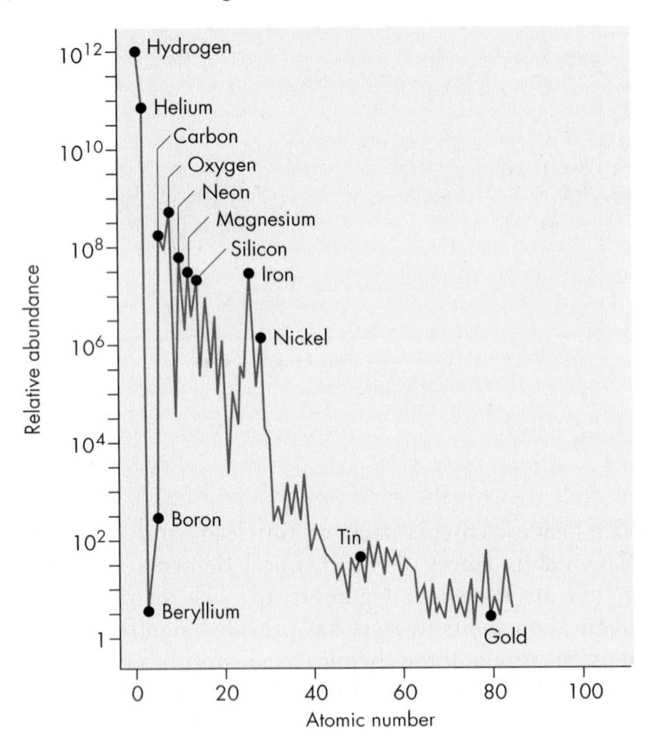

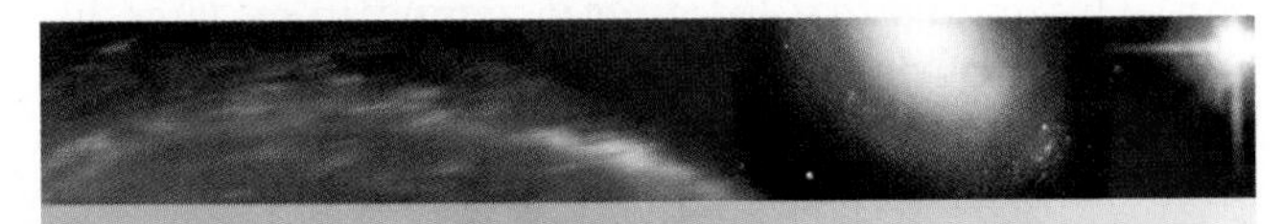

Because of the Doppler effect, the wavelengths of the spectral lines of a star depend on how rapidly the star is moving toward or away from the Earth. The Doppler shift of a star's spectral lines makes it possible to determine the speed with which the star moves toward or away from us and possibly how fast the star rotates.

FIGURE 16.16
The Radial Velocity of a Star
Only the part of a star's motion that carries it toward or away from the observer contributes to its radial velocity. The remainder of its motion is perpendicular to the line to the star and produces the proper motion of the star.

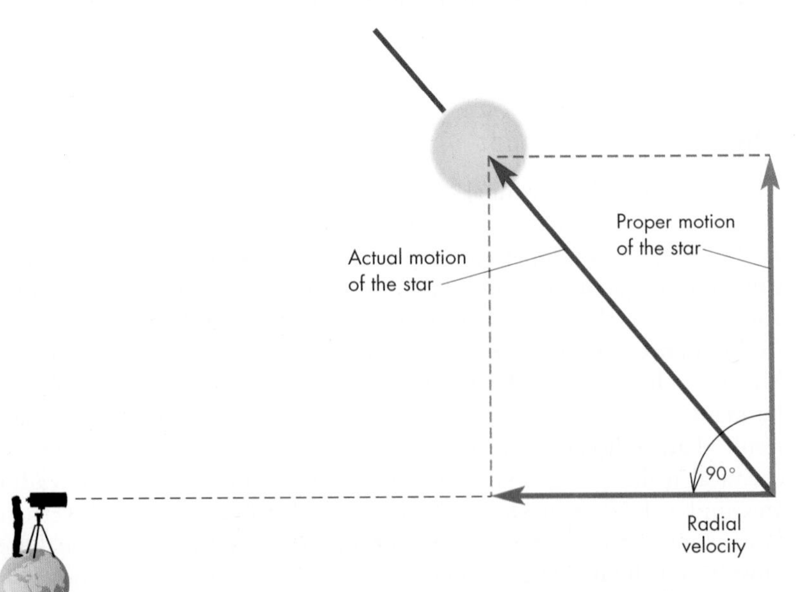

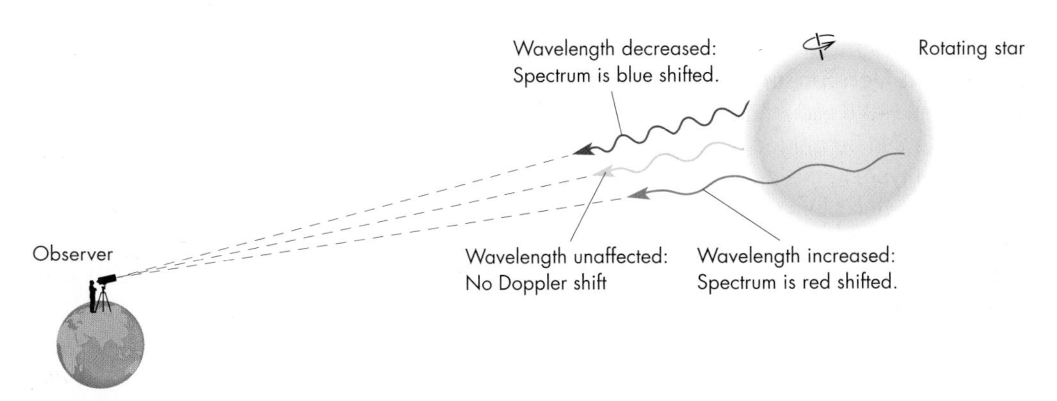

FIGURE 16.17
The Doppler Shifts of a Rotating Star
The side of the star that is moving toward us produces a blue shifted spectrum whereas the side moving away from us produces a red shifted spectrum. The part of the star nearest to us moves across the line to the star and emits a spectrum with no Doppler shift.

FIGURE 16.18
The Spectra of a Rapidly Rotating Star and a Slowly Rotating Star
The spectral lines of the rapidly rotating star are broader because of the range of Doppler shifts in the spectra emitted by different parts of its surface.

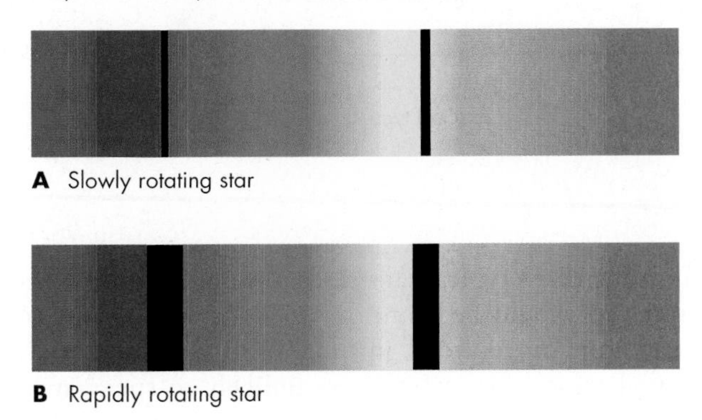

16.6 H-R DIAGRAMS

The Danish astronomer Einar Hertzsprung observed in 1905 that the luminosities of stars generally decreased with spectral type from O to M but that there were a few stars with much higher luminosities than other stars of their spectral class. In 1913 the American astronomer Henry Norris Russell independently observed the same pattern when he plotted the absolute magnitudes of stars against their spectral types. These graphs have been extremely useful for understanding the nature and evolution of stars. They are now called **Hertzsprung-Russell,** or **H-R, diagrams.** Because absolute magnitude is related to the luminosity of a star, luminosity may be substituted for absolute magnitude on the vertical axis. Similarly, temperature may be substituted for spectral type on the horizontal axis. In all cases, the same patterns appear.

A modern version of an H-R diagram, using data for a large number of stars, is shown in Figure 16.19. The best H-R diagram ever made, based on *Hipparcos* data, is shown in Figure 16.20. The most significant feature of the H-R diagram is that the stars do not uniformly fill it. There are a few areas of the H-R diagram that contain most of the stars and other regions in which few stars, if any, are located. The region of the H-R diagram that contains the most stars is called the **main sequence,** which runs diagonally from hot, luminous stars to cool, dim stars. Other well-populated regions are the giant region, which contains cool, luminous stars; the supergiant region, which contains the most luminous stars that lie at the top of the H-R diagram; and the **white dwarf** region, which contains the hot but dim stars in the lower left of the H-R diagram.

The simplest explanation for the appearance of the H-R diagram would be that stars are formed in one of four types and remain nearly steady in both spectral type and luminosity for as long as they are luminous. In reality, though, as a star ages its internal properties change, so its spectral type and luminosity change, too. This means that as a star evolves, its position in the H-R diagram changes. How these changes occur is described in Chapters 17 through 19.

The significance of the H-R diagram is that it gives important information about the evolution of stars. Well-populated regions of the diagram represent long-lived stages in the evolution of many stars. Poorly populated regions represent evolutionary stages in which few stars, if any, spend long periods of time. An analogy for using H-R diagrams to study stellar evolution would be using a satellite image to study the "behavior" of the automobiles in your town or city. The image would show many cars in garages, driveways, parking lots, and parking spaces,

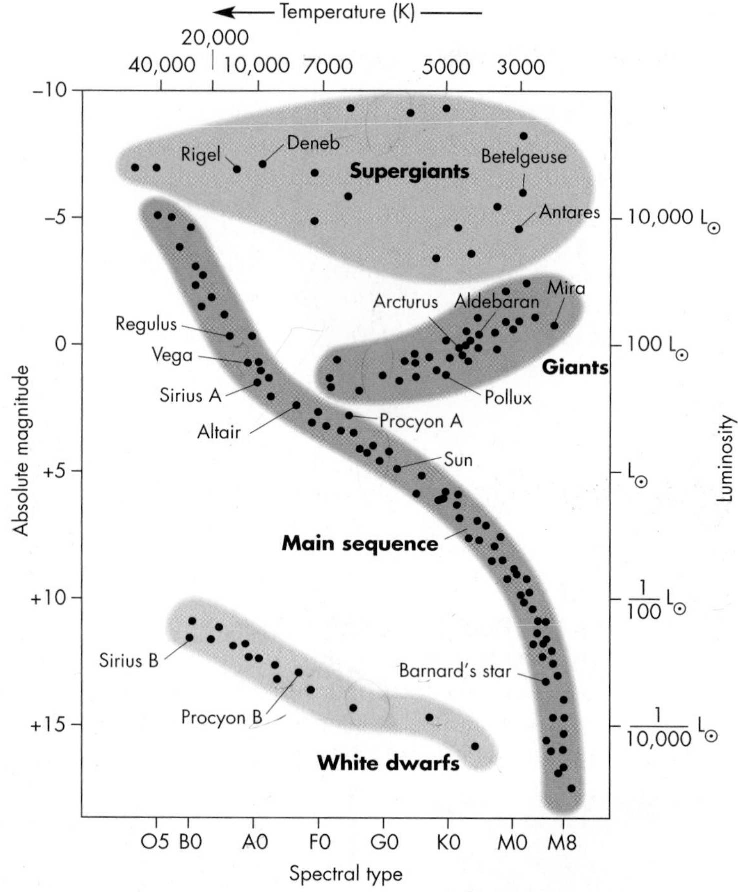

FIGURE 16.19
A Modern Version of the H-R Diagram
To make an H-R diagram, either luminosity or absolute magnitude can be plotted against either temperature or spectral type. In this diagram, luminosity is given in terms of the Sun's luminosity, $L_\odot$. The main sequence and giant regions contain most of the stars, whereas hot underluminous stars, the white dwarfs, lie below and to the left of the main sequence.

Stellar evolution and the H-R diagram

FIGURE 16.20
An H-R Diagram Based on Hipparcos Data
The graph shows temperature and luminosity for almost twenty-one thousand stars for which accurate Hipparcos distances and brightnesses are available.

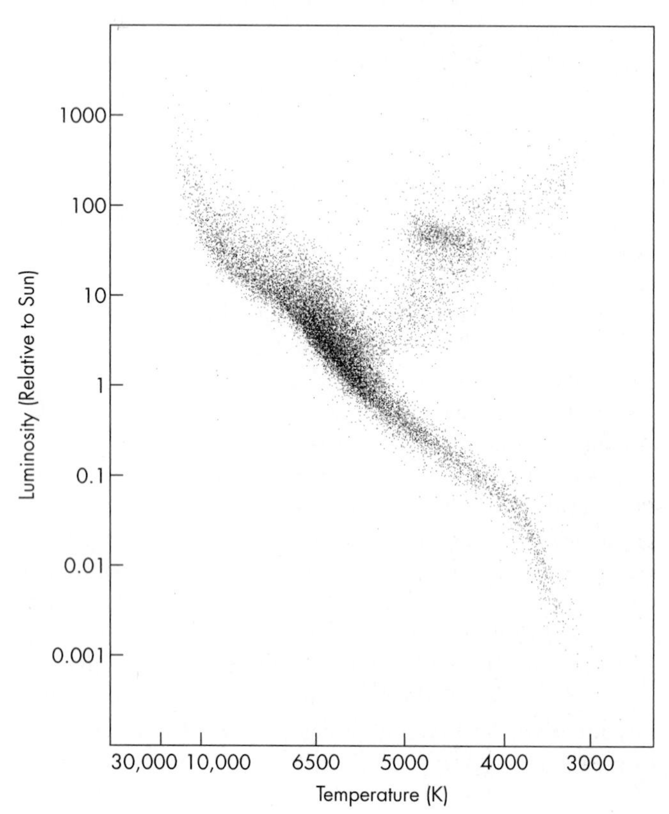

indicating that a typical car spends most of its time parked. Fewer cars would be found on roads and highways and nearly none on rooftops, in fountains, or in other places where a typical car is extremely unlikely to spend much time.

The appearance of the H-R diagram does not by itself tell us exactly how the spectral type and luminosity of a typical star change with time, but it does show that stars evolve in such a way that many of them spend large fractions of their lives as main sequence stars, giants, supergiants, and white dwarfs.

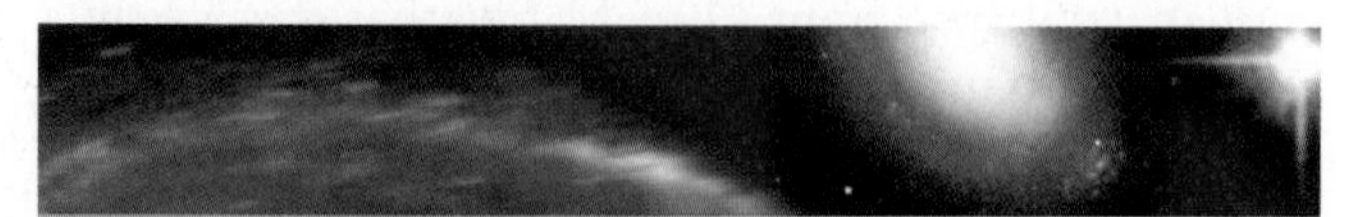

An H-R diagram is a plot of the luminosities of stars against their temperatures. Most stars are clustered within four areas of an H-R diagram. The regions within which most stars are clustered represent long-lived evolutionary stages for stars.

16.7 STELLAR MASSES

Masses of Binary Stars

Spectral types and luminosities are known for a large number of stars. Stellar masses, however, which are at least as important for understanding the workings of stars, have been found for only a few hundred members of binary star systems. The mass of a star can be determined only by calculating its gravitational attraction for a companion star (or planet). The two stars in a binary system are bound together gravitationally and revolve about their common center of mass. If the orbits of the two stars can be determined, Kepler's third law (Equation 5.4) can be used to determine their masses. That is, if the average distance between the two stars, d, is given in AU and the period of revolution of the stars, P, is given in years, then the sum of the masses of the two stars, in solar masses, is

$$M_1 + M_2 = \frac{d^3}{P^2} \tag{16.5}$$

The individual masses of the stars can be found from their distances from the center of mass by

$$\frac{M_1}{M_2} = \frac{d_2}{d_1} \tag{16.6}$$

where $d = d_1 + d_2$.

The Range of Stellar Masses

The masses of stars in binary systems range from 0.08 solar masses to 80 solar masses. It is likely that stars as massive as perhaps 150 solar masses or greater exist, although they are rare. Objects less massive than 0.08 solar masses are usually not regarded as genuine stars because nuclear fusion is not a significant energy source for them.

The most massive nonstellar objects are nearly as luminous as dim stars. Their low temperatures, however, cause most of their energy to be radiated in the infrared. These objects, which resemble stars in some respects and planets in others, are called **brown dwarfs.** Brown dwarfs are so dim that the first one wasn't discovered until 1995. Recent surveys, however, have found so many brown dwarfs that it is now thought there may be about as many brown dwarfs as stars in the galaxy. Brown dwarfs are both cooler and dimmer than the dimmest M main-sequence stars. Because of their low temperatures, their spectra are so different from M stars that three new spectral classes, L, T, and Y, were created to accommodate them. The L stars, which can be as cool as about 1500 K, show dark bands due to molecules such as iron hydride and calcium hydride. T stars, which are cooler still, show dark bands due to methane molecules. Recall that methane is one of the gases found in the atmospheres of Jupiter and Saturn. The even cooler Y stars, few of which have been identified, have broad absorption bands of ammonia in their spectra. Thus, brown stars bridge the gap between stars in planets in mass, temperature, luminosity, and spectral features.

The Mass-Luminosity Relation

The masses and luminosities of stars are related to each other. The plot in Figure 16.21 of the luminosities of individual stars versus their masses shows that larger luminosities accompany larger masses. The relationship shown

Mass and Orbital Motion

Equations 16.5 and 16.6 can be used to find the masses of the stars in a binary system if their orbital period and their distances from the center of mass of the binary system are known. Suppose the orbital period of a binary system is 10 years and the distances of the two stars from the center of mass are 2 AU and 8 AU. What is the mass of each star? The first step is to combine the two distances to find the average distance between the two stars. That is

$$d = d_1 + d_2 = 2\text{ AU} + 8\text{ AU} = 10\text{ AU}$$

Using this and the orbital period in Equation 16.5 gives

$$M_1 + M_2 = d^3/P^2 = 10^3/10^2 = 1000/100 = 10M_\odot$$

Equation 16.6 can then be used to find the ratio of the masses of the two stars from the ratio of distances.

$$\frac{M_1}{M_2} = \frac{d_2}{d_1} = 8\text{ AU}/2\text{ AU} = 4$$

Thus, star 1 is four times as massive as star 2 and, because the sum of their masses is 10 $M_\odot$, the masses of star 1 and star 2 are 8 $M_\odot$ and 2 $M_\odot$, respectively.

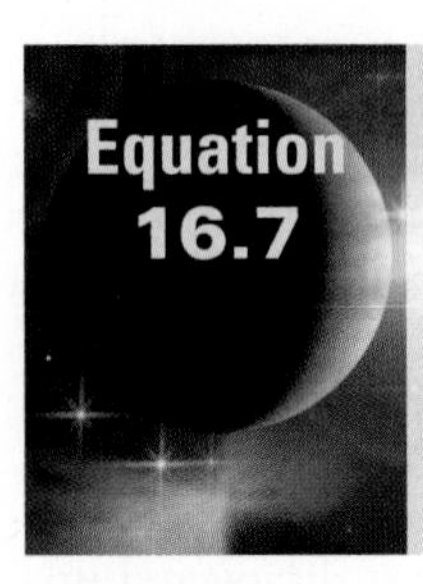

Luminosity and Mass

Equation 16.7 gives the relationship between the masses and luminosities of stars. What is the luminosity of a 5 $M_\odot$ star? Equation 16.7 gives

$$L = 5^{3.5}\ L_\odot = 280\ L_\odot$$

Notice that the mass used in Equation 16.7 must be given in terms of the Sun's mass, and the luminosity that results is in terms of the Sun's luminosity.

FIGURE 16.21
The Mass-Luminosity Relation for Main Sequence Stars
Mass and luminosity are strongly correlated. Notice that the ratio of the highest to the lowest luminosities is about one hundred million, whereas the ratio of highest to lowest mass is only about three hundred. If white dwarf stars were plotted in this figure, they would be underluminous for their masses and lie beneath the points representing the main sequence. Giant stars would be overluminous for their masses and lie above the points representing the main sequence.

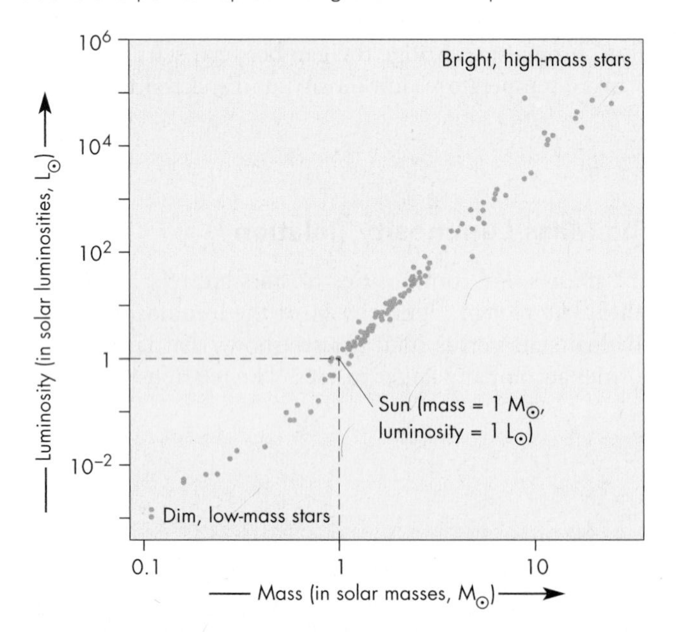

in Figure 16.21 is called the **mass-luminosity relation.** For most stars, the relationship between mass and luminosity can be approximated by

$$\frac{L}{L_\odot} = \left(\frac{M}{M_\odot}\right)^{3.5} \tag{16.7}$$

where $L_\odot$ and $M_\odot$ are the luminosity and mass of the Sun.

The mass-luminosity relation indicates that the mass of a star strongly controls the rate of energy production in the star. The relationship between stellar mass and energy production is explored in more detail in Chapters 17 and 19.

The masses of stars can be found by applying Kepler's third law to the motions of the stars in binary systems. For most stars, there is a close relationship between mass and luminosity.

Chapter Summary

- The distance of a star can be found by measuring its parallax, a small change in the position of the star when viewed from opposite sides of the Earth's orbit. If the distance to a star (d) is given in parsecs and parallax (p) is measured in seconds of arc, then $d = 1/p$. (Section 16.2)
- The motions of the stars through space cause their positions in the sky to change slowly with time. The proper motion of a star is the rate at which a star's position changes. The motion of the Sun relative to the other stars causes stars to move apart in the direction toward which the Sun is moving and to converge in the opposite direction. (16.3)
- The apparent brightnesses of stars are described by the magnitude system, in which increasingly fainter stars are assigned increasingly larger apparent

magnitudes. Each increase of one magnitude corresponds to a decrease in brightness of 2.512 times. (16.4)

- Absolute magnitude describes the intrinsic brightness of a star. The absolute magnitude of a star does not depend on the star's distance, but its apparent magnitude does. Apparent magnitude (m) and absolute magnitude (M) are related to each other and distance by

$$m = M + 5 \log (d/10). \qquad (16.4)$$

- Atoms produce dark-line spectra by absorbing radiation from a continuous source. This happens when an electron in an atom absorbs a photon and jumps to a higher energy level. When electrons jump to lower energy levels, they emit photons and produce bright-line spectra. (16.5)
- The appearance of the spectrum of a star is determined mostly by the star's temperature. Temperature controls the numbers of atoms with electrons in various energy levels and thus the strengths of absorption lines arising from those energy levels. In order of descending temperature, the stellar spectral classes are O, B, A, F, G, K, and M. (16.5)
- Because of the Doppler effect, the observed wavelengths of emission and absorption lines in a stellar spectrum depend on the motion of the star relative to the Earth. The Doppler effect makes it possible to use the observed wavelengths of spectral lines to determine how fast a star is moving toward or away from the Earth. (16.5)
- An H-R diagram is a plot of the luminosities of stars against their temperatures. In an H-R diagram the points representing stars tend to be concentrated in the main sequence, giant, supergiant, and white dwarf regions. These regions correspond to long-lived stages in the evolution of stars. (16.6)
- The masses of stars can be found by determining the orbital motion of two stars about each other in a binary system. For stars, luminosity (L) increases with mass (M) according to the mass-luminosity relation, $L = M^{3.5}$, where L and M are expressed in terms of the luminosity and mass of the Sun. (16.7)

Key Terms

absolute magnitude *381*
absorption-line *383*
antapex *380*
apex *380*
apparent brightness *380*
apparent magnitude *381*
Balmer series *385*
brown dwarf *393*
continuous spectrum *383*
dwarf *389*
emission-line *383*
energy level *385*
giant *389*
ground state *385*
Hertzsprung-Russell diagram (H-R diagram) *391*
ionization *385*
Kirchhoff's laws *383*
light year *377*
luminosity *381*
luminosity class *389*
luminosity function *383*
Lyman series *385*
magnitude *380*
main sequence *391*
mass-luminosity relation *394*
parsec *377*
proper motion *378*
radial velocity *389*
solar motion *378*
spectral class *387*
supergiant *389*
white dwarf *391*

Conceptual Questions

1. The Sun is actually much less luminous than Sirius. What effect does this fact have on the accuracy of Newton's determination of the distance of Sirius?
2. Both parallax and proper motion are changes in the apparent positions of stars. Describe how observations carried out over many years could be used to distinguish the changes caused by parallax from those due to proper motion.
3. Suppose the Sun moved through space much faster than it actually does. What effect would this have on the directions of the apex and antapex of the Sun's motion? What effect would this have on the sizes of the proper motions of nearby stars?
4. Explain why the absorption lines of an element have the same wavelengths as the emission lines of that element.
5. Why is it impossible for a gas composed of atoms that all have their electrons in the ground state to produce an emission line?
6. Make up your own mnemonic to help you remember the sequence of spectral classes.
7. Why are Balmer lines weaker in the spectra of O stars than they are in the spectra of A stars?
8. Why are Balmer lines weaker in the spectra of G stars than they are in the spectra of A stars?
9. What can be said about the radial velocity of a star that shows no Doppler shift of its spectral lines?

10. Which of the following pairs of quantities can be plotted against each other to produce an H-R diagram?
    (a) temperature and distance
    (b) luminosity and spectral class
    (c) temperature and spectral class
    (d) radial velocity and apparent magnitude
11. What is the reason that most of the stars in an H-R diagram, such as Figure 16.19, are located on the main sequence?
12. How does the brightness of a giant star compare with the brightness of a main-sequence star of the same spectral class?

Problems

1. A star has a distance of 40 pc. What is the parallax of the star?
2. A star has a parallax of 0.2 seconds of arc. What is the distance of the star?
3. A star has a parallax of 0.11 seconds of arc. Suppose we received radio signals from creatures on a planet orbiting the star and then immediately sent a radio response to the creatures. How long would it be between the time that the creatures transmitted their signal and the time that they received our signal?
4. Suppose a star has a distance of 50 pc. What would be the parallax of the star if it were measured by creatures living on Pluto? On Mercury?
5. Star A is 4 magnitudes brighter than star B. How does the apparent brightness of star A compare with that of star B?
6. Star A is 7 magnitudes brighter than star B. How does the apparent brightness of star A compare with that of star B?
7. Star A is 16 magnitudes brighter than star B. How does the apparent brightness of star A compare with that of star B?
8. Star A has an apparent magnitude of −1. Star B has an apparent magnitude of 5. How does the apparent brightness of star A compare with that of star B?
9. Star A has an apparent brightness that is 40 times as large as that of star B. What is the difference in the magnitudes of the two stars?
10. Two stars differ by 8 in absolute magnitude. What is the ratio of their luminosities?
11. A star has an absolute magnitude of 12 and a distance of 100 pc. What is the apparent magnitude of the star?
12. At what distance does a star have an apparent magnitude that is 5 magnitudes brighter (smaller) than its absolute magnitude?
13. A star is moving away from the Earth at a speed of 200 km/s. What would be the wavelength of a spectral line that would have a wavelength of 500 nm if the star were at rest?
14. In the laboratory, the wavelength of a spectral line of an atom is 450 nm. The same line in the spectrum of a star has a wavelength of 448 nm. What is the radial velocity of the star? Is the star moving toward or away from the Earth?
15. Suppose two stars obey the mass-luminosity relationship. One star is four times as massive as the other. How do the luminosities of the two stars compare?

Figure-Based Questions

1. Use Figure 16.9 to determine whether the wavelength of the α line of the Lyman series of hydrogen is longer or shorter than the wavelength of the Lyman β line.
2. Use Figure 16.5 to find the number of stars with absolute magnitude 6 compared with those of absolute magnitude 13.
3. Use Figure 16.9 to explain why the frequency of the γ line of the Lyman series of hydrogen is equal to the sum of the frequencies of the α line of the Lyman series and the β line of the Balmer series.
4. Use Figure 16.15 to find the ratio of the abundances of tin and iron.
5. Use Figure 16.21 to find the luminosity (in terms of the Sun's luminosity) of a 0.2 $M_\odot$ star and a 5 $M_\odot$ star.

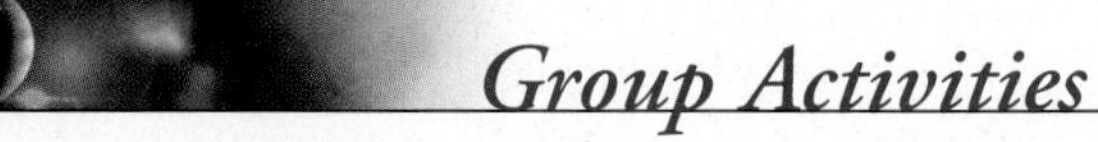

Group Activities

1. Have your partner walk ten paces away from you. With your hand, measure the angle between your partner and a distant landmark like a radio tower. Move two steps to your left and repeat the measurement. The difference in angles you measure is the parallax of your partner. Have your partner walk ten paces farther away and repeat your angular measurements. Now have your partner make the angular measurements with you at distances of ten and twenty paces. Try to deduce the relationship between parallax and distance. Compare your result with Equation 16.1.
2. Travel with your group to a dark spot on a clear, moonless night. By counting the number of stars that can be seen in one quadrant of the sky (for example, all stars between the horizon and the zenith with azimuths between due south and due west), have each member of your group estimate the total number of visible stars in the sky. If you have at least four members of your group, be sure to include all four quadrants of the sky. Average the estimates to get an overall estimate of the number of visible stars. Repeat the observations at the time of full Moon. Compare the results and have the group discuss what the results imply about the number of stars that can be seen near cities, near small towns, and far from cities and towns.

For More Information

Visit the text website at **www.mhhe.com/fix** for chapter quizzes, interactive learning exercises, and other study tools.

17

In the middle of all is the seat of the Sun. For who in this most beautiful of temples would put this lamp in any other or better place than the one from which it can illuminate everything at the same time. . . . Thus indeed the Sun as if seated on a royal throne governs his household of stars as they circle round him.

Nicholas Copernicus

Pacific Sunset

www.mhhe.com/fix

The Sun

Astronomers often refer to the Sun as a "typical star" or an "ordinary star." Compared with most stars, however, the Sun is larger, brighter, and more massive. Compared with what we are used to on the Earth, moreover, all stars, including the Sun, have such amazing properties that they are almost beyond our comprehension.

The Sun has a diameter of 1,400,000 km. How big is 1,400,000 km? Well, it's almost four times the distance to the Moon—the greatest distance any human has traveled into space. The Sun's circumference is 4 million km. Traveling in a jet at three times the speed of sound, it would take 60 days to travel that far.

The Sun is also extremely massive. It is 1000 times as massive as Jupiter and 300,000 times as massive as the Earth. The ratio of the mass of the Sun to the mass of the Earth is about the same as that of three 100 kg (220 lb) football players to a paper clip. About 99.9% of the atoms in the Sun are hydrogen and helium. However, the Sun is so massive that if all of the calcium in the Sun could be extracted, there would be more than enough to make a planet as massive as the Earth. Similarly, Earth-sized planets could also be made from the Sun's copper, silicon, vanadium, chlorine, fluorine, phosphorus, potassium, and many other elements. Of all the elements in the Sun, gold ranks about 77 in abundance. Although there isn't enough gold in the Sun to make a terrestrial planet, there is enough to gold-plate the Earth to a thickness of about 100 m.

The energy output of the Sun is 4×10^{26} watts (W). This is so much energy that if 1 second's worth of solar energy could be captured, converted to electricity, and sold at 5 cents per kilowatt hour, it would bring in

Questions to Explore

- What is the source of the Sun's energy?
- What is the structure of the Sun?
- What is the future of the Sun?
- How can we learn about the interior of the Sun?
- How is energy transported out of the Sun?
- What role do magnetic fields play in solar phenomena?
- How can we account for the sunspot cycle?

6×10^{18} dollars. This is big money, even by congressional standards. It is, in fact, about a million times as large as the gross national product of the United States. Even the small fraction of the Sun's energy output that strikes the Earth would be worth about 2 billion dollars a second. Because nearly all of the energy people have ever used is ultimately derived from sunlight, there are practical reasons to want to understand how the Sun works. We need to know how constant the Sun's energy output will be and how long the Sun's fuel supply will last.

For astronomers, another compelling reason to investigate the nature of the Sun is that it is our nearest star and the easiest to study. Astronomers who study the Sun have the luxury of abundant light, which other kinds of astronomers do not have. Although not everything that we learn about the Sun can be generalized to all the other stars, there is every reason to believe that the processes that go on in the Sun are at work in other stars as well. By learning more about the Sun, astronomers hope to improve their understanding of other stars.

Some important properties of the Sun are given in Table 17.1.

Table 17.1

The Sun

Mass	2.0×10^{30} kg
Diameter	1.4×10^{9} m
Density (relative to water)	1.4
Escape velocity	620 km/s
Rotation period (at equator)	24.7 days
Luminosity	3.8×10^{26} W
Absolute magnitude	4.8
Surface temperature	5800 K
Spectral type	G2

17.1 THE INTERNAL STRUCTURE OF THE SUN

Astronomers have been able to calculate the conditions inside the Sun by developing and solving a set of equations that describe how pressure, temperature, mass, and luminosity change with distance from the center of the Sun. The equations are based on the conservation of matter and energy and on the assumption that the Sun is in hydrostatic equilibrium, with gravity just balancing pressure differences (see Chapter 7, Section 2). To solve the equations, it is necessary to know the chemical composition of the Sun, the mechanism by which energy is carried out of the Sun (convection or radiative diffusion), and the rate of production of energy at each distance from the Sun's center. Solving the equations is straightforward, although some computer power is needed. The validity of the solutions depends on how well the composition, energy production rate, and other factors are known. For a solution to be acceptable, it must match the size and luminosity of the Sun.

The internal structure of the Sun is shown in Figure 17.1. The Sun's central temperature and density are both remarkable. The temperature is about 15.6 million K, far hotter than the core of any planet. The gas in the Sun's core is compressed by the weight of the Sun to a density of 150,000 kg/m^3, or 150 times the density of water. Both temperature and density fall smoothly from the Sun's center to its surface. Essentially all of the Sun's energy is produced in the inner 25% of the Sun's radius (about 1.5% of the Sun's volume). The concentration of energy production in the Sun's core is a consequence of the extreme temperature sensitivity of nuclear reactions that produce solar energy. At a point one-fourth the distance from the center to the surface, temperature

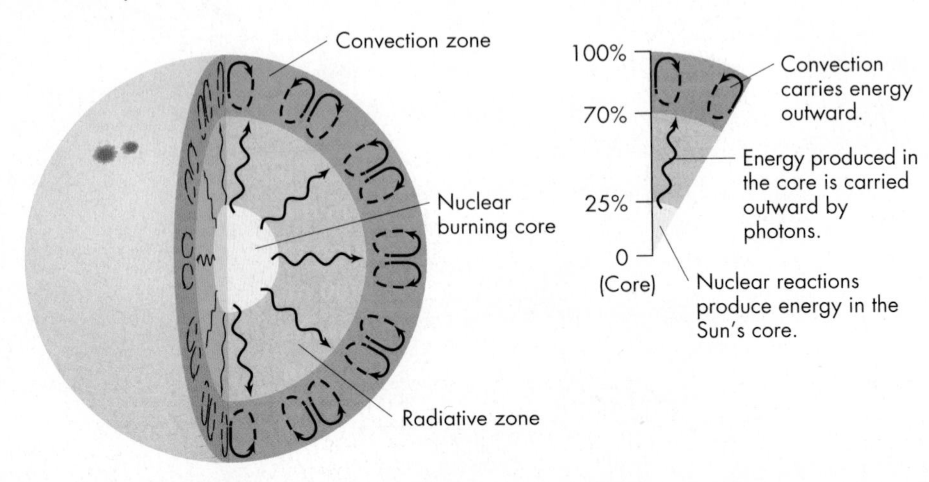

FIGURE 17.1
The Internal Structure of the Sun
Energy-producing nuclear reactions occur only within the inner 25% of the Sun's radius. The energy produced by these reactions is carried outward by photons to 70% of the Sun's radius. From that distance outward, convection carries most of the Sun's energy.

has fallen to 8 million K. This drop in temperature reduces the rate at which energy is produced to practically zero.

At the Sun's surface, hydrogen makes up 71% of each kilogram of gas. At the center, however, the hydrogen abundance is only 34%. The reason that hydrogen is less abundant at the center of the Sun is that it has been used as fuel for most of the 4.6 billion years that the Sun has been shining. During that time, a little more than half of the hydrogen at the center of the Sun has been consumed. The changes that have resulted from the gradual depletion of hydrogen in the Sun's core are described in Chapter 19.

17.2 THE SUN'S ENERGY

The Sun is enormously bright and seems to have been shining at about its present brightness for a very long time. Geologists have found rocks 3.5 billion years old that contain fossils of marine organisms. These discoveries clearly demonstrate that the Sun has warmed the Earth for at least 3.5 billion years and probably for as long as the Earth has existed. It is easy to propose ways in which the Sun could produce energy at its observed rate. The basic problem in accounting for the Sun's energy is that most proposed fuels would be exhausted far too quickly to account for the length of the Sun's luminous career.

Some Wrong Answers

Until the middle of the nineteenth century, the Sun's energy wasn't of much concern to astronomers and other scientists, because no one knew enough about physics and chemistry to appreciate how difficult it would be to account for the Sun's energy. William Herschel, probably the leading astronomer of his time, suggested that sunspots were holes in the Sun's fiery clouds, through which he could see a cool surface "lush with green vegetation." Today, this baked Alaska model of the Sun is fit for supermarket tabloids. In Herschel's time, however, it didn't seem implausible at all.

Combustion The idea incorporated in Herschel's theory, that the Sun is burning, has been around since antiquity. Combustion, a chemical reaction, can produce a great deal of energy. Massive spacecraft can be blasted into space by chemical rockets. Combustion (burning hydrogen in oxygen to produce water vapor, for example) yields about 10^7 joules (J) of energy per kilogram of fuel. If that energy were consumed at the Sun's rate, 2×10^{-4} W/kg (where 1 watt equals 1 joule per second), the Sun could remain luminous for about 50 billion seconds, or 2000 years. Two thousand years isn't long enough to account for recorded history, much less the history of the Earth. Clearly, the Sun is not on fire.

Meteoritic Impacts Another idea, proposed by Lord Kelvin in 1862, is that the Sun's luminosity is derived from meteoritic impacts. Objects falling into the Sun accelerate to about 600 km/s. When they strike the Sun, their kinetic energy is converted to heat and light. This theory seems promising, but to account for the Sun's energy, about one-tenth of an Earth mass of material would have to fall into the Sun each year. Surely we would notice this much material falling inward past the Earth. Also, if the Sun's mass increased at the rate of one-tenth of an Earth mass per year, the increased gravitational force between the Earth and the Sun would shorten the length of the year by about 1 minute per century. Such a change would be very quickly detected.

Gravitational Contraction Gravitational contraction, the most plausible nineteenth century idea about the Sun's energy, is closely related to the idea of meteoritic impacts. Instead of new material falling into the Sun, however, it involves the Sun slowly falling inward under the force of its gravity. Gravitational contraction, the ultimate source of the luminosities of Jupiter and Neptune, provides a large reservoir of energy for a massive body such as the Sun. For every meter that the Sun shrinks, enough energy is released to account for about 10 days worth of the Sun's luminosity. If the Sun were actually shrinking at this rate, about 1 km every 50 years, it would be very difficult to detect.

The problem with explaining the Sun's energy output by gravitational contraction isn't that the Sun couldn't shrink fast enough, but that it couldn't have been doing so for long enough. If the Sun began as a widely dispersed cloud of gas and then shrank to its present size, the total amount of energy released would only be enough to keep the Sun shining at its present brightness for about 20 million years. This period of time, the approximate amount of time needed for the Sun to have shrunk to nearly its present size, is called the **Kelvin-Helmholtz time,** after the two scientists who first made the calculation. When gravitational contraction was proposed as the Sun's energy source, astronomers hoped that the Kelvin-Helmholtz time was sufficient to account for the history of the solar system. We now know, however, that it is about 1/200 of the length of time that the Sun has been shining.

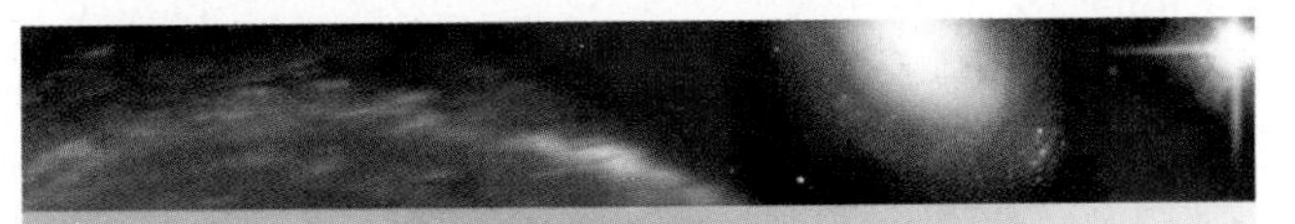

The Sun is very luminous and has been shining for billions of years. The enormous amount of energy that the Sun has produced since its formation makes it possible to rule out many energy sources, such as combustion, meteoritic impact, and gravitational contraction.

Nuclear Reactions

The vast amount of energy released in nuclear reactions began to be realized around the beginning of the twentieth century. This realization led to the proposal that fission reactions (see Chapter 7, Section 4) power the Sun. Fission is an efficient way to extract energy from matter, because during fission a small amount of matter is converted into energy, according to $E = mc^2$ (Equation 7.5). About a million times as much energy can be obtained from 1 kg of fissionable material than from 1 kg of combustible material. The fissionable isotopes in the Sun, however, are very low in abundance. Uranium, for example, makes up only about 1 in every 10^{12} nuclei in the Sun. The uranium in the Sun could only supply about 2 months' worth of sunlight if fission powered the Sun. The other fissionable isotopes in the Sun are also much too low in abundance to have produced the energy emitted by the Sun over its lifetime.

The Proton-Proton Chain What is required to explain the Sun's energy is a nuclear process involving an element that is abundant in the Sun. This process, the fusion of hydrogen into helium, was proposed in 1920 by A. S. Eddington. The details of how the process operates were worked out almost 20 years later. In fusion, as in fission, matter is converted into energy according to $E = mc^2$.

The nucleus of hydrogen, the lightest element, consists of a single proton. The nucleus of helium, on the other hand, has four nuclear particles—two protons and two neutrons. Thus, four hydrogen nuclei must combine to make one helium nucleus. It is unrealistic, however, to expect four protons to slam together and instantly make a helium nucleus. This is so improbable that it has almost certainly never happened, not even once, in the entire history of the universe. What happens instead is analogous to a multicar accident on an interstate highway. The pile-up begins when two cars collide. Other cars, one at a time, collide with the wreckage and add to the mass of disabled cars. In the same way, helium is built up by a series of reactions involving two nuclei at a time. The series of reactions is called the **proton-proton chain.** In stars more massive than the Sun, the fusion of helium occurs through a different series of reactions, the CNO cycle, which is described in Chapter 19.

The proton-proton chain begins with a reaction between two protons. The two protons must come within about 10^{-15} m of each other for a nuclear reaction to take place. However, the two protons are positively charged and strongly repel each other. This means that most collisions do not result in a reaction. Instead, the two protons deflect each other and move apart again. At room temperature there is essentially no chance that two protons will collide with sufficient energy to get close enough to each other to react.

What is required for fusion is more energetic collisions involving protons traveling at high speeds. At the center of the Sun the temperature is about 15 million K. A typical proton travels about 1 million km/hr and some travel much faster. Even at these speeds, however, the reaction of two protons is very improbable. Suppose it were possible to label a single proton at the Sun's center and follow it as it bounced off other protons many times each second. An observer would have to watch a typical proton for about 5 billion years before witnessing a fusion reaction with another proton. Improbable as a fusion is, however, there are so many protons in the Sun's core that about 10^{34} of them react each second.

Figure 17.2 shows the fusion reactions in the proton-proton chain. In the first reaction, the fusion of two protons produces **deuterium** (^{2}H), an isotope of hydrogen. The deuterium nucleus, called a deuteron, consists of one proton and one neutron, so it is about twice as massive as each of the two protons that formed it. The other products of the reaction are a positively charged electron, or positron (β^+), and a **neutrino** (ν). The energy produced by the reaction appears in several forms. One of these is kinetic energy. The deuteron and the positron are both "hot" and move more rapidly than typical particles in the gas at the center of the Sun. The deuteron quickly shares its excess kinetic energy with other particles through collisions, heating the gas. The positron itself is a kind

FIGURE 17.2
The Reactions of the Proton-Proton Chain
The symbol ν indicates neutrinos, the symbol β^+ indicates positrons, and the symbol γ indicates photons. The orange spheres represent protons, and the blue spheres represent neutrons. Although six hydrogen nuclei are needed to complete one proton-proton chain, two of them are regenerated in the final step. The net reaction, therefore, is that four hydrogen nuclei are transformed into a single helium nucleus. Energy is released in the process.

The reactions of the proton-proton chain

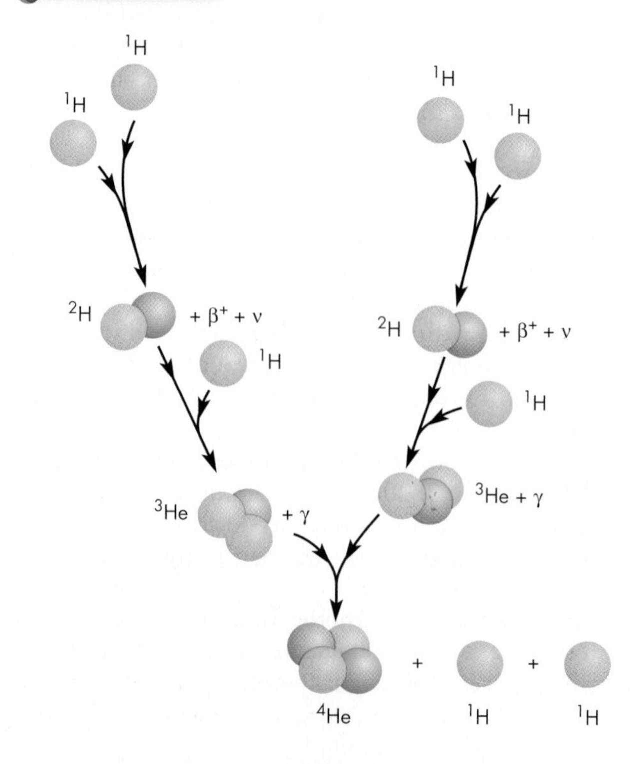

of potential energy because it is a particle of antimatter. When it collides with an electron, the positron annihilates both itself and the electron in a flash of gamma rays. The gamma rays, and those produced by the fusion reaction itself, are quickly absorbed by the surrounding gas, which is heated as a result. The fate of the neutrino is described later in this chapter.

The next reaction in the proton-proton chain is the fusion of the deuteron with a proton, which produces a helium nucleus (^{3}He) and gamma rays (γ). The ^{3}He nucleus consists of two protons and *one* neutron, whereas an ordinary helium nucleus has two protons and *two* neutrons. The reaction that produces ^{3}He from a deuteron takes place very quickly, so a typical deuteron at the center of the Sun survives for only about 4 seconds before reacting with a proton.

Usually, the final reaction in the proton-proton chain is the fusion of two ^{3}He nuclei to produce a ^{4}He nucleus and two protons. On average, a ^{3}He nucleus has to wait for 4 million years before it participates in this reaction. The net result of the chain of reactions is

$$6\,^1\text{H} \rightarrow \,^4\text{He} + 2\,^1\text{H} + 26.72 \text{ MeV} \tag{17.1}$$

Thus, although it takes six protons to make one helium nucleus, there is a net loss of only four protons because two are regenerated in the final step. Because six protons are more massive than two protons and a helium nucleus, mass is lost in the proton-proton chain and converted to energy. The energy produced by a single proton proton chain, 26.72 MeV, is tiny, about one ten-millionth of the energy required to lift a drop of water 1 cm. But there are many protons in a kilogram of hydrogen and many kilograms of hydrogen in the Sun's core. Collectively, the proton-proton chain reactions going on in the Sun are responsible for the entire luminosity of the Sun.

The Sun's Lifetime Fusion extracts energy from matter even more efficiently than fission. About 6×10^{14} J of energy are produced by the fusion of 1 kg of hydrogen into helium. This is about ten times as much energy as fission produces from 1 kg of fissionable material and more than 10 million times as much as the combustion of 1 kg of fuel. Even with such an efficient fuel, however, the Sun consumes 6×10^{11} kg of hydrogen every second. This is so much fuel that approximately 10,000,000 railroad tank cars would be needed to carry it all. Despite this rate of consumption, the Sun contains enough hydrogen to last for about 100 billion years.

This estimate is unrealistic, however, because it assumes that all of the Sun's hydrogen can be transformed into helium. In reality, only the central 10% of the Sun's mass is hot enough and dense enough for hydrogen fusion to occur at a significant rate. Thus the Sun will last for about 10 billion years before the fusion of hydrogen ceases in its core. This is more than enough time to account for the history of the Earth and the rest of the solar system. It makes the Sun a middle-aged star, now about halfway through the length of time during which it will produce nuclear energy.

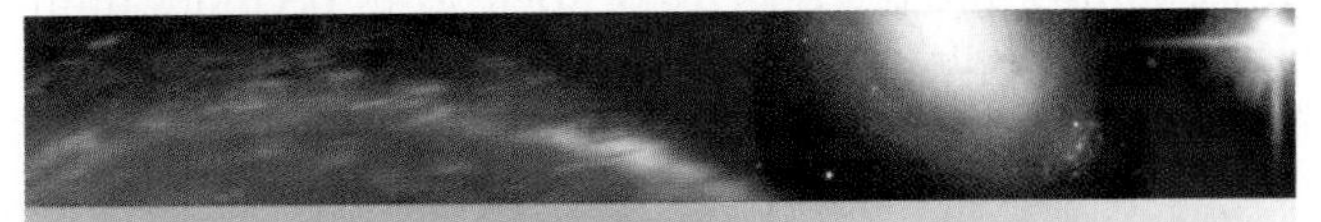

The Sun's energy is produced by the fusion of hydrogen into helium in the Sun's core. Fusion occurs through a series of nuclear reactions called the proton-proton chain. There is enough hydrogen in the Sun's core to provide energy for a luminous lifetime of about 10 billion years. The Sun is about 5 billion years old, so it has enough hydrogen in its core for about 5 billion years more.

Flow of Energy to the Surface

The energy produced by fusion in the Sun's core flows outward toward the surface. At first, the energy is carried by radiative diffusion, in which photons are emitted at one spot and absorbed at another, transporting energy between the two places. If the Sun were transparent, the photons (gamma rays) emitted by the very hot gases in the Sun's core would travel straight outward at the speed of light and escape into interplanetary space 2 seconds after being emitted. However, the Sun's gases are far from transparent. A typical photon emitted in the Sun's core travels only about 10^{-6} m before it is reabsorbed by an ion. When the photon is absorbed, its energy heats the gas that absorbs it. This hot gas then emits other photons that travel similarly short distances before they, too, are absorbed. About 10^{25} absorptions and reemissions occur before solar energy reaches the surface of the Sun.

Radiation and Convection The flow of energy out of the Sun leads to a decline in temperature with increasing distance from the center. How fast the temperature falls depends on the efficiency with which energy is transported from the hot center to the cooler surface. The efficiency with which energy is carried outward by radiation is strongly influenced by the **opacity** of the gas through which the photons flow. Opacity describes the ability of a substance to stop the flow of photons. When opacity is low (as in clear air), photons travel greater distances between emission and absorption than when opacity is high (as in cloudy, hazy, or dust-filled air). With low opacity, the transport of energy by photons is efficient, and temperature falls relatively slowly with distance. When opacity is high, on the other hand, photons travel only short distances between emission and absorption, leading to inefficient flow and a higher rate of temperature decline.

Radiative diffusion carries the Sun's energy outward from the center to about 70% of the Sun's radius. By that point, temperature has dropped from 15 million K at the center to about 1.5 million K. At this relatively low temperature the opacity of solar gases becomes quite large. As a result, the rate of temperature decline becomes steep enough that convection develops. (For more about convection, see Chapter 7, Section 5.)

The outer envelope of the Sun in which convection occurs is called the **convection zone.** Convection is a much more efficient process for transporting energy than radiative diffusion is. Although radiative diffusion still occurs in the Sun's convection zone, rising and falling currents of gas carry nearly all of the energy outward. The Sun's convection zone extends the rest of the way to the surface, where the Sun's energy is radiated into space.

The Time Delay Even though photons travel at the speed of light, the diffusion of radiation from the center of the Sun out to the convection zone is very slow. So many absorptions and reemissions occur that about 170,000 years pass before the energy produced at the Sun's center reaches the surface. Figure 17.3 shows that the energy produced in a single second is radiated from the Sun's surface over a period of more than 100,000 years. Thus the energy produced in a single second does not all reach the surface at the same time. Some energy appears after 120,000 years, other energy after 220,000 years, but most of the energy reaches the surface about 170,000 years after it is produced.

The length of time needed for energy to diffuse out of the Sun has two important implications. The first is that by observing the light emitted by the Sun, we learn almost nothing about what is going on now in the Sun's core. The fact that the Sun is now luminous tells us only that energy was generated in the Sun hundreds of thousands of years ago. The second implication is that the brightness of the Sun is very insensitive to changes (if they occur at all) in the rate at which nuclear energy is generated. Suppose energy generation were to cease in the core for a day, a year, or even a century, and then resume. By the time the energy flowed to the surface, the temporary absence of energy would be averaged over more than 100,000 years. The resulting drop in solar brightness would be too small to measure.

In some respects this is analogous to the way that an incandescent bulb responds to a 60 Hz alternating current. The current in the bulb reverses and goes to zero 120 times per second. However, it takes much longer than 1⁄120 of a second for the filament in the bulb to cool and dim very much, so the brightness of the bulb remains nearly constant.

FIGURE 17.3
The Diffusion of Energy Out of the Sun
A second's worth of solar nuclear energy reaches the surface of the Sun about 170,000 years after it is produced and is spread out over more than 100,000 years of time.

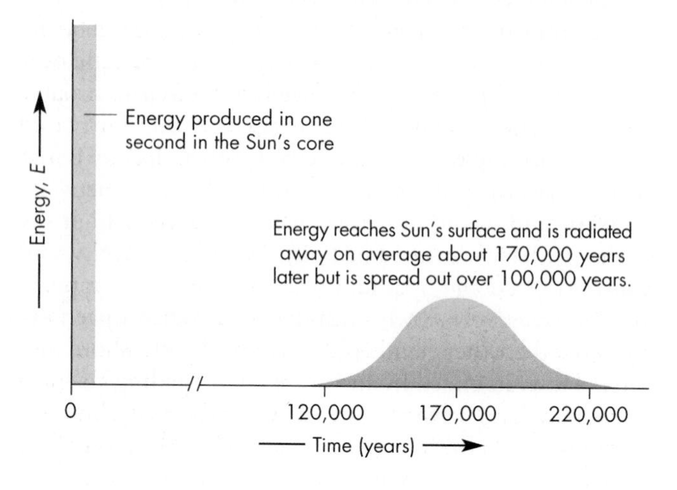
Energy produced in one second in the Sun's core
Energy reaches Sun's surface and is radiated away on average about 170,000 years later but is spread out over 100,000 years.
Energy, E
0
120,000
170,000
220,000
Time (years)

Radiative diffusion, the outward flow of photons, carries the Sun's energy outward from the core to about 70% of the Sun's radius. At that point, convection begins to occur and carries most of the energy to the surface. About 170,000 years pass between the time solar energy is produced and the time it reaches the surface.

Solar Neutrinos We could investigate the core of the Sun much more easily if photons from the core could pass directly through the Sun's gases and be detected at the Earth. (Except for scientific purposes, this would not be desirable, because most of the radiation at the core of the Sun consists of gamma rays and X rays, both of which are harmful to human tissue.) Although there is no hope of ever seeing a photon from the core, it is possible to directly observe the core by detecting the neutrinos produced there.

Neutrinos are particles produced in nuclear reactions like the proton-proton chain. Any nuclear reaction in which a proton is converted to a neutron, or vice versa, produces a neutrino. Neutrinos recently have been found to have mass, like protons and electrons. They travel nearly at the speed of light and interact weakly with matter. If a beam of neutrinos were to pass through a column of Sun-like stars over a distance of several light years, for example, only about half the neutrinos would be absorbed. The other half would emerge at the other end. Neutrinos interact so weakly with matter that nearly all of the neutrinos produced in all the nuclear reactions that have ever occurred in stars are still speeding through the universe. For neutrinos, the Sun is essentially transparent. As a result, the neutrinos produced in the Sun's core flow out of the Sun about 2 seconds later. A small fraction of them flow through the Earth about 8 minutes after they are produced. This means that solar neutrinos can tell us about what happened in the Sun's core only 8 minutes ago.

The same property that makes it possible for neutrinos to emerge from the Sun also makes it difficult to detect solar neutrinos on Earth. After all, what kind of telescope can detect particles that pass through matter? Despite this difficulty, a number of neutrino telescopes are now in operation. These telescopes rely on the fact that although neutrinos interact only weakly with any substance, they are more likely to interact with some elements (such as chlorine and gallium) than with others.

The first neutrino telescope, built by Raymond Davis in 1963, is shown in Figure 17.4. It consists of a 600,000 kg tank of cleaning fluid (C_2Cl_4) located 1.5 km deep in the Earth in the Homestake gold mine in Lead, South Dakota. The 1.5 km of rock shields the tank from cosmic rays, which could produce spurious detections. The rock, however, has no effect on neutrinos. In fact, solar neutrinos pass easily through the entire diameter of the Earth. Although 13,000 km of rock shield the telescope at night, as many solar neutrinos pass through the tank at night as during the day.

FIGURE 17.4
The First Solar Neutrino Telescope
Built by Raymond Davis in 1963, the telescope consists of a tank of cleaning fluid 1.5 km beneath the Earth's surface in the Homestake gold mine in Lead, South Dakota. The 1.5 km of Earth shields the tank from cosmic rays, which would also be detected by the telescope. About twice a day a solar neutrino reacts with a chlorine atom in the cleaning fluid to produce a radioactive argon atom. The number of argon atoms collected indicates the rate at which neutrinos reach the tank.

About 10^{16} solar neutrinos pass through each of us every second. Even more pass through the large tank containing the cleaning fluid. About once every 12 hours a solar neutrino interacts with a chlorine atom in the tank and causes a nuclear reaction in which an atom of radioactive argon is produced. The argon atoms produced by solar neutrinos are very carefully collected about every 2 months. The argon atoms are placed in a chamber where their radioactive decays can be measured to determine how much radioactive argon was produced by solar neutrinos.

The results of the Homestake experiment presented a problem for astronomers and physicists because the rate at which solar neutrinos were detected was significantly lower than predicted by models of the interior of the Sun. For his pioneering studies of solar neutrinos, Davis shared the 2002 Nobel Prize in Physics. The lower than expected rate of solar neutrino detections has been confirmed by the Super-Kamiokande experiment operated in the Kamioka mine in Japan shown in Figure 17.5. The Super-Kamiokande telescope detects neutrinos by completely different techniques and also gives the direction from which the neutrino came. The measured detections show that the Sun is the source of the neutrinos.

Two other neutrino telescopes that use gallium instead of chlorine to interact with neutrinos are SAGE and GALLEX. These experiments are especially important because they detect the relatively low-energy neutrinos produced when two protons react in the first step of the proton-proton chain. Every proton-proton chain reaction that ultimately produces helium must begin with the fusion of two protons and the emission of a low-energy neutrino. The rate at which low-energy neutrinos are produced can be calculated simply by finding out how many proton-proton chain reactions must go on each second to supply the Sun's luminosity. SAGE and GALLEX have detected low-energy solar neutrinos, but only at about two-thirds the rate predicted.

The explanation for the apparent shortage of solar neutrinos depends on the fact that the kind of neutrino produced in the Sun and captured by most neutrino telescopes is only one of three types of neutrinos. Neutrinos oscillate between these different types. So, the neutrinos produced in solar nuclear reactions turn into other varieties of neutrinos between the Sun and the Earth.

Two experiments, one in Japan and the other in the United States, have provided convincing evidence that neutrino oscillations really occur. In both experiments, a beam of neutrinos was produced by an accelerator and aimed at a detector hundreds of km away. In the U.S. experiment, the beam was produced at the Fermi National Accelerator Laboratory near Chicago and the detector was located in a mine in Soudan, Minnesota, 700 km away. The

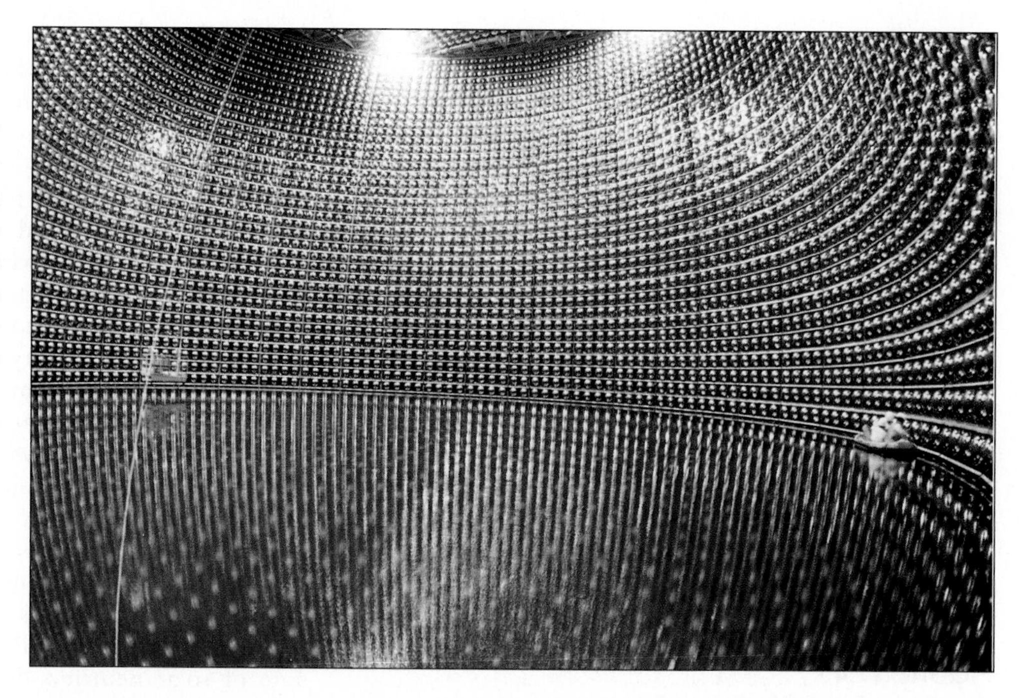

FIGURE 17.5
The Super-Kamiokande Neutrino Experiment
Super-Kamiokande consists of 50 million kg of very pure water surrounded by sensitive light detectors. When neutrinos pass through the water they cause electrons to emit feeble flashes that are measured by the detectors. About ten solar neutrinos are detected each day.

result, announced in 2006, was that only about half of the expected number of neutrino detections was seen at Soudan. The rest of the neutrinos turned into other kinds that couldn't be detected with the Soudan detector. Together with results from solar neutrino telescopes, the neutrino oscillation experiments showed that solar neutrinos are detected at a rate that is consistent with our ideas about energy generation in the Sun.

Neutrinos, produced in the proton-proton chain, pass unimpeded through the Sun's gases and escape into space. By detecting solar neutrinos, we can obtain an almost immediate view of conditions in the Sun's core. The rate at which solar neutrino telescopes measure neutrinos is consistent with ideas about energy generation in the Sun.

17.3 THE OUTER LAYERS OF THE SUN

The outer layers of the Sun are the only ones that emit radiation that reaches us here on the Earth. Light from deeper layers is absorbed by overlying gases before it can emerge from the Sun. Although the outer layers contain an insignificant fraction of the Sun's mass, by studying them astronomers have been able to discover much about the workings of the Sun and, by inference, other stars. The deepest layer that we can see is the photosphere. Above the photosphere are the chromosphere and then the corona, which extends into interplanetary space to form the solar wind.

The Photosphere

The **photosphere** is the layer that we see when we look at the visible image of the Sun. Two important properties of the photosphere can be seen in Figure 17.6. First, the photosphere has a sharp edge (or **limb**) and, second, the limb is darker than the center of the Sun's disk.

FIGURE 17.6
The Photosphere of the Sun
The photosphere is the layer that we see in visible images of the Sun. Notice that the Sun has a sharp edge (or limb) and that the edge of the Sun's disk is dimmer and redder than the center.

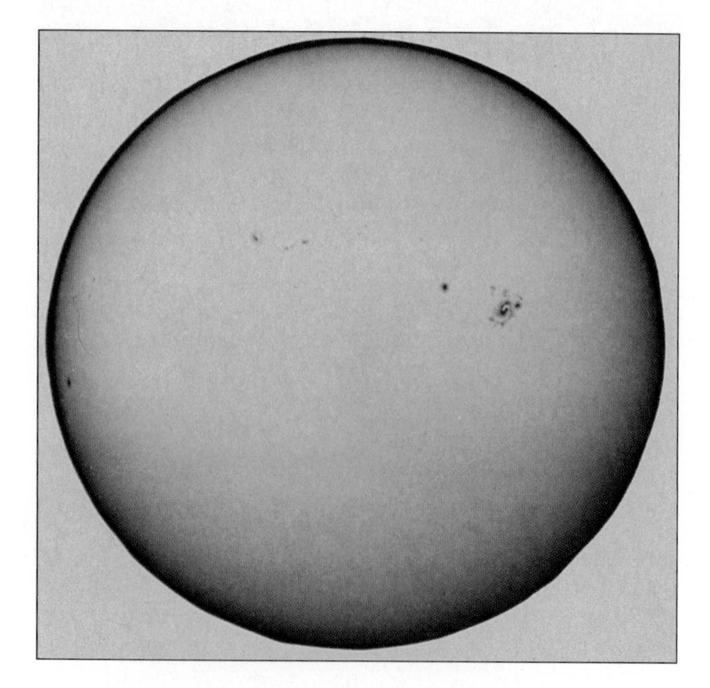

The Sharpness of the Limb The reason that the limb of the Sun is so sharply defined is that the Sun's gases go from transparent to opaque within a short distance in the photosphere. This occurs because in the photosphere density falls by 50% with every increase of 100 km in height. As shown in Figure 17.7, radiation can emerge only from layers that aren't buried too deeply. On the other hand, layers high in the photosphere have very low densities and emit very little light. It is only within a layer about 200 km thick that gas is neither buried too deeply to be seen nor is too low in density to emit much light. Because 200 km represents only about 0.03% of the Sun's radius, the edge of the Sun appears almost as sharp as if the Sun had a solid surface.

Limb Darkening The relative faintness of the Sun's limb is called **limb darkening.** In visible light, the limb is only about 40% as bright as the center of the Sun's disk. Limb darkening occurs because, within the photosphere, the Sun's temperature decreases outward. Figure 17.8 shows what we see when we look at the center and the limb of the Sun. At the center, we look straight into the Sun. The layer of gas that we see has a temperature of about 6100 K.

When we view the Sun's limb, we look obliquely through the Sun's gases. The 6100 K layer of the photosphere lies beneath enough absorbing gas that most of its light is absorbed. The layer that produces most of the light we see lies higher in the atmosphere, where the Sun's gases are cooler. The higher, cooler gas we see at the Sun's limb emits less brightly than the deeper, hotter gas seen at the center of the disk. If temperature didn't decrease outward in the photosphere, the gas we see at the limb would be just as hot and bright as the gas seen in the center of the

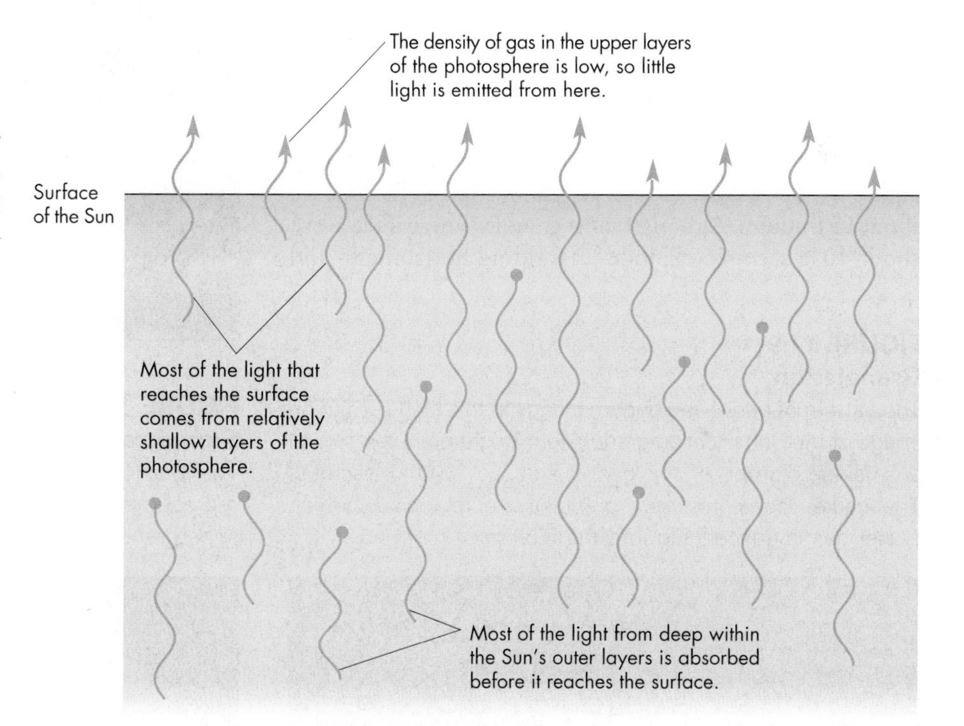
The density of gas in the upper layers of the photosphere is low, so little light is emitted from here.
Surface of the Sun
Most of the light that reaches the surface comes from relatively shallow layers of the photosphere.
Most of the light from deep within the Sun's outer layers is absorbed before it reaches the surface.

FIGURE 17.7
Light Flowing Through the Sun's Outer Layers
Nearly all of the light from shallow layers escapes, whereas much of the light from deeper layers is absorbed by overlying gas (absorptions are indicated by dots).

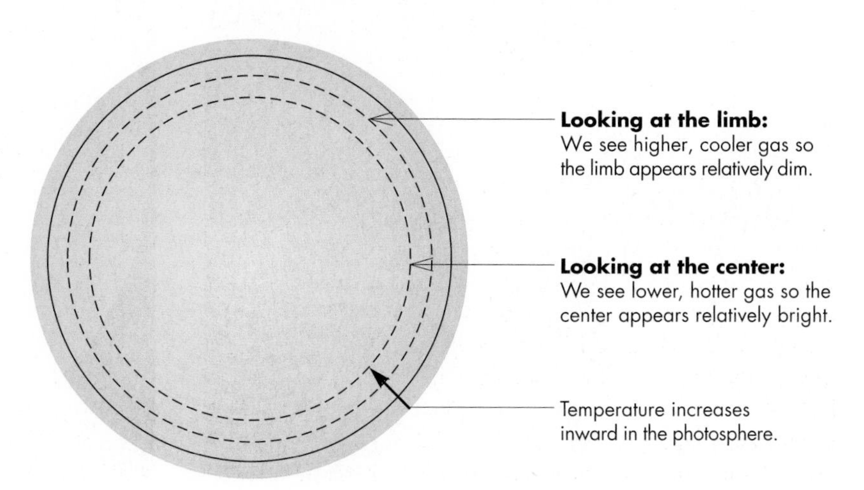
Looking at the limb:
We see higher, cooler gas so the limb appears relatively dim.
Looking at the center:
We see lower, hotter gas so the center appears relatively bright.
Temperature increases inward in the photosphere.

FIGURE 17.8
Why the Sun's Limb Appears Dark
When we look at the center of the Sun's disk, we see straight down through the layers of atmospheric gas. At the limb, we look obliquely through the gas and see a shallower layer than at the center. Because the Sun's temperature decreases with height in the photosphere, the higher layer is cooler and emits less light than the deeper layer seen at the center of the Sun's disk.

disk, and therefore limb darkening wouldn't occur. Averaged over the entire disk of the Sun, the gas that we see has a temperature of 5800 K, which is considered to be the surface temperature of the Sun.

The sharp edge of the solar disk shows that density decreases rapidly in the Sun's outer layers. Limb darkening shows that temperature decreases outward in the visible layers of the Sun.

Granulation The high-resolution picture of the Sun in Figure 17.9 shows a pattern of bright and dark markings called granulation. Granulation makes the Sun look like the pebbled surface of a basketball. The bright regions are called **granules** and they are typically about 1000 km across. Between the granules are irregular lanes that are darker than the granules because they are about 100 K cooler.

The pattern of granulation changes constantly. As Figure 17.10 shows, a typical granule has a lifetime of about 15 minutes. Although most granules simply fade away, about 40% seem to explode, expanding horizontally and pushing neighboring granules away. The Sun's granulation is caused by the convection that transports most of the Sun's energy within the outer third of the Sun's radius. The granules are rising columns of hot gas, and the dark lanes are places where cooler gas is descending into the interior.

Another convective pattern, with a scale much larger than granulation, is also present in the photosphere. This pattern, called **supergranulation,** is shown in Figure 17.11.

FIGURE 17.9
Granulation
The pattern of bright and dark markings in this high-resolution image of the Sun is called granulation. Bright granules, which are rising columns of hot gas, are about 1000 km across. The darker lanes between granules are regions in which cooler gas is descending into the interior.

FIGURE 17.10
The Changing Granulation Pattern
These six images were taken 2 minutes apart. Notice that the granule at the center of the image explodes outward as it fades from view. The size of the state of Texas is shown for comparison.

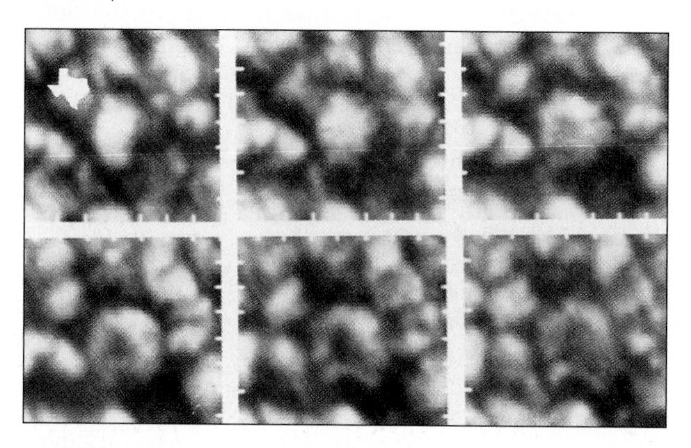

FIGURE 17.11
Supergranulation
This *SOHO* image shows a mottled bright and dark pattern, representing huge convective plumes, that are the Sun's supergranulation.

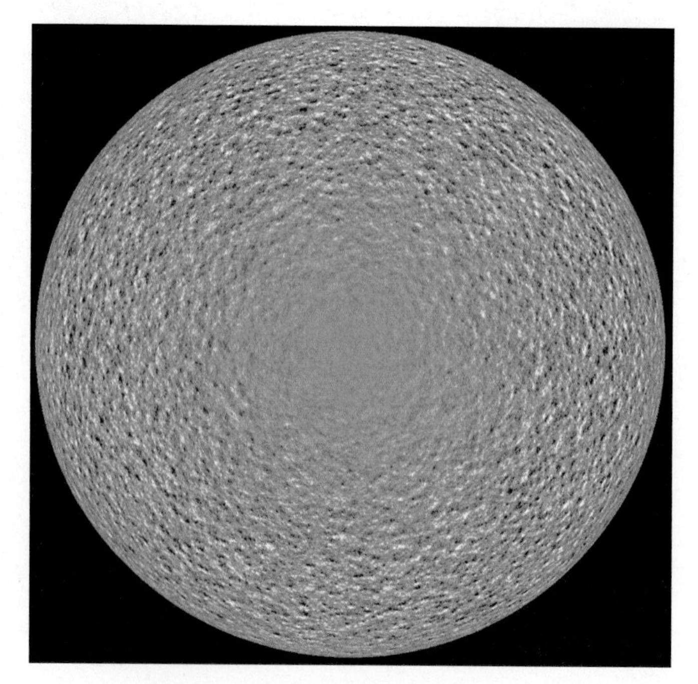

FIGURE 17.12
The Convective Patterns of Granulation and Supergranulation
Hot gases rise to the Sun's surface in the centers of supergranulation. This gas cools as it flows to the edge of the supergranule, where it sinks back into the solar interior.

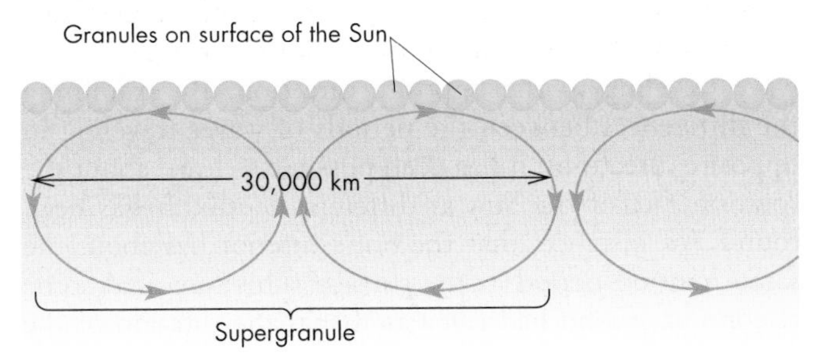

The supergranules are huge, about 30,000 km across, and contain hundreds of granules. They are the tops of convective plumes that probably extend downward 200,000 km to the bottom of the Sun's convection zone. Despite their size, supergranules are much harder to see than granules are, because supergranules carry little energy outward and are negligibly brighter than their surroundings. As Figure 17.12 shows, supergranules resemble granules in that gas rises to the surface in the center of a supergranule, spreads out, and then sinks back into the interior at the edge of a supergranule, just as it does in a granule. A typical supergranule lasts for about a day.

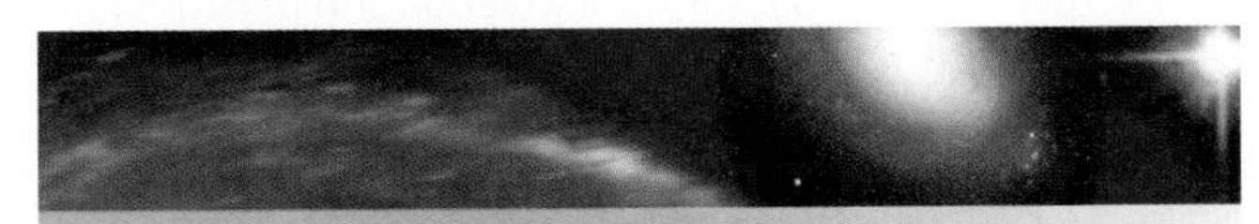

The Sun's surface is pebbled by granules, which are rising columns of hot gas. A much larger convective pattern, supergranulation, is harder to see and carries little energy outward.

Helioseismology While making detailed Doppler shift observations of the Sun's photosphere in 1960, Caltech's Robert Leighton noticed that the Sun vibrates at a variety of frequencies, like a musical instrument or the Earth after an earthquake. Upon closer examination, astronomers found that the gas in the photosphere is oscillating up and down with speeds of 100 to 300 m/s and periods of about 5 minutes. The spatial pattern of regions moving up and down at any moment is quite complex, as shown in Figure 17.13. The patterns are complex because there are millions of oscillations going on at the same time. One of the patterns, or **modes of oscillation,** is shown in Figure 17.14. Just as seismic waves in the Earth yield information about the parts of the interior through which they pass, the oscillations of the Sun have proved to be a rich source of information about its interior. The study of solar oscillations is sometimes called **helioseismology.**

FIGURE 17.13
The Oscillation Pattern of the Sun
Bright regions are moving toward the Earth and dark regions are moving away. The pattern is complex because millions of oscillation modes are occurring at the same time.

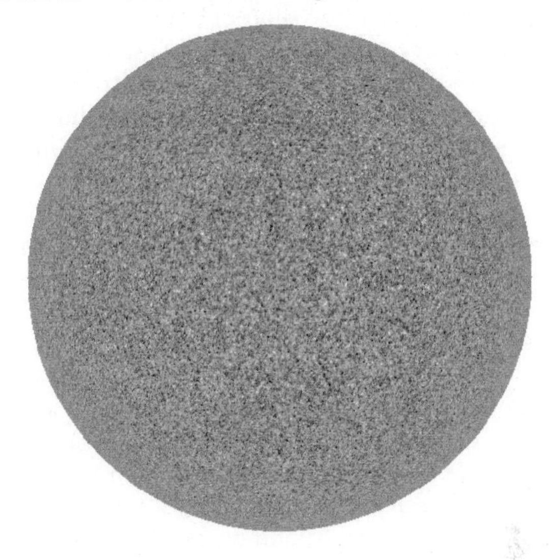

FIGURE 17.14
One of the Oscillation Modes of the Sun
Blue regions are moving outward and red regions are moving inward.

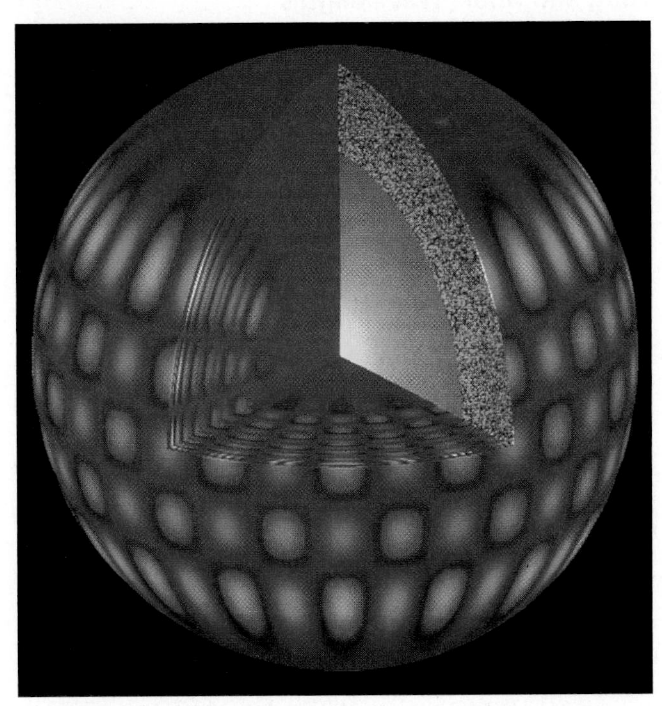

If the string of a violin or harpsichord is plucked, many different modes of oscillation are produced in the string, each mode having a different wavelength. Similarly, the violent bubbling in the Sun's convection zone excites oscillations in the Sun, producing modes with a wide range of wavelengths. The shortest modes that have been detected have wavelengths of a few thousand kilometers. These modes resonate within the outermost parts of the Sun and give very little information about the interior.

Longer wavelength modes, like those shown in Figure 17.15, penetrate much more deeply into the Sun. The period of oscillation of a particular mode depends on how long it takes the wave to penetrate to its maximum depth and return to the surface. Thus, it depends on the speed of sound in the region within which the wave resonates. Because modes of different wavelengths probe different depths in the Sun, careful analysis of the periods of the modes of oscillation reveals how the speed of sound varies with depth in the Sun. This is the same kind of information that we get from seismic waves in the Earth.

The speed of sound depends primarily on the temperature and chemical composition of the gas through which the sound wave passes. Thus solar oscillations have helped scientists understand how temperature and composition vary with depth in the Sun. Helioseismology led to the discovery of the depth at which the Sun becomes convective and also showed that helium is somewhat more abundant in the Sun than had been previously believed.

The rotation of the Sun influences the effective speed of sound within the Sun. A wave traveling in the same direction as the Sun's rotation travels faster than does a wave traveling in the opposite direction. By examining the differences between the periods of waves traveling in opposite directions, it has been possible to learn about the rotation rate of the Sun at different depths. It has been found, for instance, that the Sun's interior has about the same rotation period as the surface. This surprised some astronomers, who had calculated that the rotation of the Sun's surface layers had slowed faster than had the rotation of the interior, so that the interior should rotate faster than the photosphere.

Measurements of the solar oscillations made using the *SOHO* (*Solar and Heliospheric Observatory*) spacecraft have confirmed the presence of streams of gas moving beneath the Sun's surface. Figure 17.16 shows the pattern of streams in the Sun. One major stream lies in the

FIGURE 17.15
The Parts of the Sun Probed by Oscillations of Long and Short Wavelengths
A, The short wavelength oscillations are confined to the outer regions of the Sun, whereas **B,** the long wavelength oscillations probe the deep interior of the Sun.

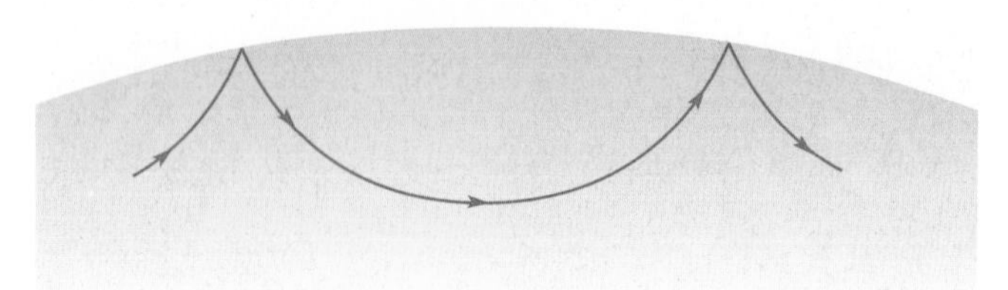

A Short wavelength oscillations probe the outer regions of the Sun.

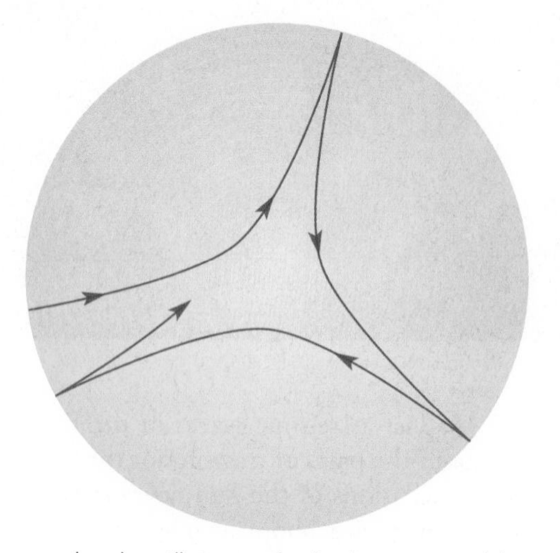

B Long wavelength oscillations probe the deep interior of the Sun.

FIGURE 17.16
Streams in the Sun
This false-color image shows streams of gas moving beneath the Sun's surface. Red indicates streams moving faster than their surroundings. Blue and green show streams moving slower than their surroundings. A very large, fast moving stream can be seen in the Sun's equatorial region. The much smaller ovals just beneath the surface in the polar regions show other fast moving streams of gas. The yellow region shows the part of the Sun buried too deeply for observations to have determined whether there are streams there.

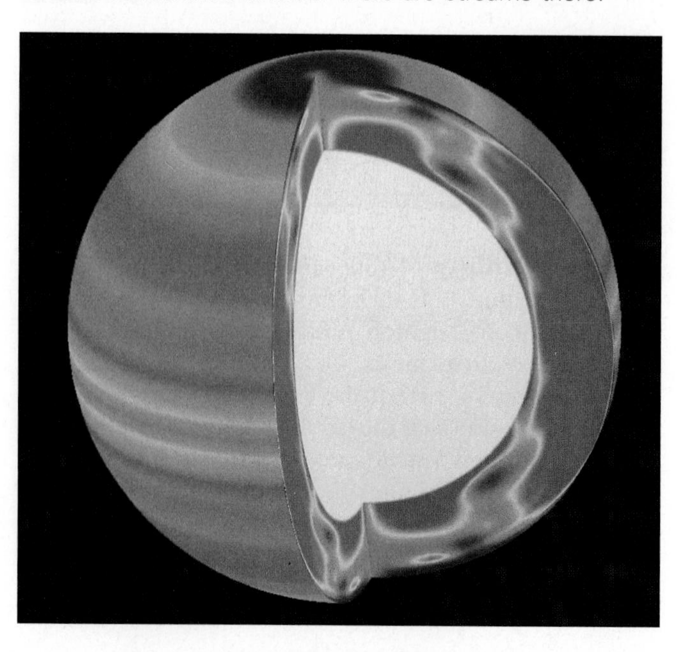

equatorial region and extends downward at least 30% of the way to the Sun's center. Other smaller streams near the poles move 10% faster than surrounding gas. All of these streams are completely submerged beneath the Sun's surface. At the Sun's surface there is a slow flow of gas from the equator to the poles. Other *SOHO* observations of solar oscillations have been used to obtain "images" of a sunspot region on the far side of the Sun. This is possible because the sunspot region interferes with waves traveling through it and disturbs the usual pattern of solar oscillations.

The Sun vibrates in a complex pattern of oscillations. Through helioseismology, the study of the pattern of oscillations, astronomers have been able to discover how temperature, composition, and rotation vary with depth in the Sun.

Sunspots and Solar Magnetism The dark, roundish markings in the photosphere are called **sunspots.** Often they are large enough to be seen without a telescope, as shown by reports of them in ancient Chinese writings. Telescopic observations of sunspots show that they range in size from pores, which are less than 2500 km across and last for less than an hour, to spots as large as 50,000 km across, or four times the diameter of the Earth. Large spots, such as the one shown in Figure 17.17, last for many months.

The inner, darkest part of a sunspot is the **umbra.** Surrounding the umbra is the **penumbra,** which consists of dark radial features that make the entire spot look something like a daisy. Larger spots form when smaller spots coalesce. Spots tend to occur in **sunspot groups** such as the ones shown in Figure 17.18. The largest sunspot groups can fill more than half a percent of the solar disk. The area around a sunspot group is called an **active region,** because flares and other events occur there.

When Galileo made the first telescopic observations of sunspots, he found that the Sun's rotation carries spots across the solar disk. (Galileo didn't look at the Sun through the eyepiece but instead used the telescope to project the Sun's image onto a screen. No one should ever look at the Sun, with or without a telescope, unless they use a special "solar filter.") Unlike the Earth, the Sun doesn't have a single rotation period. Instead, the rotation period varies with latitude. Near the equator, the rotation period is about 25 days. The rotation period lengthens from the equator to the poles, where it reaches 36 days. This phenomenon, called **differential rotation,** has important consequences for the Sun's magnetic field.

Sunspots look dark because they are cooler than the rest of the photosphere. The typical umbral temperature of a sunspot is 3800 K, about 2000 K cooler than the surrounding photosphere. Sunspots appear dark only in contrast to the brightness of the rest of the photosphere. If a large sunspot could be seen against a dark sky instead of against the photosphere, for instance, it would be about ten times brighter than the full Moon.

The relatively low temperatures of sunspots are a consequence of the very large magnetic field strengths that occur in sunspots. We are able to measure magnetic

FIGURE 17.17
A Large Sunspot
The dark inner region is the umbra. Around the umbra is the penumbra. Large sunspots can be as large as 50,000 km across and can last for many months.

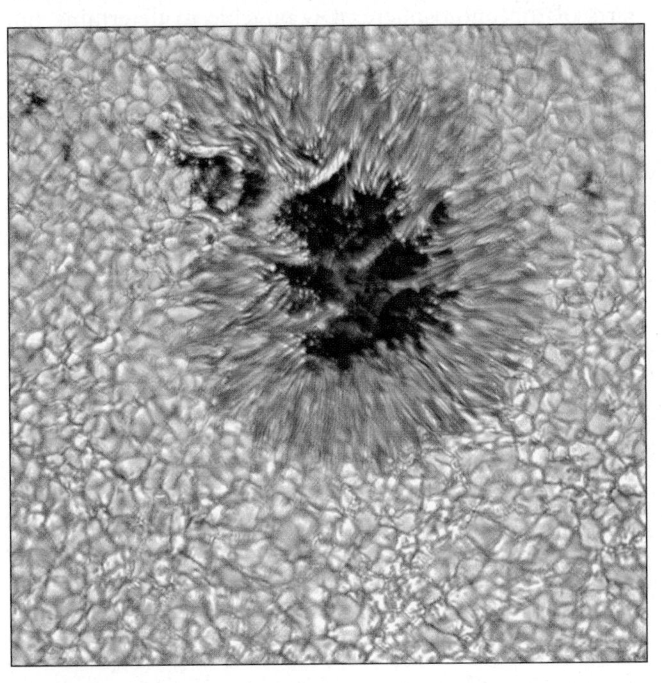

FIGURE 17.18
A Large Sunspot Group
Most sunspots are found in sunspot groups like the ones in the top and bottom center of the image.

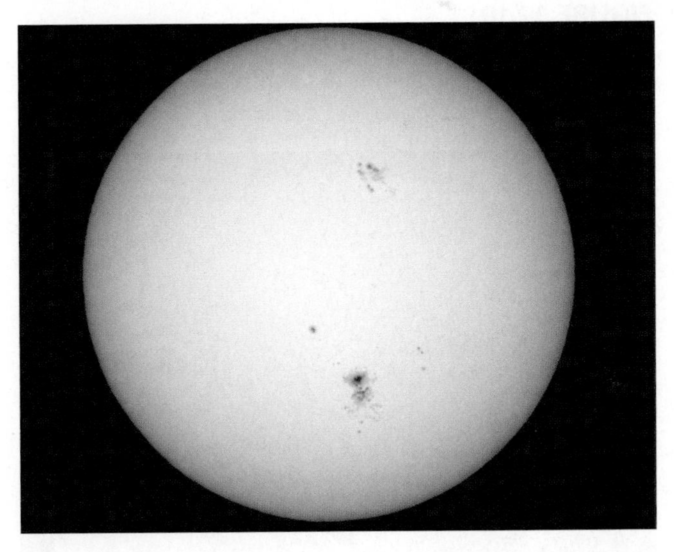

fields in sunspots and elsewhere on the Sun because of the **Zeeman effect,** in which the presence of a magnetic field causes a spectral line emitted by an atom to be split into two or more separate lines. The stronger the magnetic field, the larger the split. By examining spectra of different parts of the photosphere, it is possible to determine not only the strength of the magnetic field in that region, but also its **polarity** (that is, whether the magnetic field in that region is north or south).

A map of the strength and polarity of the Sun's magnetic fields is shown in Figure 17.19. The map shows that regions of strong magnetic fields are concentrated in sunspot groups containing areas of both north and south polarity. In a sunspot group the magnetic field can be as much as ten thousand times as strong as the magnetic field at the Earth's surface. Larger sunspots have larger magnetic field strengths.

Sunspots are cool and dark because the large magnetic fields in sunspots inhibit convection in the zone just beneath the photosphere. Ionized gases, like those in the Sun's photosphere, can flow easily along magnetic field lines, but not so easily across them. The vertical magnetic field lines in sunspots inhibit horizontal flow of the material brought to the surface by convection. This inhibits the "rolling" motion of convection and the amount of energy that convection carries to the surface. The energy that sunspots block from reaching the surface doesn't simply appear in the region around the sunspots. Instead, it is stored in the convection zone and only very gradually reaches the surface of the Sun.

FIGURE 17.19
A Map of the Sun's Magnetic Fields
Dark areas in this image have north magnetic polarity and light areas have south magnetic polarity.

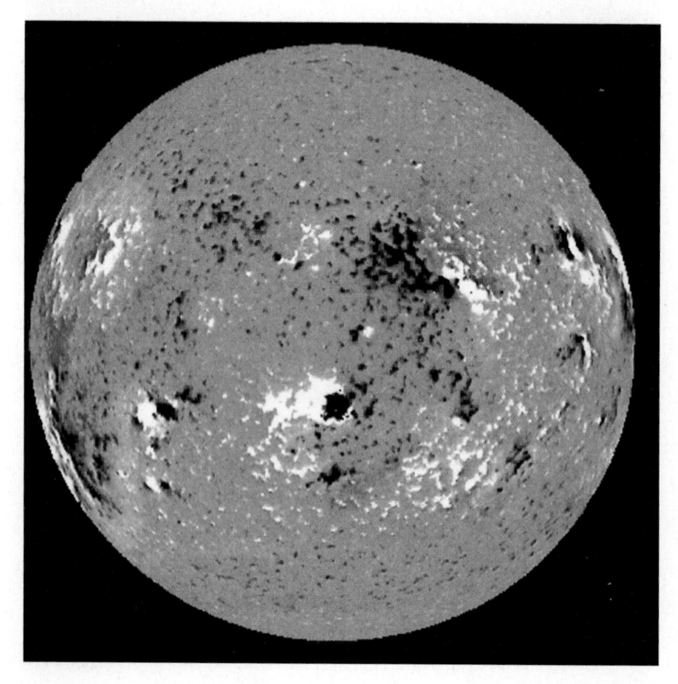

Sunspots appear dark because they are cooler than the rest of the photosphere. The large magnetic fields in sunspots inhibit convection and reduce the flow of energy to the Sun's surface. The rate at which sunspots cross the disk of the Sun varies with latitude because the rotation rate of the Sun decreases as the distance from the equator increases.

The Chromosphere

At the moment a solar eclipse becomes total and the bright photosphere is completely blocked, a reddish glow (Figure 17.20) can be seen just above the limb of the Moon. After a few seconds, the reddish glow is also eclipsed by the Moon. The glow is produced by the **chromosphere,** a tenuous region that lies just above the photosphere. It appears red because of the strength of the hydrogen Balmer α emission line in the red part of the spectrum. (For a review of emission spectra, see Chapter 16, Section 5.) Except during eclipses, the chromosphere is difficult to see. It contains so little material (only about 10^{-12} of the Sun's mass) that it is transparent at most wavelengths. We see right through it to the photosphere. At the wavelengths of strong atomic lines, however, the gases in the chromosphere are opaque enough that the chromosphere can be seen. Imaging the Sun at the wavelengths of strong atomic lines has made it possible to discover the structure of the chromosphere.

Early observations of the chromosphere during eclipses showed that it has a ragged structure that some observers compared to a burning prairie. The ragged appearance is

FIGURE 17.20
The Sun's Chromosphere Seen During a Total Solar Eclipse
The chromosphere can be seen briefly just after the Sun's bright photosphere is completely blocked by the Moon.

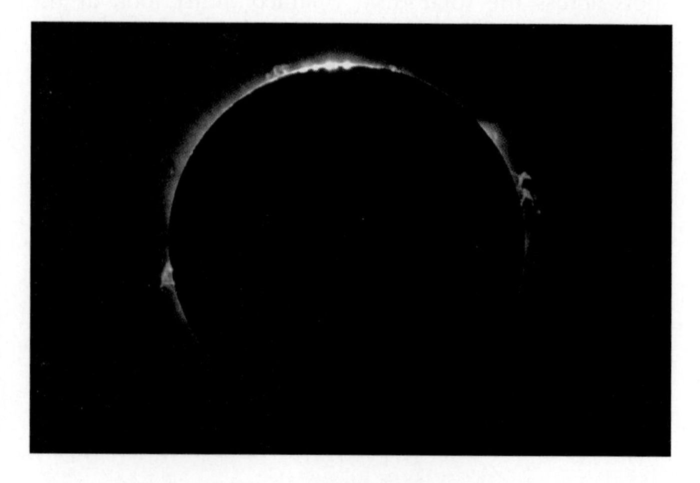

caused by **spicules,** which are jets of gas shooting upward at about 25 km/s (50,000 miles per hour). A typical spicule rises for 5 minutes before fading rapidly from sight. Although the spicules on the Sun's limb appear crowded together, this is deceptive. Figure 17.21 shows an image of the Sun at the wavelength of Balmer α. The spicules, which appear dark in this image, cover only a few percent of the Sun's surface. The spicules form a network above the edges of the photospheric supergranules. At any given moment, there are probably about one hundred thousand spicules in the chromosphere.

The best current explanation for what pushes spicules up from the base of the chromosphere involves solar oscillations. The oscillations create waves that travel through the Sun's atmosphere. Magnetic field lines in active regions provide channels through which the waves can push hot gas upward, forming spicules. This explanation is consistent with the intimate connection between spicules and magnetic fields. Even the orientation of a spicule, whether it is elongated horizontally or vertically or somewhere in between, is determined by the direction of the local magnetic field lines.

Except for the spicules, the chromosphere is mostly empty. In the cells that lie between the network of spicules, the chromosphere does not exist. Instead, the temperature rises rapidly to millions of degrees.

The chromosphere is a tenuous region of gas that lies just above the photosphere. Most of the chromospheric gas takes the form of spicules, which are rapidly rising jets of gas.

The Corona

Above the chromosphere is the corona, seen in Figure 17.22 as a ghostly glow surrounding the Moon during a total solar eclipse. Much of the light of the corona originates in the photosphere. This light is scattered toward the Earth by electrons in the corona. Some coronal light, however, consists of emission lines emitted by coronal gas. Astronomers had observed a number of these emission lines by 1935 but hadn't been able to match any of them to the spectral lines of known atoms. In desperation, some astronomers attributed the lines to a new element, coronium, which they thought existed in the corona, but not on the Earth.

It turned out that the coronal lines were hard to identify because they are emitted not by atoms or molecules but by highly ionized gases. The gases that make up the corona contain iron, nickel, and calcium (in addition to hydrogen and helium), but those ions are missing as many as 15 electrons. The ionization of the corona is a result of its very high temperature—more than 1 million K. The climb in temperature from the relatively cool chromosphere to the corona is very swift, as shown in Figure 17.23. The transition region in which temperature rises from about 4000 K to more than 10^6 K is only a few thousand kilometers thick.

Heating the Corona Astronomers do not agree about how the coronal gases are heated to such a great temperature. It is important to realize, however, that it takes relatively little energy to heat the corona. A hot gas with a low density, such as that in the corona, emits light very poorly. The rate at which energy must be supplied to the corona to balance the energy lost via coronal emission and keep the corona hot is only about one-one hundred thousandth of the energy output of the Sun. It seems likely that the mechanism that heats the corona involves magnetic

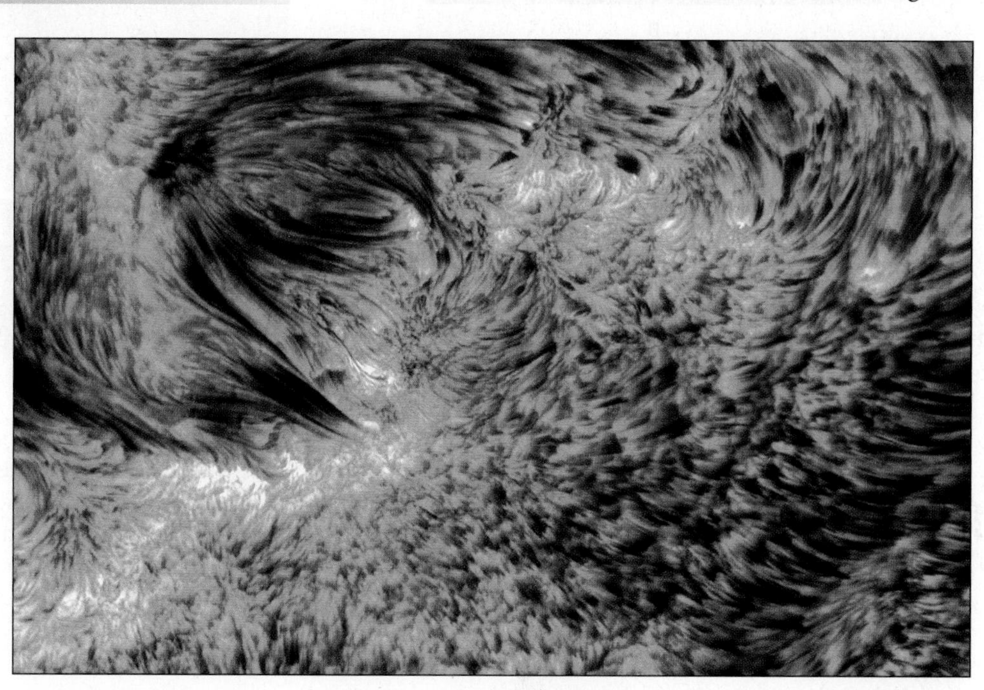

FIGURE 17.21
Spicules on the Disk of the Sun
The spicules form a network above the boundaries of photospheric supergranules.

FIGURE 17.22
The Corona During Total Eclipse
A, At the time of maximum solar activity, **B,** at the time of minimum solar activity.

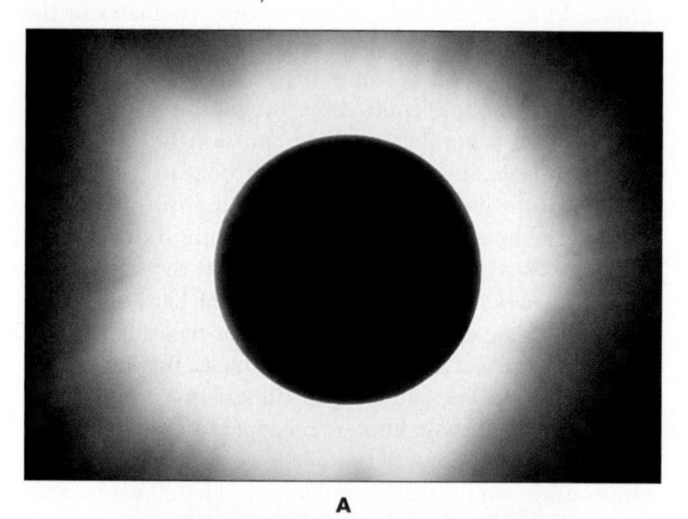

A

B

FIGURE 17.23
Temperature Variation in the Chromosphere and Corona
At the boundary between the two layers (the transition region), temperature climbs abruptly from 10,000 to 1 million K.

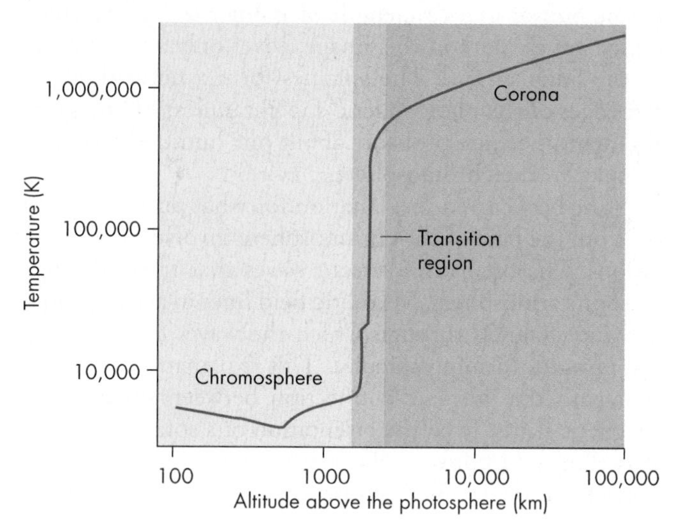

FIGURE 17.24
Magnetic Bundles in the Solar Photosphere
The inset *SOHO* image shows some of the many bundles of concentrated magnetic fields in the photosphere. The black and white spots have opposite magnetic polarities. The computer generated sunscape shows a web of field lines connecting regions of opposite magnetic polarity.

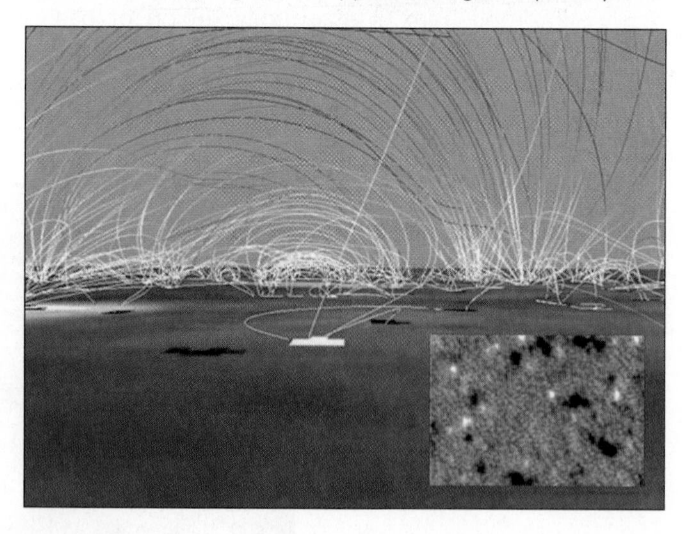

fields, which extend outward from the photosphere into the corona. Figure 17.24 shows a *SOHO* image of part of the magnetic field structure of the photosphere. The photosphere contains tens of thousands of bundles of concentrated magnetic field connected by a web of magnetic field lines. The magnetic bundles develop, drift, and merge after a typical time of 40 hours. The energy released when they merge is more than enough to heat the corona. In fact, the brightest regions of the corona lie directly above the parts of the photosphere that contain the most magnetic bundles. Although the *SOHO* data strongly support the idea that magnetic fields are responsible for the heating of the corona, astronomers do not yet understand how energy from the merging of magnetic bundles is transported to the corona thousands of kilometers above.

Coronal Structure The structure of the corona can best be seen in X-ray and ultraviolet images obtained from space, such as the one shown in Figure 17.25. Notice that the brightness of the corona is quite uneven. Bright regions cover most of the Sun but they are separated by large, dark regions known as **coronal holes.** Coronal holes appear dark because they are cooler and have lower densities than the rest of the corona. The bright parts of the corona are regions where the hot coronal gas is trapped by the Sun's magnetic field. This happens over active regions

FIGURE 17.25
An Ultraviolet *SOHO* Image of the Sun
At a temperature of several million kelvins, the corona is brightest in the X-ray and ultraviolet regions of the spectrum. Notice, however, that the brightness of the corona is uneven. The dark band in the center of the image, for instance, is a coronal hole.

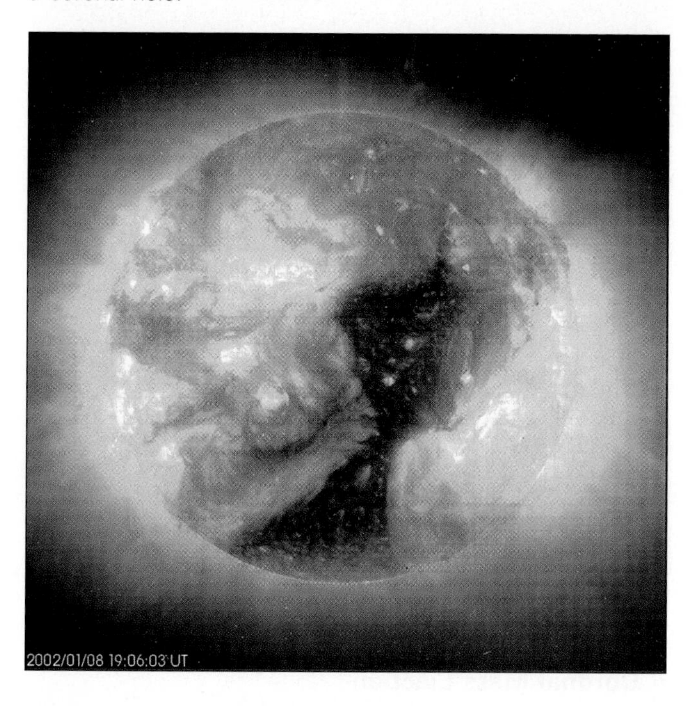

where magnetic field lines emerge from sunspots of one magnetic polarity and return through the photosphere in sunspots of the opposite polarity, as shown in Figure 17.26. Figure 17.27 shows coronal loops, which consist of hot gas lying along magnetic field lines above active regions. Away from active regions, the Sun's magnetic field is weaker and is unable to contain the hot coronal gas. Part of the energy supplied to the gas in coronal holes accelerates the gas away from the Sun, carrying the magnetic field away and releasing a high-speed stream of gas into interplanetary space.

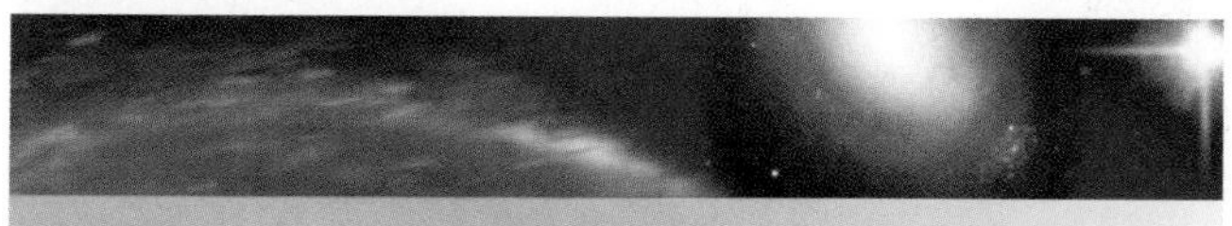

Above the chromosphere is the corona, within which the temperature is millions of kelvins. The high temperature of the corona apparently is maintained through heating by magnetic fields. The corona is densest (and brightest) above active regions, where coronal gas is trapped by magnetic fields. Away from active regions, the Sun's magnetic field can't trap coronal gas, which accelerates away from the Sun, forming dark coronal holes.

FIGURE 17.26
Open and Closed Magnetic Field Regions
In closed regions, the field lines loop outward through the corona but return to the photosphere. In open regions, the field lines are swept outward by expanding coronal gas. Coronal holes correspond to open regions.

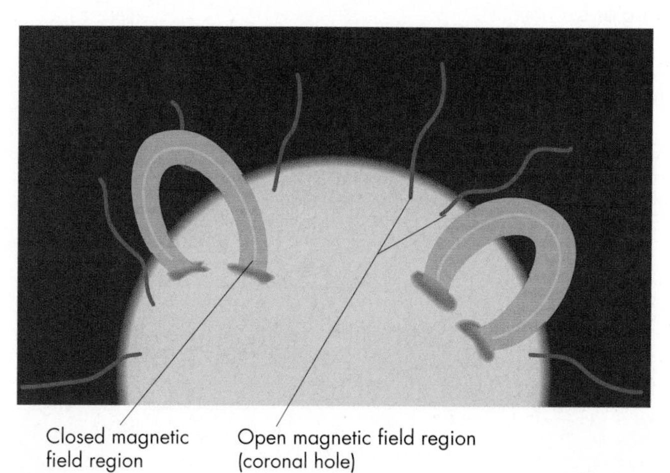

FIGURE 17.27
Coronal Loops
This image, taken at ultraviolet wavelengths by the *TRACE* spacecraft, shows numerous coronal loops. The loops consist of hot coronal gas lying along magnetic field lines that emerge from and reenter the photosphere in active regions. Typical coronal loops are dozens of times larger than the Earth.

Prominences and Flares During most total solar eclipses, bright clouds of gas, such as the ones shown in Figure 17.28, can be seen reaching upward for 50,000 km or more into the corona. These dense, relatively cool features are called **prominences** and can last for as long as 2 or 3 months. Prominences are actually cooler than the

FIGURE 17.28
Prominences on the Sun's Limb
On the left is a quiescent prominence, which can last for months. On the right is a shorter-lived, active prominence. Quiescent prominences are located away from active regions, whereas active prominences are located above active regions.

Prominences on the Sun's limb

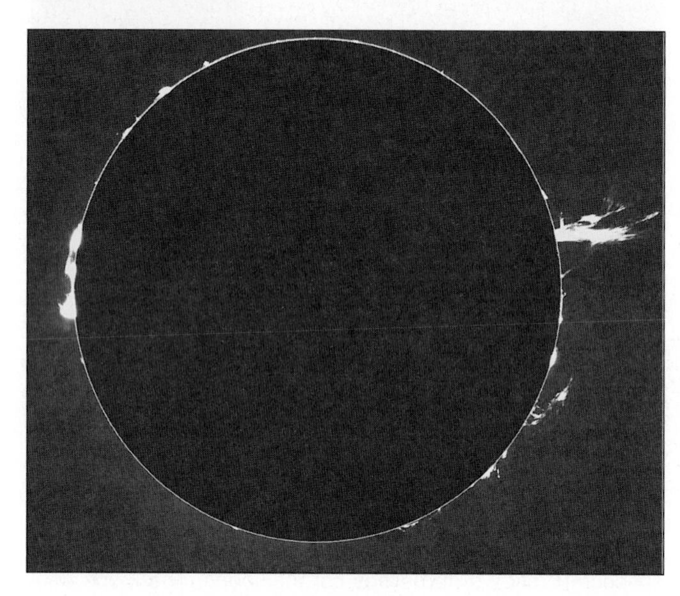

photosphere and appear as dark **filaments** when seen against the bright disk of the Sun. A more detailed picture of a prominence is shown in Figure 17.29.

Prominences seldom fade away. Instead, after rising to elevations of 50,000 km above the photosphere, they erupt. Most of them erupt upward for a few minutes to a few hours. The result of a prominence eruption often is a blast of gas moving outward through the corona and into interplanetary space at speeds as great as 1000 km/s. These events, such as the one shown in Figure 17.30, are called **coronal mass ejections** and occur on average about once or twice a day.

Solar flares, much more explosive and energetic than prominences, are abrupt releases of solar magnetic energy. These events take place in active regions where a prominence is supported against gravity by magnetic field lines. In a process not yet understood, the magnetic field structure changes abruptly. When this happens, large numbers of energetic ions and electrons are produced. Some of these ions and electrons collide with coronal gas, heating it to temperatures as high as 40 million K. For several minutes, the heated gas radiates brightly, as shown in Figure 17.31, emitting mostly X rays and ultraviolet radiation in addition to visible light. An unusually bright flare can temporarily increase the Sun's brightness by about 1%. When the X rays and ultraviolet radiation from a flare reach the Earth (8 minutes after the flare takes place), they ionize atmospheric gases, enhancing the ionosphere and disrupting

FIGURE 17.29
A Solar Prominence
The upper right of this *SOHO* image shows a prominence so large that the Earth would easily fit under its arc.

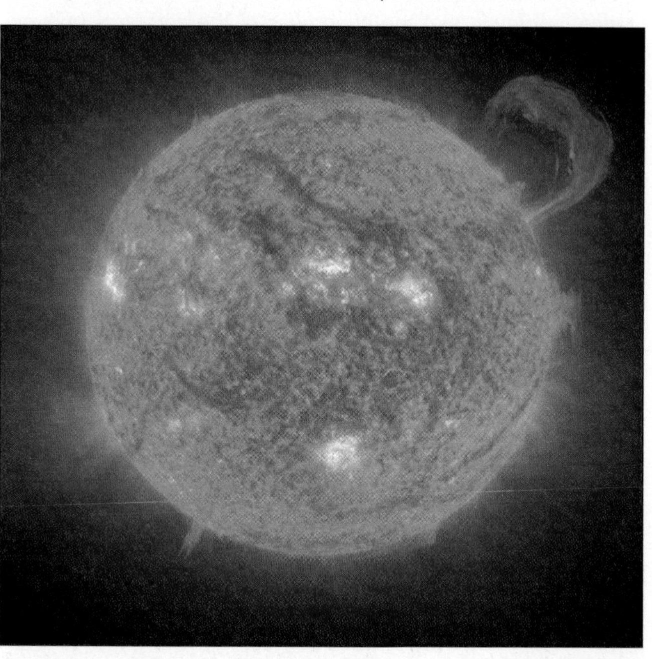

FIGURE 17.30
A Coronal Mass Ejection
A coronal mass ejection such as the one at the top of this *SOHO* image occurs when a prominence reaches a height of 50,000 km above the photosphere and erupts. These blasts of gas occur once or twice a day on average.

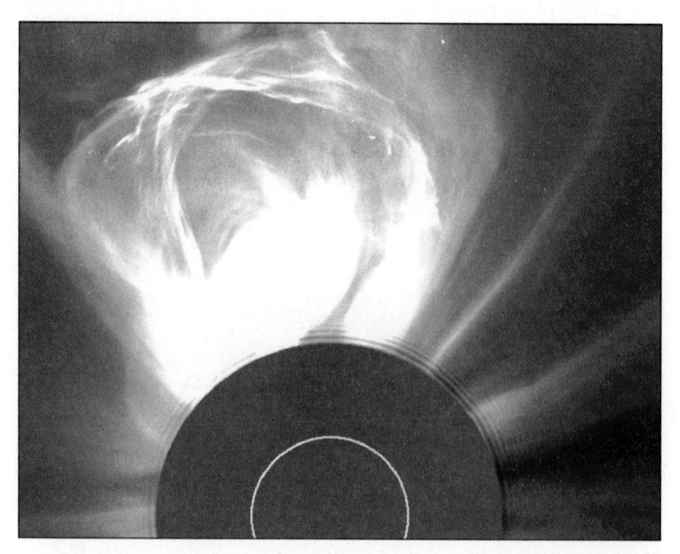

long-range radio communications. Like eruptive prominences, flares also eject large amounts of chromospheric and coronal gas into interplanetary space. Flares and coronal mass ejections often occur almost simultaneously and are different aspects of the same magnetic explosion on the Sun.

FIGURE 17.31
The Development of a Solar Flare
Solar flares are abrupt releases of solar magnetic energy. Large numbers of energetic ions and electrons collide with coronal gas and heat it to temperatures as high as 40 million K. The heated gas then radiates brightly, emitting mostly X rays and ultraviolet radiation but some visible light. These six images show the changes in a flaring region over about an hour.

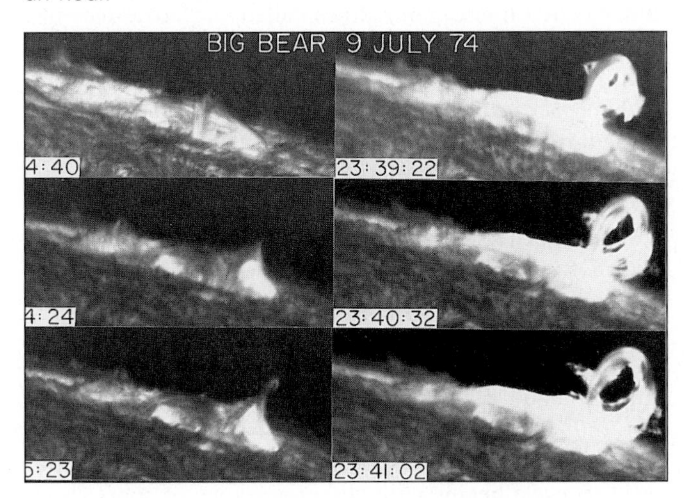

Prominences are clouds of relatively cool gas that extend upward into the corona. Large prominences often erupt, blasting gas outward into space. Flares are explosive releases of the Sun's magnetic energy. X rays, ultraviolet radiation, and energetic particles are produced by flares.

The Solar Wind

The flow of coronal gas into interplanetary space is called the solar wind. The solar wind flows away from the Sun at a speed of about 450 km/s and moves past the Earth 4 days after leaving the Sun. The magnetic field lines swept outward by the solar wind remain anchored in the Sun, which rotates every 27 days. By the time the solar wind reaches the Earth, the part of the Sun to which the field line is anchored has rotated about 50°. Thus, the magnetic field in the solar wind follows the spiral pattern illustrated in Figure 17.32. The solar wind is greatly enhanced above a coronal hole. About 4 days after a coronal hole moves across the center of the Sun's disk, we can expect a stream of gas denser than the normal solar wind to move past the Earth as fast as 700 km/s. Coronal mass ejections from eruptive prominences and blasts from flares also reach the Earth as enhancements of the solar wind.

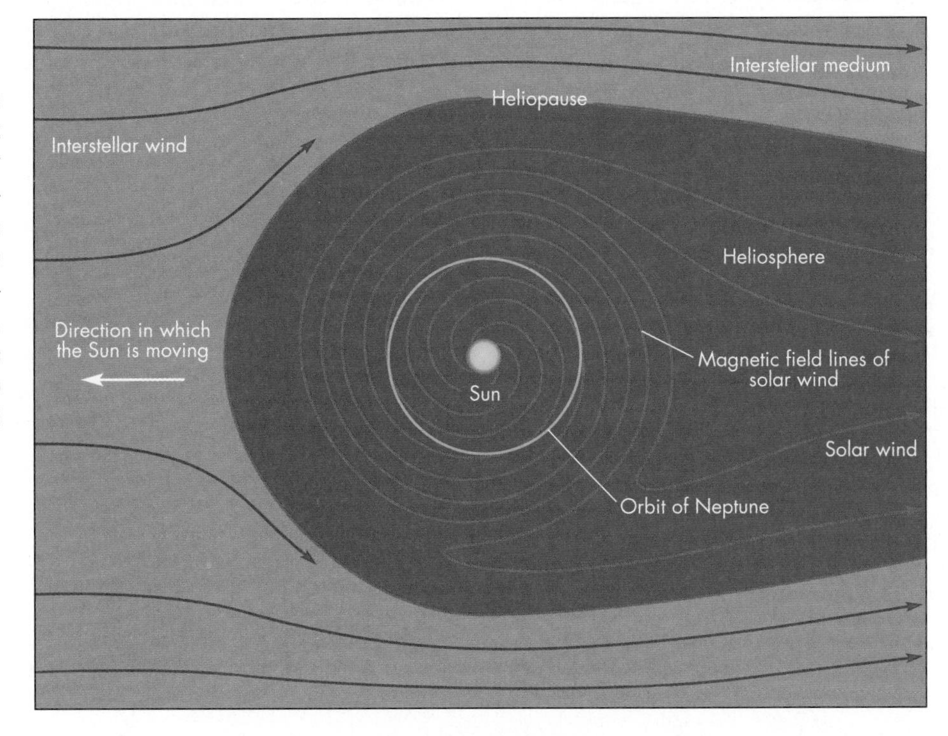

FIGURE 17.32
The Heliosphere
The region of space affected by the solar wind is called the heliosphere. At a distance of about 100 AU the solar wind begins to merge into the interstellar gas, which compresses the heliosphere in the direction that the Sun is moving and pushes the heliosphere back in the opposite direction. In the magnetosphere, the magnetic field lines are attached to the Sun, so the Sun's rotation stretches the field lines with a spiral pattern.

The effect of the solar wind on the magnetosphere of the Earth is described in Chapter 8, Section 5.

The Heliosphere No one knows for certain where the solar wind ends. Instruments on the *Voyager* spacecraft show that the solar wind continues outward to more than 90 AU from the Sun. The *Voyagers* have already passed the **termination shock,** where the speed of the solar wind drops dramatically, and are nearing the edge of the solar wind where, at a distance estimated to be about 100 AU, the solar wind merges into the interstellar gas. The boundary between the region of space dominated by the solar magnetic field and interstellar space is called the **heliopause.** The region within the heliopause is the **heliosphere** and is illustrated in Figure 17.32. Just as the Earth's magnetosphere is compressed by the solar wind on the sunward side and stretched away from the Earth on the other, the heliosphere is compressed by the interstellar gas in the direction toward which the Sun is moving and extended in the opposite direction.

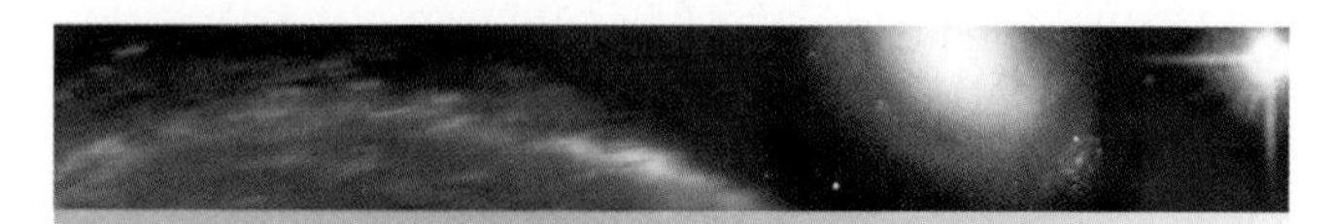

The solar wind flows outward from the Sun into interplanetary space. Astronomers think that the solar wind merges with the interstellar gas at a distance of about 100 AU.

17.4 THE SUNSPOT CYCLE

The **sunspot cycle** is a regular waxing and waning of the number of spots on the Sun. Daily records of the numbers of spots seen on the Sun have been kept since about 1850. The examination of historical records has made it possible to reconstruct the number of sunspots back as far as 1650, and there are fragmentary records from still earlier. The graph in Figure 17.33 shows annual averages of the number of spots on the Sun for the past three and a half centuries. Approximately every 11 years the number of sunspots climbs to a maximum and then declines. The height of maxima and exact amount of time between maxima vary considerably. The low maxima around 1800 and 1900 suggest that there may also be another cycle about a century long.

Sunspots and the sunspot cycle nearly disappear sometimes. Between 1640 and 1700, for instance, spots were rarely seen on the Sun. This period of time, known as the **Maunder minimum,** is of considerable interest because it coincided with a time of cold climate in western Europe and North America sometimes called the Little Ice Age. Whether the timing of the Little Ice Age and the Maunder minimum is just a coincidence is unclear.

The nature of the sunspot cycle before 1600 has been estimated from reports of observations of the aurora borealis. Auroral displays bright enough to be seen below the arctic circle are caused by solar flares. Because flares usually occur near sunspot groups, the frequency with which the aurora borealis was seen in Europe is a measure of the number of sunspots. The number of northern hemisphere auroral sightings per decade, shown in Figure 17.34, indicates that events like the Maunder minimum may have taken place during the mid–eleventh century, the early fourteenth century, and most of the fifteenth century.

The location of sunspots also varies during a sunspot cycle. A cycle begins with the appearance of spots about 30° north or south of the Sun's equator. As time passes, and individual spots come and go, the band within which sunspots are found moves steadily toward the equator. The drift of sunspot locations toward the equator, shown in

FIGURE 17.33
The Annual Average of the Number of Spots on the Sun
The data on which this graph is based go back to the time of the invention of the telescope. The number of sunspots increases and decreases in a recurring 11-year cycle, but the low maxima around 1800 and 1900 suggest there may be a 100-year cycle as well.

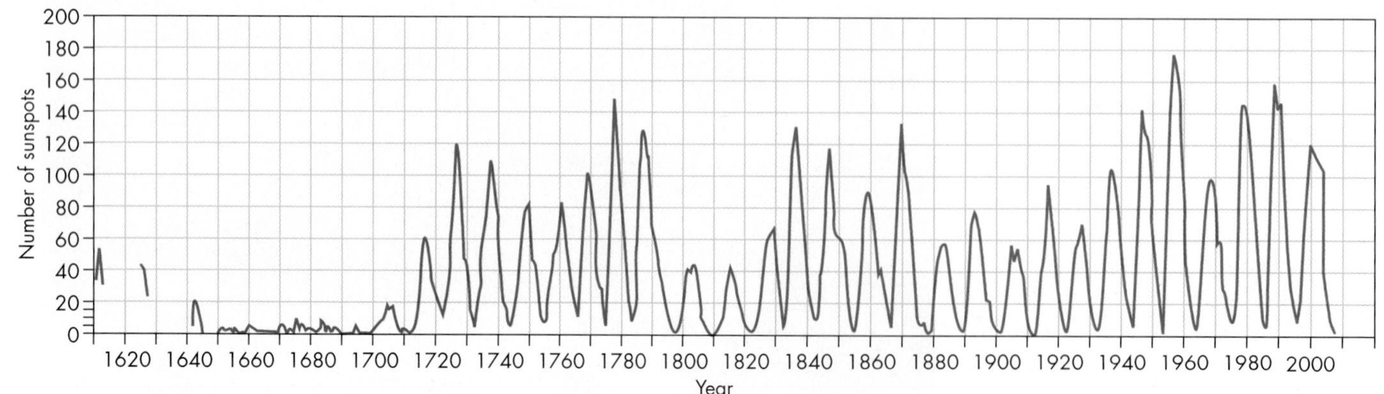

FIGURE 17.34
The Number of Sunspots Estimated from Observations of the Aurora Borealis
Solar activity, which is related to sunspot number, has been estimated from the frequency of auroral sightings before 1600. There appear to have been long periods of time, such as from 1420 to 1500, when there was little solar activity and, presumably, few sunspots.

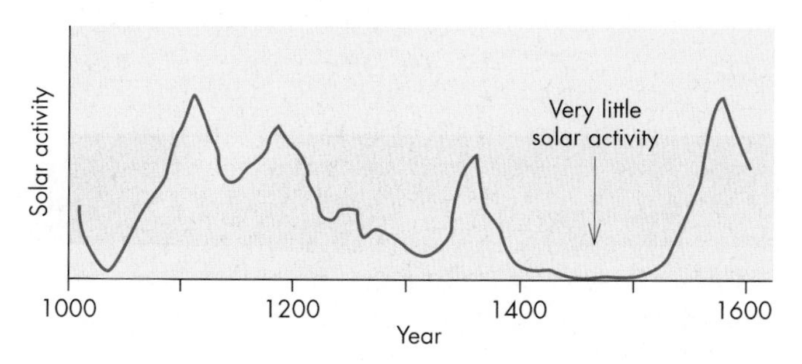

Figure 17.35, is often called the butterfly diagram. The pattern of magnetic polarities in sunspot groups is also related to the sunspot cycle. During one cycle the groups in the Sun's northern hemisphere have a preceding spot of north polarity and a following spot of south magnetic polarity. As Figure 17.36 shows, the situation is exactly reversed in the southern hemisphere, where the preceding spot has south polarity and the following spot has north polarity. During the next 11-year cycle, the northern hemisphere has preceding spots of south polarity and following spots of north polarity. The southern hemisphere again has a pattern opposite the north. The full sunspot cycle takes 22 years to complete. Each full cycle contains two 11-year cycles in which the numbers of spots vary but polarity remains the same.

A Model of the Sunspot Cycle

In 1961 Horace Babcock proposed a model that can account for many of the features of the 11-year sunspot

FIGURE 17.35
The Butterfly Diagram
This diagram shows where sunspots were found on the Sun between 1880 and 1990. At the beginning of a sunspot cycle, sunspots tend to appear around 30° north or south of the Sun's equator. As the cycle progresses, however, the band within which the sunspots are found moves toward the equator.

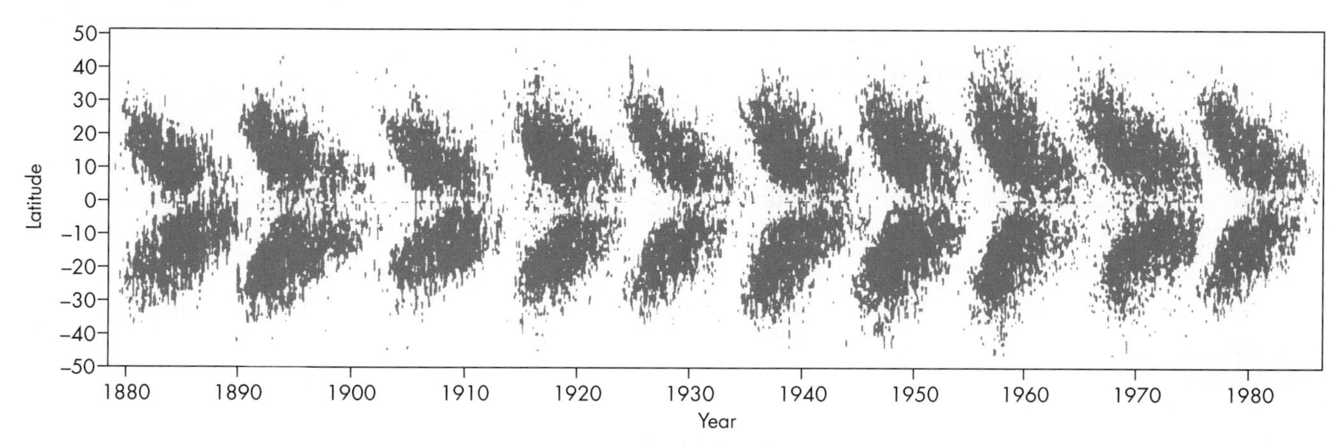

FIGURE 17.36
The Magnetic Polarity of Sunspot Groups
In cycle 1, preceding spots in the northern hemisphere have north magnetic polarity and following spots have south polarity. In the southern hemisphere, preceding spots have south polarity and following spots have north polarity. During the next 11-year sunspot cycle (cycle 2), the entire polarity pattern is the opposite from that of cycle 1.

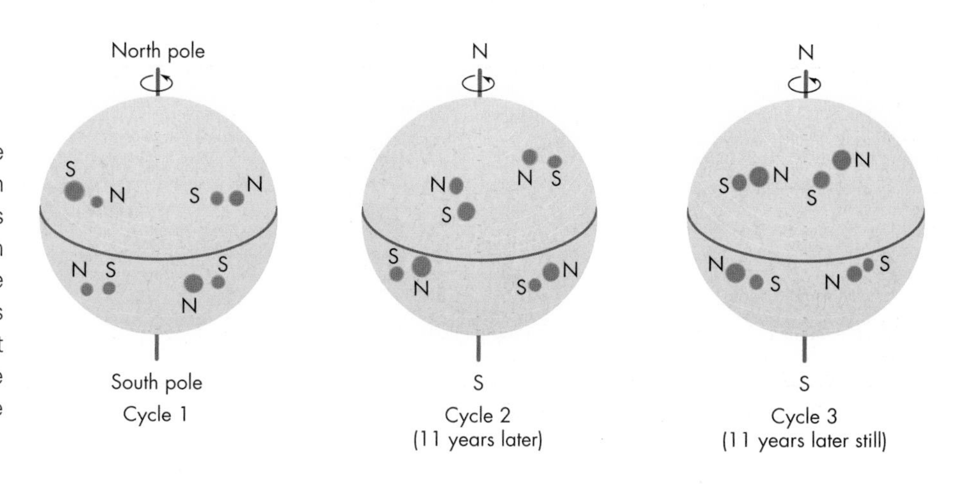

cycle and the 22-year magnetic cycle of the Sun. Some of the features of this model are shown in Figure 17.37. Babcock's model begins at sunspot minimum when the Sun has a weak magnetic field resembling that of the Earth. Magnetic field lines, which are beneath the Sun's surface near the equator, emerge at latitudes greater than 55°, extend far outward into the corona, and connect back to the opposite polar region. If the Sun rotated rigidly like a planet, the magnetic field would remain like this indefinitely. The Sun's rotation period is shortest near the equator, however, so the submerged equatorial field lines become stretched out and twisted. After about 3 years, the low-latitude magnetic field lines are wrapped around the Sun about five and a half times, greatly increasing the magnetic field strength.

Eventually, twisted ropes of magnetic field lines erupt as loops through the Sun's surface. The loops produce active regions in which the preceding and following sunspots have opposite magnetic polarity. In the other hemisphere, the polarity of active regions is just the opposite. Near the end of the cycle, the subsurface magnetic field is stretched and twisted so much that active regions and sunspot groups form near the equator. The preceding part of each active region drifts toward the equator, where areas of opposite polarity meet and cancel out. The following parts of active regions drift toward the poles where they cancel out and replace the existing global magnetic field of the Sun. The sunspot cycle ends with a weak global magnetic field opposite in magnetic polarity to the field that existed 11 years earlier. Although this picture of the solar magnetic cycle is generally accepted among astronomers, many important details remain to be worked out.

The number of sunspots increases and decreases during an 11-year sunspot cycle. However, there have been long periods of time in which very few spots have been seen on the Sun. The sunspot cycle and the pattern of magnetic polarities in sunspots have been explained by a model in which the differential rotation of the Sun stretches and twists magnetic field lines.

FIGURE 17.37
The Babcock Model of the Sunspot Cycle
A, At sunspot minimum, the Sun's magnetic field resembles that of a bar magnet (or that of the Earth). The faster rotation of the Sun's equatorial region, however, stretches and distorts magnetic field lines. The twisting of magnetic field lines strengthens the Sun's magnetic field. **B,** A few years later ropes of magnetic field erupt through the surface. Active regions and sunspot groups occur where the magnetic ropes emerge. **C,** Near the end of the cycle, the preceding parts of active regions move toward the equator where they cancel out each other. **D,** The following parts of active regions move toward the poles, where they establish a weak field that has the opposite polarity of the field at the beginning of the cycle.

Wrapping of the Sun's subsurface magnetic field

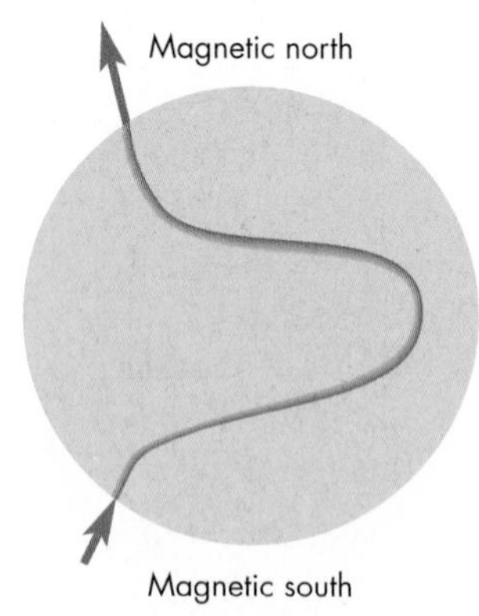

A Sunspot minimum: The Sun's rotation begins to stretch and distort magnetic field lines.

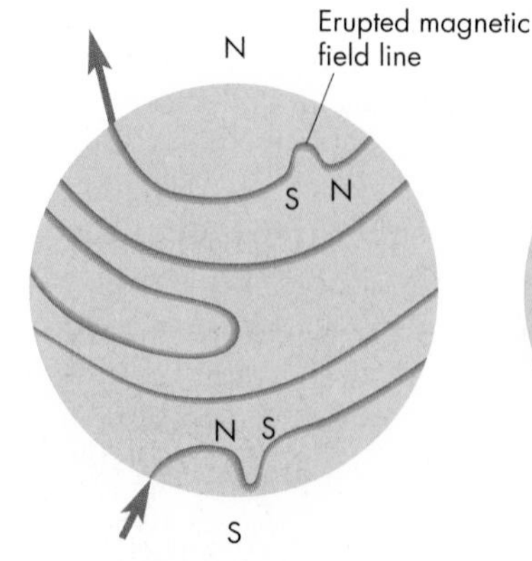

B The magnetic field lines become so twisted that they erupt through the Sun's surface.

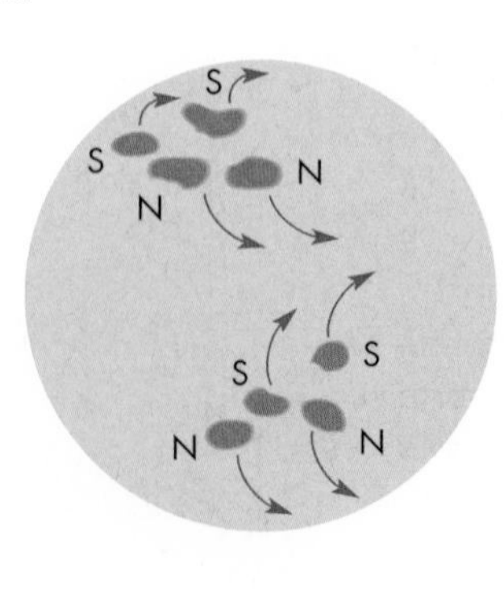

C Nearing the end of the cycle

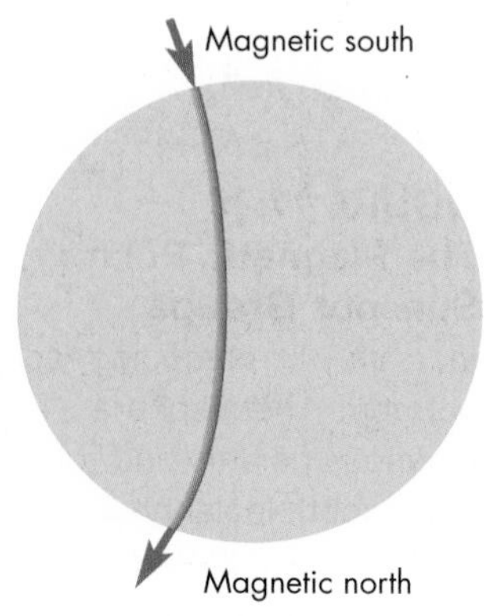

D End of the cycle: The polarity of the magnetic field is opposite what it was at the beginning of the cycle (**A**).

 SOLAR MYSTERIES

Astronomers think that they have a reasonably good understanding of the workings of stars. Yet extensive studies of the Sun, by far the best observed star, have left unanswered many important questions. Exactly how does the sunspot cycle work? How are the corona and chromosphere heated? What powers solar flares? What are spicules? Where and how does the solar wind merge into interstellar space? It is sobering to wonder how we would evaluate our understanding of other stars if we could observe them as closely as we do the Sun.

Chapter Summary

- Combustion, meteoritic impact, and gravitational contraction cannot account for the huge amount of energy that the Sun has produced during the 4.6 billion years that it has been luminous. (Section 17.2)
- The Sun's energy is produced by nuclear fusion through a sequence of reactions known as the proton-proton chain. The Sun's core contains enough hydrogen for hydrogen fusion to power the Sun for a total of about 10 billion years. (17.2)
- In the deep interior of the Sun, energy is carried outward by the flow of radiation. In the outer 30% of the Sun's radius, most of the energy is transported by convection. About 170,000 years is required to transport energy from the center of the Sun to the surface. (17.2)
- Unlike photons, solar neutrinos pass directly through the Sun's gases. By detecting neutrinos, therefore, astronomers have been able to study the core of the Sun. Because of neutrino oscillations, solar neutrinos are detected at only a fraction of the previously expected rate.
- The visible layer of the Sun is the photosphere. The sharp edge and limb darkening of the photosphere tell us that density increases swiftly inward in the photosphere and that temperature also increases inward. (17.3)
- The photosphere is pebbled by granules, which are rising convective columns of gas. An even larger convective pattern, supergranulation, is also present in the photosphere. (17.3)
- The Sun's surface vibrates in a complex pattern of oscillations. By studying these oscillations, astronomers have been able to learn about the way that temperature, composition, and rotation rate vary with depth. (17.3)
- Sunspots are dark because they are relatively cool. The low temperatures of sunspots are attributable to large magnetic fields, which inhibit convection and reduce the amount of energy reaching the solar surface. By watching sunspots move across the Sun's disk, it was discovered that the Sun's rotation rate decreases as distance from the Sun's equator increases. (17.3)
- The chromosphere is a tenuous layer that lies above the photosphere. The chromosphere consists mainly of spicules, which are rapidly rising jets of gas. (17.3)
- The outermost layer of the Sun is the corona, within which the temperature of the gas reaches millions of degrees. The corona is thought to be heated by magnetic fields. Above active regions, the corona is confined by magnetic fields and is hot, dense, and bright. In coronal holes, magnetic fields are unable to trap coronal gas, which accelerates away from the Sun. (17.3)
- Prominences are clouds of relatively dense, cool gas that reach upward into the corona. Although prominences can last for months, they often erupt, sending a blast of gas outward through the corona. Flares are sudden releases of energy stored in magnetic fields. Flares result in very hot regions of gas that emit high-energy radiation. Large numbers of energetic ions and electrons are also produced in flares. (17.3)
- The solar wind is a tenuous, hot gas that flows from the Sun into interplanetary space. At about 100 AU from the Sun, the solar wind merges with the interplanetary gas. (17.3)
- Although there have been long periods of time when sunspots were rarely seen on the Sun, the number of spots usually rises and falls in an 11-year cycle. It has been proposed that the sunspot cycle and the magnetic properties of sunspots can be explained by a model in which magnetic field lines beneath the photosphere are stretched and twisted by the differential rotation of the Sun. (17.4)

Key Terms

active region *411*
chromosphere *412*
convection zone *404*
coronal hole *414*
coronal mass ejection *416*
deuterium *402*
differential rotation *411*
filament *416*
granule *408*
heliopause *418*
helioseismology *409*
heliosphere *418*

Conceptual Questions

1. Suppose the rate at which deuterons react with protons to form ^{3}He in the Sun's core were twice as fast as it actually is. What effect would such a difference make in the rate at which the Sun's hydrogen is converted into helium?
2. The time it would take for the Sun to consume all its hydrogen (if the Sun's luminosity remained constant) is about 100 billion years. Why is this an unrealistic estimate of the lifetime of the Sun?
3. Why does it take so long for the energy produced in the Sun's core to reach the surface?
4. Why does convection develop about 70% of the way from the Sun's center to its surface?
5. Suppose the Sun's nuclear reactions were to drop by 50% next Sunday and resume their normal rate again the following Saturday. How noticeable would this temporary drop in energy production be when the energy produced next week finally reaches the Sun's surface?
6. Explain why the solar neutrino telescope in South Dakota is located 1 mile beneath the Earth's surface and why it can operate around the clock.
7. What happens to solar neutrinos as they travel between the Sun and the Earth?
8. Why is nearly all of the Sun's energy produced in the inner 1.5% of its volume?
9. Suppose the density of gas in the Sun's photosphere fell much less rapidly with increasing distance from the center than it really does. What effect would this have on the sharpness of the Sun's limb?
10. What do we mean when we say that the Sun has a temperature of 5800 K?
11. Suppose the Sun's limb were brighter than the center of the Sun's disk. What would that tell us about the temperature structure of the Sun's photosphere?
12. Describe the pattern of rising and falling gas in granules and supergranules.
13. Why do sunspots appear dark?
14. What is thought to be the reason that sunspots are cool and dark?
15. Describe a spicule.
16. Why was it difficult for astronomers to identify the elements responsible for the emission lines radiated by the corona?
17. The rate at which energy must be supplied to the corona to keep it hot is quite small. How can this be so?
18. What is the relationship between active regions and coronal holes?
19. What is the relationship between coronal holes and the solar wind?
20. How do prominences and flares compare in location and lifetime?
21. Why does the magnetic field of the solar wind follow a spiral pattern?
22. Why is the heliopause thought to be nearer the Sun in some directions than in others?
23. What evidence do we have that there is a relationship between sunspots and the temperature of the Earth?
24. What does a butterfly diagram tell us about the numbers and locations of sunspots during a sunspot cycle?
25. Describe the magnetic polarity and average location of sunspots during a full sunspot cycle of 22 years.
26. What does the differential rotation of the Sun have to do with the solar sunspot cycle?

Problems

1. Suppose the Earth produced as much energy per kilogram each second as the Sun does. What would be the total energy output of the Earth? How would the absolute magnitudes of the Earth and the Sun compare?
2. Suppose the Earth were made entirely of hydrogen and oxygen, with two atoms of hydrogen for each atom of oxygen. If this hydrogen and oxygen were converted to water, how long could the Earth produce the same luminosity as the Sun?
3. If gravitational contraction were responsible for the Sun's energy, how long would it take for the Sun's diameter to shrink by 0.1%?

Figure-Based Question

Use Figure 17.33 to find the sunspot number at the highest sunspot maximum of the nineteenth century.

Group Activity

Go to the website http://www.bbso.njit.edu/Images/daily/images/wfullb2.jpg, where today's visible image of the Sun is displayed. Sketch the image of the Sun, paying particular attention to the location of sunspot groups. Have different members of your group make such a solar sketch every day for 2 weeks. As a group, discuss how the location of the spots changes with time and try to determine the rotation period of the Sun. (Note: Galileo did this in the early 1600s.)

For More Information

Visit the text website at **www.mhhe.com/fix** for chapter quizzes, interactive learning exercises, and other study tools.

18

And God made two great lights; the greater light to rule the day, and the lesser light to rule the night; he made the stars also. And God set them in the firmament of the heaven to give light upon the earth. And to rule over the day and over the night, and to divide the light from the darkness: and God saw that it was good. And the morning and the evening were the fourth day.

Genesis

A Spitzer Space Telescope, image of three young stars and the nebula in which they formed.

www.mhhe.com/fix

The Formation of Stars and Planets

People have always wondered about the origin of the world. The mythologies of most cultures have included stories that account for the creation of the Earth, Moon, Sun, and stars. Many of these stories describe the creation of the celestial bodies and the Earth by powerful people or animals. The creation myth of the Pueblo Native Americans, for instance, centers on the creator, Awonawilona, who assumed the form of the Sun. When the Sun appeared, mists gathered and fell to form the sea in which the world floats. Warmed by the Sun, scum formed on the sea and became the Earth Mother and Sky Father. The stars were made when the Sky Father placed shining grains of maize in the sky. The Chamorro people of the Mariana Islands believed in Puntan, a being who existed before the sky and the Earth. Before he died, Puntan told his sister to use his front and back to make the Earth and the sky, his eyes to make the Sun and the Moon, and his eyebrows to make the rainbow.

Until the seventeenth century, ideas about the formation of the major celestial bodies had little or no scientific content and differed little from ancient creation myths. By the second half of the eighteenth century, however, Immanuel Kant and Pierre de Laplace had both presented the concept that is at the heart of modern theories of the formation of stars and planetary systems. Kant and Laplace proposed that both stars and planets form within rotating, contracting clouds of interstellar gas. Although this idea fell into disfavor until about 60 years ago, observations and calculations of the last

Questions to Explore

- What evidence do we have that stars form within dark interstellar clouds?
- What triggers the collapse of interstellar clouds to form stars?
- What do newly formed stars look like?
- Why do we think that the Sun was once surrounded by a nebula of gas and dust?
- How did the planets form from a thin cloud of gas and dust?
- What evidence do we have that planetary systems are forming around other stars?

half century have confirmed the intimate relationship between the formation of stars and the formation of planetary systems. We are still uncertain about many of the details of how vast interstellar clouds become stars and planets, but the general outline becomes clearer with each passing year.

This chapter describes current ideas about the formation of stars and planets from interstellar clouds.

18.1 SOME IMPORTANT CLUES

Clues from Stars

The process of star formation goes on continuously in our galaxy. On average, about 3 to 5 $M_\odot$ of interstellar gas are converted into stars each year. Of the newly formed stars, the great majority have masses smaller than the Sun's. Because the rate at which stars form declines rapidly with increasing stellar mass, star formation favors the production of the less massive stars. Also, most stars are members of binary or multiple star systems so the star formation process often produces pairs or small groups of stars rather than individual, widely separated stars. Moreover, most young stars are found in clusters bound together by their mutual gravity or in **associations,** in which the stars were born near each other but too far apart for their mutual gravity to bind them all together. The stars within associations are gradually moving apart. The grouping of young stars in clusters and associations shows that many stars may form in a single, relatively localized region called a star formation region.

Clues from the Solar System

There are many features of the solar system that give us information about the way it formed. Some of these features are:

1. The orbital planes of the planets are nearly the same and lie almost in the equatorial plane of the Sun.
2. The planets all orbit the Sun in the same direction, and most planets rotate in the same direction that they revolve about the Sun.
3. Planetary orbits are almost circular. Even Mercury's orbit about the Sun varies only about 21% from circularity.
4. The regular satellite systems of the outer planets mimic the solar system in the characteristics listed above.
5. The Sun has more than 99% of the mass of the solar system, but less than 1% of the angular momentum. In other words, the Sun rotates much too slowly to have an angular momentum comparable to that which is contained in the orbital motions of the planets.

The first three features tell us that the solar system is very flat and that the rotation and revolution of the planets is highly organized. This suggests that the material from which the planets and Sun formed also must have been flat and spun in an organized manner. The fourth feature suggests that the same process that formed the planets also occurred on the smaller scale of satellite systems. The final feature suggests that angular momentum must have been redistributed in the collapsing cloud that formed the solar system. Otherwise, the bulk of the angular momentum would be located with most of the mass—within the Sun.

Observations of the numbers of young stars of different masses show that star formation favors less massive stars over massive stars, that stars usually form in binary or multiple systems, and that many stars form at the same time in a localized region. The structure of the solar system suggests that the material from which the planets formed was flat and rotating. Also, angular momentum was redistributed in the solar system as it formed.

18.2 STAR FORMATION

Stars form inside relatively dense concentrations of interstellar gas and dust known as **molecular clouds.** These regions are referred to as "molecular" because they are cool and dense enough that most of the gas they contain consists of molecules rather than atoms or ions. Thus, gases such as H_2 (molecular hydrogen) and CO (carbon monoxide) can be found in molecular clouds. Until recently, star formation within molecular clouds was completely hidden from our view by the large quantities of small solid particles (dust) they contain. The dust also blocks our view of the stars that lie beyond the molecular clouds. This makes molecular clouds look like dark regions containing very few stars, as shown in Figure 18.1.

Although visible light can't reach us from within molecular clouds, infrared and radio radiation are less affected by the dust particles. By using infrared and radio telescopes, astronomers have been able to probe deep within molecular clouds. Many important molecules such as CO and NH_3 (ammonia) produce emission lines in the radio part of the spectrum, so radio astronomical measurements have been especially useful for gathering information about some of the steps that lead to star formation within molecular clouds.

Giant Molecular Clouds

Molecular clouds come in a wide range of sizes and masses. The ones within which most star formation occurs are the

giant molecular clouds, which are typically about 10 pc across and may contain as much as 1 million $M_\odot$ of gas and dust. The formation of stars of relatively low mass also goes on in smaller clouds with masses of about 10,000 $M_\odot$ and sizes of a few parsecs. There are thousands of giant molecular clouds in our galaxy. The nearest one is the Orion Molecular Cloud, about 450 pc away in the direction of the constellation Orion, shown in Figure 18.2. Young hot stars heat part of the gas in the Orion Molecular Cloud to produce the Orion Nebula, shown in Figure 18.3, which can be seen with the naked eye. The Orion Molecular Cloud covers many square degrees of the sky and extends far beyond the visible Orion Nebula. Figure 18.4 compares the Orion region in visible light, the radio emission from its CO molecules, and its infrared emission. The CO emission, emitted by cool gas, and the infrared emission, which traces the distribution of dust in the Orion Molecular Cloud, have many features in common that do not appear in the visible image of Orion. The white dots in Figure 18.4C are young stars, many of which are too obscured to be seen in visible light.

FIGURE 18.1
The Visible Appearance of a Molecular Cloud
The dust in the molecular cloud blocks the light from distant stars, making the region appear dark and nearly starless. This cloud, Barnard 68, is about a half a light year in size.

Upon closer examination, a giant molecular cloud consists of smaller, denser clumps of interstellar material. Each clump is a few parsecs across and contains between 10^3 and 10^4 $M_\odot$ of gas and dust. The clumps are very cold, with temperatures of only about 10 K. At the centers of many clumps are **cloud cores,** the sites at which star formation takes place. Cloud cores come in two different varieties, cold ones and warm ones. Cold cores have temperatures of about 10 K, are about 0.1 pc across, and contain between a few tenths and 10 $M_\odot$. Warm cores, which have temperatures of 30 to 100 K, are between a few tenths and 3 pc across and contain between 10 and 1000 $M_\odot$ of material.

FIGURE 18.2
The Constellation Orion
Bright stars, some of them young and hot, form the outlines of Orion's shoulders, knees, sword belt, and sword. One of the stars in Orion's sword (indicated by the arrow) is actually a star-forming region known as the Orion Nebula.

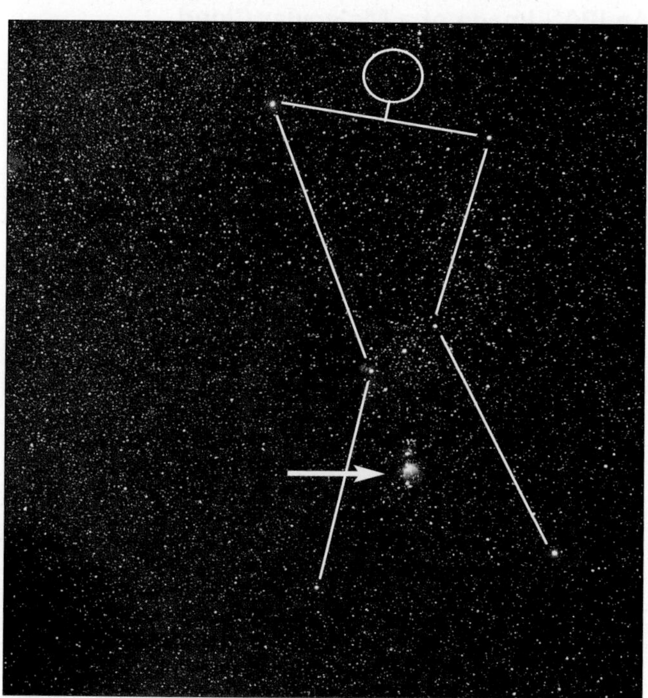

Cores and Star Formation

Two main lines of evidence suggest that cloud cores are the places where stars form. First, many cloud cores

FIGURE 18.3
The Orion Nebula
This part of the Orion Molecular Cloud is so bright it can be seen without a telescope.

FIGURE 18.4
Three Views of the Orion Star-Formation Region
A, The *Hubble Space Telescope* mosaic, taken using visible light, mainly shows stars and hot gas. **B,** The view in CO molecular emission indicates where cool gas is concentrated. **C,** The view in the infrared shows where dust is concentrated. Cool gas and dust are often located together.

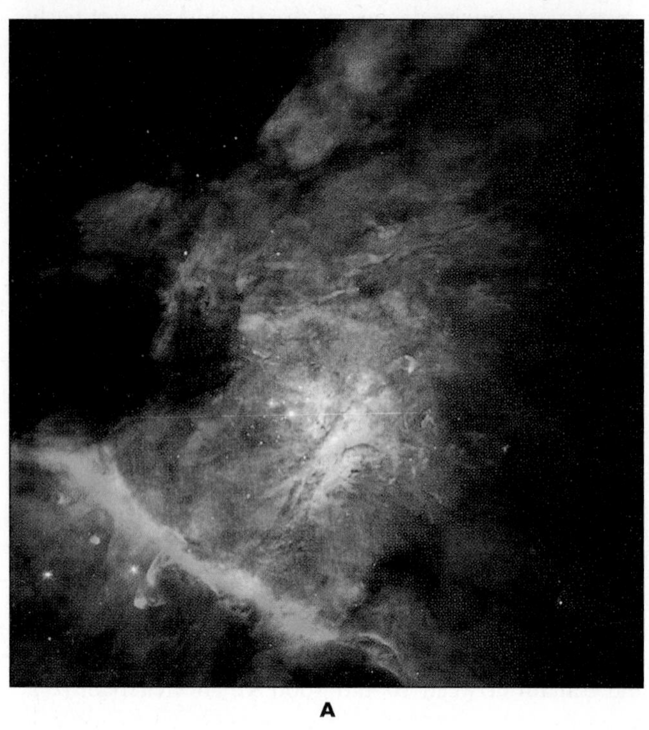

A

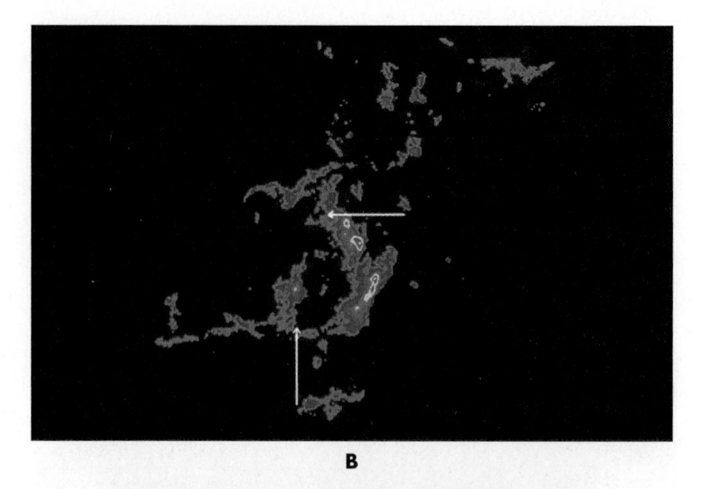

B

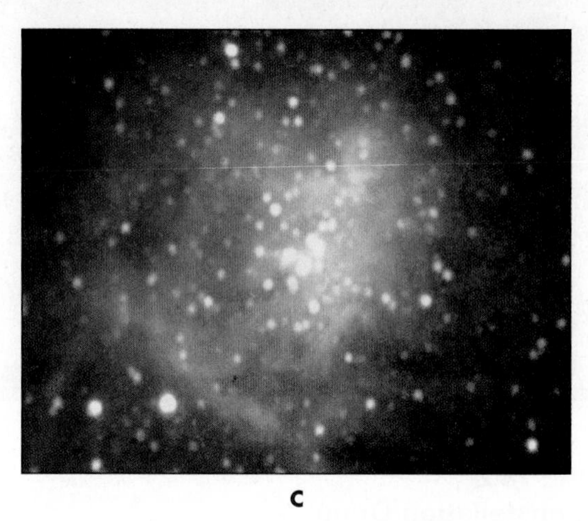

C

contain sources of intense infrared radiation, as shown in Figure 18.5. These have been identified as **protostars,** which are stars still in the process of forming. Second, young stars are frequently found in the vicinity of the cores of molecular clouds. In the case of cold cores, the young stars generally have masses of 1 $M_{\odot}$ or smaller. In the case of warm cores, some of the young stars are more massive. Warm cores are so massive that they seem to be capable of forming a cluster of young stars rather than just a single one.

Formation of Cloud Cores Astronomers do not yet agree on what it is that causes the cores of molecular clouds to form. One possible explanation is that turbulent motion of the gas in molecular clouds compresses some regions, whereas other regions of the cloud become more rarefied. The compressed regions would then become cloud cores. Another possibility is that gas ejected from a star within the cloud compresses the part of the molecular cloud that surrounds the star.

Core Collapse No matter how cloud cores form, they and the clumps of gas that surround them are susceptible

FIGURE 18.5
An IRAS Image of a Molecular Cloud
The hot spots within the cloud (indicated by white) are luminous protostars.

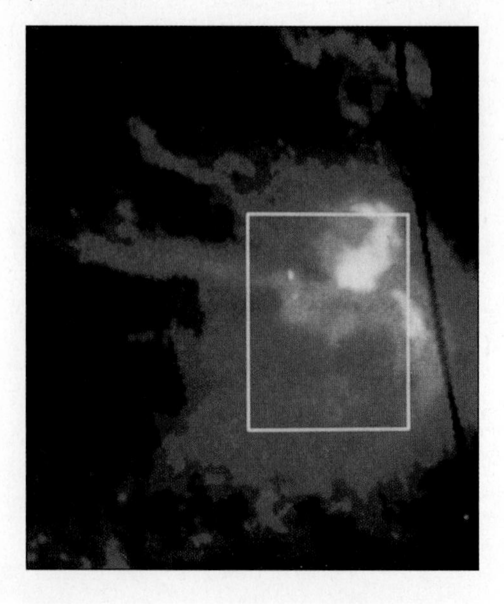

to collapse. That is, the core exerts a strong enough gravitational force on itself to overwhelm the gas pressure within the core, thus causing it to collapse under its own weight. What is puzzling to astronomers about this process is not *why* it happens but why it happens *so slowly.* Something must be retarding the collapse of cloud cores. Otherwise the formation of stars from cloud cores would occur so rapidly that the galaxy would produce far more new stars each year than it actually does.

Magnetic fields may be responsible for retarding the collapse of cloud cores. Although magnetic fields in interstellar space are very much weaker than those in the Sun or on the Earth, they nevertheless control the motions of charged particles such as protons and electrons. Just as in the Earth's magnetosphere, charged particles move freely along interstellar magnetic field lines but cannot easily cross them. On the other hand, particles that have no charge, such as atoms and molecules, can flow across field lines with no difficulty. Notice in Figure 18.6, however, that when uncharged particles cross the field lines, they collide with charged particles that are unable to move. This slows the motion of the uncharged particles toward the center of the cloud core and retards the collapse of the core.

Eventually, however, as the uncharged atoms and molecules slip across the magnetic field lines and the density of the core increases, the gravity of the core increases until it overwhelms the support provided by the magnetic field. After this point, which may take between 1 million and 10 million years to reach, the rest of core collapse takes place much more rapidly, perhaps in 100,000 years. The end result of a collapse of this kind of core is the formation of an individual star or a multiple star system.

There are other cores in which the magnetic field is never strong enough to inhibit collapse. In this case, the core collapses rapidly and breaks into fragments as it shrinks. The result is the formation of a large number of stars, a cluster, within the core. The energy emitted by the concentration of forming stars heats the gas and dust to produce a warm core.

FIGURE 18.6
The Collapse of a Core with a Magnetic Field
Ionized gas can't easily move across magnetic field lines but neutral gas can. Friction between the ionized and neutral gas makes it difficult for neutral gas to move as well. Thus, the presence of a magnetic field slows the collapse of a cloud core.

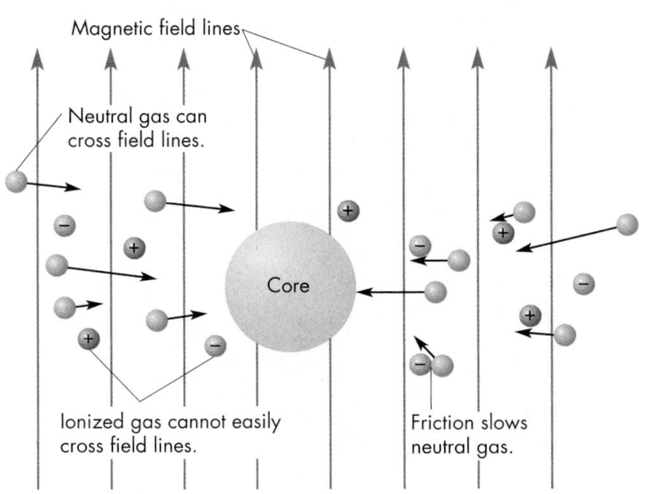

Star formation occurs in the cores of clumps of gas and dust within giant molecular clouds. Warm, massive cores seem to be where massive stars form, whereas low mass stars seem to form in cool, less massive cores. Cloud cores collapse under their own weight. The collapse is relatively slow, however, because magnetic fields make it difficult for charged particles to participate in the collapse.

18.3 PROTOSTARS

Because gravity is stronger near the center of a collapsing core than at its edges, core collapse is most rapid at the center. As a result, matter accumulates and density increases fastest at the center of the collapsing core. The dense region at the center of the core grows in mass as more and more gas becomes concentrated there. At this point the core is about to become a protostar.

The Evolution of a Protostar

The collapse of the core of an interstellar cloud releases gravitational energy. That is, the gas and dust accelerates as it falls inward, picking up speed and kinetic energy. At the same time, it is slowed by friction, which converts the kinetic energy to heat. As long as the protostar is transparent, all of the heat can be emitted as infrared radiation and the center of the cloud remains cool. Because the cloud remains cool, gas pressure, which is proportional to temperature (see Chapter 7, Section 2), doesn't increase very much and remains too small to counteract gravity and retard the collapse of the cloud. As a result, the collapse and buildup of material in the protostar goes on as fast as gravity can pull gas and dust inward toward the center of the cloud.

Eventually, the protostar becomes opaque as the material in the cloud (especially the dust, which is more effective in blocking infrared radiation) becomes more concentrated. An analogy can be made to a collection of chalk dust. Generally speaking, the more widely dispersed the chalk dust, the more transparent it is. If the chalk dust is concentrated in a pile, however, it is opaque. In the same way, as the matter in the protostar becomes more and more concentrated, it becomes less and less transparent. Once the protostar becomes opaque, infrared radiation can no longer escape to cool the protostar, so both temperature and pressure begin to rise in its center. Internal pressure soon becomes great enough to balance the weight of the protostar, so the rapid collapse of the protostar comes to a halt. This is an important milestone in the formation of a star because it marks the first time that the object at the center of the cloud core clearly resembles a star rather than an interstellar cloud.

The Growth of the Protostar When it first forms, the protostar has relatively little material—perhaps only 1% of the Sun's mass. However, the mass of the protostar grows constantly as ever more material falls onto its surface. The accretion of matter onto the protostar can't go on indefinitely or the protostar would eventually gather up the entire mass of the molecular cloud. What happens instead is that a protostar develops a strong wind (much more powerful than the solar wind) that opposes the infalling interstellar material and stops the accumulation of matter by the protostar. The wind and some of its effects are described later in this chapter. At about this point in its career, the protostar is considered to be a star because its mass is established and its future evolution is set.

The new star may not yet be visible. Eventually, though, the wind that stopped the infall of new material unveils the new star by blowing the surrounding gas and dust away.

Slow Contraction The internal structure of the protostar changes even as it gathers more mass from its surroundings. Lacking nuclear energy sources, the protostar must contract to generate the internal energy needed to balance the energy radiated away at its surface. This contraction is called Kelvin-Helmholtz contraction and is described for the case of the Sun in Chapter 17, Section 2. Contraction continues after the infall of interstellar gas and dust is halted and ceases only when the fusion of hydrogen into helium is able to supply the energy needs of the star. When fusion becomes the primary energy source for the star, the star has reached the main sequence.

The duration of the Kelvin-Helmholtz contraction phase depends on the mass of the star. For a star like the Sun, Kelvin-Helmholtz contraction goes on for about 10 million years. Less time is required for more massive stars and more time for less massive stars. Stars are cooler as they evolve toward main sequence status than they will be as main sequence stars. This means that the locations of pre–main sequence stars in the H-R diagram lie to the right of the main sequence.

Rotation and Disks

As the core of a molecular cloud collapses and a protostar develops, the rotation of the cloud core becomes more and more important. The original cloud core, at the time the collapse began, may have rotated very slowly, perhaps once every few million years. As the core shrinks, however, it rotates more and more rapidly because angular momentum must be conserved. In the equatorial plane of the core, the inertial effect of rotation partially counteracts the inward pull of gravity and slows the collapse of the cloud. Eventually, as the collapse continues and rotation becomes more rapid, the infalling gas and dust begin to accumulate in a rotating, nebular disk, as shown in Figure 18.7, rather than falling all the way to the center of the protostar.

Matter that falls inward near the rotational poles of the cloud core is unaffected by the rotation of the core and falls directly onto the protostar, as shown in Figure 18.8. At this point, the collapsing cloud consists of a central protostar, a nebular disk revolving in the equatorial plane of the protostar, and a large surrounding region of infalling gas and dust. The nebular disk lasts until the gas and dust it contains are blown away by the wind that ends the accretion of the star or evaporated by ultraviolet radiation from hot neighboring stars. This probably takes a few million years. In many cases, however, this may be

FIGURE 18.7
Formation of a Nebular Disk
The rotation of the protostar causes material falling inward near its equator to accumulate in a rotating disk.

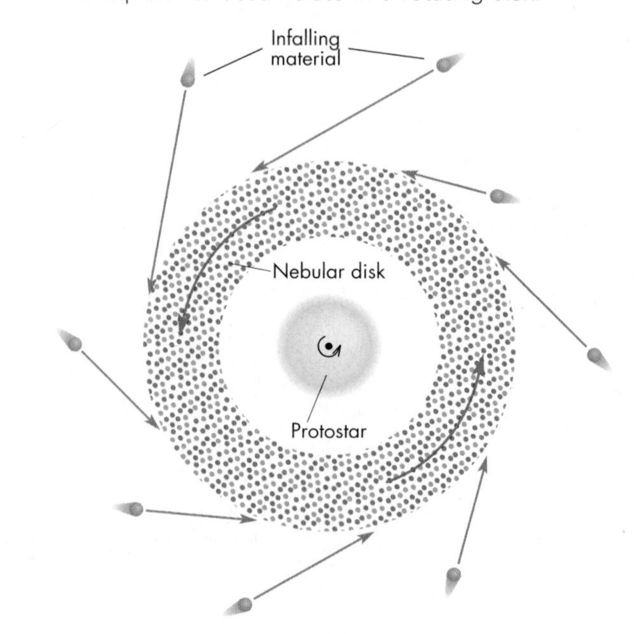

FIGURE 18.8
A Protostar After It Develops a Rotating Disk
Material falling inward in polar regions falls directly into the protostar, whereas material falling inward near the equator enters the disk.

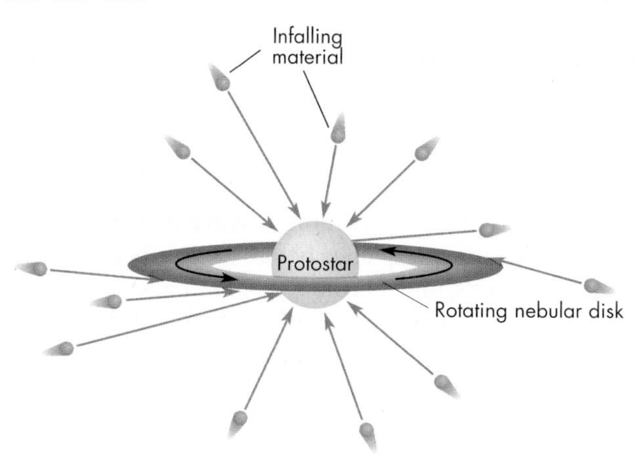

long enough for a planetary system or a second star to form within the nebular disk.

Density increases fastest at the center of a collapsing cloud core and produces a protostar. When it becomes opaque, a protostar stops collapsing and begins to accumulate infalling material. It contracts slowly until its central temperature becomes hot enough for hydrogen fusion to take place. The rotation of a cloud core causes infalling matter to accumulate in a rotating, nebular disk.

18.4 YOUNG STARS

The general picture of star formation described in the previous sections is based upon many of the properties observed for very young stars. These properties include the location of very young stars when they first appear in the H-R diagram, and the winds and disks that seem to be a universal feature of young stars.

H-R Diagrams for Young Stars

The youngest stars that we are able to see in the visible part of the spectrum are those found near molecular clouds in which other stars are still forming. Visible and infrared images of a star-forming region in Doradus are shown in Figure 18.9. Many young stars can be seen in the visible image. Some young stars, however, are still surrounded by gas and dust clouds and can be seen only in the infrared image. The H-R diagram for the stars in the Orion star-forming region is shown in Figure 18.10. Also shown are the main sequence and an approximate "birthline" where stars of different masses first become visible. The length of time during which stars are visible before they reach the main sequence decreases with increasing stellar mass. This means that stars more massive than the Sun first become visible shortly before they reach their eventual main sequence destination in the H-R diagram. Stars with masses less than or equal to the Sun's, on the other hand, blow away their surrounding gas and dust long before they achieve hydrogen fusion, so they are visible for millions of years before they reach the main sequence. Most of the youngest stars are grouped into a category called the T Tauri stars.

T Tauri Stars The **T Tauri stars** are pre–main sequence stars with masses less than 3 $M_\odot$. The H-R diagram for

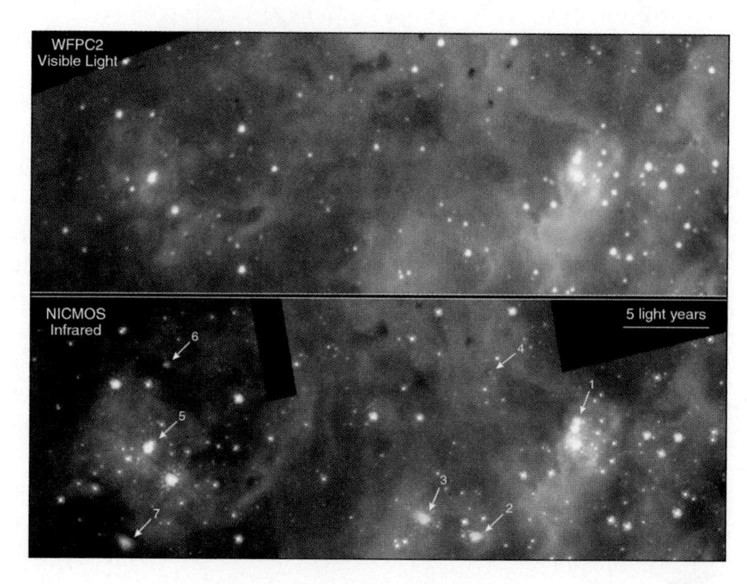

FIGURE 18.9
Visible and Infrared *Hubble Space Telescope* Images of the 30 Doradus Star-Forming Region
Most of the young stars in 30 Doradus can be seen in both visible light (top) and infrared (bottom). Stars 2, 3, and 4, however, are still surrounded by gas and dust and can only be seen in the infrared. This is because infrared radiation can pass through the dust, whereas visible light cannot.

FIGURE 18.10
The H-R Diagram for the Young Stars in the Constellation Orion
Young stars first become visible along the "birthline," which runs diagonally through the diagram. Stars become visible before they reach the main sequence.

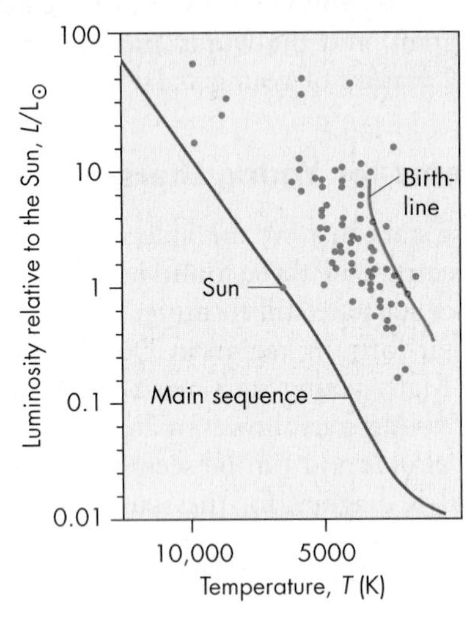

FIGURE 18.11
The H-R Diagram for T Tauri Stars
T Tauri stars are luminous, cool pre–main sequence stars.

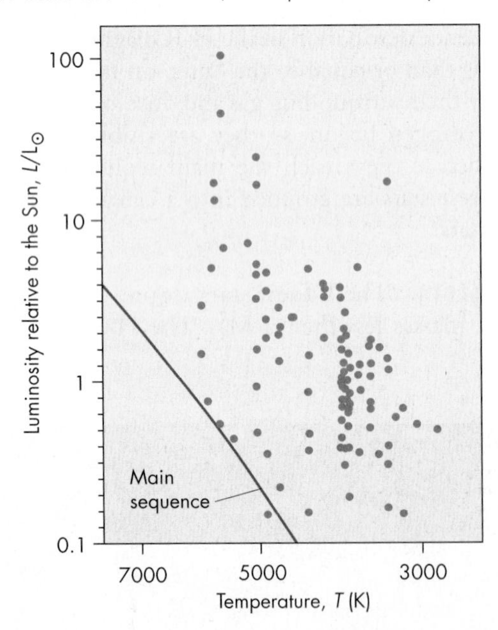

T Tauri stars is shown in Figure 18.11. The T Tauri stars are named for the first discovered star of their type. Although the T Tauri stars are cool stars with spectral types of G, K, or M, many of them show intense emission lines of the same type emitted (much more weakly) by the Sun's chromosphere. The chromospheric emission from T Tauri stars is only one indication of vigorous activity on their surfaces. They also have large dark regions, like giant sunspots. By monitoring the brightness of a T Tauri star as a spot rotates across its surface, astronomers have been able to figure out how rapidly the star rotates. A typical rotation period for a T Tauri star is about 5 days, or one-sixth of the rotation period of the Sun.

Another striking property of T Tauri stars is their strong infrared emission. The infrared radiation is emitted by dust that orbits the star. To account for the broad range of wavelengths over which there is strong infrared emission, the orbiting dust must cover a range of temperatures from warm to cool. This means, in turn, that the dust must extend over a large range of distances from the T Tauri star.

Massive Stars Stars more massive than 10 $M_\odot$ seem to reach the main sequence before they disperse their dust and gas clouds and become visible. These stars can be detected as highly luminous (more than 100 $L_\odot$) infrared objects within molecular clouds.

Winds and Disks

During the past several decades, astronomers have discovered that high-speed winds and disks are very common in young stars. In many cases, the wind and the disk interact to produce **bipolar outflows,** in which the wind blows outward only along the polar axis of the star. No wind blows outward in the equatorial plane of the star because it is blocked, at least for a while, by the gas and dust orbiting within the disk.

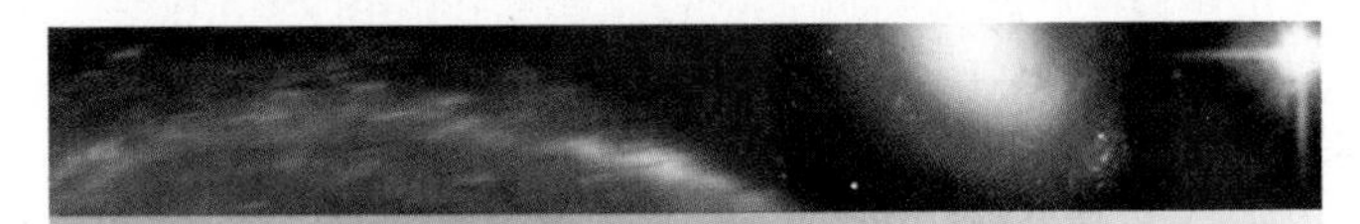

Stars become visible when they blow away the surrounding gas and dust. Most stars first become visible as T Tauri stars, which have vigorous surface activity and orbiting disks of gas and dust. The most massive stars reach the main sequence before they become visible.

What Causes the Winds? Although there is clear evidence that energetic winds from young stars thwart the infall of the molecular cloud core and blow away the surrounding gas and dust, the reason that such strong winds develop is not as clear. One hypothesis is that the initiation of the first significant nuclear fusion in the center of the star causes the strong wind to develop. When the temperature at the center of the star reaches about 10^6 K, the fusion of a deuteron and a proton (the second reaction of the proton-proton chain) becomes possible. The amount of energy produced by the fusion of deuterium is too great to be carried out of the star by the flow of radiation alone, so vigorous convection develops.

Convection and rapid rotation are the two key ingredients of a dynamo. (The Earth's core also has a dynamo, which is described in Chapter 8, Section 3.) Thus, when convection develops in a young star, the

FIGURE 18.12
***Hubble Space Telescope* Images of Disks of Dusty Material About Two Stars Like the Sun**
The disk about AU Microscopii (left) is seen edge on, whereas the disk about HD 107146 (right) is seen nearly face on. In both cases, light from the stars themselves is blocked by a device inside the telescope. Note the flatness of the dust disk about AU Microscopii. Both of the dust disks extend to distances much greater than the size of Neptune's orbit in our own solar system.

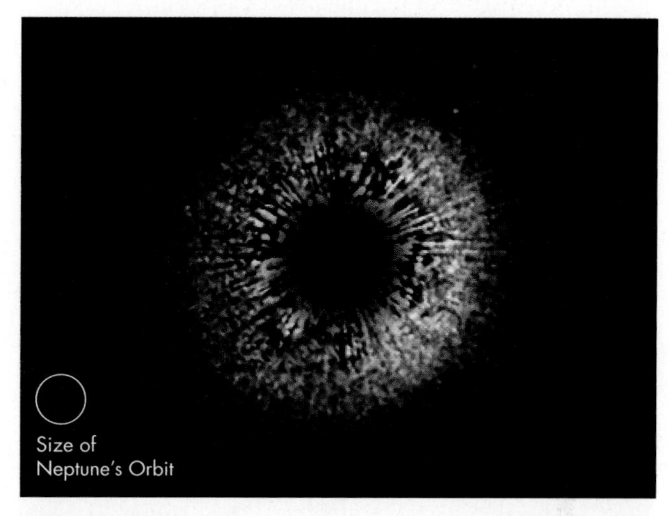

combination of convection and rapid rotation quickly builds a strong magnetic field and surface activity like that of the Sun, but much stronger. Emission lines and spots are evidence of that activity, as are winds (similar to the solar wind) that blow outward at speeds of several hundred kilometers per second. The rapid outflow of gas causes young stars to lose mass at rates of 10^{-7} $M_{\odot}$ per year. This may not sound like much but it is about 10 million times as large as the rate at which the Sun loses mass via the solar wind.

Disks Most, perhaps as many as 90% of, newly formed stars show more infrared emission than that which would normally be produced by the warm surfaces of the stars. The excess infrared emission from these stars is thought to be blackbody radiation from a swarm of relatively cool dust particles in the vicinity of the stars. At first it was thought that young stars are more or less completely surrounded by dust. However, more recent observations show that the dust is confined to a flattened disk that extends hundreds of AU from the star, as shown in Figures 18.12 and 18.13. The dust is probably mixed with several tenths of a solar mass of gas. In somewhat older disks, like the disk of AB Aurigae shown in Figure 18.14, clumps that may be forming planets can be seen. Still older disks sometimes have gaps. One possibility is that the gaps are places where a planet has formed and swept its orbit clear of gas and dust.

Within the disk, the orbital speed of the material decreases with increasing distance from the star as a consequence of Kepler's third law. The variation of orbital speed with distance leads to friction and heating. Suppose two blobs of matter orbit at slightly different distances. If the orbit of either blob is slightly eccentric, the two blobs will rub against each other from time to time. When they do, friction will cause the faster moving one (the inner one) to slow down slightly and the slower moving blob (the

FIGURE 18.13
A Dusty Disk Surrounding a Star in Orion
This *Hubble Space Telescope* infrared image shows a flattened disk of gas and dust around a young star in the Orion Nebula. The disk is about 150 AU across. About half of the young stars in Orion appear to be surrounded by disks.

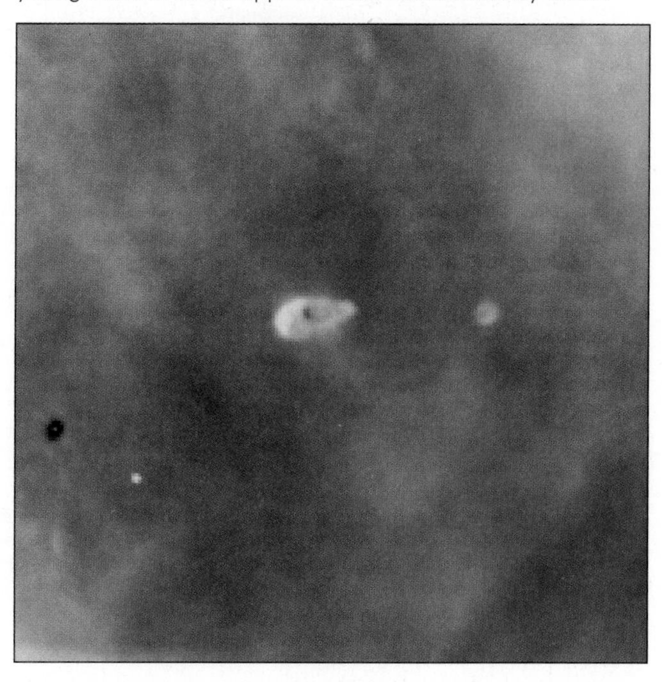

FIGURE 18.14
An Infrared *Hubble Space Telescope* Image of the Disk Around AB Aurigae
Notice that the disk shows clumps where the infrared emitting dust is concentrated. The disk is about 30 times the diameter of our solar system. The black stripes are a mask to block starlight from AB Aurigae and keep it from overwhelming the light from the disk.

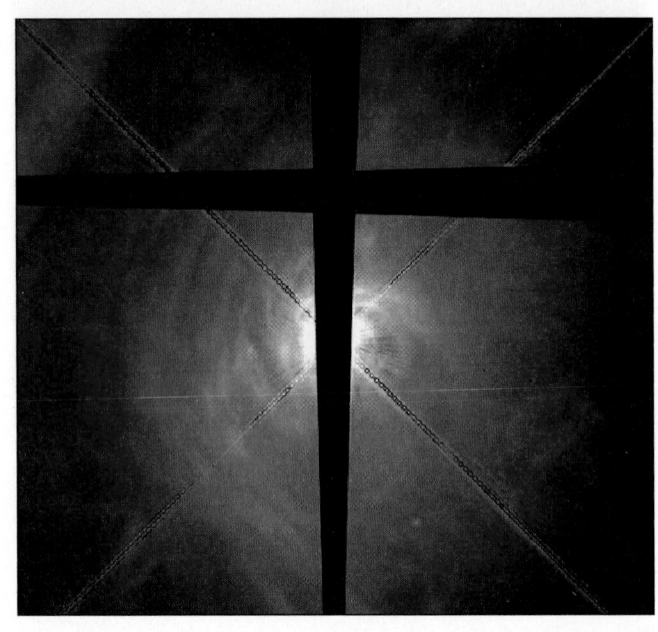

outer one) to speed up slightly. As a result, the angular momentum of the outer blob increases whereas that of the inner blob decreases. The end result of this redistribution of angular momentum is that gas and dust in the inner disk spirals inward whereas more distant gas and dust spirals outward. If this process went on for a very long time, nearly all of the mass of the disk would become concentrated in a very slowly rotating core while a small amount of orbiting material contained nearly all of the original angular momentum of the disk. This, of course, is very similar to present conditions in the solar system, in which the planets have most of the angular momentum but the Sun has nearly all the mass.

Collimation After a young star develops a wind, a battle begins between the outflowing gas in the wind and the material still falling inward from the molecular cloud. At first, as shown in Figure 18.15, the infalling material can suppress the growing wind. Eventually, the wind grows in strength and reverses the inflow along the polar axis of the star, where the amount of infalling material is least. The large amount of swirling, infalling material in the disk that lies in the star's equatorial plane is able to resist the wind. Thus, the wind blows outward only in two opposite directions. Magnetic fields probably also play a role in narrowing the wind to produce very narrow jets. If the equatorial disk is destroyed, the wind blows outward in all directions.

FIGURE 18.15
Three Stages in the Development of a Wind
A, The wind is suppressed by infalling material. **B,** Gaining strength, the wind breaks through the infalling material along the polar axis of the star. **C,** Later, the wind may blow outward in all directions and sweep away the disk of orbiting material.

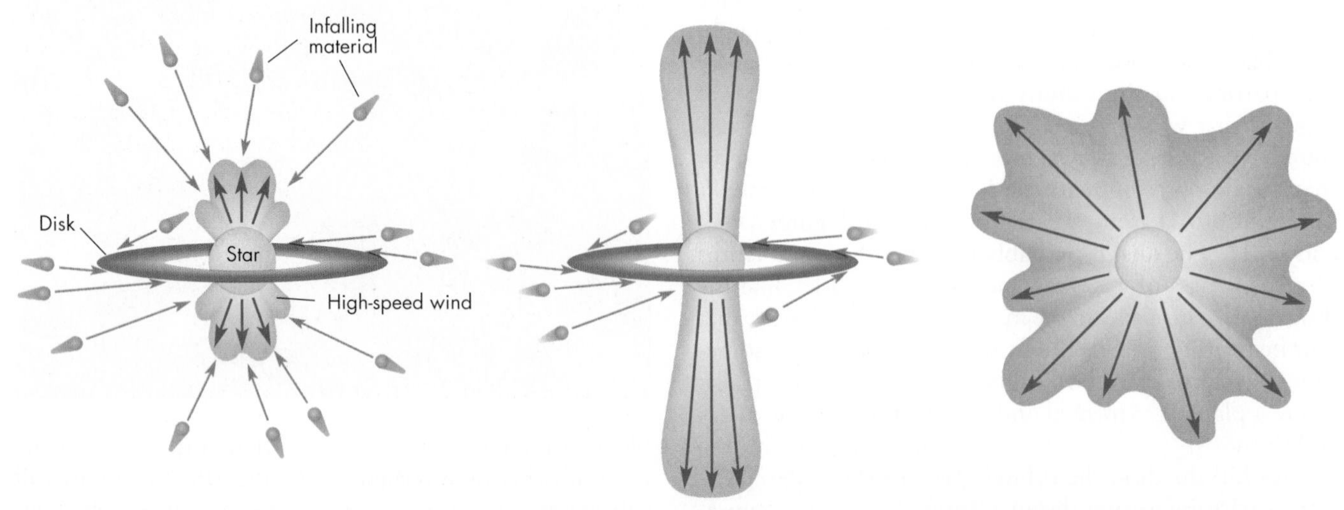
Infalling material
Disk
Star
High-speed wind
A Infalling material suppresses the wind.
B Eventually the wind becomes strong enough to break through along the polar axes.
C Later the wind sweeps away the disk and blows outward in all directions.

FIGURE 18.16
Jets of Gas from a Young Star
Two oppositely directed jets of gas from a young star can be seen in this *Hubble Space Telescope* image of part of the Carina Nebula. The star itself is still obscured by the dust that surrounds it. The bright clumps in the jets are called Herbig-Haro objects. The part of the Carina Nebula shown here is about 0.4 pc across.
Credit: NASA, ESA, N. Smith (U. California, Berkeley) et al., and The Hubble Heritage Team (STScI/AURA)

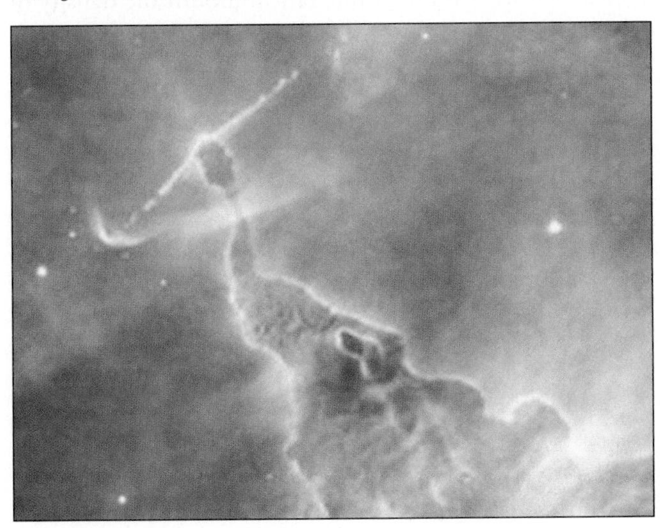

Figure 18.16 shows a wind blowing outward along the polar axis of a young star. In some cases, the jets from young stars can be traced outward for several tenths of a parsec. The bright nebulae that make this tracing possible are called **Herbig-Haro objects.** The Herbig-Haro objects, which have emission line spectra, were once thought to be nebulae surrounding protostars. However, the alignment of the Herbig-Haro objects with the jets from young stars has shown that they are actually spots where the jets collide with interstellar gas. The gas is heated by the collision to a temperature high enough to produce emission lines. Although Herbig-Haro objects are about as luminous as the Sun, they contain little material, perhaps only 10 or 20 times the mass of the Earth.

What Happens to Disks? Three processes can destroy the disk around a young star. One of these processes is the evaporation of the disk by a neighboring star. If a young star has a massive, hot star as a neighbor, its disk may absorb ultraviolet light emitted by the neighbor. When this happens, the gas and dust in the disk is heated so much that it boils away. The entire disk can be lost in as little as a few million years—far too little time for planets to form in the disk. This seems to be happening to the disks of many of the young stars near the center of the Orion Nebula shown in Figure 18.13. The second process is the dispersal of the disk by the wind that flows from the star. The third process, which sometimes occurs before evaporation or dispersal can take place, is the aggregation of the matter in the disk to form planets or a companion star. This is what happened in the disk of gas and dust that surrounded the Sun as it formed.

Young stars develop energetic winds that blow away the surrounding infalling matter. The stars also have flattened disks containing perhaps several tenths of a solar mass of gas and dust. Friction within a disk causes angular momentum to be transported from the center of the disk to the outside. At first, the wind from a young star can only blow outward along the polar axis of the star. Later it can blow away the disk of matter as well, if a planetary system or binary companion star doesn't form in the disk before it is blown away.

18.5 PLANETARY SYSTEMS

Solar system builder

Our thinking about the formation of planetary systems has been strongly dominated by efforts to account for the origin of our own solar system. After all, ours is the planetary system about which we know the most. We can hope, however, that the way the Sun and solar system formed wasn't very different from the way other stars and their planetary systems formed. In the last few years astronomers have discovered planets orbiting many nearby stars. These planets and how they were discovered are described in Chapter 27. All of the newly discovered planets are more massive than Earth and many of them orbit very close to their parent stars. Whether their proximity to their parent stars reflects a significant difference from the solar system in the way they were formed is still an open question.

The Solar Nebula

Planet formation from solar nebula

Flattening and spreading up of a collapsing interstellar cloud

The starting point for our investigation of the origin of the planets, satellites, and the small bodies of the solar system is the **solar nebula,** the disk of gas and dust that orbited the newly formed Sun.

When Did the Solar Nebula Exist? The chronology of events is probably what we know best about the solar nebula. The ages of the oldest lunar rocks and meteorites indicate that the formation of the major bodies in the solar system was complete by about 4.6 billion years ago. The length of time required to form these bodies has been calculated by studying the detailed atomic makeup of meteorites. It has been found, for instance, that chondrite meteorites contain the magnesium isotope ^{26}Mg in minerals that normally are made from aluminum. The ^{26}Mg was formed by the decay of the radioactive aluminum isotope ^{26}Al. This is remarkable because the half-life of ^{26}Al is only 700,000 years and there is no known way in which ^{26}Al could have been made within the solar system. This means the ^{26}Al must have come from somewhere else.

The most likely origin for the ^{26}Al was a supernova explosion of a massive star within the molecular cloud from which both the star and the Sun were formed. The supernova explosion could not have occurred more than a few million years earlier than the time that the meteorites formed or all of the ^{26}Al would have decayed to ^{26}Mg before it could have been incorporated in meteorites. This tells us that only a few million years elapsed between the start of the collapse of the cloud core and the time that the parent bodies of the meteorites formed.

Another radioactive isotope present in the solar nebula was ^{129}I, which decays with a half-life of 17 million years to produce ^{129}Xe, an isotope of xenon. Xenon is an inert gas that escapes unless the ^{129}I that produced it was part of a mineral within a rock. In that case, the xenon that was produced was trapped within the rock. By measuring the amounts of ^{129}Xe present in different meteorites and Earth rocks, it has been possible to calculate the interval of time during which the various objects were formed. A body that solidified early originally contained a relatively large amount of ^{129}I and should now contain a relatively large quantity of ^{129}Xe. A body that formed millions of years later, after much of the original ^{129}I had decayed, should now contain very little ^{129}Xe. Measurements of the amount of ^{129}Xe in meteorites show that most of the meteorites formed within 20 million years of each other and that planets like the Earth formed within the next 100 million years. On a cosmic timescale, the formation of the solar system took place very rapidly.

How Massive Was the Solar Nebula? The solar nebula had the same chemical composition that the outer layers of the Sun have today—that is, mostly hydrogen and helium. Only about 2% of the mass of the solar nebula was made of heavier elements, yet there is almost no hydrogen and helium in the Earth today. To yield enough silicon, oxygen, iron, magnesium, and other elements to make the Earth required about 50 Earth masses of material rich in hydrogen and helium. When similar calculations are carried out for the other planets, it turns out that the minimum mass that the solar nebula could have had was a few percent of the mass of the Sun. The solar nebula was probably more massive than that, however, because the disks that astronomers have observed around young stars have masses of a few tenths of a solar mass, and there is no reason to believe that the solar system was any different. Most of this material has been lost, probably swept out of the solar system by a T Tauri-like stellar wind.

Although a few tenths of a solar mass is certainly a large amount of material, it doesn't mean that the solar nebula was very dense. If the solar nebula extended outward to Neptune's orbit and was 1 AU thick, then the typical density was about one ten-millionth the density of air at the Earth's surface. This is much denser than interplanetary space is today, but hardly as thick as pea soup.

The abundances of isotopes in meteorites show that the solar nebula existed about 4.6 billion years ago and that only 10 to 100 million years or less were required to form the planets. The mass of the solar nebula was between a few hundredths and a few tenths of the Sun's mass.

Cooling and Condensation The solar nebula was heated by the release of gravitational energy and remained warm as long as it continued to accumulate fresh material from the infall of the cloud core. It was hottest near its center, where temperatures may have been as high as 2000 K, but temperature decreased as distance from the core increased. After the accumulation of infalling material ended, however, the solar nebula began to cool. As the gas cooled, some compounds began to condense into small particles and droplets.

The first substances to condense to solid and liquid form were those that are stable at the highest temperatures. The graph in Figure 18.17 shows that these were rare elements such as tungsten. At a slightly lower temperature, oxides such as aluminum oxide condensed. The first abundant materials to condense, at a temperature of about 1500 K, were metals such as iron and nickel and some silicate minerals. Other silicate minerals condensed as the temperature fell below 1000 K. At about 400 K, carbon and carbon-rich silicate minerals condensed. Finally, below 200 K, ices condensed, completing the sequence of condensation.

If the temperature throughout the entire solar nebula had fallen to 100 K, all the planets and other bodies would have formed from mixtures of the condensates described in the preceding paragraph. Instead, the condensation process ended when the temperature at the inner edge of the nebula had fallen to about 1000 K. In cooler regions, farther from the center, the sequence of condensation had proceeded further and more compounds had condensed. At this time, the remaining gas in the solar nebula was swept away, leaving behind only the solids and liquids that had already condensed. At the time condensation stopped, temperature and the materials already condensed varied with distance from the Sun as shown in Figure 18.18. This profile matches the actual compositions of the planets and their satellites.

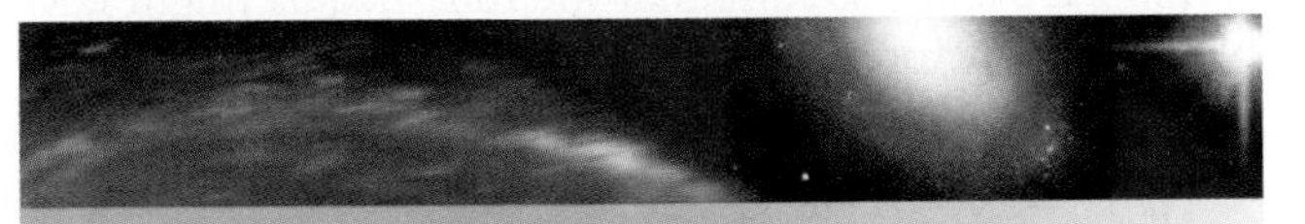

As the solar nebula cooled, solid particles began to condense. The condensation continued until the remaining solar nebula was blown away. Temperatures in the inner solar nebula remained so high that only metals, oxides, and silicate minerals condensed. Farther out, temperatures fell low enough for ices to condense, too.

FIGURE 18.17
The Sequence of Condensation in a Cooling Gas
Metals and some oxides condense at temperatures of 1500 to 2000 K. Silicates condense at temperatures of about 1000 K and carbon-rich silicates at about 400 K. Ices condense only if the temperature of the gas falls to 200 K or lower.

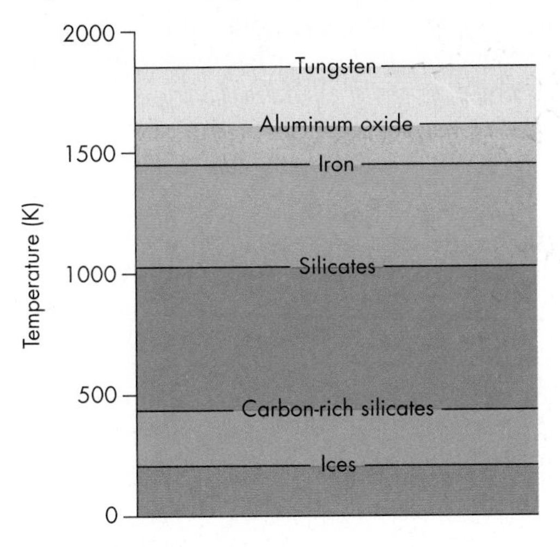

Growth of Planetesimals The next step in the development of the planetary system was the accumulation of microscopic grains of condensed material into larger bodies called **planetesimals.** Planetesimals may have ranged in size from a few millimeters to hundreds of kilometers (as big as large asteroids). At first the growth of grains into larger bodies probably was very gradual. Grains orbiting near each other had very similar orbital speeds and drifted past each other very slowly. As they slowly grazed each other, they may have stuck together. Gradually, gentle collisions built up very fluffy clumps that may have looked like dust bunnies or like the micrometeorites that have been collected in the Earth's upper atmosphere. Although the growth of kilometer-sized bodies from these fluffy clumps is not yet understood, it must have taken place on a timescale of a few centuries. Otherwise the meter-sized intermediate bodies would have drifted inward through the solar nebula and been destroyed near the Sun. As the planetesimals grew in size and mass, they began to exert gravitational influences on each other, attracting each other and enhancing the probability of a collision. The larger bodies grew by accumulating smaller ones, which broke apart when they collided with the surface of the larger body. The fragments of the collision were held on the surface of the larger body by gravity. This description may give the impression that planetesimals grew only after condensation of grains from the solar nebula had stopped. In reality, the growth of the planetesimals probably began before the condensation of new grains had finished completely.

Planet Formation Once planetesimals grew to kilometer size, further growth occurred through collisions among the planetesimals. In only a few tens of thousands of years, a collection of kilometer-sized bodies grew through collisions to produce a few bodies 1000 kilometers or more in diameter. The final development (at least for the terrestrial planets) took place as the largest body or bodies swept up the

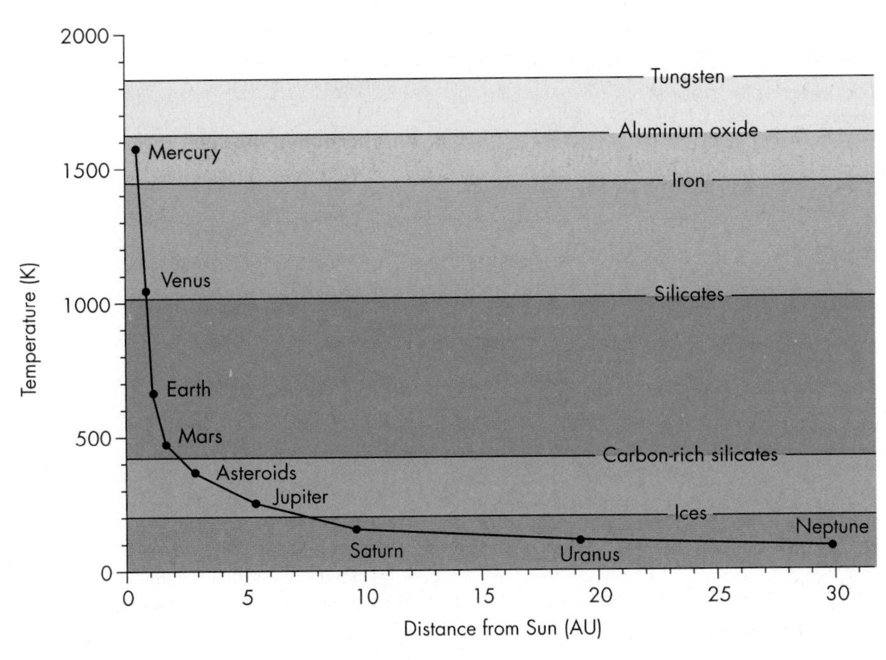

FIGURE 18.18
Condensation and the Chemical Compositions of the Planets
The temperature in the solar system at various distances is shown at the time that condensation stopped. The materials that condensed in each part of the solar system (and for each planet) are also shown. For example, at the Earth's distance, only metals, oxides, and silicates condensed. This mixture is consistent with the chemical makeup of the Earth. At Uranus's distance all materials, including ices, condensed. Thus, Uranus and its satellites have a significant ice content.

rest of the planetesimals and reached their final size. Many of the impact craters that cover the surfaces of Mercury and Mars may have been produced during the period that the remaining planetesimals were swept up. This period is referred to in Chapter 9 as the "bombardment era."

Giant planets, such as Jupiter and Saturn, could not have grown entirely by the accumulation of planetesimals. The hydrogen and helium they are made from must have been collected directly from the gas that remained in the solar nebula while the condensation of grains was taking place. Figure 18.19A, illustrates one way in which this could have happened. This possibility, called the gravitational instability model, suggests that Jupiter-sized masses of gas became gravitationally unstable and collapsed under their own weight, perhaps within unstable rings. This collapse could have taken as little as a few hundred years. Another possibility, shown in Figure 18.19B, is that cores of solid material formed at the distances of the giant planets just as they did in the inner solar system. These cores, perhaps ten times as massive as the Earth, then captured massive gaseous envelopes from the surrounding solar nebula before the gas in the solar nebula was swept away.

Neither of these hypotheses completely explains the properties of the giant planets. It is hard to see, for instance, how the collapse of a large quantity of gas could lead to a planet like Uranus or Neptune, in which a core of rock and ice makes up most of the mass of the planet. On the other hand, accumulation of gas around a rocky core would have taken so long that the solar nebula might have blown away before Jupiter and Saturn could have grown to their present masses. Probably the best way to tell which of the two models is better would be to determine with complete certainty whether Jupiter and Saturn have rocky cores containing ten or more times the mass of the Earth. If they do, then the core capture model is more likely to be correct. If not, then the gravitational instability model probably is closer to the truth.

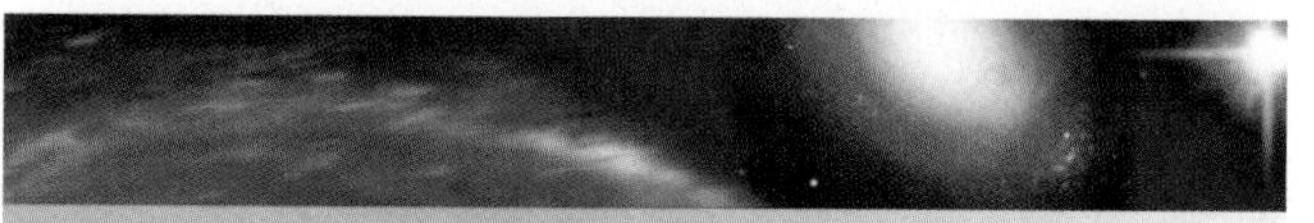

Dust particles within the solar nebula accumulated to produce planetesimals, which ranged in size from millimeters to hundreds of kilometers. These bodies collected to form the terrestrial planets and, probably, the cores of the giant planets. The cores of the giant planets, in turn, captured gas from the solar nebula to produce their massive gaseous envelopes. Alternatively, the giant planets may have formed by the gravitational collapse of parts of the solar nebula.

FIGURE 18.19
Two Models for the Formation of a Giant Planet
A, In the gravitational instability model for the formation of a giant planet, an unstable ring of gas developed in the solar nebula. A Jupiter-sized mass within this ring collapsed under its own weight to form a giant planet. **B,** A second possibility is that the core of a Jovian planet formed first and then captured gas from the surrounding solar nebula.

A Gravitational instability model:

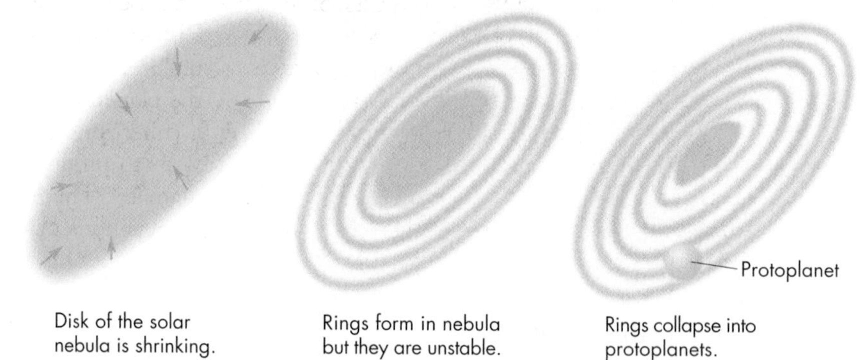
Protoplanet
Disk of the solar nebula is shrinking.
Rings form in nebula but they are unstable.
Rings collapse into protoplanets.

B An alternative model:

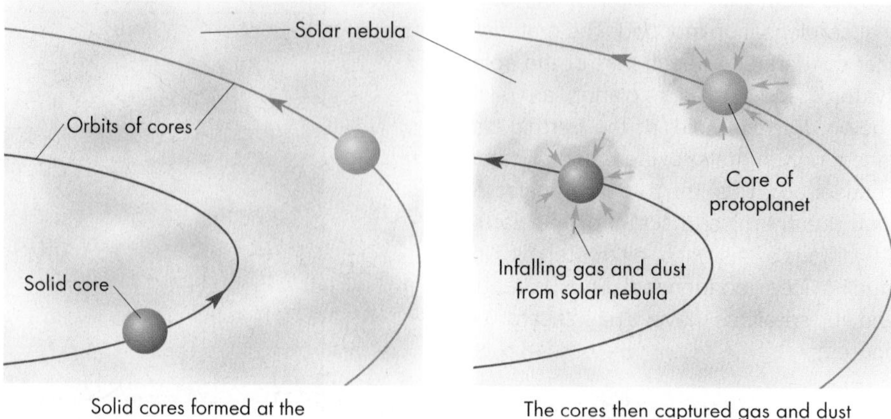
Solar nebula
Orbits of cores
Solid core
Core of protoplanet
Infalling gas and dust from solar nebula
Solid cores formed at the distances of the giant planet.
The cores then captured gas and dust from the surrounding solar nebula.

Satellite Systems The regular satellite systems of the outer planets probably formed the same way the planetary system formed. The gas that collected to form a giant planet was accompanied by additional material that orbited the planet in a disk. Taking into account the masses and chemical makeup of the satellites of the outer planets, it has been estimated that the disks were at least 1% as massive as the planets about which they swirled. The disks were heated by both the friction within them and the luminosity of the planet about which they orbited. Remember that, for a brief time, the giant planets were millions of times more luminous than they are now. As the planets dimmed, the disks cooled and material condensed within them just as it did in the planetary system. The inner satellites were formed of denser, more easily condensed material whereas the outer satellites, which formed in a cooler environment, incorporated large quantities of icy material as well.

This scenario accounts very well for the formation of the regular satellites of the giant planets, such as the Galilean satellites of Jupiter. Most of the other satellites of the giant planets are probably planetesimals that formed elsewhere in the solar system and were later captured by the giant planets. In the solar system today, the permanent capture of a small body to become a satellite of a planet is quite unlikely. Jupiter captures a comet, at least temporarily, several times per century. The most recent such capture that we know of was Comet Shoemaker-Levy 9, which crashed into Jupiter in 1994. The orbits of these captured comets are very large and the comets usually escape from orbit about Jupiter within a few orbital periods. When a small body approaches a planet, it is accelerated by the gravity of the planet so much that it reaches the planet traveling at a speed greater than escape velocity. For it to go into a permanent orbit about the planet, the small body would have to be slowed considerably. This is improbable, however, because it would require very special conditions such as a near encounter with an existing satellite.

Capture must have been much easier when the giant planets had accompanying gaseous disks. Frictional drag of the gas in a disk could have slowed an approaching body enough for it to go into orbit. Nevertheless, the timing of the capture must have been critical, because once a body was captured into orbit, frictional drag would have continued to act on the body, causing it to spiral into the planet and be destroyed. Only if capture took place just as the disk was vanishing could destruction of the captured body be avoided. It's possible that many bodies were captured by the giant planets. The early ones spiraled deep into the planet, adding to the mass of its rocky, icy core, whereas the bodies captured later became satellites.

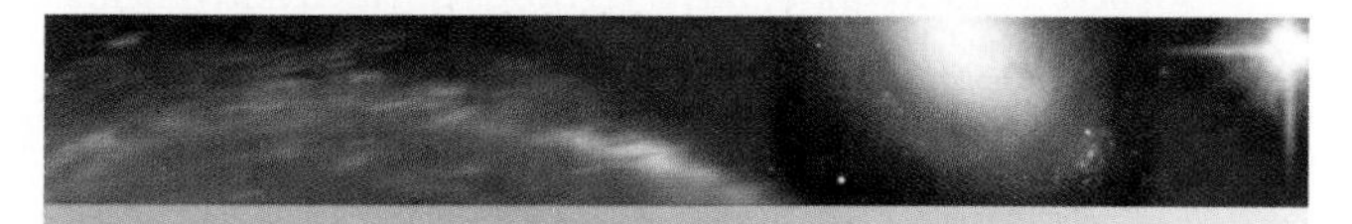

The satellite systems of the outer planets probably formed in a miniature version of the way the planets formed. Some satellites, however, may be planetesimals captured while the planets still had gaseous disks to slow down approaching bodies.

Final Collisions It now seems possible that the final stage in the formation of the solar system consisted of catastrophic collisions between planet-sized objects. One such collision, between the Earth and a Mars-sized body, may have been responsible for the origin of the Moon. Similarly, Mercury may be the remnant of a larger planet that collided with another body about one-sixth as massive as itself. The violence of the collision blasted away most of the silicate-rich mantle of the proto-Mercury but only some of its metallic core. If this theory is correct, then the high density of Mercury is a result of the destruction of the outer, less dense parts of the original planet. Yet another possibility is that a collision between Uranus and a planet twice as massive as the Earth caused the large tilt of Uranus's rotation axis. It is possible that the solar system once had more planets than there are now and that the nine we see today are the survivors of spectacular collisions that took place more than 4 billion years ago.

Chapter Summary

- The number of young stars of different masses shows that the formation of less massive stars is more common than the formation of massive ones. Binary or multiple stars form more often than single ones. Many stars form together in a star formation region. The flat shape and regular rotation of the solar system show that the material from which it formed also was flat and rotating. (Section 18.1)
- Stars form in the cores of clumps of material within giant molecular clouds. Massive stars and clusters of stars form within warm, massive cores whereas less massive stars form within cold cores. Cores collapse under their own weight. Magnetic fields within the cores may slow the collapse. (18.2)
- A protostar develops at the center of a collapsing core. The protostar is originally transparent, but

eventually becomes opaque. At this point, the protostar stops collapsing and begins to grow in mass by accumulating infalling material. It begins a period of slow contraction that ends when the star becomes hot enough for hydrogen fusion to occur. As a protostar collapses, it rotates ever more rapidly. Rotation eventually causes infalling material to accumulate in a rotating nebular disk. (18.3)

- Young stars first become visible when they shed their surrounding gas and dust. For most stars, this takes place before they reach the main sequence. These stars, called T Tauri stars, have vigorous surface activity and are orbited by disks of material. Massive stars, on the other hand, become visible only after they have already reached the main sequence. (18.4)
- The vigorous surface activity of a young star produces a wind that eventually blows away infalling material. At the same time, friction within the star's gas and dust disk causes its center to spin more slowly and carries angular momentum outward. At first the disk prevents the wind from blowing outward. This produces a collimated wind along the star's polar axis. Eventually, the disk is blown away, but sometimes this doesn't happen until after a companion star or a planetary system forms. (18.4)
- The abundances of atomic isotopes in meteorites suggest that the solar nebula existed 4.6 billion years ago and that the formation of the planets took 100 million years or less to complete. The solar nebula contained an amount of material equal to between a few hundredths and a few tenths of the mass of the Sun. (18.5)
- As the solar nebula cooled, solid particles and liquid droplets condensed from the gas. The condensation continued until the gas in the solar nebula was blown away by a strong wind from the Sun. By the time this happened, only metals, oxides, and silicates had condensed in the hot, inner part of the solar nebula. In the outer nebula, temperatures had dropped enough that ices had condensed as well. (18.5)
- After dust particles formed, they accumulated into planetesimals, which ranged up to hundreds of kilometers in diameter. These then collided with each other to produce the planets. In the outer solar system, the cores of the planets Jupiter, Saturn, Uranus, and Neptune captured gas from the solar nebula to produce their massive gaseous envelopes. Alternatively, the giant planets may have formed when parts of the solar nebula collapsed under their own weight. (18.5)
- The regular satellites of the outer solar system probably formed in a process similar to that which formed the planets. Some satellites, however, may be planetesimals captured by the planets while they still had gaseous disks. The disks were able to slow down approaching planetesimals enough for the planetesimals to be captured into orbits about the planets. (18.5)

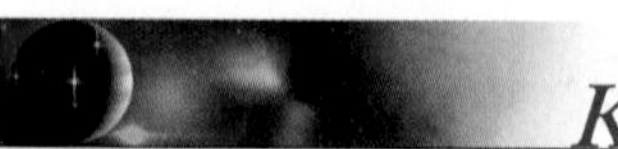

Key Terms

association *426*
bipolar outflow *432*
cloud core *427*
giant molecular cloud *427*
Herbig-Haro object *435*
molecular cloud *426*
planetesimal *437*
protostar *428*
solar nebula *435*
T Tauri star *431*

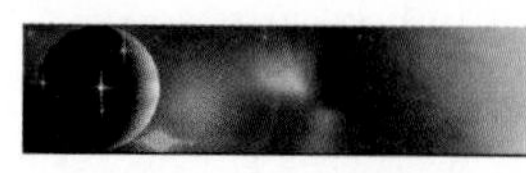

Conceptual Questions

1. What do the orbits of planets and satellites tell us about the rotation of the cloud that formed the solar system?
2. Why are radio and infrared radiation more useful than visible light in studying molecular clouds?
3. Describe the structure of a giant molecular cloud.
4. What evidence do we have that stars form in the cores of giant molecular clouds?
5. Suppose there were no magnetic fields in interstellar space. What effect might this have on the number of young stars in the galaxy?
6. Why does a collapsing cloud remain cool as long as it is transparent, but begin to heat up when it becomes opaque?
7. What halts the collapse of a protostar?
8. What happens to the shape of a protostar as it collapses? Why does this change in shape occur?
9. Where in an H-R diagram are young stars located when they first become observable using visible light? Why aren't they visible at an earlier stage of their careers?
10. What evidence do we have that young stars are surrounded by dusty disks rather than complete spherical shells of dust?
11. What role does friction play in determining the distribution of matter and angular momentum within rotating disks about young stars?

12. What is the relationship between Herbig-Haro objects and jets of gas from young stars?
13. How do we know that the formation of the solar system took place in only about 100 million years?
14. How is it possible to estimate the mass of the solar nebula?
15. Describe the steps in the process through which microscopic solid grains accumulated to form objects a few thousand kilometers in diameter.
16. What evidence do we have that the planets formed as a result of collisions of planetesimals?
17. What evidence do we have that the final stage of the formation of the solar system consisted of collisions between planet-sized objects?

Figure-Based Questions

1. Use Figure 18.17 to find the order in which solid materials would condense from a gas that was heated to a temperature of 3000 K and then allowed to cool to a temperature of 50 K.
2. Use Figure 18.18 to find which materials in the solar nebula would have condensed at a distance of 20 AU but not at 1 AU.

Group Activities

1. Divide your group into two subgroups. Have one subgroup use Figure 18.17 to develop a hypothesis of what our solar system would be like today if the gas in the solar nebula had been swept away when the temperature at the inner edge of the nebula was still much higher than 1000 K. Have the other subgroup hypothesize what the solar system would be like if the gas hadn't been swept away until the temperature at the inner edge of the nebula had fallen to 300 K. Have the two subgroups compare their results with the actual solar system and with each other. Discuss the implications of your hypotheses for what other planetary systems might be like.
2. With the help of Figure 18.2, use binoculars to observe the Orion Nebula. See how many members of your group can see the Orion Nebula without binoculars. (You should do this in winter or spring when the constellation Orion is in the sky in the evening.)

For More Information

Visit the text website at **www.mhhe.com/fix** for chapter quizzes, interactive learning exercises, and other study tools.

19

Roll on, ye stars! exult in
youthful prime,
Mark with bright curves the
printless steps of time. . . .
Flowers of the sky! ye too to
age must yield,
Frail as your silken sisters of
the field.

Erasmus Darwin

The Cat's Eye Nebula. The blue region is a *Chandra X-Ray Observatory* image. Red and purple show the *Hubble Space Telescope* image.

www.mhhe.com/fix

The Evolution of Stars

A typical star—one like the Sun—lives (that is, produces nuclear energy) for about 10 billion years. This incredible length of time is greater than 100 million human lifetimes. Astronomers have had the observational equipment necessary to carry out detailed examinations of stars and the understanding of physics required to comprehend the meaning of their observations for only about a century. Observing the Sun for 100 years covers about the same fraction of its lifetime as observing a person for 20 seconds. This might make understanding the career of a star seem impossible. After all, how much could be learned about human development and aging if a person could be observed for only 20 seconds? In that space of time, the person would breathe a few times and experience a couple of dozen heartbeats. The person might walk a few tens of meters or, if asleep, remain nearly stationary.

The task isn't nearly as hopeless, however, if more than one person is available for study. By observing a great many people for 20 seconds each, a scientist could build up a statistical picture of the behavior of typical people. About one-third of the 20-second observations would reveal the person in a horizontal, quiet state. This would lead to the conclusion that typical people spend about one-third of their lives sleeping. Similar conclusions about human activities could be made on the basis of the percentage of people observed to be eating, riding elevators, and brushing their teeth. Occasionally, important events such as births and deaths would be observed. The observation that only one in every 100 million observations recorded a death would lead to the conclusion that people live 100 million times longer than

Questions to Explore

- Why do stars evolve?
- What can be learned about stellar evolution by studying H-R diagrams?
- In what ways are all main sequence stars alike and in what ways are they different?
- What happens to a star when it finishes the main sequence part of its evolution?
- Why do some stars pulsate?
- How do stars make heavy elements?

20 seconds, or about 70 years. The observation that new people (babies) are small but most dying people are large would contribute to a theory of human growth and development. Eventually, with enough reasoning and observation, the story of people could be pieced together.

The story of stellar evolution has been pieced together in just this way. By observing a great many stars, astronomers have developed a statistical picture of the lives of typical stars. Astronomers also have observed occasional milestone transitions in the lives of stars. These observations, combined with considerable hard thinking, have brought us to the point where we have a reasonably good understanding of the evolution of stars. This chapter describes what we know, or think we know, about the lives of stars between their births (Chapter 18) and their deaths (Chapter 20).

19.1 WHY DO STARS EVOLVE?

When astronomers say that a star evolves, they mean that it changes its appearance and internal structure. Its appearance and internal structure change because the fusion reactions that supply the star's energy transform light elements into heavier ones, changing the internal chemical composition of the star. This is significant because important processes going on in stars depend on the kinds of atoms and ions that make up the star. These processes change as the chemical composition changes. As a result, as the internal structure of the star changes, its size, luminosity, and surface temperature may change as well.

In stars, there are three processes that depend on chemical composition. One is the rate at which the star produces energy. A second is the rate at which energy flows to the surface of the star, where it is radiated into space. The third is the generation of pressure, which resists gravity and prevents the collapse of the star. As these processes change, the star evolves.

Energy Generation

The two ways in which stars produce energy are fusion and gravitational contraction or collapse.

Hydrogen Fusion The temperature at the center of a star first rises to about 10 million K at the end of the star's long contraction from a fragment of a molecular cloud. At this point, the fusion of hydrogen into helium becomes the major source of energy in the star. In the Sun and in stars less massive than the Sun, the fusion of hydrogen into helium occurs by means of the proton-proton chain described in Chapter 17, Section 2. In stars more massive than the Sun, however, hydrogen fusion occurs by means of the **carbon cycle,** in which carbon, nitrogen, and oxygen nuclei act as catalysts for the production of helium from hydrogen. Figure 19.1 shows the nuclear reactions that make up the carbon cycle. For either set of reactions, hydrogen becomes less abundant and helium more abundant as time passes.

The graphs in Figure 19.2 show how the fractional amount of hydrogen varies with time in a star of 1 $M_\odot$. Initially (Graph A), the amount of hydrogen is the same throughout the star. After fusion begins, however, it occurs most rapidly at the center of the star because the temperature is highest there. The more rapidly fusion goes on, the more rapidly the composition of the star changes, so the amount of hydrogen decreases most rapidly at the center of the star (Graph B). As fusion continues over time (Graph C), more and more hydrogen is consumed in the core of the star. The number of hydrogen nuclei (protons) drops because they are being fused into helium, so the rate of collisions involving protons decreases. The rate of fusion would decrease, as a result, except that the temperature increases slightly at the same time. Collisions are more energetic at the higher temperature and are more likely to lead to fusion. Thus, rising temperature compensates for the falling number of protons to keep the rate of energy production high.

Eventually, however, all of the hydrogen in the core of a star is consumed. In many stars, this happens only after the temperature in the region just outside the core of the star has risen high enough that hydrogen fusion can occur there. Thus, core hydrogen fusion is sometimes followed by an evolutionary phase in which fusion and energy generation go on in a thin shell that surrounds the helium-rich core.

Other Nuclear Fuels High temperatures are needed to fuse hydrogen because the hydrogen nuclei, protons, are electrically charged and repel each other. Nuclei of elements other than hydrogen contain even more protons, so they have even larger electrical charges to overcome during fusion. As a result, even higher temperatures are needed to bring these other nuclei close enough for fusion to take place.

At a temperature of 100 million K, the fusion of helium into carbon becomes possible. The pair of reactions through which helium fuses into carbon is called the **triple α process** because three helium nuclei (also known as **α particles**) are needed to make each carbon nucleus. By the time the core of a star gets hot enough for the triple α process to take place, all of the hydrogen has been fused into helium. Helium is now the most abundant fuel and can supply the triple α process for a relatively long time. If the temperature in a star reaches between 500 million to 1 billion K, the carbon that results from the triple α process becomes a fuel. If the core temperature rises even higher in a star, other nuclei can be consumed in fusion reactions. It is important to remember, however, that fusion can produce energy only from fuels less massive than iron (see Chapter 7, Section 4). The fusion of iron and more massive elements can never produce energy for the star in which those reactions take place.

FIGURE 19.1
The Nuclear Reactions of the Carbon Cycle
The cycle consists of both captures of protons by nuclei and nuclear decays. An example of a proton capture is the reaction between ^{12}C and ^{1}H to produce ^{13}N. The capture of a proton by ^{15}N, however, produces ^{12}C and ^{4}He. An example of a nuclear decay is the emission of a positron (β^+) and a neutrino by ^{13}N. Notice that the cycle of reactions does not change the total number of carbon, nitrogen, and oxygen nuclei in the star. The net result of the reactions, however, is the fusion of four hydrogen nuclei into a single helium nucleus.

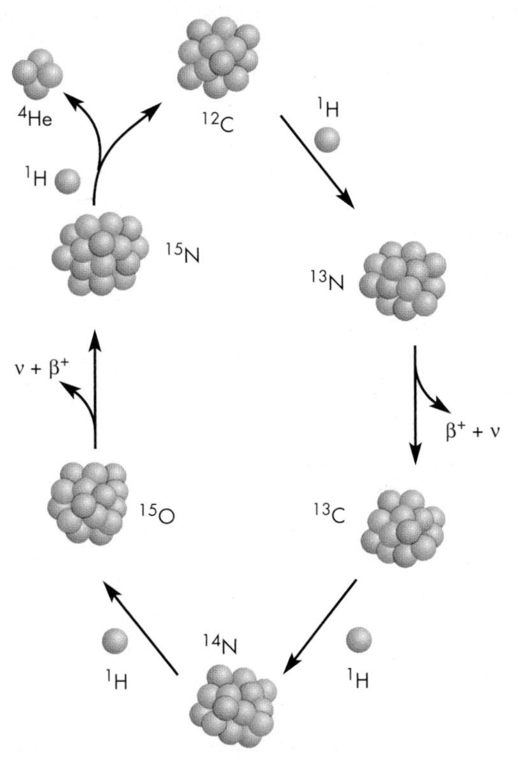

FIGURE 19.2
The Depletion of Hydrogen in the Core of a One-Solar-Mass Main Sequence Star
A, The hydrogen content throughout the star is uniform before hydrogen fusion begins. **B,** During hydrogen burning, the hydrogen content drops fastest at the center of the star because the rate of fusion reactions is greatest there. **C,** Near the end of core hydrogen burning, the hydrogen content in the core is reduced nearly to zero. Notice that the hydrogen content of the outer stellar layers is unaffected by hydrogen burning in the core.

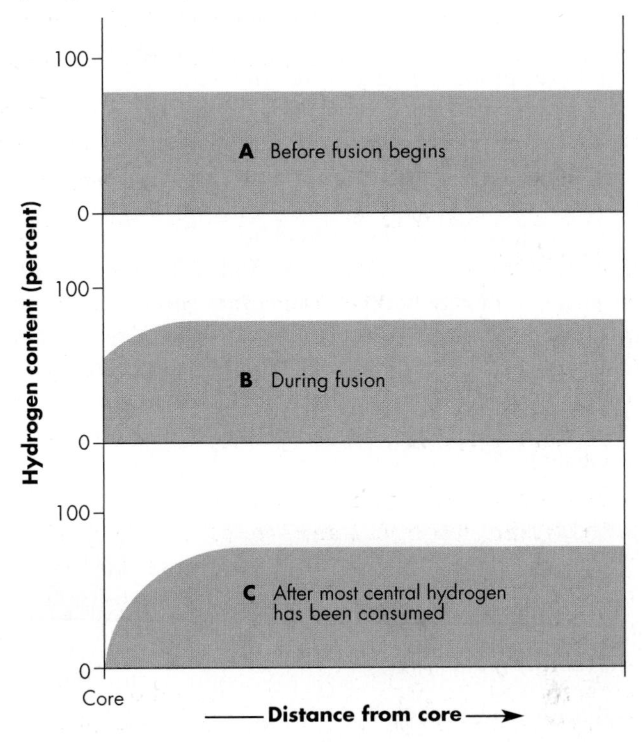

Contraction and Collapse Any time a star or a part of a star shrinks in size, gravitational energy is released. Generally, as in the case of pre–main sequence evolution, a star shrinks slowly and releases gravitational energy at a controlled rate. A star often contracts after a nuclear fuel (for example, hydrogen) has been used up in the core of the star. The energy released during contraction heats the core of the star, however, so contraction usually continues only until the core temperature rises enough for another kind of nucleus (for example, helium) to become a nuclear fuel.

The total gravitational energy that has ever been released by a star as it shrank to a given size is approximately given by

$$E = \frac{GM^2}{R} \tag{19.1}$$

Gravitational Energy

Equation 19.1 can be used to find the total energy released by the Sun as it shrank to become a main sequence star.

To calculate this, we need to use the present mass and radius of the Sun, $M_\odot = 2 \times 10^{30}$ kg and $R_\odot = 7 \times 10^8$ m, respectively. The value of the gravitational constant, G, is 6.67×10^{-11} Nm^2/kg^2. Using these values, Equation 19.1 is

$$E = \frac{GM^2}{R} = \frac{6.67 \times 10^{-11}(2 \times 10^{30})^2}{7 \times 10^8} = 4 \times 10^{41}\ \text{J}$$

At the Sun's present luminosity, this would be enough energy for 30 million years. The estimate of the Sun's Kelvin-Helmholtz contraction time, described in Chapter 17, is based on this calculation.

where G is the gravitational constant, M is the mass of the star, and R is its radius.

Because the total energy produced by gravitational contraction increases as the radius of a star decreases, the Sun would need to contract by only a factor of two in size (to half its present radius) to generate an amount of energy equivalent to that produced during the entire time the Sun shrank to its present size. If a star like the Sun were to shrink to the size of the Earth, an amount of energy equal to 3 billion years' worth of solar luminosity would be produced. If the star were to shrink to a radius of 20 km, 1000 billion years' worth of solar energy would be released. This is far more than the total energy that fusion could ever produce in a star like the Sun.

The gravitational contraction of a star usually occurs relatively slowly during the period of time before the star achieves hydrogen fusion or in the intervals of time between the exhaustion of one fuel and the ignition of another. Thus, the energy released by gravitational contraction carries the star through the periods when fusion is not an effective energy source. Sometimes, however, gravity completely overwhelms the pressure within a star. When this happens, the star collapses rapidly and releases enormous amounts of energy. The supernova explosion that is the result of such an event is described more completely in Chapter 20.

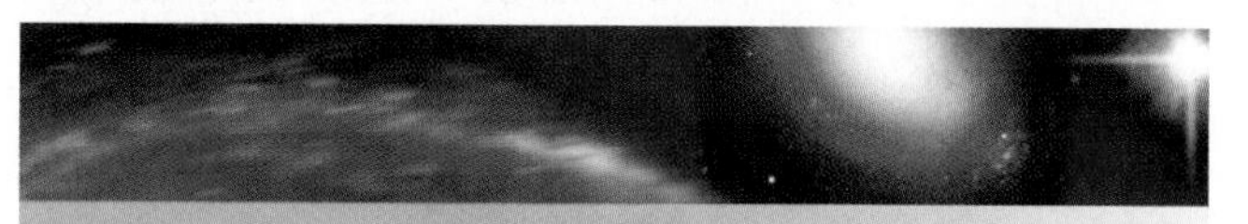

Helium, the product of hydrogen fusion, can itself become a fuel in a star if the star's central temperature becomes high enough. At still higher temperatures, nuclei such as carbon also can become fuels. In the absence of nuclear energy generation, gravitational contraction or collapse produces energy within a star.

Opacity

Opacity is the ability of stellar material to impede the flow of radiation. Recall from Chapter 17, Section 2, that when opacity is low, the gas within a star is relatively ineffective at absorbing photons. Radiation flows easily within the star, and temperature declines slowly with distance from the core. When opacity is high, on the other hand, photons flowing out of a star travel only short distances before they are reabsorbed. Under these conditions, photons diffuse slowly, so temperature declines more swiftly with distance from the center of the star. Where opacity is very high, temperature declines so rapidly that convection occurs. When this happens, convection becomes the primary mechanism by which energy is transported out of the star.

Opacity depends on chemical composition. As nuclear reactions change the chemical makeup within a star, the opacity within the star changes as well.

Opacity, the ability to impede the flow of radiation, depends on the chemical composition of the gas within a star. When opacity is low, radiation flows easily out of a star. When opacity is very high, convection replaces radiation as the mechanism that carries energy out of a star.

Equation of State

The **equation of state** of a gas is a prescription that relates the pressure of the gas to its temperature and density. The prescription depends on the chemical composition of the gas. This means that as nuclear reactions change the chemical makeup of a star, they also change the amount of pressure produced by a given temperature and density.

Normal Gases The ideal gas law, introduced in Chapter 7, Section 2, is the equation of state for the gases that people normally see or measure. When a gas obeys the ideal gas law, its pressure is proportional to the product of temperature and density (Equation 7.2). This means that the internal pressure of a star will decrease if its temperature or density decreases and increase if its temperature or density increases. As a result, the interior of a star that obeys the ideal gas law must stay hot to maintain a pressure sufficient to resist compression by gravity. Stars constantly lose energy to space, however, so they cool off unless they have internal sources of energy. As long as its nuclear energy sources last, a star can remain about the same size. When its nuclear energy runs out, however, it must produce energy through gravitational contraction.

Degeneracy Gases obey the ideal gas law only as long as the particles they contain aren't too densely packed together. A fundamental physical law called the **Pauli exclusion principle** limits the number of particles of a given kind that can be put into a cubic meter of space. As the density of a gas becomes so high that it approaches the limit set by the Pauli exclusion principle, the gas begins to behave somewhat like a liquid. The pressure of the gas increases as density increases, but the pressure is nearly independent of temperature. A high-density gas in which pressure is almost independent of temperature is called a **degenerate gas.** The particles, such as electrons or neutrons, that are responsible for the liquidlike behavior of the gas are said to be degenerate. Not all of the particles in a gas become degenerate at the same density. The electrons in a gas become degenerate at densities of about 10^9 kg/m^3. To visualize a density as large as this, imagine cramming several automobiles into a thimble. The density needed for neutrons to become degenerate is even greater, about 10^{18} kg/m^3. To achieve this density, the entire Earth would have to be packed into a cube less than 200 meters (two football fields) on a side.

When the gases inside a star become degenerate, the star no longer needs to remain hot. It can, and does, cool off. It doesn't contract, however, because its internal pressure remains high even as its temperature falls. In most cases, the star can't contract any more and has reached its final size. Thus, most stars end their careers as small, dense spheres of degenerate gas slowly cooling and radiating their remaining heat into space. These stars, called white dwarfs and neutron stars, are described in Chapter 20.

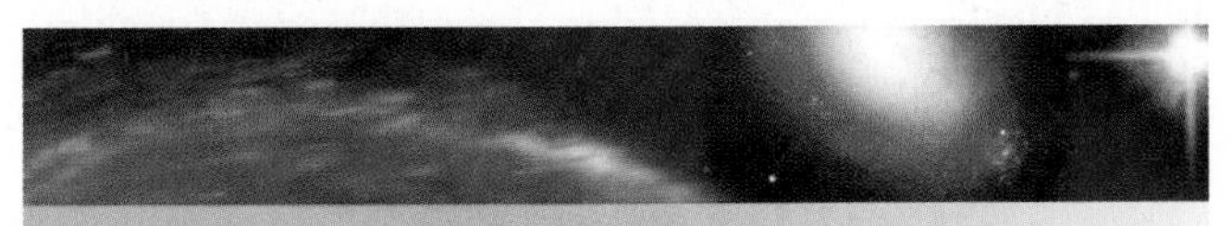

At high densities, particles in a gas are packed so tightly that they strongly resist further compression. This kind of gas, called a degenerate gas, can cool without losing its pressure.

The Vogt-Russell Theorem

The **Vogt-Russell theorem** states that the entire evolution of a star is determined by its initial mass and chemical composition. This predestination occurs because all of the important processes in the star—energy generation, energy transport, and equation of state—depend on the composition of the star. These processes determine the structure of the star, including the rates and kinds of nuclear reactions that go on in the star. These nuclear reaction rates, in turn, are what cause the composition of the star to change. From the moment a star is formed, the entire course of its evolution is fixed. If this weren't the case, if the evolution of a star depended on random, external events, astronomers would be powerless to calculate evolutionary models for stars or make predictions about stellar evolution.

The Vogt-Russell theorem applies only to single stars or those in multiple star systems that always remain far from their companion stars. Stars that orbit each other so closely that they can exchange material or interact in some fashion evolve in ways that depend on the nature of their companion as well as on their own initial mass and composition. The careers of interacting binary stars, often much more complex than those of single stars, are described in Chapter 21.

The Vogt-Russell theorem says that the entire evolution of a star, unless it has a close binary companion, is determined by its initial mass and chemical composition.

Models of Stars

Because stars usually evolve too slowly for their changes to be detectable and because their interiors are hidden from us, much of what we know about the structure and evolution of stars has been discovered through model building. If astronomers know the mass and composition of a star, they can solve a set of equations describing how pressure, temperature, mass, and luminosity vary from its center to its surface. To make a mathematical model of a star, an astronomer must know how to calculate the rate of energy generation, opacity, and equation of state throughout the star. All of these important processes depend on chemical composition.

One outcome of the modeling process is the rate at which nuclear fuels are consumed at different distances from the center of the star. This information can be used to calculate the rate at which the chemical composition of the star changes and enables astronomers to predict what the composition will be at a later time. Using the updated composition, a new model can be calculated. The new model can be used to estimate the composition at a still later time. In this way, astronomers can construct a sequence of mathematical models that follows the evolution of the star. To be sure that their model evolutionary sequences are correct, astronomers must compare the properties of their model stars to those of real stars. The way that such a comparison can be made is described in the next section of this chapter.

EVOLUTIONARY TRACKS AND STAR CLUSTERS

Astronomers follow the evolution of a sequence of models of a star by noting how the position of the star in an H-R diagram changes as its temperature and luminosity vary with time.

Changing Appearance in the H-R Diagram

Suppose we wanted to record the changes that a person undergoes as that person ages. Many measures of the aging process are possible, ranging from changes in cell structure to the ability to remember strings of numbers. Most of these measures are difficult to obtain, however, and we might decide to use two easily measured quantities—height and weight. The life of a person, recorded in a height-weight diagram, might look like Figure 19.3. The person's age is indicated at four points in the height-weight track. The track shows that the person gained in both height and weight for the first 15 to 20 years after birth. This period was followed by about 30 years during which height remained constant, but weight rose slowly. Toward the end of the person's life (ages 50 to 70), both weight and height decreased slowly.

FIGURE 19.3
A Height-Weight Diagram Describing the Life of a Person
Points on the graph represent the person's height and weight at 0, 15, 50, and 70 years of age.

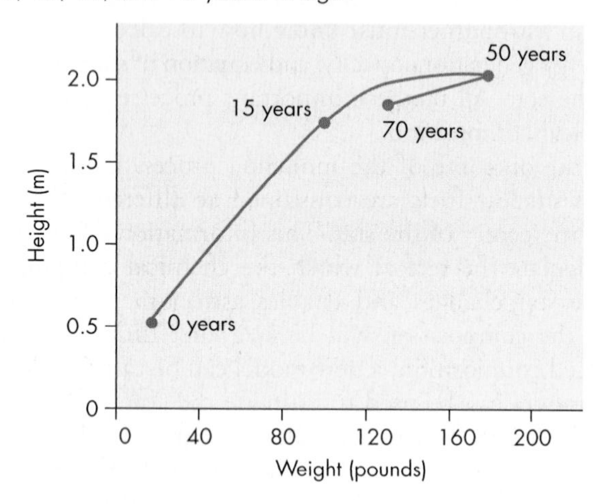

A diagram similar to the height-weight diagram is used to record the evolutionary career of a star. As a star evolves, changes in its internal structure usually are accompanied by changes in its appearance. Although no single diagram can describe the changing internal structure of a star, it is possible to record the corresponding changes in surface temperature and luminosity on an H-R diagram. The path through an H-R diagram that a star follows as it evolves is called the star's **evolutionary track.** Stars evolve so slowly compared with the lifetimes of astronomers that it is very unusual to record actual changes in a star's location in an H-R diagram. Instead, most evolutionary tracks for stars are based on model calculations that take into account the changing chemical composition of the star and its consequences.

The evolutionary track of a 1 $M_\odot$ star is shown in Figure 19.4. The pre–main sequence phase of the star's career is described in Chapter 18. As the star shrinks during that phase, it grows hotter and dimmer until fusion begins in its core. The main sequence phase of the star's life and later phases recorded in the evolutionary track are described in this chapter and the next one.

H-R Diagrams of Star Clusters

Suppose an astronomer calculated evolutionary tracks for stars of different masses. An example might be evolutionary tracks for pre–main sequence stars such as those shown in Figure 19.5. Suppose now that all of the evolutionary tracks represent stars that began to form at the same time. We could stop each evolutionary track after the stars all had a chance to evolve for the same length of time. The positions of the stars in the H-R diagram would fall on an **isochrone,**

FIGURE 19.4
An H-R Diagram Showing the Evolutionary Track of a One-Solar-Mass Star
Luminosity is shown relative to the Sun's present luminosity. The location of the star in the H-R diagram changes with time as nuclear reactions change the star's structure and appearance. The pre–main sequence stage is described in Chapter 18. The main sequence, red giant, horizontal branch, asymptotic giant branch, and planetary nebula stages are described in this chapter. The white dwarf stage is described in Chapter 20.

Stellar evolution

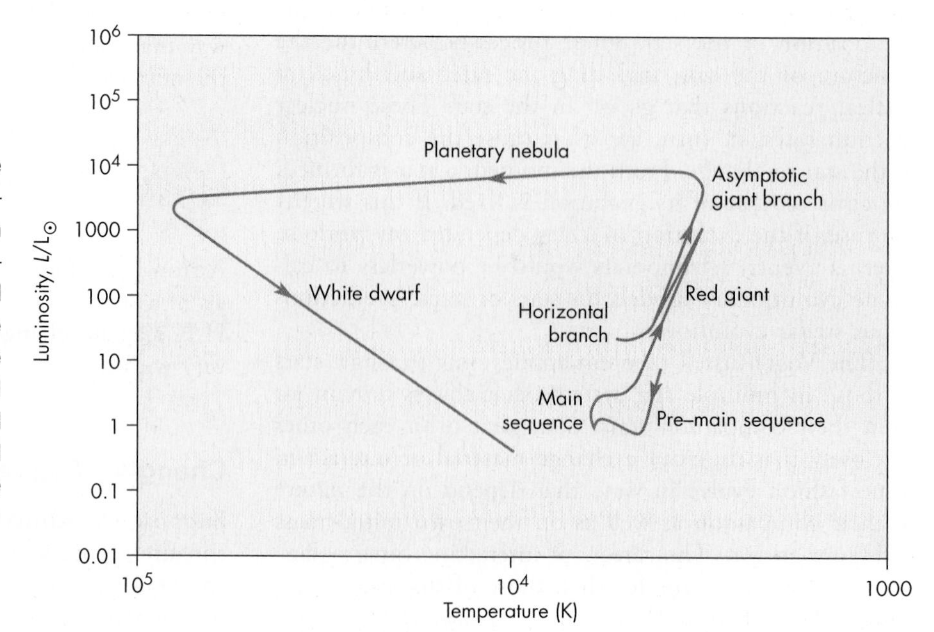

FIGURE 19.5
Evolutionary Tracks for Pre–Main Sequence Stars of Different Masses
The evolutionary tracks are labeled with the mass of the star. Isochrones for ages of 10^6 and 10^7 years are shown as dashed lines. The isochrone for 10^6 years shows that at that age stars more massive than 5 solar masses already have reached the main sequence, while less massive stars have not. The isochrone for 10^7 years shows that at that age stars more massive than 1.5 solar masses have already reached the main sequence, while less massive stars have not.

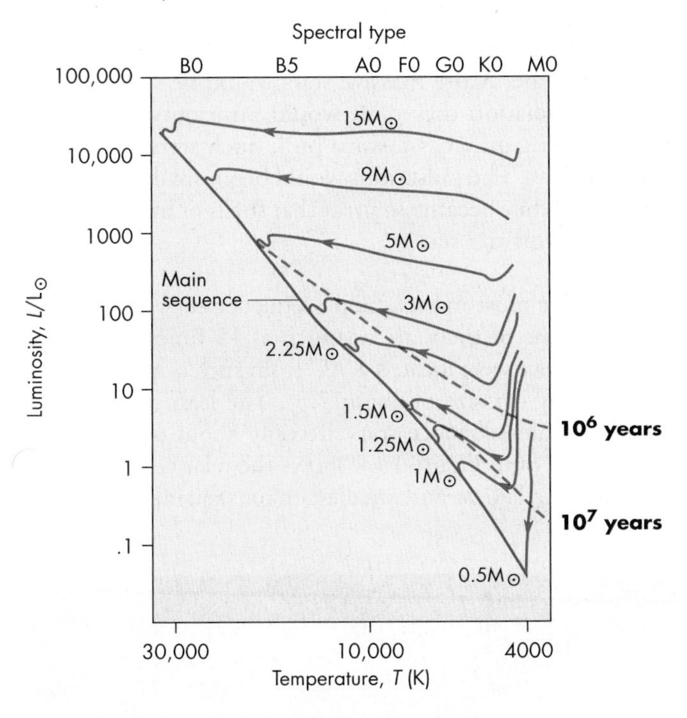

a line in an H-R diagram that shows the temperatures and luminosities of a collection of stars that have the same age but different masses. Isochrones for ages of 1 million and 10 million years are shown in Figure 19.5. Notice that for the 1-million-year isochrone (10^6 years), the most massive, hottest stars have reached the main sequence while less massive stars are still evolving toward the main sequence and lie above and to the right of the main sequence. For the 10-million-year isochrone (10^7 years), stars of intermediate mass have reached the main sequence, leaving only the low-mass stars above the main sequence.

Many approximations must be made when calculating evolutionary tracks for stars, so astronomers are usually reluctant to accept the validity of these calculations unless they can somehow be compared with the H-R diagrams of real stars. H-R diagrams for clusters of stars typically are used for these comparisons because the stars in a given cluster formed at nearly the same time within the same molecular cloud. It seems reasonable to assume, therefore, that they share a common age and chemical composition. They differ only in mass, so they make excellent standards against which to compare isochrones from evolutionary

FIGURE 19.6
The H-R Diagram for the Cluster NGC 7788, Which Is About 10 Million Years Old
The blue dots indicate the location of the stars in NGC 7788 in the H-R diagram. NGC 7788 is old enough that F0 stars almost have reached the main sequence. Stars that will become F, G, K, and M main sequence stars have not yet had time to reach the main sequence. The dashed line shows an isochrone calculated from models of stars for an age of 10^7 years. The similarity of the H-R diagram of the cluster to the isochrone shows that the cluster is about 10^7 years old.

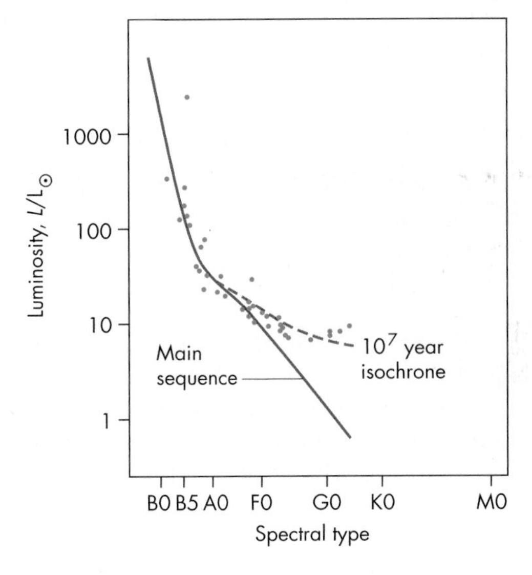

tracks. For example, Figure 19.6 shows the H-R diagram for the cluster NGC 7788 as well as an isochrone based on models of stars with ages of 10 million years. The massive stars in NGC 7788 already have reached the main sequence, while the less massive stars, which evolve more slowly, still lie above and to the right of the main sequence. The H-R diagram of the cluster strongly resembles the isochrone for 10-million-year-old stars, from which astronomers have concluded that NGC 7788 is about 10 million years old. The fact that the H-R diagrams of clusters of stars resemble isochrones of different ages has given astronomers considerable confidence that current stellar evolutionary tracks are reasonably accurate.

The evolutionary track of a star is the path the star traces through an H-R diagram as it evolves. By comparing calculated evolutionary tracks for stars of different masses to H-R diagrams of clusters of stars, which have the same age and chemical composition but different masses, astronomers have been able to check the validity of the calculated evolutionary tracks.

19.3 MAIN SEQUENCE STARS

The main sequence phase in the evolution of a star is the period of time when it is consuming hydrogen in its core. It is a period of stability during which both the structure and appearance of the star change only very gradually. The main sequence phase lasts a long time, and the star changes slowly during that time because main sequence stars consume their nuclear fuel relatively slowly. Most of the post–main sequence career of a star is spent in phases where the energy output and rate of fuel consumption are much larger than they are for the main sequence. Because stars spend the majority of their nuclear careers on the main sequence, main sequence stars make up the majority of the visible stars.

The Variety of Main Sequence Stars

Although all main sequence stars generate energy by the fusion of hydrogen into helium in their cores, they differ from one another in many important respects, such as mass, size, temperature, luminosity, and internal structure.

Mass Main sequence stars range in mass from 0.08 $M_\odot$ to perhaps 130 $M_\odot$. The lower limit on mass is due to degeneracy. As a protostar contracts, its core grows both denser and hotter. Its increasing temperature brings it steadily nearer the development of fusion in its core. Its increasing density brings it steadily nearer degeneracy. For a star more massive than 0.08 $M_\odot$, fusion begins before degeneracy is reached, so the star begins its main sequence career. For a less massive star, degeneracy develops before hydrogen fusion begins. After the star becomes degenerate, the pressure supplied by its degenerate electrons doesn't depend on temperature. The star stops contracting and begins to grow cooler and dimmer. Perhaps such an object shouldn't be called a star at all, because it never generates a significant amount of nuclear energy. The term brown dwarf is used to refer to objects that narrowly miss becoming nuclear stars.

The upper limit to the mass that a main sequence star can have is not very well known. There are probably main sequence stars as massive as 130 $M_\odot$, although they are extremely rare. More massive stars would be so luminous that the radiation they emit would push outward with a force rivaling gravity's inward pull. Such stars would begin pulsating. The pulsations would begin small but would grow until they became so great that shells of matter would be thrown off the star.

Size The most massive main sequence stars are also the largest. Some of them are as much as 15 times as large as the Sun. If we were located 1 AU from such a star, it would have an angular size of about 7.5°. The least massive and smallest main sequence stars are only about one-tenth as large as the Sun. Figure 19.7 shows the relative sizes of the Sun and the largest and smallest main sequence stars.

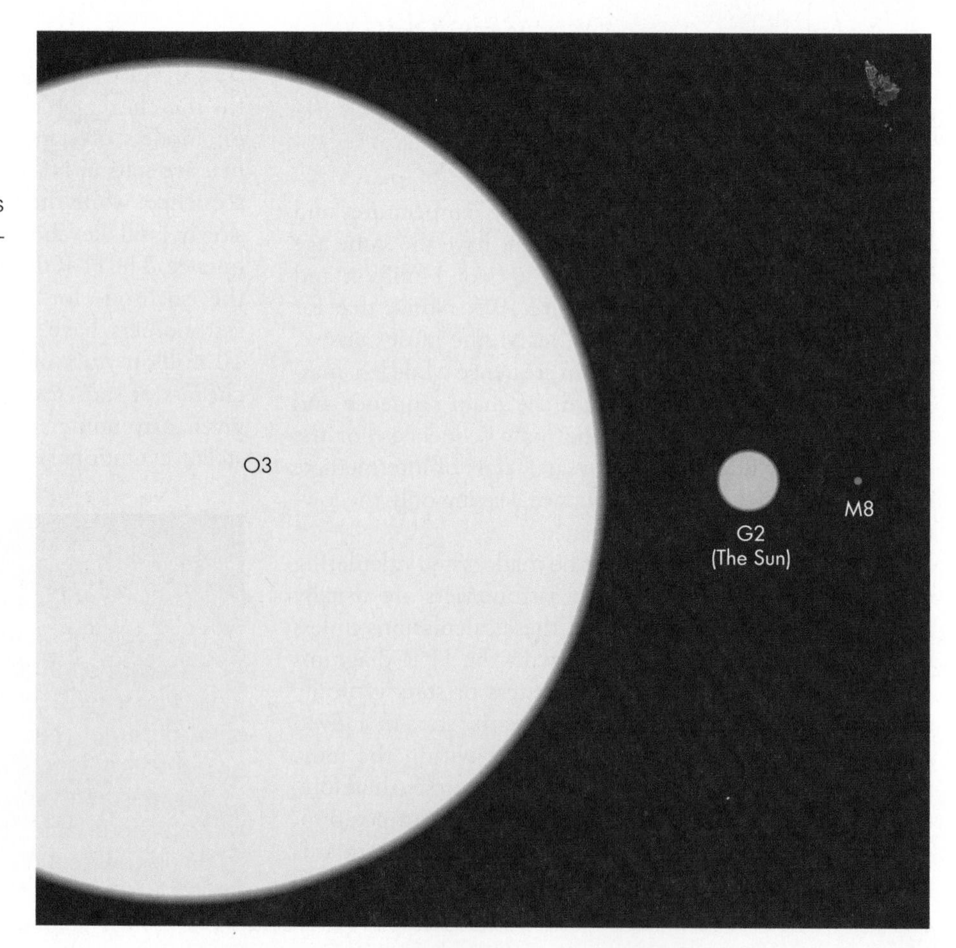

FIGURE 19.7
The Relative Sizes of Main Sequence Stars with Spectral Types of O3, G2 (the Sun), and M8
The O3 star is 15 times as large as the Sun, whereas the M8 star is one-tenth as large as the Sun.

Temperature and Luminosity Like size, the temperatures and luminosities of main sequence stars increase with increasing mass. Massive main sequence stars have surface temperatures of about 50,000 K and are spectral type O3. The main sequence stars of lowest mass have temperatures of only 2400 K and are spectral type M8. Brown dwarfs are cooler and dimmer still and fall in spectral classes L, T, and Y (see Chapter 16, Section 7). Notice that although the most massive main sequence stars are 750 times as massive and 150 times as large as the least massive main sequence stars, the most massive stars are only about 20 times as hot as the least massive stars. The Sun, which has a temperature of 5800 K and is spectral type G2, is below the middle of the range of temperatures for main sequence stars.

Main sequence stars show an extremely wide range of luminosities. The most luminous (and most massive) are more than a million times as bright as the Sun. If the Sun were replaced by one of these stars, the temperature of the Earth would rise to about 10,000 K, so the Earth and the other terrestrial planets would quickly be vaporized. The smallest, least massive main sequence stars are about 1000 times dimmer than the Sun. If one of these stars replaced the Sun, the Earth's temperature would fall to about 50 K, approximately the temperature of Neptune. The spectral types, temperatures, radii, and luminosities for main sequence stars of various masses are given in Table 19.1. In Table 19.1, masses, radii, and luminosities are given in terms of those of the Sun ($M_\odot$, $R_\odot$, and $L_\odot$).

Internal Structure The internal structure of a main sequence star depends on its mass. The more massive main sequence stars have the highest central temperatures—as great as 40 million K. This is hot enough that massive main sequence stars can use the carbon cycle to fuse hydrogen into helium. Because the rate at which fusion reactions occur is very sensitive to temperature, the massive main sequence stars consume their hydrogen at rates much faster than the Sun, which has a core temperature of only about 15 million K. The rapid consumption of hydrogen in massive main sequence stars is the source of their tremendous luminosities. Main sequence stars of low mass, in contrast, have core temperatures below 10 million K, so nuclear fusion goes on very slowly in their cores and produces only a relatively feeble output of energy.

The core of a massive main sequence star produces far too much energy for radiation alone to carry energy outward. Instead, the core of a massive star has vigorous convection. The convective region ends, however, far beneath the surface of the star. The surface presumably is smooth and without the mottled appearance shown by the Sun. The central convective region is largest for the most massive main sequence stars and grows smaller with decreasing mass until it disappears for stars a little more massive than the Sun. Low-mass main sequence stars also have convective regions, but in this case it is their surface layers that are convective, like the Sun's photosphere. The Sun has a shallow convective region, but for stars with lower masses, the depth of the surface convective zone increases with decreasing mass until, for the main sequence stars of lowest mass, it extends all the way to the core of the star. Figure 19.8 shows the internal structure for main sequence stars of several masses.

All main sequence stars consume hydrogen in their cores. Their masses, sizes, luminosities, surface temperatures, and internal structures vary widely, however. Massive main sequence stars are larger, hotter, and more luminous than the Sun, whereas low-mass main sequence stars are smaller, cooler, and dimmer than the Sun.

Table 19.1

Main Sequence Stars

$M/M_\odot$	SPECTRAL TYPE	TEMPERATURE (K)	$R/R_\odot$	$L/L_\odot$
60	O3	50,000	15	1,400,000
40	O5	40,000	12	500,000
18	B0	28,000	7	20,000
3.2	A0	10,000	2.5	80
1.7	F0	7,400	1.3	6
1.1	G0	6,000	1.05	1.2
1	G2	5,800	1	1
0.8	K0	4,900	0.85	0.4
0.5	M0	3,500	0.6	0.06
0.1	M8	2,400	0.1	0.001

FIGURE 19.8
The Internal Structures of Main Sequence Stars of 60 Solar Masses (O3), 1 Solar Mass (G2), and 0.1 Solar Mass (M8)
The dotted regions are those within which nuclear energy generation is occurring. Yellow regions are convective. The stars all are shown to be the same size in the figure, although the O3 star is really 15 times as large as the Sun and the M8 star is really only one-tenth as large as the Sun.

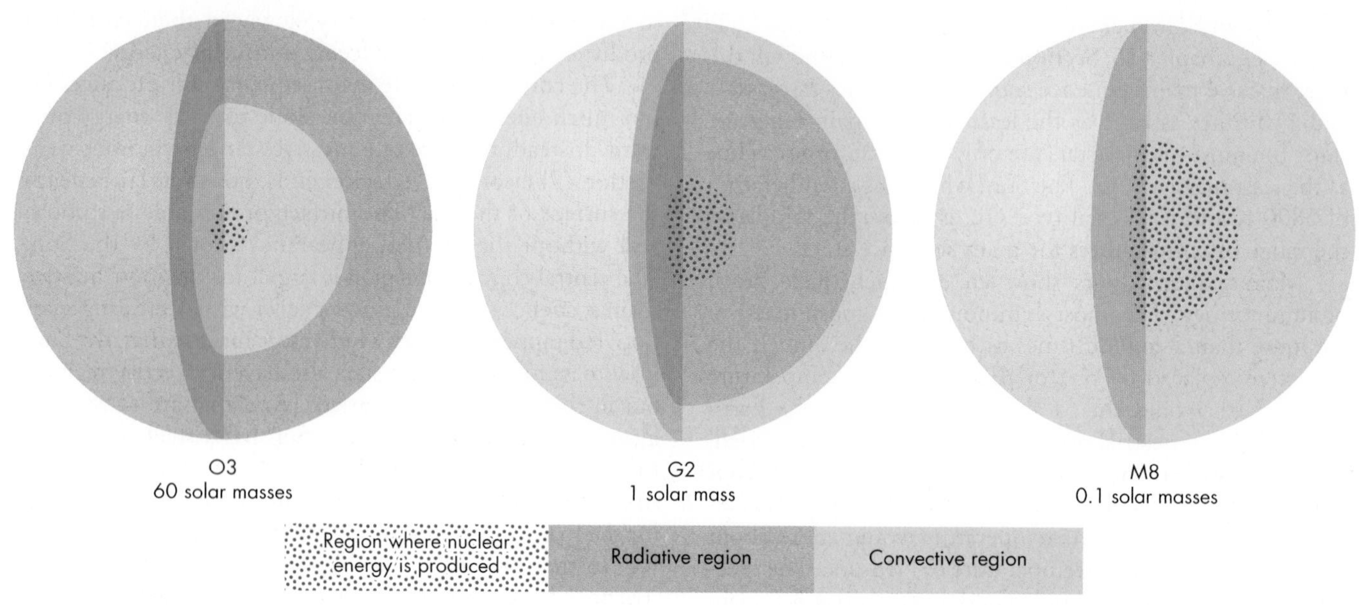

Main Sequence Lifetime

The **main sequence lifetime** of a star is the length of time that it spends consuming hydrogen in its core. Thus, the main sequence lifetime of a star depends on its internal structure and its evolution. Main sequence lifetimes are difficult to calculate accurately, but they can be estimated reasonably well using the masses and luminosities of main sequence stars. The method used to estimate main sequence lifetimes is essentially the same one that someone would use to estimate how long a fire can be kept going with a specified supply of wood.

To estimate how long the fire can be kept burning, one needs to know the amount of fuel (the size of the pile of wood) and the brightness of the fire (how much wood will be consumed each hour). The length of time that the fire will burn is directly proportional to the amount of fuel and inversely proportional to the brightness of the fire. This means that a small fire with a large supply of wood will last a long time, but a large fire with a small supply of wood will burn out quickly. Exactly the same idea can be used to estimate the lifetimes of main sequence stars. In the case of stars, the amount of fuel is the mass of hydrogen that becomes hot enough to undergo fusion. Although this varies somewhat for different stellar masses, it is approximately 10% of the entire mass of the star. This means that the amount of fuel is proportional to the mass of the star. The brightness of the fire in the case of a wood fire corresponds to the luminosity of the star, which varies relatively little during its main sequence lifetime. In either case, the wood fire or the main sequence stage of the evolution of a star, the lifetime is given by

$$t \propto \frac{M}{L} \tag{19.2}$$

For a wood fire, t is the lifetime of the fire, M is the mass of wood, and L is the luminosity (brightness) of the fire. For a main sequence star t is its main sequence lifetime, M is the mass of the star, and L is its luminosity.

For a wood fire, any combination of mass of wood and brightness of fire is possible. Recall from Chapter 16, however, that the luminosities and masses of main sequence stars are closely related to each other by the mass-luminosity relation found from observations of binary stars. The mass-luminosity relation can be expressed as

$$L \propto M^{3.5} \tag{19.3}$$

If Equations 19.2 and 19.3 are combined to eliminate L, the result is

$$t \propto M^{-2.5} \tag{19.4}$$

Main Sequence Lifetime

Equation 19.5 can be used to calculate the main sequence lifetime of a star of 10 $M_\odot$. Using $M/M_\odot = 10$, Equation 19.5 becomes

$$\frac{t}{t_\odot} = \left(\frac{M}{M_\odot}\right)^{-2.5} = 10^{-2.5} = \frac{1}{300}$$

Because the Sun's hydrogen will last for about 10 billion years, the main sequence lifetime of a 10 solar mass star is about 1/300 of 10 billion years, or 30 million years. This is only about half the time that has passed since the extinction of the dinosaurs.

In terms of the Sun's mass ($M_\odot$) and main sequence lifetime ($t_\odot$), Equation 19.4 can be rewritten as:

$$\frac{t}{t_\odot} = \left(\frac{M}{M_\odot}\right)^{-2.5} \quad (19.5)$$

Using Equation 19.5, the main sequence lifetime of a star half as massive as the Sun is found to be $(1/2)^{-2.5} = 6$ times the Sun's lifetime, or about 60 billion years. As described in Chapter 26, this is much longer than the universe has existed. This means that there hasn't been enough time for any low-mass star to have exhausted its central hydrogen. For this reason, very little effort has been spent studying the evolution of low-mass stars after their main sequence phase is completed. The main sequence lifetimes of stars with masses between 1 and 30 $M_\odot$ are graphed in Figure 19.9.

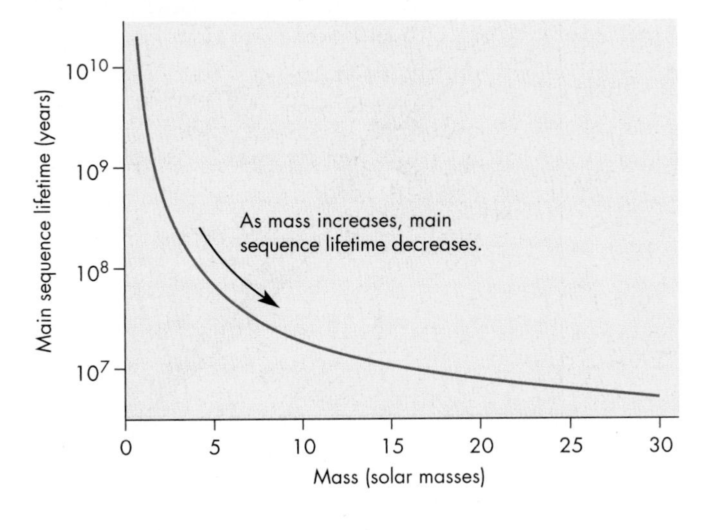

FIGURE 19.9
The Main Sequence Lifetime for Stars of Different Masses
Low-mass stars have much longer main sequence lifetimes than massive stars.

Evolution on the Main Sequence

Even though the main sequence phase of a star is a period of stability, without rapid changes in structure or appearance, gradual changes occur while a star consumes its central hydrogen. As hydrogen is depleted, main sequence stars increase in both size and luminosity. Because main sequence stars of a given mass vary in luminosity as they age, the main sequence appears as a band rather than a line in the H-R diagram. For the Sun, the increases during its main sequence lifetime are about 40% in radius and 100% in luminosity. The Sun's temperature will remain nearly constant throughout its main sequence career. The doubling of the Sun's luminosity during the 10 billion years it spends in the main sequence is a small increase compared with the luminosity changes that precede and follow the main sequence phase of its life. However, even this increase in solar energy output has had and will have consequences for the other bodies in the solar system.

We can get an idea of how serious these consequences might be by assuming that the Earth has always had and always will have the same albedo and greenhouse effect that it has today. If that were really the case, then the Earth's average temperature would have been below freezing when the Sun was beginning its main sequence life about 4.6 billion years ago. At that time the Sun was about 25% less luminous than it is today. The fact that oceans existed on the Earth almost from the Earth's beginning is evidence that the Earth couldn't have had an average temperature below freezing. One possibility is that the Earth's primitive atmosphere produced a strong greenhouse effect that raised the Earth's temperature well above the freezing point. During the next 5 billion years, the Sun will increase in brightness by about 60%. If the Earth keeps the same albedo and greenhouse effect (highly unlikely), its average temperature will rise to about 325 K, or about 125° F. This would make the tropical regions and perhaps the entire Earth uninhabitable. Nevertheless, these changes will occur so gradually that hundreds of millions of years will pass before they become a matter for great concern. We have much more reason to worry about short-term changes in the Sun's energy output.

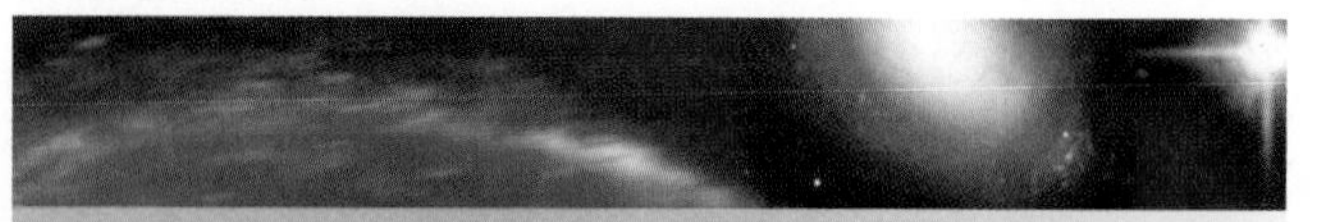

The main sequence lifetime of a star is proportional to its mass and inversely proportional to its luminosity. Luminosity increases as mass increases, however, so the most massive main sequence stars have the briefest lives. The temperature and luminosity of a star change relatively little while it is on the main sequence.

19.4 AFTER THE MAIN SEQUENCE

The production of energy by fusion at the center of a star ceases when its central supply of hydrogen has been fused into helium. Without a source of nuclear energy, the star must turn to gravitational contraction to meet its energy needs. As the star shrinks and releases gravitational energy, its interior heats up. In a relatively short time, the region around the hydrogen-depleted core becomes hot enough to ignite the fusion of hydrogen into helium in a thin shell, as shown in Figure 19.10. For a star like the Sun, a hydrogen-consuming shell develops almost at the same time that fusion stops in the core. Thus a Sun-like star has a nearly steady supply of nuclear energy. For more massive stars, there is an interval of a few hundred thousand to a few million years when the star is without nuclear energy.

After a hydrogen "burning" shell develops in a star, its structure and appearance begin to change rather quickly. (Astronomers often refer to hydrogen fusion as "burning" even though they know that fusion is a totally different process from real burning, or combustion.) The shell within which hydrogen is being converted to helium burns itself outward through the mass of the star like a prairie fire. The helium "ash" left behind settles into the core of the star, making the core denser and more massive. At the same time, the surface of the star expands and cools. As Figure 19.11 shows, the star moves upward and to the right in the H-R diagram as it becomes a red giant star. This takes about a billion years for a star like the Sun, but only a million years for a star of 9 $M_\odot$.

Because massive stars have main sequence lifetimes that are shorter than those of less massive stars, the conversion of a cluster of main sequence stars to red giants begins with the most massive stars and then proceeds down to successively less massive stars. The H-R diagram of the cluster of stars shows a main sequence that first lacks

FIGURE 19.10
The Structure of a Star When Hydrogen Shell Burning Begins
The core consists mainly of helium, but the outer layers of the star are hydrogen rich. The shell within which energy production takes place is quite thin.

The structure of a star when hydrogen shell burning begins

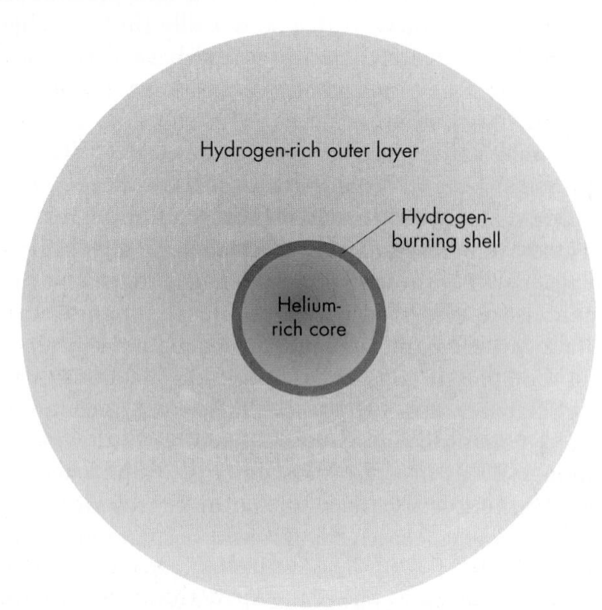

FIGURE 19.11
Evolutionary Tracks from the Main Sequence to the Red Giant Region for Stars of Different Masses
Isochrones are shown for ages of 10 million, 50 million, 100 million, and 1 billion years. Notice that a star of 15 solar masses exits the main sequence (the blue shaded region) in about 1⁄100 the time required for a star of 1.5 solar masses.

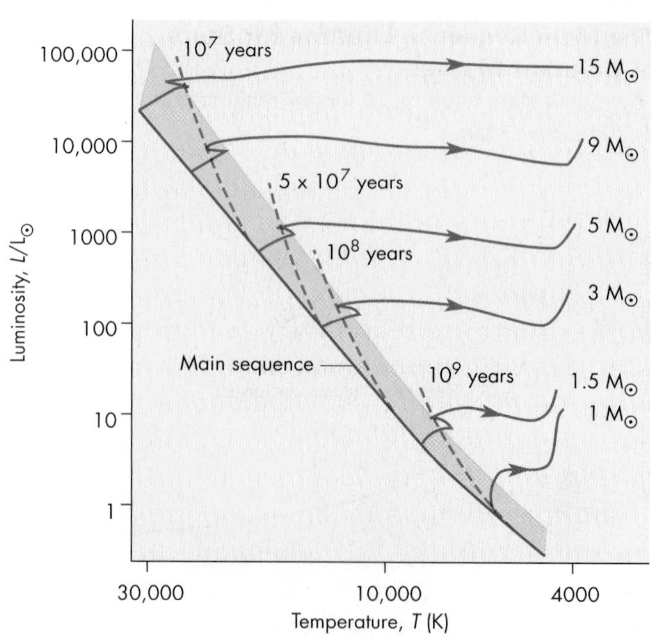

O stars (the most massive main sequence stars), then B stars, then A stars, and so on as time passes. Figure 19.12 depicts this erosion of the main sequence by comparing H-R diagrams of clusters of different ages. In each case, some of the stars that have left the main sequence can be found as red giants or as stars on their way to becoming red giants. The spectral type (or temperature) of the hottest star left on the main sequence can be used to estimate the age of a cluster of stars. For example, suppose the hottest main sequence star in a cluster has a spectral type of A0. Hotter, more massive main sequence stars already have evolved to become red giants. Because A0 stars have main sequence lifetimes of about 100 million years, the age of the cluster must be about 100 million years.

Red Giant Stars

All stars with masses greater than or equal to the Sun's expand to become red giants. What happens to a star after it reaches the red giant stage, however, depends on the mass of the particular star. For stars about as massive as the Sun, the red giant stage is terminated by a powerful internal explosion called the helium flash.

FIGURE 19.12
H-R Diagrams for Several Clusters of Different Ages
NGC 2362 is the youngest cluster shown, followed in order of increasing age by h + χ Persei, the Pleiades, M41, the Hyades, Praesepe, M3, and M67. Hot main sequence stars evolve faster than cool main sequence stars, so the hot end of a cluster's main sequence ends at cooler and cooler stars as time passes.

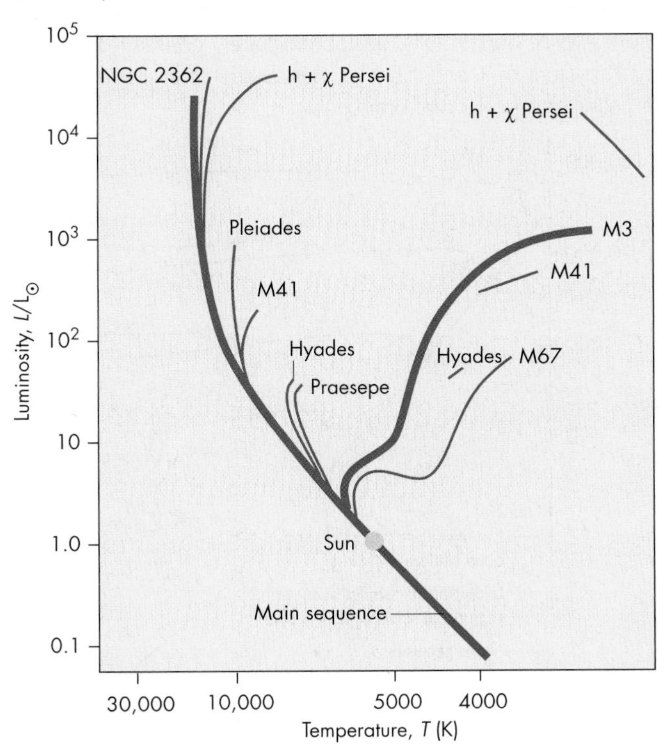

The Helium Flash The core of helium becomes more massive within a 1 M$_\odot$ red giant as the hydrogen-burning shell eats its way outward. As it becomes more massive, the core contracts and becomes hotter and denser. Before it becomes hot enough for the fusion of helium to take place, however, the core becomes so dense that its electrons become degenerate. After this happens, the degenerate electrons provide enough pressure to prevent further contraction of the core, and the temperature of the core no longer affects the internal pressure of the star.

As the hydrogen shell continues to burn, the degenerate core grows hotter and more massive. When the core temperature reaches 100 million K and the mass of the helium core reaches 0.6 M$_\odot$ (that is, when the inner 60% of the hydrogen in the star has been converted to helium), the triple α process begins to convert helium into carbon and to produce nuclear energy in the core. When this takes place, the star is nearly 1 AU in radius and is about 1000 times as luminous as the Sun.

The energy released by the fusion of helium heats the core of the star, raising its temperature. Under normal circumstances, heating the core of a star also would raise its pressure, causing it to expand and cool. (This is why nuclear reactions do not usually produce a rapid increase in the central temperature of a star.) The gas in the core of a 1 M$_\odot$ red giant, however, isn't a normal gas. It is a gas with degenerate electrons, so the temperature increase that accompanies the ignition of helium does not increase internal pressure. Increasing the temperature does, however, strongly affect the rate at which the triple α process occurs. In fact, doubling the temperature increases the rate of the triple α process by about a billion times.

The energy produced by the triple α process heats the core, raising its temperature even more. As the consumption of helium continues, temperature and the rate of energy production both rise swiftly until the temperature reaches about 300 million K. A nearly explosive consumption of helium takes place during this rapid heating, so the event is called the **helium flash.** At the peak of the helium flash, the core of the star has, for a few minutes, an energy production rate equal to 10^{14} times the luminosity of the Sun. This is roughly 100 times as great as the rate of energy output for the entire Milky Way galaxy.

One might think that the helium flash would make the star extremely luminous and conspicuous. That isn't the case, however, because the electrons stop being degenerate when the core gets hot enough. The interior of the star expands so both temperature and density decrease. Essentially none of the energy of the helium flash ever reaches the surface of the star where we might be able to see it because most of it is consumed in the expansion of the interior of the star. Thus, some of the most energetic events that occur in stars are hidden from view.

More Massive Red Giants Red giant stars more massive than about 2 M$_\odot$ are spared the helium flash.

The cores of these stars become hot enough to initiate the triple α process before the electrons there become degenerate. At this point, energy production by nuclear fusion resumes in the center of the star, this time with helium as the fuel.

After its central hydrogen is consumed, a star burns hydrogen in a shell surrounding its helium core. During this time, the star expands and cools to become a red giant. Eventually the core becomes hot enough for helium to become a fuel. This happens either explosively during the helium flash for stars as massive as the Sun or gently in more massive stars.

Core Helium Burning

After helium burning begins (either explosively or gradually) in the core of a star, the star has two sources of nuclear energy. One is the helium-burning core, and the other is the surrounding shell, in which hydrogen fusion takes place. As Figure 19.13 shows, the outer layers of such a helium-burning star are hydrogen rich, and the intermediate layers are helium rich. As helium burns in the core of the star, it decreases in abundance as the carbon abundance increases. Oxygen also is produced in the core by nuclear reactions between carbon nuclei and helium nuclei.

Horizontal Branch Stars Stars like the Sun become hotter and smaller after core helium burning begins. They evolve across the H-R diagram at roughly constant luminosity. This portion of the H-R diagram, shown in Figure 19.14, is called the horizontal branch, so stars at this point in their evolution are called **horizontal branch stars.** Many examples of horizontal branch stars can be seen in the H-R diagrams of old clusters of stars such as M3, shown in Figure 19.15.

Pulsating Stars Stars more massive than the Sun also contract somewhat and move horizontally across the H-R diagram as they grow hotter at roughly constant luminosity. While they are doing so, they become unstable and begin to pulsate. During pulsations, the surface of a star moves in and out. As its size changes, so do its temperature and luminosity. One might be tempted to think that pulsations are caused by variations in the rate at which energy is produced in a star. This is not the case, however. The rate

FIGURE 19.13
The Structure of a Star While It Burns Helium in Its Core
Energy is generated not only in the core but also in a thin hydrogen-burning shell. As time passes, the shell burns its way outward through the mass of the star, increasing the mass of the helium-rich core. In this diagram, the sizes of the core and hydrogen-burning shell are greatly exaggerated compared with the size of the outer envelope of the star.

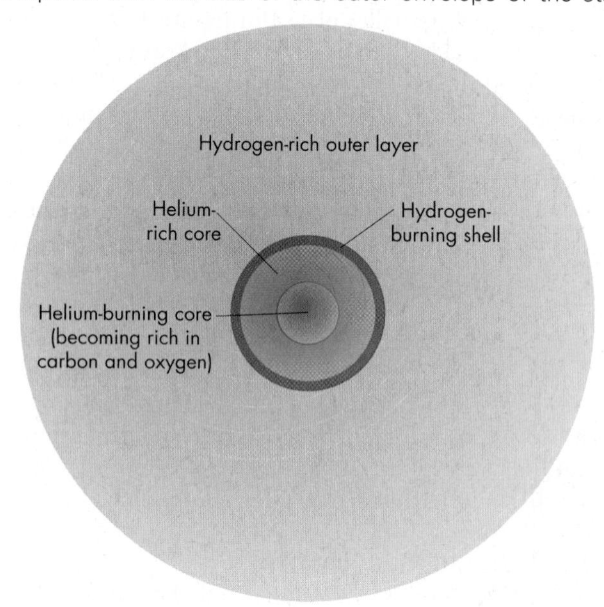

FIGURE 19.14
Evolutionary Tracks During Core Helium Burning for Stars of Different Masses
The evolutionary tracks during core helium burning are shown as purple lines. Evolutionary tracks from the main sequence to the red giant stage are shown as blue lines. Generally, the stars evolve horizontally, at approximately constant luminosity. For stars of 1 and 1.5 $M_\odot$, this phase of stellar evolution is called the horizontal branch.

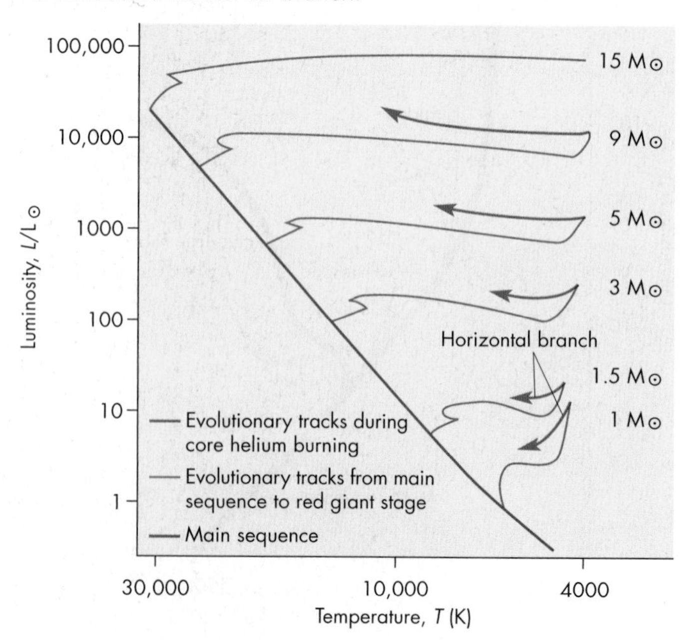

FIGURE 19.15
The H-R Diagram for the Old Cluster, M3
The region of the horizontal branch is indicated in the diagram, as are the main sequence and the red giant region.

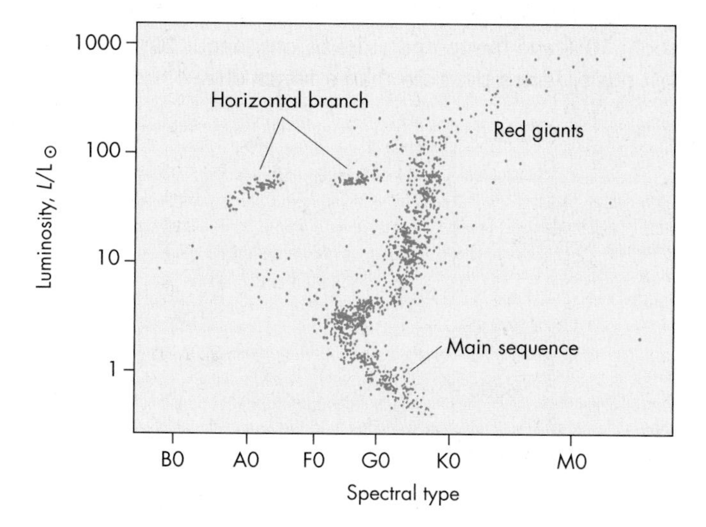
1000
100
10
1
Luminosity, L/L⊙
Horizontal branch
Red giants
Main sequence
BO
AO
FO
GO
KO
MO
Spectral type

of nuclear energy production remains constant in a pulsating star. Instead, the pulsations are caused by variations in the rate at which energy can escape from the star.

In a normal star, forces resulting from pressure balance the force of gravity. Imagine, however, a star in which pressure forces in the outer layers exceeded gravity. If this were to happen, the outer layers of the star would begin to expand, pushed outward by the excess pressure, as shown in Figure 19.16. As the star expanded, its gravity would drop, but its pressure forces would drop more quickly. When it had expanded to a certain size, pressure and gravity would once again be in balance, but the star wouldn't stop expanding. The inertia of its outward-moving layers would carry the layers beyond the balance point. By the time they had been brought to rest by gravity, their pressure would be too low to balance gravity, and they would begin to fall inward again. Now gravity would rise, but less than pressure. The outer layers of the star would fall past the balance point until eventually they were brought to rest by pressure forces. At this point, the cycle of pulsations would begin again.

A pulsating star behaves something like a weight on a spring. If the weight is pulled downward and released, it begins to oscillate around the position at which gravity and the tension in the spring are in balance. However, friction in the spring eventually causes the oscillations to die away unless the spring gets a small upward push each time it reaches the bottom of an oscillation. Similarly, a pulsating star needs an outward push each time it contracts to minimum size, or the pulsations would die out. In a star, this push is supplied by energy trapped when the outer layers of the star fall inward. The energy is released again after the star reaches its smallest size, so it pushes the outer layers back outward, overcoming frictional forces. It is helium gas that traps the energy in a pulsating star. When a pulsating star contracts, the gas beneath its surface is heated.

FIGURE 19.16
Pressure Forces and Gravity at Several Points in the Pulsation Cycle of a Pulsating Star
A, When the star is smallest, pressure forces exceed gravity so the star begins to expand. **B,** Pressure decreases as the star expands. When the star is expanding most rapidly, the two forces reach a balance. Inertia, however, expands the star's surface beyond the balance point. Gravity becomes greater than pressure forces, so eventually the expansion stops. **C,** When the star is largest, gravity is larger than pressure forces and contraction begins. **D,** When the star is contracting most rapidly, pressure and gravity are in balance again but inertia carries the stellar material inward past the balance point. Pressure becomes greater than gravity again and eventually brings the contraction to a halt at **A,** beginning the pulsation cycle over again.

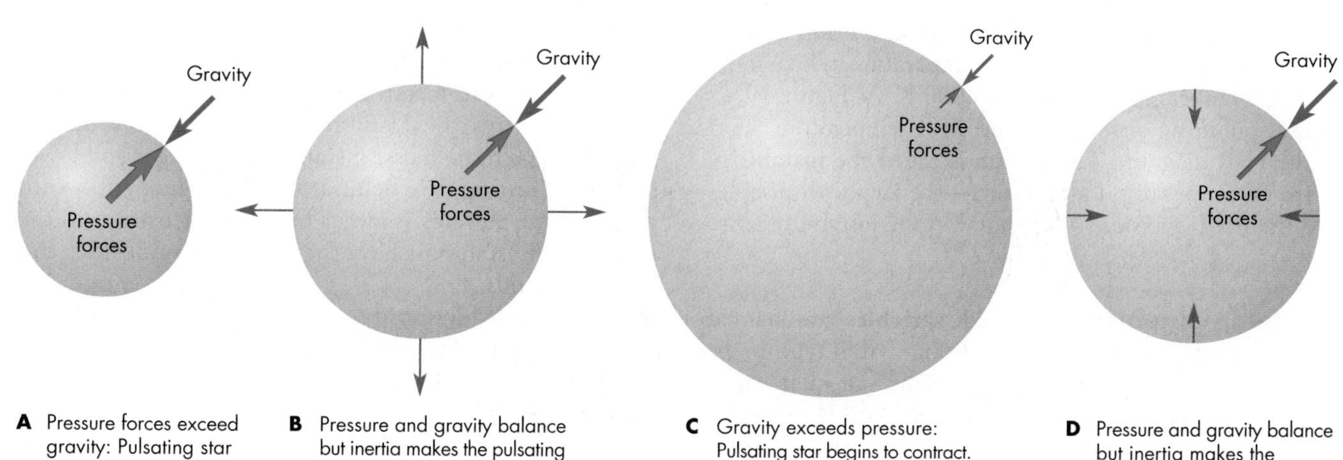
Gravity
Pressure forces
Gravity
Pressure forces
Gravity
Pressure forces
Gravity
Pressure forces
A Pressure forces exceed gravity: Pulsating star begins to expand.
B Pressure and gravity balance but inertia makes the pulsating star expand farther.
C Gravity exceeds pressure: Pulsating star begins to contract.
D Pressure and gravity balance but inertia makes the pulsating star contract farther.

FIGURE 19.17
The Location of the Instability Strip in the H-R Diagram
The Cepheid variables and the RR Lyrae stars are located in the instability strip. Notice that the instability strip covers a wide range of luminosity but a relatively small range of temperature.

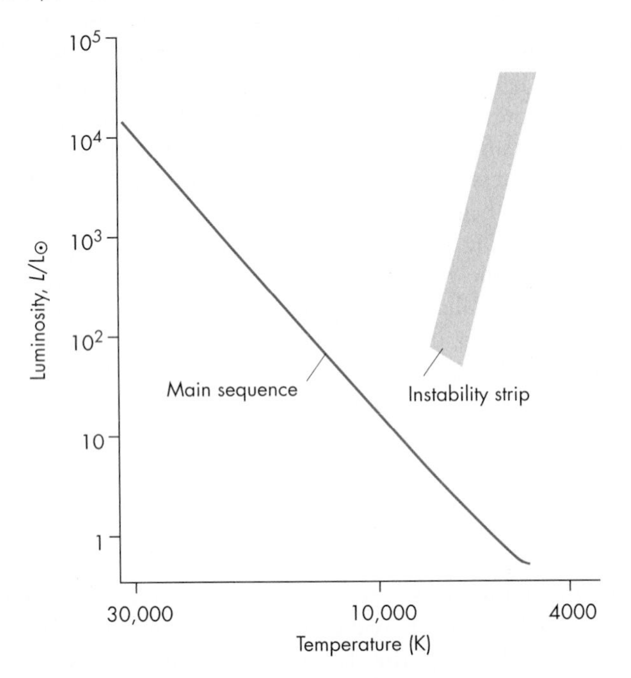

FIGURE 19.18
The Luminosity, Size, and Temperature of δ Cephei During Pulsations
Notice that the star is brightest roughly at the time when it is smallest but hottest. The size of the star varies by only about 10% and the temperature by only about 20%, but the luminosity varies by more than a factor of two.

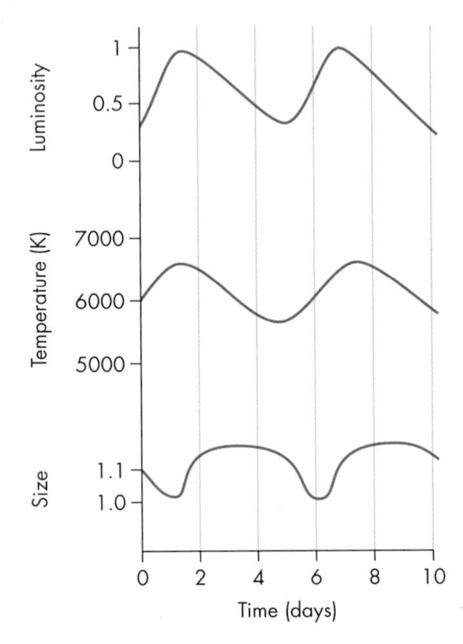

The effect of the heating, however, isn't higher temperature but rather the ionization of the helium in the gas. Ionized helium is very effective at absorbing radiation, so it absorbs radiant energy flowing outward through it. The energy is trapped, therefore, where it can supply the push that helps propel the outer layers of the star back outward again. As the star expands, electrons and helium ions recombine, which makes the gas more transparent and lets the stored energy escape.

For a star to be susceptible to pulsations, it must have a layer just beneath its surface in which helium gas is partially ionized. The existence of such a layer of partially ionized helium depends somewhat on the size and mass of the star, but mostly on its surface temperature, which must be in the range of about 5000 to 8000 K. As Figure 19.17 shows, pulsating stars are found in a temperature band in the H-R diagram. This region is called the **instability strip.** Two types of pulsating stars—the Cepheid variables and the RR Lyrae stars—are found in the instability strip.

Cepheid Variables **Cepheid variables** are named for δ Cephei, the first pulsating star of this type to be discovered. δ Cephei is a yellow giant star that varies by about a factor of two in brightness with a period of 5½ days. The variations in luminosity, size, and temperature for δ Cephei are shown in Figure 19.18. Notice that its luminosity and temperature are at maximum when its size is at minimum, and that its size is at maximum when its luminosity and temperature are at minimum. A remarkable and very useful property of the Cepheid variables is their **period-luminosity relationship,** the correlation between their pulsation periods and their luminosities. Figure 19.19 illustrates the period-luminosity relationship and shows that luminosity increases as pulsation period increases. Faint Cepheids, for instance, which are actually several hundred times as bright as the Sun, pulsate with a period of 1 day. The brightest Cepheid variables, 30,000 times as bright as the Sun, pulsate with periods of about 100 days. Figure 19.20 shows that the period-luminosity relationship of Cepheid variables arises because the more massive stars also are more luminous while they cross the H-R diagram during core helium burning. The more massive stars also are larger and lower in density during core helium burning. The period with which a star pulsates is larger for lower densities, so the massive pulsating stars have the greatest luminosities and longest periods.

The period-luminosity relationship for Cepheid variables is useful to astronomers because it provides a way to find the distances to Cepheid variables and star clusters or galaxies that contain Cepheids. If a star can be identified as a Cepheid and its pulsation period measured, then its luminosity and absolute magnitude can be found. The

FIGURE 19.19
The Period-Luminosity Relationship for Cepheid Variables
The more luminous the Cepheid, the longer the pulsation period. By measuring the pulsation period of the Cepheid, one can find the luminosity of the star. This makes it possible to determine the distance of the Cepheid from its luminosity and its apparent brightness (or absolute and apparent magnitude).

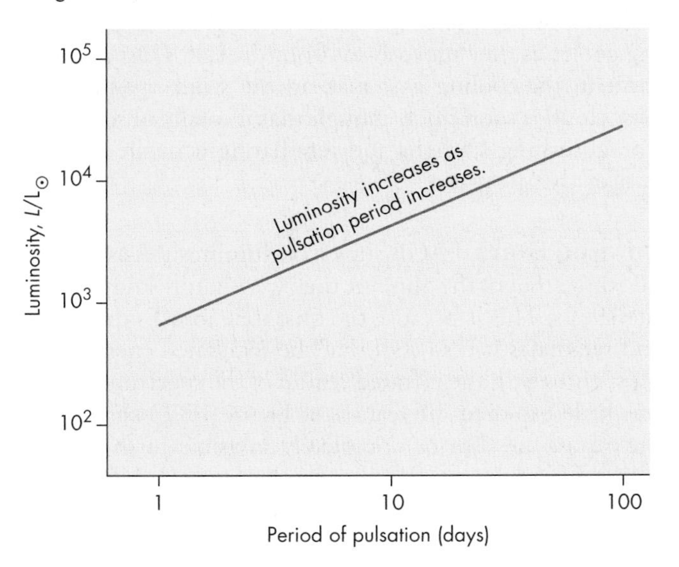

FIGURE 19.20
The Instability Strip and Evolutionary Tracks for Stars of Different Masses
More massive stars (with longer pulsation periods) intersect the instability strip at higher luminosities than do less massive stars.

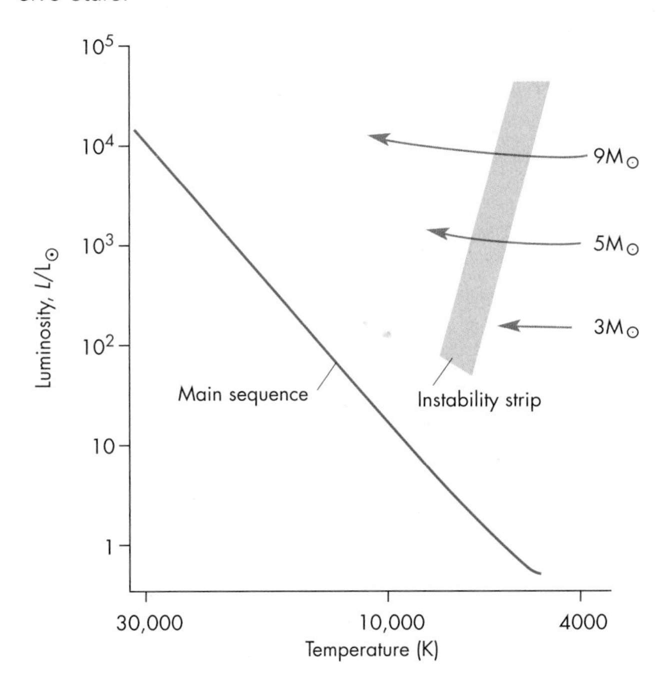

absolute magnitude then can be used with the star's apparent magnitude in Equation 16.3 to calculate its distance. One of the main tasks of the *Hubble Space Telescope* was to observe Cepheid variables and, thus, find distances to remote galaxies.

RR Lyrae Stars The dimmest, hottest stars in the instability strip are the **RR Lyrae stars.** The RR Lyrae stars are horizontal branch stars of about one solar mass that pulsate as they pass through the instability strip. They are small and dense compared with the Cepheids (although they are nearly ten times as large as the Sun and about one hundred times as luminous) and have pulsation periods between about 1.5 hours and 1 day.

After core helium burning begins, a star moves horizontally across the H-R diagram, growing hotter at constant luminosity. During this time, the star enters a phase during which it pulsates. The period-luminosity relationship for Cepheids, a type of pulsating star, allows astronomers to determine the distances to star clusters and galaxies containing Cepheids.

The Asymptotic Giant Branch

Eventually, all stars with masses greater than or equal to the mass of the Sun consume all the helium in their cores. As nuclear reactions cease in their cores, helium fusion begins in a shell surrounding the core, which now consists mostly of carbon and oxygen. The electrons in the core soon become degenerate. Just as the star did while it had a hydrogen-burning shell, the star expands, cools, and becomes more luminous. It moves upward and to the right in the H-R diagram, as shown in Figure 19.21. In fact, the star follows a track in the H-R diagram very similar to the one it followed when it became a red giant. This phase in the evolution of a star is referred to as the **asymptotic giant branch** or **AGB.** The structure of an AGB star is shown in Figure 19.22. Most of the time, the hydrogen-burning shell provides most of the energy of the star while the helium-burning shell is almost dormant. However, after enough helium has built up between the shells, the helium is consumed in an almost explosive event called a **thermal pulse.** After a thermal pulse, the star resumes its former appearance until enough helium builds up for another pulse to occur. With each thermal pulse, the mass of the degenerate carbon and oxygen core increases. For massive stars, a thermal pulse occurs in the deep interior and produces only a slight, temporary change in the star's brightness. For stars as massive as the Sun, the

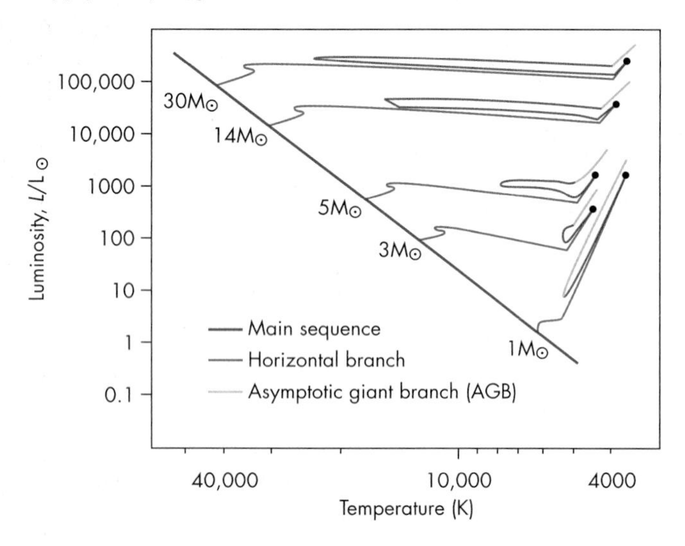

FIGURE 19.21
The Evolutionary Tracks of Stars While They Are on the Horizontal Branch and the Asymptotic Giant Branch (AGB)
This part of the evolutionary track is shown in red and yellow for each star. A star, during this stage in its evolution, develops a helium-burning shell and follows a path in the H-R diagram very similar to the one it followed when it developed a hydrogen-burning shell after exhausting its core supply of hydrogen.

thermal pulse is near enough to the surface that it causes the luminosity of the star to increase about tenfold for about a century. Because the length of time between thermal pulses is about 100,000 years, we are not very likely to witness a particular star undergo a thermal pulse.

Mass Loss A star continues to increase in both size and brightness, rising up the AGB in the H-R diagram, even as it experiences thermal pulses. A wind develops in the star during this time and blows the star's outer layers into interstellar space. Astronomers do not agree on the cause of the winds from AGB stars. However, AGB stars are so large that their surface gravities are very small. Any sort of disturbance is capable of expelling material. The outer layers of an AGB star flow outward at about 10 km/s (about 2% of the speed of the solar wind), growing cooler as they move away from the star. Dust particles form in the cooling gas, making the wind opaque. The dust cloud is soon thick enough that it totally obscures the star, absorbing starlight and reradiating it in the infrared part of the spectrum.

Infrared Stars AGB stars have luminosities as large as 10^5 times that of the Sun, yet they were hardly known until the 1960s. This is because the dust that absorbs their light and reradiates it is so cool that the reradiated energy is almost entirely in the infrared region of the spectrum, which was little explored 50 years ago. Figure 19.23 shows that these stars are dim or completely invisible in the visible region of the spectrum, whereas the Sun is bright in the visible region but very dim in the infrared. The dust cloud formed by the wind marks the surface of an infrared star. For some AGB stars, this surface has a diameter as large as 1000 AU, about 20 times the size of the solar system. The outer layers of an infrared star are very tenuous and contain only a small fraction of the star's mass. Most of the mass is concentrated in the carbon-oxygen core and the energy generating layers that surround it. Thus, the central part of the star is small and very dense, but the outer parts are enormous and have very low density.

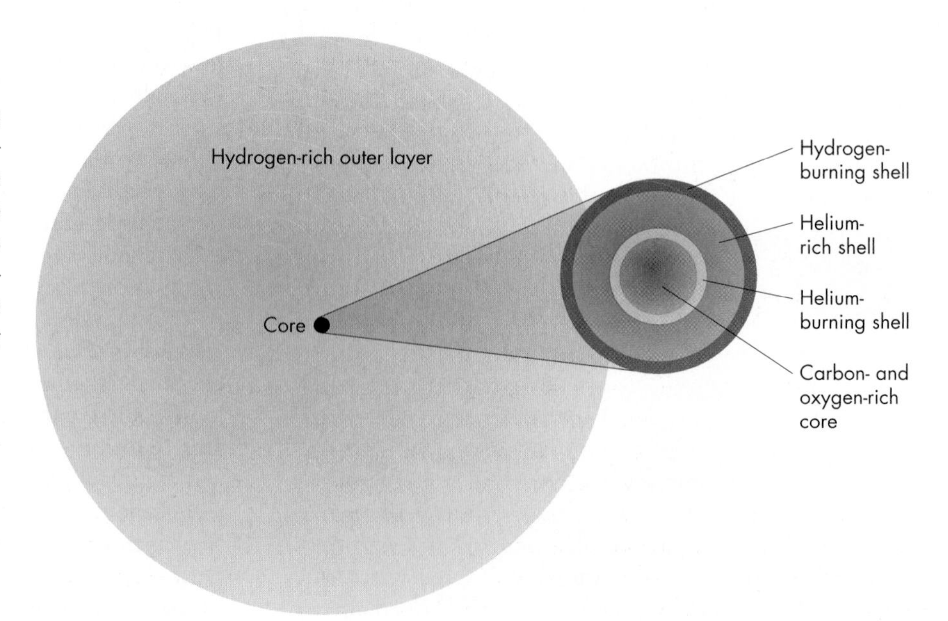

FIGURE 19.22
The Structure of an AGB Star
Energy is produced in hydrogen-burning and helium-burning shells. The degenerate carbon-oxygen core inside the helium-burning shell becomes more massive as time passes. This core is about as large as the Earth. The surface of the star is much larger than the core and may extend to hundreds or thousands of AU.

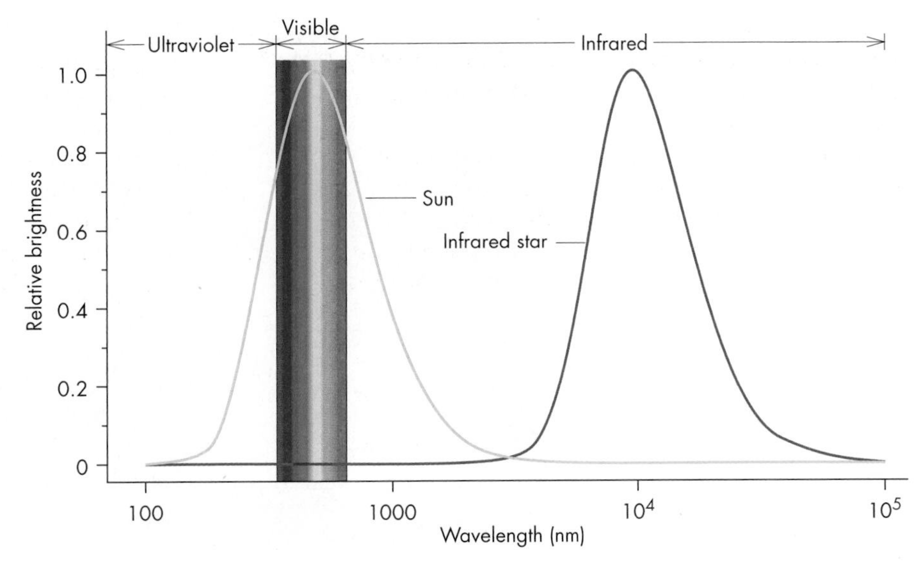

FIGURE 19.23
The Spectra of the Sun and an Infrared Star
Nearly all of the energy emitted by the infrared star is radiated by cool dust and is invisible to the human eye. Much of the energy emitted by the Sun, on the other hand, lies in the visible region of the spectrum, where it can be detected by the human eye.

The End of a Star's AGB Career As an AGB star becomes larger and more luminous, the rate at which it loses mass also increases. The rate of mass loss gradually grows to as much as 10^{-4} $M_\odot$ per year. If the Sun lost mass at a rate that large, it would be gone in only 10,000 years. Obviously, rapid mass loss can't go on for very long in any star. For stars less massive than about 8 $M_\odot$, the wind soon strips away the outer layers of the star almost down to the degenerate core. The loss of its outer layers terminates the AGB career of the star. For stars more massive than 8 $M_\odot$, the AGB ends much more spectacularly as the star explodes in a supernova detonation. Supernovae are described in Chapter 20.

A star once again becomes cool and luminous while it burns helium in a thin shell around its core. At the same time, the star develops a cool wind that carries its outer layers into space. Dust that forms in the wind obscures the star and converts its light to infrared radiation. Most stars end this phase of their lives when they have almost completely shed their outer layers.

Planetary Nebulae At the end of its AGB career, all that remains of a star is a degenerate core of carbon and oxygen surrounded by a thin shell in which hydrogen fusion takes place. The gas and dust expelled during the star's AGB career are still moving outward at tens of kilometers per second. As the debris moves away, the hot, dense core of the star becomes visible. The star now begins to move quite rapidly toward the left in the H-R diagram, growing hotter at roughly constant luminosity. Only a few thousand years are needed for its surface temperature to increase to 30,000 K. At this temperature, the star emits large amounts of ultraviolet radiation capable of ionizing hydrogen in the outward-moving matter. The heated, ionized gas begins to glow, producing a **planetary nebula** like the Ring Nebula shown in Figure 19.24. The bright blue

FIGURE 19.24
The Ring Nebula
The reddish color of the planetary nebula results from the strong red emission line of the Balmer series of hydrogen. The luminous blue star that heats and ionizes the planetary nebula can be seen at the center of the nebula.

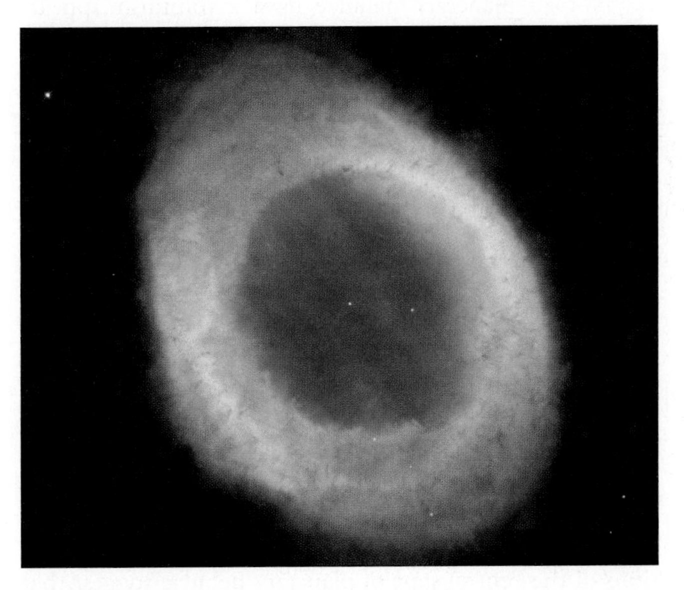

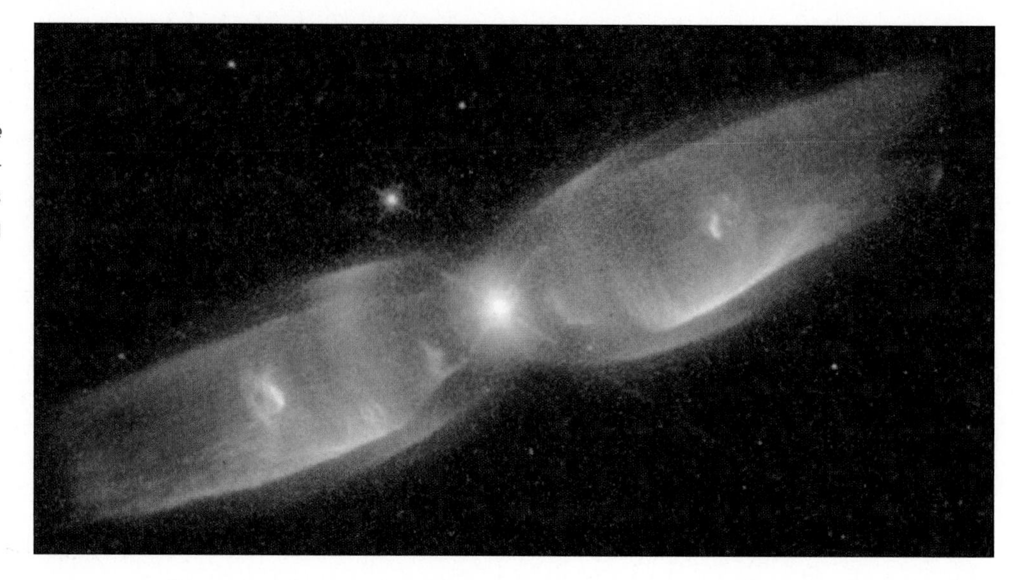

FIGURE 19.25
A Flattened Planetary Nebula
This *Hubble Space Telescope* image shows the planetary nebula M2-9. Notice that M2-9 looks quite different from the Ring Nebula shown in Figure 19.24.

star from which the nebular gas was ejected can be seen at the center of the planetary nebula. Planetary nebulae were originally named for their round appearance and have nothing to do with planets.

Not all planetary nebulae have a roundish appearance. For example, Figure 19.25 shows that the planetary nebula M2-9 has a very elongated shape. It is likely that the appearance of a planetary nebula depends on the angle from which we view it. This is because the wind from the central star of a planetary nebula probably has at least two parts, as shown in Figure 19.26. One of these is a ring, or torus. The other is a flow of gas outward in two lobes that are perpendicular to the torus. When we view the nebula along the axis of the torus, we see a roundish shape like the Ring Nebula. When we view the nebula perpendicular to the axis of the torus, we see a flattened shape like that of M2-9. One possible reason why the gas in a planetary nebula doesn't have a spherical shape may be that the star that shed the gas that became the planetary nebula is a member of a binary star system. Interaction with the companion star may cause the gas shed from an AGB star to become concentrated in a torus. Observations of the central stars of planetary nebulae suggest that many of them may actually be binary stars in close orbits around one another. The interactions of stars in close binary systems are described more fully in Chapter 21.

Two factors work together to make the evolution of the star at the center of planetary nebulae very rapid. First, the star is so luminous (100,000 times as bright as the Sun) that it consumes its hydrogen at a prodigious rate. Second, little hydrogen is left in the thin shell around the degenerate core. Thus, little fuel remains to be consumed. Some planetary nebula stars may have as little as a few millionths of a solar mass of hydrogen left to burn. These luminous, fuel-poor stars fade rapidly. Figure 19.27 shows that they make a nose dive in the H-R diagram as their luminosities fall by 90% in as little as a few decades (although most require a few thousand years to fade). Without a large supply of ionizing photons, the planetary nebula grows dark and disperses into interstellar space. The

FIGURE 19.26
A Model for Planetary Nebulae
A planetary nebula may consist of a ring and a wind perpendicular to the ring. The shape we see depends on the angle from which we view the nebula.

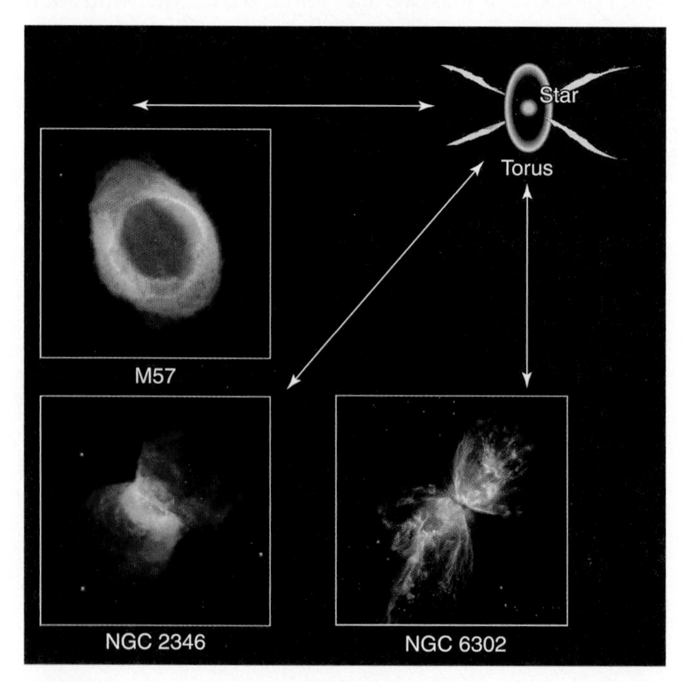

FIGURE 19.27
The Evolution of a Star After It Leaves the AGB
When the temperature of the star reaches 30,000 K, the star lights up the shell it earlier emitted, producing a planetary nebula. Within a few thousand years, the star fades dramatically in brightness and becomes a white dwarf.

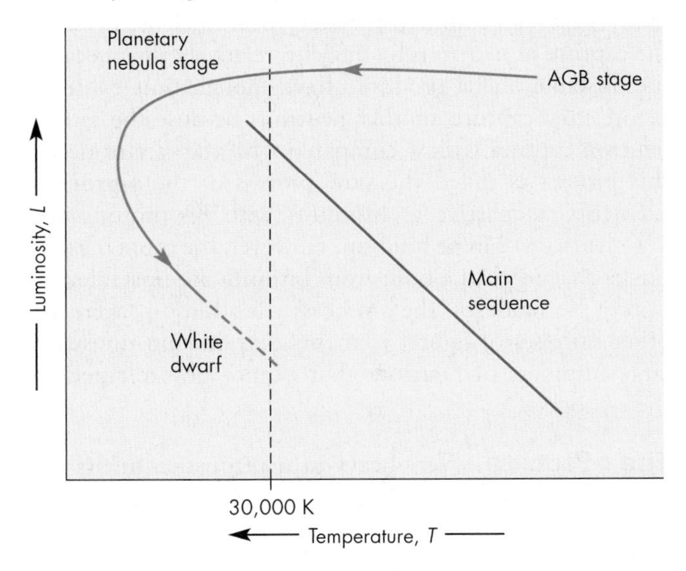

star itself fades to become a white dwarf star, described in Chapter 20.

At the end of its AGB career, a star quickly becomes hot enough to ionize the matter it had ejected when it was an AGB star. The glowing gas becomes a planetary nebula.

Very Massive Stars

The most massive stars, those with masses greater than 40 M☉, lead brief but interesting careers. They are the most luminous stars and, because of their great luminosities, consume their nuclear fuels quickly. A 120 M☉ star, for instance, is several million times as luminous as the Sun and lasts for only a few million years. After they leave the main sequence, massive stars usually become red supergiants while they consume helium in their cores. After the red supergiant stage, massive stars quickly grow hotter again and become blue supergiants. A massive star develops an iron core that eventually becomes so massive that it collapses, producing a supernova explosion.

Massive stars shed mass throughout their lives. The most massive of them can lose as much as 20% of their mass while they are main sequence stars and another 30% in later stages. For most massive stars, the mass is lost in relatively steady winds. In some cases, however, almost explosive mass loss occurs. The most spectacular example of rapid mass loss yet observed is η Carinae, which lies about 2500 pc away in the southern constellation Carina. η Carinae is normally too faint to be seen with the unaided eye. In 1837, however, η Carinae increased in brightness so that between 1837 and 1860 it was the second brightest star in the sky. After 1860 it quickly faded to about its present brightness. Later, astronomers discovered that part of the reason it faded in 1860 was that a thick cloud of dust particles formed in gas that it had ejected during its outburst. The cloud of gas and dust ejected by η Carinae, shown in Figure 19.28, contains at least 10 M☉ of material and is moving away from η Carinae at 700 km/s. It is possible that this is not the first time people have witnessed the ejection of mass from η Carinae. In 3000 B.C. the Sumerians discovered a bright new star that barely rose above the southern horizon. The "new" star was probably η Carinae during an outburst. η Carinae doubled in brightness between 1997 and 1999 and has undergone smaller changes in brightness since 1999. η Carinae is now losing mass at about 10^{-3} M☉ per year. Obviously, it can't continue to do so for very long. η Carinae is near the end of its career and is a good candidate for a supernova explosion in the relatively near future.

FIGURE 19.28
Gas and Dust Ejected by η Carinae
This *Hubble Space Telescope* image shows two lobes of gas and dust moving rapidly away from η Carinae, the bright white spot at the center. η Carinae is actually 100,000 times brighter than the lobes of gas and dust around it. Its brightness has been suppressed to show details of the gas and dust more clearly. The gas and dust cloud ejected by η Carinae is more than 500 times the size of the solar system across.

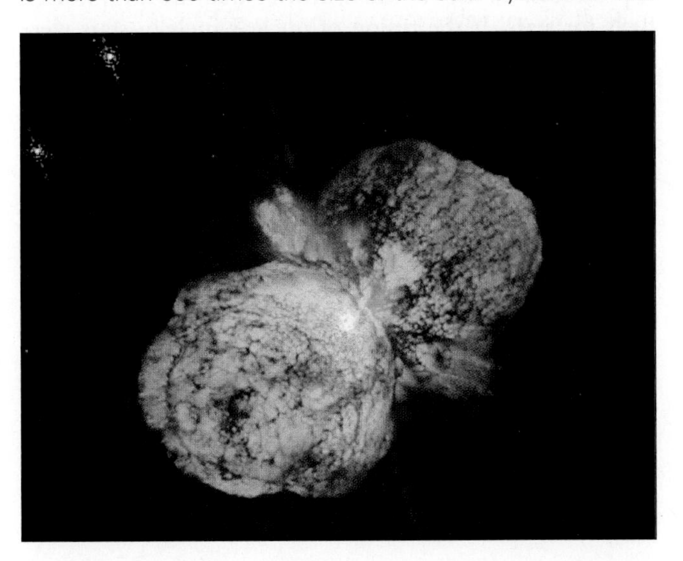

The Formation of Heavy Elements in Stars

There are more than one hundred known elements, most of which exist naturally in the universe. One of the great successes of the theory of stellar evolution has been in explaining the relative abundances of the elements and their isotopes.

Nuclear Fuels Some of the elements are formed in energy-producing fusion reactions in stars. The fusion of hydrogen into helium, and helium into carbon and oxygen, for example, is described earlier in this chapter. In massive stars, core temperatures eventually become high enough for both carbon and oxygen to become nuclear fuels. The fusion of carbon and oxygen results in a mixture of nuclei—mostly silicon, sulfur, and magnesium. One might think that the next development would be a further heating of the core of the star so that silicon and the other products of carbon and oxygen burning could become fuels. Something else occurs, however, before the required temperature (several billion kelvins) can be reached.

The radiation emitted by a gas at a temperature of 1 billion K consists mostly of very energetic gamma rays. These gamma rays are capable of disrupting nuclei and causing them to emit protons, neutrons, and α particles (helium nuclei). The emitted protons, neutrons, and α particles immediately react with other nuclei to produce new nuclei. For a few seconds a complicated network of nuclear reactions goes on all at once. The product of one reaction is the fuel for another. At the end of that time, the gas consists of elements that have 50 or 60 protons and neutrons in their nuclei, such as iron and nickel.

Reactions with Neutrons The reactions that accompany the disruption of nuclei by gamma rays are the final reactions that can produce energy in a star. As described in Chapter 7, Section 4, fusion of an element with a mass number (the number of protons and neutrons) greater than about 60 uses up more energy than it produces. Thus, elements heavier than iron cannot be fuels for stars. If the fusion of nuclear fuels in stars were the only way to produce heavy elements, there would be no nuclei with mass numbers greater than 60. Clearly this is not the case. Although elements heavier than iron and nickel are relatively rare, they do exist and are extremely important for biology and technology. These elements include copper, gold, zirconium, tin, iodine, tungsten, and uranium. They are produced by reactions between nuclei and neutrons.

The s-Process The reaction of a massive nucleus with a charged particle such as a proton or an α particle occurs only with difficulty because of the repulsive force between the positively charged particles. Neutrons, on the other hand, are electrically neutral so they can be captured relatively easily by massive nuclei. Given enough neutrons, very massive nuclei can be built up from elements like iron.

The buildup of massive nuclei by neutron capture begins during the red giant phase of a star's career but becomes more vigorous while the star is burning helium in its core. The neutrons needed for the buildup are released when neon (and other) nuclei capture α particles. During this stage of evolution, there are plenty of neutrons, as many as several hundred for every iron nucleus. Even so, the capture of neutrons by nuclei is relatively slow because the unstable nuclei produced have enough time to decay before they capture another neutron. Because the rate of neutron capture is slow compared with the rate of decay, this process is called the slow process or the **s-process.** Elements as massive as bismuth (with 83 protons and 126 neutrons) can be built up. However, the more massive nuclei that might be built from bismuth are unstable and cannot be made by the s-process. In addition, there are other isotopes of lighter elements that contain unusually large numbers of neutrons that cannot be explained by the s-process.

The r-Process Very heavy or neutron-rich nuclei can be produced only when there are so many neutrons that neutron capture is very rapid compared with decay rates. When this happens, the neutron-capture process is called the rapid process, or **r-process.** A density of neutrons large enough for the r-process to take place occurs only during the collapse of the core of a star during a supernova explosion. Supernova explosions are described in Chapter 20. During a supernova, very high densities and temperatures are produced in the contracting core. A variety of nuclear reactions that yield neutrons occur. A massive nucleus captures many neutrons each second. Between captures there is not enough time for unstable nuclei to decay. In only a few seconds, elements heavier than bismuth can be formed. These heaviest elements include radium, uranium, and plutonium.

Not only does a supernova explosion produce very heavy elements, it also propels them into space. The heavy elements gradually are mixed into the interstellar gas so that when later generations of stars form, they are enriched in heavy elements. Our own Sun formed from gas enriched by billions of years of heavy-element formation in many generations of massive stars that died as supernovae long before the Sun was born.

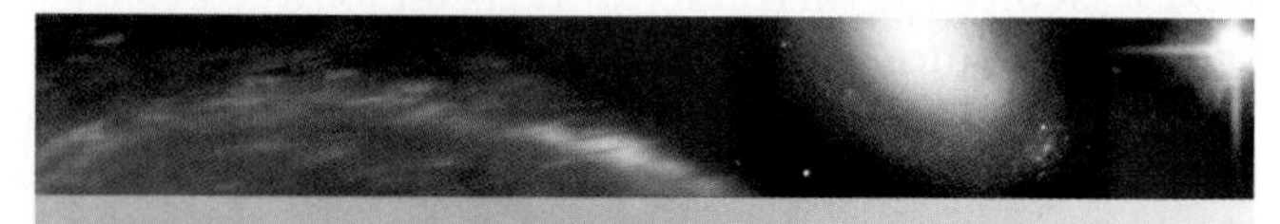

Elements as massive as iron can be made in energy-producing reactions in stars. More massive elements are made by the reaction of neutrons with nuclei. Just which kinds of elements and isotopes are made by this process depends on whether the neutrons are added rapidly (the r-process) or slowly (the s-process).

Chapter Summary

- Hydrogen can be fused into helium at a temperature of about 10 million K. Fusion of helium and heavier elements requires still higher temperatures. When a star lacks nuclear energy sources, energy can be produced by gravitational contraction or collapse. (Section 19.1)
- The opacity of a gas describes its ability to block the flow of radiation. Where opacity is low, radiation flows nearly unimpeded, so it easily carries energy outward through a star. Where opacity is high, the flow of radiant energy is inefficient. Convection currents develop and carry most of the energy outward. (19.1)
- When a gas is degenerate, its particles are compressed so much that they strongly resist further compression. As a result, the temperature of a degenerate gas can decrease without causing a similar decrease in pressure. (19.1)
- The Vogt-Russell theorem states that the initial mass and chemical composition of a star determine its entire evolution. Stars with close binary companions are exceptions to the Vogt-Russell theorem. (19.1)
- As a star evolves, its luminosity and surface temperature change. Thus, the evolution of the star can be described by its changing position in an H-R diagram. An H-R diagram of a star cluster is useful because it represents a snapshot of the evolution of a group of stars that have different masses but the same age and chemical composition. (19.2)
- Main sequence stars are those that consume hydrogen in their cores. Despite this common property, main sequence stars of different masses show a wide range of sizes, luminosities, temperatures, and internal structures. (19.3)
- Massive stars spend much less time on the main sequence than less massive main sequence stars because their greater luminosities result in more rapid consumption of their hydrogen. While on the main sequence, stars change little in temperature and luminosity. (19.3)
- After the hydrogen in the core of a star is used up, fusion begins in a thin shell surrounding its helium core. The star swells and cools to become a red giant. Rising temperature in the core of the star eventually initiates the fusion of helium. In stars like the Sun, this happens in an explosive helium flash. (19.4)
- While burning helium in its core, a star becomes hotter, moving across the H-R diagram at constant luminosity. During part of this horizontal track, the star develops pulsations that cause its size, temperature, and brightness to vary periodically. The period-luminosity relationship for Cepheid variables can be used to determine their distances. (19.4)
- When the supply of helium in the core of a star is exhausted, the shell of helium surrounding the core begins to be consumed. A star with a helium-burning shell once again swells and cools, becoming an asymptotic giant branch (AGB) star. AGB stars shed their outer layers in cool winds in which dust particles form. The dust shields the star and converts its light to infrared radiation. Eventually, for all but the most massive AGB stars, the wind strips away nearly all the outer layers of the star. (19.4)
- At the end of its AGB phase, a star quickly becomes hotter. When its surface temperature reaches 30,000 K, it ionizes the gas that it had earlier shed as a cool wind. The ionized matter glows as a planetary nebula. (19.4)
- Elements as massive as iron can be formed in energy-generating nuclear reactions in stars. More massive elements are made by the capture of neutrons by nuclei. The mix of elements and isotopes that results from neutron capture depends on whether the rate of neutron captures is rapid (the r-process) or slow (the s-process). (19.4)

Key Terms

Conceptual Questions

1. Why does the chemical composition of a main sequence star change most rapidly at its center?
2. Why can't hydrogen fusion and helium fusion go on at the same time at the center of a star?
3. Why do nuclei of elements other than hydrogen require higher temperatures to undergo fusion than does hydrogen?
4. What is the relationship between the opacity of the gas in a star and whether convection takes place within the star?
5. Suppose the temperature of a normal gas doubles while its density remains the same. What happens to the pressure of the gas?
6. Suppose the temperature of a degenerate gas doubles while its density remains the same. What happens to the pressure of the gas?
7. Suppose two single stars form at the same time. They have the same masses and chemical compositions. What can be said about the evolution of the two stars from that point forward?
8. What does the evolutionary track of a star tell us about its motion through space?
9. What is the relationship between isochrones and the HR diagrams of clusters of stars?
10. In cluster 1, the main sequence extends from spectral class O to spectral class K. In cluster 2, there are no main sequence stars cooler than spectral class G. Which cluster is older and how do we know?
11. What do all main sequence stars have in common?
12. How do the temperatures, masses, radii, and luminosities of the most massive main sequence stars compare with those of the least massive main sequence stars?
13. Why is it believed that there are no main sequence stars less massive than 0.08 $M_\odot$?
14. Describe the changes in the Sun that will occur as a result of the Sun's main sequence evolution. What effect will these changes have on the Earth?
15. Why doesn't the core of a 1 $M_\odot$ star expand when helium fusion begins to raise the temperature of the core?
16. Why doesn't the energy released during a helium flash dramatically increase the brightness of the star?
17. Describe the relationship between pressure and gravity throughout a pulsation cycle of a Cepheid variable.
18. Why are AGB stars faint in the visible part of the spectrum but very luminous in the infrared?
19. What is the origin of the gas that can be seen as a planetary nebula around a hot, post-AGB star?
20. Why must there be a hot star at the center of a planetary nebula?
21. What would the chemical composition of the universe be like if the only nuclear reactions that had ever occurred in stars had been those involved in energy production?

Problems

1. If one thousand people were observed at random times of day, about how many of them would be observed to be eating?
2. Suppose a star has the same mass as the Sun but has a diameter 5 times as large. Compare the total gravitational energy radiated away by the star with that radiated away by the Sun.
3. Suppose the Sun were to shrink to 50% of its present diameter. How would the gravitational energy released during that contraction compare with the gravitational energy released during the Sun's contraction to its present diameter?
4. Star A is 3 times as massive and 60 times as luminous as star B. How do the lifetimes of the two stars compare?
5. Main sequence star A is 5 times as massive as main sequence star B. How do the main sequence lifetimes of the two stars compare?

Figure-Based Questions

1. Suppose there is a cluster of stars in which there are no main sequence stars hotter than spectral class A0. Use Table 19.1 and Figure 19.9 to estimate the age of the cluster.
2. Use Figure 19.12 to determine the luminosity of the hottest main sequence star in the cluster h + χ Persei. Be careful when you try to read the luminosity off the horizontal scale—this is a logarithmic plot. Use Figure 16.21 to find the mass that corresponds to the luminosity of the hottest main sequence star in the cluster. Then use the mass in Figure 19.9 to find the main sequence lifetime of the star and the cluster.
3. Use Figure 19.9 to find the main sequence lifetime of a 5 $M_\odot$ star.
4. Use Figure 19.19 to find the luminosity of a Cepheid variable that has a period of 100 days.

Group Activity

Using Table 19.1, Appendix 12, and the star charts at the beginning and end of the book, have your group identify some bright stars with hot and cool surface temperatures. Have the group members decide if they can see any difference between the colors of the hot and cool stars. Try looking at the stars through binoculars to see if this makes any difference in your ability to detect the colors of bright stars.

For More Information

Visit the text website at **www.mhhe.com/fix** for chapter quizzes, interactive learning exercises, and other study tools.

20

Apparently then we have a star of mass about equal to the Sun and of radius much less than Uranus. The calculated density is 61,000 grams per cubic centimeter—just about a ton to the cubic inch. This argument has been known for some years. I think it has generally been considered proper to add the conclusion "which is absurd."

Sir Arthur S. Eddington,
The Internal Constitution of the Stars, 1926

A Planet orbiting pulsar PSR B1257+12.

www.mhhe.com/fix

White Dwarfs, Neutron Stars, and Black Holes

This chapter is about absurd stars—those that have masses comparable to the Sun, but that are as small as or smaller than planets. These objects, white dwarfs, neutron stars, and black holes, represent the final stages in the evolution of stars. Unless it has a close binary companion, a star that becomes a white dwarf or neutron star is doomed to remain in that condition forever. A star that produces a black hole disappears permanently from the rest of the universe.

Stars that become white dwarfs or neutron stars or that produce black holes could be considered dead stars because they have used up their available nuclear fuels. This language seems too strong, however, in view of the fact that many of them are quite visible and that they can exert a very strong influence on their surroundings. The interactions that these stars have with binary companions are varied and sometimes spectacular.

Questions to Explore

- How does a star get to be a white dwarf?
- What happens to stars too massive to become white dwarfs?
- How are neutron stars and supernovae related?
- What is a pulsar, and how does it produce pulses?
- What is curvature of space, and how is it an alternative to Newton's theory of gravity?
- What is a black hole, and how does a collapsing star form one?

20.1 WHITE DWARF STARS

Our knowledge of white dwarf stars began before 1850 when astronomers began to suspect that the brightest star, Sirius, had a binary companion. When the companion of Sirius (called Sirius B) was finally seen in 1862, it turned out to be only one ten-thousandth as bright as Sirius itself, as shown in Figure 20.1. Its mass, however, was 0.98 $M_{\odot}$. After the mass-luminosity relationship for stars was discovered in about 1910, it was clear that Sirius B and a few other stars were extremely underluminous for their masses. The nature of underluminous stars like Sirius B grew more mysterious in 1914 when it was found that Sirius B and another star, the companion of o Eridani, were A stars. This means that their temperatures are about 10,000 K and that they emit about eight times as much radiation from every square meter of their surfaces as the Sun does. The only way that they could be both hot and dim was to have small sizes—so they were called **white dwarf stars.**

White Dwarfs and Electron Degeneracy

After sizes and masses had been found for a few white dwarfs, astronomers like Eddington (whose words appear at the beginning of this chapter) realized that the gas within white dwarfs was much too dense to behave like an ideal gas. Shortly thereafter, in 1926, the theory of the behavior of degenerate gases was developed. In 1930, S. Chandrasekhar used the theory to calculate models for white dwarf stars. Chandrasekhar (whose 1983 Nobel prize in physics was partly for this work) showed that the pressure of degenerate electrons is capable of supporting a star against gravity if the entire star is compressed to about the size of the Earth. He also worked out several very surprising properties of white dwarf stars, including the fact that the more massive white dwarf stars are smaller than those of low mass.

FIGURE 20.1
Sirius and Its White Dwarf Companion, Sirius B
Sirius B is located just to the right and below the image of Sirius. The pattern of radial spots was caused by the telescope used to observe Sirius.

Mass-Radius Relationship Our experience, both everyday and astronomical, tells us that as the mass of an object increases, so does its size. This is true for dogs, cars, planets, and main sequence stars. For white dwarf stars, however, just the opposite is true. The more massive a white dwarf star, the smaller its size. This behavior arises because of the way the pressure in a white dwarf depends on density.

Consider an imaginary experiment designed to study the mass-radius relationship for main sequence stars. This experiment is illustrated in Figure 20.2. We begin by holding the size of a star constant while increasing its mass. Because the star has more mass in the same volume, its density increases. For a main sequence star, pressure and density are related to each other by the ideal gas law, for which pressure is directly proportional to density. Thus, increasing the density of the star also increases its internal pressure. This, in the absence of other forces, would push outward, causing the star to expand. Increasing the density of the star also increases gravity throughout the star because it forces each atom, ion, and electron nearer to its neighbors. In the absence of other forces, the increased gravity would cause the star to contract. What actually happens depends on whether the increase in pressure or the increase in gravity is larger. For a main sequence star, the increase in pressure is larger, so the star expands. As it expands, both pressure and gravity decrease until they come into balance again. When they reach this balance, the star is larger than it was before. Thus radius increases as mass increases.

The same imaginary experiment performed on a white dwarf star gives the opposite result. Increased density leads to increased pressure (just as for main sequence stars), but the pressure in a white dwarf is produced by degenerate electrons. Increased density also leads to increased gravity. For a white dwarf, however, the increase in gravity exceeds the increase in pressure, so the star contracts. As it shrinks, gravity and pressure both increase until they come into balance at a smaller size. Thus, as shown in Figure 20.3, the more massive a white dwarf is, the smaller it is. For example, a 0.5 $M_{\odot}$ white dwarf star is about 50% larger than the Earth, but a 1.0 $M_{\odot}$ white dwarf is about 90% as large as the Earth. A 1.3 $M_{\odot}$ white dwarf is only about 40% as large as the Earth—about as big as Mercury.

The Chandrasekhar Limit The mass-radius relationship shown in Figure 20.3 also shows that there is a

FIGURE 20.2
Increasing the Mass of a Main Sequence Star While Keeping Its Radius Constant
A, Before mass is added, the star is in balance, with pressure forces just equal to gravity. **B,** The effect of adding mass to the star is to increase both pressure and gravity. Pressure, however, increases more. **C,** The pressure excess causes the star to expand and both pressure and gravity to drop until both become equal again. At this point, the star stabilizes at the larger size.

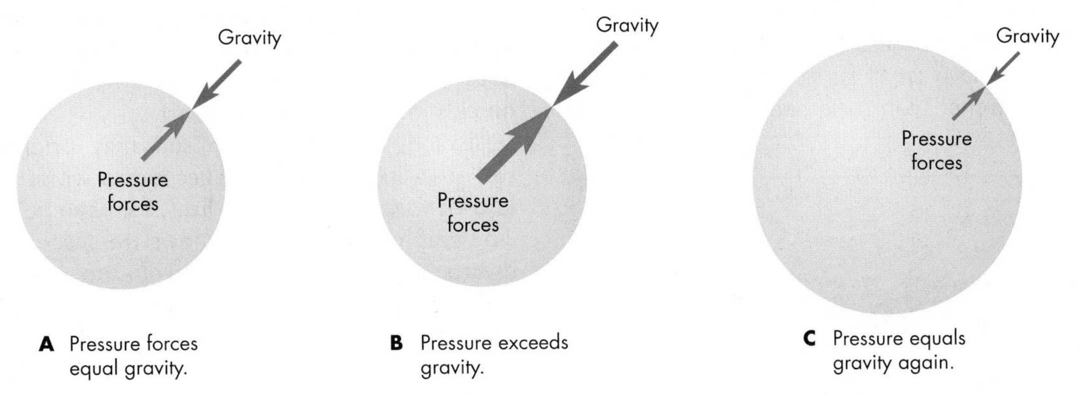

FIGURE 20.3
The Mass-Radius Relationship for White Dwarfs
Radius is given in terms of the radius of the Earth, $R_\oplus$, and mass is in units of solar masses, $M_\odot$. The more massive a white dwarf, the smaller it is. Notice that a white dwarf of 0.5 $M_\odot$ is larger than the Earth, but a white dwarf of 1 $M_\odot$ is about 90% as large as the Earth and a white dwarf of 1.3 $M_\odot$ is only about 40% the size of the Earth. The size of a white dwarf falls to zero for a mass of 1.4 $M_\odot$.

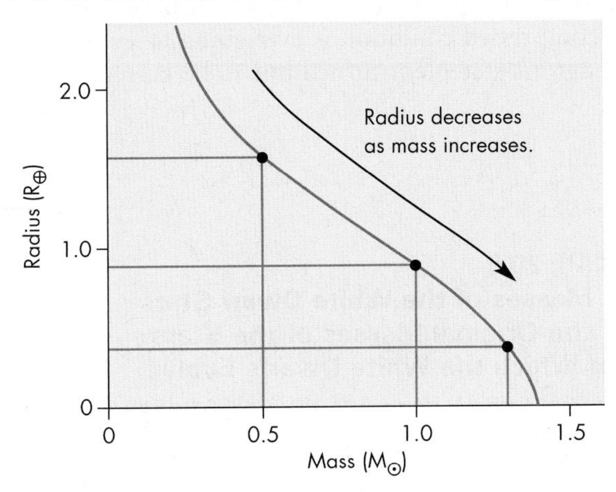

maximum mass that a white dwarf can have. This mass, about 1.4 $M_\odot$, is called the **Chandrasekhar limit.** It is the mass for which the mass-radius relationship drops to zero. In other words, a white dwarf with a mass equal to the Chandrasekhar limit, shrinks to a very small size. No star more massive than 1.4 $M_\odot$ can be supported against gravity by the pressure of its degenerate electrons. All main sequence O, B, and A stars, and some F stars, are more massive than the Chandrasekhar limit, however. If these stars are to become white dwarfs, they must reduce their masses by shedding their outer layers while they are asymptotic giant branch (AGB) stars. If they fail to lose enough mass, their final contraction cannot be stopped by their degenerate electrons when they reach about the size of the Earth. Instead, they will shrink to smaller and smaller sizes and either become neutron stars or disappear inside black holes.

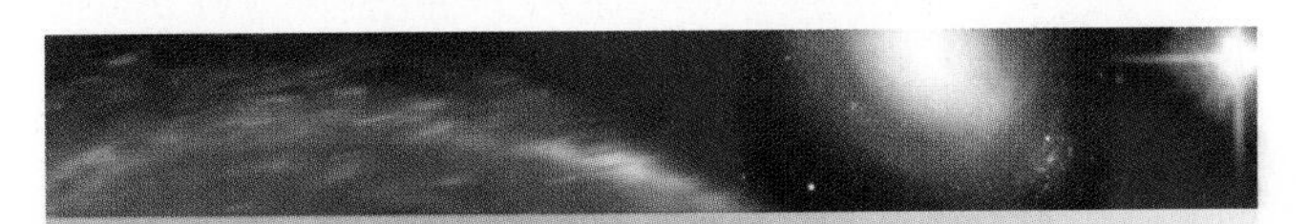

White dwarfs are planet-sized stars supported by the pressure of their degenerate electrons. The more massive the white dwarf star, the smaller its size. The greatest mass a white dwarf can have is about 1.4 $M_\odot$, the Chandrasekhar limit.

Evolution of White Dwarf Stars

Once a white dwarf star contracts to its final size, it no longer has any available nuclear fuels. It still has a very hot interior and a large supply of heat, however. As time passes, the white dwarf cools by radiating its heat into space. It keeps the same size as it cools, so it grows dimmer in much the same way that a metal ball cools and dims after it is removed from a furnace. Figure 20.4 shows that the evolutionary track of a white dwarf in the H-R diagram is downward and to the right as it grows dimmer and cooler. Massive white dwarfs are smaller and have smaller surface areas than less massive white dwarfs. This means that massive white dwarfs are less luminous for a given temperature, so their evolutionary tracks are below those of less massive white dwarf stars.

Calculations of white dwarf cooling show that white dwarfs with masses of 0.6 $M_\odot$ will decline to 0.1 $L_\odot$ in about 20 million years. Further declines in brightness take progressively longer amounts of time. The white dwarf

FIGURE 20.4
Evolutionary Tracks for White Dwarf Stars Ranging in Mass from 0.25 $M_\odot$ to 1 $M_\odot$
A white dwarf cools and grows dimmer, so it moves downward and to the right in the H-R diagram. More massive white dwarfs are smaller and dimmer than less massive white dwarfs, so the track for a 1 $M_\odot$ white dwarf lies below the tracks for less massive white dwarfs. Because they are small and have small surface areas, white dwarf stars are much dimmer than main sequence stars of the same temperature.

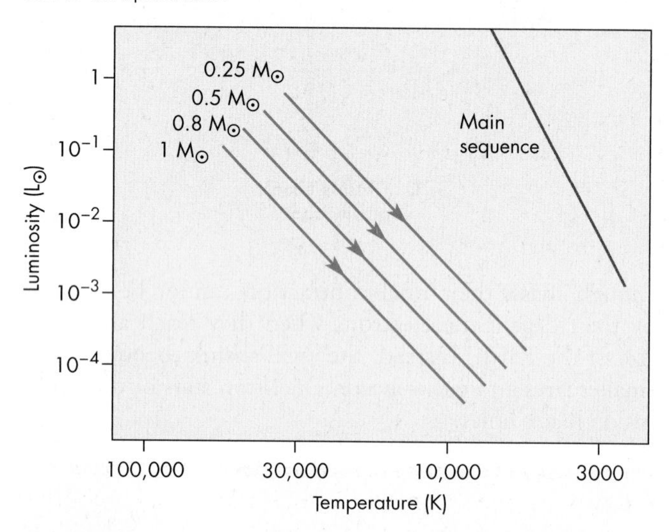

reaches 0.01 $L_\odot$ in 300 million years and 0.001 $L_\odot$ in a billion years. Six billion years after it forms, the white dwarf is only 0.0001 as bright as the Sun. At this point, the temperature and color of the white dwarf are similar to those of the Sun. It is so faint, however, that it is undetectable unless it is within a few parsecs of the Sun. If we were 1 AU from such a star, it would shine only a little more brightly than the full Moon. White dwarfs with masses greater than 0.6 $M_\odot$ have more internal heat and smaller luminosities, so they take even longer to cool and grow dim.

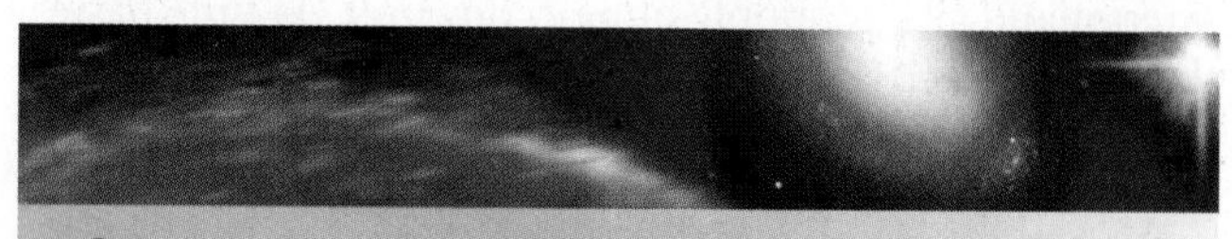

Once a star becomes a white dwarf, its size remains constant. As it cools, it grows dimmer and evolves down and to the right in the H-R diagram. After billions of years, a white dwarf is thousands of times less luminous than the Sun and very hard to detect.

What Is the Origin of White Dwarf Stars?

Most white dwarf stars evolve directly from the central stars of planetary nebulae, which are the former cores of AGB stars, as discussed in Chapter 19, Section 4. During the AGB stage, a star loses much of its mass in a cool wind. A white dwarf results if the wind strips away enough material to lower the mass of the star below the Chandrasekhar limit. What is left behind is the carbon-oxygen core of the star surrounded by a thin layer of helium-rich gas. In some cases, a thin outer layer of hydrogen-rich gas is left as well. When nuclear reactions cease at the end of the planetary nebula stage, the star quickly becomes a white dwarf.

Exactly how massive a star may originally be and still lose enough mass to become a white dwarf is not known with certainty. The limit seems to be about 8 $M_\odot$. The graph in Figure 20.5 shows the masses of the white dwarfs that result from the evolution of stars that had various masses while they were on the main sequence. Main sequence stars with masses between 2 and about 8 $M_\odot$ produce white dwarfs with masses between 0.7 and about 1.4 $M_\odot$. Main sequence stars with masses less than about 2 $M_\odot$ yield white dwarfs with masses in the range of about 0.6 to 0.7 $M_\odot$. White dwarf stars with masses less than 0.6 $M_\odot$ would evolve from main sequence stars with masses less than 1 $M_\odot$. The main sequence lifetimes of these stars are so long, however, that the universe is not yet old enough for them to have become white dwarfs. Thus, there are no white dwarf stars with masses much less than about 0.6 $M_\odot$.

About 80 years ago Eddington wrote about white dwarfs, "I do not see how a star which has once got into this compressed condition is ever going to get out of it." Eddington's statement turned out to be correct for single

FIGURE 20.5
The Masses of the White Dwarf Stars and the Original Masses of the Stars from Which the White Dwarfs Evolve
Stars that originally were more massive than about 8 $M_\odot$ can't produce white dwarfs because their degenerate cores become more massive than the Chandrasekhar limit.

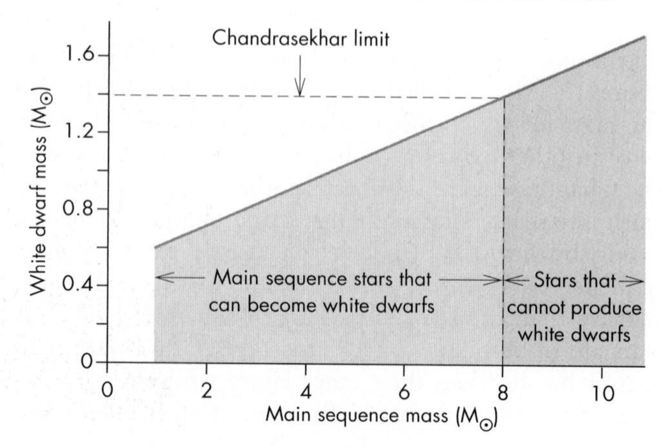

white dwarfs. For those in some binary systems, however, a spectacular way out is described in Chapter 21.

White dwarfs evolve from AGB stars that have lost most of their outer layers via cool winds. Stars as massive as the Sun yield white dwarfs of about 0.6 $M_\odot$. More massive white dwarfs evolve from stars with original masses ranging up to about 8 $M_\odot$.

20.2 NEUTRON STARS

Stars that originally are more massive than around 8 $M_\odot$ eventually collapse explosively to produce supernovae. Unless the star originally had more than about 25 $M_\odot$, what remains after the supernova explosion is a neutron star.

Supernovae

Astronomers began to recognize the spectacular nature of supernovae in 1885, when a "new star" appeared in the relatively nearby Andromeda Galaxy. Nova outbursts (described in Chapter 21) already had been seen many times in our own galaxy and in other galaxies. The explosion in the Andromeda Galaxy, however, was about ten thousand times as luminous as the brightest nova outburst. It was called a **supernova** because its luminosity was so much greater than a nova. At its brightest, the supernova in the Andromeda Galaxy shone with the luminosity of nearly 10 billion stars like the Sun. It was about one-tenth as bright as the entire Andromeda Galaxy.

Because supernovae are so bright, they can be seen even in distant galaxies. In fact, thousands have been observed in other galaxies, and many new ones are found each year. The pictures in Figure 20.6, for example, show a supernova outburst observed in the galaxy NGC 4725 in 1940. The rate at which supernovae occur in other galaxies suggests that a few per century should occur in the Milky Way. Historical records, on the other hand, show that only five supernovae have been observed in our galaxy in the past thousand years. The brightest supernova in recorded history was seen in 1006. It was reported to be dazzling to the eye and must have been at least one hundred times as bright as Venus. The Chinese recorded a supernova, visible in the daytime, in 1054, and another in 1181. Five centuries later, two supernovae were extensively observed by Tycho Brahe (1572) and Kepler (1604). The latter of these was seen only a few years before the invention of the telescope. Since then, no supernovae have been observed in the Milky Way.

We can't expect to see all the supernovae in our galaxy because distant ones are obscured by interstellar dust clouds. Even so, we are long overdue to observe another supernova within the Milky Way. It takes hundreds to tens of thousands of years for the light produced by a supernova to reach us, so we observe a supernova only long after the explosion has taken place. The light from one hundred to one thousand supernovae that already have occurred in the Milky Way is on its way to us now.

Types of Supernovae Supernovae can be divided into two general groups, called type I and type II, based

FIGURE 20.6
The Galaxy NGC 4725 During the Outburst of a Supernova (left) and 8 Months Later, After the Supernova Had Faded from View (right)
At its brightest, a supernova can shine almost as brightly as an entire galaxy.

on the kinds of emission lines that appear in their spectra. Some type I supernovae, which do not show hydrogen lines, result from the explosions of white dwarf stars in binary star systems. These supernovae are described in Chapter 21. Type II supernovae, which do show hydrogen lines, are produced by the catastrophic collapses of the cores of massive stars.

Type II Supernovae **Type II supernova** explosions take place in the centers of massive stars that have consumed the last of their nuclear fuels. As a massive star evolves, it burns heavier and heavier elements as nuclear fuels. Each nuclear burning phase is briefer than the previous one. Hydrogen lasts for tens of millions of years. Helium is consumed in a million years, carbon in a few centuries, and oxygen in months. The final nuclear fuel, silicon, burns to produce iron and nickel in only a few days. During the final nuclear burning stage, the star has an inert iron and nickel core surrounded by shells of silicon, oxygen, neon, carbon, helium, and hydrogen, as shown in Figure 20.7. The core, supported by the pressure of its degenerate electrons, is essentially a white dwarf star surrounded by the outer layers of the red giant. When the growing mass of the core exceeds the Chandrasekhar limit, its weight is too great for its degenerate electrons to support, and the core begins to contract.

As the core contracts and its density increases, **neutronization** begins to take place. That is, electrons react with the protons in the iron nuclei to produce neutrons. Each neutronization reaction also produces a neutrino. As more and more electrons react with protons, fewer and fewer are left to support the core and resist compression. As the electrons (and protons) disappear, the contraction accelerates and quickly becomes a collapse. In about a second, the core collapses from thousands of kilometers to 50 km in radius. In another few seconds, it shrinks to a radius of about 5 km. The gravitational energy released during the collapse is equal to the Sun's luminosity for tens of billions of years. Much of this energy takes the form of neutrinos, which are produced not only by neutronization, but also by gamma rays and other mechanisms.

When the central 0.6 to 0.8 $M_\odot$ of the collapsing core reaches a density equal to the density of the nuclei of atoms, the neutrons become degenerate and strongly resist compression. To achieve this density, the entire Earth would have to be compressed to a sphere 300 m (1000 feet) in diameter. The central core, which has become a **neutron star,** rebounds and pushes strongly against the rest of the infalling core, driving it outward in a shock wave, as shown in Figure 20.8. Initially, the shock wave moves outward at about one-sixth the speed of light. If nothing further happened, the shock wave would die out long before it reached the surface of the star. Instead, the neutrinos produced in the collapse reenergize the shock and help it push outward toward the surface. Normally, neutrinos pass freely through matter. At the heart of a supernova, however, the density of matter is so great that not even neutrinos can escape directly. They travel only a few meters between collisions with neutrons. Most collisions only deflect the neutrinos, which diffuse out of the core of the star. Some collisions, however, result in the absorption of neutrinos. The energy of the absorbed neutrinos heats the gas behind the shock front and helps push it outward. More than 90% of the energy of the explosion is carried

FIGURE 20.7
The Structure of the Core of a Star Just Before It Erupts as a Supernova
The core consists of layers in which nuclear fusion has reached different stages. The layers have such different sizes that they can't all be shown on the same scale. The outer part of the star consists of hydrogen and helium. Inside that is a layer in which hydrogen has been converted to helium. Closer to the center of the star are layers of helium and carbon; carbon and oxygen; oxygen, neon, and magnesium; and silicon and sulfur. At the center of the core is a region in which the fusion of silicon and sulfur has produced iron and nickel.

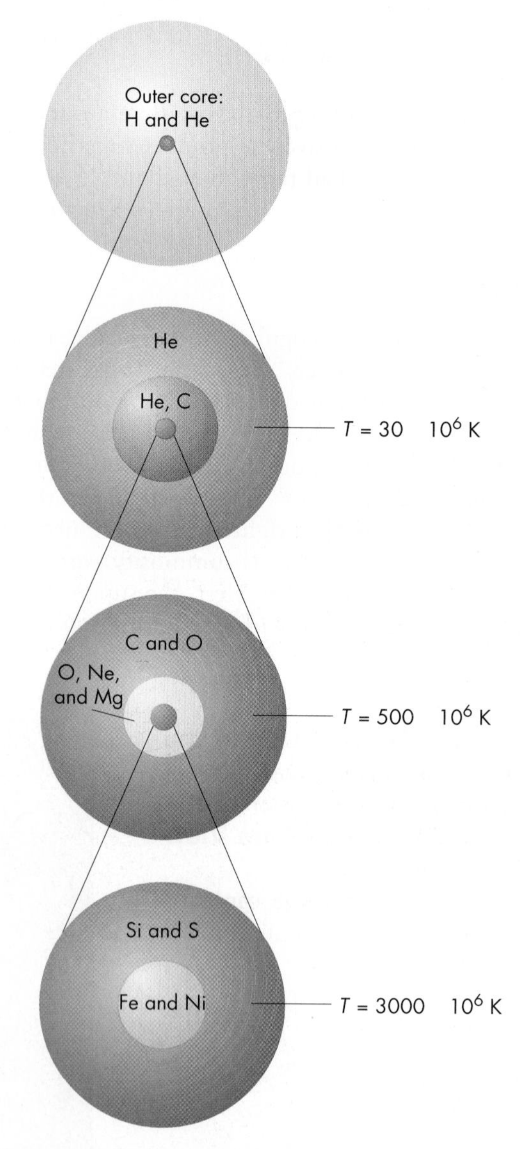

FIGURE 20.8
Steps in the Explosion of a Supernova
A, In step 1, the core of the star collapses. **B,** In step 2, the neutron-rich core rebounds and pushes infalling material outward in a shock wave. **C,** In step 3, the shock wave moves outward through the star.

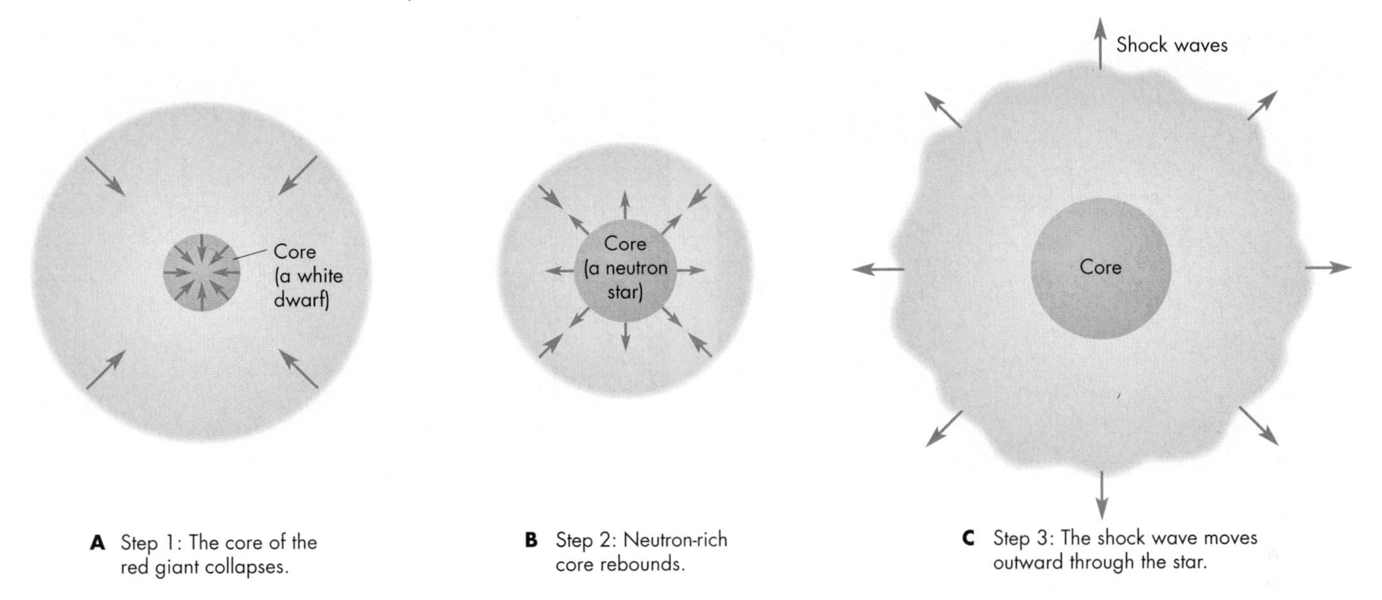

A Step 1: The core of the red giant collapses.

B Step 2: Neutron-rich core rebounds.

C Step 3: The shock wave moves outward through the star.

outward by neutrinos and escapes from the star in a few seconds. The shock front reaches the surface of the star in a few hours. Most of the matter in the star is pushed outward by the shock and is expelled from the star at a few percent of the speed of light. The ejected material carries away a small percentage of the energy of the explosion.

The visible light emitted during the explosion amounts to much less than 1% of the total energy released. At first, the brightness of the supernova increases rapidly as the shock wave reaches the surface of the star and causes the size of the star to expand. As it expands, the surface cools, and the brightness of the supernova begins to drop. The graph in Figure 20.9 shows how the brightness of a typical type II supernova varies with time. After several months, the main source of the supernova's light is the radioactive decay of nickel and cobalt nuclei produced in the explosion. The energy from these radioactive decays keeps the supernova luminous for years.

A type II supernova occurs when the degenerate core of a massive star collapses under its own weight. As the collapse proceeds, electrons and protons combine to form neutrons, producing a neutron star in the core of the star. Infalling material rebounds from the neutron star and is driven outward at great speed. When the resulting shock wave reaches the surface, it expands the star, greatly increasing its brightness.

Supernova 1987A On February 24, 1987, astronomers at several observatories in the southern hemisphere discovered the first supernova in almost 400 years that could be seen without the aid of a telescope. The supernova, a type II event that was designated SN 1987A, didn't occur in our galaxy, however. The explosion took place in the Large Magellanic Cloud, one of the very nearest galaxies beyond our own. Figure 20.10 shows part of the Large Magellanic Cloud before and during the supernova outburst. SN 1987A was bright enough that it could be studied using a variety of techniques in many parts of the spectrum. It is certainly one of the most intensely studied

FIGURE 20.9
The Brightness of a Typical Type II Supernova Before and After the Supernova Reaches Maximum Brightness
After 100 days, the supernova is only a few percent as bright as it was at maximum brightness.

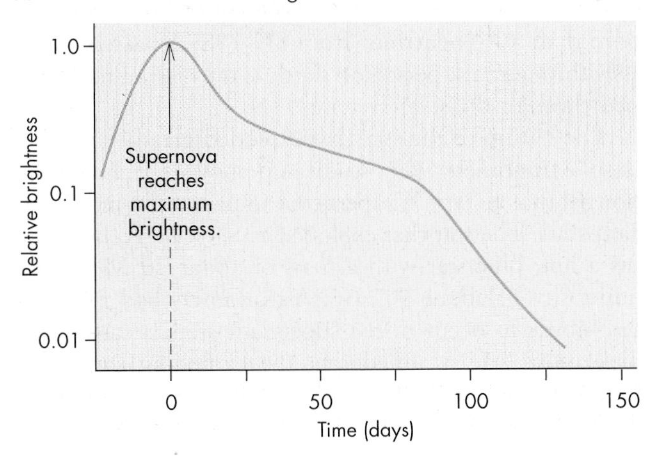

FIGURE 20.10
Supernova SN 1987A
The picture on the left shows a region of the Large Magellanic Cloud before the supernova occurred. The arrow shows the star that became the supernova. The picture on the right shows the same region after the supernova erupted.

astronomical events in history. Perhaps the most significant thing about SN 1987A, however, was that for the first time astronomers were able to find out precisely when the explosion took place and could identify the star that exploded.

Before SN 1987A, all of the supernovae that had been studied with modern observing techniques had taken place in distant galaxies. These galaxies are observed relatively infrequently, so supernovae in them usually are discovered only well after the original brightening has taken place. The time of the explosion is uncertain by hours, days, or weeks. In the case of SN 1987A, however, the explosion was detected even before the first visual observations of the brightening were made. On February 23, 1987, almost a full day before the visual detection of SN 1987A, bursts of neutrinos were observed at neutrino detectors in Europe, Japan, and the United States. The neutrinos left the core of the exploding star within a few seconds of the collapse of the core and hours before the shock wave of the explosion reached the surface. Only 19 neutrinos were detected (because neutrinos pass through matter very easily), but these 19 detections meant that a total of 10^{58} neutrinos were emitted in the explosion. The Earth is 50,000 pc (150,000 light years) from the explosion, yet more than 10^{14} neutrinos from SN 1987A passed harmlessly through each person on Earth as the blast of neutrinos sped through the solar system.

The nature of the star that exploded greatly surprised most astronomers who study supernovae. It had been thought that all type II supernovae took place in red supergiant stars. The star that exploded as SN 1987A, however, was a hot, blue star with a mass of about 20 $M_\odot$ and a luminosity of about 10^4 $L_\odot$. Astronomers had expected supernovae to occur in red supergiant stars because their calculations had assumed that the exploding star has a chemical composition similar to that of the Sun. The Large Magellanic Cloud, the site of SN 1987A, however, is deficient in elements heavier than helium. Calculations that have been carried out since SN 1987A show that a star of about 20 $M_\odot$ that is deficient in elements heavier than hydrogen becomes a red supergiant and then, shortly before it explodes, evolves leftward across the H-R diagram to become a hot star again. This means that the hot star that exploded as SN 1987A was a red supergiant until only about 30,000 years ago. Evidence in support of this scenario is a shell of gas moving outward at about 20 km/s a parsec or so from the explosion. This is probably the material lost as a cool wind by SN 1987A when it was a red supergiant. When the light from the supernova reached the shell of gas ejected during the red supergiant stage, the shell became visible, as shown in the 1994 *Hubble Space Telescope* image of the supernova in Figure 20.11. The inner, bright ring is due to ejected material, whereas the thinner, outer rings are probably related to the ejection of matter when the star was a red supergiant.

SN 1987A was a very important event for astronomers because it confirmed existing ideas about core collapse as the cause of type II supernovae. It also taught astronomers that there are many variations on the basic supernova theme.

Supernova 1987A, in the Large Magellanic Cloud, was the first supernova visible without a telescope in almost 400 years. Neutrinos from the supernova were detected, pinpointing the moment of the collapse of the core of the supernova and confirming the core collapse model of type II supernovae.

FIGURE 20.11
Rings Surrounding SN 1987A
The inner, bright ring is a shell of gas and dust that escaped from the star that became SN 1987A while it was a red supergiant. When light from the supernova reached the shell, the shell became bright. The outer rings may be related to the shell of mass lost from the red supergiant.

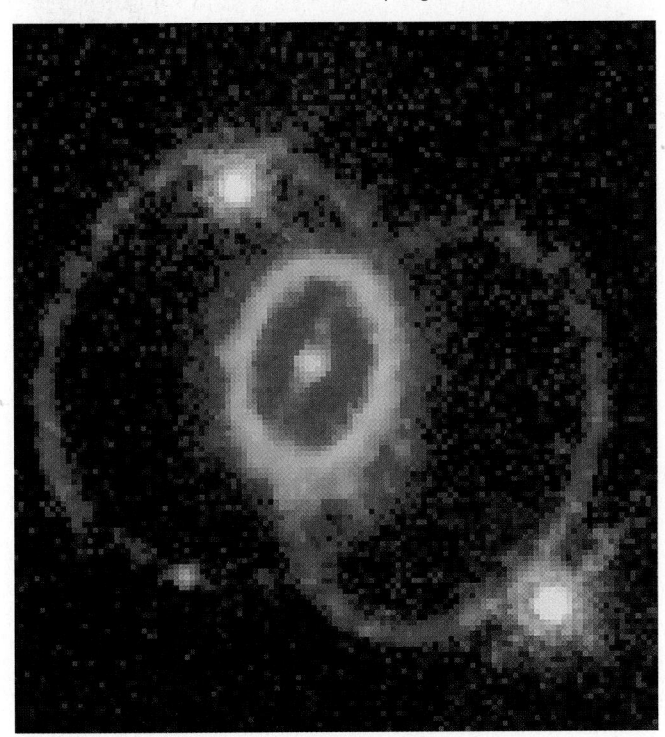

Gamma Ray Bursts

Astronomers have observed brief but intense bursts of gamma rays since the late 1960s. Until recently, however, the origin of those bursts was mysterious and controversial. Now it appears that most **gamma ray bursts** arise from the collapses of the cores of massive stars and are closely related to supernovae. The first observations of gamma ray bursts were made by Vela satellites that were put in orbit to monitor violations of the Nuclear Test Ban Treaty. Since then, thousands of gamma ray bursts have been detected. Gamma ray bursts last from a fraction of a second to a few minutes. The bursts come from random directions in the sky, showing no tendency to be located near the plane of the Milky Way. When astronomers realized that the bursts come from all directions, they also realized that the sources of gamma ray bursts weren't confined to the disk of the Milky Way, and might be coming from extremely remote places in the universe.

The question of the distances of gamma ray bursts was answered when astronomers were able to locate faint visible light and radio sources at points in the sky where gamma ray bursts had taken place. These visible and radio "afterglows," detected hours or days after the gamma ray bursts, lie in the directions of faint, distant galaxies. The great distances of the gamma ray bursts mean that they appear to be the most violent events in the universe. If a gamma ray burst releases energy uniformly in all directions, then in a few seconds it emits an amount of energy equal to tens to thousands of supernovae.

Astronomers were eager to use optical and radio telescopes to observe gamma ray bursts while they happen rather than after the gamma rays have faded. This was difficult because the gamma rays telescopes used to detect gamma ray bursts usually yielded positions only a day or so after the burst. To circumvent that difficulty, astronomers intercepted data from the Burst and Transient Source Experiment (BATSE) on the *Compton Gamma Ray Observatory*. Within seconds, they could calculate the position of the gamma ray burst and circulate the position over the Internet to observers all over the world.

On January 23, 1999, the position of a gamma ray burst was sent out. Among the recipients was a small robotic telescope in New Mexico. Only 22 seconds after the beginning of the gamma ray burst, the robotic telescope obtained an image of the part of the sky around the position of the burst. The image showed a new point of light at the position of the burst. In the next 25 seconds the point of light increased in brightness until it became about one-tenth as bright as the faintest stars visible with the naked eye. It could have been seen easily with a small pair of binoculars. It then began to fade and was 100 times fainter after 10 minutes. Four hours after the burst it was almost 100 times fainter still.

The next day, astronomers using the Keck Telescope on Mauna Kea obtained a spectrum of the fading point of light. They found that the source of the light lies at a distance of 10 billion light years. Given such a great distance, the amount of energy released is enormous. If the burst was emitted uniformly in all directions, it had about 1000 times the total energy the Sun will release in its entire lifetime. This is equivalent to converting more than one solar mass of matter into energy with perfect efficiency in a few seconds. The small portion of the energy that emerged as visible light was also enormous. If the burst had taken place in the Milky Way at a distance of 1000 pc, it would have been as bright as the Sun and turned night into day. Observations with the *Hubble Space Telescope* showed that the gamma ray burst took place in the faint galaxy shown in Figure 20.12.

The brightest gamma ray burst observed so far was seen in March 2008. This event reached naked-eye brightness for a few seconds. At a distance of 1000 pc it would have been 10 times as bright as the Sun. Its gamma rays would have seriously damaged Earth's atmosphere.

The mechanism that produces gamma ray bursts is still not completely understood, but it now appears that most gamma ray bursts occur when the core of a massive star suddenly collapses. This is also what happens in a type II supernova. In the case of a rapidly rotating, very massive star, however, a black hole rather than a neutron star is produced. The rotation of the matter surrounding the black hole produces a dense, extremely hot disk of matter deep

FIGURE 20.12
The Host Galaxy of a Powerful Gamma Ray Burst
This *Hubble Space Telescope* image shows the gamma ray burst as a bright point of light. The faint glow around the gamma ray burst appears to be an irregular galaxy. The image was obtained three days after the gamma ray burst was detected.

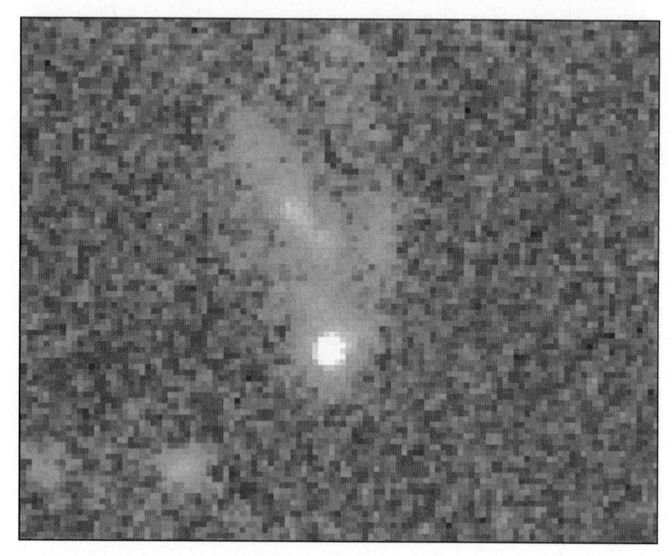

inside the star. Because of the disk, energetic particles near the black hole can't escape in all directions but instead are confined to narrow jets along the polar axis of the star. The jets travel at over 99.99% of the speed of light and quickly blast through to the surface of the star. At about the same time, the star explodes as a supernova. The gamma rays that we later detect are produced by collisions among energetic particles within the jets.

This scenario for gamma ray bursts has several important consequences. First, gamma ray bursts should be accompanied by supernovae. This has been difficult to check because most gamma ray bursts originate in extremely distant galaxies and the supernovae that accompany them are very faint. In 2003, however, an extremely bright, relatively nearby gamma ray burst was detected in the constellation Leo. About a week after the gamma ray burst was detected, astronomers observing at visible wavelengths found a brightening that proved to be a supernova. The supernova and the gamma ray burst lay in precisely the same direction. Second, the gamma rays are emitted in narrow beams, perhaps only a few degrees wide, rather than in all directions. This means that less total energy is involved than if they emitted uniformly in all directions. In fact, the total energy output of a gamma ray burst is about the same as a supernova. Finally, the small angle into which the gamma rays are emitted means that the great majority of the bursts aren't pointed at the Earth. For every one we observe, there are many hundreds that are aimed elsewhere and are invisible to us. There may be at least 1500 gamma ray bursts produced every day somewhere in the universe.

The collapse of the core of a massive star seems like a good explanation for some gamma ray bursts. Gamma ray bursts, however, come in two varieties, which differ in the length of time they can be detected. Core collapse explains the long duration bursts. Something else, perhaps the merger of a neutron star and a black hole or the merger of two black holes, may be required to explain the rest. This possibility is described in Chapter 21.

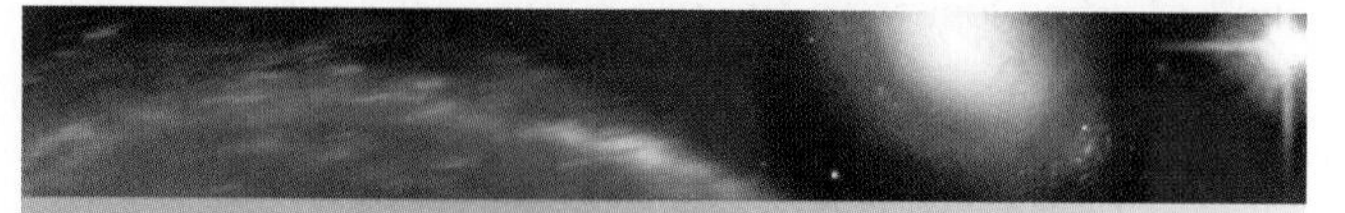

Gamma ray bursts are blasts of gamma rays that originate in enormous explosions in distant galaxies. Most gamma ray bursts are thought to originate in the collapse of the core of a massive star that also produces a supernova.

Supernova Remnants

The outer layers of a supernova, blasted outward at speeds as great as 20,000 km/s, carry a tremendous amount of energy into the surrounding interstellar gas and dust. The result is a luminous nebula called a **supernova remnant.**

Radiation from the Supernova Remnant As the outer layers of a supernova expand, they form a shell that sweeps up the interstellar gas it encounters. This slows down the shell and heats the swept-up gas to millions of kelvins. Because the swept-up material is densest just behind the shock front where the outward-moving gas strikes the interstellar material, the gas in the supernova remnant is concentrated in a thin shell. Most of the ultraviolet and visible radiation comes from relatively dense, cool pockets where the gas has been compressed. Much of the ultraviolet and visible light consists of emission lines from hydrogen atoms and other atoms and ions. Where the gas has not been compressed, it remains hot and radiates intensely in the X-ray part of the spectrum, as shown in the X-ray image of Cas A in Figure 20.13. Bright infrared emission also can be observed. This comes from interstellar dust particles trapped in the swept-up gas. The dust particles are heated by the hot gas and radiate in the infrared.

Many of the 150 or so known supernova remnants in our galaxy were detected using radio telescopes. The radio radiation from supernova remnants is emitted by the extremely energetic electrons produced in the supernova. The electrons are trapped and accelerated by magnetic fields in the supernova remnant. As they spiral around the magnetic field lines, they emit radiation by a process called **synchrotron emission.** For most supernova remnants, synchrotron emission is detectable only in the radio part of the spectrum. Figure 20.14 shows that the radio image of Tycho's supernova remnant, like most supernova remnants, appears as a thin shell.

Evolution of a Supernova Remnant Although the supernova may blast gas outward uniformly in all directions,

FIGURE 20.13
A Chandra X-Ray Observatory Image of the Supernova Remnant, Cas A
The X rays from a supernova remnant come from very hot gas within the expanding shell of gas ejected by the supernova. Cas A is the remains of a star that exploded about 300 years ago.

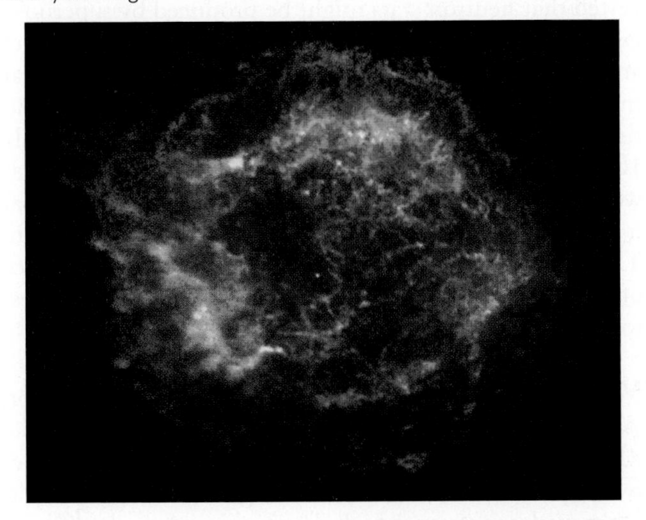

FIGURE 20.14
The Radio Frequency Image of the Remnant of Tycho's Supernova
The bright regions of the image show where the radio radiation is most intense. Most of the radio radiation is produced by synchrotron radiation. Notice that most of the radio radiation comes from a relatively thin shell of swept-up interstellar gas.

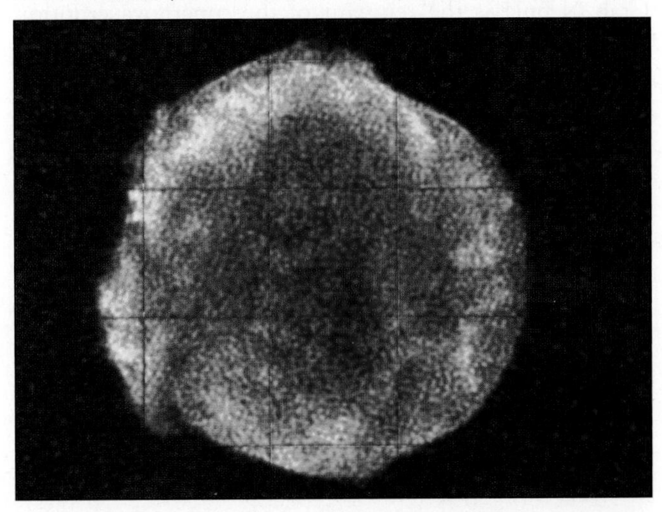

FIGURE 20.15
Part of the Cygnus Loop, an Old Supernova Remnant
The Cygnus Loop is more than 30 pc across and is estimated to be about 15,000 years old. In this false color image blue indicates emission by oxygen atoms. Red shows emission by sulfur, and green shows emission by hydrogen.

a supernova remnant soon develops structure and asymmetry. Part of the structure and asymmetry is due to the uneven distribution of interstellar gas. The remnant expands fastest in directions in which the density of surrounding gas is lowest. When small pockets of dense interstellar gas are swept up in the supernova remnant, they radiate strongly in the visible and ultraviolet and appear as bright spots. As a supernova remnant ages, it appears less round and regular. For example, Figure 20.15 shows part of the Cygnus Loop, an old, extremely irregular supernova remnant.

The expansion of the supernova remnant slows steadily as time passes. Eventually, the expansion velocity of the remnant declines to the point where it is comparable with the speeds of typical interstellar clouds. After this, the remnant merges into the interstellar gas. This only happens, however, tens or hundreds of thousands of years after the supernova explosion. The overall effect of a supernova remnant on its surroundings is to produce a region of low-density gas by sweeping out most of the original interstellar gas. The hot, low-density gas around an old supernova emits little radiation to cool itself, so it can remain hot for a long time. Another effect of a supernova remnant is to stir up the interstellar gas and increase the speeds of local interstellar clouds. Finally, supernova remnants carry the heavy elements produced in the core of the exploded star into space. These elements are mixed into the interstellar gas and enrich future generations of stars.

Not all supernova remnants have a shell-like appearance. A few have filled centers. The most conspicuous of these is the Crab Nebula, shown in Figure 20.16. The Crab Nebula is the remnant of the supernova eruption seen in 1054. It is thought that supernova remnants with filled centers differ from other supernova remnants in that they contain pulsars.

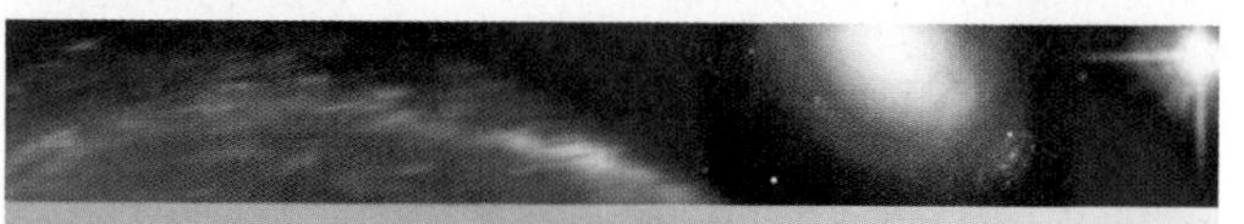

The shell of gas ejected by a supernova sweeps up and heats interstellar gas around the supernova, producing a luminous supernova remnant. Energetic electrons trapped in the magnetic field of the supernova remnant emit synchrotron radiation, which can be detected using radio telescopes. The supernova remnant merges with the interstellar gas after tens or hundreds of thousands of years.

FIGURE 20.16
The Crab Nebula
The Crab Nebula is an example of a supernova remnant with a filled center. The radiation from the filamentary structure in the nebula is due to emission lines from hot gas. The remainder of the light is produced by synchrotron radiation. The blue color of the synchrotron radiation is exaggerated. The position of the Crab Nebula pulsar is shown by the arrow.

Properties of Neutron Stars

Neutron stars

The idea of neutron stars has been around for more than 75 years. In 1933, Fritz Zwicky and Walter Baade first suggested that neutron stars might be produced by supernova explosions. A few years later, in 1939, Robert Oppenheimer and George Volkoff calculated the properties of a star made entirely of a gas of degenerate neutrons. They found that a neutron star of 0.7 $M_\odot$ would have a radius of only 10 km. Even if the surface temperature of this kind of star were 50,000 K, it would have such a small surface area that it would be about a million times fainter than the Sun. This would be much too faint to observe at the distance of the nearest stars, so neutron stars were of little interest to astronomers for almost 30 years.

Masses and Radii The exact structure of a neutron star is still not completely known. This is because, unlike white dwarfs, the equation of state (the relationship between pressure and density) for degenerate neutrons and other nuclear matter at high densities is extremely difficult to calculate. Physicists have spent a great deal of effort studying how neutrons (and protons) interact under these conditions, but many important questions remain unanswered. These uncertainties make it impossible at present to be completely certain of the structure of neutron stars.

Despite these difficulties, many models for neutron stars have been proposed. The results of various calculations differ in detail but share many important features. First, the sizes of neutron stars, like white dwarfs, decrease as mass increases. Figure 20.17 shows the mass-radius relationship for one possible equation of state. Notice that the radii of neutron stars fall in the range of 10 to 15 km for most masses. Thus, a neutron star is about as big as a major city such as Boston, Washington, D.C., or Dallas. The maximum mass that a neutron star can have depends on its equation of state. All of the calculations that astronomers have made, however, fall in the range of 1.5 to 2.7 $M_\odot$. Thus a neutron star can be more massive than a white dwarf, but a very massive star cannot become a neutron star without losing most of its mass first.

Rotation and Magnetic Fields Two things that can be safely predicted about newly formed neutron stars are that they rotate very rapidly and that they have enormous magnetic fields. A neutron star rotates rapidly because rotational angular momentum was conserved as it shrank. If the Sun, with its rotation period of 30 days, shrank to the size of a neutron star, its rotation period would have to decrease to about 10^{-3} seconds (1 millisecond, ms) for angular momentum to be conserved. It is reasonable to expect that neutron stars rotate as rapidly as hundreds or thousands of times per second.

FIGURE 20.17
The Mass-Radius Relationship for Neutron Stars
The downward slope of the line indicates that radius decreases as mass increases. Most neutron stars have radii that are between 10 and 15 km.

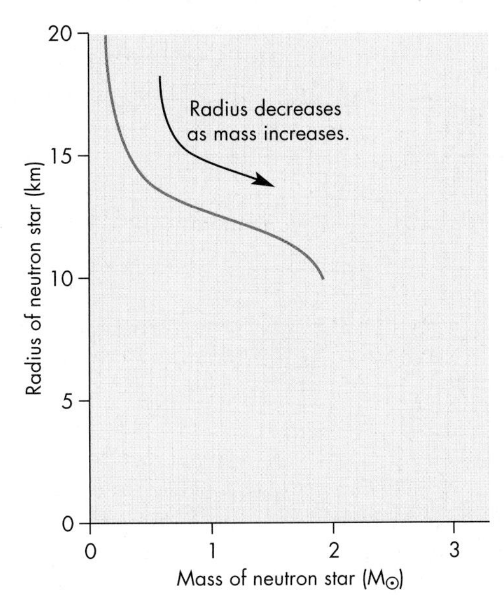

The magnetic field strength of a star also increases as its diameter decreases. If the Sun collapsed to become a neutron star, its magnetic field strength would be amplified by about 10^{10}. Thus, neutron stars should have magnetic field strengths as great as 10^8 tesla. This is much larger than the magnetic fields of ordinary stars, which usually are less than 1 tesla.

Neutron stars, which are supported by the pressure of degenerate neutrons, are about 10 to 15 km in radius. Like white dwarfs, more massive neutron stars are smaller than less massive ones. Conservation of angular momentum makes neutron stars rotate very rapidly. Large magnetic fields develop during the formation of a neutron star.

Pulsars

Discovery **Pulsars** are neutron stars from which extremely regular bursts of radio radiation are detected. They were discovered by accident in 1967 during an attempt to detect the twinkling of distant sources of radio radiation. Just as stars twinkle when we observe them through the Earth's atmosphere, radio-emitting objects with small angular sizes twinkle when observed through the solar wind. The explanation for the two effects is essentially the same. For stars, the wind blows density irregularities in the atmosphere past the observer. Light is refracted by the density irregularities, causing "seeing" (discussed in Chapter 6). For distant radio sources, the solar wind blows pockets of ionized gas past the telescope, causing the radio sources to flicker on a timescale of tenths of seconds. Antony Hewish, a British radio astronomer, designed and built a telescope to measure this flickering so that the angular sizes of the radio sources could be found. The telescope was unusual in that it recorded the strength of the signal at least ten times per second. At that time, most radio astronomical measurements consisted of averages of signal strength over many seconds or minutes. Only by making very frequent measurements could pulsars have been detected.

In August 1967, Jocelyn Bell, a research student who worked with Hewish, found a source of radio emission that flickered even when it was observed in the middle of the night, when flickering due to the solar wind was usually minimal. On November 28, a high-speed recorder was used to observe the object. The recording, reproduced in Figure 20.18, showed extremely evenly spaced pulses of radiation. The period of the pulses was 1.337 seconds and was constant to an accuracy of one part in 10 million. The object, designated CP 1919 because its right ascension was $19^h\,19^m$, was the first pulsar to be discovered. Within a year, more than two dozen were found. Today, we know of more than 1000. Although their periods range from 0.9 ms to 8.5 seconds, nearly all of them fall between 0.1 and 2.5 seconds.

What Are Pulsars? Theories to explain the pulses of radio radiation from pulsars sprang up almost as soon as the first pulsar was discovered. One early proposal was that the pulses were attempts at communication by civilizations around distant stars. More plausible explanations were that they were either white dwarfs or neutron stars. A white dwarf or neutron star could produce pulses of radiation if it pulsated like a Cepheid variable. Alternatively, it could rotate like the light in a lighthouse, bringing bright and dark parts of its surface alternately into view.

The idea that pulsars could be either pulsating white dwarfs or neutron stars was rejected because their densities could not account for the range of pulsation periods observed for pulsars. The density of a star and its pulsation period are closely related. The greater the density, the shorter the pulsation period. White dwarfs have pulsation periods of a few seconds. Thus pulsating white dwarfs couldn't account for the pulsars with the shortest periods. Pulsating neutron stars, on the other hand, could account for the fastest-pulsing pulsars but couldn't account for the majority of pulsars, which have periods of tenths of seconds or seconds.

Pulsars can't be rotating white dwarfs, either, because a white dwarf that rotated faster than about once per second

FIGURE 20.18
A Recording of Pulses from the Pulsar CP 1919
The recording shows about 20 seconds of data. The data are shown with energy increasing downward. The pulses appear as downward directed peaks that occur every 1.337 seconds.

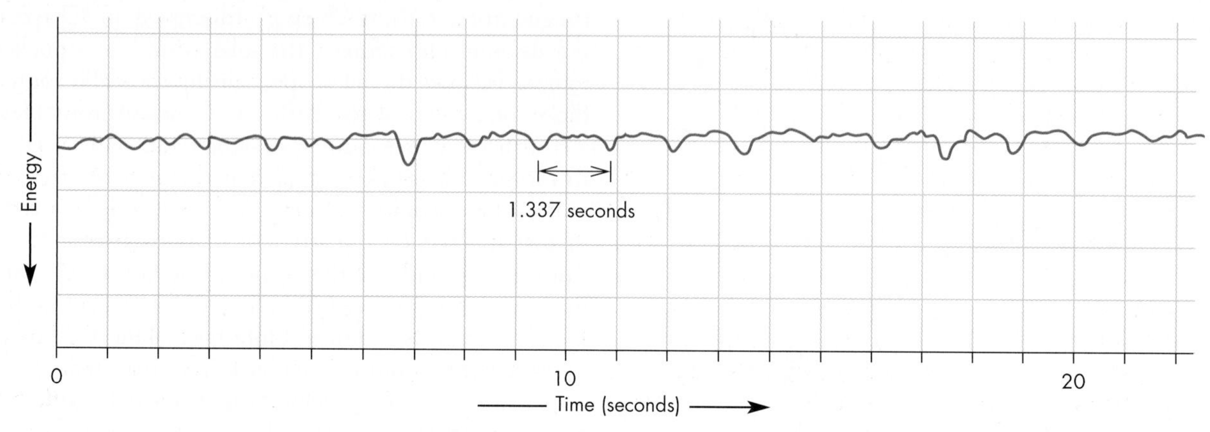

would fly apart. The gas at its equator would have to move faster than escape velocity. A neutron star, on the other hand, could rotate as fast as one thousand times per second without disintegrating. The neutron star also could rotate more slowly, of course, so rotating neutron stars can account for the entire range of observed periods of pulsars. When this was realized, neutron stars, neglected for 30 years, became a subject of intense interest to many astronomers.

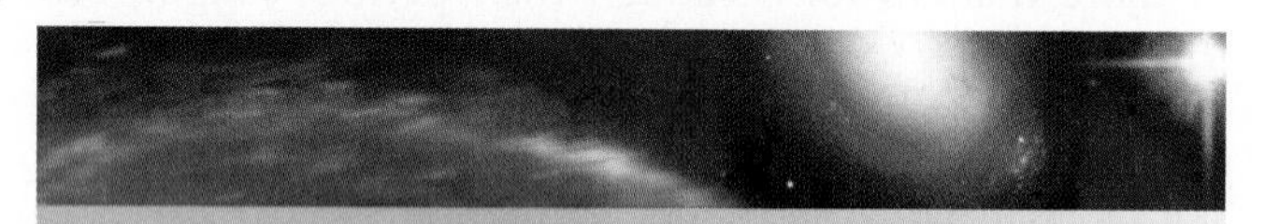

Pulsars were discovered fortuitously in 1967. They emit very regularly spaced bursts of radio radiation as often as one thousand times per second. Most pulsars have periods of tenths of seconds or seconds. Pulsars are rotating neutron stars.

What Makes Pulsars Shine? Many models have been proposed to explain the pulsed emission from pulsars. The most widely accepted of these is shown in Figure 20.19. In this model, the magnetic axis of the neutron star is tipped with respect to the rotation axis. Extremely energetic charged particles traveling along magnetic field lines near the magnetic poles produce emission that is beamed outward along the direction of the field lines. As the neutron star rotates, the beamed radiation sweeps across the Earth, producing a pulse with each rotation (or two pulses per rotation if the beams from both poles sweep past Earth).

Although this general picture is probably correct, there are many unanswered questions about the exact way in which the energetic particles produce the observed radiation. One clue about the emission process can be obtained

FIGURE 20.19
A Model of a Pulsar
The magnetic axis of the pulsar is inclined with respect to the rotation axis. Rapidly moving electrons in the regions near the magnetic poles emit radiation in a beam pointed outward. When the beam sweeps past the Earth, a pulse is seen.

A model of a pulsar

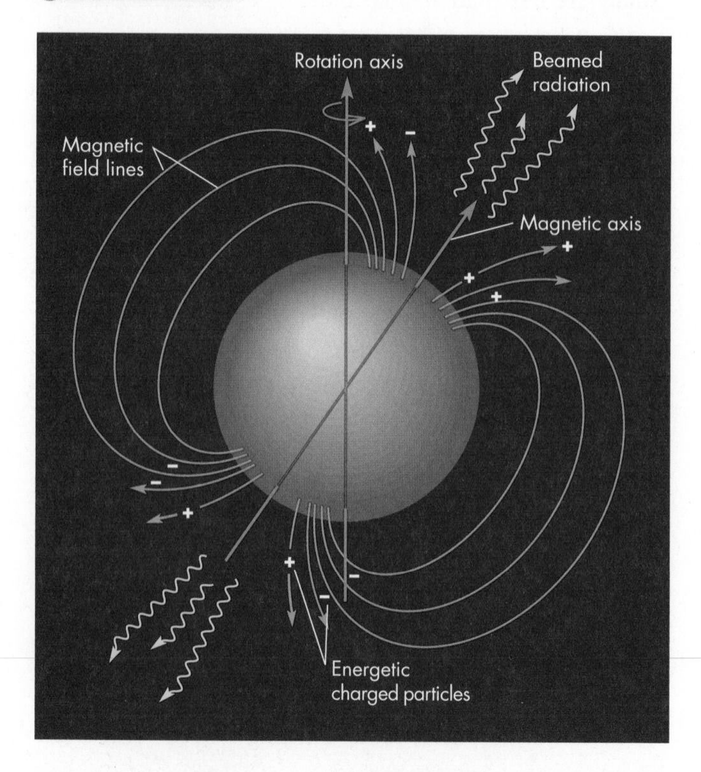

by looking closely at many successive pulses from a single pulsar. Figure 20.20 shows, for example, that there can be a tremendous variation in the pulse shape from pulse to pulse. Brief, intense micropulses often are seen. In addition, there are periods of time when the pulses completely disappear. This suggests that the radiation is emitted by bunches or clouds of energetic charged particles rather than a steady stream of particles.

Evolution of Pulsars One consequence of the pulsar model shown in Figure 20.19 is that the pulsar must spin more slowly as time passes. Periodic electromagnetic disturbances (electromagnetic radiation) can be produced by moving a magnet back and forth. This is exactly what happens as the magnetic axis of a neutron star whirls around the star's rotation axis. The frequency of the radiation is the same as the rotational frequency of the pulsar—that is, 1 Hz for a pulsar with a rotation period of 1 second and 1000 Hz for a pulsar with a period of 0.001 seconds. The energy carried away by the radiation is taken from the rotational energy of the neutron star and makes it spin more slowly. For a pulsar that has a period of about 1 second, the rate at which the rotational period increases is about 10^{-15} seconds per rotation. At this rate, the period of the pulsar would increase to 2 seconds in about 10^{15} seconds, or 30 million years.

If the mechanism by which pulses are produced continued to work throughout the evolution of the pulsar, then we should see pulsars with periods of many seconds or even minutes. In fact, we see no pulsars with periods even as long as 10 seconds. The pulse-producing mechanism must turn off after the period of the pulsar increases to more than a few seconds, so we see neutron stars as pulsars only during the first several tens of millions of years after the supernova explosions that formed them in the first place. There are, therefore, many extinct pulsars for each active pulsar that we observe.

FIGURE 20.20
One Hundred Consecutive Pulses from CP 1133+16
Individual pulses show a wide range of shapes. There are very brief, intense micropulses and periods of time when no pulses at all are seen.

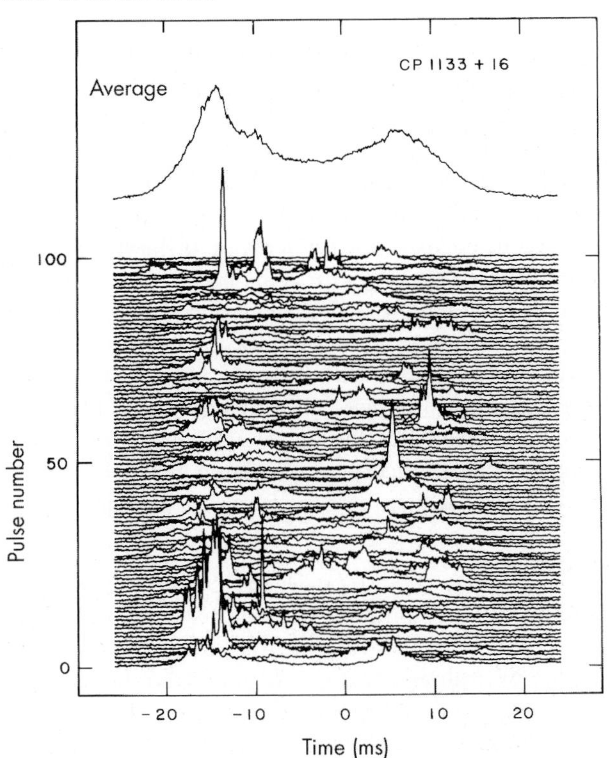

Magnetars **Magnetars** are neutron stars with magnetic fields tens to millions of times larger than those of radio-emitting pulsars. Because of its enormous magnetic field, a magnetar occasionally experiences a "starquake" in its rigid outer crust. This happens when the shifting magnetic field causes the crust to buckle and crack. The energy released during a starquake produces a powerful blast of gamma rays. The gamma rays are observed in pulses emitted every time the part of the magnetar that experienced the quake rotates into view.

Three times in the past 30 years, magnetars have produced gamma ray flares that have been among the brightest astronomical events ever observed from Earth. The brightest of these events occurred on December 27, 2004, when a blast from the magnetar SGR 1806-20 swept through the solar system. The peak of the flare lasted for 0.2 seconds. During this brief moment, SGR 1806-20 was ten times as bright as all the stars in the Milky Way galaxy combined. SGR 1806-20 released more energy during the flare than the Sun produces in 250,000 years. Although SGR 1806-20 is about 15,000 parsecs from the Earth, at the peak of the flare it was brighter than all astronomical objects in history except for the Sun and a few spectacular comets. For 0.2 seconds, the Earth received more energy from SGR 1806-20 than from the full Moon. X rays from SGR 1806-20 ripped apart atoms in the Earth's atmosphere and significantly increased the numbers of atoms and ions in the ionosphere. However, the Earth's lower atmosphere shielded us from the blast.

The flare from SGR 1806-20 may have been produced during a rearrangement of the magnetic field of the magnetar. If this is the case, the rearrangement caused the crust to break in many places all over the magnetar. This caused an explosion of energetic particles and gamma rays. As the energetic particles cooled, they emitted X rays as well as gamma rays. The flare from SGR 1806-20 was so luminous that it could have been detected if it had occurred in a galaxy tens of millions of parsecs away. Thus, magnetars may explain some of the short-duration gamma ray bursts.

The burst of gamma rays from SGR 1806-20 showed evenly spaced pulses that allowed astronomers to measure the rotation period of the magnetar. The rotation period was 7.56 seconds. Similar pulses were observed during a giant flare emitted by another magnetar, SGR 1900+14, in

1998. A comparison of the period of the pulses during the giant flare with the period measured during an earlier, smaller flare from 1900+14 showed that the magnetar is slowing down at the rate of about one second every 300 years, much faster than pulsars slow down.

This means that the magnetars we see today are very young and were produced in relatively recent supernova explosions. In fact, three of the four known magnetars are so young they are still located in or near the remnants of the supernova explosions that produced them. Because magnetars slow down their spins and fade from view so quickly, there may be many old, inactive magnetars in the galaxy, perhaps as many as 10 to 100 million.

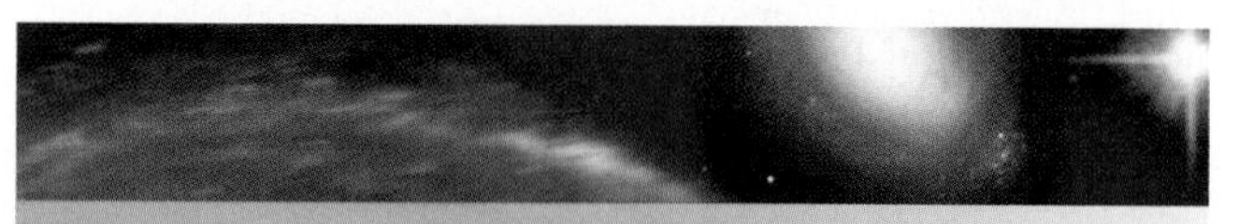

The radio radiation from pulsars is probably produced by energetic electrons near the magnetic poles, which are tipped with respect to the rotational axis. Pulsars lose rotational energy and slow down as time passes. After tens of millions of years, the period of rotation of a pulsar reaches several seconds, after which it stops emitting radio pulses. Magnetars are highly magnetized neutron stars that emit bursts of gamma rays.

Pulsars and Supernova Remnants Just as radio waves are reflected from the ionized gas in the Earth's ionosphere, the low-frequency radiation produced by a pulsar is reflected by the plasma that surrounds the pulsar. The plasma from which the radiation is reflected is pushed strongly outward and carries energy outward toward the supernova remnant around the pulsar. Electrons are accelerated to very high speeds where the radiation reflects from the plasma. As these electrons spiral around magnetic field lines, they produce synchrotron radiation throughout the region inside the expanding supernova remnant. Thus, a supernova remnant that has a pulsar at its center has a filled-in appearance (as in Figure 20.16) rather than a shell-like appearance (Figure 20.14). The idea that supernova remnants with filled centers are powered by rotational energy taken from the pulsars at their centers is strongly supported by the observed slowing of the rotation of the Crab Nebula's pulsar. The observed slowing reduces the neutron star's rotational energy at a rate of 6×10^{31} W, almost identical to the 5×10^{31} W emitted by the Crab Nebula.

Only about a dozen pulsars can be identified with supernova remnants. This may seem surprising because pulsars, rotating neutron stars, are thought to be born in supernova explosions. Why aren't more pulsars found within the expanding shells produced by supernovae? One reason is that there may be neutron stars that are oriented so that their beams don't sweep past the Earth. In that case, we can't detect them as pulsars. A more important reason is that there is a striking difference in the lengths of time that supernova remnants and pulsars can remain visible. Recall that a supernova remnant merges into the interstellar gas and can no longer be recognized about 100,000 years after the supernova explosion. A pulsar, on the other hand, remains visible until its period increases to several seconds, which takes tens of millions of years. Thus, pulsars "live" one hundred times longer than supernova remnants. Only the youngest 1% of pulsars should be accompanied by supernova remnants. For all the rest, the remnant has vanished.

The rotational energy of a pulsar is carried away by low-frequency electromagnetic radiation, which accelerates electrons near the pulsar. The synchrotron radiation from the energetic electrons is emitted in the region inside the expanding shell of the supernova remnant. The Crab Nebula is an example of a supernova remnant with a filled center.

20.3 BLACK HOLES

If a star has a mass smaller than 1.4 $M_\odot$, its contraction is halted by the pressure of degenerate electrons. After these stars contract, they become white dwarfs. Somewhat more massive stars collapse until their neutrons become degenerate and they become neutron stars. What halts the collapse of a star too massive to become a neutron star? The answer is that nothing stops the collapse. The star (or the core of a star) collapses until it forms a black hole and disappears.

The idea that stars could produce black holes was first proposed in the late 1700s by John Mitchell. Mitchell noted that an object can escape from the surface of a star only if the object has a speed greater than or equal to escape velocity (Equation 5.6). As the radius of the star decreases, its escape velocity increases. If the star gets small enough, the escape velocity at its surface becomes equal to the speed of light. Mitchell reasoned that when this happened, not even light could escape from the star.

Mitchell's thinking about black holes was based on Newton's theory of gravity. In reality, however, black holes can be understood only through another theory of gravity—Einstein's general theory of relativity. Relativity explains gravity as a consequence of the curvature of spacetime.

Spacetime

Spacetime is the combination of three space coordinates and one time coordinate that we must use to locate an event. Whether we realize it or not, most of us use the concept of spacetime every day. Suppose, for example, that you made an appointment with a friend to meet on the third floor of the Jefferson Building (at the corner of Dubuque and

FIGURE 20.21
The Four Coordinates of Spacetime Used to Locate an Event in Space and Time
To meet a friend, you must specify a time (3 P.M.) as well as three spatial coordinates (Dubuque Street, College Street, third floor).

College) at 3 P.M. These four coordinates (Dubuque, College, third floor, 3 P.M.) are shown in Figure 20.21. If any one of them is misunderstood, the meeting won't take place.

It is often useful to describe spacetime by drawing a **spacetime diagram,** such as the one in Figure 20.22. It isn't possible to draw all three spatial coordinates plus the time coordinate on a piece of paper, however, so the spacetime diagram is limited to the time coordinate plus a single space coordinate. A stationary body keeps the same spatial coordinate in a spacetime diagram. Its time coordinate changes steadily, however, so it moves straight upward in the spacetime diagram. A body that moves follows a sloping line in the spacetime diagram because its position in space changes with time. The faster the body moves, the more tilted the line describing it becomes. Lines for slow- and fast-moving bodies are shown in Figure 20.22. There is a limit to how tilted the line describing its motion can become, however, because the body can't move faster than the speed of light. Spacetime diagrams are usually drawn so that light beams follow lines that slope at 45° with respect to the horizontal axis, as shown in Figure 20.22.

Figure 20.23 shows a point, *P*, in spacetime. The lines passing through *P* at angles of 45° represent beams of light. The only part of spacetime that an observer at *P* can

FIGURE 20.22
A Spacetime Diagram
Spacetime diagrams show time along the vertical axis and one spatial coordinate along the horizontal axis. A body that remains stationary follows a vertical track in the spacetime diagram. A moving body follows a sloping line. The faster the body moves, the more steeply sloped the line it follows. Light is usually represented by a 45° line, so any track more tilted than the one for light represents a speed greater than the speed of light, which is impossible.

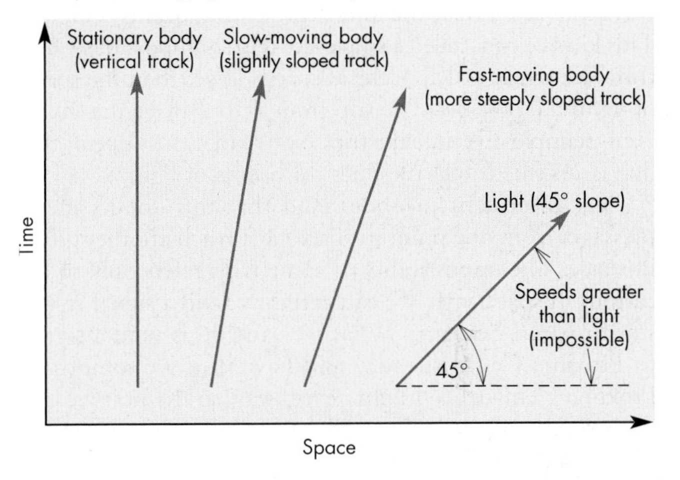

FIGURE 20.23
A Spacetime Diagram Showing Two Different Kinds of Trips
Light travels along 45° lines in the spacetime diagram. The two light beams that pass through point *P* bound the region from which *P* can receive signals and to which signals from *P* may go. Trip *BP* is a possible timelike trip. Trip *AP* is an impossible spacelike trip because it requires a velocity greater than the speed of light.

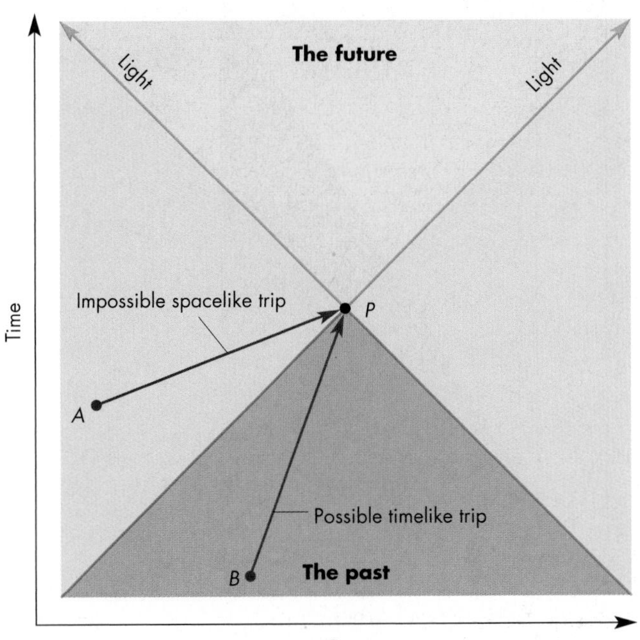

go to or send signals to is the part above P between the light beams (the future). The only part of spacetime that can reach or send signals to the observer at P is the part below P between the lines representing light beams (the past). To see that this is true, imagine that an event took place at location A in spacetime. For a message from A to reach P, that message would have to travel a greater distance in the time interval between A and P than a light beam travels. In other words, it would have to travel faster than the speed of light (notice in Figure 20.23 that the line between A and P is more tilted than the lines representing beams of light). This kind of trip, called a **spacelike trip,** is impossible. Only **timelike trips,** which occur at speeds lower than the speed of light, are possible. The trip from B to P in Figure 20.23 is an example of a timelike trip. Notice that the slope of this line is *less* tilted than the lines for beams of light.

It is important to understand that this doesn't mean messages from one point in space can't reach another point in space. The impossibility of some trips refers only to locations in spacetime. We can certainly send a signal to the star Proxima Centauri. What we can't do is send a signal to Proxima Centauri today and have it arrive tomorrow. Proxima Centauri is 4 light years away, so the best we can do is send a message that will arrive at Proxima Centauri 4 years after it is sent. Figure 20.24 shows some possible and impossible trips in spacetime between the Earth and Proxima Centauri.

FIGURE 20.24
Some Possible and Impossible Trips from the Earth to Proxima Centauri
Both the Earth and Proxima Centauri move vertically in the spacetime diagram. Possible trips are those that are less tilted than the line representing a light beam.

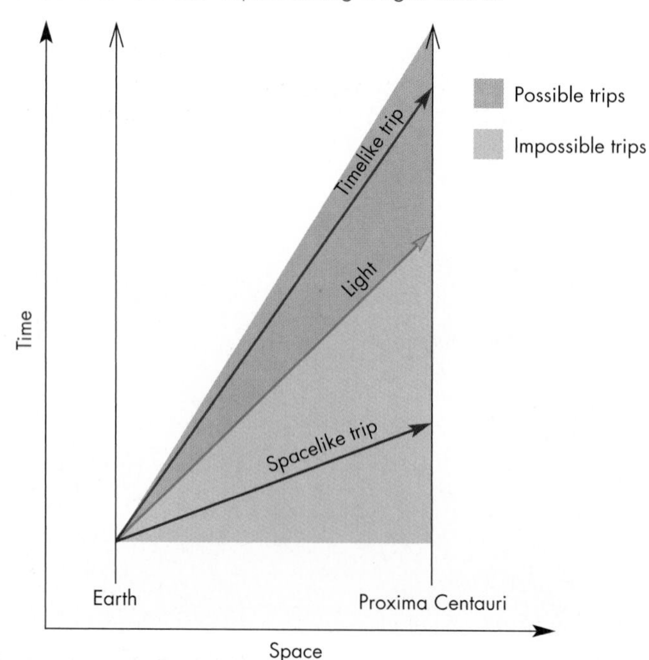

Curved Spacetime We are so accustomed to thinking of space as flat that few, if any, of us can imagine what curved three-dimensional space looks like. Thinking about curved spacetime is even more difficult. It is possible to investigate some of the properties of curved spacetime, however, by comparing two-dimensional spaces—one curved and the other flat—like the plane and the sphere shown in Figure 20.25. The surface of a sphere is an example of two-dimensional curved space. A flat piece of paper is an example of two-dimensional flat space. Each is two dimensional because two coordinates are required to locate a point in that space. In the case of the sphere, the two coordinates might be longitude and latitude. For the piece of paper, the two coordinates might be distance from the left edge of the paper and distance from the bottom.

There are real differences between the geometries of the curved and flat two-dimensional spaces. These differences begin with the difference between **geodesics** in the two spaces. A geodesic is the shortest distance between two points. In spacetime, it is also the path followed by a light beam or a freely moving object. In flat space, a geodesic is a straight line such as the line between A and B in Figure 20.25. In curved space, there are no straight lines. Geodesics are segments of great circles, such as the one between D and E in Figure 20.25. The obvious reason that airplanes follow great circle routes is that they are the shortest routes between points on the Earth's curved surface. In flat space, a triangle is defined by three straight lines, as shown by ABC in Figure 20.26. The sum of the angles of the triangle equals 180°. A triangle in curved space is defined by three great circles, as shown by DEF in Figure 20.26. Only if the triangle is much smaller than the sphere on which it is drawn will the sum of its angles approximate 180°. The larger the triangle, the more the sum of its angles exceeds 180°. In fact, the sum of the angles

FIGURE 20.25
Flat and Curved Two-Dimensional Spaces
In each space, a geodesic is the shortest distance between two points. **A,** In flat space, a geodesic is a straight line (*AB*). **B,** On a sphere, a geodesic is a segment of a great circle (*DE*).

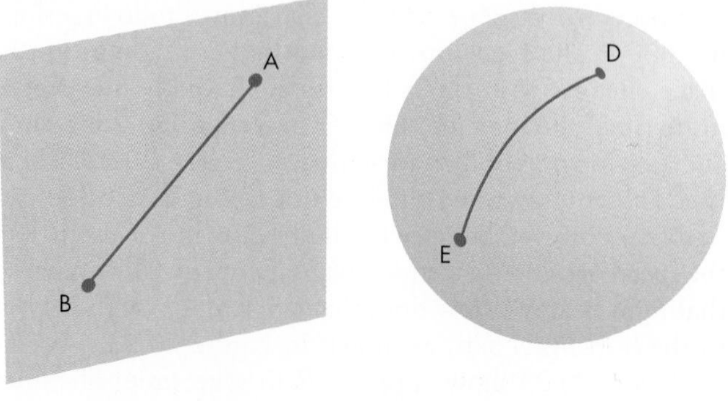

can be as much as 360°. There are many other differences, such as the areas of circles and the surfaces of spheres of a given radius, between the geometries of curved and flat spacetime.

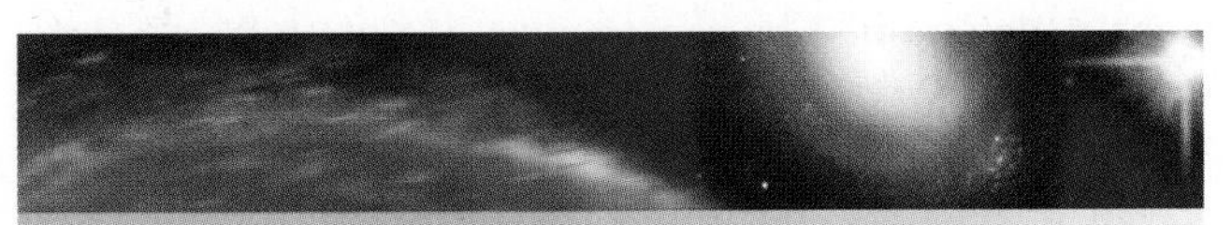

Spacetime is the three spatial coordinates and one time coordinate that are used to locate events in space and time. Spacetime may be curved or flat. A geodesic, the shortest distance between two points, is a straight line in flat spacetime but is a curved path in curved spacetime. Geometric quantities, such as the sum of the angles of a triangle, are different for flat and curved spacetime.

FIGURE 20.26
Triangles in Flat and Curved Space
A, The sum of the angles of a flat space triangle (*ABC*) is 180°. **B,** The sum of the angles of the triangle on the curved surface of the sphere (*DEF*) is more than 180°.

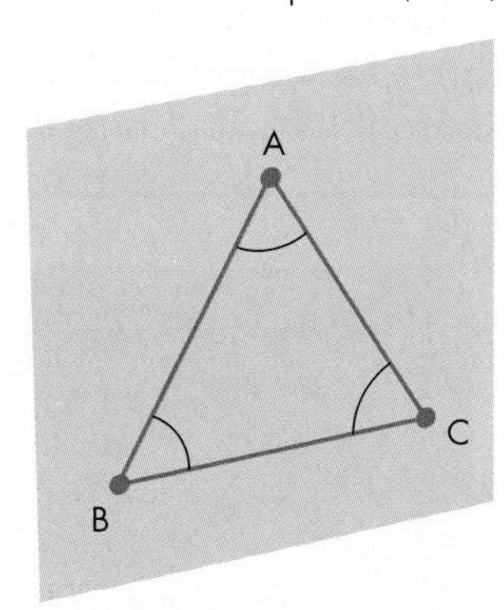

A A triangle in flat space: Angles sum to 180∞

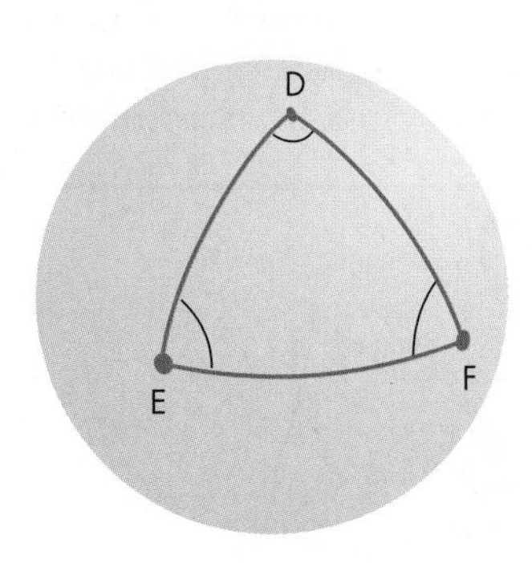

B A triangle in curved space: Angles sum to more than 180∞

What Is Gravity? Einstein's explanation of gravity is based on the idea that matter causes curvature of the spacetime in which it is located. Light and freely moving bodies follow geodesics in curved spacetime. We, however, are unable to see the curvature of spacetime. When we see light and moving bodies depart from straight-line motion, we conclude that a force, gravity, must be acting. In essence, gravity is a nonexistent force that people have invented because they are unable to perceive curved spacetime near massive objects.

Is this really possible? Could our failure to perceive the curvature of spacetime really lead to the invention of gravity? Donald Menzel, Fred Whipple, and Gerard de Vaucouleurs devised an analogy that demonstrates that such a thing could, in fact, happen. Imagine a sailor who plans to sail a triangular course around the island shown in Figure 20.27. The sailor plots a course with three equal sides and with three 60° angles and then sets out. The sailor, of course, has forgotten that the surface of the Earth is a curved space rather than a flat one. Triangles (defined by great circles) have angles that sum to more than 180°. Suppose that for the size triangle the sailor has plotted, the angles actually sum to 183°, or 61° per angle. Not realizing this, the sailor sails from *A* to *B* and turns through 60°, which is too sharp an angle. On reaching the end of the second leg, the sailor again turns through too sharp an angle. Midway through the third leg, the sailor notices that the ship is closer to the island than it is supposed to be.

In searching for an explanation for this strange turn of events, the sailor might invent a nonexistent force. He might argue, for instance, that the boat has metal on it and that a deposit of magnetic ore on the island has pulled the ship toward it. Or, perhaps there is an intense low-pressure region over the island so that the wind blows inward toward the island from all directions. Alternatively, the ocean currents flow toward the island from all directions, carrying the ship along. Each of these explanations invokes a nonexistent force to compensate for the sailor's ignorance of the curvature of the space in which the ship has been sailing.

FIGURE 20.27
A Trip Around an Island
A, In flat space, the trip would consist of three angles of 60° each. **B,** In curved space, however, the sum of the angles of a triangle is more than 180° (let's assume 183°). The correct course should consist of three angles of 61° each. If the sailor ignores the curvature of space, his course will bring the ship closer to the island than he expects. The sailor may invent nonexistent forces to explain the path of his ship.

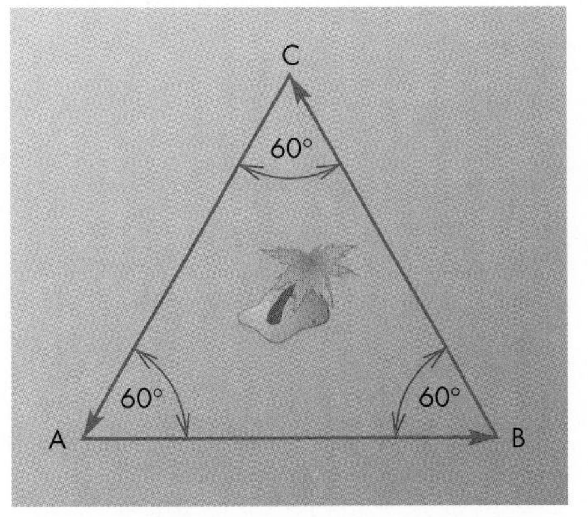

A The trip in flat space

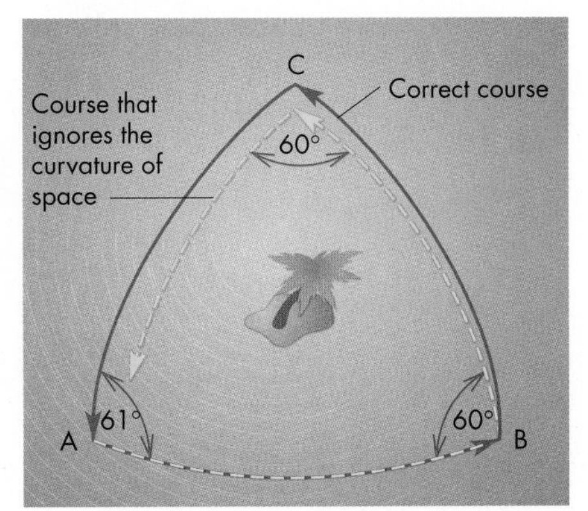

B The trip in curved space

Another way to see the difference between flat and curved space is to consider a stretched sheet of rubber as shown in Figure 20.28. A television camera is positioned over the rubber sheet. The television screen is flat, so it conceals any curvature that the rubber sheet might have. If the rubber sheet is stretched tight, however, there should be no curvature. A marble rolls across the top of the rubber sheet in a straight line at constant speed, indicating that no forces are acting on the marble. Now imagine placing a heavy ball in the middle of the rubber sheet. It pulls the sheet down and produces curvature in the rubber sheet. If a marble is rolled across the sheet, it follows the curvature of the sheet and traces a curved path. People looking at the television picture, however, can't see that the rubber sheet is curved. Instead, they see the marble travel in a curved path past the heavy ball and conclude that there must be an attractive force acting between the ball and the marble. Again, a non-existent force has been invented to explain results that would require no forces if the curvature of space could be perceived.

These analogies have concentrated on the curvature of space produced by the presence of matter. In reality, it is spacetime that is curved. The time coordinate is curved as well. What this means is that time passes more slowly in strongly curved spacetime near a massive object than it does where spacetime is flat far from the object. Clocks run more slowly, and so do biological and atomic processes. Because all of the ways of measuring time slow down in curved spacetime, we don't notice that time is passing more slowly. Only when clocks in different regions of spacetime are compared can the effect be detected.

One way to compare clocks is to send light beams from one part of spacetime to another. The frequency of the light is determined by the amount of time between wave crests. Because time passes at different rates in parts of spacetime that have different curvatures, the interval of time between wave crests and thus the frequency of the wave will be different. Suppose a beam of light is sent from a curved region of spacetime near a massive object to an observer far from the massive object. The distant observer will measure a lower frequency and longer wavelength for the light than the observer near the massive object. The change in wavelength of the light is a **gravitational redshift.** It is not a Doppler shift and has nothing to do with motion. It is a consequence of the curvature of spacetime.

Einstein's description of gravity has two advantages over Newton's gravitational theory. First, it resolves the problem of "force at a distance" (Chapter 5), which had bothered Newton and many other scientists. Newton could not explain how two objects, far from each other, could exert forces on each other. General relativity says that no forces are involved. Bodies cause curvature of spacetime. The curvature grows smaller with distance, but never completely vanishes. Other bodies move in response to that curvature. The second advantage of general relativity is that it gives correct answers for regions of spacetime close to massive objects. General relativity correctly

A

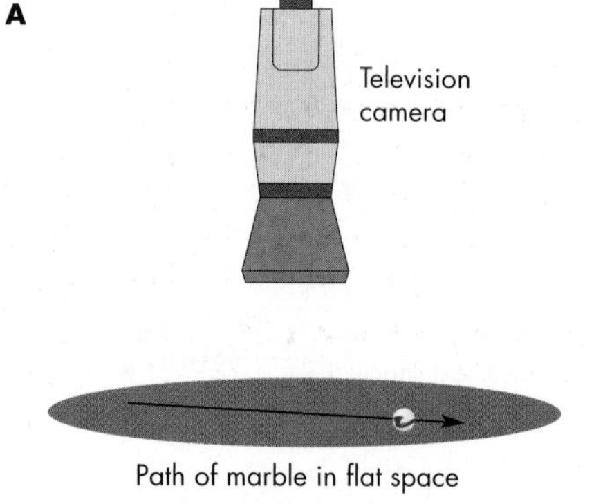

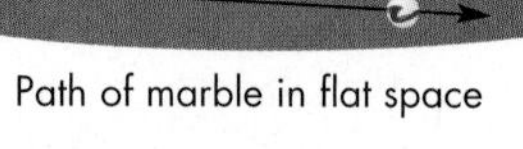

B

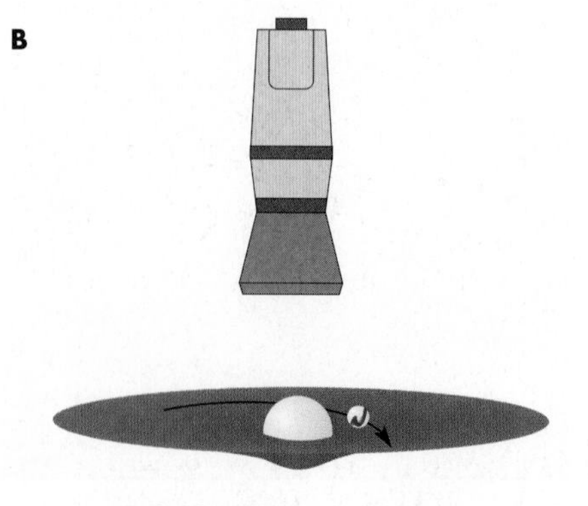

FIGURE 20.28
Paths of a Marble and the Television Image of the Paths for the Cases of Flat and Curved Space
The television image does not show the curvature of space, so someone watching the path of the marble may think that the mass that curves the space is attracting the marble.

accounts for effects, such as the bending of light near the Sun and changes in Mercury's orbit, that simply can't be adequately explained by Newtonian gravity. Nevertheless, in regions of spacetime in which curvature is small, Newton's description of gravity is extremely accurate. Thus, Newton's law of gravity continues to be used to calculate everything from the path of a missile near the Earth to the orbit of a star about the galaxy.

Einstein's explanation of gravity is that matter causes curvature of spacetime. Light and freely moving bodies move along geodesics in curved spacetime. We fail to perceive the curvature of spacetime, however, so we invent a nonexistent force, gravity, to account for the curved motion that we see.

The Formation of a Black Hole

A **black hole** is a region of spacetime from which nothing, not even light, can emerge. A black hole is the result of the collapse of the core of a star when the collapsing core is too massive to become a neutron star. Just how massive a star must be to have its core collapse to form a black hole during a supernova explosion is not well known, but it seems to be about 25 $M_\odot$.

Collapse to the Black Hole The formation of a black hole is something we are never likely to see because it is hidden beneath many solar masses of outrushing stellar material. Let's imagine, however, that we could see through the ejected material. We would see the core of the star, already no larger than a neutron star, continuing to collapse. As the core becomes more compact, its surface gravity increases. Alternatively, the curvature of spacetime near the core increases. Before the curvature gets too great, imagine laser beams shined outward in many directions from a spot on the surface of the core, as shown in Figure 20.29. Because curvature is small, the beams travel outward in nearly straight lines and escape from the core. A moment later, the curvature of spacetime is more severe and beams directed outward at large angles follow paths that curve them back into the core. Soon, only beams pointed almost directly outward can escape. Finally, the curvature becomes so great that none of the beams escape from the collapsing core. All of the beams are curved back to the core. At that moment, a black hole has formed. From that time onward, nothing can escape from the region of spacetime around the collapsing core. The boundary of the black hole is called the **event horizon,** the surface where escape velocity is equal to the speed of light. Nothing from inside the event horizon can emerge. Anything that crosses the event horizon is trapped. The radius of the event horizon is the **Schwarzschild radius,** given by

$$R_s = \frac{2GM}{c^2} \qquad (20.1)$$

If mass is given in terms of the Sun's mass, then the equation for the Schwarzschild radius becomes

$$R_s = 3\text{ km} \times \frac{M}{M_\odot} \qquad (20.2)$$

A black hole is a region of spacetime from which nothing can emerge. A black hole forms when the core of a massive star collapses. A black hole is bounded by an event horizon. Anything that crosses the event horizon becomes trapped in the black hole. The size of the event horizon is the Schwarzschild radius and is given by 3 km times the mass of the black hole in solar masses.

Falling into a Black Hole Suppose an astronaut (usually described as brave but doomed) has volunteered to fall into a black hole. The astronaut begins falling straight toward the black hole, accelerating to great speed. (Tidal forces become very great near a black hole and would stretch the astronaut out into a very thin wire. We

Equation 20.2

The Radius of a Black Hole

Equation 20.2 can be used to calculate the Schwarzschild radius of a black hole with a mass of 10 $M_\odot$. Using M = 10 $M_\odot$, the equation becomes

$$R_s = 3\text{ km} \times \frac{10\ M_\odot}{M_\odot} = 3\text{ km} \times 10 = 30\text{ km}$$

Thus, the collapse of a 10 $M_\odot$ stellar core forms a black hole with a radius of 30 km.

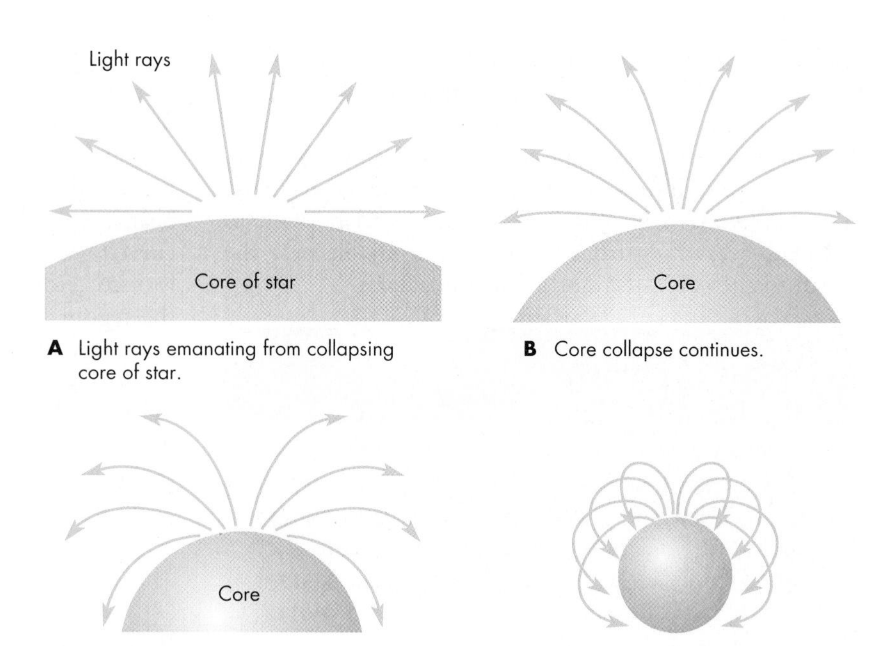
Light rays
Core of star
A Light rays emanating from collapsing core of star.
Core
B Core collapse continues.
Core
C Fewer light beams escape.
D No light escapes:
A black hole has formed.

FIGURE 20.29
The Paths of Light Rays Aimed Outward in Different Directions from the Collapsing Core of a Star
As the collapse continues, fewer of the light beams escape. When none escape, a black hole has been formed.

ANIMATION

The paths of light rays aimed outward in different directions from the collapsing core of a star

will ignore this fatal complication.) As the astronaut nears the event horizon, we see her slow down. The astronaut says, however, that she is still accelerating. The reason for the difference in the two descriptions of what is happening is that the curvature of spacetime is causing time to pass more slowly for the astronaut than for us. As far as the astronaut is concerned, she passes through the event horizon without anything special happening and accelerates rapidly toward the center of the black hole. To us, however, the astronaut continues to slow down and finally hovers just above the event horizon forever. Actually, we soon lose sight of the astronaut. The light that she emits becomes steadily more gravitationally redshifted and soon disappears. Remarkably, we disagree with the astronaut on whether she ever crosses the event horizon.

Properties of a Black Hole Nothing can ever reach us from a black hole. All we can ever tell about it is how much mass is inside it, how much angular momentum it has, and how much electrical charge it contains. These three quantities determine all the properties of the black hole. Things like what kind of material produced it or how long it has existed are impossible to tell.

It is important to realize that strange phenomena involving space and time take place only near the black hole. At great distances, the black hole has the same gravitational influence as a star of the same mass. Students often want to know what would happen if the Sun were replaced suddenly by a black hole containing 1 $M_\odot$ of material. The answer is that, at a distance of 1 AU, the curvature of spacetime would be exactly the same as before. This means that gravity would be exactly the same so the Earth would be pulled inward with exactly the same force and would orbit in exactly the same amount of time. Black holes have such strong local effects because they are so small. By approaching a black hole, it is possible to get very close to a large amount of mass. At the event horizon of a 1 $M_\odot$ black hole, an observer is within 6 km (twice the Schwarzschild radius) of an entire solar mass of material. Nowhere inside the Sun is it possible to get that close to anything like one solar mass of material. Even near the Sun's core, most of the Sun's material is much farther than 6 km away. The extreme concentration of matter within a black hole is what causes the great curvature of spacetime.

Detecting Black Holes

Because a black hole can't emit any light, it should be the ultimate in undetectable objects. One thing a black hole can do, however, is exert a strong gravitational influence on anything in its immediate vicinity. It is by observing the gravitational effects caused by a black hole that we can hope to detect its presence. Two special places where conditions are favorable for detecting the presence of black holes are in binary star systems (discussed in Chapter 21) and in the centers of galaxies (discussed in Chapter 23).

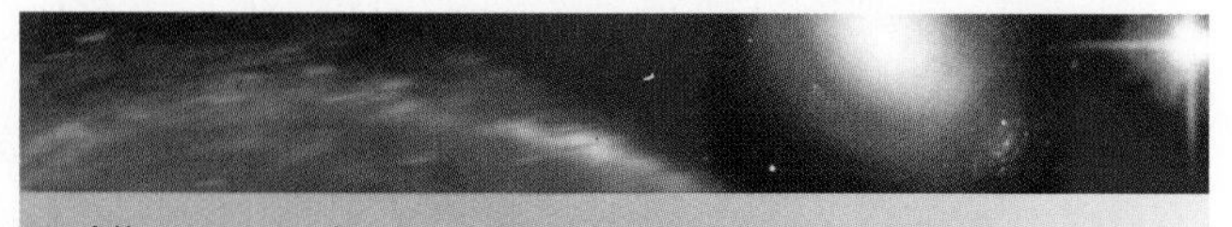

All we can know about a black hole is its mass, its angular momentum, and its electrical charge. Although black holes emit no light, we can hope to detect them by means of the strong gravitational effects that they exert on matter located near them.

Chapter Summary

- A white dwarf is a planet-sized star supported by degenerate electrons. Massive white dwarfs are smaller than low-mass white dwarfs. The Chandrasekhar limit, the greatest mass that a white dwarf can have, is about 1.4 $M_{\odot}$. (Section 20.1)
- A white dwarf keeps a constant size as it evolves. It radiates away its heat and grows cooler and dimmer simultaneously. After billions of years, a white dwarf becomes so dim it is difficult to detect. (20.1)
- White dwarf stars evolve from the cores of asymptotic giant branch stars that have lost their outer layers as cool winds. A star like the Sun will produce a white dwarf of about 0.6 $M_{\odot}$. The most massive white dwarfs originate in main sequence stars of about 8 $M_{\odot}$. (20.1)
- A type II supernova is the result of the collapse of the core of a massive star. As the core collapses, its protons and electrons combine to form neutrons. The inner core becomes a neutron star. Infalling matter rebounds from the neutron star and produces a shock wave that moves outward rapidly through the star. After a few hours, the shock wave expands the surface of the star and produces a great brightening. (20.2)
- Supernova 1987A, in the Large Magellanic Cloud, a nearby galaxy, was the first supernova visible to the naked eye in nearly 400 years. Neutrinos, which were detected almost a day before the supernova brightened, marked the time of core collapse and confirmed the idea that type II supernovae result from the collapse of the cores of stars. (20.2)
- Gamma ray bursts are brief blasts of gamma rays that originate in distant galaxies. Most gamma ray bursts are probably produced by the collapses of cores of massive stars. (20.2)
- The blast of gas ejected from a supernova sweeps up the surrounding interstellar gas and heats it to produce a luminous supernova remnant. High-energy electrons in the supernova remnant emit synchrotron radiation, which makes a supernova remnant visible using radio telescopes. After about 10,000 to 100,000 years, the supernova remnant merges into the interstellar gas. (20.2)
- Neutron stars, about 10 km in radius, are supported by degenerate neutrons. Like white dwarf stars, the more massive neutron stars are the smallest. The greatest mass that a neutron star can have is estimated to be between 1.5 and 2.7 $M_{\odot}$. A newly formed neutron star spins very rapidly and has a large magnetic field. (20.2)
- Pulsars are rotating neutron stars that emit beams of radiation. The rotation of the neutron star causes the beams to sweep past the Earth, causing us to observe pulses of radiation as often as one thousand times per second. (20.2)
- The beamed radio emission from pulsars is probably produced by energetic electrons in regions near the magnetic poles, which are tipped with respect to the rotation axis of a pulsar. Pulsars lose rotational energy as time passes and, after perhaps 10 million years, slow to periods of a few seconds. Magnetars are highly magnetized neutron stars that produce gamma ray bursts. (20.2)
- The rotation of the magnetic field of a pulsar produces low-frequency electromagnetic radiation that carries off the rotational energy of the pulsar. The radiation energizes electrons, which emit synchrotron radiation that fills in the center of the surrounding supernova remnant. (20.2)
- Spacetime is the combination of three space coordinates and one time coordinate that locates events in space and time. A geodesic, the shortest distance between two points, is a straight line in flat spacetime, but a curved path in curved spacetime. Geometric quantities, such as the sum of the angles of a triangle, depend on the curvature of spacetime. (20.3)
- According to general relativity, gravity is a consequence of the curvature of spacetime by mass. Objects and light follow geodesics in curved spacetime near massive objects. We cannot see that spacetime is curved, so we have invented the force of gravity to account for the motion of light and moving bodies. (20.3)
- A black hole forms as a result of the collapse of the core of a star when the core is too massive to become a neutron star. The black hole is bounded by its event horizon, through which nothing can emerge, not even light. The Schwarzschild radius, the radius of the event horizon, is 3 km times the mass of the black hole in solar masses. (20.3)
- All that we can ever learn about a black hole is its mass, angular momentum, and electrical charge. We can detect the presence of black holes, however, by looking for the strong gravitational influence that they have on their immediate surroundings. (20.3)

Key Terms

Conceptual Questions

1. Describe the main difference between the mass-radius relationship for main sequence stars and the mass-radius relationship for white dwarf stars.
2. A 2 $M_\odot$ core of a star contracts after using its nuclear fuels. Explain why we can be sure that the star will not become a white dwarf.
3. Suppose two white dwarf stars have the same surface temperature. Why is the more massive of the two white dwarfs the less luminous?
4. Describe the evolution of a star after it becomes a white dwarf.
5. Main sequence stars of 5 $M_\odot$ are thought to evolve into 1 $M_\odot$ white dwarfs. What happens to the 4 $M_\odot$ that do not become part of the white dwarf ?
6. What is the relationship between the central stars of planetary nebulae and white dwarf stars?
7. Suppose AGB stars did not lose mass through cool winds. What effect would this have on the number of white dwarf stars in the galaxy?
8. What is the difference between the spectra of type I supernovae and those of type II supernovae?
9. The core of a massive AGB star consists of iron and nickel surrounded by shells of successively lighter elements. What does this structure have to do with the history of consumption of nuclear fuels by the star?
10. Why does the process of neutronization reduce the ability of the degenerate electrons in the core of a massive AGB star to support the weight of the star?
11. What are the processes that reverse the infall of matter during a type II supernova explosion and blast material out of the star?
12. What is the ultimate origin of the energy released in a type II supernova?
13. How do we know that most gamma ray bursts originate far beyond the Milky Way?
14. What is thought to be the relationship between gamma ray bursts and supernovae?
15. Only a small percentage of the energy of a type II supernova is carried away by radiation and the shell of matter blasted outward. What happens to the rest of the energy released in the explosion?
16. Why are many supernova remnants bright in the radio part of the spectrum?
17. Why does the expansion of a supernova remnant slow as time passes?
18. What effect do supernova explosions have on the chemical makeup of interstellar gas?
19. What are the similarities and differences between the mass-radius relationships of white dwarfs and neutron stars?
20. What does the concept of conservation of angular momentum have to do with the rapid rotation of neutron stars?
21. Why was it possible to reject the idea that pulsars were rotating white dwarfs, pulsating white dwarfs, or pulsating neutron stars?
22. Why would no pulses be observed from a rotating neutron star if its magnetic axis and spin axis were aligned?
23. What happens to the rotation rate of a pulsar as time passes?
24. Why don't we see any pulsars with periods longer than a few seconds?
25. Why are all pulsars not found within supernova remnants?
26. How many coordinates are used in spacetime?
27. Why is a stationary body represented by a vertical line rather than a point in a spacetime diagram?
28. Why can't a body move horizontally in a spacetime diagram?
29. What is the difference between spacelike trips and timelike trips?
30. Suppose you lived in a two-dimensional world. Describe a way you could use geometry to determine whether your world was flat or curved.
31. Suppose you made a triangle in your backyard. You used a stretched string to make three geodesics to form the sides of the triangle. You then used a protractor and found that the sum of the angles of the triangle was 180°. You know that the surface of the Earth is curved, so why didn't your triangle contain more than 180°?

32. Why does Einstein's theory of relativity imply that gravity is a nonexistent force?
33. Suppose an astronaut falling toward a black hole used a watch to check her pulse rate. She notes that her pulse rate remains constant as she falls closer to the black hole. (Obviously, this astronaut is tough.) Why, if clocks run more slowly in strongly curved regions of spacetime, does the astronaut find that her pulse rate doesn't change?
34. What effect does the curvature of spacetime have on the frequency of radiation passing through it?
35. The spectra of white dwarfs usually show large redshifts. Why doesn't this imply that most white dwarfs are moving rapidly away from us?
36. Describe what we would see if we watched an object fall toward a black hole.
37. What are the only things we can tell about the nature of the material within a black hole?
38. What would happen to the distance of the Earth from the Sun and the length of the year if the Sun were instantly replaced by a 1 $M_\odot$ black hole?

Problems

1. A black hole has a Schwarzschild radius of 25 km. How much mass is contained in the black hole?
2. How large is the Schwarzschild radius of a black hole containing a mass equal to that of the Earth?
3. The visible light from a gamma ray burst has an apparent magnitude of 8.0. The distance of the galaxy in which the burst occurred is 3 billion pc. Find the absolute magnitude of the visible light. How does this compare with the absolute magnitude of the Sun? (Use Equation 16.3.)

Figure-Based Questions

1. Use Figure 20.3 to estimate the radius of a 0.9 $M_\odot$ white dwarf star.
2. A white dwarf star has a radius two-thirds as large as the Earth's. Use Figure 20.3 to find the mass of the star.
3. Use Figure 20.5 to find the mass of the white dwarf star that results from the evolution of a 6 $M_\odot$ main sequence star.
4. Use Figure 20.9 to find the brightness (relative to its maximum brightness) of a type II supernova 100 days after the time of maximum light.
5. Use Figure 20.17 to estimate the radius of a 0.5 $M_\odot$ neutron star.

Group Activity

Borrow a "chalk globe" from your instructor. Working with your group, make a triangle on the chalk globe by picking three points not too far apart and connecting the points with geodesics. You can make the geodesics by stretching a piece of string tight between a pair of points. Use a protractor to measure the angles of the triangle. Find the sum of the angles. Repeat the measurements using three points located fairly far apart on the chalk globe. Have the group discuss how the sum of the angles in your triangles compares with the sum of the angles in a plane triangle.

For More Information

Visit the text website at **www.mhhe.com/fix** for chapter quizzes, interactive learning exercises, and other study tools.

21

On with the dance! let joy

be unconfined;

No sleep till morn, when

Youth and Pleasure meet

To chase the glowing hours

with flying feet.

Byron
Childe Harold's Pilgrimage

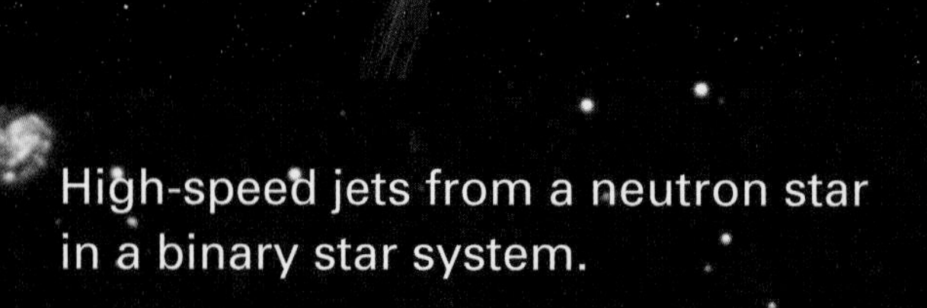

High-speed jets from a neutron star in a binary star system.

www.mhhe.com/fix

Binary Star Systems

Some of the favorite settings used by science fiction writers and "space artists" are planets in binary star systems. Many exotic and spectacular sights can be imagined. From a planet in the γ Virginis system, for example, there could be times when two suns are in the sky at once. One sun is setting. The other, 100 times dimmer, but still far too bright to look at directly, is rising. During this season, there is no night.

Sunrise on a planet in the U Cephei system would bring two suns into view. The blue sun and the dimmer, larger, yellow sun are so close together that they nearly touch. Each sun is distorted by the gravity of its companion. A tongue of gas reaches from the yellow sun to the blue sun. Multiple colored shadows make the scene even stranger.

Is this just fantasy? Could there really be planets orbiting in binary star systems? The answer is yes, there could be and, in fact, some have already been discovered. Although the gravitational forces of two stars combine to make some orbits unstable, there are two kinds of stable orbits. If the two stars orbit each other at great distances, each star could have a family of planets in small orbits. If the stars are very close together, on the other hand, there could be planets in large orbits that circle both stars at once.

In many cases, planets in binary star systems probably are completely uninhabitable. Close binary systems, such as U Cephei, are dangerous places. During the evolution of close binary systems, there are sometimes spectacular nova and supernova explosions. In other close binary systems, there are bursts and pulses of X rays and intense ultraviolet radiation. Sometimes binary

Questions to Explore

- How is it possible to tell that a point of light in the sky is actually two stars rather than one?
- How can we find masses and diameters for binary stars?
- How do binary star systems form?
- How do the stars in a binary system interact with each other?
- How do these interactions influence the evolution of the stars?
- What causes novae and type I supernovae?
- Why are binary star systems often the sources of bright, variable X-ray and ultraviolet radiation?
- Are there black holes in some binary systems?

star systems lose so much of their mass that their gravitational hold on a hypothetical planetary companion becomes so weak the planet would fly away into the depths of interstellar space. Although the scenery in binary systems might be beautiful and bizarre, there is little doubt that we are better off in a system with a single sun.

This chapter describes binary stars and the many dramatic events that can occur in binary systems. It also describes how observations of binary stars can allow us to determine stellar masses and diameters.

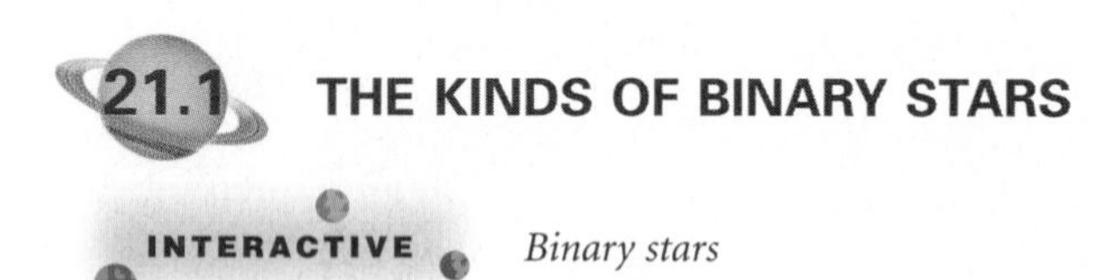

21.1 THE KINDS OF BINARY STARS

INTERACTIVE *Binary stars*

The most meaningful way of classifying binary stars is on the basis of whether the stars affect each other's evolution. When two stars are far enough apart that they evolve independently, they are called a **wide pair.** Each star in a wide pair follows the same career as a single star of the same mass. When the two stars are close enough to each other that they can transfer matter to one another at some stages of their evolution, they are called a **close pair.** Through these exchanges of mass, the stars in a close pair alter the way that each evolves.

Detecting Binary Stars

Binary stars also can be classified according to how it was learned that there are two stars rather than one. Some methods of recognizing a **binary star system** favor the discovery of wide pairs; others favor the discovery of close pairs.

Visual Binaries The existence of "double stars"—two stars that lie nearly in the same direction—has been known at least since the time of Ptolemy, who included one such pair in his catalog of stars. Probably the most famous of the naked-eye double stars are Mizar and Alcor, in the Big Dipper. After the invention of the telescope, other pairs too close together to be separated by the unaided eye were discovered. Until 1800, however, astronomers believed that all double stars were just unrelated stars that were at very different distances but happened to lie in almost the same direction. This belief led William Herschel, around 1780, to measure the separations and orientations of about 700 double stars. He thought that the brighter star was nearer and hoped that he would be able to measure the parallax of the brighter star as it shifted back and forth during a year. He intended to use the fainter star as a reference against which to detect the small parallax shifts of the brighter star. Herschel found no parallaxes but made a second set of measurements of the double stars around 1801. He found that for about 50 of the pairs, the orientation of the two stars had changed over the 2 decades between the observations. He explained the changes as orbital motion of the fainter star about the brighter. Herschel's results showed that most of the double stars were actually **visual binary stars,** or orbiting pairs of stars in which the two stars can be seen as separate images when viewed through a telescope.

For a pair of binary stars to be seen as a visual binary, they must be far enough apart that their images can be distinguished. This usually means that their separation must be many astronomical units and the pair must be relatively close to us. Because of the large sizes of their orbits, most visual binaries have orbital periods of decades or centuries. Often generations of astronomers must observe a visual binary before it completes a single orbit. For example, Figure 21.1 shows that Castor (α Geminorum), one of the visual binaries observed by William Herschel, has not yet completed its 467-year orbit since it was first observed as a double star in 1719.

Interpreting the orbital motion of a visual binary is complicated because the actual orbit may be tipped rather than viewed from directly above the orbital plane. Figure 21.2 shows how our orientation relative to the plane of the orbit affects the way the orbit appears to us. The amount of tilt and the shape and size of the actual orbit can be found by using Kepler's laws (Chapter 4, Section 5). Figure 21.3 shows

FIGURE 21.1
The Visual Binary Castor (α Geminorum)
The position of the fainter star relative to that of the brighter star has been measured for almost 3 centuries, yet the binary has not yet completed a single 467-year orbit in that time. When the first observations of Castor were made in 1719, the two stars were farthest apart and orbited each other very slowly. In 1958, they were closest together, although this isn't apparent in the diagram because Castor's orbital plane is tilted away from us.

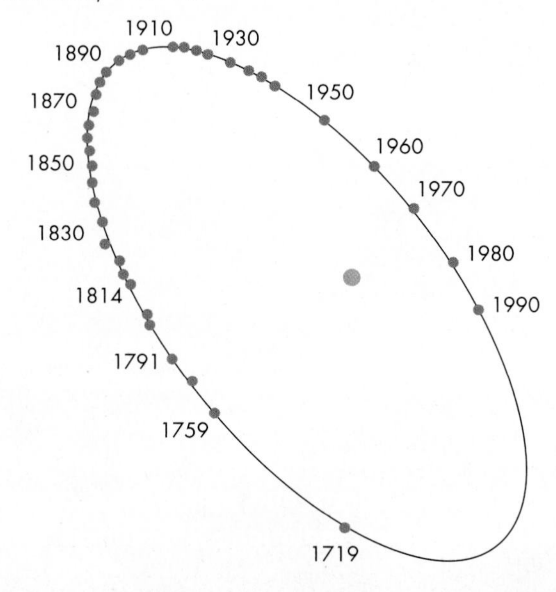

FIGURE 21.2
The Apparent Orbit of a Visual Binary
The apparent orbit of a visual binary star system depends on the orientation from which it is viewed. The left-hand diagram shows four directions from which a binary star could be viewed. The diagrams at the right show the apparent orbit as seen from each direction. **A,** When the binary is viewed from above, the apparent orbit is the true orbit. **D,** When the binary is viewed nearly edge-on and along the major axis, the apparent orbit is greatly foreshortened. In views **B** and **C,** the apparent orbit is progressively more foreshortened, appearing nearly circular in **D.**

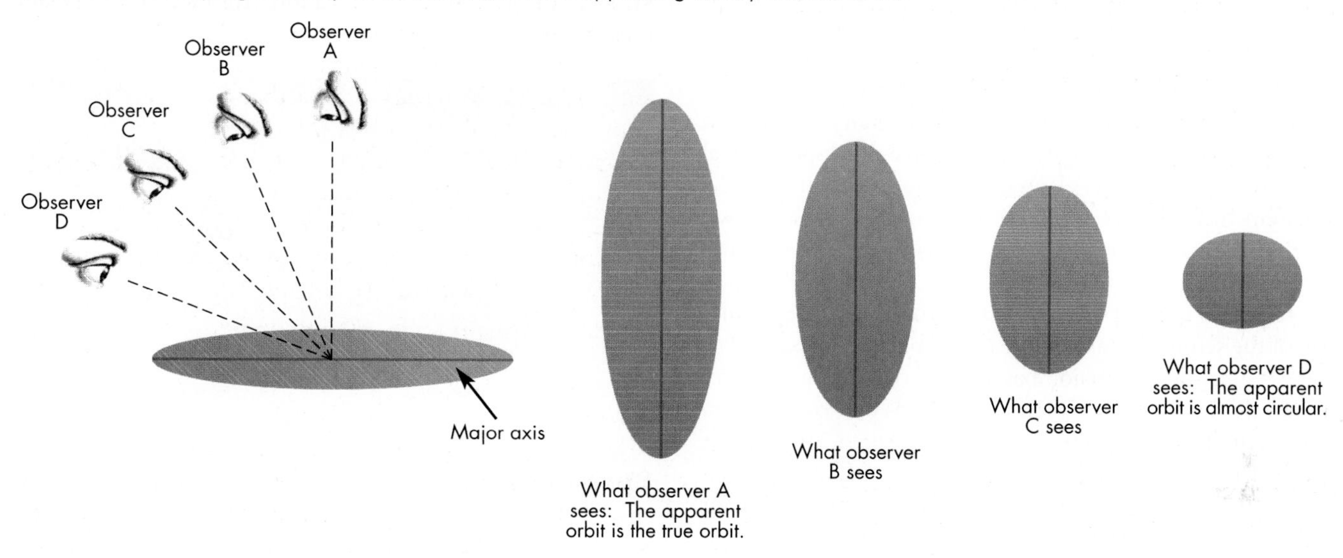

FIGURE 21.3
The True (right) and Apparent (left) Orbits of the Visual Binary α Centauri
Notice in the apparent orbit, **A,** that the brighter star is not at the focus of the ellipse and that the fainter star doesn't move fastest when it is nearest the brighter star, all of which would seem to violate Kepler's laws. Notice in the true orbit, **B,** however, that Kepler's laws are obeyed.

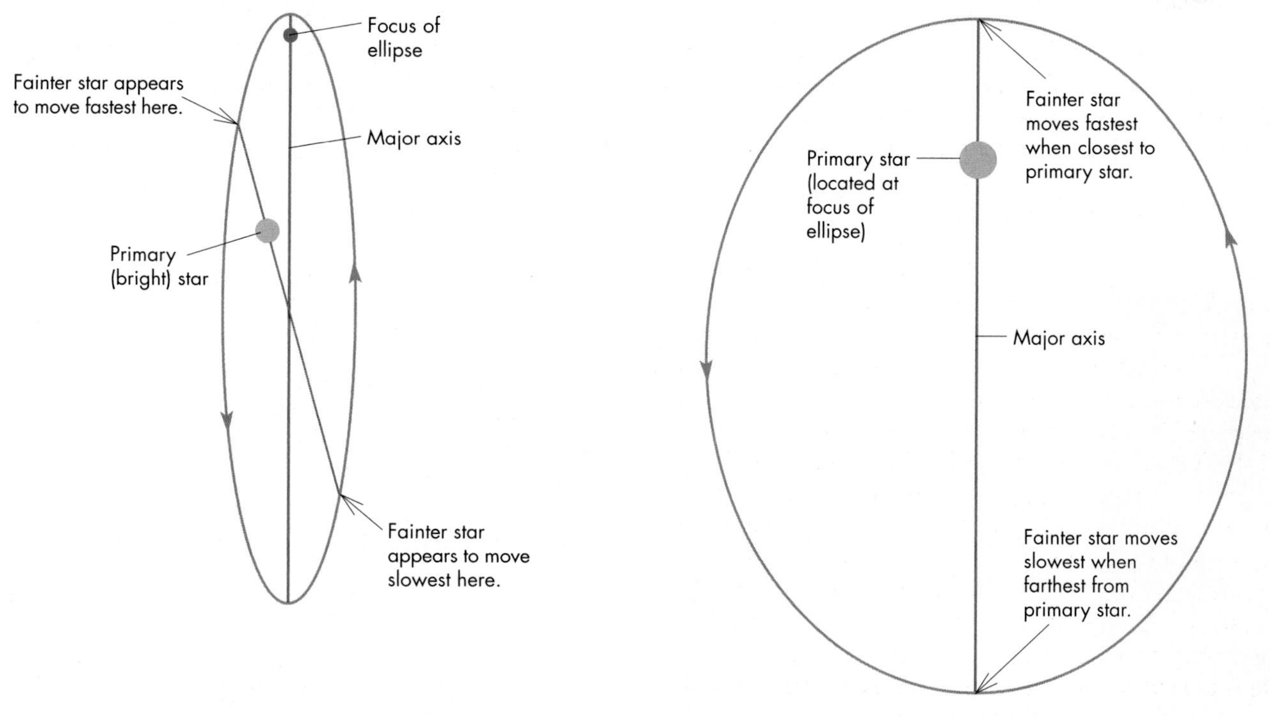

A Apparent orbit: Kepler's laws appear to be violated.

B True orbit: Kepler's laws are obeyed.

the apparent orbit of the fainter star around α Centauri. Two things tell us that the orbit we see can't be the true orbit of the star. First, the brighter star (called the primary) is not at a focus of the ellipse traced out by the fainter star (the secondary). Kepler's first law tells us that the primary star must be at a focus of the orbit of the secondary star about the primary. Second, the places where the secondary star moves fastest and slowest in its orbit are not the places where its apparent orbit brings the secondary nearest to and farthest from the primary. Kepler's second law says that the secondary must move fastest when it is nearest the primary and slowest when it is farthest away. Astronomers have developed several methods that use Kepler's laws to find the true orbit from the apparent orbit. For α Centauri, the true orbit, also shown in Figure 21.3, is tipped by 79° and is much less eccentric than the apparent orbit.

By correcting for the tilt of the true orbit, the angular size of the semimajor axis of the true orbit can be found. For α Centauri, the semimajor axis is 17.7″. For most visual binaries, the semimajor axis is a few seconds of arc or smaller. If the parallax of the binary can be measured, its distance can be found. Knowing the distance, the small angle equation (Equation 3.1) can be used to find the true semimajor axis. The small angle equation can be written $D = d \times \theta$, where D is the semimajor axis in AU, d is the distance in parsecs, and θ is the angular semimajor axis in seconds of arc. α Centauri's distance is 1.3 pc, so its semimajor axis is $1.3 \times 17.7 = 23$ AU. Thus, the two stars that make up α Centauri are about as far apart as the Sun and Uranus.

The main reason that astronomers have been willing to devote decades of effort to observing visual binaries is that in those cases for which the distance is known and the true orbit can be determined, it is possible to find the masses of the stars. Newton's form of Kepler's third law is

$$P^2 = \frac{D^3}{M_1 + M_2} \quad (21.1)$$

where P is the orbital period in years, D is the semimajor axis of the orbit in astronomical units, and M_1 and M_2 are the masses of the two stars in solar masses. This equation can be solved for the sum of the masses of the stars to give

$$M_1 + M_2 = \frac{D^3}{P^2} \quad (21.2)$$

The great majority of the several hundred masses known for stars have been found from studies of visual binaries.

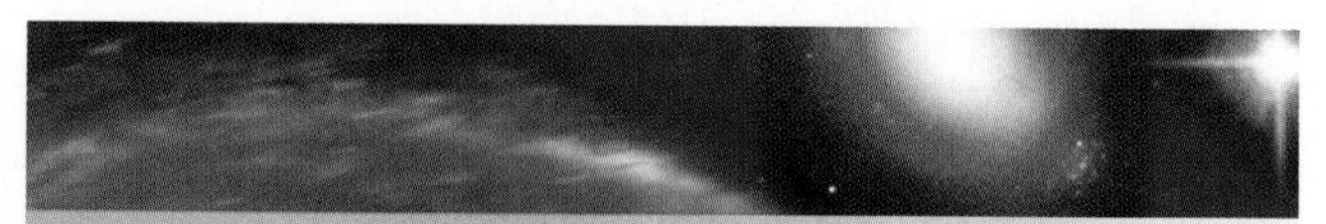

Visual binaries are those binary systems for which the images of the two stars can be distinguished. Most visual binaries are widely separated and have long orbital periods. The masses of the stars in a visual binary can be determined if the distance to the binary can be found.

Spectroscopic Binaries Some stars that appear to be single when viewed through a telescope are found to be binaries when their spectra are examined. The orbital motion of these stars, called **spectroscopic binaries,** carries them toward and then away from the Earth. The Doppler effect associated with the orbital motion causes the wavelengths of their spectral lines to shift back and forth. However, only the part of the motion of a star directed toward or away from the Earth contributes to the Doppler shift. Thus, when the two stars in a binary system are in the parts of their orbits where their motion is across our line of sight, as shown in Figure 21.4, neither star has an orbital Doppler shift. At this time, the spectral lines of the two stars are superimposed. At other times, the lines appear doubled as one star moves toward the Earth and the other away. If one of the two stars is much brighter than the other, its light dominates the spectrum and only the shifting lines of the brighter star can be seen, as shown in Figure 21.5.

Masses of Binary Stars

Equation 21.2 can be used to calculate the sum of the masses of the stars in a binary system (in solar masses) if the orbital period (in years) and semimajor axis of the orbit (in astronomical units) of one star relative to the other are known. For α Centauri, the orbital period has been observed to be 80 years and the semimajor axis is 23.2 AU. Using these values in Equation 21.2,

$$M_1 + M_2 = \frac{D^3}{P^2} = \frac{23.2^3}{80^2} = \frac{12{,}487}{6400} = 1.96\ M_\odot.$$

As the two stars orbit their center of mass, one of the stars (a K star) is always 1.23 times as far from the center of mass as the other star, a G star like the Sun. This shows that the G star is 1.23 times as massive as the K star. Thus the G star has a mass of 1.08 $M_\odot$ and the K star has a mass of 0.88 $M_\odot$.

FIGURE 21.4

Orbital Motion Reflected in the Shifting Spectral Lines of a Spectroscopic Binary

When they are at positions 1(*A*) and 1(*B*), the two stars are moving across the line of sight. Because neither of them is moving toward or away from the Earth, neither of them has redshifted or blueshifted spectral lines. Their spectral lines (three of which are indicated in spectrum 1) are superimposed. When they are at positions 2, star *A* is moving away from the Earth, and its spectral lines are redshifted. At the same time, star *B* is moving toward the Earth and has blueshifted spectral lines. At positions 3, the two stars are again moving across the line of sight, and their spectral lines again are superimposed. At positions 4, star *A* is moving toward the Earth and has blueshifted spectral lines. Star *B* is moving away from the Earth and has redshifted spectral lines. Thus, the lines of the two stars shift back and forth in the spectrum during each orbital period of the binary star system.

ANIMATION *Orbital motion reflected in the shifting spectral lines of a spectroscopic binary*

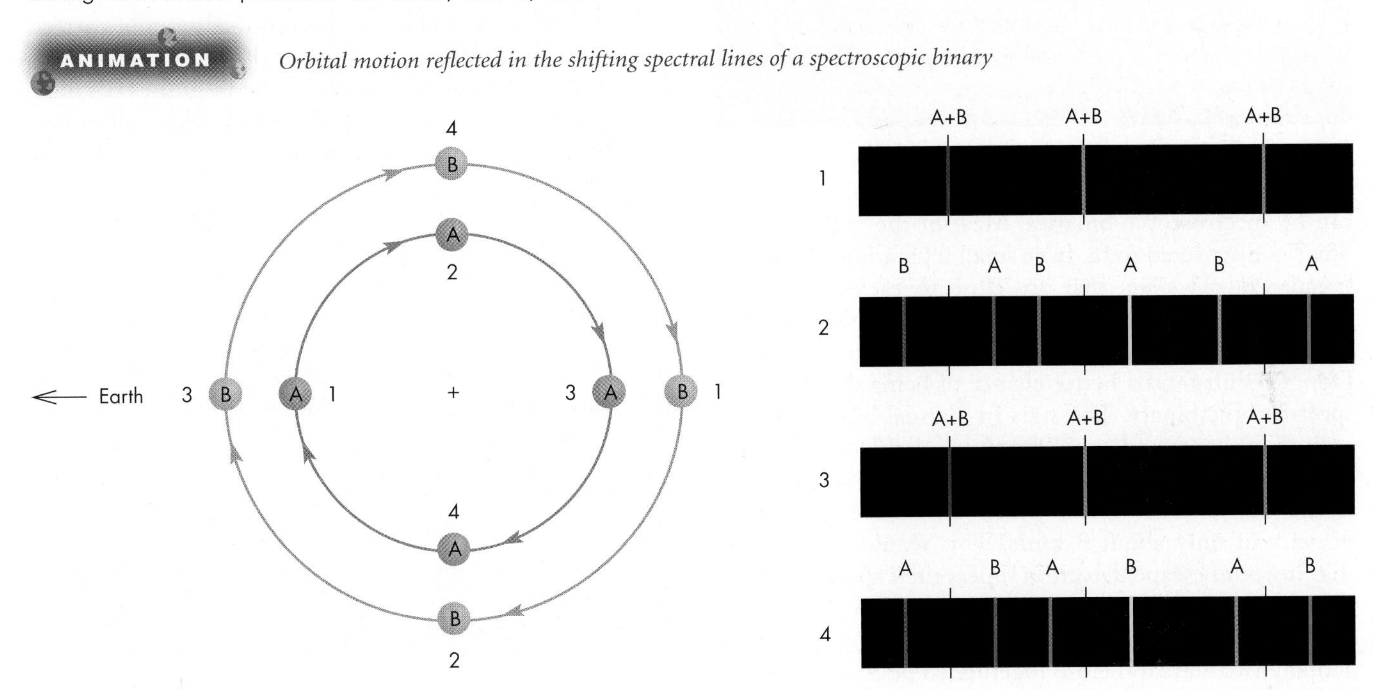

FIGURE 21.5

A Single-Line Spectroscopic Binary

Only one set of spectral lines can be seen because one of the stars is too dim for its spectrum to be visible. The single set of spectral lines shifts back and forth during each orbit, however, which is strong evidence that the star whose spectral lines can be seen has a binary companion about which it orbits. The orbit of the dim star is shown as a dashed line.

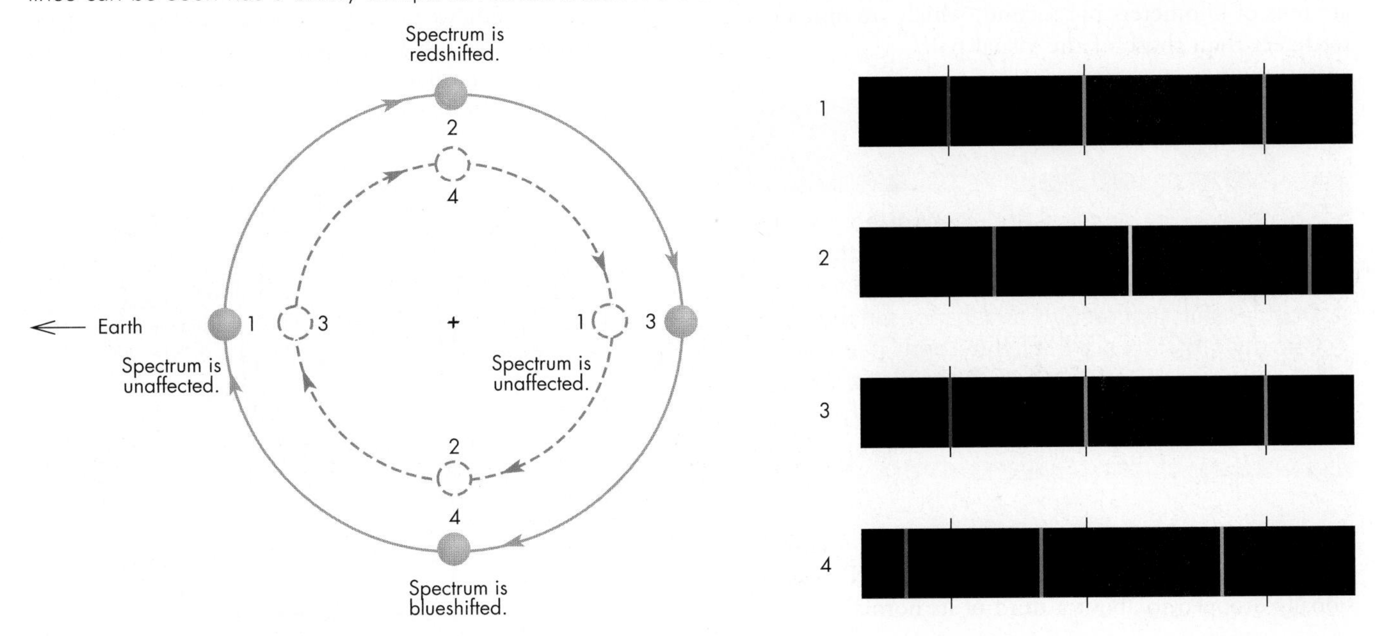

FIGURE 21.6
Part of the Castor System
Each of the components of the visual binary is itself a spectroscopic binary.

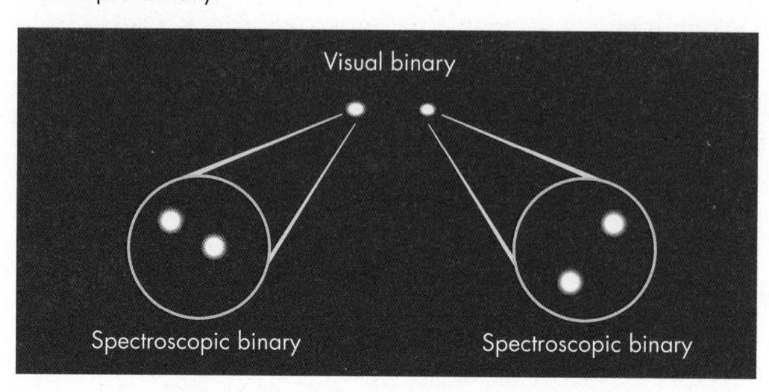

Binary pairs separated by many astronomical units can be spectroscopic binaries. Most of the spectroscopic binaries discovered so far have small separations, however, because binary pairs that are close to each other have larger orbital velocities than those of more widely separated pairs. Larger orbital velocities, in turn, lead to larger Doppler shifts and a better chance of being detected as a spectroscopic binary. The stars in a visual binary usually are too far apart and orbit too slowly for Doppler shifts to be detected easily. For example, each of the two stars in the Castor visual binary system has an average orbital velocity of only about 3 km/s. This would be so small that it would escape detection unless their spectrum were inspected very carefully. It turns out, however, that each of the "stars" that make up the Castor visual binary is actually two stars too close together to be resolved. Each of these pairs can be observed as a spectroscopic binary, as shown in Figure 21.6. The components of each spectroscopic binary are separated by only a few million kilometers (a few hundredths of an astronomical unit) and orbit with periods of a few days. Their orbital velocities are tens of kilometers per second, which are much easier to detect than those of the visual pair.

Spectroscopic binaries are recognized by the shifting of the wavelengths of their spectral lines over time. The shifting is due to variations in the Doppler shift that result from changes in the radial velocities of the stars as they orbit each other. Most spectroscopic binaries have small separations and short orbital periods, resulting in high orbital speeds and large Doppler shifts.

Eclipsing Binaries In 1667, Italian astronomer G. Montanari noticed that the star Algol (β Persei) occasionally dropped to about a third of its normal brightness. A century later, in 1783, English astronomer John Goodricke found that the dimmings of Algol were periodic. He found that every $2^d\ 20^h\ 49^m$, Algol begins a swift decline in brightness. It is dimmer than usual for about 8 hours, and then resumes its normal brightness. A graph of Algol's brightness versus time is shown in Figure 21.7. Goodricke proposed that the dimming was due to a dark star that orbited Algol and eclipsed it once during each orbit. Other astronomers, however, interpreted the brightness variations of Algol as consequences of a spotted surface. Goodricke's explanation, although correct, was ignored until the late nineteenth century. **Eclipsing binaries** are recognized as such by the brightness changes that occur when the two stars alternately block all or part of each other from our view. A graph of the way that the brightness of an eclipsing binary varies with time as the two stars eclipse each other is called a **light curve.**

Thousands of eclipsing binaries are known, most with periods of a few days and separations of less than 1 AU. Eclipsing binaries have been studied extensively because they can yield many kinds of information about the stars in the binary system. For example, Figure 21.8 shows that the shape of the light curve during eclipse is related to the relative sizes of the two stars. If the two stars are of equal size, there is only an instant when one fully covers the other. In that case, the **minimum** of the light curve, the time of minimum light, lasts only a moment. If one star is larger than the other, the minimum lasts longer.

The temperatures of each of the stars can be found from the light curve of an eclipsing binary, too. The higher

FIGURE 21.7
The Light Curve of Algol
The light curve, a graph of brightness versus time, shows periodic dips in brightness when one of the stars is eclipsed by the other. Primary minima, the deepest drops in brightness, occur when the hotter star is eclipsed. Secondary minima occur when the cooler star is eclipsed. Both a primary minimum and a secondary minimum occur once each orbital period.

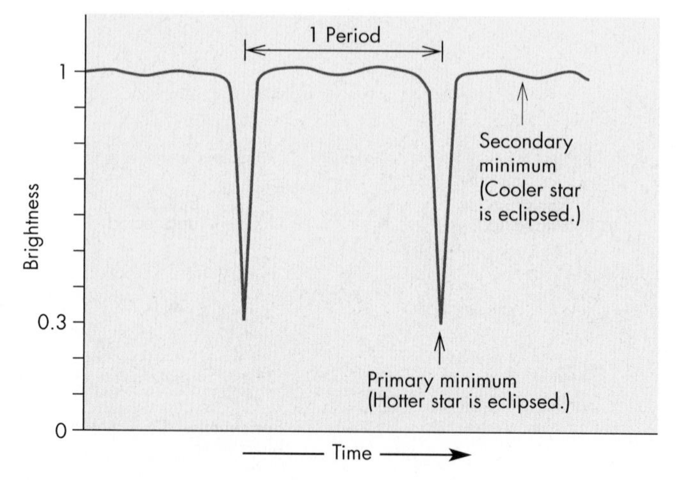

FIGURE 21.8
How the Relative Sizes of the Stars Affect the Light Curve
If the two stars have the same diameter, **A,** one blocks the other to produce a minimum for only an instant. However, when one star is smaller than the other, **B,** one of the stars is in front of the other for some time, producing a flat-bottomed minimum. The greater the difference in the sizes of the two stars, the longer the duration of minimum light.

ANIMATION *An eclipsing binary and its light curve*

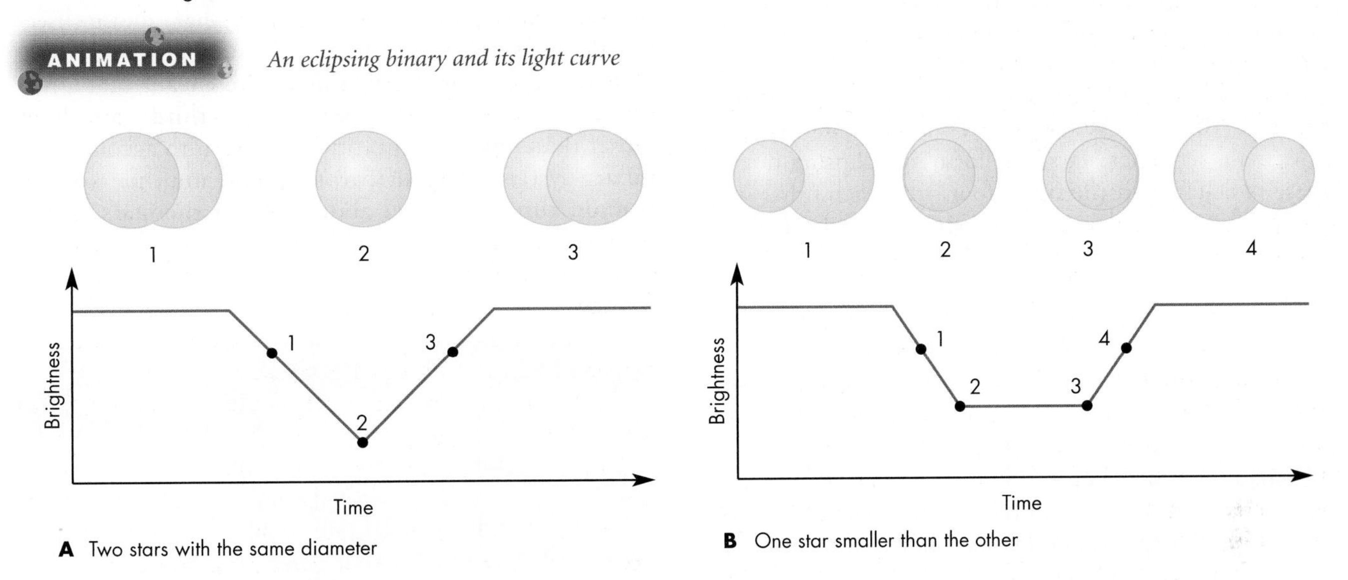

the temperature of a star, the greater the rate at which radiant energy is emitted by every square meter of the star's surface. Figure 21.9 shows the light curve of an eclipsing binary in which the two stars have different temperatures. During the two eclipses that occur in each revolution of the stars, the same amount of area is blocked on the eclipsed star. The depth of the minimum that occurs when a star is eclipsed depends on how much light the star emits in the blocked area. It is the hotter star, not necessarily the bigger or brighter star, that radiates the larger amount of energy in the blocked area and produces the deeper minimum when it is eclipsed. Other information that sometimes can be obtained from studying an eclipsing binary includes the shape of the stars (when the two stars are close to each other, tidal forces elongate them toward each other), whether the stars have dark or bright regions on their surfaces, and whether one or both of the stars is surrounded by gas or dust. For eclipsing binaries that are also spectroscopic binaries, the masses of the two stars can be measured as well.

Eclipsing binaries show periodic drops in brightness as the two stars alternately eclipse each other. Most eclipsing binaries have small separations and short periods. The temperatures and sizes of the two stars can be determined from the shape of their light curve.

FIGURE 21.9
The Depths of Eclipses and the Temperatures of the Two Stars
The same area is blocked in both eclipses, so the deeper eclipse occurs when the hotter star, which emits more light for a given area, is eclipsed. Thus, primary minimum occurs when the hotter star is eclipsed by the cooler one, and secondary minimum occurs when the cooler star is eclipsed by the hotter one.

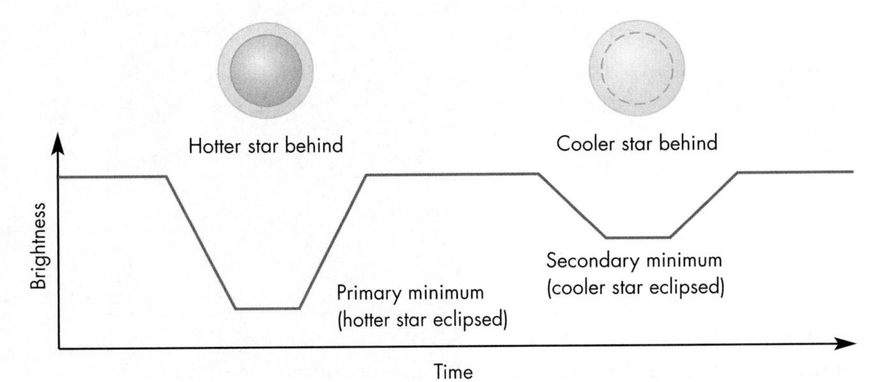

Multiple Star Systems

Many binary star systems are themselves part of larger, multiple star systems. As described in the preceding section, Castor consists of a visual binary with a period of 467 years. Each member of the visual binary is itself a spectroscopic binary with a period of a few days. In addition, the visual binary (consisting of four stars) is part of a much more widely separated visual binary, as shown in Figure 21.10. At a distance of 72″, there is another pair of M main sequence stars that form an eclipsing binary. These two stars and the other four orbit each other very slowly. Their orbital period is estimated to be perhaps 20,000 years.

A more typical multiple star system is α Centauri. Recall that it contains a visual binary with a period of 80 years. More than 2° (12,000 AU) from the bright visual binary of α Centauri is Proxima Centauri, the third member of the system. Proxima Centauri is an M main sequence star slightly closer to us than the other stars in α Centauri and is the nearest known star besides the Sun. The arrangement of stars in α Centauri, where the third star is hundreds or thousands of times farther from the closer pair than the closer pair stars are from each other, is common to most triple star systems. Quadruple systems usually consist of two close pairs widely separated from each other.

For stars like the Sun, single stars are a distinct minority compared with binary and multiple stars. Of every 100 star systems in which the primary is a K star or hotter, it has been estimated that only 30 contain single stars, 47 are binaries, and the remaining 23 are multiples, most of which are triples. The 100 star systems contain about 200 stars, so if only 30 of them are single stars, then 85% of them are in binary or multiple systems. The proportion of stars that are in binary or multiple systems may even be higher than 85%, moreover, because faint distant companions of what appear to be single stars or close binaries may have been overlooked or gone undetected.

The situation is quite different for M stars, however. Recent surveys of M stars, often in attempts to detect planets orbiting them, have shown that about three-fourths are single stars. Because M stars make up about 80% of all the stars in the galaxy, this means that a majority of all star systems, perhaps two-thirds, are single rather than binary or multiple. Taking into account that binary systems contain two stars and multiple systems contain three or more, about half of the stars in our part of the galaxy are single stars and the other half are members of binary or multiple systems.

Multiple star systems consist of close pairs orbited, at a much greater distance, by other pairs or single stars. About half of all stars are members of binary or multiple star systems.

21.2 THE FORMATION OF BINARY SYSTEMS

Many theories have been proposed to explain the formation of binary star systems. Some of the theories are applicable to the formation of close binaries, others to wide binaries. It seems likely that no single mechanism can account for the formation of all binary star systems. The problem that astronomers face today is deciding which mechanism is the most important one for binary stars with different separations. Unfortunately, none of the

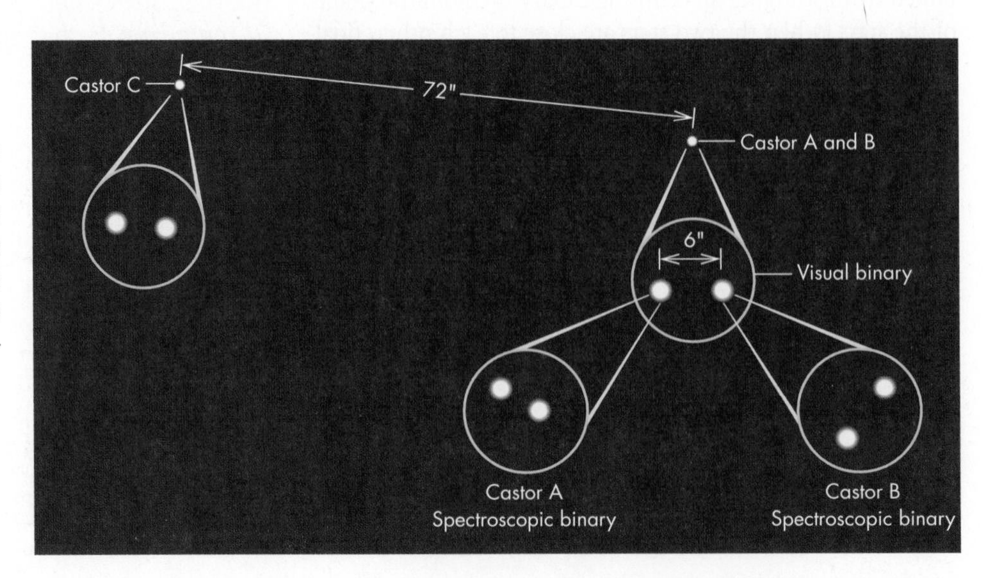

FIGURE 21.10
The Entire Castor Multiple Star System
The visual binary that contains two spectroscopic binaries (also shown in Figure 21.6) is part of a wider visual binary. The distant companion, Castor C, is also a spectroscopic binary. Thus, there are a total of six stars in the Castor system.

proposed mechanisms is well enough understood for us to be certain whether it actually works and, if so, under what conditions.

Wide Binaries

Tidal capture is one possible explanation for the origin of wide binary systems. Two stars or protostars can't capture each other, however, unless they lose energy and slow down. Otherwise, they will approach each other, pass at speeds greater than escape velocity, and then recede again without going into orbit. One way to lose energy is through tidal interactions between two cloud fragments before the fragments contract to become stars. After the two fragments capture each other, each evolves separately into a star.

Conucleation is an alternative to tidal capture. Conucleation differs from tidal capture in that when the two fragments that become the binary stars form out of a larger cloud, they are already in orbit about each other. Exactly how this might happen has not been determined.

Close Binaries

Fragmentation and **fission** are two ways that close binary star systems may form. In fragmentation, a collapsing cloud breaks into several pieces, each of which collapses to form a star. This process seems likely, however, only for systems separated by many AU. Some other process must account for binary systems separated by less than 1 AU. Fission, while similar to fragmentation, takes place in a star rather than in a collapsing cloud. In fission, the star splits into two pieces, each of which becomes a star. It turns out, however, that fission can produce only binary star systems in which one of the stars is much more massive than the other.

Another possibility is that the gas in the disks that form about single stars during their collapse can collect to form an additional star. The process through which this may happen is called **disk instability.** The disks around newly formed stars are massive and quite close to the star. If disk instability is an efficient process, it might lead to stars of roughly equal mass and very close separations.

Several mechanisms have been proposed to account for the formation of binary stars. Some of these may apply to wide binaries, others to close binaries. At present, none of the mechanisms is well enough understood for astronomers to be certain just how important it is in the formation of binary systems.

21.3 EVOLUTION OF CLOSE BINARIES

The stars in a wide binary pair evolve independently of each other. For close binaries, however, the gravitational interactions of the two stars strongly influence their evolution.

Evidence for Interaction: The Algol Paradox

There are many binary systems that show that their evolution has been influenced by their binary nature. One of these is β Persei, or Algol, an eclipsing spectroscopic binary. The two stars in Algol orbit each other every 2.9 days at a separation of 0.07 AU (10 million km). The two stars have masses of 3.7 and 0.8 $M_\odot$. The more massive star is a B8 main sequence star. The less massive star is a K0 giant. Recall from Chapter 19 that the more massive a star, the more quickly it evolves. If two stars form at about the same time, as they should in a binary star system, the more massive one should leave the main sequence and become a giant first. The paradox of the Algol system is that the more massive star is still on the main sequence and the less massive star is the one that has evolved to become a giant. This combination of masses and evolutionary statuses is inconsistent with what we know about the evolution of single stars. Somehow, in Algol and many other binary star systems, the less massive star seems to have evolved more quickly and become a giant before its more massive companion.

The resolution of the Algol paradox is that the K0 giant, now the less massive star, has not always been less massive than its companion. The interaction between the two stars has caused the star that originally was more massive to lose part of its mass to the originally less massive star. The transfer of mass between the components of a close binary star system can account not only for the Algol paradox but also for many spectacular characteristics of close binary star systems.

Equipotentials

A concept that astronomers have found very helpful in visualizing some of the interactions in a binary star system is the **equipotential.** An equipotential is a line or surface that connects all the places in the binary system that have the same potential energy (Chapter 5, Section 4). Although the equipotential may sound like an exotic concept, most of us are already familiar with another use of equipotentials—on geographic contour maps. On the surface of the Earth, potential energy is almost entirely due to gravitational energy. Therefore lines of constant altitude on a contour map, such as the one shown in Figure 21.11, are lines of constant potential energy. That is, they are equipotentials.

The equipotentials on a contour map tell us two important things about the nature of the landscape they portray. First, they tell us which way something will go if

FIGURE 21.11
A Geographic Contour Map
Lines of constant altitude (labeled in meters in this figure) in a contour map are equipotentials. Objects free to move (such as boulders or volumes of water) move downhill across equipotentials.

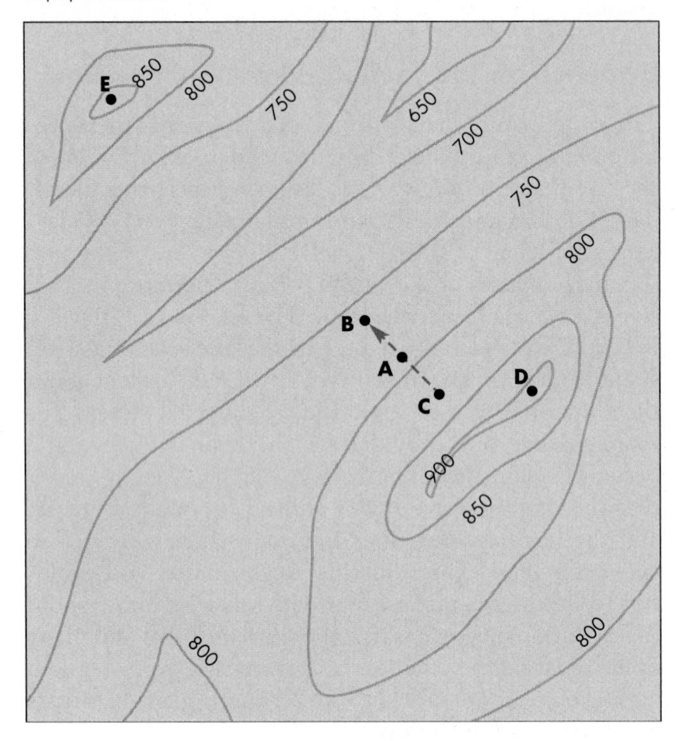

it is free to move. A boulder or a volume of water, for instance, will roll or flow downhill, in the direction of decreasing potential energy. In Figure 21.11, for instance, water will flow from point *A* to point *B*, not from point *A* to point *C*. If we want something to move uphill, in the direction of increasing potential, we must supply energy to push it upward. Along the crest of a ridge or at the top of a hill (points *D* and *E* in Figure 21.11), both of which are local maxima of potential, it is unclear which way a boulder will roll because more than one direction is downhill. At the top of a hill, all directions are downhill. A boulder on a hilltop can rest in that spot, but it is unstable. A small disturbance that pushes it slightly off the top of the hill will cause it to begin rolling downhill in that direction.

A second important thing a contour map tells us is the shape that a fluid (such as water) takes. Water flows downhill until it reaches a local minimum of potential, such as a valley or some other depression. The water then fills the depression to a constant elevation. This means that the boundary of the pond or lake that the water forms is an equipotential, a contour line of the map. Suppose there are two depressions, side by side as in Figure 21.12, and that water flows into one of them. In A, the level of the water rises to 700 m, the equipotential that encircles not only the first depression but the second as well. After this, as in B, the water flows from the first depression to the second through the low point between the two pits. When enough water has flowed into the second pit to fill it to the level of the lowest equipotential that the two depressions share (700 m), the two ponds merge. More water raises the level of the single pond, which always has an equipotential as its boundary, as shown in C.

Roche Lobes Although there are strong similarities, the equipotentials in a binary star system are more complicated than those on a contour map of the surface of the Earth. First, each of the two stars exerts its gravity on everything in the binary star system. Second, the binary system is revolving, so inertial effects are important. All of the matter in the stars is in orbital motion about the center of mass of the system. The inertia of the orbiting matter carries it away from the center of gravity. We see a similar effect when a marble is placed on a turntable. Friction may let it remain on the turntable for a spin or two, but its inertia soon carries it straight off the edge of the turntable. The equipotentials in a binary star system must be calculated taking into account the gravities of the two stars and the inertia caused by their revolution. These equipotentials are harder to visualize than those of a contour map because equipotentials in the binary system are surfaces rather than lines. We can simplify things, however, by looking at only the orbital plane in which the two stars revolve. Some of the equipotentials in the orbital plane of a binary star system are shown in Figure 21.13.

The two stars in Figure 21.13 occupy deep depressions of potential. Close to the center of either star, the gravity of that star dominates and produces equipotentials that are nearly spherical, like those of a single star. Farther from the center of either star, the equipotentials become distorted and are stretched along the line that connects the two stars. This is due to the tidal interaction of the two stars. Just as the Moon and the Earth exert tidal forces that distort each other, each star is distorted by the tidal force of its companion. A very important contour is the first one that encloses both stars in an hourglass shape. Around either star, it marks the boundary of the star's **Roche lobe,** the region in which the gravity of that star dominates. Matter within a star's Roche lobe is gravitationally bound to that star, so it tends to flow downhill toward the star. Matter outside either star's Roche lobe belongs to the binary system rather than to either star alone. $\mathbf{L_1}$, the point at the neck of the hourglass shape, corresponds to the low point between two depressions in a geographic contour map. It is the place where gas can flow from one star to the other if the first star fills its Roche lobe, just as the corresponding

FIGURE 21.12
A Contour Map with Two Depressions
A, Water fills one depression until it reaches the height of the equipotential that encircles both depressions. **B,** It then fills the second depression. **C,** Eventually, the water level rises as a single pond.

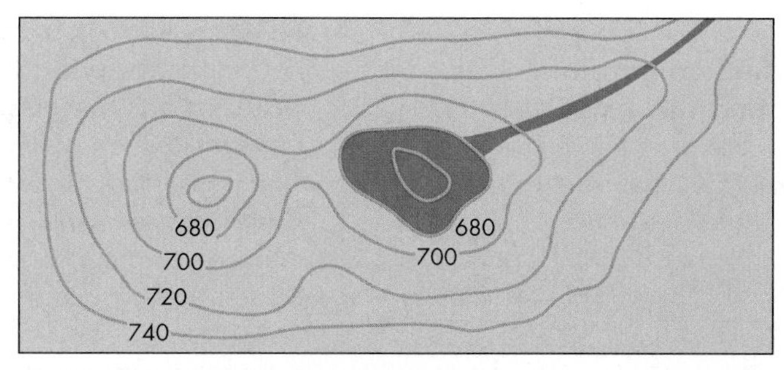

A Water fills one depression first.

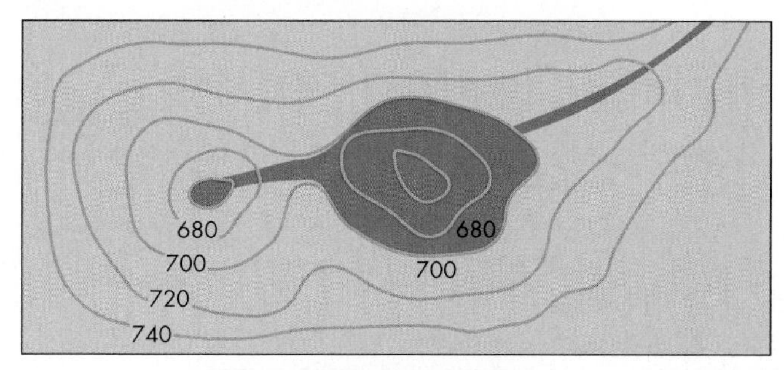

B Water spills over into the second depression.

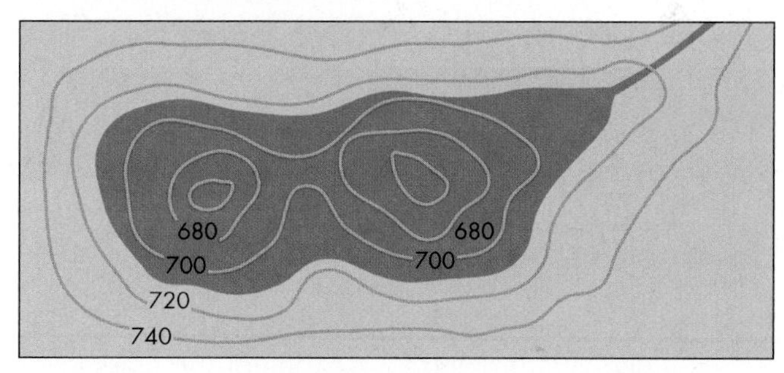

C Water level rises as a single pond.

FIGURE 21.13
Equipotentials in the Orbital Plane of a Binary Star System
The first equipotential that includes both stars defines the boundaries of the Roche lobes of the two stars. Matter flows from one star to another through L_1 and away from the binary system through L_2. The dashed line shows where the cross section of Figure 21.14 cuts through the orbital plane.

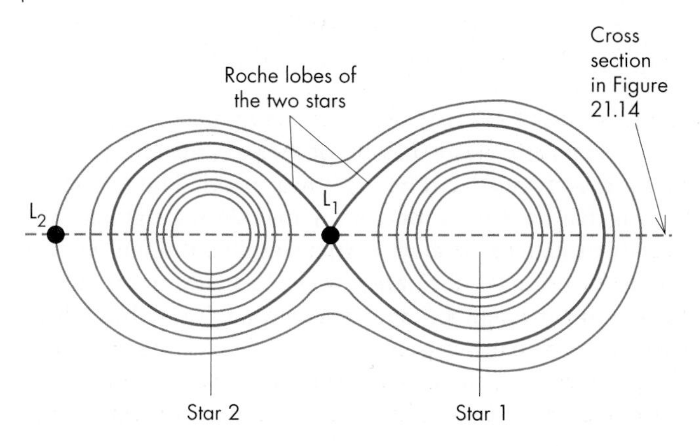

point in a contour map marks the spot where water flows into the second depression after the first fills with water.

Inertial effects become very important as the distance from the center of mass of the binary increases. Eventually, at L_2, the combined gravitational forces of the two stars are just sufficient to hold matter in orbit at that distance. Matter at L_2 is at a local maximum of potential. If it is pushed slightly outward, it will flow away, moving downhill in potential, and be lost to the binary system. Figure 21.14 shows a cross section of the potential map along the line connecting the two stars. Suppose that the right-hand star, the more massive one, expands. After it fills its own Roche lobe, it sheds material to the other star through L_1. If the Roche lobe of the less massive star also fills, the two stars become a single "pond," which grows in size until the equipotential that it fills includes L_2. Further expansion of the star then results in a loss of mass as gas flows out of the binary system through L_2.

FIGURE 21.14
The Cross Section of the Equipotential Map Along the Line Connecting the Two Stars
Notice that the stars occupy the deep depressions and that L_1 lies between the two stars. Gas that reaches L_2 can flow "downhill" away from the binary system.

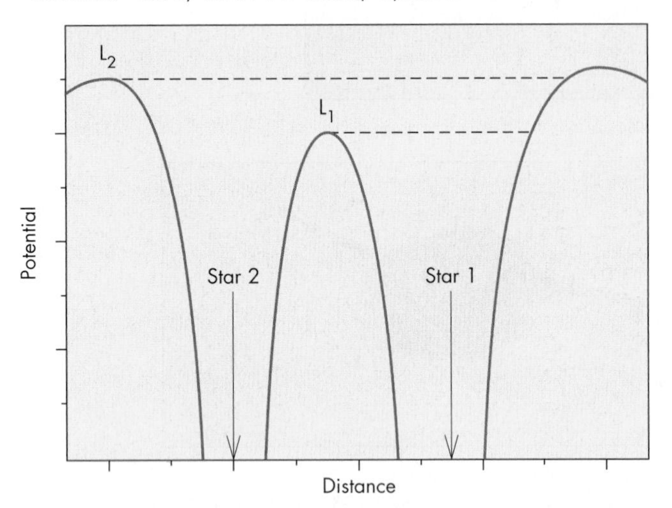

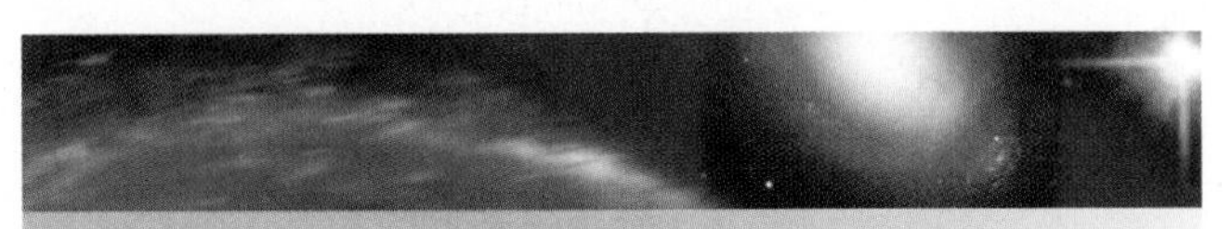

Equipotentials, analogous to lines of constant altitude in a contour map, connect places that have the same potential energy. Equipotentials indicate the directions in which matter can flow in a binary system and the shapes of the stars in the binary. A very important equipotential marks the surface of the Roche lobe of each star—the region in which its gravity dominates.

Angular Momentum and Mass Transfer The lines on a geographic contour map are relatively constant in time. Mountains and valleys gradually erode, but the change is imperceptible. The equipotentials of a binary star system, on the other hand, can change very rapidly. Rapid change occurs when mass and angular momentum are exchanged between the stars or lost from the system. As a result, the separation of the two stars and the size of the Roche lobe of each star can change as time passes.

The graph in Figure 21.15 shows how the separation of two stars with a given combined mass depends on the ratio of the masses of the two stars. When the two stars have the same mass ($M_1/M_2 = 1$), they are closest together and revolve most rapidly about their center of mass. This is because the gravitational force between the two stars depends on the product of their masses, which is greatest when the two stars have the same mass. The greater the ratio of the mass of the more massive star to the mass of the less massive star (points A and B, for instance, in

FIGURE 21.15
A Graph Showing How the Separation of the Stars in a Binary System Depends on the Ratio of the Masses of the Two Stars
The smallest separation occurs when the two stars have equal masses ($M_1/M_2 = 1$). At point **A,** M_1 is much smaller than M_2, so the stars are far apart. At point **B,** M_1 is much greater than M_2, so the stars are also far apart.

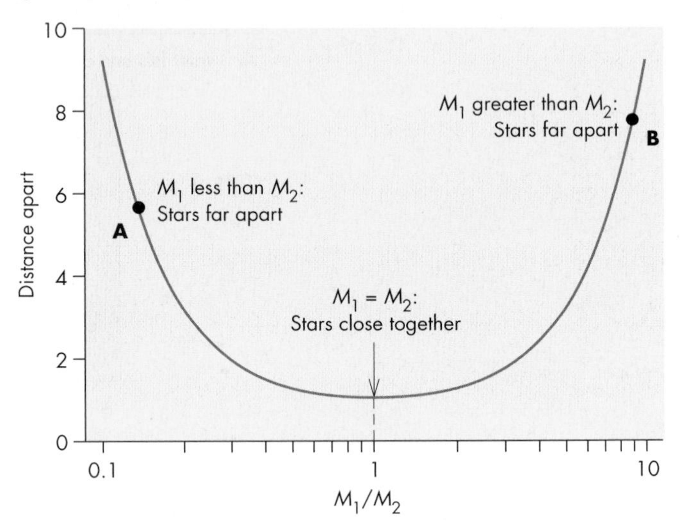

Figure 21.15), the greater the separation of the two stars. For any separation, the more massive star has the larger Roche lobe.

What happens to the separation of the stars in a binary system and the sizes of their Roche lobes when one star sheds mass to the other depends on which star is the one that fills its Roche lobe and sheds mass. If the more massive star in a binary system fills its Roche lobe and sheds mass onto its companion, then the masses of the two stars become more equal, the separation of the two stars shrinks, and the size of the Roche lobe of the mass-losing star becomes smaller, as shown in Figure 21.16. On the other hand, if the less massive star is the one that transfers mass to its more massive companion, then the masses of the stars become even more unequal, the distance between the two stars increases, and the size of the Roche lobe of the mass-losing star grows larger. Whether it is the more massive or less massive star that is shedding mass determines how rapidly mass is exchanged between the two stars.

The Timescale for Mass Transfer If the mass-losing star is the more massive of the pair, the rate at which mass is exchanged can be extremely rapid. The reason for this is that the Roche lobe of the mass-losing star becomes steadily smaller as it sheds matter onto its companion. The gas left outside the shrinking Roche lobe of the mass-losing star flows through L_1 onto the companion. Thus, shedding mass actually causes more loss of mass from the more massive star. Its Roche lobe shrinks, which leads to further mass loss, which causes its Roche lobe to continue to decrease in size, and so on. In a time that may be as short

FIGURE 21.16
A Schematic Diagram Showing How the Roche Lobes of the Stars in a Binary System Change Size as Mass Is Transferred from One Star to Another
A, If the more massive star loses mass, the size of its Roche lobe shrinks. **B,** If the less massive star loses mass, the size of its Roche lobe increases.

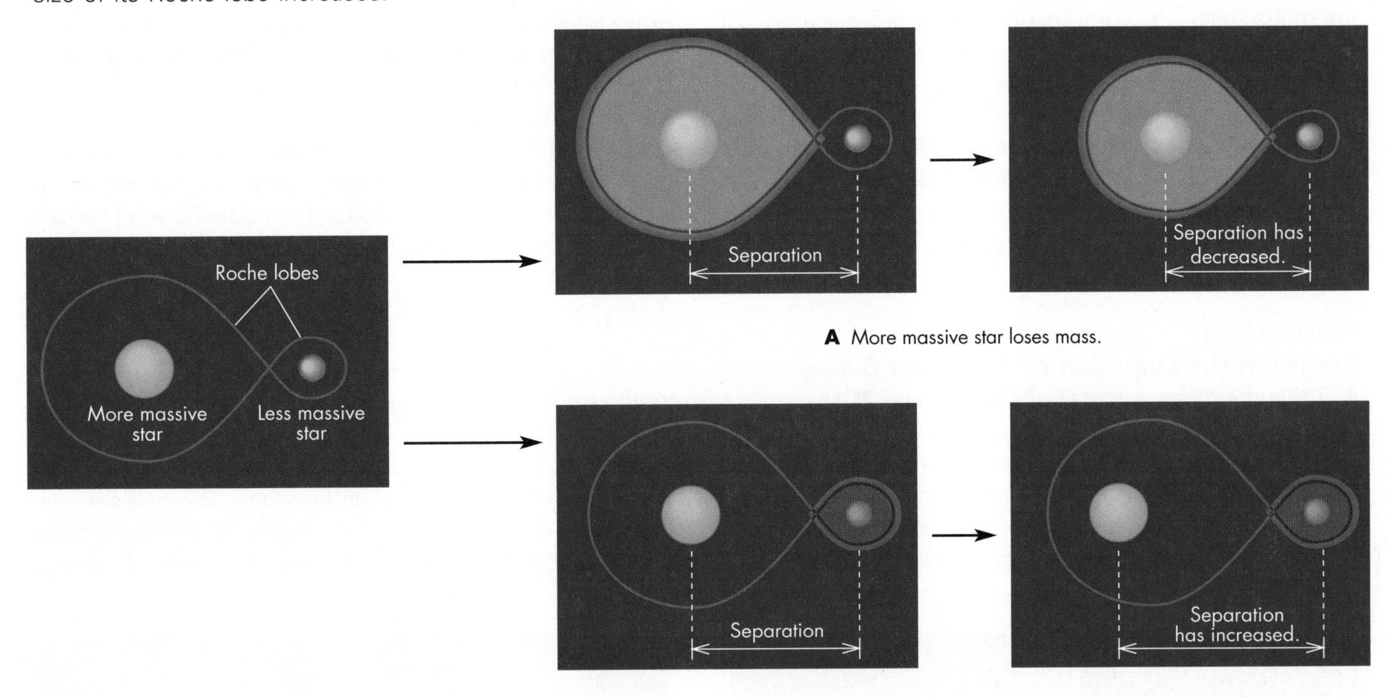

as a few days, the mass-losing star loses enough matter to make the masses of the two stars nearly equal.

When the masses of the two stars become equal, the mass-losing star is still larger than its Roche lobe and continues to shed mass onto its companion, making the companion the more massive star. The masses of the two stars grow more unequal, the distance between them increases, and the Roche lobe of the mass-losing star grows larger. Mass loss continues, at a smaller rate, until the mass-losing star again becomes smaller than its Roche lobe. This may take tens of thousands of years (an instant by stellar standards), during which time the mass ratio of the binary system is nearly reversed. This, presumably, is what has already happened to the stars in Algol.

If the star that fills its Roche lobe is the less massive member of the binary, mass exchange is much slower. As mass is exchanged, the mass ratio of the system increases and the stars draw apart. The Roche lobe of the mass-losing star grows larger. If the mass-losing star continues to expand, it adjusts to fill its Roche lobe and to transfer mass to its companion. The rate at which the star and its Roche lobe increase in size is governed by the rate of expansion of the star. The rate of expansion depends, in turn, on the star's evolutionary stage and generally is very slow. Slow mass exchange may go on for millions or billions of years.

Although most of the mass exchange between the stars in a binary goes on during episodes of rapid evolution, most of the close binaries that have been discovered are exchanging mass at slow rates. The episodes of rapid mass exchange are very brief, so we are unlikely to catch a binary star system in the act of rapid mass exchange. Because the duration of slow mass exchange is thousands of times longer than the episodes of rapid mass exchange, the overwhelming majority of interacting binaries are exchanging mass slowly at any given time.

As mass is transferred from one member of a binary system to the other, conservation of angular momentum causes the separation and the period of the system to change. The more nearly equal the masses of the two stars, the smaller the separation. If the more massive star is the one losing mass, the transfer takes place very quickly. Slow mass transfer occurs if the mass-losing star is the less massive of the pair.

The Evolution of Massive Binaries Binary star systems can form with a wide range of stellar masses and separations, so it might appear that there is an infinite variety of possible evolutionary scenarios for binary star systems. Actually, although details vary from binary system to binary system, the majority of binaries follows one of two general evolutionary paths. One of these is for massive binaries, the other for binary systems with two low-mass stars. The combination of a massive star with a low-mass companion is seldom encountered because massive stars tend to form in groups.

The story of two massive stars begins with the formation of a binary in which the stars may be separated by as much as several astronomical units. The two stars evolve independently until the more massive star exhausts its central hydrogen and expands toward the red giant stage, as shown in Figure 21.17, A. When it fills its Roche lobe, it begins to transfer mass to its companion at a rapid rate. The companion can't accommodate new material at such a large rate, so it expands, too. Soon both stars fill their Roche lobes. The matter shed by the more massive star flows into the region just outside the Roche lobes of the two stars, producing a **common envelope** around the two stars, as in Figure 21.17, B.

As the two stellar cores revolve within their common envelope, they experience friction with the gas in the envelope. The friction does two things. First, it slows the orbital speed of the stellar cores, making them spiral inward toward each other as shown in Figure 21.17, C. Second, energy

FIGURE 21.17
Stages in the Evolution of a Massive Binary
A, The more massive star fills its Roche lobe. **B,** Mass is transferred to a common envelope. **C,** The stars spiral toward each other inside the common envelope. **D,** The common envelope is ejected. **E,** The more massive star becomes a neutron star or black hole. **F,** The companion, now the more massive star, sheds matter to the compact object by means of a wind. **G,** The companion fills its Roche lobe and sheds mass to the compact object and a common envelope. The two stars either merge to form a single black hole inside the common envelope, **H,** or eject the common envelope, **H′.** If the common envelope is ejected, the normal companion evolves until it explodes, either ending the binary system, **I,** or leaving a binary with two compact objects, **I′.**

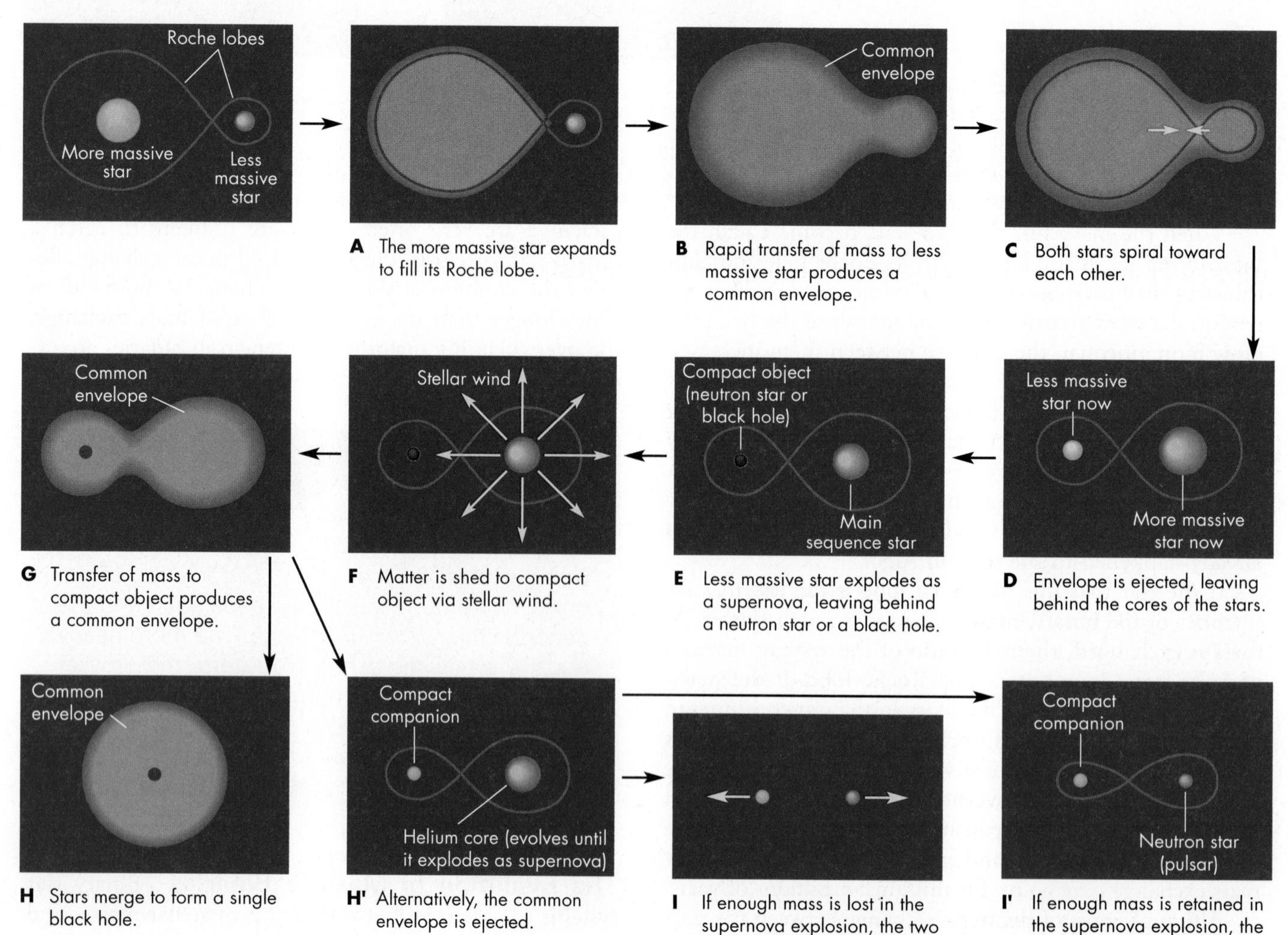

Roche lobes
More massive star
Less massive star
A The more massive star expands to fill its Roche lobe.
Common envelope
B Rapid transfer of mass to less massive star produces a common envelope.
C Both stars spiral toward each other.
Less massive star now
More massive star now
D Envelope is ejected, leaving behind the cores of the stars.
Compact object (neutron star or black hole)
Main sequence star
E Less massive star explodes as a supernova, leaving behind a neutron star or a black hole.
Stellar wind
F Matter is shed to compact object via stellar wind.
Common envelope
G Transfer of mass to compact object produces a common envelope.
Common envelope
H Stars merge to form a single black hole.
Compact companion
Helium core (evolves until it explodes as supernova)
H' Alternatively, the common envelope is ejected.
I If enough mass is lost in the supernova explosion, the two stars go their separate ways.
Compact companion
Neutron star (pulsar)
I' If enough mass is retained in the supernova explosion, the two stars remain in close orbit.

and angular momentum from the stellar cores are given to the common envelope. The common envelope eventually acquires so much rotational energy that it is ejected from the binary system, leaving only the stellar cores behind (Figure 21.17, D). The star that was less massive is now the more massive star in the binary. It has gained some mass during the period of mass transfer but looks like a more or less normal main sequence star. The star that was once the more massive is now less massive than its companion. It is the core of what was once a massive star and consists mostly of helium and heavy elements produced by nuclear reactions in the massive star. Soon the star burns the rest of its nuclear fuel, explodes as a supernova, and leaves behind a neutron star or a black hole, as shown in Figure 21.17, E.

At this point in its career, the binary system begins a long period of stability while the massive main sequence star (an O or a B star) consumes its central hydrogen. Like other O and B stars, the more massive star (which was originally less massive than its exploded companion) has a wind that blows matter outward into space (Figure 21.17, F). Part of the mass that blows away from the star is captured by its compact companion. Like water flowing into a deep pit, the captured gas is accelerated to great speeds and gains energy as it falls toward the neutron star or black hole. The amount of energy produced as it falls inward is tens of times greater than the energy that would be produced if the gas were consumed in fusion reactions. The fate of the infalling gas and the observable effects it produces are described later in this chapter.

Eventually, the main sequence star consumes its core hydrogen and begins to expand. When it fills its Roche lobe, matter begins to flow toward its compact companion through L_1 (Figure 21.17, G). Because the star is more massive than its compact companion, the separation of the stars decreases as mass is exchanged. The Roche lobe of the mass losing star shrinks, leading to very rapid mass loss. Gas pours through L_1 and falls toward the compact object at a great rate. Part of the gas shed by the more massive star spills outward into a common envelope. Within the common envelope, the core of the star and the compact object spiral toward each other. In most cases, they eventually merge, producing a single black hole within the envelope (Figure 21.17, H). The merging of the star and the compact object marks the end of the binary system.

In some cases, the common envelope is ejected before the two stars spiral into each other. All the hydrogen-rich outer layers of the mass-losing star have been lost, leaving only the helium core of the star and its compact companion (Figure 21.17, H′). Again, a wind from the helium star supplies gas that falls into the compact companion. After it uses its remaining nuclear fuel, the helium star collapses in a supernova explosion. If enough mass is blasted away by the explosion, the gravitational attraction of the binary system is weakened enough that the two components exceed escape velocity and go their separate ways, ending the binary system (Figure 21.17, I). In other cases, the binary retains enough mass to bind the stars together. When this happens, the binary system consists of a neutron star (possibly seen as a pulsar) in a close orbit about the original neutron star or black hole (Figure 21.17, I′). The stages in the evolution of a massive binary star system are summarized in Table 21.1.

Table 21.1

Evolution of a Massive Binary System (see Figure 21.17)

STAGE	WHAT HAPPENS DURING STAGE
A	Massive star expands, rapidly transfers mass to companion.
B	Common envelope forms about both stars.
C	Stars spiral toward each other, eject common envelope.
D	Formerly less massive star is now more massive star.
E	Less massive star explodes as supernova, becomes black hole or neutron star.
F	More massive star loses mass in a hot wind.
G	More massive star expands, forms common envelope.
H	Stars merge within common envelope.
H′	Common envelope ejected, more massive star explodes as supernova.
I	Stars fly apart, ending binary system.
I′	Stars, both compact objects, remain in close orbit.

In a massive binary system, the more massive star leaves the main sequence first, swells to fill its Roche lobe, and sheds mass to its companion and into a common envelope. Ejection of the common envelope leaves a close binary system in which the star that was more massive is now a neutron star or black hole and is less massive than its companion. During a long period of stability, a wind from the companion sheds matter onto the compact object. The evolution of the companion leaves either a single compact object or a binary containing a neutron star and the original compact object.

The Evolution of Low-Mass Binaries The evolution of a binary system containing two stars, each less massive than a few solar masses, has many similarities to the evolution of more massive stars. A major difference, however, is that the compact object that results from the evolution of a low-mass star is a white dwarf rather than a neutron star or black hole.

The career of a low-mass binary begins with both stars widely separated and evolving separately. The more massive star finishes hydrogen burning first and expands to fill its Roche lobe, as shown in Figure 21.18, A. Because the more massive star is the one that fills its Roche lobe, mass exchange is very rapid and a common envelope develops about the two stars (Figure 21.18, B). The stars spiral toward each other, providing energy and angular momentum to the common envelope (Figure 21.18, C). Eventually, the common envelope is ejected. What remains is the core of the mass-losing star (a white dwarf) and the second star, still less massive and on the main sequence (Figure 21.18, D). The orbital period of the binary is now a few days or less. The binary system now begins to lose angular momentum because of a stellar wind from the main sequence star or because the system emits gravitational radiation. As angular momentum is lost the stars slowly spiral toward each other and their Roche lobes shrink as shown in Figure 21.18, E. When the Roche lobe of the main sequence star becomes smaller than the star, the main sequence star begins to slowly transfer mass to the white dwarf. This continues until nuclear reactions die out in the star and it, too, becomes a white dwarf. The stages in the evolution of a low-mass binary star system are summarized in Table 21.2.

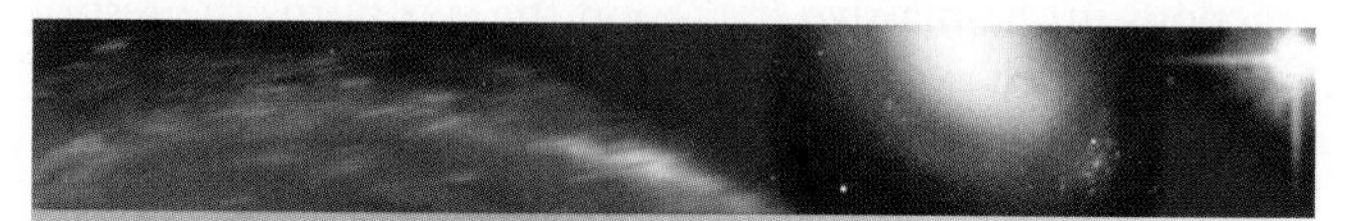

The evolution of a low-mass binary initially resembles that of a massive binary except that the more massive star becomes a white dwarf more massive than its companion. The stars spiral toward each other as the main sequence star transfers mass to its companion.

FIGURE 21.18
Stages in the Evolution of a Low-Mass Binary
A, The more massive star fills its Roche lobe. **B,** Mass is transferred to a common envelope. **C,** Friction in the common envelope brings the stars closer together. **D,** The common envelope is ejected, leaving a white dwarf that is more massive than its normal companion. **E,** The system loses angular momentum and the two stars spiral toward each other. The main sequence star transfers mass to the white dwarf.

ANIMATION *Stages in the evolution of a low-mass binary*

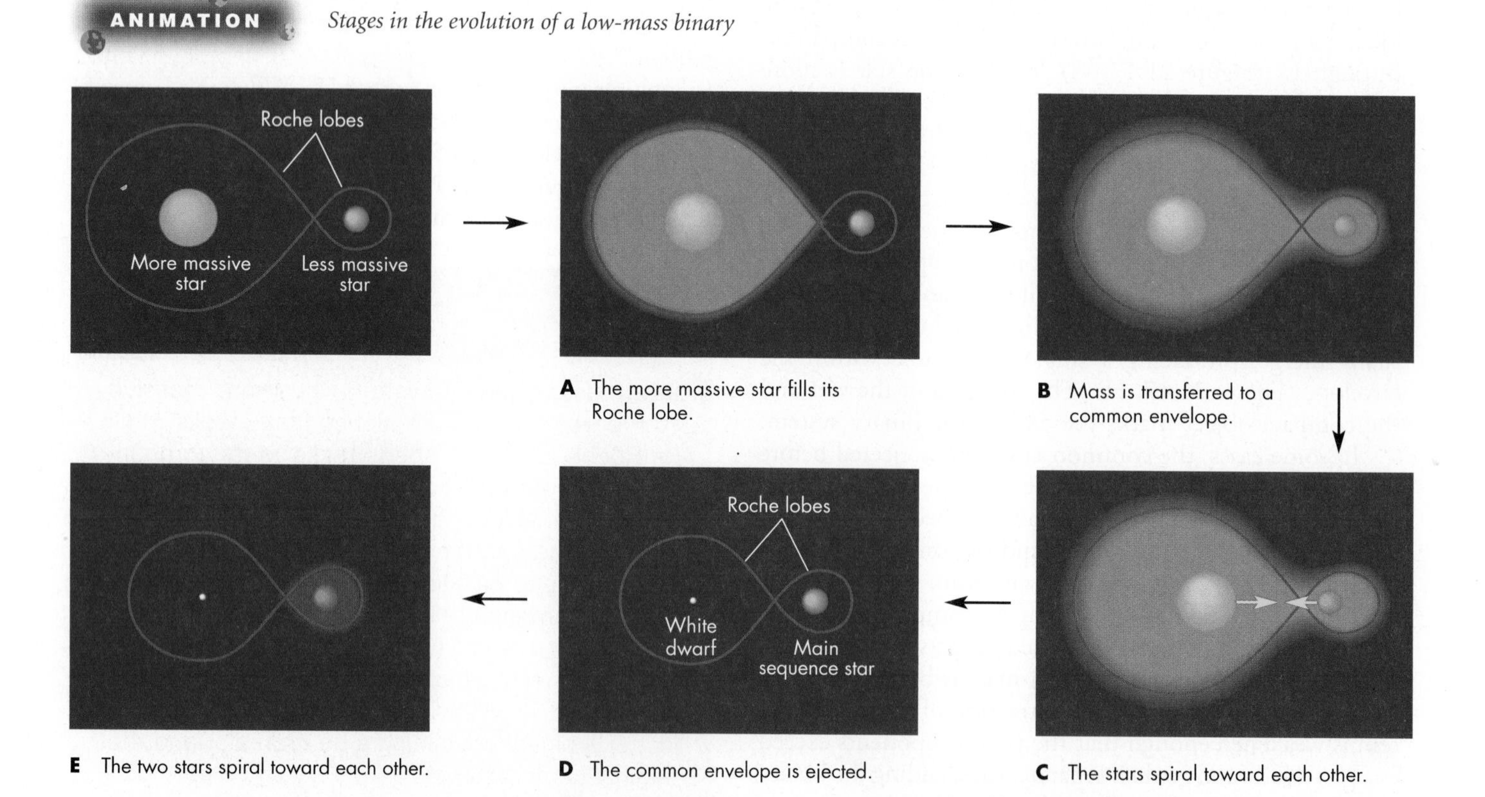

Table 21.2

Evolution of a Low-Mass Binary System (see Figure 21.18)

STAGE	WHAT HAPPENS DURING STAGE
A	Massive star expands, rapidly transfers mass to companion.
B	Common envelope forms about both stars.
C	Stars spiral toward each other, eject common envelope.
D	More massive star is now a white dwarf.
E	The two stars slowly spiral toward each other.

21.4 BINARIES WITH COMPACT OBJECTS

Binary systems that contain a compact object (a white dwarf, neutron star, or black hole) produce some of the most energetic and spectacular stellar phenomena. Many of these phenomena are caused or triggered by the transfer of mass from a star to its compact companion. The compact object is small, so gravity is very great near its surface. Matter falling toward the compact object is accelerated to high speeds, converting gravitational potential energy to kinetic energy. Large amounts of gravitational energy are released when matter falls toward the compact object after entering its Roche lobe. A kilogram of gas that falls onto a white dwarf releases about as much energy as would be produced if the gas were used as fuel for fusion reactions. Matter falling onto a neutron star or into a black hole releases one hundred times as much energy. Some of the energy that is released heats the infalling gas to 10^7 K in the case of a neutron star or black hole and 10^5 K in the case of a white dwarf. For neutron stars and black holes, the heated gas radiates strongly in the X-ray region of the spectrum. Gas falling onto a white dwarf radiates mostly in the ultraviolet.

Accretion Disks

In most cases, the gas flowing toward a compact object forms a thin disk of orbiting material called an **accretion disk** in which gas spirals inward toward the surface of the compact object. An accretion disk forms because the compact object is small compared with the distance the infalling gas must fall to reach it from L_1. As Figure 21.19 shows, the infalling gas will miss the compact object and go into orbit about it unless the infalling gas is pointed precisely at the compact object. Thus, the stream of matter falling inward from the companion forms a disk in orbit about the compact object. The infalling gas originated in the companion of the compact object and retains the direction of orbital motion of the companion. As a result, the accretion disk revolves in the orbital plane of the binary star system.

If there were no friction in the accretion disk, gas at a given distance would continue to orbit at that distance indefinitely. The friction between gas in neighboring orbits, however, causes the gas in the accretion disk to spiral inward until it finally hits the surface of the compact object or passes through the event horizon of the black hole. As the spiraling gas moves inward, gravitational energy is released. About half of this goes into orbital motion, while the rest heats the accretion disk. The release of gravitational energy is greatest in the inner accretion disk, so the gas there is hotter and radiates energy at a greater rate than in the outer disk. Intense X-ray or ultraviolet radiation from an accretion disk is the signature of a close binary system that contains a compact object.

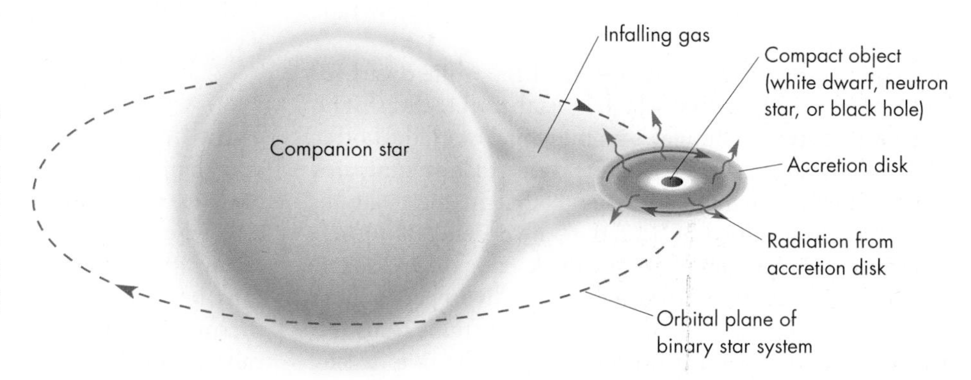

FIGURE 21.19
The Formation of an Accretion Disk
Gas shed by the companion of a black hole, neutron star, or white dwarf falls toward the compact object. The infalling gas has orbital angular momentum, however, so it misses the compact object and flows into an accretion disk around the compact object.

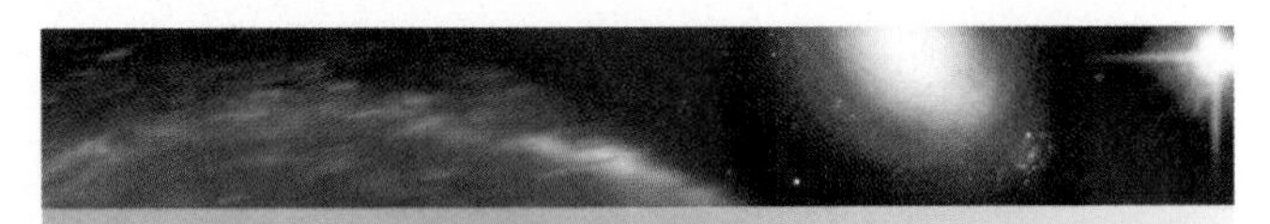

Gas falling toward a compact object releases gravitational energy that makes the gas hot enough to emit X rays and ultraviolet radiation. The angular momentum of the infalling matter often causes the matter to go into orbit about the compact object, forming an accretion disk. Friction in the accretion disk causes gas to spiral inward, releasing energy that heats the disk.

Binaries with White Dwarfs

Binary systems in which a normal star sheds its mass onto a white dwarf companion undergo large, temporary increases in brightness. In some cases, the outburst is caused by the swift, violent release of nuclear energy on the surface of the white dwarf. In other cases, a much more violent explosion results from the almost instantaneous consumption of nearly all the remaining nuclear fuel within the white dwarf.

Novae About once a decade, on average, astronomers observe a "new" naked-eye star that alters the traditional shape of a constellation. These stars, named **novae** from the Latin word for new, fade in brightness and can be seen with the naked eye only for a few weeks. Using telescopes, however, astronomers can follow their diminishing brightness for months. By comparing the positions of novae with existing star charts, astronomers have found that novae are actually old stars that increase dramatically in brightness. These spectacular brightenings occur when the normal star in a binary system sheds matter onto its white dwarf companion.

The gas shed by the normal star enters the Roche lobe of the white dwarf, spirals inward through the white dwarf's accretion disk, and eventually reaches the surface of the white dwarf. Over long periods of time, this gas builds a hydrogen-rich layer on top of the white dwarf, which is composed mostly of carbon and oxygen. What happens to the hydrogen-rich layer depends on how rapidly the layer can accumulate.

The gas that reaches the surface of the white dwarf releases gravitational energy, which heats the white dwarf. The faster that matter accumulates on the white dwarf, the hotter the white dwarf becomes. A white dwarf that accumulates matter slowly remains cool enough that its outer layers, including those added by accretion, are degenerate. When the layer of hydrogen becomes thick enough, hydrogen fusion begins at its bottom, where the hydrogen is hottest. The energy released by the fusion reactions heats the gas but doesn't expand it because the pressure in a degenerate gas doesn't depend on its temperature. The heated hydrogen burns even faster, raising its temperature and again increasing the fusion rate. A very rapid thermonuclear runaway consumes nearly all of the accreted hydrogen. Within a few hours or days, the white dwarf increases in brightness by as much as a million times, producing a nova outburst. The layer of accreted matter is blasted into space at a few thousand kilometers per second. As the shell of ejected matter expands and cools, it gradually becomes dimmer. Over months or years, the brightness of the nova declines to normal.

Novae are much more common than supernovae. About 200 novae have been recorded in the Milky Way in the last several centuries, compared with only a handful of supernovae. Taking into account those novae that have occurred in parts of the Milky Way that we can't observe, there may be as many as 50 per year in our galaxy. Novae are also much less destructive than supernovae. At the end of a nova explosion, the white dwarf is nearly the same as it was when it began accreting gas from its companion. As accretion resumes, the process that led to the nova outburst begins anew. In a few cases, more than one nova outburst has been seen from the same binary system. These binaries, known as **recurrent novae,** have smaller outbursts than other novae. This may be due to the brief interval of time, as short as a few decades, during which hydrogen can build up between thermonuclear runaways. It seems likely that more powerful novae, which have been observed to explode only once, also recur, but at intervals of tens of thousands to millions of years.

Novae are temporary brightenings of binary systems in which a normal star transfers mass to its white dwarf companion. The white dwarf accumulates an outer layer of hydrogen-rich gas that burns explosively in a thermonuclear runaway in the surface layers of the white dwarf. The nuclear energy that is released blasts the outer layer of the white dwarf into space. The same white dwarf may produce a nova explosion each time enough new matter has accumulated.

Type Ia Supernovae An outburst even more explosive than a nova occurs if the normal star sheds matter onto its white dwarf companion at a rapid rate. In that case, the gravitational energy released as gas falls onto the white dwarf keeps the layer of new material so hot it can't become degenerate. Without degeneracy, the kind of thermonuclear runaway that happens in a nova outburst can't occur. Instead, the fusion of hydrogen into helium goes on at a more or less steady rate at the base of the hydrogen-rich shell. No matter is ejected via nova explosions, so the mass of accreted material and the mass of the white dwarf gradually increase over tens of millions of years.

As the white dwarf becomes more massive, it also becomes smaller and hotter. When the temperature at its center reaches 10 billion K, carbon burning begins. Because the interior of the white dwarf is degenerate, the energy released by the sudden ignition of carbon doesn't expand the core of the star. Instead, it increases the temperature, leading to a thermonuclear runaway in which the gas in the core is almost instantly processed through a series of nuclear reactions to form elements like iron, nickel, and cobalt. The huge amount of energy released by the sudden fusion of carbon makes the interior of the white dwarf convective. Carbon-rich gas from the outer part of the white dwarf is carried by convection currents into the core of the star, where it, too, is rapidly consumed. In about 1 second, most of the carbon in the white dwarf is transported to the core and consumed. The energy released during this nuclear burning blows the white dwarf completely apart, leaving nothing behind. The explosive disruption of a white dwarf produces a **type Ia supernova.** (There are other, rarer, type I supernovae that do not seem to have anything to do with white dwarfs.)

The nuclear energy released in a type Ia supernova in 1 second is comparable to the total energy that the Sun will produce in its entire main sequence lifetime. Most of the energy is used to blast the white dwarf apart at a speed of about 10,000 km/s. Enough energy is emitted as light, however, that the supernova can outshine the galaxy in which it lies. Like type II supernovae, the radioactive decay of the isotopes ^{56}Ni and ^{56}Co is responsible for much of the light emitted by a type Ia supernova. The light produced by the radioactive decays reaches a maximum about 15 days after the explosion and then declines slowly over many years, as shown in Figure 21.20. Type Ia supernovae are extremely important because, as described in Chapter 26, they have allowed astronomers to find the distances of very remote galaxies.

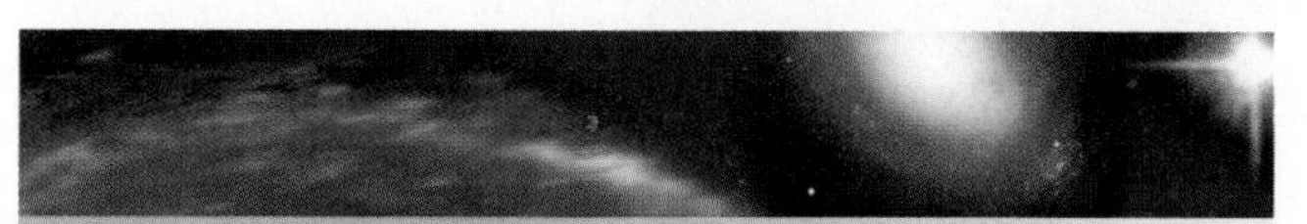

If the rate of mass transfer is high, the outer layers of a white dwarf in a binary system are so hot that they do not become degenerate. In this case, fusion of hydrogen to helium in the accreted gas goes on relatively smoothly. Eventually, the mass of the white dwarf increases so that its core becomes dense and hot enough for the almost instantaneous fusion of carbon and other elements to take place. The energy released by the fusion reactions blows the white dwarf apart in a type Ia supernova.

FIGURE 21.20
The Light Curve of a Type Ia Supernova
At maximum brightness, the supernova can be as bright as 10 billion stars like the Sun. In the year after maximum brightness, it fades to less than 1% of its maximum brightness. For comparison, the light curve of a typical type II supernova is shown in Figure 20.9.

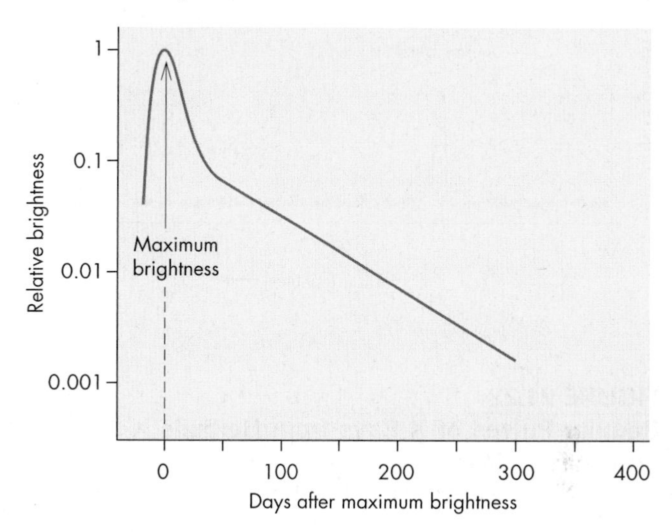

Binaries with Neutron Stars or Black Holes

Most of the 150 or so sources of bright X-ray emission in our galaxy are binary systems that contain a neutron star or black hole. Most of the X-ray radiation from most X-ray binaries is steady or flickering emission from accretion disks. In some systems, however, some of the X rays originate in different ways.

X-Ray Bursters Sporadic bursts of X rays, called **X-ray bursts,** are produced by runaway thermonuclear reactions on the surfaces of neutron stars that accrete matter from their companions. The gas that reaches the surface of the neutron star releases enough gravitational energy to heat the surface to 10 million K. At such a high temperature, the surface always emits some X-ray radiation. The gas that accretes on the neutron star is hydrogen-rich. After reaching the neutron star, it burns steadily to form helium. When the temperature and density of the helium become high enough, the fusion of helium into carbon (and then carbon into heavier elements) begins. Because the fusion reactions take place in a degenerate gas, the accumulated helium burns almost instantly. The released energy heats the surface of the neutron star and abruptly increases the emission of X rays. The graph in Figure 21.21 shows that the neutron star cools within a few minutes and resumes its normal level of X-ray emission. After a few hours, enough fresh gas accumulates on the neutron star that another X-ray burst occurs. The energy released in an X-ray burst is remarkable. In a few seconds the neutron star emits as much energy as the Sun emits in a week.

FIGURE 21.21
A Graph of the Brightness of X Rays from an X-Ray Burster
Bursts occur every few hours. During a burst, X-ray brightness increases abruptly and then returns to its normal level in less than a minute.

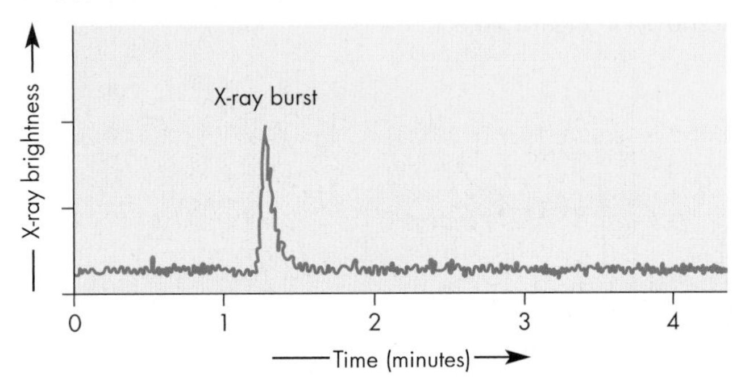

FIGURE 21.22
Regular Pulses of X Rays from Herculis X-1, an X-Ray Pulsar
Pulses of X rays are emitted by Herculis X-1 at regular intervals of 1.24 seconds. Other X-ray pulsars have periods that range from less than 1 second to about 15 minutes.

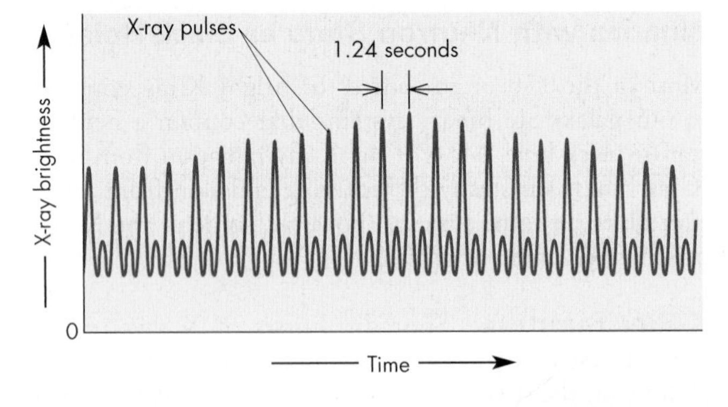

X-Ray Pulsars

X-ray pulsars show a periodic variation in their X-ray brightness, as shown in Figure 21.22. The periods of the X-ray pulsars range from less than a second to 15 minutes. The brightness variations of X-ray pulsars are produced by a process analogous to the one that produces bursts of radio emission from ordinary pulsars. The X rays are emitted by gas that falls onto a neutron star with a large magnetic field. As Figure 21.23 shows, the only place that the infalling gas can reach the surface of the neutron star is near its magnetic poles. Thus, there are hot spots on the neutron star at its magnetic poles. If the magnetic axis is tipped with respect to the rotation axis, the rotation of the neutron star sweeps the bright regions into view and then out of view again. This causes the X-ray brightness of the neutron star to brighten and then dim once or twice per rotation, depending on whether we are able to see one or both of the magnetic poles.

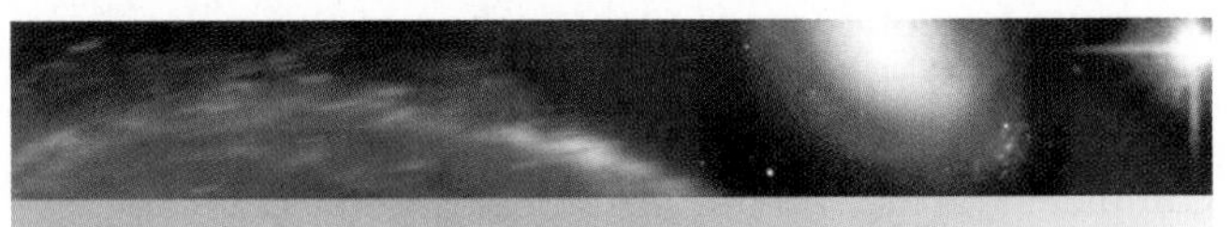

The accretion of gas onto neutron stars can produce bursts of X rays when the neutron star's surface is heated by runaway fusion in the accreted gas. Pulses of X rays are seen when an X-ray-emitting magnetic pole of the neutron star sweeps past.

Black Hole Candidates

Most of the X-ray-emitting binaries that have been detected contain neutron stars rather than black holes. Neither X-ray pulses nor X-ray bursts, which are seen in the majority of X-ray binaries, can be produced by a black hole. A minority of the X-ray binaries, however, display only steady or flickering X-ray

FIGURE 21.23
A Model for the Emission of X Rays from an X-Ray Pulsar
Infalling ionized gas can reach the neutron star only near the magnetic poles, where X-ray-emitting hot spots are formed. The rotation of the pulsar carries the hot spots into and out of view.

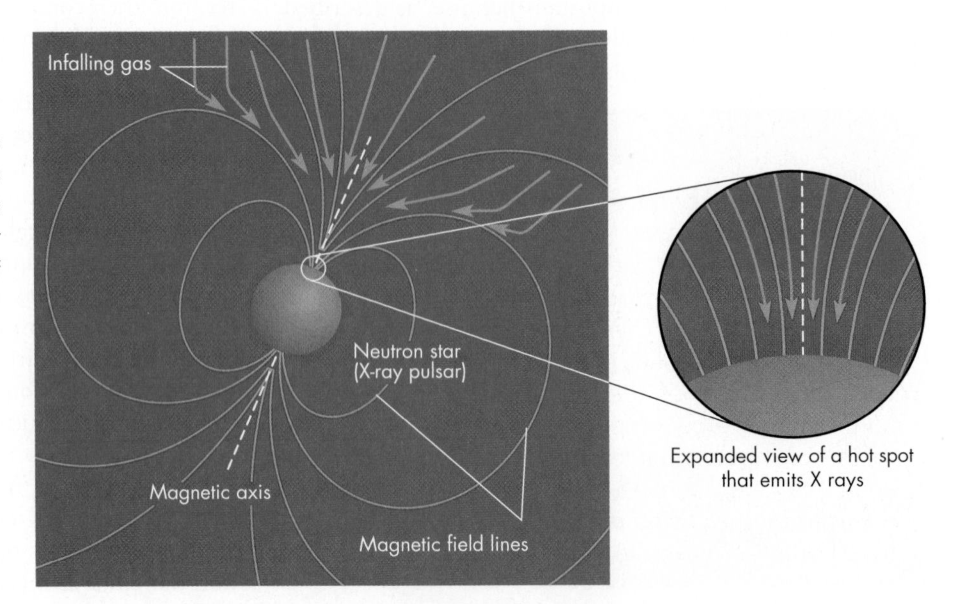

emission, which could be produced in an accretion disk around either a neutron star or a black hole. Unless pulses or bursts are seen, no one can tell whether a binary system contains a neutron star or a black hole by looking at the properties of its X-ray emission.

At present, the surest way to identify binary systems that contain black holes is to determine the mass of the compact object. The maximum mass that a neutron star can have is about 2 or 3 $M_\odot$. A compact object more massive than this must be, by elimination, a black hole. To find the mass of a compact object, its binary companion must produce a detectable spectrum. If the orbital speed of the companion can be found by measuring the Doppler shift of its spectral lines, Kepler's laws can be used to find the masses of the two stars in the binary system.

Two strong cases for black holes, Cygnus X-1 and A0620-00, are represented in Figure 21.24. If the companion of the compact object in Cygnus X-1 is what it appears to be (a normal O supergiant), then the O star has a mass of 30 $M_\odot$, and the compact object has a mass of 7 $M_\odot$. However, stars in binary systems often have traded mass back and forth and sometimes do not have masses that correspond to their spectral types. Nonetheless, it is likely that the compact object in Cygnus X-1 is too massive to be anything but a black hole. The visible star in A0620-00 is a K main sequence star likely to have a normal mass. The compact object is more massive than 3.2 $M_\odot$ and is probably a black hole.

FIGURE 21.24
Cygnus X-1 and A0620-00, Two Likely Candidates for Black Hole Binary Systems
A, Cygnus X-1 contains an O supergiant and a compact companion, which may be a black hole. **B,** A0620-00 contains an apparently normal K7 main sequence star and a compact companion likely to be a black hole.

ANIMATION *X rays from a black hole binary*

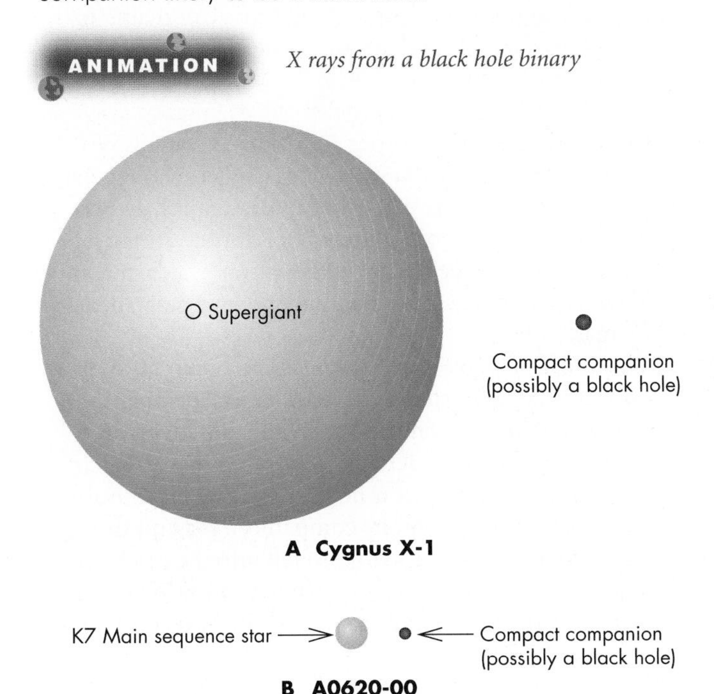

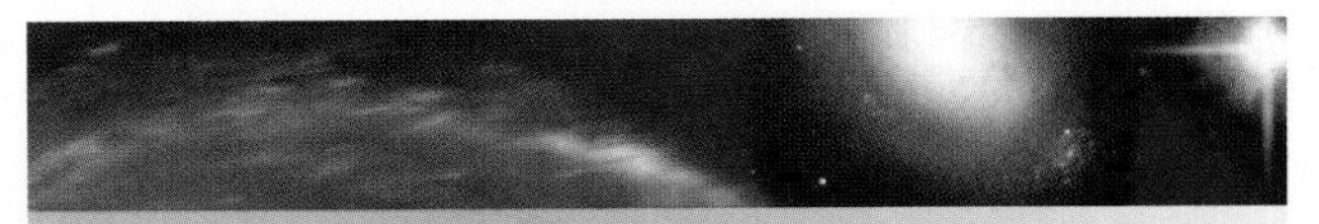

Some X-ray binaries may contain black holes rather than neutron stars. The most likely cases for black holes are those in which the mass of the compact object is too great for it to be a neutron star.

Binary and Millisecond Pulsars Most pulsars seem to be single stars. Several dozen, however, are members of binary star systems in which the other member (too dim to be seen) is a neutron star or a white dwarf. About half of the binary pulsars have rotational periods (the time between pulses) of only a few milliseconds. This is unusually rapid rotation for a pulsar, which ordinarily slows down quickly after it forms. The rapid rotation of some of the binary pulsars is a result of mass transfer from their companions. This mass transfer took place long after the neutron star formed and before the companion became a compact object. The angular momentum of the infalling material sped up the rotation of the pulsar so that it now rotates much faster than single neutron stars of the same age.

Mass exchange from a binary companion can speed up the rotation of a pulsar in a binary system and explain the millisecond pulsars in binary systems. Until recently, however, there was no good explanation for the few millisecond pulsars that are not members of binary star systems. However, a binary millisecond pulsar has been discovered that has a companion of only 0.02 $M_\odot$. The companion is losing mass at a rapid rate, probably because it is being "evaporated" by the intense bombardment of high-energy particles and radiation from the pulsar, as shown in Figure 21.25. Within a few hundred million years, the companion will have disappeared completely, leaving only a single millisecond pulsar. Astronomers speculate that the other millisecond pulsars not now in binary star systems may once have had binary companions that were disrupted long ago.

Binary pulsars have neutron star or white dwarf companions. Before the companions became compact objects, they shed matter to the pulsars, increasing their rate of spin so that many of them have rotation periods of only a few milliseconds. Single millisecond pulsars may have destroyed their companions by bombarding them with energetic particles and radiation.

FIGURE 21.25
The Disruption of the Companion of a Pulsar
A, The companion lies only about 0.01 AU (1.5 million km) from the pulsar, which bombards the companion with energetic particles, gamma rays, and X rays. **B,** The gas in the companion is heated and driven off. Eventually, the companion may be completely "evaporated" by the pulsar.

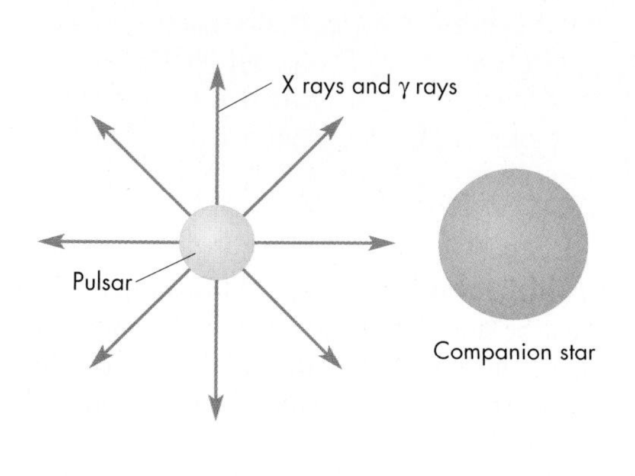

A Pulsar bombards companion star with energetic particles.

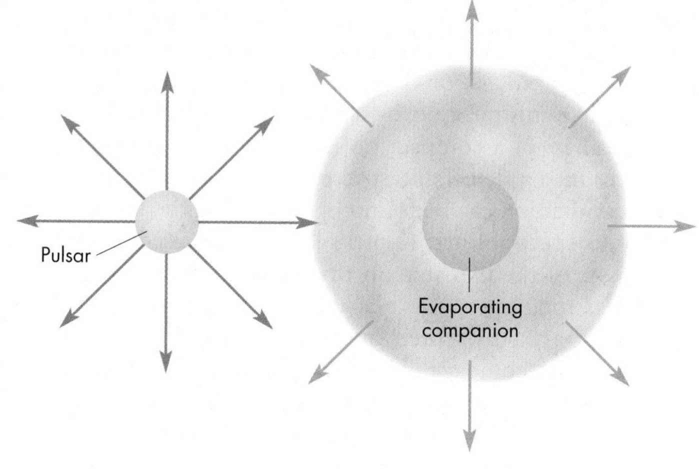

B The gas in the companion is heated and driven off.

Chapter Summary

- The stars in a visual binary are widely enough separated that their images can be distinguished. They usually have a long orbital period. If the distance to the binary is known, the masses of the stars can be calculated. (Section 21.1)
- The stars in a spectroscopic binary usually are close together and orbit each other with large velocities. As the stars orbit, their radial velocities change. These changes cause the Doppler shifts of their spectral lines to change in a periodic manner. (21.1)
- The stars in an eclipsing binary periodically eclipse each other, causing the binary to dim and then brighten at regular intervals. The light curve of the binary can be used to determine the sizes and temperatures of the stars. (21.1)
- About 50% of all stars are members of binary or multiple star systems. Multiple star systems consist of close pairs orbited at great distances by single stars or other close pairs. (21.1)
- Mechanisms have been proposed to account for the formation of both close and wide binary systems. None of these mechanisms is understood well enough for astronomers to be sure just how important it is in the formation of binary stars. (21.2)
- Equipotentials can be used to determine the directions in which gas will flow in a binary system and to find the shapes of its stars. The Roche lobes of a binary system mark the gravitational domains of each of its stars. Gas outside of the Roche lobes is common to both stars. (21.3)
- Angular momentum is conserved as matter flows from one star to the other in a binary star system. This causes the stars to grow closer together as their masses become more equal and to draw apart as their masses become more different. Rapid transfer of mass occurs if the more massive star sheds mass. Slow transfer occurs if the less massive star sheds its mass. (21.3)
- The more massive star in a massive binary system evolves first and sheds its mass to its companion and to a common envelope. When the envelope is ejected, the stars are left in a close binary in which the evolved star is now a neutron star or black hole and is less massive than its companion. A wind from the companion causes matter to fall into the compact object. The evolution of the companion results in a merger of the two stars or a system with two compact objects. (21.3)

- The initial evolution of a low-mass binary leads to a white dwarf and a less massive companion. A long period of slow mass exchange begins as the stars spiral toward one another. (21.3)
- Matter accreting onto a compact object becomes hot enough to emit X rays and ultraviolet radiation. In many cases, the accreting gas falls into orbit about the compact object, forming an accretion disk. Friction in the accretion disk causes gas to spiral inward until it reaches the compact object. (21.4)
- Binaries that contain accreting white dwarfs show temporary increases in brightness caused by runaway nuclear burning of the accreted gas on the white dwarf. Nova explosions result in the ejection of the accreted matter. (21.4)
- In cases in which the rate of accretion onto a white dwarf is large, the accreted matter remains too hot to become degenerate. The white dwarf gradually becomes more massive. Eventually, the fusion of carbon occurs explosively in the core of the white dwarf, producing a type Ia supernova. The explosion completely destroys the white dwarf. Type Ia supernovae are used to find the distances of remote galaxies. (21.4)
- X-ray bursts are produced by runaway fusion in the accreted gas on a neutron star. X-ray pulsars produce pulses of X rays when the X-ray-emitting magnetic polar regions of a neutron star sweep past the Earth. (21.4)
- Some X-ray binaries may contain black holes rather than neutron stars. A black hole is most likely in those cases where the mass of the compact object is too great to be a neutron star. (21.4)
- Binary pulsars have neutron star or white dwarf companions. Some binary pulsars have periods of only a few milliseconds. The rotation rate of these pulsars increased when their companions, before they became compact objects, transferred mass to the pulsars. Single millisecond pulsars probably had companions that they destroyed by bombarding them with high-energy particles and radiation. (21.4)

Key Terms

accretion disk *511*
binary star system *496*
close pair *496*
common envelope *508*
conucleation *503*
disk instability *503*
eclipsing binary *500*
equipotential *503*
fission *503*
fragmentation *503*
L_1 *504*
light curve *500*
minimum *500*
novae *512*
recurrent novae *512*
Roche lobe *504*
spectroscopic binary *498*
tidal capture *503*
type Ia supernova *513*
visual binary star *496*
wide pair *496*
X-ray burst *513*
X-ray pulsar *514*

Conceptual Questions

1. What is the distinction between a "wide pair" and a "close pair"?
2. Why are there few visual binaries with orbital periods of a few years or less?
3. How is it possible to tell that the orbit of a visual binary is tipped so that the apparent orbit isn't the true orbit?
4. Why are there few spectroscopic binaries in which the stars are separated by more than 10 AU?
5. Under what circumstances is the brightness of an eclipsing binary during primary minimum equal to the brightness during secondary minimum?
6. In an eclipsing binary system, star *A* is brighter, larger, and cooler than its companion, star *B*. What can be said about the depths of the minima when each of the stars is eclipsed?
7. What information in the light curve of an eclipsing binary tells about the relative sizes of the two stars?
8. The stars in a binary system are a 4 $M_\odot$ main sequence star and a 1 $M_\odot$ red giant. Explain why this binary system makes sense if the two stars are close together but is inconsistent with what we know about stellar evolution if the stars are a wide pair.
9. The surface of a lake is an equipotential while the surface of a river is not. What is the significance of this difference?
10. What is the significance of the Roche lobe of a star in a binary system?
11. Suppose a blob of gas is motionless at L_1 in a binary system. Can the blob remain at L_1? If not, is it possible to predict which way it will move?
12. Suppose the less massive star (star 2) in Figure 21.14 expands until it fills its Roche lobe. Describe what will happen as the star continues to expand.
13. Suppose the two stars in a binary system have masses of 1.5 and 2.5 $M_\odot$. What will happen to the separation

of the two stars if the 1.5 $M_\odot$ star sheds mass onto the 2.5 $M_\odot$ star?

14. What would happen to the distance between Jupiter and the Sun if the Sun shed mass onto Jupiter?
15. Explain why mass exchange can proceed very rapidly if the mass-losing star is the more massive of the pair.
16. Why do the stellar cores within a common envelope move closer together?
17. What are the two possibilities for the final state of a close, massive binary system?
18. What is the ultimate source of the energy emitted by accretion disks in binary systems?
19. What causes the gas in an accretion disk to spiral inward rather than to orbit at a constant distance from the compact object?
20. In a nova, why is the shell of hydrogen on the white dwarf consumed explosively rather than steadily?
21. Why is it possible for repeated nova explosions to occur in the same binary system?
22. What does mass exchange have to do with the occurrence of a type Ia supernova?
23. What happens on the surface of a neutron star to produce bursts of X rays?
24. Describe the evidence that Cygnus X-1 and A0620-00 are binary systems containing black holes.

Problems

1. In the box showing how to use Equation 21.2, it is claimed that if the sum of the masses of two stars is 1.96 $M_\odot$ and the ratio of the masses of the two stars is 1.23, then the masses of the two stars are 1.08 and 0.88 $M_\odot$. Show that this is true.
2. The two stars in a binary system are separated by a distance of 1 AU and have a period of 1 year. What is the sum of the masses of the two stars?
3. What is the orbital period of a binary system in which the semimajor axis is 5 AU and each of the stars has a mass of 2 $M_\odot$?
4. What is the orbital period of Proxima Centauri and α Centauri about each other if the α Centauri binary system has a total mass of 1.96 $M_\odot$, Proxima Centauri has a mass of 0.3 $M_\odot$, and the semimajor axis is 12,000 AU?

Figure-Based Questions

1. Suppose the more massive star in a binary system is five times as massive as its companion. Use Figure 21.15 to estimate how the distance between the two stars compares with the separation they would have if the more massive star were six times as massive as its companion. (Assume that the combined mass of the two stars is the same in both cases.)
2. Suppose the two stars in a binary system have masses of 9 and 1 $M_\odot$. The more massive star loses mass to its companion, and the distance between them shrinks until it is only half its original size. Use Figure 21.15 to find the new masses of the two stars.

Group Activity

The second star from the end of the handle of the Big Dipper is actually a double star. The two stars, Alcor and Mizar, appear double to the naked eye only for people with excellent vision. Working with your group, identify Alcor and Mizar. See if anyone in your group can see two stars rather than one. Use a pair of binoculars to let everyone see that there are really two stars.

For More Information

Visit the text website at **www.mhhe.com/fix** for chapter quizzes, interactive learning exercises, and other study tools.

22

Now the Milky Way is not strictly speaking a circle, but rather a belt of a sort of milky color overall (whence it got its name); moreover this belt is neither uniform nor regular, but varies in width, color, density and situation, and in one section is bifurcated. All that is very apparent even to the casual eye . . .

Claudius Ptolemy

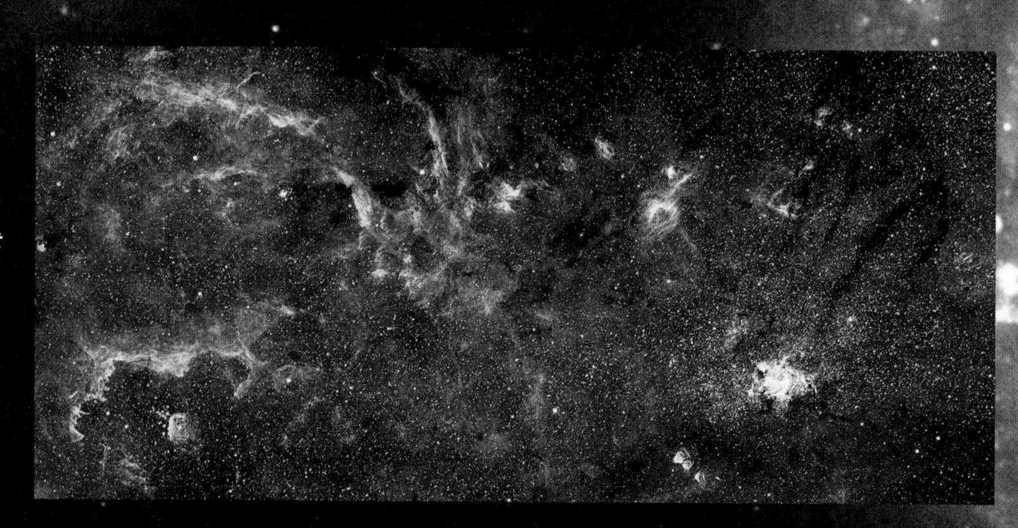

An infrared panorama of the center of the Milky Way from combined *Spitzer* and *Hubble* images.

The Milky Way

This statement in Ptolemy's Almagest is part of a detailed description of the visual appearance of the Milky Way. Ptolemy's visual description has never been surpassed, but it has been greatly extended. After photography was invented and telescopes were set up in both the northern and southern hemispheres, it became possible to obtain the composite photographic image of the visible Milky Way shown in Figure 22.1. The most striking feature of Figure 22.1 is the glow produced by stars. The stars, however, are only one of many components of the galaxy, and the expansion of astronomy into other regions of the electromagnetic spectrum has provided views of these other components. The infrared image of the galaxy shown in Figure 22.2, for instance, is much thinner than the visible view and mostly shows us emission from interstellar dust. The galactic image at a wavelength of 21 cm (in the radio part of the spectrum), shows where warm and cool interstellar gas is found (Figure 22.3). Like the dust, the warm and cool interstellar gas is concentrated in a thin band. At other radio wavelengths (Figure 22.4) we see mainly synchrotron emission from rapidly moving electrons. Tongues of emission extend far above and below the thin band of the Milky Way. The hottest gas in the galaxy emits X rays. The X-ray image of the galaxy in Figure 22.5 shows X-ray emission from all directions. In fact, the brightest parts of the X-ray sky are at the galactic poles—the regions farthest from the visible-, infrared-, and radio-bright parts of the Milky Way.

It might seem that it would be easy for astronomers, given all this information, to figure out the shape, size, and structure of the galaxy. On the

Questions to Explore

- How do we know that there is gas and dust in interstellar space?
- How were the size and shape of the Milky Way discovered?
- Why do astronomers think that much of the Milky Way consists of "dark matter"?
- Why does the galaxy have spiral arms?
- Is there a massive black hole at the center of the galaxy?
- How did the galaxy form, and how has it changed with time?

FIGURE 22.1
The Milky Way in Visible Light
This image shows the entire sky, with the galactic plane running from left to right across the center. The galactic poles, 90° from the galactic plane, are at the top and bottom of the image. Most of the light in this picture is produced by stars. Notice that there is a sprinkling of stars above and below the galactic plane but that most stars are quite close to the plane.

FIGURE 22.2
The Milky Way in the Infrared
Most of the radiation in the picture is produced by interstellar dust. Notice that almost all of the dust lies very close to the galactic plane.

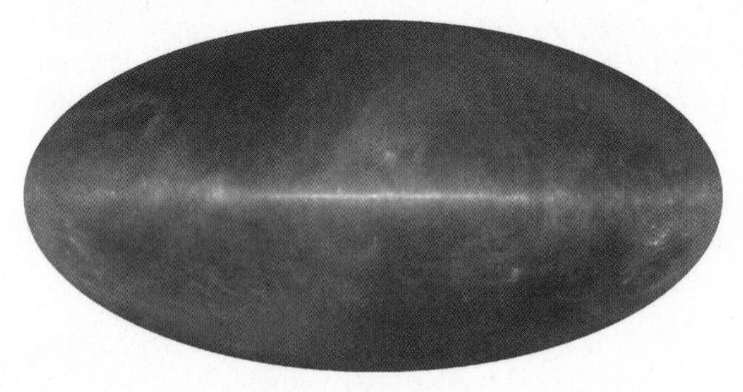

FIGURE 22.3
The Milky Way at 21 cm Wavelength in the Radio Part of the Spectrum
The emission from interstellar hydrogen atoms is strongly confined to the galactic plane. Red and white mark the brightest radiation, whereas blue indicates faint radiation.

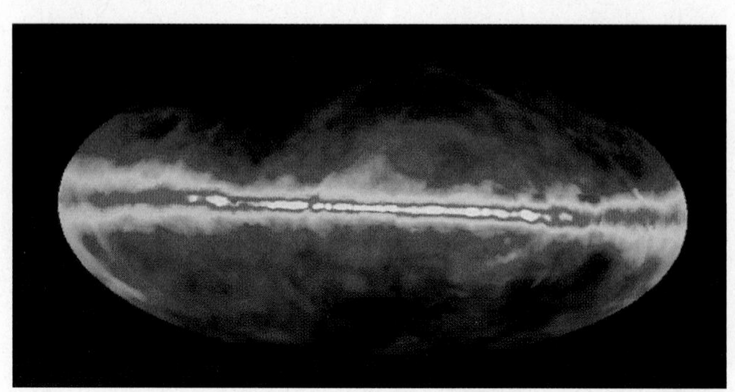

FIGURE 22.4
The Milky Way at Radio Wavelengths
Synchrotron emission from energetic electrons can be seen as tongues that reach far above and below the galactic plane. Red represents the brightest radiation and blue represents the dimmest.

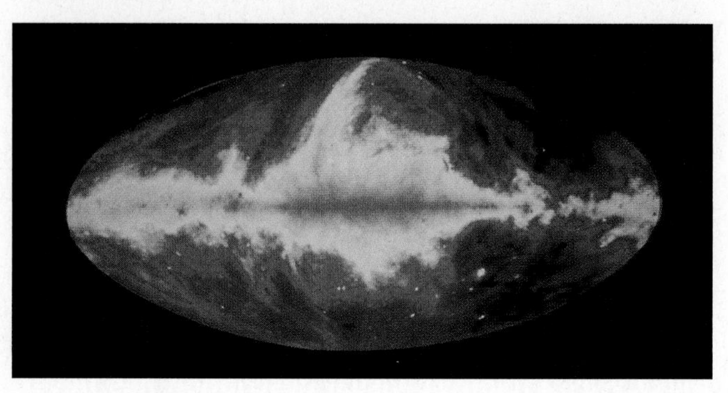

FIGURE 22.5
The Milky Way at X-Ray Wavelengths
X-ray emission from hot gas can be seen in all directions and is most intense far from the galactic plane. White represents the brightest radiation and blue represents the faintest.

contrary, the fact that we view the Milky Way from within has made it very hard to learn about its shape and form. In many ways we know more about the structure of other galaxies than we know about our own. Thus, there are still many uncertainties about the organization of the galaxy and about the way that its many components interact with each other. What we have already learned about the Milky Way is described in this chapter.

22.1 INTERSTELLAR MATTER

Glowing clouds of interstellar gas can be seen even without the use of a telescope. Concentrations of interstellar dust also can be detected quite easily. Yet until early in the twentieth century, it was generally thought that interstellar space was empty except for localized pockets of gas and dust. It is now clear that **interstellar matter is** found nearly everywhere in the galaxy. It makes up perhaps 20% of the galaxy's mass. In astronomers' attempts to determine the structure and properties of the galaxy, interstellar matter is both a curse and a blessing. It is a curse because interstellar dust prevents us from seeing visible light from most of the Milky Way. It is a blessing, however, because emission from interstellar gas can be detected from the remotest parts of the galaxy. This emission has made it possible to learn a great deal about the Milky Way.

Interstellar Gas

About 99% of the matter in interstellar space is gaseous. Like stars, most of the interstellar gas is hydrogen and helium. Interstellar gas is most conspicuous when it takes the form of a bright emission nebula—an HII region. Elsewhere, interstellar gas is much harder to detect visually.

HII Regions **HII regions** are bright nebulae that surround hot stars. Figure 22.6 shows the Orion Nebula, an HII region near enough that it can be seen without the use of a telescope. HII (pronounced "h two") regions take their name from the fact that the hydrogen gas in an HII region is ionized. A region of atomic hydrogen gas is called an HI (pronounced "h one") region.

FIGURE 22.6
The Orion Nebula
This emission nebula can be seen with the unaided eye as the middle "star" of three stars that make up Orion's sword.

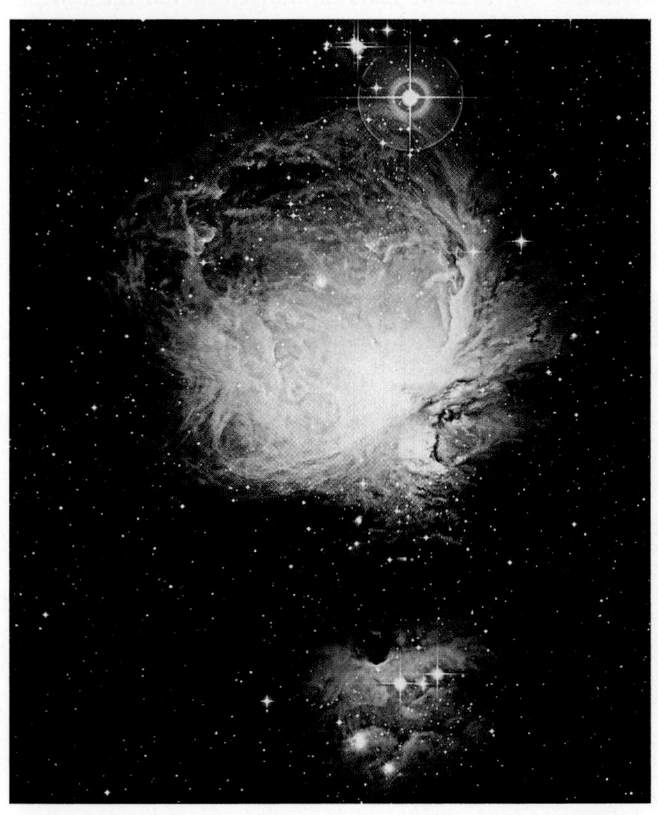

Spectra The spectrum of an HII region consists primarily of strong emission lines of hydrogen and other elements. The light emitted by the gas in an HII region is produced by a process that begins with the emission of ultraviolet radiation by the hot star in the center of the HII region. The ultraviolet photons emitted by the star strike hydrogen atoms in the surrounding gas and ionize them, producing protons and electrons. Because the gas in the nebula has a very low density (it is a more perfect vacuum than can be made in any laboratory), protons and electrons usually are far apart. Occasionally, however, an electron is captured by a proton to form an atom. As Figure 22.7 shows, the electron quickly jumps downward toward the lowest energy level, emitting photons in the process. The emitted photons make up the bright emission lines of the HII region. A particularly bright Balmer emission line of hydrogen gives an HII region a characteristic red color. After a while, the atom is ionized by another stellar ultraviolet photon, and the process begins all over again. The gas in the HII region is heated by absorbing stellar radiation and cools itself by emitting emission lines. These two processes balance each other when the temperature of the gas is about 10,000 K.

Sizes HII regions come in a variety of sizes. The main factor that determines the size of an HII region is the rate at which the central star emits ultraviolet photons energetic enough to ionize hydrogen atoms. This in turn depends primarily on the temperature and spectral type of the star. The hottest main sequence O stars emit enough ultraviolet photons each second to keep a region more than 100 pc

FIGURE 22.7
The Recombination of an Ion and an Electron
When an electron passes close to a proton, it can be captured, often into an excited energy level (level n = 4 in this case). The electron then drops to the ground state, emitting photons in the process. The photons emitted by the atoms formed when ions and electrons recombine make up an emission line spectrum.

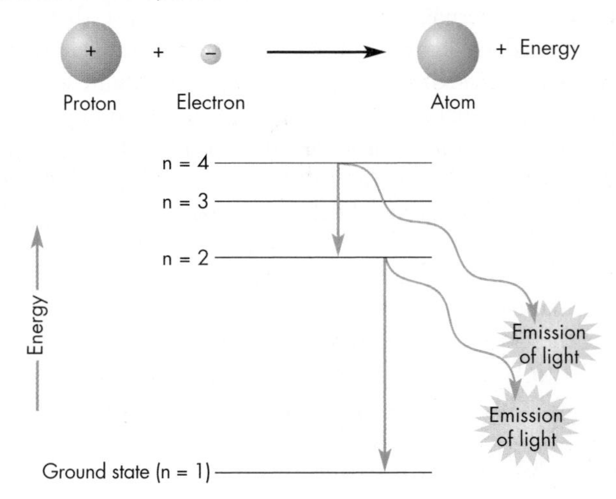

FIGURE 22.8
The Development of an HII Region
A, A hot star forms within a molecular cloud. Ultraviolet radiation from the star produces an ultracompact HII region that expands slowly in the dense gas of the cloud. **B,** After reaching the edge of the cloud, the compact HII region expands rapidly into the lower density gas outside the cloud. **C,** The rapid expansion of the classical HII region makes it look like a blister on the side of the molecular cloud.

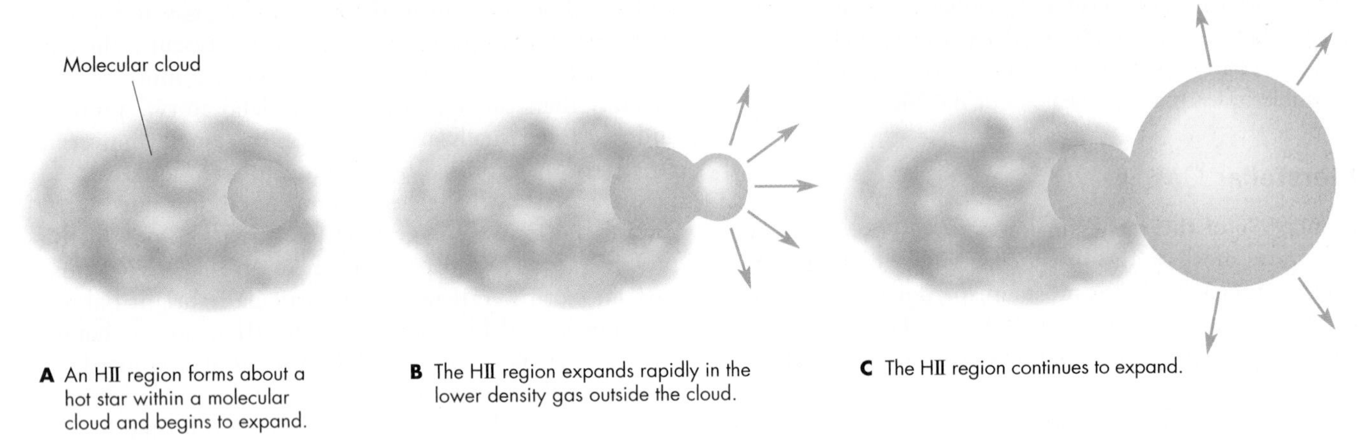

in diameter hot and ionized. The smallest conspicuous HII regions, a few parsecs in size, surround B stars. Cooler stars do not emit enough ultraviolet photons to produce detectable HII regions. Hot, massive stars form in groups, so they are often close enough together for their HII regions to overlap. When they do, they produce complex-shaped nebulae that enclose tens to thousands of hot stars.

Molecular cloud
A An HII region forms about a hot star within a molecular cloud and begins to expand.
B The HII region expands rapidly in the lower density gas outside the cloud.
C The HII region continues to expand.

Evolution of HII Regions An HII region begins to develop when a hot, massive star forms within a molecular cloud. The ultraviolet radiation from the star quickly ionizes the surrounding gas. However, because the gas within a molecular cloud is much denser than the gas in a typical part of interstellar space, the ultraviolet photons emitted by the star ionize atoms and are used up before they can travel very far from the star. The result is an ultracompact HII region, perhaps only 0.1 pc across. Dust that has been heated by light from the star emits infrared radiation that makes the HII region easily detectable in the infrared. The surrounding molecular cloud obscures the HII region so much that it can't be seen in visible light. As Figure 22.8, A, shows, the hot gas in the HII region expands outward, ionizing the surrounding molecular cloud. Within 10,000 to 100,000 years, the HII region becomes so large that it breaks through the nearest boundary of the molecular cloud, as shown in Figure 22.8, B. As it emerges from the cloud, it becomes visible as a compact HII region with a diameter of about 1 pc. The Orion Nebula (Figure 22.6) is an example of a compact HII region. As Figure 22.8, C, shows, the HII region continues to expand into the low-density interstellar gas outside the cloud and becomes a classical HII region like the Rosette Nebula shown in Figure 22.9. Because they emerge from within a molecular cloud, HII regions tend to take the shape of ionized blisters on the molecular cloud. The massive stars that produce HII regions have short main sequence lifetimes—10 million years or shorter. When the central star of an HII region grows cooler and evolves away from

FIGURE 22.9
The Rosette Nebula, a Classical HII Region
The cluster of hot young stars that keep the HII region ionized can be seen in the center of the nebula. Note the reddish color of the nebula.

the main sequence, it stops emitting copious ultraviolet photons, and the HII region fades out.

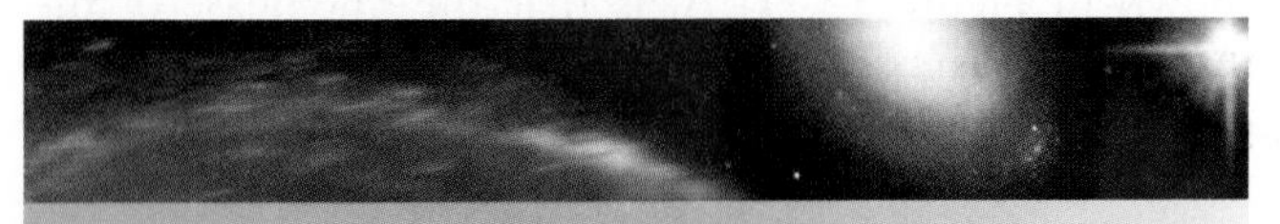

HII regions are bright nebulae that surround hot stars. The gas in an HII region is kept warm and ionized by ultraviolet radiation from the hot star. When a hot star forms within a molecular cloud, its HII region expands and eventually breaks out of the molecular cloud. When the hot star grows cooler and leaves the main sequence, its production of ultraviolet radiation declines, and the HII region fades out.

The Diffuse Interstellar Gas The realization that there is gas everywhere in interstellar space began with the work of J. Hartmann. In 1904, Hartmann obtained spectra of δ Orionis, a spectroscopic binary that lies about 400 pc from our solar system. Hartmann found two sets of spectral lines that shifted back and forth in the spectrum as the two stars orbited each other. He also found a very narrow line of ionized calcium that did not vary in wavelength at all. This line is caused by the absorption of the light of δ Orionis by interstellar calcium ions that lie between Earth and δ Orionis.

Since Hartmann's discovery, absorption lines of many other kinds of interstellar atoms and ions have been identified. Because many of these lines lie in the ultraviolet part of the spectrum, observations with orbiting observatories often have been used to investigate the diffuse interstellar gas. For many stars, the interstellar absorption lines have more than one component, demonstrating that the interstellar gas is clumpy. The different radial velocities of the clumps produce different Doppler shifts and make it possible to distinguish individual clumps of gas.

The 21 cm Line of Hydrogen Astronomers can use visual and ultraviolet absorption lines of interstellar atoms and ions to study only the interstellar gas that lies along the line of sight to bright, relatively nearby stars. Fortunately, however, radio telescopes can be used to observe interstellar hydrogen atoms in almost every direction and at great distances from the solar system. The hydrogen emission observed at radio wavelengths arises within the lowest energy level of hydrogen, as shown in Figure 22.10. Both the electron and the proton in a hydrogen atom spin. The energy of the electron in the innermost orbit of a hydrogen atom depends on whether the spin of the electron is in the same direction or the opposite direction as that of the proton. If the spins of the electron and proton are the same, the hydrogen atom has a slightly higher energy than if the two spins are different. Given enough time (about 10 million years on average), an electron that spins in the same direction as its proton does will flip its spin. When this happens, a photon is emitted. Because the energy difference is so small (about a millionth of the energy difference between the first [n = 1] and second [n = 2] energy levels of the hydrogen atom), the photon has very little energy and a very long wavelength—21 cm, about the width of this page.

FIGURE 22.10
The 21 cm Line of Hydrogen
The lowest energy level of the hydrogen atom is the n = 1 level. This level is actually two levels that differ in energy by only about one-millionth of the energy difference between the n = 1 and n = 2 levels of hydrogen. When an electron jumps from the upper to the lower level of the ground state 21 cm, radio radiation is emitted.

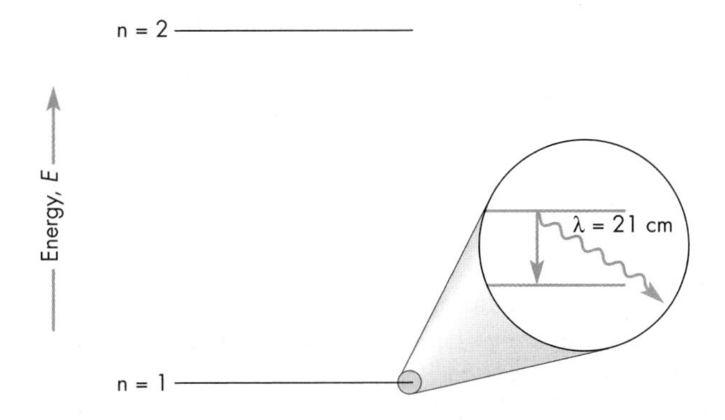

In 1944, H. C. van de Hulst suggested that the 21 cm line from interstellar hydrogen might be detected with radio telescopes. Seven years later, the 21 cm line was found by Harold Ewen and Edward Purcell. Since that time, the emission from interstellar hydrogen has been observed extensively in the Milky Way and also detected from many other galaxies.

Interstellar Molecules The molecules CH, CH^+ (ionized CH molecules), and CN were detected in the 1930s, when absorption lines of these molecules were seen in the visible spectra of stars. No further discoveries were made until 1968, when radio telescopes were used to detect interstellar water and ammonia. Since then, new molecules have been discovered by radio astronomers almost every year. More than 100 different molecules have been discovered, some with as many as 13 atoms.

Most interstellar molecules are very fragile and can be broken apart by the ultraviolet radiation that pervades most of interstellar space. As a result, they can survive only inside dark, opaque clouds where ultraviolet radiation can't penetrate. These are the molecular clouds within which star formation takes place (Chapter 18, Section 2). Even within molecular clouds, most interstellar molecules are present only in very small amounts. Hydrogen molecules make up most of the gas within molecular clouds. The next most abundant interstellar molecule is CO, but it is present at only about 20 parts per million. Even less abundant molecules are present at levels of only a few parts per trillion.

Phases of the Interstellar Gas The interstellar gas has several phases that have very different densities, temperatures, and degrees of ionization. About half of the interstellar gas exists in relatively low-temperature, high-density regions called clouds. The densest of these are the molecular clouds, with temperatures near 10 K. In clouds that have lower density than the molecular clouds, hydrogen exists in atomic form. In the inner part of our galaxy, molecular clouds are more common than atomic clouds. In the outer part of the galaxy, atomic clouds are more common. Near the Sun, the amounts of gas in atomic and molecular clouds are about equal.

Between the clouds, the interstellar gas is much warmer and much less dense. Some of the intercloud gas consists of either atomic or ionized hydrogen with a temperature of about 10,000 K. This is called the warm phase of the intercloud medium. Elsewhere, interstellar gas with a temperature of about 1 million K makes up the hot phase of the intercloud medium. The hot interstellar gas emits the X rays detected in all directions (Figure 22.5). The hot phase of the interstellar gas is produced when high-velocity shock waves move through the interstellar gas after a supernova. The shock waves sweep most gas out of the vicinity of the supernova, leaving behind a bubble of low-density, very hot gas. In a region of the galaxy where there is a cluster of young, massive stars, tens or hundreds of supernovae can occur within a few million years. The cumulative result of the supernovae is a superbubble that may be more than a thousand parsecs across. The Sun is located in a bubble possibly caused by supernovae in the Scorpius-Centaurus association of massive young stars about 10 million years ago.

At present, the fraction of the intercloud gas that exists in the hot phase is not well known. One possibility, shown in Figure 22.11, A, is that the hot phase of the intercloud gas exists only locally where supernovae have occurred. This would happen if the supernova rate for the galaxy is low or the density of interstellar gas is high enough to slow the expansion of supernova shock waves so that only small bubbles form. The other alternative, shown in Figure 22.11, B, is that the bubbles or superbubbles produced by supernovae are large enough and numerous enough that they connect with each other and fill most of interstellar space. This would happen if the supernova rate for the galaxy is large or the density of interstellar gas is low. Although it is still too early to be sure, mounting evidence favors the first possibility (Figure 22.11, A), in which the hot interstellar gas exists in only about one thousand superbubbles in the galaxy.

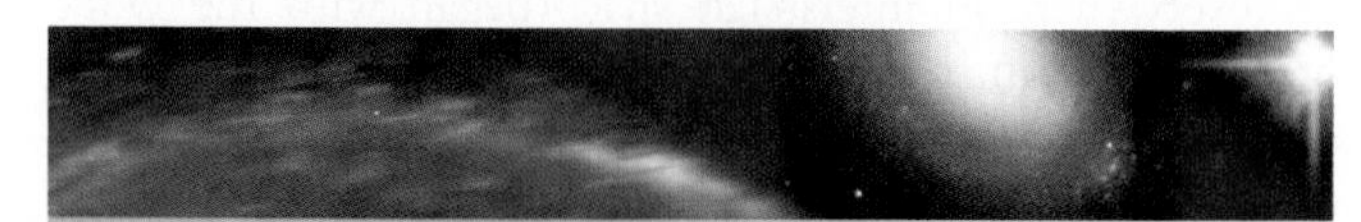

Atomic and molecular absorption and emission lines reveal that there is interstellar gas everywhere in the galaxy. The intercloud interstellar gas has a temperature of about 10,000 K, except where it has been heated to 1 million K by shock waves from supernovae. When many supernovae occur in a cluster of stars, a superbubble of hot, ionized gas is produced.

FIGURE 22.11
Two Models of the Intercloud Gas
A, If the rate of supernovae is low or the density of the interstellar gas is high, then supernovae produce only localized bubbles of hot, low-density gas. **B,** If the rate of supernovae is high or the density of interstellar gas is low, then bubbles are common enough and large enough to interconnect and constitute most of the intercloud gas.

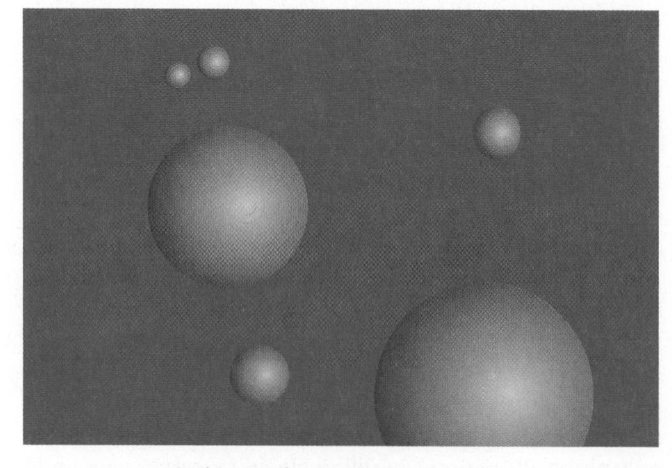

A The rate of supernovae is low or the density of the interstellar gas is high.

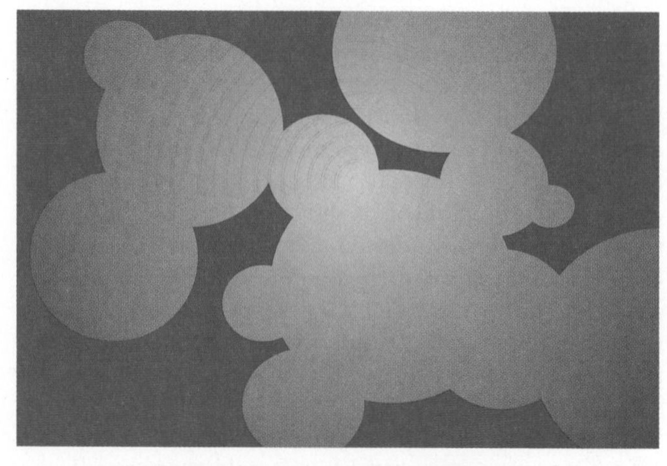

B The rate of supernovae is high or the density of the interstellar gas is low.

FIGURE 22.12
Panoramic View of the Southern Milky Way
The galactic equator runs horizontally across the center image. Dark rifts caused by interstellar dust divide the Milky Way in several places. Emission nebulae are easily identified because of their reddish color.

Interstellar Dust

Interstellar dust consists of small bits of solid material. Most of the bits are extremely small—no more than a millionth of a meter in size. Although one would need a microscope to see a single interstellar dust particle, large collections of interstellar dust particles can be seen easily. Sometimes the dust in interstellar clouds obscures our view of the stars behind it. From the Earth, these regions appear to be holes in the galaxy in which few if any stars can be seen. The dust in other interstellar clouds shines brightly by reflecting the light that falls on it. This process is known as **scattering.** Interstellar particles also emit radiation in the infrared part of the spectrum. Although this infrared radiation is concentrated near the galactic plane (Figure 22.2), it can be seen in nearly every direction in the sky.

Dark Nebulae On a dark summer night, the Milky Way arches overhead as a pale glowing band, part of which is shown in Figure 22.12. Even to the unaided eye, dark rifts and splotches can be seen against the glow. In fact, the Inca saw shapes resembling animals in the splotches and called them "dark constellations." The most conspicuous of the dark splotches is the Great Rift in the constellations Aquila and Cygnus. Figure 22.13 shows a small dark splotch in the constellation Scorpius. Until the twentieth century, astronomers generally interpreted the dark regions of the Milky Way as places devoid of stars. Astronomers thought they were tunnels through the star swarms of the Milky Way. We now realize that they are actually caused by relatively nearby interstellar clouds in which the dust particles almost completely block distant stars from view. The clouds in which dust obscures bright regions of the sky are called **dark nebulae.** Even though the interstellar clouds that produce dark nebulae are often referred to as "dust clouds," they actually contain far more gas than dust. The relatively small amount of dust they contain, however, is more effective than their gas in blocking the light of background stars.

FIGURE 22.13
A Small Rift in the Milky Way
The rift is produced by foreground dust that blocks our view of more distant gas and star swarms.

For an interstellar cloud to produce a dark nebula, it must lie between the Earth and either a dense star swarm or a bright nebula. This happens frequently for bright nebulae, because the gas that produces a bright nebula is accompanied by dust. Many nebulae are combinations of glowing background gas and beautiful, extremely intricate,

FIGURE 22.14
The Horsehead Nebula in Orion
A distinctive cloud of gas and dust blocks light from the more distant emission nebula. Notice that faint stars are more numerous in the upper half of the picture than in the lower half, where the light from distant stars is obscured by the dust in the cloud of which the Horsehead Nebula is a part. (Color image creation by Jean-Charles Cuillandre, The Canada-France-Hawaii Telescope Corp. Copyright © 2001 CFHT)

FIGURE 22.15
A *Hubble Space Telescope* Image of Dust Pillars in the Eagle Nebula
The pillars are dense clouds of gas and dust as much as ⅓ pc in height. The pillars contain denser globules in which stars are forming. The pillars are being eroded by ultraviolet radiation from hot stars off the top edge of the image.

wisps and lanes of dust that give the nebula its shape and, often, its name. The Horsehead Nebula in Orion, for instance, is shown in Figure 22.14. Its shape is due to a tongue of dust that blocks our view of an emission nebula. Dust also gives the dark columns in the Eagle Nebula, shown in Figure 22.15, their remarkable appearance. In IC 2944 (Figure 22.16), dust forms roundish globules that can be seen against the bright background. The dark lanes of the Great Nebula in Carina (Figure 22.17) are produced by elongated clouds of foreground dust. Dark nebulae range in size from much less than 1 pc to more than 10 pc across.

A closer look at a dark nebula shows that stars aren't completely missing in the direction of the nebula. There are often bright stars that lie between Earth and the nebula. Faint stars can be seen as well, but in smaller numbers than in other directions. The dust within the nebula doesn't completely prevent light from distant stars from reaching us. Instead, all the stars behind the nebula suffer about the same reduction in brightness. For most dark nebulae, about 75% of the starlight is lost. In some nebulae, however, more than 95% of the light is stopped by the dust.

Reflection Nebulae Clumps of interstellar dust also become visible if they lie near a bright star. In that case, the dust particles scatter the starlight toward us and produce a **reflection nebula.** Reflection nebulae can be found next to emission nebulae, as in the case of the Trifid Nebula in Figure 22.18, and near young star clusters such as the Pleiades, shown in Figure 22.19. One way to distinguish a reflection nebula from an emission nebula is by its color. Reflection nebulae are bluish in color for much the same reason that the Earth's sky is blue. Because the dust in the reflection nebula scatters blue light better than red light, the light from the nebula is bluer than the star that illuminates the nebula. Emission nebulae, as described earlier in this chapter, are reddish.

FIGURE 22.16
Globules in IC 2944
The globules, roughly spherical concentrations of gas and dust, are visible in the *Hubble Space Telescope* image as dark, roundish features against the bright background of the emission nebula.

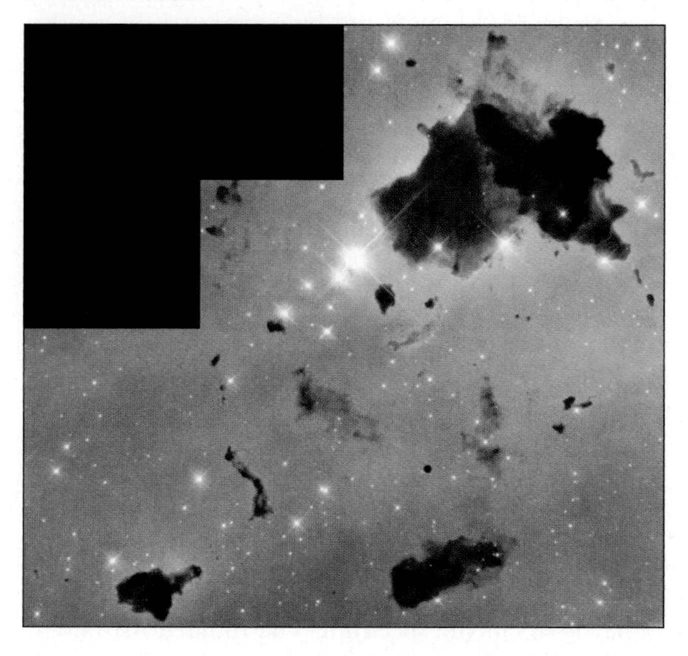

FIGURE 22.17
The Carina Nebula
Most of the pattern of bright and dark features is produced by dust that obscures the bright background gas. In the absence of dust, the emission nebula in the background would be smooth and featureless.

FIGURE 22.18
The Trifid Nebula
The reddish part of the nebula is due to emission by hydrogen atoms. The blue part of the nebula is due to reflection by dust particles, which reflect blue starlight better than red starlight.

FIGURE 22.19
Reflection Nebulae in the Pleiades
Like the Pleiades, many young clusters of stars are surrounded by dust that produces reflection nebulae.

Emission and reflection nebulae also have very different spectra. The spectrum of an emission nebula is dominated by bright emission lines, whereas that of a reflection nebula resembles the spectrum of the star whose light is scattered by the dust.

Clouds of interstellar dust can be seen as dark nebulae that block the light of the stars that lie behind them, and as reflection nebulae, which shine by scattering the light of nearby stars.

Diffuse Interstellar Dust Only a small fraction of the galaxy is filled with dark nebulae. Is there dust in the rest of interstellar space? The answer to this question was discovered by Robert Trumpler in 1930 as a result of his study of clusters of stars. He measured the angular sizes and distances of one hundred clusters. Surprisingly, he found that cluster size tended to increase with distance. He also found that the stars in distant clusters were redder than stars of the same spectral type in nearer clusters. Trumpler concluded that interstellar space is loosely filled with dust particles that both obscure and redden light that passes through them. The reason that distant clusters seemed larger was that Trumpler had overestimated their distances. This had happened because interstellar dust made the stars in the clusters look dimmer, and therefore more distant, than they would appear if there were no interstellar dust.

Reddening and Extinction The dimming of starlight by interstellar dust is called **extinction.** On average, interstellar dust extinguishes visible light that passes through it by about one magnitude (or a factor of 2.5 in brightness) for each 1000 pc. This means that only 40% of the light from a star at 1000 pc is able to penetrate the interstellar dust. For a star 2000 pc away, only 16% of the light gets through. At 5000 pc, only one photon in a hundred is left. At 10,000 pc, only one photon in 10,000. Because the Milky Way is many tens of thousands of parsecs across, most of our galaxy is obscured from our view by interstellar dust.

The interstellar dust extinguishes blue light more strongly than red light. Only about 2.5% of the blue light reaches us from a star at a distance of 3000 pc, while 6% of the red light gets through. This effect, called **interstellar reddening,** alters the color of any celestial body viewed through the interstellar dust. The stopping power of interstellar dust decreases with increasing wavelength (red light has a longer wavelength than blue light) so that interstellar space is nearly transparent at infrared and radio wavelengths. As a result, most of what we know about the more distant parts of the galaxy has been discovered in those parts of the electromagnetic spectrum.

Absorption or Scattering? The two ways that interstellar particles can extinguish the light of distant stars are by absorbing it and by scattering it in another direction. Which of these ways is more important depends on what interstellar dust particles are made of and how large they are, both of which are not completely known. It appears, however, that both scattering and absorption are important but that the particles scatter a little more light than they absorb.

The Galactic Glow Starlight scattered by interstellar dust brightens the galaxy in the same way that sunlight scattered by molecules in the Earth's atmosphere brightens our sky. The galaxy is filled with this diffuse light, which can't be directly attributed to any star. About 75% of the brightness we see in a typical direction along the Milky Way is due to stars both bright and faint. Most of the rest is due to the starlight that has been scattered to us by interstellar dust.

The Infrared Perspective Starlight absorbed by interstellar dust heats the dust and causes it to radiate in the infrared part of the spectrum. The Infrared Astronomy Satellite (IRAS) surveyed the sky in the infrared and found that emission from interstellar dust can be seen in nearly every direction. The dust emission is particularly easy to see in directions away from the band of the Milky Way. The wispy appearance of the dust emission, shown in Figure 22.20, has led astronomers to call it "infrared cirrus." Most of the dust that produces infrared cirrus is very cold—about 20 to 30 K—because most of it lies far from any star and intercepts only a small amount of starlight.

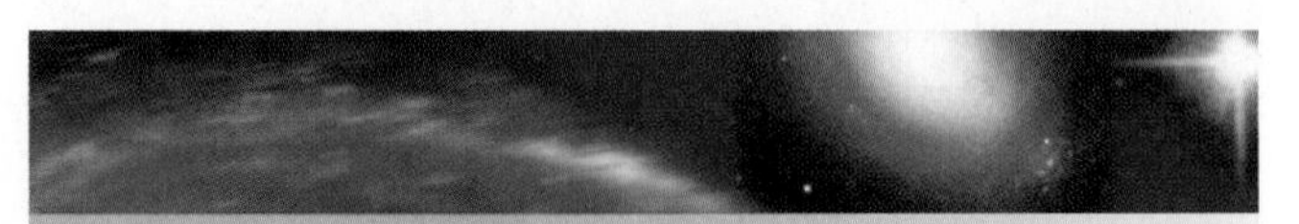

The diffuse interstellar dust both dims and reddens the light from distant stars. Part of the interstellar extinction is due to scattering of light into other directions and part is due to absorption. The scattered light appears as a general glow in the galaxy. Absorbed starlight heats interstellar particles, which then emit infrared radiation. This infrared radiation can be seen in nearly all directions.

What Are the Interstellar Particles Like? The way that interstellar dust particles absorb and scatter light shows that the particles must be very small. The largest

FIGURE 22.20
Infrared Cirrus
Thin, wispy clouds of interstellar dust particles emit infrared radiation when they are heated by starlight. Infrared radiation from interstellar dust can be seen in nearly all directions in the sky. This image covers a 40° by 20° region centered on the galactic plane.

of them are probably about 6×10^{-7} meters in diameter. (This is about the same size as a bacterium.) Despite their small sizes, however, they are large enough to contain hundreds of millions of atoms. Interstellar particles are made of a variety of different substances. Some of them are made of the same types of silicate materials that form the rocks of the Earth. Others are made of carbon, mostly in the form of soot. Most of the interstellar silicon, aluminum, magnesium, calcium, iron, and other heavy elements is locked up in the particles. As Figure 22.21 shows, the rocky or sooty cores of many interstellar dust particles are covered by a layer of ices.

FIGURE 22.21
An Interstellar Dust Particle
The core of the particle is made of either silicate or carbon soot. A mantle of water, ammonia, methane, and other icy material slowly grows around the core as interstellar atoms and molecules stick to its surface.

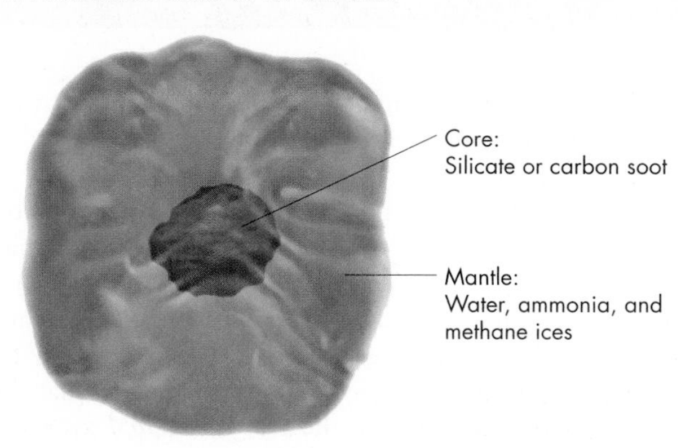

The Origin of Interstellar Dust Most interstellar dust particles form in the atmospheres of cool giant and supergiant stars. In stars in which there is more oxygen than carbon (the normal situation for stars), silicate minerals condense in the cooling gas that flows out of the stars. For stars in which there is more carbon than oxygen, carbon compounds such as soot condense instead. After the particles flow into interstellar space, they may acquire coatings of other compounds when interstellar atoms and molecules strike their surfaces and stick. In time, layers of ice can build up in this way.

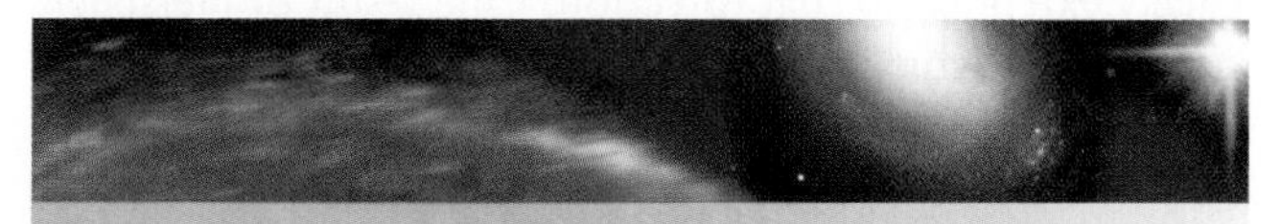

Interstellar dust particles are extremely small. They condense in the gas that flows from cool giant stars into interstellar space. While in interstellar space, the particles may acquire icy coatings.

22.2 THE SHAPE AND SIZE OF THE GALAXY

Galileo's first glimpse of the Milky Way through his telescope confirmed the idea, originally expressed by the Greek philosopher Democritus, that the visible glow is made up of swarms of stars. For much of the four centuries since Galileo, astronomers have been trying to discover the way that the stars are arranged in the galaxy. In other words, they have tried, and are still trying, to determine the shape, size, and structure of the Milky Way.

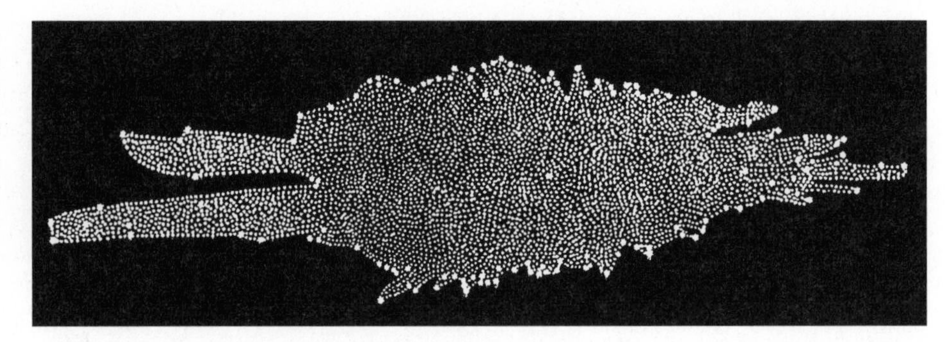

FIGURE 22.22
Herschel's Model of the Galaxy
A cross section of Herschel's star counts apparently showed that the galaxy consists of a flattened, irregular swarm of stars with the Sun near the center.

Star Counting

The earliest attempts to determine the shape and size of the Milky Way were based on the simple method of counting faint stars in various directions. If we find fewer faint stars in one direction than in another, we may conclude that the stars "thin out" faster in the direction in which fewer faint stars are seen. The validity of this conclusion depends, however, on our ability to see all of the stars in a given direction, whatever their distances.

Herschel's "Gaging the Heavens" The first person to use star counts to determine the shape of the galaxy was William Herschel, in 1785. Herschel counted all the stars that he could see with his telescopes in hundreds of different regions of the sky. He assumed that the stars were uniformly distributed throughout space within the galaxy and absent from the region beyond the galaxy. The number of stars seen in a given direction (which ranged from just one to many hundreds) determined how far the galaxy extended in that direction. One result of Herschel's star counts, which he called "gages," was the cross section of the Milky Way shown in Figure 22.22. Herschel found the galaxy to be flattened along the Milky Way. In some directions, it extended to about five hundred times the distance of Sirius, the brightest star, which corresponds to about 1500 pc. Herschel explained that the galaxy looks like a band of light because the Sun is situated near the midplane of a wide but thin layer of stars, as shown in Figure 22.23. He argued that bright stars, those within the inner circle, could be seen in all directions with equal probability. Fainter stars, which combine to make up the glow of the Milky Way, can be seen only when we look along the layer (as at points X and Y in Figure 22.23) and are missing when we look out of the layer (as at points M and N).

Kapteyn's Universe Early in the twentieth century, Jacobus Kapteyn used star counts to produce a model for the galaxy often referred to as the "Kapteyn Universe." His results located the Sun about 600 pc from the center of a flattened galaxy. From the center of the Kapteyn Universe, the stars thin out in all directions. The density of stars falls to 1% of the central density at 600 pc for directions that lie perpendicular to the Milky Way and at 3000 pc for directions in the Milky Way.

FIGURE 22.23
Herschel's Explanation of the Appearance of the Milky Way
If the galaxy is a thin disk, faint stars (those that lie outside the inner circle) will appear only in those directions in which we look along the disk. As a result, the collective glow of the faint stars will produce a band of light that runs completely around the sky. Bright stars (those within the inner circle) will be seen in all directions. In this drawing, the disk is viewed from the side.

The Milky Way

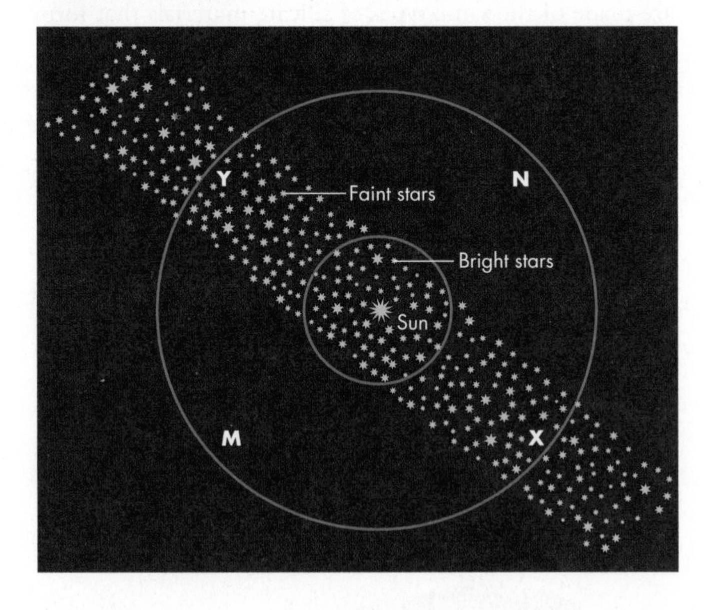

Apart from the result that the galaxy is flattened, neither Herschel's nor Kapteyn's model resembles the actual shape and size of the Milky Way because neither astronomer realized that light from distant stars is dimmed by interstellar dust. As a result, we can see only those stars that lie within a few thousand parsecs of the Sun. Interstellar dust almost completely obscures the rest of the galaxy, which extends much farther than the relatively small regions that Herschel and Kapteyn believed they had mapped out.

Attempts to determine the shape of the galaxy by counting stars in various directions led to the conclusion that the Sun is near the center of the Milky Way, and that the Milky Way extends only a few thousand parsecs from the Sun. These models were incorrect, however, because they did not take into account the extinction of starlight by interstellar dust.

The Modern View of the Galaxy

The modern view of the size, shape, and structure of the Milky Way began to emerge in 1915 with Harlow Shapley's study of the locations of globular clusters.

Globular Clusters The clusters of stars associated with star formation regions are loose collections that may contain a few thousand stars. A **globular cluster,** on the other hand, is a tight, spherical collection of as many as several hundred thousand stars. Two globular clusters, M4 and Omega Centauri, are shown in Figure 22.24. Although they look very compact, globular clusters have diameters between 20 and 100 pc, so the stars they contain are, on average, about a parsec apart. The highly distinctive appearance of globular clusters makes them easy to identify, even at great distances from the solar system. The approximately 150 globular clusters that have been found probably make up two-thirds to three-fourths of the total number in the Milky Way.

The Size of the Galaxy Even before Shapley began his work, astronomers knew that the globular clusters weren't distributed uniformly around the sky. Instead, all but a few of the clusters are found in the half of the sky centered on the brightest, thickest part of the Milky Way in the constellation Sagittarius. Figure 22.25 shows a region in Sagittarius. The region covers only 2% of the sky but has about one-fifth of the known globular clusters.

Shapley concentrated on finding the distances to as many globular clusters as he could. For those in which he was able to identify RR Lyrae variable stars (see Chapter 19, Section 4), he used the fact that all RR Lyrae stars have nearly the same luminosity. He compared the apparent brightnesses with the luminosities of the RR Lyrae stars to find their distances. For other clusters, he estimated distances from the brightnesses of the brightest stars or assumed that the sizes of all globular clusters were the same (not a very good assumption) so that he could calculate distance from angular size. After he had distances for many globular clusters, he mapped out their positions in the galaxy, as shown in Figure 22.26. He found that the globular clusters are centered on a point about 15,000 pc, or 15 kpc (a **kiloparsec,** abbreviated kpc, is 1000 pc), from the Sun in the direction of Sagittarius.

Shapley decided that the distribution of the globular clusters is centered on the center of the galaxy. This meant that the Sun is very far from the center and that the galaxy

FIGURE 22.24
Two Globular Clusters, M4 and Omega Centauri
Both M4, **A,** and Omega Centauri, **B,** lie in the southern sky. M4 can be seen from anywhere in the United States. Omega Centauri can be seen from the extreme southern United States.

A

B

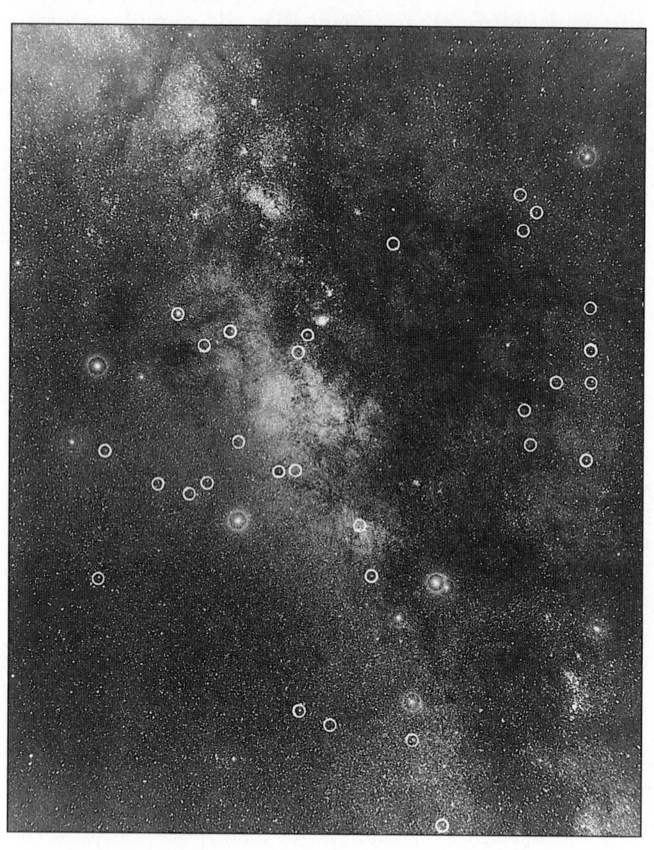

FIGURE 22.25
The Concentration of Globular Clusters in the Constellation Sagittarius
Although this picture covers only 2% of the sky, it contains about one-fifth of the known globular clusters. The globular clusters are circled.

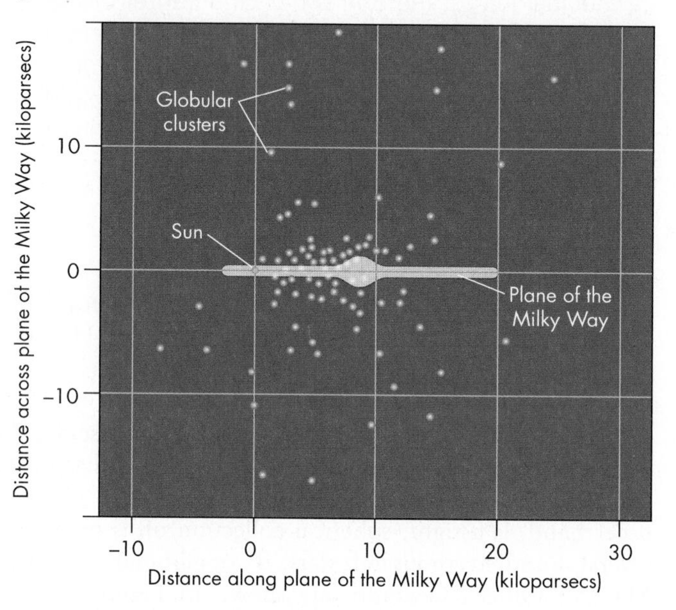

FIGURE 22.26
The Distribution of Globular Clusters in the Galaxy
The distribution of globular clusters is centered in the plane of the galaxy about 8000 pc from the Sun. The modern value for the distance from the center of the distribution of globular clusters to the Sun is about half as large as the value originally calculated by Harlow Shapley from his studies of globular clusters.

is very large. It is much larger, in fact, than the models proposed by Herschel and Kapteyn. Because he was unaware of the extinction of starlight by interstellar dust, Shapley overestimated the distance to the center of the galaxy by nearly a factor of two. The distance is thought to be between 8 and 8.5 kpc.

Galactic Coordinates The galactic coordinate system has been developed to describe the directions of celestial objects with respect to the Sun's location in the galaxy. As shown in Figure 22.27, the **galactic equator** is an imaginary line that runs around the sky in the middle of the glowing band of the Milky Way. Angular distance from the galactic equator is defined as **galactic latitude,** just as latitude on Earth is angular distance from the Earth's equator. **Galactic longitude,** analogous to longitude on the Earth, is the angular distance along the galactic equator between the galactic center and the point on the galactic equator that lies nearest to the direction to the object. The center of the galaxy has galactic latitude 0° and galactic longitude 0°. The galactic poles are found at galactic latitudes 90° and −90°.

The Shape of the Galaxy The shape of the Milky Way is somewhat hard to pin down because the galaxy thins out rather than ending with abrupt boundaries. In addition, the shape depends on the kind of objects being observed. Figure 22.28 shows the way in which the stars in the galaxy are distributed. Most of the stars are located in a disk 40 kpc in diameter and about 2 kpc thick. This **galactic disk** has about the same shape as a pizza. The star density (the number of stars per cubic parsec) is greatest in the central plane of the galactic disk and falls off above and below the plane. The center of the Milky Way, which is called the **galactic nucleus,** is surrounded by the **galactic bulge,** a somewhat flattened collection of stars about 6 kpc across. The bulge reaches far above and below the galactic disk. The Milky Way's bulge, like that of many other galaxies, is elongated to form a bar four or five times longer than it is wide. The bar points about 45° away from the Sun. Finally, the disk and bulge are surrounded by the **galactic halo,** which consists of globular clusters and stars with very low abundances of elements other than hydrogen and helium. The galactic halo seems to be roughly spherical and to extend 30 to 40 kpc from the center of the galaxy. It may reach even farther. There are a few distant globular clusters and some small galaxies that seem to be

FIGURE 22.27
The Galactic Coordinate System
Galactic longitude is measured from the direction of the galactic center around the galactic plane. Galactic latitude is measured north or south of the galactic plane.

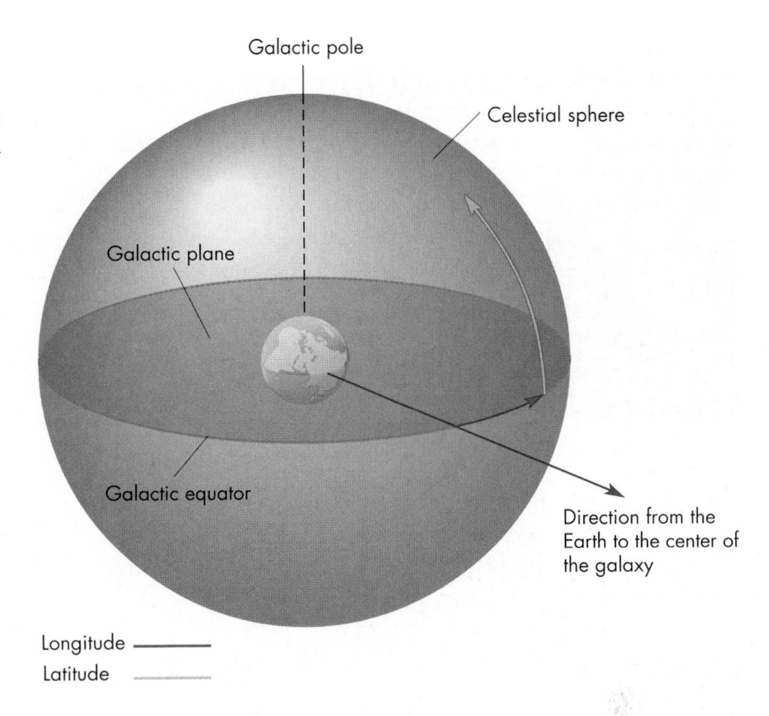

FIGURE 22.28
The Shape of the Milky Way
The galaxy consists of a thin disk with a bulge at its center. Both the galactic disk and the galactic bulge are surrounded by a much larger galactic halo.

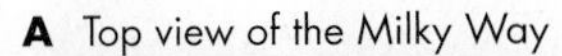

A Top view of the Milky Way

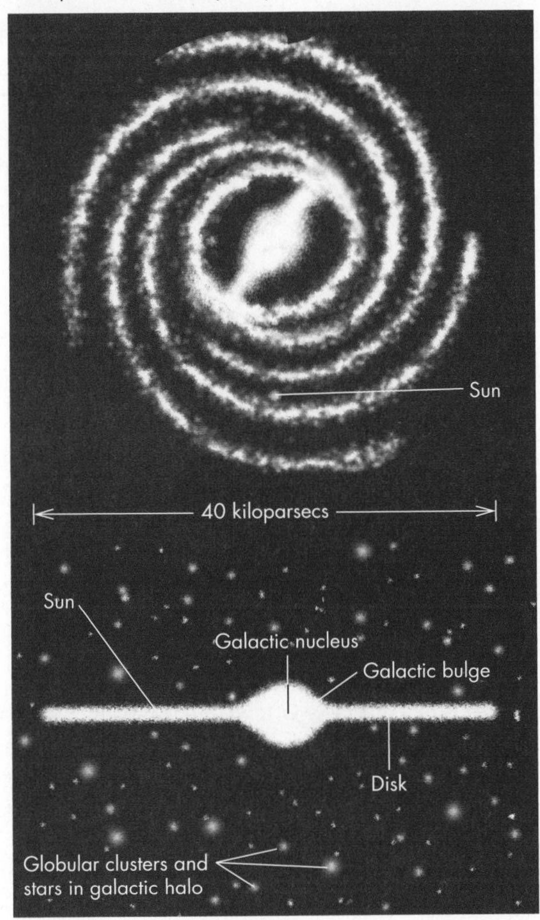

B Side view of the Milky Way

in orbit around the Milky Way and may define the outer limits of the galaxy.

The modern view of the Milky Way emerged from the study of globular clusters, which are distributed about the center of the galaxy. The galaxy consists of a flattened disk about 40 kpc in diameter, a central bulge, bar, and a spherical halo that may extend outward to more than 40 kpc. The Sun lies between 8 and 8.5 kpc from the center of the galaxy.

22.3 THE ROTATION OF THE MILKY WAY

All of the stars, gas, and dust in the Milky Way orbit about the center of the galaxy. The organized orbital motion of these objects is what astronomers mean when they say that the galaxy is rotating.

The Orbit of the Sun

The orbital speed of the Sun is about 220 km/s (or 1 pc every 4400 years). The distance from the Sun to the center of the galaxy is about 8.5 kpc. Because we have approximate values for both the Sun's orbital speed and distance,

Orbital Period, Distance, and Speed

Equation 22.1 can be used to find the orbital period of the Sun about the center of the galaxy. The Sun's orbital distance is $d = 8.5$ kpc = 8500 pc. The Sun's orbital speed is v = 220 km/s = $\frac{1}{4400}$ pc/yr. Using these values, Equation 22.1 yields

$$P = \frac{2\pi d}{v} = \frac{2\pi \times 8500 \text{ pc}}{\frac{1}{4400} \text{ pc/yr}} = 53.400 \times 4400 \text{ yr}$$
$$= 230 \text{ million yr}$$

This means that the Sun has only moved through about one-quarter of a single orbit since the dinosaurs became extinct 65 million years ago.

we can estimate the length of time it takes for the Sun to complete one revolution. If the Sun's orbit is circular, then its orbital period is just the circumference of its orbit divided by its orbital speed, or

$$P = \frac{2\pi d}{v} \qquad (22.1)$$

where P is orbital period, d is distance from the galactic center, and v is orbital speed.

Although the Sun's orbital period of 230 million years is long compared with the evolutionary timescale of life on Earth, it is only about 5% of the age of the Sun and perhaps 2% of the age of the galaxy. Thus, the Sun has already completed 20 trips about the galaxy and the galaxy had completed at least as many rotations by the time the Sun formed.

The Mass of the Inner Galaxy Given the Sun's orbital speed and its distance from the center of the galaxy, it is possible to calculate the mass of material needed to supply the gravitational force that keeps the Sun in orbit. Recall that Kepler's third law, as modified by Newton (Equation 5.4), relates orbital period and distance to the sum of the masses of the two bodies in orbit about each other. The mass of the galaxy is very much greater than the mass of the Sun, so the sum of the two masses is essentially the mass of galactic material that lies inside the Sun's orbital distance. Because orbital period depends on orbital distance and speed, Kepler's third law can be rewritten in a form that is convenient for the study of the Milky Way and other galaxies. This form is

$$M = \frac{v^2 d}{G} \qquad (22.2)$$

where M is mass, v is orbital speed, d is distance from the galactic center, and G is the gravitational constant.

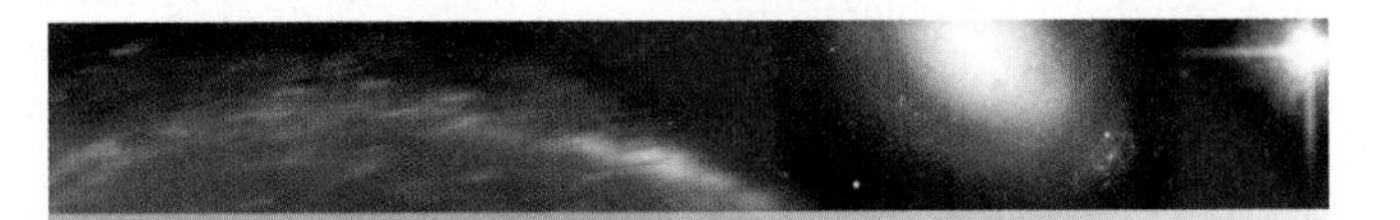

The speed and distance of the Sun can be used to find the orbital period of the Sun, about 230 million years. The mass of material inside the Sun's orbit, found from Kepler's third law, is 100 billion $M_\odot$.

The Rotation Curve of the Galaxy

The **rotation curve** of the galaxy is a plot of orbital speed versus distance from the galactic center. The importance

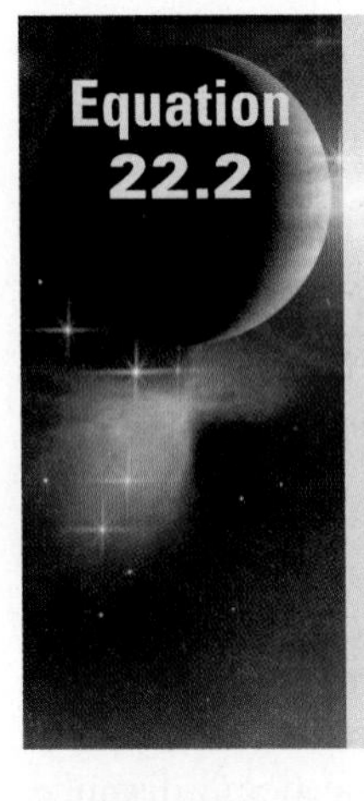

Mass, Distance, and Orbital Speed

Equation 22.2 can be used to find the mass that lies inside the Sun's orbit about the center of the galaxy. It is important in using Equation 22.2 to use the same units of measurement for G, v, and d. For G = 6.67×10^{-11} m^3/s^2 kg we need orbital speed in m/s and orbital distance in m. Converting 220 km/s gives $v = 2.2 \times 10^5$ m/s. Converting 8.5 kpc gives $d = 2.6 \times 10^{20}$ m. With these values, Equation 22.2 gives the mass inside the Sun's orbit as

$$M = \frac{v^2 d}{G} = \frac{(2.6 \times 10^5 \text{ m/s})^2 \times 2.4 \times 10^{20} \text{ m}}{6.67 \times 10^{-11} \text{ m}^3/\text{s}^2 \text{ kg}}$$
$$= 2.5 \times 10^{41} \text{ kg}$$

The Sun's mass is 2×10^{30} kg, so the mass inside the Sun's orbit is about 10^{11} times the mass of the Sun. That is, there is 100 billion $M_\odot$ of material within 8.5 kpc of the center of the galaxy.

FIGURE 22.29
Wheel-like and Planetlike Rotation
A, A wheel rotates so that speed increases with distance from its center. **B,** For planetlike rotation, speed decreases with increasing distance. Neither of these patterns resembles the rotation of the galaxy.

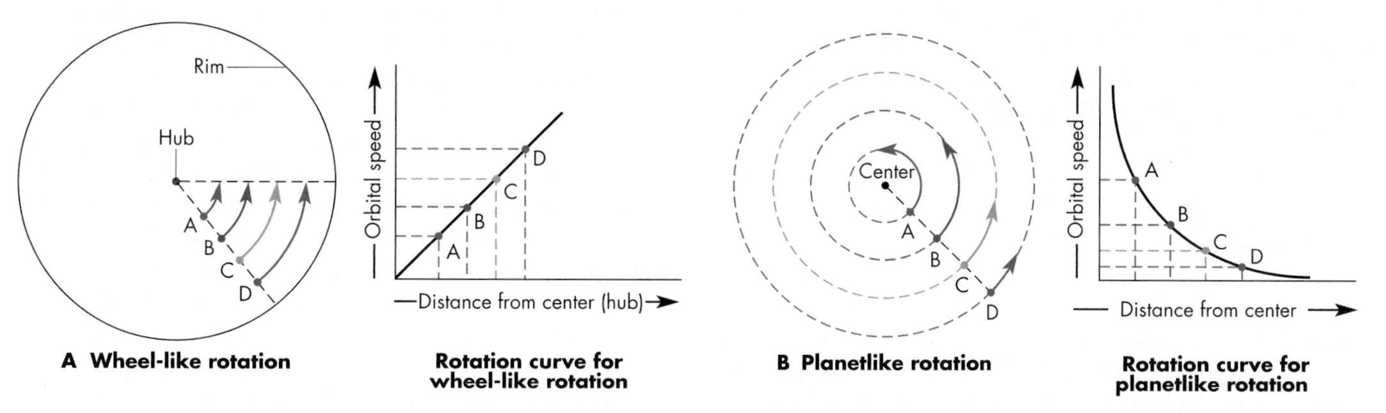

of the galactic rotation curve is that it can be used to find the distribution of mass within the galaxy.

Two simple kinds of rotation curves, one for wheel-like rotation and one for planetlike rotation, are shown in Figure 22.29. When a wheel turns, all its parts complete one rotation in the same length of time. The distance a part of the wheel travels during one rotation, however, depends on its distance from the center of the wheel. Parts close to the hub travel shorter distances than parts close to the rim. To complete the rotation in the same length of time, the rim must rotate faster than the hub. In other words, wheel-like rotation requires that orbital speed increase linearly with distance from the center. Planetlike rotation occurs when the mass responsible for orbital motion is concentrated toward the center of rotation, as with the Sun in the solar system. In this case, orbital speed falls with distance.

To determine the rotation curve of the galaxy, astronomers have had to measure the orbital speeds of objects at various distances from the galactic center. Because of the extinction of light by interstellar dust, most of this work has been carried out by using radio telescopes to measure the Doppler shifts of emission lines from interstellar atoms and molecules. For the part of the galaxy inside the Sun's orbit, the rotation curve has been found from the 21 cm line of interstellar atomic hydrogen. For the parts of the galaxy outside the Sun's orbit, the rotation curve has been found by several methods, including measuring the radial velocities of CO gas in molecular clouds. This method can be used to determine the rotation curve out to about 20 kpc (more than twice the Sun's distance). There seem to be few stars beyond that distance.

Combining the results for the inner and outer galaxy, the rotation curve for the Milky Way is essentially flat between 1 kpc and 20 kpc, as shown in Figure 22.30. The radial velocities of distant globular clusters suggest that the rotation curve of the galaxy remains approximately flat out to distances greater than 50 kpc from the galactic center.

FIGURE 22.30
The Rotation Curve of the Milky Way
Each point is shown with its uncertainty in distance and speed. The rotation curve is flat (or possibly rises slightly) to more than twice the Sun's orbital distance.

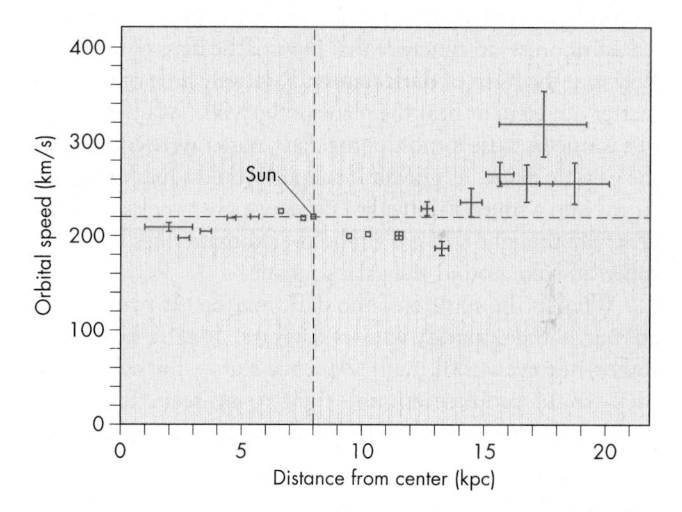

The values for the distance and rotation speed of the Sun recommended by the International Astronomical Union and used throughout this chapter are 8.5 kpc and 220 km/s. However recent measurements suggest that the Sun's orbit about the center of the Milky Way may be somewhat different. Teams of radio astronomers using the Very Long Baseline Array (VLBA) in the United States and the VERA array in Japan have measured the positions of masers in star formation regions in the Milky Way with enough precision to find parallaxes to an accuracy of about 30 millionths of a second of arc. Such parallaxes make it possible to determine accurate distances and investigate the rotation of the Milky Way as far as several kiloparsecs from the Sun. Accurate proper motions (see Chapter 16, Section 3) have been measured as well. With distances, proper motions, and radial velocities, it is possible to tell precisely where the masers and the star formation regions that include them are located and how fast they are moving relative to the Sun. Analysis of the motions of the masers that have been measured so far yields a distance and rotation speed for the Sun of 8.4 kpc and 250 km/s. The faster rotation speed of the Sun implies that the Milky Way is significantly more massive than had been previously thought.

Dark Matter and the Mass of the Galaxy According to Equation 22.2, a flat rotation curve for the galaxy means that the mass of the galaxy increases as the distance from the galactic center increases. That is, if the mass within the Sun's galactic orbit is 10^{11} $M_{\odot}$, then the mass within twice the Sun's orbit is 2×10^{11} $M_{\odot}$. And, if the rotation curve really remains flat to a distance of 50 kpc (six times the Sun's distance), the total mass of the Milky Way is about 6×10^{11} solar masses. The speeds with which nearby galaxies appear to be orbiting the Milky Way suggest that the mass of the Milky Way may be as great as 10^{12} $M_{\odot}$. The disturbing thing about these results is that little of the matter beyond the Sun's orbital distance is luminous enough to be seen. This has led astronomers to conclude that most of the mass of the Milky Way is in the form of **dark matter.** Relatively little of this dark matter can lie in or near the plane of the Milky Way. We know this is true because if most of the dark matter were confined to the galactic plane, its gravitational pull would squash interstellar gas into a much thinner layer than we observe. Instead, it is generally thought that the undiscovered matter lies in a dark, spherical halo around the galactic plane.

Dark matter

What is the nature of the dark matter? At present, the answer is that nobody knows for sure. It can't be normal stars—not even cool main sequence stars—for collectively they would produce enough light to be seen. It can't be dust, or there would have to be so much dust surrounding the galaxy that we couldn't see distant stars and globular clusters. Some possibilities that can't be ruled out are old, dim white dwarfs, brown dwarfs, Jupiter-like planets, black holes, and unknown kinds of elementary particles.

The possibility that old white dwarfs, brown dwarfs, or other bodies of similar mass are important components of the dark mass of the galaxy has been boosted by observations carried out in the last decade or so. Several independent groups of astronomers have searched for **microlensing events.** Microlensing events occur when a dim, relatively nearby body passes almost directly between the Earth and a distant star. The gravity of the nearby body focuses the light of the distant star, making the distant star temporarily brighter. How long the microlensing event lasts depends on the mass of the nearby object. An event due to a white dwarf would last for several months. An event due to a brown dwarf would last for several weeks and the event due to a Jupiter-like planet would last for several days.

Searches for microlensing events have been made using distant stars in the central region of the Milky Way and the Large and Small Magellanic Clouds, two nearby galaxies. Hundreds of microlensing events have been detected so far. The brightness of the distant star during one of these microlensing events is shown in Figure 22.31.

The number of microlensing events observed so far indicates that objects with masses between about 0.1 and 1 $M_{\odot}$ make up about 20% of the dark matter in the Milky Way. Objects less massive than 0.1 $M_{\odot}$ make up less than 20%. No multi-year-long microlensing events caused by massive black holes have yet been detected, indicating that massive black holes make up little of the dark matter. It appears that most of the dark matter in the Milky Way must be accounted for by other kinds of dark matter such as elementary particles.

Whatever the nature of the dark matter, there are strong indications that it is a very important constituent of the universe. Other galaxies have rotation curves like that of the Milky Way, implying that they too have dark

FIGURE 22.31
A Microlensing Event
The graph shows the brightness of a star in the Large Magellanic Cloud versus time. During the microlensing event the star became more than five times as bright as normal. Observations were made in both red and blue light because microlensing causes equal brightness increases in both the red and the blue. A variable star, which might be mistaken for a microlensing event, usually looks different in the red and blue. The 2-month duration of the microlensing event suggests that the body causing the microlensing has a mass a few tenths that of the Sun.

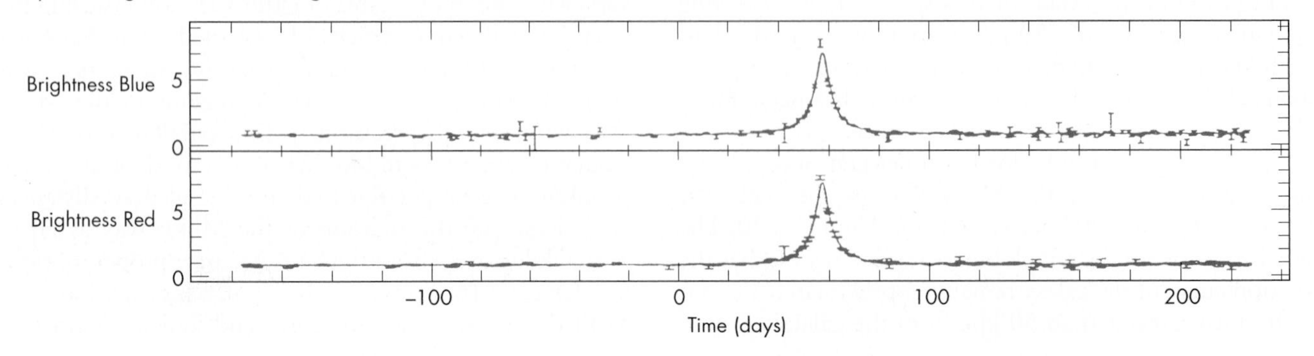

matter. Clusters of galaxies also appear to consist mainly of matter that can't be seen. The subject of dark matter is discussed again in Chapters 23, 25, and 26.

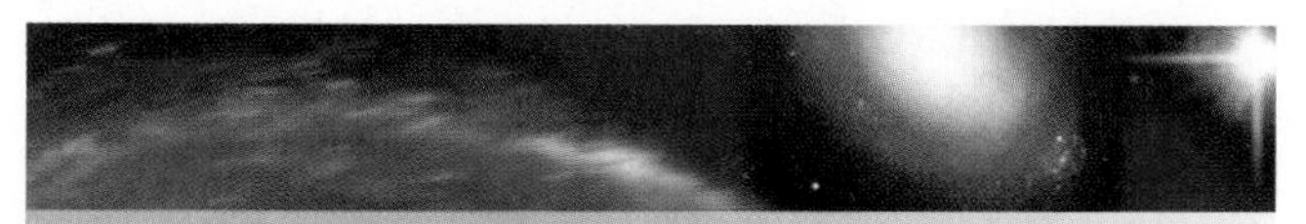

Measurements of the velocities of interstellar clouds show that rotational speed is essentially constant between 1 and 20 kpc from the center of the galaxy and perhaps to much greater distances. This implies that the mass of the galaxy increases with increasing distance. The galaxy may contain as much as six times as much material as lies within the Sun's orbit. Much of this is dark matter, the nature of which is not yet known.

22.4 THE SPIRAL STRUCTURE OF THE GALAXY

Flattened galaxies like the Andromeda Galaxy, shown in Figure 22.32, have **spiral arms.** The Milky Way—a flattened galaxy—also has a spiral structure, although the actual shape of the spiral arms and the way that they connect with each other is still not well known.

Observations of Spiral Structure

It is far from obvious, in looking at the Milky Way, that the galaxy has spiral structure. The problem is that we ourselves are part of the galaxy. Our situation is somewhat like that of a member of a marching band who wants to find out just what word or words the band is spelling out. The band member's view shows other band members in a confusing array of distances and directions. To make the comparison to our view of the galaxy more accurate, imagine that the band is performing in a fog that prevents the band member from seeing the more distant parts of the formation just as interstellar dust prevents us from seeing remote parts of the Milky Way. The only way for the band member to discover the shape of the band formation (short of asking the band director) would be to make a map, estimating the distances to fellow band members in different directions. With luck, the map would show a word or part of a phrase. With even more luck, it might be possible to guess what the rest of the formation looks like.

Optical Observations In trying to determine the structure of the Milky Way, the first step is to figure out which objects are part of the "band." This can be done by observing other galaxies to see what kinds of objects trace out their spiral arms. These turn out to be young objects such as HII regions, clusters of O and B stars, clouds of interstellar hydrogen, and molecular clouds. When HII regions and O and B stars in the Milky Way were mapped in the early 1950s, the map showed parallel sections of the three spiral arms indicated in Figure 22.33. The nearest

FIGURE 22.32
The Andromeda Galaxy
Many flattened galaxies, like the Andromeda Galaxy, have spiral arms. The two bright elliptical objects on each side of the Andromeda Galaxy are companion galaxies, which orbit about the Andromeda Galaxy.

FIGURE 22.33
Spiral Arms near the Sun
Observations of young clusters and HII regions show parts of three spiral arms within a few kiloparsecs of the Sun. Interstellar dust restricts the map, which is based on visual observations, to the region within about 5 kpc of the Sun. This region is too small to show the overall spiral structure of the galaxy or the way that the local arms are connected to each other. To see beyond a few kiloparsecs, astronomers use radio or infrared telescopes. The numbers around the outside of the figure show galactic longitude.

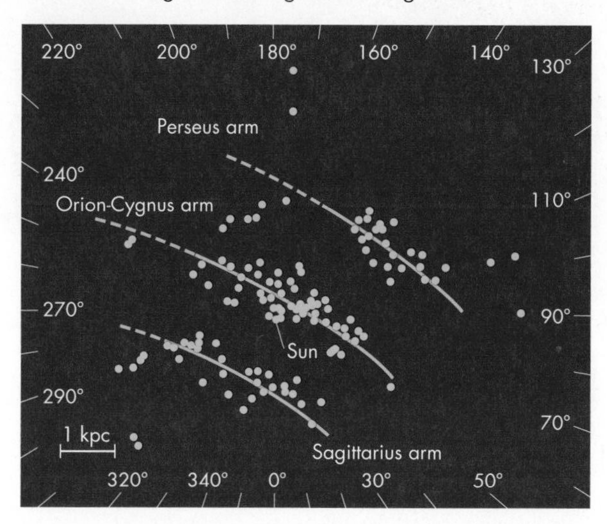

spiral arm, which may include the Sun, is the Orion arm, so-named because much of what we can see of it lies in the constellation Orion. About 2 kpc farther from the center of the galaxy is the Perseus arm. The Sagittarius arm is found about 2 kpc inside the Orion arm. An obvious problem with this optical map of the spiral arm structure of the galaxy is that it does not extend far enough from the Sun to see the overall spiral structure. This is because interstellar dust limits optical observations in most directions to those HII regions and hot stars that lie within about 5 kpc of the Sun.

Radio Observations Because interstellar space is nearly transparent to radio waves, observations of the radio emission lines of atomic hydrogen (at 21 cm) and carbon monoxide can be used to plot spiral structure to much greater distances than the limits of optical observations. Given the rotation curve of the inner galaxy, the individual bumps in the 21 cm line in a given direction can be associated with clouds at particular distances, as shown in Figure 22.34. Observations at many galactic longitudes then can be used to map the hydrogen clouds in the galaxy. The same thing can be done for CO emission from molecular clouds.

There are, however, some problems with using the radio emission to map spiral structure. First, in the directions toward and away from the galactic center, clouds at all distances are moving parallel to the Sun. The radial velocity for all of them is the same—zero. Second, if a cloud has an orbit that isn't perfectly circular, its speed and radial velocity are different from those given by the galactic rotation curve at the cloud's distance from the center of the galaxy. As a result, the cloud will be misplaced on any map of the galaxy that assumes circular orbits. Because of these problems, the interpretation of radio observations of spiral structure is complicated and controversial. Some astronomers take the extreme view that essentially nothing about spiral structure can be learned from the observations of radio lines of interstellar atoms and molecules. Other astronomers are willing to proceed with caution. A map of the galaxy, based mostly on observations of CO from molecular clouds, is shown in Figure 22.35. If this map is correct, the galaxy has a messy, complicated spiral structure. Even though the map traces spiral structure at distances as great as 20 kpc from the Sun, it is not complete enough to tell which pieces of spiral arms connect with each other to form a pattern.

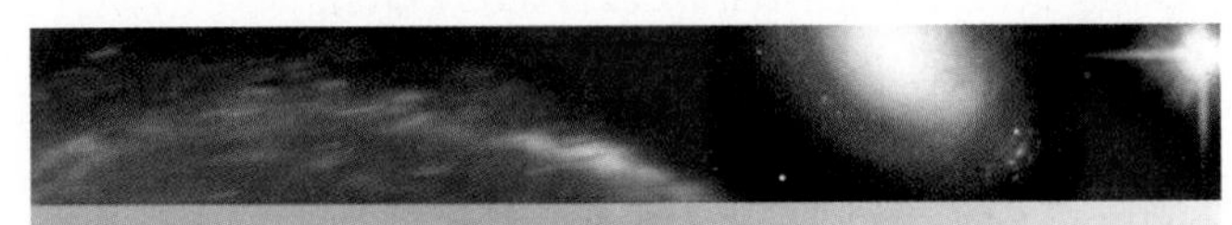

Maps of both young stars and interstellar clouds show that the Milky Way has spiral structure. The overall picture of the spiral pattern, however, is not well known.

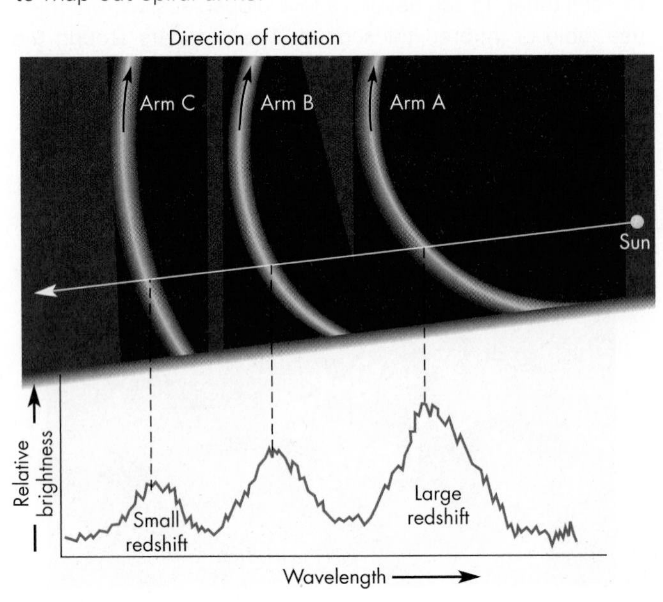

FIGURE 22.34
21 cm Observation of Spiral Structure
Bumps in the line profile of 21 cm hydrogen emission can be identified with clouds of gas at various distances. These clouds can be connected to those seen in other directions to map out spiral arms.

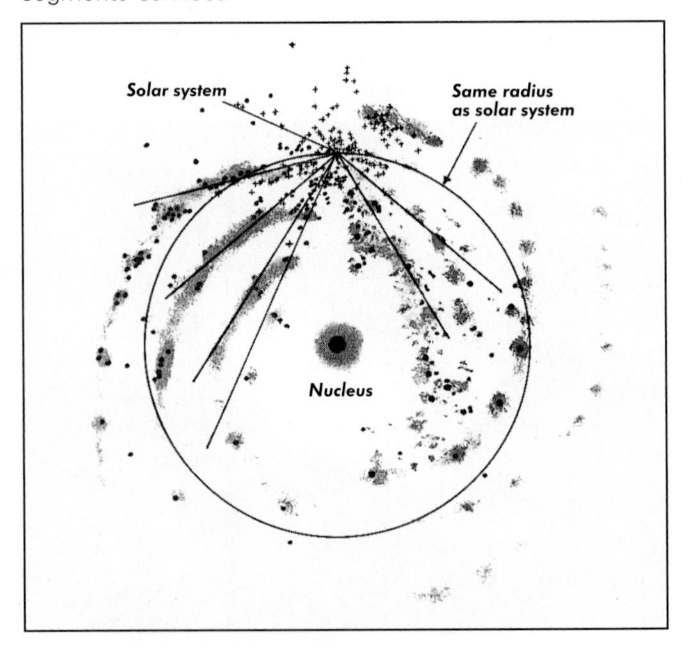

FIGURE 22.35
The Spiral Structure of the Galaxy
Although molecular clouds outline spiral arms, the map is incomplete. It is unclear, moreover, how the different arm segments connect.

Why Is There Spiral Structure?

After it was established that the Milky Way, like many other galaxies, has spiral arms, astronomers sought an explanation for the spiral structure. Two features of spiral arm structure provided important clues about the nature of spiral arms.

First, the spiral arms of other galaxies are outlined by very luminous stars and the HII regions they produce. (These are the O and B stars that were first used to map the spiral structure in the Milky Way.) Such stars live for only a few million years, so if we see them in a spiral arm, we can conclude that they were born there. The question is, why does star formation occur simultaneously along an entire spiral arm of a galaxy?

The second feature of spiral structure is often called the "winding dilemma." Because gas and stars over a wide range of distances from the center of the galaxy orbit at the same speed, orbital period increases steadily outward from the central part of the galaxy. Why, then, don't the spiral arms become very tightly wound up? Imagine a string of gas clouds in a spiral arm. At first, the clouds are lined up radially in the galaxy as in Figure 22.36. As time passes, the inner clouds race ahead of the outer ones. By the time the cloud at 3 kpc has completed half a revolution, the cloud at 8.5 kpc, the Sun's distance, has revolved through only about one-sixth of its orbit. By the time the cloud at the Sun's distance has completed only a little more than one orbit, it is about to be passed by the cloud orbiting at 3 kpc. Within only a few galactic rotations, the spiral arms should be tightly wound. Because the galaxy has completed about 40 rotations since it formed, the spiral arms should be so tightly wound that they should resemble the groove in a phonograph record. In fact, the spiral arms of the Milky Way and other galaxies are rarely wound around more than once or twice.

The winding dilemma is a problem, however, only if a spiral arm always consists of the same gas clouds. This is called a "material arm." The dilemma is avoided if the spiral structure is a wave phenomenon. If a spiral arm is like the crest of a water wave, then it is made up of different matter at different times as the wave travels through the galaxy. The wave nature of the spiral arms is the essential feature of the **density wave theory,** which has succeeded in explaining many features of spiral galaxies.

Density Waves Density waves occur frequently on crowded interstate highways. Imagine that there is a slow-moving truck in the right lane of the interstate, as in Figure 22.37. If traffic is heavy, a knot of cars develops around the truck. Drivers approaching the knot slow down. Because they are traveling slower, the cars spend more time in the vicinity of the truck than they do in an uncongested part of the interstate. In other words, there are more cars around the truck than elsewhere. The greater density of cars makes other drivers more cautious, so they slow down. By slowing down, they maintain the higher

FIGURE 22.36
The Winding Dilemma
Stars and interstellar gas clouds near the center of the galaxy complete an orbit in less time than material farther from the center. If spiral arms consisted of the same material at all times, they would become tightly wound in only a few galactic rotations.

The winding dilemma

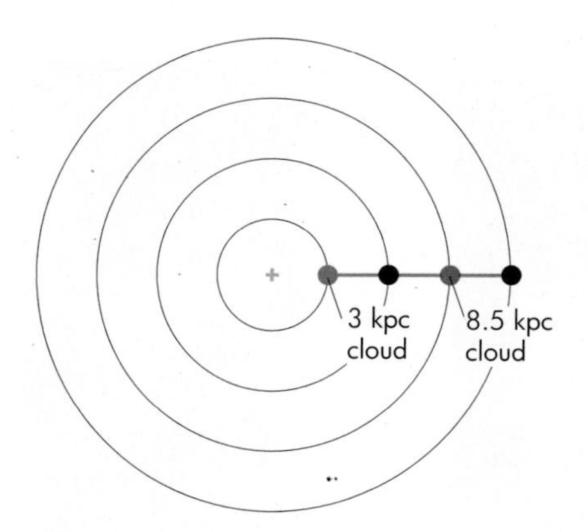

A A string of gas clouds lines up radially.

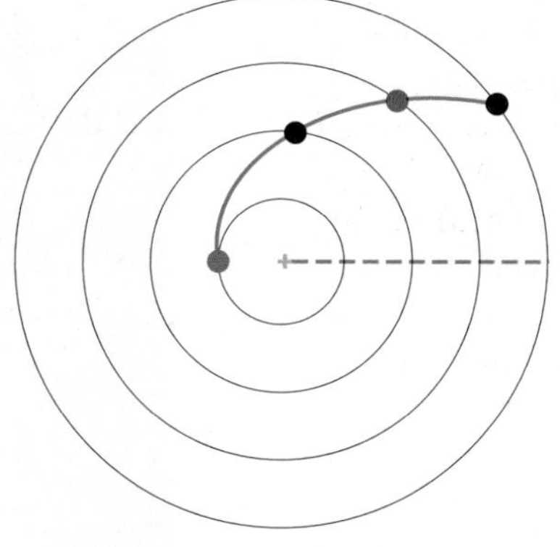

B The 3 kpc cloud completes half of a revolution in the time the 8.5 kpc cloud completes one-sixth of a revolution.

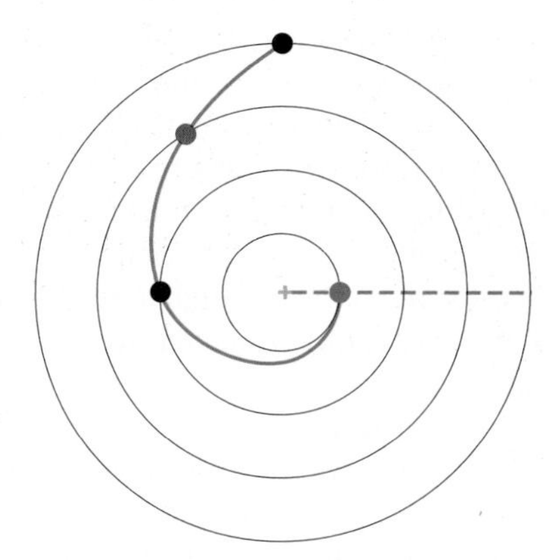

C The 3 kpc cloud will pass the 8.5 kpc cloud in little more than one orbit.

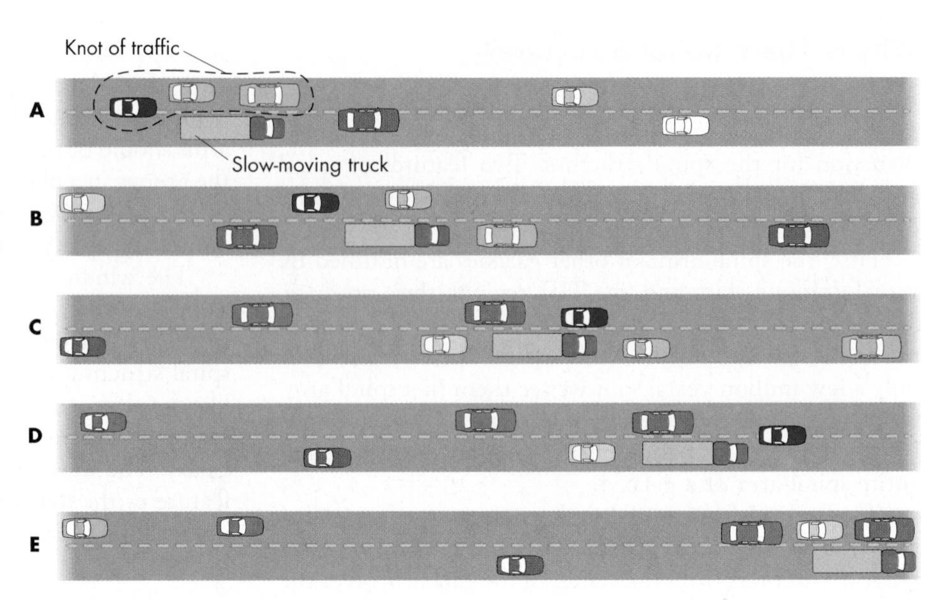

FIGURE 22.37
A Density Wave on a Highway
A slow-moving truck causes a knot of traffic that moves along the highway at the speed of the truck. Individual cars approach the traffic knot, slow down as they move carefully through the knot, and then resume speed as they leave the knot. As a result, the traffic knot consists of different cars at different times.

density around the truck. A given car slows down, works its way through the knot, and then resumes its normal speed. The knot of traffic moves down the highway at the speed of the truck. It contains different cars at different times and is a density wave.

A situation even more like that in a galaxy sometimes occurs under very heavy traffic conditions. In this case, a knot of traffic can develop even without a slow-moving truck or any other permanent obstacle. After the knot develops, it is self-perpetuating. That is, drivers slow down because there is a knot. By slowing down, they contribute to the high density of cars and maintain the knot. Eventually, they come out of the knot and resume normal speed. In this highway analogy, the density wave occurs because of the caution of drivers. In the case of a galaxy, the density wave is a consequence of gravity.

The Density Wave Theory The density wave theory of spiral structure explains that spiral arms exist because they exert a gravitational influence on stars and gas that orbit the galaxy. If the spiral arms didn't exist, a star or gas cloud would feel only a gravitational force directed toward the center of the galaxy. In response to that gravitational force, the star or cloud could orbit the center of the Milky Way in a circular path at constant speed. The spiral arms of the galaxy, however, are regions where the density of matter is higher than it is between the arms. The gravitational force felt by a star or cloud orbiting the galaxy is more complicated than in the case of a smooth galaxy without spiral arms. The orbit of a star or cloud isn't circular, and the star or cloud doesn't move at constant speed. One important result is that a star or gas cloud moves more slowly while it is in a spiral arm than it does elsewhere in its orbit. This means that matter piles up in a spiral arm, increasing the density there. This increase in density is what produces the gravitational disturbance caused by the spiral arm. The whole picture is self-consistent. Because of the gravitational influence of the spiral arm, matter orbits more slowly while in the spiral arm, as shown in Figure 22.38. Where orbital speed decreases, the density of matter increases, producing the extra gravitational force caused by the spiral arm. The direction in which a star or gas cloud moves also changes as it passes through a spiral arm. This change of direction has a very important consequence for gas clouds.

FIGURE 22.38
Orbital Speed near a Spiral Arm
Stars and gas slow down as they approach the spiral arm and then speed up again as they move away from the spiral arm. Stars and gas move more slowly near the spiral arm, much as the cars in Figure 22.37 move more slowly near the traffic knot.

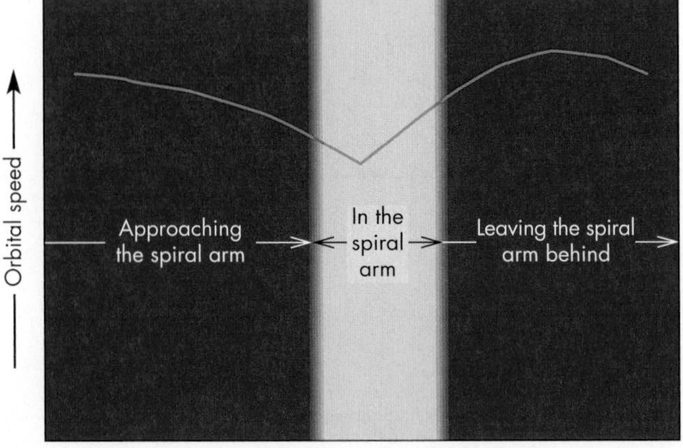

The spiral arms don't wind up because they aren't material arms. Instead, the arms contain different stars and gas at different times. The entire spiral pattern rotates with the same orbital period—about 500 million years. At the Sun's distance, the pattern moves at about half the speed of the stars and gas. Stars and gas clouds catch up with the arms from behind, work their way through the arms, then resume normal speed when they emerge. This is analogous to the way that cars pass through a traffic knot on a highway.

Star Formation and Spiral Arms It is very difficult to see spiral structure in a picture of a spiral galaxy taken in yellow light. This is because a picture taken in yellow light emphasizes yellow stars, like the Sun. The spiral arms don't stand out because there are only a few percent more solar-type stars in a spiral arm than in a region of similar size between the spiral arms. The concentration of older stars like the Sun doesn't respond more strongly to the increased gravitational attraction of the spiral arms because older stars have a fairly wide range of orbital eccentricities. In a given region of the galaxy, the older stars, while all moving in the same general direction, have a range of orbital speeds and vary in the precise direction in which they are moving. Each of them feels the influence of the spiral arm, but the effect is somewhat different for each individual star. This means that the effect of the spiral arm is blurred.

The effect of the spiral arm is not blurred for interstellar gas clouds (and the dust they contain), which traverse the galaxy on orbits nearly parallel to each other. All the gas clouds in a given region of the galaxy speed up or slow down and change direction in the same way in response to a spiral arm. Figure 22.39 shows the paths of gas clouds in a spiral galaxy. As the gas clouds enter a spiral arm, they turn sharply so that their paths are closer together. The gas is compressed, increasing its density by about six to eight times. This is enough to cause molecular clouds to collapse, triggering star formation. A burst of star formation takes place while the gas cloud passes through a spiral arm. Massive stars, which lead short lives, die (perhaps as supernovae) before they can emerge from the spiral arm. This is why hot, massive stars highlight the spiral arms. Stars like the Sun, on the other hand, live much longer. They emerge from the spiral arm and make many trips around the galaxy before they die.

How Do Spiral Arms Get Started? The spiral structure of the galaxy, once established, may be self-perpetuating. It may survive many, perhaps an indefinite number, of galactic revolutions. The problem is, how does spiral structure get established? At present, nobody knows for sure, although there have been many suggestions. One idea is that relatively small irregularities in a galaxy can grow into a full spiral pattern. Another is that spiral structure is produced by the tidal influence of a companion galaxy. The origin of spiral arms is probably the biggest remaining mystery about the spiral structures of the Milky Way and other galaxies.

FIGURE 22.39
The Motion of Gas Clouds in a Spiral Galaxy
The direction of orbital motion of clouds changes abruptly when they reach a spiral arm. As a result, the gas density rises by a large factor. The lines in this drawing represent the motions of individual clouds. Note how the direction of motion changes when a cloud encounters a spiral arm.

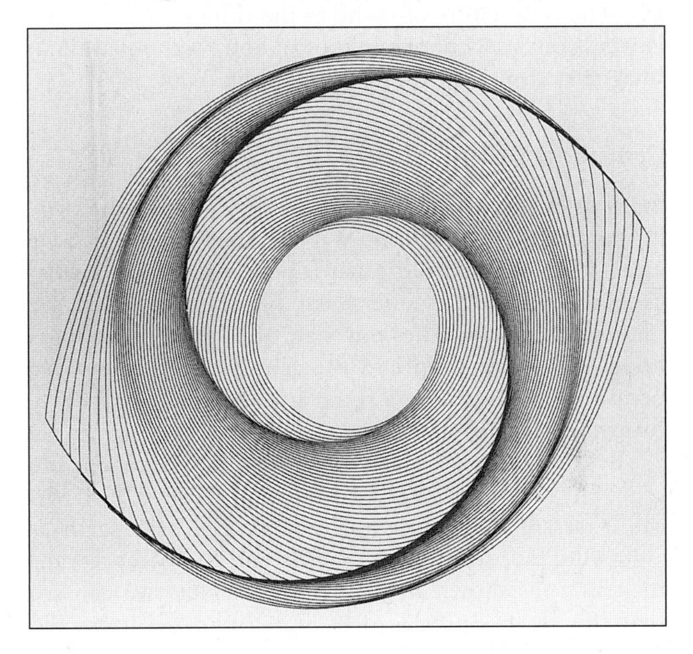

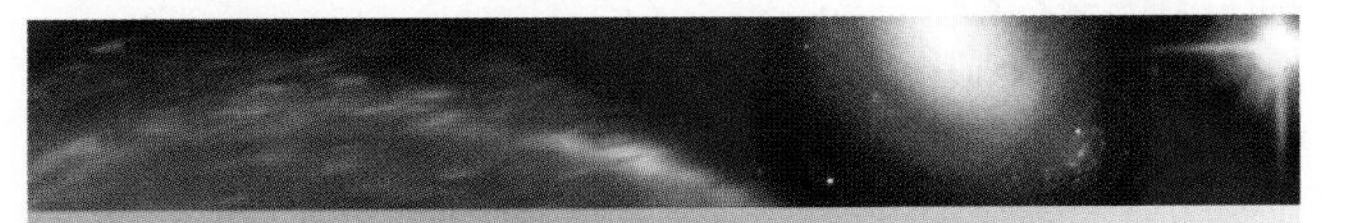

Spiral arms are crests of density waves, which contain different material at different times. The gravity caused by the extra matter in spiral arms modifies the orbits of stars and interstellar gas. The compression of interstellar clouds that enter spiral arms causes star formation. Hot stars mark the spiral arms because they do not live long enough to emerge from the spiral arm in which they are born.

22.5 THE CENTER OF THE GALAXY

If astronomers had to rely on only visible light, we would know almost nothing about the center of the Milky Way. So much interstellar dust lies between the Earth and the center that only one visible photon in a thousand billion makes it through. Fortunately, radio, infrared, and X-ray radiation reaches us from the center and tells us about the conditions in the core of the Milky Way. Although the center of the galaxy has been extensively observed for decades, it is still mysterious. There are features near the center that do not have counterparts elsewhere in the galaxy. The true nature of many of the things seen near the center is unclear. Enough is known, however, to permit an imaginary trip to the center of the Milky Way.

Crowded Stars

Starting from the Sun, our trip takes us toward the richest part of the Milky Way. Within a few kiloparsecs, we enter a part of the galaxy that is poorly known to us because it is obscured by interstellar dust. Looking back, the same dust now obscures our view of our solar neighborhood. As we travel toward the center of the galaxy, we occasionally pass through spiral arms studded with newly formed clusters of stars.

By the time we reach 1 kpc from the center, we are well within the galactic bulge (Figure 22.28). In the infrared, the bulge can be seen from the Earth as a glowing ellipse, about 2° across and brightest toward its center, as shown in Figure 22.40. Now we can see that the glow is produced by stars, many of them red giants, which become more crowded as we move inward. In the solar neighborhood, there is less than one star per cubic parsec. At 100 pc from the center, we find that the density of stars has risen to 100 per cubic parsec. There are several clusters of young, very massive, extremely bright stars. At 10 pc from the center, there are several thousand stars per cubic parsec. When we are just a few parsecs from the center, we are dazzled by hundreds of thousands of stars in each cubic parsec. The stars are only light weeks rather than light years apart. Starlight is bright enough to read by. This is a dusty region, so much of the starlight is absorbed by interstellar particles and then reradiated in the infrared region of the spectrum.

FIGURE 22.40
An Infrared View of the Milky Way
In this *Spitzer Space Telescope* image, red indicates emission from warm dust and blue indicates stars. The center of the galaxy is located in the whitish region. The image is about 2° across.

The Gaseous Ring In addition to the stars, we also see a thin ring of gas and dust that rotates about the center between 2 and 8 pc from the center. The ring is turbulent and very clumpy. It rotates at a speed of 110 km/s, which means that the gravitational force keeping it in orbit is due to 10 million $M_{\odot}$ of matter located closer to the center than the ring. Most of the 10,000 $M_{\odot}$ of gas in the ring consists of atoms and molecules. At the inner edge of the ring, however, the gas is ionized, apparently by ultraviolet radiation from the central 2 pc of the galaxy. The ring may be an accretion disk in which infalling gas accumulates and then slowly spirals inward.

When we cross the inner boundary of the ring, at 2 pc from the center, we find ourselves in a cavity in which the gas density is ten to one hundred times lower than in the ring. Because the gas in the ring slowly spirals inward, it should fill up the cavity with gas in about 100,000 years. The low gas density in the cavity implies that something, perhaps an explosion as violent as a supernova outburst, must have expelled gas from the cavity quite recently.

The Black Hole at the Center Although the region inside the gaseous ring has relatively little gas, we can see that it isn't completely empty. Streamers of hot gas form a spiral-like pattern shown in Figure 22.41. These streamers may be material falling inward from the ring. The streamers, which originally may have been small clouds of gas, have been stretched out by tidal forces during the 10,000 years that they have been falling inward. As the gas fell inward, its rotational velocity increased from 150 km/s at 0.7 pc from the center, to 260 km/s at 0.3 pc, to 700 km/s at 0.1 pc. Stars even closer to the center are revolving at even greater speeds. Such large rotational speeds at such small distances from the center could occur only if a few million solar masses of material lie within the inner 0.1 pc of the galaxy.

As we try to locate that mass, our attention turns to **Sgr A***, a small, bright source of radio radiation shown in Figure 22.42. Sgr A* is located near the spot where streamers of infalling gas merge. There are several reasons to believe that Sgr A* is a massive black hole that marks the exact center of the galaxy. Its size, 1 AU or

FIGURE 22.41
Streamers of Gas Inside the Rotating Ring near the Center of the Galaxy
The streamers, shown in blue, converge near the point labeled Sgr A*, shown in more detail in Figure 22.42. Purple indicates a disk of dust and gas.

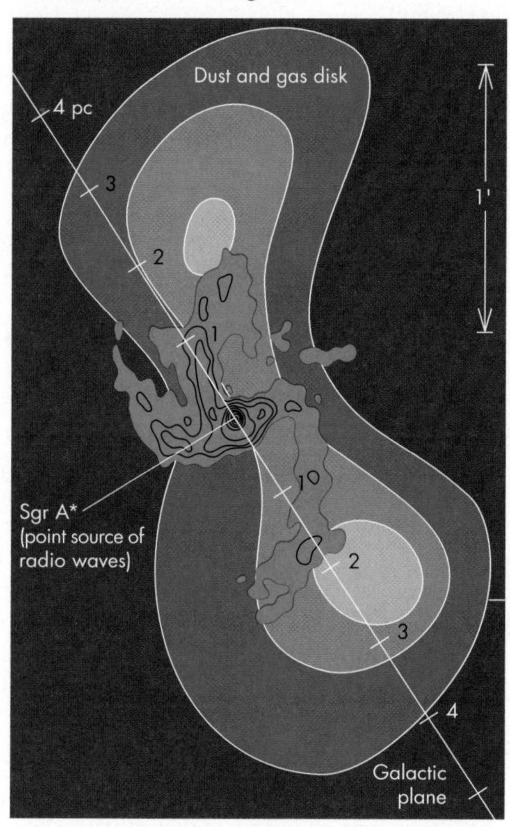

FIGURE 22.42
Sgr A*
The point where the streamers come together is the location of Sgr A*. Sgr A* is a massive black hole that lies at the exact center of the galaxy.

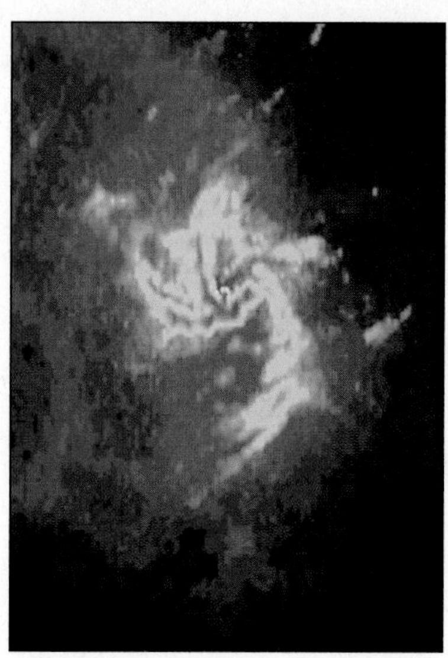

smaller, is about ten times as large as the Schwarzschild radius of a black hole containing a few million $M_\odot$. One important clue about the nature and location of Sgr A* is that it appears to be motionless. Precise measurements of the direction of the radio emission have shown that Sgr A* is moving across the line of sight at a speed less than 15 km/s. Unless Sgr A* is at the center of the galaxy, it would have to be orbiting at hundreds of km/s and would be moving much more rapidly across the line of sight.

Another reason to believe Sgr A* is a massive black hole is that astronomers have measured the orbits of stars that revolve about it. Figure 22.43 shows the orbits of a group of stars that lie within 0.02 pc of the center of the galaxy. The stars are moving at great speed and complete their orbits in only a few decades. The orbital sizes and periods of the stars orbiting Sgr A* can be used in Newton's version of Kepler's third law (Equation 5.4) to find its mass. The result is 2.6 million solar masses. The orbit of one of the stars carried it within only

FIGURE 22.43
Stars Orbiting Sgr A*
The orbits of the six stars carry them very close to Sgr A*. The orbital periods are only a few decades. The mass required to keep the stars in orbit is 2.6 million solar masses, strongly supporting the idea that Sgr A* is a black hole.

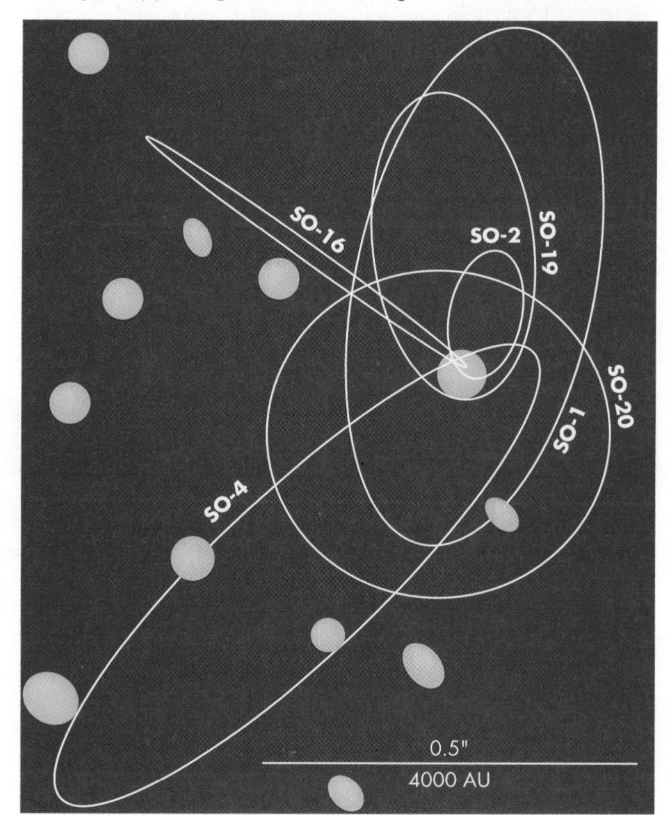

60 AU of Sgr A*. Given the great mass of Sgr A* and the fact that it is no bigger than the solar system, it almost certainly must be a black hole.

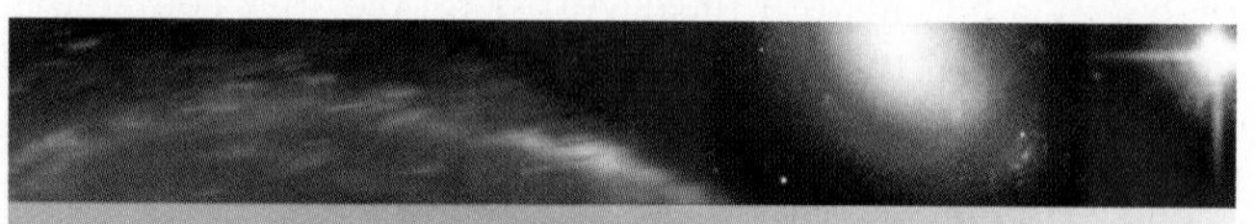

Although the galactic center is obscured by interstellar dust, it can be mapped in the infrared and radio parts of the spectrum. The region near the center contains a great concentration of stars, a rotating ring of gas extending from 2 to 8 pc from the center, and streamers of gas that may be falling into a central black hole of several million solar masses. The orbits of the stars near galactic center strongly support the idea that there is a massive black hole there.

22.6 THE HISTORY OF THE GALAXY

Many of the most important clues about the origin and development of the galaxy have been found in the spatial distributions, chemical compositions, and ages of the stars in the galaxy.

Stellar Populations

Population I and II orbits

A **stellar population** is a group of stars that resemble each other in spatial distribution, chemical composition, and age. Exactly how many stellar populations there are in the galaxy is controversial, mostly because astronomers disagree where one population ends and another begins. Although some astronomers argue that there are as many as six distinguishable stellar populations, the more prevalent view is that the galaxy can be described well enough with only the three populations listed in Table 22.1. In Table 22.1, the column marked "height" gives the distance from the galactic plane within which half of the population can be found. "Metals" refers to those atoms that are anything but hydrogen and helium. Table 22.1 gives the abundance of metals in each population relative to the abundance of metals in the Sun.

As we look from the thin-disk population to the thick-disk population to the halo population, there are two trends that deserve notice. First, the older the population, the lower the amount of heavy elements it contains. The youngest stars are somewhat richer in heavy elements than the Sun. Older stars (for example, those found in globular clusters), are up to 30 times poorer in heavy elements than the Sun. These stars consist almost entirely of hydrogen and helium. Second, height above the galactic plane increases as the age of the population increases, so the older the population, the less it is confined to the region of the galactic plane. The oldest stars in the galaxy can be found well out in the halo of the Milky Way, many kiloparsecs from the galactic plane. The halo population stars move on orbits highly inclined to the galactic plane. They pass through the galactic plane twice on each orbit (some of them can be found in the solar neighborhood) but spend most of their time far from the plane. The orbits of the youngest stars (the thin-disk population) have only very small inclinations, so they move entirely in the galactic plane.

Galactic Evolution

The fundamental things that the stellar populations tell us are the shape and chemical makeup of the gas in the galaxy at the time that stars were formed.

Stars as Fossil Gas Clouds The stars that form in a cloud of gas immediately share the orbit and the chemical composition of the gas cloud. Although the orbit of a star could be changed if it had a near collision with another star, stellar collisions are extremely rare. This means that a star, for its entire lifetime, continues to move through the galaxy in the same orbit as the cloud from which it

Table 22.1

Stellar Populations in the Milky Way

POPULATION	EXAMPLES	AGE (BILLIONS OF YEARS)	HEIGHT (PARSECS)	METALS
Thin disk	O and B stars T Tauri stars	0 to 5	250	11
Thick disk	G and K stars Planetary nebulae	5 to 14	700	1/3
Halo	Globular clusters RR Lyrae stars	11	2500	1/30

formed. Thus, the spatial distribution and orbits of the stars in each population preserve a record of the distribution and orbits of the gas clouds that filled the galaxy at the time they were formed. By comparing the populations, we learn that as time has passed, the gas in the galaxy has become increasingly flattened toward the galactic plane. Thus, the Milky Way was much more spherical at the time the halo population formed than it is today.

Although nuclear reactions gradually modify the chemical composition in the interior of a star, in most cases the surface layers of the star retain the composition of the gas cloud from which the star formed. Again comparing the populations, we find that interstellar gas clouds have become progressively richer in heavy elements over the life of the galaxy. Our theories of the evolution of the galaxy have had to take into account its steadily changing shape and composition.

The Birth of the Galaxy The galaxy began as a cloud a few hundred kiloparsecs across. The cloud was made almost entirely of hydrogen and helium. As the cloud moved past other clouds, their gravitational attractions started the protogalaxy spinning. However, the rotation of the protogalaxy was far from organized. Currents and cloudlets still moved in all directions. One of the earliest things to happen in the protogalaxy was that spheres of gas containing about 10^6 $M_\odot$ began to collapse under their own gravity to form clusters of stars. These clusters, the globular clusters, are the oldest recognizable objects in the galaxy. H-R diagrams of globular clusters show that the stars now leaving the main sequence have main sequence lifetimes of about 11 billion years. This is taken to be a lower limit on the age of the galaxy. The globular clusters typify the halo population of the Milky Way.

Gas Cloud Collisions Although collisions between stars have played an insignificant role in determining the shape of the galaxy, quite the opposite is true for collisions of gas clouds. From the earliest times, collisions between clouds of gas have taken place. These collisions have steadily compressed the interstellar gas toward the galactic plane. To see how this has happened, imagine that two clouds stick when they collide. Two clouds that have different orbital inclinations will stick and move on an orbit with intermediate inclination, as shown in Figure 22.44. As time passes, cloud–cloud collisions steadily eliminate the clouds with the greatest inclination and those moving in the direction opposite to the majority of clouds. The gas in the galaxy has, through collisions, become flatter and flatter until most of it is now in a layer only about 100 pc thick. The way that the galaxy has become flatter is similar to the flattening that took place in the solar nebula and to the way in which planetary rings become flat and thin through the collisions of orbiting particles.

FIGURE 22.44
Gas Cloud Collisions
When gas clouds collide, they stick together and share their angular momentum. As a result of collisions with other gas clouds, clouds that had unusual inclinations, eccentricities, and directions of motion merged with other clouds and lost their distinctive orbits. Gas cloud collisions have steadily flattened the disk in the galaxy.

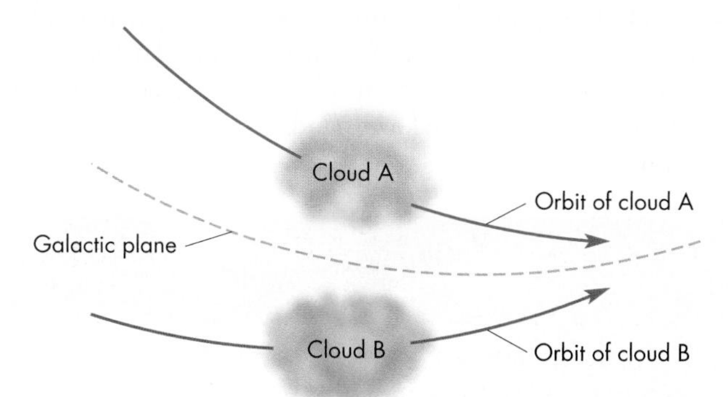

A Two clouds of gas with different orbital eccentricities

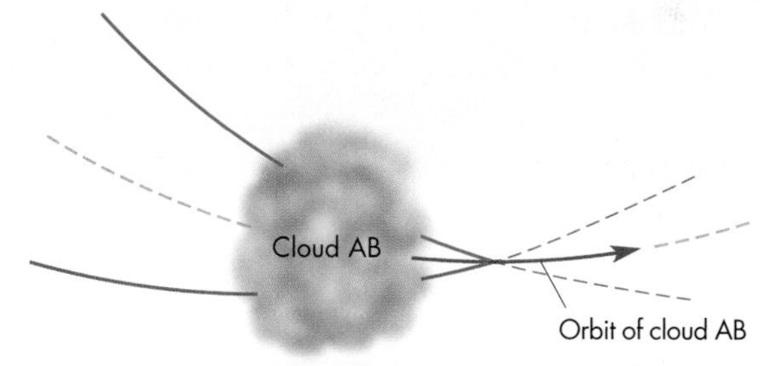

B The two clouds collide and stick together; the orbit of the merged cloud is much closer to the galactic plane.

Chemical Evolution The nuclear reactions in stars create heavy elements from hydrogen and helium. In the case of massive stars, these heavy elements are eventually dispersed into interstellar space through supernova explosions and stellar winds. The mixing of heavy elements made in stars with the original gas has steadily enriched the interstellar gas. Newly formed stars have benefited from previous generations of stars and have become richer in metals over the history of the galaxy.

The Galaxy Today At present, the cool gas in the galaxy is confined to a very thin layer, within which star formation goes on in spiral arms. Star formation is so efficient in the Milky Way that interstellar gas would soon be depleted if there were no fresh supply of gas. This supply apparently comes from the galactic halo, because clouds of gas have been observed falling rapidly from the halo toward the galactic plane. Much of the infalling matter may be gas shot into the halo in "chimneys." The chimneys form when superbubbles produced by numerous supernovae in a young cluster become so large that they extend into the galactic halo. When the hot gas that breaks out of the galactic disk in chimneys has had time to cool, it falls back into the galactic plane to replenish the interstellar gas.

Astronomers gradually have pieced together information about the size, mass, luminosity, shape, and rotation of the Milky Way galaxy. Our current picture of the Milky Way is illustrated by the artist's conception shown in Figure 22.45. If we could move out of the swirling mass of stars, gas, and dust and look down on the Milky Way from a point far above the galactic plane we would see an SBc galaxy—a large, barred spiral with two main spiral arms and several secondary arms and spurs.

The galaxy began more than 11 billion years ago as a gaseous spheroid. Over time, collisions between gas clouds flattened the galaxy. Nuclear reactions in stars increased its heavy-element content. By examining the stellar populations—stars that formed at different times—we can learn how the Milky Way has changed in shape and composition since it formed.

FIGURE 22.45
An Artist's Conception of the Milky Way

Chapter Summary

- HII regions are bright nebulae produced when ultraviolet radiation from hot stars ionizes the surrounding interstellar gas. An HII region develops when hot stars form inside molecular clouds. The HII region grows in size, eventually breaking out of the cloud. When the central star of an HII region stops emitting large numbers of ultraviolet photons, the HII region fades out. (Section 22.1)
- The presence of diffuse interstellar gas is revealed by interstellar absorption and emission lines. Between clouds, the interstellar gas is warm except where it has been heated by supernovae. Superbubbles are large regions of hot interstellar gas produced when many supernovae occur within a short time in a cluster of stars. (22.1)
- Interstellar dust clouds appear either as dark nebulae, which obscure the stars behind them, or as reflection nebulae, which scatter the light of nearby stars toward us. (22.1)
- Interstellar dust blocks starlight both by scattering it away and by absorbing it. The scattered light contributes to the general glow of the galaxy, whereas absorbed starlight heats the particles and causes them to radiate in the infrared. (22.1)
- Interstellar dust particles are very tiny. They form in the gas that flows into interstellar space from cool giant and supergiant stars. While in interstellar space, the dust particles may develop coatings of icy material. (22.1)
- Initially, counts of stars in various directions led astronomers to conclude that the Milky Way is a small galaxy centered near the Sun. This result turned out to be wrong, however, because it failed to take interstellar absorption into account. (22.2)
- The center of the galaxy is located between 8 and 8.5 kpc from the Sun. The galaxy consists of a flat disk a few kpc thick and 40 kpc across, a spheroidal bulge, and a spherical halo that may extend to more than 40 kpc from the galactic center. (22.2)
- The orbital period of the Sun is about 230 million years. According to Kepler's third law, the mass

of material inside the Sun's orbit must be about 100 billion $M_{\odot}$. (22.3)

- The rotational speed of the Milky Way is constant between 1 and 20 kpc from the center, and may be constant to a much greater distance. This shows that the mass of the galaxy increases with distance from the center and may total as much as 600 billion $M_{\odot}$. Most of the mass of the galaxy consists of dark matter of an unknown nature. (22.3)
- The spiral structure of the Milky Way has been found by mapping the locations of young stars and interstellar clouds. However, the overall structure of the galaxy's spiral pattern is still not well known. (22.4)
- The spiral arms of the Milky Way are density waves, which contain different material at different times. The gravitational force of the increased density of matter in spiral arms modifies the orbits of stars and interstellar dust. Interstellar clouds that enter spiral arms are compressed and form stars. Hot stars are found mainly in spiral arms because they don't live long enough to emerge from the spiral arm in which they form. (22.4)
- Radio and infrared observations have revealed the nature of the center of the Milky Way. Many stars are concentrated near the center, which also contains a ring of rapidly rotating gas, orbiting stars, and streamers of gas that may be falling into a central black hole of several million solar masses. (22.5)
- Since the galaxy formed more than 11 billion years ago collisions between gas clouds have steadily flattened it. Nuclear reactions in stars have steadily enriched the galaxy in heavy elements. The orbits and compositions of the different populations of stars show the shape and composition of the galaxy at the time that the stars formed. (22.6)

Key Terms

dark matter *538*
dark nebulae *527*
density wave theory *541*
extinction *530*
galactic bulge *534*
galactic disk *534*
galactic equator *534*
galactic halo *534*
galactic latitude *534*
galactic longitude *534*
galactic nucleus *534*
globular cluster *533*
HII region *523*
interstellar matter *523*
interstellar reddening *530*
kiloparsec (kpc) *533*
microlensing event *538*
reflection nebula *528*
rotation curve *536*
scattering *527*
Sgr A* *544*
spiral arm *539*
stellar population *546*

Conceptual Questions

1. Why must the star responsible for an HII region be a hot star rather than a cool star?
2. Why are the HII regions around O stars generally larger than those around B stars?
3. Describe the evolution of an HII region.
4. Why wouldn't you expect to detect strong 21 cm emission from an HII region?
5. What heats the hot phase of the interstellar gas?
6. What would most bright nebulae look like if there were no interstellar dust?
7. What is the relationship between the color of a reflection nebula and the color of the star that illuminates it?
8. What fraction of the light from a star is able to penetrate 3000 pc through a typical region of interstellar dust?
9. What property of interstellar dust has made it easier for astronomers to study distant parts of the galaxy at infrared wavelengths rather than in the visible part of the spectrum?
10. What prediction would you make about how the color of the diffuse light in the galaxy compares with the average color of the bright stars in the galaxy?
11. What is the reason that the rocky material in an interstellar dust particle forms a core and the icy material forms an outer layer rather than vice versa?
12. What was the principal reason that the measurements of Herschel and Kapteyn didn't reveal the correct shape and size of the galaxy?
13. How did Herschel explain the appearance of the Milky Way in terms of the distribution of stars in the galaxy?
14. Suppose an observer on a hypothetical planet maps the distribution of globular clusters in the sky. The observer finds that globular clusters are evenly distributed throughout all parts of the sky. Where is the observer's planet?
15. What is the evidence that most of the matter in the galaxy exists in an as-yet undetected form?
16. What kinds of objects are found preferentially in the spiral arms of a galaxy?
17. Why can't observations in visible light reveal the entire spiral pattern of the galaxy?
18. How do we know that the arms of the galaxy can't be "material arms"?

19. Why is rotation speed smaller in a spiral arm than in the region between spiral arms?
20. About how often do the stars in the Sun's part of the galaxy catch up with and pass through a spiral arm?
21. According to the density wave theory, why are O and B stars found almost exclusively in spiral arms?
22. Describe the evidence that there is a massive black hole in the center of the galaxy.
23. Why do astronomers suspect that there have been violent explosions at the center of the galaxy?
24. Why are the young objects in the galaxy confined to a thin disk, but the oldest objects occupy a nearly spherical volume of space?
25. How have the shape and chemical composition of the galaxy changed as the galaxy evolved?

Problems

1. Suppose the Sun's orbital speed were 440 km/s. What would be the Sun's orbital period?
2. Suppose the Sun's orbital speed were 440 km/s. What would be the mass required to keep the Sun in its orbit?
3. At the same distance from the center, the rotation speed of galaxy A is three times as large as the rotation speed of galaxy B. What can be said about the masses of the two galaxies?
4. The gas at a distance of 0.1 pc from the center of the galaxy has an orbital speed of 700 km/s. Use Equation 22.2 to find the mass of the material in the inner 0.1 pc of the galaxy.
5. Use the small angle equation to find the angular diameter of Sgr A* if it has a diameter of 13 AU and a distance of 8.5 kpc.

Figure-Based Question

Use Figure 22.36 to estimate how many revolutions the cloud at 3 kpc makes during a single revolution of the cloud at 8.5 kpc.

Group Activities

1. With your group, travel to a dark spot on a clear, moonless night. Trace the path of the Milky Way across the sky. Have members of your group point out dark rifts in the Milky Way to one another. (You should do this in summer or autumn when the summer Milky Way is high in the sky in the evening.)
2. Take a drive with members of your group on a busy highway. See if members of the group (not the driver) can spot moving knots of traffic that are analogous to density waves. Note the speed of the car while outside the knot and while in the knot. See if you can estimate how much denser traffic is in the knot than outside. Afterward, discuss how your observations are related to the density wave theory of spiral structure.
3. Using Figure 22.36 as a guide, have members of your group line up and hold a long piece of string. Then have the group walk in circles of different sizes, all moving at about the same speed. Notice how the string (the spiral arm) gets thoroughly wound up before even a couple of orbits have been completed.

For More Information

Visit the text website at **www.mhhe.com/fix** for chapter quizzes, interactive learning exercises, and other study tools.

23

We know our immediate neighborhood rather intimately. With increasing distance, our knowledge fades, and fades rapidly. Eventually, we reach the dim boundary—the utmost limits of our telescopes. There, we measure shadows, and we search among ghostly errors of measurement for landmarks that are scarcely more substantial.

Edwin Hubble, 1936

A *Hubble* image of the barred spiral NGC 1300.

www.mhhe.com/fix

Galaxies

The "neighborhood" about which Hubble wrote has grown ever larger throughout human history. The shadows have been pushed back to reveal the solar system and then the galaxy. Beyond the Milky Way is the realm of the galaxies, a region so vast that our studies become explorations not only of space, but also of time. The telescope becomes, in effect, a time machine because the light we observe from distant parts of the universe takes so long to reach us. Light from the Sun takes only about 8 minutes to reach us. Light from Proxima Centauri, the star closest to the solar system, travels for 4 years before falling on the Earth. The most recent information we have about Proxima Centauri, therefore, is always 4 years old. The farther we look into space, the further back in time we see.

The ability to look back in time by looking outward into space becomes important when it permits us to view objects as they were when they were only a fraction as old as they are today. This happens, however, only at very great distances. The Andromeda Galaxy is 2 million light years away, so we see it as it was 2 million years ago. The Andromeda Galaxy is over 10 billion years old, however, so while it may have changed in detail during the last 2 million years, it is fundamentally the same object today that it was then. The telescopes and detectors that we have today make it possible for us to see not just millions, but billions of light years into space. By doing so, we see far enough back in time that we can begin to learn about the ways in which galaxies, clusters of galaxies, and even the universe have changed over time. In effect, we can see them evolve.

Questions to Explore

- What are the different kinds of galaxies?
- How have astronomers determined the distances to galaxies?
- What makes astronomers think that galaxies contain vast amounts of dark matter?
- How do we think galaxies formed?
- Why do nearly all galaxies have redshifted spectra?
- Why do the redshifts of galaxies increase with their distances?

In Chapters 23 through 26 we explore the observations that permit us to see ever deeper back into time. As we do, we see a universe that was more and more different from the one in which we live. In this chapter, we learn about galaxies as they are today—near us both in space and in time. In Chapter 24, we look farther and see very strange sorts of galaxies that no longer exist today. In Chapter 26, we make use of observations that carry us back almost to the birth of the universe.

23.1 THE DISCOVERY OF GALAXIES

People have observed **galaxies** since prehistory. Two galaxies, the **Magellanic Clouds,** are conspicuous objects in the skies of the southern hemisphere. Another, the Andromeda Nebula, was described by an Arab astronomer more than a thousand years ago. After the telescope was invented, many more galaxies were discovered. They were lumped together with other glowing patches in the sky as nebulae, from the Latin word for mist or cloud. At first, these nebulae were considered nuisances by astronomers who often mistook them for the comets they were seeking. In 1781, one comet hunter, Charles Messier, compiled a list of 103 nebulae as an aid to other comet hunters. We still refer to nebulae in that list by their Messier numbers. The Andromeda Nebula, for example, is M31.

Island Universes

Christopher Wren suggested, in the seventeenth century, that the nebulae might be star systems that were so distant the light of the individual stars blurred into a milky, continuous glow. He called these star systems "island universes," a term that had the same meaning that "galaxies" has today. William Herschel held this view at first, too, because he was able to resolve some of Messier's objects into individual stars. He assumed that other nebulae were star clusters that he was unable to resolve into stars because they were too far away. In 1790, however, Herschel discovered a "nebulous star." The nebulous star was a planetary nebula in which he clearly saw a single star surrounded by a nebula. The fact that one star could be seen, but the nebula could not be resolved into stars, convinced Herschel that some nebulae, at least, could not be explained simply as unresolved starlight. When William Huggins showed in 1864 that the spectra of some nebulae consist of emission lines from hot thin gas, the island universe model for nebulae fell into disfavor.

The discovery that nebulae tend to avoid the galactic plane, as shown in Figure 23.1, seemed to contradict the island universe model, too. If the nebulae were distant collections of stars, located far from the Milky Way, it seemed an amazing coincidence that they would be found in all directions in the sky except along the galactic plane. In 1885 a point of light in the Andromeda Nebula increased in brilliance until it became about one-tenth as bright as the entire Andromeda Nebula. Astronomers were reluctant to believe, if the Andromeda Nebula were composed of many stars, that a single one of them could become as bright as millions or billions of others. By the turn of the twentieth century, most astronomers thought that there was only one island universe—our own Milky Way.

In about 1917, however, the island universe theory became popular again. The reemergence of the island universe theory began when Heber Curtis discovered that novae occur in spiral nebulae. These novae showed light curves like novae in the Milky Way, but were very much fainter, indicating that the spiral nebulae were very distant. Curtis also found that nebulae seen edge-on, as in Figure 23.2, have dark lanes that bisect them. Curtis realized that if the plane of the Milky Way also has obscuring matter, then the reason nebulae seem to avoid the Milky Way is that

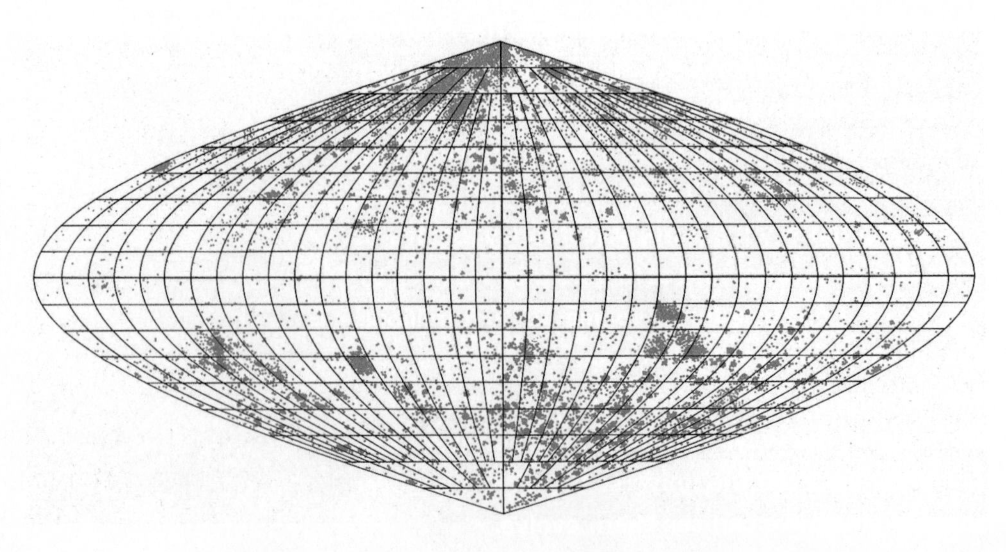

FIGURE 23.1
The Zone of Avoidance
In this map of the distribution of galaxies in the sky, the galactic plane runs horizontally across the center. Relatively few galaxies can be found near the galactic plane.

FIGURE 23.2
Dust in the Spiral Galaxy, NGC 4565
Observations of dust lanes in other galaxies convinced many astronomers that dust within our own galaxy prevents us from seeing galaxies in directions near the galactic plane, producing the zone of avoidance. NGC 4565 is seen edge-on.

obscuring matter blocks our view of them when we look along the galactic plane.

At about the same time as Curtis's discoveries of novae and dark lanes in nebulae, Adrian van Maanen began comparing photographs of spiral nebulae taken years apart. He found that some features changed position in a way that suggested the nebula rotated with a period of about 100,000 years. This was a serious problem for the island universe theory. If the spiral nebulae were distant galaxies, they would have to be so large that to revolve in 100,000 years, their outer parts would have to travel faster than the speed of light. In the early 1920s, ideas about spiral nebulae were very unsettled.

Hubble Proves Nebulae Are Galaxies

The nature of spiral nebulae was resolved in 1923 when Edwin Hubble (after whom the *Hubble Space Telescope* is named) discovered Cepheid variables in the Andromeda Nebula. The period-luminosity relationship for Cepheids (Chapter 19, Section 4) allowed him to find the distance to the Andromeda Nebula. This distance was so great that it showed, once and for all, that the Andromeda Nebula and other spiral nebulae were far beyond the Milky Way and were huge collections of stars. They were, in fact, distant galaxies like our Milky Way. It turned out that the swift rotation periods van Maanen had calculated were due to measurement errors. Hubble's discovery produced a vastly expanded vision of the universe. Instead of a single galaxy surrounded by nothing, the Milky Way became one of countless galaxies that stretched outward to the limits of observation.

Until early in the twentieth century, the nature of the spiral nebulae was controversial. Proof that they are distant galaxies of stars was supplied in 1923, when Edwin Hubble identified Cepheid variables in the Andromeda Nebula. He showed that the Andromeda Nebula is far beyond the Milky Way.

23.2 THE KINDS OF GALAXIES

No two galaxies look exactly alike. Nevertheless, most galaxies can be classified into a fairly simple system. Several classification systems have been invented, but the one most frequently used to classify galaxies was originated by Edwin Hubble and is called the Hubble classification. Hubble found that nearly all galaxies have regular shapes that take one of two basic forms—elliptical or spiral. The Hubble classification system is summarized in Figure 23.3.

Elliptical Galaxies

Elliptical galaxies, designated E, are nearly featureless and elliptical in shape. Their brightness decreases from their

FIGURE 23.3
The Hubble Classification of Galaxies
Most galaxies fit into one of the classes in the Hubble system.

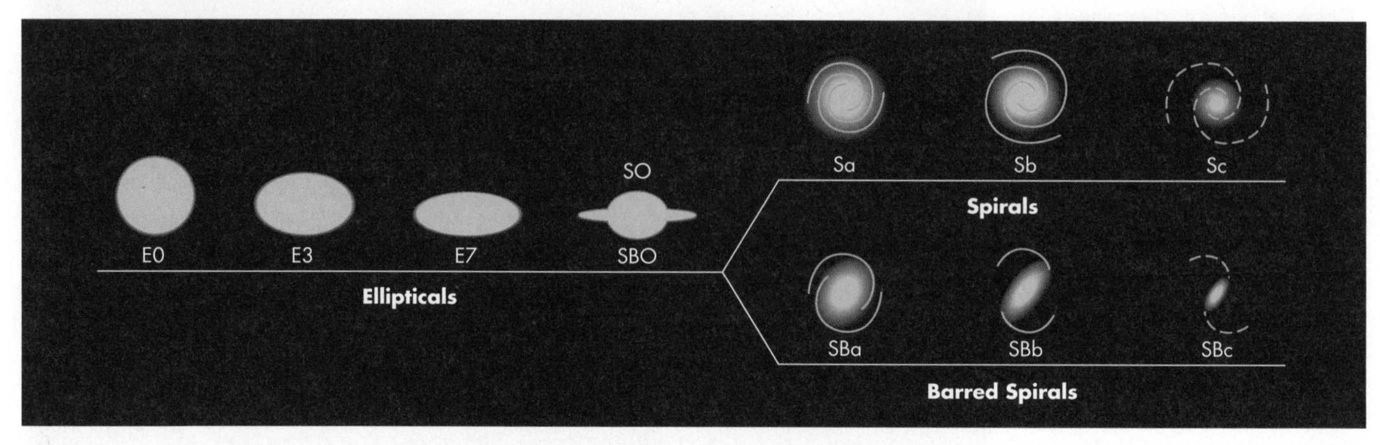

centers outward, but they have no structure otherwise. The ellipticals are classified according to how flattened they look. The most spherical ones, such as the galaxy shown in Figure 23.4, are classified E0, while the most flattened ones are designated E7, as shown in Figure 23.5.

Astronomers once thought that elliptical galaxies were flattened because they rotated rapidly, in much the same way that Jupiter and Saturn are flattened by their rapid rotation. It turns out, however, that elliptical galaxies have very little angular momentum. Overall, they hardly rotate at all. This doesn't mean that the stars within an elliptical galaxy aren't orbiting about the center of the galaxy. They are. It means, instead, that about as many stars are orbiting in one direction as in the opposite direction, as shown in Figure 23.6. If the stellar orbits were completely random, an elliptical galaxy would have a spherical shape. Instead, there are more stars in some orbital planes than in others. It is probable that the true shapes of elliptical galaxies are triaxial ellipsoids, with three unequal axes, as shown in Figure 23.7. A single elliptical galaxy can look quite different when viewed from different directions (which, of course, we can't do). Thus, the classification of an elliptical galaxy isn't an intrinsic property of the galaxy but depends on how the galaxy is oriented to our view.

Elliptical galaxies span a wide range of sizes and luminosities. The smallest of them, called dwarf spheroidal

FIGURE 23.4
An E0 Galaxy
The galaxy NGC 3379 is classified as an E0 galaxy because it is an elliptical that looks almost circular.

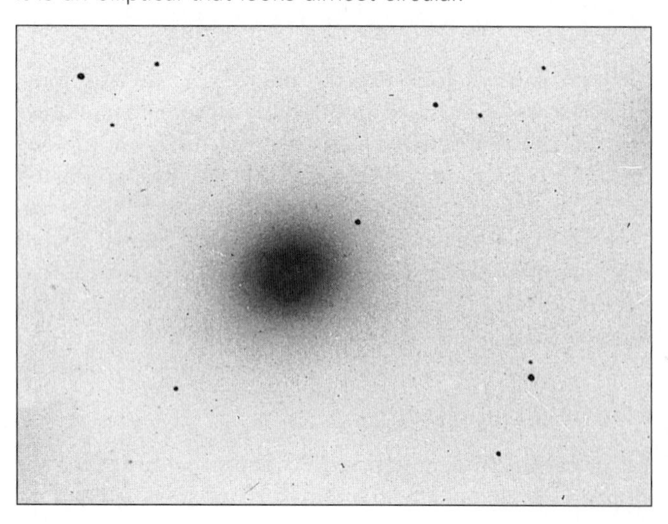

FIGURE 23.5
An E7 Galaxy
The galaxy NGC 3115 is among the most elongated-looking elliptical galaxies.

galaxies, have diameters of a few kiloparsecs or smaller. Some of them are only a few hundred thousand times brighter than the Sun. At the other extreme are the cD galaxies, which are giant ellipticals found at the centers of some clusters of galaxies. The bright central regions of these galaxies are surrounded by dim envelopes that can be as large as 2 million pc across. This makes the cD galaxies the largest known galaxies. They are as bright as several hundred billion Suns.

As with their luminosities and sizes, the elliptical galaxies span a wide range of masses. Some of the dwarf spheroidal galaxies contain only about $10^{6}{}_{\odot}$ M while cD ellipticals can be as massive as 10^{14} $M_{\odot}$ This is a hundred times as massive as the Milky Way.

FIGURE 23.6
Stellar Motions in an Elliptical Galaxy
The orbits of stars in the galaxy are approximately random. In a small region of the galaxy, there are about equal numbers of stars moving in all directions.

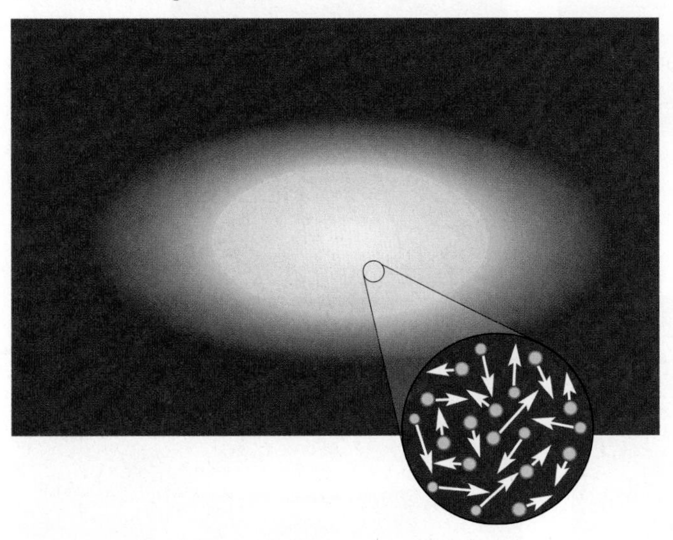

Spiral Galaxies

About 80% of all observed galaxies have flattened disks. Most of these are classified as **spiral galaxies** because they have spiral arms. There are two different types of spiral galaxies, the **barred spirals** and the **normal spirals.** The barred spirals (Figure 23.8) are designated SB and have a bright bar that crosses the nucleus. The spiral arms begin at the ends of the bar. Normal spirals (Figure 23.9) are designated S and lack a bar. The spiral arms for normal spirals begin at the nucleus. Both kinds of spirals are divided into types a, b, or c according to the appearance of the spiral arms and the brightness of the nucleus. Sa and SBa galaxies have large nuclei and tightly wound spiral arms. Sc and SBc galaxies have loosely wound spiral arms and small nuclei or no obvious nucleus at all. Sb and SBb galaxies fall between these two extremes. Galaxies with flat disks, like spirals, but that have no spiral arms, are designated S0. Spiral galaxies, like ellipticals, come in a range of sizes, masses, and luminosities, but these ranges are not as wide as they are for ellipticals. Spiral galaxies are some of the most "photogenic" celestial objects, as illustrated by Figures 23.10 and 23.11.

Elliptical and spiral galaxies differ in two important ways. First, there is much less interstellar gas and dust in elliptical galaxies than in spirals. Second, the stars in spiral galaxies have a wide range of ages, whereas elliptical galaxies have few, if any, young stars. These two differences are related because stars form from interstellar gas and dust clouds. The small amount of interstellar material in elliptical galaxies results in a small or nonexistent

FIGURE 23.7
The Shapes of Elliptical Galaxies
Elliptical galaxies are probably triaxial ellipsoids, which have different shapes when viewed from different directions. Thus, the apparent shape of an elliptical galaxy depends on how it is oriented relative to Earth.

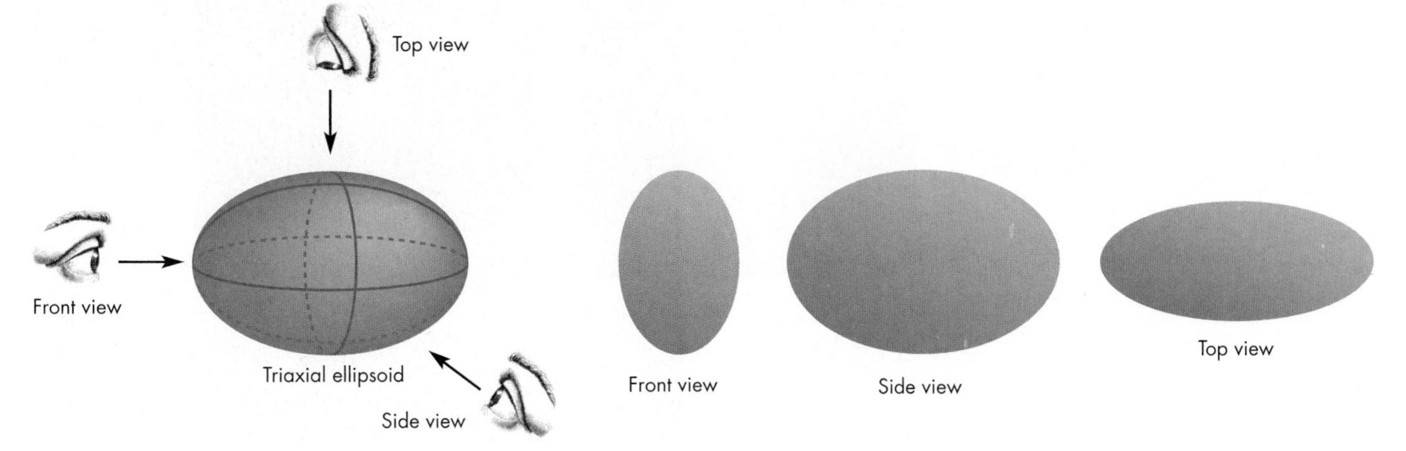

FIGURE 23.8
Barred Spirals
A, The nucleus of a barred spiral is crossed by a bar. Spiral arms begin at the ends of the bar. The tightness of the arms and the brightness of the nucleus determine whether a barred spiral is classified **B,** SBa; **C,** SBb; or **D,** SBc.

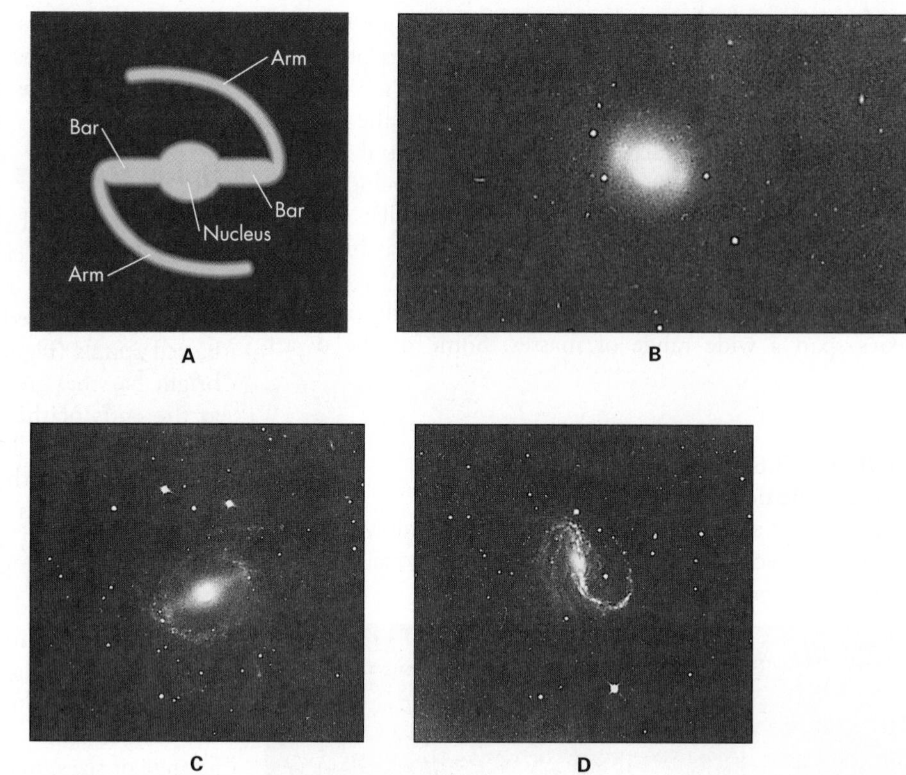

FIGURE 23.9
Normal Spirals
A, The spiral arms of a normal spiral begin at the edge of the nucleus. The tightness of the arms and the brightness of the nucleus determine whether a normal spiral is classified **B,** Sa; **C,** Sb; or **D,** Sc.

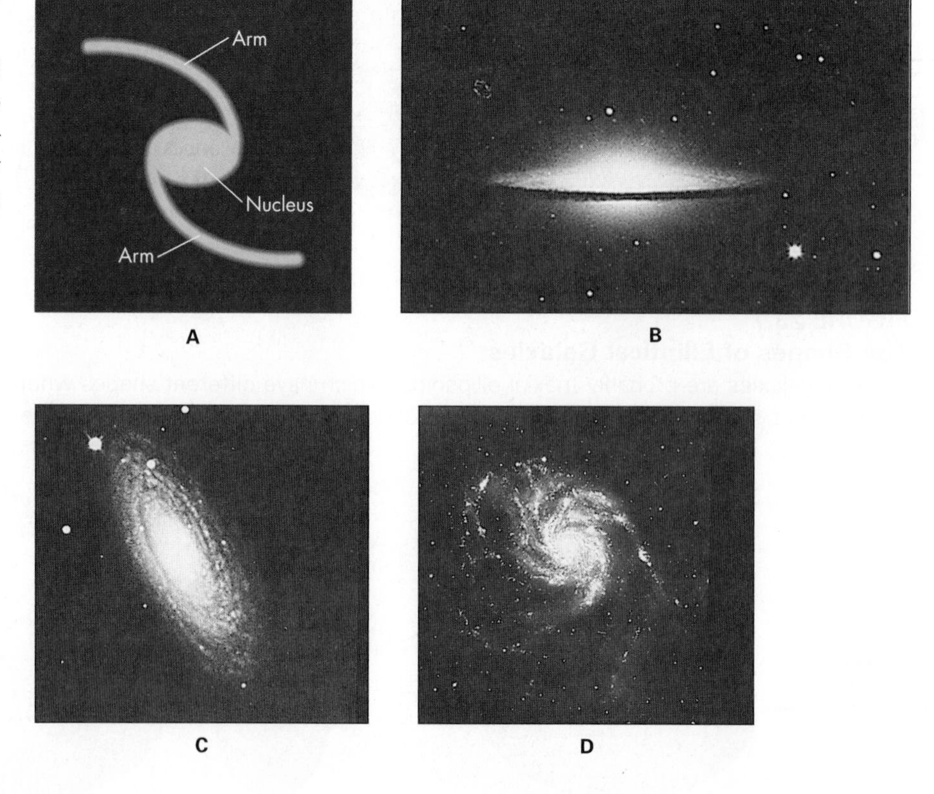

FIGURE 23.10
The Spiral Galaxy M74
M74, an Sc galaxy, is located in the constellation Pisces at a distance of 10 million parsecs.

FIGURE 23.11
The Barred Spiral Galaxy M83
M83, an SBc galaxy, is located in the constellation Hydra at a distance of 5 million parsecs.

rate of star formation. The interstellar gas in spiral galaxies seems to have been consumed slowly in star formation, but in ellipticals most interstellar material was used up in an early burst of star formation.

Irregular Galaxies

A small percentage of the galaxies is neither elliptical nor spiral and is designated *Irr,* for **irregular galaxy.** Some of these, such as the Large Magellanic Cloud, shown in Figure 23.12, may be spirals that are so patchy in brightness their overall spiral structure isn't apparent. Other irregular galaxies simply have very bizarre appearances. Many irregular galaxies contain even larger fractions of gas and dust than spiral galaxies do.

The Hubble classification system divides galaxies into different classes according to their shapes. Smooth-looking spheroidal galaxies are designated ellipticals. Spiral galaxies are those in which spiral arms extend from the nucleus or from a bar that crosses the nucleus. Ellipticals have much less interstellar gas and dust than spirals and contain few if any young stars. A small percentage of the galaxies lack overall structure and are designated irregulars.

FIGURE 23.12
The Large Magellanic Cloud
The LMC, one of the nearest galaxies to the Milky Way, is an irregular galaxy. It may have spiral structure, but the patchiness of the galaxy prevents us from seeing whatever pattern it may have.

Why Are There Different Kinds of Galaxies?

Our ideas about why there are different kinds of galaxies have changed greatly since the 1920s. Many astronomers, including Hubble himself, regarded the Hubble classification system as an evolutionary sequence. Hubble proposed that galaxies formed as elliptical systems, which then expanded. Next, a bar developed as matter was expelled from the central elliptical region. The spiral arms formed from the ends of the bar and grew larger and more open as time passed. In some galaxies, the bar remained; in others it went away.

We now recognize that the different kinds of galaxies don't represent a simple evolutionary sequence. Instead, some of the differences among galaxies reflect the differing conditions under which they are formed. Other differences are the result of encounters between galaxies. The *Hubble Space Telescope* is being used to obtain images, such as Figures 23.13 and 23.14, showing galaxies near the time they formed.

Galaxy Formation Galaxies are thought to have formed from the collapses of clouds of gas that were about ten times bigger than the galaxies are today. These gas clouds were set rotating by gravitational tugs from other gas clouds. Spiral galaxies seem to have retained most of the angular momentum that they had as gas clouds.

FIGURE 23.13
The Hubble Ultra Deep Field
The Hubble Ultra Deep Field is a one million second (11½ days) exposure with the *Hubble Space Telescope.* The region shown in the image is about one-tenth the diameter of the Moon and lies in the constellation Fornax. Many of the 10,000 galaxies seen in the Hubble Ultra Deep Field have very unusual shapes.

Distant Galaxies in the Hubble Ultra Deep Field
Hubble Space Telescope • Advanced Camera for Surveys
NASA, ESA, R. Windhorst (Arizona State University) and H. Yan (Spitzer Science Center, Caltech)
STScI-PRC04-28

FIGURE 23.14
Very Young Galaxies in the Hubble Ultra Deep Field
The faint red galaxies that are circled in the Hubble Ultra Deep Field are seen as they were when they were only a few hundred million years old. A few of these galaxies are shown in more detail in the panels on the right. Many of these young galaxies are irregular and misshapen. The reason the young galaxies appear red is that their motion away from us causes them to be redshifted. The significance of their redshifts are discussed in Chapter 23, Section 3, and Chapter 26.

Elliptical galaxies, on the other hand, lost more than 90% of the angular momentum they once had.

How can this fundamental difference between spiral and elliptical galaxies have developed? One idea that has gained considerable acceptance in the last decade requires that most of the material in a galaxy exists in the form of dark matter. If the luminous matter (stars and clusters of stars) in a protogalaxy were able to transfer its angular momentum to the dark matter, it could collapse to form an elliptical galaxy. This transfer could take place if the stars in the protogalaxy were clumped together rather than smoothly distributed. Imagine a clump of stars moving through the protogalaxy, as shown in Figure 23.15. As the clump moves, it deflects the matter through which it passes. The deflected matter converges as a wake behind the clump. The extra matter in the wake then exerts a gravitational force on the clump of stars, slowing down the clump and causing it to settle toward the center of the cloud. The clumps merge to form the elliptical galaxy. The angular momentum originally in the stellar clumps has

FIGURE 23.15
The Formation of an Elliptical Galaxy
A, As a massive star clump moves through a protogalaxy, the matter it passes is deflected by the clump's gravity and converges behind it. **B,** The gravitational pull of the deflected matter then slows the clump and causes it to settle to the center of the protogalaxy. **C,** The angular momentum of the resulting elliptical galaxy has been transferred to the surrounding (mostly dark) halo of matter.

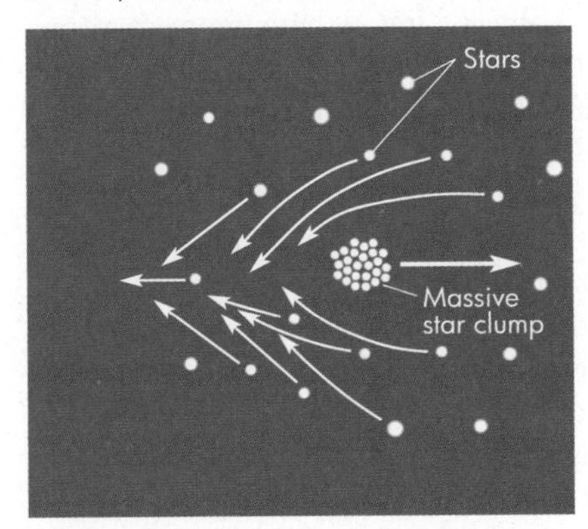

A Matter converges in a wake behind the massive star clump.

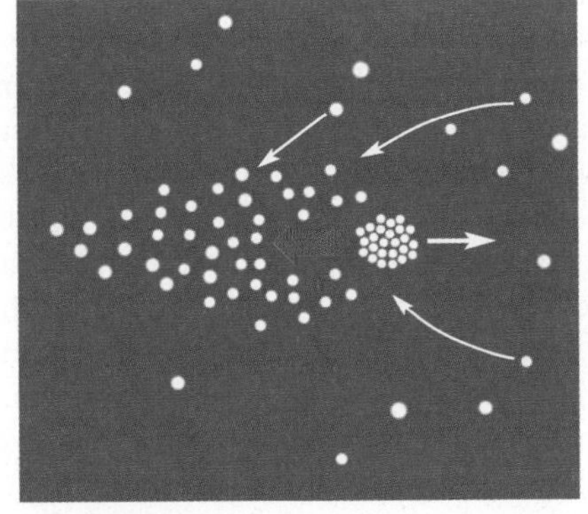

B The matter in the wake exerts a gravitational pull on the star clump.

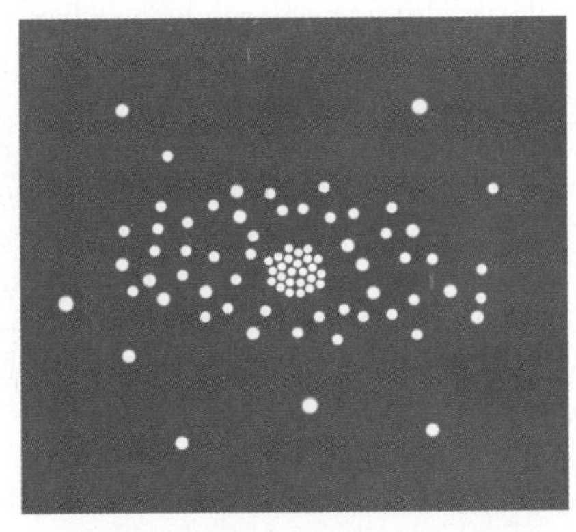

C The star clump eventually slows down and settles toward the center of the cloud.

FIGURE 23.16
M51 and NGC 5195
A bridge of stars extends from M51 (right part of the picture) to NGC 5195 (the smaller galaxy at the left).

been transferred to the (mostly) dark matter that lies in a wide halo about the elliptical galaxy. If this scenario for the formation of an elliptical galaxy is correct, then spiral galaxies must have formed from collapsing collections of stars that were smooth rather than clumpy.

But why should some galaxies have formed from clumpy clouds and other galaxies from smooth clouds? The answer to this question, which has to do with the density of the region in which the galaxy formed, is described in Chapter 25 as an aspect of the formation of galaxies in clusters.

Galaxy Collisions An example of two galaxies that collide as they orbit one another is shown in Figure 23.16. Galaxies are much more likely to collide or have near collisions than the stars that they contain. The main reason for this is that galaxies are much closer together, relative to their sizes, than are stars. The nearest star to the Sun is about 30 million solar diameters away. In contrast, the Milky Way has several neighboring galaxies within about one Milky Way diameter. Collisions and near collisions among galaxies can lead to unusual-looking or very irregular galaxies and galaxy mergers.

Even when two galaxies collide, the stars in the two galaxies almost never run into each other. Instead, the orbits of the stars (as well as the gas clouds) in each galaxy are severely disturbed by the passing galaxy. Low-speed collisions between galaxies in orbit about each other can sometimes produce long tails of matter that point away from the two galaxies, as in Figure 23.17. When one galaxy is struck almost in its center by another massive galaxy, the result can be a ring of matter that moves outward from the target galaxy. Figure 23.18 shows an example of a ring galaxy. In any of these kinds of collisions or near collisions, one or both of the galaxies involved is likely to have its structure greatly altered. The chaotic appearance of such a galaxy probably would lead to classification as an irregular.

FIGURE 23.17
NGC 4676, Galaxies with Tails
The tails, which point away from the galaxies, consist of stars that were pulled away from the galaxies as the galaxies collided at high speed.

Galaxies with tails

The repeated collisions between two galaxies in orbit about each other will eventually cause the two galaxies to spiral into one another and merge. Even if the galaxies were originally spirals, the result of the merger is likely to be a galaxy that lacks spiral arms and looks more like an elliptical galaxy. Ellipticals are the most common form of galaxy in very crowded regions where collisions are common. This is consistent with the idea that ellipticals can result from galaxy mergers. It is possible that S0 galaxies can form from mergers, too.

In regions where the density of galaxies is very high, a single galaxy might be able to merge with more than one nearby companion. The growth of a galaxy by devouring multiple companions is called **galactic cannibalism.** The most likely perpetrators of galactic cannibalism are the giant elliptical cD galaxies found in the centers of clusters. Some cD galaxies have been found to have more than one nucleus. While cannibalism is going on, some of the stars in the galaxy being devoured can be flung outward into very large orbits. These stars then make up the very diffuse extended envelopes that characterize cD galaxies. Closer to home, the Andromeda Galaxy has been found to have a double nucleus, as shown in Figure 23.19. One of the cores presumably is the nucleus of a galaxy in the process of being consumed.

FIGURE 23.18
AM0644-741, a Ring Galaxy
This galaxy, which consists of a bright center and a surrounding ring, was produced when the galaxy was struck nearly in its center by another galaxy millions of years ago.

There are different types of galaxies because they formed in different ways or because they have interacted with other galaxies. Clumpy protogalaxies are thought to have evolved into ellipticals, while smooth protogalaxies evolved into spirals. Galactic collisions can produce unusual-looking galaxies and also can cause the merger of two or more galaxies.

The Rotation of Galaxies

The greater the orbital speed of a star or gas cloud in a galaxy, the greater the amount of mass that lies within its orbit about the center of the galaxy. Thus rotation can be used to determine the distribution of mass in a galaxy. Studies of the rotation of galaxies have shown that many galaxies are made mostly of dark matter and that there are black holes in the cores of many galaxies.

Dark Matter in Galaxies The rotation curves of 23 spiral (Sb) galaxies are shown in Figure 23.20. These graphs, which show how rotational velocity depends on distance from the center of the galaxies, can be used with Equation 22.2 to calculate the distributions of mass in those galaxies. Most of the curves in Figure 23.20 rise sharply in the first 2 to 3 kpc from the center of the galaxy, then remain flat. Some are flat to more than 40 kpc from the nucleus. The Milky Way, also a spiral galaxy, has a rotation curve (Figure 22.30) similar to those in Figure 23.20; it is flat out to at least 20 kpc from the center. It turns out that the rotation curves of nearly all other spiral galaxies are flat, too.

The flat shapes of these curves indicate that mass increases with distance from the nucleus. The bright matter (the stars) in these galaxies, however, is concentrated near the nucleus, so dark matter in some form must account for the increase of mass with distance. Observations of 21-cm radiation from interstellar atomic hydrogen have shown that rotation curves of these galaxies remain flat far beyond the limits of the optical images of the galaxies. Thus, the

FIGURE 23.19
The Double Nucleus of the Andromeda Galaxy
This *Hubble Space Telescope* image of the core of the Andromeda Galaxy shows two distinct nuclei about 1.5 pc apart. The fainter spot is thought to be the true center of the galaxy because it is the center of the pattern of rotation. The brighter spot may be the core of a small galaxy recently destroyed by the Andromeda Galaxy.

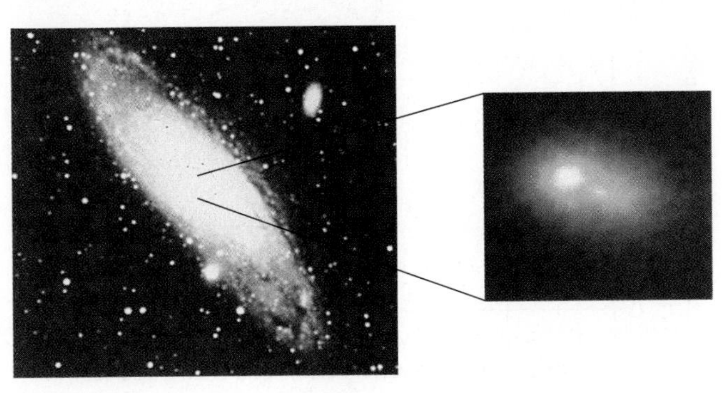

FIGURE 23.20
Rotation Curves for Spiral Galaxies
The rotation curves are flat far beyond the bright parts of the galaxies. Some flat rotation curves extend to 40 or 50 kpc. The flatness of the rotation curves means that the masses of the galaxies increase as the distances from their nuclei increase. Notice that the rotation curves are offset vertically from one another so that they don't overlap.

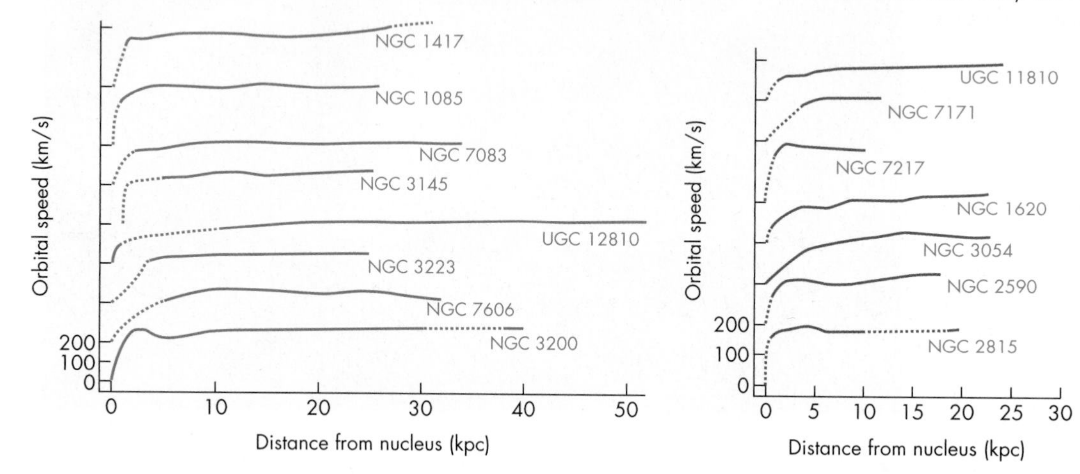

rotation curves suggest that other galaxies (like the Milky Way) must be made *mostly* of dark matter and that dark matter is widespread in the universe.

Black Holes in the Centers of Galaxies The very rapid orbital speeds very close to the centers of the galaxies suggest that massive black holes exist there. Determining the rotation curve near the center of a galaxy is difficult, however, even for the nearby Andromeda Galaxy. For one thing, we have to look through the galactic bulge to see the nuclear region, so the spectrum of the nuclear region of a galaxy is contaminated by light from stars in the bulge. Another, more serious, problem is that it is hard to obtain observations with high enough angular resolution to isolate the stars very close to the center. Even for the Andromeda Galaxy, each second of arc equals 3.4 pc. At a superb observing site such as Mauna Kea, the best "seeing" still blurs an image to about a third of a second of arc. This isn't good enough to determine structure and velocities inside the central parsec of a galaxy. The *Hubble Space Telescope* and ground-based telescopes with adaptive optics, however, can be used to obtain higher angular resolution images of the cores of galaxies than ever before.

Several groups of astronomers have determined the rotation curves of the Andromeda Galaxy and more than two dozen other galaxies to within the innermost second of arc. The results for the Andromeda Galaxy, shown in the graph in Figure 23.21, indicate that the stars revolve at speeds greater than 100 km/s within about 2 pc of the center. This corresponds to a central mass of 10^7 to 10^8 $M_\odot$. The core of the galaxy does not seem bright enough for all the mass to exist in the form of luminous stars, however. Instead, a massive black hole seems to be required to explain why so much mass is concentrated at the center of the Andromeda Galaxy. Another relatively nearby galaxy, the Sombrero Galaxy, is shown in Figure 23.22. Rotation near the center of the Sombrero Galaxy is even higher than in the Andromeda Galaxy and requires a 10^9 $M_\odot$ black hole. Other galaxies have rotation curves near their centers that require black holes of millions to billions of solar masses. It appears that black holes lurk at the centers of many galaxies. After the masses of the black holes at the centers of nearby galaxies were determined, astronomers found that the masses of the black holes were related to the combined masses of the stars in the central bulge—either the entire elliptical galaxy or the nuclear bulge of a spiral galaxy. The central bulge is,

FIGURE 23.21
Rotation in the Center of the Andromeda Galaxy
Rapid rotation within a few parsecs of the center of the Andromeda Galaxy shows that there are 10 to 100 million $M_\odot$ of matter within the central parsec. This mass may lie within a black hole.

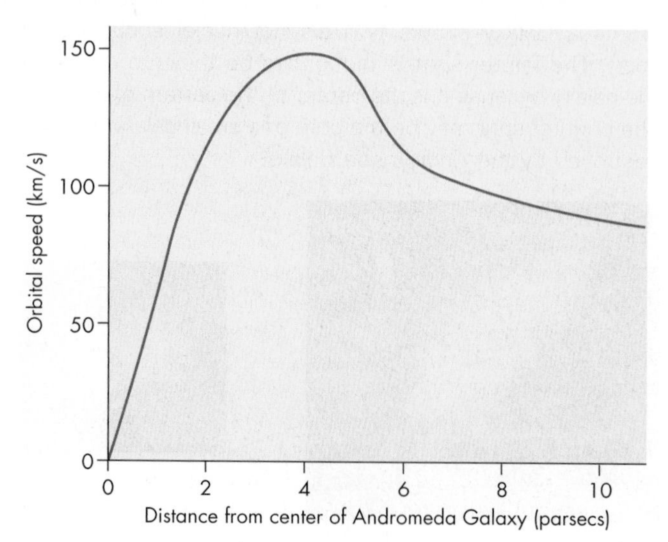

FIGURE 23.22 M104, the Sombrero Galaxy
Like the Andromeda Galaxy, M104 has rapid rotation in its center and may contain a black hole, which may be even more massive than the one believed to be in the nucleus of the Andromeda Galaxy.

on average, about 700 times as massive as the black hole. A possible explanation of this relationship is that when a black hole grows to a certain point, it terminates star formation in the central part of its parent galaxy. This may happen as a result of the energetic activity, described in Chapter 24, that accompanies supermassive black holes.

The rotation of a galaxy can be used to investigate the distribution of matter within the galaxy. The flat shapes of the rotation curves of many spiral galaxies suggest that they contain large amounts of dark matter. Rapid rotation near the centers of some galaxies suggests that they contain massive black holes.

23.3 THE COSMIC DISTANCE SCALE

One of the most important, although difficult, tasks that astronomers have undertaken in the last century has been the determination of the distances to the galaxies. This work is of special importance because galaxies and clusters of galaxies mark the basic structure of the universe. Thus our ability to answer fundamental questions about the origin and fate of the universe depends on how well we know the distances of galaxies.

The most reliable astronomical distances are those that have been found by measuring parallaxes. However, only celestial objects nearer than a few hundred parsecs show large enough parallaxes for their distances to be found by this method. The nearest galaxies lie one hundred times farther away than the limit of parallax observations, so their distances must be found by less direct methods. The strategy for finding the distances to galaxies has been to proceed stepwise. In each step, objects such as HII regions in galaxies for which distances have already been determined are compared with the same kind of objects in more distant galaxies or clusters of galaxies. The relative brightnesses or sizes of the near and distant objects give the relative distances. The use of different kinds of objects to step outward in distance is summarized in Figure 23.23.

Primary Distance Indicators

The first step is to find the distances to **primary distance indicators,** for which we know the size or brightness by observing them in the Milky Way. These can then be compared with the same kinds of objects in nearby galaxies. The best primary indicators are Cepheid variables. The period-luminosity relationship for Cepheids makes it possible to find their luminosities (or absolute magnitudes) in any galaxy near enough that individual Cepheids can be recognized and their light curves measured. By comparing their apparent brightnesses (or apparent magnitudes) with their luminosities, astronomers can find the distance to the Cepheids and the galaxy containing them.

Novae also have been used as primary distance indicators. Although novae vary considerably in their luminosity at greatest brightness, the bright novae dim faster than fainter ones, so that they all reach about the same luminosity 15 days after maximum brightness. By comparing the apparent brightness of a nova with its expected luminosity 15 days after maximum, its distance can be found.

Unfortunately, primary distance indicators aren't bright enough that they can be identified in any but the nearest galaxies. At a distance of 10 **megaparsecs** (abbreviated Mpc, 1 megaparsec equals 1000 kpc or 10^6 pc), a nova 15 days

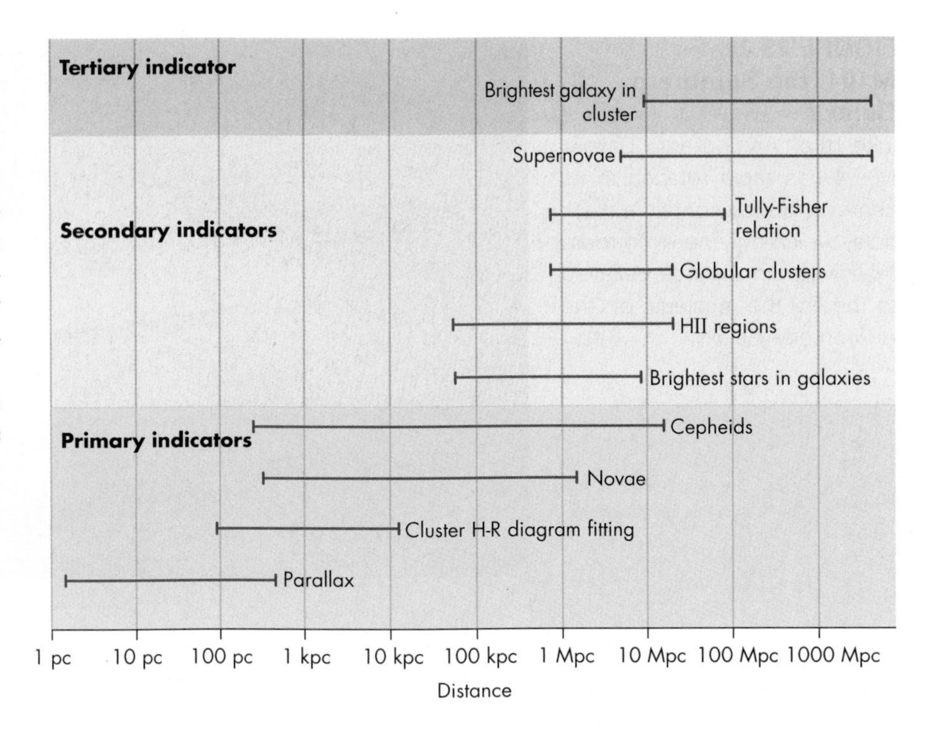

FIGURE 23.23
Objects Used to Determine Cosmic Distances
Each kind of object or technique can be used for a range of distances from the nearest object of that type to the greatest distance at which that kind of object can be seen. For instance, Cepheid variable stars can be used as indicators for distances between a few hundred parsecs and a few tens of megaparsecs, whereas the Tully-Fisher relation can be used for distances between about 1 Mpc and 100 Mpc.

after maximum would have an apparent magnitude of 25. This is a hundred million times fainter than the dimmest stars that can be seen with the naked eye. Relatively few galaxies are close enough that their distances have been found through primary indicators. These galaxies, however, provide the information needed to take the next step in determining the cosmic distance scale. One of the principal missions of the *Hubble Space Telescope* was to observe primary indicators at greater distances than ever before and to extend the range of distances obtained using primary indicators out to tens of megaparsecs. Hubble observations of Cepheid variables have yielded the distances to galaxies 20 Mpc away.

Secondary Distance Indicators

Secondary distance indicators are those for which brightnesses or sizes are known because they have been found in nearby galaxies. The distances of the nearby galaxies are known through primary indicators. Without the primary indicators, the properties of secondary indicators wouldn't be known well enough for them to be used in distance determinations. Secondary indicators are brighter than primary indicators and can be seen at considerably greater distances.

One important secondary indicator is the size of the largest HII region in a galaxy. The largest HII regions are the ones that surround the hottest, most massive, most luminous main sequence stars. The size of the largest HII region increases as the total luminosity of a galaxy increases. When the actual size of the largest HII region is known and its angular size is measured, the distance to the HII region and galaxy can be found through the small angle equation.

Another secondary indicator is the average brightness of the globular clusters in a galaxy. The average brightnesses of the globular clusters are almost the same for all nearby galaxies, although the galaxies themselves range over a factor of 1000 in brightness. In any galaxy for which enough globular clusters can be found to determine their average brightness, the distance to the galaxy can be determined.

A powerful secondary indicator makes use of the Tully-Fisher relation, a relationship between the luminosity of a spiral galaxy and the range of velocities over which its interstellar gas emits 21 cm hydrogen radiation. The more luminous a galaxy, the broader its 21 cm line. Using the Tully-Fisher relation, astronomers can estimate the luminosity of any spiral galaxy for which the 21 cm emission is strong enough to be measured using radio telescopes. By comparing the estimated luminosity with the galaxy's visual brightness, the distance to the galaxy can be found. The Tully-Fisher relation is the secondary indicator of choice because it can be used to measure galaxies as distant as 100 Mpc, whereas other secondary indicators can be used only to about 25 Mpc.

The most powerful secondary indicator is the type Ia supernova. Type Ia supernovae have been seen at distances as large as 3000 Mpc. The results of distance determinations using type Ia supernovae will be described in Chapter 26.

Tertiary Distance Indicators

Beyond about 25 Mpc, individual stars and HII regions can't be identified. Instead, entire galaxies must be used as distance indicators. This has been done by comparing distant galaxies with what are thought to be similar galaxies within the 25 Mpc limit of secondary indicators. For instance, two spiral galaxies are considered to have the same luminosity if they have the same Hubble classification and if their spiral arms are equally sharp and well defined. This method, while approximate, can be used to find galaxy distances out to hundreds of megaparsecs.

At even greater distances, the brightest galaxy in a cluster of galaxies can be used as a distance indicator. The luminosity of the brightest galaxy shows little variation from one cluster of galaxies to another. Thus, by comparing the luminosity of the brightest galaxy with its apparent brightness, the distance of the galaxy can be found and, with it, the distances to the other galaxies in the cluster. This method can be used to find the distances of galaxies to more than 1000 Mpc (that is, more than 3 billion light years away).

The distances to galaxies have been determined in a stepwise fashion in which relatively nearby objects are compared with the same kind of objects in more distant galaxies. Distances to galaxies as far away as 3000 Mpc have been determined through this method.

Hubble's Law

One of the most significant discoveries about galaxies is **Hubble's law,** a remarkable and extremely profound relationship between the speed with which a galaxy is moving away from us (its **recession velocity**) and its distance. Hubble's law is intimately related to both the history and the future of the universe. A description of the cosmological importance of Hubble's law is continued in Chapter 26.

Slipher's Radial Velocities The first radial velocity of a star was measured in 1868. By the turn of the twentieth century, the measurement of stellar radial velocities was routine. Galaxies are much fainter than the bright stars, so it was much more difficult for astronomers to obtain spectra good enough to determine the radial velocities of galaxies. The first measurement of the radial velocity of a galaxy was made by V. M. Slipher in 1913. He discovered that the Andromeda Galaxy was moving toward the Sun at the astonishing speed of 300 km/s, or about 15 times faster than the radial velocity of a typical star. It was hard to understand how something moving as rapidly as 300 km/s could remain in the Milky Way, so Slipher's result convinced many astronomers that the Andromeda Galaxy must be an "island universe" far beyond the Milky Way.

By 1917, Slipher had obtained spectra for 25 spiral galaxies. The effort expended in obtaining the spectra is almost unimaginable today, when bigger telescopes and modern detectors make it possible to obtain spectra of galaxies in minutes. Using a 60-cm (24-inch) refractor and spectrograph, Slipher had to expose photographic plates for 20, 40, or even as long as 80 hours to get a single spectrum. Thus each spectrum required many nights of observing. What Slipher found was that of his 25 galaxies, 21 showed redshifted spectral lines, indicating that they were moving away from us. Several had radial velocities of about 1100 km/s (2.5 million miles per hour). Slipher's interpretation of the radial velocities of spiral galaxies was influenced by the unfortunate spatial distribution of the galaxies he had observed. Most of them lie in one part of the sky. Of the much smaller number he had observed in the opposite part of the sky, several had negative radial velocities, indicating that they were approaching us. Slipher proposed that we were moving toward the blueshifted galaxies and away from the redshifted ones. That is, he suggested that the pattern of radial velocities was a consequence of the motion of the Sun through space at a speed of 700 km/s. He thought that as more galaxies were observed, they would confirm the pattern of negative velocities in one part of the sky and positive radial velocities in the opposite.

Velocities and Distances Even before the distances to galaxies were established by Hubble in 1923, astronomers had found that the recession velocities of galaxies were related to some of their other properties. The galaxies with the smallest angular sizes and faintest magnitudes are those moving away the fastest. After 1923, several astronomers compared Hubble's distances and Slipher's list of radial velocities and noted a tendency for the more distant galaxies to have greater recession velocities. Convincing evidence of this relationship, however, wasn't presented until 1929, when Hubble plotted recession velocity against distance for 24 galaxies for which he judged he had dependable distances. What he found is shown in the graph in Figure 23.24. Even with such a small collection of data, Hubble found that, on average, the recession velocity of a galaxy is directly proportional to its distance. This relationship between recession velocity and distance is called Hubble's law and is written

$$v = \mathrm{H}d \qquad (23.1)$$

FIGURE 23.24
Hubble's Plot of the Recession Velocities and Distances of Galaxies
Hubble found that the speed with which a galaxy recedes is proportional to its distance. This relationship is called Hubble's law.

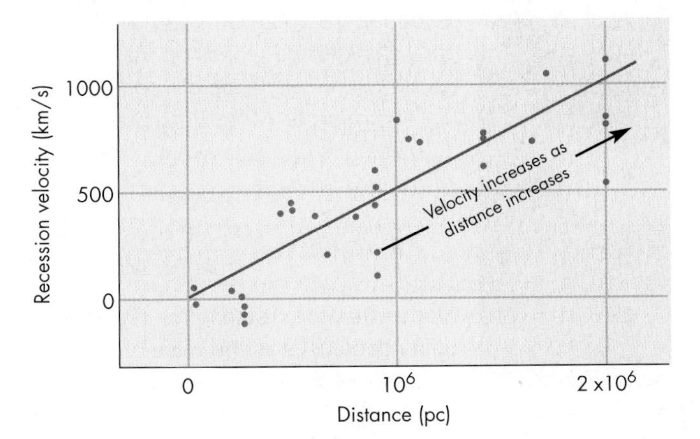

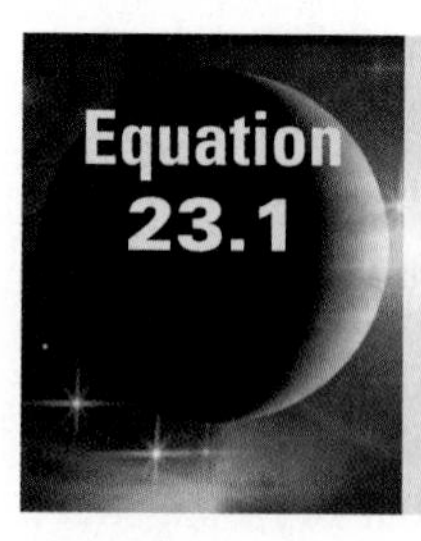

Distances Determined from Hubble's Law

Equation 23.1 can be used to find the recession velocity of a galaxy at a given distance. For instance, if Hubble's constant has a value of 71 km/s per Mpc, then Equation 23.1 gives the recession velocity of a galaxy at a distance of 400 Mpc as

$$v = Hd = 71 \text{ km/s per Mpc} \times 400 \text{ Mpc} = 28{,}800 \text{ km/s}$$

where v is recession velocity, d is distance, and H is **Hubble's constant.**

Hubble's constant is the slope of the straight line that describes the relationship between recession velocity and distance. Hubble calculated that H has the value of 500 km/s per Mpc. This means that the speed with which the galaxies are receding increases by 500 km/s for each additional megaparsec of distance from the Sun. It turns out, however, that Hubble had systematically underestimated the distances to galaxies by a factor of 5 to 10. The Virgo cluster of galaxies, for example, which Hubble placed at a distance of 2 Mpc, is actually 20 Mpc away. Underestimating the distances of the galaxies is equivalent to overestimating the value of H. Thus, today's estimates of the value of Hubble's constant range between 50 and 85 km/s per Mpc. The best estimate of H is 71 km/s per Mpc with an uncertainty of about 3 km/s per Mpc.

Using Hubble's Law The relationship between recession velocity and distance has been shown to extend far beyond the distance limits of Hubble's original observations. The graph in Figure 23.25 shows a modern representation of Hubble's law. To find the distances to very distant galaxies and other objects, astronomers often assume that Hubble's law can be extrapolated beyond the greatest distance for which distances can actually be measured. By reorganizing the equation for Hubble's law, distance (d) is related to recession velocity (v) and Hubble's constant (H) by

$$d = v/\text{H} \qquad (23.2)$$

The Expanding Universe At first, it may seem incredible that virtually all of the galaxies are moving away from the Earth, the Sun, and the Milky Way. After all, as astronomical thinking matured, it became clear first that the Earth isn't in the center of the solar system and then that the Sun is far from the center of the galaxy. The organized recession of the galaxies, which looks the same in all directions, seems to place our galaxy in a very special place—the center of the universe. However, the real message contained in Hubble's law isn't that we are in the center of the universe, but rather that the universe is expanding.

The relationship between Hubble's law and the expansion of the universe can be visualized by using the analogy of a lump of raisin bread dough put in a warm oven to rise, as in Figure 23.26. As the dough rises and expands in size, it carries along the raisins, which represent galaxies. If the size of the raisin bread universe doubles in 1 hour, then every distance in the bread doubles. Two raisins that are originally 1 cm apart (*A* and *B*) are 2 cm apart at the end of the hour. They have separated at a speed of 1 cm/hr. *A* and *C*, which begin 5 cm apart, end up 10 cm apart and have separated at 5 cm/hr. An observer on raisin *A* would find that all other raisins are moving away and that their speeds are directly proportional to their distances. In other words, the observer would see Hubble's law. There is nothing special about raisin *A*, however, because an observer on any other raisin, say *D*, also would see all other raisins receding according to Hubble's law. The same argument applies in the real universe. Just as

Hubble's Law

Equation 23.2 can be used to find the distance of a galaxy if its recession velocity can be measured. Suppose Hubble's constant has a value of 71 km/s per Mpc. Equation 23.2 then gives the distance of a galaxy with a recession velocity of 15,000 km/s as

$$d = v/\text{H} = \frac{15{,}000 \text{ km/s}}{71 \text{ km/s per Mpc}} = 210 \text{ Mpc}$$

Notice that the distance for a given recession velocity decreases as the value of Hubble's constant increases. If H is assumed to be 100 km/s per Mpc, for example, then the distance of a galaxy with a recession velocity of 15,000 km/s is

$$d = v/\text{H} = \frac{15{,}000 \text{ km/s}}{100 \text{ km/s per Mpc}} = 150 \text{ Mpc}$$

That is, for large values of Hubble's constant, the galaxies are nearer and the universe is smaller than for small values of Hubble's constant.

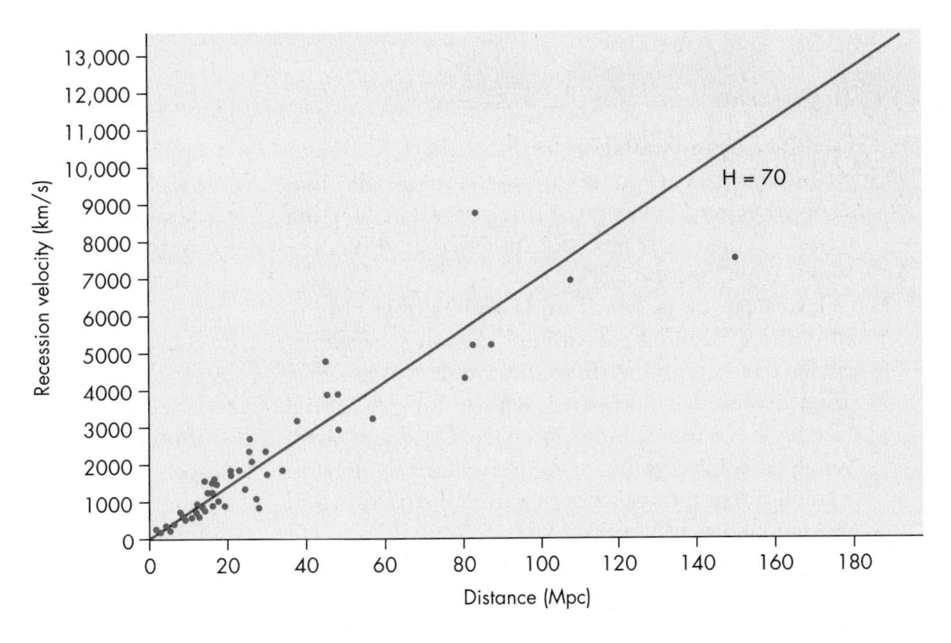

FIGURE 23.25
Hubble's Law
The modern determination of Hubble's law extends to a distance of over 100 Mpc. In this graph, Hubble's constant is taken to be 70 km/s per Mpc. The scatter of points about the line is mostly due to uncertain distances.

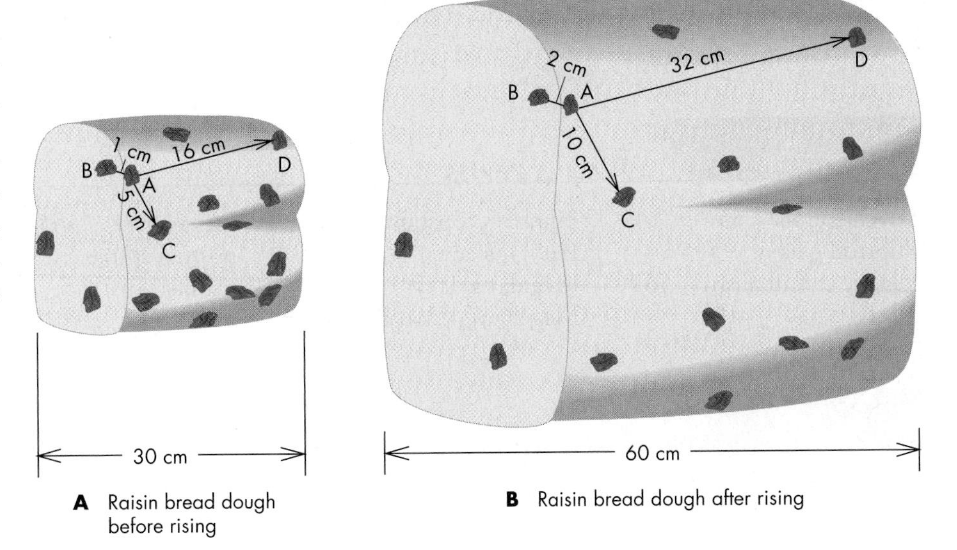

FIGURE 23.26
The Expansion of the Universe
As the universe (here represented by a loaf of raisin bread) expands, the expansion carries galaxies (represented by the raisins) away from each other at speeds that are proportional to their distances from each other. It doesn't matter within which galaxy an astronomer resides; the other galaxies all appear to be moving away.

we see the Virgo cluster of galaxies receding at 1100 km/s as part of Hubble's law, an observer in the Virgo cluster would see the Milky Way receding at 1100 km/s. That observer would record the motion and distance of the Milky Way as a data point in a plot that would show the same Hubble's law that we measure. Neither we nor the hypothetical observer in the Virgo cluster is at the center of the universe. All observers see Hubble's law in the pattern of galaxies receding from them.

Although Hubble's law is sometimes explained in terms of Doppler shifts and the velocities of galaxies, this is misleading because it leaves the impression of galaxies flying apart in a stationary universe. Actually, it is space that is expanding, carrying the galaxies along. It is more realistic to think of the galaxies as stationary in expanding space just as the raisins are stationary within the expanding dough. The redshifts of the galaxies are properly described as expansion redshifts rather than Doppler shifts because they arise from the expansion of the universe.

Hubble's law is a linear relationship between the speed with which a galaxy is moving away from us and the distance of the galaxy. Using Hubble's law, the recession velocity of a galaxy can be used to calculate its distance. Hubble's law is a result of the expansion of the universe.

Chapter Summary

- In 1923 Edwin Hubble showed that the Andromeda Nebula is a galaxy—a distant collection of stars. This resolved a longstanding controversy about whether spiral nebulae were galaxies or objects within the Milky Way. (Section 23.1)
- Galaxies can be classified using the Hubble system. Smooth, spheroidal galaxies are classified as ellipticals. Galaxies with flat disks and spiral arms are classified as spirals. Compared with ellipticals, spiral galaxies have much more interstellar gas and dust. Spirals also have young stars, but ellipticals do not. A small fraction of galaxies lack overall structure and are classified as irregular. (23.2)
- The different appearances of galaxies result from differences in the way in which they formed or from their interactions with other galaxies. It is believed that elliptical galaxies evolved from clumpy protogalaxies and spirals from smooth protogalaxies. Collisions between galaxies can alter their appearances and also can merge galaxies into larger galaxies. (23.2)
- The distribution of matter in a galaxy can be found by measuring the rotation curve of the galaxy. Many spiral galaxies have flat rotation curves, which suggests that they contain a considerable amount of dark matter. Rapid rotation near the centers of some galaxies suggests that massive black holes may reside there. (23.2)
- The distances to galaxies have been determined through a stepwise process that compares nearby objects with the same kind of objects in more distant galaxies. Distances as large as 3000 Mpc have been found through this method. (23.3)
- Galaxies are receding from us with speeds proportional to their distances. The relationship between speed and distance is called Hubble's law and is a result of the expansion of the universe. Hubble's law can be used to find distances to galaxies for which recession velocities can be measured. (23.3)

Key Terms

barred spirals *557*
elliptical galaxy *555*
galactic cannibalism *562*
galaxy *554*
Hubble's constant *568*
Hubble's law *567*
irregular galaxy *559*
Magellanic Clouds *554*
megaparsec *565*
normal spirals *557*
primary distance indicator *565*
recession velocity *567*
secondary distance indicator *566*
spiral galaxies *557*

Conceptual Questions

1. What is meant by the statement that "a telescope is a time machine"?
2. How old is the most recent information we can ever have about the dwarf planet Pluto?
3. What is the Hubble classification of an elliptical galaxy that appears spherical in shape?
4. What is the Hubble classification of a barred spiral with tightly wound arms?
5. How can the stars in an elliptical galaxy be orbiting the center of the galaxy if the galaxy has essentially no angular momentum?
6. What is the relationship between the sizes of the nuclei of spiral galaxies and how tightly the spiral arms of those galaxies are wound?
7. Why is a galaxy much more likely to collide with another galaxy than a star is to collide with another star?
8. What is believed to be the way that ring galaxies are formed?
9. What evidence do we have that there is dark matter in spiral galaxies?
10. What evidence do we have that there is a massive black hole in the center of the Andromeda Galaxy?
11. What are some of the objects that serve as primary distance indicators in determining the cosmic distance scale?
12. Why can't primary distance indicators be used to find the distances of very distant galaxies?
13. Suppose we were to discover that all of the galaxies are actually ten times farther away than previously thought. What effect would this have on the value of Hubble's constant?
14. Describe how Hubble's law is explained as a consequence of the expansion of the universe.
15. Suppose we lived in a universe in which the greater the distance to a galaxy, the faster that galaxy were approaching us. How could such a relationship between distance and velocity be explained?

Problems

1. Using a value for Hubble's constant of 71 km/s per Mpc, find the distance of a galaxy that is receding at a velocity of 1600 km/s.
2. Using a value for Hubble's constant of 71 km/s per Mpc, find the recession velocity of a galaxy at a distance of 40 Mpc.
3. At 2 pc from the center of the Andromeda Galaxy, gas revolves at 100 km/s. Use Kepler's third law (Equation 22.2) to find the mass within the central 2 pc of the Andromeda Galaxy.

Figure-Based Questions

1. Estimate the Hubble classification for the elliptical galaxy shown in Figure 23.7 if the galaxy were viewed from the front, the side, and the top.
2. Use Figure 23.20 to find the orbital velocity of matter in the galaxy NGC 3200 at distances of 10, 20, 30, and 40 kpc from the center. Use Equation 22.2 to find the amount of matter within each of these distances.

Group Activities

1. After reading Chapter 23, Section 1, divide your group into two subgroups. After preparing for a couple of minutes, have one subgroup present the arguments, as of about 1920, that the spiral nebulae were parts of the Milky Way. Have the other subgroup present the arguments, as of about 1920, that the spiral nebulae were distant galaxies. Have the entire group discuss which case seemed best at the time.
2. Divide into two subgroups. After preparing for a couple of minutes, have one subgroup present an argument that Hubble's law implies we are at the center of the universe. Have the other subgroup present an argument that Hubble's law implies that the universe is expanding. Have the entire group discuss the relative merits of the two arguments.

For More Information

Visit the text website at **www.mhhe.com/fix** for chapter quizzes, interactive learning exercises, and other study tools.

24

"I can't believe that," said Alice.
"Can't you?" the Queen said in a pitying tone.
"Try again: draw a long breath and shut your eyes."
Alice laughed. "There's no use trying," she said, "one can't believe impossible things."
"I daresay you haven't had much practice," said the Queen.
"When I was your age, I always did it for half-an-hour a day.
Why, sometimes I've believed as many as six impossible things before breakfast."

Lewis Carroll

www.mhhe.com/fix

The nucleus of the active galaxy Centaurus A.

Quasars and Other Active Galaxies

The discovery of quasars, or at least the discovery that quasars were a totally new and unexpected kind of astronomical phenomenon, marked a kind of revolution in astronomy. The discovery of quasars can be directly attributed to the development of radio astronomy during the 1940s. Radio astronomers began to compile lists of strong sources of radio radiation much as Hipparchus and Ptolemy had compiled lists of sources of visible light—stars—2000 years earlier. At first, it was thought that most of the strong radio sources also were stars. Gradually astronomers realized that this was wrong. The bright radio sources are actually nebulae (such as the Crab Nebula) and galaxies. The second brightest radio source in the sky, Cygnus A, shown in Figure 24.1, was found to be a distorted-looking galaxy that radiates more energy in the radio region alone than the entire Milky Way produces at all wavelengths.

When astronomers attempted to identify other strong radio sources, they found that many of them appeared pointlike when viewed in visible light. Thus, they were called quasars, for quasi-stellar radio sources. Later, the term quasar was extended to similar objects that do not emit intense radio radiation. Spectra of quasars have strong, broad emission lines, but the wavelengths of the emissions do not correspond to those of laboratory measurements of known atoms or ions. The emission lines in different quasars, moreover, were found to be at different wavelengths, as though each quasar had a different, unknown chemical composition.

The puzzle of the quasar emission lines was solved by Maarten Schmidt in 1963. He found that the emission lines in the quasar 3C 273 were due

Questions to Explore

- What is the significance of the large redshifts of quasars?
- Why are quasars so luminous yet so small?
- Why do the components of some quasars seem to be moving apart faster than the speed of light?
- How are the different kinds of quasarlike objects related to each other?
- What is the nature of the jets of matter that flow out of the cores of active galaxies?
- Why are quasars found only at great distances from us?
- What effect can a gravitational lens have on the appearance of a quasar?

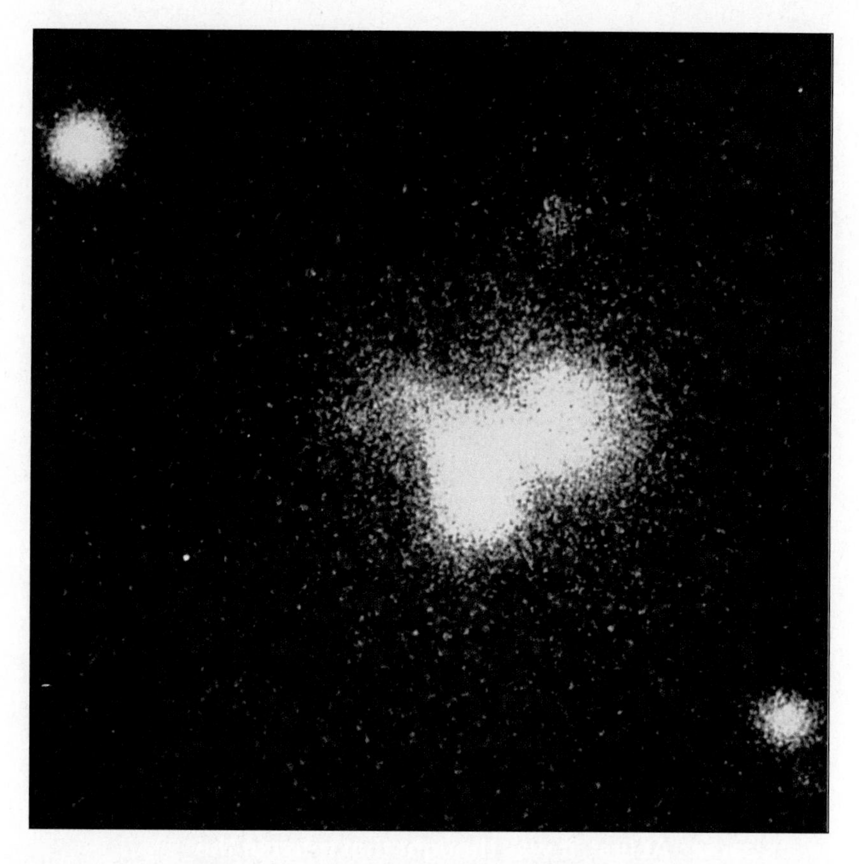

FIGURE 24.1
The Visual Image of Cygnus A
This distorted-looking galaxy is the second strongest radio source in the sky. Its distance is about 250 Mpc.

to familiar elements, including hydrogen. The entire pattern of spectral lines, however, was redshifted by 15.8%, as shown in Figure 24.2. In other words, each of the emission lines in 3C 273 was found at a wavelength 15.8% greater than it would have if measured in a laboratory. The redshift of a quasar (or any other object) is denoted by the symbol "z" and is defined as the change in the wavelength of a spectral line divided by the wavelength measured in the laboratory. Astronomers frequently observe stars and galaxies in which the spectral lines are redshifted or blueshifted. The main obstacle to identifying the emission lines of quasars was that the optical images of quasars are so small that they were assumed to be stars, for which the greatest redshifts or blueshifts are only a tiny fraction of that observed for 3C 273. Astronomers soon found that the spectra of other quasars were redshifted, too, and that some were redshifted much more than 3C 273.

As they searched for quasars, astronomers also found many new kinds of strange objects that seem to be related to quasars. The unusual properties of quasars forced astronomers to consider such things as how components of quasars can seem to move faster than the speed of light and how the images of quasars can be distorted by gravitational lenses. To account for quasars, astronomers were forced to propose explanations that included massive black holes and other possibilities that they had never had to consider before. This chapter is about the exotic properties of quasars and their relatives as well as the ways in which we can account for those properties.

24.1 QUASARS

There are many kinds of active galaxies, some of which have been known since early in the twentieth century. However, it was the discovery of quasars that showed, very dramatically, that some galaxies have nuclei that put out enormous amounts of energy. Many of our ideas about other kinds of active galaxies, including the explanation of their energy output in terms of massive black holes, have arisen through comparisons with quasars.

The Meaning of Quasar Redshifts

After the large redshifts of quasars were discovered, several explanations for the redshifts were proposed. Nearly all

FIGURE 24.2
The Spectrum of the Quasar 3C 273
The three emission lines of hydrogen in the spectrum of the quasar are redshifted by about 16% relative to the wavelengths they have in the laboratory. (The laboratory wavelengths of the hydrogen lines are indicated below the comparison spectrum.)

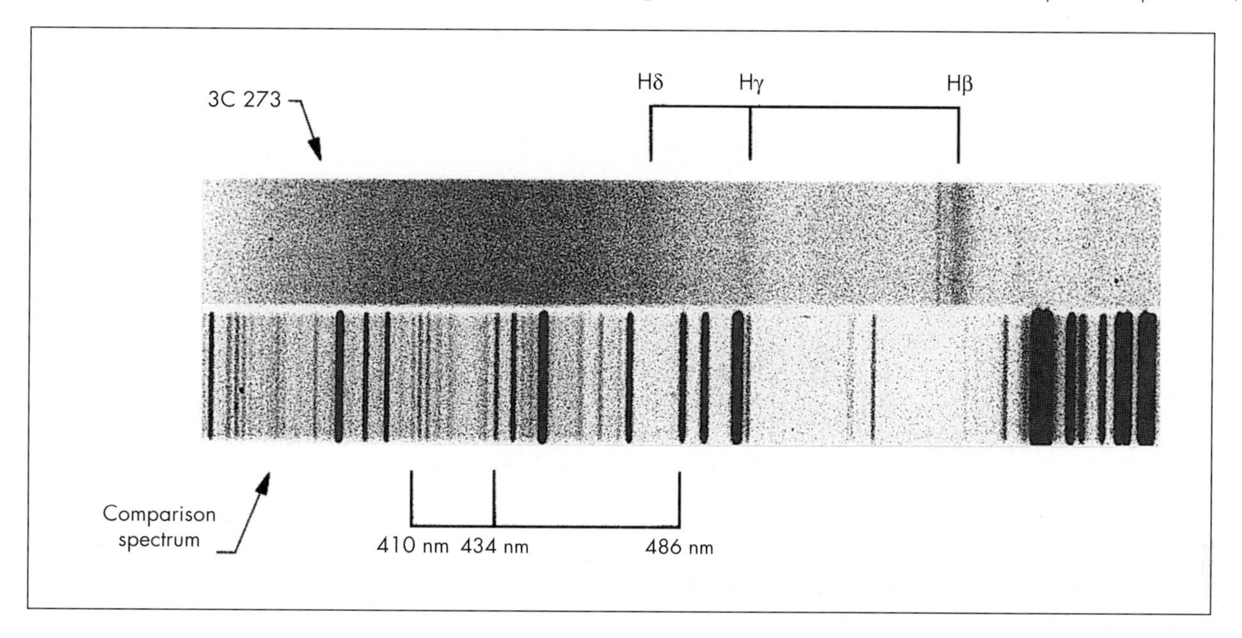

astronomers now agree, however, that the light from quasars is redshifted because quasars are very distant and receding rapidly from us as part of the expansion of the universe. Their light has been traveling for a very long time to reach us. Thus we see them as they were when the universe was only a fraction as old as it is now.

To calculate the distance to a quasar with a given redshift and the quasar's **lookback time** (the length of time since the light we are now seeing left the quasar), we must know how rapidly the universe was expanding during the time the quasar's light has been traveling to us. This means that we need to know the present value of Hubble's constant and whether the expansion has slowed or sped up with time. To get an idea about distances and lookback times for quasars, we can adopt the expansion rate history that is most consistent with current data. This and other possibilities for the history (and future) of the universe are described in Chapter 26. If Hubble's constant is 70 km/s per Mpc (the best current estimate is 71 km/s per Mpc with an uncertainty of about 3 km/s per Mpc) the graphs in Figures 24.3 and 24.4 show how present distances and lookback times depend on redshift. Using Figures 24.3 and 24.4, we see that a quasar with a redshift of 2 is now 5180 Mpc (16.9 billion light years) away. We see it as it was when the universe was only about a quarter of its present age (about 10.2 billion years ago). The most distant quasars discovered so far have redshifts of about 6.4. A quasar with a redshift of 6.4 is now 8300 Mpc (27 billion light years) away. We see it as it was 12.6 billion years ago when the universe was only 6% as old as it is today.

The large redshifts of quasars, a result of the expansion of the universe, show that quasars are extremely distant. The great distances of quasars mean that we are seeing them as they were in the remote past, when the universe was only a fraction of its present age. By observing quasars, we can sample conditions in the universe long ago.

Properties of Quasars

Although individual quasars have a great range of properties, they all have great luminosities, unusual spectra with broad emission lines, and small sizes.

Luminosity The very brightest quasars are as luminous as 10^{15} stars like the Sun. This is equivalent to the output of about ten thousand Milky Way galaxies. Such tremendous luminosities are rare, however, and the numbers of quasars increase as luminosity decreases. To be considered a quasar, the nucleus of a galaxy must have a minimum luminosity of about 10^{11} $L_{\odot}$ or about one galactic luminosity. At such a brightness, the quasar (the nucleus of a galaxy) would be as bright as the rest of its host galaxy. Below this limit, there are dimmer objects

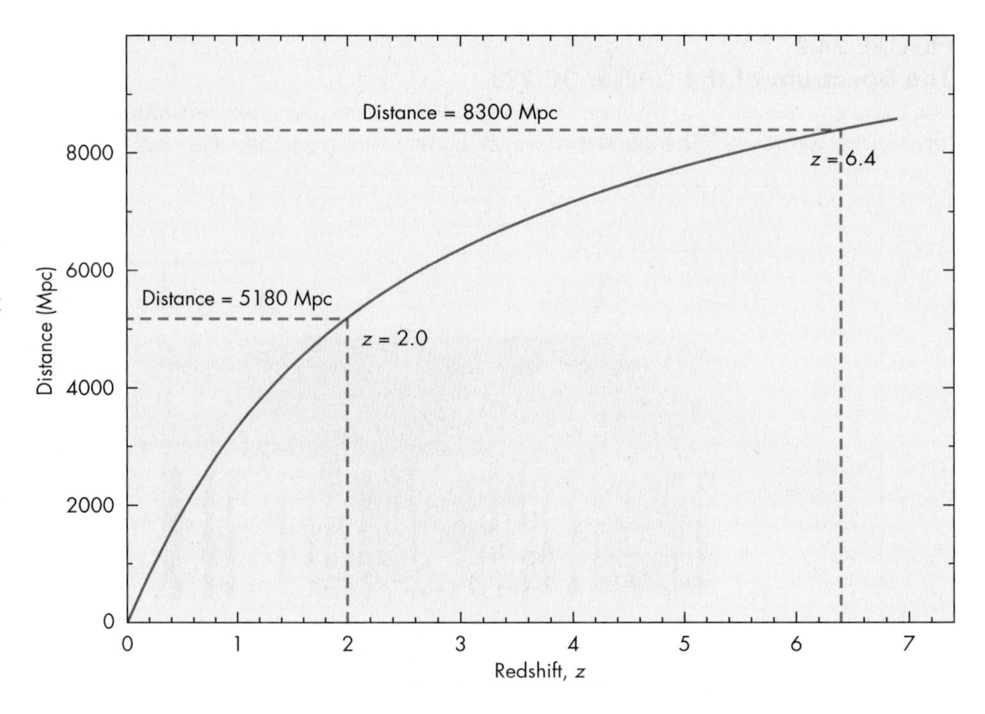

FIGURE 24.3
Distance Versus Redshift (*z*)
The shape of this graph depends on the present value of Hubble's constant and how it has changed with time. In this figure, Hubble's constant is assumed to be 70 km/s per Mpc. The distances of quasars with redshifts of 2 and 6.4 are about 5180 and 8300 Mpc, respectively.

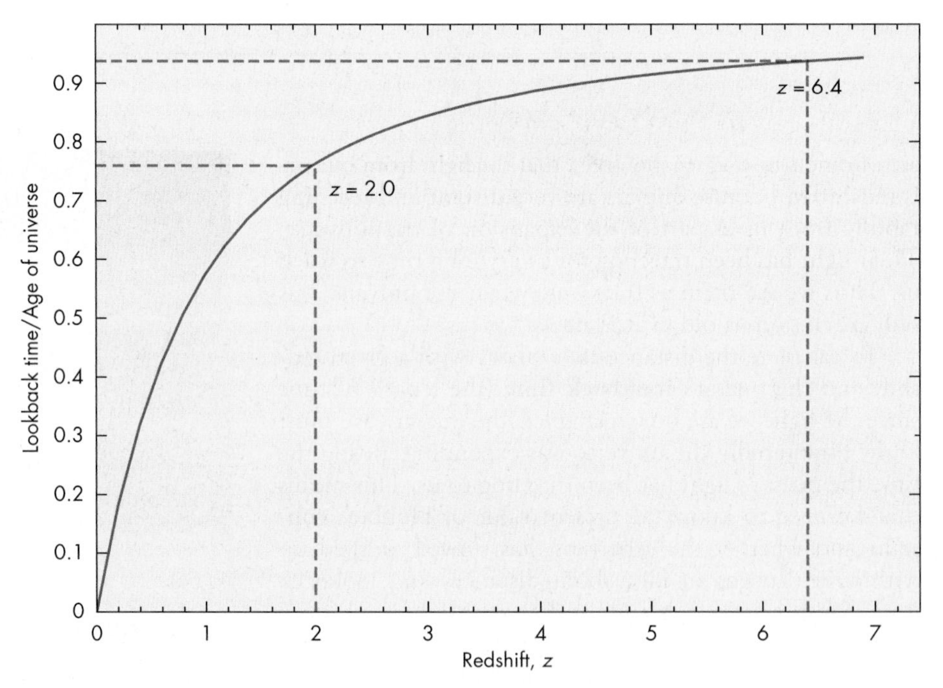

FIGURE 24.4
Lookback Time Versus Redshift (*z*)
The remotest quasars have redshifts of 6.4. They have lookback times greater than 0.9, so we see them as they were when the universe was less than 10% of its present age. The lookback time of a quasar with a redshift of 2 is 0.76.

that otherwise resemble quasars and may actually be low-luminosity quasars. These active galaxies are described later in this chapter.

Spectrum The spectrum of a quasar is made up of continuous radiation together with broad *and* narrow emission lines. The conditions needed to produce these three kinds of radiation are different enough that each probably arises in a different region of the quasar.

Continuous Radiation Most of the luminous objects in the universe are bright in only a limited region of the spectrum. Planets, for example, emit significant radiation only in the infrared. Stars like the Sun are bright only in the visible part of the spectrum and in adjacent parts of the ultraviolet and infrared. Quasars are unusual in that they have almost uniform brightness over a range of as much as 10^{11} in wavelength. This covers the region from X rays to radio waves. Over this entire range, the power emitted is approximately constant. That is, the energy emitted in the interval between 1 and 10 nm is the same as the energy emitted between 300 and 3000 nm, which is the same as that between 6 and 60 μm in the infrared, and so forth.

The unusual continuous spectra of quasars make it easy to distinguish them from most stars and galaxies. Quasars are unusually blue because they emit a large amount of blue and ultraviolet radiation compared with their visible output. One of the most productive ways to find quasars is to search for objects that look like stars but are very blue. At magnitude 22 (2.5 million times fainter than can be seen by the unaided eye), about 95% of the "blue stellar objects" turn out to be quasars. It is even easier to distinguish quasars from stars at X-ray wavelengths. Most stars are only feeble emitters of X rays, so most of the point sources observed by X-ray detectors on satellites turn out to be quasars or closely related objects.

A closer look at the continuous spectrum of a quasar shows that it actually has several separate components as shown in Figure 24.5. Over most of the spectrum of a quasar, the bulk of the continuous radiation is contributed by a featureless component that extends from the X-ray region to the far infrared. This component is probably due to synchrotron radiation from energetic electrons moving through magnetic fields. For most quasars, the brightness of the synchrotron radiation fades rapidly with increasing wavelength in the radio part of the spectrum. In fact, most quasars are completely undetectable as radio sources, even though it was intense radio radiation that originally alerted astronomers to quasars. Only a small minority of quasars is "radio loud." A possible explanation for why some quasars are radio loud while others are "radio quiet" is described later in this chapter.

A second component in the spectra has been called the "blue bump." This component peaks near the border between the visible and ultraviolet parts of the spectrum. It probably originates as blackbody radiation from dense gas at a temperature of about 10,000 K. A third component, brightest in the infrared, is probably due to blackbody radiation from warm dust.

Quasars have luminosities that typically are much larger than ordinary galaxies. Most of the radiant output of a quasar takes the form of continuous radiation that covers nearly the entire electromagnetic spectrum. A quasar spectrum has several components, including synchrotron radiation and radiation from gas and dust.

FIGURE 24.5
The Continuous Spectrum of a Typical Quasar
The spectrum has several components, including synchrotron radiation and thermal (blackbody) radiation from hot gas. The blackbody radiation makes up the "blue bump" in the visible and ultraviolet parts of the spectrum.

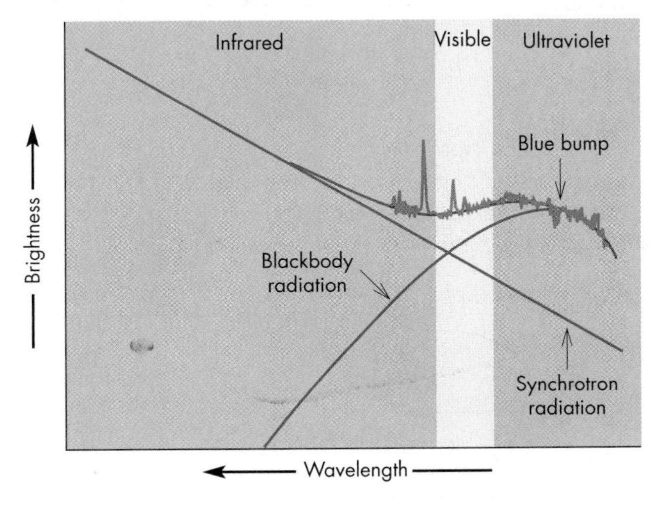

Emission Lines In addition to their continuous spectra, quasars show many strong emission lines. The emission lines from a typical quasar are shown in Figure 24.6. These lines are emitted by gas that is heated when it absorbs ultraviolet radiation and X rays from the source of continuous radiation in the quasar. The high-energy radiation ionizes the gas that absorbs it, producing a mixture of ions and electrons. This is similar to what happens to the gas in an HII region or a planetary nebula (Chapters 19 and 22).

The emission lines cool the gas in quasars so effectively that, despite their proximity to extremely intense ultraviolet radiation, the regions that produce the emission lines have temperatures of only about 10,000 K—about the same as in the HII regions that surround single stars.

The pattern of emission lines from a gas ionized by ultraviolet radiation and X rays depends on the intensity of the ultraviolet radiation and X rays, the chemical composition of the gas, and the density of the gas. Although the calculations are difficult, astronomers have been able to

FIGURE 24.6
Emission Lines in the Spectrum of a Typical Quasar
Strong lines of hydrogen and other elements can be seen.

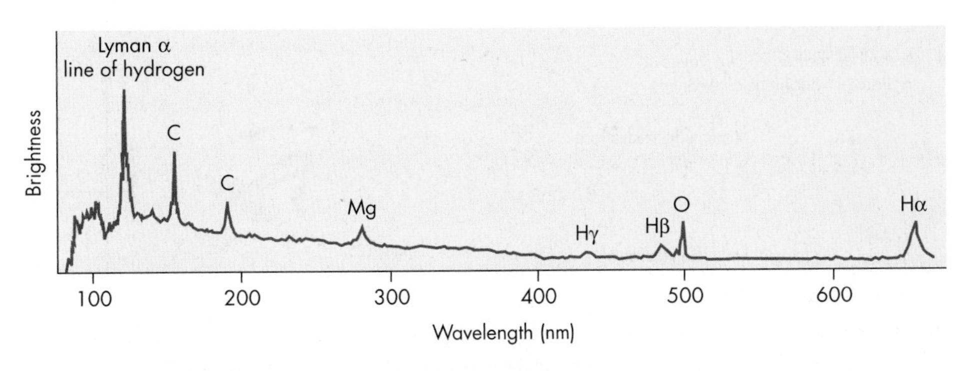

use the strengths of the emission lines from various energy levels of different elements to deduce the conditions in the emission-line-producing regions of quasars. Actually, the situation is complicated by the presence of both broad and narrow emission lines in the spectra of quasars. The patterns of the two types of lines show that they originate in distinct regions called the narrow line region and the broad line region.

The **narrow line region** in a quasar produces emission lines of hydrogen, helium, carbon, nitrogen, and other abundant elements. The density in a typical narrow line region is quite small—at most one atom per cubic meter. About 10^5 to 10^7 $M_\odot$ of gas are needed to account for the strengths of the narrow emission lines in quasars.

The pattern of lines produced in the **broad line region** requires a density of about one thousand atoms and ions per cubic meter. At such high densities, the gas emits very efficiently. As soon as the electron in an atom reaches the ground state, the atom is reexcited by a collision or is reionized. This means that a single atom produces photons at a rapid rate. The efficiency with which the emission lines are produced is so great in broad line regions that only a few solar masses of gas are needed to account for the strengths of the broad emission lines from quasars.

Compared with the narrow line region, the broad line region of a quasar has a higher density and requires more high-energy ultraviolet radiation and X rays. This suggests that the broad line region lies closer to the center of the quasar than the narrow line region because the center is where density is high and where high-energy radiation is produced. In both regions, the emission is thought to be produced by large numbers of small clouds immersed in much hotter gas. The clouds in the broad line region move much faster than those in the narrow line region, thus producing larger Doppler shifts and broader lines.

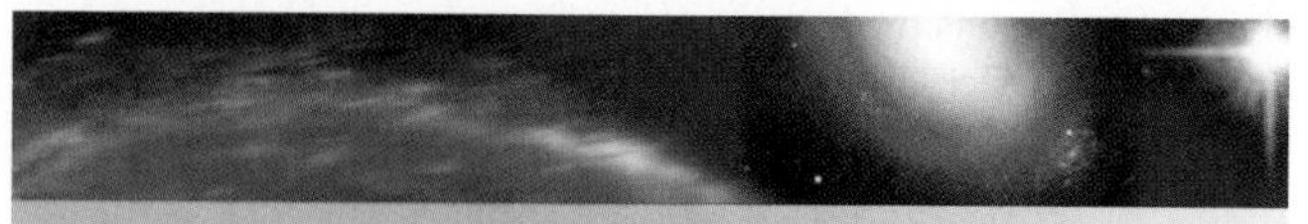

Quasar spectra contain both broad and narrow emission lines. The broad lines originate nearer the center of the quasar than do the narrow lines.

Sizes Quasars appear pointlike not only because of their great distances, but also because the regions of quasars that emit most of the radiation are only a few parsecs or less in size. The reason astronomers believe the emitting regions in quasars are small is that they vary in brightness on timescales as short as a few years or less.

Rapid variations in brightness imply small sizes because we can't observe a significant change in the brightness of an object in less time than it takes a beam of light to cross it. Figure 24.7 shows two astronomical bodies that have the shape of thin spherical shells. One of the shells is 1 light year across, the other is 2 light years across. Suppose the entire shell making up the smaller object brightens instantly and simultaneously. Because the object is far from the Earth, it will be many years before we learn

FIGURE 24.7
Brightness Changes and Sizes
A, If a shell of matter 1 light year in diameter brightens all at once, radiation from the near side of the brightened shell reaches us a year before the radiation from the far side of the brightened shell. Although the shell brightened all at once, we see the brightness increase gradually over a period of a year—the time it takes light to cross the shell. **B,** For a shell of matter 2 light years in diameter, light takes 2 years to cross the shell, so we see the shell brighten over a period of 2 years.

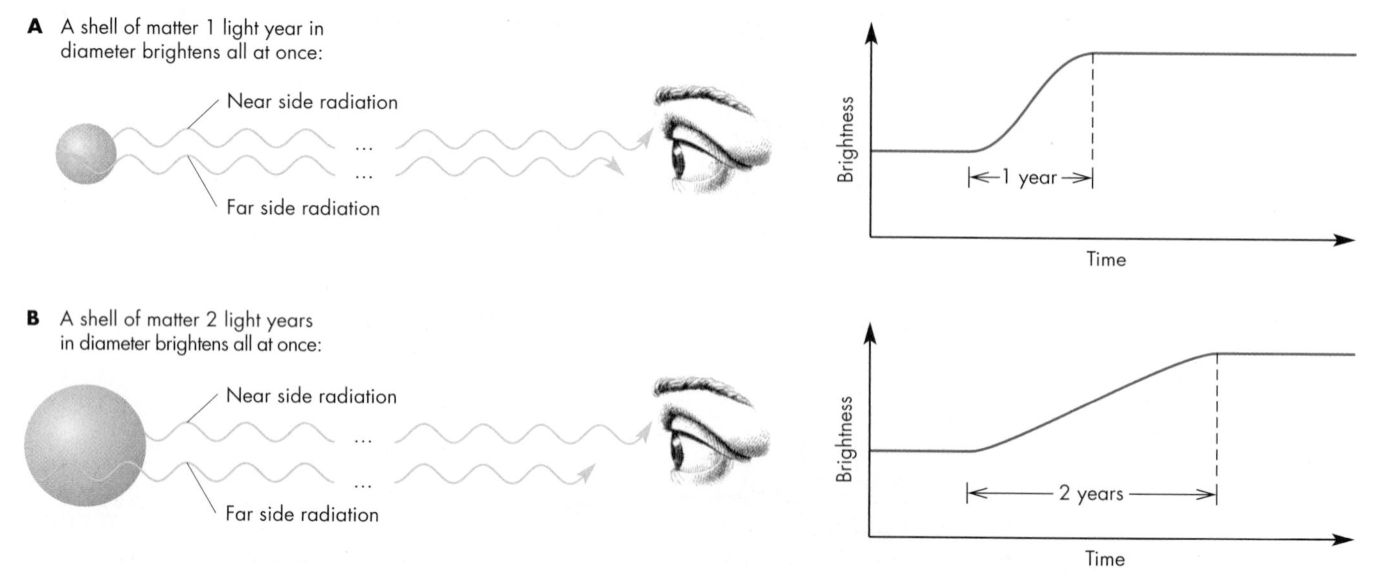

of the brightening. We first see evidence of the brightening when light from the nearest point on the shell reaches us. At that moment, we are still seeing "older," dim radiation from more distant parts of the shell. The "newer," bright radiation from those parts has not yet had a chance to reach us. As time passes, we see brightened radiation from more of the shell. The depth from which we see brightened radiation increases at the rate of 1 light second per second. Because the shell is 1 light year across, a full year will be required for light from the entire brightened shell to reach us. During that time, we will see the object brighten steadily. Although the entire object brightened instantly, we see the brightening occur over 1 year, the time it took the light to cross the shell. If the shell 2 light years across also brightened instantly, we would see its brightness increase over 2 years. In either case, the time required for brightness changes to occur is an upper limit on the size of the emitting shell. The size could actually be much smaller than the timescale on which its brightness variations occur, but it can't be larger.

For quasars, the most rapid brightness variations occur for X rays. Variations in half a day or less are common, and some quasars have been found to vary dramatically in brightness in less than an hour. Thus, the X-ray–emitting regions of quasars must be half a light day or less in diameter, not much larger than the diameter of Neptune's orbit. This means that the enormous X-ray output from quasars is produced in a region as small as or smaller than our solar system.

For many quasars, the continuous radiation at wavelengths other than X rays is also variable, but on timescales of months to years. Broad emission lines also seem to vary on this timescale, but narrow emission lines from quasars don't seem to vary in brightness at all. A comparison of the timescales of variation indicates that the X-ray–emitting region is smallest and, probably, nearest the core of the quasar. Next in size is the rest of the region that produces the continuous spectrum. The broad line region is probably somewhat larger, and the narrow line region is larger still.

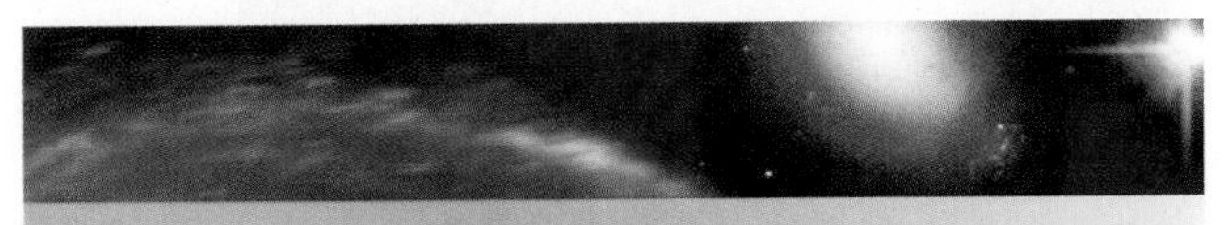

An astronomical body can't appear to vary in brightness more rapidly than the time required for a beam of light to cross it. Because quasars vary in brightness on timescales as short as days, their energy must be produced in regions of space no larger than the size of the solar system.

Superluminal Motion Most of the radio emission from a quasar has an angular size that is much too small to be resolved by a single radio telescope. When quasars are examined at high angular resolution using interferometers, they often show multiple components, separated by a few milliarcseconds (thousandths of a second of arc). In dozens of cases, these closely separated components have been observed to move progressively farther apart as time passes. The series of radio maps in Figure 24.8

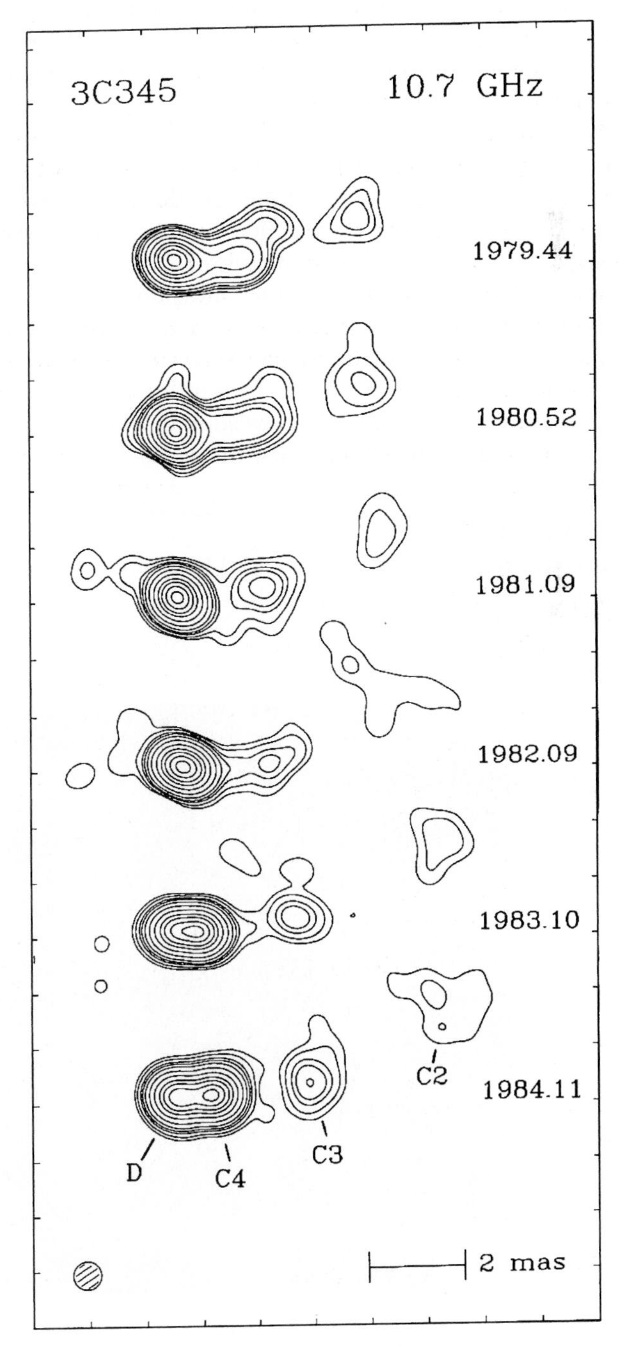

FIGURE 24.8
A Series of Radio Maps of the Quasar 3C 345
This series of contour maps shows that the components separated steadily during the 5 years in which observations were made. The speed of separation is about seven times the speed of light. The bar at the bottom shows 2 milliarcseconds (mas).

shows that several of the components of the quasar 3C 345 moved apart at a steady rate during a 5-year period. Although the angle between the components increased at a very small rate—less than 1 milliarcsecond per year—3C 345 is so distant that any detectable increase in the separation implies that they are moving apart at very high speeds. In fact, 3C 345 has a redshift of 0.6, so its distance is about 1900 Mpc, and the rate at which the components are separating is about seven times the speed of light. Apparent motions at speeds greater than the speed of light are called **superluminal motions.** Superluminal motion is common for quasars. Some quasars have components that appear to move apart as fast as ten times the speed of light.

At first glance, superluminal motion seems to violate the well-established law of physics (from special relativity) that nothing can move faster than the speed of light. It is probable, however, that the superluminal motions of quasars don't really mean that components are separating at speeds greater than the speed of light. Instead, the effect is a kind of optical illusion that occurs when we see matter moving almost directly toward us at very high speed. Suppose there are two bright components of a quasar (*A* and *B*) and that they are side by side at a distance of 1 billion light years from us, as shown in Figure 24.9. This means that a billion years after components *A* and *B* emit their light, we see them next to one another with zero angular separation. Now suppose component *B* moves at a speed barely less than the speed of light in a direction 37° from the direction to the Earth. After 5 years, it has moved almost 5 light years through space. This motion brings it 4 light years closer to the Earth and carries it 3 light years across our line of sight. At the end of the 5 years, it emits light again. This time the light takes 4 years less than a billion years to reach the Earth because component *B* is closer to us than it was earlier. The light arrives only 1 year after the light emitted when *A* and *B* were close together because the light travel time to the Earth has decreased as component *B* moves closer to the Earth. In essence, the component has almost kept up with the light it emits in our direction. It appears to us that *A* and *B* have moved apart by 3 light years in a single year, so we conclude that the components are separating at three times the speed of light.

The graphs in Figure 24.9 show that the speed with which components *A* and *B* appear to move apart depends not only on how fast component *B* is actually moving, but also on the angle at which it moves. If it is moving directly toward us (an angle of 0°), there is no motion across the line of sight, and we will detect no superluminal motion. If the motion is almost at a right angle to the direction to the Earth (an angle of 90°), its distance to us doesn't decrease as it moves. We see the components separating, but only at the actual speed of the moving component. However, at angles between 0° and 90°, we see the components appear to separate at speeds faster than they actually move. For some angles, the apparent speed can be much greater than the actual speed of separation.

FIGURE 24.9
An Explanation of Superluminal Motion

Component *B*, moving at nearly the speed of light, almost keeps up with the light it emits in our direction. Light it emits when it is 5 light years from component *A* arrives only a year after the light it emits when it is next to component *A*. We see it appear to move across the line of sight much faster than it would if it weren't moving toward us.

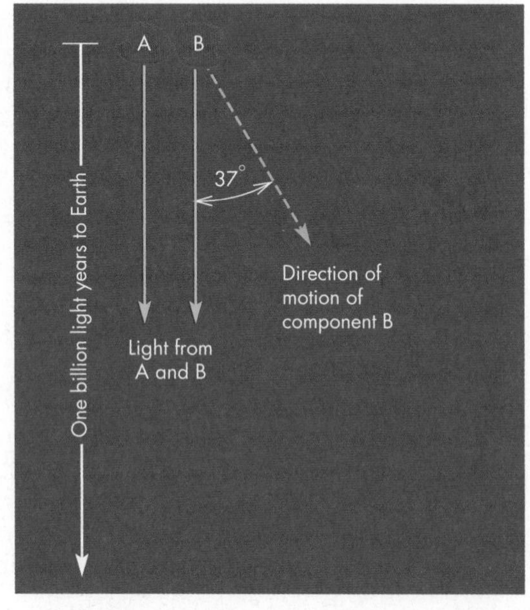

A Light from quasar components A and B is observed on Earth 1 billion years later.

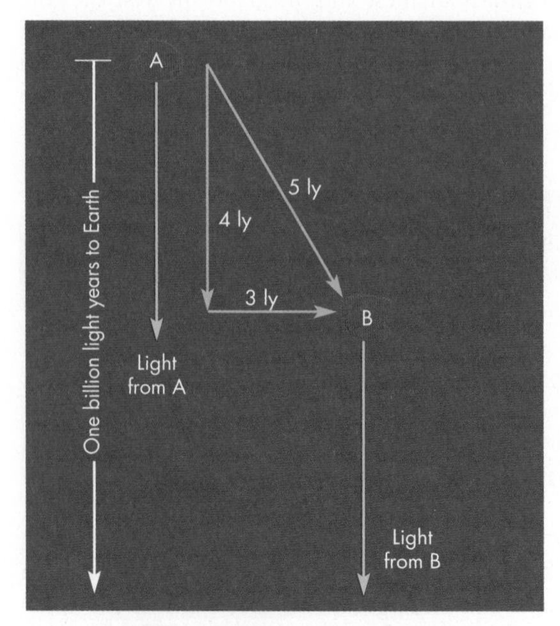

B Five years later: Light from A still takes 1 billion years to reach Earth; light from B takes 4 years less.

FIGURE 24.10
Superluminal Speed for Components Moving at 90% and 98% of the Speed of Light
In both cases, the apparent superluminal speed depends on the angle to the line of sight. For an actual speed of 90% of the speed of light, a maximum apparent speed (2 c) occurs when the component is moving at an angle of 25° from directly at the Earth. For an actual speed of 98% of the speed of light, the maximum apparent speed (almost 5 c) occurs when the component is moving at an angle of 12° from directly at the Earth.

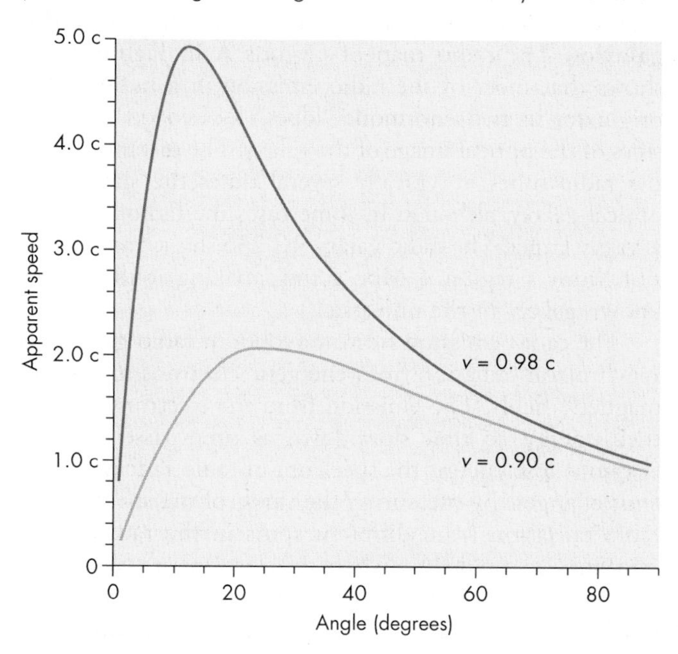

In Figure 24.10, for example, when component B is moving at a speed of 0.90 c at an angle of 25°, its apparent speed is about 2 c. When component B is moving at a speed of 0.98 c at an angle of 12°, its apparent speed is almost 5 c. To produce superluminal motions as rapid as those observed in quasars, the moving component must be traveling at a large fraction of the speed of light and also must be moving almost directly at us. How such motion could originate, some of its consequences, and why so many quasars have components moving almost directly at the Earth are described later in this chapter.

High-resolution radio images of quasars often show multiple components that appear to be moving apart faster than the speed of light. This superluminal motion can happen when a component of a quasar moves almost directly at us at almost the speed of light. The component almost keeps up with the light it emits, magnifying the rate at which it appears to move across the sky.

24.2 THE ACTIVE GALAXY ZOO

A number of kinds of objects share some of the properties of quasars. These close relatives are called **active galaxies.** The quasarlike central regions within active galaxies are called **active galactic nuclei.** Although active galaxies have been known for about a century, astronomers have shown great interest in them only since it became clear, in the 1960s, that they resemble the enigmatic quasars.

Seyfert Galaxies

In the early 1940s, Carl Seyfert identified a number of spiral and barred spiral galaxies that have bright, pointlike nuclei, as shown in Figure 24.11. These galaxies are called **Seyfert galaxies,** and they come in two types, Seyfert 1 and Seyfert 2 galaxies. The nuclear regions of Seyfert galaxies emit continuous radiation over a broad region of the spectrum. Seyfert galaxy nuclei, however, are not as luminous as quasars. For Seyfert 1 galaxies, the resemblance to quasars is strengthened by the existence of both narrow and very broad emission lines. The Seyfert 2 galaxies, on the other hand, have only narrow emission lines.

Seyfert galaxies are thought to be low-luminosity counterparts of quasars. Because they are less luminous, they can't be seen to such great distances as quasars and, for that reason, they have much smaller redshifts. Although the galaxy surrounding the active nucleus of a Seyfert galaxy is quite apparent, this is only because of the

FIGURE 24.11
***Hubble Space Telescope* Image of the Seyfert Galaxy NGC 7742**
The nucleus of this Seyfert galaxy is nearly as bright as the rest of the galaxy. If the galaxy were at a much greater distance, only the nucleus would be detectable and would be considered a quasar.

relative proximity of the Seyfert galaxy. At the distance of a typical quasar, the galaxy surrounding the active nucleus of a Seyfert galaxy would be too faint to be seen. In that case, the bright nucleus would be classified as a quasar. Thus astronomers can't always distinguish clearly between quasars and Seyfert galaxies.

The most striking difference between the two types of Seyfert galaxies is the absence of broad emission lines in Seyfert 2 galaxies. Although the two types are otherwise similar, Seyfert 1 galaxies tend to be more luminous and have much stronger X-ray emission than the Seyfert 2 galaxies. The differences between Seyfert 1 and Seyfert 2 galaxies could arise in several ways. First, the two types might simply be fundamentally different kinds of objects that just happen to resemble each other in many ways. This kind of coincidence, however, seems unlikely.

A better possibility is that the two types of Seyfert galaxies differ because clouds of gas and dust near the nucleus of a Seyfert 2 galaxy hide the X-ray–emitting region and the broad emission line region from view. In this model, the narrow emission line region lies outside the obscuring material and looks the same for both types of Seyfert galaxies. Two different theories have been proposed to account for the dust that obscures the broad line regions of Seyfert 2 galaxies. In one explanation, all Seyfert galaxies begin as dusty Seyfert 2 galaxies. As the dust is blown away by outflowing gas and radiation, the broad line region emerges into view, and the galaxy becomes a Seyfert 1 galaxy. This idea is supported by the fact that essentially all Seyfert 2 galaxies, but only about 30% of Seyfert 1 galaxies, show large amounts of infrared radiation from warm dust clouds. In the second explanation, the dust in all Seyfert galaxies is distributed in the form of a torus, or doughnut, like that shown in Figure 24.12. In those cases where we view the torus from the side, the broad line region is blocked from view. Only when we view the torus from above can we see the broad line region within it.

Radio Galaxies

A few percent of luminous elliptical galaxies are strong sources of radio radiation and are classified as **radio galaxies.** The radio map of Cygnus A in Figure 24.13 shows that most of the radio emission in a radio galaxy originates in two enormous "lobes" located on opposite sides of the optical image of the galaxy. The distance across the radio lobes is typically several times the size of the optical galaxy, although in some cases the radio structure is vastly larger. The radio galaxy 3C 236 shows radio emission from a region 4 Mpc across, making it the largest known galaxy in the universe.

The radio emission from the lobes of radio galaxies is synchrotron radiation from energetic electrons moving in magnetic fields. The emission from the electrons reduces their energy, so they slow down as time passes. As the electrons lose energy, the spectrum of radio radiation they emit changes. By measuring the shape of the spectrum of radio radiation from different spots in the radio lobes, astronomers can deduce where the electrons are young and energetic and where the electrons flow as they age. For luminous radio galaxies, the electrons are youngest and most energetic at the brightest spots in the radio lobes, located on the outer edges of the radio lobes shown in Figure 24.14.

FIGURE 24.12
The Nucleus of a Seyfert Galaxy Inside an Obscuring Torus
Viewed in the plane of the torus, the broad line region within the torus can't be seen. In this case, the galaxy is classified as a Seyfert 2. Viewed from above the torus, both the broad and narrow line regions can be seen, so the galaxy is classified as a Seyfert 1.

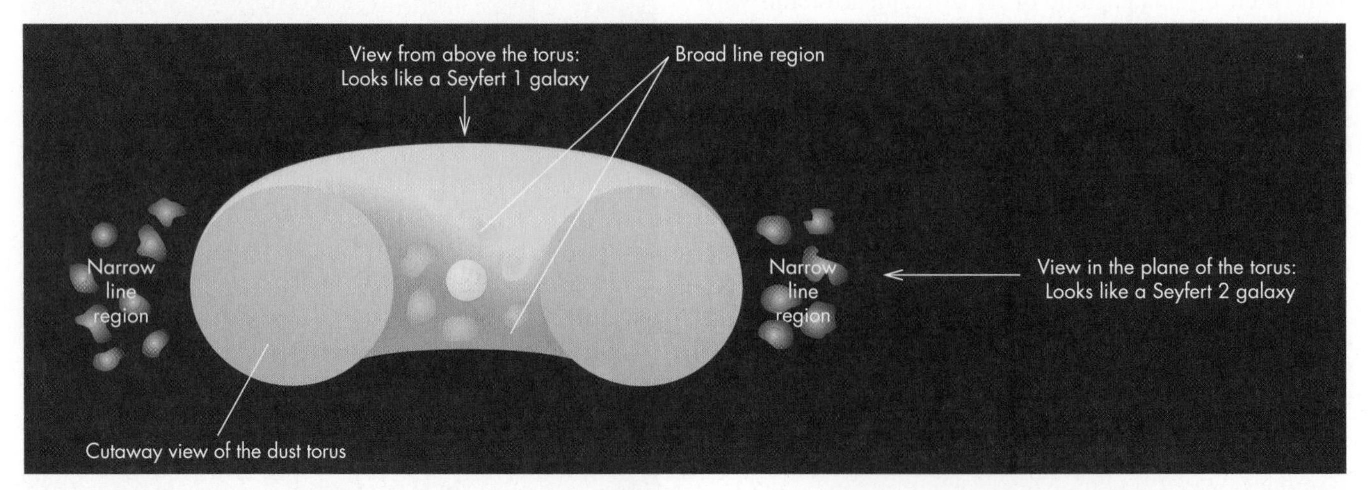

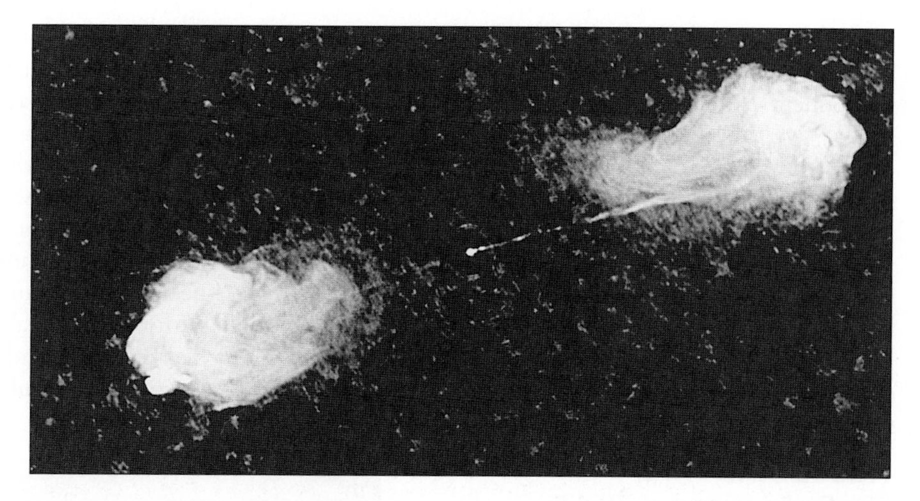

FIGURE 24.13
A Radio Map of the Radio Galaxy Cygnus A
Each lobe of radio wave–emitting material extends about 50 kpc away from the central galaxy, which lies about midway between the two lobes. Compare this view of Cygnus A with the visual image in Figure 24.1.

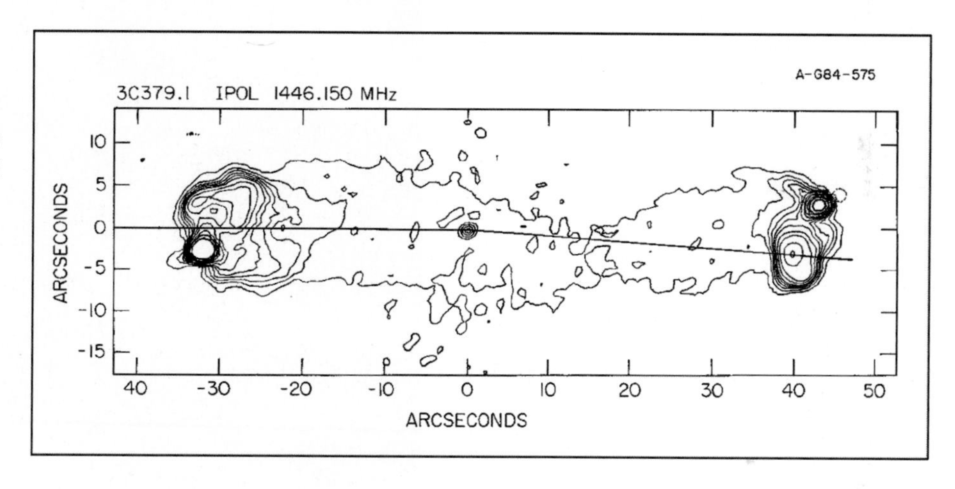

FIGURE 24.14
A Radio Contour Map of the Radio Galaxy 3C 379.1
The bright "hot spots" in each lobe are where jets of material from the central galaxy strike intergalactic gas.

The ages of the electrons increase steadily with diminishing distance from the central galaxy, indicating that the electrons flow back toward the galaxy from the outer edges of the radio lobes.

Radio Jets The inward flow of electrons (and ions) in the lobes of radio galaxies raises the question of how the electrons and the energy they carry get into the outer parts of the lobes of radio galaxies. The answer is that the electrons are injected into the lobes by tight beams of matter that stretch outward from the very center of the optical galaxy. In some radio galaxies, the beam, or **jet,** of matter can be seen in radio maps like the one in Figure 24.15. A great deal of what we know about the properties of radio jets has come from comparing the observed appearance of jets with the results of computer models of energetic jets. Figure 24.16 shows one of the models that strongly resemble observed jets. The best models are for low-density jets of extremely energetic electrons moving at nearly the speed of light. These models mimic the behavior of real jets in many ways, including the way they end in a bright hot spot at the outside of the radio lobe and then flow back toward the galaxy in a cocoon that surrounds the jet.

Beaming Because there are two radio lobes in a radio galaxy, it seems reasonable to suppose that we should see two, oppositely directed, jets—one to feed each lobe. Sometimes this is the case. In the more powerful radio galaxies, however, one of the jets is often much fainter than the other. In some cases, one jet is so faint it can't be seen. When two jets of unequal brightness are observed, the brighter jet is on the near side of the galaxy (the side toward us) and the dimmer jet is on the far side. This means that the brighter jet is pointed toward us and the dimmer jet is pointed away.

The beam that points toward us is the brighter one for reasons similar to the one used to explain superluminal motion. A luminous jet of gas moving nearly toward us at great speed almost keeps up with the light that it emits. In the example used to explain superluminal motion (Figure 24.9), light emitted over 5 years by component *B* would be received by us in a single year because component *B* was moving nearly toward us at very close to the speed of light. Thus, we receive light at five times greater a rate from component B than we would if it were at rest, making it appear five times as bright as it would if it were motionless. A luminous object moving away from us is dimmed by the same factor because of its motion, as shown in Figure 24.17. The brightness of an object moving across the line of sight is nearly unaffected by its motion. The effect of motion is to concentrate the radiation of the object in a beam that points in the direction in which it moves.

The effect is small for motion at low speeds. The effect can be quite large, however, for motion near the speed of light, like the jets of radio galaxies. In fact, the jet pointed toward us can be brightened tenfold, while the jet pointed away from us becomes ten times fainter. The amount of brightening and dimming depends on both how directly the jets point toward and away from us and how close the speed of the jets is to the speed of light. We can interpret radio galaxies with nearly equal jet brightnesses as those for which the directions of the jets are nearly at right angles to the line of sight. Radio galaxies with very unequal jets are those for which the beams point nearly toward and away from us. This means that a given radio galaxy would look very different when viewed from

FIGURE 24.15
The Radio Image of 3C 175
A jet of material can be seen extending from the central galaxy to one of the radio lobes. The jet feeding the other lobe can't be seen.

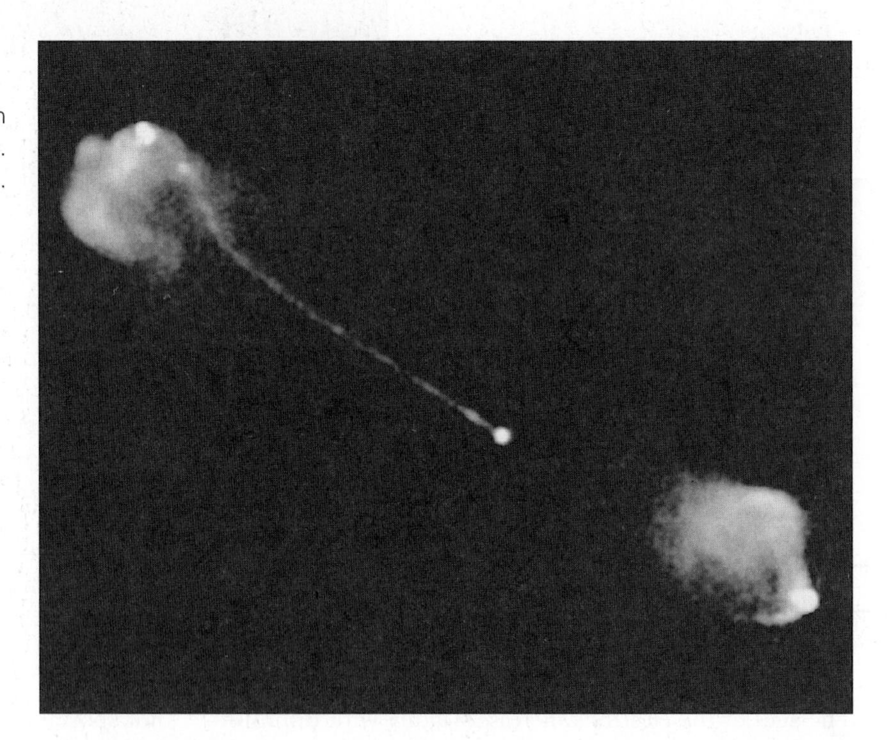

FIGURE 24.16
A Model of a Jet
Calculations of the flow of matter in a high-speed, low-density jet show that the jet remains tightly collimated until it strikes the intergalactic gas, where a hot spot is produced.

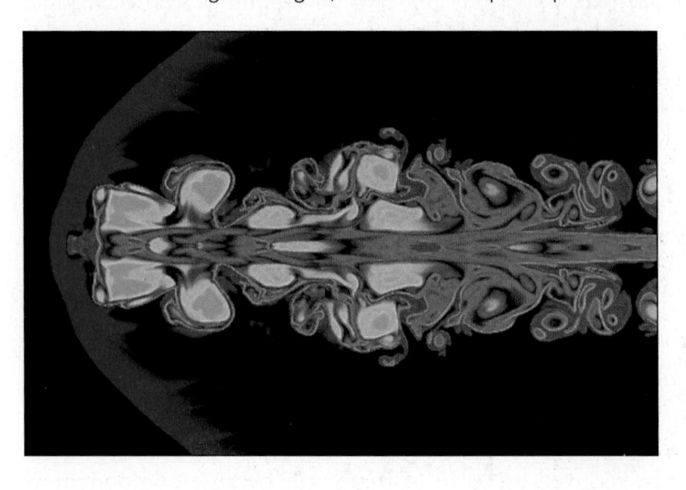

FIGURE 24.17
Viewing a High-Speed Jet
If the jet is pointing at us, we see its brightness enhanced. Viewed from behind, the jet is dimmed. From the sides, the jet is about as bright as it would be if it were motionless.

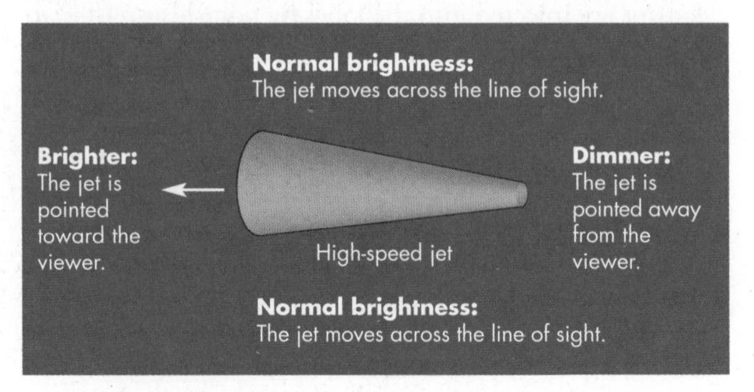

different angles. The idea that the direction from which an active galaxy is viewed controls the way it appears to us is one that is discussed again later in this chapter.

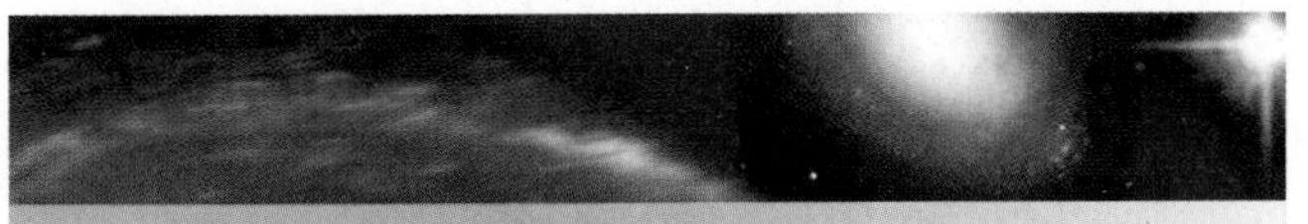

Jets of matter shot outward from the centers of galaxies feed huge lobes that emit radio radiation. In some cases, only a single jet can be seen even though two radio lobes are present. This happens because a high-speed jet pointed at us appears much more luminous than a jet pointed away from us.

Blazars

Other active galactic nuclei, called **blazars,** resemble quasars because they are small in angular size and show strong variations in brightness. Unlike quasars, however, blazars have weak emission lines or none at all. Blazars were discovered as astronomers searched for optical counterparts of radio sources. In several cases, the visible objects that were found had already been categorized as variable stars. Observations quickly showed that they weren't stars at all, but rather the nuclei of distant elliptical galaxies. One of these "variable stars," BL Lacertae, was the first of its type of active galactic nucleus to be identified. As the prototype of the group, BL Lacertae's name has been used (in modified form) for the entire class of objects.

Two different explanations for the weakness or absence of emission lines in blazars have been proposed. One theory is that blazars are like quasars except that blazars have little or no gas in their cores. The shortage of gas means that there are few if any atoms to produce emission lines. The other explanation is that the blazars have emission lines of normal strength, but that these lines are overwhelmed by very strong continuous radiation. The evidence now seems to favor the second explanation. For one thing, there are some blazars for which emission lines can't be seen except when the brightness of the blazar is at a minimum. At those times, emission lines like those of quasars can be seen. In addition, a number of blazars show superluminal motion like that seen in some quasars. Superluminal motion implies energetic jets of matter that point nearly at us. Radiation from a jet moving in our direction from the center of a blazar is beamed and boosted in brightness just as it is from a radio galaxy. It is believed that it is the boosting of brightness of the beamed radiation from the jet in a blazar that overwhelms the emission lines and makes it difficult to detect them.

IRAS Galaxies

All galaxies emit infrared radiation from the stars and dust they contain. In most cases, the infrared emission is modest compared with the visual output of the galaxies. The Infrared Astronomy Satellite (IRAS), however, found tens of thousands of galaxies that emit the bulk of their energy in the infrared. There seem to be two distinct, although possibly related, kinds of IRAS galaxies—the starburst galaxies and the most luminous IRAS galaxies.

Starburst Galaxies **Starburst galaxies** tend to be spiral galaxies in which a very large number of stars have recently formed. The ultraviolet radiation from the many young stars heats the dust grains in the galaxy. The heated dust, in turn, emits large amounts of infrared radiation. In some starburst galaxies, star formation is confined to the galactic nucleus, but in other cases it is spread throughout the disk of the galaxy. What triggers rapid star formation in a starburst galaxy is not completely understood, but it is likely that the interaction between two galaxies is a factor. A majority of starburst galaxies, including M82, shown in Figure 24.18, are interacting with other galaxies, so they have disturbed appearances. Star formation may occur when gas from one galaxy flows into the other. Another possibility is that unusually strong spiral density waves are

FIGURE 24.18
The Starburst Galaxy M82
The bright central region of M82 is the location of vigorous star formation. The distorted appearance of the galaxy is probably a consequence of a gravitational interaction with a neighboring galaxy.

produced in a galaxy by the gravitational disturbance of the galaxy with which it is interacting.

The Most Luminous IRAS Galaxies The most luminous IRAS galaxies emit 10^{12} to 10^{13} $L_\odot$ in the infrared. This is one hundred to one thousand times the infrared output of a normal spiral galaxy and comparable to the luminosities of quasars. The great luminosities of these galaxies suggest that they are dust-shrouded quasars. The dust absorbs the radiation produced by the nucleus of the quasar and emits it in the infrared. Like starburst galaxies, the most luminous IRAS galaxies are peculiar in appearance because they are pairs of interacting or merging galaxies. To some astronomers, this suggests that the quasar phenomenon is a consequence of the strong interaction of two galaxies. Starbursts and quasarlike activity seem to go on under similar conditions (and sometimes in the same infrared galaxies), but the relationship between the two phenomena is not yet understood.

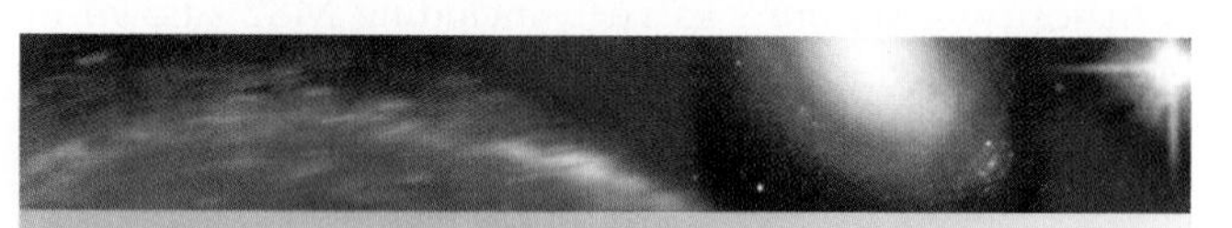

There are many kinds of galaxies that have quasarlike activity in their nuclei. Seyfert galaxies have bright, pointlike nuclei and strong emission lines. Radio galaxies have enormous radio-emitting lobes fed by jets from the cores of their central galaxies. Blazars are small and bright, like quasars, but show no emission lines. Luminous IRAS galaxies resemble dust-shrouded quasars.

MASSIVE BLACK HOLES AND ACTIVE GALAXIES

Ever since quasars were discovered, astronomers have tried to explain why quasars are so small yet so luminous. Equally mysterious have been the variability of quasars, the nature of jets in active galaxies, and the manner in which the various kinds of active galaxies are related to each other. Over the years, a model that can account for many aspects of the active galaxy phenomenon has gained general acceptance among astronomers. The fundamental feature of this model is the existence of massive black holes in the centers of quasars and other active galaxies.

Massive Black Holes

There is no instance in which a black hole actually has been observed in the center of a galaxy. In fact, because black holes emit no radiation, no direct detection of a remote black hole is possible. The gravitational influence of a massive black hole can be detected, however, and it can be very great. The evidence for black holes consists of observations that show that there must be large amounts of matter in the nuclei of active galaxies and that the matter must lie within a small volume of space. Astronomers have concluded that active galactic nuclei contain black holes by eliminating all other concentrated forms of matter from consideration. Essentially, the best evidence that there are massive black holes in the centers of active galaxies is that nothing else can produce the striking effects that occur there.

Size and Luminosity Quasar luminosities equivalent to one hundred average galaxies are common, and luminosities as great as ten thousand galaxies have been observed. The rapid brightness variations of quasars demonstrate, however, that this energy must be produced within a region a few light days across or smaller. With the exception of massive black holes, nothing seems capable of producing so much energy in such a small region of space. To show how difficult it is to account for such concentrated energy production, imagine trying to account for the energy of the most luminous quasars as starlight. The most luminous stars, the O stars, are about 500,000 times as luminous as the Sun. It would take about a billion O stars to produce the luminosity of the brightest quasars. If these stars were confined to a sphere a few light days across, they would be only about 1 AU apart. Within a short time, they would collide and merge, causing the system of stars to collapse and form a black hole.

Rate of Mass Consumption Black holes produce energy from matter very efficiently. Calculations have shown that matter falling into a black hole can emit an amount of energy equivalent to converting as much as 30% of its mass into energy. For comparison, nuclear fusion converts less than 1% of the masses of nuclei into energy. Despite the efficiency of a black hole, however, a large quantity of matter must be drawn into a black hole each year to account for the tremendous luminosity of a quasar. To produce as much energy as one hundred typical galaxies, for instance, a black hole would have to consume about 1 $M_\odot$ each year.

At this rate of consumption of matter, the black holes in active galactic nuclei have become very massive, regardless of their original masses. We know that quasars and other active galaxies have existed for nearly the entire history of the universe, more than 10 billion years. To produce one hundred galactic luminosities for that length of time, a black hole would have had to accumulate 10 billion $M_\odot$ of infalling material. Black holes in more typical active galactic nuclei would have had to collect 100 million $M_\odot$ of matter. If, on the other hand, there have been periods of time when little or no matter fell into a black hole and it was inactive, the mass of the black hole would now be smaller than if it had always been active.

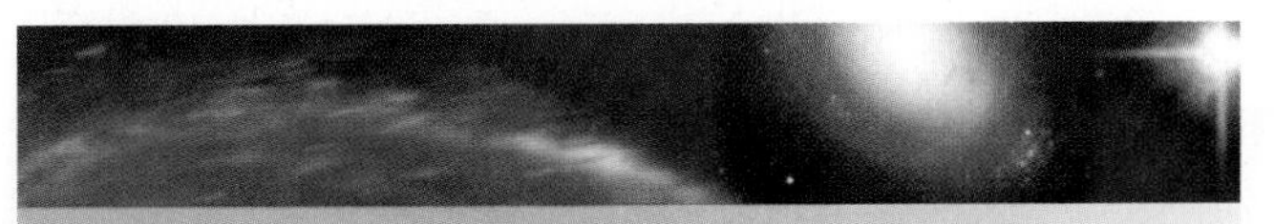

Astronomers have concluded that quasars are powered by massive central black holes partly because all other alternatives are less likely. It is difficult to imagine how normal stars or other energy sources could produce the great luminosity of a quasar in such a small volume of space. Black holes are very efficient in wringing energy from matter falling into them.

Luminosity and Mass The radiation produced in the nucleus of the galaxy pushes the surrounding gas outward with a force that depends on the brightness of the radiation. Unless there is enough mass in the nucleus that its gravity can overcome the radiation force, the gas and other matter will be blown away and no energy would be produced from infalling matter. For a given central mass, the maximum luminosity that could exist without blowing away the surrounding material is called the **Eddington luminosity.** Expressed in terms of the Sun's mass, $M_\odot$, and luminosity, $L_\odot$, the Eddington luminosity (L_{Edd}) is

$$L_{Edd} = 30{,}000\left(\frac{M}{M_\odot}\right) L_\odot \qquad (24.1)$$

where M is the mass of the black hole. If Equation 24.1 is solved for $M/M_\odot$, it gives the minimum mass a body of a given luminosity must have for it not to blow away all the matter that surrounds it. This mass, in terms of the Sun's mass, is

$$\frac{M}{M_\odot} = \frac{1}{30{,}000}\frac{L}{L_\odot} \qquad (24.2)$$

The mass of the central black hole of a quasar could be even bigger than that given by the Eddington luminosity, but it can't be smaller. The lower the luminosity of an active galaxy, the lower the mass of the black hole needed to produce its energy. Because luminosity depends on the rate at which matter falls into a black hole, the Eddington luminosity also can be thought of as a limit on the rate at which a black hole of a given mass can accrete fresh material.

The radiation emitted by an active galactic nucleus pushes outward on the surrounding gas. The black hole in an active galactic nucleus must be massive enough that its gravity overcomes the outward force of its radiation, or infalling matter will be blown away. A typical quasar must have at least several hundred million solar masses of material to continue to attract the gas that surrounds it.

Accretion Disks

If angular momentum is absent, matter can flow directly into a black hole. In more realistic situations, however, where rotation of the surrounding galactic material is taken into account, the infalling matter forms an accretion disk within which matter spirals inward to the black hole. Energy is released by friction within the disk and then carried to the surface of the disk, where it is radiated away. The structure of the disk depends on the rate at which energy is produced and the efficiency with which it can be transported to the surface to be released.

For black holes accreting matter at a relatively slow rate, the amount of energy released is small enough that it can be quickly carried to the surface of the disk and

Equation 24.2

The Eddington Luminosity

Equation 24.2 can be used to find the minimum mass that the nucleus of a quasar with a luminosity of 10^{13} $L_\odot$ must have to attract surrounding gas rather than blow the gas away. Using 10^{13} $L_\odot$ for L, Equation 24.2 gives

$$\frac{M}{M_\odot} = \frac{1}{30{,}000}\frac{L}{L_\odot} = \frac{1}{30{,}000}10^{13} = 3 \times 10^{8}$$

Thus the quasar's black hole must contain at least 300 million $M_\odot$ to remain active by consuming surrounding gas.

FIGURE 24.19
Accretion Disks in Active Galactic Nuclei
A, If the rate of accretion of matter is small, the disk remains thin and has a relatively cool interior. **B,** If the rate of accretion is rapid, the disk has a hot interior and swells in thickness. The "spikes" represent jets of matter moving outward, perpendicular to the accretion disk.

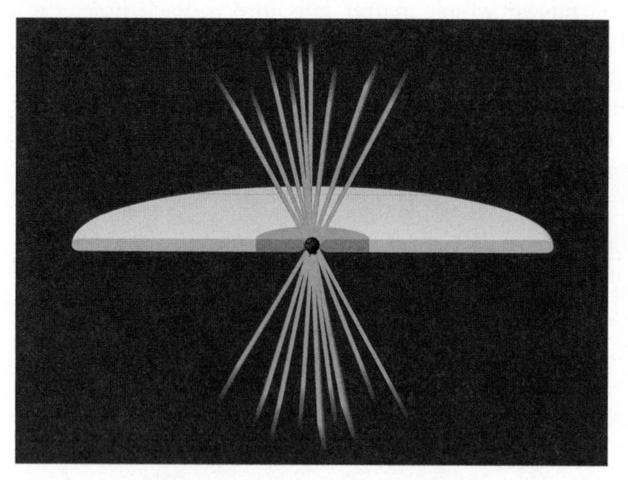

A Thin disk: Slow accretion of matter

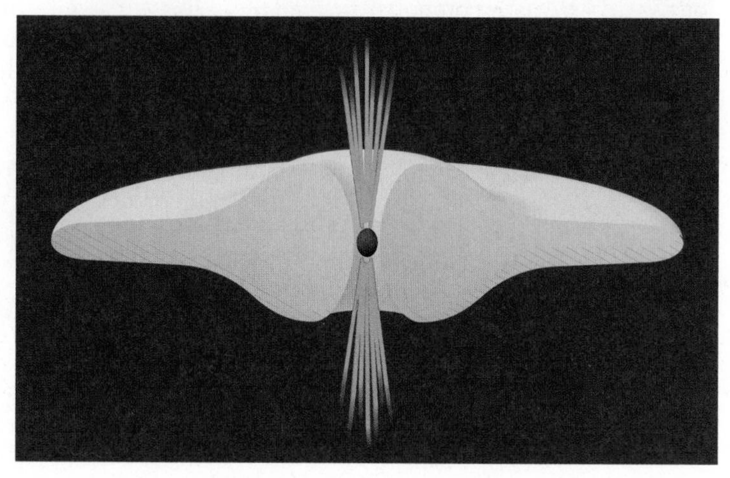

B Thick disk: Rapid accretion of matter

radiated away. The interior of the disk remains relatively cool, pressure remains low, and the disk is thin, as shown in Figure 24.19, A. When matter is being accreted at nearly the limit set by the Eddington luminosity, on the other hand, the rate of energy production is enormous, particularly for a volume of space as small as the accretion disk. As a result, the inside of the disk becomes extremely hot, increasing pressure in the disk so that it swells to a thickness proportional to the mass accretion rate. The disk takes on the appearance of a torus (or doughnut), as shown in Figure 24.19, B. The temperature at the surface of the torus is like that of a hot star, so that the torus radiates most of its energy in the ultraviolet part of the spectrum. This radiation is responsible for the "blue bumps" seen in the spectra of quasars.

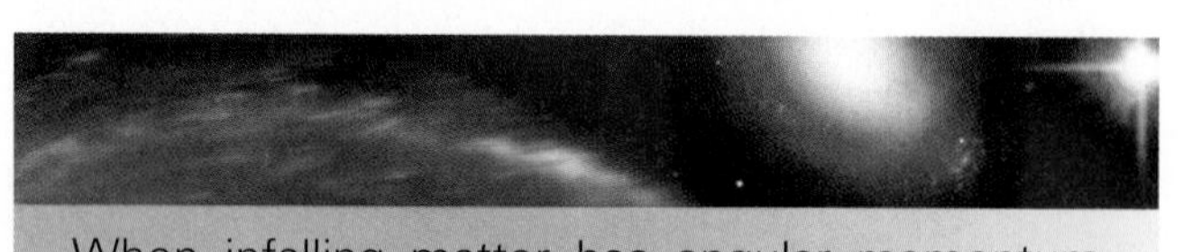

When infalling matter has angular momentum and can't fall directly into a black hole, an accretion disk is formed. The surface of an accretion disk is as hot as a star and emits visible and ultraviolet radiation that creates the "blue bumps" in the continuous spectrum of a quasar.

Jet Formation Jets from active galactic nuclei form tight beams that travel at high speed. No explanation for the acceleration of the jet material has yet been completely worked out, but one of the most attractive possibilities involves the effect of magnetic fields in the accretion disk. Figure 24.20 shows that if the magnetic field lines are tilted with respect to the rotating disk, then ions and electrons will move along the field lines and will be flung outward like beads on a rotating wire. The outflows produced in this way point in opposite directions that correspond to

FIGURE 24.20
Acceleration of Jet Material
Magnetic field lines embedded in the accretion disk rotate with the disk. Ionized gas moving along a field line is accelerated outward like beads on a rotating wire.

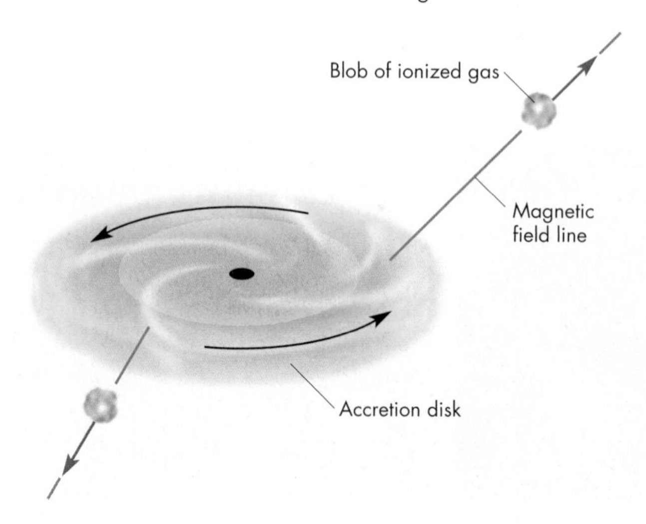

the orientations of the field lines emerging from the accretion disk. Some calculations show that most of the energy and angular momentum of the accretion disk could be carried away by oppositely directed jets.

The jets produced by this mechanism or any of the others that have been proposed are not very well collimated (confined to a narrow range of directions). Within a few parsecs of the center of an active galaxy, however, the jet has been narrowed to a very tight angle. The collimation of the jet must take place because the jet is confined by the pressure of its surroundings. If the surrounding pressure is too small, the jet will expand in angle and become too broad. If the surrounding pressure is too great, the jet may be squeezed shut and greatly disrupted. The nature of the pressure that confines and collimates the jets from active galactic nuclei isn't well understood. One possibility is that the collimation and confinement may be due to magnetic fields generated by the rapid flow of the ionized, conducting gas in the jet. The magnetic field produced by the jet, shown in Figure 24.21, will focus the jet, directing it into a narrow angle.

How Are Active Galaxies Related? Astronomers who study quasars and active galaxies are in a situation similar to that of anthropologists who study prehuman evolution. Many different kinds of early humanlike creatures have been discovered. A given creature resembles others in some ways but not in others. The question is, how are all the different types related to each other? Did some evolve into others? Are there some that are actually not closely related to the other types but just happen to look similar? With quasars and active galaxies, we want answers to the same kinds of questions. We want to know if there is a general scheme that coherently organizes all of the different kinds of active galaxies. If so, what are the parameters that describe that scheme? Are some kinds of active galaxies different stages in the development of the same galaxies?

Astronomers have proposed a number of ways in which different kinds of active galaxies might be related. There is now a general picture in which two parameters—luminosity and viewing angle—distinguish the different kinds of active galactic nuclei from each other. The picture is far from perfect because many of the relationships implied by the unified picture aren't completely consistent with the observed properties of active galactic nuclei. It is difficult to see how some objects fit within the scheme. In fact, the picture is still so preliminary that major parts of it may prove not to be true at all. We may find out that some of the pieces have been crammed into the puzzle in places where they don't fit. Nevertheless, the different types of active galaxies have too much in common for us to believe that they are completely unrelated and share no common fundamental properties.

FIGURE 24.21
Confinement of a Jet
The outward-flowing, ionized gas in the jet produces an azimuthal magnetic field that pinches the jet. This magnetic field confines the jet to a narrow range of angles.

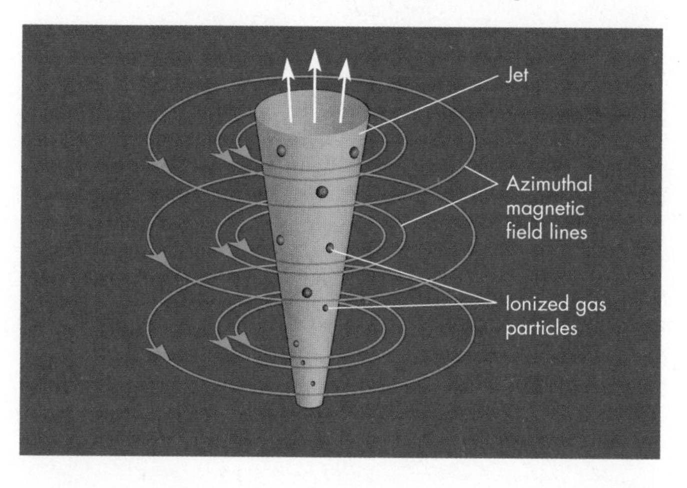

Viewing Angle The direction from which we view an active galactic nucleus can affect what we see in two different ways. First, the inner parts of the active galactic nucleus may be blocked from view by obscuring matter. Figure 24.22 shows that a dust torus lying outside the regions that produce continuous radiation and the broad line region would obscure these regions when viewed near the plane of the torus but leave them open to view when viewed near the axis of the torus. This probably accounts for the apparent difference between Seyfert 1 and Seyfert 2 galaxies (see Figure 24.12). It may also explain the difference between radio-loud quasars and powerful radio galaxies. Viewed near the axis of the torus, the nucleus looks like a quasar. Viewed near the plane of the torus, however, the nucleus looks like a radio galaxy because the bright continuous radiation and broad lines of the quasar can't be seen.

Viewing angle can affect what we see when energetic jets are present, too. If we view a jet pointed nearly in our direction, radiation from the jet will be beamed toward us, making the jet brighter than when it is viewed from other directions. This will result in bright radio emission and optical radiation so bright that it swamps the emission line and thermal radiation produced in the nucleus. Superluminal motion and rapid variability are other possible consequences of viewing a jet almost head-on. The blazars and quasars seem to be the best examples of active galaxies that have jets pointed in our direction. Viewed from other directions, the jets have reduced brightness. In cases where there are two oppositely directed jets, both jets may be visible, as in the case of radio galaxies.

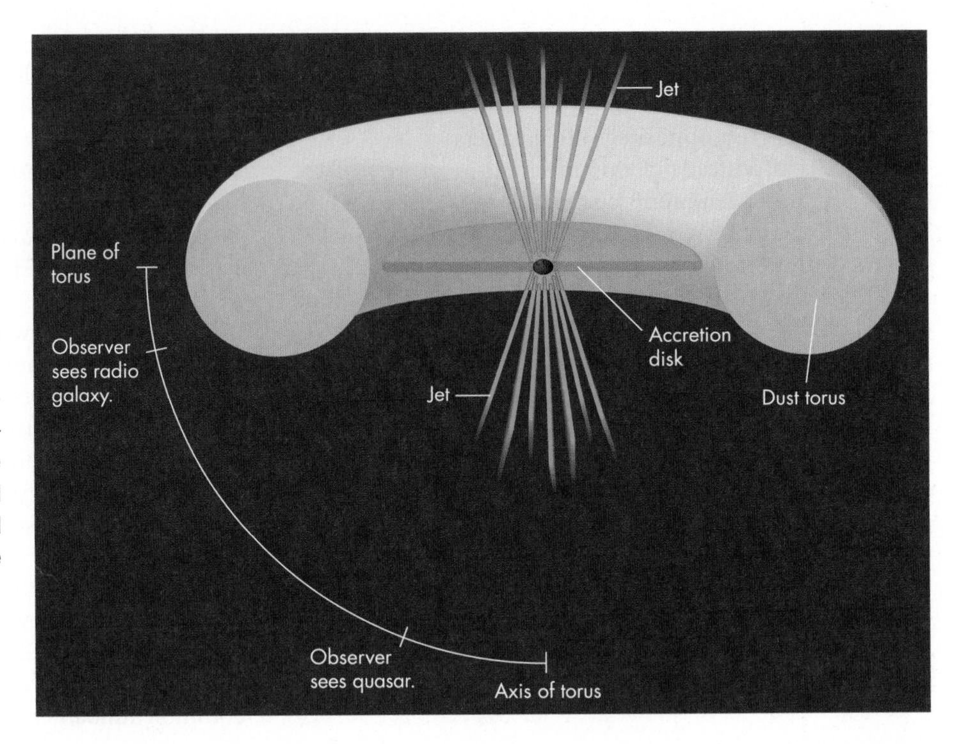

FIGURE 24.22
How the Appearance of Active Galactic Nuclei Depends on Viewing Angle
When viewed near the plane of the torus, a dust ring, or torus, obscures the regions of the nucleus that produce the continuous radiation and broad lines. The jets, furthermore, point across the line of sight. From this angle, the active galaxy appears as a radio galaxy. Seen from near the axis of the torus, the continuous radiation and broad line regions can be seen, and the jets point toward and away from us. The jet pointing toward us is boosted in brightness, and the galaxy looks like a quasar.

Luminosity Different kinds of active galaxies also differ in the amount of radiation emitted by the active nucleus. All the forms of radiant energy ultimately depend on the rate at which the central black hole accretes matter. In descending order of luminosity, the sequence of brightness is quasars, blazars, and Seyfert galaxies. Figure 24.23 shows where different types of active galaxies fall in a plot of viewing angle versus the rate at which matter is accreted by the central black hole. Although many of the different types of active galactic nuclei can be found in this diagram, not all kinds can be conveniently located. The radio-quiet quasars and bright IRAS galaxies, for instance, don't fit very easily into such a simple diagram, although they clearly seem to be related to the rest of the active galactic nuclei.

FIGURE 24.23
The Differences Between the Types of Active Galaxies
Both luminosity and viewing angle affect the way that an active galaxy appears to us and, thus, how we classify it. Viewing angle is measured from the axis of the dust torus.

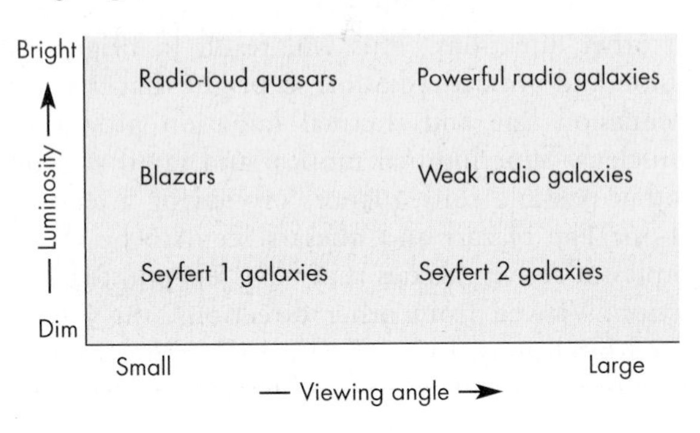

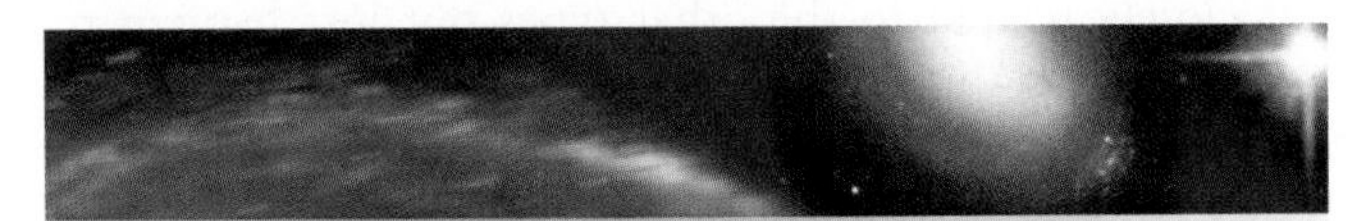

It may be possible to distinguish the many types of active galaxies by their luminosities and the angle from which we view them. The viewing angle controls which parts of the nuclear region we are able to see and also the extent to which the brightnesses of components moving rapidly toward us are increased.

The Host Galaxy

Astronomers have been able to find a number of Seyfert galaxies, blazars, and radio galaxies that are near enough and for which the active nucleus is faint enough that the surrounding host galaxy can be seen. For distant, luminous quasars, it has been difficult to study the surrounding galaxy. For one thing, the most luminous quasars are thousands of times brighter than typical galaxies. Thus the faint fuzz is all but lost in the glare of the central, ultraluminous nucleus. In addition, with increasing distance, the host galaxy shrinks in angular size, so it becomes harder to study its structure or even to see it at all. Finally, the night sky isn't completely black. For distant quasars, the light from the host galaxy is fainter than the night sky and is difficult to detect at all. Observations from space with the *Hubble Space Telescope,* however, have produced the sharp images, like those shown in Figure 24.24, that are needed to clarify many things about the host galaxies of active galactic nuclei.

Galaxy Type Seyfert galaxies are nearly always spirals, blazars are found in elliptical galaxies, and powerful radio galaxies are also ellipticals. Among the quasars, there is evidence that the fuzz around radio-loud quasars originates in elliptical galaxies, whereas the faint fuzz around radio-quiet quasars comes from spiral galaxies. This suggests that radio-quiet quasars are the high-luminosity counterparts of Seyfert galaxies. At present, there is no generally accepted explanation for why active nuclei in elliptical galaxies produce strong radio radiation but those in spiral galaxies do not. Perhaps there is a difference between the magnetic-field configurations in the centers of spiral and elliptical galaxies. If fields in the centers of spirals are weak or disordered, magnetic confinement of a jet would be less effective. The jet would spread out and lead to a low-velocity outflow incapable of producing energies sufficient for synchrotron radiation. Another possibility is that black holes formed earlier in elliptical galaxies than they did in spirals. Because they had more time to accrete matter and to develop in a more crowded environment, black holes in ellipticals are systematically more massive than black holes in spirals. It is not clear, however, just how this would result in a difference in radio radiation.

Interactions In many (perhaps most) cases, nuclear activity in galaxies seems to be triggered by interactions with other galaxies. Seyfert galaxies and the host galaxies of quasars, for instance, often have odd shapes, as if they have been disturbed by recent interactions with other galaxies. An unusually large fraction of quasars, moreover, have nearby companion galaxies with which they may be interacting. There are many ways in which an interaction with another galaxy could either bring material in the active galaxy close to its center or result in the flow of fresh material inward from the other galaxy. The effects of interactions with other galaxies are particularly easy to see in the luminous IRAS galaxies, which may hide quasar-like activity beneath a large amount of dust. Virtually all of the luminous IRAS galaxies are interacting with or in the process of merging with a companion. Presumably, the interaction that enhanced the flow of gas to the center of the IRAS galaxy also led to a wave of star formation and the production of interstellar dust. Perhaps a radio-quiet quasar emerges when the dust disperses.

For many types of active galactic nuclei, the host galaxy can be clearly observed. For some quasars, the surrounding galaxy can be seen as well. Both spiral and elliptical galaxies are hosts to active nuclei. The appearances of many active galaxies are due to interactions with other galaxies.

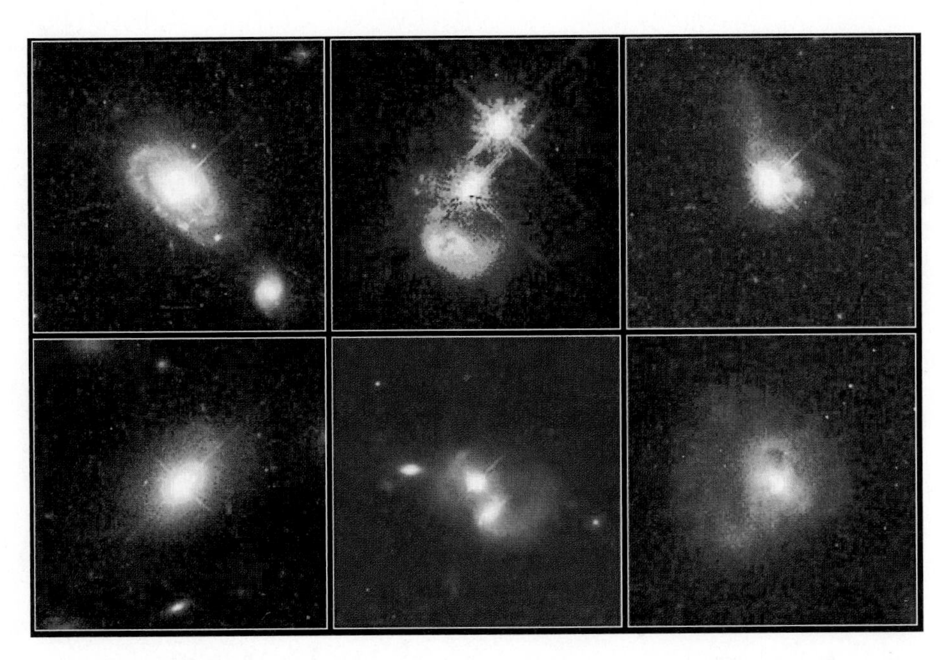

FIGURE 24.24
The Host Galaxies of Quasars
In this set of *Hubble Space Telescope* images the quasars appear as bright spots immersed in galaxies. Some of these galaxies appear to be disrupted by collisions with other galaxies.

24.4 EVOLUTION OF QUASARS

One of the major goals of astronomy is to discover how the universe and its components have changed or evolved with time. A fruitful way to study this evolution is by using the telescope as a time machine. By looking at a typical sample of the universe as it was long ago (and now far away) and by comparing the properties of that sample with the present properties of a typical sample of the universe, we can learn about the evolution that has taken place. To do this, however, we must observe very luminous objects because we must look to great distances to see far back in time. Quasars, much more luminous than normal galaxies, are luminous enough to allow us to use them to investigate more than 80% of the history of the universe. Because quasars are galaxies, we can study them to learn about the evolution of galaxies and perhaps, at very large redshifts, the formation of galaxies.

Evidence for Evolution

If quasars have changed over time, then their properties (such as brightness and the number in a given volume of space) should change with redshift. There are several ways in which we can investigate quasar evolution. One of the most straightforward is the $\boldsymbol{V/V_{max}}$ **test.**

The V/V_{max} Test Suppose, as part of a survey of quasars, that we measure the brightness of one of them. The survey has a detection limit. That is, because of instrumental sensitivity, sky brightness, and other factors, we will be unable to detect quasars fainter than a certain limiting brightness. Knowing that the apparent brightness of a quasar falls with the square of its distance, we can calculate how far away the quasar could be and still be detectable. For example, if the measured brightness of the quasar is one hundred times the detection limit for the survey, we would still be able to see it if it were at ten times its actual distance. We can then calculate the volume of space out to the distance of the detection limit (V_{max}) and compare that with the volume of space out to the actual distance of the quasar (V), as shown in Figure 24.25.

If objects of a given kind are uniformly distributed through space, we would expect to find half of them in the outer half of the maximum volume ($V/V_{max} > 0.5$) and half in the inner half of the maximum volume ($V/V_{max} < 0.5$). If we were to survey white dwarf stars, for example, we would expect to find some of them very close to us, with small values of V/V_{max}, whereas others would be so faint that they would barely be detectable. These faint white dwarfs would have values of V/V_{max} only slightly less than one. For the entire collection of

FIGURE 24.25
V/V_{max} for a Dim Quasar and a Bright Quasar
A, The dim quasar is almost as distant as it could be and still be seen. The volume of space out to the distance of the quasar (V) is almost as large as the volume out to the maximum distance at which an identical quasar could be detected (V_{max}). **B,** The bright quasar could be seen even if it were much more distant than it is. In this case V is only a small fraction of V_{max}.

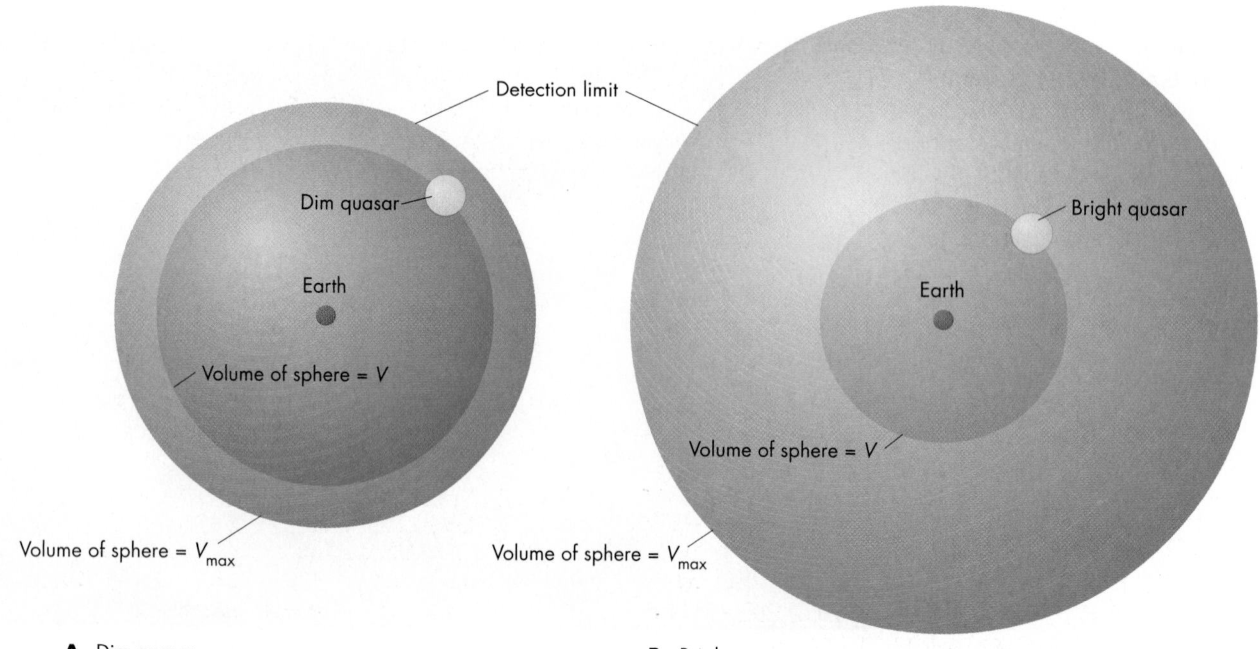

white dwarfs, however, the average value should be about 0.5 because white dwarfs are more or less uniformly distributed throughout the Milky Way. An average value V/V_{max} that is significantly smaller than 0.5 indicates that kind of object is concentrated near us. This could happen, for instance, if we were looking at planets with rings. The only four that we know about (Jupiter, Saturn, Uranus, and Neptune) are all very much brighter than the present detection limit for giant planets and have very small values of V/V_{max}. On the other hand, if the average value of V/V_{max} for some type of astronomical object is significantly greater than 0.5, then we can conclude that such objects tend to avoid our region of space.

Quasars tend to avoid our region of space. The graph in Figure 24.26 shows that the great majority of the quasars are in the outer half of the surveyed volume (V/V_{max} greater than 0.5), indicating that they are more common at great distances than they are nearby. Actually, it is time rather than distance that affects the distribution of quasars. To see very many of them, we need to look to very great distances, which means that we need to look very far back in time. What Figure 24.26 really shows is that quasars were more common in the past than they are today. This is powerful evidence that quasars have evolved and that the universe has changed during the more than 10 billion years that quasars have existed.

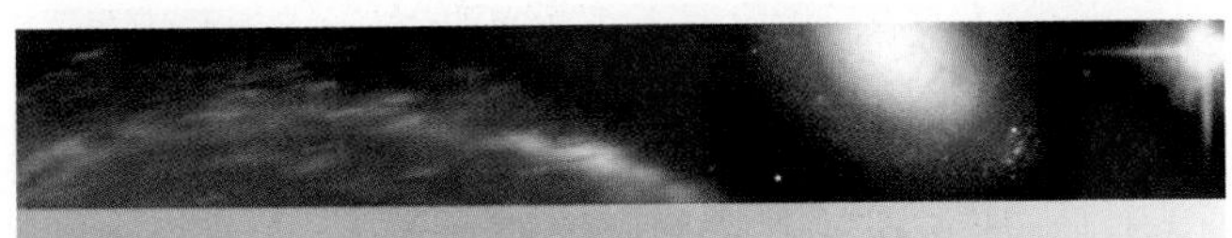

When we apply the V/V_{max} test to quasars, we find that most of them are nearly as distant as they could be and still be detected. This shows that they were more common in the past than they are today and suggests that quasars have evolved and the universe has changed with time.

Luminosity Functions When we describe the evolution of quasars, we must take into account their great range of luminosities. It isn't enough to just count the quasars that can be seen at various distances, because the faint ones would be undetectable at the greater distances. Instead, we need to determine the space density (the number that can be found in a given volume of space) as a function of luminosity. This is often expressed as the number in a cubic megaparsec brighter than a given absolute magnitude. Such a distribution is called a **luminosity function.** The luminosity functions for quasars about 10 billion years ago (seen at redshift of $z = 2$), quasars about 6 billion years ago (seen at $z = 0.5$), and Seyfert galaxies, which may be the modern counterparts of quasars, are shown in Figure 24.27. Brightness increases

FIGURE 24.26
***V/V_{max}* for a Survey of Quasars**
Most of the quasars have V/V_{max} of 0.5 or greater, indicating that they are nearly as distant as they could be and still be seen. This means that there were more of them long ago than there are today. Thus, the V/V_{max} test shows that quasars have evolved over time.

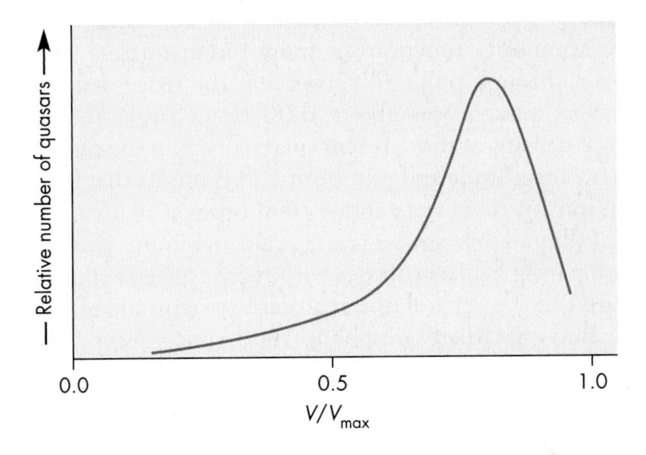

FIGURE 24.27
The Luminosity Functions for Seyfert Galaxies and Quasars
Space density is the number of Seyfert galaxies or quasars per cubic Mpc brighter than a given absolute magnitude. The vertical line at an absolute magnitude of −27 connects the space densities of ancient ($z = 2.0$) quasars (*A*), and more recent ($z = 0.5$) quasars (*B*), and shows that ancient quasars were about one thousand times more numerous. The vertical line at an absolute magnitude of −24 connects the space densities of recent quasars (*C*) and Seyfert galaxies (*D*) and shows that recent quasars were about one thousand times more numerous than Seyfert galaxies for a given brightness (absolute magnitude).

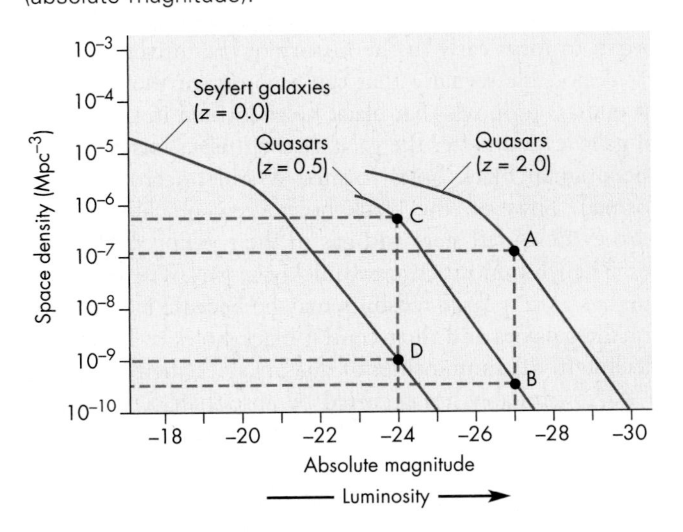

to the right in Figure 24.27, so the luminosity functions show that bright quasars and Seyfert galaxies are less numerous than dim ones.

One way to interpret these luminosity functions is to look along a vertical line, such as the one at absolute magnitude of -27 in Figure 24.27. The fact that the luminosity function for ancient quasars (which passes through point *A*) is almost a factor of 1000 higher than the luminosity function for more recent quasars (which passes through point *B*) shows that the space density of ancient quasars was about 1000 times larger than the space density of more recent quasars having the same absolute magnitude and luminosity. This means that for any luminosity, there were about 1000 times as many quasars 10 billion years ago as there were 6 billion years ago. Comparing Seyfert galaxies with recent quasars (points *C* and *D* on the vertical line at absolute magnitude of -24), we find that there are about 1000 times fewer Seyfert galaxies for a given brightness than there were quasars 4 billion years ago. The comparison of luminosity functions suggests that quasars steadily declined in brightness as time passed and, as a result, their luminosity function slid slowly horizontally toward the left in the diagram. The brightness of an average quasar declined by about 2.5 magnitudes (a factor of 10 in brightness) between 10 and 6 billion years ago and then dropped another 3 magnitudes or so between 6 billion years ago and today. The quasars of today, according to this evolutionary picture, are so dim that they don't overwhelm their host galaxy. We see them as Seyfert galaxies.

When Did Quasars Form? Although the space density of quasars of a given luminosity increases with redshift (and lookback time) to $z = 2$, this trend apparently does not continue at higher redshifts. There have been many searches for quasars with redshifts greater than 3, but relatively few have been found. The space density for quasars seems to have reached a peak at about $z = 2.5$, when the universe was about one-sixth of its present age. Some luminous quasars, however, can be found at redshifts greater than 6, showing that at least some quasars began to form early in the history of the universe.

A possible scenario that can account for the evolution of quasars proposes that black holes formed in the centers of galaxies soon after the galaxies themselves formed. How the original black holes formed is still uncertain. Once formed, however, the black holes grew quickly in mass as they devoured stars and gas in their young host galaxies. Their luminosities grew quickly as well. The dearth of quasars at very large redshifts may be because it took time for the galaxies and their central black holes to form. The decline in the luminosities of quasars since their heyday at $z = 2.5$ probably has occurred because their central black holes have gradually depleted their surroundings of available gas. If this view is correct, then there should be massive black holes in the centers of many galaxies. In most cases, these black holes may be accreting matter at such a small rate that they emit only weak versions of the continuous radiation and emission lines that characterize quasars.

Where Are the Dead Quasars?

There are two types of evidence that massive black holes—dormant quasars—can be found at the centers of many galaxies that are not active enough to be classed as quasars or even Seyfert galaxies. The first is that many elliptical and spiral galaxies have regions in their nuclei that emit weak, broad emission lines and weak synchrotron radiation. These galaxies may contain "starved monsters," or black holes that accrete only a very small supply of the infalling gas that powers them.

Also, several dozen nearby galaxies, including the Milky Way, have been found to have massive bodies, very likely black holes, at their centers. The evidence for black holes in nearby galaxies is discussed in Chapter 23, Section 2. If the black holes in nearby galaxies are, in fact, dead quasars, they must be accreting matter at a very slow rate to keep them from emitting conspicuous quasarlike radiation. This is probably because the black holes long ago depleted the gas in their vicinities by accreting it or by blowing it outward due to the pressure of the intense radiation emitted by their accretion disks.

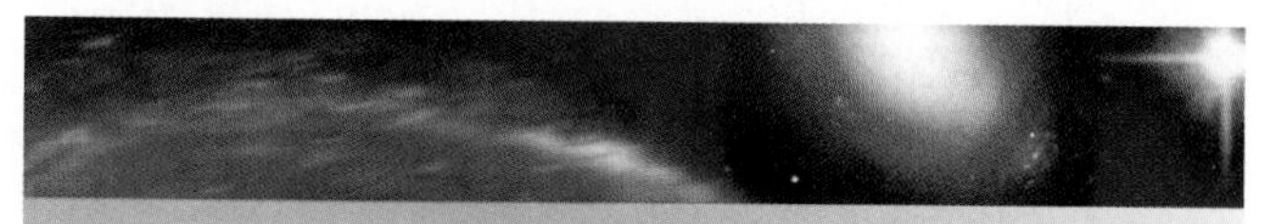

A quasar luminosity function describes the density in space of quasars of different luminosities. Comparison of luminosity functions for quasars at different redshifts suggests that quasars have dimmed by more than one hundred times since they formed. The existence of weak quasarlike activity in the cores of many galaxies suggests that dormant quasars (massive black holes) may reside in many normal galaxies today.

24.5 QUASARS AS PROBES OF THE UNIVERSE

Because the light from quasars passes through billions of parsecs of space before we see it, the light can serve as a probe of the matter that lies between us and the quasar. The existence of intervening matter has been detected through both absorption lines and gravitational lensing.

Absorption Lines

If the light from a quasar passes through a galaxy or an intergalactic gas cloud, the atoms, molecules, and ions in

FIGURE 24.28
The Visible Spectrum of the Quasar UM 402
The redshift of UM 402 ($z = 2.86$) places the Lyman α emission line of hydrogen from the quasar at a wavelength of about 470 nm. At shorter wavelengths, the spectrum is broken up by numerous absorption lines in the Lyman α forest.

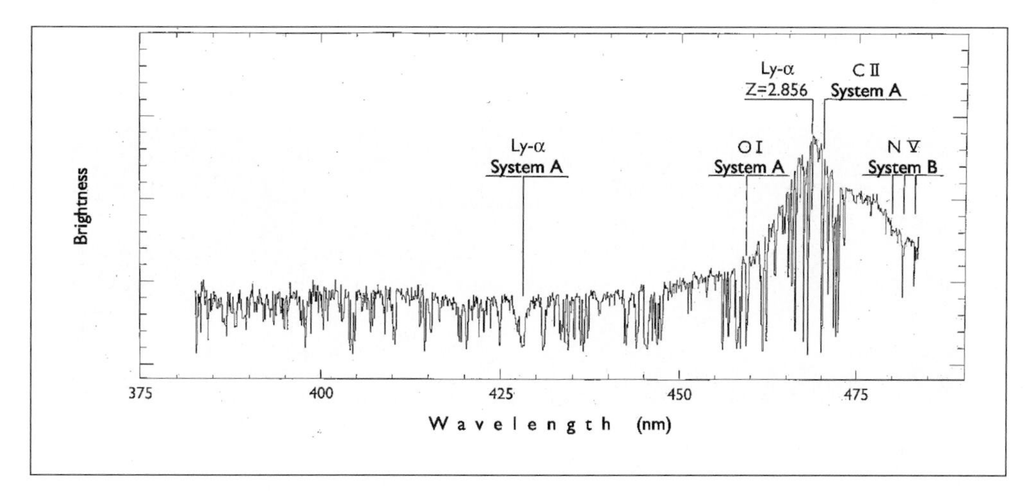

the galaxy or cloud can absorb the quasar's light and produce absorption lines. If the light from the quasar passes through more than one galaxy or intergalactic cloud, multiple absorption lines are produced.

Although absorption lines of many elements have been found in the spectra of quasars, the Lyman α absorption lines of hydrogen are by far the most common. The spectral region covering wavelengths just shorter than the Lyman α emission line of a quasar is often broken up by many narrow Lyman α absorption lines that make up what has been called the **Lyman α forest,** shown in Figure 24.28.

Some of the Lyman α absorption lines are formed by gas clouds that also produce absorption lines of other elements. This gas may lie within galaxies, where stars have processed hydrogen into heavier elements and then ejected the enriched gas back into space. For other Lyman α absorption lines, however, there are no absorption lines that correspond to other elements, probably because there are not enough atoms of the other elements in the cloud to produce absorption lines. In some cases, the heavy elements are at least a thousand times less abundant in the absorbing gas cloud than in the Sun. This gas probably lies in intergalactic space and has never been a part of any galaxy.

The clouds that produce Lyman α absorption lines in the spectra of quasars seem to have evolved in a way similar to that of the quasars. That is, the number of absorption lines produced by the clouds increases with increasing redshift. This implies that the clouds responsible for the absorption lines were more numerous in the past than they are today. Thus the quasar absorption lines suggest that the amount of primordial material left in intergalactic space has decreased as the universe grew older. Presumably this has happened because intergalactic material gradually has been gathered into galaxies.

Absorption lines in the spectra of quasars arise when the light from a quasar passes through clouds of gas that lie between the quasar and us. Some of the lines appear to be produced by intergalactic clouds that were more numerous in the time of quasars than they are today.

Gravitational Lenses

One of the predictions of Einstein's general relativity is that light can be deflected when it passes near matter. This prediction was confirmed in 1919 when the deflection of starlight by the Sun was observed during a total eclipse. The deflection of a beam of light passing near a concentration of mass is shown in Figure 24.29. The angle through which a beam is deflected can be quite large for beams that pass close to massive objects. Just as an optical lens is made by shaping a piece of glass so that its thickness varies from center to edge, a **gravitational lens** results when the mass of a galaxy (or cluster of galaxies) decreases from the center outward. The background object can be shifted in direction, its appearance distorted, and its brightness enhanced by passing through a gravitational lens. Gravitational lenses are responsible for the microlensing events discussed in Chapter 22, Section 3, and the lensing by clusters of galaxies described in Chapter 25.

The first gravitational lens was discovered in 1979 when observers found two quasars, only 6 seconds of arc apart in the sky, that had the same redshifts and similar absorption and emission lines. Radio observations show

FIGURE 24.29
The Gravitational Deflection of Light
Light from a distant quasar is bent by a cluster of galaxies near which it passes. The more massive the cluster of galaxies and the nearer the beam passes to the center of the cluster, the greater the angle θ through which the beam is bent.

Gravitational lensing

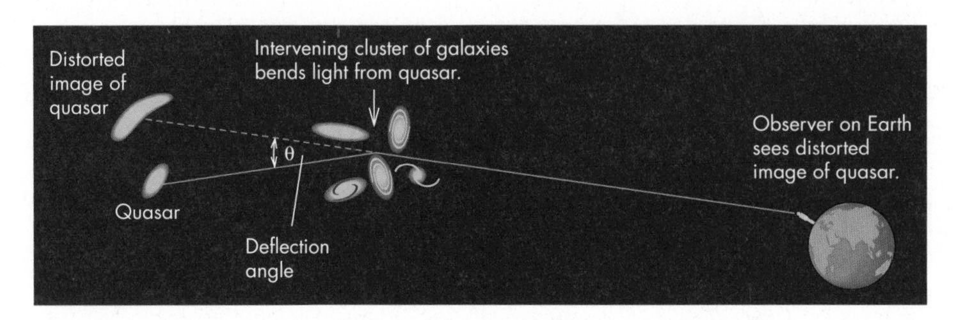

FIGURE 24.30
A Radio Map of an Einstein Ring
If a background galaxy or quasar lies almost perfectly in line with a galaxy, the image of the background source of radiation can be distorted into a ring.

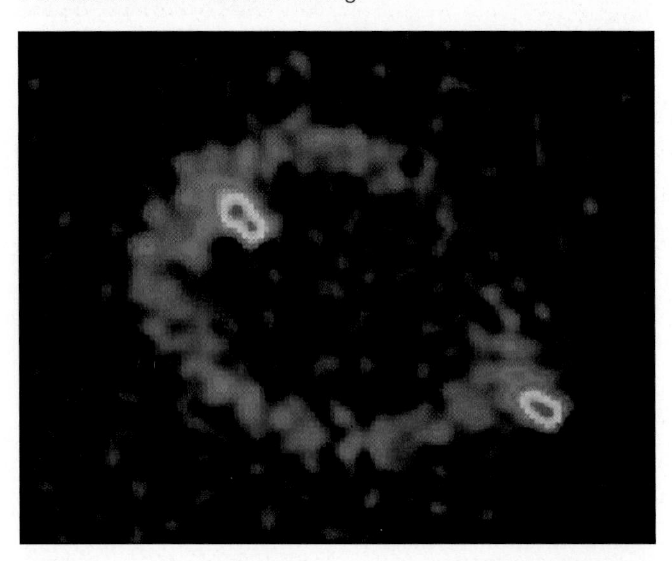

that the two quasars both have jets that point in nearly the same directions and are at nearly the same distances from the cores of the quasars. The resemblance of the two quasars was immediately recognized as too great to be coincidental. They are two images of the same quasar. After a search, the lensing galaxy, a giant elliptical, was found near one of the images. Since the initial discovery, more pairs of gravitationally lensed quasar images have been found. When a quasar lies almost precisely behind the gravitational lens, the quasar is distorted into a ring or near ring rather than a pair of images. This effect, called an **Einstein ring,** is shown in Figure 24.30.

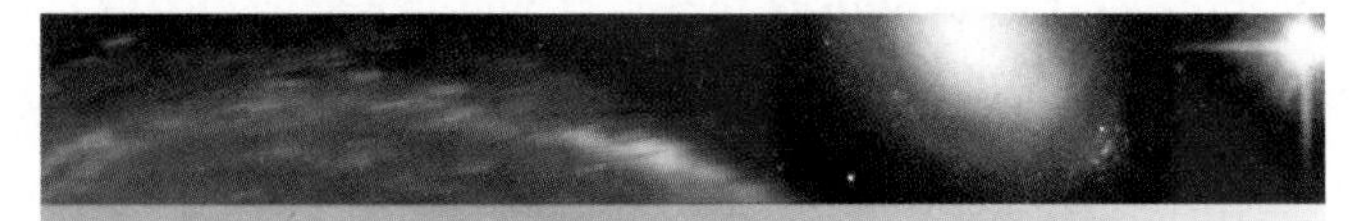

The masses of galaxies produce gravitational lenses that distort the images of background quasars and galaxies. A gravitational lens can produce multiple images of a distant quasar or a bright ring around the massive galaxy.

Chapter Summary

- The large redshifts of quasars are a result of the expansion of the universe. Quasars are very distant, so the light we are now receiving has traveled a long time to reach us. We see quasars as they were when the universe was many billions of years younger than it is today. (Section 24.1)
- A typical quasar is much more luminous than an entire galaxy. Most of the radiant output of a quasar is continuous radiation that is a combination of synchrotron radiation and emission from gas and dust. (24.2)
- Ionization of gas near the center of a quasar by ultraviolet radiation and X rays produces emission lines. The broad and narrow emission lines are produced in different regions of the quasar. (24.2)

- Quasars can vary in brightness in as little as half a day. Variations this fast show that the radiation the quasars emit must be produced in a region as small as or smaller than our solar system. (24.2)
- In many cases, radio maps of quasars show components that seem to be separating faster than the speed of light. This superluminal motion is probably due to an illusion that occurs when light is emitted by something moving almost directly at us at a speed only slightly less than that of light. (24.2)
- Radio galaxies have huge lobes of radio emission fed by high-speed jets of matter from the heart of the central galaxy. In some cases, radio galaxies have two lobes, but only a single jet can be seen. This occurs because a jet pointed at us looks brighter and a jet pointed away looks dimmer than either would if they were viewed from the side. (24.2)
- There are many kinds of active galactic nuclei, including Seyfert galaxies, radio galaxies, blazars, and starburst galaxies. All of these share some properties with quasars but are different in other important respects. (24.2)
- Quasars probably produce so much energy in such small regions because they contain massive black holes that accrete surrounding gas. Black holes produce energy very efficiently from the matter that falls into them. (24.3)
- The intense radiation in quasars would blow away infalling gas if the gravity in quasars weren't large enough to overcome the outward force of radiation. As a result, the cores of quasars must contain 100 million $M_{\odot}$ or more of matter. (24.3)
- Because it has angular momentum, matter approaching a black hole forms an accretion disk. Friction in the accretion disk releases considerable energy, which flows to the surface of the disk and is then radiated away. The surface of the accretion disk is about as hot as a star. (24.3)
- It may be possible to distinguish the different kinds of active galactic nuclei on the basis of luminosity and the angle from which we view them. Viewing angle determines whether the center of the active nucleus is hidden by an obscuring torus and whether jets of outflowing matter point nearly at us. If they do, they appear increased in brightness. (24.3)
- The host galaxies of many active galactic nuclei, including the nearer quasars, can be detected. The host galaxies are often distorted because of interactions with other galaxies. The host galaxies of distant quasars can't be seen because they appear too small and faint. (24.3)
- The density of quasars is larger at great distances than it is in our part of the universe. This shows that quasars were more numerous in the past than they are today. (24.4)
- A luminosity function gives the density of quasars of different brightness. Comparing luminosity functions for distant and relatively near quasars suggests that quasars have grown about one hundred times dimmer since they formed. Weak quasarlike activity in nearby galaxies may indicate that it is common for galaxies to have dormant quasars in their centers. (24.4)
- Absorption lines in the spectra of quasars are produced when light from the quasar passes through clouds of gas between the quasar and us. Some of the absorbing clouds appear to be intergalactic clouds, which were more numerous in the time of quasars than they are today. (24.5)
- The images of distant quasars and galaxies are sometimes distorted by gravitational lensing caused by intervening galaxies. The lens can produce multiple images of the background galaxy or quasar or arcs or rings of light around the massive intervening galaxy. (24.5)

Key Terms

active galactic nucleus *581*
active galaxy *581*
blazar *585*
broad line region *578*
Eddington luminosity *587*
Einstein ring *596*
gravitational lens *595*
jet *583*
lookback time *575*
luminosity function *593*
Lyman α forest *595*
narrow line region *578*
quasar *573*
radio galaxy *582*
Seyfert galaxy *581*
starburst galaxy *585*
superluminal motion *580*
V/V_{max} test *592*

Conceptual Questions

1. What is it about the colors of quasars that makes it easy to distinguish them from most stars and galaxies?
2. Compare the narrow line and broad line regions of quasars with respect to density and the amounts of material required to produce the emission lines they emit.
3. What evidence do we have that quasars are a few parsecs or smaller in size?
4. Suppose a quasar is observed to increase tenfold in brightness in 4 years. Approximately how large can the quasar be?
5. Describe how we can account for the superluminal motions of quasars without requiring components of quasars that are actually moving apart at speeds greater than the speed of light.
6. Describe the differences between Seyfert 1 and Seyfert 2 galaxies.
7. How can viewing angle account for the differences between the two types of Seyfert galaxies?
8. How do electrons and the energy they carry get into the lobes of radio galaxies?
9. What is the relationship between jets and the lobes of radio galaxies?
10. How could a radio galaxy be observed to have two radio-emitting lobes but only a single jet pointing at one of the lobes?
11. How are blazars similar to quasars, and how are they different?
12. How can viewing angle account for the differences between quasars and blazars?
13. How do we know that if black holes power quasars, then those black holes must be very massive?
14. Explain why, if objects of a particular kind are uniformly distributed through space, we should expect to find about half of them with a V/V_{max} less than 0.5 and half with a V/V_{max} greater than 0.5.
15. What is the significance of the observation that the average value of V/V_{max} for quasars is greater than 0.5?
16. How have quasars evolved over the past 10 billion years?
17. What evidence do we have that there was once a time when there were few, if any, quasars?
18. What evidence do we have that there may be dead quasars in the centers of normal galaxies today?
19. Under what circumstances can a gravitational lens produce an Einstein ring rather than multiple images of a distant quasar?

Problems

1. What is the Eddington luminosity of a quasar with a mass of 1 billion $M_{\odot}$?
2. Suppose a quasar has a luminosity of 10^{12} $L_{\odot}$. What is the smallest mass that the quasar could have and still continue to pull surrounding gas into itself?

Figure-Based Questions

1. Use Figure 24.3 to find the redshift of a galaxy at a distance of 3000 Mpc.
2. Use Figure 24.3 to find the distance of a galaxy that has a redshift of 1.0.
3. Use Figure 24.4 to find the ratio of lookback time to the age of the universe for a galaxy with a redshift of 1.0.
4. Use Figure 24.4 to find the redshift of a galaxy that has a ratio of lookback time to the age of the universe of 0.6.
5. Use Figure 24.8 to find the rate (in milliarcseconds per year) at which components *D* and *C2* of the quasar 3C 345 appear to be moving apart. How long would it take for them to reach a separation of 0.1 second of arc?
6. Use Figure 24.10 to find the apparent speed of a component of a quasar that is moving at 90% of the speed of light in a direction that makes an angle of 20° to the line of sight to the quasar.
7. Suppose two components of a quasar seem to be separating at four times the speed of light but that the real speed with which they are moving apart is 98% of the speed of light. Use Figure 24.10 to find the angle that the motion of one of the components of the quasar makes to the line of sight to the quasar. (There are two answers to this question.)

8. If the general picture of active galaxies presented in Figure 24.23 is correct, what would we call a low-luminosity active galaxy that we see at a large viewing angle?
9. If the general picture of active galaxies presented in Figure 24.23 is correct, what would we call an intermediate-luminosity active galaxy that we see at a small viewing angle?
10. Use Figure 24.26 to find the value of V/V_{max} for which the greatest number of quasars are found.
11. Use Figure 24.27 to find the ratio of the space densities of old quasars and more recent quasars for an absolute magnitude of −24.

Group Activity

Working as a group, try to come up with at least three explanations for the apparent superluminal motions of quasars. Ask your instructor to discuss with you the problems with the various explanations you come up with.

For More Information

Visit the text website at **www.mhhe.com/fix** for chapter quizzes, interactive learning exercises, and other study tools.

25

Our knowledge is a torch of smoky pine
That lights the pathway but one step ahead
Across a void of mystery and dread.

George Santayana
O World, Thou Choosest Not the Better Part

The galaxy cluster Abell 50740.

www.mhhe.com/fix

Galaxy Clusters and the Structure of the Universe

Two of the recurring themes in science are the search for organization and structure in the universe and the explanation of that structure once it is found. The earliest Greek astronomers sought the organization of the solar system by studying the Sun, the Moon, and the planets. Two thousand years later, astronomers attempted to discover the structure of the Milky Way. Early attempts failed because astronomers couldn't see enough of the Milky Way for the pattern to be clear. They only succeeded when they were able to use new kinds of instruments and techniques to extend the range of their observations far enough that the spiral pattern became visible.

This chapter is about the structure of the universe on a much larger scale than the Milky Way. As with the Milky Way, our knowledge of the structure of the universe has emerged only as new telescopes and techniques have pushed the limits of the observed universe steadily farther outward. In the first few decades of the twentieth century, astronomers learned that the galaxies were grouped in clusters. They began to wonder whether the clusters were organized into superclusters, the superclusters into still larger structures, and so on. Only in the past 40 years have observations permitted astronomers to see enough of the universe that the organization of the galaxies into structures larger than clusters became apparent. The entire pattern is not yet clear, so observational work is under way to clarify the picture of the large-scale structure. Similarly, astronomers are developing and testing new theories to explain the emerging observational picture.

Questions to Explore

- To what extent are galaxies found in clusters rather than in isolation?
- Are galaxies in clusters different from galaxies not in clusters?
- How did clusters of galaxies form?
- What evidence do we have that there is dark matter between the galaxies in clusters?
- How are galaxies and clusters of galaxies distributed in space?
- Why are there enormous voids in which there are few galaxies?

25.1 CLUSTERS OF GALAXIES

Although some galaxies are found alone, without any companions, most are found in **clusters of galaxies.** Clusters of galaxies range from small groups that contain barely enough mass to hold themselves together to dense clusters with thousands of galaxies. The tendency of galaxies to group together can be seen in Figure 25.1, which shows the distribution of over a million bright galaxies in the northern sky. Notice that most of the galaxies lie in a web that encloses less densely populated regions.

Classifying Clusters

Astronomers have invented a number of ways to describe clusters of galaxies. The **richness** of a cluster is a measure of the number of galaxies it contains. Richness is found by counting the number of bright galaxies in the cluster. Only the bright galaxies are counted, rather than all the visible galaxies in a cluster, because faint galaxies can be seen only in nearby clusters. If richness were measured by counting all the visible galaxies, then a nearby cluster would appear to be richer than an identical cluster too distant for its faint galaxies to be detected. For all clusters, the total number of galaxies is sure to be considerably larger than the number bright enough to be seen.

Clusters of galaxies can be classified by their shapes, too. **Regular clusters,** for instance, appear round. The galaxies in regular clusters are concentrated in the center of the cluster. **Irregular clusters,** on the other hand, lack any overall symmetry. They may be made up of several clumps of galaxies. It turns out that the richest clusters tend to be regular, whereas less rich clusters tend to be irregular, so these two classification schemes are related.

The Local Group The Milky Way is part of the **Local Group,** a small cluster of galaxies shown in Figure 25.2. We know of about 50 members of the Local Group. There may be additional members (perhaps many) that have not been found because they are obscured by interstellar dust in our galaxy or because they are just too dim. New members of the Local Group are discovered regularly. In fact, the nearest known neighbor of the Milky Way, the Canis Major Galaxy, wasn't discovered until 2004. The Canis Major Galaxy is located only about 13 kpc from the center of the Milky Way. It is so close to the Milky Way that it is being pulled apart by tides and has already been stretched into long streams of stars that loop several times around the Milky Way. The stars in the Canis Major Galaxy will eventually be absorbed into the disk of the Milky Way. The Canis Major Galaxy escaped detection until 2004 because it lies behind the plane of the Milky Way and was obscured by dust. A number of other faint dwarf galaxies that are satellites of the Milky Way have been discovered in the past few years. Some contain only a few million stars. Astronomers suspect that dozens of other faint companions of the Milky Way still await discovery.

The size of the Local Group is difficult to define. A few galaxies that lie within 1 or 2 Mpc of the Milky Way may be members of the Local Group or just passersby. The diameter of the Local Group is usually considered to be roughly 1 Mpc. Except for the few small galaxies just outside its boundaries, the Local Group is nearly alone. The space between the Local Group and the nearest clusters of galaxies, 3 to 5 Mpc away, is almost devoid of other galaxies.

The Local Group contains all of the main types of galaxies. Two spiral galaxies, the Milky Way and the Andromeda Galaxy, shown in Figure 25.3, dominate the Local Group. Together they emit almost ten times as much light as all the other galaxies in the Local Group combined. Most of the visible mass (in the form of stars and HII regions) in the Local Group is found in these two galaxies. The Milky Way and the Andromeda Galaxy are approaching each other at 300 km/s. It is likely that the two galaxies will have a close encounter in about 3 billion years and merge a few billion years after that. The third brightest galaxy in the Local Group, M33, is also a spiral. M33, shown in Figure 25.4, is barely visible to the unaided eye. It is seen nearly face-on so we have a good view of its spiral structure.

FIGURE 25.1
A Million Galaxies
This plot, covering the entire sky, shows the directions of over a million bright galaxies. The distribution of galaxies has a pattern of filamentary concentrations of galaxies separated by patches with many fewer galaxies.

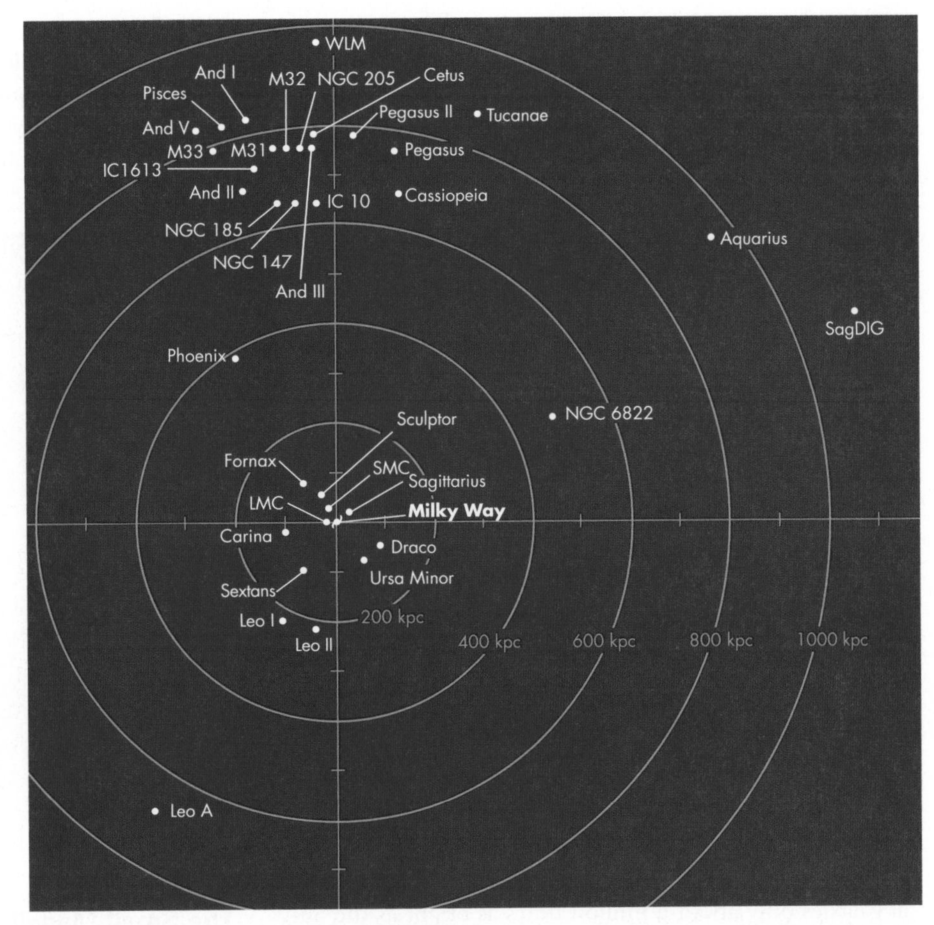

FIGURE 25.2
The Local Group of Galaxies
The Local Group consists mainly of small groups around the Milky Way and the Andromeda Galaxy. The cluster is about 1 Mpc across.

FIGURE 25.3
The Andromeda Galaxy M31 and Its Companions
M31 and the Milky Way, both spiral galaxies, are the dominant members of the Local Group. The two bright spots above and below M31 are its satellite galaxies, M32 and NGC 205. The satellites of M31 are the brightest and largest of the elliptical galaxies in the Local Group.

FIGURE 25.4
M33, a Spiral Galaxy in the Local Group
M33 is the smallest and dimmest of the three large spirals in the Local Group. It lies at a distance of about 800 kpc.

FIGURE 25.5
The Leo I Dwarf Elliptical Galaxy
This member of the Local Group is about 1 kpc across and lies at a distance of 200 kpc. It is one of a dozen or so known galaxies that are satellites of the Milky Way.

Although there are more than a dozen elliptical galaxies in the Local Group, they are all relatively small and dim. Unlike many richer clusters, the Local Group does not contain any large elliptical galaxies. No elliptical galaxy in the Local Group is more than 1% as bright as the Andromeda Galaxy. The brightest and largest of these elliptical galaxies, shown in Figure 25.3, are both companions of the Andromeda Galaxy. The rest, like Leo I (Figure 25.5), are dwarf elliptical galaxies only about a million times as bright as the Sun and a few kiloparsecs in diameter. The other galaxies in the Local Group are irregulars. The two brightest irregulars are the Large Magellanic Cloud (Figure 25.6) and the Small Magellanic Cloud (Figure 25.7). Both of the Magellanic Clouds are companions of the Milky Way. Other, dimmer, irregular galaxies are distributed throughout the Local Group. One of these, NGC 6822, is shown in Figure 25.8.

As is the case within the Milky Way, luminous matter seems to make up only a small fraction of the total mass of the Local Group. Given the velocities with which they are moving, the gravity of the galaxies of the Local Group would not be great enough to hold the cluster together. The fact that the Local Group has remained clustered together for many billions of years suggests that there must be a large amount of dark material within the boundaries of the cluster. The stability of clusters other than the Local Group also seems to depend on the presence of large amounts of unseen matter.

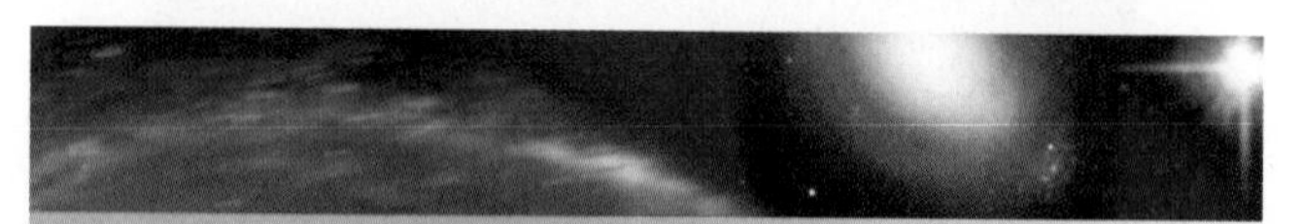

Galaxies cluster together. Clusters of galaxies are classified by the number of bright galaxies they contain and by the regularity of their appearance. The Milky Way is a member of a small cluster of galaxies called the Local Group.

FIGURE 25.6
The Large Magellanic Cloud
The Large Magellanic Cloud is an irregular galaxy about 50 kpc from the Milky Way. The reddish regions are locations of active star formation. The Large Magellanic Cloud contains about 10 billion $M_{\odot}$ of material.

FIGURE 25.7
The Small Magellanic Cloud
The Small Magellanic Cloud is an irregular galaxy about 60 kpc from the Milky Way. The Small Magellanic Cloud is only about half as large and contains only about one-tenth the mass of the Large Magellanic Cloud. Compared with the Large Magellanic Cloud, there is little star formation going on in the Small Magellanic Cloud.

FIGURE 25.8
NGC 6822, an Irregular Galaxy in the Local Group
NGC 6822 lies at a distance of about 600 kpc and is 2 kpc across. NGC 6822 is far from the other galaxies of the Local Group and does not appear to have been disrupted by the gravitational attraction of other galaxies. The reddish regions within and around NGC 6822 are HII regions.

The Virgo Cluster The Virgo cluster, about 20 Mpc away, is the nearest great cluster of galaxies. An image of part of the Virgo cluster is shown in Figure 25.9. The cluster, so large that it covers an area more than 10° across, contains more than one thousand galaxies. Fifteen of these galaxies are conspicuous enough to have made Charles Messier's original list of nebulae. The brightest members of the Virgo cluster include both spiral galaxies, such as M100 (Figure 25.10), and giant ellipticals, such as M87 (Figure 25.11). A massive black hole may lurk at the center of M87. As Figure 25.12 shows, *Hubble Space Telescope* images reveal a spiral-shaped disk of gas about 150 pc across in the core of M87. Spectra obtained from parts of the disk 20 pc on either side of the center of M87 show that the orbital speed of gas at that distance is 550 km/s. An unseen concentration of 2 to 3 billion $M_\odot$ of matter, likely to be in the form of a massive black hole, is required to explain such rapid orbital motion.

More galaxies have been discovered in the Virgo cluster than in any other cluster of galaxies, but this is mostly because Virgo is much closer than most other clusters. Most of the galaxies in the Virgo cluster are dwarf ellipticals that would not have been found if the cluster were farther away. The relatively small number of bright galaxies in Virgo means that it is classified as "poor" rather than "rich." Furthermore, it lacks symmetry and a central concentration, so it is also classified as irregular.

The Virgo cluster is moving away from the Local Group at 1000 km/s, as part of the expansion of the universe. It would be receding at 1200 to 1300 km/s, however, if it weren't for the gravitational attraction between the

FIGURE 25.9
The Central Part of the Virgo Cluster
Both spiral and giant elliptical galaxies are found in the central part of the Virgo cluster. The two giant ellipticals that mark the center of the Virgo cluster are M84 (on the right) and M86. (Color image creation by Jean-Charles Cuillandre, The Canada-France-Hawaii Telescope Corp. Copyright © 2001 CFHT.)

FIGURE 25.10
M100, a Large Spiral Galaxy in the Virgo Cluster
Long-exposure images of M100 show that there is a tongue of matter extending from M100 toward the small elliptical galaxy (top left) with which it is interacting.

FIGURE 25.11
M87, a Giant Elliptical Galaxy in the Virgo Cluster
Most of the points of light around M87 are globular clusters. There are four thousand globular clusters in M87.

Virgo cluster and the Local Group, which pulls the two together at a speed of 200 to 300 km/s. Even though the average speed at which the Virgo cluster is receding is about 1000 km/s, the radial velocities of individual galaxies vary widely. The range of radial velocities is partly due to the orbital motions of the galaxies about the center of mass of the cluster. It also arises, however, because some galaxies are still streaming inward as the cluster adds new galaxies to itself.

Other clusters are sprinkled through space with typical separations of tens of megaparsecs. A typical cluster of galaxies, the Fornax cluster, is shown in Figure 25.13. The Fornax cluster is located about 20 Mpc from the Local Group.

The Coma Cluster The Coma cluster, shown in Figure 25.14, lies at a distance of about 90 Mpc and is an extremely rich, regular cluster of galaxies. It is richer than 95% of the known clusters of galaxies and contains several hundred bright galaxies. Unlike the Virgo cluster, which contains many bright spiral galaxies, there are virtually no spiral galaxies in the central region of the Coma cluster. Instead, the region contains about equal numbers of elliptical and S0 galaxies. The only spirals in the Coma cluster are located at the fringes of the cluster.

Our sky would look very different if the Earth were located in the core of the Coma cluster. Galaxies are much closer together in Coma than in the Local Group, so in many directions there would be galaxies as close or closer to us than the Magellanic Clouds. On a dark night

FIGURE 25.12
Evidence for a Massive Black Hole in M87
Hubble Space Telescope images of M87 show that there is a spiral-shaped disk of gas orbiting in the center of the galaxy. Two regions about 20 pc on either side of the center of M87 have spectra that show that gas in the regions is orbiting the center of M87 at a speed of 550 km/s. Two to three billion $M_\odot$ of matter is required to keep the gas in orbit. The mass may be concentrated in a massive black hole at the center of M87.

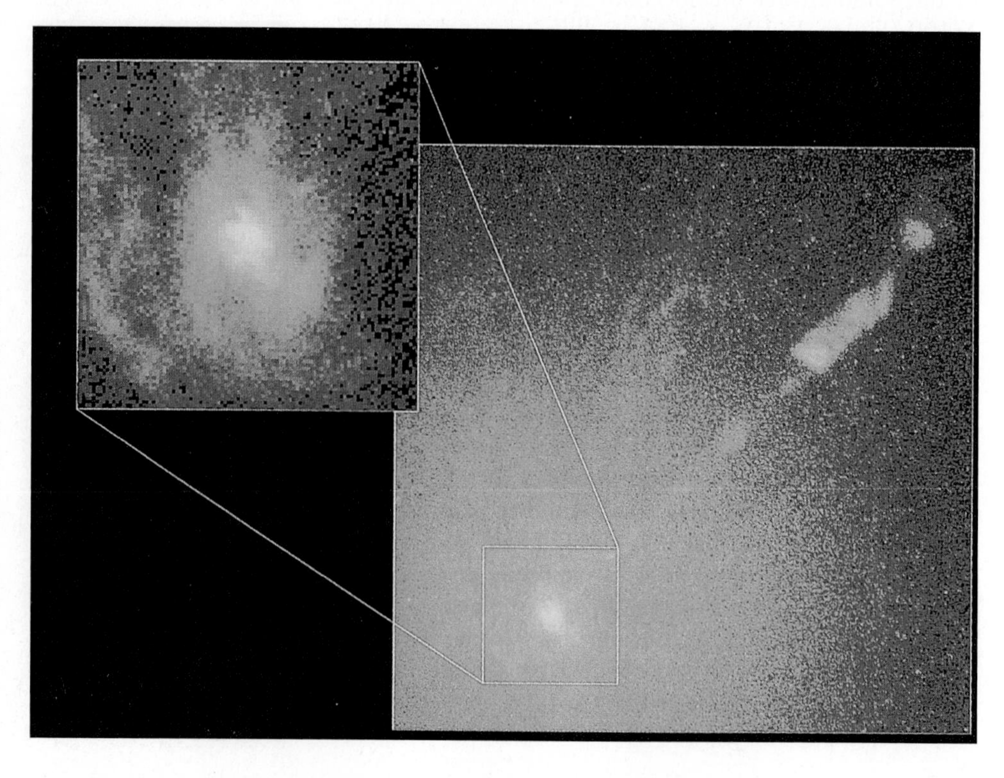

FIGURE 25.13
The Fornax Cluster
This cluster is a little more distant than the Virgo cluster and contains fewer galaxies. Notice that there are both bright ellipticals and bright spirals in the Fornax cluster. NGC 1365, at lower right, is a splendid example of a barred spiral.

we would be able to see hundreds of bright galaxies even without a telescope. The brightest galaxies, the giant ellipticals, would be pale balls of light, some of which would be brighter than Mars.

The Coma cluster is so close to us that most telescopes are unable to capture the entire cluster in a single image. We can get an idea of the overwhelming number of galaxies in a rich cluster, however, by looking at a cluster even more distant than the Coma cluster. Figure 25.15 shows the rich cluster CL 0939+4713. Nearly all of the bright objects in the direction of CL 0939+4713 are galaxies rather than foreground stars in the Milky Way.

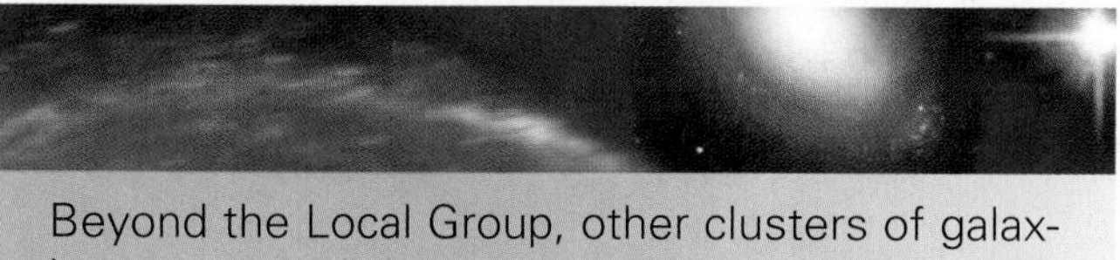

Beyond the Local Group, other clusters of galaxies are spread through space with typical separations of tens of megaparsecs. The Virgo cluster, at a distance of 20 Mpc, and the unusually rich Coma cluster, at 90 Mpc, are examples of clusters of galaxies much larger and richer than the Local Group.

FIGURE 25.14
The Central Region of the Coma Cluster, a Very Rich Cluster of Galaxies
The two largest and brightest galaxies in the picture are giant elliptical cD galaxies. Few spirals, if any, can be found in the central region of the Coma cluster. The bright object in the upper right is a foreground star in the Milky Way.

FIGURE 25.15
The Cluster CL 0939+4713
Most of the galaxies seen in this rich, distant cluster are spirals. Apparently, clusters contained more spirals in the past than they do today.

Galaxies in Clusters

The formation and evolution of galaxies has proceeded differently in the low-density regions of the universe between galaxy clusters than in the clusters themselves. The result has been a striking difference between the mix of types of galaxies found in clusters and in low-density regions. The graph in Figure 25.16 shows the relative numbers of elliptical and spiral galaxies in regions of different density. In low-density regions between clusters about 80% of the galaxies are spirals and irregulars. As the density of galaxies increases, the percentage of spiral and irregular galaxies decreases steadily whereas that of elliptical galaxies increases steadily. In rich clusters, only about 10% of the galaxies are spirals and irregulars whereas about 90% are elliptical and S0 galaxies.

There is no question that the density of the region in which a galaxy exists has played an important role in determining the appearance of the galaxy today. The question is whether it is the formation or the evolution of the galaxy (or both) that depends on density.

FIGURE 25.16
The Fraction of Elliptical and Spiral Galaxies in Regions with Different Galaxy Densities
Spiral and irregular galaxies dominate low-density regions, whereas elliptical galaxies dominate high-density regions.

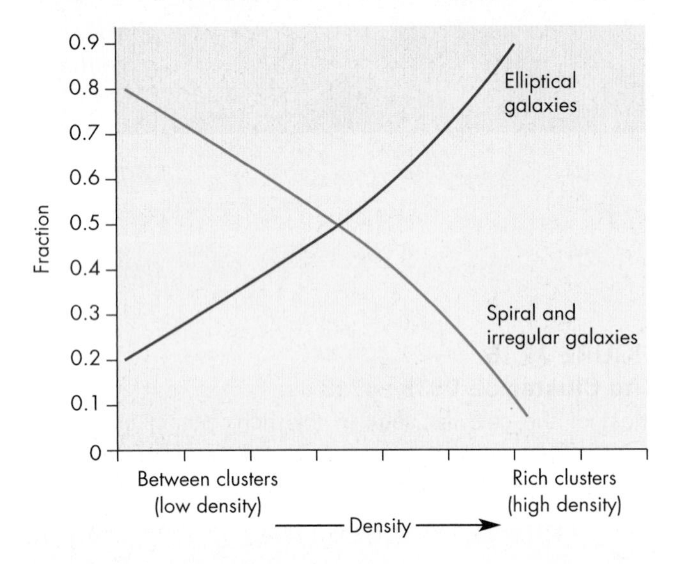

The Formation of Clusters of Galaxies An important question about clusters of galaxies is how they formed. One idea is that very large clumps of gas collected to form protoclusters, which then fragmented to produce individual galaxies. This is sometimes called the **pancake model** of cluster formation, because the collapse of a very large region of gas was more likely to lead to a flattened, pancakelike shape than a roughly spherical shape. At the other extreme is the idea that individual galaxies formed first and then collected into clusters. This is sometimes referred to as the **hierarchical clustering model** of cluster formation.

In the pancake model, illustrated in Figure 25.17, masses of about 10^{15} $M_{\odot}$, about as massive as a rich cluster of galaxies, began to collapse. The collapse was faster in one direction than the other two, so the result was a collapsing pancake. Within the pancake, smaller regions with high density and masses of about 10^{12} $M_{\odot}$ began to collapse. These became individual galaxies. One serious problem for the pancake model is that the time it would take for a pancake to collapse and for individual galaxies to form is longer than the age of the universe. This means that individual galaxies should still be forming today.

The almost total absence of very young galaxies is thus a problem for the pancake model. Another serious problem is accounting for small clusters of galaxies (such as the Local Group) and galaxies that lie between clusters. According to the pancake model, such galaxies must have formed as part of a massive cluster, yet such isolated galaxies are so distant from massive clusters that it is hard to see that they could ever have been members of massive clusters.

The hierarchical clustering model for cluster formation is illustrated in Figure 25.18. The hierarchical clustering model is preferred over the pancake model because of both the failings of the pancake model and the successes

FIGURE 25.17
The Pancake Model of the Formation of Clusters of Galaxies
In the pancake model, regions of gas about as massive as clusters of galaxies began to collapse into flattened shapes. Within the pancakes, smaller regions of gas later collapsed to form individual galaxies.

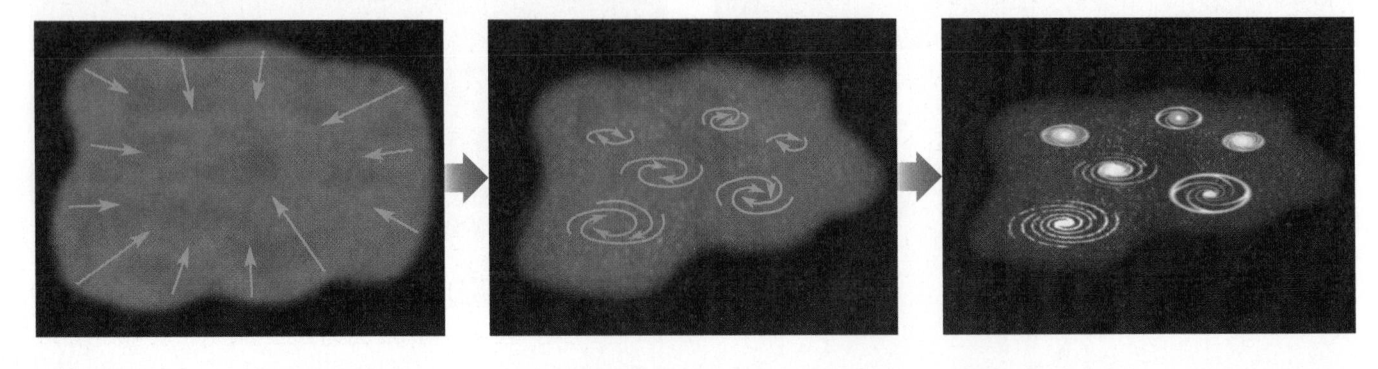

FIGURE 25.18
The Hierarchical Clustering Model of the Formation of Clusters of Galaxies
In the hierarchical clustering model, the first objects to collapse were relatively small clouds that formed clusters of stars. These clouds collected into larger masses that formed galaxies. The galaxies then gathered to form clusters of galaxies, a process that is still occurring.

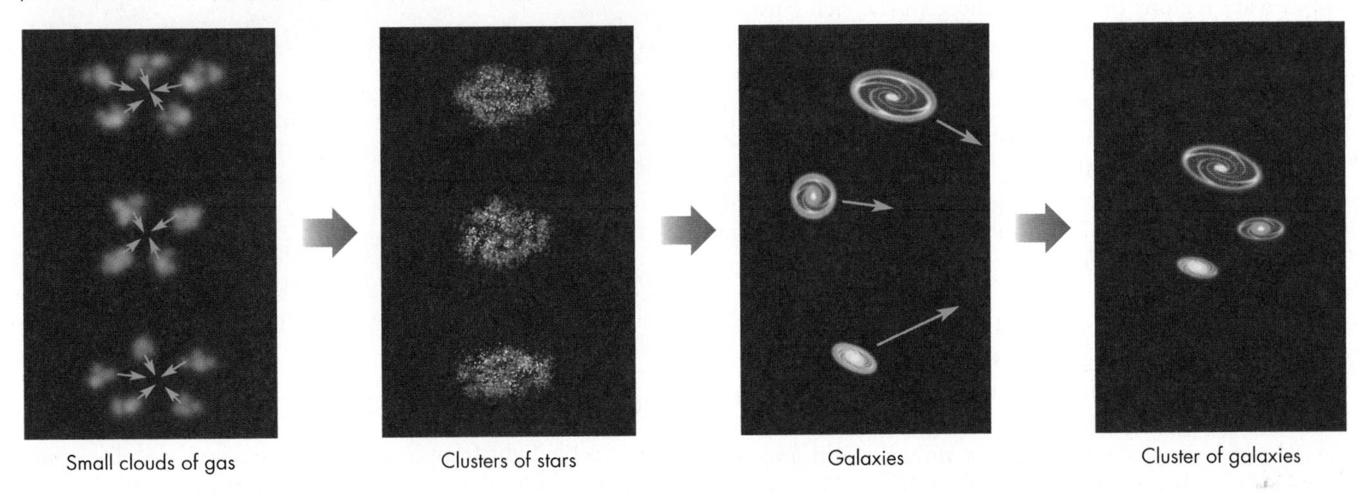

of the hierarchical clustering model. One of these successes is an explanation for why there are few young galaxies. The reason is that the era of galaxy formation occurred very early in the history of the universe, before the clusters formed. The observation that clusters of galaxies still seem to be in the process of gathering new galaxies into themselves is consistent with the hierarchical clustering model because the timescale for cluster formation, longer than the timescale for galaxy formation, is long enough that cluster formation should not have ended yet.

In the hierarchical clustering model, there were small, medium, and large regions of greater than average density in the early universe. Small regions began to collapse almost immediately, forming clouds or clusters about as massive as globular clusters. *Hubble Space Telescope* images, shown in Figures 23.14 and 25.19, may show some of these clouds about to merge to become galaxies. Larger regions, containing many of these clouds, collapsed more slowly to form galaxies. As individual galaxies collected under the influence of their gravitational attraction for each other, clusters of galaxies began to form.

Two theories of the formation of clusters of galaxies are the pancake model, in which clusters formed first and the hierarchical clustering model, in which galaxies formed first and then collected to form clusters. At the present time, the hierarchical clustering model seems to have the fewest serious shortcomings.

Density and Galaxy Formation It seems likely that galaxy formation proceeded differently in high-density regions where clusters formed and in low-density regions in between clusters. The evidence that density played a role in galaxy formation is found in the different mixes of galaxies in rich clusters and in regions between rich clusters.

FIGURE 25.19
"Pre-galactic Blobs"
Many of the small blue objects in this *Hubble Space Telescope* image are distant clouds of gas and stars that may be about to merge to form galaxies. The "blobs" support the hierarchical clustering model of galaxy cluster formation.

Only 5 to 10% of all galaxies belong to rich clusters. Therefore, most galaxies lie between clusters or are members of small clusters. Even though only 20% of the galaxies in low-density regions of the universe are elliptical and S0, the *total* number of elliptical and S0 galaxies in low-density regions outnumber those in the rich clusters. Galaxy collisions in the low-density regions of the universe are so rare that the elliptical and S0 galaxies in those regions couldn't have resulted from collisions between spiral galaxies, as has happened in high-density regions. Thus, evolutionary changes in the structure of galaxies can't be the only thing that depends on density. Instead, the different mix of galaxies in different density regions may, at least in part, reflect the way that galaxy formation depends on density.

Recall from Chapter 23 that a protogalaxy collapsed to form a disk and became a spiral galaxy if its matter was distributed smoothly and became an elliptical galaxy if it was clumpy. Imagine two protogalaxies, one forming in a low-density region of the universe and another forming in a high-density region. As the protogalaxies collapse, the clumps in them collapse to form clumps of stars. Because gravity is stronger where density is high, the clumps will collapse in the high-density protogalaxy before they collapse in the low-density one. It is through the clumps of stars that angular momentum is transferred from the luminous to the dark matter in the galaxy. Thus, galaxies that formed in high-density environments were more likely to develop stellar clumps, lose angular momentum to dark matter, and collapse to form elliptical galaxies. Spiral galaxies, formed from smoothly collapsing matter, were the likely outcome of galaxy formation in small groups and in the regions between clusters.

Galaxy Evolution in Clusters Being in a rich cluster seems to influence the evolution of galaxies, too, because collisions are much more frequent in clusters than in the regions between clusters. The collisions distort the shapes of the galaxies and lead to mergers of small galaxies with large ones. The cD galaxies found at the centers of clusters probably began as normal ellipticals. As they moved through the cluster, they pulled less massive galaxies into a wake behind them (this is analogous to what happened to clumps of stars in protogalaxies). The mass in the wake retarded the motion of the elliptical, robbing it of angular momentum and causing it to sink toward the center of the cluster. In the crowded core of the cluster, the elliptical galaxies devoured other, smaller galaxies and eventually grew to be the diffuse giants that they are today.

The large numbers of S0 galaxies found in rich clusters may also result from galaxy collisions. When two spiral galaxies collide at high speed (the typical speeds of galaxies are larger in rich clusters than elsewhere), the stars in the galaxies pass each other without harm. The gas clouds, on the other hand, collide with each other and are left behind as the galaxies continue along their paths. The galaxies, now devoid of interstellar gas, stop forming stars. Their spiral arms fade out and they soon become S0 galaxies.

Rich clusters are dominated by elliptical and S0 galaxies whereas most single galaxies and those in small clusters are spirals. This difference probably arose partly from the different densities of the regions in which rich clusters and other galaxies formed and partly from collisions that produced giant ellipticals and S0 galaxies, destroying spirals in clusters.

Other Matter in Clusters

The visible light from a cluster shows only the stellar component of its galaxies and fails to point out other important components of the cluster, such as hot gas and dark matter.

Hot Gas and X Rays Many clusters of galaxies are strong sources of X rays. Figure 25.20 shows that although some of the X rays come from individual galaxies, most

FIGURE 25.20
X Rays from a Cluster of Galaxies
The X-ray emission, shown in purple, is superimposed on a composite of infrared and visible images of the cluster of galaxies. The galaxies in the cluster are seen as colored dots. Although some X rays come from individual galaxies, most of the X rays come from hot intergalactic gas.

of the radiation comes from intergalactic gas that fills the entire cluster. The temperature of the gas is 10 to 100 million K. This may seem very hot, but at that temperature, the ions move at about 1000 km/s, about the same speed as that of the individual galaxies in the cluster.

The X-ray emission from clusters of galaxies contains emission lines of iron and other heavy elements. These elements can only be produced by fusion in stars or supernovae. Thus some of the hot gas in clusters of galaxies must have been ejected from stars within the galaxies in the cluster. However, there is too much hot gas in clusters for it all to have come from stars. Most of it must have existed as intergalactic gas since the time that the cluster formed. The mass of hot gas in a typical cluster is comparable to, or larger than, the mass of the galaxies.

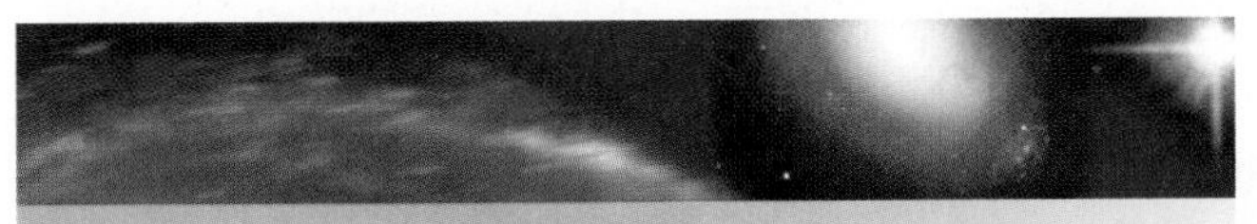

Strong X-ray radiation from many clusters of galaxies indicates that those clusters are filled with large amounts of hot, low-density gas. Most of the hot gas must have been present as intergalactic gas since the time that the cluster formed.

Dark Matter in Clusters The first evidence for large amounts of dark matter in clusters of galaxies was found by comparing the results of two different methods for calculating the masses of clusters of galaxies. The first method, essentially the same one used to find the masses of elliptical galaxies, is to assume that the orbits of the galaxies in a cluster are randomly oriented. The spread of velocities is then used to find the average orbital speed. This, with the size of the cluster, gives the mass that contributes to the gravity of the cluster. The second method is to count the galaxies and multiply by the mass of a typical galaxy. The resulting mass is then doubled to account for the hot gas in the cluster, yielding the cluster's visible and X-ray–emitting mass.

These two methods for determining the mass of a cluster of galaxies were first applied to a cluster of galaxies (the Coma cluster) by Fritz Zwicky over 70 years ago. He found that the first method yielded far more mass than the second. A similar discrepancy between the gravitational mass and visible mass of a cluster of galaxies has now been found in many other clusters. For a typical cluster, the gravitational mass is about ten times larger than the visible mass. This means that 90% of the matter in the cluster exists in a form that emits neither visible light nor X rays.

This conclusion has been reinforced through observations of gravitational lensing by clusters of galaxies. Since 1980, numerous high-redshift galaxies lensed by clusters of galaxies have been discovered. When clusters of galaxies form the gravitational lens, the background galaxies are elongated into arcs by the mass of the foreground cluster, as shown in Figures 25.21 and 25.22. The amount of distortion of the distant galaxies can be used to investigate the way that the mass of the cluster varies with distance from the center. All of the mass in the cluster—not

FIGURE 25.21
Arcs in a Cluster of Galaxies
This *Hubble Space Telescope* picture shows several arcs that are images of the blue galaxy near the center of the picture. The images of the galaxy are stretched into arcs by the gravitational lens caused by the matter in the cluster of galaxies.

FIGURE 25.22
Gravitational Lensing of Distant Galaxies
This *Hubble Space Telescope* image shows more than one hundred concentric arcs surrounding a massive cluster of galaxies. The arcs are images of distant background galaxies that would be too faint to be detected were it not for gravitational lensing.

FIGURE 25.23
Dark Matter in the Bullet Cluster
This composite image of the galaxy cluster 1E 0657-56, the Bullet Cluster, shows the region of hot, X-ray–emitting gas in red and the region of gravitational lensing dark matter in blue. When the Bullet Cluster was formed by the collision of two clusters of galaxies, the normal matter in the two clusters was halted by collisions. The dark matter wasn't slowed and now can be seen in two regions on opposite sides of the normal matter. The Bullet Cluster is about a billion Mpc away.

just the luminous mass—contributes to the gravitational lens. This means that gravitational lensing can be used to investigate the amount of dark matter in a cluster of galaxies and the way in which the dark matter is distributed. The results of these investigations yield masses for clusters of galaxies that are about the same as the masses derived from the velocities of the galaxies in the clusters. This is far larger than the amount of luminous mass that can be found. Thus dark matter exists not only in galaxies but also between galaxies within a cluster.

Further evidence for dark matter in clusters of galaxies can be seen in Figure 25.23, which shows the galaxy cluster 1E 0657-56, also known as the "Bullet Cluster." The Bullet Cluster was formed when two clusters of galaxies collided. The gas clouds of normal matter in the two clusters collided with each other and were brought to a halt. This hot gas emits X rays and is shown as red regions in Figure 25.23. Dark matter particles don't interact strongly with each other or with normal matter so the clouds of dark matter weren't slowed down as the clusters collided. Instead, they passed right through the region where the collision of the galaxy clusters occurred. The presence of the clouds of dark matter was discovered through the gravitational lensing they produce. The regions of dark matter are shown in blue in Figure 25.23. The dark matter in the Bullet Cluster makes up far more mass than the galaxies and the hot, X-ray–emitting gas combined.

The velocities of galaxies in clusters are so large that clusters must contain about ten times more dark matter than luminous matter. The masses of clusters of galaxies act as gravitational lenses that distort the images of background galaxies. The degree of distortion can be used to study dark matter in clusters of galaxies. The results support the conclusion that most of the mass in clusters is invisible.

25.2 SUPERCLUSTERS AND VOIDS

Just as galaxies are gathered into clusters, the clusters themselves are gathered into superclusters. The superclusters are not randomly distributed in space. Instead, on the biggest scale that has been measured, the superclusters form a pattern of sheets that surround regions in which few galaxies can be seen.

The Local Supercluster

The Milky Way and the Local Group of galaxies are located on the fringe of the Local Supercluster, sketched in Figure 25.24. The Local Supercluster is also sometimes called the Virgo Supercluster because it is roughly centered on the Virgo cluster of galaxies. In addition to the Virgo cluster, the Local Supercluster contains several dozen other groups or clusters of galaxies within a region of space about 40 Mpc across. The clouds of galaxies in the Local Supercluster (the light-colored regions in Figure 25.24) occupy only a small fraction of the space needed to enclose the supercluster. The space between the clouds of galaxies in the Local Supercluster is almost completely empty. Only about % of the galaxies in the supercluster lies outside of the main clouds of galaxies.

Other Superclusters

Beyond the Local Supercluster, stretching away to the limit of observations, are other superclusters. A schematic picture showing the locations of the nearest superclusters is shown in Figure 25.25. The figure is schematic because it ignores the three-dimensional structure of space and compresses all the superclusters into a plane. The superclusters appear as elongated features connected by filamentary links. This appearance is somewhat deceptive, however, because many of the striking features in the map extend as sheets above and below the map. For example, the line of galaxies that stretches left and right from the densest part

FIGURE 25.24
The Local Supercluster
The supercluster is centered approximately on the Virgo cluster, which contains most of the mass in the supercluster. The Milky Way (marked "Earth") is located far from the center. The region around the supercluster is nearly empty.

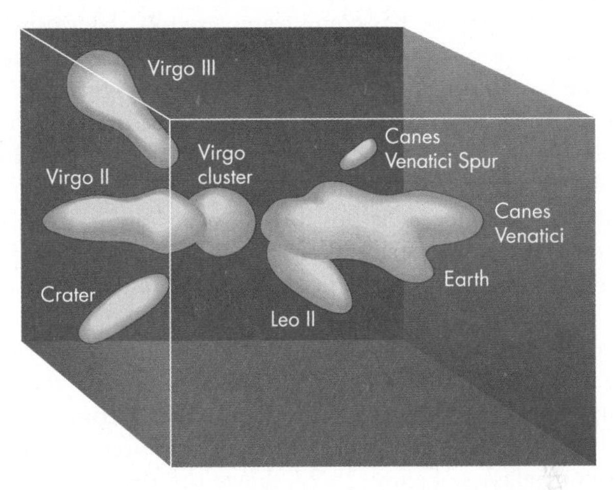

FIGURE 25.25
A Schematic Map of Relatively Nearby Superclusters
The sectors near the center of the map are obscured by dust in the Milky Way. The region that stretches left and right from the densest part of the Coma Supercluster (at the top of the picture) is part of a large sheet of galaxies known as the "Great Wall."

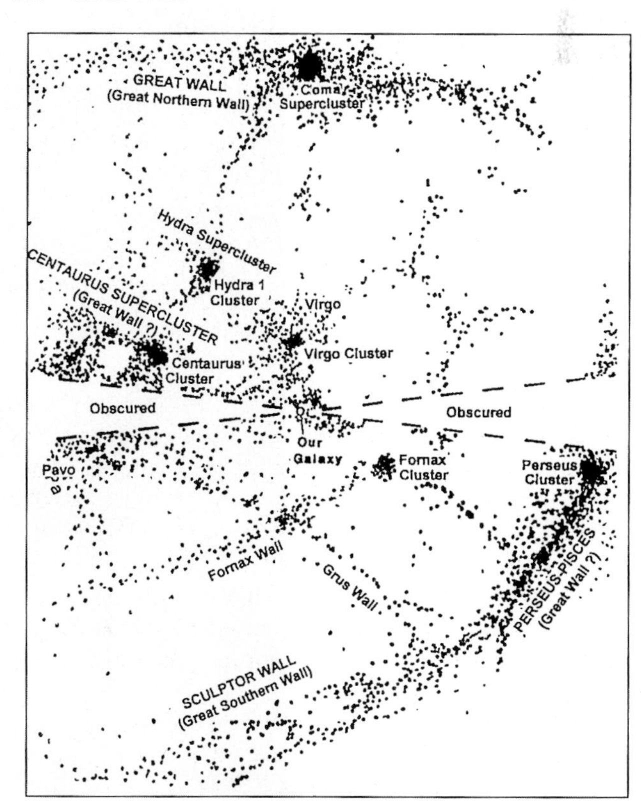

of the Coma Supercluster also reaches far above and below the map. This feature has been called the "Great Wall."

Voids

The best information that we have about the distribution of galaxies in space has come from carefully determining the locations of galaxies in slices of the universe centered on the Earth. In the 12° wide slice shown in Figure 25.26, astronomers at Harvard and Arizona measured the radial velocities of more than a thousand galaxies. Using Hubble's law, in which the radial velocities are proportional to distance, it is possible to locate the galaxies in three-dimensional space. For H = 70 km/s per Mpc, the slice reaches outward to about 180 Mpc. The galaxies are arranged in thin loops that outline huge, almost completely empty regions called **voids.** Some of the voids are as large as 300 Mpc across. Astronomers have carried out a number of projects to examine the structure of the universe by determining the distances to fainter and more distant galaxies. Figure 25.27 shows the distribution of 70,000 galaxies in a 2° wide slice that extends outward to about 750 Mpc, or 2.5 billion light years. The pattern of loops and voids is visible even on this large scale. The Sloan Great Wall, the largest structure ever detected in the universe, stretches 1.4 billion light years across Figure 25.27.

Whether the galaxies are arranged in ropes or sheets can be determined by comparing the patterns of galaxies in adjacent slices. If the pattern is one of ropes, the two maps should be quite different. If the pattern is sheets, then the adjacent slices should cut through the sheets in almost the same places, producing similar maps. The similarity of the maps for two adjacent slices leaves no doubt that the overall distribution of galaxies and clusters of

FIGURE 25.26
A Map of the Slice
The galaxies form a pattern of long, thin bands that surround voids. It is possible, however, that the voids may contain just as much dark matter as the more luminous regions of the slice.

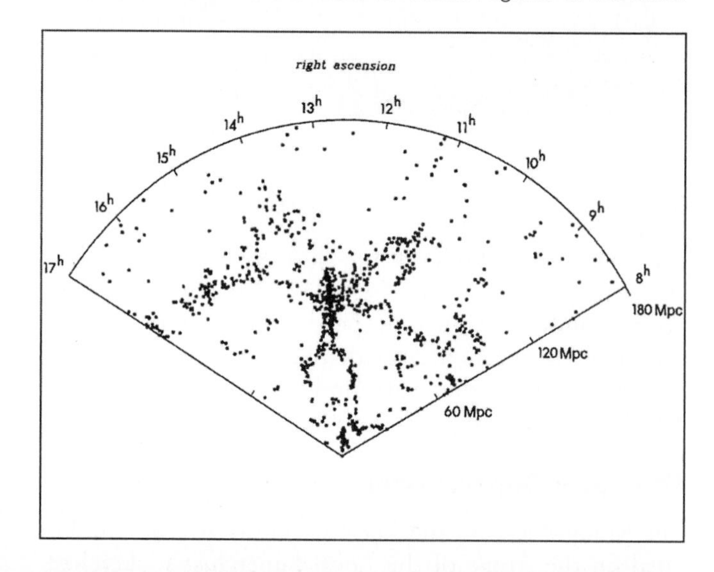

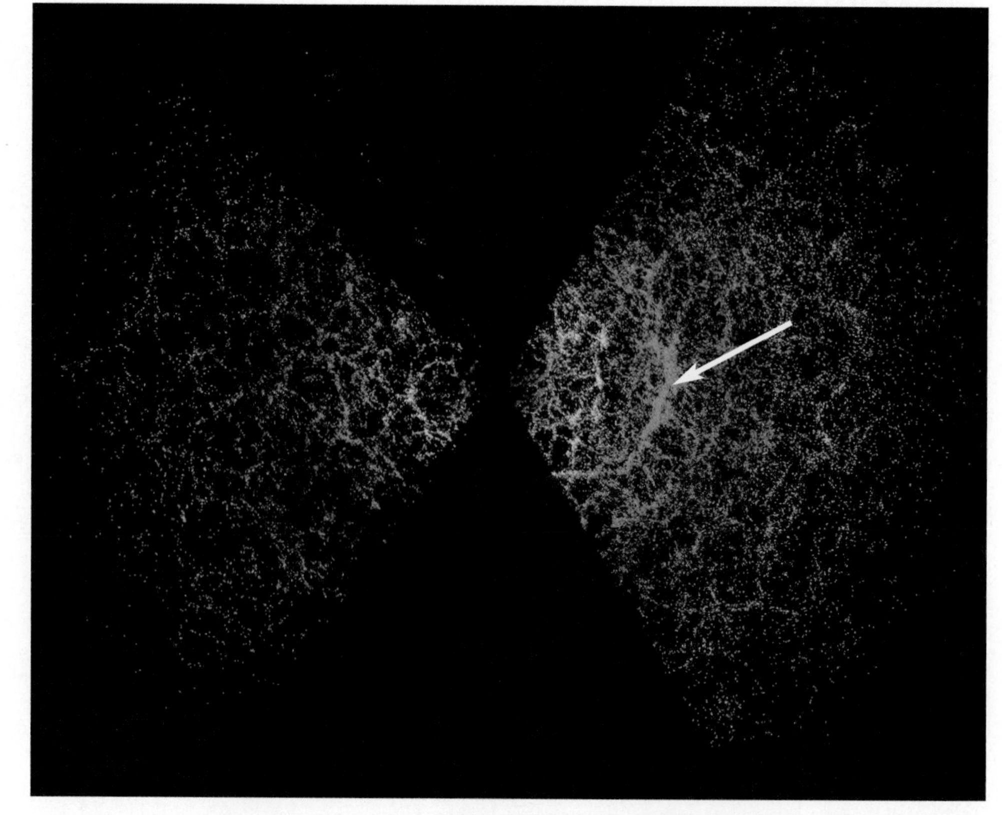

FIGURE 25.27
The Sloan Great Wall
This map shows the locations of 70,000 galaxies in a slice of the universe recorded by the Sloan Digital Survey. The slice extends outward to about 2.5 billion light years. The Sloan Great Wall, located about a billion light years away from the Milky Way, is indicated by an arrow.

galaxies is in sheets. Whether the interconnected sheets look more like a foam of soapsuds, a honeycomb, or a sponge is not yet known.

Why Are There Voids and Sheets?

One of the great uncertainties about voids and sheets is just how much they tell us about the actual distribution of matter in the universe. It appears that, at least within clusters of galaxies, 90% or more of the matter in the universe is invisible. It may be that what we observe as voids aren't really empty after all. Instead, these areas may have about the same density of dark matter as the sheetlike regions that contain the luminous matter in the universe. Perhaps the galaxies formed where the density of matter (bright and dark) is just slightly higher than elsewhere. If this is the case, then the great contrast between the voids and the sheets may not accurately reflect the actual distribution of matter in the universe.

At the present time, many groups of astronomers are investigating the formation of the pattern of voids and sheets by modeling the development of galaxies in the early universe. Most of these models include the presence of large amounts of dark matter, often in the form of unknown fundamental particles. A number of very different models have been able to reproduce the general pattern of the distribution of luminous matter in the universe. It still seems too early to decide whether any of these models are correct or whether different ideas will be required to explain the large scale appearance of the universe.

The Great Attractor

Another clue about the distribution of matter in the universe can be found in the motions of clusters and galaxies in response to the gravitational attraction of major clumps of matter. It appears that everything in our part of space, including the Local Group, the Virgo Supercluster, and even the distant Hydra-Centaurus Supercluster, is being pulled toward a great concentration of mass that lies beyond the Hydra-Centaurus Supercluster. This mass, which has been called the **Great Attractor,** appears to be a great concentration of rich clusters at a distance of about 100 Mpc. Figure 25.28 shows ACO 3627, one of the clusters of galaxies that make up the Great Attractor.

FIGURE 25.28
ACO 3627, a Distant Cluster of Galaxies
ACO 3627 appears to be part of a concentration of matter known as the Great Attractor.

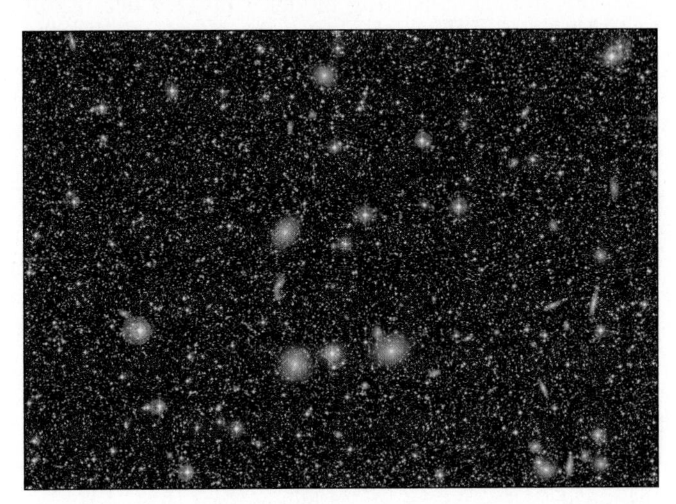

Clusters of galaxies are gathered into superclusters. Clusters and superclusters are arranged in patterns of sheets that surround voids, within which few galaxies can be found. The pattern of motions of galaxies in our part of the universe shows that there is a large concentration of mass, called the Great Attractor, which pulls us in the direction of the constellation Centaurus.

Chapter Summary

- Galaxies tend to be gathered in clusters, which differ in the number of bright galaxies they contain and in the regularity of their appearances. The Milky Way is located in a small cluster called the Local Group. (Section 25.1)
- Beyond the Local Group, other clusters of galaxies are scattered throughout space, typically separated by tens of megaparsecs. The Virgo cluster and the Coma cluster, much richer clusters than the Local Group, are among the nearest great clusters of galaxies. (25.1)
- The fraction of elliptical and S0 galaxies in a cluster increases as the density of the cluster increases. The dominance of these galaxies in rich clusters probably arose partly from a tendency of galaxies forming in dense regions to become clumpy and evolve into ellipticals. Collisions, which are more likely in rich clusters, also could have produced giant ellipticals and S0 galaxies while destroying spirals. (25.1)
- Two theories of the formation of clusters of galaxies are the pancake model, in which clusters formed

first and then fragmented to form galaxies, and the hierarchical clustering model, in which clusters collected from galaxies that had already formed. The hierarchical clustering model is favored because of its successes and the shortcomings of the pancake model. (25.1)
- X rays from many clusters show the presence of hot, tenuous gas between the galaxies. There is too much hot intergalactic gas for it to have come from stars and galaxies, so it must have been present in the cluster since the cluster formed. (25.1)
- The velocities of the galaxies in clusters are too large for the clusters to hold themselves together unless they contain about ten times more dark matter than luminous matter. The images of distant quasars and galaxies are sometimes distorted by gravitational lensing by intervening galaxies and clusters of galaxies. The distortion of these images can be used to study the distribution of matter in clusters of galaxies. The results support the conclusion that most of the matter in clusters is invisible. (25.2)
- Galaxy clusters are themselves grouped into superclusters. Clusters and superclusters are arranged in sheets that surround voids within which there are few, if any galaxies. A large concentration of mass (the Great Attractor) beyond the Hydra-Centaurus Supercluster is pulling the galaxies in our part of the universe toward itself. (25.2)

Key Terms

cluster of galaxies *602*
Great Attractor *615*
hierarchical clustering model *608*
irregular cluster *602*
Local Group *602*
pancake model *608*
regular cluster *602*
richness *602*
voids *614*

Conceptual Questions

1. What are the ways in which clusters of galaxies can be classified?
2. What evidence do we have that there is more dark matter than luminous matter in the Local Group of galaxies?
3. Explain why the Virgo cluster is considered a "poor" cluster even though it contains at least a thousand galaxies.
4. Compare the Virgo and Coma clusters with respect to distance, richness, and the types of galaxies found in their central regions.
5. Why is there a lower proportion of spiral galaxies in dense clusters than in clusters with lower density?
6. In what fundamental way are the pancake model and the hierarchical clustering model for the formation of clusters of galaxies different?
7. What is the best way to detect hot gas in clusters of galaxies?
8. What have astronomers been able to learn about dark matter in clusters of galaxies by studying the gravitational lensing of distant galaxies by clusters of galaxies?
9. What is the relationship between clusters of galaxies and superclusters?
10. Describe the structure of space on a scale that includes superclusters and voids.

Group Activities

After reviewing the section on galaxy formation, divide into two subgroups and have a debate on the relative merits of the pancake model and the hierarchical clustering model for the formation of galaxies.

For More Information

Visit the text website at **www.mhhe.com/fix** for chapter quizzes, interactive learning exercises, and other study tools.

26

You should neither seek to know what end the Gods have given to you and to me, Leuconoe, for such knowledge is forbidden to us, nor should you consult the Babylonian astrological calculations. It is better just to accept whatever will happen. We cannot know whether Jupiter bestows many more winters to us, or whether this is the last one, which is now wearing out the Tyrrhenian Sea against the craggy shore. Be wise. Spend your time on domestic chores; time is brief so restrain from far-off hopes. While we speak, envious time will have fled; seize the day, putting as little faith as possible in tomorrow.

Horace
Tu Ne Quaesieris

Distant galaxies.

Cosmology

One of the most remarkable developments in the last half century has been the series of discoveries and calculations that form the core of *cosmology*—the study of the universe as a whole. In that time we have gone from a position of almost total ignorance about the history and fate of the universe to one in which we understand the broad details of the story of the universe from almost the first instant to the present. Also, while individual men and women are no better able to predict their fates today than they were 2000 years ago in the time of Horace, we are much better informed about the fate of the entire universe. If our current ideas about the universe prove to be correct, we may soon know the entire story of the universe from its first moments into its remote future.

Questions to Explore

- What does Hubble's law tell us about the beginning of the universe?
- How can we tell how old the universe is?
- Can space itself be curved? What does that mean? How could we tell?
- What important events happened early in the history of the universe?
- What will the future of the universe be like?

26.1 HUBBLE'S LAW REVISITED

Recall from Chapter 23 that Hubble's law implies that the universe is expanding. The expansion of the universe implies that there was a moment long ago when everything that we can now see in the universe was located very close together. At that moment, the history of the universe began.

The Expansion Age of the Universe

To see how Hubble's law implies that the universe had a beginning, imagine that you live in the middle of a vast, flat desert with a roommate and a Jeep. One morning before you get up, your roommate takes the Jeep and heads for town. When you arise, you notice your roommate and the Jeep are gone and wonder when they left. You look through your binoculars and spot the Jeep just passing a landmark 20 km from home. You also notice that the speedometer reads 10 km/hr. (These are *very* good binoculars.) To estimate the time that your roommate has been traveling, you just divide the distance of the landmark (20 km) by the speed (10 km/hr). The answer is 2 hours. Notice that the accuracy of this estimate depends on how constant the speed of the Jeep has been during the trip. If your roommate started the trip traveling faster than 10 km/hr and has slowed down to 10 km/hr, then the trip took less than 2 hours. If your roommate started slowly and speeded up, it took longer than 2 hours.

This same idea can be applied to galaxies and clusters of galaxies. To find how long a galaxy has been traveling away from us (actually, the length of time that the expansion of the universe has been carrying the galaxy and us away from each other), we just divide the distance to the galaxy by the speed with which the expansion of the universe is carrying the galaxy away from us. That is, we estimate that

$$t = \frac{d}{v} \qquad (26.1)$$

where t is the time since we were together with the galaxy, d is the distance to the galaxy, and v is its speed of recession. This will be an accurate estimate only if the recession speed of the galaxy has always been the same. The box "Time, Speed, and Distance" shows that a galaxy at 100 Mpc has been receding from the Milky Way for 13.8 billion years.

Now suppose we do a calculation like the one in the box "Time, Speed, and Distance" for a galaxy at 200 Mpc. The result is the same—13.8 billion years. This is not a coincidence. The galaxy at 200 Mpc is twice as far away as the galaxy at 100 Mpc. But, according to Hubble's law, it is also receding twice as fast. The greater speed exactly compensates for the greater distance. In fact, because Hubble's law states that recession speed is proportional to distance, the same value for t is calculated for every galaxy and cluster of galaxies in the universe. This estimate of the expansion age of the universe, called the **Hubble time,** is given by

$$t = \frac{d}{v} = \frac{d}{\mathrm{H}d} = \frac{1}{\mathrm{H}} \qquad (26.2)$$

where H is Hubble's constant. In Equation 26.2, Hubble's law is used to substitute Hd for v. Approximately 1 Hubble time ago, all of the matter in the universe was located together in an arbitrarily dense state from which it then began to expand. This beginning of the expansion of the universe is usually called the **Big Bang.**

Actually, the Hubble time is only an estimate of the expansion age of the universe. The two are the same only if the expansion of the universe has always gone on at the same rate. If the expansion originally occurred at a faster rate, then the expansion has gone on for less than the Hubble time. If the early expansion occurred at a slower

Time, Speed, and Distance

Equation 26.1 can be used to calculate the length of time that the Milky Way and another galaxy have been receding from each other. Suppose there were a galaxy at a distance, d, of 100 Mpc and a recession speed, v, of 7100 km/s. (This implies that Hubble's constant has a value of 71 km/s per Mpc.) To use Equation 26.1, we must first convert the distance of the galaxy from megaparsecs to kilometers. The conversion factor is 1 Mpc = 3.09 $\times 10^{19}$ km, so the distance to the galaxy is 100 Mpc = 3.09×10^{21} km. With these values for d and v, Equation 26.1 becomes

$$t = \frac{d}{v} = \frac{3.09 \times 10^{21}\ \text{km}}{7100\ \text{km/s}} = 4.35 \times 10^{17}\ \text{s}$$

Because there are 3.16×10^7 seconds in a year, this time is equal to 13.8 billion years.

The Hubble Time

Equation 26.2 gives the relationship between Hubble's constant, H, and the Hubble time, t, which is an estimate of the length of time that the universe has been expanding. The preceding box showed that the Hubble time is 13.8 billion years if Hubble's constant is 71 km/s per Mpc. Equation 26.2 can be used to find the Hubble time if Hubble's constant is 142 km/s per Mpc. According to

$$t = \frac{1}{H}$$

the larger Hubble's constant is, the smaller the Hubble time. If Hubble's constant is 142 km/s per Mpc (twice as large as 71 km/s per Mpc), then the Hubble time is half as long as the Hubble time for H = 71 km/s per Mpc. That is,

$$t = \frac{1}{2} \times 13.8 \text{ billion years} = 6.9 \text{ billion years}$$

rate, then the universe has taken more than the Hubble time to reach its present size.

When students first encounter the concept of the Big Bang, it is common for them to get the mistaken impression that the Big Bang was like an ordinary explosion—only bigger. They picture the Big Bang as the explosive expansion of the matter in the universe into pre-existing space. The correct picture, much harder to grasp, is that the Big Bang was the rapid expansion of spacetime itself. Rather than resembling a bomb going off in a room, the Big Bang was more like the rapid expansion of the room. Matter was just carried along with expanding spacetime.

Hubble's law implies that there was a beginning of the universe—a unique moment when the universe was compressed in a dense state. The expansion of the universe from this dense state is called the Big Bang. The Hubble time, an estimate of how long ago the Big Bang took place, is calculated assuming that expansion has always gone on at the present rate.

26.2 COSMOLOGICAL MODELS

Although cosmological models—descriptions of the universe and its history—have been invented throughout human history, meaningful mathematical and physical cosmological models have only been possible since Einstein's theory of general relativity (Chapter 20) was proposed in 1915. General relativity, a theory of gravity that describes the relationship between matter and spacetime, is the cornerstone of modern models of the universe. Cosmological models based on general relativity present us with two different alternatives for the nature of the universe. These two classes of cosmological models differ in the curvature of space in the universe.

Curvature of Space

We are so used to thinking of space as nothingness that it can be very hard to understand that space is real and has distinct properties. One of these properties is curvature, discussed in Chapter 20 in the context of spacetime near black holes. Although local spatial curvature certainly occurs wherever there is matter, the curvature in cosmological models ignores these local effects and concentrates on the overall curvature of the entire universe. Because of the great difficulty that nearly everyone has in imagining three-dimensional curved space, it is helpful to explore the properties of curvature in two-dimensional spaces, which we can more easily visualize.

Positive Curvature The surface of a ball is an example of a two-dimensional, positively curved space. The ball is a three-dimensional object, but its surface is two-dimensional because a creature located at a point on the ball has only two choices about where to go without leaving that space. A creature in Figure 26.1 can decide to go north or south (choice 1) and east or west (choice 2). Any direction other than due north, south, east, or west is a combination of the two original choices. The creature can't go in toward the center of the ball or out from the ball's surface without making a three-dimensional trip. If the ball is large and the creature is small, the curvature of the ball may be hard to detect, just as it is difficult for someone on the surface of the Earth to tell that the Earth isn't flat. To make the analogy to our own universe more complete, imagine that light must also follow the curvature of the surface of the ball. This means that the creature can see the rest of the positively curved universe by peering outward with a telescope but can't see anything that isn't on the surface of the ball.

The two-dimensional, positively curved space has several important properties. First, it is *finite*. That is, there

FIGURE 26.1
A Two-Dimensional, Positively Curved Space
The surface of a sphere is two-dimensional because a creature on the surface can travel only in two directions (north-south or east-west) or combinations of these directions. The arrow on the right indicates a direction of travel that is a combination of east and north.

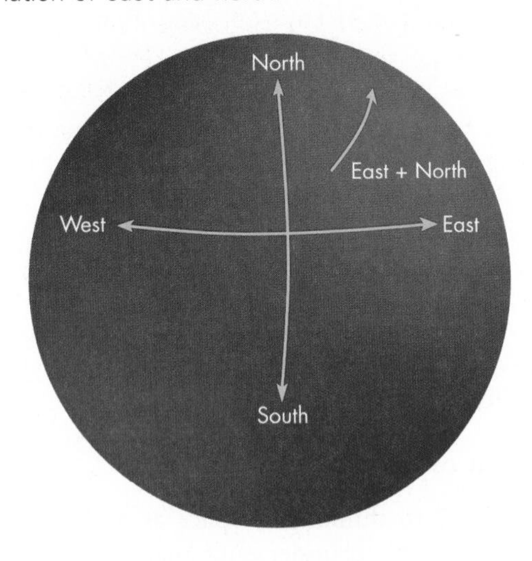

is only a certain amount of area on the surface of the ball. If there are two-dimensional galaxies on the surface of the ball, there are only a finite number of them. The total mass on the surface of the ball is finite as well. Despite being finite in area, the space is *unbounded*. A traveler can journey forever without encountering a wall that marks the boundary of the universe. The traveler will begin to follow its original track after going once around the entire space, but no universe, positively curved or otherwise, can have a boundary. If boundaries existed, the space wouldn't be an entire universe. It would only be a piece of one. Finally, the two-dimensional, positively curved space has *no center*. There is no point on the *surface* of the ball that is unique and that could be the center. The ball has a center, but a two-dimensional creature on the surface of the ball would be unable to make the three-dimensional trip required to reach the center of the ball. Also, the creature couldn't see the center of the ball or even, probably, imagine where it might be.

Can the two-dimensional space expand? Certainly. Just imagine a balloon being inflated. All points in the space get farther apart. A creature at any point sees that the expansion obeys Hubble's law. However, the point about which the universe is expanding is the center of the ball and is not a part of the two-dimensional universe. A two-dimensional creature can't look around and find the spot where the expansion began.

The geometry of two-dimensional, positively curved space is different in many respects from the Euclidean geometry taught in most high-school geometry classes. To begin with, there are no straight lines. Instead, a geodesic, the shortest distance between two points, is part of a great

FIGURE 26.2
A Geodesic in Two-Dimensional, Positively Curved Space
A geodesic, the shortest distance between two points, can be formed by a taut string held at the two points. On the surface of a sphere, a geodesic is a great circle.

FIGURE 26.3
A Triangle and a Circle in Two-Dimensional, Positively Curved Space
A triangle formed by three geodesics has angles that add up to greater than 180°. A circle is formed by marking off all the places that have the same distance from a point. The circumference of a circle, C, in positively curved space is less than 2π times the radius of the circle, r.

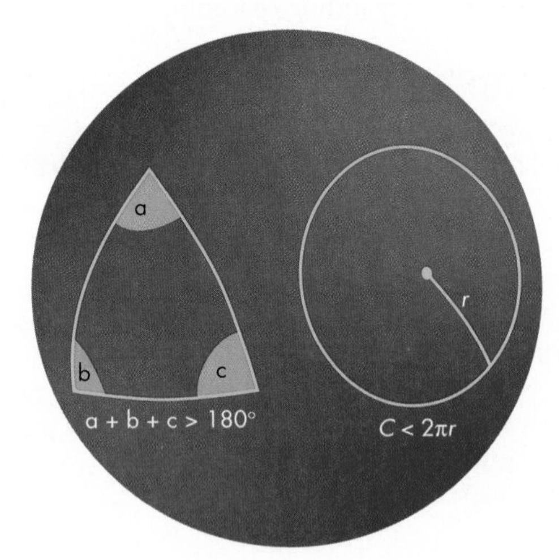

circle, as shown in Figure 26.2. Light travels along great circles in positively curved space. Figure 26.3 shows that, just as in flat space, a triangle can be constructed from three geodesics. In positively curved space, however, the sum of the angles of the triangle is greater than 180°.

Suppose the two-dimensional creatures draw a series of ever larger circles by marking off all the places at a given distance from a specified point, as shown in Figure 26.4. For small circles the circumference of the circle is related to its radius by the familiar $C = 2\pi r$, where C is circumference and r is radius. As the circles get larger, however,

FIGURE 26.4
Three Circles in Two-Dimensional, Positively Curved Space
For relatively small circles such as the circle with radius r_1 and circumference C_1, circumference increases with increasing radius. The circle with radius r_2 has the largest circumference possible in the positively curved space. For larger circles, such as the one with radius r_3, circumference actually decreases as radius increases.

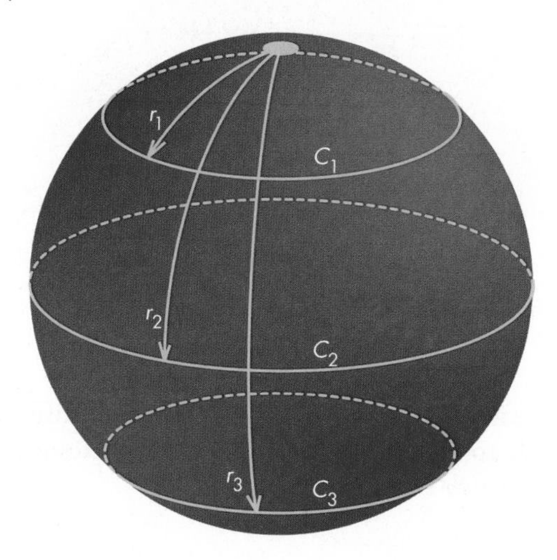

their circumferences fall short of $2\pi r$ by an ever larger amount. Eventually, the circumferences get extremely small for circles with radii that stretch almost halfway around the universe. Through measurements of the geometrical properties of their universe, it might be possible for the two-dimensional creatures to discover that they live in a universe with positive curvature.

Zero Curvature In some respects it is easier for us to imagine the properties of two-dimensional space with zero curvature—flat space—because our senses tell us that we live in its three-dimensional counterpart. A part of a two-dimensional flat space is shown in Figure 26.5. The flat space shown in Figure 26.5 is artificially bounded. It actually extends indefinitely to the right and left as well as up and down. The space is two-dimensional because there are only two choices that can be made about where to go. All trips on the sheet are combinations of up-down and right-left. Traveling in or out of the page would require a third dimension not available to the two-dimensional creatures in this space.

One important difference between two-dimensional flat space and two-dimensional positively curved space is that flat space has an *infinite* area. Light rays beamed outward in any direction travel forever and never return to the place where they originated. The flat universe contains an infinite number of galaxies and has infinite mass. Like the positively curved universe, the flat universe has *no boundary*. At no point will a traveler encounter a wall that marks the edge of the universe, for there is no edge.

Also like the positively curved universe, the flat universe has *no center*. To see that this must be so, ask yourself how someone finds the center of a room. This is usually done by finding the spot midway between the north and south walls as well as midway between the east and west walls. In other words, the center is defined only in terms of distances from the boundary of the room (the walls). If there is no boundary, there can be no center. Although it is hard to visualize an already infinite space getting larger, we can at least imagine a part of the space expanding. We need only think of a rubber sheet being stretched in all directions, as in Figure 26.6. Every two-dimensional galaxy will recede from every other galaxy, and Hubble's law will be obeyed.

The familiar geometrical properties of flat space are different from those of two-dimensional, positively curved space. In flat space, a geodesic is a straight line. A triangle made from three straight lines, as in Figure 26.7, has angles that total 180°. A circle, no matter what its radius, always has a circumference equal to $2\pi r$.

Negative Curvature The third kind of space has negative curvature. Although an entire two-dimensional, negatively curved space can't actually be drawn, we can approximate part of such a space in the middle of the saddle-shape

FIGURE 26.5
A Section of Two-Dimensional Flat Space
This section of space is artificially bounded. In reality, it goes on indefinitely. A creature located at any point in this space can travel only in two directions or combinations of directions (up or down, left or right). A trip in or out of the page is impossible because it would be a three-dimensional trip.

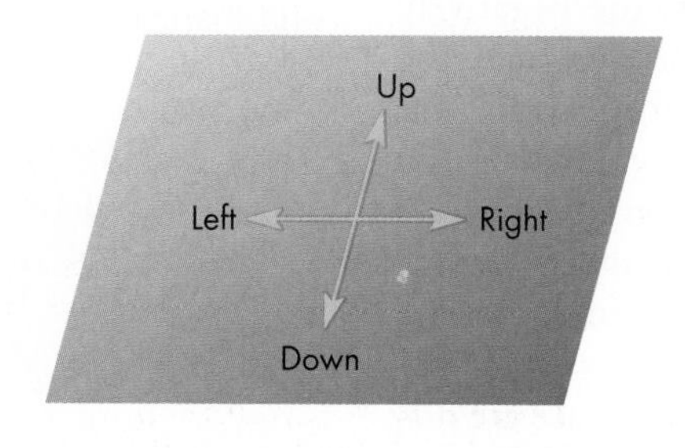

A Possible trips in two-dimensional flat space

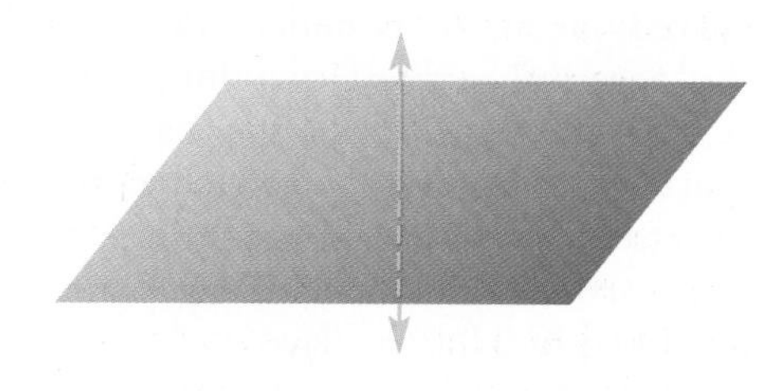

B Impossible trips in two-dimensional flat space

FIGURE 26.6
Expanding Two-Dimensional Flat Space
The same segment of this infinite rubber sheet is shown at two times during expansion. As the sheet is stretched in all directions, astronomers in all of the galaxies on it grow farther apart and observe every other galaxy moving away from them according to Hubble's law.

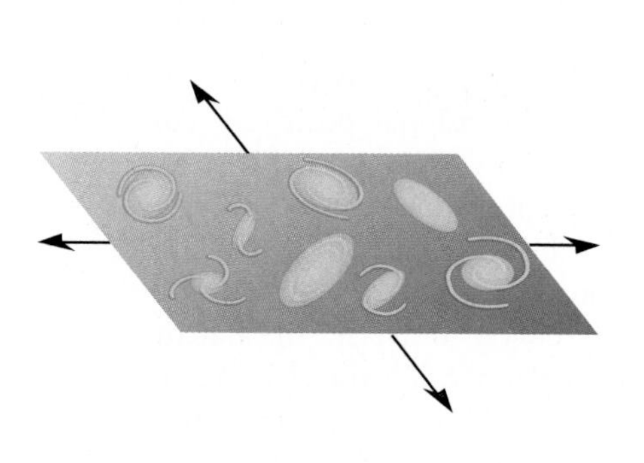

A Expanding two-dimensional flat space

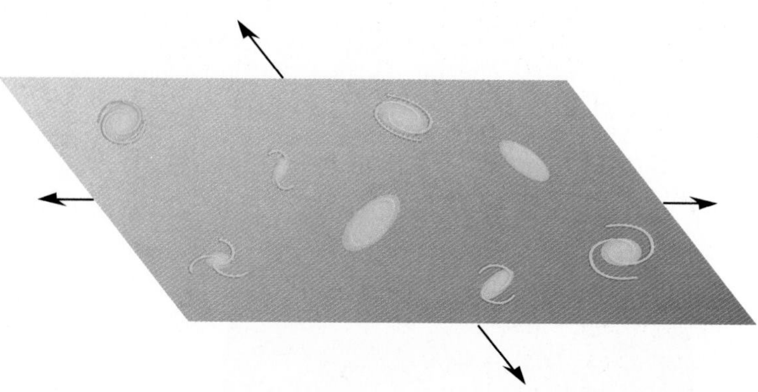

B Later during the expansion: The galaxies are farther apart.

FIGURE 26.7
Circles and Triangles in Two-Dimensional, Flat Space
In flat space, all circles have a circumference of $2\pi r$ no matter how large the radii they have and all triangles have angles that add up to 180°.

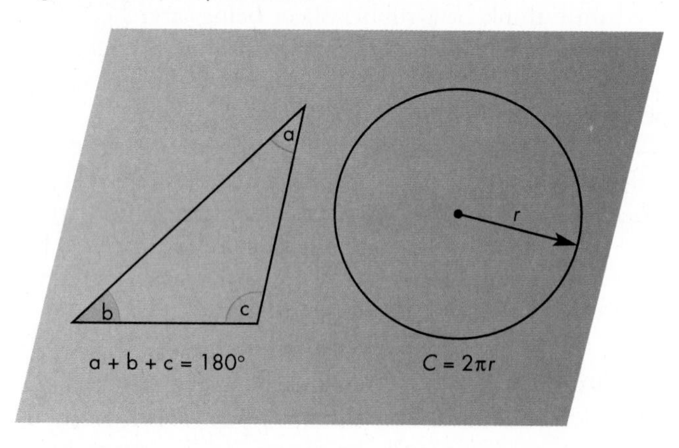

surface shown in Figure 26.8. Like flat space, negatively curved space is *infinite* in area and goes on forever in all directions. It has an infinite number of galaxies and an infinite mass. Like all other universes, the negatively curved one has no boundary. Because it has no boundary, the negatively curved two-dimensional universe has no center—all points in the space are the same. As with flat space, expansion of a negatively curved universe involves the stretching of an already infinite curved sheet. Expansion carries two-dimensional galaxies apart in a manner described by Hubble's law.

Triangles and circles in negatively curved space are different from those of either flat or positively curved space. The angles of a triangle in negatively curved space add up to less than 180°. The circumference of a circle, however, is always greater than $2\pi r$. For small circles, this difference is impossible to see, but as the radii of the circles get larger, the discrepancy steadily increases.

Testing Curvature

Suppose you were a two-dimensional creature living in a two-dimensional universe and you wanted to know what type of universe you lived in. How could you figure out whether your universe had positive, zero, or negative curvature? You could find out whether circles have ratios of circumference to radius that remain a constant 2π with increasing radius (flat space), decrease with increasing radius (positively curved space), or increase with increasing radius (negatively curved space). You could also send

FIGURE 26.8
Circles and Triangles in Two-Dimensional, Negatively Curved Space
The region near the center of a saddle approximates two-dimensional negatively curved space. In this space, circles have circumferences that exceed $2\pi r$ and triangles have angles that add up to less than 180°. Only for large circles and triangles, however, are these discrepancies apparent.

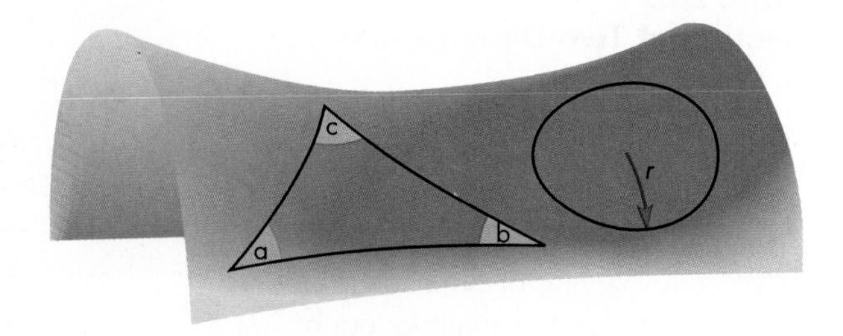

Triangle: a + b + c < 180°
Circle: Circumference (C) > $2\pi r$

friends to distant places with lasers and use the lasers to form a triangle. The number of degrees in the sum of the angles of the triangle could tell you the type of space in which you live. However, the circles and triangles you construct would have to be big enough that the curvature of space is noticeable. Small circles and triangles look the same in all three kinds of space. In reality, unless your universe was highly curved, you couldn't actually draw big enough circles and triangles to carry out the test.

One related test, though, might be possible. You could count the number of galaxies within circles of different radii. This process is similar to measuring the circumferences of circles, but it doesn't actually require you to draw the circles yourself. If space is flat and the density of galaxies is uniform, the number should increase with distance according to πr^2, the equation for the area of a circle in flat space. The graph in Figure 26.9 shows that for positively curved space, the number of galaxies increases less rapidly than πr^2 and for negatively curved space, it increases more rapidly. In principle, this test should allow you, if you use powerful enough telescopes, to determine the curvature of space. In reality, however, there might be serious complications. For one thing, the density of galaxies changes with time if your universe is expanding or contracting. This means that at great distances and lookback times there would be more or fewer galaxies than there would be in the absence of expansion or contraction. This could confuse a test for curvature that counted galaxies at different distances. Another complication is that galaxies evolve with time. The appearance or disappearance of some galaxies at certain distances and lookback times could also confuse the test. Similar tests of the curvature of our own universe have also suffered from these complications.

FIGURE 26.9
Testing Curvature in Two-Dimensional Space
The number of galaxies within circles of different distances is proportional to the area of those circles. For flat space, the number increases as πr^2. For positively curved space, the number rises more slowly with distance. For negatively curved space, the number rises more rapidly. If the universe is expanding or if galaxies evolve, this simple test becomes much more complex.

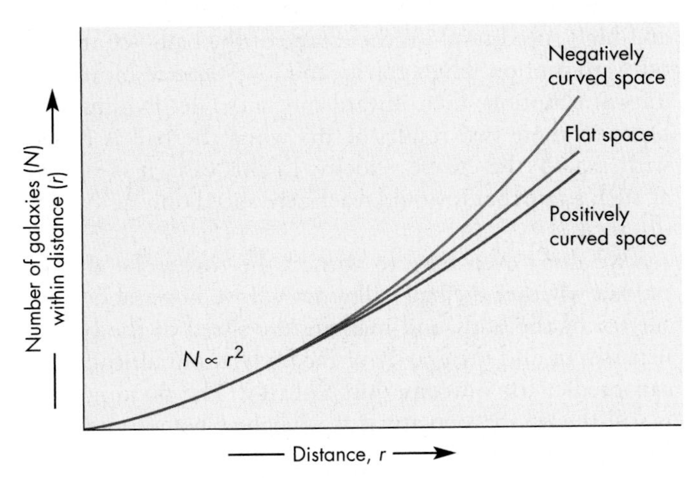

Three-Dimensional Universes

What, then, can we say about our own three-dimensional universe? General relativistic models of our universe show that the same three possibilities exist for our universe as for two-dimensional ones. The universe can be made of space with positive, zero, or negative curvature. Unfortunately, it is impossible for most people to imagine what that curved space looks like. How can space, already three-dimensional, be curved? Curved into what? To be able to visualize the answers to these questions, we would have to be able to imagine a fourth spatial dimension into which our entire universe curves. This is as impossible for most of us as it would be for two-dimensional creatures to be able to visualize a third dimension into which their own two-dimensional universe curves.

Fortunately, however, we don't have to be able to visualize the curvature of three-dimensional space to be able to understand its properties. We can safely extrapolate from the properties of two-dimensional space. Thus three dimensional positively curved space is finite in volume, just as two-dimensional positively curved space is finite in area. This means that there is a finite number of galaxies and a finite mass in our universe if it is positively curved. There is, however, no boundary. There is also no center. Even though there is no center, the universe is expanding. Every point is getting farther from every other point. It is incorrect to think of a three-dimensional, positively curved universe as a ball expanding into a much larger room. After all, a ball is a three-dimensional flat space object.

Three-dimensional flat space (zero curvature) is the kind of space that our senses tell us we live in. Like two-dimensional flat space, it is infinite in volume, contains an infinite number of galaxies and an infinite mass, has no boundary, and has no center. Three-dimensional, negatively curved space is also infinite in volume, has no boundary, and has no center.

The same kinds of geometrical tests that applied to two-dimensional universes would also be able to tell us the curvature of our universe, if we could carry them out. Like two-dimensional creatures, we could measure the angles of enormous triangles or the circumferences and areas of huge circles. We could also try to count the galaxies in volumes of space within increasing distances. Any of these, in principle, could tell us the kind of universe in which we live. In practice, however, the universe is much too large for us to make triangles and circles large enough to detect the effects of curvature.

Astronomers have tried to use galaxy counts to determine the curvature of space in the universe. These studies have been unable to determine the curvature of space so far, however, primarily because changes in the luminosities of galaxies over billions of years produce much larger effects than curvature. This doesn't mean that we can't determine the curvature of space and the fundamental nature of our universe, but the best way to do so requires measuring the history of the expansion of the universe.

The three possibilities for the curvature of space in the universe are negative, flat, and positive. If the universe has positive curvature, it is finite in volume. Otherwise its volume is infinite. For all three curvatures, however, the universe has no boundary and no center. Geometrical tests, such as determining the volumes of large spheres, can in principle be used to determine the curvature of space. In practice, such tests have proved unworkable.

Curvature and Density The basic postulate of general relativity is that matter and energy determine the curvature of space. As a result, it is hardly surprising that the average density of the universe and the spatial curvature of the universe are intimately related. Essentially, if the average density of the universe is large, the universe has positive spatial curvature. If density is small, the curvature of the universe is negative.

The dividing line between these two alternatives, flat spatial curvature, occurs if the universe has exactly the **critical density.** The critical density, ρ_c, is extremely small—equivalent to only 10^{-26} kg/m^3 or about ten hydrogen atoms per cubic meter. Because of the great significance of the critical density, estimates of the actual average density of matter and energy in the universe, ρ, are usually expressed in terms of the critical density using the parameter $\Omega_0 = \rho/\rho_c$. Thus, if the average density of the universe is less than the critical density (ρ less than ρ_c), then Ω_0 is less than one, and the universe is infinite with negative curvature. If ρ is greater than ρ_c, then Ω_0 is greater than one, and the universe is finite and has positive curvature. If ρ equals ρ_c, then Ω_0 is exactly equal to one, and the universe is infinite and flat. The symbol Ω_0 (note the subscript "0") refers to the present ratio of density to the critical density. The ratio at any past or future time is referred to simply as Ω.

It is very important to realize that the density that determines the value of Ω_0 includes all the forms of matter and energy in the universe, not just luminous matter. Thus, density includes electromagnetic radiation, neutrinos, and the dark matter that makes up much of galaxies and clusters of galaxies. Density may also include forms of energy that have not been identified in laboratory experiments and that may have properties much different from the forms of matter and energy with which we are familiar.

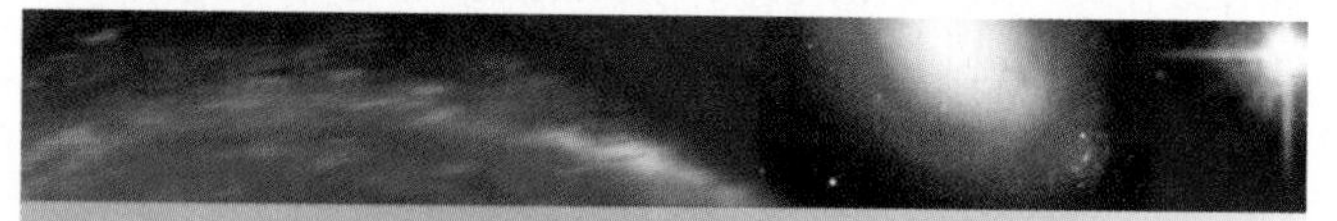

The curvature of space is closely related to the density of the universe. By determining the average density of the universe, it would be possible to calculate the curvature of space. Density includes all the forms of matter and energy in the universe.

The History and Future of Expansion The matter and energy in the universe determine not only the curvature of space but also how the expansion of the universe changes with time. We will look at some exotic possibilities later in this section, but first we will examine expansion if the density of the universe is made up only of normal matter (dark and bright) and energy. In this case, the density today determines not only how rapidly the expansion of the universe is decelerating but also whether the expansion will eventually come to an end. The expansion is slowing because of the self-gravitation of the universe.

The expansion of the universe can be compared with the flight of a ball that is tossed upward from the surface of a massive body. No matter how vigorously the ball is tossed upward, it is steadily slowed by the gravity of the body. If it is tossed upward with a small initial speed, the ball will reach a maximum height, at which time its speed slows to zero and it begins to fall back toward the body's surface, as shown in Figure 26.10. A much more energetic toss produces a large enough initial velocity that, although constantly slowed by gravity, the ball moves outward forever and escapes from the body.

What information do we need to predict, just after the ball is released, whether it is coming back or not? The two pieces of information we need are the speed of the ball and the strength of the body's gravity. Low gravity and high speed result in the escape of the ball—continual outward motion. High gravity and low speed result in first outward motion, then inward motion. The dividing line between these two results occurs when the ball is tossed with exactly the escape velocity. In this case, it is slowed at such a rate that it would reach zero speed only at infinite distance.

We don't even have to witness the toss to be able to predict whether the ball will return. If we hover above the surface of the body and measure the speed of the ball as it passes us and the gravity of the body at our altitude, we can predict the outcome just as easily. The dividing line is still the escape velocity at the height where we observe the flight of the ball.

FIGURE 26.10
Balls Thrown Upward from a Massive Body
For a weak throw (speed less than escape velocity), the ball slows as it rises, reaches a maximum height, and then falls back. For stronger throws (those with speed greater than or equal to escape velocity), the ball still slows as it rises, but nevertheless escapes from the body.

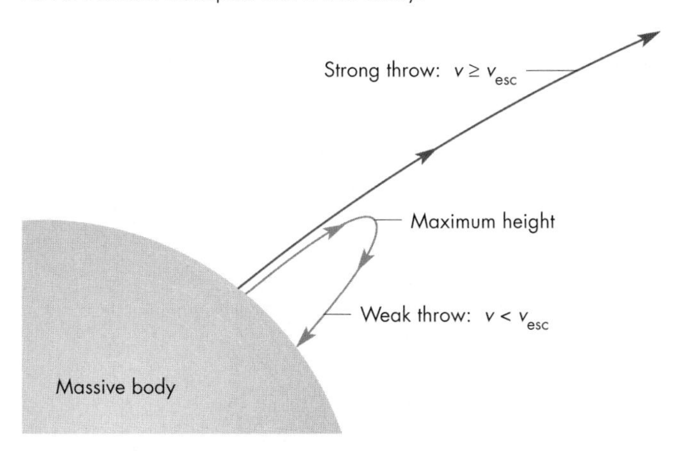

For the universe, the factors that correspond to speed and gravity are Hubble's constant and the average density of the universe. Hubble's constant describes the expansion rate of the universe. The average density tells how far apart adjacent atoms are and how strongly they attract each other gravitationally. In other words, it describes the self-gravity and curvature of the universe. The alternatives for the universe are the same as those for the ball tossed upward from the massive body. Slow expansion (a small value for Hubble's constant) and large self-gravity (high average density) lead to expansion that slows rapidly enough that it eventually halts as the universe reaches its maximum size. After that, the universe begins to contract at a rate that increases with time, as in Figure 26.11. On the other hand, a large value of Hubble's constant and a low density lead to a universe in which expansion always slows but never reaches zero. The universe expands forever.

The dividing line between the alternatives of expansion-contraction and unending expansion occurs for the same value of density that gives a flat universe, the critical density. Thus for $\Omega_0 < 1$ the universe is negatively curved and will expand forever. The flat universe, $\Omega_0 = 1$, will also expand forever. For $\Omega_0 > 1$, the universe is positively curved and will eventually contract.

Astronomers and physicists have developed a number of theories about forms of energy that behave very differently from normal matter. These forms of energy are often called **"dark energy"** by astronomers. The most famous form of dark energy is the energy associated with the **cosmological constant.** The cosmological constant was invented by Albert Einstein in 1917. Einstein used his equations of general relativity to make cosmological models. He found that the models could not stay the same

FIGURE 26.11
The Future of Expansion
For dense, positively curved models, the universe eventually begins to contract. Lower density, flat or negatively curved universe models expand forever. Because flat and negatively curved universes are infinite in size (but grow larger) it is simpler to talk about how a typical scale factor (say, the distance between clusters of galaxies) changes with time than to talk about the size of the universe.

ANIMATION *The future of expansion*

INTERACTIVE *Cosmology and the Cosmos' Fate*

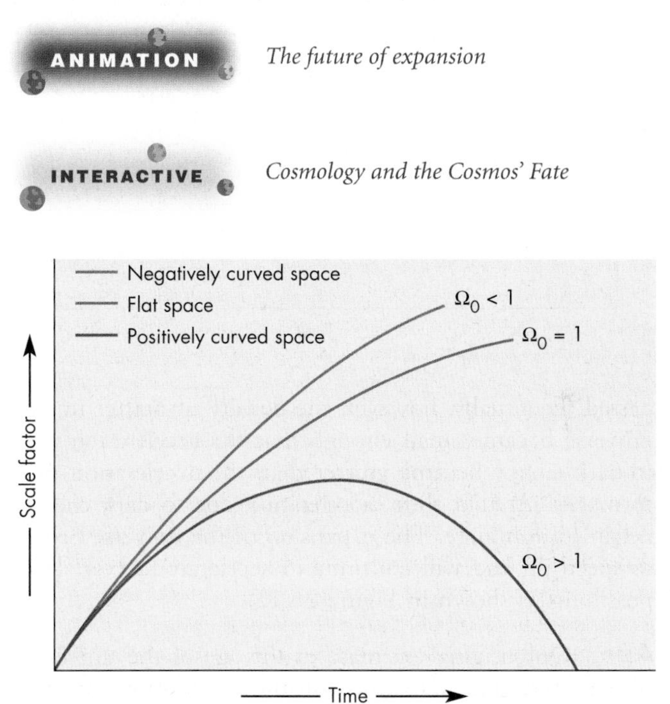

size, but needed to be either expanding or contracting. At the time Einstein made this discovery, the work of Edwin Hubble was some years in the future and it was not yet known that the universe is, indeed, expanding. To keep his models from predicting an expanding or contracting universe, Einstein proposed the existence of a self-repelling property of space to just balance the attractive force of gravity. After Hubble discovered that the universe is expanding, Einstein called the cosmological constant "the biggest blunder of my life."

Now it seems that Einstein was right after all and that dark energy really exists. If this is the case, then the history and future of expansion cannot be deduced from the density of matter alone. Instead, there are two competing factors at work. Matter has the effect of slowing expansion. As time passes and the universe grows larger, the density of matter drops and the deceleration it produces drops as well. Dark energy has the effect of accelerating expansion. Because the cosmological constant is a property of space itself, its density remains constant and the acceleration it produces stays the same as time passes. This means that if there is dark energy, then the early, high-density universe was dominated by matter. During that time, the expansion of the universe slowed as time

FIGURE 26.12
Dark Energy and the Expanding Universe
The early history of the universe was dominated by matter and expansion slowed. After density became small enough, dark energy began to dominate and produce acceleration of the expansion.

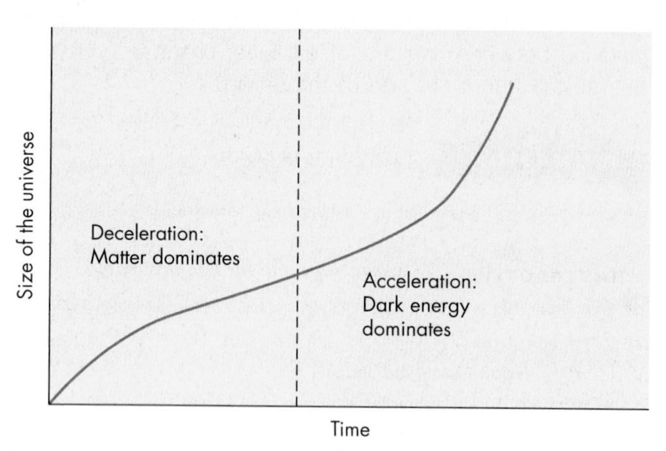

passed. Eventually, however, the density of matter in the universe became small enough that the acceleration due to dark energy became greater than the deceleration due to matter. At that time, acceleration due to dark energy began to dominate. The expansion of the universe began to speed up and will continue to accelerate forever. This possibility is shown in Figure 26.12.

Age Our original estimate of the age of the universe, the Hubble time, was based on the assumption that the expansion of the universe has always gone on at the same rate. This led to the result that the age is given by $1/H$, where H is Hubble's constant. If the rate of expansion has been variable, however, the Hubble time may not be a very good estimate.

Again, let us first look at the case in which normal matter (both dark and luminous) and energy make up the density of the universe. If so, the self-gravity of the universe has always slowed the expansion and the present expansion rate is lower than the expansion rate in the past. This means that less time was needed for the universe to reach its present size than if expansion had always gone on at the present rate. The actual age of the universe must be less than $1/H$, but by how much depends on the average density of the universe.

The graph in Figure 26.13 shows the size of the universe versus time for three values of Ω_0. For $\Omega_0 = 0$, there is no matter in the universe and never has been. In that case, the actual age of the universe is the Hubble time, $1/H$. If the density of the universe is very small (Ω_0 only slightly greater than zero) and the deceleration is slight, then the actual age of the universe approaches $1/H$. If the density of the universe equals the critical density ($\Omega_0 = 1$), the universe is flat and its age is given by $2/3H$. Denser universes (with positive curvature) are younger than $2/3H$.

FIGURE 26.13
The Age of the Universe If the Universe Contains No Dark Energy
The age of the universe is equal to the Hubble time only for empty universes ($\Omega_0 = 0$) in which there is no deceleration of expansion. For denser universes, age falls short of the Hubble time. For a universe in which space is flat and the density is equal to the critical density ($\Omega_0 = 1$), the actual age is equal to two-thirds of the Hubble time.

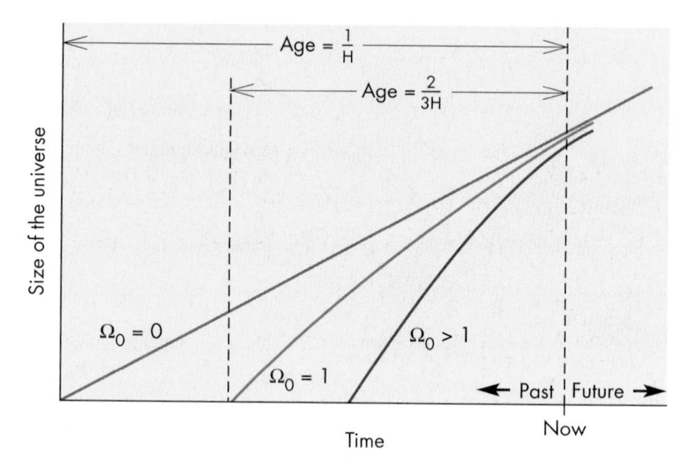

The situation is even more complicated if the universe contains dark energy. In that case, the universe first slowed and then accelerated. Depending on how long ago dark energy began to dominate and the acceleration of expansion started, the present rate of expansion may be either greater than or less than the average rate of expansion in the past. The Hubble time may be either an overestimate or an underestimate of the age of the universe.

Whether the expansion of the universe will continue forever depends on the curvature of space and thus on the density of the universe. Only if the density of the universe is low and space is negatively curved or flat will expansion continue forever. If density is high, contraction will eventually take place. The accuracy with which the Hubble time approximates the age of the universe also depends on density. Unless average density is low, the actual age of the universe will be significantly less than the Hubble time. If dark energy exists, the rate of expansion may be increasing.

Distant Supernovae

The history of the expansion of the universe depends on the density of normal matter and energy and on the acceleration due to dark energy. By determining how the rate of expansion of the universe has changed with time, it should be possible to discover the density of the universe, the contribution of the dark energy, and the curvature of space. This involves looking very far back in time and, thus, outward in space. In order to learn the history of expansion, several groups of astronomers have been observing very distant (and, therefore, ancient) type Ia supernovae. The reason that type Ia supernovae were chosen is that their light curves obey a relationship between peak luminosity and the time it takes for the supernova to fade. This means that by measuring the light curve of a distant type Ia supernova, astronomers can determine its intrinsic brightness. From intrinsic brightness and apparent brightness, the distance of the supernova can be determined.

Type Ia supernovae occur only about once every few hundred years in a given galaxy, so in order to detect many supernovae astronomers need to monitor a huge number of galaxies. In fact, one of the teams monitors almost 100,000 galaxies. This is enough to detect dozens of supernovae during every observing period. Once the supernovae are detected, their light curves are measured and their instrinsic brightnesses obtained. Using the *Hubble Space Telescope* or large telescopes like the Keck telescopes, astronomers also measure the redshifts of the galaxies in which the supernovae occur. Astronomers have found many supernovae so distant that they have redshifts as large as 1.76. Their light has been traveling toward us for 10 billion years—about 70% the age of the universe.

A surprising result of the studies of distant supernovae is that they are fainter than would be expected if the expansion of the universe were constant with time. This is because they are more distant than they would be if the expansion had been constant. We estimate the distance of a galaxy with a supernova by measuring its redshift and dividing by Hubble's constant (see Equation 23.2). If the expansion of the universe has been accelerating, then the present value of Hubble's constant is greater than it was in the past. If we use the present value of Hubble's constant, then we underestimate the distance to the galaxy and overestimate its apparent brightness. So the faintness of distant supernovae is a strong indication that dark energy exists and is causing the expansion of the universe to accelerate.

If the expansion of the universe is accelerating, then there should have been an early era during which the normal matter and energy in the universe dominated the rate of expansion and the expansion was decelerating. This means that if we can observe extremely distant supernovae, we can see the universe as it was before acceleration began. In that case, the present value of Hubble's constant would be an underestimate of its average value since the time the supernova outburst occurred. Our estimate of its distance based on the present value of Hubble's constant would be an overestimate and the supernova would appear brighter than expected. This is exactly what has been observed for the most distant supernovae—those with redshifts greater than about 1.0. Supernovae seen at a redshift of 1.7 appear about twice as bright as they would if Hubble's constant really had been constant.

The analysis of the redshifts and brightnesses of distant supernovae must take both dark energy and the density of normal matter and energy into account. The result of the analysis is that normal matter and energy supply about 28% of the density required for a flat universe. Dark energy supplies about 72% of the density required for a flat universe. Thus, Ω is very close to 1. If Ω is 1, then the universe is now and always has been flat. Using the present value of Hubble's constant of 71 km/s per Mpc, the age of the universe is 13.7 billion years. This is very close to the Hubble time—the age of the universe if the expansion rate had been constant. Thus, the effects of deceleration and acceleration have, so far, just about canceled each other out. Acceleration started 4.5 billion years ago, about the time the Sun and solar system formed. To see the time when acceleration began, we need to look outward to a redshift of 0.43. Once acceleration began, it will continue forever.

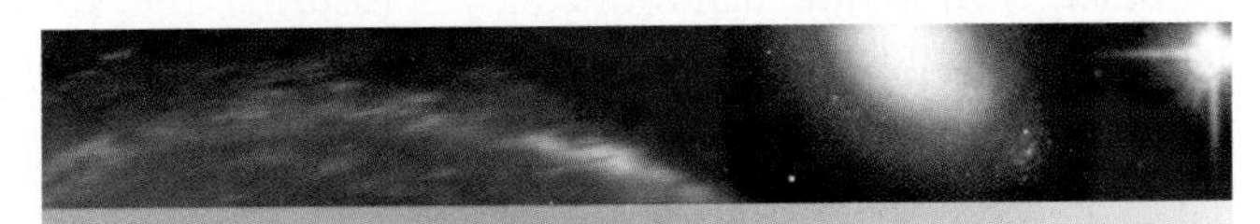

Studies of ancient type Ia supernovae show that they are fainter than would be expected if the expansion rate of the universe had been constant. Extremely ancient and distant supernovae, are brighter than expected. The best explanation for this is that dark energy exists. In the early universe, matter dominated and expansion slowed. About 4.5 billion years ago, dark energy began to dominate and expansion accelerated.

26.3 THE BIG BANG

Most of what we know about the early stages of the expansion of the universe, the Big Bang, has been learned since the 1940s. A key step was the realization that temperature during the Big Bang was very high. The high temperature and the great energy associated with it were responsible for a number of events that determined the basic properties of the universe we live in. We are reasonably certain about events and processes that took place between a tiny fraction of a second after the expansion began and millions of years later, when the universe had cooled and grown enormously. This section describes the "standard Big Bang," beginning at the point where we are quite sure about what happened. More uncertain ideas about what happened at still earlier times are described later in the chapter.

When Did the Big Bang Occur?

We can estimate the age of the universe by dating the oldest objects we can find. This method provides a lower limit on the age of the universe because the universe must be older than these ancient objects. The oldest objects in the vicinity of the Sun are white dwarf stars. Because white dwarfs dim with age, the least luminous are the oldest. These appear to be about 10 billion years old. However, white dwarf stars near the Sun lie in the galactic plane and could not have formed until the galaxy assumed nearly its present form. This means that the galaxy and the universe could be quite a bit older than the dim white dwarfs near the Sun.

Farther from the Sun are the globular clusters, probably as old as the galaxy itself. The H-R diagrams of a globular cluster can be used to find the age of the globular cluster by determining the brightness and color of the cluster's hottest main sequence stars. (To review how this can be done, refer to Figure 19.12.) In order to find the brightness of the hottest main sequence stars in a globular cluster, the distance to the cluster must be known. Distances to globular clusters are determined by measuring the apparent magnitudes of the RR Lyrae stars in the clusters. Because all RR Lyrae stars have about the same absolute magnitude, the difference between absolute and apparent magnitude gives the distance to the RR Lyrae star and the cluster containing it. The H-R diagrams of globular clusters show they are about 11 to 12 billion years old. The universe is older than this by the amount of time needed for galaxies to begin to form after the expansion of the universe began.

Taken together, the age estimate from white dwarfs and globular clusters suggest that the universe is more than 11 billion years old.

The minimum age of the universe (11 to 12 billion years) found from the ages of the oldest objects we can reliably date can be compared with age estimates for models for the history of the expansion of the universe. One such estimate, the Hubble time (see Figure 26.13), is the age of the universe if it has very low density. Using our current estimate of Hubble's constant (71 km/s per Mpc), the Hubble time is 13.8 billion years. The discovery that the expansion of the universe is accelerating as a result of dark energy leads to a model of the universe in which space is flat, but the contribution of matter to the density of the universe is small. In that case, the expansion of the universe first decelerated and then accelerated (see Figure 26.12). The age of the universe is 13.7 billion years, only slightly less than the Hubble time. This age is probably consistent with the ages of globular clusters and is another piece of evidence that favors the reality of dark energy.

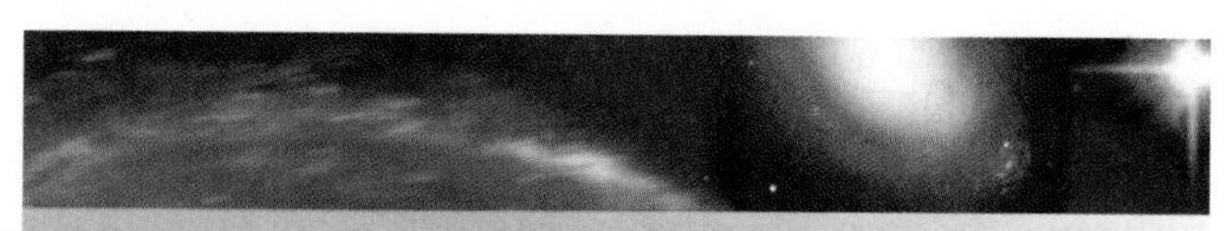

The age of the universe can be estimated from the ages of the oldest objects in the universe. The ages of white dwarf stars and globular clusters suggest that the universe is more than 11 billion years old. Cosmological models suggest the universe has been expanding for about 13.7 billion years.

What Happened?

Many of the significant events of the early Big Bang took place extremely rapidly—in many cases, in a tiny fraction of a second. It is important, however, to keep two things in mind when thinking about the early universe. The first is that at the extreme densities and temperatures that existed then, the rates of interactions among photons and particles of matter were very great. Even a tiny amount of time was long enough for many interactions of a given kind to take place. The second is that the timescale on which the universe changed significantly lengthened steadily as the expansion proceeded. Tremendous changes occurred within the first second of the expansion. A few minutes later, however, the universe hardly changed at all during a given 1-second interval.

The First Second We will begin following the expansion of the universe 1 microsecond (10^{-6} s) after it began. At this time, the temperature of the universe was 10 trillion K (10^{13} K), about a million times hotter than the center of the Sun. Although any time traveler who traveled to that moment would perish instantly (the universe would remain lethal for hundreds of thousands of years to come), we can at least imagine what an indestructible traveler might have seen. The glow of radiation would have been intense and uniform in all directions. The traveler would have been in a bright fog that limited vision to a small fraction of a centimeter.

The high-energy gamma rays that made up most of the radiation were so energetic that they were capable of **pair production,** the creation of matter from radiation. In pair production, illustrated in Figure 26.14, the collision of two photons yields a particle and its **antimatter** counterpart. For example, one set of products of pair production is an electron and a positron, which is as massive as an electron but has a positive charge. Other possible pairs are a proton and antiproton, a neutrino and antineutrino, and many others. Pair production is one of the most spectacular consequences of the equivalence of matter and energy as expressed by Einstein's famous equation, $E = mc^2$. For two photons to produce a particle, antiparticle pair, the photons must have energy at least equal

FIGURE 26.14
Pair Production and Annihilation
A, In pair production, two photons produce a pair of matter, antimatter particles. **B,** In annihilation, matter and antimatter particles destroy each other, producing photons. Pair production only can occur where there are energetic photons, whereas annihilation occurs whenever a particle and its antiparticle collide.

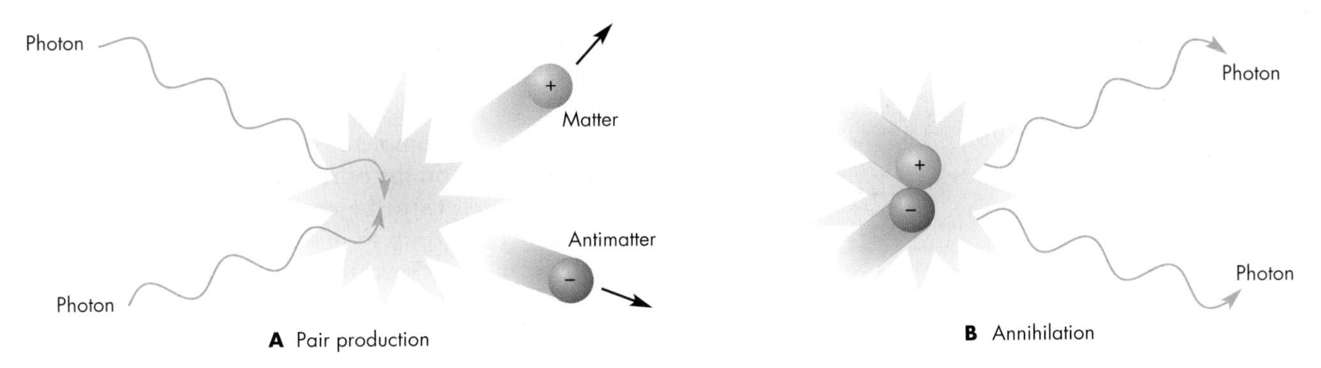

to mc^2, where m is the combined masses of the particle and antiparticle and c is the speed of light. A microsecond after the universe began, the gamma ray photons present had enough energy to produce protons, neutrons, electrons, and their antiparticles.

At the same time that pair production was occurring, particles and their antiparticles were destroying each other through **annihilation,** in which a particle and its antiparticle collide and disappear in a burst of radiation. Annihilation is the opposite of pair production and converts matter into photons. At 1 microsecond, the universe was a dense mixture of radiation and matter. Pairs of particles appeared from nowhere and then disappeared again in a flash of gamma rays, as shown in Figure 26.14.

As the universe grew older, temperature decreased and the energies of photons decreased as well. After the first microsecond, the gamma rays that dominated the universe lacked enough energy to make proton, antiproton pairs or neutron, antineutron pairs. Less massive electron, positron pairs could still be produced. Although protons and neutrons could no longer be made, they continued to annihilate with their antiparticles, destroying most of the matter that had been built up from radiation during the first microsecond. By the time the universe was 1 second old, the temperature had dropped to about 10 billion K. The photons present at 1 second lacked even enough energy to make electron, positron pairs, so pair production ceased entirely. The annihilation of electrons and positrons continued, however, swiftly reducing the abundance of electrons and positrons in the universe.

One consequence of the end of pair production and the reduction in the numbers of electrons and positrons was the liberation of the neutrinos that were abundant in the early universe (as they are today). Neutrinos interact only very weakly with matter. At the high density of the universe before it was 1 second old and with the large abundance of electrons and positrons, neutrinos interacted often enough with electrons and positrons that they were "coupled" to the matter in the universe. Neutrinos traveled only short distances between interactions with matter. At about 1 second, however, the rate of interactions between matter and neutrinos was decreasing extremely rapidly. At this time neutrinos practically stopped interacting with the rest of the universe. Since the first second, the neutrinos that existed at that time have traveled freely through the universe, moving at or near the speed of light. Only an insignificant fraction of them have ever interacted with matter again. There are about 1 billion primordial neutrinos in every cubic meter of the universe. About 10^{18} of them pass through each of us every second. There is no danger, however, because the odds are so small that any one of them will interact with the matter within us.

During the first second, the universe was so hot that photons had enough energy to make matter and antimatter through pair production. Matter and antimatter also destroyed each other through the process of annihilation. After 1 second, photons lacked sufficient energy for pair production, so no more matter and antimatter were produced. The amount of matter and antimatter in the universe decreased because the process of annihilation continued. At this time, the neutrinos in the universe stopped interacting with matter and have moved freely through the universe ever since.

Formation of Helium One second after it began to expand, the universe was a sea of high-energy radiation (gamma rays and X rays) with a smattering of neutrons,

FIGURE 26.15
Abundances of Various Isotopes During the Time of Nuclear Reactions
This graph shows the fraction of the matter in the universe that consisted of various particles and nuclei. The graph covers the first day after expansion began. For several hundred seconds, a series of nuclear reactions made helium from hydrogen. When the universe had cooled and expanded to the point where the reactions stopped, it contained a mixture of isotopes of H, He, Li, and Be.

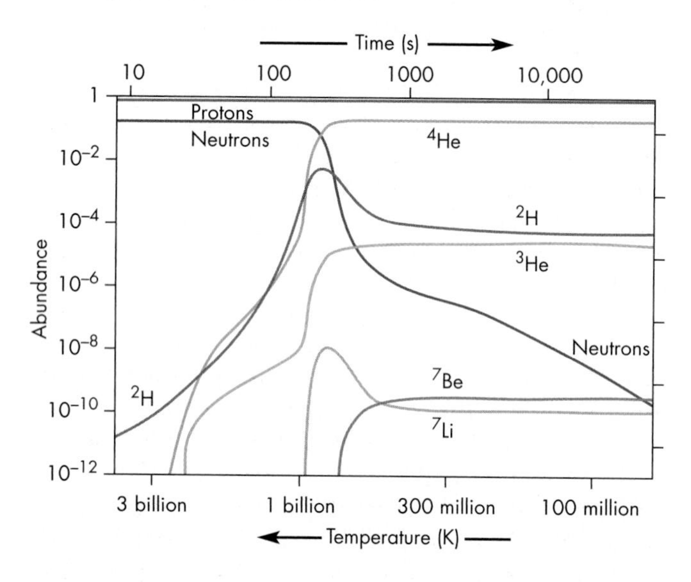

protons, and electrons. Neutrons, slightly more massive than protons, can be produced by the reaction of a proton and an electron. Neutrons, however, spontaneously decay into an electron and proton with a half-life of 13 minutes. The interactions between electrons, protons, and neutrons resulted in about one neutron for every five protons. The protons and neutrons collided with each other and reacted to form **deuterons,** which are hydrogen nuclei containing both a proton and a neutron. At first, however, deuterons could not survive. They were constantly bombarded by photons that had sufficient energy to turn them back into protons and neutrons again. Not until 100 seconds after the expansion began, when the temperature had dropped to 1 billion K (10^9 K), did the energy of a typical photon decline to the point where it could no longer disrupt a deuteron. After that time, the abundance of deuterons climbed swiftly, as shown in Figure 26.15.

Deuterons react easily with protons. After deuterons became stable, reactions with protons initiated a series of dozens of nuclear reactions that led to more and more massive nuclei. For about 200 seconds, the temperature remained high enough for nuclear reactions to change the chemical makeup of the universe from entirely hydrogen (protons) into a more complex mixture that included protons (^{1}H), deuterons (^{2}H), two isotopes of helium (^{3}He and ^{4}He), and small amounts of lithium (^{7}Li) and beryllium (^{7}Be). The way that these nuclei changed with time is shown in Figure 26.15. One of the most important clues we have about conditions in the early universe is the relative abundances of these isotopes in primordial material that has not been altered in composition by nuclear burning in stars. If the density of protons and neutrons was high at the time nuclear reactions in the Big Bang began, then the rate of nuclear reactions would have been high. One consequence would have been the almost complete consumption of deuterons. If density was low, on the other hand, there was not enough time for deuterons to be entirely consumed or for ^{3}He to be entirely used up to make ^{4}He. Figure 26.16 shows the abundances of several key isotopes that would have resulted from different densities during the nuclear reactions in the early Big Bang.

FIGURE 26.16
Primordial Abundances and the Density of the Universe
The abundances of the isotopes of hydrogen, helium, and lithium produced during the first few minutes of the Big Bang depend on the density of protons and neutrons at the time of the nuclear reactions. The density of the universe today (shown in the figure) also depends on the density at the time the nuclear reactions were taking place. The horizontal shaded regions show current estimates of the abundances of the isotopes ^{4}He, ^{2}H, ^{3}He, and ^{7}Li relative to protons. The vertical shaded region shows the range of densities required to explain the abundances of these isotopes. This range, which includes about 2 to about 5 $\times$ 10^{-28} kg/m^3, is much less than the critical density (10^{-26} kg/m^3).

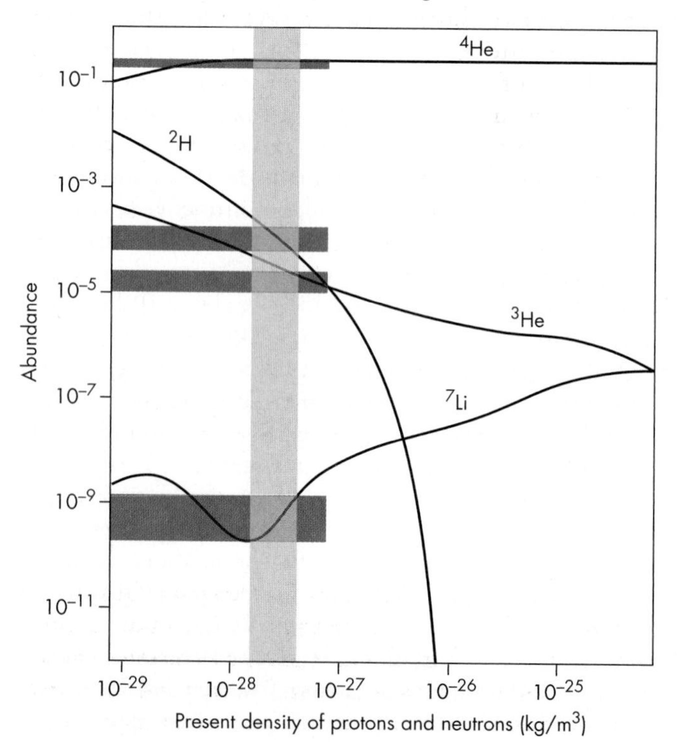

Astronomers have been able to measure the primordial abundances of several of these isotopes by examining the spectra of very old stars. These studies show that ^{4}He makes up 24% of primordial matter (about 1 kg in every 4), ^{3}He makes up about one part in 10^4, deuterium about one part in 10^5, and ^{7}Li about one part in 10^9. These are the kind of abundances that would have resulted from a low-density environment rather than a high-density environment. In fact, the most probable value for the density of the universe at the time nuclear reactions were taking place was only a few percent of the critical density. Thus, the abundances of the isotopes produced during the first few minutes of the Big Bang indicate that matter in the form of electrons and nuclei falls far short of providing the self-gravity to produce a positively curved or flat universe. The "dark" matter in the universe probably can't be electrons and nuclei, but must take the form of exotic, undiscovered, types of particles.

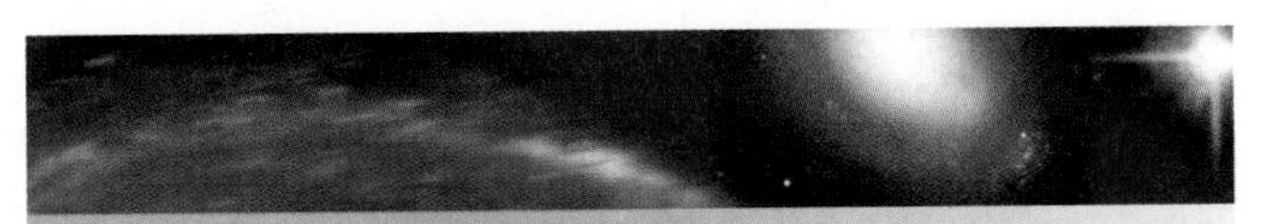

From about 100 to 300 seconds after expansion began, conditions were right for nuclear reactions to fuse hydrogen into helium. The abundances of isotopes produced during that time show that the density of electrons, protons, neutrons, and nuclei alone was too low for the universe to have positive curvature.

The Radiation Era Radiation dominated the universe during the first few minutes after expansion began and for a long time afterward. That is, if the radiant energy in a cubic meter of space was expressed as mass according to $E = mc^2$, it would have had a density much larger than that of matter. This stage in the development of the universe, therefore, is known as the **radiation era.**

During the radiation era, the universe was still too hot (1 million K after the first few years) for atoms to be stable. Whenever an electron and a proton combined to form an atom, the atom was almost immediately torn apart by an energetic photon or by a collision with another particle. The gas in the universe was very opaque, because free electrons are very efficient at absorbing and scattering radiation. The light in the universe never got very far from the place where it was emitted before it was destroyed or scattered again. An observer would have been surrounded by a bright mist that made it impossible to see more than a short distance.

Things were gradually changing, however. Both temperature and density were dropping. Collisions became less violent. The radiation that flooded the universe gradually shifted from gamma rays and X rays (high energy, short wavelengths) to ultraviolet and visible light (lower energy, longer wavelengths). The surrounding mist became dimmer and redder as time went by and the distance to which it would have been possible to see slowly increased. While this was happening, the relative importance of radiation and matter was also changing. Although the density of both matter and radiation was falling, the density of radiation was falling faster. About 380 thousand years after expansion began, when the temperature had fallen to a few thousand kelvins, two events happened almost simultaneously. Matter began to dominate the universe, and the universe became transparent.

The Clearing of the Universe The universe changed from opaque to transparent because the temperature became low enough for atoms to survive. Collisions and radiation, which had ripped atoms apart as soon as they formed, no longer had enough energy to disrupt atoms. Atoms absorb and scatter radiation far less efficiently than an ionized gas of electrons and ions. As soon as atoms formed, therefore, radiation could no longer be absorbed or scattered as soon as it was produced. Light and other forms of electromagnetic radiation could then travel through space with almost complete freedom. The moment when light broke free from matter is known as the **decoupling epoch** or the **recombination epoch,** because electrons and nuclei combined to form atoms. Figure 26.17 illustrates how it was much easier for light and other forms of radiation to travel through space after recombination than it was before.

The decoupling epoch has special significance for our efforts to explore the early universe by looking deep into space and far back into time. This is because all the light emitted before the decoupling epoch was destroyed almost immediately after it was emitted. Thus it is impossible to study the first 380 thousand years of expansion using electromagnetic radiation. In contrast, nearly all of the radiation emitted during and after the decoupling epoch is still present in the universe. In fact, by looking far enough into space, we can still see the glowing gas that made up the universe at the time that radiation broke free from matter. The discovery of that glow, called the **cosmic microwave background (CMB),** was one of the most significant scientific findings of the twentieth century.

About 380 thousand years after expansion began, ions and electrons combined. When this happened, the universe became transparent to radiation. All the radiation that existed before this time was destroyed. Most of the radiation emitted during and after recombination is still present in the universe.

FIGURE 26.17
The Universe Before and After Recombination of Nuclei and Electrons
A, Before recombination, free electrons absorbed and scattered radiation, making the universe opaque. **B,** When the universe had cooled enough for atoms to form, the universe became transparent, making it possible for radiation to travel great distances.

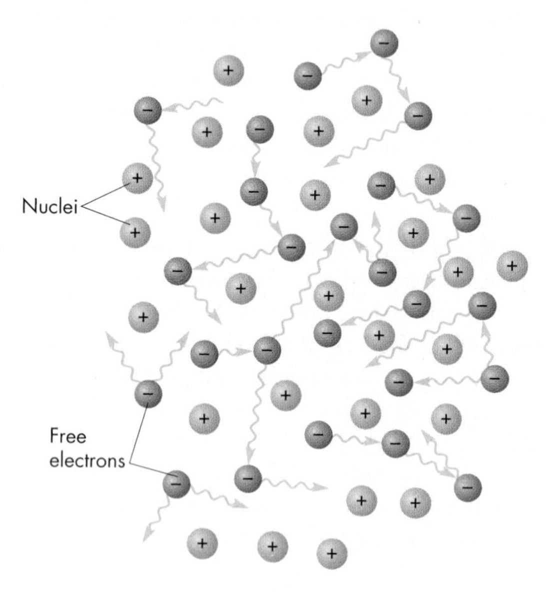

A Before recombination: The universe was opaque.

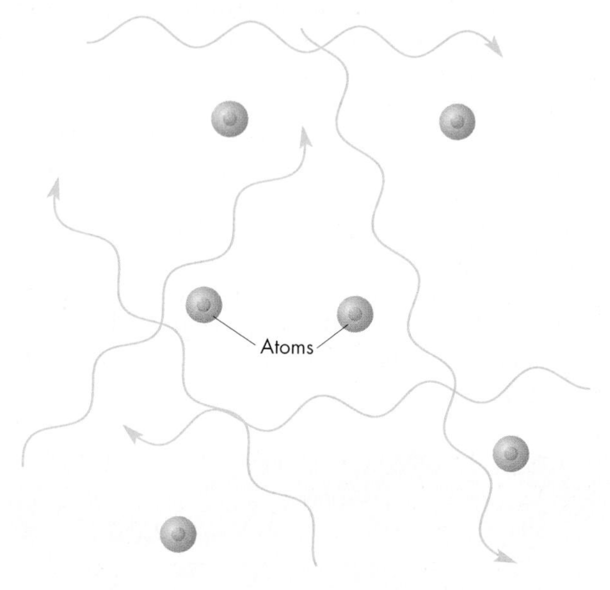

B After recombination: The universe was transparent.

The Cosmic Microwave Background

Like many other important scientific discoveries, the cosmic microwave background (abbreviated CMB) was found by accident. By the time of its discovery, however, its general properties had been understood for more than a decade.

What Should the CMB Look Like? The cosmic microwave background is the electromagnetic radiation emitted by the universe at the epoch of decoupling. At that time the universe had a temperature of about 3000 K, so the radiation that was present was mostly in the visible and infrared parts of the spectrum. Does this mean that the CMB should be sought with optical and infrared telescopes? The answer is no. To see the CMB, we need to observe the universe as it was about four hundred thousand years after expansion began. This means that we need to use the telescope as a time machine and look backward nearly to the beginning of the universe.

The farther we look into space, however, the greater the redshift we observe. To see far enough to observe the CMB, we need to view parts of the universe that have a redshift of about 1000. Thus, the CMB now has 1000 times the wavelength it had when it was emitted. To detect it, infrared and radio telescopes (hence the "microwave" in the name CMB) are needed rather than optical telescopes. Instead of showing the spectrum emitted by the universe when its temperature was 3000 K, the CMB appears today with the spectrum characteristic of a body with a temperature of 3 K.

Where should we look to see the CMB? The answer is that it can be seen in all directions. We can see the CMB in any direction in which we can look outward to a redshift of 1000. In any direction in the sky that isn't blocked by some foreground object, we should see a glow like that emitted by a blackbody with a temperature of about 3 K. Why do we have to look so far to see the CMB? Why can't we see the radiation emitted at the time of decoupling by the matter in our part of the universe? The reason is that the CMB emitted by the matter that would ultimately form the Milky Way is long gone. It left our part of the universe at the speed of light billions of years ago and now forms part of the CMB for observers in remote parts of the universe, as shown in Figure 26.18.

Discovery When George Gamow and other astronomers first realized in the 1940s that the early universe had been very hot, they also realized that there should be surviving radiation from the early universe. They predicted that the radiation should have a temperature of between 5 and 50 K but made no effort to search for it. A team of astronomers at Princeton University refined the calculations in 1964 and concluded that the surviving radiation should have a temperature of only a few kelvins. They began to build a radio telescope to search for it.

FIGURE 26.18
The Origin of the Cosmic Microwave Background for Us (the Milky Way) and Creatures in a Very Distant Galaxy
The spherical surfaces show the places where the CMB observed today originated for each location. Just as the matter that became the distant galaxy radiated light that became part of our CMB, so did the matter that formed the Milky Way emit radiation that is now being received at the distant galaxy as part of the CMB that astronomers in that galaxy observe.

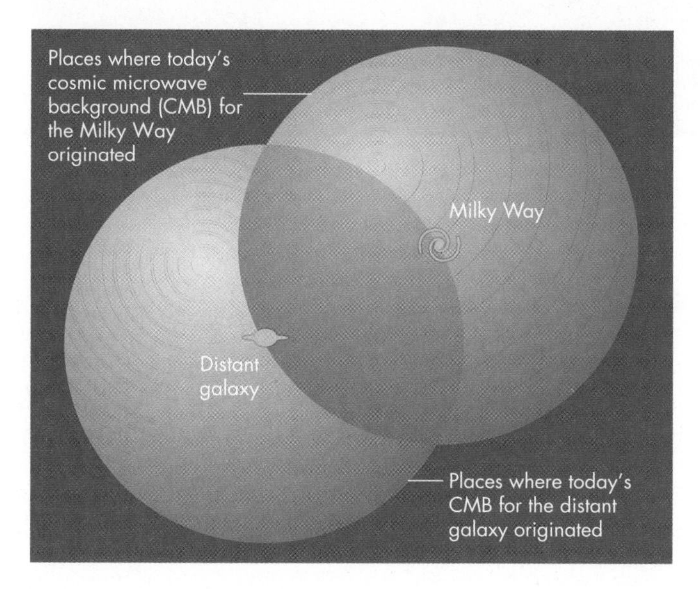

FIGURE 26.19
The Discoverers of the Cosmic Microwave Background
Arno Penzias (right) and Robert Wilson (left) are shown near the radio antenna that they used to discover the CMB. Penzias and Wilson shared the 1978 Nobel Prize in Physics for their discovery.

Before they could complete their equipment, however, the CMB was found by Arno Penzias and Robert Wilson of the Bell Telephone Laboratories in New Jersey. Penzias and Wilson used a small horn-shaped antenna, shown in Figure 26.19, that had been built for satellite communications. Although they had carefully measured and accounted for many sources of noise, including those of the Earth's atmosphere, the Sun, and the Milky Way, they were unable to explain a source of noise that seemed to be of equal strength in all directions in the sky. They consulted a colleague, who was aware of the work of the Princeton group. When Penzias and Wilson talked with the Princeton astronomers, the two groups realized that the source of noise was the cosmic microwave background. Since its original discovery, the spectrum and brightness of the CMB have been measured and found to correspond to a blackbody with a temperature of 2.74 K.

Isotropy of the CMB When they discovered the cosmic microwave background, Penzias and Wilson found that its brightness is remarkably **isotropic.** That is, its brightness is nearly the same in all directions. Numerous careful measurements of the brightness of the CMB have confirmed that it is uniform to a level of a few hundred parts per million. The departures from uniformity are very important, however, because they tell us about conditions in the early universe. Although the fluctuations in the CMB have been measured by many experiments, the most complete set of measurements was carried out by the *Wilkinson Microwave Anisotropy Probe* (*WMAP*), a satellite that was launched in 2001. Figure 26.20 shows the *WMAP* map of the CMB across the entire celestial sphere. Fluctuations can be seen on all scales from much less than a degree to many degrees. On average, the bright regions are about a degree across.

The fluctuations in the CMB were produced by sound-like waves that rippled back and forth across the early universe. The bright spots are about as large as the maximum distance the waves could have traveled between the Big Bang and the epoch of decoupling. The speed of the waves and the structure they produced depend on the density of normal matter at the time the CMB was emitted. Analysis of the structure of the CMB shows that normal matter made up only a few percent of the density of the universe

FIGURE 26.20 Fluctuations in the Cosmic Microwave Background
This map, based on *Wilkinson Microwave Anisotropy Probe* (*WMAP*) data, shows the brightness of the CMB across the entire celestial sphere, shown as two hemispheres. Warmer regions are shown in red, cooler regions in blue. The fluctuations in the CMB are extremely small—only about 1/100,000th of the CMB's average brightness.

at the time the CMB was emitted. The contribution of normal matter to the density of the universe should be similarly small today. The sizes of the fluctuations also depend on the curvature of space and show that space was and is very nearly flat. A complete analysis of the *WMAP* data shows that space is flat ($\Omega = 1$), normal matter makes up about 5% of the density of the universe, dark matter makes up 23%, and dark energy (72%) dominates the universe. Other *WMAP* results are that the age of the universe is 13.7 billion years, the Hubble constant has a value of 71 km/s per Mpc, and the first stars formed when the universe was only about 400 million years old. Light from the first stars lit up the universe and ended the "dark ages" that had lasted since the time of decoupling. The close agreement between the results of *WMAP* and those from the study of distant supernovae, the analysis of the elements produced in the first few minutes of the Big Bang, and the large-scale distribution of galaxies in space (Chapter 25) gives astronomers some confidence that our current description of the universe—dominated by dark matter and dark energy—is correct.

The cosmic microwave background (CMB) is the light emitted at the time when the universe became transparent. It is seen in all directions but comes to us from such great distance that it is redshifted by a factor of about 1000. The CMB is redshifted to such an extent that it is observed as radio waves. The spatial structure of the CMB shows that space is flat and normal matter contributes only a few percent of the density of the universe.

26.4 INFLATION

Despite its great successes, the standard model of the Big Bang has some difficulties. To overcome these difficulties, astronomers have developed a modification of the standard model in which the early universe underwent a brief period of extremely rapid and enormous expansion. This expansion is often referred to as **inflation.**

Problems with the Standard Big Bang

The standard model of the Big Bang has three main difficulties, which have come to be known as the horizon problem, the flatness problem, and the structure problem.

The Horizon Problem The cosmic microwave background that comes to us from opposite sides of our sky originated in regions of the universe that are now very distant from each other. These regions were one thousand times closer together at the epoch of decoupling, when the CMB originated, than they are today. Nevertheless, they were millions of light years apart at that time, when the universe was only four hundred thousand years old. Nothing, not even light, could have traveled from one region to the other during the history of the universe up to that time. There was no way in which either region could have ever received any information about the other or could have sent energy to or received energy from the other region. If one of the regions had been hotter than the other there would not have been time for energy to flow from the hotter region to the cooler region to equalize the temperature between the two regions. Why, then, were temperature and density so similar across the region from which we now receive the CMB? In the standard Big Bang model, there is no explanation for the extreme uniformity of the early universe.

The Flatness Problem The universe is flat. This is remarkable because there was a strong tendency in the earliest stages in the expansion of the universe for curvature to grow. If the early universe had been precisely flat ($\Omega = 1$), it would have remained flat forever. However, if the universe were slightly positively curved (Ω very slightly greater than 1), then Ω would have grown rapidly with time. In a short time the universe would have reached its maximum size and contracted again.

On the other hand, if the universe had been slightly negatively curved (Ω slightly less than 1), Ω would have fallen quickly. Thus, the universe today would be strongly negatively curved and would have a very low density. Only if the early universe were almost perfectly flat and Ω were almost exactly equal to 1 could the actual density be as close to the critical density as it is today. For Ω to be as close to 1 as it now is, the value of Ω one second after expansion began must have been different from one by only 1 part in 10^{15}. Earlier than one second after expansion began, Ω must have been equal to one to an even greater accuracy. The standard model of the Big Bang cannot explain extreme flatness of the early universe. It would seem reasonable that the initial density of the universe could have been quite different from the critical density, leading to a universe of either extremely large positive or negative curvature. Thus, with the standard Big Bang, there is no way to account for the flatness of the universe in which we find ourselves.

The Structure Problem Even though the universe is quite uniform on the very largest scales, it has complicated structure and is highly nonuniform on smaller scales, such as the sizes of clusters of galaxies. This structure developed from irregularities in the universe at the decoupling epoch. These irregularities can be seen in the brightness fluctuations in the cosmic background radiation. The question is, where did that structure originate? There is no way that the standard model of the Big Bang can explain why there were density fluctuations in the early universe. According to the standard model, they must simply have been there all along, even at the moment expansion began.

The standard model of the Big Bang can only explain three important features—the uniformity of widely separated regions of the early universe, the early flatness of the universe, and the presence of density structure—by saying, "That's just the way it was." Each of these features can be adequately explained, however, as a natural result of a period of inflation in the very early universe.

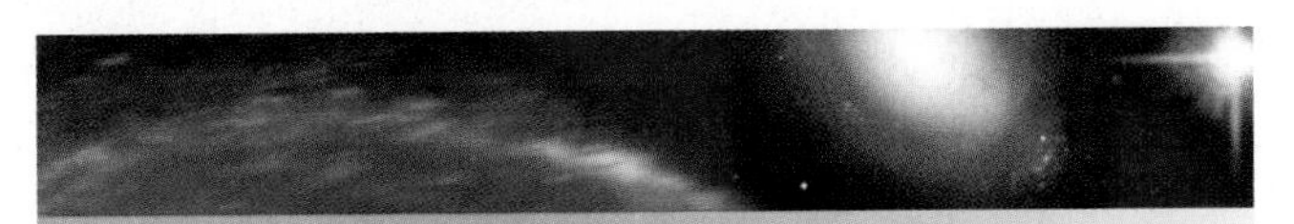

The standard model of the Big Bang has three serious problems. It cannot account for the extreme isotropy of the CMB, the flatness of the universe, and the development of large-scale structure in the universe.

How Inflation Solves the Problems

The inflationary model for the early universe proposes that, starting about 10^{-34} seconds after expansion began, the rate of expansion began to increase rapidly with time. Figure 26.21 shows that during the next 10^{-34} seconds the universe doubled in size and continued to double in size during each succeeding 10^{-34} seconds until the inflationary epoch ended at a time of about 10^{-32} seconds. Because 10^{-32} seconds contains one hundred intervals each 10^{-34} seconds long, there was time for one hundred

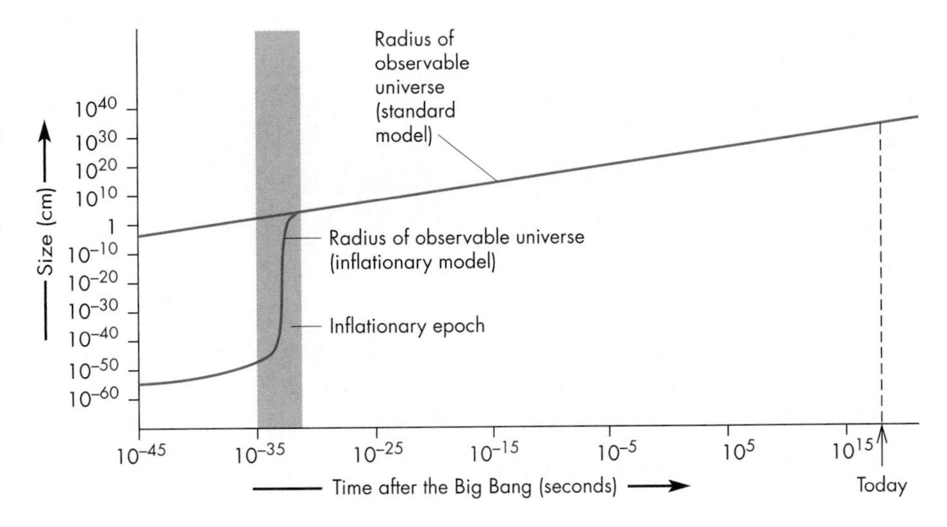

FIGURE 26.21
The Radius of the Observable Universe Versus Time
In the inflationary model, the universe became enormously larger about 10^{-34} seconds after expansion began. The period of inflation is indicated by the vertical shaded strip.

doublings of the size of the universe in the first 10^{-32} seconds. While inflation was going on, the universe grew in size by at least a factor of 10^{25} and perhaps much more depending on exactly how long inflation lasted.

Inflation provides an easy answer to the horizon problem. The region from which we are now receiving the cosmic background radiation was extremely small just before inflation began. Even though the universe was very young, the region was small enough that there had been enough time for energy to flow from hotter places to cooler places and make the temperature throughout the region uniform. During inflation, this very tiny region expanded to become the entire part of the universe that we can now observe. Thus the uniformity of the CMB and the general sameness of the observable universe on very large scales were established before inflation when the entire region was much smaller than the nucleus of an atom.

Inflation answers the flatness problem, too. During inflation, the curvature of the universe was rapidly driven toward flatness in much the same way that the curvature of a balloon flattens when the balloon is inflated. Figure 26.22 shows that as the balloon gets larger, it looks flatter and flatter to a creature on its surface. As the universe expanded, it became flatter and the value of Ω rapidly approached 1. In fact, Ω became so nearly equal to one at the end of inflation that the actual density and critical density should still be the same to 1 part in 10,000.

Inflation explains the origin of the structure that later became galaxies and clusters as a consequence of the very tiny original size of the portion of the universe that we can see. On large scales, things behave predictably according to the laws of physics. On small scales, however, such as those that apply to individual particles, behavior can only be described by probabilities rather than certainties. That is, no one can predict exactly when a given radioactive nucleus will decay. All we can say for sure is that there is a certain probability that it will decay within a given interval of time. Before inflation, the part of the universe that we can observe was so small that density fluctuations appeared and disappeared in a random manner that can only be described by probabilities. At the instant inflation began, the existing fluctuations were inflated to great sizes and became the fluctuations in the CMB and the seeds of large-scale structure in the universe.

Inflation solves the problems of the isotropy of the CMB and flatness by proposing that the entire observable universe was extremely tiny before inflation took place. Large-scale structure in the universe developed from random density fluctuations in the early, very compact universe.

Why Inflation?

Inflation can provide natural answers to the problems of the standard model of the Big Bang. But what caused the epoch of inflation? To answer this question, we need to see what conditions would lead to very rapid, very brief expansion. In a flat universe (which the universe would have become shortly after inflation began), the expansion rate is proportional to the square root of density. For a universe dominated by normal matter, density drops as the universe expands and both density and the expansion rate fall as time increases. But density includes all forms of matter and energy. If the early universe were dominated by a form of energy whose density didn't fall as the universe expanded, the expansion rate would have remained high.

FIGURE 26.22
Inflation and the Flatness Problem
As a sphere is inflated, a small region of its surface looks steadily flatter to an observer located on the surface of the sphere.

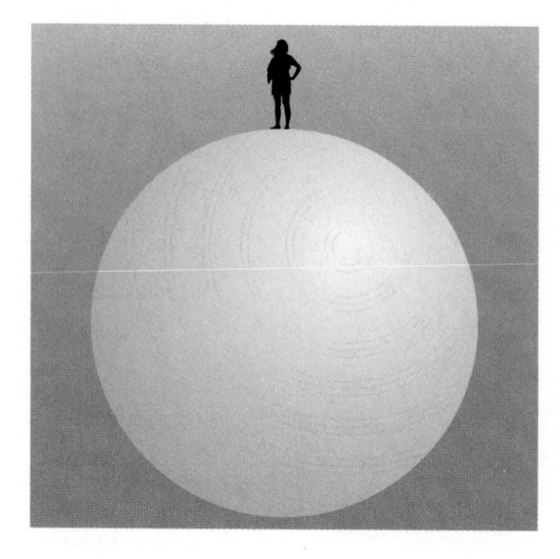

To account for inflation, we need a form of energy whose density remained high while the universe got vastly larger and then dropped enough that matter and dark energy dominated the universe and inflation ended.

No form of energy yet detected would have behaved in such a way, but physicists have predicted the existence of just such a kind of energy. This energy is a **scalar field,** which physicists have proposed to explain the properties of elementary particles. Although most people aren't familiar with scalar fields, they have properties that are similar to those of more familiar gravitational, magnetic, and electric fields. A field has an energy density throughout space. The energy density of the field contributes to the density of the universe and, thus, the expansion rate. The energy densities of the kinds of fields we are familiar with drop as the universe expands and lead to a decrease in the expansion rate. In that way they behave similarly to normal matter. A scalar field has an energy density that would have decreased very slowly as the universe expanded until the energy density of the scalar field dropped to a certain point. After that, the energy density of the scalar field fell swiftly.

So inflation can be explained if the early universe had one or more kinds of scalar fields with large energy densities. No experiments conducted so far have detected scalar fields, but physicists propose that several kinds exist. It is possible that scalar fields will be detected in experiments using particle accelerators that will be developed in the relatively near future.

The most widely accepted explanation for inflation depends on scalar fields that fill the universe. During the first moments of the expansion, the density due to scalar fields remained high and caused the expansion rate to increase in an epoch of inflation.

26.5 THE FATE OF THE UNIVERSE

In the last half century, astronomers have learned about many of the important stages in the development of the universe. After the first second, we think we have a good general idea of the history of the universe. Although we aren't yet certain about the future of the universe, the choices have been narrowed to only two.

The Choices

The two choices for the fate of the universe are expansion forever and expansion leading eventually to contraction. We can sketch the future for both eternal expansion or expansion leading to contraction.

Contraction and Beyond

The present expansion of the universe will have to slow down and come to a halt if expansion is ever to give way to contraction. We can be sure that this isn't going to happen for at least several tens of billions of years. In that time, the universe will have changed greatly from the way it is today. Today's luminous stars will have long since become very dim white dwarfs, neutron stars, or black holes. As our stars die, they will be replaced by new stars, although not at a rate that can sustain the number of luminous stars. The reason for this is that the raw ingredients for stars, interstellar gas and dust, will diminish as time goes by. The universe will darken as the number of luminous stars decreases and galaxies become dimmer and dimmer.

On the day the universe reaches its maximum size, the dim galaxies will pause for a moment and then begin to fall toward each other with increasing speed. If there is anyone in the universe to carry out observations, they will see the opposite of Hubble's law gradually establish itself. At first only the nearest galaxies will show blueshifts. The more distant galaxies will still be seen as they were in the past, before contraction began. Eventually, though, the observers will see a pattern of blueshifts increasing with distance. The implication of that observation will be that all the galaxies are rushing toward an appointment with each other in what has been called the "Big Crunch."

As the galaxies rush together, first the clusters and then individual galaxies begin to overlap. As the stars accelerate, they rush through interstellar gas at such great speeds that they are eroded by friction. The friction heats the gas enough to make it glow. As temperature rises, the gas becomes ionized and opaque, duplicating the conditions that occurred billions of years ago at the time of decoupling. In a few hundred thousand years, the density and temperature of the universe rise as the collapse repeats the initial stages in expansion in reverse. Eventually, however, the density and temperature of the universe become so extreme that we cannot predict what will happen next. One intriguing possibility is that at a maximum stage of compression the universe will rebound and begin another cycle of expansion and contraction, perhaps with different amounts of matter or even different physical laws. Perhaps our current cycle is but one in an eternal series of cycles of expansion and contraction. This possibility was first described, remarkably, by Edgar Allen Poe in 1848. In the speculative essay "Eureka," he wrote, "[T]he majestic remnants of the tribe of Stars flash, at length, into a common embrace. . . . Are we not, indeed, more than justified in entertaining a belief—let us say, rather in indulging a hope—that the processes we have here ventured to contemplate will be renewed forever, and forever, and forever; a novel Universe swelling into existence, and then subsiding into nothingness, at every throb of the Heart Divine."

Continued Expansion

The alternative to this "death by fire" in the Big Crunch is "death by ice," in which the universe just fades away. In this case, the expansion continues as the nuclear fires in galaxies go out. Binary stars and galaxies in clusters gradually spiral into each other, emitting gravitational waves because of their motion. The stars in the clusters gradually shrink together for the same reason, leaving most of the masses of the clusters in massive black holes.

We tend to think of matter as permanent and imagine that things like protons last forever. Yet this may not be the case. It is likely that on a timescale of 10^{32} years, the protons in the universe will decay into radiation and particles that annihilate each other to produce radiation. Gradually, all the matter in the universe will turn into radiation. Even black holes may not be permanent; instead, they may radiate away their mass. A black hole with the mass of a galaxy could radiate away its mass in 10^{100} years. Eventually, all the mass in the universe disappears. All that remains are gravitational waves, neutrinos, and cosmic radiation that reddens and weakens as the expansion of the universe continues for eternity.

FIGURE 26.23
The Future of the Universe If the Density of Dark Energy Is Small
The scale factor of the universe for negative ($\Omega_0 < 1$) or flat ($\Omega_0 = 1$) curvature increases forever. If the universe has a density greater than the critical density and space is positively curved ($\Omega_0 > 1$), the expansion eventually gives way to contraction. After the contraction, a new cycle of expansion and contraction may occur.

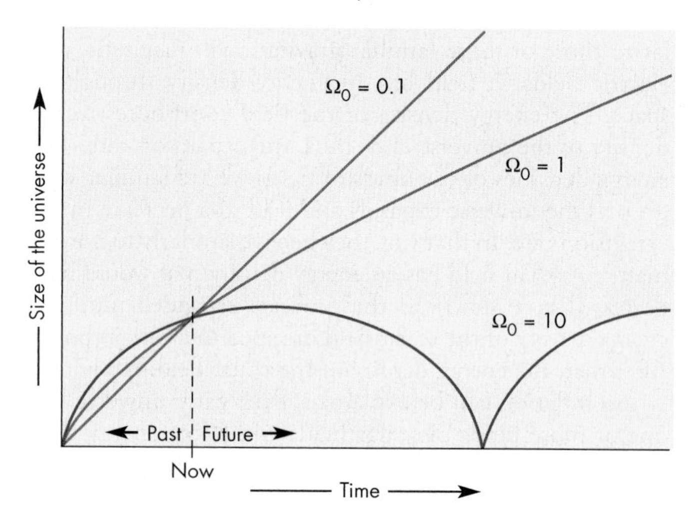

Which Will It Be?

If the density of dark energy were small or zero, then the choice between expansion forever and eventual contraction would depend only on the curvature of space and the density of the universe relative to the critical density. If density exceeded the critical density (Ω_0 greater than 1) then space would be positively curved and expansion would slow until the universe reached a maximum size. After that, it would begin to contract, as shown in Figure 26.23. If space were negatively curved or flat, on the other hand, then Ω_0 and the self-gravity of the universe would never be able to bring the expansion to a halt. In that case, the expansion would continue forever.

The discovery that dark energy dominates the universe makes such simple relationships among density, curvature, and the future of expansion obsolete. The expansion of the universe appears to be accelerating. Whether expansion will continue forever depends on the still uncertain nature of dark energy. One possibility is that dark energy is a property of space and exerts a constant pressure as space expands. If that is the case, then expansion will continue forever and the universe will just fade away, as described in the previous section. However, it may be that dark energy is exerted by exotic, as yet undiscovered, particles that grow farther apart as space expands. In that case the force causing acceleration could change with time. One possible outcome would be a "Big Rip" in which the expansion of space eventually becomes so rapid that all matter—from stars to atoms—is torn apart. Another possibility is that the force could reverse its direction, halt the expansion of the universe and eventually cause the "Big Crunch" described earlier in this section. Thus, the ultimate fate of the universe seems to depend on the exact nature of the dark energy that dominates the universe. The best information so far on the nature of dark energy has come from *WMAP* results, which indicate that dark energy behaves like the cosmological constant. That is, that the amount of dark energy in a cubic meter of space has remained the same throughout the history of the universe and will continue to remain constant. In this case, the universe will continue to expand forever, without either a "Big Rip" or a "Big Crunch." Experiments now underway or in the planning stage should refine our knowledge of the nature of dark energy and make it possible to forecast the future of the universe with certainty.

The two choices for the future of the universe are eventual contraction and continued expansion. For eventual contraction, the universe will eventually reach a maximum size and then begin to contract. In this case, the early stages of the Big Bang will run in reverse until conditions exceed our ability to predict what will happen. A new cycle of expansion and contraction is one possibility. If continued expansion occurs, the matter in the universe will eventually decay and the radiation in the universe will redden and dim forever. The choice between the two alternatives depends on the nature of dark energy.

Chapter Summary

- The expansion of the universe implies that the expansion began when the universe was very dense and hot. This early state of expansion is called the Big Bang. The time since the Big Bang can be estimated by the Hubble time, which assumes that the expansion has always gone on at the present rate. (Section 26.1)
- The space in the universe may be either positively curved, negatively curved, or flat. If it is positively curved, the universe is finite in volume. Otherwise it is infinite. For all three cases, the universe has no center and is unbounded. In principle, geometrical tests can be used to determine the curvature of the universe. In practice, however, these are unworkable because the universe has changed with time. (26.2)
- The curvature of the universe can be tested by measuring the density of the universe. If the density is equal to the critical density (ρ_c), space is flat. If the density is less than ρ_c, space is negatively curved. If it is greater than ρ_c, then space is positively curved. (26.2)
- If the universe is negatively curved or flat, it will continue to expand forever. If the universe has positive curvature, contraction will eventually occur. The accuracy of the Hubble time as an estimate of the age of the universe also depends on curvature. Only for a low density and negative curvature is the actual age as great as the Hubble time. (26.2)
- Studies of ancient supernovae show that the expansion of the universe originally slowed but later began to accelerate. Dark energy is responsible for the acceleration. (26.2)
- The ages of the oldest objects in the universe suggest that the Big Bang occurred more than 11 billion years ago. Estimates of how long the universe has been expanding yield 13.7 billion years. (26.3)
- During the first second of expansion, the universe was so hot that photons had sufficient energy to make matter and antimatter. At the same time, matter and antimatter continually destroyed each other through annihilation. After the first second, photons no longer had enough energy to make matter and antimatter, but annihilation continued to reduce the amount of matter in the universe. When this happened neutrinos were free to travel throughout the universe without interacting with matter. (26.3)
- From about 100 to 300 seconds after expansion began, conditions were right for nuclear reactions to produce helium from hydrogen. The abundances of isotopes made at that time show that the density of protons, neutrons, and helium was too low for the universe to be flat or have positive curvature unless the universe is mostly composed of dark matter and dark energy. (26.3)
- After 380 thousand years, the universe cooled enough to permit atoms to form from nuclei and electrons. When this occurred, the universe became transparent to radiation, so radiation could begin to travel freely through the universe. The moment when light broke free from matter is known as the decoupling epoch or the recombination epoch. The radiation emitted before recombination is all gone, but most of the radiation emitted since that time is still present in the universe. (26.3)
- The cosmic microwave background is the light emitted at the time the universe became transparent. It is almost equally bright in all directions but must be observed at infrared and radio wavelengths because it comes from very remote parts of the universe and is strongly redshifted. Fluctuations in the cosmic microwave background show that space is flat and that normal matter contributes little to the density of the universe. (26.3)
- The standard model of the Big Bang cannot account for the isotropy of the cosmic microwave background, the near flatness of the universe, and the origin of structure in the universe. (26.4)
- The theory of inflation proposes that the early universe increased in size by an enormous factor. This solves the problems of isotropy of the cosmic background radiation and the flatness of the universe because the part of the universe that we can see was once extremely compact. The structure in the universe arose from random fluctuations in that compact region of space. (26.4)
- Inflation is believed to have been produced by scalar fields that caused density to remain high as the universe grew. (26.4)
- The two alternatives for the future of the universe are continued expansion and eventual contraction. Contraction would lead to a dense state in which the early Big Bang is played out in reverse. At some point, conditions would become so extreme that we can't predict what will happen, although one possibility is another cycle of expansion and contraction. If expansion continues forever matter will decay, and the radiation in the universe will become redder and dimmer forever. Alternatively, dark energy may cause an eventual contraction of the universe. (26.5)

Key Terms

annihilation *631*
antimatter *630*
Big Bang *620*
cosmic microwave background (CMB) *633*
cosmological constant *627*
cosmology *619*
critical density *626*
dark energy *627*
decoupling epoch *633*
deuteron *632*
Hubble time *620*
inflation *636*
isotropic *635*
pair production *630*
radiation era *633*
recombination epoch *633*
scalar field *639*

Conceptual Questions

1. Under what conditions would the expansion age of the universe be the actual length of time that has passed since the Big Bang?
2. Suppose astronomers were to discover that the value of Hubble's constant were twice as large as it is now thought to be. What effect would this discovery have on the expansion age of the universe?
3. Suppose we use Hubble's law to find that the galaxies at a distance of 200 Mpc were adjacent to the Milky Way 12 billion years ago. How long ago were the galaxies at 400 Mpc adjacent to the Milky Way?
4. What is meant by the statement that the surface of a ball has no center?
5. Compare flat space universes with positively curved universes with respect to whether they are finite, whether they are bounded, and whether they have centers.
6. Describe two geometrical tests that could be used to distinguish positively curved universes from flat universes.
7. Describe two geometrical tests that could be used to distinguish positively curved universes from negatively curved universes.
8. Suppose you lived in a two-dimensional universe. Describe how you could use counts of distant galaxies to learn the curvature of space in your universe.
9. What is the curvature of the universe if its density is less than the critical density?
10. What is the value of Ω_0 if the universe is flat?
11. What combination of Hubble's constant and density can produce a universe in which expansion will continue forever?
12. What is the relationship between the value of Ω_0 and the extent to which the expansion age of the universe differs from the actual age of the universe?
13. How do ancient supernovae contribute to the idea that expansion is accelerating?
14. Why did Einstein invent the cosmological constant?
15. How does dark energy affect the expansion of the universe?
16. How can globular clusters be used to place a lower limit on the age of the universe?
17. Why did pair production stop when the universe was 1 second old?
18. What was the origin of the primordial neutrinos that are still present in the universe?
19. Why were conditions right for the production of helium from hydrogen only between 100 and 300 seconds after the universe began?
20. What happened to all of the radiation that was produced during the first 380 thousand years after the universe began?
21. The cosmic microwave background was emitted by gas at a temperature of about 3000 K. Why, then, is the CMB brighter in the radio part of the spectrum than at visible wavelengths?
22. What is meant by the statement that the CMB has a temperature of 2.74 K?
23. What do fluctuations in the brightness of the CMB tell us about the universe?
24. Why is the isotropy of the CMB a problem for the standard Big Bang model?
25. Why is the present flatness of the universe a problem for the standard Big Bang model?
26. Describe how inflation solves the flatness problem of the standard Big Bang.
27. What is the fate of the universe if it has positive curvature and there is little or no dark energy?

Problems

1. Suppose a galaxy at a distance of 300 Mpc has a recession speed of 21,000 km/s. If its recession speed has been constant over time, how long ago was the galaxy adjacent to the Milky Way?
2. What is the Hubble time if Hubble's constant is 40 km/s per Mpc?
3. If the Hubble time is 20 billion years, what is the value of Hubble's constant?

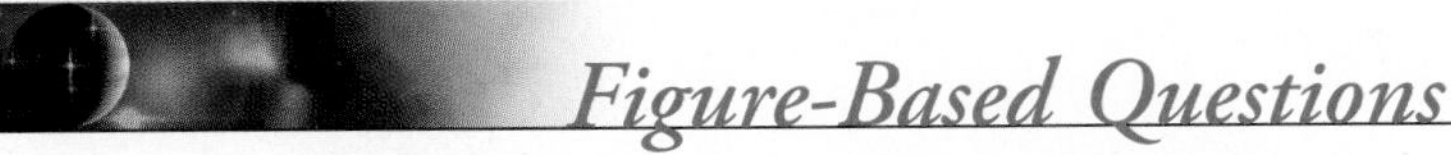

Figure-Based Questions

1. Use Figure 26.15 to find the two most abundant particles or nuclei in the universe when it was 100 seconds old and when it was 1000 seconds old. What happened between these two times to produce a change in the makeup of the universe?
2. Suppose ^{3}He and ^{7}Li nuclei were equally abundant in the universe. Use Figure 26.16 to estimate the present density of protons and neutrons in the universe.

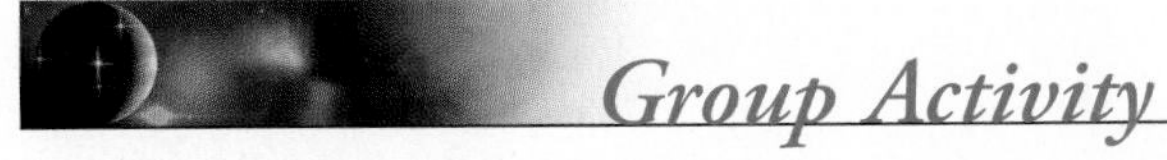

Group Activity

In the late 1940s, a group of astronomers proposed the steady-state theory of cosmology. The crucial feature of the steady-state theory was that the universe has always looked about the same as it does today. The theory proposed that new matter is spontaneously created from nothing and collected into new galaxies at just the rate required to fill the gaps left behind as existing matter is carried apart by the expansion of the universe. Working as a group, try to come up with some ways that the steady-state theory could be tested.

For More Information

Visit the text website at **www.mhhe.com/fix** for chapter quizzes, interactive learning exercises, and other study tools.

27

We cast this message into the cosmos. It is likely to survive a billion years into the future, when our civilization is profoundly altered and the surface of the Earth may be vastly changed. Of the 200 billion stars in the Milky Way galaxy, some—perhaps many—may have inhabited planets and spacefaring civilizations. If one such civilization intercepts Voyager and can understand these recorded contents, here is our message: "This is a present from a small distant world, a token of our sounds, our science, our images, our music, our thoughts and our feelings. We are attempting to survive our time so we may live into yours. We hope someday, having solved the problems we face, to join a community of galactic civilizations. This record represents our hope and our determination, and our good will in a vast and awesome universe."

Jimmy Carter, President of the United States
June 16, 1977

Alien Life Forms by William K. Hartmann.

Life in the Universe

People have wondered for thousands of years about whether there might be life on worlds beyond the Earth. The plurality of inhabited worlds was a common theme in the writings of the Greeks and Romans. After Copernicus showed that the Earth was but one of many planets orbiting the Sun, it was a short step to the belief that other planets in the solar system might be inhabited or that there might be inhabited planets orbiting other stars. Giordano Bruno wrote in the late sixteenth century that there were Earthlike planets orbiting the infinitely many stars and that those planets were inhabited. Bruno was a genuine heretic who also believed that the stars and planets had souls. When he traveled to Italy to convert the Pope to his peculiar brand of Christianity he was captured, tried before the Inquisition, and burned at the stake in 1600. In the seventeenth and eighteenth centuries, many writers, including Cyrano de Bergerac and Voltaire, wrote about life on other planets and the Moon. The idea that there were other inhabited worlds was also shared by prominent scientists such as Herschel and Newton. Later, in the late nineteenth and early twentieth centuries, the idea that Mars was inhabited was popularized by Schiaparelli and Lowell. The ideas of Schiaparelli and Lowell are discussed more fully in Chapter 11.

Today, the idea that there are other inhabited worlds is widespread. Other planets are frequently used as locales for science fiction novels and movies. There have also been many popular movies about extraterrestrials (sometimes good, sometimes bad)

Questions to Explore

- What is life?
- What substances are essential for terrestrial life? Is it possible that life elsewhere could be based on different substances?
- How and when did life begin on Earth? Is it likely that life originated elsewhere in the solar system?
- What evidence do we have that there are other planetary systems?
- What can be said about the likelihood that there is life elsewhere in the galaxy? Is it likely there is intelligent life?
- What efforts have been made to communicate with creatures on other habitable planets?

visiting the Earth. One of the most widely held opinions about unidentified flying objects is that they are piloted by extraterrestrials. The evidence for life elsewhere in the universe, however, is as yet unconvincing to most scientists. Even so, we are gradually learning enough about other planetary systems and the origin of life that we may eventually be able to estimate with some confidence how common life is in the universe. There are many groups of scientists who are even now searching for signals from creatures on planets orbiting other stars.

27.1 LIFE

A unique feature of the Earth is the **life** it supports. With human activities producing significant environmental changes, it is more important than ever for us to learn about how life developed on Earth and the range of conditions that could yield habitable planets.

What Is Life?

The question "What is life?" doesn't have an easy answer. Yet people generally have little difficulty distinguishing living organisms from nonliving objects. Perhaps the best way to define life is to determine the general properties that all terrestrial organisms have in common. A list of those properties would include

1. Organization. All terrestrial organisms are composed of **cells,** which are organized collections of molecules enclosed in a wall-like membrane. (Viruses, which most scientists consider to be alive, have protein coats rather than membranes for walls.)
2. Metabolism. Organisms can use energy from their environment to maintain their internal organization.
3. Regeneration and growth. Organisms can fully or partially repair parts of themselves and can grow.
4. Response to stimulation. Organisms react to their environment through motion, behavior, and metabolic change.
5. Reproduction and evolution. The property that most clearly divides living organisms from nonliving objects is the ability to reproduce and to evolve through a series of changes from one generation to the next.

The ability to reproduce isn't sufficient, by itself, to define life. There are, for example, liquid-filled spheres enclosed by membranes, like those shown in Figure 27.1,

FIGURE 27.1
Protocells
Certain molecules, similar to fats, in a water solution spontaneously form hollow spheres that resemble cells. Although the protocells can grow and divide, they do not evolve and thus cannot be considered to be alive.

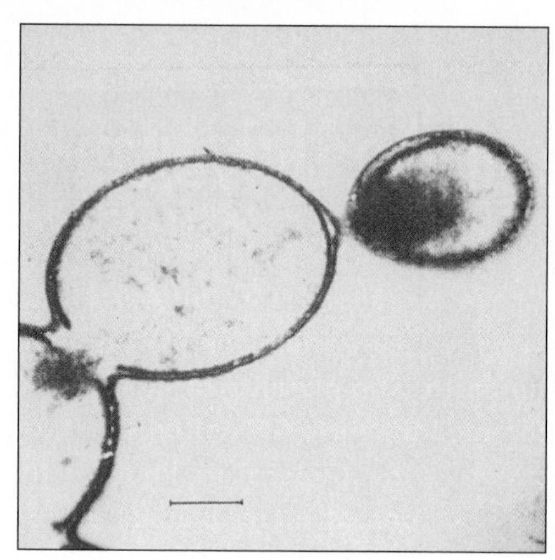

that can grow and subdivide, but they are no more alive than soap bubbles. Being alive depends on the ability to change, reproduce, and pass the changes down to a new generation. This requires a genetic system through which changes can be preserved. The changes that are preserved tend to be the ones that enhance the chances of survival of individual organisms. Fossils show that evolution has been going on as long as there has been life on the Earth.

Life on Earth

All life on Earth is based on organic molecules, which are molecules containing carbon atoms or chains of carbon atoms. The carbon atoms combine with hydrogen, oxygen, nitrogen, and sulfur atoms to produce a relatively small number of different molecules, including **amino acids** and **nucleotides.** Chains of amino acids form **proteins,** which act as structural materials for the bodies of organisms. Long chains of nucleotides make up **nucleic acids,** including deoxyribonucleic acids (DNA) and ribonucleic acids (RNA). DNA stores the hereditary information of an organism. In other words, DNA specifies the characteristics that are passed on to the offspring of that organism. The information in DNA is transferred to RNA, which then makes proteins. The resulting protein depends on the section of the DNA molecule from which information was transferred to RNA. The pattern of proteins from which an organism is made determines the characteristics of the organism. Thus, an organism's characteristics are encoded in its DNA. By passing its DNA along to the cells of its offspring, the organism causes its offspring to inherit its characteristics. Although we should not expect DNA and RNA to exist in extraterrestrial organisms, the definition of life requires that some kind of molecules play the same role in those organisms that DNA and RNA do in terrestrial life.

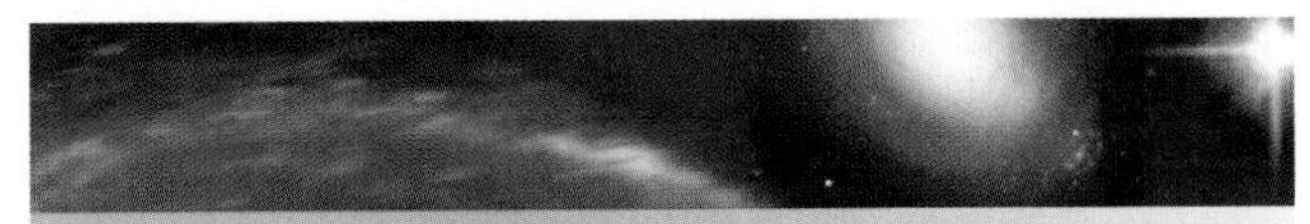

Life is distinguished by its ability to reproduce and evolve. Terrestrial life is based on carbon atoms and water, but it seems possible that there may be life elsewhere based on substances other than carbon and water.

Other Possibilities

There may be life-forms in the universe that are very different from those of Earth. It seems possible that a genetic system capable of regulating reproduction and heredity could evolve in an environment quite different from that of the early Earth (see Chapter 8, Section 6). If so, that life would have to be based on substances other than carbon and water, essential ingredients of life on Earth.

One feature of carbon that makes it essential for life on Earth is its ability to combine with hydrogen (and other atoms) to form large, complex molecules. It does this by forming long chains of connected carbon atoms to which the other atoms are attached. Silicon is another element, about one-quarter as abundant as carbon, that shares many of the chemical properties of carbon. The chemical bond between silicon atoms, however, is only about half as strong as that between carbon atoms. In fact, bonds between silicon atoms and oxygen atoms or between silicon atoms and hydrogen atoms are actually stronger than the bonds between silicon atoms. This makes it difficult for silicon atoms to form long chains or rings as carbon atoms do. While it may be possible for life to be based on silicon, it seems much less likely than carbon-based life.

Water plays a crucial role in terrestrial life. Life began in water and water is still the medium in which molecules in cells move freely about. No known plant or animal can grow and reproduce in an environment in which water is absent. Water has a number of properties that make it beneficial to life. For one, water is an excellent solvent. Water can dissolve more substances than any other solvent, bringing these substances together to react with one another. Another important property of water is its large heat capacity—the amount of heat needed to raise the temperature of a kilogram of a substance by 1 kelvin. This means that water stabilizes body temperature and prevents organisms from getting too hot or too cold. Water is also liquid over a wide range of temperatures, protecting organisms from freezing or boiling. Finally, water has a high heat of vaporization—the heat needed to change a liquid into a gas once the boiling temperature has been reached. This means that a small amount of evaporation can cool an organism and prevent it from overheating.

Are there liquids on other planets that play the same role that water plays on Earth? Two substances that have been proposed are ammonia and methyl alcohol, both of which are liquid at temperatures far below the freezing point of water. Ammonia, however, is liquid over a much smaller range of temperatures than water and has a much lower heat of vaporization. Methyl alcohol is liquid over a wider temperature range, but has a lower heat capacity and heat of vaporization than does water. It may be possible for life based on ammonia or methyl alcohol to exist in the universe. It seems likely, however, that the superior properties of water mean that most extraterrestrial forms of life are based on water just as most of them are probably based on carbon.

THE ORIGIN AND EVOLUTION OF LIFE ON EARTH

The fossil record shows that life on Earth began relatively soon after the planet cooled and formed a solid surface. Figure 27.2 shows fossils of ancient bacteria. The oldest fossil bacteria date from more than 3.5 billion years ago. This is about three-fourths of the age of the Earth and the origin of life may be considerably more ancient than that.

An early step in the development of life on Earth was the formation of complex organic compounds. Whether those organic compounds originated elsewhere and fell to Earth or formed on Earth is still uncertain. Astronomical

FIGURE 27.2
Fossil Bacteria
These 0.01-mm-diameter fossil bacteria were found in the Bitter Springs formation in Australia. These fossils are about 850 million years old and resemble fossil bacteria that are as much as 3.5 billion years old.

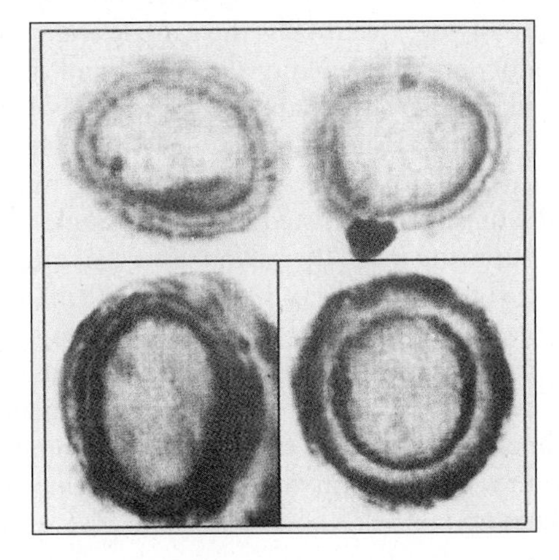

evidence suggests that the formation of complex organic molecules, including amino acids, can occur in a wide range of environments. As described in Chapter 22, complex organic molecules have been detected in interstellar clouds. Also, complex organic molecules, including more than 70 amino acids, have been found in carbonaceous chondrite meteorites. It seems likely that complex organic molecules existed in primordial interplanetary bodies. It is possible that these molecules survived the impacts of interplanetary bodies with the Earth and accumulated on Earth's surface.

It is also possible that complex organic molecules could have formed on the Earth from simpler compounds. A number of interesting experiments have shown how complex organic molecules could have formed on the early Earth. In the 1950s, chemists and biologists experimented with mixtures of gases such as hydrogen, methane, ammonia, and water vapor. When such gases were energized by electrical discharges (like lightning), solar radiation, heat, or shock waves, amino acids were produced. It is now thought that the early atmosphere of Earth was more likely rich in carbon dioxide and nitrogen than in hydrogen, methane, and ammonia. Experiments with mixtures of gases rich in carbon dioxide and nitrogen have also produced complex organic compounds, though in much smaller amounts than experiments with hydrogen-rich gases. In either case, however, as sunlight, lightning, volcanic eruptions, and other energy sources steadily produced complex organic molecules, the oceans must have become a dilute, warm soup of the basic ingredients of life.

The next step in the development of life was probably the gathering of proteins and other molecules from the primeval soup into roundish clusters that resembled cells. This may have occurred in extreme environments such as hydrothermal vents on the ocean floor, where very hot water provides a steady supply of complex molecules and energy. Gradually, the cell-like clusters became more complex. Some clusters were better able to collect additional molecules and energy from their surroundings. These clusters would have lasted longer and sometimes grown large enough to divide into daughter clusters that resembled their parent. The daughters, too, were able to grow and to divide. Eventually, the genetic ability to regulate reproduction and heredity developed and life began. It seems likely that this happened when RNA evolved and began to act as the informational molecule in cells. DNA is thought to have evolved later. For most of its history, life on Earth consisted of simple, single-celled organisms. Approximately 700 million years ago, multicellular life evolved. Life on land evolved about 400 million years ago and mammals appeared about 200 million years ago. *Homo sapiens* appeared about five hundred thousand years ago, and developed civilization only a few thousand years ago. Technological civilization, capable of communicating with creatures elsewhere in the universe, developed in the last half-century.

LIFE ELSEWHERE IN THE SOLAR SYSTEM

Nearly all the planets and satellites in the solar system have now been examined by space probes from the Earth. None of the other solar system bodies seem hospitable to life. There is still a chance, however, that there may be microbial life on Mars that began when conditions on Mars were more favorable than they are today. There is also a chance that there may be life on Jupiter's satellite Europa, the only solar system body other than the Earth where there is known to be liquid water near the surface. The existence on Earth of life in environments with very high or low temperature and where conditions are very acidic or alkaline suggests that the search for life in the solar system should not be confined to the most benign environments. Life even thrives in rocks as far as 5 km beneath the Earth's surface.

The possibility that life originated on Mars billions of years ago was strengthened in 1996 by a close examination of a meteorite thought to have come from Mars. The meteorite, shown in Figure 27.3, is made of igneous rock that solidified about 4.5 billion years ago at the time that Mars formed. About 3.6 billion years ago globules of carbonate minerals were deposited in cracks in the rock. The carbonate minerals may have been deposited when liquid water seeped into the cracks. The impact of an asteroid or comet on Mars 16 million years ago ejected the rock from Mars into interplanetary space. About thirteen thousand years ago the rock fell into the Antarctic ice fields as a meteorite.

FIGURE 27.3
A Meteorite Containing Evidence of Possible Ancient Life on Mars
The meteorite, about 10 cm across and having a mass of 2 kg, is believed to have been blasted from Mars during an impact 16 million years ago. It contains organic and inorganic compounds and structures that suggest to some scientists that life existed on Mars 3.6 billion years ago.

It was collected in Antarctica in 1984, and its origin on Mars was recognized in 1993.

Two years of intense study of the meteorite led to the announcement in 1996 of several kinds of evidence that life once existed in the carbonate deposits located in the cracks in the meteorite. One piece of evidence was the discovery of easily detectable amounts of organic compounds concentrated in the vicinity of the carbonate globules. The kinds and amounts of organic compounds are consistent with those that would result from the fossilization of primitive microorganisms. Another piece of evidence was the discovery of inorganic compounds like iron sulfides that can be produced by bacteria and other terrestrial organisms. The most dramatic evidence, however, is tiny structures in the carbonate globules that resemble microscopic fossils of ancient terrestrial bacteria. Some of the structures are egg-shaped while others, like the one shown in Figure 27.4, are tubular. The team of scientists who examined the meteorite suggests that the structures may be the remains of ancient organisms that helped form the carbonate globules and were fossilized like the organisms fossilized in limestone on the Earth.

After the announcement of possible evidence of life in the meteorite, other scientists carried out tests. Some of them reported that the meteorite shows no evidence of life. They found that the carbonate deposits probably were formed by a very hot gas rather than water passing through the rock. They also suggest that the organic compounds may have entered the meteorite while it lay for thousands of years in the Antarctic snow rather than much earlier on Mars. Finally, they reported that the egg-shaped and tubular structures appear to be crystals rather than fossils. The original investigators and other scientists then presented additional studies supporting the hypothesis that the globules, organic compounds, iron sulfides, and tubular structures are, in fact, evidence of ancient life on Mars. It is not now clear what the scientific community will conclude about what the meteorite has to say about life on Mars. It is quite possible that the evidence will remain inconclusive and that the question of ancient Martian life will only be answered by future missions to Mars.

FIGURE 27.4
A Possible Martian Fossil
The tubular structure in the center of the image resembles terrestrial bacteria and was found in 3.6-billion-year-old carbonate deposits in a meteorite believed to have come from Mars. The structure is about 100 nm in length. This is about 1/100th of the diameter of the period at the end of this sentence.

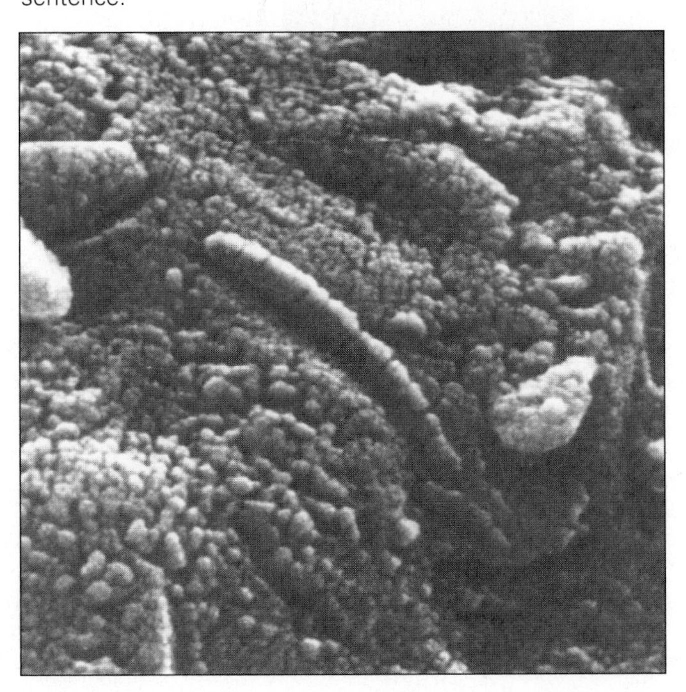

People have speculated about life on Mars for centuries. The possibility that Europa may bear life has been considered seriously only since astronomers realized, after looking at *Voyager* images, that there is an ocean of liquid water beneath Europa's icy crust. *Galileo* images have shown that the icy surface is cracked and distorted. Blocks of ice appear to have shifted about as though they were slipping across an underlying layer of water. The cracks, caused by tidal flexing of Europa, may last for centuries before they are filled in with ice. This means that sunlight may be able to pass through the cracks into the ocean below and produce complex organic compounds. If there is life on Europa, the cracks may provide enough light for photosynthesis to occur in Europa's oceans. It doesn't seem likely that we will be able to really explore the possibility of life on Europa until future space probes can penetrate through Europa's icy crust and explore its oceans.

Life on Earth developed relatively soon after the Earth formed. We do not have any conclusive evidence that this happened elsewhere in the solar system, though there is a chance that life developed on Mars, Europa, or elsewhere.

LIFE IN OTHER PLANETARY SYSTEMS

Extra solar planets

Astronomers think that planetary systems form as a common by-product of the formation of stars. Current ideas about the formation of planetary systems are described in Chapter 18. If our current ideas are correct, then planetary systems should be extremely numerous in the galaxy. Only a half-century or so ago, however, it was believed that our own solar system was the result of a catastrophic, highly unusual event involving the Sun. The most widely accepted theory was that the Sun was struck a glancing blow, or was narrowly missed, by a passing star. Material pulled from the Sun and the other star went into orbit about the Sun, cooled, and produced the planetary system.

If such a catastrophe were required to produce the solar system, then planetary systems would be very rare because the stars are very small compared to the distances between them. Given the speeds with which they move with respect to one another, huge amounts of time are required for stars to cover the distances that separate them. This means that collisions and near-collisions between stars are very unusual events. Probably only a few collisions have taken place in the disk of the galaxy during its entire history. The catastrophic theory of the formation of the solar system led to the conclusion that ours might be the only planetary system in the galaxy. The situation is certainly very different today. Astronomers have now detected planets around many nearby solar-type stars. The implication is that planetary systems are common in the universe. How common life is in these planetary systems, however, is still an unanswered question.

Finding Other Planetary Systems

Astronomers have used a number of different methods to detect the presence of planets in orbit about other stars. All of the methods favor the detection of giant planets like Jupiter because giant planets are more massive, larger, and brighter than terrestrial planets. As a result, giant planets are much easier to detect than terrestrial planets.

Doppler Shifts Some of the efforts to detect planets orbiting other stars have looked for the gravitational influence of a planet on the star it orbits. This technique is based on the fact that planets don't actually orbit stars. Instead, a planet and its parent star orbit their center of mass, which lies between the star and the planet. The star is much more massive than the planet, however, so the star moves on a much smaller orbit about the center of mass than does the planet. As the star orbits, it moves first toward and then away from us, causing a change in the Doppler shift of its spectral lines. Figure 27.5 shows

FIGURE 27.5
The Position and Doppler Shift of a Star Orbiting Its Common Center of Mass with a Planet
In **A,** the planet is moving away from us and the star is approaching us, so the spectral lines of the star are blueshifted. In **B,** the planet is approaching us and the star is moving away from us, so the spectral lines of the star are redshifted. The position of the center of mass is marked with an ×. Both the shift in position and the Doppler shift of the spectral lines are very greatly exaggerated.

The position and Doppler shift of a star orbiting its common center of mass with a planet.

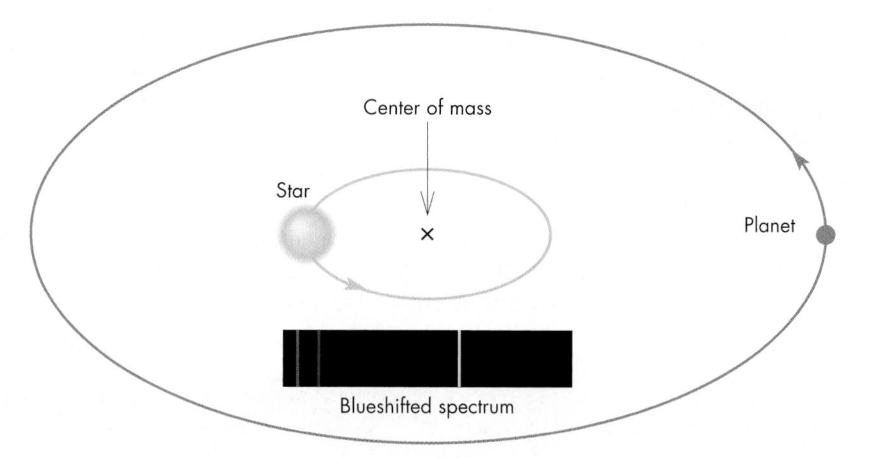

A The planet moves away from Earth; the star moves toward Earth.

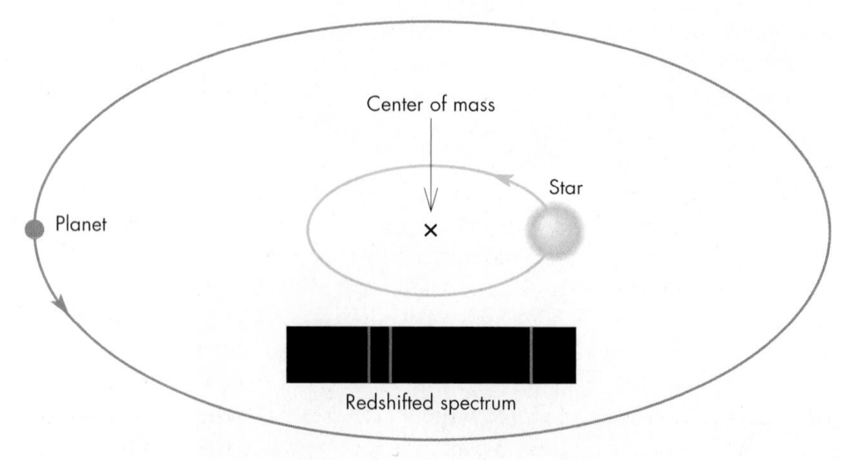

B The planet moves toward Earth; the star moves away from Earth.

how the position and Doppler shift of the spectral lines of a star vary as it and a companion planet orbit their common center of mass. It is important to note that the Doppler shift of the spectral lines of a star shows only the part of the star's orbital motion that carries the star directly toward and away from the Earth. If we observed a star from directly above its orbital plane, it would show no variations in Doppler shift at all. Because we don't know the angle from which we observe the orbital motion of a star and its planet, we can use the Doppler method only to determine lower limits for the masses of planets orbiting stars. That is, if we see Doppler-shift variations for a given star, we can be certain only that the variations are caused by a planet more massive than a certain lower limit.

To see how small the changes in Doppler shifts caused by planets might be, consider the case of Jupiter and the Sun. Jupiter orbits the center of mass of the Jupiter-Sun system at a distance of 5.2 AU and at a speed of 13 km/s. The Sun, a thousand times as massive as Jupiter, orbits at a distance of 0.005 AU, only about 1% of the orbital size of Mercury. The orbital speed of the Sun is only 13 m/s, about 30% faster than a world-class sprinter can run. This is only 40 billionths of the speed of light and produces a shift in wavelength of one part in 20 million. However, planets more massive than Jupiter and nearer their parent stars will have Doppler shifts that are greater than those of Jupiter and can be more easily detected.

Astronomers have carefully examined many nearby stars for periodic variations in Doppler shift. In 1995, astronomers in Switzerland and the United States reported the first discoveries of giant planets orbiting stars like the Sun. Since then the number of planets detected by the Doppler shifts they produce has risen to about 300, some in systems with as many as five planets.

Transits A second way to find a planet orbiting another star is to observe the slight dips in brightness of the star that occur as the planet passes in front of it. This kind of event is called a **transit** of the star by the planet. To see what such a transit would be like, consider again the case of Jupiter and the Sun. Jupiter is about 1/10th as large as the Sun, so the disk of Jupiter has about 1/100th the area of the Sun's disk. So a transit of the Sun by Jupiter would produce a 1% dip in the Sun's brightness. The dip would last for about a day—the length of time it would take Jupiter to cross the Sun's disk. The dips would occur at intervals of Jupiter's orbital period, about 12 years. Because the Sun and Jupiter are both very small compared with the distance between them, the chances that a given observer would be aligned perfectly enough to see a transit are extremely small. The chances of detecting a transit, however, are much better for Jupiter-sized planets or larger orbiting very close to their parent stars. The interval between transits will be shorter and the probability of being aligned perfectly enough to see the transit is larger.

Transits by orbiting planets have been discovered for dozens of stars. Most of the transiting planets have been found by repeatedly observing a star field with thousands of stars. Careful analysis can distinguish transits from the many other causes of stellar brightness variations. Because the chances that transits can be detected are better for planets orbiting very close to their parent stars, most of the planets detected so far are in small orbits. Also, large planets produce deeper transits than do small planets, so the transit method favors giant planets. To detect Earth-like planets orbiting at distances comparable to Earth's orbital distance it is necessary to examine many stars with a telescopic system that can detect small brightness variations. Two space missions, *Corot* (European Space Agency) and *Kepler* (NASA) have been launched to do just that. Both *Corot* and *Kepler* repeatedly image about 100,000 stars. The two missions should result in the detections of a large number of giant planets and a few tens to a hundred terrestrial planets with masses and sizes comparable to Earth's.

Microlensing As described in Chapter 22, Section 3, microlensing events occur when a dim, relatively nearby body passes almost directly between the Earth and a distant star. Hundreds of microlensing events, such as the one shown in Figure 22.31, have been observed. In a handful of cases, the microlensing event has been found to have structure that reveals the presence of a planet.

Imaging Most of the planets that have been detected by Doppler shifts, transits, and microlensing have such small orbits that when viewed from Earth they are lost in the glare of their parent stars and cannot be seen as separate points of light. By blocking the light of the star, however, astronomers have succeeded in imaging visible and infrared radiation from a number of planets with orbits comparable to or larger than the solar system. Figure 27.6 shows three planets orbiting HR 8799.

Pulsar Planets There are two pulsars for which the careful timing of the intervals between pulses revealed that the pulsar was orbiting its center of mass with one or more planets. In one case, the pulsar has three planets, in the other case, one planet. It is difficult to understand how a planet could have survived the violence that is believed to accompany the formation of a pulsar, and pulsar planets are still mysterious. It is likely, however, that they have little or nothing in common with the planets that have been found orbiting stars like the Sun.

Planets Orbiting Stars Like the Sun The stars about which planets have been discovered are mostly main sequence stars ranging from a little hotter and more luminous than the Sun to M stars about half as hot and only a few percent as luminous as the Sun. Most of the stars lie within 100 pc of the Sun and are either barely

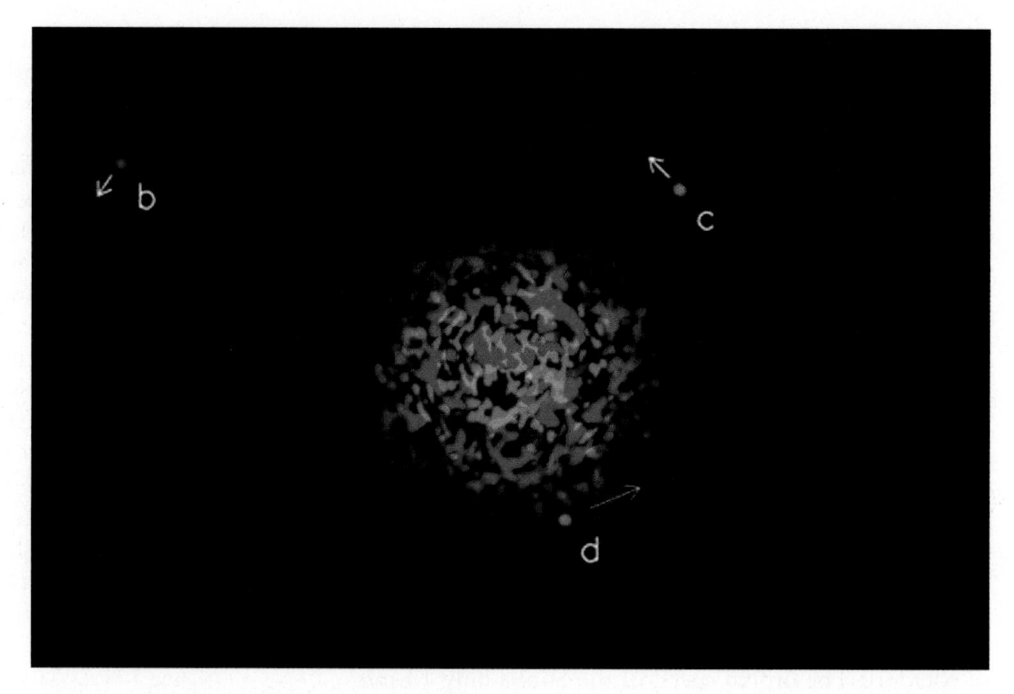

FIGURE 27.6
An Infrared Image of Three Planets of HR 8799.
The three planets orbit HR 8799 at distances of 68 AU (b), 38 AU (c), and 24 AU (d). Their masses are estimated to be about ten times the mass of Jupiter. The multicolored blur marks the location of the artificially obscured star. HR 8799 is about 50% more massive than the Sun.

visible to the naked eye or too faint to be seen except with the aid of a telescope. The orbital distances of the planets discovered as of 2009 are summarized in Figure 27.7. The planets range in distance from 0.01 AU to 700 AU. Orbital periods range from less than a day to 900 years, although some of the planets discovered by imaging have even longer periods that have not yet been determined. The smallest planet discovered so far has a mass about twice that of Earth.

Most of the newly discovered planetary systems bear little resemblance to our own. About a third of the planets are giant planets in orbits within about 0.3 AU of their stars. This is closer than the orbit of Mercury in our own solar system and leads to orbital periods ranging from a few days to a few weeks. Current theories of the formation of planetary systems predict that a giant planet cannot form so close to its parent star. Instead, giant planets are thought to form far enough from their stars that ices can condense to solid form. One possible explanation for the existence of a giant planet in a small orbit is that it formed farther from its star and then slowly spiraled inward. This could have happened if gravitational interactions with debris in the protoplanetary disk robbed the planet of some of its angular momentum. Presumably, if there were any giant planets that formed inside the original orbit of the planet, they would already have spiraled all the way into the star and been destroyed. Additional information about giant planets in small orbits can be learned if they also transit their parent star. An example is the planet orbiting HD 209458. The planet is 63% as massive as Jupiter and orbits every 3.5 days. Whenever it passes in front of HD 209458 it blocks 1.7% of the star's surface and causes a corresponding drop in brightness. Although the planet is less massive than Jupiter, its diameter is 1.6 times as great as Jupiter's. This is because the intense starlight falling on the planet heats it to much higher temperatures than Jupiter's and swells the planet.

FIGURE 27.7
The Orbital Distances of the New Planets
This graph shows the number of planets discovered orbiting other stars for various ranges of distance from the stars. For comparison, Mercury orbits the Sun at 0.4 AU, Venus at 0.7 AU, and Earth at 1 AU.

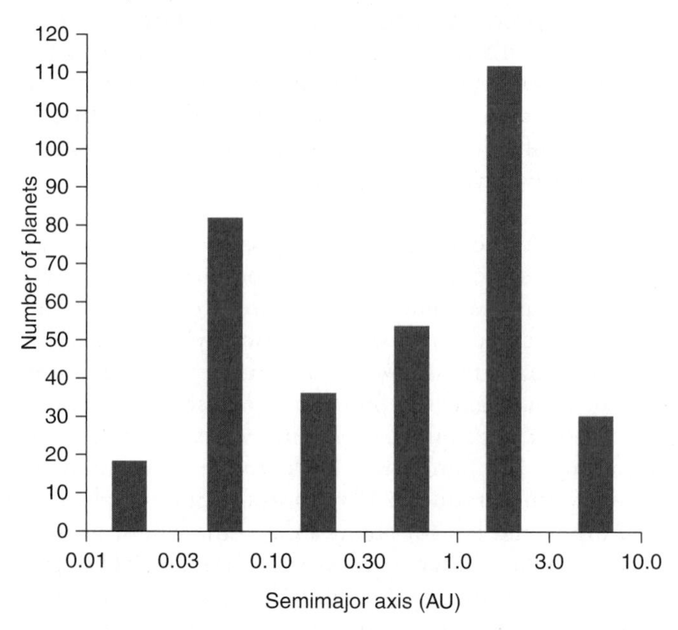

About half of the known planets have orbits larger than Earth's, but many of these have orbits unlike those of the giant planets in the solar system because of their large eccentricities. Almost half have eccentricities of 0.3 or larger and several have eccentricities greater than 0.9. Such eccentricities are much larger than that of Mercury (0.21), which is the largest of any planet in the solar system. The giant planets in the solar system have eccentricities of 0.056 or smaller. The effect of large orbital eccentricity is a large range of orbital distance. For example, 16 Cygni B's planet, which has an orbital eccentricity of 0.69 and a semimajor axis of 1.7 AU (about the same as Mars's) ranges between 0.5 AU and 2.9 AU from its star. Gravitational interactions with this planet would have long ago disrupted or expelled any other planets that occupied the region between 0.5 AU and 2.9 AU. In our solar system all of the terrestrial planets orbit in or near this region. Thus, 16 Cygni B is unlikely to have terrestrial planets in orbits similar to those of Earth. Many of the new planets have orbits that would have swept the terrestrial planet zone clear of planets.

About a dozen planets have orbits that resemble those of the giant planets in the solar system in that they are roughly circular and lie between the distances of Mars and Neptune. Most of these planets are considerably more massive than Jupiter, but they are probably similar to Jupiter and Saturn in general appearance and properties.

In most cases, only single planets have been found, but for about 10% of the stars, multiple planets have been discovered. There could be multiple giant planets orbiting many of the other stars as well. The planets found so far may simply be the ones that are near enough their parent stars and massive enough to produce relatively large Doppler shifts. Of the multiple planet systems, there are several that resemble the solar system in having multiple giant planets beyond 1 AU. One of these is the three-planet system of υ Andromedae. One of the planets, only 0.06 AU from υ Andromedae and with an orbital period of 4.6 days, was discovered early in the search for extrasolar planets. Later, two more planets with orbital distances of 2.0 and 4.0 AU were found. A comparison of our solar system and the υ Andromedae system is shown in Figure 27.8.

Although few of the planets discovered so far are likely to be habitable, it is likely that some of their stars also have terrestrial planets in orbits like those of Venus, Earth, and Mars. The Doppler shifts due to terrestrial planets at distances near 1 AU are too small to be detected with present techniques and instruments. Another intriguing possibility is that some of the giant planets may have satellite systems like that of Jupiter. For those planets, like the ones orbiting ι Horologii or HD 177830, that orbit near 1 AU the temperatures on their satellites might be in the range in which liquid water could exist. If Jupiter's satellite, Europa, were orbiting ι Horologii's planet, for example, it probably would have oceans and a thick atmosphere and might be habitable.

The discoveries so far have shown that planetary systems are common, but have given us only a lower limit on the fraction of stars that have planets. The fact that a given star has, so far, not been found to have planets does not show that there are no planets orbiting it. Of the planets discovered using Doppler shifts or transits, few have orbital periods longer than 10 years. This is partly because distant planets cause smaller Doppler shifts than closer planets do and are less likely to transit their parent stars. Another factor is that astronomers seeking planets cannot be sure that the pattern of Doppler shifts they see is due to a planet until they have observed the star for at least one full orbital period. Even for planets with relatively short orbital periods, it may be necessary to average data over several orbital periods to be sure that the observed pattern is real. This means that as time passes more planets are sure to be discovered. Many of these will have distances and masses like those of the giant planets in our solar system. Additional multiplanet systems will be found as well.

Scientists are now developing plans to build arrays of optical/infrared telescopes in space. The first such arrays would be used to detect the light reflected from giant planets in other solar systems. Later arrays would be sensitive enough to search for terrestrial planets. The arrays would have high enough angular resolution to distinguish details such as oceans, continents, and mountain ranges on planets orbiting other stars.

FIGURE 27.8
The Planetary System of υ Andromedae and the Inner Solar System
The three planets of υ Andromedae span the range of distances of the terrestrial planets in our solar system. The inner planet lies well inside the orbital distance of Mercury. The outer planet orbits between the distances of Mars and Jupiter. Notice that the middle and outer planets have much greater orbital eccentricities than any of the terrestrial planets in our solar system.

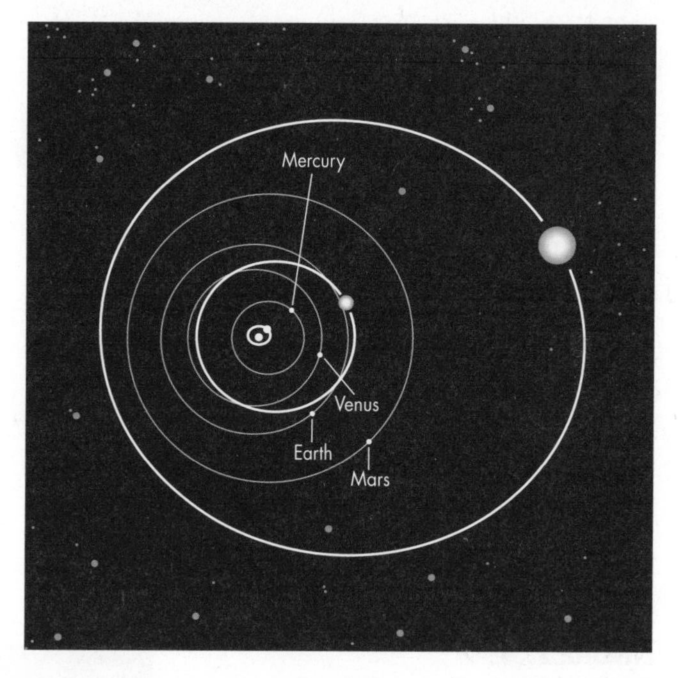

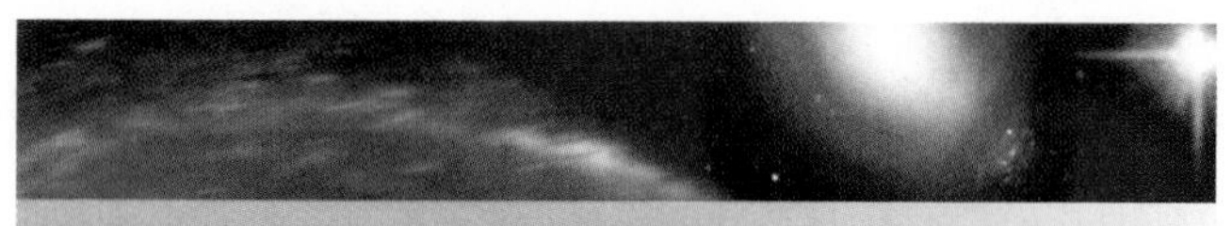

Observations have discovered many planets orbiting stars like the Sun. Studies now under way are likely to lead to the discovery of numerous additional planetary systems. It now appears that a significant fraction of single stars like the Sun are accompanied by planetary systems.

Habitable Planets

Certain conditions seem essential for life similar to that on the Earth to have developed on another planet. Life on Earth probably began in the sea. Thus, the distance of a life-bearing planet from its star must be one that leads to liquid surface water rather than ice or steam. The planet cannot be so small that it retains no atmosphere nor so large that it is a gas giant like Jupiter and Saturn. The star about which the planet orbits and the climate on the planet must be stable long enough for life to develop.

Some of these ideas have been combined in the concept of the **habitable zone** of a star. The habitable zone is the range of distances from a star within which liquid water can exist at the surface of an Earthlike planet. Astronomers have used results of stellar evolution calculations to determine the luminosity of stars of different masses throughout their evolution. They have also used planetary climate models to calculate what planets with the size, mass, and composition of the Earth would be like at various distances from stars of different luminosities. They find that the inner edge of the habitable zone is determined by the distance at which a planet would undergo a runaway greenhouse effect and become like Venus. The outer radius of the habitable zone is the distance at which a dense carbon dioxide atmosphere would begin to condense as dry ice. When this happens, the greenhouse warming of the planet by carbon dioxide diminishes and the temperature of the planet soon drops below the freezing point of water. The habitable zone for the Sun is shown in Figure 27.9.

As the Sun has grown brighter during its main sequence phase, the habitable zone has moved farther from the Sun. Figure 27.10 shows how the habitable zone of the Sun has changed and will continue to change throughout the main sequence lifetime of the Sun. When the Sun reached the main sequence, the inner edge of the habitable zone of the solar system was located at about 0.8 AU (outside the orbit of Venus). The inner edge of the habitable zone has moved outward to 0.95 AU as the Sun brightened during the past 4.6 billion years. The outer edge of the habitable zone has remained at about 1.4 AU (just inside the orbit of Mars). In the future, as the Sun continues to brighten, the habitable zone will move outward, passing the Earth in about 1.5 billion years. The most important result of these studies is that

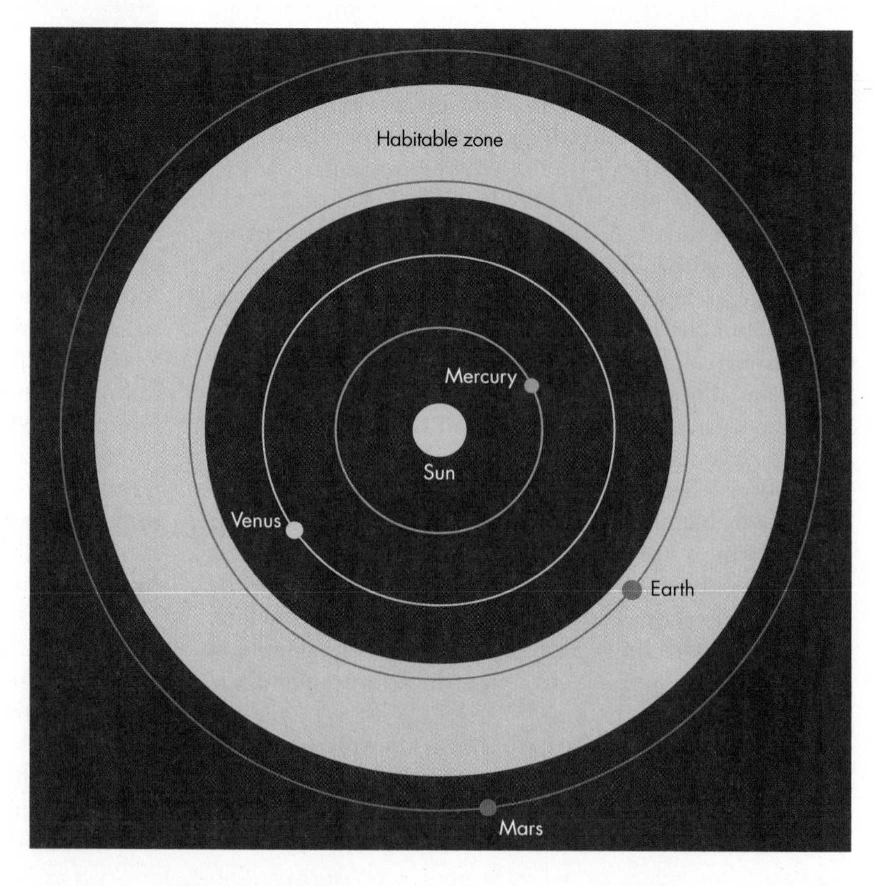

FIGURE 27.9
The Habitable Zone of the Sun
The habitable zone is the range of distances from a star at which water would exist in liquid form on a planet like the Earth. The habitable zone of the Sun begins just inside the orbital distance of the Earth and extends nearly to the orbital distance of Mars. Thus, the Earth is the only planet that lies within the Sun's habitable zone.

FIGURE 27.10
How the Habitable Zone of the Sun Has Changed Since the Sun Formed
As the Sun has grown more luminous, its habitable zone has moved farther from the Sun. In the distant future, the habitable zone will move past the Earth, leaving the Earth in a Venuslike condition. At an even more distant time, the habitable zone will encompass Mars. This does not mean, however, that life will begin on Mars at that time.

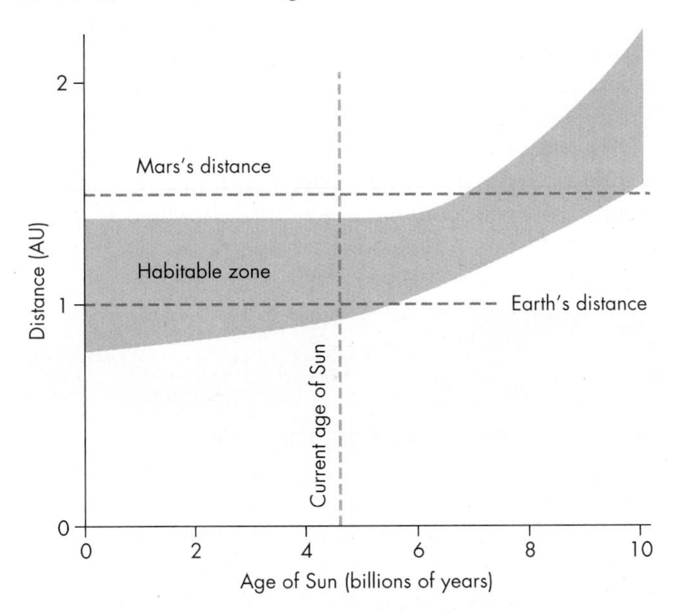

the habitable zone for a star like the Sun is always at least several tenths of an AU in width. This is wide enough that it is likely that a large fraction of planetary systems around stars like the Sun will have one or more planets within the habitable zone. Stars somewhat more massive or less massive than the Sun also have habitable zones that are at least several tenths of an AU in width. Thus, it seems likely that habitable planets are relatively common in planetary systems. Furthermore, the discovery that there is liquid water beneath the icy crust of Jupiter's satellite Europa suggests that habitable planets may exist outside of habitable zones around stars.

Intelligent Life

This is as far as astronomy, even astronomical speculation, can take us. Any further conclusions about extraterrestrial life require deductions about the origin, development, and durability of life that are still quite uncertain. Given our current ideas about the origin of life on Earth, however, it appears that the origin of life may be a common and spontaneous development whenever conditions are favorable for a long enough period of time. This would imply that life has originated on many planets and at many times throughout the history of the galaxy.

Even if it were true, however, that the development of life is common, that wouldn't necessarily imply that the galaxy is now filled with life-bearing planets. In order for there to be life on a planet, it is also necessary for conditions to permit the continued existence of life on that planet once life begins. For example, life might not be expected to exist on a planet that suffers frequent bombardments by large interplanetary bodies. In fact, it has been speculated that life on Earth may have originated and then been obliterated, perhaps several times, by impacts during the first few hundred million years of the Earth's history.

To see how the number of inhabited planets depends on how long life survives on a typical planet, suppose that one-tenth of the stars in the galaxy were orbited by planets on which life developed. Suppose also that the average time that life lasts on a planet were 1 million years and that the average star and its planetary system were 10 billion years old. This would mean that life existed for only 1/10,000th of the history of a typical planetary system. Life would now exist on only 1 in 10,000 planets on which it has ever existed. Life would be extinct on the other 9,999 planets out of 10,000. Because only one star in ten is assumed to have planets on which life developed, we would expect only 1 in every 100,000 stars to be accompanied by a life-bearing planet right now. There are about 100 billion stars in the Milky Way, so there would be about one million inhabited planets.

On the other hand, suppose that life lasts for 1 billion years, or one-tenth of the age of a typical planet. This would mean that one-tenth of the planets on which life ever developed are inhabited right now. It would also mean that one in every one hundred stars is now orbited by a life-bearing planet and that there are one billion inhabited planets in the Milky Way. Now, none of these assumptions about how often life arises or how long it lasts may actually be true. The point is that the number of life-bearing planets in the galaxy today depends directly on how long life lasts once it develops. Life has lasted for about 3 billion years on Earth, but we don't really know whether that means life is very durable or that we've been lucky.

Even if life is common in the galaxy, it will be very difficult for us to discover life outside the solar system unless intelligent life often develops. This is because the easiest way to search for life in the universe (and the only way available to us right now) is to look for signals from intelligent creatures on other planets. The search for interstellar communication is discussed in the next section of this chapter.

There is a considerable range of opinions about the inevitability of the development of advanced intelligence once life begins. Some scientists have argued that intelligence confers an evolutionary advantage that leads to more and more intelligent creatures. Other scientists have pointed out that all plants and most animals show no sign of intelligence. A low level of intelligence has evolved many times but its evolution has nearly always stopped at some moderate level and stayed there. There are few examples of progressive growth in brain size in the fossil record. It is also not clear that intelligence has survival value. Many species of creatures closely related to humans (such as

Australopithecus and *Gigantopithecus*) have developed and then become extinct. At the present time it seems very difficult to choose between these two points of view.

The Drake Equation In 1961, astronomer Frank Drake devised an equation that, if we knew the values of all of its terms, could be used to calculate the number of extraterrestrial civilizations in the Milky Way galaxy with which we might potentially establish communication. This equation, now known as the **Drake Equation,** is given by

$$N = R_* f_p n_e f_l f_i f_c L \tag{27.1}$$

where R_* is the number of stars that form in the galaxy each year, f_p is the fraction of stars that have planets, n_e is the average number of planets in a planetary system that have the potential to support life, f_l is the fraction of such planets on which life actually develops, f_i is the fraction of inhabited planets on which intelligent life develops, f_c is the fraction of planets bearing intelligent life on which those intelligent beings have an interest in and the capability to establish interstellar communication, and L is the average length of time that such a civilization maintains efforts to communicate.

Because the values of so many of the terms in the Drake Equation are completely unknown at the present time, the equation is more useful in helping us organize our thinking about life elsewhere in the galaxy than it is for actually calculating the number of potential communicating partners. The term R_* is very likely to be about 10 to 20 stars per year. The term f_p was essentially unknown only 20 years ago, but recent discoveries of other planetary systems give us some confidence that as many as half of the stars may have planetary systems. The number of potentially habitable planets per planetary system (n_e) is less certain. Potentially habitable planets would be impossible in most of the planetary systems discovered so far. But solar systems like our own would be much more difficult to detect than those with giant planets orbiting within a few AU of their parent stars, so there could be many of them in nearby space still awaiting detection. The terms f_l and f_i, as described in the preceding section, have been estimated by different scientists to have values ranging from essentially zero to nearly one. Estimates of the terms f_c and L are little better than guesses. We really know nothing about the motivation of nonhuman civilizations to establish communications (f_c). The length of time a typical civilization lasts and tries to establish communication (L) probably depends on many factors ranging from the probability that civilizations can survive the destructive capability of the same technology that enables them to communicate to the possibility that establishing communication with other civilizations gives them the knowledge required to survive for an extremely long time.

Pessimists Versus Optimists Obviously, it is difficult to make sensible statements about extraterrestrial life, particularly intelligent life. Even so, a number of scientists have used the Drake Equation in attempts to estimate how common life is in the universe and how often life develops into a civilization that might attempt to communicate with us. The estimates have ranged very widely. Extreme optimists have argued that as many as 1% of all stars may have planets bearing potential partners in communication. This implies that the nearest planet bearing intelligent life may be no more than a few parsecs away. In contrast, extreme pessimists have argued that life is so rare and its development so precarious that we live on the only planet in the Milky Way on which intelligent life has ever developed. Even given the billions of galaxies in the universe, extreme pessimists believe that life is so rare that the nearest planet bearing intelligent life is extremely distant.

The truth lies somewhere between the optimistic and pessimistic extremes. The safest conclusion is that the occurrence of life and intelligent life elsewhere is highly uncertain.

The habitable zones of stars like the Sun are large enough that there are likely to be habitable planets in many planetary systems. Our current ideas about the origin of life on Earth suggest that life may have begun on many planets in the galaxy. Uncertainty about the durability of life and the inevitability of intelligence makes it difficult to make reliable estimates of how common life and intelligent life are in the universe.

27.5 INTERSTELLAR COMMUNICATION

Because there may be intelligent creatures elsewhere, perhaps within our galaxy, many scientists have considered how communication with those beings might be achieved. The opening of communications with intelligent beings from another planetary system would be one of the most important events in human history.

Travel

Despite the apparent ease with which spaceships cross interstellar space in science fiction movies, interstellar travel is very far beyond our abilities now and may remain impossible forever. The main obstacle to interstellar travel is the immense distances involved. The fastest moving spacecraft ever launched from Earth, the *Voyagers,* left the planetary system traveling at about 20 km/s. At that speed, less than 1/100th of 1% of the speed of light, it will take the *Voyagers* about 60,000 years to reach the distance of the nearest star. Because the *Voyagers* have become interstellar probes, scientists took advantage of the opportunity to use them as very slow interstellar messengers. Recordings of images and

FIGURE 27.11
The Cover of the Recording Attached to the *Voyager* Spacecraft
The cover presents information from which it would be possible to discover when and where the spacecraft was launched. The recording itself contains images and sounds of the Earth. It is hoped that the recording will be played by intelligent creatures who intercept *Voyager* as it drifts among the stars hundreds of thousands or millions of years from now.

sounds of the planet Earth were attached to the spacecraft. One of the recordings is shown in Figure 27.11. The design on the cover of the recording contains a "map" that could be decoded to tell whoever intercepts the spacecraft when and where the spacecraft was launched.

If we could find a way to shorten the trip by traveling at nearly the speed of light, there would still be serious problems. For one thing, the energy expenditure required to accelerate either an inhabited spacecraft or a robot probe to nearly the speed of light is so enormous that no civilization is likely to use such an inefficient means of communication. Another serious problem is that interstellar atoms, molecules, and dust particles would be struck by the spacecraft at such speeds that they would be deadly. There may not be a practical way to prevent these energetic particles from striking the spacecraft and its inhabitants.

There is an interesting argument that has been used to show that the obstacles to interstellar travel must be difficult enough that no civilization has ever, in the entire history of the galaxy, overcome them. The argument goes like this: Suppose that interstellar travel is possible, if only at speeds of 1000 km/s, or 0.3% of the speed of light. (This is still 50 times as fast as the *Voyagers* left the solar system.) Suppose a civilization arose, mastered the technology of interstellar flight, and began to send out spacecraft to colonize the galaxy. It would take about 1000 years to travel from one planetary system to its nearest neighbors. Suppose that after every trip, the civilization spent 1000 years building spacecraft to continue the colonizing effort. Thus, the wave of colonization would spread through the galaxy at an average speed of 0.15% of the speed of light and would fill the galaxy in about 100 million years. Obviously, this is a very long time, but it is only 1% of the age of the galaxy. This means that there has been plenty of time for this to happen. The fact that we aren't part of a vast galactic empire and have no firm evidence that the Earth has ever been visited by extraterrestrials may indicate that no interstellar civilization has ever developed in the galaxy, perhaps because interstellar travel is too difficult. Another explanation—one offered by extreme pessimists—is that intelligent life has never arisen in the galaxy and we are completely alone.

Radio Communication

Although other means of communication have been proposed, including robot spacecraft and lasers, most of the ideas about communication with extraterrestrials involve the transmission and reception of radio signals. One reason that radio communication is generally considered more likely than communication in other parts of the spectrum such as the visible and infrared, is that the stars likely to have habitable planets are bright in the visible and infrared wavelengths, making it difficult for even a powerful laser to be seen. Another reason is that sophisticated equipment and signal processing techniques for radio communication have already been developed for radio communications on the Earth. Earlier in this chapter we saw that the number of life-bearing planets in the galaxy is proportional to the length of time that life survives once it begins on a planet. In the same way, the number of civilizations with whom we might communicate is proportional to the length of time that the average civilization possesses and uses the kind of technology that might make it possible for us to communicate with it. That technology involves the use of large radio telescopes and powerful radio transmitters.

There are two reasons why the length of time a typical civilization uses powerful radio technology might be relatively brief. Pessimists have pointed out that the development of radio technology on Earth coincided closely with the development of potent weapons of mass destruction such as nuclear bombs and chemical and biological agents. It may be that most or all civilizations destroy themselves shortly after developing the technology needed to communicate. In this case, the number of potential partners in communication would be very small and our chances of establishing communications would be slim. Another possibility is that civilizations develop superior technologies and abandon the use of radio transmitters after a relatively brief time. On the other hand, if civilizations survive and maintain radio technology (perhaps for purposes of interstellar communication), there may be many of them and they may be near enough that we can find them.

One important consequence of the dependence of our chances of contacting other civilizations on the length of time they use radio technology is that we are very likely to be the junior partner in any successful attempt at communication. We have had the ability to carry out interstellar communications for only about 60 years. If civilizations either perish or give up radio technology after an average of less than 60 years, then there will be, at best, only a few such civilizations in the galaxy. Our chances of finding one are almost zero. On the other hand, if civilizations last much longer than ours has so far, then there may be many of them. They would almost certainly have much more powerful radio transmitters than we do. We should attempt the less demanding activity of listening rather than transmitting signals in hopes of a reply. Nearly all of the attempts at interstellar communications carried out so far have been listening rather than signaling. These attempts are collectively called **SETI,** the Search for Extraterrestrial Intelligence. Scientists involved in SETI have proposed that a highly sensitive array of radio telescopes be built to search for faint signals from very distant civilizations. One possibility for such an array is shown in Figure 27.12. To isolate the array from radio signals from Earth, advocates have suggested that it be built on the back side of the Moon or in space in the outer solar system. As a first step, an array of 350 six-meter telescopes is now under construction in northern California. The array, will be used exclusively to search for interstellar signals.

Attempts at Communication At our present level of technology, our large radio telescopes could detect signals from creatures using similar technology on planets anywhere in the galaxy. A very big problem, however, is that to receive signals, we have to be looking in the right direction at the right time and have to be observing at the correct frequency.

What Frequency to Observe? It seems probable that any creatures attempting to communicate across interstellar distances will transmit signals with a very narrow range of frequencies. This is because by packing a given amount of power into a narrow frequency range (or bandwidth), the signals will be much brighter and can be detected much farther away than if that amount of power is spread over a wider bandwidth. The ideal bandwidth may be somewhere between 0.001 Hz and 0.1 Hz.

There is no sure way of knowing what frequency extraterrestrials might use to try to communicate with us. There is, however, a part of the radio spectrum within which communication should be easiest. This is because within that part of the spectrum there is less naturally occurring radio emission to compete with an artificial signal. Figure 27.13 shows the radio part of the spectrum. At frequencies below approximately 10^9 Hz (or 1 GHz [gigaHertz]), the galaxy is very bright because of synchrotron emission by energetic electrons. Above about 10 GHz, the Earth's atmosphere becomes steadily brighter due to molecules of water vapor and oxygen. Although this would not be a problem for radio transmitters and receivers in space, it would be a problem for habitable planets with atmospheres containing water vapor and oxygen. Thus, the quietest part of the radio spectrum is between 1 and 10 GHz. It is in that part of the spectrum that most efforts to detect interstellar signals have concentrated.

Even though we can identify 1 to 10 GHz as the best region of the spectrum in which to search for signals, the number of frequencies to search in that region of

FIGURE 27.12
A Proposed Array of Radio Telescopes to Search for Interstellar Signals
The proposed array would consist of at least a thousand 30-m telescopes working together to try to detect extremely faint interstellar communications.

FIGURE 27.13
The Most Favorable Part of the Radio Spectrum for Interstellar Communications
Background noise, which reduces the ability of a telescope to detect faint signals, is at a minimum for frequencies between about 1 and 10 GHz. Below 1 GHz, the sky is bright because of emission from the galaxy. Above 10 GHz, molecules in the Earth's atmosphere (or the atmosphere of any Earthlike planet) make the sky bright.

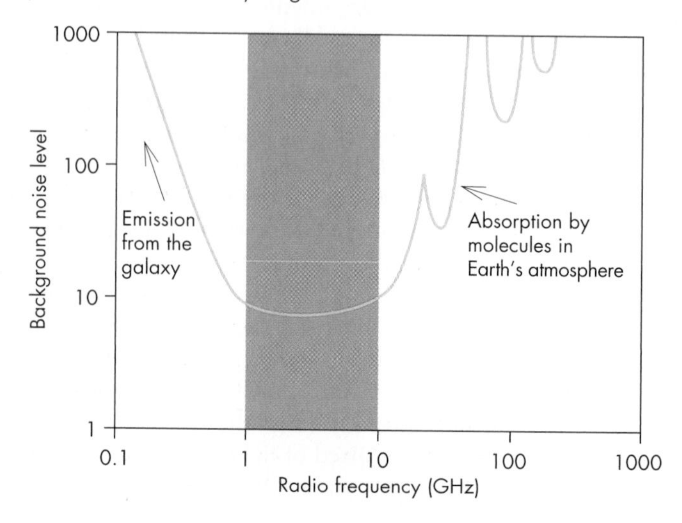

the spectrum is immense. There are 9 billion "channels," each 1 Hz wide, in the region between 1 and 10 GHz. There are 9 trillion channels, each 0.001 Hz wide, in the region. How can we possibly pick the same channel to listen to that the extraterrestrials chose to transmit? There have been some ingenious suggestions. Most of these suggestions make use of the fact that there is one frequency that both we and the extraterrestrials know about. That is the frequency, 1.420 GHz (a wavelength of 21 cm), at which interstellar hydrogen atoms emit. Any civilization capable of building radio telescopes is sure to be aware of the emission line of interstellar hydrogen. Some scientists have suggested observing at 1.420 GHz. Others have suggested 4.461 GHz, which is 1.420 GHz times the universal number π, or 3.860 GHz, which is 1.420 GHz times the number e.

In any case, most scientists have concluded that it is prudent to search as much of the 1 to 10 GHz region as possible. They have built signal processors capable of measuring as many as hundreds of millions of narrow, adjacent channels simultaneously. Given so much data, it would be impossible for scientists to personally examine every channel to see if it contains a signal. Instead, sophisticated computer programs have been used to search for signals in the data.

Enormous computer power is needed to search the data for possible signals. In one project, called SETI@home, that computer power is supplied by millions of personal computers in homes around the world. Each computer is supplied with special software and then given data from the Arecibo Observatory to examine. Each set of data covers a particular time and frequency interval. The program has performed over 2 million hours of computing since it was begun in 1999.

One interesting test of this radio method of searching for extraterrestrial signals was an attempt to detect radio transmissions from the *Pioneer 10* spacecraft. At the time of the attempt, *Pioneer 10* had already left the planetary system and was about 70 AU from the Earth. Figure 27.14 shows one thousand frequency channels during successive 0.7 second intervals of time. The snowy appearance of the figure is due to noise (random static). The weak signal from *Pioneer 10* shifted in frequency as time passed because the rotation of the Earth caused the Doppler shift of the signal to vary. Thus, the *Pioneer 10* signal, like a true extraterrestrial signal, appears as a diagonal line.

FIGURE 27.14
A Signal from Interstellar Space
This figure shows the brightness in one thousand adjacent frequency channels during successive 0.7-second intervals of time. The snowy appearance is due to noise. The signal, produced by *Pioneer 10* when it was beyond the orbit of Neptune, appears as a diagonal line. The signal from *Pioneer 10* changed in frequency, causing the line to slope, because of the changing Doppler shift due to the rotation of the Earth.

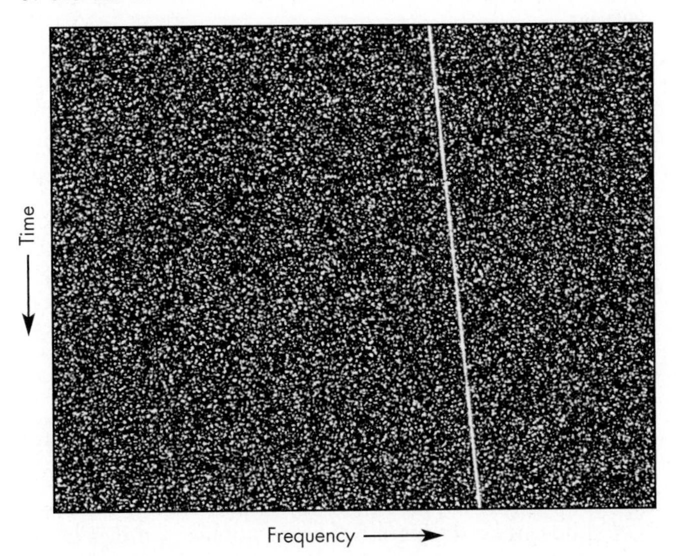

Where to Look? Just as the strength of a transmitted signal is greater the narrower the frequency range used, so the signal is stronger the smaller the angle it is transmitted into. This means that in order to detect a signal from creatures in another planetary system we probably have to point the radio telescope directly at the source of the signals.

Attempts to detect extraterrestrial signals have adopted at least three different strategies for selecting where to look. One choice has been to concentrate on high-priority targets, which are usually chosen to be nearby single stars like the Sun. The first attempt to detect interstellar radio signals, in 1960, observed two nearby Sunlike stars. This attempt, called Project Ozma after the Princess of the Land of Oz, was carried out at a frequency of 1.420 GHz.

Another strategy is to sweep the sky listening for signals from planets orbiting more remote stars. This can be done by pointing a telescope at a fixed altitude and azimuth and letting the rotation of the Earth carry different targets into view. Both selecting high-priority targets and surveying the sky require dedicated telescope time. This has meant that searches for interstellar communications have been carried out either using relatively small radio telescopes or during relatively infrequent allocations of observing time on large, more sensitive, radio telescopes. To increase the amount of observing time available to SETI on large telescopes, a third strategy has been used. This strategy, sometimes called the "piggyback" mode, simply uses whatever frequency and target are selected for other observing programs on the telescope. It then looks for extraterrestrial signals in the data acquired for the other program. The piggyback mode is used to acquire the data used in the SETI@home project.

None of the searches so far carried out has been successful. However, this suggests only that relatively nearby planets aren't bristling with radio telescopes beaming recognizable signals at the Earth.

Signals from Earth There have also been intentional and unintentional signals sent into space from the Earth. In 1974, an intentional signal was sent from the Arecibo Observatory toward a globular star cluster. Unintentional signals come from radio and television transmissions, some of which escape from the atmosphere and spread into space. It is sobering to realize that our first contact with extraterrestrials may be their reception of rock music or situation comedies. Such signals fill a volume of space that expands away from the Earth at the rate of one light year per year.

Light Travel Time Radio signals travel at the speed of light. Even the nearest stars are so distant that it takes several years for their light to reach us. This means that a radio conversation with creatures on a planet orbiting one of the nearest stars would be a slow one. Years or decades would elapse between a question and its answer. In the likely case that the nearest intelligent life is more distant, even longer lapses would occur. Communication under such circumstances could hardly be a dialogue. Instead, each message would have to be self-contained and as clear as possible. The reception of such a communication would be like finding a message in a bottle cast into the sea long ago.

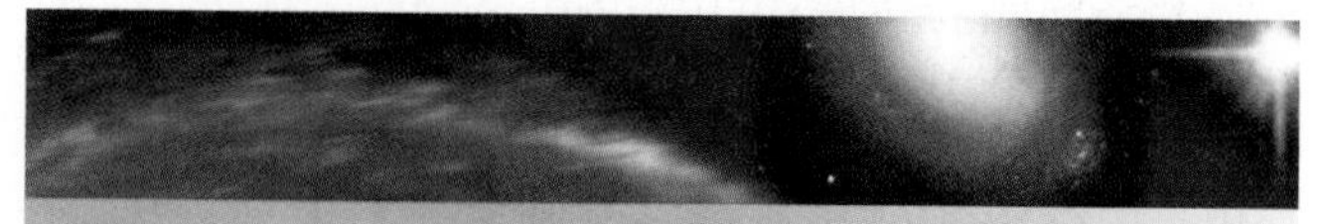

Radio signals seem to be the best way for civilizations in different planetary systems to communicate. There is considerable uncertainty, however, about what frequency to use and in which directions to search. There have been numerous attempts to detect signals from other civilizations. None has yet been successful. If the nearest civilizations are more than a few light years away, the light travel time of the signals will make it difficult to carry out a dialogue.

Chapter Summary

- Life can be distinguished from nonliving things by its ability to reproduce and evolve. All terrestrial life is based on carbon atoms and water, but it may be possible that there are extraterrestrial life-forms based on substances other than carbon and water. (Section 27.1)
- Life on Earth began within a billion years or less after the Earth formed. We have no conclusive evidence that life originated elsewhere in the solar system though it seems possible that life may have developed on Mars and Europa. (27.2, 27.3)
- Astronomers have discovered planets orbiting hundreds of nearby stars. We can now be confident that many single stars like the Sun have planetary systems. (27.4)
- Stars like the Sun have habitable zones large enough that they are likely to encompass habitable planets in many planetary systems. Our current ideas about the origin of life on Earth suggest that life may have begun on many planets in the galaxy. We are now too uncertain about the durability of life and the inevitability of evolution of intelligence to make reliable estimates of how common life and intelligent life are in the universe. (27.4)
- Radio communication using powerful transmitters and large radio telescopes seems to be the best way for civilizations in different planetary systems to communicate with each other and with us. In order to detect signals from extraterrestrials, however, we need to know what frequency to use and in what direction or directions to search. There have been many attempts, without success, to detect radio signals from other civilizations. The distances between us and the nearest civilizations are likely to make it difficult to carry out a conversation. (27.5)

Key Terms

amino acid *646*
cell *646*
Drake Equation *656*
habitable zone *654*
life *646*
nucleic acid *646*
nucleotide *646*
protein *646*
SETI *658*
transit *651*

Conceptual Questions

1. Describe why, in order to be considered living, an object must contain DNA or another substance that has the same capabilities.
2. What properties of water are crucial to life on Earth? Discuss the advantages and disadvantages of ammonia and methyl alcohol versus water as liquids on which life might be based.
3. The oceans of the early Earth probably contained high concentrations of organic molecules. Where did those molecules come from?
4. How does the nebular theory differ from the catastrophic theories of the origin of the solar system with respect to its predictions about the rarity of other planetary systems?
5. Why have attempts to find planets beyond the solar system concentrated on giant planets like Jupiter rather than terrestrial planets?
6. What kind of observations were used to discover planets orbiting nearby stars like the Sun?
7. Discuss the ways in which the recently discovered planetary systems are similar to and different from the solar system.
8. Why does the habitable zone of a star change its location as time passes?
9. What is the relationship between the length of time that life endures on a typical planet and the number of inhabited planets in the galaxy?
10. Why does it follow that if our attempts at interstellar communication are likely to succeed, we are very likely to be the junior partners in the conversation?
11. Suppose someone suggested that the "logical" frequencies at which to search for interstellar signals are precisely 1 or 10 GHz. Present an argument that there is nothing special or logical about those frequencies.
12. If we discover life on other planets, how is human civilization likely to be affected? (There is no "correct" answer to this question, but it is an interesting subject for speculation.)

Problems

1. What would the angle between the Sun and the Earth be if the solar system were viewed from a distance of 1 pc?
2. The Sun is about 300,000 times as massive as the Earth. Calculate the radius of the Sun's orbit about its common center of mass with the Earth. How big would the radius of the Sun's orbit appear if the solar system were viewed from a distance of 1 pc?
3. At what speed does the Sun orbit its common center of mass with the Earth? (For this problem, you need to look up the Earth's orbital speed in Appendix 5.)
4. The planet orbiting 16 Cyg B has an orbital distance of 1.7 AU. If 16 Cyg B has the same mass as the Sun, what is the orbital period of the planet?
5. Use the Drake Equation and your own estimates of its terms to estimate the number of communicating civilizations in the galaxy. (There is no correct answer or incorrect answer to the problem.)
6. How long will it take the *Voyagers* to travel 10 pc through interstellar space?
7. Suppose scientists built a signal processor capable of simultaneously looking for interstellar signals in 100 million channels, each 1 Hz wide. How many times would the processor need to be used to examine the entire spectrum of a candidate star between 1 and 10 GHz?
8. Approximately how long would it take radio signals from the Earth to reach Neptune? How long would it take radio signals to reach the Earth from the nearest star, Proxima Centauri?

Figure-Based Questions

1. Use Figure 27.9 to find the width (in AU) of the Sun's habitable zone.
2. Use Figure 27.10 to find the age of the Sun when the Sun's habitable zone moves beyond the Earth's orbit. Given the age of the Sun (4.5 billion years), how long will it be before the Earth ceases to be in the Sun's habitable zone?
3. Use Figure 27.10 to find how long will it be before Mars is within the Sun's habitable zone.
4. Use Figure 27.10 to find how long Mars will be within the Sun's habitable zone.

Group Activities

1. Divide your group into two subgroups. After preparing for a few minutes, have one group present reasons why the discovery of life elsewhere in the universe would be "good" for humanity. Have the other group present reasons why the discovery of life elsewhere would be "bad" for humanity. Have the entire group discuss the two cases and try to come to a consensus.
2. Divide your group into two subgroups. After preparing for a few minutes, have one group present reasons why the initiation of communications with creatures elsewhere in the universe would be "good" for humanity. Have the other group present reasons why starting such communications would be "bad" for humanity. Have the entire group discuss the two cases and try to come to a consensus.
3. Divide your group into two subgroups. Have each subgroup make up a "picture" like that in Figure 27.11 that could tell extraterrestrial creatures about humanity and life on Earth. Then ask each subgroup to try to interpret the other subgroup's message.

For More Information

Visit the text website at **www.mhhe.com/fix** for chapter quizzes, interactive learning exercises, and other study tools.

Appendixes

Appendix 1

Mathematical and Physical Constants

pi (π)	3.1415926
degrees in radian	57.296
seconds of arc in radian	206264.8
speed of light in vacuum (c)	2.9979×10^{8} m/s
gravitational constant (G)	6.6726×10^{-11} $m^3/kg\ s^2$
Planck's constant (h)	6.6261×10^{-34} W s^2
mass of electron (m_e)	9.1094×10^{-31} kg
mass of hydrogen atom (m_H)	1.6735×10^{-27} kg
Stefan-Boltzmann constant (σ)	5.6705×10^{-8} $W/m^2\ K^4$
Boltzmann constant (k)	1.3805×10^{-23} Ws/K

Appendix 2

Astronomical Constants

astronomical unit (AU)	1.496×10^{11} m
parsec (pc)	3.086×10^{16} m
light year (ly)	9.460×10^{15} m
mass of Earth (M_{Earth})	5.974×10^{24} kg
equatorial radius of Earth (R_{Earth})	6378 km
solar mass ($M_\odot$)	1.989×10^{30} kg
solar radius ($R_\odot$)	6.960×10^{8} m
solar luminosity ($L_\odot$)	3.847×10^{26} W
solar day	86,400 s
sidereal month	2.361×10^{6} s
synodic month	2.551×10^{6} s
sidereal year	3.156×10^{7} s

Appendix 3

Conversion Factors

DISTANCE

1 nm	=	10^{-9} m
1 micrometer	=	10^{-6} m
1 cm	=	10^{-2} m
	=	0.3937 inches
1 m	=	3.28 feet
1 km	=	0.6214 mile
1 AU	=	1.4960×10^{8} km
	=	92.96 million miles
1 pc	=	206,265 AU
	=	3.26 light years (ly)
1 kpc	=	10^{3} pc
1 Mpc	=	10^{6} pc

MASS

1 kg	=	10^{3} g

TEMPERATURE

T(K)	=	T(°C) + 273.15
T(°C)	=	$[T(°F) - 32]\frac{5}{9}$
T(°F)	=	$\frac{9}{5}T(°C) + 32$

TIME

1 sidereal month	=	27.3217 solar days
1 solar month	=	29.5306 solar days
1 sidereal year	=	365.2564 solar days

Appendix 4

Periodic Table of the Elements

9 **F** Fluorine 19.00	
9	Atomic number
19.00	Atomic mass

1 1A	2 2A	3 3B	4 4B	5 5B	6 6B	7 7B	8 8B	9 8B	10 8B	11 1B	12 2B	13 3A	14 4A	15 5A	16 6A	17 7A	18 8A
1 **H** Hydrogen 1.008																	2 **He** Helium 4.003
3 **Li** Lithium 6.941	4 **Be** Beryllium 9.012											5 **B** Boron 10.81	6 **C** Carbon 12.01	7 **N** Nitrogen 14.01	8 **O** Oxygen 16.00	9 **F** Fluorine 19.00	10 **Ne** Neon 20.18
11 **Na** Sodium 22.99	12 **Mg** Magnesium 24.31											13 **Al** Aluminum 26.98	14 **Si** Silicon 28.09	15 **P** Phosphorus 30.97	16 **S** Sulfur 32.07	17 **Cl** Chlorine 35.45	18 **Ar** Argon 39.95
19 **K** Potassium 39.10	20 **Ca** Calcium 40.08	21 **Sc** Scandium 44.96	22 **Ti** Titanium 47.88	23 **V** Vanadium 50.94	24 **Cr** Chromium 52.00	25 **Mn** Manganese 54.94	26 **Fe** Iron 55.85	27 **Co** Cobalt 58.93	28 **Ni** Nickel 58.69	29 **Cu** Copper 63.55	30 **Zn** Zinc 65.39	31 **Ga** Gallium 69.72	32 **Ge** Germanium 72.59	33 **As** Arsenic 74.92	34 **Se** Selenium 78.96	35 **Br** Bromine 79.90	36 **Kr** Krypton 83.80
37 **Rb** Rubidium 85.47	38 **Sr** Strontium 87.62	39 **Y** Yttrium 88.91	40 **Zr** Zirconium 91.22	41 **Nb** Niobium 92.91	42 **Mo** Molybdenum 95.94	43 **Tc** Technetium (98)	44 **Ru** Ruthenium 101.1	45 **Rh** Rhodium 102.9	46 **Pd** Palladium 106.4	47 **Ag** Silver 107.9	48 **Cd** Cadmium 112.4	49 **In** Indium 114.8	50 **Sn** Tin 118.7	51 **Sb** Antimony 121.8	52 **Te** Tellurium 127.6	53 **I** Iodine 126.9	54 **Xe** Xenon 131.3
55 **Cs** Cesium 132.9	56 **Ba** Barium 137.3	57 **La** Lanthanum 138.9	72 **Hf** Hafnium 178.5	73 **Ta** Tantalum 180.9	74 **W** Tungsten 183.9	75 **Re** Rhenium 186.2	76 **Os** Osmium 190.2	77 **Ir** Iridium 192.2	78 **Pt** Platinum 195.1	79 **Au** Gold 197.0	80 **Hg** Mercury 200.6	81 **Tl** Thallium 204.4	82 **Pb** Lead 207.2	83 **Bi** Bismuth 209.0	84 **Po** Polonium (210)	85 **At** Astatine (210)	86 **Rn** Radon (222)
87 **Fr** Francium (223)	88 **Ra** Radium (226)	89 **Ac** Actinium (227)	104 **Rf** Rutherfordium (257)	105 **Db** Dubnium (260)	106 **Sg** Seaborgium (263)	107 **Bh** Bohrium (262)	108 **Hs** Hassium (265)	109 **Mt** Meitnerium (266)	110 **Ds** Darmstadtium (269)	111 **Rg** Roentgenium (272)	112	(113)	114	(115)	116	(117)	(118)

58 **Ce** Cerium 140.1	59 **Pr** Praseodymium 140.9	60 **Nd** Neodymium 144.2	61 **Pm** Promethium (147)	62 **Sm** Samarium 150.4	63 **Eu** Europium 152.0	64 **Gd** Gadolinium 157.3	65 **Tb** Terbium 158.9	66 **Dy** Dysprosium 162.5	67 **Ho** Holmium 164.9	68 **Er** Erbium 167.3	69 **Tm** Thulium 168.9	70 **Yb** Ytterbium 173.0	71 **Lu** Lutetium 175.0
90 **Th** Thorium 232.0	91 **Pa** Protactinium (231)	92 **U** Uranium 238.0	93 **Np** Neptunium (237)	94 **Pu** Plutonium (242)	95 **Am** Americium (243)	96 **Cm** Curium (247)	97 **Bk** Berkelium (247)	98 **Cf** Californium (249)	99 **Es** Einsteinium (254)	100 **Fm** Fermium (253)	101 **Md** Mendelevium (256)	102 **No** Nobelium (254)	103 **Lr** Lawrencium (257)

Metals

Metalloids

Nonmetals

The 1–18 group designation has been recommended by the International Union of Pure and Applied Chemistry (IUPAC) but is not yet in wide use. In this text we use the standard U.S. notation for group numbers (1A–8A and 1B–8B). No names have been assigned for elements 112, 114, and 116. Elements 113, 115, 117, and 118 have not yet been synthesized.

Appendix 5

Orbital Properties of the Planets and Dwarf Planets

NAME	SEMIMAJOR AXIS, IN AU	SEMIMAJOR AXIS, IN 10^6 km	SIDEREAL PERIOD	SYNODIC PERIOD, IN DAYS	AVERAGE ORBITAL SPEED, IN km/s	ORBITAL ECCENTRICITY	INCLINATION OF ORBIT, IN DEGREES
Planet							
Mercury	0.3871	57.9	87.97 days	115.9	47.9	0.206	7.00
Venus	0.7233	108.2	224.7 days	583.9	35.0	0.007	3.39
Earth	1.0	149.6	365.3 days	—	29.8	0.017	0.00
Mars	1.523	227.9	687.0 days	780.0	24.1	0.093	1.85
Jupiter	5.202	778.3	11.86 years	398.9	13.1	0.048	1.31
Saturn	9.539	1,427	29.46 years	378.1	9.6	0.056	2.49
Uranus	19.19	2,870	84.01 years	369.7	6.8	0.046	0.77
Neptune	30.06	4,497	164.8 years	367.5	5.4	0.010	1.77
Dwarf Planet							
Ceres	2.77	413.7	4.60 years	466.7	17.9	0.080	10.59
Pluto	39.53	5,900	248.5 years	366.7	4.7	0.248	17.15
Haumea	43.1	6,450	283 years	366.5	4.5	0.195	28.22
Makemake	45.8	6,850	310 years	366.4	4.4	0.159	28.96
Eris	67.7	10,210	560 years	365.9	3.4	0.442	44.19

Appendix 6

Physical Properties of the Planets and Dwarf Planets

NAME	EQUATORIAL DIAMETER, IN km	EQUATORIAL DIAMETER, IN EARTH DIAMETERS	MASS, IN EARTH MASSES	DENSITY, RELATIVE TO WATER	ACCELERATION OF GRAVITY AT SURFACE, RELATIVE TO EARTH	ESCAPE VELOCITY, IN km/s	SIDEREAL ROTATION PERIOD, IN DAYS	ALBEDO	AXIAL TILT, IN DEGREES
Planet									
Mercury	4,878	0.38	0.055	5.4	0.38	4.3	58.65	0.11	2
Venus	12,104	0.95	0.815	5.2	0.91	10.4	243.02	0.65	177.3*
Earth	12,756	1.0	1.0	5.5	1.0	11.2	1.00	0.37	23.5
Mars	6,787	0.53	0.107	3.9	0.38	5.0	1.03	0.15	25.2
Jupiter	142,980	11.2	317.9	1.3	2.54	59.6	0.41	0.52	3.1
Saturn	120,540	9.5	95.2	0.7	1.08	35.6	0.44	0.47	26.7
Uranus	51,120	4.0	14.5	1.3	0.91	21.3	0.72	0.40	97.9*
Neptune	49,530	3.9	17.1	1.6	1.19	23.8	0.67	0.35	29.6
Dwarf Planet									
Ceres	950	0.074	0.00016	2.1	0.03	0.5	0.38	0.09	4
Pluto	2,300	0.18	0.002	2.0	0.06	1.2	6.39	0.6	122.5*
Haumea	1,400	0.11	0.0007	3	0.04	0.8	0.16	0.8	—
Makemake	1,500	0.12	0.0007	2	0.05	0.8	—	0.8	—
Eris	2,400	0.19	0.0025	2	0.06	1.3	—	0.86	—

*Retrograde rotation

Data for Appendix 6 from National Space Science Data Center.

Appendix 7

Properties of the Major Satellites of the Planets and Dwarf Planets

Satellite of Earth

NAME OF SATELLITE	AVERAGE DIAMETER, IN km	MASS, IN UNITS OF MASS OF MOON (7.35×10^{22} kg)	DENSITY, RELATIVE TO WATER	ORBITAL SEMIMAJOR AXIS, IN 10^3 km	ORBITAL SEMIMAJOR AXIS, IN UNITS OF RADIUS OF PLANET	ORBITAL PERIOD, DAYS	ALBEDO
Moon	3,476	1.0	3.34	384.4	60.2	27.32	0.12

Satellites of Mars

NAME OF SATELLITE	AVERAGE DIAMETER, IN km	MASS, IN UNITS OF MASS OF MOON (7.35×10^{22} kg)	DENSITY, RELATIVE TO WATER	ORBITAL SEMIMAJOR AXIS, IN 10^3 km	ORBITAL SEMIMAJOR AXIS, IN UNITS OF RADIUS OF PLANET	ORBITAL PERIOD, DAYS	ALBEDO
Phobos	25	1.4×10^{-7}	2.0	9.4	2.8	0.32	0.07
Deimos	14	3.3×10^{-8}	1.7	23.5	6.9	1.26	0.08

Major Satellites of Jupiter

NAME OF SATELLITE	AVERAGE DIAMETER, IN km	MASS, IN UNITS OF MASS OF MOON (7.35×10^{22} kg)	DENSITY, RELATIVE TO WATER	ORBITAL SEMIMAJOR AXIS, IN 10^3 km	ORBITAL SEMIMAJOR AXIS, IN UNITS OF RADIUS OF PLANET	ORBITAL PERIOD, DAYS	ALBEDO
Amalthea	170	—	—	181	2.5	0.50	0.05
Thebe	100	—	—	222	3.1	0.68	0.05
Io	3,640	1.21	3.53	422	6.0	1.77	0.6
Europa	3,122	0.66	3.01	671	9.5	3.55	0.6
Ganymede	5,262	2.03	1.94	1,070	15.1	7.15	0.4
Callisto	4,820	1.46	1.83	1,883	26.6	16.69	0.2
Himalia	185	—	—	11,460	160	251	0.03

Major Satellites of Saturn

NAME OF SATELLITE	AVERAGE DIAMETER, IN km	MASS, IN UNITS OF MASS OF MOON (7.35×10^{22} kg)	DENSITY, RELATIVE TO WATER	ORBITAL SEMIMAJOR AXIS, IN 10^3 km	ORBITAL SEMIMAJOR AXIS, IN UNITS OF RADIUS OF PLANET	ORBITAL PERIOD, DAYS	ALBEDO
Prometheus	100	—	—	139	2.3	0.61	0.5
Pandora	90	—	—	142	2.3	0.63	0.7
Janus	190	—	—	151	2.5	0.69	0.9
Epimetheus	120	—	—	152	2.5	0.69	0.8
Mimas	400	5.2×10^{-4}	1.14	186	3.1	0.94	0.5
Enceladus	500	1.1×10^{-3}	1.00	238	4.0	1.37	1.0
Tethys	1,050	1.0×10^{-2}	1.00	295	4.9	1.89	0.9
Dione	1,120	1.4×10^{-2}	1.50	377	6.3	2.74	0.7
Rhea	1,530	3.4×10^{-2}	1.24	527	8.7	4.52	0.7
Titan	5,150	1.83	1.88	1,222	20.3	15.95	0.2
Hyperion	300	2.7×10^{-4}	1.5	1,481	24.6	21.27	0.3
Iapetus	1,440	2.6×10^{-2}	1.02	3,561	59	79.33	0.05 to 0.5
Phoebe	220	—	—	12,954	215	550	0.08

Appendix 7

Properties of the Major Satellites of the Planets and Dwarf Planets—cont'd

Major Satellites of Uranus

NAME OF SATELLITE	AVERAGE DIAMETER, IN km	MASS, IN UNITS OF MASS OF MOON (7.35×10^{22} kg)	DENSITY, RELATIVE TO WATER	ORBITAL SEMIMAJOR AXIS, IN 10^3 km	ORBITAL SEMIMAJOR AXIS, IN UNITS OF RADIUS OF PLANET	ORBITAL PERIOD, DAYS	ALBEDO
Cressida	80	—	—	62	2.4	0.46	—
Juliet	95	—	—	64	2.5	0.49	—
Portia	135	—	—	66	2.6	0.51	—
Belinda	80	—	—	75	2.9	0.62	—
Puck	160	—	—	86	3.4	0.76	0.07
Miranda	470	9.0×10^{-4}	1.20	129	5.1	1.41	0.3
Ariel	1,160	1.8×10^{-2}	1.67	191	7.5	2.52	0.4
Umbriel	1,170	1.6×10^{-2}	1.40	266	10.4	4.14	0.2
Titania	1,580	4.8×10^{-2}	1.71	436	17.1	8.71	0.3
Oberon	1,520	4.1×10^{-2}	1.63	583	22.8	13.46	0.2
Caliban	100	—	—	7,230	283	580	—
Sycorax	190	—	—	12,179	476	1,283	—

Major Satellites of Neptune

NAME OF SATELLITE	AVERAGE DIAMETER, IN km	MASS, IN UNITS OF MASS OF MOON (7.35×10^{22} kg)	DENSITY, RELATIVE TO WATER	ORBITAL SEMIMAJOR AXIS, IN 10^3 km	ORBITAL SEMIMAJOR AXIS, IN UNITS OF RADIUS OF PLANET	ORBITAL PERIOD, DAYS	ALBEDO
Thalassa	80	—	—	50	2.0	0.31	0.06
Despina	150	—	—	53	2.1	0.33	0.06
Galatea	160	—	—	62	2.5	0.43	0.06
Larissa	190	—	—	74	3.0	0.55	0.06
Proteus	420	—	—	118	4.8	1.12	0.06
Triton	2,700	0.29	2.05	355	14.3	5.88	0.80
Nereid	340	—	—	5,510	223.0	360.10	0.40

Satellite of Pluto

NAME OF SATELLITE	AVERAGE DIAMETER, IN km	MASS, IN UNITS OF MASS OF MOON (7.35×10^{22} kg)	DENSITY, RELATIVE TO WATER	ORBITAL SEMIMAJOR AXIS, IN 10^3 km	ORBITAL SEMIMAJOR AXIS, IN UNITS OF RADIUS OF PLANET	ORBITAL PERIOD, DAYS	ALBEDO
Charon	1,200	2.6×10^{-2}	2.0	19	8.3	6.39	0.5

Appendix 8

Orbital Properties of Selected Asteroids

ASTEROID NAME	YEAR OF DISCOVERY	DIAMETER, IN km	SEMIMAJOR AXIS, IN AU	ORBITAL PERIOD, IN YEARS	ORBITAL ECCENTRICITY	ORBITAL INCLINATION, IN DEGREES
Main-belt Asteroids						
Pallas	1802	520	2.77	4.61	0.180	34.8
Vesta	1807	500	2.36	3.63	0.097	7.1
Hygiea	1849	430	3.14	5.59	0.136	3.8
Davida	1903	340	3.18	5.67	0.171	15.9
Interamnia	1910	330	3.06	5.36	0.081	17.3
Europa	1858	310	3.10	5.46	0.119	7.4
Euphrosyne	1854	250	3.16	5.58	0.099	26.3
Juno	1804	240	2.67	4.36	0.218	13.0
Bamberga	1892	240	2.68	4.41	0.285	11.1
Earth-crossing Asteroids						
1973 NA	1973	6	2.43	3.78	0.64	68.0
Phaethon	1983	6	1.27	1.43	0.89	22.0
Toro	1948	5	1.37	1.60	0.44	9.4
Daedalus	1971	3	1.46	1.77	0.62	22.2
Ra-Shalom	1978	3	0.83	0.76	0.44	15.8
Geographos	1951	2	1.24	1.39	0.34	13.3
Adonis	1936	1	1.88	2.57	0.76	1.4
Aten	1976	1	0.97	0.95	0.18	18.9
Icarus	1949	1	1.08	1.12	0.83	22.9
Apollo	1932	1	1.47	1.78	0.56	6.4
Asteroids at Jupiter's Distance and Beyond						
Hektor	1907	220	5.15	11.7	0.03	18.3
Agamemnon	1919	170	5.20	11.9	0.07	21.8
Chiron	1977	180	13.72	50.8	0.38	6.9

Appendix 9

Orbital Properties of Selected Comets

NAME OF COMET	YEAR OF NEXT PERIHELION	ORBITAL PERIOD, IN YEARS	PERIHELION DISTANCE, IN AU	ORBITAL ECCENTRICITY	ORBITAL INCLINATION, IN DEGREES
Giacobini-Zinner	2008	3.3	1.03	0.71	31.9
Encke	2015	11.9	0.34	0.85	11.9
Schwassman-Wachmann 1	2019	15.0	5.45	0.10	9.8
Halley	2061	76.0	0.59	0.97	162.2
Swift-Tuttle	?	120	0.96	0.96	113.6
Ikeya-Seki	2845	880	0.01	1.00	141.9
Hale-Bopp	4377	2380	0.91	1.00	89.4

Appendix 10

Properties of Selected Meteor Showers

NAME OF METEOR SHOWER	DATE OF MAXIMUM	DURATION, IN DAYS	RATE OF VISIBLE METEORS AT MAXIMUM, IN NUMBER PER HOUR
Quadrantids	January 3	0.5	85
Lyrids	April 22	2	15
η Aquarids	May 5	6	30
Perseids	August 13	5	100
Orionids	October 22	5	20
Taurids	November 3	30	15
Leonids	November 17	4	12
Geminids	December 14	3	95

Appendix 11

The Constellations

NAME OF CONSTELLATION	ABBREVIATION	TRANSLATION	APPROXIMATE RIGHT ASCENSION, IN HOURS*	APPROXIMATE DECLINATION, IN DEGREES	AREA, IN SQUARE DEGREES
Northern Constellations					
Andromeda	And	Princess of Ethiopia	1	+40	722
Auriga	Aur	Charioteer	6	+40	657
Camelopardalis	Cam	Giraffe	6	+70	757
Canes Venatici	CVn	Hunting Dogs	13	+40	465
Cassiopeia	Cas	Queen of Ethiopia	1	+60	598
Cepheus	Cep	King of Ethiopia	22	+70	588
Cygnus	Cyg	Swan	21	+40	804
Draco	Dra	Dragon	17	+65	1,083
Lacerta	Lac	Lizard	22	+45	201
Leo Minor	LMi	Little Lion	10	+35	232
Lynx	Lyn	Lynx	8	+45	545
Lyra	Lyr	Lyre, Harp	19	+40	287
Perseus	Per	Rescuer of Andromeda	3	+45	615
Ursa Major	UMa	Big Bear	11	+50	1,280
Ursa Minor	UMi	Little Bear	15	+70	256
Equatorial Constellations					
Aquarius	Aqr	Water Bearer	23	−15	980
Aquila	Aql	Eagle	20	+5	653
Aries	Ari	Ram	3	+20	441

*Right ascension is often given in hours, where 1 hour equals 15 degrees. One hour of right ascension can be divided into 60 minutes and 1 minute into 60 seconds of right ascension. Right ascension is measured eastward around the celestial equator starting at the vernal equinox.

Appendix 11

The Constellations—cont'd

NAME OF CONSTELLATION	ABBREVIATION	TRANSLATION	APPROXIMATE RIGHT ASCENSION, IN HOURS	APPROXIMATE DECLINATION, IN DEGREES	AREA, IN SQUARE DEGREES
Equatorial Constellations					
Bootes	Boo	Herdsman	15	+30	907
Cancer	Cnc	Crab	9	+20	506
Canis Major	CMa	Big Dog	7	−20	380
Canis Minor	CMi	Little Dog	8	+5	183
Capricornus	Cap	Goat	21	−20	414
Cetus	Cet	Whale	2	−10	1,231
Coma Berenices	Com	Berenice's Hair	13	+20	387
Corona Borealis	CrB	Northern Crown	16	+30	179
Corvus	Crv	Crow	12	−20	184
Crater	Crt	Cup	11	−15	282
Delphinus	Del	Dolphin	21	+10	189
Equuleus	Equ	Little Horse	21	+10	72
Eridanus	Eri	River Eridanus	3	−20	1,138
Fornax	For	Furnace	3	−30	398
Gemini	Gem	Twins	7	+20	514
Hercules	Her	Hero	17	+30	1,225
Hydra	Hya	Water Snake (female)	10	−20	1,303
Leo	Leo	Lion	11	+15	947
Lepus	Lep	Rabbit	6	−20	290
Libra	Lib	Scales	15	−15	538
Monoceros	Mon	Unicorn	7	−5	482
Ophiucus	Oph	Serpent-bearer	17	0	948
Orion	Ori	Hunter	5	+5	594
Pegasus	Peg	Winged Horse	22	+20	1,121
Pisces	Psc	Fish	1	+15	889
Piscis Austrinus	PsA	Southern Fish	22	−30	245
Pyxis	Pyx	Ship's Compass	9	−30	221
Sagitta	Sge	Arrow	20	+10	80
Sagittarius	Sgr	Archer	19	−25	867
Sculptor	Scl	Sculptor	0	−30	475
Scutum	Sct	Shield	19	−10	109
Serpens	Ser	Serpent	17	0	637
Sextans	Sex	Sextant	10	0	314
Taurus	Tau	Bull	4	+15	797
Triangulum	Tri	Triangle	2	+30	132
Virgo	Vir	Maiden	13	0	1,294
Vulpecula	Vul	Little Fox	20	+25	268

Appendix 11

The Constellations—cont'd

NAME OF CONSTELLATION	ABBREVIATION	TRANSLATION	APPROXIMATE RIGHT ASCENSION, IN HOURS	APPROXIMATE DECLINATION, IN DEGREES	AREA, IN SQUARE DEGREES
Southern Constellations					
Antlia	Ant	Air Pump	10	−35	239
Apus	Aps	Bird of Paradise	16	−75	206
Ara	Ara	Altar	17	−55	237
Caelum	Cae	Chisel	5	−40	125
Carina	Car	Ship's Keel	9	−60	494
Centaurus	Cen	Centaur	13	−50	1,060
Chamaeleon	Cha	Chameleon	11	−80	132
Circinus	Cir	Compass	15	−60	93
Columba	Col	Dove	6	−35	270
Corona Australis	CrA	Southern Crown	19	−40	128
Crux	Cru	Southern Cross	12	−60	68
Dorado	Dor	Swordfish	5	−65	179
Grus	Gru	Crane	22	−45	366
Horologium	Hor	Clock	3	−60	249
Hydrus	Hyi	Water Snake (male)	2	−75	243
Indus	Ind	Indian	21	−55	294
Lupus	Lup	Wolf	15	−45	334
Mensa	Men	Table (Mountain)	5	−80	154
Microscopium	Mic	Microscope	21	−35	210
Musca	Mus	Fly	12	−70	138
Norma	Nor	Square, Level	16	−50	165
Octans	Oct	Octant	22	−85	291
Pavo	Pav	Peacock	20	−65	378
Phoenix	Phe	Phoenix	1	−50	469
Pictor	Pic	Painter, Easel	6	−55	247
Puppis	Pup	Ships' Stern	8	−40	673
Reticulum	Ret	Net	4	−60	114
Scorpius	Sco	Scorpion	17	−40	497
Telescopium	Tel	Telescope	19	−50	252
Triangulum Australe	TrA	Southern Triangle	16	−65	110
Tucana	Tuc	Toucan	0	−65	295
Vela	Vel	Ship's Sail	9	−50	500
Volans	Vol	Flying Fish	8	−70	141

Appendix 12

Stars Nearer Than Four Parsecs

NAME OF STAR	RIGHT ASCENSION, IN HOURS AND MINUTES	DECLINATION, IN DEGREES AND MINUTES	DISTANCE, IN PARSECS	SPECTRAL TYPE	APPARENT VISUAL MAGNITUDE	ABSOLUTE VISUAL MAGNITUDE
Sun	—	—	—	G2	−26.7	4.8
Proxima Cen	14 30	−62 41	1.30	M5	11.1	15.5
α Cen A	14 40	−60 50	1.33	G2	0.0	4.4
B				K0	1.3	5.7
Barnard's star	17 58	+4 42	1.83	M4	9.6	13.2
Wolf 359	10 56	+7 01	2.39	M6	13.4	16.6
BD+36 2147	11 03	+35 58	2.52	M2	7.5	10.5
L726-8 A	1 39	−17 57	2.63	M6	12.4	15.3
B				M6	13.2	16.1
Sirius A	6 45	−16 43	2.63	A1	−1.4	1.5
B				white dwarf	8.4	11.3
Ross 154	18 50	−23 50	2.93	M3	10.5	13.1
Ross 248	23 42	+44 10	3.17	M5	12.3	14.8
ϵ Eri	3 33	−9 28	3.27	K2	3.7	6.2
Ross 128	11 48	+0 48	3.32	M4	11.1	13.5
L789-6	22 39	−15 18	3.40	M5	12.3	14.7
BD+43 44 A	0 18	+44 01	3.45	M1	8.1	10.4
B				M3	11.1	13.4
ϵ Ind	22 03	−56 47	3.46	K5	4.7	7.0
61 Cyg A	21 07	+38 45	3.46	K5	5.2	7.5
B				K7	6.0	8.3
BD+59 1915 A	18 43	+59 38	3.49	M3	8.9	11.2
B				M3	9.7	12.0
τ Cet	1 44	−15 56	3.49	G8	3.5	5.8
Procyon A	7 39	+5 14	3.50	F5	0.4	2.7
B				white dwarf	10.7	13.0
CD-36 15693	23 06	−35 51	3.52	M1	7.3	9.6
G51-15	8 30	+26 47	3.63	M6	14.8	17.0
GJ 1061	3 36	−44 31	3.70	M5	13.0	15.2
L725-32	1 13	−17 00	3.74	M4	12.0	14.2
BD+5 1688	7 27	+5 14	3.78	M3	9.9	12.0
CD-39 14192	21 17	−38 52	3.87	M0	6.7	8.7
Kapteyn's star	5 12	−45 01	3.87	M0	8.8	10.9
Kruger 60 A	22 28	+57 42	3.97	M3	9.8	11.9
B				M4	11.3	13.3

Appendix 13

The Brightest Stars

NAME OF STAR	COMMON NAME	RIGHT ASCENSION, IN HOURS AND MINUTES	DECLINATION, IN DEGREES AND MINUTES	DISTANCE, IN PARSECS	SPECTRAL TYPE	APPARENT VISUAL MAGNITUDE	ABSOLUTE VISUAL MAGNITUDE
—	Sun	— —			G2	−26.72	+4.8
α CMa	Sirius	6 43	−16 39	2.7	A1	−1.47	+1.4
α Car	Canopus	6 23	−52 40	30	F0	−0.72	−3.1
α Boo	Arcturus	14 13	+19 27	11	K2	−0.06	−0.3
α Cen	Rigel Kentaurus	14 36	−60 38	1.3	G2	−0.01	+4.4
α Lyr	Vega	18 35	+38 44	8	A0	+0.04	+0.5
α Aur	Capella	5 13	+45 57	14	G8	+0.05	−0.7
β Ori	Rigel	5 12	−8 15	250	B8	+0.14	−6.8
α CMi	Procyon	7 37	+5 21	3.5	F5	+0.37	+2.6
α Ori	Betelgeuse	5 52	+7 24	150	M2	+0.41	−5.5
α Eri	Achernar	1 36	−57 29	20	B5	+0.51	−1.0
β Cen	Hadar	14 00	−60 08	90	B1	+0.63	−4.1
α Aql	Altair	19 48	+8 44	5.1	A7	+0.77	+2.2
α Tau	Aldebaran	4 33	+16 25	16	K5	+0.86	−0.2
α Cru	Acrux	12 24	−62 49	120	B1	+0.90	−4.5
α Vir	Spica	13 23	−10 54	80	B1	+0.91	−3.6
α Sco	Antares	16 26	−26 19	120	M1	+0.92	−4.5
α PsA	Fomalhaut	22 55	−29 53	7.0	A3	+1.15	+1.9
β Gem	Pollux	7 42	+28 09	12	K0	+1.16	+0.8
α Cyg	Deneb	20 40	+45 06	430	A2	+1.26	−6.9
β Cru	Beta Crucis	12 45	−59 24	150	B0.5	+1.28	−4.6

Appendix 14

Principal Members of the Local Group of Galaxies

NAME OF GALAXY	RIGHT ASCENSION, IN HOURS AND MINUTES	DECLINATION, IN DEGREES AND MINUTES	HUBBLE TYPE	DISTANCE, IN Mpc	DIAMETER, IN kpc	VISUAL APPARENT MAGNITUDE	APPROXIMATE LUMINOSITY, IN MILLIONS OF SOLAR LUMINOSITIES
IC10	0 20	+59 18	Irr	0.7	1	11	300
NGC 147	0 33	+48 30	E5	0.7	2	10	100
NGC 185	0 39	+48 20	E3	0.7	2	9	150
NGC 205	0 40	+41 41	E5	0.8	5	8	300
Andromeda VIII	0 42	+40 37	E	0.8	14	9	150
M32	0 43	+40 52	E2	0.8	2	8	300
M31	0 43	+41 16	Sb	0.8	40	3	25,000
SMC	0 57	−72 50	Irr	0.06	5	2	600
IC 1613	1 05	+02 07	Irr	0.7	3	9	100
M33	1 34	+30 40	Sc	0.8	16	6	3,000
LMC	5 23	−69 45	Irr	0.05	0.6	1	2,000
NGC 3109	10 03	−26 09	Irr	1.38	5	10	100
Milky Way	17 46	−29 00	SBc	0.01	40	—	20,000
NGC 6822	19 45	−14 48	Irr	0.5	2	9	200

Appendix 15

The Brightest Galaxies Beyond the Local Group

NAME OF GALAXY	GALAXY TYPE	DIAMETER, IN kpc	APPROXIMATE LUMINOSITY, IN MILLIONS OF SOLAR LUMINOSITIES	DISTANCE, IN Mpc	RECESSION SPEED, IN km/s
NGC 300	Sc	15	3,000	2.4	+625
NGC 55	Sc	25	8,000	3.1	+115
NGC 2403	Sc	15	6,000	3.6	+300
NGC 3031 (M81)	Sb	15	15,000	3.6	+124
NGC 3034 (M82)	Irr	7	4,000	3.6	+410
NGC 253	Sc	25	20,000	4.2	+500
NGC 1313	SBc	8	8,000	5.2	+260
NGC 4736 (M94)	Sab	10	20,000	6.9	+345
NGC 5128 (Cen A)	S0	20	50,000	6.9	+250
NGC 5236 (M83)	SBc	20	30,000	6.9	+275
NGC 4826 (M64)	Sab	15	15,000	7.0	+350
NGC 5457 (M101)	Sc	50	50,000	7.6	+370
NGC 2903	Sc	30	20,000	9.4	+470
NGC 4258 (M106)	Sb	60	40,000	10	+520
NGC 5194 (M51)	Sbc	40	70,000	11	+540
NGC 5055 (M63)	Sbc	25	35,000	11	+550
NGC 6744	Sbc	35	50,000	13	+660
NGC 1291	SBa	20	60,000	15	+510
NGC 4594 (M104)	Sab	35	90,000	17	+870
NGC 4472 (M49)	E1/S0	30	140,000	22	+820

Appendix 16

Properties of Selected Clusters of Galaxies

NAME OF CLUSTER OF GALAXIES	ESTIMATED DISTANCE, IN Mpc	DIAMETER, IN Mpc	NUMBER OF KNOWN GALAXIES	REDSHIFT	RECESSION SPEED, IN km/s
Virgo	18	4	2,500	0.004	+1,200
Fornax	20	2	350	0.005	+1,400
Cancer	46	2	150	0.016	+4,800
Pegasus I	47	1	100	0.013	+3,700
Coma	95	8	800	0.022	+6,700
Hercules	150	0.3	300	0.034	+10,300
Ursa Major I	220	3	300	0.051	+15,400
Leo	280	3	300	0.065	+19,500
Corona Borealis	300	3	400	0.072	+21,600
Bootes	560	3	150	0.131	+39,400
Ursa Major II	590	2	200	0.137	+41,000
Hydra	870	—	—	0.201	+60,600

Appendix 17

Logarithmic Graphs

Many of the graphs in this book use logarithmic scales in order to show several powers of 10 on a single graph. One of these logarithmic graphs, Figure 7.12, is reproduced below. The key to using a logarithmic graph is realizing that equal distances along an axis represent equal ratios. That is, on the horizontal axis (which shows mass) in Figure 7.13 the distance from 0.1 to 1 is 15 mm. The distance from 1 to 10 is also 15 mm. Any two points on the horizontal axis that are separated by 15 mm have values that are in the ratio 10 to 1.

The value of the point midway between 1 and 10 (7.5 mm from each) is the square root of 10, or 3.16. This is because the ratio of 3.16 to 1 is equal to the ratio of 10 to 3.16. The value of the point one-third of the way from 1 to 10 (5 mm from 1 and 10 mm from 10) is the cube root of 10, or 2.16. The value one-third of the way from 1.0 to 1.9 is equal to the cube root of 10 times 0.1, or 0.216.

Logarithmic graphs can be made easier to read if values between consecutive powers of 10 are shown. In Figure 7.12 the positions corresponding to 2, 3, 4, 5, 6, 7, 8, and 9 are shown as small tick marks between the larger tick marks representing 1 and 10.

As an example of using logarithmic graphs, we can find the mass and accretional energy per kilogram (both relative to the Earth) for Uranus. Reading straight downward from the dot representing Uranus, we find that the point on the horizontal axis representing the mass of Uranus lies about midway between 10 and the next small tick mark, which corresponds to 20. The ratio of 20 to 10 is 2, and the square root of 2 is 1.41, so the mass of Uranus is about 1.41 times 10 or 14 Earth masses. Reading horizontally across from the dot representing Uranus, we find that the point on the vertical axis (also a logarithmic scale) is the third small tick mark above 1. This small tick mark corresponds to 4, so the accretional energy per kilogram of Uranus was 4 times as large as that of the Earth.

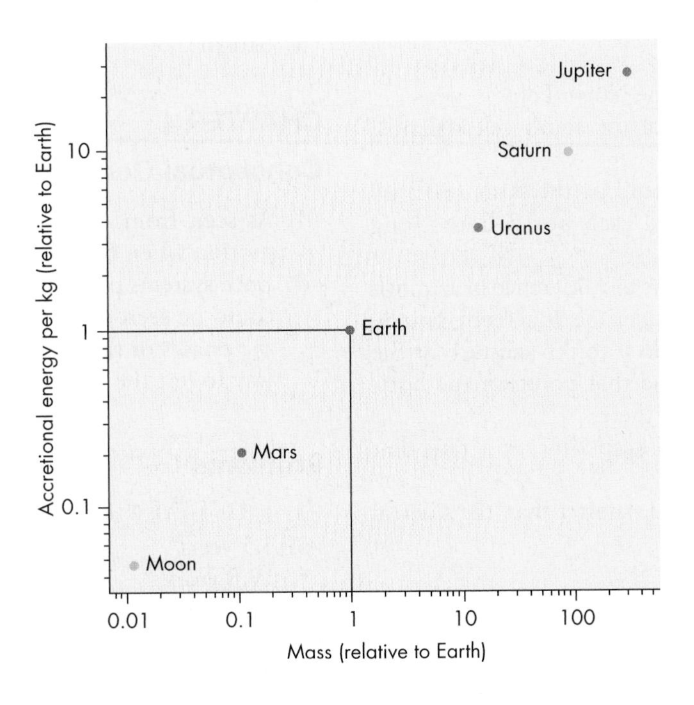

Appendix 18

Answers to Odd-Numbered Questions

CHAPTER 1

Conceptual Questions

1. The motions are seen more easily when there is a fixed object for comparison.
3. 90° N. All of the lines of constant longitude cross at the pole.

Problem

1. 84′, 5040″

Figure-Based Questions

1. 40°, 90°
3. 0°, 135°

CHAPTER 2

Conceptual Questions

1. Diurnal motion is from east to west. East is on our left as we face south and on our right as we face north.
3. 0 hours
5. White (The observer is at the North Pole.)
7. 90° S, where the altitude of the south celestial pole is 90°
9. The azimuth of sunrise would be the same (90°) all year. Day and night would each be 12 hours long throughout the year.
11. One way would be to measure the difference in azimuth between the summer and winter solstitial rising points. Halve the difference and add it to the azimuth of the summer solstitial rising, find that point on the horizon, and put a pile of rocks there.
13. The watch would need to keep time at a rate that varied throughout the year.
15. The synodic month would be shorter than the sidereal month.

Figure-Based Questions

1. 20°, 45°
3. Full, 3rd quarter, waning crescent, new, waxing gibbous

CHAPTER 3

Conceptual Questions

1. One way would be to have two kinds of years—one with 13 months = 325 days and another with 14 months = 350 days. There would have to be 1 long year for every 4 short years.
3. During some eclipses the curvature of the Earth's shadow would be more pronounced than during other eclipses.
5. Mars could be seen at gibbous or full phase. Venus could be seen at either new and crescent or full and gibbous phases, depending on whether its deferent was smaller or larger than that of the Sun.
7. Yes, in fact there would be little difference in the apparent motion of the planet, although the planet would appear dimmest instead of brightest during retrograde.

Problems

1. 11 synodic periods = 12 years
3. 62,000 m = 62 km
5. Its angular diameter is reduced to ⅓.
7. 3.9

Figure-Based Question

1. Saturn

CHAPTER 4

Conceptual Questions

1. As seen from Mars, the Earth undergoes retrograde motion when the Earth passes Mars.
3. Both systems predicted that Mars, Jupiter, and Saturn could be seen only at gibbous and full phases. Thus, the phases of the outer planets could not be used as a way to test the superiority of either system.

Problems

1. 1.5 years, 1.5 years
3. 3.5 years
5. 19.8 years
7. 0.94 AU
9. The star's diameter would be 2 AU, much larger than the Sun.
11. 0 cm, this is a circle.
13. 0.59 years
15. 0.79 AU

Figure-Based Questions

1. 0.47 AU, 1 AU, the angle is 28°, the sine of 28° is 0.47
3. About 52

CHAPTER 5

Conceptual Questions

1. Advocates of the impetus theory said that there was a constant force acting on the body. Galileo said that there was no force acting on the body.
3. It accelerates because the direction of its motion changes constantly as it orbits.
5. There are several methods. One would be to remove a shoe and throw it horizontally. The person would experience a force in the opposite direction as the shoe is thrown, accelerate briefly, and then glide to the shore.
7. The Earth is 80 times as massive as the Moon. This more than compensates for the fact that the Moon is one-fourth the size of the Earth.

Problems

1. 5.0 m/s^2
3. The acceleration of A is ¼ the acceleration of B.
5. 2 m/s^2 to the south.
7. The mass of A is 1/20 the mass of B.
9. The force would be the same.
11. It becomes 4 times as great.
13. 490 N
15. 8.3 km/s
17. 0.32 as massive as the Sun
19. 6.7 km/s, 67 km/s
21. 7.6 km/s

CHAPTER 6

Conceptual Questions

1. The photoelectric effect shows that light behaves like a stream of particles. The operation of a radio interferometer, in which electromagnetic radiation interferes like waves, shows that electromagnetic radiation behaves like waves.
3. To make drawing, consult Figure 6.19. Replace the lens with a mirror and draw rays as if reflected from the mirror rather than refracted.
5. Astronomers seek transparency, clear skies, good seeing, and dark skies.

Problems

1. The wavelength of A is one-fourth the wavelength of B.
3. 60 m
5. 7.5×10^8 Hz, this is a radio wave
7. 1/100
9. 300 km/s away from each other
11. A has twice the energy of B.
13. 36 times as great
15. The focal length of the f/10 mirror is twice as long.
17. 2

Figure-Based Questions

1. 3×10^{22} Hz
3. 50%

CHAPTER 7

Conceptual Questions

1. Outward moving atoms in the lower atmosphere strike other atoms and are deflected downward before they can escape. Collisions are rare enough in the exosphere that an outward moving atom is unlikely to strike another atom before escaping.
3. Because the weight of overlying gas drops with increasing altitude
5. The body will emit more energy than it absorbs, so it will cool. Its temperature will fall and the rate at which it emits energy will fall until it reaches thermal equilibrium.
7. The number of protons determines the chemical element, so the two atoms are different elements.
9. Radioactive atoms decay into other atoms at a predictable rate. By comparing the relative numbers of radioactive atoms and the atoms they decay into, it is possible to tell how long decays have been occurring.
11. a. Radiation, b. Conduction, c. Convection

Problems

1. 5.2 km/s
3. 0.22 times as great
5. It would become 4 times as large.
7. 27.5 km
9. Estimated area = 2 m^2, force = 2×10^5 N
11. The temperature of A is one-fourth the temperature of B.
13. It will become 2.1 times as great.
15. 0.011 times as bright
17. 3 minutes
19. About 1/180,000

Figure-Based Questions

1. 0.21
3. The answer found by counting squares is 1, 0.6, 0.3.
5. ^{26}Al, ^{40}K

CHAPTER 8

Conceptual Questions

1. When the Earth is moving directly toward or directly away from the star, the telescope would be pointed

directly at the star. At other times, the telescope would have to point slightly ahead of the direction to the star. The amount of tilt would be maximum when the Earth was moving at right angles to the direction to the star.

3. The circulation is clockwise, because air flowing in from the south is deflected westward, and air flowing in from the north is deflected eastward.
5. Observers in Florida are farther from the north magnetic pole.

Problems

1. 199 pounds, the person's mass would be the same
3. Into the ocean floor, about 0.15%
5. 16%
7. About 4000 km, 200 million years ago
9. The amount emitted by the planet would double (so that the same amount as before would escape into space). Temperature would become 1.2 times as high.

Figure-Based Questions

1. 5500 kg/m^3, 11,500 kg/m^3, 1.6 and 3.3 times as large as the densities of typical surface rocks
3. 280 K, 0.9 or 0.003 or 0.0003 or less than 0.00001 atmosphere

CHAPTER 9

Conceptual Questions

1. The Moon's synodic period would be longer than it actually is because the Earth would move farther in its orbit during a sidereal month, and thus the Moon would have to move farther in its orbit to bring it back to the same orientation relative to the Earth and Sun. (If the Earth's orbital period were 225 days, the synodic period of the Moon would be 31.1 days.)
3. A total solar eclipse.
5. Lunar samples are more abundant in elements that form compounds that condense at high temperatures but less abundant in water, elements that are easily vaporized, and iron.
7. The crater density has steadily increased.
9. The surface of Io must be extremely young.
11. The average density of the Moon is too low for a large amount of a dense material like iron to be present.

Problems

1. 3470 km
3. 2.5 seconds
5. 240 million years
7. 70 to 140 km in diameter

Table-Based Questions

1. 1st
3. 2009

CHAPTER 10

Conceptual Questions

1. Nights are 88 Earth days long and there is no atmosphere to insulate the surface.
3. Because Mercury has a high density despite little internal compression.
5. The cloud droplets mostly scatter (reflect) light rather than absorb it, so after many scatters some sunlight penetrates through the clouds.
7. Red sunlight penetrates through the clouds more easily than blue sunlight, so everything looks redder (orangish) than it would in unfiltered sunlight.
9. Midocean ridges, trench systems, and mountain ranges would be visible.
11. Earth's carbon dioxide is mostly incorporated in carbonate minerals.

Problems

1. 1.5
3. 1.6 years, it is the same length of time

Figure-Based Question

1. Both planets have broad plains (the ocean bottoms in the case of the Earth) and regions of mountainous terrain; no mountain chains, long trenches, or oceanic ridge systems are found on Venus; no circular chasms, such as the one on the southern edge of Aphrodite Terra, are found on Earth.

CHAPTER 11

Conceptual Questions

1. The ejecta looks as if it were fluid enough to flow away from the impact. This might have happened because subsurface ice was melted by the impact.
3. The northern plains of Mars were formed by volcanic flows about 3.0 billion years ago. The flows wiped out any evidence of earlier craters.
5. They formed from very fluid material, perhaps pyroclastic flows.
7. The orbital eccentricity of Mars is much larger than that of the Earth, so the difference in solar heating between perihelion and aphelion is much greater for Mars.
9. Oxidized iron compounds give the surface a rusty color.
11. Season winds blow dust particles on and then off Martian soil and rock.

13. Water is known to exist in the atmosphere and polar caps; it is believed to exist as subsurface ice.
15. Carbon dioxide may have been removed from the atmosphere to form carbonaceous rocks. Water vapor condensed and became permanent ice.
17. The consensus was that the soil of Mars is chemically active but not biologically active.

Problems

1. 0.38 AU, 0.66 AU
3. 100 km, about one-sixth the diameter of Olympus Mons

Figure- and Table-Based Questions

1. 3.6 billion years
3. Between 12 and 155 km

CHAPTER 12

Conceptual Questions

1. The inertia of rapidly moving regions near the equator “lifts” them against gravity producing a larger equatorial than polar diameter.
3. The great abundance of hydrogen means that hydrogen compounds rather than oxygen compounds dominate. Most oxygen is in the form of water molecules.
5. Both planets have numerous circular cloud features with hurricane-like circulation patterns.
7. Generally, dark features are lower and hotter than light features. The Great Red Spot is an exception.
9. The blue regions are deep in the atmosphere. The light we see is sunlight reflected from atmospheric molecules, which reflect blue sunlight better than red sunlight.
11. They are denser than they would be if they were made from only hydrogen and helium.
13. The excess energy is energy stored from the formation of Jupiter. In Saturn’s case the excess energy may come from the settling of helium toward the center of the planet.
15. The rings are very thin. When viewed edge-on, they are much harder to see than when we see them from above or below.
17. *Voyager* showed that there is much more structure (ringlets) in the rings than previously thought and that spokes form in the rings.
19. It is torn to pieces by tidal forces.

Problem

1. 91%

Figure-Based Questions

1. 10 km/s just before it encountered Jupiter
3. ½

CHAPTER 13

Conceptual Questions

1. He had to see whether the new object moved with respect to the stars.
3. He calculated that there was an unknown planet near the position where other astronomers later found Neptune.
5. Because the axis of Uranus lies in its orbital plane there are long periods of time when one pole or the other nearly points at the Sun. During those times, one hemisphere is always in daylight and the other in darkness.
7. Little heat flows out of Uranus to drive convention and cause atmospheric features.
9. Such models are consistent with the gravitational field measurements made by the *Voyager* spacecraft.
11. Their magnetic field axes are more tilted than the other planets and are also displaced from their centers.
13. The rings produced temporary drops in the light of stars when the rings passed between the stars and the Earth.
15. Small dust particles are efficient at forward scattering and the regions between the rings appear bright when viewed looking backward toward the Sun.
17. At perihelion, Pluto is closer to the Sun than Neptune is.
19. Using Charon’s orbital period and distance, Kepler’s third law was used to calculate the mass of Pluto.
21. The scattered disk objects have orbits that are larger, more eccentric, and more inclined to the ecliptic.

Problems

1. 8.9 m/s^2, 11.7 m/s^2
3. 586 years

Figure-Based Question

1. 1 atmosphere

CHAPTER 14

Conceptual Questions

1. A regular satellite
3. Masses were determined by how much the gravities of the satellites affected the motion of the spacecraft.
5. If the satellite is large, orbits close to a massive planet, and has an eccentric orbit.
7. There are no impact craters on Io.
9. Ice in the surface layers of Europa may flow to reduce the heights of crater walls and fill in craters. Also, there may be geyserlike flows of water that cover old craters and then freeze.
11. They have surface cracks and smooth regions that appear to have been flooded by watery material from within.
13. Both atmospheres are made mostly of nitrogen. Titan’s atmosphere is colder and has a higher pressure.

15. The orange color is due to molecules formed in reactions that are triggered by solar ultraviolet radiation.
17. Coronae resemble the ridge and trough systems in the younger, brighter regions of Ganymede. The rolling, cratered terrain resembles regions of Umbriel and several other satellites.
19. They are plumes of dark material ejected in geysers and blown downwind by the thin atmosphere of Triton.

Problems

1. 1900 kg/m^3
3. We are seeking the synodic period of revolution of Phobos, given by $1/S = 1/P$ (Phobos) $- 1/P$ (Mars), where P (Phobos) is 7.7 hours and P (Mars) is the rotation period of Mars $=$ 24.6 hours. Inserting these values gives $S =$ 11.2 hours.

CHAPTER 15

Conceptual Questions

1. A meteoroid is a small body in interplanetary space. When it enters the Earth's atmosphere, it produces a meteor. The parts of it that reach the Earth's surface become meteorites.
3. They are traveling along parallel paths from the direction of the radiant.
5. They appear to be unaltered samples of material from the early solar system.
7. It tells how long it has been since the meteorite broke off from a larger parent body.
9. It is too faint to be seen with the unaided eye.
11. When the nucleus comes within about 3 AU of the Sun, it is warmed by sunlight to the point where water and other ices evaporate and flow outward carrying dust particles with them.
13. Dust is pushed away from the Sun by solar radiation. The solar wind pushes the ionized gas in the plasma tail away from the Sun.
15. Their orbits are altered by the gravity of passing stars.
17. They strike the Sun or planets, are eroded by many passages near the Sun, or are fragmented by passing too near the Sun or a planet.
19. No comet has ever been found to have a hyperbolic path when it first entered the planetary system. Thus, none of them have speeds as great as escape velocity.
21. An iridium-rich layer of sediment was laid down at the time that widespread extinction took place at the end of the Cretaceous period. Iridium is abundant in meteorites but rare in the Earth's surface rocks.

Problems

1. 3.0 years, 6.0 years
3. Twice each 1.12-year-long orbital period

CHAPTER 16

Conceptual Questions

1. Because Sirius is brighter than Newton assumed it was, it must be more distant than he calculated to have the apparent brightness it does.
3. The directions of the apex and antapex would stay the same. Nearby stars would have larger proper motions because their directions relative to the Sun would change at a greater rate.
5. For emission to occur, the electron must jump to a lower energy level. An electron in the ground state can't jump to a lower level.
7. Essentially all the hydrogen in an O star is in ionized form rather than atomic form and only hydrogen atoms can produce Balmer lines. In the cooler A star, there is more atomic hydrogen.
9. The radial velocity is 0 km/s.
11. Most stars spend most of their luminous lifetimes as main sequence stars.

Problems

1. 0.025 seconds of arc
3. 37 years
5. 40 times as bright
7. Star A is 2.5 million times as bright.
9. 4 magnitudes
11. 17
13. 500.33 nm
15. The more massive is 128 times as luminous.

Figure-Based Questions

1. The energy of the Lyman α line is less than the energy of the Lyman β line, so the wavelength of the Lyman α line is longer.
3. The energy of the Lyman γ line is the same as the combined energies of the Lyman α and Balmer β lines, so, because the energy of a photon is proportional to its frequency, the frequency of the Lyman γ line is equal to the sum of the frequencies of the Lyman α and Balmer β lines.
5. About 0.005; about 400

CHAPTER 17

Conceptual Questions

1. It would have no effect at all, because the rate of energy production is limited by the rate at which protons react with protons, which is much slower than the rate at which deuterons react with protons.
3. Photons in the Sun travel such short distances that many absorptions and reemissions must occur before radioactive diffusion can carry the Sun's energy to the surface.

5. Not noticeable at all because the 1-week deficit would be spread over more than 100,000 years.
7. They turn into other kinds of neutrinos.
9. It would make the Sun's limb much fuzzier than it is.
11. It would tell us that temperature increases with height in the photosphere.
13. They are cooler than the rest of the Sun's surface.
15. A spicule is a rapidly rising jet of gas in the chromosphere.
17. A small rate of energy input is sufficient to keep the corona hot because its gases are very inefficient at radiating away energy to cool the corona.
19. Gas in coronal holes streams away from the Sun to produce the solar wind.
21. Because the magnetic field lines in the solar wind are attached to the Sun, which rotates and stretches the field lines into a spiral pattern.
23. The Maunder minimum, a period within which few sunspots were seen, coincided with a time of cold climate.
25. The cycle begins with a few spots at about 40° N and S latitudes. The leading spots in groups in the northern and southern hemisphere have different magnetic polarity. Over the next 11 years the average latitude of spots shifts toward the equator. After about 11 years new spots begin to appear at 40° N and S latitude. This time, however, the polarity of the spots in the two hemispheres is different than it was during the first 11 years of the cycle.

Problems

1. 1.2×10^{21} W; absolute magnitude = 18.6
3. 35,000 years

Figure-Based Question

1. 130

CHAPTER 18

Conceptual Questions

1. The cloud was flattened and had organized rotation.
3. Giant molecular clouds are about 10 pc across and contain up to 1 million $M_\odot$. They contain cold clumps a few parsecs across. At the center of the clumps are cloud cores, which are the sites of star formation.
5. Without magnetic fields, there might be many more stars formed each year.
7. Collapse stops when temperature and pressure in the cloud become great enough to balance the weight of the protostar.
9. They are located above and to the right of the main sequence when they first become visible. They aren't visible earlier because they were surrounded by clouds of dust.
11. Friction removes angular momentum from the inner part of the disk and gives it to the outer part of the disk.
13. Radioactive iodine decays into xenon. The amount of xenon in a rock tells how much radioactive iodine was left in the solar nebula when the rock formed. Measurements of the amounts of xenon in meteorites and the Earth show that they all formed within about 100 million years.
15. At first, grains accumulated by gently running into one another and sticking. Later, larger bodies accumulated because their gravity attracted nearby grains and small bodies.
17. Large impacts have been successful in accounting for the origin of the Moon and the iron-rich nature of Mercury.

Figure-Based Question

1. Tungsten first, then aluminum oxide, iron, silicates, carbon-rich silicates, and ices

CHAPTER 19

Conceptual Questions

1. The temperature is highest at the center, so nuclear reactions proceed most rapidly there.
3. Their nuclei have greater positive charges than hydrogen, so a higher temperature is required for the nuclei to have enough energy to overcome the repulsive electrical force between the nuclei.
5. The pressure doubles.
7. Their evolution should be identical.
9. Because the stars in a cluster have nearly the same age, they fall on an isochrone in an H-R diagram.
11. They have hydrogen fusion taking place in their cores.
13. Less massive stars become degenerate and stop heating up before hydrogen fusion can begin.
15. At the time core helium fusion begins, the core of a 1 $M_\odot$ main sequence star is degenerate, so the energy released by helium increases the temperature without increasing pressure and expanding the core.
17. When the star is smallest, pressure dominates gravity so the star begins to expand. When the star reaches average size, pressure and gravity balance, but inertia carries the surface outward until maximum size is reached and gravity dominates pressure. The star begins to contract again until it reaches minimum size and the cycle begins anew.
19. The gas in the planetary nebula was shed from the star while it was an AGB star.
21. There would be no elements heavier than iron.

Problems

1. About 40 of them (this assumes a typical person spends 1 hour per day eating)
3. It would be twice as large.
5. The main sequence lifetime of B is 56 times as long.

Figure-Based Questions

1. 1.7×10^8 years
3. 10^8 years

CHAPTER 20

Conceptual Questions

1. For main sequence stars, radius increases as mass increases. For white dwarf stars, radius decreases as mass increases.
3. The more massive white dwarf is dimmer because it is smaller and has less surface to radiate energy.
5. The rest of the star's mass is blown into space as a cool wind while the star is an asymptotic giant branch star.
7. There would be fewer white dwarf stars, because stars originally more massive than 1.4 $M_\odot$ couldn't become white dwarfs.
9. The iron in the core of an AGB star was produced by a long series of nuclear reactions. The shells surrounding the iron core represent the products of earlier series of reactions. The earliest nuclear products are farthest from the core.
11. Infalling material rebounds from the neutron core of the star. Also, and more important, neutrinos are absorbed by the outward moving gas. The energy from the absorbed neutrinos pushes material outward.
13. The galaxies in which several gamma ray bursts occurred have been identified.
15. Most of the energy is released in the form of neutrinos.
17. The expansion is slowed by sweeping up surrounding interstellar gas.
19. In both cases, as mass increases radius decreases. For a given mass, a neutron star is much smaller than a white dwarf.
21. A white dwarf rotating as fast as a pulsar would break up. A pulsating white dwarf can't pulsate as fast as the fastest pulsars. A pulsating neutron star can't pulsate as slowly as the slowest pulsars.
23. The rotation rate slows as time passes.
25. Pulsars "live" much longer than supernova remnants, so the remnants that once surrounded most pulsars can't be seen.
27. The object maintains the same location in space (represented horizontally) but changes its vertical position as time passes.
29. Spacelike trips require faster than light motion and are impossible. Timelike trips require speeds slower than the speed of light and are possible.
31. The difference between 180° and the sum of the angles in a triangle in curved space increases as the size of the triangle increases. Only if the triangle is reasonably large can the difference be measured accurately enough to provide a real test of curvature.
33. Her pulse rate remains constant, because to measure her pulse she compares her biological clock to a mechanical clock. All clocks run more slowly in highly curved regions of spacetime, so both her clocks slow at the same rate and her pulse rate remains constant.
35. The redshifts of white dwarf spectra are mostly gravitational redshifts rather than Doppler shifts.
37. All we can measure are the mass, electric charge, and angular momentum of the material within a black hole.

Problems

1. 8 $M_\odot$
3. −34.4, 39 magnitudes brighter

Figure-Based Questions

1. 1.0 R_{Earth}
3. 1.2 $M_\odot$
5. 14 km

CHAPTER 21

Conceptual Questions

1. The stars in a wide pair are too far apart to exchange matter with each other. The stars in a close pair are close enough to do so.
3. If the binary doesn't appear to obey Kepler's laws, the orbit must be tipped.
5. Primary and secondary minimum are equally deep if the two stars have the same temperature.
7. If the minima have flat bottoms, the two stars have unequal sizes.
9. Water flows toward lower potentials, so there is little flow of water across a lake while water flows rapidly in a river.
11. The blob can't remain at L_1. The tiniest push will send it one way or the other. Whether it flows into star 1 or star 2 can't be predicted.
13. This will make the masses of the star more unequal, so the separation of the two stars will increase.
15. If the mass-losing star is more massive, the separation of the two stars decreases, and the mass-losing star continues to fill its Roche lobe even as it loses mass.
17. The two stars can exceed escape velocity and go their separate ways, form a close binary consisting of two compact objects (neutron stars or black holes), or merge.
19. Friction in the disk causes matter to spiral inward.

21. A nova outburst can occur whenever enough fresh material is accumulated for fusion to begin.
23. X-ray bursts occur whenever freshly accumulated gas becomes hot enough for helium fusion to occur.

Problems

1. $1.08\ M_\odot + 0.88\ M_\odot = 1.96\ M_\odot$, $1.08\ M_\odot/0.88\ M_\odot = 1.2$
3. 5.6 years

Figure-Based Question

1. If the more massive is 6 times as massive as its companion, the stars are 1.3 times farther apart than if the more massive star is 5 times as massive as its companion.

CHAPTER 22

Conceptual Questions

1. The star must emit a large amount of ultraviolet radiation. Cool stars don't emit much of their radiation in the ultraviolet.
3. An HII region forms about a young, hot star in a molecular cloud and begins to expand. When the HII region reaches the edge of the molecular cloud, it begins to expand rapidly in the lower density region outside the cloud. When the star leaves the main sequence and becomes cooler, the HII region fades out.
5. The hot phase is heated by shock waves from supernovae.
7. The reflection nebula is bluer than the star that illuminates it.
9. The dust is nearly transparent in the infrared.
11. The core forms at relatively high temperatures in outflows from cool stars. The icy outer layer can only form at the much lower temperatures of interstellar space.
13. He explained that the galaxy is flat, so we can only see distant stars (that form the glow of the Milky Way) when we look along the disk rather than out of it.
15. More matter is required to account for the speed of rotation of the galaxy than can be found in visible forms such as stars and nebulae.
17. Interstellar dust prevents us from seeing far enough at visible wavelengths to discern the entire spiral pattern.
19. In the vicinity of the spiral arms, the gravity of the material in the arms has an effect on the speeds of orbiting matter.
21. Stars are formed in spiral arms. O and B stars have such short lives that they don't survive long enough to leave the spiral arm in which they form.
23. The low density in the innermost 2 pc of the galaxy suggests that explosions periodically blast matter out of the central region.
25. The galaxy has become flatter and richer in heavy elements as time passes.

Problems

1. 110 million years
3. The mass of A is nine times the mass of B.
5. 0.0017 seconds of arc

Figure-Based Question

1. About 4

CHAPTER 23

Conceptual Questions

1. Because time is required for the light from an object to reach us, we always see objects as they were in the past.
3. E0
5. The stars orbit in random directions, so even though each star has angular momentum, the sum of all the angular momentum of the stars is nearly zero.
7. Galaxies are much closer together in terms of their sizes than are stars.
9. The rotations of spiral galaxies are generally much too rapid to be accounted for by the matter we can see.
11. Some primary distance indicators are Cepheid variables and novae.
13. Hubble's constant would be one-tenth of its actual value.
15. The relationship would imply that the universe was contracting and that the galaxies would all come together at a specified time in the future.

Problems

1. 22.5 Mpc
3. 4 million $M_\odot$

Figure-Based Question

1. E3, E2, E6

CHAPTER 24

Conceptual Questions

1. Quasars are extremely blue in color.
3. Quasars vary in brightness in a few years or less.
5. Superluminal motions will occur if components of quasars move almost directly at us at speeds just less than the speed of light.

7. The broad line region of a Seyfert galaxy is within a torus of dust that blocks the broad line region, when viewed from near the plane of the dust, resulting in the classification of the galaxy as a Seyfert 2. Viewed from above the torus, the broad line region can be seen and the galaxy is classified as Seyfert 1.
9. The lobes are regions in which electrons in the jets are slowed and emit synchrotron radiation.
11. Blazars are small, bright, and show brightness variations. Unlike quasars, they have weak emission lines or no lines at all.
13. The great luminosities of quasars would blow surrounding matter away unless the black holes were massive enough to attract surrounding gas.
15. This shows that most quasars are very distant and are seen as they were in the remote past.
17. Despite many searches, few quasars with redshifts greater than three have been found. This shows that there were few quasars in the early universe.
19. An Einstein ring is produced if the distant quasar lies directly behind the galaxy or galaxies producing the gravitational lens.

Problem

1. $3 \times 10^{13}\ L_{\odot}$

Figure-Based Questions

1. 0.9
3. 0.6
5. 0.4 milliarcseconds/y, 250 years
7. 7° or 23°
9. Blazar
11. The space density of old quasars was about 5 times as large.

CHAPTER 25

Conceptual Questions

1. Clusters can be classified according to richness, a measure of the number of galaxies in the cluster, and according to whether their shapes are regular or irregular.
3. Most of the galaxies in the Virgo cluster have luminosities too low for them to be seen at the distances of other clusters, so many galaxies have been found in the Virgo Cluster only because it is much closer than other clusters.
5. The matter in low-density regions was distributed smoothly and formed few of the clumps needed for the formation of elliptical galaxies.
7. By observing X rays from the space between the galaxies in the cluster.
9. Clusters are gathered into superclusters just as galaxies are gathered into clusters.

CHAPTER 26

Conceptual Questions

1. If the expansion has neither speeded up nor slowed down since the Big Bang.
3. 12 billion years
5. Flat universes are infinite, positively curved universes are finite. Neither flat universes nor positively curved universes have boundaries or centers.
7. One way would be to measure the sum of the angles of an extremely large triangle. In a positively curved universe, the sum is greater than 180°. In a negatively curved universe, the sum is less than 180°. Another way is to measure the ratio of circumference to radius for extremely large circles. In a positively curved universe, the ratio is less than 2π. In a negatively curved universe, the ratio is greater than 2π.
9. Negative
11. A large value of Hubble's constant and a small value of density.
13. They are fainter and more distant than expected if expansion were constant or slowing down. This means that the present value of Hubble's constant, which we use to estimate distances, is greater than it has been in the past. Thus, expansion is accelerating.
15. Dark energy would cause the expansion rate to increase with time.
17. After 1 second, the radiation in the universe lacked the energy to produce pairs of particles.
19. Before 100 seconds, energetic gamma rays disrupted deuterons as soon as they formed. After 300 seconds, protons lacked sufficient energy to react to form deuterons.
21. To see the universe at the time the CBR was emitted, we need to look far back in time and so far out in space that the radiation from those parts of the universe is redshifted from visible light to radio radiation.
23. They tell us that space is flat and that normal matter made up only a few percent of the density of the universe at the time the CBR was emitted.
25. Because the departure from flatness has increased greatly since the Big Bang. This means that the early universe must have been flat to one part in 10^{15}.
27. It will eventually contract.

Problems

1. 14 billion years
3. 49 km/s per Mpc

Figure-Based Question

1. Protons and neutrons at 100 s, protons and helium nuclei at 1000 s, fusion of hydrogen into helium took place between those two times.

CHAPTER 27

Conceptual Questions

1. DNA or a similar substance is necessary so that the offspring of an organism can inherit the organism's characteristics.
3. The two possibilities are that the organic molecules fell to Earth in primordial solar system bodies or were formed from the gases in Earth's atmosphere by lightning, sunlight, heat, or shock waves.
5. It is much easier to detect Jupiter-like planets than terrestrial planets.
7. They are similar to the solar system in that they contain one or more giant planets. They are different in that in many planetary systems the giant planets are much closer to the star than Jupiter is to the Sun and that many of the giant planets have highly elliptical orbits.
9. The longer life endures on a typical planet, the greater the number of inhabited planets in the galaxy.
11. 1 and 10 GHz seem special to us only because our unit of frequency, 1 Hz (or one cycle per second), depends on the basic unit of time being defined as 1 second and we use a base ten counting system (because we have ten fingers). Creatures elsewhere would almost certainly have a different basic unit of time and possibly a different counting system. To them, the frequency we describe as 1 GHz might be described as, say, 1.7934 Gzornies and wouldn't be special at all.

Problems

1. 1 second of arc
3. 0.1 m/s
5. No correct or incorrect answer
7. 90

Figure-Based Questions

1. 0.5 AU
3. 2.5 billion years from now

Glossary

Note: This glossary includes definitions of the key terms and other important terms relevant to the study of astronomy. Page numbers indicate where the key terms are boldfaced in the text.

A

aberration of starlight The angular shift in the apparent direction of a star caused by the orbital motion of the Earth. (p. 154)

absolute magnitude The apparent magnitude a star would have if it were at a distance of 10 parsecs. (p. 381)

absolute zero The lowest possible temperature. At absolute zero, atoms and other particles stop moving. (p. 134)

absorption line A dark line superimposed on a continuous spectrum when a gas absorbs light from a continuous source that is hotter than the absorbing gas. (p. 383)

acceleration The rate of change of velocity. An acceleration may involve a change of speed, direction of motion, or both. (p. 83)

acceleration of gravity The acceleration of a body, equal to 9.8 meters per second per second, caused by the force of gravity near the surface of the Earth. (p. 89)

accretion The growth in the mass of a body by the infall of matter gravitationally attracted to the body. (p. 145)

accretional heating The heating of a body by the impacts that occur as it grows by adding infalling material. (p. 145)

accretion disk A disk of gas and dust spiraling inward toward a star or toward the nucleus of a galaxy. (p. 511)

achondrite A stony meteorite lacking chondrules. (p. 351)

active galactic nucleus The nucleus of an active galaxy. (p. 581)

active galaxy A galaxy whose nucleus is unusually bright and small. Seyfert galaxies, BL Lacertae objects, and quasars are examples of active galaxies. (p. 581)

active region A region of the Sun's surface layers that has a large magnetic field and in which sunspots, flares, and prominences preferentially occur. (p. 411)

adaptive optics A system for modifying the shape of the mirror of a telescope to compensate for atmospheric seeing and to produce sharp images. (p. 118)

aerosol Liquid droplets and solids suspended in the atmosphere of a planet or satellite. (p. 171)

aesthenosphere A layer of plastic, deformable rock located in the upper mantle of a planet directly below the lithosphere. (p. 163)

albedo The ratio of the light reflected in all directions by a surface to the light incident on it. A perfectly reflecting surface has an albedo of 1, a perfectly absorbing surface has an albedo of 0. (p. 141)

alpha particle The nucleus of a helium atom, consisting of two protons and two neutrons. (p. 444)

altitude The angular distance between the direction to an object and the horizon. Altitude ranges from 0° for an object on the horizon to 90° for an object directly overhead. (p. 9)

amino acid A carbon-based molecule from which protein molecules are assembled. (p. 646)

Amor asteroid A member of a class of asteroids having orbits that cross the orbital distance of the Earth. (p. 357)

angular momentum The momentum of a body associated with its rotation or revolution. For a body in a circular orbit, angular momentum is the product of orbital distance, orbital speed, and mass. When two bodies collide or interact, angular momentum is conserved. (p. 89)

annihilation The mutual destruction of a matter–antimatter pair of particles. The charges on the two particles cancel and the mass of the particles is entirely converted to energy. (p. 631)

annular eclipse A solar eclipse in which the Moon is too far from the Earth to block the entire Sun from view and a thin ring of sunlight appears around the Moon. (p. 187)

antapex The direction in the sky away from which the Sun is moving. Because of the Sun's motion, nearby stars appear to converge toward the antapex. (p. 380)

antimatter A type of matter that annihilates ordinary matter on contact. For every particle, there is a corresponding antimatter particle. For example, the antimatter counterpart of the proton is the antiproton. (p. 630)

apex The direction in the sky toward which the Sun is moving. Because of the Sun's motion, nearby stars appear to diverge from the apex. (p. 380)

aphelion The point in the orbit of a solar system body where it is farthest from the Sun. (p. 69)

Apollo asteroid A member of a class of asteroids having orbits that cross the orbital distance of the Earth. (p. 357)

apparent brightness The observed brightness of a celestial body. (p. 380)

apparent magnitude The observed magnitude of a celestial body. (p. 381)

apparent solar time Time kept according to the actual position of the Sun in the sky. Apparent solar noon occurs when the Sun crosses an observer's meridian. (p. 27)

arachnoid A circular feature on the surface of Venus connected to other similar features by a web of fractures. (p. 224)

ascending node The point in the Moon's orbit where it crosses the ecliptic from south to north. (p. 184)

association A group of stars whose gravity is insufficient to hold it together but has not yet had time to disperse. (p. 426)

asteroid A small, planetlike solar system body. Most asteroids are rocky in makeup and have orbits of low eccentricity and inclination. (p. 353)

asteroid belt The region of the solar system lying between 2.1 and 3.3 astronomical units (AU) from the Sun. The great majority of asteroids is found in the asteroid belt. (p. 354)

astrology A pseudoscience that holds that people and events are influenced by the configurations of the Sun, Moon, and planets with respect to each other and the stars. (p. 39)

astronomical unit (AU) The average distance between the Earth and the Sun. (p. 64)

asymptotic giant branch (AGB) The portion of the H-R diagram occupied by enormous, cool stars with helium-burning shells. (p. 459)

Aten asteroid An asteroid having an orbit with semimajor axis smaller than 1 AU. (p. 357)

atom A particle consisting of a nucleus and one or more surrounding electrons. (p. 132)

atomic number The number of protons in the nucleus of an atom. Unless the atom is ionized, the atomic number is also the number of electrons orbiting the nucleus of the atom. (p. 142)

aurora australis Light emitted by atoms and ions in the upper atmosphere near the south magnetic pole. The emission occurs when atoms and ions are struck by energetic particles from the Sun. (p. 173)

aurora borealis Light emitted by atoms and ions in the upper atmosphere near the north magnetic pole. The emission occurs when atoms and ions are struck by energetic particles from the Sun. (p. 173)

autumnal equinox The point in the sky where the Sun appears to cross the celestial equator moving from north to south. This happens on approximately September 22. (p. 23)

azimuth The angular distance between the north point on the horizon eastward around the horizon to the point on the horizon nearest to the direction to a celestial body. (p. 10)

B

Baily's beads Points of light around the limb of the Moon just before and just after a total eclipse of the Sun. Baily's beads are caused by sunlight shining through valleys on the Moon's limb. (p. 185)

Balmer series A series of absorption or emission lines of hydrogen seen in the visible part of the spectrum. (p. 385)

barred spirals A spiral galaxy in which the nucleus is crossed by a bar. The spiral arms start at the ends of the bar. (p. 557)

basalt An igneous rock often produced in volcanic eruptions. (p. 159)

belt A dark region of clouds located at a given latitude on one of the giant planets. (p. 270)

Big Bang The explosive event at the beginning of the universe. The expansion produced the Big Bang that continues today. (p. 620)

binary accretion theory A theory of the origin of the Moon that holds that the Moon and the Earth formed at about the same time out of the same swarm or cloud of material. (p. 202)

binary star system A pair of stars that orbit each other under their mutual gravitational attraction. (p. 496)

bipolar outflow Relatively narrow beams of matter ejected in opposite directions by a protostar. (p. 432)

blackbody An object that is a perfect absorber of radiation falling on it. (p. 138)

blackbody radiation The electromagnetic radiation emitted by a blackbody. The spectrum and intensity of blackbody radiation are controlled by the temperature of the blackbody. Many stars and other celestial bodies approximate blackbodies. (p. 138)

black hole A region of space from which no matter or radiation can escape. A black hole is a result of the extreme curvature of space by a massive compact body. (p. 489)

blazar A type of active galaxy named for BL Lacertae, the first of the type discovered. Blazars show rapid, unpredictable variations in brightness. (p. 585)

bow shock The region where the solar wind is slowed as it impinges on the Earth's magnetosphere. (p. 172)

broad line region The high-density region in a quasar where broad emission lines are formed. (p. 578)

brown dwarf A star with too low a mass for nuclear fusion to begin in its core. (p. 393)

C

capture theory The theory of the origin of the Moon that holds that the Moon formed elsewhere in the solar system and then was captured into orbit about the Earth. (p. 202)

carbonaceous chondrite A stony meteorite that contains carbon-rich material. Carbonaceous chondrites are thought to be primitive samples of material from the early solar system. (p. 351)

carbon cycle The series of reactions by means of which massive stars fuse hydrogen into helium. (p. 444)

Cassini division A conspicuous 1800-kilometer-wide gap between the outermost rings of Saturn. (p. 285)

celestial equator The circle where the Earth's equator, if extended outward into space, would intersect the celestial sphere. (p. 18)

celestial horizon The circle on the celestial sphere that is 90° from the zenith. The celestial horizon is approximately the boundary between the Earth and sky. (p. 9)

celestial mechanics The part of physics and astronomy that deals with the motions of celestial bodies under the influence of their mutual gravitational attraction. (p. 294)

celestial sphere An imaginary sphere surrounding the Earth. The celestial bodies appear to carry out their motions on the celestial sphere. (p. 8)

cell An organized collection of molecules enclosed in a wall-like membrane. (p. 646)

Centaur A small solar system body whose orbit lies between the orbits of Jupiter and Neptune. (p. 310)

central force A force directed at the center of motion of a body. Gravity is the central force that accounts for the orbital motion of solar system bodies. (p. 82)

centripetal acceleration The acceleration toward the center of motion, that causes the path of an orbiting body to continually bend away from a straight-line path. (p. 86)

centripetal force The central force that produces centripetal acceleration. (p. 86)

Cepheid variable A member of a class of yellow pulsating stars that vary in brightness as they expand and contract. The period of a Cepheid is related to its luminosity. (p. 458)

Chandrasekhar limit The maximum mass, about 1.4 solar masses, that a white dwarf star can have. (p. 471)

charge coupled device (CCD) An array of photosensitive electronic elements that can be used to record an image falling on it. (p. 114)

chondrite A meteorite containing chondrules. (p. 350)

chondrule A small, spherical body embedded in a meteorite. Chondrules are composed of iron, aluminum, and magnesium silicate rock. (p. 350)

chromosphere The part of the Sun's atmosphere between the photosphere and the corona. (p. 412)

circle A curve on which all points are equidistant from the center. (p. 93)

circular speed The speed that causes an orbiting body to have a circular orbit rather than an elliptic one. (p. 91)

circumpolar stars A body is close enough to a celestial pole that its diurnal circle is always above the horizon. Circumpolar stars neither rise nor set. (p. 12)

classical KBO A Kuiper Belt Object with an orbital perihelion distance far enough from Neptune that it does not experience large perturbations due to Neptune. (p. 309)

close pair A binary system in which the two stars are close enough together that they transfer matter to one another during some stages of their evolution. (p. 496)

cloud core The dense part of a molecular cloud where star formation takes place. (p. 427)

cluster of galaxies A group of galaxies held together by their mutual gravitational attraction. (p. 602)

collision fragment A satellite that probably is a fragment of a larger satellite broken apart by a collision with a meteoroid. (p. 316)

coma A spherical gaseous region that surrounds the nucleus of a comet. The coma of a comet may be 100,000 kilometers or more in diameter. (p. 360)

comet A small, icy body in orbit about the Sun. When a comet is near the Sun, it displays a coma and a tail. (p. 359)

common envelope A stage in the evolution of a close pair of stars in which matter shed by one of the stars fills the region just outside the Roche lobes of the two stars. (p. 508)

conduction The transfer of heat by means of direct collisions between adjacent atoms, molecules, or ions. (p. 146)

conic section One of four kinds of curves (circle, ellipse, hyperbola, and parabola) that can be formed by slicing a right circular cone with a plane. (p. 93)

conjunction The appearance of two celestial bodies, often a planet and the Sun, in approximately the same direction. (p. 32)

conucleation A possible explanation for the origin of a wide binary pair of stars in which the two cloud fragments that become the stars are already in orbit about one another when they form. (p. 503)

constellation One of 88 regions into which the celestial sphere is divided. (p. 4)

continuous spectrum A spectrum containing neither emission nor absorption lines. (p. 383)

convection The process of energy transport in which heat is carried by hot, rising and cool, falling currents or bubbles of liquid or gas. (p. 146)

convection zone The outer part of the Sun's interior in which convection occurs. (p. 404)

coordinates The numbers used in a coordinate system. Longitude and latitude are examples of coordinates. (p. 6)

coordinate system A system in which numbers are used to give the location of a body or event. The longitude-latitude system is an example of a coordinate system used to locate things on the Earth's surface. (p. 6)

core The innermost region of the interior of the Earth or another planet. (p. 163)

Coriolis effect The acceleration that a body experiences when it moves across the surface of a rotating body. The acceleration results in a westward deflection of projectiles and currents of air or water when they move toward the Earth's equator and an eastward deflection when they move away from the equator. (p. 156)

corona **(A)** The outermost layer of the Sun's atmosphere. Gases in the corona are tenuous and hot. (p. 185) **(B)** A circular feature on the surface of Venus. Coronae appear to be collapsed volcanic domes and can be as much as several hundred kilometers across. **(C)** A type of surface feature of Uranus's satellite Miranda. Coronae consist of parallel ridges and troughs producing a striped appearance. Coronae have sharp boundaries. (pp. 224, 337)

coronal hole A low-density, dim region in the Sun's corona. Coronal holes occur in regions of open magnetic field lines where gases can flow freely away from the Sun to form the solar wind. (p. 414)

coronal mass ejection A blast of gas moving outward through the Sun's corona and into interplanetary space following the eruption of a prominence. (p. 416)

cosmic microwave background (CMB) Radiation emitted by the universe about 400,000 years after the universe began to expand. Because the CMB comes to us from extremely remote parts of the universe, it is highly redshifted. (p. 633)

cosmic ray Extremely energetic ions and electrons that travel through space almost at the speed of light. Most cosmic rays come from great distances and may be produced in supernovae and pulsars. (p. 352)

cosmic ray exposure age The length of time that has passed since a meteorite broke off from a larger body and became exposed to radiation damage from cosmic rays. (p. 352)

cosmological constant A self-repelling property of space first proposed by Einstein. (p. 627)

cosmology The study of the universe as a whole. (p. 619)

crater A roughly circular feature on the surface of a solar system body caused by the impact of an asteroid or comet. (p. 194)

crater density The number of craters of a given size per unit area of the surface of a solar system body. (p. 198)

crater saturation The maximum crater density a solar system body can have. Once saturation is reached, new craters can only be produced by eradicating old ones. (p. 198)

crescent phase The phase of the Moon at which only a small, crescent-shaped portion of the near side of the Moon is illuminated by sunlight. Crescent phase occurs just before and after new Moon. (p. 29)

critical density The value that the average density of the universe must equal or exceed if the universe is closed. If the density of the universe is less than the critical density, the universe will continue to expand forever. (p. 626)

crust The outermost layer of the interior of a planet or satellite. (p. 163)

C-type asteroid One of a class of very dark asteroids whose reflectance spectra show no absorption features due to the presence of minerals. (p. 358)

D

dark energy A form of energy that may be causing the expansion of the universe to accelerate. (p. 627)

dark matter Matter that cannot be detected or has not yet been detected by the radiation it emits. The presence of dark matter can be deduced from its gravitational interaction with other bodies. (p. 538)

dark nebula A dense, interstellar cloud containing enough gas and dust to block the light of background stars. The dimming of background stars gives the appearance of a region with no stars. (p. 527)

declination The angular distance of a celestial body north or south of the celestial equator. Declination is analogous to latitude in the terrestrial coordinate system. (p. 19)

decoupling epoch The time about a million years after the expansion of the universe began when the universe became transparent and light could, for the first time, travel great distances before being absorbed or scattered. The cosmic background radiation was produced at the decoupling epoch. (p. 633)

deferent One of the circles on which a planet moved according to the Ptolemaic model of the solar system. (p. 51)

degenerate gas A gas in which a type of particle (electrons or neutrons) are as tightly packed as permitted by the Pauli exclusion principle. In a degenerate gas, temperature has essentially no influence on pressure. (p. 446)

degree A unit used to measure angles. There are 360° in a circle. (p. 8)

density The mass of a body divided by its volume. (p. 135)

density wave theory A theory to account for the spiral arms of spiral galaxies. According to the density wave theory, spiral arms are the crests of waves moving through a galaxy like water waves move through water. (p. 541)

descending node The point in the Moon's orbit where it crosses the ecliptic from north to south. (p. 184)

detector A device used to measure light once it has been brought into focus by a telescope. (p. 113)

deuterium An isotope of hydrogen. The nucleus of a deuterium atom is a deuteron. (pp. 142, 402)

deuteron A nucleus of deuterium, an isotope of hydrogen. A deuteron contains one proton and one neutron. (p. 632)

diamond ring The last of Baily's beads, which seems to shine with special brilliance just before a solar eclipse becomes total. (p. 185)

differential rotation Rotation in which the rotation period of a body varies with latitude. Differential rotation occurs for gaseous bodies like the Sun or for planets with thick atmospheres. (p. 411)

differentiation The gravitational separation of the interior of a planet into layers according to density. When differentiation occurs inside a molten body, the heavier materials sink to the center and the light materials rise to the surface. (p. 175)

diffraction The change in the direction of a wave after passing an obstacle or through an aperture that has about the same size as the wavelength of the wave. (p. 112)

direct motion The eastward apparent motion of a solar system body with respect to the stars. Direct motion is interrupted by regular episodes of retrograde (westward) motion. (p. 32)

disk instability A possible explanation for the origin of a close binary pair of stars in which one star forms within the disk of gas and dust orbiting another, newly formed star. (p. 503)

dispersion The separation of white light according to wavelength. Dispersion produces a rainbowlike spectrum. (p. 109)

diurnal Daily. (p. 18)

diurnal circle The circular path that a celestial body traces out as it appears to move across the sky during an entire day. Diurnal circles are centered on the north and south celestial poles. (p. 18)

Doppler effect The change in the frequency of a wave (such as electromagnetic radiation) caused by the motion of the source and observer toward or away from each other. (p. 105)

Drake Equation An equation that, if we knew the values of all its terms, could be used to calculate the number of extraterrestrial civilizations in the Milky Way galaxy with which we might potentially establish communication. (p. 656)

dust tail A comet tail that is luminous because it contains dust that reflects sunlight. The dust in a comet tail is expelled from the nucleus of the comet. (p. 361)

dwarf A main sequence star. (p. 389)

dwarf planet A solar system body that does not orbit a planet and whose gravity is strong enough to make it almost spherical yet that does not dominate the region near its orbit by clearing the region of smaller bodies. (p. 309)

dynamo A process in which electric currents within a rotating, convective body produce a magnetic field. (p. 161)

E

eccentricity A measure of the extent to which an orbit departs from circularity. Eccentricity ranges from 0.0 for a circle to 1.0 for a parabola. (p. 69)

eclipse The obscuration of the light from the Sun when the observer enters the Moon's shadow or the Moon when it enters the Earth's shadow. Also, the obscuration of a star when it passes behind its binary companion. (p. 184)

eclipse seasons The times, separated by about 5½ months, when eclipses of the Sun and Moon are possible. (p. 189)

eclipse track The path of the Moon's shadow across the Earth during a solar eclipse. (p. 188)

eclipse year The interval of time (346.6 days) from one passage of the Sun through a node of the Moon's orbit to the next passage through the same node. (p. 190)

eclipsing binary Binary star systems for which the orbital plane of the stars lies so nearly in the line of sight that two stars alternately pass in front of one another, causing eclipses. (p. 500)

ecliptic The plane of the Earth's orbit about the Sun. As a result of the Earth's motion, the Sun appears to move among the stars, following a path that is also called the ecliptic. (p. 22)

Eddington luminosity The maximum luminosity that a body could emit without driving away surrounding material. (p. 587)

Einstein ring The ring or near ring into which the image of a distant quasar is distorted if the quasar lies directly behind a galaxy or cluster of galaxies producing a gravitational lens. (p. 596)

electromagnetic spectrum The range of frequency or wavelength of all possible electromagnetic radiation. The electromagnetic spectrum includes (in order of increasing wavelength) gamma ray, X ray, ultraviolet, visible, infrared, and radio. (p. 104)

electromagnetic wave A periodic electrical and magnetic disturbance that propagates through space and transparent materials at the speed of light. Light is an example of an electromagnetic wave. (p. 102)

electron A low-mass, negatively charged particle that can either orbit a nucleus as part of an atom or exist independently as part of a plasma. (p. 142)

element A substance that cannot be broken down into a simpler chemical substance. Oxygen, nitrogen, and silicon are examples of the approximately 100 known elements. (p. 133)

ellipse A closed, elongated curve describing the shape of the orbit that one body follows about another. (pp. 69, 93)

elliptical galaxy A galaxy having an ellipsoidal shape and lacking spiral arms. (p. 555)

emission line A narrow, bright region of the spectrum. Emission lines are produced when electrons in atoms jump from one energy level to a lower energy level. (p. 383)

energy flux The rate at which a wave carries energy through a given area. (p. 102)

energy level Any of the many energy states that an atom may have. Different energy levels correspond to different distances of the electron from the nucleus. (p. 385)

epicycle One of the circles upon which a planet moved according to the Ptolemaic (geocentric) model of the solar system. The center of the epicycle moved on a larger circle, called the deferent. (p. 51)

equant In the Ptolemaic system, the point from which the motion of the epicycle around the deferent is uniform. (p. 51)

equation of state The relationship among pressure, density, and temperature for a gas or fluid. The ideal gas law, for which pressure is proportional to the product of temperature and density, is an example of an equation of state. (p. 446)

equator The line around the surface of a rotating body that is midway between the rotational poles. The equator divides the body into northern and southern hemispheres. (p. 7)

equatorial jet The high-speed, eastward, zonal wind in the equatorial region of Jupiter's atmosphere. (p. 271)

equatorial system A coordinate system, using right ascension and declination as coordinates, used to describe the angular location of bodies in the sky. (p. 19)

equipotential A line or surface of equal potential energy. On the Earth, a line of equal elevation is approximately an equipotential. (p. 503)

escape velocity The speed that an object must have to achieve a parabolic trajectory and escape from its parent body. (p. 92)

event horizon The boundary of a black hole. No matter or radiation can escape from within the event horizon. (p. 489)

evolutionary track The path in an H-R diagram followed by the point representing the changing luminosity and temperature of a star as it evolves. (p. 448)

exosphere The outer part of the thermosphere. Atoms and ions can escape from the exosphere directly into space. (p. 135)

extinction The dimming of starlight due to absorption and scattering by interstellar dust particles. (p. 530)

F

filament A dark line on the Sun's surface when a prominence is seen projected against the solar disk. (p. 416)

fireball An especially bright streak of light in the sky produced when an interplanetary dust particle enters the Earth's atmosphere, vaporizing the particle and heating the atmosphere. (p. 347)

fission **(A)** A nuclear reaction in which a nucleus splits to produce two less massive nuclei. (p. 143) **(B)** A possible explanation for the origin of a close binary pair of stars in which a star splits into two pieces, each of which becomes a star. (p. 503)

fission theory A theory for the origin of the Moon in which the Moon consists of matter that was flung from the primitive Earth because of the Earth's rapid rotation. (p. 201)

focal length The distance between a mirror or lens and the point at which the lens or mirror brings light to a focus. (p. 110)

focal plane The surface where the objective lens or mirror of a telescope forms the image of an extended object. (p. 111)

focal point The spot where parallel beams of light striking a lens or mirror are brought to a focus. (p. 110)

focus (foci) One of two points from which an ellipse is generated. For all points on the ellipse, the sum of the distances to the two foci is the same. (p. 69)

force A push or a pull. (p. 84)

fragmentation A possible explanation for the origin of a close binary pair of stars in which a collapsing cloud breaks into several pieces, each of which becomes a star. (p. 503)

frequency The number of oscillations per second of a wave. (p. 102)

full phase The phase of the Moon at which the bright side of the Moon is the face turned toward the Earth. (p. 28)

fusion A nuclear reaction in which two nuclei merge to form a more massive nucleus. (p. 143)

G

galactic bulge A somewhat flattened distribution of stars, about 6 kiloparsecs in diameter, surrounding the nucleus of the Milky Way. (p. 534)

galactic cannibalism The capture and disruption of one galaxy by another. (p. 562)

galactic disk A disk of matter, about 30 kiloparsecs in diameter and 2 kiloparsecs thick, containing most of the stars and interstellar matter in the Milky Way. (p. 534)

galactic equator The great circle around the sky that corresponds approximately to the center of the glowing band of the Milky Way. (p. 534)

galactic halo The roughly spherical outermost component of the Milky Way, reaching to at least 30 to 40 kiloparsecs from the center. (p. 534)

galactic latitude The angular distance of a body above or below the galactic equator. (p. 534)

galactic longitude The angular distance, measured eastward around the galactic equator, from the galactic center to the point on the equator nearest the direction to a body. (p. 534)

galactic nucleus The central region of the Milky Way. (p. 534)

galaxy A massive system of stars, gas, and dark matter held together by its own gravity. (p. 554)

gamma ray The part of the electromagnetic spectrum having the shortest wavelengths. (p. 104)

gamma ray burst A brief blast of gamma rays that originates in a distant galaxy. (p. 477)

geocentric Centered on the Earth. In a geocentric model of the solar system, the planets moved about the Earth. (p. 51)

geodesic The path in spacetime followed by a light beam or a freely moving object. (p. 486)

giant A star larger and more luminous than a main sequence star (dwarf) of the same temperature and spectral type. (p. 389)

giant impact theory The theory of the origin of the Moon that holds that the Moon formed from debris blasted into orbit when the Earth was struck by a Mars sized body. (p. 202)

giant molecular cloud An unusually large molecular cloud that may contain as much as 1 million $M_{\odot}$. (p. 427)

gibbous phase The phase of the Moon at which the near side of the Moon is more than half illuminated by sunlight. Gibbous phase occurs just before and after full Moon. (p. 28)

globular cluster A tightly packed, spherically shaped group of thousands to millions of old stars. (p. 533)

granule A bright convective cell or current of gas in the Sun's photosphere. Granules appear bright because they are hotter than the descending gas that separates them. (p. 408)

gravitational lens A massive body that bends light passing near it. A gravitational lens can distort or focus the light of background sources of electromagnetic radiation. (p. 595)

gravitational potential energy The energy stored in a body subject to the gravitational attraction of another body. As the body falls, its gravitational potential energy decreases and is converted into kinetic energy. (p. 91)

gravitational redshift The increase in the wavelength of electromagnetic radiation that occurs when the radiation travels outward through the gravitational field of a body. (p. 488)

gravity The force of attraction between two bodies generated by their masses. (p. 83)

Great Attractor A great concentration of mass toward which everything in our part of the universe apparently is being pulled. (p. 615)

great circle A circle that bisects a sphere. The celestial equator and ecliptic are examples of great circles. (p. 7)

greatest elongation The position of Mercury or Venus when it has the greatest angular distance from the Sun. (p. 64)

Great Red Spot A reddish elliptical spot about 40,000 km by 15,000 km in size in the southern hemisphere of the atmosphere of Jupiter. The Red Spot has existed for at least 3½ centuries. (p. 270)

greenhouse effect The blocking of infrared radiation by a planet's atmospheric gases. Because its atmosphere blocks the outward passage of infrared radiation emitted by the ground and lower atmosphere, the planet cannot cool itself effectively and becomes hotter than it would be without an atmosphere. (p. 170)

ground state The lowest energy level of an atom. (p. 385)

H

HII region A region of ionized hydrogen surrounding a hot star. Ultraviolet radiation from the star keeps the gas in the HII region ionized. (p. 523)

habitable zone The range of distances from a star within which liquid water can exist on the surface of an Earthlike planet. (p. 654)

half-life The time required for half of the atoms of a radioactive substance to disintegrate. (p. 144)

heliocentric Centered on the Sun. In the heliocentric model of the solar system, the planets move about the Sun. (p. 61)

heliopause The boundary of the heliosphere, where the solar wind merges into the interstellar gas. (p. 418)

helioseismology A technique used to study the internal structure of the Sun by measuring and analyzing oscillations of the Sun's surface layers. (p. 409)

heliosphere The region of space dominated by the solar wind and the Sun's magnetic field. (p. 418)

helium flash The explosive consumption of helium in the core of a star when helium fusion begins in a degenerate gas in which pressure doesn't rise as energy is produced and temperature increases. (p. 455)

Herbig-Haro object A clump of gas illuminated by a jet of matter streaming away from a young star. (p. 435)

Hertzsprung-Russell diagram (H-R diagram) A plot of luminosities of stars against their temperatures. Magnitude may be used in place of luminosity and spectral type in place of temperature. (p. 391)

hierarchical clustering model A model for the formation of clusters of galaxies in which individual galaxies form and then begin to collect into clusters. (p. 608)

horizon system A coordinate system, using altitude and azimuth as coordinates, used to locate the positions of objects in the sky. (p. 9)

horizontal branch star A star that is undergoing helium fusion in its core and hydrogen fusion in a shell surrounding the core. (p. 456)

Hubble's constant (H) The rate at which the recession speeds of galaxies increase with distance. Current estimates of Hubble's constant range from 50 to 100 kilometers per second per megaparsec. (p. 568)

Hubble's law The linear relationship between the recession speeds of galaxies and their distances. The slope of Hubble's law is Hubble's constant. (p. 567)

Hubble time An estimate of the age of the universe obtained by taking the inverse of Hubble's constant. The estimate is only valid if there has been no acceleration or deceleration of the expansion of the universe. (p. 620)

hydrostatic equilibrium The balance between the inward directed gravitational force and the outward directed pressure force within a celestial body. (p. 136)

hyperbola A curved path that does not close on itself. A body moving with a speed greater than escape velocity follows a hyperbola. (p. 93)

I

ideal gas law The equation of state for a low-density gas in which pressure is proportional to the product of density and temperature. (p. 8)

igneous rock A rock formed by solidification of molten material. (p. 159)

impetus A theory of motion, developed in the fourteenth and fifteenth centuries, that motion could continue only so long as a force was at work. (p. 59)

index of refraction The ratio of the speed of light in a vacuum to the speed of light in a particular substance. The index of refraction, which always has a value greater than 1.0, describes how much a beam of light is bent on entering or emerging from the substance. (p. 108)

inertia The tendency of a body at rest to remain at rest and a body in motion to remain in motion at a constant speed and in constant direction. (p. 83)

inertial motion Motion in a straight line at constant speed followed by a body when there are no unbalanced forces acting on it. (p. 81)

inferior planet A planet whose orbit lies inside the Earth's orbit. (p. 64)

inflation A brief period of extremely rapid and enormous expansion that may have occurred very early in the history of the universe. (p. 636)

infrared The part of the electromagnetic spectrum having wavelengths longer than visible light but shorter than radio waves. (p. 107)

instability strip A region of the H-R diagram occupied by pulsating stars, including Cepheid variables and RR Lyrae stars. (p. 458)

intercrater plain Smooth portions of the surface of Mercury that lie between and around clusters of large craters. (p. 212)

interferometry The use of two or more telescopes connected together to operate as a single instrument. Interferometers can achieve high angular resolution if the individual telescopes of which they are made are widely separated. (p. 124)

interstellar matter Gas and dust in the space between the stars. (p. 523)

interstellar reddening The obscuration, by interstellar dust particles, of blue starlight more strongly than red starlight. (p. 530)

ion An atom from which one or more electrons has been removed. (p. 133)

ionization The removal of one or more electrons from an atom. (p. 385)

ionosphere The lower part of the thermosphere of a planet in which many atoms have been ionized by ultraviolet solar photons. (p. 170)

iron meteorite A meteorite composed primarily of iron and nickel. (p. 351)

irregular cluster A cluster of galaxies that lacks a symmetrical shape and structure. (p. 602)

irregular galaxy A galaxy having an amorphous shape and lacking symmetry. (p. 559)

isochrone Lines in an H-R diagram occupied by stars of different masses but the same age. (p. 448)

isotopes Nuclei with the same number of protons but different numbers of neutrons. (p. 142)

isotropic Looking the same in all directions. (p. 635)

J

jet A narrow beam of gas ejected from a star or the nucleus of an active galaxy. (p. 583)

K

Kelvin-Helmholtz time The time it would take a star to contract from infinite diameter down to the main sequence while radiating away the gravitational energy released during contraction. (p. 401)

Kelvin temperature scale A temperature scale (like Fahrenheit and Celsius) in which 0 K is defined as absolute zero and 273.15 K is defined as the melting point of ice. (p. 134)

Kepler's laws of planetary motion Three laws, discovered by Kepler, that describe the motions of the planets around the Sun. (p. 68)

kiloparsec (kpc) A unit of distance, equal to 1000 parsecs (pc), often used to describe distances within the Milky Way or the Local Group of galaxies. (p. 533)

kinetic energy Energy of motion. Kinetic energy is given by one-half the product of a body's mass and the square of its speed. (p. 91)

Kirchhoff's laws Three "laws" that describe how continuous, bright line, and dark line spectra are produced. (p. 383)

Kuiper Belt Object (KBO) A trans-Neptunian body in a stable orbit with a semimajor axis less than about 50 AU. (p. 309)

L

L_1 The point between two stars in a binary system where matter may flow from one star to the other. (p. 504)

latitude The angular distance of a point north or south of the equator of a body as measured by a hypothetical observer at the center of a body. (p. 7)

lava Molten rock at the surface of a planet or satellite. (p. 159)

leap year A year in which there are 366 days. (p. 28)

life The property of a body that has (1) organization, (2) metabolism, (3) regeneration and growth, (4) response to stimulation, and (5) reproduction and evolution. (p. 646)

light The visible form of electromagnetic radiation. (p. 102)

light curve A plot of the brightness of a body versus time. (p. 500)

light-gathering power A number, proportional to the area of the principal lens or mirror of a telescope, that describes the amount of light that is collected and focused by the telescope. (p. 111)

light year The distance that light travels in a year. (p. 377)

limb The apparent edge of the disk of a celestial body. (p. 406)

limb darkening The relative faintness of the edge of the Sun's disk (limb) compared with the center of the Sun's disk. (p. 407)

line of nodes The line connecting the two nodes of the Moon's orbit around the Earth. (p. 189)

lithosphere The rigid outer layer of a planet or satellite, composed of the crust and upper mantle. (p. 163)

Local Group The small cluster of galaxies of which the Milky Way is a member. (p. 602)

local hour angle The angle, measured westward around the celestial equator, between the meridian and the point on the equator nearest a particular celestial object. (p. 20)

longitude The angular distance around the equator of a body from a zero point to the place on the equator nearest a particular point as measured by a hypothetical observer at the center of a body. (p. 7)

long-period comet A comet with an orbital period of 200 years or longer. (p. 362)

lookback time The length of time that has elapsed since the light we are now receiving from a distant object was emitted. (p. 575)

luminosity The rate of total radiant energy output of a body. (p. 381)

luminosity class The classification of a star's spectrum according to luminosity for a given spectral type. Luminosity class ranges from I for a supergiant to V for a dwarf (main sequence star). (p. 389)

luminosity function The distribution of stars or galaxies according to their luminosities. A luminosity function is often expressed as the number of objects per unit volume of space that are brighter than a given absolute magnitude or luminosity. (pp. 383, 593)

Lyman α forest The large number of absorption lines seen at wavelengths just longer than the wavelength of the Lyman α line of hydrogen in the spectrum of a quasar. The Lyman α forest is caused by absorption by gas clouds lying between the quasar and the Earth. (p. 595)

Lyman series A series of absorption or emission lines of hydrogen lying in the ultraviolet part of the spectrum. (p. 385)

M

Magellanic Clouds Two irregular galaxies that are among the nearest neighbors of the Milky Way. (p. 554)

magma Molten rock within a planet or satellite. (p. 159)

magnetar A highly magnetized neutron star that emits bursts of gamma rays. (p. 483)

magnetopause The outer boundary of the magnetosphere of a planet. (p. 172)

magnetosphere The outermost part of the atmosphere of a planet, within which a very thin plasma is dominated by the planet's magnetic field. (p. 172)

magnetotail The part of the magnetosphere of a planet stretched behind the planet by the force of the solar wind. (p. 172)

magnitude A number, based on a logarithmic scale, used to describe the brightness of a star or other luminous body. Apparent magnitude describes the brightness of a star as we see it. Absolute magnitude describes the intrinsic brightness of a star. (p. 380)

main sequence The region in an H-R diagram occupied by stars that are fusing hydrogen into helium in their cores. The main sequence runs from hot, luminous stars to cool, dim stars. (p. 391)

main sequence lifetime The length of time that a star spends as a main sequence star. (p. 452)

major axis The axis of an ellipse that passes through both foci. The major axis is the longest straight line that can be drawn inside an ellipse. (p. 69)

mantle The part of a planet lying between its crust and its core. (p. 163)

maria A dark, smooth region on the Moon formed by flows of basaltic lava. (p. 193)

mascon A concentration of mass below the surface of the Moon that slightly alters the orbit of a spacecraft orbiting the Moon. (p. 197)

mass A measure of the amount of matter a body contains. Mass is also a measure of the inertia of a body. (p. 83)

mass-luminosity relation The relationship between luminosity and mass for stars. More massive stars have greater luminosities. (p. 394)

mass number A measure of the mass of a nucleus given by the total number of protons and neutrons in the nucleus. (p. 142)

Maunder minimum A period of few sunspots and low solar activity that occurred between 1640 and 1700. (p. 418)

mean solar time Time kept according to the average length of the solar day. (p. 27)

megaparsec (Mpc) A unit of distance, equal to 1 million parsecs, often used to describe the distances of objects beyond the Local Group. (p. 565)

meridian The great circle passing through an observer's zenith and the north and south celestial poles. (p. 18)

mesopause The upper boundary of the mesosphere layer of the atmosphere of a planet. (p. 170)

mesosphere The layer of a planet's atmosphere above the stratosphere. The mesosphere is heated by absorbing solar radiation. (p. 170)

metallic hydrogen A form of hydrogen in which the atoms have been forced into a lattice structure typical of metals. In the solar system, the pressures and temperatures required for metallic hydrogen to exist only occur in the cores of Jupiter and Saturn. (p. 279)

metamorphic rock A rock that has been altered by heat and pressure. (p. 160)

meteor A streak of light produced by a meteoroid moving rapidly through the Earth's atmosphere. Friction vaporizes the meteoroid and heats atmospheric gases along the path of the meteoroid. (p. 346)

meteorite The portion of a meteoroid that reaches the Earth's surface. (p. 347)

meteoroid A solid interplanetary particle passing through the Earth's atmosphere. (p. 346)

meteor shower A temporary increase in the normal rate at which meteors occur. Meteor showers last for a few hours or days and occur on about the same date each year. (p. 347)

microlensing event The temporary brightening of a distant object that occurs because its light is focused on the Earth by the gravitational lensing of a nearer body. (p. 538)

micrometeorite A meteoritic particle less than 50 millionths of a meter in diameter. Micrometeorites are slowed by atmospheric gas before they can be vaporized, so they drift slowly to the ground. (p. 349)

Milky Way The galaxy to which the Sun and Earth belong. Seen as a pale, glowing band across the sky. (p. 4)

mineral A solid chemical compound. (p. 159)

minimum The time of minimum light in a light curve. (p. 500)

minor planet Another name for asteroid. (p. 353)

minute of arc A unit of angular measurement equal to one-sixtieth of a degree. (p. 9)

mode of oscillation A particular pattern of vibration of the Sun. (p. 409)

molecular cloud A relatively dense, cool interstellar cloud in which molecules are common. (p. 426)

momentum A quantity, equal to the product of a body's mass and velocity, used to describe the motion of the body. When two bodies collide or otherwise interact, the sum of their momenta is conserved. (p. 83)

M-type asteroid One of a class of asteroids that have reflectance spectra like those of metallic iron and nickel. (p. 358)

N

narrow line region The low density region in a quasar where narrow emission lines are formed. (p. 578)

neap tide An unusually low high tide and unusually high low tide that occur when the tidal forces of the Sun and Moon act at right angles to one another. (p. 96)

neutral gas A gas containing atoms and molecules but essentially no ions or free electrons. (p. 133)

neutrino A particle with no charge and probably no mass that is produced in nuclear reactions. Neutrinos pass freely through matter and travel at or near the speed of light. (p. 402)

neutron A nuclear particle with no electric charge. (p. 142)

neutronization A process by which, during the collapse of the core of a star, protons and electrons are forced together to make neutrons. (p. 474)

neutron star A star composed primarily of neutrons and supported by the degenerate pressure of the neutrons. (p. 474)

new comet A comet that has entered the inner solar system for the first time. (p. 362)

new phase The phase of the Moon in which none or almost none of the near side of the Moon is illuminated by sunlight, so the near side appears dark. (p. 29)

nodes The points in the orbit of the Moon where the Moon crosses the ecliptic plane. (p. 31)

normal spirals A galaxy in which the spiral arms emerge from the nucleus. (p. 557)

north celestial pole The point above the Earth's North Pole where the Earth's polar axis, if extended outward into space, would intersect the celestial sphere. The diurnal circles of stars in the northern hemisphere are centered on the north celestial pole. (p. 18)

north circumpolar region The region of the northern sky within which the diurnal circles of stars do not dip below the horizon. The size of the north circumpolar region varies with the latitude of the observer. (p. 18)

nova An explosion on the surface of a white dwarf star in which hydrogen is abruptly converted into helium. (p. 512)

nuclei *See* nucleus.

nucleic acid A long chain of nucleotides. DNA and RNA are nucleic acids. (p. 646)

nucleotide The class of organic molecules of which nucleic acids are composed. (p. 646)

nucleus **(A)** The massive, positively charged core of an atom. The nucleus of an atom is surrounded by one or more electrons. A nucleus missing one or more accompanying electrons is called an ion. Nuclei consist of protons and electrons. (p. 142) **(B)** An irregularly shaped, loosely packed lump of dirty ice several kilometers across that is the permanent part of a comet. (p. 359)

number density The number of particles in a given volume of space. (p. 135)

O

objective The main lens or mirror of a telescope. (p. 110)

oblate A departure from spherical shape of a body in which the body's polar diameter is smaller than its equatorial diameter. (p. 156)

Oort cloud The region beyond the planetary system, extending to 100,000 AU or more, within which a vast number of comets orbit the Sun. When comets from the Oort cloud enter the inner solar system, they become new comets. (p. 363)

opacity The ability of a substance to absorb radiation. The higher the opacity, the less transparent the substance is. (p. 403)

opposition The configuration of a planet or other body when it appears opposite the Sun in the sky. (p. 32)

orbit The elliptical or circular path followed by a body that is bound to another body by their mutual gravitational attraction. (p. 31)

orbital resonance A condition that occurs when two bodies have orbital periods that have the ratio of small integers, such as 1:2, 2:3, etc. (p. 306)

organic molecule A molecule containing carbon. (p. 260)

outflow channel A Martian valley with few tributaries probably formed by the sudden melting and runoff of subsurface water. (p. 243)

outgassing The release of gas from the interior of a planet or satellite. (p. 175)

ozone A molecule consisting of three oxygen atoms. Ozone molecules are responsible for the absorption of solar ultraviolet radiation in the Earth's atmosphere. (p. 170)

P

pair production A process in which gamma rays are transformed into a particle and its antiparticle (such as an electron and a positron). (p. 630)

pancake model A model for the formation of clusters of galaxies in which protoclusters form first and then fragment into individual galaxies. (p. 608)

parabola A geometric curve followed by a body that moves with a speed exactly equal to escape velocity. (p. 93)

parsec (pc) The distance at which a star has a parallax of 1 second of arc. At a distance of 1 parsec, an AU fills an angle of 1 second of arc. (p. 377)

patera A type of Martian volcano that resembles shield volcanos, but has even more gentle slopes. (p. 240)

Pauli exclusion principle A physical law that limits the number of particles of a particular kind that can be placed in a given volume. A gas in which that limit is reached is degenerate. (p. 446)

penumbra The outer part of the shadow of a body where sunlight is partially blocked by the body. (pp. 186, 411)

perihelion The point in the orbit of a body when it is closest to the Sun. (p. 69)

period-luminosity relationship The relationship between the period of brightness variation and the luminosity of a Cepheid variable star. The longer the period of a Cepheid is, the more luminous the Cepheid. (p. 458)

perturbation A deviation of the orbit of a solar system body from a perfect ellipse due to the gravitational attraction of one of the planets. (p. 294)

photon A massless particle of electromagnetic energy. (p. 106)

photosphere The visible region of the atmosphere of the Sun or another star. (p. 406)

pixel A "picture element," consisting of an individual detector in an array of detectors used to capture an image. (p. 114)

planet One of the eight major bodies in orbit around the Sun. (p. 4)

planetary nebula A luminous shell surrounding a hot star. The gas in a planetary nebula was ejected from the star while it was a red giant. (p. 461)

planetesimal A primordial solar system body of intermediate size that accreted with other planetesimals to form planets and satellites. (p. 437)

planetology The comparative study of the properties of planets. (p. 147)

plasma A fully or partially ionized gas. (p. 133)

plasma tail A narrow, ionized comet tail pointing directly away from the Sun. (p. 361)

plate A section of the Earth's lithosphere pushed about by convective currents within the mantle. (p. 165)

plate tectonics The hypothesis that the features of the Earth's crust, such as mountains and trenches, are caused by the slow movement of crustal plates. (p. 164)

plume A rising column of gas over a hot region in the interior or atmosphere of a body. (p. 321)

polarity The property of a magnet that causes it to have north and south magnetic regions. (p. 412)

precession The slow, periodic conical motion of the rotation axis of the Earth or another rotating body. (p. 49)

pressure The force exerted per unit area. (p. 136)

primary distance indicator A type of object, such as a Cepheid variable, for which we know the size or brightness by observing them in the Milky Way. (p. 565)

prime meridian The circle on the Earth's surface that runs from pole to pole through Greenwich, England. The zero point of longitude occurs where the prime meridian intersects the Earth's equator. (p. 8)

primeval atmosphere The original atmosphere of a planet. (p. 175)

prograde motion The eastward (normal) revolution of a solar system body. (p. 32)

prominence A region of cool gas embedded in the corona. Prominences are bright when seen above the Sun's limb, but appear as dark filaments when seen against the Sun's disk. (p. 415)

proper motion The rate at which a star appears to move across the celestial sphere with respect to very distant objects. (p. 378)

protein A large molecule, consisting of a chain of amino acids, that makes up the bodies of organisms. (p. 646)

proton A positively charged nuclear particle. (p. 142)

proton-proton chain A series of nuclear reactions through which stars like the Sun produce energy by converting hydrogen to helium. Named because the first reaction in the series is the reaction of one proton with another. (p. 402)

protostar A star in the process of formation. (p. 428)

pulsar A rotating neutron star with beams of radiation emerging from its magnetic poles. When the beams sweep past the Earth, we see "pulses" of radiation. (p. 481)

Q

quarter phase The phase of the Moon in which half of the near side of the Moon is illuminated by the Sun. (p. 29)

quasar A distant galaxy, seen as it was in the remote past, with a very small, luminous nucleus. (p. 573)

R

radial velocity The part of the velocity of a body that is directed toward or away from an observer. The radial velocity of a body can be determined by the Doppler shift of its spectral lines. (p. 389)

radiant The point in the sky from which the meteors in a meteor shower seem to originate. (p. 348)

radiation era The period of time, before about 1 million years after the expansion of the universe began, when radiation rather than matter was the dominant constituent of the universe. (p. 633)

radiative transfer The transport of energy by electromagnetic radiation. (p. 146)

radioactivity The spontaneous disintegration of an unstable nucleus of an atom. (p. 144)

radio galaxy A galaxy that is a strong source of radio radiation. (p. 582)

rays Long, narrow light streaks on the Moon and other bodies that radiate from relatively young craters. Rays consist of material ejected from a crater at the time it was formed by an impact. (p. 195)

recession velocity The rate of movement of a galaxy away from the Milky Way caused by the expansion of the universe. (p. 567)

recombination epoch The time, about 1 million years after the expansion of the universe began, when most of the ions and electrons in the universe combined to form atoms. (p. 633)

recurrent nova A binary system in which the white dwarf star undergoes repeated nova outbursts. (p. 512)

reflectance spectrum The reflectivity of a body as a function of wavelength. (p. 358)

reflection The bouncing of a wave from a surface. (p. 108)

reflection nebulae A cloud of interstellar gas and dust that is luminous because the dust it contains reflects the light of a nearby star. (p. 528)

reflectivity The ability of a surface to reflect electromagnetic waves. The reflectivity of a surface ranges from 0% for a surface that reflects no light to 100% for a surface that reflects all the light falling on it. (p. 108)

reflector A telescope in which the objective is a mirror. (p. 110)

refraction The bending of light when it passes from a material having one index of refraction to another material having a different index of refraction. (p. 108)

refractor A telescope in which the objective is a lens. (p. 110)

regolith The surface layer of dust and fragmented rock, caused by meteoritic impacts, on a planet, satellite, or asteroid. (p. 196)

regular cluster A cluster of galaxies that has roughly spherical symmetry. (p. 602)

regular satellites Regularly spaced satellites with nearly circular orbits that form miniature "solar systems" about their parent planets. (p. 316)

resolution The ability of a telescope to distinguish fine details of an image. (p. 112)

resonance The repetitive gravitational tug of one body on another when the orbital period of one is a multiple of the orbital period of the other. (p. 286)

resonant KBO A Kuiper Belt Object in an orbital resonance with Neptune. Most resonant KBOs are, like Pluto, in a 2:3 orbital resonance with Neptune and orbit the Sun twice for every three orbits of Neptune. (p. 310)

retrograde motion The westward revolution of a solar system body around the Sun. (p. 32)

retrograde rotation The westward rotation of a solar system body. (p. 216)

richness A measure of the number of galaxies in a cluster. The more galaxies there are, the greater the richness. (p. 602)

right ascension Angular distance of a body along the celestial equator from the vernal equinox eastward to the point on the equator nearest the body. Right ascension is analogous to longitude in the terrestrial coordinate system. (p. 19)

rille A lunar valley, probably the result of volcanic activity. (p. 197)

Roche distance The distance from a planet or other celestial body within which tidal forces from the body would disintegrate a smaller object. (p. 288)

Roche lobe The region around a star in a binary system in which the gravity of that star dominates. (p. 504)

rock A solid aggregation of grains of one or more minerals. (p. 159)

rotation curve A plot of the speed of revolution of the stars and gas in a galaxy versus distance from the center of the galaxy. (p. 536)

r-process The process of building up massive nuclei in which neutrons are captured at a rate faster than the newly produced nuclei can undergo radioactive decay. (p. 464)

RR Lyrae star A member of a class of giant pulsating stars, all of which have pulsation periods of about 1 day. (p. 459)

runoff channel One of a network of Martian valleys that probably were formed by the collection of widespread rainfall. (p. 243)

S

saros The length of time between one member of a series of similar eclipses and the next (6585⅓ days). (p. 190)

scalar field A form of energy that has been proposed as the cause of inflation in the early universe. (p. 639)

scarp A cliff produced by vertical movement of a section of the crust of a planet or satellite. (p. 212)

scattering The redirection of light in random directions when it strikes atoms, molecules, or solid particles. (p. 527)

scattered disk object A trans-Neptunian body with a relatively large, highly eccentric orbit. The perihelion distances of scattered disk objects are close enough to Neptune's orbit that they will eventually experience an encounter with Neptune and move into the planetary system. (p. 310)

Schwarzschild radius The radius of the event horizon of a black hole. (p. 489)

seafloor spreading The splitting of the oceanic crust where magma forces the existing crust apart, creating new ocean floor. (p. 165)

secondary atmosphere The atmosphere that forms after a planet has lost any original atmosphere it had. (p. 175)

secondary distance indicator A type of object for which we know the size or brightness because objects of that type have been found in nearby galaxies. (p. 566)

second of arc A unit of angular measurement equal to $1/60$ of a minute of arc or $1/3600$ of a degree. (p. 9)

sedimentary rock A rock formed by the accumulation of small mineral grains carried by wind, water, or ice to the spot where they were deposited. (p. 159)

seeing A measure of the blurring of the image of an astronomical object caused by turbulence in the Earth's atmosphere. (p. 118)

seismic wave Waves that travel through the interior of a planet or satellite and are produced by earthquakes or their equivalent. (p. 162)

seismometer Sensitive device used to measure the strengths and arrival times of seismic waves. (p. 162)

semimajor axis Half of the major axis of an ellipse. Also equal to the average distance from the focus of a body moving on an elliptical orbit. (p. 69)

SETI The search for extraterrestrial intelligence. (p. 658)

Seyfert galaxy A barred or normal spiral galaxy with a small, very bright nucleus. (p. 581)

Sgr A* A small, bright source of radio emission, possibly the accretion disk of a black hole, that probably marks the exact center of the Milky Way. (p. 544)

shield volcano A broad, gently sloped volcano built up by the repeated eruption of very fluid lava. (p. 223)

short-period comet A comet with an orbital period shorter than 200 years. (p. 362)

sidereal clock A clock that marks the local hour angle of the vernal equinox. (p. 20)

sidereal day The length of time (23 hours, 56 minutes, 4.091 seconds) between successive appearances of a star on the meridian. (p. 26)

sidereal month The length of time required for the Moon to return to the same apparent position among the stars. (p. 30)

sidereal period The time it takes for a planet or satellite to complete one full orbit about the Sun or its parent planet. (p. 62)

silicate A mineral whose crystalline structure is dominated by silicon and oxygen atoms. (p. 159)

sinuous rille A winding lunar valley possibly caused by the collapse of a lava tube. (p. 197)

smooth plains Widespread sparsely cratered regions of the surface of Mercury possibly having a volcanic origin. (p. 212)

solar constant The solar energy received by a square meter of surface oriented at right angles to the direction to the Sun at the Earth's average distance (1 AU) from the Sun. The value of the solar constant is 1372 watts per square meter. (p. 140)

solar day The amount of time that passes between successive appearances of the Sun on the meridian. The solar day varies in length throughout the year. (p. 26)

solar flare An explosive release of solar magnetic energy. (p. 416)

solar motion The motion of the Sun with respect to the nearby stars. (p. 378)

solar nebula The rotating disk of gas and dust, surrounding the newly formed Sun, from which planets and smaller solar system bodies formed. (p. 435)

solar wind The hot plasma that flows outward from the Sun. (p. 172)

solidification age The amount of time that has passed since a meteorite solidified from the molten state. (p. 352)

south celestial pole The point above the Earth's South Pole where the Earth's polar axis, if extended outward into space, would intersect the celestial sphere. The diurnal circles of stars in the southern hemisphere are centered on the south celestial pole. (p. 18)

spacelike trip A path in spacetime that would require motion at a speed faster than the speed of light. (p. 486)

spacetime The combination of three spatial coordinates and one time coordinate that we use to locate an event. (p. 484)

spacetime diagram A diagram showing one spatial coordinate against time, in which the paths of bodies and beams of light can be plotted. (p. 485)

spectral class A categorization, based on the pattern of spectral lines of stars, that groups stars according to their surface temperatures. (p. 387)

spectrograph A device used to produce and record a spectrum. (p. 114)

spectroscopic binary A pair of stars whose binary nature can be detected by observing the periodic Doppler shifts of their spectral lines as they move about one another. (p. 498)

spectroscopy The recording and analysis of spectra. (p. 114)

spicule A hot jet of gas moving outward through the Sun's chromosphere. (p. 413)

spiral arm A long, narrow feature of a spiral galaxy in which interstellar gas, young stars, and other young objects are found. (p. 539)

spiral galaxy A flattened galaxy in which hot stars, interstellar clouds, and other young objects form a spiral pattern. (p. 557)

spokes Dark, short-lived radial streaks in Saturn's rings. (p. 287)

spring tide Unusually high, high tide and unusually low, low tide that occur when the tidal forces of the Sun and Moon are aligned. This occurs at full Moon and new Moon. (p. 96)

s-process The process of building up massive nuclei in which neutrons are captured at a rate slower than the newly produced nuclei can undergo radioactive decay. (p. 464)

standard time The time kept throughout one of Earth's approximately 15-degree-wide time zones. (p. 27)

star A massive gaseous body that has used, is using, or will use nuclear fusion to produce the bulk of the energy it radiates into space. (p. 4)

starburst galaxy A galaxy in which a very large number of stars have recently formed. (p. 585)

Stefan-Boltzmann law The relationship between the temperature of a blackbody and the rate at which it emits radiant energy. (p. 139)

stellar occultation The obstruction of the light from a star when a solar system body passes between the star and the observer. (p. 301)

stellar parallax The shift in the direction of a star caused by the change in the position of the Earth as it moves about the Sun. (p. 67)

stellar population A group of stars that are similar in spatial distribution, chemical composition, and age. (p. 546)

stony-iron meteorite A meteorite made partially of stone and partially of iron and other metals. (p. 352)

stony meteorite A meteorite made of silicate rock. (p. 350)

stratopause The layer of Earth's atmosphere at the boundary between the stratosphere and the mesosphere. (p. 170)

stratosphere The region of the atmosphere of a planet immediately above the troposphere. (p. 169)

S-type asteroid One of a class of asteroids whose reflectance spectra show an absorption feature due to the mineral olivine. (p. 358)

subduction The process through which lithospheric plates of a planet or satellite are forced downward into the mantle. (p. 166)

summer solstice The point on the ecliptic where the Sun's declination is most northerly. The time when the Sun is at the summer solstice, around June 21, marks the beginning of summer. (p. 22)

sunspot A region of the Sun's photosphere that appears darker than its surroundings because it is cooler. (p. 411)

sunspot cycle The regular waxing and waning of the number of spots on the Sun. The amount of time between one sunspot maximum and the next is about 11 years. (p. 418)

sunspot group A cluster of sunspots. (p. 411)

supergiant An extremely luminous star of large size and mass. (p. 389)

supergranulation The pattern of very large (15,000 to 30,000 km in diameter) convective cells in the Sun's photosphere. (p. 408)

superior planet A planet whose orbit lies outside the Earth's orbit. (p. 64)

superluminal motion The apparent separation of components of a quasar at speeds faster than the speed of light. (p. 580)

supernova An explosion in which a star's brightness temporarily increases by as much as 1 billion times. Type I supernovae are caused by the rapid fusion of carbon and oxygen within a white dwarf. Type II supernovae are produced by the collapse of the core of a star. (p. 473)

supernova remnant The luminous, expanding region of gas driven outward by a supernova explosion. (p. 478)

synchronous rotation Rotation for which the period of rotation is equal to the period of revolution. An example of synchronous rotation is the Moon, for which the period of rotation and the period of revolution about the Earth are both 1 month. (p. 182)

synchrotron emission Electromagnetic radiation, usually observed in the radio region of the spectrum, produced by energetic electrons spiraling about magnetic field lines. (p. 478)

synodic month The length of time (29.53 days) between successive occurrences of the same phase of the Moon. (p. 31)

synodic period The length of time it takes a solar system body to return to the same configuration (opposition to opposition, for example) with respect to the Earth and the Sun. (p. 32)

T

T Tauri star A pre–main sequence star, less massive than about three solar masses, showing intense emission lines. (p. 431)

temperature A measure of the average energies of the particles in a system. For a gas, temperature is a measure of the motions of the particles. (p. 134)

terminal velocity The speed with which a body falls through the atmosphere of a planet when the force of gravity pulling it downward is balanced by the force of air resistance. (p. 347)

termination shock The point where the solar wind slows significantly as it interacts with interstellar gas. The termination shock is located on a surface whose distance from the Sun varies with direction and solar wind speed. (p. 418)

terrae The light-colored, ancient, heavily cratered portions of the surface of the Moon. (p. 193)

terrestrial planet A rocky planet located in the inner solar system. (p. 207)

thermal equilibrium The condition in which a body or a portion of a body gains energy (by generating it or absorbing it) at the same rate at which energy is transported away from it. (p. 140)

thermal pulse The rapid consumption of helium in a shell within an asymptotic giant branch star. (p. 459)

thermosphere The layer of the atmosphere of a planet lying above the mesosphere. The lower thermosphere is the ionosphere. The upper thermosphere is the exosphere. (p. 170)

tidal capture A possible explanation for the origin of a wide binary pair of stars in which two cloud fragments tidally interact with and capture one another. (p. 503)

tidal force The differences in gravity in a body being attracted by another body. (p. 94)

tidal heating The frictional heating of the interior of a satellite as it is flexed and released by a variable tidal force due to its parent planet. (p. 317)

tides Distortions in a body's shape resulting from tidal forces. (p. 94)

timelike trip A path in spacetime that can be followed by a body moving slower than the speed of light. (p. 486)

transform fault The boundary between two of the Earth's crustal plates that are sliding past each other. (p. 167)

transit An event that occurs when a smaller body passes between a larger body and an observer causing the apparent brightness of the larger body to drop. (p. 651)

Trans-Neptunian Object (TNO) A small, icy solar system body in an orbit beyond Neptune. (p. 304)

transverse velocity The part of the orbital speed of a body perpendicular to the Sun between the body and the Sun. (p. 69)

triple α process A pair of nuclear reactions through which three helium nuclei (alpha particles) are transformed into a carbon nucleus. (p. 444)

Trojan asteroid One of a group of asteroids that orbit the Sun at Jupiter's distance and lie 60° ahead of or behind Jupiter in its orbit. (p. 357)

tropical year The interval of time, equal to 365.242 solar days, between successive appearances of the Sun at the vernal equinox. (p. 27)

tropopause The upper boundary of the troposphere of the atmosphere of a planet. (p. 169)

troposphere The lowest layer of the atmosphere of a planet, within which convection produces weather. (p. 169)

type Ia supernova An extremely energetic explosion produced by the abrupt fusion of carbon and oxygen in the interior of a collapsing white dwarf star. (p. 513)

type II supernova An extremely energetic explosion that occurs when the core of a massive star collapses, probably producing a neutron star or black hole. (p. 474)

U

ultraviolet The part of the electromagnetic spectrum with wavelengths longer than X rays, but shorter than visible light. (p. 107)

umbra **(A)** The inner portion of the shadow of a body, within which sunlight is completely blocked. (p. 186) **(B)** The dark central portion of a sunspot. (p. 411)

universe All the matter and space there is. (p. 59)

V

Van Allen belts Two doughnut-shaped regions in the Earth's magnetosphere within which many energetic ions and electrons are trapped. (p. 172)

vector A quantity that has both direction and magnitude. Velocity is a vector, whereas speed is not. (p. 83)

velocity A physical quantity that gives the speed of a body and the direction in which it is moving. (p. 83)

vernal equinox The point in the sky where the Sun appears to cross the celestial equator moving from south to north. This happens approximately on March 21. (p. 19)

visual binary star A pair of stars orbiting a common center of mass in which the images of the components can be distinguished using a telescope and which have detectable orbital motion. (p. 496)

Vogt-Russell theorem The concept that the original mass and chemical composition of an isolated star completely determine the course of its evolution. (p. 447)

voids Immense volumes of space in which few galaxies and clusters of galaxies can be found. (p. 614)

V-type asteroid The asteroid Vesta, which is unique in having a reflectance spectra resembling those of basaltic lava flows. (p. 358)

V/V_{max} test A statistical method used to determine whether quasars have changed over time. (p. 592)

W

waning crescent The Moon's crescent phase that occurs just before new Moon. (p. 30)

wave A regular series of disturbances that moves through a material medium or through empty space. (p. 102)

wavelength The distance between crests of a wave. For visible light, wavelength determines color. (p. 102)

waxing crescent The Moon's crescent phase that occurs just after new Moon. (p. 29)

weight The gravitational force exerted on a body by the Earth (or another astronomical object). (p. 89)

white dwarf A small, dense star that is supported against gravity by the degenerate pressure of its electrons. (pp. 391, 470)

wide pair A binary star system in which the components are so distant from one another that they evolve independently. (p. 496)

Wien's law The relationship between the temperature of a blackbody and the wavelength at which its emission is brightest. (p. 138)

winter solstice The point on the ecliptic where the Sun has the most southerly declination. The time when the Sun is at the winter solstice, around December 22, marks the beginning of winter. (p. 22)

X

X ray The part of the electromagnetic spectrum with wavelengths longer than gamma rays but shorter than ultraviolet. (p. 107)

X-ray burst Sporadic burst of X rays originating in the rapid consumption of nuclear fuels on the surface of the neutron star in a binary system. (p. 513)

X-ray pulsar A neutron star from which periodic bursts of X rays are observed. (p. 514)

Y

year The length of time required for the Earth to orbit the Sun. (p. 22)

Z

Zeeman effect The splitting of a spectral line into two or more components when the atoms or molecules emitting the line are located in a magnetic field. (p. 412)

zenith The point on the celestial sphere directly above an observer. (p. 9)

zero point The point from which the coordinates in a coordinate system are measured. For example, the vernal equinox is the zero point of right ascension and declination in the celestial coordinate system. (p. 7)

zodiacal constellations The band of constellations along the ecliptic. The Sun appears to move through the 12 zodiacal constellations during a year. (p. 22)

zodiacal light The faint glow extending away from the Sun caused by the scattering of sunlight by interplanetary dust particles lying in and near the ecliptic. (p. 366)

zonal winds The pattern of winds in the atmosphere of a planet in which the pattern of wind speeds varies with latitude. (p. 271)

zone A light region of clouds located at a given latitude on one of the giant planets. (p. 270)

zone of convergence According to plate tectonics, a plate boundary at which the crustal plates of a planet are moving toward one another. Crust is destroyed in zones of convergence. (p. 166)

zone of divergence According to plate tectonics, a plate boundary at which the crustal plates of a planet are moving away from one another. Crust is created in zones of divergence. (p. 166)

References

CHAPTER 2

Allen, Clabon Walter. *Astrophysical Quantities,* ed 3. London: Athlone Press, 1973.

Marshack, Alexander. *The Roots of Civilization: The Cognitive Beginnings of Man's First Art, Symbol, and Notation,* ed 1. New York: McGraw-Hill, 1971.

CHAPTER 3

Boorstin, Daniel Joseph. *The Discoverers.* New York: Random House, 1983.

Christianson, Gale E. *This Wild Abyss: The Story of the Men Who Made Modern Astronomy.* New York: Free Press, 1978.

Couch, Carl J. *Constructing Civilizations.* Greenwich, Conn.: JAI Press, 1984.

Dicks, D. R. *Early Greek Astronomy to Aristotle.* Ithaca, N.Y.: Cornell University Press, 1970.

Gillings, Richard J. *Mathematics in the Time of Pharaohs.* New York: Dover, 1982.

Hoyle, Sir Fred. *Astronomy.* New York: Crescent Books, 1962.

Lockyer, Sir Norman. *The Dawn of Astronomy: A Study of the Temple-Worship and Mythology of the Ancient Egyptians.* New York: Macmillan, 1897.

Neugebauer, Otto. *A History of Ancient Mathematical Astronomy.* New York: Springer-Verlag Publishing, 1975.

———. *Astronomy and History: Selected Essays.* New York: Springer-Verlag, 1983.

O'Neil, William Matthew. *Early Astronomy from Babylonia to Copernicus.* Sydney: Sydney University Press, 1986.

Pannekoek, Anton. *A History of Astronomy.* New York: Interscience Publishers, 1961.

Ptolemy, Claudius. *Ptolemy's Almagest.* Translated and annotated by G. J. Toomer. London: Duckworth, 1984.

———. *Tetrabiblos.* Edited and translated by F. E. Robbins. Cambridge, Mass.: Harvard University Press, 1940.

Ronan, Colin A. *The Astronomers.* New York: Hill and Wang, 1964.

Tompkins, Peter. *Secrets of the Great Pyramid.* New York: Harper and Row, 1971.

Van Helden, Albert. *Measuring the Universe: Cosmic Dimensions from Aristarchus to Halley.* Chicago: University of Chicago Press, 1985.

CHAPTER 4

Adamczewski, Jan, and Piszek, E. J. *Nicholas Copernicus and His Epoch.* Philadelphia: Copernicus Society of America, 1974.

Bienkowska, Barbara. *The Scientific World of Copernicus on the Occasion of the 500th Anniversary of His Birth.* Boston: D. Reidel Publishing Co., 1973.

Christianson, Gale E. *This Wild Abyss: The Story of the Men Who Made Modern Astronomy.* New York: Free Press, 1978.

Copernicus, Nicholaus. *On the Revolutions of the Heavenly Spheres.* Translated by Alistair Matheson Duncan. New York: Barnes and Noble, 1976.

Kuhn, Thomas S. *The Copernican Revolution: Planetary Astronomy in the Development of Western Thought.* Cambridge, Mass.: Harvard University Press, 1957.

O'Neil, William Matthew. *Early Astronomy from Babylonia to Copernicus.* Sydney: Sydney University Press, 1986.

Pannekoek, Anton. *A History of Astronomy.* New York: Interscience Publishers, 1961.

Ronan, Colin A. *The Astronomers.* New York: Hill and Wang, 1964.

CHAPTER 5

Christianson, Gale E. *This Wild Abyss: The Story of the Men Who Made Modern Astronomy.* New York: Free Press, 1978.

Kuhn, Thomas S. *The Copernican Revolution: Planetary Astronomy in the Development of Western Thought.* Cambridge, Mass.: Harvard University Press, 1957.

Newton, Sir Isaac. *Sir Isaac Newton's Mathematical Principles of Natural Philosophy, and His System of the World.* Berkeley: University of California Press, 1962.

Pannekoek, Anton. *A History of Astronomy.* New York: Interscience Publishers, 1961.

Taton, Reni and Wilson, Curtis, eds., *Planetary Astronomy from the Renaissance to the Rise of Astrophysics.* New York: Cambridge University Press, 1989.

CHAPTER 6

Tipler, Paul A. *Physics,* ed 2. New York: World Publishing, Inc., 1982.

CHAPTER 8

Garland, George David. *Introduction to Geophysics: Mantle, Core, and Crust.* Philadelphia: Saunders College Publishing, 1971.

Hartmann, William K. *Moons and Planets,* ed 2. Belmont, Cal.: Wadsworth Publishing Co., 1983.

Hoskin, Michael A. *Stellar Astronomy: Historical Studies.* Chalfont St. Giles, Bucks, England: Science History Publishing, 1982.

Kaula, William M. *An Introduction to Planetary Physics: The Terrestrial Planets.* New York: John Wiley & Sons, Inc., 1968.

Press, Frank, and Siever, Raymond. *Earth,* ed 2. San Francisco, Cal.: W. H. Freeman, 1978.

States, Robert J., and Gardner, Chester. "Thermal Structure of the Mesopause Region (80–105 km) at 40° N Latitude. Part I: Seasonal Variations." *Journal of the Atmospheric Sciences* 57 (2000):66–77.

Wetherill, George W. "Formation of the Earth." In George W. Wetherill, Arden L. Albee, and Francis G. Stehli, eds., *Annual Review of Earth and Planetary Sciences,* vol. 18. Palo Alto, Cal.: Annual Reviews, Inc., 1990:205–256.

CHAPTER 9

Allen, Clabon Walter. *Astrophysical Quantities.* London: Athlone Press, 1973.

Cadogan, Peter H. *The Moon—Our Sister Planet.* New York: Cambridge University Press, 1981.

Canup, Robin M. "Dynamics of Lunar Formation." In Geoffrey Burbidge, Roger Blandford, and Allan Sandage, eds., *Annual Review of Astronomy and Astrophysics,* vol. 42. Palo Alto, Cal.: Annual Reviews, Inc., 2004:441–475.

Hartmann, William K. *Moons and Planets,* ed 2. Belmont, Cal.: Wadsworth Publishing Co., 1983.

Herodotus. *The History.* Translated by David Grene. Chicago: University of Chicago Press, 1987.

Kaula, William M. *An Introduction to Planetary Physics: The Terrestrial Planets.* New York: John Wiley & Sons, Inc., 1968.

King, Elbert A. *Space Geology: An Introduction.* New York: John Wiley & Sons, Inc., 1976.

Kopal, Zdenek. *The Moon in the Post-Apollo Era.* Boston: D. Reidel Publishing Co., 1974.

Kopal, Zdenek, and Carder, Robert W. *Mapping of the Moon: Past and Present.* Boston: D. Reidel Publishing Co., 1974.

Link, Frantisek. *Eclipse Phenomena in Astronomy.* New York: Springer Publishing, 1969.

Melchior, Paul J. *The Tides of the Planet Earth.* New York: Pergamon Press, 1978.

Mitchell, Samuel A. *Eclipses of the Sun,* ed 5. New York: Columbia University Press, 1951.

Neugebauer, Otto. *A History of Ancient Mathematical Astronomy.* New York: Springer-Verlag Publishing, 1975.

Ottewell, Guy. *The Astronomical Calendar 1986.* Greenville, S.C.: Department of Physics, Furman University, 1986.

Plutarch. *The Lives of the Noble Grecians and Romans.* Translated by John Dryden. Chicago: Encyclopedia Brittanica, 1952.

Rudaux, Lucien, and de Vaucouleurs, G. *Larousse Encyclopedia of Astronomy,* ed 2. Revised by Colin A. Ronan and Henry C. King. New York: Prometheus Press, 1967.

Taylor, Stuart Ross. *Lunar Science: A Post-Apollo View: Scientific Results and Insights from the Lunar Samples.* New York: Pergamon Press, 1975.

CHAPTER 10

Burgess, Eric. *Venus, An Errant Twin.* New York: Columbia University Press, 1985.

Caldwell, John J. "Planetary and Satellite Atmospheres." In Stephen P. Maran, ed., *The Astronomy and Astrophysics Encyclopedia.* New York: Van Nostrand Reinhold, 1992:506–508.

Cattermole, Peter John. *Planetary Volcanism: A Study of Volcanic Activity in the Solar System.* New York: Halstead Publishing, 1984.

Christon, Stephen P. "Mercury, Magnetosphere." In Stephen P. Maran, ed., *The Astronomy and Astrophysics Encyclopedia.* New York: Van Nostrand Reinhold, 1992:427–430.

Encrenaz, Therese. "Water in the Solar System." In Roger Blandford, John Kormendy, and Ewine van Dishoeck, eds., *Annual Review of Astronomy and Astrophysics,* vol. 46. Pal Alto, Cal.: Annual Reviews, Inc., 2008:57–88.

Fimmel, Richard O., Colin, Lawrence, and Burgess, Eric. *Pioneer Venus.* Washington, D.C.: Scientific and Technical Information Branch, National Aeronautics and Space Administration, 1983.

Golombek, Matthew P. "Planetary Tectonic Processes, Terrestrial Planets." In Stephen P. Maran, ed., *The Astronomy and Astrophysics Encyclopedia.* New York: Van Nostrand Reinhold, 1992:544–546.

Harris, Alan W. "Planetary Rotational Properties." In Stephen P. Maran, ed., *The Astronomy and Astrophysics Encyclopedia.* New York: Van Nostrand Reinhold, 1992:540–542.

Head James W., and Campbell, Donald B. "Venus Volcanism: Initial Analysis from Magellan Data." *Science* 252:5003 (12 April 1991):276–288.

Hood, Lon L. "Planetary Interiors, Terrestrial Planets." In Stephen P. Maran, ed., *The Astronomy and Astrophysics Encyclopedia.* New York: Van Nostrand Reinhold, 1992:528–530.

Hubbard, William B. *Planetary Interiors.* New York: Van Nostrand Reinhold Publishing, 1984.

Hunten, D. M., et al., eds. *Venus.* Tucson: University of Arizona Press, 1983.

Hunten, Donald M. "Venus: Atmosphere." In Lucy-Ann McFadden, Paul R. Weissman, and Torrence V. Johnson, eds., *Encyclopedia of the Solar System,* ed 2. San Diego, Cal.: Academic Press, 2007:139–148.

Ip, Wing Huen. "Mercury and the Moon, Atmospheres." In Stephen P. Maran, ed., *The Astronomy and Astrophysics Encyclopedia.* New York: Van Nostrand Reinhold, 1992:418–420.

Newcott, William. "Venus Revealed." *National Geographic* 183:2 (February 1993):36–59.

Phillips, Roger J., and Arvidson, Raymond E. "Impact Craters on Venus: Initial Analysis from Magellan." *Science* 252:5003 (12 April 1991):288–297.

Prinn, Ronald G., and Fegley, Bruce, Jr. "The Atmospheres of Venus, Earth, and Mars: A Critical Comparison." In George W. Wetherill, Arden L. Albee, and Francis G. Stehli, eds., *Annual Review of Earth and Planetary Sciences,* vol. 15. Palo Alto, Cal.: Annual Reviews, Inc., 1987.

Sandwell, David T., and Schubert, Gerald. "Evidence for Retrograde Lithospheric Subduction on Venus." *Science* 257:5071 (7 August 1992):766–770.

Saunders, R. Stephen. "The Surface of Venus." *Scientific American* 263:6 (December 1990):60–65.

Saunders, R. Stephen, and Arvidson, Raymond E. "An Overview of Venus Geology." *Science* 253:5003 (12 April 1991): 249–252.

Saunders, R. Stephen, and Carr, Michael H. "Venus." In Michael H. Carr, ed., *The Geology of the Terrestrial Planets.* Washington, D.C.: Scientific and Technical Information Branch, National Aeronautics and Space Administration, 1984:57–77.

Saunders, R. Stephen, and Doyle, Jim. "The Exploration of Venus: A Magellan Progress Report." *Mercury* 20:5 (September 1991):130–144.

Seiff, Alvin. "Venus, Atmosphere." In Stephen P. Maran, ed., *The Astronomy and Astrophysics Encyclopedia.* New York: Van Nostrand Reinhold, 1992:942–944.

Solomon, Sean C., and Head, James W. "Venus Tectonics: Initial Analysis from Magellan." *Science* 252:5003 (12 April 1991): 297–312.

Smrekar, Suzanne, E., and Stofan, Ellen R. "Venus: Surface and Interior." In Lucy-Ann McFadden, Paul R. Weissman, and Torrence V. Johnson, eds., *Encyclopedia of the Solar System,* ed 2. San Diego, Cal.: Academic Press, 2007:149–168

Strom, Robert G. "Mercury." In Lucy-Ann McFadden, Paul R. Weissman, and Torrence V. Johnson, eds., *Encyclopedia of the Solar System,* ed 2. San Diego, Cal.: Academic Press, 2007:117–138.

Strom, Robert G. "Mercury." In Michael H. Carr, ed., *The Geology of the Terrestrial Planets.* Washington, D.C.: Scientific and Technical Information Branch, National Aeronautics and Space Administration, 1984:13–55.

———. "Mercury: A Post-*Mariner 10* Assessment." *Space Science Reviews* 24:1 (September 1979):3–70.

———. "Mercury, Geology and Geophysics." In Stephen P. Maran, ed., *The Astronomy and Astrophysics Encyclopedia.* New York: Van Nostrand Reinhold, 1992:424–427.

Vilas, Faith, Clark, R., and Matthews, Mildred Shapley, eds., *Mercury.* Tucson: University of Arizona Press, 1988.

Zuber, Maria T. "Venus, Geology and Geophysics." In Stephen P. Maran, ed., *The Astronomy and Astrophysics Encyclopedia.* New York: Van Nostrand Reinhold, 1992:944–948.

CHAPTER 11

Anderson, Don L., et al. "Meteorological Results from the Surface of Mars: *Viking 1 and 2.*" *Journal of Geophysical Research* 82:28 (30 September 1977):4524–4546.

Baker, Victor R. "The Channels of Mars." In Duke B. Reiber, ed., *The NASA Mars Conference.* San Diego: Univelt 1988: 75–90.

Biemann, K., et al. "The Search for Organic Substances and Inorganic Volatile Compounds in the Surface of Mars." *Journal of Geophysical Research* 82:28 (30 September 1977):4641–4658.

Binder, Alan B., et al. "The Geology of the *Viking* Lander 1 Site." *Journal of Geophysical Research* 82:28 (30 September 1977): 4439–4451.

Blasius, Karl R., and Cutts, James A. "Geology of the Valles Marineris: First Analysis of Imaging from the *Viking 1* Orbiter Primary Mission." *Journal of Geophysical Research* 82:28 (30 September 1977):4067–4091.

Briggs, Geoffrey, et al. "Martian Dynamical Phenomena During June–November 1976: *Viking* Orbiter Imaging Results." *Journal of Geophysical Research* 82:28 (30 September 1977): 4121–4149.

Carr, Michael H. "Mars." In Michael H. Carr, ed., *The Geology of the Terrestrial Planets.* Washington, D.C.: Scientific and Technical Information Branch, NASA, 1984:207–263.

———. "The Volcanism of Mars." In Duke B. Reiber, ed., *The NASA Mars Conference.* San Diego, Cal.: Published for NASA and the American Astronautical Society by Univelt, Inc., 1988:55–73.

———. Carr, Michael H. "Water on Mars." *Nature* 325:6108 (5 March 1987):30–35.

Carr, Michael H. "Mars: Surface and Interior." In Lucy-Ann McFadden, Paul R. Weissman, and Torrence V. Johnson, eds., *Encyclopedia of the Solar System,* ed 2. San Diego, Cal.: Academic Press, 2007:315–330.

Carr, Michael H., and Schaber, Gerald G. "Martian Permafrost Features." *Journal of Geophysical Research* 82:28 (30 September 1977):4039–4054.

Carr, Michael H., et al. "Some Martian Volcanic Features as Viewed from the *Viking* Orbiters." *Journal of Geophysical Research* 82:28 (30 September 1977):3985–4015.

Catling, David C., and Loevy, Conway. "Mars Atmosphere: History and Surface Interactions." In Lucy-Ann McFadden, Paul R. Weissman, and Torrence V. Johnson, *Encyclopedia of the Solar System,* ed 2. San Diego, Cal.: Academic Press, 2007: 301–314.

Cattermole, Peter John. *Planetary Volcanism: A Study of Volcanic Activity in the Solar System.* New York: Halstead, 1989.

Connerney, J. E. P., et al. "Tectonic Implications of Mars Crustal Magnetism." *Proceedings of the National Academy of Sciences* (18 October 2005):14970–14975.

Covault, Craig. "Analysis Points to Meteorite Coming from Mars." *Aviation Week and Space Technology* 117:22 (29 November 1982):18–21.

Fanale, Fraser P. "The Water and Other Volatiles of Mars." In Duke B. Reiber, ed., *The NASA Mars Conference.* San Diego, Cal.: Published for NASA and the American Astronautical Society by Univelt, Inc., 1988:157–174.

Golombek, Matthew P., and McSween, Harry Y., Jr. "Mars Landing Site Geology, Minerology, and Geochemistry." In Lucy-Ann McFadden, Paul R. Weissman, and Torrence V. Johnson, eds., *Encyclopedia of the Solar System,* ed 2. San Diego, Cal.: Academic Press, 2007:331–348.

Hess, S. L., et al. "Meteorological Results from the Surface of Mars: *Viking 1* and *2.*" *Journal of Geophysical Research* 82:28 (30 September 1977):4559–4574.

Horowitz, Norman H. "The Biological Question of Mars." In Duke B. Reiber, ed., *The NASA Mars Conference.* San Diego, Cal.: Published for NASA and the American Astronautical Society by Univelt, Inc., 1988:177–185.

Horowitz, Norman H., et al. "*Viking* on Mars: The Carbon Assimilation Experiments." *Journal of Geophysical Research* 82:28 (30 September 1977):4659–4662.

Hoyt, William Graves. *Lowell and Mars.* Tucson: The University of Arizona Press, 1976.

Hubbard, William B. *Planetary Interiors.* New York: Van Nostrand Reinhold, 1984.

Klein, H. P. "The *Viking* Biological Investigation: General Aspects." *Journal of Geophysical Research* 82:28 (30 September 1977):4677–4680.

Krasnopolsky, Vladimir A., and Feldman, Paul D. "Detection of Molecular Hydrogen in the Atmosphere of Mars." *Science* 294 (30 November 2001):1914–1917.

Leovy, Conway. "The Meteorology of Mars." In Duke B. Reiber, ed., *The NASA Mars Conference.* San Diego, Cal.: Published for NASA and the American Astronautical Society by Univelt, Inc., 1988:133–154.

Levin, Gilbert V., and Straat, Patricia Ann. "A Reappraisal of Life on Mars." In Duke B. Reiber, ed., *The NASA Mars Conference.* San Diego, Cal.: Published for NASA and the American Astronautical Society by Univelt, Inc., 1988:187–207.

———. "Recent Results from the *Viking* Labeled Release Experiment on Mars." *Journal of Geophysical Research* 82:28 (30 September 1977):4663–4667.

Malin, Michael C., and Edgett, Kenneth S. "Evidence for Recent Groundwater Seepage and Surface Runoff on Mars." *Science 288* (30 June 2000):2330–2335.

Martin Marietta Corp. *Viking Mars Expedition 1976.* Denver, Colo.: Martin Marietta Corp., 1978.

Moore, Henry J., et al. "Surface Materials of the *Viking* Landing Sites." *Journal of Geophysical Research* 82:28 (30 September 1977):4497–4523.

Mutch, Thomas A., et al. "The Geology of the *Viking* Lander 2 Site." *Journal of Geophysical Research* 82:28 (30 September 1977):4452–4467.

Neukum, G., et al. "Recent and Episodic Volcanic and Glacial Activity on Mars Revealed by the High Resolution Stereo Camera" *Nature* 432 (23 December 2004):971–979.

Oyama, Vance I., and Berdahl, Bonnie J. "The *Viking* Gas Exchange Experiment Results From Chryse and Utopia Surface Samples." *Journal of Geophysical Research* 82:28 (30 September 1977):4669–4676.

Picardi, Giovanni, et al. "Radar Soundings of the Subsurface of Mars." *Science* 310 (23 December 2005):1925–1928.

Plescia, J. B. "Recent Flood Lavas in the Elysium Region of Mars." *Icarus* 88:2 (December 1990):465–490.

Powers, Robert M. "Far Lonelier Than Barsoom." *The* Viking *Mission to Mars.* Denver, Colo.: Martin Marietta Corp., 1975.

Soffen, Gerald A. "The Viking Project." *Journal of Geophysical Research* 82:28 (30 September 1977):3959–3970.

Squyres, Steven W. "The History of Water on Mars." In George Wetherill, Arden L. Albee, and Francis G. Stehli, eds., *Annual Review of Earth and Planetary Sciences,* vol. 12. Palo Alto, Cal.: Annual Reviews, Inc., 1984:83–106.

CHAPTER 12

Burns, Joseph A. "Planetary Rings." In J. Kelly Beatty and Andrew Chaikin, eds., *The New Solar System,* ed 3. Cambridge, Mass.: Sky Publishing Corp., 1990:153–170.

Conrath, B. J., et al. "Thermal Structure and Dynamics of the Jovian Atmosphere 2: Visible Cloud Features." *Journal of Geophysical Research* 86:A10 (30 September 1981): 8769–8775.

Flasar, F. Michael, et al. "Thermal Structure and Dynamics of the Jovian Atmosphere 1: The Great Red Spot." *Journal of Geophysical Research* 86:A10 (30 September 1981):8759–8767.

Hanel, R. A., et al. "Albedo, Internal Heat, and Energy Balance of Jupiter: Preliminary Results of the *Voyager* Infrared Investigation." *Journal of Geophysical Research* 86:A10 (30 September 1981):8705–8712.

Hubbard, W. B., Burrows, A., and Lunine, J. I. "Theory of Giant Planets." In Geoffrey Burbidge, Allan Sandage, and Frank H. Shu, eds., *Annual Review of Astronomy and Astrophysics,* vol. 40. Palo Alto, Cal.: Annual Reviews, Inc., 2002:103–136.

Hubbard, William B. "Interiors of the Giant Planets." In J. Kelly Beatty and Andrew Chaikin, eds., *The New Solar System,* ed 3. Cambridge, Mass.: Sky Publishing Corp., 1990:131–138.

———. *Planetary Interiors.* New York: Van Nostrand Reinhold, 1984.

Hubbard, William B., and Stevenson, D. J. "Interior Structure of Saturn." In Tom Gehrels and Mildred Shapley Matthews, eds., *Saturn.* Tucson: University of Arizona Press, 1984:47–87.

Ingersoll, Andrew P. "Atmospheres of the Giant Planets." In J. Kelly Beatty and Andrew Chaikin, eds., *The New Solar System,* ed 3. Cambridge, Mass.: Sky Publishing Corp., 1990:139–152.

Ingersoll, Andrew P., et al. "Interaction of Eddies and Mean Zonal Flow on Jupiter as Inferred from *Voyager 1* and *Voyager 2* Images." *Journal of Geophysical Research* 86:A10 (30 September 1981):8733–8743.

Ingersoll, Andrew P., et al. "Structure and Dynamics of Saturn's Atmosphere." In Tom Gehrels and Mildred Shapley Matthews, eds., *Saturn.* Tucson: University of Arizona Press, 1984:195–238.

Marley, Mark S., and Fortney, Jonathan J. "Interiors of the Giant Planets." In Lucy-Ann McFadden, Paul R. Weissman, and Torrence V. Johnson, eds., *Encyclopedia of the Solar System,* ed 2. San Diego, Cal.: Academic Press, 2007:403–418.

Morrison, David. *Voyages to Saturn.* Washington, D.C.: Scientific and Technical Information Branch, NASA, 1982.

Owen, Tobias, and Terrile, Richard J. "Colors on Jupiter." *Journal of Geophysical Research* 86:A10 (30 September 1981): 8797–8814.

Porco, Carolyn C., and Hamilton, Douglas P. "Planetary Rings." In Lucy-Ann McFadden, Paul R. Weissman, and Torrence V. Johnson, eds., *Encyclopedia of the Solar System,* ed 2. San Diego, Cal.: Academic Press, 2007:503–518.

Prinn, Ronald G., et al. "Composition and Chemistry of Saturn's Atmosphere." In Tom Gehrels and Mildred Shapley Matthews, eds., *Saturn.* Tucson: University of Arizona Press, 1984:88–149.

Shan, Lin-Hua, and Goertz, C. K. "On the Radial Structure of Saturn's B Ring." *The Astrophysical Journal* 367:1 (20 January 1991):350–360.

Smith, Bradford A., et al. "A New Look at the Saturn System: The *Voyager 2* Images." *Science* 215:4532 (29 January 1982):504–537.

Smoluchowski, R. "Origin and Structure of Jupiter and Its Satellites." In Tom Gehrels and Mildred Shapley Matthews, eds., *Jupiter: Studies of the Interior, Atmosphere, Magnetosphere and Satellites.* Tucson: The University of Arizona Press, 1976:3–21.

Stevenson, D. J., and Salpeter, E. E. "Interior Models of Jupiter." In Tom Gehrels and Mildred Shapley Matthews, eds., *Jupiter: Studies of the Interior, Atmosphere, Magnetosphere and Satellites.* Tucson: The University of Arizona Press, 1976:85–112.

———. "The Phase Diagram and Transport Properties for Hydrogen-Helium Fluid Planets." *The Astrophysical Journal Supplement Series* 35:2 (October 1977):221–237.

Terrile, Richard J., et al. "Infrared Images of Jupiter at 5-Micrometer Wavelength During the *Voyager 1* Encounter." *Science* 204: 4396 (1 June 1979):1007–1008.

Van Allen, James A. "Magnetospheres, Cosmic Rays, and the Interplanetary Medium." In J. Kelly Beatty and Andrew Chaikin, eds., *The New Solar System,* ed 3. Cambridge, Mass.: Sky Publishing Corp., 1990:29–40.

West, Robert A. "Atmospheres of the Giant Planets." In Lucy-Ann McFadden, Paul R. Weissman, and Torrence V. Johnson, eds., *Encyclopedia of the Solar System,* ed 2. San Diego, Cal.: Academic Press, 2007:383–402.

CHAPTER 13

Broadfoot, A. L., et al. "Ultraviolet Spectrometer Observations of Neptune and Triton." *Science* 246:4936 (15 December 1989):1459–1466.

Burns, Joseph A. "Planetary Rings." In J. Kelly Beatty and Andrew Chaikin, eds., *The New Solar System,* ed 3. Cambridge, Mass.: Sky Publishing Corp., 1990:153–170.

Conrath, B., et al. "Infrared Observations of the Neptunian System." *Science* 246:4936 (15 December 1989):1454–1459.

Cruikshank, Dale P., and Morrison, David. "Icy Bodies of the Outer Solar System." In J. Kelly Beatty and Andrew Chaikin, eds., *The New Solar System*, ed 3. Cambridge, Mass.: Sky Publishing Corp., 1990:195–206.

Hubbard, William B. "Interiors of the Giant Planets." In J. Kelly Beatty and Andrew Chaikin, eds., *The New Solar System*, ed 3. Cambridge, Mass.: Sky Publishing Corp., 1990:131–138.

———. *Planetary Interiors*. New York: Van Nostrand Reinhold, 1984.

Ingersoll, Andrew P. "Atmospheres of the Giant Planets." In J. Kelly Beatty and Andrew Chaikin, eds., *The New Solar System*, ed 3. Cambridge, Mass.: Sky Publishing Corp., 1990:139–152.

Miner, Ellis D. *Uranus: The Planet, Rings and Satellites*. New York: Ellis Horwood, 1990.

Morbidelli, Alessandro, and Levinson, Harold F. "Kuiper Belt: Dynamics." In Lucy-Ann McFadden, Paul R. Weissman, and Torrence V. Johnson, eds., *Encyclopedia of the Solar System,* ed 2. San Diego, Cal.: Academic Press, 2007:589–604.

Ness, Norman F., et al. "Magnetic Fields at Neptune." *Science* 246:4936 (15 December 1989):1473–1478.

Pannekoek, Anton. *A History of Astronomy*. New York: Interscience Publishers, 1961.

Smith, Bradford A., et al. "*Voyager 2* at Neptune: Imaging Science Results." *Science* 246:4936 (15 December 1989):1422–1449.

Stern, Alan S. "Pluto." In Lucy-Ann McFadden, Paul R. Weissman, and Torrence V. Johnson, eds., *Encyclopedia of the Solar System,* ed 2. San Diego, Cal.: Academic Press, 2007:541–556.

Tegler, Stephen C. "Kuiper Belt Objects: Physical Studies." In Lucy-Ann McFadden, Paul R. Weissman, and Torrence V. Johnson, eds., *Encyclopedia of the Solar System,* ed 2. San Diego, Cal.: Academic Press, 2007:605–620.

CHAPTER 14

Broadfoot, A. L., et al. "Ultraviolet Spectrometer Observations of Neptune and Triton." *Science* 246:4936 (15 December 1989):1459–1466.

Burath, B. J., et al. "*Cassini* Visual and Infrared Mapping Spectrometer Observations of Iapetus: Detection of CO." *The Astrophysical Journal* 622 (1 April 2005):L149–L152.

Buratti, Bonnie J., and Thomas, Peter C. "Planetary Satellites." In Lucy-Ann McFadden, Paul R. Weissman, and Torrence V. Johnson, eds., *Encyclopedia of the Solar System,* ed 2. San Diego, Cal.: Academic Press, 2007:365–382.

Cattermole, Peter John. *Planetary Volcanism: A Study of Volcanic Activity in the Solar System*. New York: Halstead, 1989.

Conrath, B., et al. "Infrared Observations of the Neptunian System." *Science* 246:4936 (15 December 1989):1454–1459.

Collins, Geoffrey, and Johnson, Torrence V. "Ganymede and Callisto." In Lucy-Ann McFadden, Paul R. Weissman, and Torrence V. Johnson, eds., *Encyclopedia of the Solar System,* ed 2. San Diego, Cal.: Academic Press, 2007:449–466.

Coustenis, Athena. "Titan." In Lucy-Ann McFadden, Paul R. Weissman, and Torrence V. Johnson, eds., *Encyclopedia of the Solar System,* ed 2. San Diego, Cal.: Academic Press, 2007:467–482.

Cruikshank, Dale P., and Brown, Robert Hamilton. "Satellites of Uranus and Neptune, and the Pluto-Charon System." In Joseph A. Burns and Mildred Shapley Matthews, eds., *Satellites*. Tucson: The University of Arizona Press, 1986:836–873.

Cruikshank, Dale P., and Morrison, David. "Icy Bodies of the Outer Solar System." In J. Kelly Beatty and Andrew Chaikin, eds., *The New Solar System*, ed 3. Cambridge, Mass.: Sky Publishing Corp., 1990:195–206.

Griffith, Caitlin A., McKay, Christopher P., and Ferri, Francesca. "Titan's Tropical Storms in an Evolving Atmosphere." *The Astrophysical Journal* 687 (1 November 2008):L41–L44.

Hubbard, William B. *Planetary Interiors*. New York: Van Nostrand Reinhold, 1984.

Johnson, Torrence V. "The Galilean Satellites." In J. Kelly Beatty and Andrew Chaikin, eds., *The New Solar System*, ed 3. Cambridge, Mass.: Sky Publishing Corp., 1990:171–188.

Lopes, Rosaly M. C. "Io: The Volcanic Moon." In Lucy-Ann McFadden, Paul R. Weissman, and Torrence V. Johnson, eds., *Encyclopedia of the Solar System,* ed 2. San Diego, Cal.: Academic Press, 419–430.

Malin, Michael C., and Pieri, David C. "Europa." In Joseph A. Burns and Mildred Shapley Matthews, eds., *Satellites*. Tucson: The University of Arizona Press, 1986:689–717.

McKinnon, William B., and Kirk, Randolph L. "Triton." In Lucy-Ann McFadden, Paul R. Weissman, and Torrence V. Johnson, eds., *Encyclopedia of the Solar System,* ed 2. San Diego, Cal.: Academic Press, 2007:483–502.

McKinnon, William B., and Parmentier, E. M. "Ganymede and Callisto." In Joseph A. Burns and Mildred Shapley Matthews, eds., *Satellites*. Tucson: The University of Arizona Press, 1986:718–763.

Miner, Ellis D. *Uranus: The Planet, Rings and Satellites*. New York: Ellis Horwood, 1990.

Morrison, David, et al. "The Satellites of Saturn." In Joseph A. Burns and Mildred Shapley Matthews, eds., *Satellites*. Tucson: The University of Arizona Press, 1986:764–801.

Nash, Douglas B., et al. "Io." In Joseph A. Burns and Mildred Shapley Matthews, eds., *Satellites*. Tucson: The University of Arizona Press, 1986:629–688.

Owen, Tobias. "Titan." In J. Kelly Beatty and Andrew Chaikin, eds., *The New Solar System*, ed 3. Cambridge, Mass.: Sky Publishing Corp., 1990:189–194.

Peale, S. J. "Orbital Resonances, Unusual Configurations and Exotic Rotation States Among Planetary Satellites." In Joseph A. Burns and Mildred Shapley Matthews, eds., *Satellites*. Tucson: The University of Arizona Press, 1986: 159–223.

Prockter, Louise M., and Pappalardo, Robert T. "Europa." In Lucy-Ann McFadden, Paul R. Weissman, and Torrence V. Johnson, eds., *Encyclopedia of the Solar System,* ed 2. San Diego, Cal.: Academic Press, 431–448.

Schubert, Gerald, et al. "Thermal Histories, Compositions and Internal Structures of the Moons of the Solar System." In Joseph A. Burns and Mildred Shapley Matthews, eds., *Satellites*. Tucson: The University of Arizona Press, 1986: 224–292.

Smith, Bradford A., et al. "The Galilean Satellites and Jupiter: *Voyager 2* Imaging Science Results." *Science* 206:4421 (23 November 1979):927–950.

———. "The Jupiter System Through the Eyes of *Voyager 1*." *Science* 204:4396 (1 June 1979):951–972.

———. "A New Look at the Saturn System: The *Voyager 2* Images." *Science* 215:4532 (29 January 1982):504–537.

———. Smith, Bradford A., et al. "*Voyager 2* at Neptune: Imaging Science Results." *Science* 246:4936 (15 December 1989):1422–1449.

Soderblom, L. A., et al. "Triton's Geyser-Like Plumes: Discovery and Basic Characterization." *Science* 250:4979 (19 October 1990):410–415.

Squyres, Steven W., and Croft, Steven K. "The Tectonics of Icy Satellites." In Joseph A. Burns and Mildred Shapley Matthews, eds., *Satellites*. Tucson: The University of Arizona Press, 1986: 293–341.

Stevenson, D. J., et al. "Origins of Satellites." In Joseph A. Burns and Mildred Shapley Matthews, eds., *Satellites*. Tucson: The University of Arizona Press, 1986:39–88.

Thomas, P., et al. "Small Satellites." In Joseph A. Burns and Mildred Shapley Matthews, eds., *Satellites*. Tucson: The University of Arizona Press, 1986:802–835.

Tyler, G. L., et al. "*Voyager* Radio Science Observations of Neptune and Triton." *Science* 246:4936 (15 December 1989):1466–1473.

CHAPTER 15

Bailey, M. E., Clube, S. V. M., and Napier, W. M. *The Origin of Comets*. Oxford, N.Y.: Pergamon Press, 1990.

Brandt, John C. "Comets." In J. Kelly Beatty and Andrew Chaikin, eds., *The New Solar System*, ed 3. Cambridge, Mass.: Sky Publishing Corp., 1990:217–230.

Brandt, John C., and Hodge, Paul W. *Solar System Astrophysics*. New York: McGraw-Hill Book Company, Inc., 1964.

Brown, M. E., Schaller, E. L., Roe, H. G., Rabinowitz, D. L., and Trujillo, C. A. "Direct Measurement of the Size of 2003 UB 313 From the *Hubble Space Telescope*." *The Astrophysical Journal* 643 (20 May 2006):L61–L63.

Brown, M. E., Trujillo, C. A., and Rabinowitz, D. L. "Discovery of a Planetary-Sized Object in the Scattered Kuiper Belt." *The Astrophysical Journal* 635 (10 December 2005):L97–L100.

Burke, John G. *Cosmic Debris: Meteorites in History*. Berkeley: University of California Press, 1986.

Chapman, Clark R. "Asteroids." In J. Kelly Beatty and Andrew Chaikin, eds., *The New Solar System,* ed 3. Cambridge, Mass.: Sky Publishing Corp., 1990:231–240.

Connelly, H. C. "From Stars to Dust: Looking into a Circumstellar Disk Through Chondritic Meteorites." *Science* 307 (7 January 2005):75–76.

Degewig, Johan, and Tedesco, Edward F. "Do Comets Evolve into Asteroids? Evidence from Studies." In Laurel L. Wilkening and Mildred Shapley Matthews, eds., *Comets*. Tucson: The University of Arizona Press, 1986:665–695.

Donn, Bertram, and Rahe, Jurgen. "Structure and Origin of Cometary Nuclei." In Laurel L. Wilkening and Mildred Shapley Matthews, eds., *Comets*. Tucson: The University of Arizona Press, 1986:56–82.

Everhart, Edgar. "Evolution of Long- and Short-Period Orbits." In Laurel L. Wilkening and Mildred Shapley Matthews, eds., *Comets*. Tucson: The University of Arizona Press, 1986: 659–664.

Fraundorf, P., et al. "Laboratory Studies of Interplanetary Dust." In Laurel L. Wilkening and Mildred Shapley Matthews, eds., *Comets*. Tucson: The University of Arizona Press, 1986:383–409.

Gladman, Brett, Kavelaars, J. J., Petit, Jean-Marc, Morbidelli, Alessandro, Holman, Mathew J., and Loredo, T. "The Structure of the Kuiper Belt: Size Distribution and Radial Extent." *The Astronomical Journal* 122 (August 2001):1051–1066.

Hartman, William K. "Small Bodies and Their Origins." In J. Kelly Beatty and Andrew Chaikin, eds., *The New Solar System*, ed 3. Cambridge, Mass.: Sky Publishing Corp., 1990:251–258.

Heide, Fritz. *Meteorites*. Chicago: University of Chicago Press, 1964.

Keller, H. Uwe. "The Nucleus." In Walter F. Huebner, ed., *Physics and Chemistry of Comets*. Berlin, Heidelberg: Springer-Verlag, 1990:13–68.

King, Elbert A. *Space Geology: An Introduction*. New York: John Wiley and Sons, 1976.

Kohoutek, L. "Atmospheric Heights of Telescopic Meteors." In Lubor Kresak and Peter M. Millman, eds., *Physics and Dynamics of Meteors*. Dordrecht-Holland: D. Reidel Publishing Company; New York: Springer-Verlag, 1968:143–146.

Kowal, Charles T. *Asteroids: Their Nature and Utilization*. Chichester, England: Ellis Horwood; New York: Halsted Press, 1988.

Kresak, Lubor. "Comet Discoveries, Statistics, and Observational Selection." In Laurel L. Wilkening and Mildred Shapley Matthews, eds., *Comets*. Tucson: The University of Arizona Press, 1986:56–82.

Luu, Jane X., and Jewitt, David C. "Kuiper Belt Objects: Relics from the Accretion Disk of the Sun." In Geoffrey Burbidge, Allan Sandage, and Frank H. Shu, eds., *Annual Review of Astronomy and Astrophysics,* vol. 40. Palo Alto, Cal.: Annual Reviews, Inc., 2002:63–101.

Oort, Jan H. *Orbital Distribution of Comets*. In Walter F. Huebner, ed., *Physics and Chemistry of Comets,* 1990.

Pannekoek, Anton. *A History of Astronomy*. New York: Interscience Publishers, 1961.

Reinhard, Rudeger. "The Halley Encounters." In J. Kelly Beatty and Andrew Chaikin, eds., *The New Solar System,* ed 3. Cambridge, Mass.: Sky Publishing Corp., 1990:207–216.

Rickman, Hans, and Huebner, Walter F. "Comet Formation and Evolution." In Walter F. Huebner, ed., *Physics and Chemistry of Comets,* 1990.

Schorghofer, Norbert. "The Lifetime of Ice on a Main Belt Asteroid." *The Astrophysical Journal* 682 (20 July 2008):697–705.

Shoemaker, Eugene M., and Shoemaker, Carolyn S. "The Collision of Solid Bodies." In J. Kelly Beatty and Andrew Chaikin, eds., *The New Solar System,* ed 3. Cambridge, Mass.: Sky Publishing Corp., 1990:259–274.

Weissman, Paul R. "Dynamical History of the Oort Cloud." In Laurel L. Wilkening and Mildred Shapley Matthews, eds., *Comets*. Tucson: The University of Arizona Press, 1986: 637–658.

Wood, John A. "Meteorites." In J. Kelly Beatty and Andrew Chaikin, eds., *The New Solar System,* ed 3. Cambridge, Mass.: Sky Publishing Corp., 1990:241–250.

The following references are from *Asteroids II,* edited by Richard P. Binzel, Tom Gehrels, and Mildred Shapley Matthews. Tucson: The University of Arizona Press, 1989:

Bell, Jeffrey F., Davis, Donald R., Hartmann, William K., and Gaffey, Michael J. "Asteroids: The Big Picture." Pages 921–945.

Binzel, Richard P. "An Overview of the Asteroids." Pages 3–18.

Davis, Donald R., Weidenschilling, Stuart J., Farinella, Paolo, Paolicchi, Paolo, and Binzel, Richard P. "Asteroid Collisional History: Effects on Sizes and Spins." Pages 805–826.

French, Linda M., Vilas, Faith, Hartmann, William K., and Tholen, David J. "Distant Asteroids and Chiron." Pages 468–486.

Gaffey, Michael J., Bell, Jeffrey F., and Cruikshank, Dale P. "Reflectance Spectroscopy and Asteroid Surface Mineralogy." Pages 98–127.

Gradie, Jonathon C., Chapman, Clark R., and Tedesco, Edward F. "Distribution of Taxonomic Classes and the Compositional Structure of the Asteroid Belt." Pages 316–335.

Greenberg, Richard, and Nolan, Michael C. "Delivery of Asteroids and Meteorites to the Inner Solar System." Pages 778–804.

Lipschutz, Michael E., Gaffey, Michael J., and Pellas, Paul. "Meteoritic Parent Bodies: Nature, Number, Size and Relation to Present-Day Asteroids." Pages 740–777.

Matson, Dennis L., Veeder, Glenn J., Tedesco, Edward F., and Lebofsky, Larry A. "The IRAS Asteroid and Comet Survey." Pages 269–281.

McFadden, Lucy-Ann, Tholen, David J., and Veeder, Glenn J. "Physical Properties of Aten, Apollo and Amor Asteroids." Pages 442–467.

Scott, E. R. D., Taylor, G. J., Newsom, H. E., Herbert, F., Zolensky, M., and Kerridge, J. F. "Chemical, Thermal and Impact Processing of Asteroids." Pages 701–739.

Shoemaker, Eugene M., Shoemaker, Carolyn S., and Wolfe, Ruth F. "Trojan Asteroids: Populations, Dynamical Structure and Origin of the L4 and L5 Swarms." Pages 487–523.

Tholen, David J., and Barucci, M. Antonietta. "Asteroid Taxonomy." Pages 298–315.

Trujillo, Chadwick A., and Brown, Michael E. "The Radial Distribution of the Kuiper Belt." *The Astrophysical Journal* 554 (10 June 2001):L95–L98.

Trujillo, Chadwick A., Jewitt, David C., and Luu, Jane X. "Population of the Scattered Kuiper Belt." *The Astrophysical Journal* 529 (1 February 2000):L103–L106.

Veeder, Glenn J., Tedesco, Edward F., and Matson, Dennis L. "Asteroid Results from the IRAS Survey." Pages 282–289.

Weissman, Paul R., A'Hearn, Michael F., McFadden, L. A., and Rickman, H. "Evolution of Comets into Asteroids." Pages 880–920.

Wetherill, George W. "Origin of the Asteroid Belt." Pages 661–680.

CHAPTER 16

Basri, Gibor. "Observations of Brown Dwarfs." In Geoffrey Burbidge, Alan Sandage, and Frank H. Shu, eds., *Annual Review of Astronomy and Astrophysics,* vol. 39. Palo Alto, Cal.: Annual Reviews, Inc., 2001:485–520.

Hoskin, Michael A. *Stellar Astronomy: Historical Studies.* Chalfont St. Giles, Bucks, England: Science History Publishing, 1987.

Jura, M. "Other Kuiper Belts." *The Astrophysical Journal* 603 (10 March 2004):729–737.

Kuhn, Thomas S. *The Copernican Revolution; Planetary Astronomy in the Development of Western Thought.* Cambridge, Mass.: Harvard University Press, 1957.

Matsumura, Soko, and Pudritz, Ralph E. "The Origin of Jovian Planets in Protostellar Disks: The Role of Dead Zones." *The Astrophysical Journal* 598 (20 November 2003): 645–656.

Menzel, Donald Howard, Whipple, Fred Lawrence, and de Valcouleurs, Gerard Henri. *Survey of the Universe.* Englewood Cliffs, N.J.: Prentice-Hall, 1970.

Mihalas, Dimitri, and Routly, Paul McRae. *Galactic Astronomy.* San Francisco, Cal.: W. H. Freeman and Co., 1968.

Neugebauer, Otto. *Astronomy and History: Selected Essays.* New York: Springer-Verlag, 1983.

O'Neil, William Matthew. *Early Astronomy: From Babylonia to Copernicus.* Portland, Ore.: International Specialized Book Services (distributor), 1986.

Pannekoek, Anton. *A History of Astronomy.* New York: Interscience Publishers, 1961.

Russell, Henry Norris, Dugan, Raymond Smith, and Steward, John Quincy. *Astronomy: A Revision of Young's Manual of Astronomy.* Boston, Mass.: Ginn, 1926–1927.

van Leeuwen, F. "Validation of the New Hipparcos Reduction." *Astronomy and Astrophysics* 474:653–664.

CHAPTER 17

Aschwanden, Markus J., Poland, Arthur I., and Rabin, Douglas M. "The New Solar Corona." In Geoffrey Burbidge, Alan Sandage, and Frank H. Shu, eds., *Annual Review of Astronomy and Astrophysics,* vol. 39. Palo Alto, Cal.: Annual Reviews, Inc., 2001:175–210.

Brandt, John C., and Hodge, Paul W. *Solar System Astrophysics.* New York: McGraw-Hill, 1964.

Foukal, Peter. *Solar Astrophysics.* New York: John Wiley and Sons, 1990.

Noyes, Robert W. "The Sun." In J. Kelly Beatty and Andrew Chaikin, eds., *The New Solar System,* ed 3. Cambridge, Mass.: Sky Publishing Corp., 1990:15–28.

Pannekoek, Anton. *A History of Astronomy.* New York: Interscience Publishers, 1961.

Van Allen, James A. "Magnetospheres, Cosmic Rays, and the Interplanetary Medium." In J. Kelly Beatty and Andrew Chaikin, eds.,*The New Solar System,* ed 3. Cambridge, Mass.: Sky Publishing Corp., 1990:29–40.

Zirin, Harold. *Astrophysics of the Sun.* New York: Cambridge University Press, 1988.

CHAPTER 18

Bergin, Edwin A., and Tafalla, Mario. "Cold Dark Clouds: The Initial Conditions for Star Formation." In Roger Blandford, John Kormendy, and Ewine van Dishoeck, eds., *Annual Review of Astronomy and Astrophysics,* vol. 45. Palo Alto, Cal.: Annual Reviews, Inc., 2007:339–396.

Bertout, Claude. "T Tauri Stars: Wild as Dust." In Geoffrey Burbidge, David Layzer, and John G. Phillips, eds., *Annual Review of Astronomy and Astrophysics,* vol. 27. Palo Alto, Cal.: Annual Reviews, Inc., 1989:351–395.

Blum, Jurgen, and Wurm, Gerhard. "The Growth Mechanisms of Macroscopic Bodies in Protoplanetary Disks." In Roger Blandford, John Kormendy, and Ewine van Dishoeck, eds., *Annual Review of Astronomy and Astrophysics,* vol. 46. Palo Alto, Cal.: Annual Reviews, Inc., 2008:21–56.

Brandt, John C., and Hodge, Paul W. *Solar System Astrophysics.* New York: McGraw-Hill, 1964.

Cameron, A. G. W. "Formation and Evolution of the Primitive Solar Nebula." In David C. Black and Mildred Shapley Matthews, eds., *Protostars and Planets II.* Tucson: The University of Arizona Press, 1985:1073–1099.

———. "Origin of the Solar System." In Geoffrey Burbidge, David Layzer, and John G. Phillips, *Annual Review of Astronomy and Astrophysics,* vol. 27. Palo Alto, Cal.: Annual Reviews, Inc., 1989:441–472.

Cassen, Patrick, Shu, Frank H., and Terebey, Susan. "Protostellar Disks and Star Formation." In David C. Black and Mildred Shapley Matthews, eds., *Protostars and Planets II*. Tucson: The University of Arizona Press, 1985:448–483.

Elmegreen, Bruce G. "Molecular Clouds and Star Formation: An Overview." In David C. Black and Mildred Shapley Matthews, eds., *Protostars and Planets II*. Tucson: The University of Arizona Press, 1985:33–58.

Glassgold, A. E. "Symposium Summary: Fragmentation and Star Formation in Molecular Clouds." In E. Falgarone, F. Boulanger, and G. Duvert, eds., *Fragmentation of Molecular Clouds and Star Formation; Proceedings of the 147th Symposium of the International Astronomical Union, Held in Grenoble, France, June 12–16, 1990.* Dordrect, Holland: Kluwer Academic Publishers, 1991:379–386.

Harvey, Paul M. "Observational Evidence for Disks Around Young Stars." In David C. Black and Mildred Shapley Matthews, eds., *Protostars and Planets II*. Tucson: The University of Arizona Press, 1985:484–492.

Lada, Charles J., and Lada, Elizabeth A. "Embedded Clusters in Molecular Clouds." In Geoffrey Burbidge, Roger Blandford, and Allan Sandage, eds., *Annual Review of Astronomy and Astrophysics,* vol. 41. Palo Alto, Cal.: Annual Reviews, Inc., 2003:57–115.

Larson, Richard B. "Some Processes Influencing the Stellar Initial Mass Function." In E. Falgarone, F. Boulanger, and G. Duvert, eds., *Fragmentation of Molecular Clouds and Star Formation; Proceedings of the 147th Symposium of the International Astronomical Union, Held in Grenoble, France, June 12–16, 1990.* Dordrect, Holland: Kluwer Academic Publishers, 1991: 261–273.

Mundt, Reinhard. "Highly Collimated Mass Outflows from Young Stars." In David C. Black and Mildred Shapley Matthews, eds., *Protostars and Planets II.* Tucson: The University of Arizona Press, 1985:414–433.

Myers, Philip C. "Molecular Cloud Cores." In David C. Black and Mildred Shapley Matthews, eds., *Protostars and Planets II.* Tucson: The University of Arizona Press, 1985:81–103.

Palla, Francesco. "Theoretical and Observational Aspects of Young Stars of Intermediate Mass." In E. Falgarone, F. Boulanger, and G. Duvert, eds., *Fragmentation of Molecular Clouds and Star Formation; Proceedings of the 147th Symposium of the International Astronomical Union, Held in Grenoble, France, June 12–16, 1990.* Dordrect, Holland: Kluwer Academic Publishers, 1991:331–344.

Poignant, Roslyn. *Oceanic Mythology: The Myths of Polynesia, Micronesia, Melanesia, and Australia.* London: Hamlyn, 1967.

Rydgren, A. Eric, and Cohen, Martin. "Young Stellar Objects and Their Circumstellar Dust: An Overview." In David C. Black and Mildred Shapley Matthews, eds., *Protostars and Planets II.* Tucson: The University of Arizona Press, 1985:371–385.

Schwartz, Richard D. "The Nature and Origin of Herbig-Haro Objects." In David C. Black and Mildred Shapley Matthews, eds., *Protostars and Planets II.* Tucson: The University of Arizona Press, 1985:405–413.

Shu, Frank H., Adams, Fred C., and Lizano, Susana. "Star Formation in Molecular Clouds: Observation and Theory." In Geoffrey Burbidge, David Layzer, and John G. Phillips, eds., *Annual Review of Astronomy and Astrophysics,* vol. 25. Palo Alto, Cal.: Annual Reviews, Inc., 1987.

Thompson, Rodger. "High Mass Versus Low Mass Star Formation." In David C. Black and Mildred Shapley Matthews, eds., *Protostars and Planets II.* Tucson: The University of Arizona Press, 1985:434–447.

Zinner, E., and Gopen, C. "Aluminum-26 in H4 Chondrites: Implications for Its Production and Its Usefulness As a Fine-Scale Chronometer for Early Solar System Events." *Meteorit. Planet. Sci.* 37 (2002):1001–1013.

CHAPTER 19

Balick, Bruce, and Frank, Adam. "Shapes and Shaping of Planetary Nebulae." In Geoffrey Burbidge, Allan Sandage, and Frank H. Shu, eds., *Annual Review of Astronomy and Astrophysics,* vol. 40. Palo Alto, Cal.: Annual Reviews, Inc., 2002:439–486.

Baud, B., and Habing, H. J. "The Master Strength of OH/IR Stars, Evolution of Mass Loss and the Creation of a Superwind." *In Astronomy and Astrophysics* 127:78–83.

Clayton, Donald D. *Principles of Stellar Evolution and Nucleosynthesis; With a New Preface.* Chicago: University of Chicago Press, 1983.

Cox, John P., and Giuli, Thomas. *Principles of Stellar Structure.* New York: Gordon and Breach, 1968.

De Marco, Orsola, Bond, Howard E., Harmer, Diane, and Fleming, Andrew J. "Indications of a Large Fraction of Spectroscopic Binaries Among Nuclei of Planetary Nebulae." *The Astrophysical Journal* 602 (20 February 2004):L93–L96.

Motz, Lloyd. *Astrophysics and Stellar Structure.* Waltham, Mass.: Ginn, 1970.

Schwarzschild, Martin. *Structure and Evolution of the Stars.* Princeton: Princeton University Press, 1958.

Soker, N. "Observed Planetary Nebulae As Descendants of Interacting Binary Systems." *The Astrophysical Journal* 645 (1 July 2006):L57–L60.

CHAPTER 20

Baud, B., and Habing, H. J. "The Master Strength of OH/IR Stars, Evolution of Mass Loss and the Creation of a Superwind." *Astronomy and Astrophysics* 127:73–83.

Bethe, H. A. "Supernova Mechanisms." *Reviews of Modern Physics* 62:4 (October 1990):801–866.

Clayton, Donald D. *Principles of Stellar Evolution and Nucleosynthesis; With a New Preface.* Chicago: University of Chicago Press, 1983.

Cox, John P., and Giuli, R. Thomas. *Principles of Stellar Structure.* New York: Gordon and Breach, 1968.

D'Antona, Francesca, and Mazzitelli, Italo. "Cooling of White Dwarfs." In Geoffrey Burbidge, David Layzer, and Allan Sandage, *Annual Review of Astronomy and Astrophysics,* vol. 28. Palo Alto, Cal.: Annual Reviews, Inc., 1990:139–181.

Davies, P. C. W. *Space and Time in the Modern Universe.* New York: Cambridge University Press, 1977.

Eddington, Sir Arthur Stanley. *The Internal Constitution of the Stars.* Cambridge, England: Cambridge University Press, 1926.

Hurley, K., et al. "An Exceptionally Bright Flare from SGR 1806-20 and the Origins of Short-Duration Gamma-ray Bursts." *Nature* 434 (28 April 2005):1098–1103.

Iben, Icko, Jr., and Renzini, Alvio. "Asymptotic Giant Branch Evolution and Beyond." In Geoffrey Burbidge, David Layzer, and John G. Phillips, eds., *Annual Review of Astronomy and Astrophysics,* vol. 21. Palo Alto, Cal.: Annual Reviews, Inc., 1983:271–342.

Lang, Kenneth R., and Gingerich, Owen, eds., *A Source Book in Astronomy and Astrophysics,* 1900–1975. Cambridge, Mass.: Harvard University Press, 1979.

Liebert, James. "White Dwarf Stars." In Geoffrey Burbidge, David Layzer, and John G. Phillips, eds., *Annual Review of Astronomy and Astrophysics,* vol. 18. Palo Alto, Cal.: Annual Reviews, Inc., 1980:363–398.

Menzel, Donald Howard, Whipple, Fred Lawrence, and de Vaulcouleurs, Gerard Henri. *Survey of the Universe.* Englewood Cliffs, N.J.: Prentice-Hall, 1970.

Meszaros, P. "Theory of Gamma-Ray Bursts." In Geoffrey Burbidge, Allan Sandage, and Frank H. Shu, eds., *Annual Review of Astronomy and Astrophysics,* vol. 40. Palo Alto, Cal.: Annual Reviews, Inc., 2002:137–169.

Motz, Lloyd. *Astrophysics and Stellar Structure.* Waltham, Mass.: Ginn, 1970.

Payne-Gaposchkin, Cecilia, and Haramundanis, Katherine. *Introduction to Astronomy,* ed 2. Englewood Cliffs, N.J.: Prentice-Hall, Inc., 1970.

Schwarzschild, Martin. *Structure and Evolution of the Stars.* Princeton: Princeton University Press, 1958.

Shapiro, Stuart L., and Teukolsky, Saul A. *Black Holes, White Dwarfs, and Neutron Stars: The Physics of Compact Objects.* New York: John Wiley and Sons, 1983.

Taylor, J. H., and Stinebring, D. R. "Recent Progress in the Understanding of Pulsars." In Geoffrey Burbidge, David Layzer, and John G. Phillips, *Annual Review of Astronomy and Astrophysics,* vol. 24. Palo Alto, Cal.: Annual Reviews, Inc., 1986:285–327.

Weidemann, Volker. "Masses and Evolutionary Status of White Dwarfs and Their Progenitors." In Geoffrey Burbidge, David Layzer, and John G. Phillips, eds., *Annual Review of Astronomy and Astrophysics,* vol. 28. Palo Alto, Cal.: Annual Reviews, Inc., 1990:103–137.

Weiler, Kurt W., and Sramek, Richard A. "Supernovae and Supernova Remnants." In Geoffrey Burbidge, David Layzer, and John G. Phillips, eds., *Annual Review of Astronomy and Astrophysics,* vol. 26. Palo Alto, Cal.: Annual Reviews, Inc., 1988:295–341.

Woosley, S. E., and Weaver, Thomas A. "The Physics of Supernova Explosions." In Geoffrey Burbidge, David Layzer, and John G. Phillips, eds., *Annual Review of Astronomy and Astrophysics,* vol. 24. Palo Alto, Cal.: Annual Reviews, Inc., 1986:205–253.

The following references are from *The Astronomy and Astrophysics Encyclopedia,* edited by Stephen P. Maran. New York: Van Nostrand Reinhold; Cambridge, England: Cambridge University Press, 1992:

Bekenstein, Jacob D. "Gravitational Theories." Pages 296–299.

Chevalier, Roger A. "Supernova Remnants, Evolution and Interaction with the Interstellar Medium." Pages 900–902.

Clark, David H. "Supernovae, Historical." Pages 886–887.

Dickel, John R. "Supernova Remnants, Observed Properties." Pages 902–904.

Fesen, Robert A. "Supernovae, General Properties." Pages 883–886.

Krauss, Lawrence M. "Neutrinos, Supernova." Pages 489–491.

Manchester, Richard N. "Supernova Remnants and Pulsars, Galactic Distribution." Pages 895–898.

Price, Richard H. "Black Holes, Theory." Pages 87–88.

Rankin, Joanna M. "Pulsars, Observed Properties." Pages 567–570.

Stinebring, Daniel R. "Pulsars, Binary." Pages 564–566.

Watson, M. G. "Black Holes, Stellar, Observational Evidence." Pages 85–87.

Woosley, Stanford E. "Supernovae, Type II, Theory and Interpretation." Pages 892–895.

CHAPTER 21

Aitken, Robert Grant. *The Binary Stars.* New York: D. C. McMurtrie, 1918.

Bowers, Richard L., and Deeming, Terry. *Astrophysics,* vol. 1. Boston: Jones and Bartlett, 1984.

Heintz, Wulff D. *Double Stars.* Boston: D. Reidel Publishing Co., 1978.

Lada, C. J. "Stellar Multiplicity and the Initial Mass Function: Most Stars are Single." *The Astrophysical Journal* 640 (20 March 2006):L63–L66.

Livio, Mario, and Truran, James W. "Type I Supernovae and Accretion-Induced Collapses from Cataclysmic Variables?" *The Astrophysical Journal* 389:2 (20 April 1992):695–703.

Pannekoek, Anton. *A History of Astronomy.* New York: Interscience Publishers, 1961.

Payne-Gaposchkin, Cecilia. *The Galactic Novae.* New York: Interscience Publishers, 1957.

Payne-Gaposchkin, Cecilia, and Gaposchkin, Sergei. *Variable Stars.* Cambridge, Mass.: Published by The Observatory, 1938.

Payne-Gaposchkin, Cecilia, and Haramundanis, Katherine. *Introduction to Astronomy,* vol. 2. Englewood Cliffs, N.J.: Prentice-Hall, 1970.

Ritter, Hans, et al. "The White Dwarf Mass Distribution in Classical Nova Systems." *The Astrophysical Journal* 20:376 (20 July 1990):177–185.

Shapiro, Stuart L., and Teukolsky, Saul A. *Black Holes, White Dwarfs, and Neutron Stars: The Physics of Compact Objects.* New York: John Wiley and Sons, 1983.

Tohline, Joel E. "The Origin of Binary Stars." In Geoffrey Burbidge, Allan Sandage, and Frank H. Shu, eds., *Annual Review of Astronomy and Astrophysics,* vol. 40. Palo Alto, Cal.: Annual Reviews, Inc., 2002:349–385.

The following references are from *The Astronomy and Astrophysics Encyclopedia,* edited by Stephen P. Maran. New York: Van Nostrand Reinhold; Cambridge, England: Cambridge University Press, 1992:

Andersen, Johannes. "Binary Stars, Eclipsing, Determination of Stellar Parameters." Pages 68–71.

Backer, Donald C. "Pulsars, Millisecond." Pages 566–567.

Batten, Alan H. "Binary and Multiple Stars, Types and Statistics." Pages 54–55.

De Greve, Jean-Pierre. "Binary Stars, Semidetached." Pages 77–79.

Durisen, Richard H. "Binary and Multiple Stars, Origin." Pages 52–54.

Fesen, Robert A. "Supernovae, General Properties." Pages 883–886.

Hertz, Paul L. "Star Clusters, Globular, X-Ray Sources." Pages 687–689.

Lewin, Walter H. G. "X-Ray Bursters." Pages 959–962.

Lippincott, Sarah Lee. "Binary Stars, Astrometric and Visual." Pages 55–58.

Nomoto, Ken'ichi. "Supernovae, Type I, Theory and Interpretation." Pages 890–892.

Pryor, Carlton. "Star Clusters, Globular, Binary Stars." Pages 667–668.

Ritter, Hans. "Binary Stars, Cataclysmic." Pages 61–63.
Savonije, GertJan. "Binary Stars, X-Ray, Formation and Evolution." Pages 83–85.
Spruit, Henk C. "Accretion." Pages 1–2.
Stinebring, Daniel R. "Pulsars, Binary." Pages 564–566.
Taam, Ronald E. "Stellar Evolution, Binary Systems." Pages 832–835.
Underhill, Anne B. "Binary Stars, Spectroscopic." Pages 79–80.
Warwick, Robert S. "X-Ray Sources, Galactic Distribution." Pages 962–964.
Watson, M. G. "Black Holes, Stellar, Observational Evidence." Pages 85–87.
Webbink, Ronald F. "Binary Stars, Theory of Mass Loss and Transfer in Close Systems." Pages 81–83.

CHAPTER 22

Allen, D. A. "The Stellar Cluster." In Donald C. Backer, ed., *American Institute of Physics AIP Conference Proceedings 155: The Galactic Center; Proceedings of the Symposium Honoring C. H. Townes, Berkeley, CA, 1986.* New York: American Institute of Physics, 1987:1–7.

Backer, D. C., and Sramek, R. A. "Proper Motion of the Compact, Nonthermal Radio Source in the Galactic Center." In Donald C. Backer, ed., *American Institute of Physics AIP Conference Proceedings 155: The Galactic Center; Proceedings of the Symposium Honoring C. H. Townes, Berkeley, CA, 1986.* New York: American Institute of Physics, 1987:163–165.

Benjamin, R. A., et al. "First GLIMPSE Results on the Stellar Structure of the Galaxy." *The Astrophysical Journal* 630 (10 September 2005):L49–L53.

Bok, Bart J., and Bok, Priscilla F. *The Milky Way,* ed 5. Cambridge, Mass.: Harvard University Press, 1981.

Dyson, J. E., and Williams, D. A. *Physics of the Interstellar Medium.* Manchester, England: Manchester University Press, 1980.

Eisenhauer, F., Schodel, R., Genzel, R., Ott, T., Tecza, M. Abuter, R., Eckart, A., and Alexander, T. "A Geometric Determination of the Distance to the Galactic Center." *The Astrophysical Journal 597* (10 November 2003):L121–L124.

Fich, Michel, and Tremaine, Scott. "The Mass of the Galaxy." In Geoffrey Burbidge, David Layzer, and Allan Sandage, eds., *Annual Review of Astronomy and Astrophysics,* vol. 209. Palo Alto, Cal.: Annual Reviews, Inc., 1991:409–445.

Freeman, Ken, and Bland-Hawthorn, Joss. "The New Galaxy: Signatures of Its Formation." In Geoffrey Burbidge, Allan Sandage, and Frank H. Shu, eds., *Annual Review of Astronomy and Astrophysics,* vol. 40. Palo Alto, Cal.: Annual Reviews, Inc., 2002:487–537.

Geballe, T. R. "The Ionized Gas in the Galactic Center." In Donald C. Backer, ed., *American Institute of Physics AIP Conference Proceedings 155: The Galactic Center; Proceedings of the Symposium Honoring C. H. Townes, Berkeley, CA, 1986.* New York: American Institute of Physics, 1987:30–38.

Gehz, A. M., et al. "Stellar Orbits Around the Galactic Center Black Hole." *The Astrophysical Journal* 620 (20 February 2005):744–757.

Genzel, R., and Townes, C. H. "Physical Conditions, Dynamics, and Mass Distribution in the Center of the Galaxy." In Geoffrey Burbidge, David Layzer, and John G. Phillips, eds., *Annual Review of Astronomy and Astrophysics,* vol. 25. Palo Alto, Cal.: Annual Reviews, Inc., 1987:377–423.

Gilmore, G., King, I., and Van der Kruit, P. *The Milky Way as a Galaxy: Nineteenth Advanced Course of the Swiss Society of Astrophysics and Astronomy.* Roland Buser and Ivan King, eds., Sauverny-Versoix, Switzerland: Geneva Observatory, 1989.

Gusten, R. "Atomic and Molecular Gas in the Circumnuclear Disk." In Donald C. Backer, ed., *American Institute of Physics AIP Conference Proceedings 155: The Galactic Center; Proceedings of the Symposium Honoring C. H. Townes, Berkeley, CA, 1986.* New York: American Institute of Physics, 1987:19–29.

Gusten, R., et al. "Hat Creek Aperture Synthesis Observations of the Circum-Nuclear Ring in the Galactic Center." In Donald C. Backer, ed., *American Institute of Physics AIP Conference Proceedings 155: The Galactic Center; Proceedings of the Symposium Honoring C. H. Townes, Berkeley, CA, 1986.* New York: American Institute of Physics, 1987:103–105.

Honma, M., and Sofue, Y. "Rotation Curve of the Galaxy." *Publ. Astron. Soc. Japan* 49 (1997):453–460.

Hoskin, Michael. "The Milky Way from Antiquity to Modern Times." In Hugo van Woerden, Ronald J. Allen, and W. Butler Burton, eds., *The Milky Way Galaxy: Proceedings of the 106th Symposium of the International Astronomical Union Held in Groningen, The Netherlands, 30 May–3 June, 1983.* Dordrecht, Holland: D. Reidel Publishing Co., 1985:11–24.

Hoskin, Michael A. *William Herschel and the Construction of the Heavens.* New York: Norton, 1963.

Kapteyn, Jacobus Cornelius. "First Attempt at a Theory of the Arrangement and Motion of the Sidereal System." In Kenneth R. Lang and Owen Gingerich, eds., *A Source Book in Astronomy and Astrophysics 1900–1975.* Cambridge, Mass.: Harvard University Press, 1979:542–549.

Lebofsky, M. J., and Rieke, G. H. "The Stellar Population at the Galactic Center." In Donald C. Backer, ed., *American Institute of Physics AIP Conference Proceedings 155: The Galactic Center; Proceedings of the Symposium Honoring C. H. Townes, Berkeley, CA, 1986.* New York: American Institute of Physics, 1987:79–82.

Lo, K. Y. "The Galactic Center Compact Nonthermal Radio Source." In Donald C. Backer, ed., *American Institute of Physics AIP Conference Proceedings 155: The Galactic Center; Proceedings of the Symposium Honoring C. H. Townes, Berkeley, CA, 1986.* New York: American Institute of Physics, 1987:30–38.

Maiz-Apellaniz, Jesus. "The Origin of the Local Bubble." *The Astrophysical Journal* 560 (20 October 2001):L83–L86.

McGinn, M. T., et al. "Stellar Kinematics in the Central 10 PC of the Galaxy." In Donald C. Backer, ed., *American Institute of Physics AIP Conference Proceedings 155: The Galactic Center; Proceedings of the Symposium Honoring C. H. Townes, Berkeley, CA, 1986.* New York: American Institute of Physics, 1987:87–90.

Mihalas, Dimitri, and Routly, Paul McRae. *Galactic Astronomy.* San Francisco, Cal.: W. H. Freeman and Co., 1968.

Norman, Colin A., and Ikeuchi, Satoru. "The Disk-Halo Interaction: Superbubblers and the Structure of the Interstellar Medium. *The Astrophysical Journal* 345:1 (1 October 1989): 372–383.

Oort, J. H. "The Galactic Center." In Geoffrey Burbidge, David Layzer, and John G. Phillips, eds., *Annual Review of Astronomy and Astrophysics,* vol. 15. Palo Alto, Cal.: Annual Reviews, Inc., 1977.

Osterbrock, Donald E. *Astrophysics of Gaseous Nebulae.* San Francisco, Cal.: W. H. Freeman and Co., 1974.

Pannekoek, Anton. *A History of Astronomy.* New York: Interscience Publishers, 1961.

Reid, M. J., et al. "Trigonometric Parallaxes of Massive Star-Forming Regions. VI. Galactic Structure, Fundamental Parameters, and Noncircular Motions." *The Astrophysical Journal* 700 (20 July 2009):137–148.

Reynolds, R. J., Tufte, S. L., Haffner, L. M., Jaehnig, K., and Percival, J. W. "The Wisconsin Hα Mapper (WHAM): A Brief Review of Performance Characteristics and Early Scientific Results." Publications of the Astronomical Society of Australia 15 (1998):14–18.

Shu, Frank H. *The Physical Universe: An Introduction to Astronomy.* Mill Valley, Cal.: University Science Books, 1982.

Spitzer, Lyman. *Diffuse Matter in Space.* New York: Interscience Publishers, 1968.

Spitzer, Lyman, Jr. *Searching Between the Stars.* New Haven, Conn.: Yale University Press, 1982.

The following references are from *The Astronomy and Astrophysics Encyclopedia,* edited by Stephen P. Maran. New York: Van Nostrand Reinhold; Cambridge, England: Cambridge University Press, 1992:

Barnes, Joshua E. "Galaxies, Disk, Evolution." Pages 236–238.

Blanco, Victor M. "Galactic Bulge." Pages 196–197.

Blitz, Leo. "Galactic Structure, Interstellar Clouds." Pages 202–204.

Bruhweiler, Frederick C. "Interstellar Medium, Local." Pages 377–378.

Burton, W. B. "Galactic Structure, Large Scale." Pages 204–208.

Carlberg, Ray. "Galaxies, Spiral, Nature of Spiral Arms." Pages 268–270.

Cox, Donald P. "Interstellar Medium, Hot Phase." Page 376.

Cudworth, Kyle M. "Galactic Structure, Globular Clusters." Pages 200–202.

Dalgarno, Alexander. "Interstellar Clouds, Chemistry." Pages 343–345.

Edmunds, Michael G. "Galaxy, Dynamical Models." Pages 280–282.

FitzGerald, M. Pim Vatter. "Galactic Structure, Optical Tracers." Pages 211–218.

Hesser, James E., and Bolte, Michael J. "Star Clusters, Globular." Pages 663–667.

Hills, Jack G. "Missing Mass, Galactic." Pages 445–446.

Ho, Paul T. P. "Galactic Center." Pages 197–200.

Israel, Frank P. "HII Regions." Pages 300–301.

Kahn, Franz D. "HII Regions, Dynamics and Evolution." Pages 301–303.

Leisawitz, David. "Interstellar Medium, Dust, Large Scale Galactic Properties." Pages 366–368.

Magnani, Loris. "Interstellar Medium, Dust, High Galactic Latitude." Pages 364–366.

Mathis, John S. "Interstellar Medium, Dust Grains." Pages 362–364.

Morris, Mark. "Interstellar Medium, Galactic Center." Pages 369–371.

Reid, M. J., et al. "Trigonometric Parallaxes of Massive Star-Forming Regions. VI. Galactic Structure, Fundamental Parameters, and Noncircular Motions." *The Astrophysical Journal* 700 (20 July 2009):137–148.

Reynolds, Ronald J. "Interstellar Medium." Pages 352–354.

Roberts, William W., Jr. "Galactic Structure, Spiral, Interstellar Gas, Theory." Pages 218–221.

Robinson, Brian. "Galactic Structure, Spiral, Observations." Pages 221–224.

Shuter, William L. H. "Interstellar Medium, Galactic Atomic Hydrogen." Pages 368–369.

Turner, Barry E. "Interstellar Medium, Molecules." Pages 378–381.

Williams, David A. "Interstellar Extinction, Galactic." Pages 350–352.

Wyse, Rosemary, F. G. "Galactic Structure, Stellar Populations." Pages 227–229.

CHAPTER 23

Bell, Eric F. "Galaxy Bulges and Their Black Holes: A Requirement for the Quenching of Star Formation." *The Astrophysical Journal* 682 (20 July 2008):355–360.

Dressler, Alan, and Richstone, Douglas O. "Stellar Dynamics in the Nuclei of M31 and M32: Evidence for Massive Black Holes." *The Astrophysical Journal,* 324:2 (15 January 1988):701–713.

Gallagher, John S., III, and Hunter, Deidre A. "Structure and Evolution of Irregular Galaxies." In Geoffrey Burbidge, David Layzer, and John G. Phillips, eds., *Annual Review of Astronomy and Astrophysics,* vol. 22. Palo Alto, Cal.: Annual Reviews, Inc., 1984:37–74.

Haring, Nadine, and Rix, Hans-Walter. "On the Black Hole Mass—Bulge Mass Relation." *The Astrophysical Journal* 604 (1 April 2004):L89–L92.

Ho, Luis C. "A Relation between Distance and Radial Velocity Among Extra-Galactic Nebulae." In Kenneth R. Lang and Owen Gingerich, eds., *A Source Book in Astronomy and Astrophysics, 1900–1975.* Cambridge, Mass.: Harvard University Press, 1979:725–728.

———. "Nuclear Activity in Nearby Galaxies." In Roger Blandford, John Kormendy, and Ewine van Dishoeck, eds., *Annual Review of Astronomy and Astrophysics,* vol. 46. Palo Alto, Cal.: Annual Reviews, Inc., 2008:475–539.

———. *The Realm of Nebulae.* New York: Dover Publications, Inc., 1958.

Hodge, Paul W. *Galaxies.* Cambridge, Mass.: Harvard University Press, 1986.

Hubble, Edwin P. "Extra-Galactic Nebulae." In Kenneth R. Lang and Owen Gingerich, eds., *A Source Book in Astronomy and Astrophysics, 1900–1975.* Cambridge, Mass.: Harvard University Press, 1979:716–724.

Kormendy, John. "Evidence for a Central Dark Mass in NGC 4594 (The Sombrero Galaxy)." *The Astrophysical Journal* 335:1 (1 December 1988):40–56. Chicago: University of Chicago Press, 1988.

———. "Evidence for a Supermassive Black Hole in the Nucleus of M31." *The Astrophysical Journal* 325:1 (1 February 1988):128–141.

———. "Observations of Galaxy Structure and Dynamics." In J. Binney, J. Kormendy, and S. White, authors, and L. Martinet and M. Mayor, eds., *Morphology and Dynamics of Galaxies; Twelfth Advanced Course of the Swiss Society of Astronomy and Astrophysics.* Sauverny-Versoix, Switzerland: Geneva Observatory, 1982:113–288.

Rowan-Robinson, Michael. *The Cosmological Distance Ladder: Distance and Time in the Universe.* New York: W. H. Freeman and Co., 1985.

Rubin, Vera C. "Weighing the Universe: Dark Matter and Missing Mass." In James Cornell, ed., *Bubbles, Voids, and Bumps in Time: The New Cosmology.* Cambridge, England: Cambridge

University Press (by the Smithsonian Astrophysical Observatory), 1989.

Shu, Frank H. *The Physical Universe: An Introduction to Astronomy.* Mill Valley, Cal.: University Science Books, 1982.

Slipher, Vesto M. "A Spectrographic Investigation of Spiral Nebulae." In Kenneth R. Lang and Owen Gingerich, eds., *A Source Book in Astronomy and Astrophysics, 1900–1975.* Cambridge, Mass.: Harvard University Press, 1979:704–707.

Smith, Robert W. *The Expanding Universe: Astronomy's "Great Debate" 1900–1931.* Cambridge, England: Cambridge University Press, 1982.

de Valcouleurs, G. "The Extragalactic Distance Scale. I. A Review of Distance Indicators: Zero Points and Errors of Primary Indicators." *The Astrophysical Journal* 223:2 (15 July 1978): 351–363.

The following references are from *The Astronomy and Astrophysics Encyclopedia,* edited by Stephen P. Maran. New York: Van Nostrand Reinhold; Cambridge, England: Cambridge University Press, 1992:

Barnes, Joshua E. "Galaxies, Disk, Evolution." Pages 236–238.

Baum, William A. "Galaxies, Elliptical." Pages 240–241.

Fabbiano, Giuseppina. "Galaxies, X-Ray Emission." Pages 276–278.

Freedman, Wendy L. "Hubble Constant." Pages 308–310.

Gottesman, Stephen T., and Hunter, James H., Jr. "Galaxies, Barred Spiral." Pages 229–232.

Hunter, Deidre A. "Galaxies, Irregular." Pages 252–254.

Keel, William C. "Galaxies, Binary and Multiple, Interactions." Pages 232–234.

Lake, George. "Galaxies, Formation." Pages 246–248.

Merritt, David. "Galaxies, Elliptical, Dynamics." Pages 241–243.

Mould, Jeremy. "Galaxies, Dwarf Spheroidal." Pages 238–240.

Quinn, Peter J. "Galaxies, Elliptical, Origin and Evolution." Pages 243–246.

Tully, R. Brent. "Distance Indicators, Extragalactic." Pages 173–175.

Walterbos, Rene A. M. "Andromeda Galaxy." Pages 21–23.

CHAPTER 24

Barthel, Peter D. "Is Every Quasar Beamed?" *The Astrophysical Journal* 336:2 (15 January 1989):606–611.

Blandford, R. D. "Physical Process in Active Galactic Nuclei." In R. D. Blandford, H. Netzer, and L. Woltjer, authors, T. J. L. Courvoisier and M. Mayor, eds., *Active Galactic Nuclei: Saas-Free Advanced Course 20, Lecture Notes 1990, Swiss Society for Astrophysics and Astronomy.* Berlin: Springer-Verlag, 1990:161–275.

Blank, David L., and Soker, Noam. "Evolutionary Sequence of Seyfert Galaxies." In H. Richard Miller and Paul J. Wiita, eds., *Lecture Notes in Physics 307: Active Galactic Nuclei; Proceedings of a Conference Held at the Georgia State University, Atlanta, Georgia, October 28–30, 1987.* Berlin: Springer-Verlag, 1988:427–429.

Cohen, Marshall H. "Radio Sources: Small Scale Structure." In H. Richard Miller and Paul J. Wiita, eds., *Lecture Notes in Physics 307: Active Galactic Nuclei; Proceedings of a Conference Held at the Georgia State University, Atlanta, Georgia, October 28–30, 1987.* Berlin: Springer-Verlag, 1988:290–301.

Fried, J. W. "Host Galaxies of Quasars." In G. Giuricin, F. Mardirossian, M. Mezzetti, and M. Ramella, eds., *Structure and Evolution of Active Galactic Nuclei; International Meeting Held in Trieste, Italy, April 10–13, 1985.* Dordrecht, Holland: D. Reidel Publishing Co., 1986:309–314.

Hewitt, J. N., et al. "Unusual Radio Source MG1131+0456: A Possible Einstein Ring." *Nature* 333:6173 (9 June 1988):537–540.

Kochanek, Christopher S. "The Theory and Practice of Radio Ring Lenses." In Y. Mellier, B. Fort, and G. Soucail, eds., *Lecture Notes in Physics 360: Gravitational Lensing; Proceedings of a Workshop Held in Toulouse, France, September 13–15, 1989.* Berlin: Springer-Verlag, 1990:244–253.

Lawrence, A. "Classification of Active Galaxies and the Prospect of a Unified Phenomenology." *Publications of the Astronomical Society of the Pacific* 99:615 (May 1987):309–334.

Longair, Malcolm. "The New Astrophysics." In Paul Davies, ed., *The New Physics.* Cambridge, England: Cambridge University Press, 1989.

Marshall, Herman L. "The Luminosity Function of Quasars and Low Luminosity Active Galactic Nuclei." In G. Giuricin, F. Mardirossian, M. Mezzetti, and M. Ramella, eds., *Structure and Evolution of Active Galactic Nuclei; International Meeting Held in Trieste, Italy, April 10–13, 1985.* Dordrecht, Holland: D. Reidel Publishing Co., 1986:627–632.

Myers, Steven T., and Spangler, Steven R. "Synchrotron Aging in the Lobes of Luminous Radio Galaxies." *The Astrophysical Journal* 291:1 (1 April 1985):52–62.

Osterbrock, Donald E. "Emission-Line Spectra and the Nature of Active Galactic Nuclei." In H. Richard Miller and Paul J. Wiita, eds., *Lecture Notes in Physics 307: Active Galactic Nuclei; Proceedings of a Conference Held at the Georgia State University, Atlanta, Georgia, October 28–30, 1987.* Berlin: Springer-Verlag, 1988:1–16.

Porcas, R. W., et al. "First VLBI Hybrid Maps of 0957+561 A and B." In J. M. Moran, J. N. Hewitt, and K. Y. Lo, eds., *Lecture Notes in Physics 330: Gravitational Lenses; Proceedings of a Conference Held at the Massachusetts Institute of Technology, Cambridge, Massachusetts, in Honour of Bernard F. Burke's 60th Birthday, June 20, 1988.* Berlin: Springer-Verlag, 1989:82–83.

Schmidt, Maarten. "Space Distribution and Luminosity Function of Quasars." In H. Richard Miller and Paul J. Wiita, eds., *Lecture Notes in Physics 307: Active Galactic Nuclei; Proceedings of a Conference Held at the Georgia State University, Atlanta, Georgia, October 28–30, 1987.* Berlin: Springer-Verlag, 1988:408–416.

Tyson, Anthony. "Mapping Dark Matter with Gravitational Lenses." *Physics Today* 45:6 (June 1992):24–32.

Walker, R. C., Benson, J. M., and Unwin, S. C. "The Radio Morphology of 3C 120 on Scales from .5 Parsecs to 400 Kiloparsecs." *The Astrophysical Journal* 316:2 (15 May 1987): 546–572.

Weedman, Daniel W. "Evolutionary Connection of Seyfert Galaxies and Quasars." In G. Giuricin, F. Mardirossian, M. Mezzetti, and M. Ramella, eds., *Structure and Evolution of Active Galactic Nuclei; International Meeting Held in Trieste, Italy, April 10–13, 1985.* Dordrecht, Holland: D. Reidel Publishing Co., 1986: 215–226.

Weedman, Daniel W. *Quasar Astronomy.* Cambridge, England: Cambridge University Press, 1986.

Woltjer, J. "Phenomenology of Active Galactic Nuclei." In R. D. Blanford, H. Netzer, and L. Woltjer, authors, T. J. L. Courvoisier and M. Mayor, eds., *Active Galactic Nuclei: Saas-Free Advanced Course 20, Lecture Notes 1990, Swiss Society for Astrophysics and Astronomy.* Berlin: Springer-Verlag, 1990:1–55.

The following references are from *The Astronomy and Astrophysics Encyclopedia,* edited by Stephen P. Maran. New York: Van Nostrand Reinhold; Cambridge, England: Cambridge University Press, 1992:

Antonucci, Robert. "Active Galaxies and Quasistellar Objects, Blazars." Pages 6–7.

Barthel, Peter. "Active Galaxies and Quasistellar Objects, Interrelations of Various Types." Pages 12–15.

Blandford, Roger D. "Active Galaxies and Quasistellar Objects, Accretion." Pages 5–6.

Cohen, Marshall H. "Active Galaxies and Quasistellar Objects, Superluminal Motion." Pages 17–19.

Eilek, Jean A. "Active Galaxies and Quasistellar Objects, Jets." Pages 15–17.

Heckman, Timothy M. "Galaxies, Nuclei." Pages 261–263.

Konigl, Arieh. "Jets, Theory of." Pages 391–393.

Krolik, Julian H. "Active Galaxies and Quasistellar Objects, Central Engine." Pages 7–9.

Mushotzky, Richard. "Active Galaxies and Quasistellar Objects, X-Ray Emission." Pages 19–21.

Peacock, John A. "Gravitational Lenses." Pages 292–295.

Peterson, Bradley M. "Quasistellar Objects, Absorption Lines." Pages 571–572.

Reichert, Gail A. "Active Galaxies, Seyfert Type." Pages 2–4.

Rieke, George H. "Active Galaxies and Quasistellar Objects, Infrared Emission and Dust." Pages 10–12.

Shields, Gregory A. "Active Galaxies and Quasistellar Objects, Emission Line Regions." Pages 9–10.

Smith, Malcolm G. "Quasistellar Objects, Statistics and Distribution." Pages 578–580.

Spruit, Henk C. "Accretion." Pages 1–2.

Stockton, Alan. "Quasistellar Objects, Host Galaxies." Pages 572–574.

Wills, Beverley J. "Quasistellar Objects, Spectroscopic and Photometric Properties." Pages 575–578.

CHAPTER 25

Bahcall, Neta A. "Clusters of Galaxies." In Geoffrey Burbidge, David Layzer, and John G. Phillips, eds., *Annual Review of Astronomy and Astrophysics,* vol. 15. Palo Alto, Cal.: Annual Reviews, Inc., 1977:505–540.

Dressler, A. "The Evolution of Galaxies in Clusters." In Geoffrey Burbidge, David Layzer, and John G. Phillips, eds., *Annual Review of Astronomy and Astrophysics,* vol. 22. Palo Alto, Cal.: Annual Reviews, Inc., 1984:185–222.

Geller, Margaret J. "Mapping the Universe: Slices and Bubbles." In James Cornell, ed., *Bubbles, Voids, and Bumps in Time: The New Cosmology.* Cambridge, England: Cambridge University Press, 1989.

Giovanelli, Riccardo, and Haynes, Martha P. "Redshift Surveys of Galaxies." In Geoffrey Burbidge, David Layzer, and John G. Phillips, eds., *Annual Review of Astronomy and Astrophysics,* vol. 29. Palo Alto, Cal.: Annual Reviews, Inc., 1991:499–541.

Forman, W., and Jones, C. "Hot Gas in Clusters of Galaxies." In William R. Oegerle, Michael J. Fitchett, and Laura Danly, eds., *Clusters of Galaxies; Proceedings of the Clusters of Galaxies Meeting, Baltimore 1989 May 15–17.* Cambridge, England: Cambridge University Press, 1990:257–277.

Haynes, Martha P. "Evidence for Gas Deficiency in Cluster Galaxies." In William R. Oegerle, Michael J. Fitchett, and Laura Danly, eds., *Clusters of Galaxies; Proceedings of the Clusters of Galaxies Meeting, Baltimore 1989 May 15–17.* Cambridge, England: Cambridge University Press, 1990:177–193.

van den Bergh, Sidney. "Updated Information on the Local Group." *Publications of the Astronomical Society of the Pacific* 112 (April 2000):529–536.

West, Michael J. "Cosmogony and the Structure of Rich Clusters of Galaxies." In William R. Oegerle, Michael J. Fitchett, and Laura Danly, eds., *Clusters of Galaxies; Proceedings of the Clusters of Galaxies Meeting, Baltimore 1989 May 15–17.* Cambridge, England: Cambridge University Press, 1990:65–103.

Whitmore, Bradley C. "The Effect of the Cluster Environment on Galaxies." In William R. Oegerle, Michael J. Fitchett, and Laura Danly, eds., *Clusters of Galaxies; Proceedings of the Clusters of Galaxies Meeting, Baltimore 1989 May 15–17.* Cambridge, England: Cambridge University Press, 1990:139–167.

Zucker, D. B., et al. "A New Milky Way Dwarf Satellite in Canes Venatici." *The Astrophysical Journal* 643 (1 June 2006): L103–L106.

The following references are from *The Astronomy and Astrophysics Encyclopedia,* edited by Stephen P. Maran. New York: Van Nostrand Reinhold; Cambridge, England: Cambridge University Press, 1992:

Bertschinger, Edmund. "Dark Matter, Cosmological." Pages 170–171.

Binggeli, Bruno. "Virgo Cluster." Pages 950–952.

Chincarini, Guido, Scaramella, Roberto, and Vettolani, Paolo. "Superclusters, Dynamics and Models." Pages 877–881.

Dressler, Alan. "Galaxies, Properties in Relation to Environment." Pages 263–265.

Fairall, Anthony P. "Superclusters, Observed Properties." Pages 881–883.

Giuricin, Giuliano. "Galaxies, Local Supercluster." Pages 256–259.

Hodge, Paul. "Galaxies, Local Group." Pages 254–256.

Jones, Christine. "Clusters of Galaxies, X-Ray Observations." Pages 102–104.

Lapparent-Gurriet, Valerie de. "Voids, Extragalactic." Pages 952–953.

Oemler, Augustus, Jr. "Clusters of Galaxies, Component Galaxy Characteristics." Pages 97–99.

Saslaw, William C. "Cosmology, Clustering and Superclustering." Pages 150–151.

West, Michael J. "Clusters of Galaxies." Pages 95–97.

CHAPTER 26

Bromm, Volker. "The First Sources of Light." *Publications of the Astronomical Society of the Pacific* 115 (February 2004): 103–114.

Cowan, John J., Thielemann, Friedrich-Karl, and Truran, James W. "Radioactive Dating of the Elements." In Geoffrey Burbidge, David Layzer, and John G. Phillips, eds., *Annual Review of Astronomy and Astrophysics,* vol. 29. Palo Alto, Cal.: Annual Reviews, Inc., 1991:447–497.

Ellis, G. F. R. "Alternatives to the Big Bang." In Geoffrey Burbidge, David Layzer, and John G. Phillips, eds., *Annual Review of Astronomy and Astrophysics,* vol. 22. Palo Alto, Cal.: Annual Reviews, Inc., 1984:157–184.

Filippenko, Alexie V. "Einstein's Biggest Blunder? High-Redshift Supernovae and the Accelerating Universe." *Publications of the*

Astronomical Society of the Pacific 113 (December 2001): 1441–1448.

Freedman, Wendy L., et al. "Final Results from the Hubble Space Telescope Key Project to Measure the Hubble Constant." *The Astrophysical Journal* 553 (20 May 2001):47–72.

Frieman, Joshua A., Turner Michael S., and Huterer, Dragan. "Dark Energy and the Accelerating Universe." In Roger Blandford, John Kormendy, and Ewine van Dishoeck, eds., *Annual Review of Astronomy and Astrophysics,* vol. 46. Palo Alto, Cal.: Annual Reviews, Inc., 2008:385–432.

Gudmundsson, Einar H., and Gunnlaugur, Bjornsson. "Dark Energy and the Observable Universe." *The Astrophysical Journal* 565 (20 January 2002):1–16.

Gunn, James E. "Expanding the Universe: Space Telescopes and Beyond in the Next Twenty Years." In James Cornell, ed., *Bubbles, Voids, and Bumps in Time: The New Cosmology.* Cambridge, England: Cambridge University Press, 1989:147–182.

Guth, Alan, and Steinhardt, Paul. "The Inflationary Universe." In Paul Davies, ed., *The New Physics.* Cambridge, England: Cambridge University Press, 1989:34–60.

Guth, Alan H. "Starting the Universe: The Big Bang and Cosmic Inflation." In James Cornell, ed., *Bubbles, Voids, and Bumps in Time: The New Cosmology.* Cambridge, England: Cambridge University Press, 1989:105–146.

Harrison, Edward Robert. *Cosmology: The Science of the Universe.* New York: Cambridge University Press, 1981.

———. *Darkness at Night: A Riddle of the Universe.* Cambridge, Mass.: Harvard University Press, 1987.

Hu, Wayne, and Dodelson, Scott. "Cosmic Microwave Background Anisotropies." In Geoffrey Burbidge, Allan Sandage, and Frank H. Shu, eds., *Annual Review of Astronomy and Astrophysics,* vol. 40. Palo Alto, Cal.: Annual Reviews, Inc., 2002:171–216.

Leibundgut, Bruno. "Cosmological Implications from Observations of Type Ia Supernovae." In Geoffrey Burbidge, Alan Sandage, and Frank H. Shu, eds., *Annual Review of Astronomy and Astrophysics,* vol. 39. Palo Alto, Cal.: Annual Reviews, Inc., 2001:67–98.

Levi, Barbara Goss. "COBE Measures Anisotropy in Cosmic Microwave Background Radiation." *Physics Today* 45:6 (June 1992):17–20.

Loeb, Abraham, and Barkana, Rennan. "The Reionization of the Universe by the First Stars." In Geoffrey Burbidge, Allan Sandage, and Frank H. Shu, eds., *Annual Review of Astronomy and Astrophysics,* vol. 39. Palo Alto, Cal.: Annual Reviews, Inc., 2001:19–66.

Longair, Malcolm. "The New Astrophysics." In Paul Davies, ed., *The New Physics.* Cambridge, England: Cambridge University Press, 1989:94–208.

Miller, Christopher J., Nichol, Robert C., Genovese, Christopher, and Wasserman, Larry. "A Nonparametric Analysis of the Cosmic Microwave Background Power Spectrum." *The Astrophysical Journal* 565 (1 February 2002):L67–L70.

Narlikar, J. V., and Padmanabhan, T. "Inflation for Astronomers." In Geoffrey Burbidge, David Layzer, and John G. Phillips, eds., *Annual Review of Astronomy and Astrophysics,* vol. 29. Palo Alto, Cal.: Annual Reviews, Inc., 1991: 325–362.

Ohanian, Hans C. *Gravitation and Spacetime.* New York: Norton, 1976.

Peebles, P. J. E. *Physical Cosmology.* Princeton, N.J.: Princeton University Press, 1971.

Ress, Martin. *New Perspectives in Astrophysical Cosmology.* Cambridge, England: Cambridge University Press, 2000.

Sandage, Allan. "Observational Tests of World Models." In Geoffrey Burbidge, David Layzer, and John G. Phillips, eds., *Annual Review of Astronomy and Astrophysics,* vol. 26. Palo Alto, Cal.: Annual Reviews, Inc., 1988: 561–630.

Sciama, D. W. *Modern Cosmology.* Cambridge, England: Cambridge University Press, 1971.

Shu, Frank H. *The Physical Universe: An Introduction to Astronomy.* Mill Valley, Cal.: University Science Books, 1982.

Silk, Joseph. *The Big Bang,* ed 3. New York: W. H. Freeman and Company, 2001.

Weedman, Daniel W. *Quasar Astronomy.* Cambridge, England: Cambridge University Press, 1986.

Wright, E. L. "A Cosmology Calculator for the World Wide Web." *Publications of the Astronomical Society of the Pacific* 118 (December 2006):1711–1715.

The following references are from *The Astronomy and Astrophysics Encyclopedia,* edited by Stephen P. Maran. New York: Van Nostrand Reinhold; Cambridge, England: Cambridge University Press, 1992:

Bertschinger, Edmund. "Dark Matter, Cosmological." Pages 170–171.

Boughn, S. P., and Partridge, R. B. "Background Radiation, Microwave." Pages 42–45.

Brandenberger, Robert H. "Cosmology, Inflationary Universe." Pages 158–159.

Clayton, Donald D. "Cosmology, Cosmochronology." Pages 153–156.

Freedman, Wendy L. "Hubble Constant." Pages 308–310.

Pagel, Bernard E. J. "Cosmology, Big Bang Theory." Pages 147–149.

Ryden, Barbara. "Cosmology, Galaxy Formation." Pages 156–158.

Trimble, Virginia, and Maran, Stephen P. "Cosmology. Observational Tests." Pages 162–165.

Wagoner, Robert V. "Cosmology, Nucleogenesis." Pages 160–162.

CHAPTER 27

Black, David C. "Planetary Systems, Formation, Observational Evidence." In Stephen P. Maran, ed., *The Astronomy and Astrophysics Encyclopedia.* New York: Van Nostrand Reinhold, 1992:542–544.

———. "Completing the Copernican Revolution: The Search for Other Planetary Systems." In Geoffrey Burbidge and Allan Sandage, eds., *Annual Review of Astronomy and Astrophysics,* vol. 33. Palo Alto, Cal.: Annual Reviews, Inc., 1995:359–380.

Dawkins, Richard. *The Blind Watchmaker.* New York: W. W. Norton & Company, 1987.

Gaudi, B. S., Seager, S., and Mallen-Ornelas, G. "On the Period Distribution of Close-In Extrasolar Giant Planets." *The Astrophysical Journal* 623 (10 April 2005):472–481.

Gould, A., et al. "Microlens OGLE-2005-BLG-169 Implies That Cool Neptune-Like Planets Are Common." *The Astrophysical Journal* 644 (10 June 2006):L37–L40.

Goldsmith, Daniel, and Owen, Tobias. *The Search for Life in the Universe.* Menlo Park, Cal.: The Benjamin Cummings Publishing Company, Inc., 1980.

The following references are from Heidmann, J., and Klein, M. J., eds., *Bioastronomy: The Search for Extraterrestrial Life—The Exploration Broadens.* Berlin: Springer-Verlag, 1991:

Calvin, William H. "The Antecedents of Consciousness: Evolving the "Intelligent" Ability to Simulate Situations and Contemplate the Consequences of Novel Courses of Action," Pages 311–318.

Donnelly, C., Bowyer, S., Herrick, W., Werthimer, D., Lampton, M., and Hiatt, T. "The Serendip II SETI Project: Observations and RFI Analysis," Pages 223–228.

Tarter, Jill, and Klein, Michael J. "SETI: On the Telescope and On the Drawing Board," Pages 229–234.

Whitmire, D. P., Reynolds, R. T., and Kasting, J. F. "Habitable Zones for Earth-Like Planets Around Main Sequence Stars," Pages 173–178.

Kasting, James F., and Catling, David. "Evolution of a Habitable Planet." In Geoffrey Burbidge, Roger Blandford, and Allan Sandage, eds., *Annual Review of Astronomy and Astrophysics,* vol. 41. Palo Alto, Cal.: Annual Reviews, Inc., 2003: 429–463.

Klein, Michael J., and Gulkis, Samuel. "The Impact of Technology on SETI," Pages 203–208.

Kuchner, Marc J. "Volatile-Rich Earth-Mass Planets in the Habitable Zone." *The Astrophysical Journal* 596 (10 October 2003): L105–L108.

Lineweaver, Charles H., and Grether, Daniel. "What Fraction of Sun-Like Stars Have Planets?" *The Astrophysical Journal* 598 (1 December 2003):1350–1360.

Mandell, Avi M., and Sigurdsson, Steinn. "Survival of Terrestrial Planets in the Presence of Giant Planet Migration." *The Astrophysical Journal* 599 (20 December 2003): L111–L114.

The following references are from Marx, George, ed., *Bioastronomy—The Next Steps.* Dordrecht: Kluwer Academic Publishers, 1988:

Brown, Robert A. "Systematic Aspects of Direct Extrasolar Planetary Detection," Pages 117–123.

Brown, Ronald D. "Exotic Chemical Life," Pages 179–185.

Burke, B. F. "Optical Interferometry and the Detection of Evidence of Life," Pages 139–152.

Calvin, W. H. "Fast Tracks to Intelligence (Considerations from Neurobiology and Evolutionary Biology)," Pages 237–245.

Campbell, B., Walker, G. A. H., and Yang, S. "A Search for Planetary Mass Companions to Nearby Stars," 83–90.

Fraknoi, A. G. "What If We Succeed? (A Media Viewpoint)," Pages 417–424.

Linscott, I., Duluk, J., Burr, J., and Peterson, A. "Artificial Signal Detectors," Pages 319–335.

Oro, J. "Constraints Imposed by Cosmic Evolution Towards the Development of Life," Pages 161–165.

Pfleiderer, Mircea. "Some Biological Implications on Drake's Formula," Pages 279–280.

Schwartzman, D., and Rickard, L. J. "Being Optimistic About SETI," Pages 305–312.

Terrile, R. J. "Direct Imaging of Extra-Solar Planetary Systems With a Low-Scattered Light Telescope," Pages 125–130.

Oro, J., Miller, Stanley L., and Lazcano, Antonio. "The Origin and Early Evolution of Life on Earth." In George W. Wetherill, Arden L. Albee, and Francis G. Stehli, eds., *Annual Review of Earth and Planetary Physics,* vol. 18. Palo Alto, Cal.: Annual Reviews, Inc., 1990:317–356.

Ponnamperuma, Cyril, and Cameron, A. G. W. *Interstellar Communication: Scientific Perspectives.* Boston: Houghton Mifflin Company, 1974.

Press, Frank, and Siever, Raymond. *Earth,* ed 2. San Francisco, Cal.: W. H. Freeman, 1978.

Rampino, M. R., and Caldeira, K. "The Goldilocks Problem: Climatic Evolution and Long-Term Habitability of Terrestrial Planets." *Annual Reviews of Astronomy and Astrophysics,* vol. 32. Palo Alto, Cal.: Annual Reviews, Inc., 1994:83–114.

Raven, Peter H., and Johnson, George B. *Biology,* ed 2. St. Louis: Times Mirror/Mosby College Publishing, 1989.

Sagan, Carl, and Shckovskii, I. S. *Intelligent Life in the Universe.* San Francisco: Holden Day, Inc., 1966.

Udalski, A., et al. "A Jovian-Mass Planet in Microlensing Event OGLE-2005-BLG-071." *The Astrophysical Journal* 628 (1 August 2005):L109–L112.

Udry, Stephane, and Santos, Nuno C. "Statistical Properties of Exoplanets." In Roger Blandford, John Kormendy, and Ewine van Dishoeck, eds., *Annual Review of Astronomy and Astrophysics,* vol. 45. Palo Alto, Cal.: Annual Reviews, Inc., 2008:397–439.

Zucker, Shay, and Mazeh Tsevi. "Derivation of the Mass Distribution of Extrasolar Planets with Maxlima, a Maximum Likelihood Algorithm." *The Astrophysical Journal* 562 (1 December 2001):1038–1044.

Credits

PHOTOGRAPHS

Design Element for Chapter Openers: © Stocktrek/age fotostock RF

Chapter 1

Opener: Face of the Earth ™ ARC Science Simulations, © 2007. www.arcscience.com; **1.1–1.4:** © Matt Neely; **1.8:** © Hulton Archive/ Getty; **1.15–1.17:** Matt Neely; **1.18:** © Getty RF.

Chapter 2

Opener: STS-47 Crew, NASA; **2.1:** © Roger Ressmeyer/Corbis; **2.14:** © Goeff Dore Photography; **2.15:** © The Solstice Project/photo by Karl Kernberger; **2.16:** Courtesy E.C. Krupp, Griffith Observatory; **2.23a, b:** © Matt Neely.

Chapter 3

Opener: © Getty RF; **3.1:** © North Wind Picture Archives; **3.3 (Sapphos):** © Bettmann/Corbis; **3.3 (The Great):** © Hulton Archive/Getty; **3.3 (Archimedes):** © Hulton Archive/Getty; **3.3 (Claudius):** © Time Life Pictures/Getty.

Chapter 4

Opener: © National Geographic/Getty; **4.2 (All):** © North Wind Picture Archives; **4.10:** © Hulton Archives/Getty; **4.16:** © The Granger Collection; **4.17:** Galileo Galilei, Istoria e dimostrazioni intorno alle macche solari . . ., 1613, pp. 72–83; **4.18:** Courtesy Yerkes Observatory.

Chapter 5

Opener: © Brand X Pictures/PunchStock RF; **5.10:** © Soren Hald/ Getty; **5.13a, b:** Courtesy John S. Neff.

Chapter 6

Opener: © Corbis RF; **6.22:** Courtesy Smithsonian Institution Astrophysical Observatory; **6.23a, b:** Courtesy of Dr. Jerom Kristian, © The Observatories of the Carnegie Institution in Washington; **6.25, 6.26:** © Robert Ressmeyer/Corbis; **6.27, 6.28:** © European Southern Observatory (ESO); **6.30:** NOAA.NGDC, DMSP Digital Archive; **6.31a, b:** © France Hawaii Telescope Corporation, 1996. Photo provided by Dr. Oliver Lai; **6.32:** AURA/NOAO/National Science Foundation; **6.33:** Courtesy Roger Smith/NOAO/ AURA/NSF; **6.34:** © Brand X Pictures/PunchStock RF; **6.35:** STS-82 Crew/STSci/NASA; **6.36:** Courtesy D. Crisp and WFPC2 Team/JPL/Caltech, HST, STSI, NASA; **6.37:** NASA/JPL-Caltech/ S. Willner (Harvard-Smithsonian Center for Astrophysics; **6.39 (X-ray):** Courtesy D. Burrows (PSU) et al., Optical: P. Challis (CfA) et al., NASA; **6.40:** © David Parker/SPL/Photo Researchers; **6.41:** Photo by Dave Finley, NRAO/AUI; **6.43a, b:** NRAO/AUI; **6.44c:** © John D. Fix.

Chapter 7

Opener: NASA; **7.13a:** © John D. Fix; **7.13b:** © Richard Megna/ Fundamental Photographs; **7.13c:** © Corbis RF.

Chapter 8

Opener: © NASA/Corbis RF; **8.3:** © AP Wide World Photos; **8.8:** NOAA/NASA; **8.9:** © Sylvain Grandadam/Photo Researchers; **8.11:** © National Geographic Society; **8.18:** World Ocean Floor Panorama, Bruce C. Heezen and Marie Tharp, 1977. Copyright © by Marie Tharp 1977/2003. Reproduced by permission of Marie Tharp Maps, LLC 8 Edward St., Sparkill, NY 10976; **8.25:** © Stocktrek/ Frank Rosotto; **8.26:** NASA; **8.29:** © Alaska Division of Tourism; **8.30:** © StockTrek/Getty RF; **8.31:** Courtesy L.A. Frank, University of Iowa, NASA.

Chapter 9

Opener: © Corbis RF; **9.3:** NASA/Johnson Space Center; **9.5:** Courtesy Rob Scott; **9.6:** Courtesy Andreas Gada; **9.7:** Solar Eclipse, July 1991, © Roland & Marjorie Christen, taken with Astro-Physics 130mm StarFire refractor.; **9.8:** Courtesy Jeff Schroeder/NASA; **9.13:** © Jean-Pierre Haignere/CNES; **9.18:** Courtesy Ewen A. Whitaker; **9.19:** Courtesy Ewen A. Whitaker; **9.20a, b:** © Jay L. Inge, National Geographic Society; **9.21:** NASA; **9.22:** © The Observatories of the Carnegie Institution of Washington; **9.23, 9.24:** NASA; **9.26a–c:** Don E. Wilhems and Donald E. Davis/US Geological Survey.

Chapter 10

Opener: © Brand X Pictures/PunchStock RF; **10.1 (All):** Courtesy JPL; **10.4, 10.5:** MESSENGER, NASA, JHU APL, CIW; **10.6:** NASA; **10.7–10.9:** NASA/Johns Hopkins University Applied Physics Laboratory/Carnegie Institution of Washington; **10.11:** NASA/Brown RPIF; **10.19a–d:** Courtesy Soviet Academy of Sciences/U.S.S.R/ GSFC/NASA; **10.20:** JPL; **10.21:** NASA/USGS; **10.22:** Courtesy Calvin J. Hamilton; **10.23:** JPL; **10.24:** NASA; **10.25–10.28:** JPL; **10.29:** NASA; **10.30:** © Calvin Hamilton; **10.31, 10.32:** JPL; **10.33a, b:** Courtesy David T. Sandwell and Gerald Schubert.

Chapter 11

Opener: Pat Rawlings/NASA; **11.1 (series: Ten photos):** Courtesy Lowell Observatory; **11.2:** Courtesy Steve Lee (Univ. of Colorado)/Jim Bell (Cornell University)/Mike Wolf (Space Science Institute)/NASA; **11.3:** NASA/Brown RPIF; **11.4:** GSFC/NASA; **11.5:** NASA/JPL; **11.6, 11.7:** NASA/Brown RPIF; **11.9:** Adapted from S.P. Neszaros, NASA Technical Momorandum 56207: Photographic Catalog of Selected Planetary Size Comparisons, NASA, 1985: 2-61; **11.10:** Courtesy NASA/JPL/Malin Space Science Systems; **11.11:** NSSDC/ NASA; **11.12:** Courtesy Calvin J. Hamilton; **11.13:** NASA/JPL/ Malin Space Science Systems; **11.14:** GSFC/NASA; **11.15:** Courtesy Calvin J. Hamilton; **11.16:** NASA/JPL/University of Arizona; **11.17:** NASA/JPL/Malin Space Science Systems; **11.18:** Courtesy

Calvin J. Hamilton; **11.19:** NASA/JPL/University of Arizona; **11.20, 11.22:** NASA/JPL/Malin Space Science Systems; **11.23:** JPL; **11.24:** © Calvin Hamilton; **11.25 (top, bottom):** JPL; **11.26:** NASA/JPL; **11.27:** NASA/JPL/CORNELL; **11.28:** Courtesy HiRISE, MRO, LPL (U. Arizona, NASA; **11.29:** Courtesy NASA/JPL/CORNELL; **11.30, 11.31:** NASA/JPL; **11.32:** Courtesy MRO-HiRISE, NASA, JPL, Univ. Arizona; **11.33:** Courtesy Phoenix Mission team, NASA, JPL-CalTech, U. Arizona, Texas A&M University; **11.34:** Courtesy Marco Di Lorenzo, Kenneth Kremer, Phoenix Mission, NASA, JPL, UA, Max Planck Inst., Spaceflight; **11.37:** Courtesy J. Bell (Cornell), M. Wolf (Space Science Inst.), Hubble Heritage Team (STSci/AURA, NASA; **11.38:** JPL/Malin Space Science Systems/NASA; **11.39 (left, right):** NASA; **11.40:** Courtesy Harvard College Observatory; **11.41:** Courtesy Planetary Science Team, Los Alamos Laboratory; **11.42:** NASA/JPL/University of Arizona.

Chapter 12

Opener: © Alan Gutierrez; **12.3:** Courtesy Joseph A. Burns/NASA/The Planet Saturn, AFOD Alexander, (Faber and Faber); **12.4:** NASA and The Hubble Heritage Team (STSCi/AURA); **12.5 (left, right):** NASA; **12.6:** Courtesy Reta Beebe, Amy Simon (New Mexico State University), and NASA; **12.7:** NASA/JPL/Space Science Institute; **12.10:** JPL; **12.11:** NASA/JPL/Cornell University; **12.12:** The Galileo Project, JPL, NASA; **12.13:** JPL; **12.14:** NASA/JPL/Space Science Institute; **12.15 (left):** Courtesy Richard Terrile/JPL; **12.15 (right):** Courtesy Richard Terrile/JPL; **12.16:** Courtesy David Dunn and Lawrence Molnar/The University of Iowa/The University of Iowa; **12.20:** Galileo Project/NASA; **12.23:** JPL/NASA; **12.24:** Lowell Observatory; **12.25:** NASA and The Hubble Heritage Team (STScI/AURA); **12.27:** NASA; **12.29, 12.30:** Courtesy M. Belton (NOAO)/J.Burns (Cornell)/Galileo Project; **12.31, 12.34:** NASA/JPL/Space Science Institute; **12.35:** JPL.

Chapter 13

Opener: © StockTrek/Getty RF; **13.2a, b:** Voyager 2/NASA; **13.4:** Courtesy Lawrence Sromovsky, University of Wisconsin-Madison/W.M. Keck Observatory; **13.6:** NASA/Brown/RPIF; **13.7, 13.8:** JPL; **13.12:** Courtesy Erich Karkoschka/University of Arizona/NASA; **13.14, 13.15:** JPL; **13.16:** NASA; **13.17, 13.18:** JPL; **13.21:** Lowell Observatory; **13.22:** Courtesy of James Christy, US Naval Observatory; **13.23:** AURA/STSci.

Chapter 14

Opener: © Stockbyte/Getty RF; **14.3:** Viking Project/JPL/NASA; **14.4:** Courtesy HiRise, MRO, LPL (U. Arizona), NASA; **14.5:** JPL/Malin Space Science Systems; **14.6:** NASA/Brown/RPIF; **14.7:** Galileo Project/JPL/NASA; **14.8:** NASA/JPL/University of Arizona; **14.9:** Galileo Project, SSI, University of Arizona; **14.10:** Galileo Project/JPL/NASA; **14.11:** Courtesy © NASA enhanced by Tom Stack & Associates; **14.13:** Galileo Project/JPL/NASA; **14.15:** NASA/JPL/University of Arizona; **14.16:** Galileo Project/JPL/NASA; **14.17:** NASA/JPL/University of Arizona/University of Colorado; **14.18, 14.19:** Galileo Project/JPL/NASA; **14.20:** NASA; **14.21:** Galileo Project/JPL/NASA; **14.22:** Courtesy Calvin J. Hamilton; **14.23:** Galileo Project/JPL/NASA; **14.24:** Cassini Imaging Team, NASA, JPL, SSI; **14.25, 14.26:** Cassini Imaging Team, SSI, JPL, ESA, NASA; **14.27, 14.28:** NASA/JPL/Space Science Institute; **14.29, 14.30:** Cassini Imaging Team, SSI, JPL, ESA, NASA; **14.31:** NASA/JPL/Space Science Institute; **14.32:** ESA/NASA/JPL/University of Arizona; **14.33:** Courtesy Huygens raw images: ESA, NASA, JPL, University of Arizona. Image processing and panorama: René Pascal, Bonn, Germany; **14.34:** ESA/NASA/JPL/University of Arizona; **14.35:** NASA/JPL/USGS; **14.36:** JPL; **14.37:** Astrogeology Team, USGS, The Voyager Project; **14.38–14.40:** JPL; **14.41:** NASA/JPL; **14.42, 14.45–14.47:** JPL.

Chapter 15

Opener: © Brand X Pictures/PunchStock RF; **15.1:** © Jerry Schad/Photo Researchers; **15.3:** Courtesy Chen Huang Ming 2001; **15.3:** Courtesy Pl Jenniskens/NASA; **15.5:** Courtesy Donald Brownlee/University of Washington; **15.6–15.8:** © John A. Wood; **15.12:** Courtesy NEAR John Hopkins University Applied Physics Lab/NASA; **15.13:** NASA/JPL; **15.14–15.16:** Courtesy NEAR Project, NLR, JHUAPL, Goddard SVS, NASA; **15.18:** Detail from the Bayeux Tapestry, 11th century; by special permission of the City of Bayeux. Photo by J. Kelly Beatty/Sky Publishing Corporation/New Track Media; **15.19:** © Anglo-Australian Obs/David Malin Images; **15.21:** NASA/JPL/UMD; **15.22:** © Ronald Royer/SPL/Photo Researchers; **15.24:** © Wally Pacholka/AstroPics.com; **15.27:** Courtesy D. Jewitt & J. Luu, University of Hawaii; **15.28:** Galileo Project, Brown University, JPL, NASA; **15.29:** © MSSSO, ANU/SPL/Photo Researchers, Inc; **15.30:** Space Telescope Science Institute; **15.31:** © Matt Neely; **15.32:** Courtesy David Roddy, Carl Zeller/USGS, Flagstaff; **15.33:** NASA and Science Graphics; **15.34:** Courtesy Walter Alvarez/University of California Berkeley.

Chapter 16

Opener: © Photodisc/Getty RF; **16.2b, c:** Courtesy Yerkes Observatory; **16.10:** From Richard Herzbert, *Atomic Spectra & Atomic Structure*, Copyright © Prentice Hall, Dover Publications. Reprinted by permission.; **16.11, 16.20:** NASA/JPL.

Chapter 17

Opener: © Corbis RF; **17.4:** Courtesy Brookhaven National Laboratory; **17.5:** Courtesy The Institute for Cosmic Ray Research/University of Tokyo; **17.6:** NOAO; **17.9:** © Dr. Thomas Berger/Lockheed Martin Solar and Astrophysics Laboratory; **17.10:** © Alan Title/Lockheed Solar Observatory; **17.11:** SOHO (ESA & NASA); **17.13:** NASA; **17.14:** NOAO; **17.16:** NASA; **17.17:** Courtesy T. Rimmele (NSO), M. Hanna (NOAA)/AURA/NSF; **17.18:** Courtesy Ray Gralak; **17.19:** Courtesy NSO/Kitt Peak data used here are produced cooperatively by NSF.NOAO, NASA/GSFC, and NOAA/SEL; **17.20:** National Solar Observatory; **17.21:** Courtesy SST, Royal Swedish Academy of Sciences, LMSAL; **17.22a:** Courtesy NOAO; **17.22b:** Courtesy NOAO; **17.24:** Courtesy Stanford-Lockheed Institute for Space Research (Palo Alto, CA)/NASA Goddard Space Flight Center; **17.25:** SOHO (ESA & NASA); **17.27:** Courtesy M. Aschwanden et al. (LMSAL), TRACE, NASA; **17.28:** Courtesy Institute for Astronomy, University of Hawaii; **17.29 & 17.30:** Courtesy SOHO-EIT Consortium/ESA/NASA; **17.31:** Courtesy Big Bear Solar Observatory, New Jersey Institute of Technology.

Chapter 18

Opener: © Stocktrek/age fotostock RF; **18.1:** FORS Team, **8.2:** -meter VLT Antu, ESO; **18.2:** © Dennis di Cicco, Sky and Telescope Magazine/New Track Media; **18.3:** © Anglo-Australian Obs/David Malin Images; **18.4a:** Courtesy AURA/StSci; **18.4b:** Courtesy Ronald J. Maddalena, Mark Morris, J. Moscowitz, and P. Thaddeus/National Radio Astronomy

Observatory; **18.4c:** © Anglo-Australian Obs/David Malin Images; **18.5:** NOAO; **18.9:** Courtesy N. Walborn (STSci)/R. Barba (La Plata Observatory)/NASA; **18.12 (left, right):** NASA, ESA, J.E. Krist (STScI/JPL); D.R. Ardila (JHU); D.A. Golimowski (JHU); M. Clampin (NASA/Goddard); H.C. Ford (JHU); G.D. Illingworth (UCO-Lick); G.F. Hartig (STScI) and the ACS Science Team; **18.13:** Courtesy O'Dell/Rice University/NASA; **18.14:** Courtesy C.A. Grady and B. Woodgate (NASA Goddard Space Flight Center), F. Bruhweiler and A. Boggess (Catholic University of America), P. Plait and D. Lindler (ACC, Inc., Goddard Space Flight Center), M. Clampin (Space Telescope Science Institute), and NASA; **18.16:** NASA, ESA, N. Smith (University of California Berkeley), and the Hubble Heritage Team, (STScI/AURA) and Courtesy CTIO image N. Smith and NOAO/AURA/NSF.

Chapter 19

Opener (X-ray): NASA/CXC/SAO; Optical: NASA/STScI; **19.24:** Hubble Heritage Team (AURA.STScI) NASA; **19.25:** Courtesy Bruce Balick (University of Washington), Vincent Icke (Leiden University, The Netherlands), Garrelt Ellema (Stockholm University), and NASA; **19.26 (top, bottom left):** Hubble Heritage team (AURA/STScI/NASA); **19.26 (bottom right):** NASA, ESA, and the Hubble SM4 ERO Team; **19.28:** NASA.

Chapter 20

Opener: © Stocktrek Images/Getty RF; **20.1:** Courtesy McDonald Observatory; **20.6a, b:** © Mount Wilson and Las Campanas Observatories; Carnegie Observatory; **20.10a, b:** © Anglo-Australian Obs/David Malin Images; **20.11:** Space Telescope Science Institute; **20.12:** HST GRB Collaboration, STIS, HST, NASA; **20.13:** CXC/SAO/Rutgers/J. Hughes/NASA; **20.14:** Remote Sensing Division, Naval Research Laboratory; **20.15:** Courtesy J. Hester (ASU), NASA; **20.16:** Courtesy European Southern Observatory.

Chapter 21

Opener: © Stocktrek Images/Getty RF.

Chapter 22

Opener: Credit for Hubble Image: NASA, ESA, and Q.D. Wang University of Massachusetts, Amherst) Credit for Spitzer Image: NASA, JPL, and S. Stolovy (Spitzer Science Center/Caltech); **22.1:** © Axel Melinger; **22.2:** Courtesy Edward L. Wright; **22.3:** © Christine Jones/Smithsonian/Astrophysical Observatory; **22.4:** © C. Haslam et al., Max Planck Institut for Radio Astronomy, SkyView; **22.5:** © Michael Freyberg/Max-Planck-Institut fŸr extraterrestrische Physik; **22.6:** © Anglo-Australian Obs/David Malin Images; **22.9:** © Royal Observatory, Edinburgh/AAO/SPL/Photo Researchers; **22.12:** © John P. Gleason/Celestial Images; **22.13:** Courtesy Roff/Lowell Observatory; **22.14:** Courtesy Canada-France-Hawaii Telescope/J.-C. Cuillandre/Coelum © 2001 CFHT; 22.15: AURA/STSci; **22.16:** NASA and The Hubble Heritage Team (STSci/AURA); **22.17–22.19:** © Anglo-Australian Obs/David Malin Images; **22.20:** National Space Science Data Center/NASA Goddard Space Flight Center; **22.22:** © Dr. Jeremy Burgess/SPL/Photo Researchers,; **22.24a:** Courtesy Patrick Seitzer, University of Michigan; **22.24b:** T2KA, KPNO 0.0-m Telescope, NOAO, AURA, NSF; **22.25:** Courtesy Harvard College Observatory; **22.32:** JPL; **22.39:** From H.C.D. Visser in *Astronomy and Astrophysics*, 88, 159–174, 1980. Reproduced with permission. Copyright © The European Southern Observatory (ESO); **22.40:** Courtesy Susan Stolovy (SSC/Caltech) et al., JPL-Caltech, NASA; **22.42:** NRAO; **22.45:** NASA/JPL-Caltech/Robert Hurt (SSC).

Chapter 23

Opener: NASA, ESA, and The Hubble Heritage Team (STScI/AURA) Acknowledgment: P. Knezek (WYN); **23.2:** Courtesy Bruce Hugo and Leslie Gaul, Adam Block (KPNO Visitor Program), NOAO, AURA, NSF; **23.4, 23.5, 23.8b–d, 23.9b–d:** © The Observatories of the Carnegie Institute of Washington; **23.10:** Courtesy Gemini Observatory-GMOS Team; **23.11:** © FORS Team, 8.2-meter VLT, European Southern Observatory; **23.12:** © Anglo-Australian Obs/David Malin Images; **23.13:** NASA, ESA, S. Beckwith (STScI) and the HUDF Team; **23.14:** NASA, ESA, R. Windhorst (Arizona State University) and H. Yan (Spitzer Science Center, Caltech); **23.16:** Courtesy Todd Boroson (NOAO), AURA, NOAO, NSF; **23.17:** NASA/G. Llingworth/M. Clampin; **23.18:** NASA, ESA, and The Hubble Heritage Team (AURA/STScI); **23.19 (left, right):** NASA; **23.22:** NASA and The Hubble Heritage Team (STScI/AURA).

Chapter 24

Opener: NASA, ESA, and The Heritage Team (STScI/AURA). Acknowledgment: J. Gallagher (University of Wisconsin), M. Mountain, (STScI) and P. Puxley (National Science Foundation); **24.1:** Courtesy Palomar/California Institute of Technology; **24.2:** Courtesy Maarten Schmidt; **24.11:** AURA/STSci/NASA; **24.13, 24.15:** NRAO/AUI; **24.16:** Courtesy M. Norman, NCSA, UIUC; **24.18:** NOAO; **24.24:** Courtesy J. Bahcall (Institute for Advanced Studies)/M. Disney, University of Wales/NASA; **24.30:** Courtesy Jacqueline Hewitt.

Chapter 25

Opener: NASA, ESA, and The Hubble Heritage Team (STScI/AURA) Acknowledgment: J. Blakeslee (Washington State University); **25.1:** Courtesy Two Micron All Sky Survey, a joint project of the University of MA and the Infrared Processing and Analysis Center/CA Institute of Technology, funded by NASA/National Science Foundation/T. H. Jarrett, J. Carpenter, & R. Hurt. Perha; **25.3:** Courtesy T.A. Rector and B.A.Wolpa/NOAO/AURA/NSF; **25.4–25.7:** © Anglo-Australian Obs/David Malin Images; **25.8:** Courtesy Local Group Galaxies Survey Team, NOAO, AURA, NSF; **25.9:** © Canada-France-Hawaii Telescope/J.-C. Cuillandre/Coelum (c) 2001 CFHT; **25.10, 25.11:** © Anglo-Australian Obs/David Malin Images; **25.12:** Space Telescope Science Institute; **25.13:** © Anglo-Australian Obs/David Malin Images; **25.14:** Courtesy Omar Lopez-Cruz & Ian Shelton/NOAO/AURA/NSF; **25.15:** Courtesy A. Dressler (CIW)/NASA; **25.19:** Courtesy Rogier Windhorst and Sam Pascarelle (Arizona State University) and NASA; **25.20:** Courtesy X-ray: NASA/CXC/ESO/P. Rosati et al.; Optical: ESO/VLT/P. Rosati et al.; **25.21:** AURA.STSci; **25.22:** Courtesy Richard Ellis (Cambridge University) and NASA; **25.23:** Courtesy Composite Credit: X-ray: NASA/CXC/CfA/M.Markevitch et al.; Lensing Map: NASA/STScI; ESO WFI; Magellan/U.Arizona/D. Clowe et al. Optical: NASA/STScI; Magellan/U.Arizona/D. Clowe et al.; **25.27:** © Astrophysical Research Consortium (ARC) and Sloan Digital Sky Survey (SDSS) Collaboration, http://www.sdss.org; **25.28:** Courtesy 2P2 Team, WFI, MPG/ESO 2.2-m Telescope, La Silla, ESO.

Chapter 26

Opener: NASA, ESA, J. Blakeslee and F. Ford (John Hopkins University); **26.19:** Courtesy of AT&T Archives and History Center; **26.20:** WMAP Science Team, NASA.

Chapter 27

Opener: © William K. Hartmann; **27.1:** Courtesy Sidney Fox; **27.2:** Courtesy J.W. Schopf; **27.3, 27.4:** NASA; **27.6:** Courtesy C. Marois et al./NRC Canada & Keck Observatory; **27.11, 27.12, 27.14:** NASA.

LINE ART

Chapter 6

6.27: From Own Concept Study by Roberto Gilmozzi and Philippe Diericky, European Southern Observatory. Reprinted with permission of EOS.

Chapter 13

13.12: Reprinted with permission from the *Annual Review of Astronomy and Astrophysics,* Volume 17 © 1979 by Annual Reviews. *www.annualreviews.org.*

Chapter 15

15.11: Image courtesy of the Minor Plant Center, Smithsonian Astrophysical Observatory; **15.25:** Oort Comet Cloud illustration. Copyright © 1990 by Sky Publishing Corp.

Chapter 16

16.10: From Atomic Spectra and Atomic Structure, by Richard Herzberg, Dover Publications, 1937; **16.21:** Hipparcos Satellite, (ESA) European Space Agency.

Chapter 17

17.35: Courtesy of Dr. Jim O'Donnell, Royal Observatory Greenwich.

Chapter 22

22.43: From Sky & Telescope, April 2003. Copyright © 2003 Sky Publishing Corp.

Chapter 24

24.8: Courtesy of John Biretta, Space Telescope Science Institute.

Chapter 25

25.28: Reprinted with permission of Astrophysical Research Consortium (ARC) and the Sloan Digital Sky Survey (SDSS) Collaboration, *http://www.sdss.org.*

Index

C

E

F

H

M

S

T

U

V

W

X

Y

Z

The Night Sky in Spring

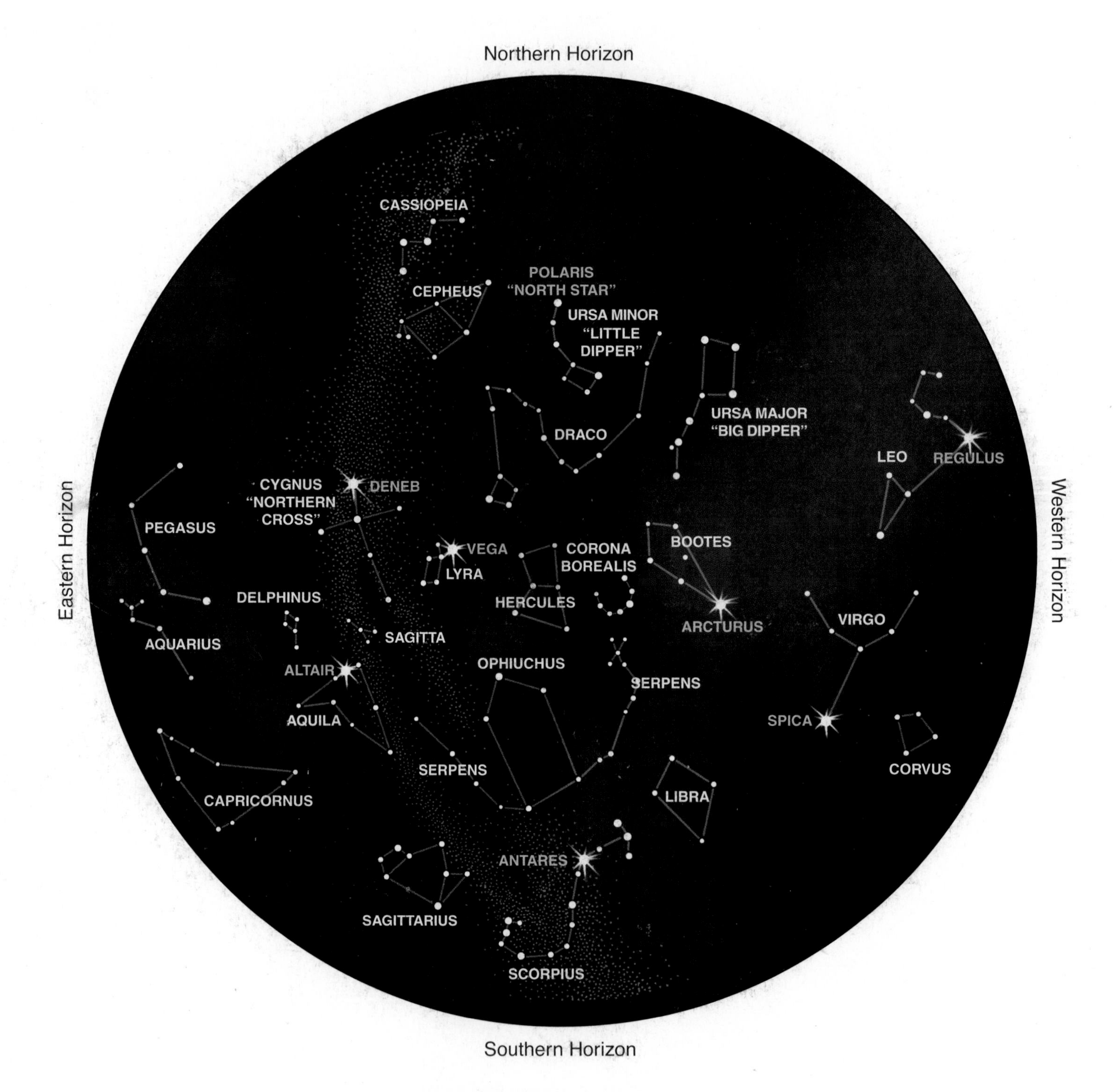
Northern Horizon
Eastern Horizon
Western Horizon
Southern Horizon
CASSIOPEIA
CEPHEUS
POLARIS
"NORTH STAR"
URSA MINOR
"LITTLE
DIPPER"
URSA MAJOR
"BIG DIPPER"
DRACO
LEO
REGULUS
CYGNUS
"NORTHERN
CROSS"
DENEB
PEGASUS
VEGA
LYRA
CORONA
BOREALIS
BOOTES
DELPHINUS
HERCULES
ARCTURUS
VIRGO
AQUARIUS
SAGITTA
ALTAIR
OPHIUCHUS
SERPENS
AQUILA
SPICA
SERPENS
CORVUS
CAPRICORNUS
LIBRA
ANTARES
SAGITTARIUS
SCORPIUS

The Night Sky in Summer